HISTORY OF INSECTS

HISTORY OF INSECTS

by

*N.V. Belayeva, V.A. Blagoderov, V.Yu. Dmitriev, K.Yu. Eskov, A.V. Gorokhov,
V.D. Ivanov, N.Yu. Kluge, M.V. Kozlov, E.D. Lukashevich, M.B. Mostovski,
V.G. Novokshonov, A.G. Ponomarenko, Yu.A. Popov, L.N. Pritykina, A.P. Rasnitsyn,
D.E. Shcherbakov, N.D. Sinitshenkova, S.Yu. Storozhenko, I.D. Sukatsheva,
V.N. Vishniakova, Peter Vršanský and [†]V.V. Zherikhin*

Edited by

ALEXANDR P. RASNITSYN AND DONALD L.J. QUICKE

KLUWER ACADEMIC PUBLISHERS
DORDRECHT / BOSTON / LONDON

Library of Congress Cataloging-in-Publication data is available.

ISBN 1-4020-0026-X

Published by Kluwer Academic Publishers,
P.O.Box 17, 3300 AA Dordrecht, The Netherlands

Sold and distributed in North, Central and South America
by Kluwer Academic Publishers,
101 Philip Drive, Norwell, MA 02061, U.S.A.

In all other countries, sold and distributed
by Kluwer Academic Publishers, Distribution Center,
P.O. Box 322, 3300 AH Dordrecht, The Netherlands

Printed on acid-free paper

Typeset by Newgen Imaging Systems Pvt. Ltd., Chennai, India.

Printed and bound in Great Britain by MPG Books Ltd.,
Bodmin, Cornwall.

This book is dedicated to the three people
most responsible for the present state of palaeoentomology,

Andrey Vassilievich Martynov,
Boris Borisovich Rohdendorf and
Frank Morton Carpenter

We also extend our dedication to
Vladimir Vassilievich Zherikhin, whose untimely
death in December 2001,
was a great loss to
palaeoentomology

Contents

Preface

Insects are not dinosaurs – and they probably pose us more strange puzzles and unexpected questions. A million extant species, that is several times more than all other living taxa together, is still a very conservative estimate, and their real number is for sure many times more. They are incomparably diverse in terms of their size, structure and way of life – and yet they are all small – by our standard at least – why? And they practically ignore the cradle of life, the sea – again, why? Of course, some survive and even reproduce in salt water, but nevertheless very few of them are specialised for marine life. Some insects have developed highly elaborate forms of sociality, far surpassing all achievements reached by vertebrates (including ourselves), at least in terms of interdependence of individuals. Once again, why have they developed this way of life? The history of insects is also full of unexpected discoveries as well as of painful gaps, though these omissions are not overwhelming nor necessarily senseless. There is a special branch of palaeontology called taphonomy whose aim is to learn about the burial pattern of the past organisms: what governs their chance of becoming fossilised, and how this chance depends on features of both the organisms themselves and on their environments. Depending on how deep the taphonomical background in a particular research field, it is even possible to glean information from the very gaps in the fossil record. For instance, lake deposits are known to be favourable for insect burial in contrast to deposits left by running waters. If a group of insect fossils has a poor record in spite of being aquatic (judging from its morphology: otherwise habits are unknown), it most likely occurred in streams, and almost certainly so in cases where the few available fossils are worn and incomplete in contrast to other aquatic fossils collected there.

Palaeontology is an important, though not the only, way to trace the history of insects: phylogenetics is another and is equally important, and it is noteworthy that it is largely independent of the fossil record in its sources and inferences. Both ways are widely used and are mutually helpful and exert controls on one another. Using the words of Willi Hennig, they provide mutual illumination. There are even quantitative tools in progress aimed at helping to assess how well the fossil record and various phylogenetic hypotheses agree with one another (e.g. the ghost range method, RASNITSYN 2000a).

Phylogenetic research is popular and widespread, mainly in the form of cladistics (particularly since its fundamentals were rediscovered by the English speaking audience in HENNIG 1966a). However, this is not the case for palaeoentomology, because palaeoentomologists were and are still few in number, and are widely scattered, except for the Moscow research group with its associates, which has been the most prolific and productive such group in the world since the 1960s. So it is not completely by chance that the present book is written mainly by those Russian scientists, but importantly it attempts to cover all the world's material and all the available information.

ACKNOWLEDGEMENTS

For A.P. Rasnitsyn, preparing of various parts of this book was supported in part by grants: by the International Science Foundation, by the Leverhulme Trust to D.L.J. Quicke and M.G. Fitton; by the Royal Society Joint Project with the FSU to APR and E.A. Jarzembowski; by ESF Project 'Fossil Insects Network'; by RFFI grants 95-04-11105, 98-04-48518; by the Smithsonian Institution and California Academy of Sciences; and by various help, including sharing unpublished information, rendering material, advice, publications, shelter and working facilities from many friends and colleagues, of which APR would like to mention specially, besides the co-editor and all co-authors of the present book, also Dr. H.H. Basibuyuk, Dr J.M. Carpenter, Mr. J. Cooper, Mr R. Coram, Dr. H. Dathe, Dr. E.A. Jarzembowski, Dr. C.C. Labandeira, Mr. D. Kohls, Dr. J. Kukalov-Peck, Dr. X. Martínes-Delclòs, Dr. L. Masner, Dr. A. Nel, Mr. L. Pribyl, Dr W.J. Pulawski, Dr. A.J. Ross, Dr. J.M. Rowland, Dr. W. Zessin.

For V.D. Ivanov, this study was supported by the RFFI grants N 99-04-49564, 00-15-97934, by the Federal program 'Universities of Russia' (project N 3917), and by project BR-7 of the program 'Bioraznoobrazie'.

N.Yu Kluge's contribution was supported by the Federal programme for support of leading scientific schools, RFFI grant N 00-15-97934.

Participation of M.B. Mostovski was helped by Dr. J. Ansorge (Institut für Geologische Wissenschaften, Greifswald), Dr. B.R. Stuckenberg (Natal Museum, South Africa), Dr. E.A. Jarzembowski (Maidstone Museum and University of Reading, UK), Mr. R. Coram (University of Reading, UK), Dr. X. Martínes-Delclòs (Universitat de Barcelona, Spain), Dr. A. Arillo (Universidad Complutense, Madrid, Spain), and was supported in part by the International Science Foundation, the Palaeontological Society (USA, grants nos. RG0-638 (B) and RG0-822-7) and the Geologists' Association (UK), ESF Project "Fossil Insects Network".

Studies on terrestrial and aquatic insect palaeoecology (†V.V. Zherikhin and N.D. Sinitshenkova, respectively) have been partially supported by the European Science Foundation (project "Fossil Insects") and the Federal Scientific and Technical Programs (programmes "Evolution of the Biosphere" and "Co-evolution of Ecosystems").

Unless stated otherwise, illustrations are provided by the authors of the respective chapters. M.K. Emelyanova (Palaeontological Institute RAS, Moscow) helped in the formatting and the enhancing of many insect restorations and photographs. Thanks also go to Mrs. I.V. Kistchinskaya (Kishchinskaya) (Data+, Ltd.) for help in preparing the maps of the fossil sites (see Figs. 3–5) with the use of the ArcView GIS software programme.

Alexandr P. Rasnitsyn and Donald L.J. Quicke

Contributors

Natalya V. Belayeva, Chair of Entomology, the Moscow State University, Vorob'evy Gory, Moscow, 119899 Russia, nvb@3.entomol.bio.msu.ru.

Vladimir A Blagoderov, Palaeontological Institute RAS, Moscow 117868 Russia, currently Department of Entomology, American Museum of Natural History, Central Park West at 79th Street, New York, NY 10024-5192, USA, vblago@amnh.org.

Viktor Yu. Dmitriev, Palaeontological Institute RAS, Moscow 117868 Russia, admin@paleo.ru.

Kirill Yu. Eskov, Palaeontological Institute RAS, Moscow 117868 Russia, afranius@newmail.ru.

Andrey V. Gorokhov, Insect Taxonomy Laboratory, Zoological Institute RAS, University Embarkment 1, St Petersburg 199034 Russia, orthopt@zin.ru.

Vladimir D. Ivanov, St Petersburg State University, University Embarkment 7, St Petersburg 199034 Russia, vladi@vdi.usr.pu.ru.

Nikita Ju. Kluge, St Petersburg State University, University Embarkment 7, St Petersburg 199034 Russia, kluge@ent.bio.pu.ru.

Mikhail V. Kozlov, Laboratory of Ecological Zoology, University of Turku, Turku FIN 20500 Finland, mikoz@mailhost.utu.fi.

Elena D. Lukashevich, Palaeontological Institute RAS, Moscow 117868 Russia, elukashevich@hotmail.com.

Mikhail B. Mostovski, Palaeontological Institute RAS, Moscow 117868 Russia, phorids@hotmail.com.

Viktor G. Novokshonov, Perm State University, Bukireva 15, Perm 614600 Russia, Larisa.Zhuzhgova@psu.ru.

Alexandr G. Ponomarenko, Palaeontological Institute RAS, Moscow 117868 Russia, aponom@paleo.ru.

Yuri A. Popov, Palaeontological Institute RAS, Moscow 117868 Russia, elena@advizer.msk.ru.

Ludmila N. Pritykina, retired from Palaeontological Institute RAS, Moscow 117868 Russia, rasna@online.ru.

Donald L.J. Quicke, Department of Biology, Imperial College at Silwood Park, Ascot, Berkshire SL5 7PY UK, d.quicke@ic.ac.uk.

Alexandr P. Rasnitsyn, Palaeontological Institute RAS, Moscow 117868 Russia, rasna@online.ru.

Dmitry E. Shcherbakov, Palaeontological Institute RAS, Moscow 117868 Russia, prosbole@hotmail.com.

Nina D. Sinitshenkova, Palaeontological Institute RAS, Moscow 117868 Russia, vzher@aha.ru.

Sergey Yu. Storozhenko, Biological and Pedological Institute, Far East Branch RAS, Vladivostok 690022 Russia, entomol@online.marine.su.

Irina D. Sukatsheva, Palaeontological Institute RAS, Moscow 117868 Russia, rasna@online.ru.

Valentina N. Vishniakova, retired from Palaeontological Institute RAS, Moscow 117868 Russia, rasna@online.ru.

Peter Vršanský, Palaeontological Institute RAS, Moscow 117868 Russia, ambasvk@stonline.sk.

†Vladimir V. Zherikhin, Palaeontological Institute RAS, Moscow 117868 Russia.

1
Introduction to Palaeoentomology

1.1
Scope and Approach

A.P. RASNITSYN

1.1.1. COVERAGE

This book tries in general to cover the whole extent of the history of insects in time and space in detail (Fig. 1). The exception is the Quaternary Period which is only occasionally considered here in depth because Quaternary palaeoentomology is a specialised field with its own, rather specific methods and approaches. Quaternary palaeoentomology has recently been dealt with in considerable detail by ELIAS (1994), and this might excuse our omission here. We understand that the selection of which subjects to present and discuss here reflects our personal views and tastes, and that many important subjects may have been missed, either by mistake or ignorance. We hope that our readers will help us to make any future editions better than the present one.

The geochronological scale employed here is shown in Fig. 2, and many insect fossil sites are displayed on the maps (Figs. 3–5). The sites/site groups numbered and labelled there are the same as those catalogued in Appendix (Chapter 4).

The class Insecta is taken here in its narrow sense, that is, not including the entognathous orders (Acerentomida = Protura, Campodeida = Diplura, Podurida = Collembola). Although the relationships of these three orders is not fully resolved, there is growing evidence that they may form a monophyletic group together with the myriapods (REMINGTON 1955, HANDSHIN 1958, MELNIKOV 1974a,b, RASNITSYN 1976, MELNIKOV & RASNITSYN 1984, D. Shcherbakov, submitted) as explained in some detail below.

1.1.2. PHYLOGENETIC APPROACH

It may come as a surprise to many members of the western scientific community that cladistics has not been universally accepted as the one and only method to be applied to classifying organisms by entomologists and other systematists of the former Soviet Union. That is not to say that cladistics and parsimony analysis are not regarded as being very useful tools, or that groups should not be clustered on the basis of shared derived characters (**synapomorphies**) wherever possible. Rather, many believe that phylogenetic inferences are exactly the same as all other scientific statements in that they can never be proved nor rejected, but are for ever destined to persist as more or less likely hypotheses. Indeed, the inaccessibility of the final verification of a scientific statement has become almost a truism in post-Popperian times, and yet the final falsification of one statement is nothing more than the verification of an alternative statement, that is, the falsifying result is neither by chance nor due to unconsidered influences or circumstances. As a result there is no single best method for inferring phylogenies, such as the outgroup approach comparison praised by some authors. Contrarily, many methods are good, each in its own place, and their application can be efficiently regulated in the form of presumption. Thus much of the work described in this book relies on making presumptions. This is not a trivial matter because of the often incomplete nature of the evidence that scientists in general, and palaeontologists in particular, have to deal with. A **presumption** is a statement based on observations that a particular kind of result occurs more commonly in particular circumstances, and thus it is to be considered as most likely irrespective of the existence of confirmatory evidence, but not, of course, in the presence of reasonably sound contrary evidence (RASNITSYN 1988a, 1992a, 1996, RASNITSYN & DLUSSKY 1988). For instance, we should consider any similarity between organisms as inherited from a common ancestor and not gained independently (as homoplasy), unless and until strong contrary evidence is presented. This has been termed the **presumption of cognisability of evolution**, and it is equivalent to the auxiliary principle of HENNIG (1966a). Another type of presumption that is particularly relevant to palaeontology (the **palaeontological presumption**) is that of two apparently closely related groups, the one entering the fossil record earlier should be considered as ancestral unless and until sound contrary evidence is presented. Likewise, we can recognise a **biogenetic presumption**, that is, a transformation series should be polarised in agreement with the ontogenetic succession of the respective character states; the **outgroup presumption**, that a character state found only within a group should be considered apomorphic in respect to that found both within and outside the group, and of course, the **presumption of parsimony**, i.e. that the most likely cladogram is that one necessitating the least number of homoplasies (i.e. the most parsimonious one). All of these presumptions are accepted below unless there is convincing evidence to the

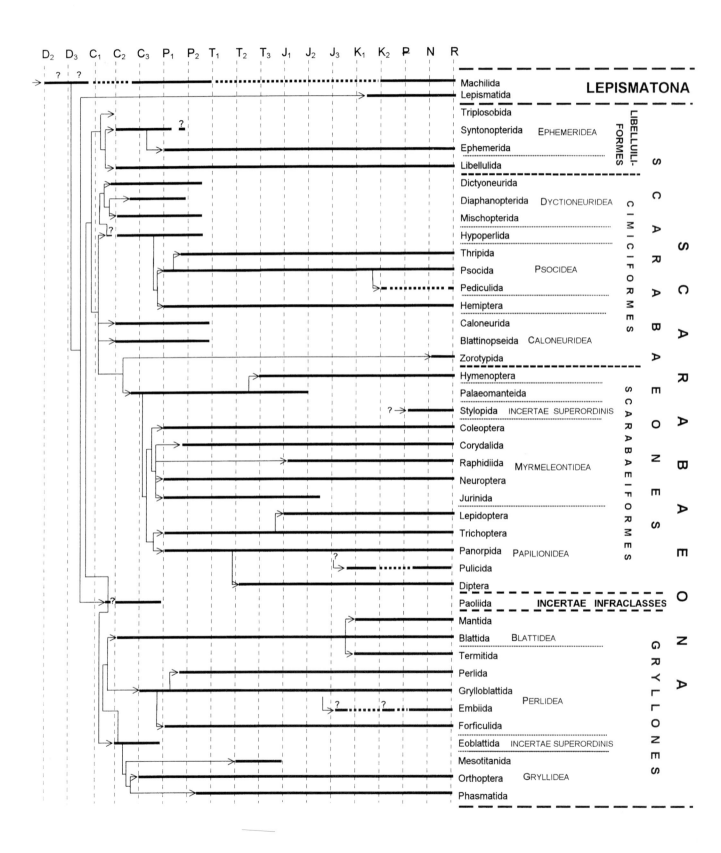

contrary. For a more thorough discourse on the use of presumptions in phylogenetics see RASNITSYN & DLUSSKY (1988) and RASNITSYN (1996).

For the most part, the authors in this volume try to follow the cladistic approach to phylogeny reconstruction (HENNIG 1966a, etc.). In short, we appreciate that a clade should be considered definable by a **synapomorphy**, that is, an advanced character state hypothesised to have been acquired by the first member of the clade. The **clade** itself as the main notion in phylogenetics can be defined as a part of the tree-like figure representing relationships between groups of organisms, which is formed by cutting off a single ancestral line. **Symplesiomorphy** (similarity in a character state inherited from a more distant common ancestor) has no value in phylogenetics, as well as any kind of the **homoplasy** (independently gained similarities, including convergence and reversals, that is the convergence with an ancestral state of a character).

Another minor deviation here is that phylogenies based extensively on fossil information, in comparison with modern phylogenetic reconstructions prepared by students of living insects, pay relatively less attention to synapomorphies of detailed internal anatomy. Such characters are usually simply unknown for the phylogenetically most interesting extinct groups, and even living representatives have sometimes been explored very unevenly for these systems.

A more important deviation results from understanding that the cladistic concept is a methodological claim rather than an ontological one. By this it is meant that cladistics makes available a very efficient method for the identification and, in more advanced versions, for the calculation of relatedness, that result in enhanced objectivity and reproducibility of the results. At the same time, the basic hypotheses that either implicitly or explicitly form the basis of cladistics are not necessarily correct ontologically, that is, they do not necessarily agree with what we can observe in nature. Indeed, amongst the most important such hypotheses, there is the claim that each taxonomic group appears as a result of a divergence event accompanied by the gain of one or more apomorphy(ies), and that these groups are each destined to either die out, or to disappear in the next divergence event resulted in appearance of two (neither more nor less) new lines, each marked with their own apomorphies. Biological theory is aware of no mechanisms possible to secure these rules and to prevent, say the ancestral line not dying out when it bears a daughter line, but survives and to give birth to two or more further daughter lines. In other words, a polychotomy (the case of a line giving birth to more than two daughter lines) is not necessarily a case of incomplete knowledge: indeed it well might be effectively correct. Equally, no mechanisms are known to prevent a line from giving birth to only one, and not to two daughter lines. Of course, it is a rare occasion when we have evidence for one or another of the possibilities discussed above. Nevertheless, this makes it evident that a cladistic concept is an hypothesis and not the final truth, and that a cladistic result does not mark the end of a study but rather is the material for further testing by different methods (RASNITSYN 1996).

The most powerful of these tests are various ways of comparison of the cladist's cladogram with the palaeontological succession of fossils and their characters, e.g. tracing of the morphological transitions formed by fossils (e.g. the ghost range method; RASNITSYN 2000a). Of course, these methods do not give us any ultimate truth either, but mutual hypothesis testing and correction is always beneficial.

It is clear from the above discussion that the principles of taxonomy followed here differ somewhat from those employed by many, if not all, current cladists (see WILEY 1981, QUICKE 1993). This particularly concerns the interrelationship between the taxonomic system employed and the hypothesised system of relatedness between the organisms concerned (as derived from cladograms). The cladistic demand is to construct the first system so as to make it isomorphic with the second, as if the ultimate goal were to prepare the genealogical system of organisms. Thus, for most cladists only monophyletic taxa are appreciated as legitimate, with monophyly being taken in its narrowest sense, referring to a taxon which includes all the descendants of its single, oldest member. This kind of taxon has the single ancestral line and no descendant lines beyond its limits.

Because of the existence of another, older and broader understanding of the term monophyly which is of no less use or popularity, a new term was coined for the cladistic version of monophyly, viz. **holophyly** (ASHLOCK 1971). This latter term is subordinate to **monophyly sensu lato**, which implies a taxon equally with the single ancestral line beyond its limits, but with no limitation imposed on its descendant lines. The second subordinate term of monophyly *s. l.* which is complementary to holophyly is **paraphyly**. A taxon is paraphyletic if it has a single ancestral line but has one or more descendant lines beyond its limits. So monophyly *s. l.* is definable through possession of only a single ancestral line beyond its limits. A taxon with more than one ancestral line is called polyphyletic irrespective of the presence of descendant lines. It is only monophyly **sensu lato** that is used throughout the present book, with the alternative notion, whenever employed, being termed holophyly.

Thus the purist cladistic approach to taxonomy has been largely abandoned here for several reasons (for details see RASNITSYN 1996). First of all, reflection of genealogy is not the ultimate goal of taxonomy, even for many cladists, for the cladogram coupled with necessary explanations (technically termed scenarios) reflects genealogy much better than any nomenclatural system. Implicit here is the belief that the system must reflect the full balance of similarities and dissimilarities out of the totality of all possible characters including not only morphological, physiological, behavioural, aesthetic and so on, but also those that are already explored and those that are as yet completely unknown. Because of the obvious predestination, which history (as it is materialised in the realised type of organisation) imposes on the structure, functions and further evolution of an organism, history (in form of a genealogical scheme, a cladogram) has been hypothesised to be best correlated with the above mentioned total balance of similarities and dissimilarities. This inference, which has much truth, albeit not all the truth, was the starting point of cladistics.

Fig. 1 Phylogeny and system of the insects (Class Insecta) at order level. Time scale intervals are abbreviated as follows: D_2, D_3 – Middle and Late (Upper) Devonian, C_1, C_2, C_3 – Early (Lower), Middle and Late (Upper) Carboniferous, P_1, P_2 – Early (Lower) and Late (Upper) Permian, T_1, T_2, T_3 – Early (Lower), Middle and Late (Upper) Triassic, J_1, J_2, J_3 – Early (Lower), Middle and Late (Upper) Jurassic, K_1, K_2 – Early (Lower) and Late (Upper) Cretaceous, Ᵽ – Palaeogene (Early Tertiary, viz. Palaeocene + Oligocene + Eocene), N – Neogene (Late Tertiary, viz. Miocene and Pliocene), R – present time (Holocene). The name columns are, left to right, orders, superorders, cohorts, infra- and subclasses (except subclass Lepismatona which is out of order to save the space). The customary names of taxa are given in the titles of the respective chapters. Arrows show ancestry, thick bars indicate known temporal durations of taxa (questioned in the case of debatable records, dashed for long gaps in the fossil record)

Era	Period	Epoch[1]		Age	Myr[2]
Cenozoicum			Q[3]		
	Neogene	Pliocene	L	Piacentian	1,75
			E	Zanclian	
					5,3
		Miocene	L	Messinian	
				Tortonian	11,0
			M	Serravallian	
				Langhian	15,8
			E	Burdigalian	
				Aquitanian	23,0
	Paleogene	Oligocene	L	Chattian	
			E	Rupelian	33,7
		Eocene	L	Priabonian	37,0
			M	Bartonian	
				Lutetian	46,0
			E	Ypresian	53
		Paleocene	L	Thanetian	
			E	Danian	65
Mesozoicum	Cretaceous		L	Maestrichtian	
				Campanian	
				Santonian	
				Coniacian	
				Turonian	
				Cenomanian	96
			E	Albian	
				Aptian	
				Neocomian[4] L Barremian	
				Hauterivian	
				E Valanginian	
				Berriasian	135
	Jurassic	Malm	L	Tithonian	
				Kimmeridgian	
				Oxfordian	154
		Dogger	M	Callovian	
				Bathonian	
				Bajocian	
				Aalenian	175
		Lias	E	Toarcian	
				Pliensbachian	
				Sinemurian	
				Hettangian	203
	Triassic		L	Rhaetian	
				Norian	
				Carnian	230
			M	Ladinian	
				Anisian	240
			E	Olenekian	
				Induan	250
Paleozoicum	Permian		L	Tatarian	
				Kazanian	
				Ufimian	275
			E	Kungurian	
				Artinskian	
				Sakmarian	
				Asselian	295
	Carboniferous		L	Gzhelian	
				Kasimovian	305
			M	Moscovian	
				Bashkirian	315
			E	Serpukhovian	
				Visean	
				Tournaisian	355
	Devonian				

Carboniferous in Western Europe

Carboniferous	L	Stephanian	C	
			B	
			A	
		Westphalian	D	
			C	
			B	
			A	
		Namurian	C	Yeadonian
			B	Marsdenian
				Kinderscoutian
			A	Alportian
				Chokierian
				Arnsbergian
				Pendleyian
	E	Visean		
		Tournaisian		

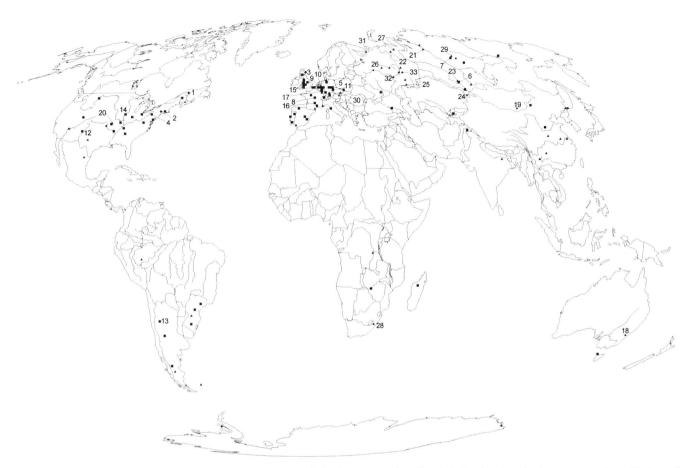

Fig. 3 Palaeozoic insect fossil sites: (crosses are Devonian, squares – Carboniferous, triangles – Permian). Numbered are sites/site groups catalogued in Appendix (Chapter 4), as follows. Devonian: 1 – Gaspé, 2 – Gilboa, 3 – Rhynie. Carboniferous: 4 – Allegheny Series, 5 – Bitterfeld/Delitz, 6 – Cheremichkino, Zavyalovo, Zheltyi Yar etc., 7 – Chunya, 8 – Commentry, 9 – Gulpen, 10 – Hagen-Vorhalle, 11 – Horní Suchà and Vrapice, 12 – Madera Formation, 13 – Malanzán Formation, 14 – Mazon Creek, 15 – Middle Coal Measure, 16 – Montceau-Les-Mines, 17 – Saar Basin. Permian: 18 – Belmont, 19 – Bor-Tologoy, 20 – Elmo, 21 – Fatyanikha, 22 – Inta Formation, 23 – Kaltan, Erunakovo Formation, Ilyinskaya Formation etc., 24 – Karaungir, 25 – Kargala mines, 26 – Kityak, 27 – Lek-Vorkuta Formation, 28 – Middle Beaufort Series, 29 – Nizhnyaya Tunguska, 30 – Obora, 31 – Soyana, 32 – Tikhiye Gory, 33 – Tshekarda. Because of small scale, this and the following maps often show groups of localities rather than individual fossil sites. Compiled by K.Yu. Eskov

The first and most important problem is that even in theory, good correlation between a cladogram and the total balance of similarities and dissimilarities is possible only under the condition of essentially equal evolutionary rates, which is not generally held. For example, the birds are usually considered to form a class of their own, in spite of the fact that they are monophyletic with crocodiles (in terms of living reptiles) because they have been deviating (at least morphologically and physiologically) from the common ancestor of both at a much greater rate in comparison with the crocodiles.

Another problem arises when we appreciate that the cladistic system (that is one that is isomorphic with the cladogram) can be constructed only from taxa whose origin can be justified by the possession of at least one synapomorphy. In fact we could consider synapomorphy as ultimate evidence that a taxon has appeared, and a divergence event as a heuristic method permitting us to locate the proper place of the synapomorphy on the cladogram. Alternatively, we can use synapomorphy as a heuristic method to determine the sequence of past divergence events so forming a cladogram. Irrespective of this dichotomy, considerable uncertainty results from the absence of a precise correlation between the divergence event and the gain of a defining synapomorphy. As discussed above, one can take place without the other: a population of an ancestor can evolve into another species due to the

Fig. 2 Geochronological chart of the insect range of during Earth history (compiled by V.Yu. Dmitriev). Columns are the geochronological units of successively lower ranks, as indicated in the upper row, except for the last column which shows the dates of the respective boundaries in million years. The Russian geochronological scale is taken as the basis of the chart because palaeoentomological information using this version of the scale is the particularly abundant. The West European scale is added for the Carboniferous where the discrepancy is particularly deep. Notes: [1] epoch subdivisions are abbreviated as follows: E – Early, M – Middle, L – Late (equivalent to the stratigraphical subdivisions L – Lower, M – Middle, U – Upper). [2] after ODIN (1994). [3] Quaternary. [4] Neocome is taken here including the Barremian (not a universal practice); in Asia, two large insect assemblages are often described as originating from the early and later parts of the Neocomian, although the position of their borderline is unknown: correlation of this boundary with that of Valanginian and Hauterivian is quite arbitrary

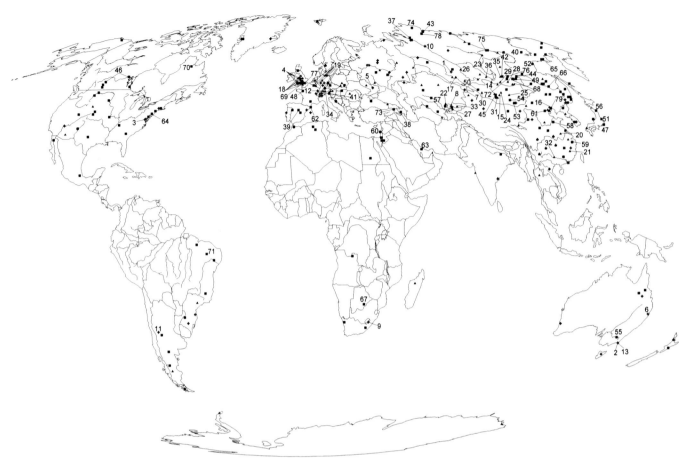

Fig. 4 Mesozoic insect fossil sites: (crosses are Triassic, triangles – Jurassic, squares – Cretaceous). Numbered are sites/site groups catalogued in Appendix (Chapter 4), as follows. Triassic: 1 – Babiy Kamen', 2 – Brookvale, 3 – Cow Branch, 4 – English Rhaetian, 5 – Garazhovka, 6 – Ipswich, 7 – Kenderlyk, 8 – Madygen Formation, 9 – Molteno Formation, 10 – Pelyatka, 11 – Potrerillos, 12 – Vosges, 13 – Wianamatta. Jurassic: 14 – Bakhar, 15 – Bayan-Teg, 16 – Beipiao, 17 – Chernyi Etap etc., 18 – English Lias, 19 – German Lias, 20 – Hansan Formation, 21 – Jiaoshang Formation, 22 – Karatau, 23 – Kempendyay, 24 – Khotont, 25 – Khoutiyn-Khotgor, 26 – Kubekovo, 27 – Kyzyl-Kiya, Sagul, Shurab, 28 – Mogzon, 29 – Novospasskoye, 30 – Oshin-Boro-Udzur-Ula, Zhargalant, 31 – Shar Teg, 32 – Shiti, 33 – Sogyuty, 34 – Solnhofen etc., 35 – Uda, 36 – Ust-Baley, Iya etc. Cretaceous: 37 – Agapa, 38 – Agdzhakend, 39 – Álava, 40 – Arkagala, 41 – Austrian amber, 42 – Baissa, 43 – Begichev Formation, 44 – Bolboy etc., 45 – Bon-Tsagan, Khurilt etc., 46 – Canada amber, 47 – Choshi, 48 – French amber, 49 – Glushkovo Formation, 50 – Gurvan-Eren Formation, 51 – Iwaki, 52 – Khetana, 53 – Khutel Khara, 54 – Khutuliyn, 55 – Koonwarra, 56 – Kuji, 57 – Kzyl-Zhar, 58 – Laiyang Formation, 59 – Laocun, 60 – Lebanese amber, 61 – Lushangfeng Formation, 62 – Montsec, 63 – Near East amber (deep sea sample), 64 – New Jersey amber, 65 – Obeshchayushchiy, 66 – Obluchye, 67 – Orapa, 68 – Polovaya, 69 – Purbeck, 70 – Redmond, 71 – Santana, 72 – Semyon, 73 – Shavarshavan, 74 – Taimyr Lake retinites, 75 – Timmerdyakh, 76 – Turga, 77 – Wealden, 78 – Yantardakh, 79 – Yixan Formation. Compiled by K.Yu. Eskov

gain of different synapomorphies, while the residual population remains essentially unaffected. And conversely, a synapomorphy can be gained in the course of so called phyletic evolution, that is without any divergence events. What is worse, there is no evolutionary mechanism known that would make any of the above processes rare. We should consider such cases as fairly common, and yet each of them makes impossible precise reconstruction of the cladogram and, as a result, the delimitation of the respective taxa.

One more problem with the cladistic taxon deserves mention. It is equally rooted in that the taxon can be legitimised by either a gained synapomorphy, or by the occurrence of divergence: both sorts of event are important but only by their absence or presence, not by any of their other features. That is why any synapomorphy and any divergence event is of equal value in the ranking of taxa. Therefore subordinated ranks must either be as numerous as the number of succeeding

divergence events (conservatively counting a divergence event each million years gives hundreds of subordinate ranks), or such ranking should be abandoned and replaced by simple numbering (cf. HENNIG 1981). Alternatively, taxonomic decisions can be arbitrary. Consequently, the claimed consistency of cladistic taxonomy is delusory, and its other advantages are equally problematic, thus permitting a chance for other taxonomic principles.

One alternative to the **cladistic** approach is the **phenetic** one which claims superiority because the classification relies on weighting of the raw similarity (for details see SOKAL & SNEATH 1973). This approach has its own advantages and deficiencies. At present it has been practically abandoned and can be recommended only as an auxiliary, albeit useful, tool of taxonomic study.

Another alternative was termed **phylistic**. It is similar to traditional taxonomy but can be explicated as follows (RASNITSYN 1996). Both

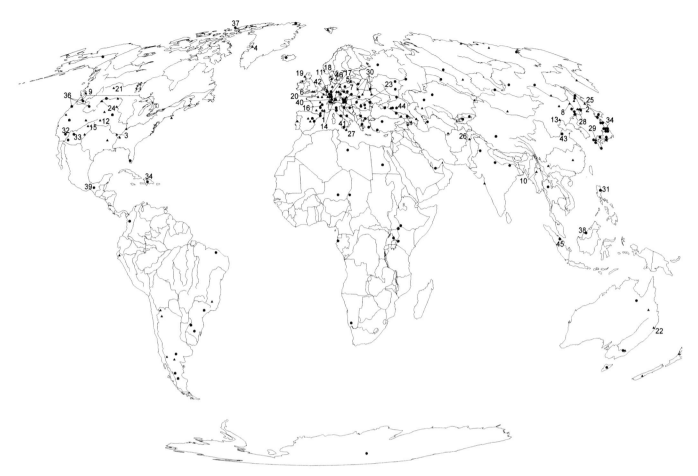

Fig. 5 Cainozoic insect fossil sites: (triangles – Palaeogene, circles – Neogene). Numbered are sites/site groups catalogued in Appendix (Chapter 4), as follows. Palaeogene: 1 – Aix-en-Provence, 2 – Amgu, 3 – Arkansas amber, 4 – Atanikerdluk, 5 – Baltic amber, 6 – Bembridge Marl, 7 – Bitterfeld amber, 8 – Bol'shaya Svetlovodnaya, 9 – British Columbia amber, 10 – Burmese amber, 11 – Eckfelder Maar, 12 – Florissant, 13 – Fushun, 14 – Geiseltal, 15 – Green River, 16 – Menat, 17 – Messel, 18 – Mo Clay, 19 – Mull Is., 20 – Oise amber, 21 – Pascapoo Formation, 22 – Redbank Plain, 23 – Romanian amber, 24 – Ruby River, 25 – Sakhalin amber, 26 – Salt Range, 27 – Sicilian amber, 28 – Tadushi, 29 – Ube, 30 – Ukrainian amber. Neogene: 31 – Bandoc, 32 – Barstow, 33 – Bonner Quarry, 34 – Dominican amber, 35 – Kamiwada, 36 – Latah, 37 – Meigen I., 38 – Merit-Pila, 39 – Mexican amber, 40 – Oeningen, 41 – Radoboj, 42 – Rott, 43 – Shanwang, 44 – Stavropol, 45 – Sumatra resin, 46 – Willershausen. Compiled by K.Yu. Eskov

similarity and relatedness are involved in a taxonomic system's organising principles, but their application is not arbitrary. In fact, the total balance of similarities and differences in all, even unavailable, characters is considered as the ultimate but directly unattainable goal to be reflected in the system. Raw similarity is used there as a heuristic method to construct (delimit) taxa while relatedness (monophyly) works as a method to test the resulting taxon. It is believed that the history (monophyly) is at least as suggestive concerning the ultimate goal of taxonomy as is raw similarity, and therefore, in case of contradiction (for example, when a taxon seems to be polyphyletic), it is recommended to re-test the whole case on a wider basis (using more material and deeper analysis). Technically, the taxon is termed a **monophyletic continuum**, with monophyly defined as above (i.e. holophyly + paraphyly) while the continuum can be understood as a chain of taxa, either simple or branching, with every neighbouring pair of links being more similar to each other than to members of other continua. Thus taxa can be delimited by tracing the area of the least similarity, that is along hiati. Metaphorically, a taxon is a cloud of characters in a space, with the cloud's integrity being of primary importance, and the particular

characters of secondarily so ("Scias Characterem non constituere Genus, sed Genus Characterem", that is "Know, the character does not constitute the genus, but the genus the character"; Linné 1751, #169).

1.1.3. NOMENCLATURE

Since Rohdendorf's proposal (Rohdendorf 1977, Rasnitsyn 1982), the thorough typification of all insect taxa including those of the highest rank is possibly one of the most striking features of the Moscow palaeoentomological school, and this practice is generally retained here. There are several reasons for this (Rasnitsyn 1996). The first is the old, but still valid, observation that a consistent order in the formation of taxonomic names makes the work of taxonomists easier. This applies equally to names of any rank, and the urgency for progress in the unification of higher taxon names can be still be appreciated even by entomologists (e.g. Boudreaux 1979), not to mention other zoologists (e.g. Starobogatov 1991), and particularly by botanists who have even changed their Code respectively (ICBN 1980). The second reason is based on another observation, namely that typification works well when applied to taxa of lower rank, and therefore should also be of help in respect to higher taxa.

Both of these reasons are clear and have been well known for centuries, indeed consistent typification of names of insect orders was first proposed 220 years ago; LAICHARTING (1781). Nevertheless, resistance to the proposed changes is strong, because a considerable cost would have to be paid in exchange for the advantages promised by comprehensive typification. The cost is the necessity to abandon many customary descriptive names and to learn new, typified names, and this will concern many more people than do routine nomenclatural changes that typically affect only tens of, and rarely hundreds of, taxonomists. Nevertheless, the cost is worth the price, and it will be paid sooner or later, for the cost of refusal is also high, and additionally is accumulating, with a consequential detrimental effect on taxonomic practice. This is because, contrary to the ICZN (p. xiii), nomenclature is not neutral in respect to taxonomy.

The present book is not the proper place to discuss details of the problem, for which see RASNITSYN (1996). Only those aspects that seem more important concerning the issue of the typification of higher rank names will be briefly considered here.

"The name-bearing type provides the objective standard of reference by which the application of the name it bears is determined, no matter how the boundaries of the taxon may change" (ICZN #61a). The type principle is thus a method of introducing a taxon into the system. Indeed, although in practice we introduce it by means of a taxon diagnosis rather than by direct comparison with the type, the diagnosis must always agree with the characters of the type, and in cases of uncertainty only such comparison can finally resolve the issue. At the same time, application of the type principle is a rather sophisticated job, though the principle is nevertheless universally applied to lower ranked taxa, thus indicating that other possible ways to introduce a taxon into the system are not efficient. This is because the taxon-continuum (see above), being a cloud of subordinate taxa with highly unstable boundaries and content, cannot be effectively worked with unless its name is pinned by a type as a tag.

When we refuse to typify higher rank taxa, we should imply that it is because they either differ cardinally from that of lower rank and thus can be introduced in the system in another way, or they are so unimportant that any rigorous system of formation and application of their names is superfluous. The first alternative does not seem real, for there is no essential differences between lower and higher rank taxa that has been reported yet. Higher rank taxa are not particularly well characterised to be introduced just by referring to their characters. Nor do they exhibit especially high integrity which could permit us to consider them, unlike lower ranked taxa, as true individuals: this would make it possible to introduce the taxon into the system referring to its untypified (attributed to the individual as a whole) name.

As a result the inference appears inescapable that the common resistance to the thorough typification of taxonomic names in zoology rests on the belief, probably unconscious, that the names of the higher ranked taxa are of only secondary importance, and therefore that the problems borne by arbitrariness in their application may be neglected comparing the labour which should be otherwise spent on learning new names. This myopic arrogance in respect to higher taxa is rather natural as a result of ongoing specialisation of taxonomists for lesser and lesser ranked taxa, coupled with the lowering prestige of taxonomy itself whose needs and demands meet the decreasing respect of other scientists and laymen. And yet such an approach is treating taxonomy itself, because the longer the disregard of higher level taxonomy lasts, the deeper will be its disorganising effect on lower level taxonomy. It is our appreciation that this

danger is a real one which has forced us (the Moscow school) to continue using typified names of higher taxa, in spite of understanding that this will hardly increase the rating of this book, at least in short term.

In spite of a full understanding of the ultimate advantages of thorough typification, the painful proximate disadvantages of our hard adherence to its ultimate goal have forced us to soften our attitude in a way approved by botanists. They recently made exception for a short list of traditional, non-typified names that can be legitimately used, along with the typified ones, for a while at least (e.g. Umbelliferae together with Apiaceae) (ICBN 1980). Similarly it was agreed among the contributors of the present volume, that several non-typified names should be permitted to be used herein, viz. Insecta, Odonata, Hemiptera, Coleoptera, Neuroptera, Trichoptera, Lepidoptera, Diptera, Hymenoptera and Orthoptera. All other names are left typified (see Fig. 1).

Another topic worth discussing in respect to nomenclature is the use of parataxa. This problem is not unique to palaeontology but is particularly important therein. Parataxa are those groups created in direct violation of particular, taxonomy-dependent principles of nomenclature, as discussed in detail by RASNITSYN (1996). The violation is inevitable when we have to classify insufficiently known organisms – whether it is due to their incomplete preservation, as is normal in palaeontology, the stage of development, the life cycle or the sex at hand, bears no taxonomically important characters (as in some fungi and parasitic worms, as well as in some insects with pronounced sexual dimorphism).

The following kinds of parataxa have been proposed: taxon *incertae sedis*, form taxon, and collective taxon. A taxon *incertae sedis* is a group that is properly housed at some taxonomic levels but not at some others. For instance, a genus *incertae sedis* is that for which only the encompassing order and not the family is identified as yet. The form taxon is created for material with a particular type of taxonomic deficiency: it can be treated as a seemingly normal taxon when compared to other formal taxa of the same kind, e.g. genera of 'Fungi Imperfecti' (a group of fungi with the sexual stage of the life cycle unknown) or form taxa of the detached beetle elytra (Chapter 2.2.1.3.2.1). However, they cannot be properly compared with normal taxa (orthotaxa). The weakest form of the parataxon is the collective group which is a collection of species differing from each other and possible to be assigned to an orthotaxon of any rank, but which are impossible to be either distributed among its subtaxa or to be organised in a system of parataxa of their own. For example, *Carabilarva* Ponomarenko is the assemblage of fossil beetle larvae that can be reasonably attributed to the family Carabidae but not to any of its particular subtaxa (RASNITSYN 1996). In fact, a collective name is just a generic name for species *incertae sedis*.

1.2
Special Features of the Study of Fossil Insects

A.P. RASNITSYN

When working with fossils, palaeoentomologists are faced with problems and difficulties that are rarely, if ever, completely unknown to neontologists. However, the routine problems are essentially different in the two fields. Although differences in the material they are working and how it is handled are the most apparent, it is the differences in the way the observed results are interpreted that are of primary importance.

(a) Handling fossils. Palaeoentomologists not only deal with insects that died a long time ago but they are also enclosed, at least partly, in some medium (usually a rock matrix or fossil resin), and so they are not free. Exceptions are rare and mostly concern Quaternary fossils (for which see ELIAS 1994). Inclusions in fossil resin are perhaps more analogous with living insects enclosed in a special medium like the Canada balsam. However, the preparation methods for fossils (grinding and polishing) are quite specific: descriptions and useful references can be found in GRIMALDI (1993). No universally standard methods have as yet been developed in palaeoentomology, and those that have been proposed reflect personal experience, at least to an extent. It is no wonder therefore that material is treated in a slightly different manner depending on both the students and their institution. For example, for less polymerised, and therefore fragile resins (those other than true amber, e.g., the retinite common in the Cretaceous deposits), it is usual that resin pieces are enclosed permanently in a medium with appropriate optical properties, either liquid (e.g., mineral oil – but strictly avoiding oil of plant origin, for it will slowly but irreversibly dissolve the resin!) or solid (e.g. Canada balsam, various plastics, etc.). These methods have their own shortcomings. Using media that hardens like balsam is always time-consuming and inappropriate for mass processing of material. Further, both liquid or hardening media can sometimes have unfavourable or even disastrous consequences for the fossil because of chemical interactions. For example, the first crop of inclusions in Cretaceous fossil resin from the North Siberia collected in 1970–71 by Moscow palaeoentomologists has been seriously damaged by use of castor oil as an immersion medium, and both HEIE (1967) and GRIMALDI et al. (1997) describe very sad consequences of embedding specimens in plastic. KLEBS (1910) recommended keeping Baltic amber specimens either embedded in modern natural resins or submerged in water with a small amount of camphor; both of these methods were used for decades without any evident alterations in the state of the amber. Light is dangerous for fossil resins because it appears to result in more rapid oxidation; accordingly, specimens exposed in museums under a strong light are most endangered. Theoretically the best conservation method, preventing resin oxidation, would be to keep it in a vacuum (or, rather in atmosphere of inert gas like nitrogen or argon), but such conditions are only used for mineralogical specimens.

One potential issue that needs to be mentioned specially is that a liquid preservative might penetrate into an inclusion via cracks of the resin, even if it does not necessarily interact with the resin chemically. Penetration of a liquid medium into the resin also modifies its optical features – sometimes to the better but often to the worse! Media with optical properties similar to those of amber, make cracks less visible and so enhance visibility. In the opposing direction, when penetrating the inclusion a medium can cancel all the surface effects thus leaving only the survived fragments of the cuticle and internal structures of the fossil visible. Sometimes this is for the better, but usually it's not because the cast of the fossil on the inner resin surface is often preserved much more complete and detailed than the real fossil tissues: this is particularly true for the wing membrane which is the first to become invisible in the affected fossils.

For all of the above reasons, and until an optimal way of handling the inclusions is developed, we have used a palliative approach at the

Arthropoda Laboratory, Palaeontological Institute RAS in Moscow since 1973, in which the main material is stored dry and only preliminarily ground to permit the most general identification of the fossil(s) inside. Only when needed for closer study is the resin piece ground and polished better, and then it is observed through a drop of thick sugar solution covered with a microscope cover slip. The commonest device is that shown by GRIMALDI (1993: figs. 2a, b), except that we usually use plasticine rather than a mixture of paraffin and mineral oil as recommended by Grimaldi, though his material may well be superior. Sugar syrup has moderately good optical properties: it is used instead of mineral oil because it hardly penetrates through the fine cracks, and can easily be washed out.

One particular danger with insects enclosed in fossil resins, particularly amber fossils, is the existence of numerous forgeries. GRIMALDI et al. (1994a) describe the main kinds and diagnostics of amber forgeries.

Rock material differs from the ordinary entomological specimens more seriously than do resin inclusions. In many cases these fossils resemble real insects flattened against the rock though sometimes as if they have been chemically treated to make them more or less transparent. Close examination, however, shows that this analogy is incomplete. The fossil is available for observation only when the rock is split and the split usually runs through the fossil or very close to it. If the fossil is only partly exposed, the overlaying rock matrix should be removed with either manual or mechanical instruments: this is a time consuming job and one that requires considerable skill. When the fossil is covered by a thin layer of rock matrix (but is still traceable by the rock relief, or can be seen through particularly thin layer of the matrix), it may be made more clearly visible by applying some liquid, usually alcohol (95 or less per cent): however, alcohol (particularly at lower concentration) must not be used when the matrix contains salts (a common feature of rocks collected in arid environments). Viewing under polarised light can be of a kind of substitute when alcohol cannot be employed.

A fossil is rarely ever split precisely along one of its external surfaces: normally it splits through the inside of the body, so we see the fossil insect preserved on both halves of the split rock piece, and so we see each half from inside. If the insect has buried in the dorsoventral position, and we can ignore any internal structures preserved in the fossil, it is more or less equivalent to observing the dorsal surface *from below* on one half of the rock, and the ventral surface *from above* on another half. So the two parts of the fossil complement each other and are not at all simple duplicates of one another (see Fig. 344), and so both have to be studied separately. The two halves of the fossil are commonly termed the obverse and reverse, or positive and negative, or the part and counterpart, in each case implying that the former term corresponds to the dorsal side of the fossil, and the latter one the ventral side. Of course, this distinction cannot be applied if the fossil is buried in a strictly lateral position. Even with dorsoventrally preserved fossils the usage is not straightforward, however, it is standard, to call the **part** (=**obverse**, **positive impression**) that half that shows the wing veins as fluting, like in the normal insect wing seen from above (that is, with vein SC concave, R convex etc., for details see Chapter 2.2c). Respectively, the half of the rock showing the veins in reverse position (like in the wing seen from below) is termed the **counterpart** (or **reverse**, or **negative impressions**). Hence the obverse fossil half (the part) is in fact the one showing the *ventral* body side (or internal structures) *from above* (in dorsal view), while the reverse fossil half (the counterpart) displays the *dorsal* body side (again unless some other internal structures) *from below* (in ventral view).

Unfortunately, additional uncertainty appears due to a tendency to call the part (obverse) the half that displays surface sculpture such as tubercles, pits, keels, etc., in their normal and not inverted form. This distinction is in any case difficult because the surface sculpture of fossils is generally a matter of interpretation. Sculpture is only seen in a natural way when the split runs strictly along the external surface of the cuticle: otherwise we can see either the internal cuticular surface with the sculpture inverted, or the cast of this internal surface either in the matrix or internal insect tissue with seemingly normal relief (not mirrored) though representing anything but the real cuticular surface. We rarely know beforehand if a particular surface has convex (tubercular, ridged) or concave (pitted, furrowed) sculpture, or both, so its interpretation primarily depends on correct interpretation of *which* side of the fossil we can see and in *what view*. That is why the first thing that we need to understand is which half of the fossil we are observing. Is it the part or the counterpart? This information then helps us to identify what is in fact surface sculpture. Contrary to the common expectation, fossils in rocks often display an excellent state of preservation (see Figs. 49 and 50) and may preserve several layers of internal structures. In these cases identification of the precise position of the split plane is particularly important because it greatly affects one's visual picture of the insect. An important discussion concerning the visual features of insect structure depending on the view and position of the split is presented by PONOMARENKO (1969) using the beetle elytron as an example.

(b) Collecting fossil insects. For insects that are directly embedded in rocks, the main collecting methods are generally the same as for other invertebrate (or fish, or plant) macrofossils, that is, splitting tons of rocks with a hammer. There is again some palaeontological speciality, though this time it is not very striking. The most promising substrates for fossil insects are the finest-grained rocks of lacustrine origin. Importantly, the huge diversity of insects, even in the past, means that very many fossils should be collected from any site to minimally characterise it. That is why, whenever possible we spend weeks and months at each fossil site, rather than just a few hours or days like other palaeontologists used to do. A very concise description of collecting methods, mainly orientated to lay people, is given (in Russian) by MARTYNOVA (1953).

Collecting fossiliferous resins is an essentially different job, even if one ignores the special field of industrial amber 'mining' (see, e.g., POINAR 1992a). Resin pieces accumulated in unconsolidated deposits (sands or, usually, lenses of fossil wood chips within sand deposits) can be collected by sifting followed by floating the residue in a saturated (or nearly so) salt solution on which the less dense resin pieces float and the fossil wood sinks (ZHERIKHIN & SUKATSHEVA 1973, PIKE 1993). The most recent detailed review of methods and techniques for collecting fossil insects, and their preparation, conservation and study is given by PEÑALVER (1996). Insect extraction from unconsolidated Quaternary sediments requires special techniques as have been described by I. WALKER & PATERSON (1985), COOPE (1986), BIDASHKO (1987), and ELIAS (1994).

(c) Interpretation of observations. Science is the matter of interpretation, not just observation. Even the first and most elementary step in handling the observational results is purely interpretative, when we try to distinguish the 'real' data obtained (those reflecting, not necessarily in a direct way, the features of the very object of study), and the 'mistakes' (the results primarily due to external influences, either of environmental nature, or caused by particular methods applied, or else borne by our personal idiosyncrasies, etc.). Palaeontology is a particularly interpretative field compared with neontology because the

possible sources of mistakes are much more numerous and diverse, and also because the observable features are more limited, and their generalisation involves more hypothetical elements. Naturally, it is vitally important for the palaeontologist to be aware of both sources of mistakes and of the rules that help produce safer and more reliable hypotheses.

Palaeontology has its own theory of mistakes called taphonomy: its application to studying fossil insects is discussed at length below (Chapter 1.4). Considered here in a cursory way are some problems specific to the generalisation of the palaeoentomological observations.

When dealing with living insects, an entomologist has, at least in principle, the possibility of studying every aspect of their structure, function, and relationships. In fact this possibility is only potential for the absolute majority of insect species, and the more so for infraspecific entities. That is why all our generalisations are hypothetical to an extent even concerning the extant insects. When dealing with fossils, we also sometimes have a possibility to see fine structures available at the electron microscope level, or to study the food of long extinct insects, or to observe their developmental stages, as is exemplified in the following pages. Unfortunately, these are particularly uncommon occasions, at least at the present state of knowledge in palaeoentomology. Generally speaking, the differences between palaeontology and neontology are each quantitative rather than qualitative, but in their totality they do indeed change the quality.

Palaeoentomological inferences are primarily based on the following types of primary data: gross insect morphology permitting both comparative and functional interpretation, the co-occurrence of fossil taxa (including their numerical relationships), and the available geological data indicative of the landscape, biotope, climate and sometimes even of the seasonal environment. Examples of this sort are scattered throughout this whole book, particularly in the chapters discussing insect taphonomy (Chapter 1.4) and ecology (Chapters 3.2 and 3.3). That is why only some more general features are briefly discussed below, as well as a few important ones that are not considered elsewhere in sufficient detail.

The fossil record presents us with an highly generalised and biased picture of past life and environments which lacks many important details and entire fragments (Chapter 1.4). We have practically no chance of discovering rare organic forms (either taxa or intraspecific forms or parts), so what we can see are all typical in some sense of a particular spot in time and space. When we have a minimal amount of material, it gives us a picture averaged over a long time interval (hundreds and thousands of years) and not a momentary portrait of the local population. So the chance to make any inference from our data concerning short duration environmental deviations (ranging from days to tens of years) or of population oscillations, is negligible. This makes palaeontological observations less detailed but more reliable. Another important aspect of palaeoentomology is that its observations are limited in many details at its lowermost taxonomically discernible level, and this is high by neontological standards. Insect fossils rarely demonstrate diagnostic features commonly used to distinguish extant species, such as fine genitalic structures, details of the colour pattern, not to mention the sound production, identity of the host animal or plant, specific seasonal or circadian activity time, and other characters relevant to taxonomy of some living groups. As a result, what we call species would in many cases be species groups or even genera by neontological standards. Of course, neontological practice is not guaranteed against this sort of mistake either, for many, maybe even most, living species as defined by taxonomists are probably actually bunches of sibling species!

The reverse danger is equally real in both palaeontology and neontology, that is the possibility of mistaking various intraspecific forms (developmental stages, opposite sexes, alternating generations, seasonally or permanently co-occurring genotypically or just phenotypically different forms) as belonging to different taxa. When dealing with living insects we can occasionally or intentionally observe the transformation of a caterpillar into a pupa and further into a butterfly, or catch a pair *in copula*, or study the progeny of particular parents in breeding experiments. Palaeoentomologists either lack these possibilities, or they have them only exceptionally rarely (except for some groups, mainly within dipterans, which are regularly found buried *in copula* in fossil resins). That is why our inferences on the conspecificity of morphologically different fossils are all indirect, based on their co-occurrence, and appear only when we anticipate the possibility of their conspecificity (no such idea would ever appear for example with mayfly nymphs and adults unless we were aware of the life cycle of living mayflies).

Most fossil insects lived in environments different from those they become buried in, so the fact that their fossils co-occur in a fossil site does not necessarily imply their co-occurrence in life. Lacustrine insects are apparently an exception, some of them have been identified as immatures and adults of one and the same species, and no wonder that one of the best developed criteria of conspecificity in the insect fossil record concerns developmental stages of such lake-dwelling insects. These are most explicitly formulated by SINITSHENKOVA (1987: 86–87) who singles out the following three criteria: (i) co-occurrence of the fossils, (ii) taxonomic compatibility (the taxonomic characters available indicate both to belong to one and the same higher taxon and not to different subtaxa), (iii) compatible body size. It should be stressed here that these criteria are not to be taken as hard rules, at risk of making serious mistakes. Rather, these statements are to be considered as agreeing with the above (Chapter 1.1c) definition of presumption, that is, they are to be held true until and unless any serious contrary arguments appear. With proper care, the same or similar rules can be probably applied successfully to infer conspecificity of different sexes and, possibly, other morphologically distinct forms. However, accumulated experience does not seem rich enough to further develop the above recommendations and to detail their application to various particular situations. For the present, the case of immature and adult lacustrine insects is paradigmatic for identification of conspecificity, and the examples of this sort look the best explaining the rules (presumptions), their application and limitations. One of such example is presented in more detail below.

PRITYKINA (1977) described a new dragonfly, *Hemeroscopus baissicus* Pritykina (Hemeroscopidae, see Fig. 98), based on thousands of nymphs and hundreds of adults from the Early Cretaceous locality Baissa in Transbaikalia. The nymphs and adults are well matching in size and co-occurred perfectly. Indeed, Zherikhin (unpublished) has later estimated density (as the number of specimens per m²) of the dominant insect groups in 28 succeeding beds in the Baissa section, and found that in 12 of these test samples at least two *H. baissicus* individuals were present. Of these 12 samples, nymphs and adults are found together in 11 samples, what means the probability of their occasional association $P = 6.0 \times 10^{-7}$ (correlation coefficient $r = 0.92$; $\chi^2 = 23.5$). PRITYKINA (1977) has identified only 6 other dragonfly species in the Baissa assemblage, all are very rare (known from a total of just 7 specimens altogether), including a single wing of a member of Aeschnidiidae (the family also involved in the case, see below) which is not compatible with the nymphal *H. baissicus* because of its size

(estimated hind wing length 40 mm as compared to the body length of 46–54 mm of the older nymphs (not counting the paraprocts)). Pritykina has drawn attention to the unusual combination of adult and immature traits in this species, and considered this as sufficient reason to establish the new family Hemeroscopidae. Quite similar was the case of another dragonfly, *Sona nectes* Pritykina (Sonidae), described by the same student from the Early Cretaceous Gurvan-Eren Formation in West Mongolia (PRITYKINA 1986). The only difference is that *S. nectes* is described based on considerable, though far from huge, material (17 adults and almost 300 nymphs) and that the only co-occurring dragonfly species, a libelluloid *Eocordulia cretacea* Pritykina, is not only rare (three fragmentary wings found), but also is clearly taxonomically incompatible.

Contrary to the above inferences, BECHLY *et al.* (1998) have attributed the adult *H. baissicus* to Hemeroscopidae placed basally in the libelluloid clade of dragonflies, adult *S. nectes* to Proterogomphidae (gomphoid clade), and both immature *S. nectes* and *H. baissicus* to Sonidae (basal in the advanced Libellulina and supposedly a synonym of the Aeschnidiidae). These conclusions are based solely on phylogenetic hypotheses which are incompatible with the combined nymphal and adult character sets of Pritykina's species (for further discussion see Chapter 2.2.1.1.2e). As to the evidence derived by Pritykina from the co-occurrence of nymphal and adult fossils, BECHLY *et al.* (1998) disregards them by reference to a Turkish Oligocene locality where one libellulid species is found as 3 nymphs and 50 adults, and one lestid species is represented by one wing and more than 200 nymphs. This case does indeed exemplify that the co-occurrence criterion is not at all easy and straightforward in use. However, the Turkish example is appreciably different from the East Asian one in that both the morphological and ecological differences between the dragonflies (libelluloids in Turkey and all East Asian species) and damselflies (lestids) are much more deep and reliable than between the relevant dragonfly taxa. The close association of the Turkey immatures and adults displaying characters of either dragon- or damselflies is therefore beyond doubt, so their contradictory numerical ratios should be ascribed to unknown details of their biotopic preferences. The evidence is quite the reverse for the East Asian fossils: both morphology and, by inference and by analogy with living dragonflies, ecology are much more similar, implying more similar taphonomic properties and hence more indicative co-occurrence of the adult and immature fossils. That is why the Turkish example can hardly affect the strength of the co-occurrence evidence, even in the case of *S. nectes*, and decidedly cannot cast any serious doubt on the interpretation of *H. baissicus* which is additionally supported by the huge amount of material involved.

While the above discussion illustrates some of the problems that palaeoentomologists meet with in their routine work, it far from exhausts them. Indeed, all the above cases call for further taphonomic analysis of the assemblages involved. The nymphs of both *H. baissicus* and *S. nectes* clearly developed in the same water body where they were buried, as their perfect preservation state and the presence of succeeding nymphal instars implies. They should consequently have produced adults that dwelled nearby and so become buried there not only in number, but also in good condition (i.e. not much worn as might be caused by long transport towards a remote burial site). In contrast, those species known solely as adults may have developed, at least in part, in remote water bodies and so arrived more or less worn, thus opening yet another way to analyse the fossil data (for which see Chapter 1.4). Other approaches cannot be excluded either, for any co-ordination in the characters known for the dragonfly nymphs and adults (e.g., rooted in their common ecological preferences, functional or purely developmental correlation) can be of much help in the examination of their relationships. The same is true whenever dealing with any fossils where there is a suspicion that different individuals represent different forms of one and the same species.

1.3
Concise History of Palaeoentomology

A.P. RASNITSYN

Although insects entombed in Baltic amber were known to people since antiquity, scientific study of fossil insects began comparatively late. The first research using binomial names was published a year after Linné's death (BLOCH 1779; Fig. 6) and dealt with insects included in the copal (relatively young, at most a million years old, fossil resin). The first

description of a mass of fossil material occurred still some sixty years later (GERMAR 1837, 1839). It can be taken with better reasons as the starting point of the history of palaeoentomology.

During more than a century and a half of the history of palaeoentomology, not only have the names of students changed, but so too their favourite styles and the aims of their research. The first and longest was the faunistic stage, when a set of various fossil insects collected from a site was the usual unit of research material. The students described these fossils without any trouble even though they usually belonged to quite different orders: they felt themselves as universal in palaeoentomology despite being specialists when working as neoentomologists. For example, Germar (Fig. 7) himself clearly preferred beetles when dealing with living insects, and yet he described practically any fossil insects. In fact, it was considered as possible, and even normal, to describe fossil insects while not even being an entomologist. There were even quite a few geologists, especially palaeobotanists, among the authors of palaeoentomological papers, both in the 19th and 20th centuries (e.g. Oswald Heer, Fig. 8, Christoph Giebel, Fig. 9, Friedrich Goldenberg, Charles Brongniart, Fig. 10, Paul Guthörl, Mikhail Zalessky etc.).

Publications by the universal palaeoentomologists were intended mostly to reveal the insect world of the past. Also they tried to correlate deposits and to reconstruct the past environments using insects, or, especially in the post-Darwinian time, to improve ideas about insect

Fig. 6 Marcus Elieser Bloch (1723–1799) (courtesy Deitsches Entomologisches Institut Eberswalde, Germany)

Fig. 7 Ernst Friedrich Germar (1786–1859), professor of mineralogy in Halle, Germany (courtesy Deitsches Entomologisches Institut Eberswalde, Germany)

Fig. 8 Oswald Heer (1809–1883), professor in Zurich, Switzerland (courtesy Deitsches Entomologisches Institut Eberswalde, Germany)

Fig. 9 Christoph Gotfried Andreas Giebel (1820–1881), professor in Halle, Germany (courtesy Deitsches Entomologisches Institut Eberswalde, Germany)

phylogeny. These aims seemed to be of subordinate importance for most students. Rather they were simply interested in inquiring into a wide direction of study, coupled with the lack of specialisation of students in particular insect groups, has resulted in rather superficial or even incorrect descriptions and drawings in many publications which not uncommonly fail to meet current taxonomic standards. Nevertheless it was quite a large piece of work which was carried out at the first stage of palaeoentomology.

The Non-Professional palaeoentomologist period came to an end and was replaced, quite logically, by the Time of the *Great Universals*, who summarised the results accumulated before them, and who made inescapable the occurrence of the next stage in the development of palaeoentomology. The first of these Great Universals was Samuel Hubbard Scudder (1837–1911; Fig. 11), based at Boston, USA, and who was also known as a prominent orthopterist and lepidopterist. He described many North American fossil insects and thus laid the foundations of palaeoentomology in the New World. Further, he catalogued all known fossil insects and the respective literature (SCUDDER 1890a, 1891), creating the necessary basis for the prolific activity of the another Great Universal, a Viennese entomologist, Anton Handlirsch (1865–1935; Fig. 12). He was well known for his taxonomic accounts on living sphecid wasps, but he later shifted to the field of palaeoentomology. His voluminous compendium of all hitherto known fossil insects (HANDLIRSCH 1906–1908, 1937, 1939) was a milestone in the field and, albeit outdated, it still rests as one of most frequently used and cited palaeoentomological publications. His descriptions of fossil insects have not been forgotten either (although they are not easy to use

now because at least in part they tend to be vague or even incorrect); his review chapters on insect palaeontology, taxonomy, etc. in the famous KÜKENTAL's and SCHRÖDER's 'grossbuchs' are also still in use.

Two younger Great Universals were Andrey Vassilievich Martynov (1879–1938; Fig. 13) from Leningrad/St Petersburg and Moscow, Russia, and Frank Morton Carpenter (1902–1994; Fig. 14) from Harvard University, Cambridge, Massachusetts, USA. The first began his career as a specialist in caddisflies and amphipod crustaceans, but he was also interested in the taxonomy and phylogeny of insects as a whole. He has proposed a classification system for the Insecta (MARTYNOV 1925a, 1938c), which was generally accepted worldwide for several decades and still gains wide appreciation. He felt the necessity to test it palaeontologically, and this was the primary cause of Martynov's turn towards the study of fossil insects. Frank M. Carpenter was initially a specialist in Neuroptera and also became more interested in studying fossils than living insects. His long and prolific activity in the field has been crowned with another of the great milestones in palaeoentomology, his two volumes on insects in 'Treatise on Invertebrate Palaeontology' (F. CARPENTER 1992a).

The main merit of both Martynov and Carpenter was that they established new and much raised standards in describing and illustrating fossils. Illustrations are being emphasised here because in palaeoentomology, the drawings and photographs generally play a more important role than the descriptions themselves. Especially reliable were Carpenter's wing drawings. Because of his prolific and very long activity in this science, Carpenter's style became the paradigm.

Fig. 10   Charles Brongniart (1859–1899), entomologist in Paris, France (courtesy Deitsches Entomologisches Institut Eberswalde, Germany)

Fig. 11   Samuel Hubbard Scudder (1837–1911), entomologist in Cambridge, USA (courtesy Frank M Carpenter)

Fig. 12   Anton Handlirsch (1865–1935), professor in Wien, Austria (from HANDLIRSCH 1937)

Martynov's descriptions and figures were also changing gradually towards being more accurate in comparison with his predecessors, especially in his later years, and this influenced other palaeoentomologists, especially his students. His main merit, however, was different: it was he who realised deeply that, similar to the study of the living insects, universalism in palaeoentomology was not the best strategy, because this did not provide a sufficiently high level of study. Simple specialisation of lone palaeoentomologists on a particular insect group could hardly improve the situation, however. Because of the still very low level of knowledge of the past insect worlds, such specialists would regularly encounter enigmatic fossils beyond their regular domain and so he/she would fail to overcome the temptation to describe them. Thus, effective specialisation is possible only within a group of palaeoentomologists, and Martynov undertook efforts to organise such a group. He left his beloved Leningrad and Zoological Institute, USSR Academy of Sciences, to join the palaeontological Institute in Moscow, also USSR Academy of Sciences, where the Arthropoda laboratory had been established in January, 1936. Sadly, he was happy to lead the laboratory for only fourteen months before his death, and twenty long years after this, the team there, headed by Boris Borissovich Rohdendorf (1904–1977; Fig. 15), remained quite small, including only three students. However,

Fig. 14 Frank Morton Carpenter (1902–1994), professor in Cambridge, USA (courtesy Frank M. Carpenter)

Fig. 13 Andrey Vassilievich Martynov (1879–1938), professor in Leningrad and Moscow, USSR (by Streblov) (LEONOVA & ROZANOV 2000)

since 1955 the laboratory began growing, and eventually comprised up to 12 palaeoentomologists in the staff during the mid-nineteen sixties, subsequently stabilising at more or less this level.

It was natural that faunistically and stratigraphically oriented research persisted in palaeoentomology in and after the time of the Great Universal palaeoentomologists had begun. This style of work could even dominate, as can be exemplified by its characteristic representatives. Among them were an especially important Australian student, Robin John Tillyard (1881–1937; Fig. 16), who developed much descriptive work on the Late Triassic insects of Queensland and on Early Permian ones from Elmo, Kansas, and the legendary Theodore Dru Alison Cockerell (1866–1948; Fig. 17), an author of almost four thousand (indeed, not four hundred!) publications, many of which include at best average quality descriptions of hundreds of fossil insects of almost any age and provenance. Paul Guthörl, also warrants mention, as he devoted tens of publications to insects and other fossils from the Permian and Carboniferous of the Saar Coal Basin.

Nevertheless the general situation has changed considerably, for a palaeoentomologist who specialises in the study of a particular insect group becomes perfectly predisposed for its phylogenetic research. Phylogenetically oriented and palaeontologically based publications

appeared already during the nineteen thirties (ZEUNER 1939) and then slowly accumulated (ROHDENDORF 1946, MARTYNOVA 1948), and since the mid-sixties, the Moscow palaeoentomological team (with its associates) alone has published more than a dozen such books (ROHDENDORF 1964, SHAROV 1968, PONOMARENKO 1969, 1988a, RASNITSYN 1969, 1980, POPOV 1971, ROHDENDORF & RASNITSYN 1980, SUKATSHEVA 1982, SINITSHENKOVA 1987, NOVOKSHONOV 1997a, STOROZHENKO 1998) and a lot more articles. This work, coupled with the growing popularity of phylogenetics after the rediscovery of the Hennig's phylogenetic systematics by an English-reading audience (HENNIG 1966), has stimulated other students to contribute to insect phylogeny using fossils (for example, the abundant publications by Jarmila Kukalová-Peck, André Nel and others). Eventually this completely changed the appearance of palaeoentomology between 1960 and 2000, thus delimiting a period that can be identified as its 'Phylogenetic Stage'.

Although phylogenetic as well as purely descriptive publications still dominate the last several decades, a new direction of research has manifested itself during the nineteen seventies and eighties, this being the integrated analysis of the composition, ecology, biocoenology, biogeography and dynamics of entire insect assemblages of various areas and ages. These varied considerably in their scope, ranging from studies of local assemblages from a particular fossil site (KALUGINA 1980a) to regional (RASNITSYN 1985a,b, 1986b) and global (ZHERIKHIN 1978, 1980a, KALUGINA 1980b) works. Another interesting form of wide-scale palaeoecological and palaeobiocoenetical research using fossil insects are studies of particular interorganismic interactions through

Fig. 15 Boris Borissovich Rohdendorf (1904–1977), professor in Moscow, USSR (LEONOVA & ROZANOV 2000)

Fig. 16 Robin John Tillyard (1881–1937), Fellow of Royal Society, Nelson in New Zealand and Canberra in Australia (E[VANS] 1946)

time and space, such as arthropod-plant interactions (HUGHES & SMART 1967, A. SCOTT 1977, A. SCOTT & TAYLOR 1983, LABANDEIRA & PHILLIPS 1992, 1996a,b, LABANDEIRA *et al.* 1994a, 1997, 2000; see Chapters 2.3.5, 3.2, 3.4 for more references). Particularly worthy of mention are two sets of studies. Firstly, analyses of pollen contents of the guts of Palaeozoic and Mesozoic insects and similar studies (KRASSILOV & RASNITSYN 1982, 1997, KRASSILOV *et al.* 1997a,b, KRASSILOV, RASNITSYN & AFONIN 1999, RASNITSYN & KRASSILOV 1996a,b, 2000), and secondly reconstructions of the historical development of insect communities populating particular media, like fresh water (WOOTTON 1972, KALUGINA 1980a) and terrestrial environments (ZHERIKHIN 1980a). Also noteworthy are numerous publications analysing the past taxonomic dynamics of insects (ZHERIKHIN 1978, DMITRIEV & ZHERIKHIN 1988, RASNITSYN 1988b, 1989a, DMITRIEV *et al.* 1994a,b, 1995, LABANDEIRA & SEPKOSKI 1993, JARZEMBOWSKI & ROSS 1996, A. ROSS *et al.* 2000, ALEKSEEV *et al.* 2001).

Fig. 17 Theodore Dru Alison Cockerell (1866–1948), professor in Boulder, USA (courtesy Frank M. Carpenter)

1.4
Pattern of Insect Burial and Conservation

†*V.V. ZHERIKHIN*

1.4.1. GENERAL

Palaeontological data are strongly biased and their interpretation is by no way simple and easy. The most important thing is not so much the fact that we have not discovered many extinct taxa yet as that we shall never find any direct evidence for existence of majority of them. If this basic incompleteness of the fossil record was incidental and irregular the very scientific value of palaeontology might be questionable. Fortunately, the biases are not random and can be estimated and so corrected for. Pattern of the fossil record constitutes the subject of the discipline called *taphonomy* (from the Greek *taphos*, a tomb) by EFREMOV (1950), and its importance for palaeontology is comparable with that of the theory of errors in statistics. Palaeontological data not evaluated taphonomically are often misleading and useless for any considerations besides a simple registration of the existence of the fossil.

The term *burial*, used here to mean removal from biological turnover, is a necessary condition for the conservation of organic remains. Normally organic production is consumed or decays within ecosystem, and burial is a relatively rare event. Its probability depends on many circumstances varying in space and time and differs strongly between different taxa in the same environment as well as between different environments for the same taxon.

Burial most often takes place in the bottom sediments of a water body. Other sedimentary environments like aeolian deposits, soils, etc., are less stable in time and much less protective. However, even after the remains have been buried they are by no means guaranteed to survive destruction. Unconsolidated sediment layers may be reworked by burrowing organisms, or they may be washed away, or blown out if the water body dries up. Later the remains may be crushed or dissolved during the sediment lithification, or the lithified rock may be metamorphosed resulting in the loss of embedded fossils. If all these threats have been avoided, fossiliferous rock will be exposed some time or other and eroded with the loss of the bulk of its fossils though some of them may be re-deposited into clastic rock masses. Finally, when an explorer is fortunate enough to discover a fossil site before its natural ruination, some remains may still get overlooked or damaged irretrievably during collecting.

The total set of these events may be described as a step-by-step sieving of primal biological information resulting in consecutive selective biases and losses at each stage. The following principal stages may be recognised (Fig. 18): 1, *community or biocoenosis* (all organisms inhabiting certain habitat); 2, *thanatocoenosis, or death assemblage* (all corpses in the habitat; the organisms predated upon or which drifted away have been removed while others may be transported in from the outside) (WASMUND 1926); 3, *taphocoenosis, or burial assemblage* (all organic remains entombed in unconsolidated sediments; the corpses consumed by scavengers, decayed, mechanically removed or destroyed have been excluded) (QUENSTEDT 1927); 4, *oryctocoenosis, or fossil*

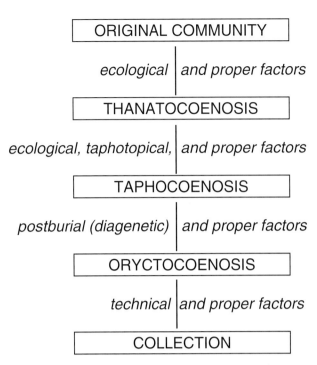

Fig. 18 The main stages and factors of generalised taphonomic process

assemblage (all fossils preserved inside the rock after its lithification; those crushed by pression or crystallisation, chemically dissolved, etc., have been lost) (EFREMOV 1950); 5, *collection* (all fossils in hands; the specimens overlooked or badly damaged during collecting have been abolished). Corresponding environments may be designated as: 1, *biotope (habitat)*, 2, *thanatotope* (death environment) (WASMUND 1926), 3, *taphotope* (burial environment); 4, *oryctotope* (fossilisation environment) and 5, *depository*. There is no sharp border between tapho- and oryctocoenosis because at the earliest diagenetic stages an oryctocoenosis remains nearly identical with the burial assemblage and gradually becomes more and more different. That is why the term "taphocoenosis" is often used in a broad sense including the early oryctocoenoses. Consequently, the line between *subfossils* (i.e. components of a taphocoenosis) and true *fossils* constituting an oryctocoenosis is purely conventional. Some kinds of remains are produced by living organisms like those parts which may be lost during the life (e.g. pollen grains and fallen leaves of plants, moult skins, hairs, feathers, etc. of animals) and the traces of activity or 'Lebensspuren' (tracks, trails, burrows, feeding damages, nests, etc.). In these cases the special terms *merocoenosis* (part assemblage) and *ichnocoenosis* (trace assemblage) are available (DAVITASHVILI 1964). The German word 'Lagerstätte' (pl. Lagerstätten) is now internationally accepted for the "rock bodies unusually rich in palaeontological information, either in quantitative or qualitative sense" (SEILACHER et al. 1985: 5). They are broadly divided into the *concentration Lagerstätten*, originated due to an accumulation of large amounts of remains, and *conservation Lagerstätten*, connected with conditions that are especially favourable for preservation (SEILACHER et al. 1985).

The factors affecting consecutive assemblages may roughly be arranged into five principal classes (ZHERIKHIN 1997a) (see Fig. 18): 1, *autotaphonomical*, *proper* or *intrinsic* (characteristics of organisms which may be important at any stage of the taphonomical process) and *allotaphonomical*, including 2, *ecological or preburial* (environmental factors operating within bio- and thanatotopes including both biotic and abiotic agents), 3, **taphotopical or burial** (related to burial environments), 4, **post-burial** (depending on diagenetic, metamorphic, and hypergenetic rock alteration), and 5, **technical** (depending on methods of collecting, conservation and studying of fossils). Each class may be further subdivided into taphonomically **positive** (i.e. favouring to preservation) and **negative** factors. The most important general positive autotaphonomical factors are: presence of mechanically and chemically resistant skeletal parts, aquatic (especially marine) or near-water mode of life, high abundance, high ecological diversity as well as wide geographical distribution. Soft tissues are capable of fossilisation only in conditions that are either especially protective against decay or that permit very rapid mineralisation. The ecological factors differs principally between the terrestrial and aquatic environments. In water bodies, the low activity of consumers, especially scavengers, and the absence of agents removing or destroying living and dead organisms (e.g. currents and wave action) are positive, while in terrestrial thanatotopes the disturbing factors (winds, rainfalls, landslides, occasionally also predators and scavengers) may play a positive role in transporting living or dead organisms to water bodies. The main positive taphotopic factors are an absence of both aggressive chemicals and sediment disturbance by physical or biological agents. On the other hand, episodic rapid sedimentation connected, for example, with flooding or volcanic ash-fall may be positive. Taphotopic conditions preventing decay by rapid dehydration or antibiotic effects favour exceptional preservation. In contrast, postburial factors mainly have negative effects. The technical factors are strongly connected with personal experience of collectors and curators; small or badly preserved remains are often overlooked. Taphonomical studies are focused mainly on taphotopical and post-burial factors while autotaphonomical, ecological and technical factors are relatively rarely discussed in detail.

One particularly important concept concerns taphonomical **autochthonicity** vs. **allochthonicity** (POTONIÉ 1910). In the simplest case the biotope, thanatotope, taphotope, and oryctotope all coincide spatially. However, individuals can die somewhere out of their natural habitat, corpses may be displaced by water, wind, gravitation, or scavengers, and, finally, fossils may be re-deposited. When the taphotope corresponds to the biotope of the organism, the burial is **autochthonous**, otherwise it is **allochthonous** (NAUMANN 1858). Depending on the oryctotope coinciding or not with the taphotope, the burial may be either **primary** or **secondary**. Sometimes autochthonicity is treated as *in situ* burial only. This strict concept is admissible for sedentary organisms; however, it is impossible to distinguish, for example, between a dragonfly nymph that shifted actively to its future burial place and died there and another one which died elsewhere within the same pool and was then displaced by a current. That is why here the notion of autochthonicity is treated in a broad sense, covering all organisms buried within the same zone of the water body which they inhabited. For fossil insects autochthonicity is rarely absolutely certain and mostly should be inferred as more or less probable. When there are important reasons to believe that the organism inhabited a different zone of the same water body, it is called to be **subautochthonous**, and if one wishes to emphasise the autochthonicity in the strict sense, the term **euautochthonous** should be used (POTONIÉ 1958). Terrestrial fossils are nearly always allochthonous but if the organism lived just near the shoreline, the word **hypoautochthonous** is available (KRASSILOV 1972a). Fossils occurring together may all originate from the same habitat (in which case the assemblage is referred to as **monotopic**) or from two or more different biotopes (**polytopic** assemblage) (ILYINSKAYA 1958). Additionally, the

concept of autochthonicity *vs.* allochthonicity is applicable to the components of mero- and ichnocoenoses.

Besides the fossil assemblages themselves, taphonomy also deals with presentday thanato- and taphocoenoses (**actuapalaeontology**) and the fate of dead organisms under controlled environments (i.e. **experimental taphonomy**). We should also differentiate between the taphonomy of taxa and taphonomy of communities (**syntaphonomy** after OCHEV 1997). The same factors can have different effects on taphonomy of taxa and syntaphonomy. For instance, any addition of allochthonous organisms increase the number of represented taxa, but produces syntaphonomical biases.

It should be stressed that while such biases as fragmentation, partial dissolution or deformation of fossils are unfortunate for biological studies, they may provide useful information about the physical and chemical environments of tapho- and oryctotopes (M. WILSON 1988b).

An important aspect of modern taphonomy is **molecular taphonomy** which concentrates on the chemical preservation and alteration of fossils (EGLINGTON & LOGAN 1991). Besides the chemical investigation of morphologically preserved fossils, molecular methods are also applied to dispersed organic matter and other biogenic materials preserved in sediments (**chemofossils**).

Quantitative probabilistic approaches are of great potential value in taphonomy but unfortunately, the corresponding methods are not sufficiently advanced yet. Such studies may be centred on the representativeness of either certain fossil samples or of the total fossil record of a taxonomic or ecological unit. The gap analysis method developed by C. MARSHALL (1991) is worthy of mention in this connection. This method allows one to calculate the confidence intervals around the probable time of origin of a taxon on the basis of the frequency and distribution of its fossils and seems to be important for testing phylogenetic hypotheses. This approach is often used in palaeontology but usually in an informal and rather intuitive manner. Methods of quantitatively estimating the representativeness of fossil samples are still at a very early stage of development and testing (e.g. RASNITSYN & PONOMARENKO 1967, RASNITSYN & KHOVANOV 1972, HOLTZMAN 1979, SAPUNOV 1990).

1.4.2. INSECT TAPHONOMY

Though fossil insects are often considered to be rare, they are in fact widespread in a wide range of oryctocoenoses. Research papers dealing specifically with insect taphonomy are not numerous but their number has been increasing in the last few decades (e.g. M. WILSON 1980, 1988a, ZHERIKHIN 1980a, 1992, 1997a, ZHERIKHIN & SUKATSHEVA 1989, 1992, MARTINELL & MARTÍNEZ-DELCLÒS 1990, HENWOOD 1992a,b, 1993a,b, MARTÍNEZ-DELCLÒS & MARTINELL 1993, KOHRING & SCHLÜTER 1995, LUTZ 1997, PONOMARENKO 1997, MCCOBB et al. 1998, PEÑALVER 1998, RUST 1998). In particular, actuapalaeontology and experimental taphonomy of insects are still at the very early stage of development. Numerous occasional observations and reflections are scattered in palaeontological and entomological literature but these have not yet been reviewed.

There are two main modes of insect conservation as well as a number of rarer and special ones as discussed below.

1.4.2.1. Direct burial in sedimentary deposits

This kind of burial prevails over the insect fossil record. The burial happens when a dead (or still living) insect or a part of its body

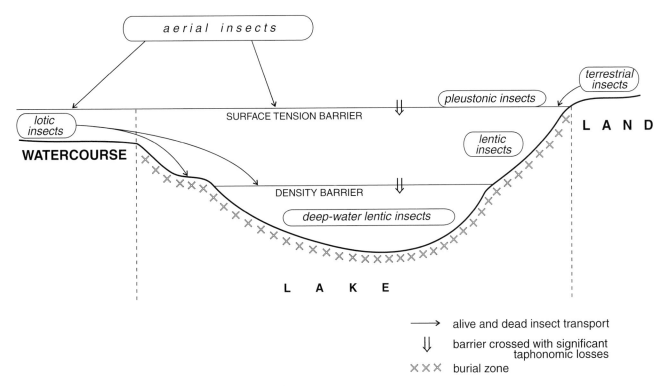

Fig. 19   The principal burial barriers for terrestrial and aquatic insects

becomes overlain by bottom sediment. For an insect which died on the land, three *pre-burial stages* of the taphonomic process are absolutely necessary to achieve burial: it has to get to a water body, to pass through *surface tension barrier*, and then to be overlain by sediments (Fig. 19). Only the former stage is excluded if an insect dies on the water surface and two former stages if it died under water. An additional pre-burial physical barrier occurs within the water column if there is a sudden change in water density, and consequently one more stage, passing through the *density barrier*, becomes necessary. Another additional stage appears when an allochthonous insect has to be transported to the taphotope from outside by water currents. Ichnofossils can be directly buried when the activity traces have been made on or in sediments themselves.

Insect remains occur in all main types of sedimentary deposits (clastic, chemical, biogenic, or mixed) as well as in volcanic-sedimentary rocks. The frequency of finds decreases with increasing rock grain size from mudstone and siltstone to sandstone and further to gritstone and psephite. Insects are represented mainly either by *compression fossils* with preserved (organic or mineralised) cuticle but flattened and often more or less distorted (Fig. 20) or by *impressions* in the rock matrix lacking cuticle (Fig. 21). Occasionally slightly or undeformed *three-dimensional fossils* occur either as cavities or as mineralised inclusions in a rock.

1.4.2.1.1. Autotaphonomical factors

In general, insect fossils are rarer than those of plants or marine molluscs, but are commoner than other non-marine arthropods, suggesting that their fossilisation potential should be moderately high. A number of autotaphonomical characteristics of insects in general affect the probability of their conservation in sediments.

The **chitinous exoskeleton**, which is both mechanically durable and chemically resistant, is taphonomically positive. Chitinous parts of a dead insect may be exposed to the environment for several years before burial without being seriously destroyed while soft tissues decay in a short time. They are not subjected to chemical dissolution unlike shells or bones but they can be decomposed both aerobically and anaerobically by chitinolytic bacteria (D. JOHNSON 1931, ZOBELL & RITTENBERG 1938, KOPP & MARKIANOVA 1950, LEAR 1961, SEKI & TAGA 1963, 1965, OKAFOR 1966, HOOD & MYERS 1974, YAMAMOTO & SEKI 1979, WARNES & RANDLES 1980, LAKSHMANAPERUMALSAMY 1983, PLOTNICK 1986, GOODAY *et al.* 1990, SEKI *et al.* 1990, R. MILLER 1991, MIYAMOTO *et al.* 1991). There are some records of chemically preserved chitin in fossil insect remains (LENGERKEN 1922, ABDERHALDEN & HEYNS 1933, R. MILLER 1991) but recent studies show that chitin rarely can be chemically recognised even in excellently morphologically preserved fossils (STANKIEWICZ *et al.* 1997a,b, 1998a, McCOBB *et al.* 1998). In Quaternary remains, only 130 thousand year old, the content of chitin is significantly lower than in fresh cuticle (R. MILLER *et al.* 1993). Normally fossil cuticle is chemically altered by lipid polymerisation, coalified, or replaced by a mineral substance. Too few analyses have been made to ascertain which tapho- and oryctotopic conditions favour to chemical preservation of chitin. Differential cuticle sclerotisation is an important source of biased representation of different insect taxa and instars in the fossil record, and soft-bodied insects with very thin cuticle are rare as fossils.

Another positive feature connected with the presence of the chitinous exoskeleton is **development with ecdysis**. An exuvium can be entombed, at least in quiet environments, just so easily as a dead insect. Consequently, during its life each individual potentially produces the number of complete fossils equal to the number of moults plus one. Though this potential is rarely realised in the full measure,

Fig. 21 Grylloblattidan *Ideliopsina nana* Storozhenko (Ideliidae) from the Middle or Late Triassic of Dzhayloucho in Kirghyzstan (holotype, photo by A.P. Rasnitsyn), preserved in barely lithified mud subject to plastic deformation (arrow shows the supposed direction of tension); right fore wing 24 mm long

Fig. 20 Unidentified broad-headed bug (Alydidae) from the Late Cretaceous of Obeshchayushchiy in Russian Far East (PIN 3901/900, photo by D.E. Shcherbakov); body 16.7 mm long

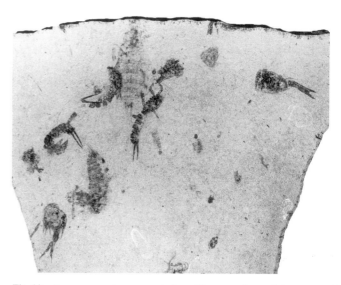

Fig. 22 A merocoenosis composed of moulting casts of aquatic immatures of the heptageniid mayfly *Ephemeropsis melanurus* Cockerell (dark, wide in the uppermost left and lowermost right corners; note their shrivelled condition indicating drying-out), coptoclavid beetle *Coptoclava longipoda* Ping (dark, slender, numerous), and hemeroscopid dragonfly *Hemeroscopus baissicus* Pritykina (pale, wide) in the Early Cretaceous of Baissa in Siberia (PIN 3064/6642, photo by D.E. Shcherbakov); slab 92 mm high as shown

moult skins of aquatic or shore-inhabiting immatures often numerically dominate in fossil assemblages, indicating that they are in fact mainly merocoenoses (Fig. 22). If so, mass mortality events have to be less taphonomically important for insects than is generally believed for the majority of animals. Skins may be recognised on the basis of a dorsal fissure at the anterior part of the body or by a deformed head and thorax (Fig. 23). Many fragmentary fossils should also be fragments of skins but it is often difficult to prove. Occasional finds of specimens fossilised during moulting provide direct evidence of multiplication of the number of fossils by moults (Fig. 24). The number, manner, and place of moulting are important for interpretation of quantitative data. For example, in the Early Cretaceous lacustrine assemblage of Baissa, Transbaikalia, Russia, *Hemeroscopus* dragonfly nymphs clearly dominate over another large autochthonous aquatic predator, *Coptoclava* beetle larvae, normally being two to ten times and occasionally even 100 times more frequent. However, in both taxa exuviae constitute the bulk of the material and because the moult number in dragonflies is normally several times more than in beetles their relative abundance

seems to be biased (ZHERIKHIN 1997a). A high rate of exuviae production strongly biased the distribution of size cohorts in fossil samples in comparison with the living population (HARTNOLL & BRYANT 1990). The number of moults generally decreases in advanced insect orders, and this point may be of some importance for the accurate comparison of insect abundance in the Palaeozoic and younger times.

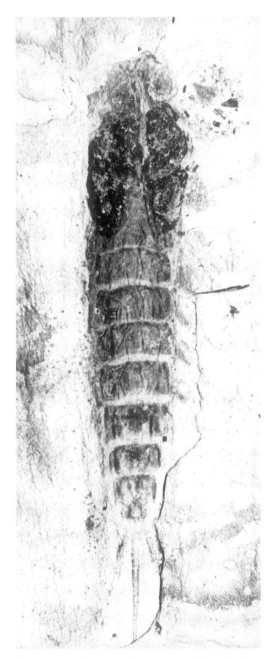

Fig. 24 A moulting larva of the coptoclavid beetle *Coptoclava longipoda* Ping; Early Cretaceous of Baissa in Siberia (PIN 3064/6546, photo by D.E. Shcherbakov); fossil 17 mm long

Fig. 23 Nymphal moulting cast of the heptageniid mayfly *Ephemeropsis melanurus* Cockerell; note the clearly visible fissure at the head and thorax; Early Cretaceous of Baissa in Siberia (PIN 3064/3333, photo by D.E. Shcherbakov); body 47 mm long

Insects are nearly omnipresent in non-marine habitats (except for ice shields) being occasionally even the only metazoans as in some hypersaline lakes (COLLINS 1980). This **ubiquity** is reflected in the fossil record by the presence of remains in a variety of palaeoenvironments. In this respect the insects generally evolved in a taphonomically positive direction because their ecological diversity increased greatly since their first appearance in the Devonian (Chapter 3.2.2.1). The comparative taphonomical potential of different environments is discussed below (Chapter 1.4.2.4).

The tremendous **abundance** of insects is of great positive importance. The actual degree of rarity is grossly exaggerating in the fossil record. For this reason both the appearance and disappearance of a taxon in the fossil record should respectively indicate an increase or decrease in abundance (other taphonomically important things being equal) rather than origin or extinction. Many taxa are unknown in the fossil state over a long time interval and would be considered as extinct if there were no known living relicts. Some of such 'living fossils' are widely known, like coelacanthid fish; among the insects the anaxyelid sawflies, ironomyiid flies, eccoptarthrid and jurodid beetles as well as the order Grylloblattida may be mentioned. This fact is largely ignored in computations of taxonomic dynamics making them inaccurate. For extinct taxa we know the time of disappearance from the fossil record, not of extinction, and to make the data comparable we should take living taxa missing as fossils over the Neogene or the entire Cainozoic as disappeared as well ('practically extinct': ZHERIKHIN 1978). Note also that any statements about 'mass extinction' at a strictly definite time level are taphonomically naïve even when some dramatic changes are doubtless (as at the famous iridium layer at the Mesozoic/Cainozoic boundary): the last dinosaur bone found is almost certainly not a bone of the last dinosaur that ever existed, and in any case, this cannot be proved in any way.

One of the most taphonomically profitable insect features is the **capacity of flight**. Flying organisms can easily fall into water, especially when they are small and their active flight ability is restricted. The spiders are no less common and widespread in nature than are terrestrial insects but they are far less well represented in the fossil record (ZHERIKHIN 1980a). Winged terrestrial insects are much more abundant (usually by 3 to 4 orders of magnitude) in oryctocoenoses than are flightless ones, whereas in actual communities the ratio might be reversed if not only flightless species but also flightless ontogenetic stages are taken into account. Even such abundant and highly active flightless insects as worker ants are virtually absent as fossils in sedimentary rocks (F. CARPENTER 1930a) but common in fossil resins; the same is true for the wingless morphs of the aphids. According to KENWARD's (1975) actualistic observations, aerial insects predominate absolutely over wingless ones in taphocoenoses even in such small water bodies as drainage ditches. On the other hand, the strongest and most manoeuvrable fliers like butterflies, large moths and many wasps are also underrepresented in the fossil record, probably because they are less affected by accidental wind rush actions. An exceptional case, that of dragonflies which are quite common in many oryctocoenoses, may be explained by their close connection to water bodies. Burial probability also decreases for rarely and unwillingly flying species. Thus frequent but not especially strong flight seems to be the most advantageous, and the dominance of beetles, bugs, leafhoppers, flies, etc., in most oryctocoenoses corroborate this opinion. Insects with different flight capacities also tend to be buried at different distances from the shoreline, and consequently the near-shore and off-shore tapho- and oryctocoenoses reflect the same original community in a different way (M. WILSON 1980).

Insects constitute an important component of the aerial zooplankton both inland and over the sea as aircraft and ship trap studies demonstrate (GLICK 1939, J. FREEMAN 1945, C.G. JOHNSON 1957, 1969, L. TAYLOR 1960, HOLZAPFEL 1978, RAINEY 1983, FARROW 1984, ASHMOLE & ASHMOLE 1988, PEDGLEY 1990). In particular, extensive aircraft and ship trap studies demonstrate that insects regularly occur over sea, sometimes at a very long distance from the land (GRESSITT & NAKATA 1958, YOSHIMOTO & GRESSITT 1959, 1960, 1961, 1963, GRESSITT et al. 1960, 1961, 1962, YOSHIMOTO, GRESSITT & MITCHELL 1962, HARRELL & YOSHIMOTO 1964, CLAGG 1966, HARRELL & HOLZAPFEL 1966, HOLZAPFEL & HARRELL 1968, HOLZAPFEL & PERKINS 1969, GUILMETTE et al. 1970, HOLZAPFEL et al. 1970, 1978, BOWDEN & JOHNSON 1976, HARDY & CHENG 1986, SPARKS et al. 1986, WOLFF et al. 1986, DELETTRE 1990, DIOLI 1992, IRWIN & ISARD 1992, S. JOHNSON 1992, PECK 1994). Observations on nival aeolian ecosystems also confirm a large-scale insect transfer by air currents, often far outside the altitudinal limits of their natural habitats (KAISILA 1952, MANI 1968, J. EDWARDS 1987, FURNISS & FURNISS 1972, J. EDWARDS & BANKO 1976, PAPP 1978, PAPP & JOHNSON 1979, SPALDING 1979, ASHMOLE et al. 1983, L. PRICE 1985, ASHMOLE & ASHMOLE 1988). Their abundance varies greatly and generally decreases with increasing height and distance from land but some specimens have been detected up to 3000 m altitude (occasionally even up to 6000 m), and at distances of hundreds kilometres from the nearest land (HOLZAPFEL 1978). A find of European aphids at Spitsbergen (ELTON 1925) make it clear that the distance of passive transport can be as great as 1300 km. The volume of arthropod fallout is as much as 20 specimens per m^2 per day even at the barren Anak Krakatau Island where it is constituted exclusively by alien fauna transported across the sea (THORNTON et al. 1988). The quantity of insects carried out to coastal seas may be so great that sometimes they form a significant additional food source for marine life (CHENG & BIRCH 1977, 1978; LOCKE & COREY 1986). Insect flight behaviour is managed by a complex set of genetic, physiological and environmental factors (C. WILLIAMS 1958, C.G. JOHNSON 1963, 1969, DRAKE et al. 1995, GATEHOUSE 1997). Usually the aerial plankton is dominated by small homopterans (especially aphids) and dipterans. However, not all components of the aerial plankton are equally well represented as fossils. Some wingless arthropods such as minute insect larvae, mites and ballooning spiders abundant in aerial drift (GLICK 1939, MEDLER & GHOSH 1968, MUMCUOGLU & STIX 1974, FARROW 1982, WASHBURN & WASHBURN 1984, MARGOLIES 1987) are nearly absent in sedimentary oryctocoenoses. It is not clear why the representation of wingless animals in the air drift and in the fossil record differs so dramatically. Perhaps, most of them are of too small a size to be detected even on fine-grained rocks where objects 1 mm or less in size are rarely discernible. However, if so, they should occur in microscopic palynological samples which is likely not the case. According to HESPENHEIDE (1977) the size of airborne insects decreases continuously with altitude above ground; thus the probability to falling into a water body may be higher for larger winged insects from the near-ground air layer. More actualistic observations on insect fallout are necessary to explain this pattern definitely.

In any case, actualistic data on the aerial plankton indicates that flying insects approach pollen grains in respect of their capacity of aerial dispersion and, consequently, they should represent the next most important source of palaeontological information about watershed areas after palynology. Consequently, oryctocoenoses are often highly heterotopic and exhibit not only strictly local near-water communities but also a collective faunal portrait of the surroundings including sometimes rather remote habitats.

On the other hand, terrestrial flightless insect taxa, ontogenetic stages or casts, etc. are strongly underrepresented except for inhabitants of a narrow area just along the water line. The insects had evolved active flight by the end of the Early Carboniferous and this was probably the most taphonomically substantial milestone in their evolutionary history. In early winged insects, archemetaboly (gradual metamorphosis with more than one flying stage (Chapter 2.2c) was widespread (see Figs 71, 356, 388). Correspondingly, their subadults and late instar nymphs occurred rather frequently. The extinction of archemetabolous orders, except for mayflies, during the Permian will have had a negative taphonomic effect in addition to the secondary evolution of flightlessness common in many extant taxa. Any circumstances conducive to the development of flightlessness such as ectoparasitic specialisation, windy environments like mountain tops and small oceanic islands, etc., just serve to reduce the completeness of the fossil record.

If high ecological diversity is a positive feature, the general **ecological distribution** of insects proves to be rather unhelpful in that this is the major negative factor determining the fragmentary nature of their fossil record. The most profitable taphotopes are shallow seas and, secondarily, inland standing waters. But the insects are predominantly terrestrial, and aquatic taxa constitute only a minor part of their total diversity. Moreover, many of the aquatic taxa inhabit highly dynamic environments such as watercourses or the breaker zones of lakes where most organic remains are rapidly destroyed. About 1400 living species of 14 insect orders are connected with marine habitats (including coastal debris) but there are very few true marine taxa (CHENG 1982). As a rule, the 'marine' insects in fact live in brackish coastal sites, often under a rather unstable and potentially destructive physical regime. As regards the quiet parts of lakes, their entomofauna is only moderately diverse, and no higher taxa are restricted to them. The terrestrial insects populate taphonomically unequal biotopes including the most disadvantageous ones, and their mode of life often hinders burial even at more fortunate sites. The burial probability is particularly low for soil and litter dwellers, nidicoles, troglobionts, parasites, desert and alpine fauna, etc.

Any analysis of palaeontological data might take these points very seriously. The earliest insects were almost certainly terrestrial and probably cryptobiotic. The oldest positive evidence of insect invasion into aquatic habitats is dated by the Early Permian (though perhaps some Late Carboniferous wingless insects were aquatic or semiaquatic (Chapter 3.3.2)), and no highly diverse lacustrine assemblages are known before the Triassic. In the Late Cretaceous a dramatic decline of the lentic fauna is evident extending in time up until the Oligocene. Some marine insects such as marine water striders occur in the fossil state (N. ANDERSEN 1998) but there is no ground to suppose that at any time in the past the insects have been better represented in any marine communities than they are nowadays. The same is true for other taphonomically favourable habitats.

Changes in habitat preferences of a taxon will make its fossil record incomparable at different time intervals as illustrated by the fossil record of the stoneflies (SINITSHENKOVA 1985b). The number of Mesozoic stonefly taxa is roughly three times more than in the Palaeozoic and nearly six times that of the Cainozoic. If these figures were taken uncritically, they could be interpreted as evidence for a flourishing of the order in the Mesozoic and a subsequent dramatic decline. However, almost all living stoneflies are lotic, and the same seems to be true for the entire Cainozoic and Late Cretaceous fauna, whilst some Late Permian and many Mesozoic (especially Jurassic) taxa were lentic, and their nymphs are common in lacustrine deposits. Thus the representativeness of Mesozoic stonefly assemblages is disproportionally high. Similar, and often more complicated, phenomena are widespread in the history of other orders. They may, in particular, be responsible for the majority of apparent incidences of recurrence, or the so-called Lazarus effect, when a taxon re-appears in the fossil record after a long time of absence or near-absence. The psyllomorph homopterans provide perhaps the best example of the Lazarus effect among the insects. Several extinct psyllomorph families are common in the Late Permian, Triassic and Jurassic (BEKKER-MIGDISOVA 1985, and undescribed materials) but few specimens have been found in the Early Cretaceous and no fossils are known from the Late Cretaceous or the Palaeocene. Some living families appear in the Eocene when they are extremely rare but since the Oligocene, fossil finds again become frequent. Psyllomorph ecological evolution in the Late Mesozoic and Early Cainozoic is obscure but the distributional pattern suggests that in the Cretaceous and Early Tertiary they were restricted to some extremely taphonomically unfavourable habitats, perhaps arid or alpine, and then re-colonised other environments.

Taphonomical biases connected to habitat preference may impart to fossil assemblages an image differing radically from the primal community type. For instance, desert oryctocoenoses are often dominated by inhabitants of near-water oases and may be easily misinterpreted. There are some localities in Italy, dated by the Messinian stage of the Miocene when the Mediterranean Basin had temporarily lost its connection with the Atlantic and so turned into a vast hot depression with hypersaline lakes (the Messinian salinity crisis: HSÜ et al. 1977). Nevertheless, oryctocoenoses are rich in dragonflies, craneflies, fungus gnats, march flies, hoverflies and other taxa characteristic for mesic environments (CAVALLO & GALLETTI 1987, GENTILINI 1984, 1989).

The **composite** and **water-repellent exoskeleton** of insects can also be of negative taphonomic importance. Even if each sclerite is rather durable, it is connected with others by a soft membrane. When transported by current, insect bodies often become disarticulated, and isolated sclerites may be deposited separately owing to their different weight and hydrodynamic properties. A dead body may disintegrate before burial also without any mechanical influences as a consequence of decay of the softer membranes; this process depends upon the level of bacterial activity and the time of exposure to it. MARTINELL & MARTÍNEZ-DELCLÒS (1990) have found that the most fragile articulation points are located between the head and prothorax, between the head and the bases of antennae, between the antennal segments, between the thoracic segments and legs, between the articulated parts of the legs, between the pterothorax and wings and between the metathorax and abdomen. Different skeletal parts are unequally durable, and the strongest (especially elytra and tegmina) are overrepresented in drift assemblages (Fig. 25). It should be noted that even the strongest sclerites are fragile in comparison to shells, wood, pollen, or bones and are nearly incapable of re-deposition. This is a negative point for fossil persistence in time but fortunate for dating because any doubt whether the fossil is synchronous to the fossiliferous sediment or not is easily rejected (unless we are dealing with a fossiliferous pebble, cobble, erratic block, or xenolith).

Fig. 25 Isolated beetle elytra accumulated by water flow in a fluvial sandstone; Oligocene of Kenderlyk in Kazakhstan (PIN 1341/5–7, photo by D.E. Shcherbakov); left elytron 7 mm long

The water-proofing ability of insect integument inhibits the submergence of an insect that falls on to a water surface even after its death, and a floating corpse may be consumed, destroyed or cast ashore. However, the thin epicuticular hydrophobic layer is liable to microbial decay. Actualistic experiments in aquaria are described by MARTINELL & MARTÍNEZ-DELCLÒS (1990) and MARTÍNEZ-DELCLÒS & MARTINELL (1993). They demonstrate that insects that have fallen on to a water surface while still alive and then killed by asphyxia can float for from several days up to two weeks or more and then sink rather slowly. Besides the water-repellent cuticle, low specific weight and sprawled wings increase the time before sinking. Floating time depends on the entry of water into the tracheal system, the rate of decomposition, the development of micro-organisms on the corpse, and on mechanical actions (waves and rain). The authors interpret the effect of fungal, algal and microbial growth mainly as a result of a consequent increase in the weight of the corpse, but perhaps some microorganisms also destroy the epicuticular layer. The time that an insect body remains floating increases by up to several months if an insect is placed on to water after its death; however, more observations are necessary to estimate the effect of different thanatotopic factors on the epicuticular layer and on the corpse's buoyancy. T. WAGNER et al. (1996) have demonstrated experimentally that the wettability of insect wings depends also on their surface sculpture, pubescence, scaling, etc.

The taphonomical effect of **small body size** seems to be equivocal. It reduces the probability of preservation in coarse-grained sediments and hinders detection during collecting but the most important negative thing is a decreasing probability of passing through the surface film of water. On the other hand, smaller size facilitates wind transfer and, as discussed before, moderately small (body length 1 mm or more) flying insects are well represented as fossils indicating that the surface tension barrier is not insuperable for them. For instance, in the Early Cretaceous lacustrine deposits of Baissa in Siberia such small-sized forms as winged aphids (see Fig. 189) constitute about one third of all terrestrial insect fossils.

To summarise the above, the main general autotaphonomic features of the insects are as follows: high abundance and high ecological diversity, flight ability, moults, chitinous integument (all positive), prevalence of terrestrial mode of life, composite articulated exoskeleton, water-repellent cuticle and, partly, minute size (negative). Besides these, each insect taxon has its own taphonomically important combination of characters. Only some aspects may be briefly mentioned here.

Any **connection to a near-water habitat** such as oviposition preference, feeding or hibernation sites, etc., raise the burial probability. Mass **migrations** may favour mass mortality even when the distance of migration is short. For instance, migration of ladybirds (Coccinelidae) to hibernation places is occasionally accompanied by an impressive concentration of dead individuals at lake and sea shores (SCHÄFER 1962, TURNOCK & TURNOCK 1979, LEE 1980, KLAUSNITZER 1989); SUCHÝ (1991: 77) describes stripes 10–20 cm wide and several centimetres thick formed by dead beetles along several kilometres of a beach.

Circadian activity pattern may be important, at least at large lakes and coastal seas. The shore night breeze should blow flying insects off of the land favouring their accidental drowning, while the opposing daytime ('sea') breeze is not of benefit to entombment. This effect may explain, in particular, why nocturnal moths occur rather frequently in some oryctocoenoses buried in marine sediments (e.g. in the Miocene of Stavropol, North Caucasus, as well as in the Palaeocene Fur

Formation in Denmark: KOZLOV 1988, RUST 1998) while the diurnal butterflies are extremely rare. Some other insect groups with presumably nocturnal activity are also frequent in the same deposits (e.g. ichneumonid wasps of the subfamily Ophioninae in Stavropol). The **polarised light reflected by water surfaces** is attractive for many insects including aquatic bugs and beetles (SCHWIND 1989) which occasionally bring them into the sea. This misorientation is the most probable explanation for a relatively high rate of those taxa in some localities of a marine origin (e.g. in the Jurassic of Solnhofen, Germany and the Miocene of Stavropol, Russia).

Many aspects of **swarming behaviour** seem to be influential, including the time and place of swarming, swarm size, etc. A near- or above-water swarming is taphonomically profitable but the mass nuptial flights of ants and termites is probably at least equally favourable owing to their rather weak flight ability and especially to their wings breaking off: wings broken after the flight are carried by wind and surface runoff forming a merocoenoses. More advanced sociality reinforces this trend due to increasing production of alate individuals.

The **swimming ability** of terrestrial insects might affect their burial probability in small ponds as well as in the near-shore zone but hardly in the central parts of larger water bodies. This problem is discussed by MARTINELL & MARTÍNEZ-DELCLÒS (1990), MARTÍNEZ-DELCLÒS & MARTINELL (1993), and RUST (1998). Swimming ability may differ significantly even between closely related taxa (e.g., between subfamilies of grasshoppers: LOCKWOOD et al. 1989) but observations are sporadic and concern too few species.

The **mode of oviposition** and of **adult emergence** may be important, especially for those insects with aquatic larvae. Oviposition into water and underwater eclosion are most profitable. For instance, teneral caddisfly adults still lacking fully developed wings occur commonly in some lacustrine oryctocoenoses indicating the autochthonicity of fossils (Fig. 26). Mass burial at some bedding planes (Fig. 27) is most probable for species with mass synchronous emergence.

The **disarticulation pattern** differs in different taxa depending on their morphological peculiarities. In ant fossils, the basal abdominal segment is more or less fused with the thorax and is associated with the latter more often than with the remaining part of the abdomen (RUST 1998). The compact abdomen of bugs and beetles is often completely preserved unlike the long and slender abdomen of dragonflies (Figs. 28, 29). In those taxa that possess a well-defined clavus of the forewings (roaches, mantises, true bugs, auchenorhynch homopterans, lower hymenopterans), the wing breaks easily along the claval furrow so that fossil forewings are often incomplete, with the clavus missing; isolated clavi also occur frequently (Fig. 30). LUTZ (1984b) observed this breakage of cockroach wings in actualistic experiments.

1.4.2.1.2. Ecological factors

Ecological factors affect the fossil record mainly indirectly, through modification of animal behaviour or their abundance and temporal and spatial distribution patterns, both of living and dead individuals. Ecological factors seem to be even more diverse and variable than autotaphonomical characteristics, so it is nearly impossible to discuss many of them in a general way. From the taphonomical point of view, ecological factors can be divided into those affecting organisms in their lifetime, those affecting mortality patterns and those acting after death.

Ecological effects are often highly complicated, depending on a complex set of variables that may, at least partly, operate in opposite

Fig. 27 Mass burial of adult *Plutopteryx* stoneflies indicating synchronous mass emergence; Jurassic of Bayan-Teg in Mongolia (PIN 4023/389–403, photo by D.E. Shcherbakov); aggregation 40 mm wide

Fig. 26 A teneral caddisfly specimen; Early Cretaceous of Baissa in Siberia (PIN 3064/8152); 13.8 mm long excluding antenna

directions to one another. These effects are insufficiently studied from the taphonomical point of view, and available data are often equivocal. The following review of ecological factors is surely very incomplete, but its main purpose is to provide the reader with some general ideas about their importance and complex nature.

1.4.2.1.2.1. Ecological factors affecting organisms in their life-time

This group of ecological factors includes various environmental circumstances affecting the abundance and spatial distribution of populations in any directly or, more commonly, indirectly taphonomically important way.

Any increase in abundance is taphonomically positive because the more abundant the organisms, the more probable their entry into the fossil record. That is particularly so when other morphological or biological characteristics are not highly taphonomically favourable. In many terrestrial and lotic insects, high population density provokes more or less chaotic migratory behaviour and 'random' dispersal by flight or passive drift, with a consequential increasing probability of their accidental death in standing waters. Other environmental factors inducing random dispersal include local disturbance, food depletion, oxygen deficiency, etc. Some environmental stresses inspire non-random migrations which are taphonomically positive when they lead

Fig. 28 Abdomen of a naucorid bug; Late Triassic of Madygen in Kyrghyzstan (PIN 2087/11, photo by D.E. Shcherbakov); abdomen 4.5 mm long

to increasing insect density in coastal or still-water habitats, but negative when they operate in the opposite direction. However, such factors which are positive for the taphonomy of the taxon, are often syntaphonomically negative because they result in the formation of mixed, highly polytopic death assemblages.

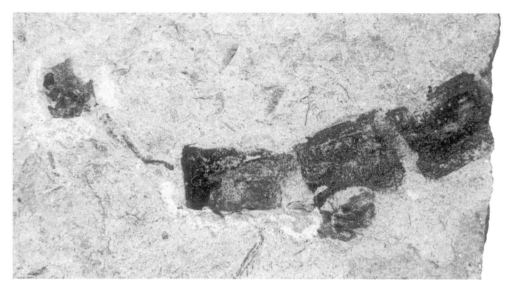

Fig. 29 A disarticulated abdomen of an adult *Hemeroscopus* dragonfly; Early Cretaceous of Baissa in Siberia (PIN 1989/2640, photo by D.E. Shcherbakov); abdomen 29 mm long as preserved

A few factors deserve special discussion in this respect.

Fluctuations in insect abundance and of their spatial distribution are profoundly influenced by **weather**. Wind conditions are crucial for the aerial transfer of insects, and atmospheric fronts surely affect insect flight activity though the data are somewhat controversial, perhaps due to the differences between different insect taxa and different climatic areas (WELLINGTON 1954, GLICK & NOBLE 1961, C.G. JOHNSON 1967, 1969, L. TAYLOR 1967, R. BERRY & TAYLOR 1968, HOLZAPFEL & HARRELL 1968, HURST 1969, TOMLINSON 1973, HAESELER 1974, LANDIN & SOLBRECK 1974, RAINEY 1978, 1983, FARROW 1984, SPARKS *et al.* 1986, DRAKE & FARROW 1988, 1989, PEDGLEY 1990, BERGER 1992, S. JOHNSON 1992, PEDGLEY & REYNOLDS 1992, DRAKE & GATEHOUSE 1995, GATEHOUSE 1997). After strong storms, mass occurrences of insects, both dead and still living, have been noticed in beach-drifts by different authors (SAUNDERS 1836, MAPLETON 1879, SCHWARZ 1931, PLATONOFF 1940, HOWDEN 1977, BURKE *et al.* 1991) indicating that thousands of individuals have been transported out to sea, sometimes from quite remote areas. Precipitation and temperature are also of great importance. In particular, during seasonal or incidental droughts, many insects migrate to wetter habitats where the vegetation is less affected by water stress.

In running waters the main long-distance insect dispersal mechanism is a downstream **drift**. This phenomenon is widely known and has been discussed repeatedly in hydrobiological papers (e.g. T. WATERS 1965, N. ANDERSON & LEHMKUHL 1968, WENINGER 1968, H. HYNES 1970, 1975, BOURNAUD & THIBAULT 1973, R. WILSON & BRIGHT 1973, KROGER 1974, NEVEU & ECHAUBARD 1975, ADAMUS & GAUFIN 1976, NEVEU 1980, IRVINE 1985, OTTO & SJÖSTRÖM 1986, RESH *et al.* 1988, STATZNER *et al.* 1988, MACKAY 1992, LANCASTER *et al.* 1996). Available data indicate that the total volume of insect drift in rivers is considerable (though highly variable in time and space) and that a large quantity of drifting insects reaches both lakes and seas (DENDY 1944, LINDBERG 1949, D. WILLIAMS 1980, LOCKE & COREY 1986). However, the actual taphonomic role of insect input from running waters into lacustrine and coastal marine oryctocoenoses seems to be surprisingly low (ZHERIKHIN 1980a, 1985). Indeed, in the Cainozoic where almost all families and many genera are identical with those of the present-day (so that the risk of misinterpretation of their ecology is minimal) and the diversity of palaeo-environments represented in the fossil record is maximal (see below), finds of lotic insect larvae are extremely rare and their percentage is consistently much less than 1% of the total insect number. Moreover, throughout the insect fossil record, finds of aquatic insect stages in marine deposits are exceptional. Consequently the common finds of aquatic stages of some taxa related to modern lotic groups in the Mesozoic and Palaeozoic lacustrine deposits have to be explained rather by a switch in ecology of the group than by the action of river drift (SINITSHENKOVA 1987).

In both running and standing waters passive insect drift on floating algae, higher aquatic vegetation or driftwood may be of some importance. Some authors (e.g. LUTZ 1990, 1991a, PONOMARENKO 1996) postulate its importance for a transport of near-shore lacustrine insects to deep-water areas of large lakes (see Chapter 3.3 for further discussion).

The importance of **environmental dynamics** should be stressed. During the course of hydrological and physiographic evolution of a sedimentary basin and its surroundings, the pattern of both aquatic and terrestrial communities changes in time and space. As a result a number of different fossil assemblages occur within the same sequence corresponding to different stages of basin and landscape evolution. This makes the scanning of successive oryctocoenoses equivalent to sampling along an environmental gradient. Hence the resulting diversity of faunas and palaeoenvironments represented in the fossil record of the same sedimentary sequence increases greatly. Short-time course fluctuations such as a reduction of the shore vegetation area during severe droughts may facilitate the burial of members of less hygrophilic communities than usually. Over larger time intervals secondary succession in a coastal habitat after a fire, wind-fall or other disturbance produces the same effect. Finally, climatic fluctuations with a period of tens or hundreds of thousands years induce faunal migrations equivalent to a spatial shift measuring hundreds and occasionally several thousand kilometres, as documented in particular by Quaternary faunal migrations. Because the natural samples from different communities are preserved separately in different layers, this

Fig. 30 Roach fore wings with clavus nearly detached (above) and missing (below) (holotypes of, respectively, *Ignaroblatta sibirica* Becker-Migdissova and *Archaeotiphites captiosus* Becker-Migdissova (described as Archimylacrididae) from the Late Carboniferous of Zheltyi Yar in S. Siberia; wings 30 and 25 mm long as preserved, respectively, from ROHDENDORF *et al.* 1961)

effect is positive from the point of view of both auto- and syntaphonomy. However, it can lead to stratigraphic confusion. PONOMARENKO & POPOV (1980) stressed that if similar faunas existed for several millions of years or more, the diachronous sediments corresponding to the same stages of lake evolution in different basins may be easily miscorrelated. ZHERIKHIN & KALUGINA (1985) pointed out that the same problem also arises in dating terrestrial assemblages and that physiographic changes induced by erosion cycles can imitate climatic changes very closely (for example, regional changes in drainage may be misinterpreted as having been caused by fluctuations in precipitation).

1.4.2.1.2.2. Mortality factors

The **place of death** (***thanatotope***) is of great taphonomical importance. The most favourable situation exists when the thanatotope coincides directly with a taphotope of a high preservation potential.

These are firstly the bottom areas at still parts of standing water bodies; however, even here the necrocoenoses are vulnerable to various destructive agents. Other taphonomically favourable death environments are sporadic (see Chapter 1.4.2.1.2.3 for further detail). Less favourable is the situation when the thanato- and taphotope coincide but the preservation potential of the latter is relatively low. If the thanatotope is different from any potential taphotope (as, for instance, usually in subaerial environments) the components of the necrocoenosis have to be removed from it and transported to the future burial place. The probability of transport differs between different environments, and various post-mortem disturbances or barriers can impede bodies reaching the taphotope or even make it absolutely impossible.

Mortality rate is another important factor. Since the early days of palaeontology when the catastrophe theory appeared, the importance of **mass mortality events** was continuously stressed by many authors. Impressive pictures of terrible local overkill episodes come to mind easily when one looks on the most spectacular fossil sites with their high concentrations of remains. A number of different catastrophic taphonomical scenarios proposed for different organisms and environments can be found in the general taphonomical literature cited above. The origin of some fossiliferous deposits was indeed connected with different mass mortality factors, but their general taphonomical importance was surely overestimated for a long time. The continuous accumulation of remains resulting in Lagerstätten formation is quite possible in different environments as a routine process without any extraordinary incidents. Important in this respect is the existence of different kinds of permanently or periodically operating **natural traps**. Some kinds of traps catch mainly living organisms and kill them, while others collect mainly dead bodies. Even a high concentration of fossils at some bedding surfaces is not necessarily connected with any catastrophic events. For instance, in many insect taxa with synchronous emergence and short-lived adults, mass mortality is a normal and not a catastrophic pattern. In palaeoecology and studies on fossil populations the difference between mass mortality and continuous accumulation is especially important. An occasional drying up, winter-kill or suffocation event provides a 'snapshot' of ancient populations adequately reflecting their age structure, size cohort ratio, sex ratio, etc., whilst the accumulated assemblages better represent the mortality pattern than the structure of the living population. It should be also remembered that aquatic insect oryctocoenoses are often in fact merocoenoses formed under a low rate of mortality.

Any **water body** acts as an incidental trap for terrestrial (especially aerial) insects. This effect is strengthened by regular local atmospheric processes resulting from details of the local topography. Water bodies act as a kind of a suction trap owing to katabatic air currents produced by the temperature contrast between water and land surface (PALM 1949, NORLIN 1964, 1967, GUSEV, 1971). The sharpness of this contrast (and accordingly the trap efficiency) depends strongly on the volume and shape of the lake body (limnion). Small or shallow ponds are less effective traps than are larger and deeper lakes. This is why aerial insects are scarcely represented in the oryctocoenoses of small lakelets, and is well demonstrated for the non-marine Jurassic deposits in Siberia (ZHERIKHIN 1985) where winged insects, and in particular small-sized ones, occur only rarely in the lentils of fine-grained sediments deposited in small oxbow lakes within the fluviatile sequences (even the winged stages of mayflies and stoneflies are nearly absent in spite of the high abundance of their nymphs), whereas the sediments of the larger lakes yield rich terrestrial insect assemblages dominated by small dipterans and homopterans. The shallow-water zone of large lakes also provides a comparatively low efficiency trap compared with off-shore zones so that the share of terrestrial insects in lacustrine deposits increases away from the shore (M. WILSON 1980, PONOMARENKO 1997, ZHERIKHIN et al. 1999). In semiarid and arid regions the water surface itself is probably positively attractive for many aerial insects (DUVIARD & ROTH 1973). The attractiveness of shallow waters depends probably also from the bottom colour. This phenomenon is well known and widely used in the so-called yellow (Moericke) traps used in modern insect collecting, but there are no actualistic comparative observations on the insect fall for natural water bodies of different colours. The trapping effect also occurs at sea and is amplified by the strong attractive effect of the water surface (depending on polarisation of the reflected light: SCHWIND 1989) for many aquatic bugs and beetles which often become disoriented during their flight and come down to the sea but cannot survive in sea water. The role of breezes in burial selectivity has been discussed above. Other

mechanisms of living insect entry to water traps, such as drifting and rainwater, are likely to be of minor importance.

Only some of the insects that fall on to water die there. The **killing potential** of water traps depends on various factors, e.g. surface area and water chemistry. Extremely high concentration of dead insects have been observed at hypersaline, inland desert ponds and lakes (RUSTAMOV 1947, KARTASHEV & KRYZHANOVSKY 1954, GARYAINOV 1974, 1978) as well as at temporary hypersaline, littoral ponds (CAUSSANEL et al. 1997). TUROV (1950) described dead insects and vertebrates accumulating en mass at the coast of an island in the hypersaline Kara-Bogaz Gulf of Caspian Sea (but see below on hypersaline lakes as an unfavourable taphotope). Strongly mineralised thermal waters kill their occasional visitors very rapidly; also the mineralisation rate of organic matter in this kind of environment is extremely high (LIVINGSTONE & LIVINGSTONE 1865, HILTON 1946, PIERCE 1951).

Oil seepage also produces efficient killing traps (VINOGRADOV & STAL'MAKOVA 1938). Although this phenomenon is uncommon in natural environments, subfossil localities formed in natural surface oil and asphalt deposits are discovered in Europe (ŁOMNICKI 1894, SCHILLE 1916), Caucasus (BOGACHEV 1939, BURCHAK-ABRAMOVICH & DZHAFAROV 1955), North (PIERCE 1947, S.E. MILLER 1983, GUST 1992) and South America (K. BLAIR 1927, KUGLER 1927, CHURCHER 1966). Actualistic observations on this type natural traps are scarce (BARASH et al. 1970, PIERCE 1949). Fresh tar seems to be attractive for some insects (SAYLOR 1933); an additional and highly selective attraction mechanism should be the high concentration of vertebrate carrion at such sites.

Weather conditions affect both the place of death and the mortality rate of insects as well as the transport of their dead bodies or parts. The role of catastrophic storms in the accumulation of flying insects in water bodies and at their shorelines has been mentioned above; this mass mortality mechanism may be of a great importance. Besides, in permanently stratified (meromictic) lakes, strong winds can cause uplifting of toxic hypolimnion waters resulting in suffocation and mass mortality of aquatic organisms dwelling the shallow water zone; this mechanism was postulated to explain the origin of some lacustrine insect oryctocoenoses (e.g. PONOMARENKO & KALUGINA 1980, PONOMARENKO 1997). Other important weather events favouring the mass transport to water of terrestrial insects, dead or alive, are floods (BONESS 1975, KISELYOV 1981, NAZAROV 1984, BONESS & STARÝ 1988, RÖDEL & KAUPP 1994, TOWNSEND 1994) and strong rains (NAZAROV 1984) as well as ice break-up (HUDON 1994). Precipitation also effects fluctuations of water level, and muddy areas that are temporarily exposed after large floods or during droughts seem to be ideal places for trapping insects and their subsequent conservation. Even when shallow pools remain at such areas, many aquatic insects using them as refugia can be killed by warming. The unstable playa-lake model is suggested for some important insect Lagerstätten, e.g. for the Eocene Green River Series in USA (EUGSTER & SURDAM 1973, EUGSTER & HARDIE 1975, SURDAM & WOLFBAUER 1975, BOYER 1982). Recently D. SMITH (1999) demonstrated an important similarity between the Green River oryctocoenoses and modern thanatocoenoses of a seasonally drying playa lake in Arizona. At the same time she found that the oryctocoenosis of another well-known North American, Early Tertiary locality, Florissant, that originated in a much more stable lake environment, differs considerably. PANFILOV (1968) supposed that in the Late Jurassic Karatau Lagerstätte many terrestrial insects and their fragments had settled to the muddy bottom in areas that were only temporarily covered by lake waters after strong rains. This model may well be correct for some other localities, but hardly for Karatau;

in general, muddy traps are likely to be of limited importance in insect taphonomy. Some fossiliferous layers in different localities demonstrate evidences of temporary drying up such as sediment cracks, shrivelled insect exuvia, etc. (see Fig. 22), but this is not a common occurrence. Large-sized insect fossils rarely demonstrate any traces of the insect having crawled on to mud or other attempts to escape; this suggests that they had probably died before contact with the mud surface. Perhaps the preservation potential of muddy areas in general is not high because dead insects can easily be destroyed before being overlain by sediments carried by the next flooding.

Volcanic eruptions are infrequent but their taphonomic importance is disproportionally great because volcanic deposits are often highly resistant to erosion and thus over-represented in the rock record. Insect fossils occur in many volcano-clastic sequences though they are usually restricted to a few, thin, fine-grained layers in each (ZHERIKHIN 1997). SCOBLO (1968) stressed that insects are the only fossil organisms regularly occurring in Jurassic tuffite deposits in Western Transbaikalia where other animal remains are only occasional, and he established a special facies type of 'insect shale' in this area. This opinion is confirmed in a number of Jurassic, Cretaceous and Cainozoic localities in Siberia and the Russian Far East (personal observations). In the course of a catastrophic eruption, hot toxic gases and volcanic ash-fall cause mass mortality of diverse living organisms, including insects, over large areas. Many insects can be killed directly in the air, and this may explain, for instance, an unusually high share of strong fliers buried in ash layers such as the hymenopterans (RASNITSYN 1980). Hot ash-fall can kill aquatic insects as well, first of all by oxygen depletion. SINITZA (1993) notes the presence of caddisfly larvae in their cases buried in the Late Jurassic or Early Cretaceous tuffite deposits of Khutulyin-Khira in Mongolia. Normally nearly all fossil cases are buried empty, and the *in situ* conservation of larvae in this site was almost surely connected with lethal effect of ash. COCKERELL (1908a) called the well-known Florissant Lagerstätte in Colorado "a Miocene Pompeii". However, he has probably somewhat underestimated the importance of ash-falls in the formation of this important insect site. Actualistic observations on insects killed by ash-falls are rare. CHANEY (1938) observed abundant plant fragments and dead insects in fresh subaerial ash deposits in Alaska, and KRIVOLUTZKAYA & NECHAEV (1963) describe insect mortality during an ash-fall on the Kuril Islands in the North Pacific. Volcanic ash demonstrates an insecticidal effect for a long time after its fall, probably due to mechanical damage of cuticle with consequent desiccation or infection (J. EDWARDS & SCHWARTZ 1981, FYE 1983). If so, cuticular abrasion by ash particles may have an additional positive taphonomical effect facilitating penetration of the corpse through the surface tension barrier. Probably, this effect is selective depending on insect size and cuticle thickness but available data are scarce. While catastrophic eruptions can be regarded as taphonomically positive phenomenon, a continual volcanic activity should have a negative effect because of strong numerical impoverishment of the fauna (e.g. HUTCHESON 1992).

The **biotic factors of mortality** are taphonomically important as well. The main taphonomical effect of **predation** is usually negative because many individuals become lost for fossilisation. However, predators can also facilitate the burial of the most resistant undigested remains in their faeces, pellets and regurgitates. The importance of this source of palaeontological information is well demonstrated by both palaeontological and actualistic data, especially for small vertebrates (MELLETT 1974, MAYHEW 1977, POPLIN 1986, KUSMER 1990, EMSLIE & MESSENGER 1991). In regard to insects this mode of burial is

discussed in more detail below (Chapter 1.4.2.2.2). The indirect taphonomical effects of predation should be of considerable importance as well. It is easy to observe that numerous insects inhabiting shores, and especially species that escape predation by jumping, such as shore bugs, leafhoppers, grasshoppers, flea beetles, etc., can fall incidentally into water when disturbed by a potential enemy. Many insectivores consume their prey only partially, and the wings are among the least edible parts. Mammals often tear off the wings before devouring an insect, and the wings remain largely unswallowed when a bird is pecking at a prey insect. Small fish and some carnivorous insects (e.g. vespid wasps) also leave wings uneaten. There are no published quantitative data on wings lost in natural habitats where they may be seen occasionally, but their total production should be considerable, especially in places where insect predation is high. In Northern Siberia I have personally observed numerous horsefly wings (up to 14 wings/dm^2) on a tent after a week of a pied wagtail (*Motacilla alba*) feeding. STEFANESCU (1997) during his ten-days-long studies of lepidopteran migration across the Mediterranean Sea observed birds repeatedly attacking moths and butterflies and twice saw isolated noctuid wings, probably belonging to individuals who had been preyed upon; it should be stressed that his observations were made under a low abundance of migrating insects (the total number of flying lepidopterans observed was less than 50). Some predatory insects, both terrestrial and aquatic, as well as other invertebrates, suck out their prey leaving the integument practically undamaged. Such 'non-destructive' predation could have a positive taphonomic effect but its actual importance is unknown. In general, evolution of prey capture, feeding behaviour and digestive physiology of predators as well as the evolution of defence behaviours of prey may change the taphonomical role of predation from clearly negative to more or less positive, and *vice versa*. Mass mortality induced by **parasites** and **pathogens** may be of some positive importance because, in this case, the killing organisms usually do not destroy the host's body.

For parasitic insects, the biological and behavioural **characteristics of hosts** are often more taphonomically important than their own. Clearly, when a parasite's remains are attached to the fossilised remains of the host, the taphonomical properties of the latter are crucial (Chapter 1.4.2.2). A particularly interesting case is the occurrence of the remains of two species of supposed pterosaur ectoparasites in the Early Cretaceous of Transbaikalia (PONOMARENKO 1976a, RASNITSYN 1992b, RASNITSYN & ZHERIKHIN 2000). One of these species, *Saurophthirus longipes* Ponomarenko, is represented by a dozen or so specimens from two different localities (Chapter 2.2.1.3.4.5), and the large chewing louse *Saurodectes* is also found at one of them (Chapter 2.2.1.2.4.2). The Late Jurassic *Strashila incredibilis* Rasnitsyn may be a pterosaur ectoparasite as well (Chapter 2.2.1.3.4.5). The frequency of these finds is quite low; in Baissa, where the majority of specimens have been found, *Saurophthirus* constitutes less than 0.1% of the total terrestrial insect fossils. It must be remembered, however, that from the Cainozoic sedimentary deposits there has never been even a single recorded find of a wingless ectoparasite of any bird or mammal (including bats). Hence it can be suggested with a good degree of confidence that the pterosaurs should differ from the bats in their hair covering, or rubbing behaviour, or both.

1.4.2.1.2.3. Post-mortem ecological factors

Any destructive agents, both abiotic and biotic, affecting corpses in the thanatocoenoces are generally negative. The ecological factors can also create different physical barriers for burial (negative) or facilitate their crossing (positive). Unfortunately, the role of ecological factors in insect taphonomy is poorly studied and the available data are often controversial in some important respects.

The post-mortem factors depend on the place of death and differ principally between the subaerial and subaquatic death environments. First of all, **surface tension** creates an important barrier for terrestrial and aerial insect burial, preventing corpse submersion. For aquatic insects, both autochthonous and transported, this factor is of minor importance except for the floating exuviae of the pupae and final instar nymphs. However, the problem is still insufficiently studied and, as discussed above, the actualistic experimental data do not match very well with the composition of oryctocoenoses.

Wind (mostly indirectly, through its induced wave action) and **rain** are also important mechanisms permitting the surface tension barrier to be overcome and thus reducing the time that dead insects remain afloat (NAZAROV 1984, MARTÍNEZ-DELCLÒS & MARTINELL 1993). According to MARTÍNEZ-DELCLÒS & MARTINELL (1993) all insects in experimental aquaria have sunk after a strong rain. NAZAROV (1984) stressed that the mud and soil particles which adhered to rain-transported insects are also important in this respect.

Water density would also seem *a priori* to have potentially significant effect, with the probability of sinking of a dead insect (and the more so for isolated wings) decreasing from fresh to brackish and saline and further to hypersaline waters. Here again the observed palaeontological pattern is somewhat difficult to explain because, in fact, small insects with thin cuticle and low specific weight are not rare in some deposits formed in brackish and marine environments. LUTZ (1997) has analysed the relative abundance of 'heavy' (beetles and bugs) and 'light' (hymenopterans and dipterans) insects in selected Tertiary oryctocoenoses in relation to the supposed electrolyte concentration in the basin at burial. He found that the share of the 'heavy-bodied' forms increases with increasing 'salinity'. He also argued that in the permanently stratified (meromictic) lakes there is an important density barrier at the upper border of hypolimnion connected with the **chemocline**. Besides the increasing water density, a **bacterial film** that exists here acts to trap dead insects preventing their downshift to the bottom. However, LUTZ's paper is based mainly on a compilation of published data which are too often incomplete and non-quantitative. My personal observations on numerous, mostly Mesozoic localities of lacustrine origin disagree with the Lutz's model which is most probably oversimplified and ignores too many other taphonomical factors. The only way to test it accurately would be a multivariate analysis of a representative set of quantitative data for different localities, because it is practically impossible to select for comparison localities that differ only in the presence/absence of a chemocline. Actualistic observations could be useful as well but unfortunately the effect of water density has never been investigated experimentally.

Water current is the most important physical factor in the running waters. Its negative effect is obvious because many dead insects become disarticulated and abraded, especially in rapid and highly turbulent flows. Drifted fossil assemblages are composed mainly of the most durable insect sclerites such as beetle elytra (see Fig. 25). More delicate fossils such as membranous wings can predominate within the alluvial sequences in the most fine-grained mudstone accumulated in the stillest backwater environments. However, even in this case intact remains are virtually absent in a large (more than 1,000 specimens) collection from Kempendyay, Jakutia. In the sections of Jurassic oxbow lakes in Siberia and West Mongolia, the more coarse-grained

layers indicating increasing influence of river flow, contain more remains of terrestrial insects but only as fragments (mainly isolated homopteran wings). In contrast, the oryctocoenoses of the most fine-grained layers include more rare but much better preserved and often intact terrestrial insect fossils (ZHERIKHIN 1985, ZHERIKHIN & KALUGINA 1985).

Weathering is the dominant factor affecting organic remains in subaerial environments. Both physical and chemical weathering are taphonomically destructive and make subaerial thanatotopes extremely unfavourable for long term preservation of any remains including the most resistant ones (such as the vertebrate bones). There are no detailed observations on the time taken for the destruction of insect remains in natural environments; it seems that the most durable insect exoskeletons (e.g. those of large beetles) can persist for several years.

The effect of **fire** should be specially noted. Wildfires are a regular phenomenon in some terrestrial ecosystems, especially in those where a large amount of undecayed dry organic matter accumulate, and there is evidence of fires in the palaeontological record (SANDER & GEE 1990). Obviously, fire would normally have a strongly negative taphonomical effect, destroying many organic remains in the burned thanatotope. However, some types of the organic matter rapidly become coalified producing a characteristic coal variety called fusain. Fusainised remains are resistant to further decay and thus have a high preservation potential. Occasionally they demonstrate an excellent state of preservation including very delicate structures. Fusainised insect remains occasionally occur in small numbers together with other fossils (e.g. JARZEMBOWSKI 1995b: fig. 2). Some deposits rich in fusainised plant remains probably also contain rather numerous insect fossils. GRIMALDI *et al.* (2000) mention a rather rich insect assemblage discovered recently in the Late Cretaceous of New Jersey, USA, together with abundant fusainised plant remains. Undescribed beetle remains have been collected in similar Late Cretaceous deposits in Sweden (B.A. Korotyaev, pers. comm.). The insect fossils are three-dimensional and display fine morphological structures beautifully preserved. Such assemblages are of great interest because they probably reflect mainly the litter fauna which is otherwise rarely represented in the fossil record. Hence, layers that are rich in fusainised plant debris deserve much more attention by palaeoentomologists.

Rapid mineralisation of exposed organic remains can be occasionally induced by strongly mineralised soil solutions in extra-arid regions (FERSMAN 1924), but this mechanism is exceptional and its taphonomical importance is likely to be negligible.

Post-mortem biotic factors are diverse. Living or dead organisms rarely create simple mechanical taphonomical barriers or traps. Dead logs in watercourses may work as traps and current shelters (E. KELLER & SWANSON 1979) but the importance of such traps in insect taphonomy is probably insignificant because insects are rarely preserved in channel deposits. Algal mats on the bottom can probably operate as rather effective traps for dead bodies transported there by currents but this is a purely speculative idea and special observations are necessary to estimate the importance of algal traps. As mentioned above, bacterial growth possibly reinforces the barrier effect of the chemocline but here again no direct observations are available. The negative effect of dense terrestrial and especially **shore vegetation** is more obvious. NAZAROV (1984) demonstrated that in the Quaternary, the taxonomic content of insect oryctocoenoses becomes strongly impoverished with increasing abundance of the shore plants such as sedges. In such deposits there are very few terrestrial insects other than those species that inhabit the sedge growth (e.g. weevils of the genus *Notaris*). Nazarov explained this pattern as the result of the sedge "brush" trapping dead insects carried by rain water. ZHERIKHIN (1985) confirmed this observations for Mesozoic oryctocoenoses with abundant semiaquatic horsetails and pointed out that dense vegetation in shallow-water can also catch both living and dead insects floating on the water surface together with other debris and foam, and prevent their sinking.

The **benthic scavenger activity** is commonly regarded as one of the most taphonomically destructive ecological factors and the origin of the conservation Lagerstätten is often believed to be connected with environments with low or even no benthic metazoan activity (e.g. J. GALL 1983, SEILACHER *et al.* 1985, BEHRENSMEYER & HOOK 1992). Consequently anoxic environments are often considered as the most taphonomically favourable, and near-bottom anoxia has been postulated for many fossil sites including important insect Lagerstätten (e.g. PONOMARENKO & KALUGINA 1980, LUTZ 1990, FREGENAL-MARTÍNEZ *et al.* 1992, MARTILL 1993, PEÑALVER 1998). Although the anoxic model may well be correct for some sites, it seems to be far from universal. In fact anoxia is ineffective as a long-time preservation mechanism, and additional mechanisms are necessary (ALLISON 1988). On the other hand, the destructive effect of bottom surface scavengers is probably comparable with that of predators and terrestrial scavengers, in that it often leaves at least the most durable parts undamaged. This is illustrated by the common presence of traces of locomotion of worm-like benthic organisms (see Fig. 465) and invertebrate coprolites (Fig. 31) in the fossiliferous lacustrine marls in Baissa, one of the most spectacular Lagerstätten of Early Cretaceous insects. The coprolites are often concentrated at fossil remains suggesting that scavengers were feeding on the dead insects but nevertheless it had not been fatal for fossil preservation. In a number of localities in the Late Jurassic or Early Cretaceous lacustrine deposits of the Glushkovo and Ukurey Formations in Transbaikalia, Russia, beautifully preserved insect remains including delicate intact midges and gnats occur together with numerous tadpole shrimps and their coprolites, the latter occasionally containing insect fragments. It should be stressed that in these localities, as well as in Baissa, evidence of benthic life and well-preserved insect fossils can be seen on the same bedding surface. Most probably, the negative taphonomical effect of the surface (epifaunal) benthos is related to its abundance and perhaps also its taxonomic composition. Burrowing benthic (infaunal) organisms may be more destructive, indeed (Chapter 1.4.2) but their distribution is controlled by oxygen availability within the sediment layer rather than in the near-bottom waters and hence near-bottom anoxia is not necessary to reduce their activity.

Microbial activity is another important but insufficiently studied ecological factor. Here again, the negative taphonomic effect is the most evident but surely not the only point. The negative effect of partial decay was demonstrated by ALLISON (1986) who found that in laboratory experiment the probability of disarticulation of crustacean exoskeleton depends more on the state of decay than on the distance of transport by currents; this is probably true for insects as well. Gas production during the course of internal tissue decay is a negative factor because it increases the corpse's buoyancy, while microbial decay of the epicuticle should be taphonomically positive as discussed above. The development of microbial and fungal growth on a dead insect body may reduce the time necessary for its sinking (MARTÍNEZ-DELCLÒS & MARTINELL 1993) but if they are gas-producing they may

Fig. 31 Invertebrate coprolites covering nymphal *Hemeroscopus baissicus* Pritykina; Early Cretaceous of Baissa in Siberia (PIN 3064/10579, photo by A.P. Rasnitsyn); 6 visible nymphal segments 22 mm long

increase the insect's buoyancy by producing bubbles (LUTZ 1990). In water bodies the bacterial consumption of dissolved oxygen (both direct, through microbial breathing, and indirect, through oxidation of methane and other microbial metabolites) can suppress benthic life including destructive predators and scavengers. Micro-organisms also strongly affect sedimentation and the processes within sediments but this is a taphotopical rather than an ecological question (see below).

1.4.2.1.3. Taphotopical factors

This and the following group of factors generally attract much attention in taphonomy and they are what the bulk of taphonomical papers deal with. However, in insect taphonomy the effect of taphotopic environments is poorly studied. Actualistic observations on the fate of insects buried in recent sediments are scarce. Moreover, there are striking differences between the most common subfossil and fossil insect assemblages which are difficult to explain. Lacustrine sites rich in intact insect remains are quite exceptional in the Quaternary record (e.g. Shiobara in Japan) which is usually strongly dominated either by beetle fragments or by head capsules of larval chironomids. Hence no good subfossil model sites are known that can help to elucidate the origin of older insect Lagerstätten, and only general speculative conclusions can be made.

From a formal viewpoint taphotopical factors operate from the beginning of the burial process, i.e. after the formation of the first sediment layer overlying the remains. Accordingly, all preburial barriers have to be crossed up to this time point. However, surface bottom sediments are still not completely excluded from ecosystem turnover and thus it is not always easy to strictly delimit taphotopical and preburial ecological factors. The same is true for the delimitation between taphotopical and diagenetic factors: diagenetic processes (e.g. density increase and chemical alteration) occur even in fresh soft sediments while the final lithification takes places only after thousands of years or more. Usually the most dynamic microenvironments occur in the upper few centimetres of the sediment layers. Any aggressive agents affecting buried remains or traces are taphonomically negative, and many of them are basically identical with the post-mortem ecological factors discussed above (physical disturbances, predator, scavenger and decomposer activity). However, it does not necessarily mean that any taphonomically favourable thanatotope is equally favourable as a taphotope and *vice versa*.

The **sedimentation rate** is one of the most important taphotopical factors. When it is very low, the probability of remains being destroyed before burial increases. On the other hand, extremely rapid sedimentation often suggests catastrophic physical events (mud flows, lahars, etc.) which themselves are often taphonomically destructive (except for volcanic ash-falls). Hence, other things being equal, moderately rapid and continuous sedimentation is taphonomically optimal. No sedimentary sequences are really continuous, and all of them contain innumerable stratigraphic hiatuses of various duration reflecting repeating sediment disturbances. The rate and continuity of sedimentation depends on a complex set of physiographical, hydrological and hydrochemical conditions. In particular, in low energy waters the ratio between the areas of drainage and sedimentation is important, and hence the average sedimentation rate in lakes (which are smaller in relation to their drainage basin area) is significantly (by about an order of magnitude) higher than in the sea (SLY 1978). On the other hand, the continuity of deposition in the lakes (even in their open-water areas with a low rate of disturbance) is generally low in comparison with marine sediments. The comparative effects of clastic, chemical and biogenic sedimentation on insect taphonomy are poorly known; in general, the latter two often suggest a quieter sedimentation environment compared with the former.

Thermal conditions seem to be influential because in the Quaternary assemblages the remains from interglacial deposits (NAZAROV 1984) and warmer climatic areas (BIDASHKO 1987) generally demonstrate poorer cuticle preservation than do glacial assemblages. However, it seems to be largely a result of more intensive biological decomposition and thus should be considered as a biotic rather than a direct physical phenomenon.

Any **physical disturbances** of sediment (e.g. by currents, wave action or desiccation) are taphonomically negative, though their effects can vary greatly. As a result, a high energy environment is of a low preservation potential even when dead insects are concentrated in large quantities (e.g. at the swash zone of a beach as discussed above). This is a good example of a discrepancy between the taphonomical potential of the same environment as the thanato- and taphotope.

The **mechanical composition** of sediments is important as well. Even when insect remains occur in coarse-grained deposits, their preservation state is generally poor. This is partially due to mechanical deformation by coarse mineral particles and partially to more rapid decomposition because of higher rates of fluid flow and of exchange of dissolved gases between the water and the sediment; further deterioration occurs in diagenesis (see below).

Data on insect preservation in relation to environmental **chemistry** are scarce and equivocal. In general, any factors favouring rapid **mineralisation** should be positive but their effects on insect remains in fresh sediments are virtually unexplored except for the aforementioned observations on thermal springs. Occasionally remains become mineralised even before their burial (Fig. 32). The mineralisation rate depends strongly on biotic (microbial) modifications of sediment chemistry and are considered below together with other microbial processes. Rapid sediment **cementation**, and in particular a syngenetic (very early diagenetic) formation of concretions in anaerobic conditions, favours the conservation organic remains inside them (BLOM & ALBERT 1985, K. MÜLLER 1985, BAIRD *et al.* 1986, PARK 1995).

Aggressive chemical environments are likely to be of relatively minor importance to insect taphonomy because the insect exoskeleton is resistant to the commonest destructive factor, which is high acidity (except for the rare case of calcified remains). On the other hand,

Fig. 32 A mineralised nymphal skin of a modern stonefly from a spring with strongly mineralised water; Chita Region in Siberia (photo by D.E. Shcherbakov); skin 20 mm long as preserved

hypersaline environments are probably unfavourable for insect conservation. GARYAINOV (1974, 1978) failed to find insect remains in shore sediments of a hypersaline lake where nevertheless a high concentration of dead insect bodies occurred at the ground surface. Here again the thanato- and taphotopical properties of the same environment are likely to be incongruent. The mechanism of insect destruction in hypersaline sediments is unstudied; perhaps, this is a physical rather than chemical process because salt crystallisation can mechanically destroy fragile insect exoskeletons. Indeed, salt deposits contains diverse insect microfossils (setae, scales, cuticle fragments) which may be the products of such a process (MANI 1945, 1947, GEORGE 1952). However, there are some rich fossil insect assemblages in saliniferous deposits (e.g., in the Oligocene of France and Nakhichevan' in Azerbaidzhan: QUIEVREUX 1935, ERMISCH 1936, BOGACHEV 1940). Garyainov's observations are restricted to the shallowest near-shore zone and so they are not necessarily applicable to deep-water hypersaline environments.

The role of **biotic factors** operating in the taphotope is great. The sedimentation process and character of sediments are under strong biological control, and biogenic components (bioclasts) often constitute an important part of the total sediment volume. In deep water areas where the transport of terrigenic particles is normally low to negligible, bioclasts can prevail over other components. Biological evolution affects sedimentation in different ways, and some type of sediments are restricted to certain geological time intervals. For instance, diatomite, one of the most favourable sediment types for insect conservation, appears in the marine record since the Jurassic but in non-marine sequences no earlier than in the latest Cretaceous.

Moreover, bottom surface deposits themselves are never biologically inert. In particular, **microbial activity** seems to be the key factor of early chemical transformation of buried organic matter (BERNER 1971, 1976, 1980, 1981). The negative taphonomical effect of microbial decomposition is clearly evident, including chemical degradation and physical thinning of cuticle, disarticulation of sclerites, etc. Even the most resistant parts may be somewhat affected by decay and maceration; in particular, occasionally a doubling of wing veins occurs owing to an exfoliation of the wing with some displacement of one wing wall in relation to another. However, in a taphotope the decay is less destructive than in thanatotopes because sclerites which have become disarticulated after membrane decomposition can at most be slightly displaced within the fluid mud and so can be preserved near one another (unless additional physical sediment disturbances occur). Sometimes such disarticulated remains are even more suitable for morphological studies than are intact fossils. Ponomarenko (in ARNOLDI et al. 1977) notes that the enlarged hind coxae in intact Mesozoic trachypachid ground beetles are often unclear or nearly indiscernible against the abdominal background but they are clearly visible when the abdomen is separated from the metathorax.

There are few detailed actualistic observations on arthropod decay though, as discussed before, chitinolytic bacteria, which are of a special interest in insect taphonomy as potential decomposers, are widespread in diverse terrestrial, freshwater and marine environments. Chemical degradation of chitin and cuticular proteins in arthropod exoskeletons in seminatural subaquatic environments proceeds rather rapidly, with major loss in the first few weeks (STANKIEWICZ et al. 1998b). Some experimental studies indicate that arthropod cuticle in fresh water degrades less rapidly than in marine conditions (STANKIEWICZ et al. 1998b, BRIGGS et al. 1998a); however, these interesting data are based on few observations and need confirmation. The discovery of

chemically preserved chitin in insect remains from Oligocene lake deposits of Enspel (STANKIEWICZ et al. 1997a) and a high scatter in the data on the chitin contents in subfossil beetle remains (STANKIEWICZ et al. 1998a) indicate that the taphotopic environment rather than age controls the long term chemical preservation of cuticle. Unfortunately, the relative conservation potential of different taphotopes is unknown. When some quantities of chemically preserved chitin are recognisable, its absolute content correlates positively with cuticle thickness (BRIGGS et al. 1998b). The chemical composition of studied cuticle samples demonstrates obvious differences between materials from different deposits and even within the same deposit, which could hardly be explained solely by different original composition or selective preservation, and indicates rather complex chemical transformations including secondary re-polymerisation of biomolecules (BRIGGS et al. 1998a, STANKIEWICZ et al. 1998d, BRIGGS 1999). Scanning electron microscopy of fossil cuticle demonstrates a microstructural pattern suggesting different decomposition rates of different components, similar to those noted in actualistic experiments (McCOBB et al. 1998). In the best chemically preserved samples the microstructural preservation is also exceptional, showing the chitinous fibres still enclosed in a proteinaceous matrix quite like those in fresh modern cuticle (STANKIEWICZ et al. 1997a). However, there is no simple correlation between the chemical and macrostructural preservation of the fossils, and strongly chemically altered remains can show remarkably perfect state of morphological preservation (e.g., McCOBB et al. 1998).

It should be noted that the available chemical and microstructural data are restricted to material from a few insect fossil sites, mainly those where the remains demonstrate visually well-preserved and evidently weakly altered cuticle. However, visual observations indicate that the preservation state is extremely variable, including numerous obviously specific cases which are still awaiting detailed investigation. In particular, insect cuticle is often well recognisable under polarised light, even when under normal lightening the remains look just like an impression lacking any trace of organic matter. Moreover, sometimes (e.g., in the Triassic deposits at Garazhovka, Ukraine) some insect specimens are either extremely unclear or even practically invisible whereas under polarised light their morphology is rather well observable. Hence even a high degree of organic matter degradation is not necessarily taphonomically fatal.

At a number of sites, thick cuticle is apparently coalified (it should be noted, however, that it was never studied chemically and the actual state of the cuticle is unknown): this is mainly the case in some volcanoclastic deposits as well as some coal-bearing strata. Both differ in some significant taphonomical features suggesting that the coalification occurred under strongly different taphotopical conditions. This may be illustrated by two contrasting Cretaceous localities in the Russian Far East, namely some tuffite beds of the Emanra Formation (described in detail in GROMOV et al. 1993) and the Arkagala Formation containing economically important coal layers. Both insect oryctocoenoses are numerically quite rich but are strongly dominated by beetles and caddis cases. Other insects with strong cuticle (e.g., cockroaches, auchenorhynchan homopterans and large-sized aculeate hymenopterans) occur occasionally in relatively small numbers but the membranous (non-elytrised) wings are never observable, and insects with thin exoskeletons are completely missing.

In the Emanra Formation, coalified remains are restricted to several quite thin (up to few centimetres thick) layers of the finely laminated fine-grained tuffite; there are other fossiliferous beds within the sequence as well, but with a different preservation pattern. Intact beetle remains predominate suggesting that the burial was probably rapid and the intensity of decay in the taphotope was low or negligible. In this case the coalification was probably caused directly by a hot ash fall, and the thanatotope stage was probably missing because of a catastrophic burial pattern resulting in practically instantaneous sediment accumulation. In other volcanoclastic deposits similar effects are usually much less pronounced, suggesting that although the burial rate was about as high as in the Emanra Formation the ash temperature was significantly lower. For example, in the Late Cretaceous Ola Formation in the same region and in the Upper Jurassic Uda Formation in Transbaikalia, the coalification is less obvious, insects with thin cuticle are quite abundant, and membranous wings are well preserved. In fact, most fine-grained ash fractions usually reach ground surface at a considerable distance from the eruption point, so that the ash temperature would be expected to decrease significantly during the air transport. Indeed, coarse-grained ash deposits probably formed near the volcanoes occasionally contain coalified beetle remains and caddis cases like in the Emanra Formation but these are poorly preserved in the coarse rock matrix (e.g., in the Cretaceous Arzamasov Formation in Russian Far East: ZHERIKHIN 1978).

Unlike the Emanra deposit, in the Arkagala Formation the preservation pattern is rather uniform throughout the section where insect remains occur in several rather thick layers (up to 1–1.5 m, with weak if any lamination both under and below economical coal layers). In one layer the material seems to be somewhat less coalified and only in this layer several less sclerotised objects have been collected (e.g., orthopteran and neuropteran wings). Intact beetles are very rare (no more than one or two per cent), and fragments probably belonging to the same beetle specimen are sometimes at a distance of 2 to 3 centimetres from each other. Even the abdominal segments are often disarticulated. This pattern suggests a complete decomposition of the membranous parts, and consequent disintegration of the exoskeleton into separated sclerites under a low sedimentation rate in a forested swamp environment. Other Mesozoic coal-bearing deposits often demonstrate a similar pattern of disarticulation and selective coalification of the most sclerotised skeletal parts only, but for unknown reasons the Arkagala oryctocoenosis is exceptional in respect of its high insect abundance. However, there is some evidence that the conditions of superficially similar remains may differ in different deposits. For example, the apparently coalified beetle elytra from the Lower Cretaceous Ognevka Formation (Syndasko Bay, Taimyr Peninsula, North Siberia) are unrecognisable under polarised light. The Palaeozoic situation seems to be different from the Mesozoic as indicated by numerous records of insect wings from Carboniferous and Permian coal measures. In general, the effects of different palaeoenvironments (in particular, the taphotopic conditions) on coal petrography are well studied (TEICHMÜLLER 1989), and comparative studies of insect preservation in genetically different coal-bearing deposits should make an important contribution to insect taphonomy.

The chemical studies of subfossil and fossil insects referred to above concern mainly the pattern of chitin degradation; other potentially important aspects of chemical preservation are even less investigated. This is true in particular for insect pigments. A contrasting pattern of dark and pale parts is often clearly visible on insect fossils including the oldest known ones (e.g., Figs. 108, 125, 136, 196, 254, 272, 319, 357, 376, 428, 436). However, this is probably owing to differential cuticle density and thickness rather than chemically preserved melanins or products of their degradation. Even in subfossil remains,

natural colours other than black or dark brown are usually lost or at least they are strongly modified. However, occasionally at least some traces of supposedly natural red and yellow colours are recognisable. I have seen subfossil ladybird beetle (Coccinelidae) elytra still with a red background (contrary to ELIAS 1994 who describes them as usually being so discoloured as to retain only a distinct black and pale pattern). Some Early Cretaceous buprestid beetle elytra (e.g., from Baissa in Transbaikalia and Oyun'-Khaya in Yakutia, Siberia) show the pale spots as distinctly yellow in freshly exposed specimens, like in many present-day species: this colour disappearing gradually over several months or even longer (e.g., with the holotype of *Metabuprestium oyunchaiense* Alexeev). A Miocene rhipiphorid beetle from Chon-Tuz, Kyrgyzstan, collected in 1936, still displayed a pale orange abdomen like in modern *Metoecus*, though much less intense in colour, by the end of the 1960's. During mechanical preparation of an unidentified beetle elytron from the Palaeocene Tadushi Formation (Russian Far East), I was surprised to discover a distinct light red apical spot. The red colour disappeared over the following quarter of an hour, surely because of rapid pigment oxidation, and only a pale area of the rock matrix colour has persisted at its place. In all those cases, carotinoid pigments were probably preserved in fossils but degraded upon exposure to the atmosphere or light.

Another interesting case was observed repeatedly during fossil collecting at Baissa. Freshly exposed specimens of some insects (mainly large-sized dragonfly nymphs but occasionally also adult mayflies and backswimmers (Notonectidae)), rather often show more or less distinctly pink-coloured eyes, which fade out gradually over several days after the rock splitting. Moulted skins never show this feature. Possibly it could be related to rhodopsin conserved in the rock which degrades after contact with air. This unique colour conservation was observed exclusively in finely laminated light-grey marl layers and never in dark or yellowish laminated marls, massive marls of any colours or any other rocks occurring in the section including the bituminous shale. Interestingly, I never saw any insect remains, either fossil or even subfossil, showing a green pigment colour such as is so common in living katydids, grasshoppers or leafhoppers; probably the green insect pigments are much less resistant to chemical degradation than are red or yellow ones. Such sporadical observations on coloured fossils indicate that they could provide interesting additional information on the taphotopic conditions. More attention should be paid to the colour changes in freshly exposed specimens, and development of special techniques for rapid colour conservation could enlarge the field of insect palaeobiochemistry.

One more notable aspect is the preservation of structural metallic coloration depending on optical interference in the fine structures of the endocuticle. This type of coloration is rather common in modern insects, especially in some beetles, lepidopterans and odonatans. Subfossil beetle remains often demonstrate metallic colours as well, either natural or somewhat altered (ELIAS 1994), but there are strong differences between materials from different sites. An alteration may be caused by mechanical deformations, or structural changes in the course of chemical degradation, or both. Further ultrastructural studies should elucidate the relative role of possible alteration mechanisms, and then the preservation of metallic colours probably could be used as an indicator of taphotopical environments. Some authors (e.g., LUTZ 1990) believe that the metallic colours which often occurred on fossil beetle remains from some localities may be natural or only slightly altered. This opinion needs confirmation by ultrastructural examination of the cuticle because of the possibility of its diagenetic compression. It should be noted that this peculiar type of preservation is almost completely restricted to very few localities where the preservation state of fossil remains is quite exceptional in many respects (Geiseltal, Messel, Enspel), and that only one of these sites (Enspel) has the unique chemical and microstructural preservation of insect cuticle already been demonstrated (STANKIEWICZ et al. 1997a). In other deposits, metallic coloration occurs very rarely. I have seen a unique ground beetle specimen from a marl layer in Baissa, with a metallic blue cuticle. Its cuticle was extremely thin suggesting significant compression or degradation and hence hardly could preserve its original colour unaltered. Unfortunately, unlike any other fossils from the same locality and layer, the cuticle rapidly became exfoliated and fragmented after its contact with the air and desiccation, and so could not be conserved. In the Palaeocene lacustrine Tadushi Formation (Russian Far East) an unidentified chalcidoid wasp fossil displays a metallic green coloration, similar to that which is so characteristic of its extant relatives. Interestingly, this fossil was found in the same deposit with the above-mentioned pigmented beetle elytron.

Microbial activity controls not only the organic matter decomposition but also general chemistry of sediments and porous waters, both directly and indirectly (REDFIELD 1958, S. KUZNETSOV et al. 1962, S. KUZNETSOV 1970, SIEBURTH 1979, SUESS 1979, EHRLICH 1981, SCHINK 1989). It also plays an important role in early diagenesis including the formation of concretions in surface sediments (BERNER 1968, TYLER & BUCKNEY 1980, RAISWELL 1987, 1988), as well as in mineral replacement of organic matter (S. KUZNETSOV 1970, BERNER 1981, LUCAS & PREVOT 1985, WILBY 1993). In general, decomposition in anaerobic environments is slower and less destructive than in aerobic ones, though an oxygen deficiency is neither a necessary nor sufficient condition for good preservation. Some fossiliferous rocks are interpreted as deposited by bottom cyanobacterial mats (e.g., the Jurassic lithographic limestone of Germany: K. BARTHEL 1970, 1972, K. BARTHEL et al. 1990), and exceptional fossil preservation including structural preservation of soft tissues is often a result of fossilisation of bacteria rather than of the tissue itself (WÜTTKE 1983, MARTILL 1987, WILLEMS & WÜTTKE 1987, J. GALL 1990). Thicker and more massive algal stromatolites in freshwater deposits may contain abundant caddisfly cases (BERTRAND-SARFATI et al. 1966, SINITZA 1993), but usually few if any other insect remains. Those finds are unusual in two respects: firstly, the cases are three-dimensionally preserved suggesting an extremely rapid calcification of their soft silky base, and, secondly, the cases which are potentially easily transportable by bottom currents are here surely euautochthonous (buried exactly at the feeding places of algophagous caddisworms).

Insect mineralisation is best documented for three-dimensionally preserved petrified remains occurring occasionally in diverse palaeoenvironments. However, even in those unusual fossils the replacing minerals have only been identified precisely in a few cases, and their records in the old palaeontological literature might be taken with reservation as they were never proved by detailed chemical analysis. Data on mineral composition of more ordinary compressed fossils are also quite scarce. Nevertheless, a number of minerals has been identified in both three-dimensional and compressed remains, including different varieties of calcite (calcium carbonate), gypsum (calcium sulphate), apatite (calcium phosphate), quartz (silicon dioxide), pyrite (iron sulphide) and limonite (iron hydroxide) as well as analcite (sodium aluminosilicate) and other zeolites, siderite (iron carbonate), celestite (strontium sulphate), and others (ZEUNER 1931, FRANZ 1942, LEAKEY 1952, PALMER et al. 1957, BRITTON 1960, DOBERENZ et al.

1966, DORF 1967, BACHMAYER et al. 1971, RUNDLE & COOPER 1971, GRIMALDI & MAISEY 1990, PARK 1995, DUNCAN et al. 1998, PEÑALVER 1998). The composition of authigenic minerals provides valuable information on taphotopic environments; in particular, iron oxide minerals indicate oxidising, and siderite and pyrite reducing environments. Mineralisation can occur at various stages of the taphonomical process from the thanatotope before burial as discussed above but especially in the taphotope, and microbial processes often play an important role in this process, though these are not always well studied. Commonly, only hard skeletal parts are mineralised while soft insect tissues, especially internal organs, are preserved quite exceptionally (e.g., PALMER et al. 1957), indicating that mineralisation usually occurs some time after death when they are already decomposed. A few deposits containing mineralised insects were carefully investigated taphonomically, including the Late Palaeocene-Early Eocene marine London Clay in England with pyritised insect fossils (ALLISON 1988), and early diagenetic concretions in the Miocene lacustrine Barstow Formation in California, USA, with a remarkably rich suite of minerals being involved, dominated by celestite and quartz (PARK 1995). In the London Clay wood-boring insect taxa dominate numerically, and their mineralisation could occur at least partially in a peculiar microtaphotope inside pyritised wood. However, this was probably not the case for other deposits with pyritised insects such as the Miocene (Tarkhanian) marine beds at the Kerch Peninsula (Crimea, Ukraine).

Besides their chemical activity, bacterial and algal films stabilise the bottom surface reducing the level of physical disturbance (E. HOLLAND et al. 1974, RHOADS et al. 1978, YINGST & RHOADS 1978, HENLE 1992).

In general, the effects of microbial activity in the taphotope seem to be one of the most essential fields in insect taphonomy. Future special studies in this direction can resolve a number of unanswered problems, and broad comparative investigations of hydrochemically, thermally and trophically different water bodies are highly desirable. However, it is quite possible that the principal types of standing waters in the past differed from present-day basins in important hydrochemical and trophic characteristics, and thus actualistic models might be of a limited use. In particular, the Cainozoic deep-water lacustrine bituminous shale often contains well-preserved insects while in similar facial environments in the Mesozoic, the state of insect preservation is usually poor.

Infaunal benthos activity is generally taphonomically negative but its consequences depend strongly on both the abundance and taxonomic composition of the benthos. FISHER (1982) and MCCALL & TEVESZ (1982) presented useful reviews of actualistic data on invertebrate activity in freshwater sediments. The most important taphonomically negative actions of the benthos are scavenging, direct sediment disturbance (bioturbation), and effects indirectly increasing sediment erodability (increasing porosity and destruction of bacterial and algal films). A further taphonomically negative consequence of bioturbation is that it keeps organic remains in the surface zone where degradation processes are most intensive (HART 1986). Possible positive effects are limited and include mainly sediment stabilisation by invertebrate pellets. However, the relations between the invertebrate fauna and microbial processes in sediments are still poorly studied, especially in non-marine environments, and thus the existence of additional indirect taphonomically positive phenomena cannot be excluded. Large-sized filter-feeders and especially the bivalve molluscs are probably the most destructive macroinvertebrates in fresh water taphotopes, and the effect of the smaller burrowing benthos (e.g., oligochaetes and chironomids) varies strongly depending on their density and spatial distribution. It should be stressed that bottom anoxia is not a necessary condition for a low benthos activity; for instance, in the areas of low biological productivity the sediment disturbances induced by benthic invertebrates are minimal even in oxygenated waters (COHEN 1984). The ichnological record indicates that colonisation of inland waters by infaunal benthos was a very slow process (BUATOIS *et al.* 1998), and consequently its taphonomical role should be different at different geological time intervals. In particular, in pre-Permian non-marine deposits, burrows are extremely rare, and bioturbation depth seems to increase in the Jurassic.

The **subaerial taphotopes** are generally much less taphonomically profitable than subaquatic ones because of intense abiotic and biotic weathering and erosion except for some highly specific situations (e.g., permafrost). Subfossil insects occur (occasionally in large numbers) in diverse subaerial palaeoenvironments including buried soils, tar seeps, archaeological sites, ice shields, and in the organic deposits that fill crevices, caves and vertebrate burrows. Those deposits are unstable in time and are virtually restricted to the Quaternary. The sediments of small temporary pools and springs, despite being deposited under water, resemble subaerial burial places in this respect and are virtually absent from the fossil record before the Neogene. Palaeosols and aeolian deposits themselves are much more stable and occur widely in the geological record since the Palaeozoic but are unfavourable for long-time conservation of any organic remains. Pre-Quaternary insects are represented in those environments exclusively by ichnofossils which are occasionally abundant and highly diverse. Karst deposits constitute a notable exception as they occur sporadically at different stratigraphical levels and occasionally contain insect fossils including some remarkably preserved three-dimensional remains suggesting highly specific taphotopical conditions with rapid (probably microbial) mineralisation. No detailed studies on taphotopical factors affecting insect preservation in subaerial environments are available. Diverse insect deposits of subaerial origins are discussed in more detail below (Chapter 1.4.2.4).

1.4.2.1.4. Postburial factors

The process of fossilisation includes various physical (filling by sediments, flattening, crushing, distortion) and chemical (various kinds of organic matter degradation and mineralisation) changes of organic remains. In fact, physical and chemical changes begin very early on, and thus the border between taphotopical and strictly diagenetic factors is by no means sharp. The diagenetic factors are conventionally restricted here to that stage of the taphonomic process when the direct effects of living organisms on sediment alterations are negligible and, accordingly, the earliest diagenetic processes dominated by biotic factors are considered above as taphotopical ones. However, after this stage there may be a rather long time interval when biological activity in sediments is already negligible but the sediment is still not lithified and the overlaid deposits are not very thick. An early lithification (e.g., due to formation of early diagenetic concretions) can exclude the remains from biological turnover even in surface sediments. If lithification is slow, the accumulation of overlying sediments is the most important factor controlling the degree of direct biotic effects. The depth of burial necessary to stop them varies widely but in bottom sediments they are rarely significant below 10–15 cm under the bottom surface. All subsurface processes, both taphotopical and oryctotopical

(or diagenetic in the strict sense), are generally regarded as being **early diagenesis**. It includes mechanical sediment compaction and consolidation as well as cementation and diverse other chemical modifications. This is the last stage of the taphonomic process when some changes (in particular, rapid cementation) can still be taphonomically favourable. Later on, any further modifications of the fossiliferous rock can only deteriorate the preservation state of enclosed fossils, and hence the less change that occurs the better the taphonomical result. Diagenetic changes continue in a rock deep in the lithosphere (**late diagenesis** or, according to the classification generally accepted in Russian geological literature, the **kata-** and **metagenesis**). If those changes are induced by outside influences, a special term, **metamorphism**, is used (with additional subdivisions in relation to the cause of changes, e.g., the **dynamic**, **thermal**, or **contact metamorphism**). When a rock becomes exposed, it undergoes **hypergenesis**, including oxidation, washing, biogenic weathering, formation of a weathered crust (regolith) and soil, etc.

The taphonomical literature dealing with diagenetic factors is quite extensive but few special studies on insects are available. After the isolation of the sediment layer from direct biological activity, the only significant taphonomically positive type of changes is probably a rapid sediment cementation (including local cementation within the layer leading to formation of concretions) that prevents further physical deformation of enclosed organic remains. The degree of deformation of insect fossils varies strongly but usually they demonstrate at least some evidence of compression. Its degree can vary between different insects in the same layer (mainly due to different cuticle thickness) as well as between different layers in the same sequence (due to different lithification rate). Often the remains are completely flattened, with both dorsal and ventral (or right and left side) structures superimposed and observable at the same slab; sclerotised internal structures (e.g., tracheae, proventricular armature, spermathecae, eggs) and gut contents are also occasionally observable. Such interposition of different structures leads sometimes to serious misinterpretations of the fossil's morphology. Strong compression affects the sculpture of the cuticle, or even causes its complete disappearance. As a result, conspecific specimens from different sites or even layers may have a very different general appearance; this is another important source of mistakes. Occasionally nearly undeformed three-dimensional fossils occur either as mineralised inclusions (Fig. 33) or as cavities in a rock (Fig. 34). In the former case, rapid mineralisation of the remains (usually owing to taphotopical bacterial activity as discussed above) is necessary while in the latter the rate of sediment lithification is the crucial factor of preservation. Three-dimensional cavities, although they maintain the original shape more exactly, are often less convenient for study, especially when filled in by sediment or crystals which is in fact rather common. If, however, there are no internal fillings, fine structural details are easily observable and latex or plastic replicas of such cavities appear as very exact copies of the original shape and sculpture of the insect (Fig. 35). No actualistic observations on insect deformation under the overlying sediment pressure are available. Quaternary assemblages show that there is no simple universal correlation between the sediment age and the degree of insect deformation; however, some authors (e.g., NAZAROV & KOVALYUKH 1993, ZINOVYEV 1997) believe that the pattern of preservation of beetle cuticular microsculpture can be used for dating of Quaternary insect faunas. The importance of cuticle thickness and sclerotisation is evident: the strongest and most sclerotised beetles from Quaternary deposits are nearly undeformed even when the insect-bearing layer is overlain by

Fig. 34 A chafer (Coleoptera: Scarabaeidae: Melolonthinae) preserved as an undeformed three-dimensional cavity in bog iron ore; Holocene of Protopopovo, Kemerovo Region in Siberia (photo by A.P. Rasnitsyn); elytra 11 mm long

Fig. 33 A three-dimensionally preserved pyritised weevil (Coleoptera, Curculionidae, ?Molytinae); Early Miocene (Tarkhanian) of Kerch in Ukraine (287/738, photo by D.E. Shcherbakov); 4.4 mm long

younger sediments several metres thick (Fig. 36). Soft-bodied insect larvae and the silky bases of caddis cases are the entomological objects least resistant to the pressure: even in those oryctocoenoses in which other insects are preserved more or less three-dimensionally they are usually completely flattened, and their three-dimensional preservation is a good indicator of very early (taphotopical) mineralisation (Fig. 37). Insect fossils are rarely strongly crushed; this suggests that the compression under increasing pressure of overlying sediments is usually rather gradual. This gradual flattening can distort the original body shape, especially when it was strongly convex to begin with.

Any displacement of disarticulated parts is only possible in a fluid mud and ceases after sediment lithification. However, other deformations are still possible even in a lithified rock under dynamic metamorphism. If the fossiliferous rock still retains some plasticity, tectonic disturbances can lead to more or less obvious distortion of fossils due to deformation of the rock matrix. In some cases this distortion can be quite strong, resulting in bizarrely shaped fossil remains, as in the Triassic deposits at Dzhayloucho in Kyrgyzstan (see Fig. 21). A method of restoration of the original shape of distorted fossils was proposed by Sdzuy (1962; see also Ponomarenko 1969). When the rock is very hard, similar disturbances induce crushing, sometimes with

a slight but well-observable displacement of fragments (Fig. 38). However, under a high pressure any rock flows, and hence plastic deformations are in principle possible at any lithification stage, especially in the areas of active volcanism and plate collision zones.

The nature of the rock matrix also considerably affects the preservation state. Other things being equal, the more fine-grained is the rock is, the better the preservation of delicate structures. Hence micrites of different nature (mudstone, micritic limestone and tuffite, diatomite, etc.) often contain excellently preserved insects, while in sandstone the preservation state is always poor. Poorly sorted rocks are unfavourable for insect preservation as well.

As discussed above, the original chemical composition of cuticle is normally altered, even at the earliest diagenetic stage in the taphotope, and this process would continue further during the lithification. In particular, mineral replacement of degraded organic matter under chemical processes in the rock is a common phenomenon widely discussed in taphonomical literature. However, the diagenetic alteration of arthropod cuticle in oryctotopic conditions remains largely uninvestigated. In particular, thermal diagenesis can be strongly destructive. Insect remains in the porcellanites (burnt rocks affected by coal seam fires) usually lack any recognisable cuticular traces and their morphology is unclear in many detail (e.g., in the porcelain jaspers from Chernovskie Mines, Transbaikalia: Sinitshenkova 1998a). On the other hand, moderately thermally metamorphosed rocks can contain

Fig. 35 A plastic replica of a *Ceutorhynchus* weevil (Coleoptera, Barididae: Ceutorhynchinae) preserved undeformed as a three dimensional cavity in a rock; Oligocene of Izarra in Spain (photo by D.E. Shcherbakov); weevil 2.5 mm long

well preserved insects as, for example, in the Triassic Mine Group in Japan where the state of carbonaceous materials suggests heating up to more than 160°C (AIZAWA 1991). MONETTA & PEREYRA (1986) describe insect remains from metamorphosed Carboniferous rocks in Argentina.

The presence of crystals inside internal cavities of some insect fossils as well as occasional mineral pseudomorphs associated with insect remains indicate the influences of dissolved minerals transported by waters slowly soaking inside the rock. Their effects on insect preservation are unexamined except for obvious obscuring of morphological details. There are some records of insect inclusions in crystals formed in rock cavities (KAWALL 1976, TILLYARD 1922a); this most rare and unusual mode of conservation is again mainly unstudied.

The hypergenesis of exposed fossiliferous rocks is surely a strongly negative taphonomical factor. Rates of weathering and ruination vary widely for different rocks and under different conditions of exposure. At this stage abiotic factors usually dominate though biological weathering can be important as well (e.g., by root penetration into rock layers). Sometimes destruction is so rapid that it can be observed during fossil collecting. In particular, salt crystallisation in freshly excavated wet mudstone under rapid drying occurs not infrequently in arid regions and can damage some specimens severely. Usually, however, destruction is rather slow and gradual, and in many localities more or less weathered but still recognisable remains can be seen on rock surfaces exposed for several years or longer. Direct biological destruction of fossils is probably negligible in most cases; however, I have observed mould growth on some insect remains collected from

Fig. 36 Well-preserved undeformed subfossil remains of *Stephanocleonus fossulatus* Fischer (Coleoptera, Curculionidae); the Edoma Formation, Late Pleistocene of Krestovka in Yakutia-Sakha Rep., Russia (photo by D.E. Shcherbakov); elytra 9.3, head and pronotum 5 mm long

Fig. 38 Roach forewing damaged due to deformation of the hard rock matrix (holotype of *Phyloblatta aliena* Becker-Migdissova, Phyloblattidae, from the Late Carboniferous of Zheltyi Yar in S. Siberia; wing 30 mm long as preserved, from ROHDENDORF *et al.* 1961)

Fig. 37 Three-dimensionally preserved caddis cases *Indusia tubulata* Brongniart; the *Indusia* Limestone, Early Miocene of Condailly Pit in St.-Gerand-le-Puy, France (PIN 4615/1, photo by D.E. Shcherbakov); aggregation 65 mm high

Oligocene diatomite at Bol'shaya Svetlovodnaya River in Russian Far East and packed under a high air humidity.

At this stage **contamination** of fossiliferous rocks by recent remains is possible (Chapter 1.4.3).

Some of the most resistant organic remains from exposed sediments can survive rock ruination and be **re-deposited** in younger sediments. Unlike vertebrate bones, shells and fossil woods, direct insect re-deposition seems to be extremely rare and its taphonomical effect is negligible. S. Kuzmina (pers. comm.) has observed exposed beetle fragments in the contemporary alluvium of the Lena River in Yakutia, that belong to species that have not been discovered in the present-day fauna of the area but are represented as subfossils in nearby exposed Quaternary deposits. This proves that the most mechanically resistant insect remains can potentially be re-deposited after several thousand years. A re-deposition of mineralised insect fossils in younger sediments is also possible; for instance, pyritised beetle remains originating from the London Clay deposits were collected repeatedly on a modern beach (VENABLES & TAYLOR 1963) and hence their secondary burial in beach deposits cannot be excluded. However, the only re-deposition mechanism of a real importance for insect taphonomy is their indirect re-deposition in containers (Chapter 1.4.2.2).

The most taphonomically important general consequence of sediment and rock erosion is that the rock record itself is strongly selective. This aspect is discussed in some detail below (Chapter 1.4.2.4.5).

1.4.2.1.5. Technical factors

The technical factors of fossil collecting are rarely discussed specifically in the taphonomic literature, and collections are often regarded as being at least roughly equivalent to oryctocoenoses. AMITROV (1989) is one of the very few authors who has discussed the problems of collection biases in relation to oryctocoenoses and of comparability between collections. He recognised three basic divisions of technical biases including statistical differences between collections, the differences connected with collecting methods and working conditions, and differences connected with uneven fossil distribution within a deposit. The biases of the former type can be estimated by standard methods of statistics; the latter two groups are more difficult for a formal analysis and need a brief discussion. However, the collecting methods and working conditions might be considered separately, because the nature of biases they produce are quite different in these two cases.

The availability of fossiliferous rocks for sampling fundamentally limits the possibilities of collecting. Only a very small part of surviving fossiliferous deposits are exposed and thus potentially available for careful examination. In this respect badlands are the most favourable areas, while in the heavily vegetated regions usually only a few small outcrops scattered mainly along watercourses are available. That is why such regions as Central Asian, Mongolian and North Chinese deserts or the North American Big Basin region seem to be so rich in fossils. The rocks unavailable in natural exposures can also occasionally be revealed by artificial exposures, mines, etc., and in this case the probability of their discovery depends on the level of industrial activity in the region. Not only present-day, but also past mining industry may be useful in this respect. For example, at Kargala (Orenburg Region, Russia) rich collections of Late Permian insects have been made from the abandoned copper mines and their slag-heaps in the 1930s, while the mines themselves were exploited most intensively between 1840s and 1860s and exhausted finally in 1913 (CHERNYKH 1997). On the other hand, some fossil sites can be completely destroyed by commercial pits. Fossils occur in drilling cores as well. Materials extracted by drilling are often highly valuable for their detailed stratigraphy, especially when many boreholes from the same area are available. However, their representativeness is always low (at least in regard to macrofossils) and they can hardly give more than a general idea on the dominant faunal components even when some layers rich in fossils are represented in the cores. Any deposits from the areas presently covered by sea or by ice shields are quite rarely obtainable for palaeontological sampling. This shortage of data is sometimes of major importance. For instance, during the large-scale Quaternary marine regressions vast areas of present-day continental shelf became exposed; those ancient coastal areas could have played an important role as faunal refugia in glaciated regions, but their biota is virtually unknown except for few sporadic materials from sea drilling (e.g., S.S. ELIAS *et al.* 1992a, 1996). Even the information on bottom sediments of modern large lakes is fragmentary (PICARD & HIGH 1981). The above-mentioned dissimilarity between the commonest Quaternary and older insect assemblages may be explained, at least partially, by this underrepresentation of open-water taphocoenoses in the Quaternary materials.

Techniques of collecting and studying fossils have changed historically and they have become more and more sophisticated; accordingly, more sources of palaeontological information have become available. However, in palaeoentomology the use of such recently introduced sources is still very limited. At present, in fossil insect studies new techniques are more involved with obtaining additional information (e.g., microstructural and palaeobiochemical) from the "traditional" sources (macrofossils) than with adding new ones (e.g., microfossils and chemofossils, see Chapter 1.4.2.3). There is nothing comparable with the revolutionary development of palynology and dispersed cuticle studies in palaeobotany, except perhaps the rapidly growing use of some microfossils (mainly the larval head capsules of chironomids) in Quaternary palaeoentomology. That does not mean that there has been no significant progress in insect collecting practices; however, this progress is mainly due to a growth in the general knowledge of distribution patterns of insect fossils (the collective scientific experience) and not to the development of new technology. Taphonomical observations are in fact the most powerful tool for minimising information loss, and making the collections more and more representative and comparable. Hence the most important negative technical factor in palaeoentomology is at present the scarcity of palaeontological information in standard courses of palaeontology for students, and in general textbooks on fossils for amateur collectors. Usually they are illustrated by a few photographs of Solnhofen dragonflies or other spectacular large-sized intact insect fossils, while much more common and widespread fragmentary or small-sized remains are rarely mentioned at all. As a result of this misleading educational practice, not only amateurs but also many professional palaeobotanists and vertebrate palaeontologists pay no attention in the field to even completely preserved aphids, thrips, and other small objects as well as to isolated wings and elytra, simply because they have never seen them and have no idea about their importance. My own experience tells me that when five to ten insect specimens have been collected from a fossil site by a geologist or palaeontologist who previously had no contacts with palaeoentomologists, this site should be a very rich insect Lagerstätte. More information on fossil insects might be included into educational programs and made available to the public to reduce irrevocable losses of invaluable palaeoentomological information. Unfortunately, no popular books on fossil insects are available at present.

1.4.2.2. Indirect burial in fossil containers

Insect fossils or ichnofossils quite often occur in or on other fossil objects which can be termed *fossil containers.* In this case, a thanato-, mero- or ichnocoenosis was formed primarily on or inside other organisms, dead or while still alive, and then those organisms or their parts might be buried to secure the conservation of insect remains or traces. Hence the probability and state of insect preservation depends both on their own properties and on the properties of the container which may be quite different. So the taphonomical process is more complicated here than in the case of a direct burial. In particular, some container types are much more mechanically resistant than the insect bodies themselves and therefore long-distance insect transport

inside a container is a common phenomenon in insect taphonomy. The evolution of container-producing organisms and their co-evolutionary relationships with the organisms buried in the containers are important in that the container taphonomy is more variable in time than the taphonomy of direct burial in sediments.

When a fossiliferous rock is being destroyed, its fragments (pebbles, nodules, erratic boulders, crystals, etc.) can be regarded as ***secondary containers*** of fossils which have been primarily buried directly in sedimentary rocks in accordance with the burial pattern described above (Chapter 1.4.2.1). In this case only the taphonomy of containers and not the taphonomy of contained objects needs some additional considerations. Occasionally the primary containers can be re-deposited in secondary containers as well.

1.4.2.2.1. Fossil resins

A number of plants produce liquid secretions that can operate as sticky traps for diverse small objects including insects. Often those secretions are quite limited, or they are water-soluble (e.g. polysaccharide gums) and thus cannot be fossilised. However, there is a category of plant secretions that is of great taphonomical significance, namely resin. This provides the most important kind of fossil container in palaeoentomology because of its wide distribution in the fossil record and the very peculiar state of insect preservation it achieves.

Ecological, taxonomic, physiological and biochemical aspects of resin secretion are discussed in a number of papers (e.g., HOWES 1949, NICHOLAS 1963, 1973, PIGULEVSKY 1965, WEISSMANN 1966, PRIDHAM 1967, LANGENHEIM 1969, 1990, CLAYTON 1970, THOMAS 1970, DELL & MCCOMB 1978, FAHN 1979). The resins are produced by parenchyma cells, mainly inside specialised ducts and cavities in different tissues and organs of a number of modern conifers and angiosperms. The resin can be excreted and accumulated either schizogenically (in intercellular spaces), or lysigenically (with destruction of the secretory cells), or with a combination of the two. Resin production varies greatly between different plant taxa from negligible up to very large quantities, occasionally with fluxes several dozens kilograms each. It often increases considerably with mechanical or thermal plant injury; some bacterial and fungal infections also induce a high levels of resin secretion. Chemically the resins are composed by a complex mixture of diverse terpenoids, i.e. mono-, di-, tri- and polymers of the isoprene unit

$$\begin{array}{c} CH_3 \\ | \\ -CH_2-C=CH-CH_2- \end{array}$$

Chemical composition varies widely both between and within plant species as well as between different organs of the same plant.

Fossilised resins are known since the Late Palaeozoic but before the Early Cretaceous they are only represented either by internal bodies in plant tissues or by extremely small external particles. The evolution of resin production and chemistry and the possible role of insect/plant co-evolution in this process are discussed briefly in Chapter 3.2.3.1. From the Early Cretaceous onwards fossil resins become widespread and diverse in respect of their physical properties and chemical composition (TSCHIRCH & STOCK 1936, LANGENHEIM & BECK 1968, LANGENHEIM 1969, VÁVRA & VYCUDILIK 1976, SAVKEVICH 1980, SCHLEE 1984, 1990, K. ANDERSON *et al.* 1992, K. ANDERSON & BOTTO 1993, LAMBERT *et al.* 1993, VÁVRA 1993, K. ANDERSON 1994, 1995, KRUMBIEGEL & KRUMBIEGEL 1996a,b, C. BECK 1999, KOSMOWSKA-CERANOWICZ 1999,

ZHERIKHIN & ESKOV 1999). It should be noted that some occasional records scattered in the geological literature have been overlooked in recent reviews (e.g., MCLACHLAN & MCMILLAN 1976, on the latest Jurassic or earliest Cretaceous finds in South Africa, SAINFELD 1952 for the Oligocene of Tunisia, R. JOHNSTON 1888 for the Palaeogene of Tasmania, and PISHCHIKOVA 1992 for the Miocene of Crimea Peninsula, Ukraine). These records, often old and forgotten, merit much more attention from palaeoentomologists. It seems that all fossil resins (except those represented by internal resin bodies or only by extremely small fragments) contain some organic inclusions, though their content is very variable, and it is sometimes difficult to collect the resin in sufficient quantity to find even a single inclusion.

Fossil resins are often collectively called "amber" though this word is commonly associated in mind with the well-known ornamental stone and may be misleading when applied to a brittle substance that rather more resembles colophon. It is therefore better to make a distinction between true **amber** (which is mechanically durable and potentially suitable for ornamentals, e.g., the Baltic, Saxon and Ukrainian succinite, the Rumanian, Sicilian, Burmese and Sakhalinian rumanite-type amber as well as Mexican and Dominican amber) and other **amber-like resins**. However, some of the latter are traditionally referred to as amber in both popular and scientific literature (e.g., the Lebanese, Canadian and New Jersey 'amber'). Those fossil resins other than the true amber are sometimes collectively designated as **retinites** or **resinites** though those terms are not clearly defined. In the old mineralogical literature a great number of specific names were proposed for the resins of different origin, chemistry and physical properties (e.g., succinite, gedanite, glessite, beckerite and krantzite for different Baltic resins, burmite and harcine for Burmese resins, simetite for the Sicilian amber, rumanite for the Rumanian amber, chemawinite and cedarite for Cretaceous Canadian resins, ajkaite for a Cretaceous resin from Hungary, walchowite for a Czechian Cretaceous resin, plaffeite for a Early Palaeogene resin from Switzerland, etc.; see TSCHIRCH & STOCK 1936, for a review). However, in modern mineralogy the resins are no longer regarded as minerals because of their unstable chemical composition and thus their nomenclature is not ordered. Subfossil resins are usually called **copal**. This term was incorrectly applied to some fossil resins as well, and, to the contrary, some Quaternary resins (e.g., from Columbia) are sometimes wrongly designated as amber.

Following the works of LANGENHEIM & BECK (1965) and LANGENHEIM (1966, 1969) there have been repeat attempts to determine the botanical sources of fossil resins on the basis of their chemistry. The results have often been equivocal because of both diagenetic alterations of the original resin and the biochemical evolution of producers. Thus its botanical origin hardly can provide a firm basis for the universal fossil resin classification, and a purely chemical approach to resin classification is more likely to be consistent (K. ANDERSON et al. 1992, K. ANDERSON & BOTTO 1993, K. ANDERSON 1994, 1995, C. BECK 1999); unfortunately the classifications available at present seem to be preliminary and require further elaboration.

To be fossilised in resin, a living or dead object has to be trapped by fresh liquid plant resin and then the resin flow containing it has to be buried in a sedimentary deposit. The former stage is highly specific in respect of the autotaphonomical, ecological, and taphotopical factors influencing the burial selectivity and preservation probability.

For a review of the fauna of different fossil resins see POINAR (1992a) and references herein; some other important recent publications are by KOHRING & SCHLÜTER (1989), GRIMALDI (1996a,b), WICHARD & WEITSCHAT (1996), AZAR (1997), A. ROSS (1998)

WEITSCHAT & WICHARD (1998), POINAR & POINAR (1999), ALONSO et al. (2000), GRIMALDI et al. (2000), and RASNITSYN & ROSS (2000). Diverse aspects of inclusion taphonomy are specially discussed by KORNILOVICH (1903), LENGERKEN (1913), BRUES (1933), PETRUNKEVITCH (1935), VOIGT (1937), SCHLÜTER & KÜHNE (1974), MIERZEJEWSKI (1976, 1978), ZHERIKHIN & SUKATSHEVA (1989, 1992), POINAR & HESS (1985), HENWOOD (1992a,b, 1993a,b), PIKE (1993), GRIMALDI et al. (1994b), J. AUSTIN et al. (1997), and STANKIEWICZ et al. (1998c), and a general review of fossil resin deposit formation is given by DIETRICH (1979). Unfortunately, there are surprisingly few actualistic observations on animal inclusions in modern resins (SKALSKI 1975, ZHERIKHIN & SUKATSHEVA 1989, 1992).

1.4.2.2.1.1. Autotaphonomical factors

The burial in resin is highly selective, and its selectivity is quite peculiar and very different from, and sometimes even opposite to, the selectivity pattern discussed above under the direct burial in sediments (Chapter 1.4.2.1).

Firstly, the **ecological selectivity** is unusual. The assemblages of resin inclusions are strongly dominated by terrestrial insects, mainly by the species inhabiting trunks and branches of resin-producing trees or at least visiting them regularly. Aquatic taxa are represented by flying adults and occasionally by nymphal exuviae left by final instar nymphs moulted at near-water vegetation (LARSSON 1978, WICHARD & WEITSCHAT 1996). Finds of aquatic stages are quite exceptional and their presence is difficult to explain; some of them can belong to insects dwelling in aquatic microhabitats on trees (phytotelmata) (ANDERSEN 1999, POINAR & POINAR 1999). The assemblages reflect nearly exclusively the fauna of densely forested areas, and differences between different inclusion assemblages depend strongly on the original forest type (usually either riverine or swampy forests) (ZHERIKHIN & SUKATSHEVA 1992, ZHERIKHIN & ESKOV 1999). The distribution within the forest is also influential. In particular, actualistic data show that the inhabitants of the herb layer are strongly underrepresented even in resin samples collected from isolated trees at a meadow, while many species dwelling in the bush understorey and tree species other than the resin producers are not rare (ZHERIKHIN & SUKATSHEVA 1989, 1992). The soil and litter fauna is generally scarcely represented, except for those species inhabiting debris accumulations in tree holes and under bark or moss growth on trunks and branches. However, resin flows may occasionally reach the trunk base and impregnate the neighbourhood together with associated organisms. Subterranean resin excretion by roots can also occur though the available data are equivocal (HENWOOD 1993c). Some examples of supposed soil inhabitants enclosed in the Baltic amber, as well as a large sublinear group of worker ants supposedly covered suddenly by a fallen resin drop, are referred to by BACHOFEN-ECHT (1949). A similar ant group found in Dominican amber was studied in detail by WAGENSBERG & BRANDÃO (1998); those authors found that the ants had been included simultaneously and that their antennae are not randomly oriented indicating that at the time of the accident a large group of workers was moving in an orderly fashion along a route. DUBOIS & LAPOLLA (1999) found that the ant fauna of the subfossil Colombian copal more closely resembled the much older assemblage of the Dominican amber than the present-day fauna of Columbia suggesting that different resin assemblages are biased in a similar manner, probably mainly in connection with differences in the nesting and foraging area between different ant taxa.

The burial in amber is strongly **size-selective** because only relatively small objects can be trapped by resin. The upper size limit of the inclusions is different in different resins: some of them (e.g., the Mexican and Dominican amber) occasionally contain even small vertebrates (with a body length of up to about 10cm) while in others (e.g., in the Siberian Cretaceous resins) inclusions more than 5 to 6 mm long are virtually absent. The maximal inclusion size correlates positively with the size of resin pieces. However, even in those resins where relatively large-sized organisms occasionally occur, they are clearly strongly underrepresented simply because most of them are able to free themselves from a resin trap. In this respect the inclusion assemblages differ radically from the sedimentary rock oryctocoenoses.

There are also other ways for an insect to escape resin trapping including **autotomy**. In particular, isolated cranefly legs are not rare in resins suggesting that the insect itself has evaded death at the price of the lost appendage. Similarly, finds of numerous moth scales in some pieces of resin (e.g., SCHLÜTER 1978) indicate that the scaly wing covering also helps to avoid capture. To the contrary, the effect of autotomy on the probability of direct burial in sediments is likely to be negligible.

On the other hand, some autotaphonomical characteristics significant for the direct burial in sediments are of little if any importance for burial in resins. In particular, this concerns the thickness and mechanical durability of cuticle: the conservation potential of resin is so high that the most soft-bodied insects (e.g., fly larvae) are often perfectly preserved. Flight ability is also much less important than in the case of direct burial. Flightless insects and other arthropods are quite common as the resin inclusions, and especially the workers of ants and termites which are virtually absent in compression fossil assemblages whilst they are among the numerical dominants in many resins, both fossil and modern (ZHERIKHIN & SUKATSHEVA 1989, 1992). However, to be enclosed, the flightless insects have to visit the resin-producing tree species at least occasionally. For example, while aphid alates represented in resin assemblages can easily belong to species inhabiting other hosts than the resin producer, the apterous morphs found most probably dwelled on the producer. An interesting analysis of probable host relations of the Baltic amber aphids from this point of view is given by HEIE (1967).

All things being equal, those taxa trophically or topically connected with the resin-producing tree species should be over-represented in the resin assemblages. Theoretically it is possible that some of the producer plant inhabitants are adapted to resin excretion in a way that reduces the probability of them becoming entrapped (e.g., highly tolerant to resin toxicity). This could explain, for instance, the absence of some important present-day conifer inhabitants in the rich and well-studied Baltic amber fauna. However, many taxa not represented in the Baltic amber (such as the xyelid sawflies, ipine bark beetles and lachnid aphids) are discovered in modern resin samples. Thus their absence there might be regarded as taphonomically significant (ZHERIKHIN & SUKATSHEVA 1989, 1992). On the other hand, modern conifer resins often contain some herbivorous taxa living on angiosperm hosts (e.g., the weevil *Betulapion simile* Kirby developing on birch, and leaf-beetles of the genera *Phratora* Chevr. and *Chrysomela* L. feeding on willows and alder) so that even repeated finds in a resin do not necessarily mean any close direct relation with the resin producer. However, they should indicate at least a high abundance of the insect in the habitat and consequently, in the case of herbivores, a significant role of their host plants in the vegetation. The dominant Baltic amber bark-beetle genus, *Phloeosinites* Hagedorn, is closely allied to modern *Phloeosinus* Chapius which lives on Cupressaceae, a family which was common in the Eocene 'amber'

forests as indicated by numerous inclusions in the Baltic amber (CZECZOTT 1961, LARSSON 1978). Both Cupressaceae and *Phloeosinites* are represented in the Baltic amber by a number of species and trophic connections between them seem to be quite probable. However, when there are numerous finds of the herbivorous taxa with an *a priori* low burial probability (e.g., sedentary flightless aphids of the genus *Germaraphis* Heie, relatively large-sized weevils of the genus *Electrotribus* Hustache and longhorn beetles of the genus *Notorrhina* Redtb., and the stick-insect larvae of the genus *Pseudoperla* Pictet, which are both flightless and relatively large-sized, in the Baltic amber) their most probable host plant is the resin-producing tree.

The probability of burial in a resin depends also on both the insect's **behaviour** and **physiology**. Resin is a repellent for some insects and an attractant for others, and some of its components are toxic. Hence it seems that BRUES' (1933) attempt to compare the Baltic amber fauna with the insects trapped by tanglefoot in a modern forest may be somewhat inaccurate because of differences in the attractive/repulsive effects of the two sticky substances. Insect reactions to resins are relatively well-studied for a number of forest pest species where they are often both species-specific and host-specific (see Chapter 3.2.3 for a brief discussion and references) but the information on other taxa is scarce. It should be noted that the bioxylophages (wood-borers infesting living trees) do not seem to be very abundant in both fossil and modern resins. In modern forests they usually avoid healthy trees with a high level of resin production but can attack them intensively during outbreaks; the absence of their mass burial in fossil resins may indicate rarity of such extreme pest outbreaks in natural environments.

Some bees (especially Meliponini and Anthidiini) are **resin gatherers** that use fresh resin in their nests as a combination sticky cementing material and fungicide (WILLE & MICHENER 1973, PASTEELS 1977, WILLE 1983, FROHLICH & PARKER 1985, HOWARD 1985, MESSER 1985). Resin collecting is described also for some digger wasps of the subfamily Pemphredoninae which close entrance to their tunnels by resin plugs (MUDD & CORBEL 1975). This mode of behaviour probably favours burial, and in particular, the stingless meliponine bees are common in many fossil (Sicilian, Apennine, Dominican, and Mexican amber) and subfossil resins (American, African and South Asian copal). POINAR (1992b) examined 750 specimens of *Proplebeia dominicana* Wille & Chandler in Dominican amber and found that 21 of them had rather large "resin balls" on their hind tibiae but none contained any pollen in their corbiculae. Meliponine workers change their tasks during their lifetime, and resin gathering is usually their last duty (WILLE 1983). It seems that pollen-collecting individuals very rarely become trapped by resin, and stingless bees are represented in the resin taphocoenoses mainly by resin collectors. The predominance of specimens lacking any pollen load may be further explained by their becoming entrapped during fights between the resin gatherers which are quite frequent and often desperate (HOWARD 1985). It should be noted, however, that GRIMALDI et al. (1994b), in a controversy with Poinar's above-mentioned results, mentions pollen clumps as not uncommon on the hind legs and abdomens of stingless bees in Dominican amber.

As well as stingless bees, pemphredonine wasps are also not rare in diverse fossil resins (though in this case their nesting in wood may be at least equally important as resin collecting). This is not the case of the anthidiine bees but they are usually too large to be trapped easily. A different strategy of resin use is described in some reduviid bugs of the subfamily Apiomerinae which cover their legs in fresh sticky resin to help them catch their prey (USINGER 1958), and POINAR (1992b) did discover resin clumps on the fore legs of several apiomerine nymphs in the Dominican amber.

Other behavioural features may influence the probability of becoming entrapped as well. Occasionally I have observed a worker ant making contact with a fresh drop of very liquid resin. After its antenna had touched the drop for the first time, the insect stopped immediately and began to clean it; very soon it smeared a leg, tried to clean it, dirtied another leg, and after ten minutes or so had become partly coated with resin. Similar field observations of insects on tree trunks can no doubt uncover other behavioural patterns either favouring or preventing burial.

In the present day, some dipterans (e.g., several genera of the syrphid and anthomyiid flies) develop directly in resin fluxes (MAMAEV 1971), and their puparia occur commonly in modern resin samples (ZHERIKHIN & SUKATSHEVA 1989); strangely, they are unrecorded from any fossil resins. It can mean either that this habit is very recent or that the damaged resin pieces have a very low preservation potential.

LARSSON (1978) suggested that the **seasonal activity** pattern affects the probability of burial in parts of the world with a seasonal climate. For example, he supposed that the absence of alate morph of the most common Baltic amber aphid genus, *Germaraphis*, may be because the winged generation was restricted to a season of minimal resin secretion. The possible role of phenology on the composition of the amber fauna was also discussed by SKALSKI (1992). **Circadian activity** should also be of some importance because of the positive correlation between ambient temperature and the trapping ability of resin.

Host biology is important in the case of permanent ectoparasites. For instance, the fleas which occasionally occur in both Baltic and Dominican amber belong to very few genera which probably parasitised arboreal hosts. The unique finds of lice eggs on mammal hairs in the Baltic amber (VOIGT 1952) and scale insects on a twig inclusion in New Jersey Cretaceous resin (GRIMALDI *et al.* 2000) represent rare examples of a 'container-in-container' burial.

As can be seen from the above, the pattern of insect burial in resins differs radically from those in sedimentary rocks and hence, both types of oryctocoenoses are complementary rather than comparable. Unfortunately, they practically never occur in the same deposits (see below). Some important taphonomic differences occur between the oryctocoenoses of different resins as well (e.g., in the share of aquatic taxa as discussed below). However, in comparison with sedimentary rock assemblages the resin faunas are much more uniform, which is not surprising because they reflect the fauna of much less diverse habitats and because the differences in the trapping potential between the resins are relatively small.

1.4.2.2.1.2. Ecological factors

The ecological factors affecting insects are of a limited importance from the viewpoint of their burial in resins. The thanatotope and primary taphotope usually coincide in this case. As a rule, living insects become trapped so that death and burial are the same event. Occasionally a resin flux can cover a dead insect on or under the bark; in this case the insect corpse before burial is exposed for the same principal mechanical and biological destructive influences as usually in terrestrial environments. A few mortality factors can be of some (but generally minor) importance to burial probability. In particular, spider web fragments, not rare in fossil resins, are occasionally accompanied by insect remains which seem to be 'sucked out', but those insects usually belong to species commonly occurring in the same resin (e.g., WUNDERLICH 1988). LARSSON (1978) suggested that some aquatic

insect stages could be accidentally enclosed in resin during catastrophic flooding of forested areas, but this is surely a very rare phenomenon.

The ecological factors affecting the resin production are taphotopical in relation to inclusions and thus are considered below.

1.4.2.2.1.3. Taphotopical factors

Because the resin not only kills but also envelopes trapped insects, isolating them from external influences, it might be regarded as the primary taphotope of the objects enclosed; in fact, the isolation from both abiotic and biotic external factors is much more effective than that provided by a sediment layer of similar thickness. Even when an inclusion is partially exposed (Fig. 39) it can be fossilised though in this case the preservation state of the fossil will be poor because of diagenetic oxidation (see below).

Resin availability in the habitat depends on the **distribution and abundance of resin producers** as well as on **resin production**. Other things being equal, the greater the role of resin producers in the local and regional vegetation and mean resin production of each producer individual, the higher is the probability of insect burial. The **resin producer anatomy and physiology** can also affect the probability. HENWOOD (1993b) stresses that though there is a large literature on resin secretion (first of all in connection with tree tapping for industrial resin extraction), the available information is too often equivocal and few definitive conclusions can be made. In particular, resin excretion is often associated with mechanical damage or pathogen action (which

Fig. 39   An ant enclosed in a small drop of modern *Callitris* (Cupressaceae) resin, Launceston in Tasmania (photo by D.E. Shcherbakov); ant head and thorax 1 mm long

are not necessarily mutually exclusive). However, some trees (e.g., modern Araucariaceae) produce external resin drops even in the absence of any obvious injuries, and healthy plants often produce significantly more resin than sick or weakened individuals. The model of 'succinosis' (a peculiar pathological state of forest stands provoking an abnormally high resin excretion) proposed by CONWENTZ (1890) to explain the origin of the enormous Baltic amber deposits is considered as unnecessary by almost all modern authors. Weather and edaphic conditions surely affect resin production as well as do wildfires and diverse biological agents, and their effects seem to differ not only between tree species but even between populations (and probably genetic strains) of the same species. Interactions between different factors might further complicate the resulting pattern. Individual resin production can vary widely. For instance, GIANNO & KOCHUMMEN (1981) estimate the annual resin production of tapped *Dipterocarpus kerrii* trees between 0.5 and 40 l. HENWOOD (1992c) has failed to obtain fresh resin by slashing of *Hymenaea courbaril* trees in Panama during two-month-long studies though this species is well-known as a resin producer and old hardened resin fluxes were observed (though not very commonly) in the fieldwork area.

A wide range of modern conifers and angiosperms produce resins, and resin secretion is also documented for several extinct Mesozoic and Palaeozoic gymnosperm groups. However, their resin production is often limited and thus of no real importance for insect taphonomy (though microscopic organic inclusions are reported from very small Triassic resin particles; G. POINAR et al. 1993). Interestingly, no reasonably large external resin fluxes are documented for any extinct plant orders. Theoretically, some insects inclusions can be found even in small and dispersed resin pieces as well as in resin fillings of internal channels and cavities in a wood, but the probability of such finds is extremely low. In fact, only external resin secretion is of any real importance, and resin clumps below 1.5 to 2 mm are of no practical interest.

The contents of inclusions is highly variable: for instance, in one sample of a Cenomanian resin from Yukhary Agdzhakend in Azerbaidzhan weighing about 179 g 100 insects and insect fragments have been found whereas in a resin from the Begichev Formation (Albian-Cenomanian) of Taimyr, Siberia, the average concentration of inclusions is as low as one specimen per 1 kg of resin (ZHERIKHIN 1978). Based upon my own experience, a fossil resin sample about 1 kg is normally sufficient for finding at least one or a few inclusions. Variations in the frequency of inclusions suggest that there could be differences in the **trapping potential of fresh resin**. There are few comparative observations on inclusion content in different modern resins. The preliminary results of modern resin sampling suggest some interspecific differences (e.g., insects seem to be more common in spruce resin in comparison with that of larch, pers. obs.) but the dispersion of the data is high for all kinds of resins examined so that no statistically robust conclusions can be drawn. Inclusion frequency probably depends on a complex set of factors. A low frequency can sometimes be due to actual low insect abundance or short season of activity as in the case of modern larch resin from Khatanga River Basin, Taimyr, Siberia (ZHERIKHIN & SUKATSHEVA 1989). However, more often, inclusion frequency depends primarily on the trapping potential of the resin, and in particular on its viscosity, rate of hardening and frequency of repeating resin flows. In different fossil resins the highest content of inclusions occurs as a rule in pieces composed by numerous sequential resin layers (Schlauben of German authors: KLEBS 1910, BACHOFEN-ECHT 1949) and especially in stalactite-like

ones (ZHERIKHIN & SUKATSHEVA 1973, PIKE 1993). This is not surprising for two reasons: firstly, the total trapping surface increases manyfold in comparison with solid pieces of the same volume and, secondly, surface inclusions become well-protected by overlaying resin flows. When the resin is not extremely rich in inclusions, pieces of any other morphologies yield so few inclusions that their investigation is often unnecessary (ZHERIKHIN & SUKATSHEVA 1973). In both cases, the resin might be very fluid, i.e. either quite freshly exposed, or secondarily softened under direct sun rays. The physical properties of the resin have some effects on the selectivity of burial as well. HENWOOD (1993c) stresses that more liquid resins should be more effective as traps for smaller insects while more viscous ones can probably catch larger objects. Resin viscosity, and the time taken for it to harden, depends on its chemical composition, and most of all, on the rate of loss of volatile components. Very fluid resins collect more debris and can reach the forest floor easily where so many soil and litter particles can be entombed that the resin often becomes dirty and opaque. In particular, such opaque 'earthy amber' is quite common in the Neogene of South Asia (TSCHIRCH & STOCK 1936, DURHAM 1956, HILLMER et al. 1992a,b, SCHLEE & PHEN 1992). It contains insect inclusions which, however, are poorly visible in natural resin lumps and can be discovered only when it is broken or cut into thin sections. Other taphonomical aspects connected with resin chemistry (its attractive and repellent action and toxicity for insects) were mentioned above.

Amber inclusions have been famous for ages as one of the most spectacular types of fossils illustrating the remarkably high **conservation potential of natural resins**. HENWOOD (1992b) attributes this potential to a strong antibiotic effect of resins, and to rapid tissue dehydration which in combination can suppress or even halt the decay processes in enclosed organisms. Anaerobic conditions are probably of minor if any importance because even polymerised amber is probably not completely impermeable to air (though occasional pyritisation of inclusions would indicate an oxygen-deficient microenvironment). The dehydration process in resins has not been studied in detail but seems to be complex rather than based on simple osmosis (HENWOOD 1992b). The comparative importance of antibiotic and dehydratative properties of the resin is unclear, but both almost certainly vary in relation to resin chemistry, physical environments (e.g., temperature) and autotaphonomical characteristics of the enclosed object. Indeed, the preservation state of inclusions varies considerably between both different resins and different insect taxa. Broken inclusions often demonstrate that there is nothing more than a very fine cast impregnated by extremely thin and fragile cuticle remains (Fig. 40) whereas in other cases well-preserved cuticle and internal skeletal parts with remains of musculature can be seen (Fig. 41). Inclusions that appear somewhat decomposed are not uncommon in some resins (e.g., in the Cenomanian resins from Azerbaidzhan). More often insects entombed in different resins are partly (or, more rarely, completely) covered by a white clouding that obscures morphological structures (Fig. 42). Similar clouding can also be found in pieces that lack inclusions and are common in some deposits including the Baltic amber where the proportion of partially (the so-called 'Bastard') or entirely whitish pieces (the so-called 'knochiger Bernstein' or 'bone amber') is high (TSCHIRCH & STOCK 1936). Some other fossil resins are almost always more or less cloudy (e.g., the Cenomanian resins from France and Azerbaidzhan: Fig. 43). The clouding is formed by innumerable extremely small bubbles and indicates an emulgation of fresh liquid resin with water or other fluids (SAVKEVICH 1970). Thus, clouding around inclusions was probably caused by fluids exuding from the

Fig. 40 A broken limoniid cranefly in Baltic amber demonstrating preservation of cuticle as a thin, partially destroyed film (PIN 964/807, photo by D.E. Shcherbakov); wing 6.3 mm long as preserved

Fig. 41 A click-beetle (Coleoptera, Elateridae) in Baltic amber with preserved musculature (PIN 964/808, photo by D.E. Shcherbakov); 5.3 mm long as preserved

corpse, and so indicates some decay of freshly embedded objects. MIERZEJEWSKI (1978) estimated the frequency of clouding associated with different inclusions in the Baltic amber and found no correlation with either body size or cuticle thickness. SCHLÜTER & KOHRING (1974) emphasise that the clouding is often located at one side of inclusion; they suppose that this is a result of resin homogenisation under heating at the sun while the emulged areas survive only on the shadowed side. Sometimes clouding probably disappears during diagenesis (see below).

Chemical analysis indicates that organic matter is altered not only in amber inclusions but also in insect remains enclosed in young (about 2 thousand years old) subfossil copal; however the presence of chitin and proteins is documented in the latter, and hence major degradation must occur later in the course of resin diagenesis (STANKIEWICZ et al. 1998c). There are some records of non-racemised amino-acids from fossil resin inclusions (BADA et al. 1994, X. WANG et al. 1995, H. POINAR et al. 1996) suggesting remarkably long term chemical preservation. In this connection recent attempts to isolate DNA fragments from resin inclusions should be mentioned. There are several reports on DNA retrieval

from insect and plant inclusions in the Dominican and Baltic amber as well as from a beetle preserved in the Early Cretaceous Lebanese resin (CANO et al. 1992a,b, 1993, DESALLE et al. 1992, 1993, H. POINAR et al. 1993, DESALLE 1994, GRIMALDI et al. 1994b, HOWLAND & HEWITT 1994, POINAR 1994a, 1999a, PAWLOWSKI et al. 1996a,b). However, those results could not been replicated independently in careful and extensive studies by J. AUSTIN et al. (1997). Those authors used several extraction methods and different PCR conditions and have found that DNA sequences can be sometimes amplified, but the results are not reproducible and, moreover, DNA fragments can be detected in amber pieces lacking any macroscopic inclusions. The same results were obtained from subfossil copal materials. Accordingly, it seems that in fact natural resin is poor environment for DNA preservation and previous records are based probably on experimental artefacts (LINDAHL 1997, WALDEN & ROBERTSON 1997, STANKIEWICZ et al. 1998c). No studies of DNA degradation in insect corpses embedded in modern resins are available.

Fig. 42 A scirtid beetle in Baltic amber showing milky clouding on one-side (PIN 964/808, photo by D.E. Shcherbakov); beetle 2.1 mm long

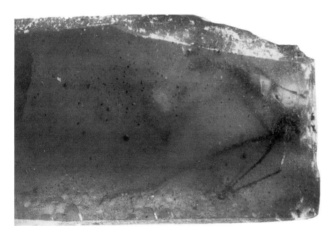

Fig. 43 A nematoceran fly preserved in a dull amber-like resin; Late Cretaceous (Cenomanian) of Yukhary Agdzhakend in Azerbaidzhan (photo by D.E. Shcherbakov); amber block 5.7 mm wide

All of the above data indicate that both chemical degradation and decay are probably normal for resin-embedded insects. On the other hand, recent ultrastructural studies show that not only the structural preservation of sclerotised skeletal parts is often perfect but some inclusions contain mummified soft tissues including muscles, parts of digestive, excretory and reproductive systems, dorsal aorta and nerve ganglia (HENWOOD 1992b, GRIMALDI et al. 1994b). Even some intracellular structures can sometimes be observed such as cell nuclei, mitochondria and perhaps ribosomes and lipid droplets (POINAR & HESS 1982, 1985, HENWOOD 1992a). No quantitative data are available but it seems that the specimens with well-preserved internal anatomy are much less rare in the Dominican than in Baltic amber (GRIMALDI et al. 1994b). Inclusions in the Cretaceous Siberian (pers. obs.) and New Jersey (GRIMALDI et al. 2000) resins are usually empty but ovarian tissue has been discovered in a weevil from the Early Cretaceous Lebanese resin (CANO et al. 1993).

If the resin is the original taphotope, it itself might in turn be buried; otherwise the resin flows will be lost together with all inclusions. Exposed resins are subject to both oxidation and bacterial decomposition (LANGENHEIM 1969), as well as occasional bioerosion by resinicolous insect larvae (MAMAEV 1971), and deeply weathered old flows can be often seen on tree trunks. The time of preservation of exposed resin was never determined accurately and probably varies with resin composition and the physical environment (e.g., temperature, humidity and lightning) but is unlikely normally to exceed a dozen or so years.

Resin taphotopes can be subaquatic as well as subaerial. Subaerially resins occur in subfossil soils and palaeosols, subaquatically – in diverse fluvial, lacustrine, swamp, and coastal marine deposits. The general mechanisms of burial are the same as discussed above for direct burial in sediments but, of course, the autotaphonomical resin characteristics are peculiar in many respects. **Resin fossilisation** is a slow process surely depending on taphotopic and diagenetic conditions but its early stage is poorly studied. In particular, there are no comparative data on chemical alteration of resin in different taphotopes. Anaerobic conditions are probably important because oxidation of resin is destructive; the role of other factors is unclear.

No resin producers are known among true aquatic plants, either extant or extinct. Hence **autochthonous resin burial** (i.e. directly within the resin-producing tree stands) occurs mainly in subaerial environments. Subfossil autochthonous deposits in soils seem to be rather common and widespread; in particular, the high-quality copal that was gathered for the varnish industry from soil in different parts of the world for many decades suggesting that total amount of resin buried in soils is extremely high (TSCHIRCH & STOCK 1936, HOWES 1949, B. THOMAS 1969, WHITMORE 1980, SCHLÜTER & GNIELINSKI 1987). HENWOOD (1993a) assumes that the contribution of subterranean resin production by roots to the forming of soil resin deposits is also significant, but there are no conclusive data to support this suggestion. On the contrary, the available data on the insect fauna of copal indicate that the resin was often originally exposed well above the ground level. Resins in soil are affected by decay and chemical degradation but those processes are virtually uninvestigated. There is some evidence of differences in resistance to decay between the resins of different plants (LANGENHEIM 1969). At least some resins can be preserved for hundreds and even thousands of years in deep soil sections. In particular, kauri copal in New Zealand was collected for industrial purposes in the XIX and early XX centuries at the sites where its producer, Agathis, had already disappeared, well before the time of European colonisation (TSCHIRCH & STOCK 1936, B. THOMAS 1969). LAMBERT et al. (1993) refer the ^{14}C data indicating an age of up to 37,000 years for some kauri copal deposits; the geological information in the paper is scarce so that it is difficult to say whether the material studied has originated actually from the soil deposits but that is quite possible. The soil profile overlying resin deposits can exceed 2 m in depth (HOWES 1949), and resins sometimes occur at several different levels within the profile (TSCHIRCH & STOCK 1936). Subfossil copal is obviously harder than old exposed resins and is covered with a more or less distinct surface crust indicating that the resin is chemically altered. Most probably, polymerisation and other chemical processes in buried resins vary in connection with soil chemistry, but this problem is poorly studied. Resins are occasionally accompanied by the remains of logs but the rate of wood decay in forest soils is generally much higher. The inclusion assemblages contain more terrestrial and fewer aquatic insects because the resin samples represent the fauna of the interior parts of forest stands (ZHERIKHIN & SUKATSHEVA 1989, 1992).

Some modern conifers (e.g., Scotch pine, *Pinus sylvaticus*, and bald cypress, *Taxodium distichum*) inhabit swampy habitats and fossil resins occur rather widely in coal-bearing lacustrine and swamp deposits of some areas (e.g., in the Palaeogene of Siberia, Kazakhstan, and Russian Far East: ZHERIKHIN & ESKOV 1999). Such deposits should be also treated as **autochthonous in the broad sense** (though resin itself has been secreted above the water level) or as **parautochthonous** (if the plant materials with the resin have been transported by lake currents). Resin conservation pattern in subaquatic sediments is even less documented by actualistic observations than for subaerial ones. Subfossil resins occasionally occur autochthonously or parautochthoniously in peat deposits; some of them have been named (e.g. phylloretine, fichtelite, quickbornite: TSCHIRCH & STOCK 1936) and briefly described, but the available information is scarce. They are associated with tree (mostly pine) logs and often resemble more or less weathered modern resin flows suggesting a low degree of polymerisation. Fossil subaquatic autochthonous deposits are restricted to coal-bearing strata and not infrequently they contain resin in a high concentration. The state of resin polymerisation in coals varies but a high degree of polymerisation is probably due to later diagenetic changes rather than taphotopic ones. The fauna is rather rich in aquatic insects that developed in standing waters (ZHERIKHIN & SUKATSHEVA 1989, 1992). On the contrary, in parautochthonous deposits, resins only occur sporadically as dispersed pieces and hence are not of practical use for palaeoentomology. Fossil wood in this case is also sporadic.

Allochthonous primary deposits contain resins transported on tree trunks, branches, or bark, that has of fallen directly into water from undermined banks. The most common mechanism of their formation is transport by river flow, and the highest resin concentration occurs in this case in abandoned channels, especially in deltaic environments where tree remains are accumulated (Fig. 44). Allochthonous primary deposits are widespread in the fossil record but few data are available on subfossil deposits of this type, and actualistic observations are practically absent. Unlike autochthonous deposits, allochthonous ones are polytopic and contain resin transported from different parts of the river basin. However, there are no good indicators of the transport distance because the resin carried on tree remains shows no clear evidence of rolling or other mechanical actions. In allochthonous primary deposits the resin is associated with abundant wood and/or bark remains; its concentration is often high. The fauna is enriched by lotic insects that emerged from rivers (ZHERIKHIN & SUKATSHEVA 1989, 1992).

Secondary allochthonous deposits are formed when primary resin deposits (either auto-, or allochthonous) become eroded and the resin is re-deposited; this process is possible at the taphotope as well as at the oryctotope stage. Resin concentration may be very high because of

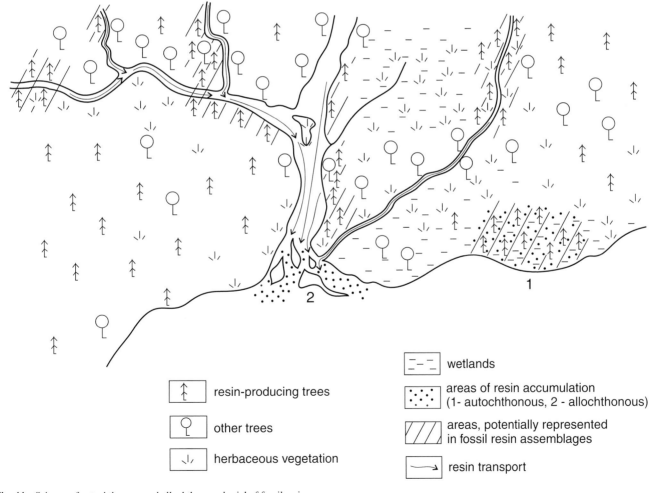

resin-producing trees

other trees

herbaceous vegetation

wetlands

areas of resin accumulation (1- autochthonous, 2 - allochthonous)

areas, potentially represented in fossil resin assemblages

resin transport

Fig. 44   Scheme of autochthonous and allochthonous burial of fossil resin

secondary sorting; unlike primary deposits, wood and other plant remains are usually scarce. The secondary resin assemblages should be highly polytopic but it is difficult to recognise the materials originated from different habitats. The only way to do it is an analysis of correlation between the presence of certain species in the same resin piece (the *syninclusions* after KOTEJA 1996); however, such analysis needs large databases and is highly time-consuming.

Mixed allochthonous deposits are formed when both primary and secondary buried resins occur together in fluvial sediments. In particular, the Baltic amber deposits are probably of this type. SAVKEVICH (1970, 1983) suggests that the origin of the succinite-type amber is related to some decay in primary autochthonous soil deposits and subsequent re-deposition by rivers. However, in the Baltic Sea area it is accompanied with gedanite, a retinite-type resin with a much lower degree of polymerisation (known in jewellery as "the rotten amber" because of its brittleness). Probably, as with that of buried insect cuticle, the resin polymerisation is a slow complex process including some alteration and degradation of the original chemicals and subsequent re-polymerisation of its products. There are also intermediate resin types in the Baltic deposits called collectively the gedano-succinites. Most probably, the gedanite was originally buried allochthonously in fluvial deposits, and the gedano-succinites have been washed out from forest soils in a rather short time after their primary autochthonous burial, too early to produce true high-quality amber. Unfortunately, there are no comparative data on inclusion assemblages in different varieties of the Baltic fossil resins. If the above hypothesis is correct, it may be predicted that the gedanite fauna should be richer in lotic taxa in comparison with the succinite fauna (ZHERIKHIN & SUKATSHEVA 1989, 1992).

In general, special comparative studies on subfossil resins buried in different taphotopic environments are necessary to elucidate the taphotopical processes in more detail.

1.4.2.2.1.4. Postburial factors

Resin containers protect the enclosed objects from the majority of diagenetic factors. Nevertheless, some changes seem to occur due to interactions between the inclusion and the resin during its fossilisation. Unfortunately, very few comparative observations on the state of preservation of inclusions in subfossil and fossil resins are available. Diagenetic alterations of resins themselves are more evident but also insufficiently studied. Resins, even the least polymerised ones, are highly resistant to aggressive inorganic chemicals. The main trend of their alteration during diagenesis is probably a gradual polymerisation. The outer layer of resin lumps turns out to be more altered and forms a distinct surface crust; this crust differs in different resins in colour, thickness and physical properties but the factors influencing its formation and characters are poorly known. Polymerisation is normally so slow that some fossil resins retain clear chemical fingerprints of their botanical origin (e.g., the Mexican and Dominican amber was almost certainly produced by an extinct species of the leguminous tree genus, *Hymenaea*: LANGENHEIM 1966, LANGENHEIM & BECK 1965, 1968, C. BECK 1999). However, in other cases there may be major differences between the chemical compositions of the original resin and its products fossilisation because of diagenetic processes (SAVKEVICH 1970, 1980, 1983, K. ANDERSON *et al.* 1992, VÁVRA 1993, KOLLER *et al.* 1997). There have been some attempts to use the degree of chemical alteration to date fossil resins, but the results often contradict with other evidences. In particular, LAMBERT *et al.* 1985, postulated on this basis a wide range of ages (from the Late Eocene up to Early Miocene) for the Dominican amber originating from different mines. However,

microfossils from the amber-bearing sediments indicate that none of them are older than Early Miocene (ITURRALDE-VINCENT & MACPHEE 1996). Probably, the regular pattern of diagenetic 'maturation' is too dependent upon different diagenetic processes to be useful for dating with any certainty. It is quite possible that rather young, and even Quaternary resins can be occasionally polymerised to a degree comparable with much older materials in other areas, particularly in tectonically active or volcanic regions.

In addition to its polymerisation, resin chemistry changes due to interactions with sediments (SAVKEVICH 1983). In particular, its sulphur content is occasionally high as a result of hydrogen sulphide diffusion and subsequent binding with oxygen (KOSMOWSKA-CERANOWICZ *et al.* 1996).

Slow changes in resin composition probably only have a limited effect on inclusions. STANKIEWICZ *et al.* (1998c) have found some differences between the chemical pattern of copal and Dominican amber inclusions, suggesting continuous slow changes connected with a gradual cuticle degradation and increasing polymerisation of resin components. Inclusions that are either partially exposed or situated within the external crust are always poorly preserved, and their cuticle is completely destroyed. Sometimes resin inclusions are pyritised, as is common in the case of French Cenomanian resins (SCHLÜTER 1978), but pyritisation also occurs occasionally in other resins as well, including Baltic amber (BARONI-URBANI & GRAESER 1987, and pers. obs.). However, it is not clear when during diagenesis and under what special conditions inclusions become pyritised. Occasionally concentric fractures can be seen around inclusions, indicating the action of pressure of overlaying sediments; sometimes these have even been misinterpreted as traces of insect moving in fresh liquid resin (e.g., BACHOFEN-ECHT 1949).

In two cases fossil resins undergo much more rapid changes that also affect their inclusions, specifically when they are heated or oxidised. **Heating** under high pressure occurs as a result of thermal metamorphism and leads to a very high degree of polymerisation. Naturally heated resins are distributed mainly in regions of faulting (e.g., Carpathians, Apennines, Minor Caucasus, Sakhalin Island). WUNDERLICH (1999) discusses the presence of heated specimens in the Dominican amber. The heated resins are especially hard because of their high degree of polymerisation and they can be considered as true amber. They have been described under different names (e.g., the rumanite, simetite, delatynite, schraufite, burmite, etc.) but their similarity in many respects justifies their collective designation as the rumanite-type amber (SAVKEVICH 1980, KOSMOWSKA-CERANOWICZ 1999, ZHERIKHIN & ESKOV 1999). All rumanite-type ambers have a similar infrared spectrum but they still demonstrate some chemical differences that are probably partially connected with original resin composition. The characteristics of the Sakhalinian amber as described by ZHERIKHIN (1978) are generally applied to all rumanite-type ambers, and allow us to recognise it easily. Because of the high degree of polymerisation, this type is more resistant to organic solvents and oxidation than any other fossil resins. The colour of resin varies from yellow to dark-red but red varieties predominate; freshly exposed specimens often demonstrate a greenish iridescence that gradually disappears. The milky varieties are rare, probably because the microscopic cavities fill up with softened resin. The external crust is thin, variable in the colour, often fissured in many places, with the paler internal substance occasionally squeezed out, indicating that it has melted under high pressure. Metamorphosed pieces are typically flattened and may be misinterpreted as pebbles. Other common evidence of melting include the presence of irregular internal reddish films (probably originating as a result of deformation of original laminated structure of fluxes) and elongate gas bubbles arranged into linear series. However, these features are not universal suggesting that the conditions of metamorphism are not identical. In particular, burmite and Late Cretaceous amber from Gorchu in Azerbaidzhan often shows internal fractures that are secondarily filled with the calcite. These fractures indicate that the resin has probably been only somewhat softened and crushed by high pressure and has not melted, perhaps because of lower temperature or higher melting point, the latter varying widely between different resins. Nevertheless, both amber varieties demonstrate typical rumanite-type infrared spectrum suggesting a high degree of polymerisation (SAVKEVICH 1980). The condition of organic inclusions in rumanite-type amber is also peculiar: they are more or less obviously deformed (Fig. 45) and often pale, looking somewhat like specimens that have been boiled in KOH, with their internal cavity usually being filled with resin. A number of deformed inclusions in the Fushun amber and simetite are illustrated by HONG (1981) and SKALSKI & VEGGIANI (1990); again, the deformation is less pronounced in the burmite. DLUSSKY (1988) proposed a method for the graphic reconstruction of original body shape for deformed inclusions. In experiments, inclusions in Cretaceous Taimyr retinite were found to become deformed and cleared after artificial heating under pressure. Milky clouding around inclusions is extremely rare in rumanite-type amber, surely because of secondary resin filling of microcavities.

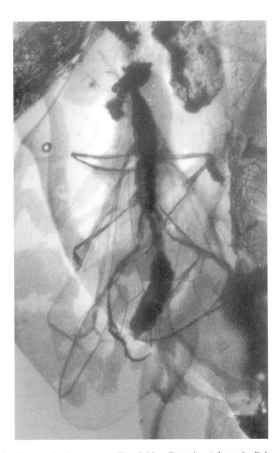

Fig. 45 Undescribed queen ant (Formicidae: Ponerinae) from the Palaeocene Sakhalin amber distorted due to the high temperature plastic deformation of the amber in the course of diagenesis: note the compressed thorax and legs (specimen 3387/116, photo by A.P. Rasnitsyn); body 3 mm long as preserved

Resin **oxidation** occurs as a katagenetic process in exposed resin-bearing strata. Oxidised samples can be recognised easily from their dark (usually reddish) colour and abundant fractures; often only a more or less thick outer layer is affected. The outer parts of long-exposed resin-bearing deposits always have their resin more or less oxidised. However, the oxidation pattern in nature is unexplored, and only data from resins kept in collections are available. The rate of oxidation is rather low and differs for different resin varieties. In general, the more polymerised is the resin, the slower its oxidation, though the process is accelerated by the action of light. Even the least polymerised retinites do not demonstrate any obvious signs of oxidation for several decades when they are protected from light. On the other hand, old (80 to 100 years) specimens of more polymerised succinite clearly become oxidised, at least for a several millimetre thick outer layer. The quality of inclusions deteriorates slowly up to the complete destruction of the specimen by crumbling. In exposed deposits, fossil resins are probably affected by other destructive agents as well (e.g., fluctuations of temperature) but their effects are poorly studied.

Resins are only moderately mechanically resistant but enough so to permit re-deposition from eroded deposits. Their low specific gravity (ranging from 0.965 to 1.119 but usually between 1.03 and 1.09: TSCHIRCH & STOCK 1938) facilitates their transport by water currents with minimal abrasion. Nevertheless, strong currents, wave and other mechanical actions can grate resin into small particles as indicated by the occurrence of sands that are enriched by amber grains (CLEVE-EULER & HESSLAND 1948). Mass re-deposition is documented for the Baltic amber in different areas of Europe (BACHOFEN-ECHT 1949, LARSSON 1978); in some cases repeated re-deposition (up to five or even six times) has been suggested (MEYN 1876). In some areas fossil resins are known only in re-deposited state like, for example, the simetite in Sicily (SKALSKI & VEGGIANI 1990) and the supposedly Cretaceous resins from the north of European Russia (ZHERIKHIN & ESKOV 1999, and references herein). LARSSON (1978) even assumed extremely long distance resin transport from North America to Iceland and the Kola Peninsula by the Gulf Stream but this hypothesis seems to be implausible. Not only true amber but also fragile retinites occur re-deposited in Quaternary sediments, sometimes in large quantities (YUSHKIN 1973, ZHERIKHIN & SUKATSHEVA 1973, ZHERIKHIN & ESKOV 1999). Sometimes the re-deposited resin is strongly oxidised (e.g., in the Quaternary of Gubina Gora near Khatanga, Taimyr, North Siberia) suggesting a long term exposition but this is not a universal feature. Nor is the morphology of re-deposited resins diagnostic though some pieces can be seen to have been rolled or abraded. Re-deposition creates significant problems connected with the age and geographic origin of re-deposited materials. In particular, some authors believe that the Ukrainian, Belorussian and Polish amber deposits in Quaternary sediments have originated due to a re-deposition of Baltic amber (e.g., BACHOFEN-ECHT 1949, LARSSON 1978). However, studies on amber-bearing sediments in Eastern Poland and Western Belorussia indicate the Ukrainian Shield as the area of transport (KOSMOWSKA-CERANOWICZ et al. 1990). In the Quaternary, re-deposited resins can be recognised relatively easily on the basis of their fossilisation state; in older strata it is often difficult to say if the resin is contemporaneous with the resin-bearing sediments or older. For instance, the Burmese amber occurs in Eocene deposits but its fauna contains a number of taxa not found elsewhere in the Cainozoic. Perhaps, it was re-deposited from more ancient (Cretaceous) strata but a long period of survival of Cretaceous relics in an island refugium is another possible explanation (ZHERIKHIN & ROSS 2000). Another intriguing unresolved

Fig. 46 Sakhalinian amber (dark pieces) and a modern conifer resin (pale pieces) collected together in a mixed assemblage from recent beach deposits at the Sea of Okhotsk coast in Starodubskoe (photo by D.E. Shcherbakov); wooden chip with the amber piece impressed (centre) 33 mm of maximum length

problem is the origin of the so-called Saxonian amber occurring in the Early Miocene of Germany but containing an insect fauna that is very similar to that of the Baltic amber (WEITSCHAT 1997, RÖSCHMANN 1999a). Unlike the Burmese amber which is accompanied only by thin coal layers, in Saxony the amber occurs in brown-coal deposits so that a re-deposition seems to be less probable, and the problem of extremely high faunistic similarity with the Baltic amber is still unresolved. In secondary deposits, resins of very different age can co-occur in mixed assemblages, and this is well documented by observations. KOSMOWSKA-CERANOWICZ *et al.* (1996) have found that the true amber on the Baltic coast of Poland is accompanied by so-called 'young amber' or 'colophon' with an age ranging between 525 ± 60 and more than 60,000 years according to ^{14}C dating. At the Okhotsk Sea coast of Sakhalin Island, Palaeocene Sakhalinian amber occurs together with a soft pale resin of a recent origin (Fig. 46). Similar mixed assemblages might co-occur in older deposits as well, but are hardly recognisable. Chemically different resins occurring together could well be produced synchronously by different plants or (in mixed deposits) processed in different taphotopes. Any significant differences in their age can be argued only on the basis of their faunas, but it suggests that each piece with inclusions should be examined chemically, which unfortunately is a rather expensive and time-consuming procedure.

1.4.2.2.1.5. Technical factors

The availability of fossiliferous rocks for sampling is even more important as a constraint in the case of fossil resins than for compression fossils. The unexposed resin-bearing deposits are out of practical interest from the viewpoint of inclusions because the probability of finding any inclusions in very small quantities of resin, as are available from boreholes, is extremely low. Fossil resins are only rarely so common in certain deposits that their records from drilling cores are regular (e.g., at the West Siberian Plain the Late Cretaceous Pokur Formation was informally known for a long time as 'the amber

formation'). However, even in such cases a representative collection of inclusions can hardly be made except for the cases when inclusion rate is unusually high.

The effects of collecting **techniques** are discussed by PIKE (1993) who has demonstrated that resin lumps collected by means of floatation, contain on average more inclusions than those collected by hand because in this last case, a significant part of smaller fluxes (i.e. those that are richer in inclusions) gets overlooked. However, the most serious technical problem is connected with the search for inclusions. Even when the resin itself is transparent, its outer crust obscures its contents, and the smallest inclusions are particularly difficult to observe. A minimal representative sample of resin should include several tens of kilograms, that is many thousands of pieces, and it is practically impossible to remove the crust from them before the search for inclusions (especially in the case of the fragile retinite-type resins which should be treated individually to circumvent their mechanical ruination). Hence some biases connected with selective losses of the smallest inclusions are practically inevitable, even for an experienced collector, as the time required for sample examination would become unrealistic. The problem can be partially resolved by means of a strong transmitted light and screening the fluxes immersed into water. The loss increases considerably in dark-coloured, cloudy 'earthy' resins but even in transparent laminated pieces (richest in fossils) the internal light refraction still obscures some inclusions. In cloudy resins inclusions can be detected under X-rays, but probably only when they are pyritised (SCHLÜTER 1975, 1978). At present no simple effective methods for locating inclusions have been developed, and only relatively small resin samples can be examined carefully. Moreover, in many cases the available collections have been composed by amateurs, often either inexperienced or uninterested in some kinds of inclusions (e.g., in the smallest ones or in the most common taxa such as small nematoceran dipterans or ants). As a result, the majority of the world's collections of inclusions are more or less strongly biased.

Further problems arise because of the commerce in amber. True amber has been used for jewellery for centuries, and large quantities are available on the commercial market; and copal has also been commercially important for a long time as a varnish constituent. Consequently, inclusion collections often contain a large number of purchased specimens, and this is particularly true in the case of many Baltic and Dominican amber collections, as well as the only available large collection of the burmite which is housed in The Natural History Museum, London (ZHERIKHIN & ROSS 2000). However, commercial materials are rarely labelled exactly, and as a result, in some old collections even copal and amber materials were mixed so that some taxa described allegedly from the Baltic amber in fact originated from Quaternary copal of doubtful geographic origin (see CROSSKEY 1966, HENNIG 1966, NEL & PAICHELER 1993). Modern chemical techniques exclude such great mistakes but do not allow us to distinguish between, say, the Baltic amber from different parts of the Baltic Sea region or even between Baltic, Saxonian and Ukrainian succinite. There are some data on faunistic differences between the Baltic amber from the Samland Peninsula and other areas (e.g., from Lithuania and Denmark: KLEBS 1910, LARSSON 1978) and therefore the absence of precise data labels in many museum collections is distressing.

1.4.2.2.2. Other primary fossil containers

Not only resins but also some other fossil organic substances may contain insect remains or traces of their activity. For all of them the taphonomical properties of both insects and the fossiliferous substance are important.

Wood has a high fossilisation potential due to its mechanical resistance and low rate of decay in most ecosystems, except the tropical rainforests where it rapidly becomes destroyed by termites. Dead logs and wood fragments may be easily transported by water and accumulated in burial places. The sedimentary environments of wood burial vary widely as well as the modes of its fossilisation including coalification, diverse types of mineralisation (pyritisation, sideritisation, silicification, etc.), and sometimes mummification (BREZINOVA 1977). Standing trees may even occasionally be buried by mudflows or volcanic ashfalls and preserved over geological time (the so-called 'fossil forests'). Many insects are wood-borers and traces of their activity are easily observable in wood. These insects relatively rarely attack live wood but are often abundant in exposed dead logs, both standing and fallen. However, the probability of preservation of strongly destroyed wood decreases, and hence the early stages of decomposition seem to be the most taphonomically favourable from palaeoentomological viewpoint. Some insects (mayfly nymphs, chironomid larvae) make mines in submerged wood but the majority of wood-borers are terrestrial. Wood compression in sediments can obscure both its structure and insect traces, and rapid mineralisation of wood can produce false traces which may be misinterpreted as insect borings (FISK & FRITZ 1984).

Insect borings occur frequently in both subfossil and fossil wood and often contain faecal pellets (see Chapters 2.3.3 and 2.3.5) but there are few published records of other organic remains in these burrows. Studies of subfossil (600–4200 yr BP) conifer (mainly pine) trunks and stumps from a peat bog at Piilonsuo, Finland (KOPONEN & NUORTEVA 1973, LEKANDER et al. 1975) demonstrate that not only beetle remains, but also delicate objects such as the larval skins of beetles, cranefly, and hoverfly larvae occur in tunnels. Besides the wood borers and their inquilines several species of parasitic wasps were represented as well as some inhabitants of bracket-fungi. The shallow surface pits made on the submerged wood by larvae of the aquatic leaf beetle, *Donacia*, have been also found, sometimes with cocoons of the beetle inside them. The number of remains is significant: tunnels in one pine log about 3 m long have yielded no less than 43 remains of the bark beetle, *Pityogenes quadridens* Hartm., and at least four of the darkling beetle, *Hypophloeus linearis* F., as well as fragmentary remains of the parasitic wasp, *Cosmophorus cembrae* Ruschka, larvae of *Hypophloeus*, the checkered beetle *Thanasimus formicarius* L., and the longhorn beetle *Acanthocinus* sp. A 10 cm thick section of a pine stump provided seven remains of larvae from three longhorn beetle species. Besides the wood borers, their commensals and parasites, numerous remains of diverse terrestrial and aquatic insects have been found in peat filling of the holes and crevices in logs and stumps; the same species occur in number in the peat deposits outside the logs as well.

Up to now, Piilonsuo is the only locality where the content of tunnels in wood has been investigated in detail. Occasional records of other similarly preserved Quaternary finds (e.g., BUCKLAND et al. 1974, HARDING & PLANT 1978, SCHOTT 1984, KLINGER 1988) suggest that this mode of preservation is not unique and perhaps not even very rare. The pre-Quaternary records are scarce. The state of preservation of the cossonine weevil, *Rhyncolus kathrynae* Sleeper, described from a burrow in a pine branch of the Pliocene age from Fernley, Nevada, USA, is nearly the same as in the Quaternary (SLEEPER 1968). KOLBE (1888) described under the name *Anthribites rechenbergi*, tunnels and a three-dimensionally preserved pupa of an anthribid beetle from the Oligocene of Germany; unfortunately, the condition of the pupa was not characterised in detail. ROSELT & FEUSTEL (1960) and DUDICH (1961) recorded "mummified" insect remains found in supposed sphecid wasp nests in lignitised wood from the Late Tertiary of Germany and Hungary. NESSOV (1995) mentions phosphatised cockroach oothecae found in fissures of a large sideritised tree branch in the Late Cretaceous lagoon deposits of Uzbekistan. Available data indicate that insects represented in wood burrow contents include the xylobionts (wood inhabitants, i.e. the wood borers, their predators, parasitoids and commensals), the prey of wood-nesting carnivores, and incidental remains, mostly those washed into burrows in submerged wood.

Insect remains in wood borings are probably not so extremely rare as could be concluded from the published record, but rather hard to detect and quite fragile. Searching for them provides many technical problems. Because of the scarcity of available information it is impossible to say at present which types of wood preservation are best suited for palaeoentomological investigations.

Traces of insect damage are common on different plant organs such as **leaves**, **stems**, **seeds**, etc. They are often easily recognisable on fossil plant remains and provide a valuable source of information on insect/plant relationships in ancient ecosystems (see Chapter 2.3.5). For a discussion of plant taphonomy see KRASSILOV (1972a).

There are some records of sedentary scale insect females and whitefly puparia on fossil leaves and twigs both in sedimentary rocks (PAMPALONI 1902, ZEUNER 1938, RIETSCHEL 1983) and in fossil resins (GRIMALDI et al. 2000). Fossils of this type are inconspicuous and are probably being largely overlooked by collectors so that their rarity may be exaggerated.

Insect borings occur also in **vertebrate skeletal parts**. The preservation potential of bones is rather high but only in non-acid environments. To be bored by insects, bone has to be exposed for some time; consequently, insect traces mainly occur on disarticulated skeletal parts. Two types of insect traces in subfossil and fossil bones have been described in detail: the supposed dermestid beetle pupation chambers (originated in subaerial environments when the carrion is dried) and mayfly nymph burrows (originated in subaquatic environments) (Chapter 2.3.3). Preserved carrion fly puparia have been repeatedly recorded from subfossil vertebrate remains (e.g., GAUTIER & SCHUMANN 1973, GAUTIER 1975, COOPE & LISTER 1987, HEINRICH 1988, VERVOENEN 1991, LISTER 1993, GERMONPRÉ & LECLERCQ 1994). They occur mainly in skull fillings or inside horns of large mammals (mammoths, woolly rhinoceroses, bisons) but occasionally also in bird bones (PIERCE 1945a, GAGNÉ 1982). The only pre-Quaternary record is a reference of fly puparia from the interior of *Lophiodon* bones from the Eocene of Geiseltal, Germany (WEIGELT 1935).

One more rare and special case of fossil containers is **parasitic insect burial on or inside their hosts**. There are some records of supposed stylopised insects with remains of their parasite (LUTZ 1990, KINZELBACH & POHL 1994). HANDSCHIN (1947) described a braconid pupa found inside a phosphatised fly puparium from Quercy in France. In this case the taphonomy of the host is of crucial importance from an autotaphonomical point of view, but parasite preservation requires special taphotopic conditions providing exceptional conservation state.

Insect remains in fossilised vertebrate **faeces (coprolites) and pellets** are not uncommon, though their published records are not numerous (e.g., COCKERELL 1925a, L. MEDVEDEV 1976, GIRLING 1977, HIURA 1982, TVERDOKHLEBOVA 1984, KUSCHEL 1987, SCHMITZ 1991, ANSORGE 1993). It is a possibility that, in some papers the connection of insect fossils with coprolites is not noticed while in other cases their presence in coprolites has not been recognised. A number of pellets with insect wings representing supposedly pterosaur regurgitates occur in the Late Jurassic of Karatau, Kazakhstan (Fig. 47). The faeces of actively swimming predators with preserved food remains permit us to compare dominance pattern of local oryctocoenoses with those of lake ecosystems in general. For example, coprolites of the sturgeon-like fish, *Stychopterus*, in the lacustrine Early Cretaceous of Western Mongolia largely contain fragments of corixid bugs, indicating their wide distribution and abundance in the lakes, while in the coprolites of the same opportunistic predator from other localities, other organisms (mayfly nymphs, caddis cases, crustaceans, etc.) predominate (PONOMARENKO 1986a). A predator can sometimes transport prey remains from taphonomically unfavourable environments to more suitable ones. In the Oligocene lacustrine deposits at Bol'shaya Svetlovodnaya (Russian Far East), the incomplete remains of a burrowing mayfly nymph discovered in a fish coprolite belong to a species that supposedly inhabited a lake tributary (MCCAFFERTY & SINITSHENKOVA 1983). Insect fragments also occur occasionally in the stomach contents of fossil insectivorous vertebrates (FRANZEN 1984, RICHTER & STORCH 1980, RICHTER 1988, HABERSETZER & STORCH 1989), but such finds suggest an exceptional preservation state of the predator's remains, and the physiology of digestion may be important as well. STORCH (1978) who has found only very few fragments of insect cuticle among plant material in the stomach of the fossil pangolin, *Eomanis*, supposes that insect tissues have been digested rapidly by this specialised insectivore. Insect remains in human coprolites from archaeological sites indicate a significant role of

Fig. 47 A supposed pterosaur regurgitate with odonatan wings from the Karabastau Formation, Late Jurassic of Karatau in Kazakhstan (PIN 2904/3, photo by D.E. Shcherbakov); regurgitate 27 mm wide

some insects (e.g., grasshoppers and termites) in the human diet in the past (H. HALL 1977, STIGER 1977).

Vertebrate coprolites should also occasionally contain remains of coprophagous insects. However, the probability of their preservation might be low because the specialised coprophagous insects can only colonise faeces in terrestrial environments, and the fragile, insect-burrowed faecal mass can be buried without complete destruction only exceptionally. Thus, there are only a very few records of insect remains from subfossil coprolites restricted to Quaternary subaerial environments (e.g., WAAGE 1976, HOLLOWAY 1990). Insect trace fossils in intact coprolites are also extremely rare (CHIN & GILL 1996); the only exceptions are the dung balls constructed by dung beetles occurring regularly in some palaeosols (Chapter 2.3.6).

1.4.2.2.3. Secondary fossil containers

There are few records of insects re-deposited inside secondary containers, that is in fragments of lithified and then eroded sediments. Two main mechanisms of re-deposition of large rocks fragments are water and glacial erosion, and both participate in insect re-deposition in secondary containers. At outcrops of insect-bearing rocks along modern rivers, rather large rock fragments containing insect remains may commonly be seen in recent fluvial sediments, illustrating the possibility of re-deposition. The Early Triassic fluvial deposits at Zalazna in the Kirov Region, European Russia, contain pebbles of underlying Late Permian rocks with plant and insect remains. In the Netherlands Early Jurassic insects have been collected from erratic boulders transported by Quaternary glaciers (HUCKE & VOIGT 1967).

The time interval between primary burial and the re-deposition event can vary widely. In the Early Cretaceous sequence at Baissa in Transbaikalia a rich insect fauna is preserved in lacustrine sediments, particularly in marl layers. A thick conglomerate lens in the lower part of the section contains marl fragments with the same insect assemblage. In this case the lake sediments have not yet fully consolidated having probably been disturbed by a high-energy flow entering the lake, and partially re-deposited together with abundant pebbles transported from outside. In this case the re-deposition event occurred relatively soon (probably no more than few thousand years) after primary deposition, still in the time of existence of the same lake.

A special case of fossil containers is that of fossil-bearing crystals. Organic inclusions in crystals are extremely rare, and the mechanism of their embedding is poorly known. In any case, the crystallisation should occur in sediments during diagenesis so that the crystals are always secondary containers of the fossils. The most intriguing example of insect inclusions has been described by KAWALL (1876) who has found several small lepidopteran larvae inside a quartz crystal from the Urals. Unfortunately, this remarkable specimen cannot be traced now. Another interesting find is a wing embedded in a selenite crystal described by TILLYARD (1922a). In both cases the age of the fossils is uncertain. MARTÍNEZ-DELCLÒS (1996) mentions insect inclusions in salt crystals in the Late Miocene (Messinian) of Elba, Italy. However, in this case at least, any re-deposition is improbable because the container is water-soluble.

Finally, not only directly buried fossils but also fossil containers can be re-deposited in secondary containers; in particular, the Sakhalinian amber in recent beach deposits is accompanied by coal pebbles containing the same type of resin (ZHERIKHIN 1978). The coal is similar to the coals from the Palaeocene Lower Due Formation, and this was an important age indicator which helped in the location of the original amber deposits.

1.4.3. PRODUCTS OF THE TAPHONOMICAL PROCESS: INSECT FOSSILS AND ICHNOFOSSILS IN DIFFERENT PALAEOENVIRONMENTS AND MODES OF THEIR PRESERVATION

As pointed out above, in any environment the taphonomical potential depends on a complex set of taphonomical factors acting in different, and often opposite, directions. The following brief review of insect-bearing deposits is aimed at summarising the relative palaeoentomological importance of different palaeoenvironments which is not easily seen from the foregoing characteristics of taphonomical factors. For a more general summary of terrestrial organisms occurrence in different depositional environments see BEHRENSMEYER & HOOK (1992). The distribution of insect microfossils and chemofossils in different palaeoenvironments is poorly known, and the available data are discussed separately (Chapter 1.4.3.5).

1.4.3.1. Marine deposits

Marine deposits, and especially those accumulated in comparatively shallow epicontinental seas, are widespread. Marine sedimentary sequences are generally relatively complete and, because of a low mean sedimentation rate, often represent rather long time intervals (millions of years). Accordingly, the palaeontological record of skeletal organisms in the marine deposits is more complete than in any others.

Insects can be buried in marine sediments both directly and in containers. Insect trace fossils occur rarely and are represented nearly exclusively by bored wood and feeding traces on other plant organs. The alleged trichopteran cases from the marine Neogene of North Caucasus are in fact polychaetan tubes (VIALOV 1972).

As a rule, the frequency of insect finds in marine deposits is very low. Only few insect Lagerstätten of marine genesis are known; interestingly, none of them are older than the Jurassic although occasional insects occur in more ancient marine deposits as well. The best known examples are: German Liassic deposits connected with the so-called *Posidonia* Shales (Toarcian) of Germany and adjacent territories, the Late Jurassic Solnhofen lithographic limestone of Bavaria, the so-called Mo Clay of Denmark and northern Germany (latest Palaeocene-Early Eocene), the London Clay of England of the same age, the Miocene (Burdigalian-Early Helvetian) of Radoboj in Croatia, and the Miocene (Chokrakian-Karaganian) deposits of Stavropol, North Caucasus, Russia (for the references see Chapter 4.1). Other note worthy insect faunas originate from the latest Triassic–Liassic deposits of England, the Liassic of Switzerland (Aargau), the Eocene (Monte Bolca) and Late Miocene (e.g., Monte Castellaro) of Italy, and the Miocene (Tarkhanian) of Kerch Peninsula, Crimea, Ukraine. In all these cases the fossiliferous rocks are fine-grained but their composition varies widely, including the clastic sediments, organic-rich black shales, algal laminites, and occasionally diatomites with numerous volcanic ash layers (in the Mo Clay: RUST 1999a). In the latter case, insects are remarkably common and often represented by well-preserved articulated fossils. Depositional environments range from shallow-water sea bays (e.g., Monte Castellaro: GENTILINI 1989) up to deep-water offshore areas (e.g., the Mo Clay: BONDE 1974). The lateral insect distribution in marine strata is extremely uneven, and outside the principal Lagerstätten the same beds may contain few if any insects (e.g., in the *Posidonia* Shales). Oryctocoenoses from different facial environments are very different in taxonomic composition and hardly comparable with each other; however, the faunas of different localities within the same beds are rather similar and may include a number of common species (again, e.g., in the *Posidonia* Shales: ANSORGE 1996).

There are few truly marine modern insects, and their fossil finds are also scarce. These include rare marine water striders in the Cainozoic (N. ANDERSEN 1998), and perhaps some extinct taxa (e.g., some Chresmodidae, Chapters 2.2.2e, 3.3.5). Consequently, the insect assemblages in marine deposits are almost entirely allochthonous and highly polytopic, containing nearly exclusively aerial adults transported by winds from a wide range of terrestrial habitats (except for the case of fossil containers). Due to their ability to fly, insects occur in marine deposits relatively frequently in comparison with other terrestrial organisms. With few exceptions (e.g., Solnhofen), marine orycto-coenoses are dominated by small and medium-sized insects. The absence of aquatic larvae and their transportable trace fossils (the trichopteran or chironomid tubes), indicates that the role of watercourse transport is negligible. Adult aquatic insects are sometimes rather common: while dragonflies occur occasionally in open sea sediments, the repeating finds of adult aquatic beetles and bugs (e.g., in Solnhofen and Stavropol) should indicate a relatively proximity of land. Mass migrations may play an important role in some cases. RUST (2000) points that an unidentified lepidopteran species is remarkably common in the Fur Formation (the upper part of the Mo Clay), with more than 100 specimens collected (and as many as 14 together in the same slab). He interprets this pattern as evidence for a mass migration across the sea, and the same model cannot be excluded for the most common odonatans in Solnhofen and other marine deposits.

Insect remains in marine deposits are usually dismembered, and most commonly represented by isolated wings; however, in some localities (e.g., in Solnhofen), intact fossils predominate. They are preserved as compression fossils or, rarely, as three-dimensional cavities (e.g., in the Tarkhanian of Kerch) or pyritised remains (in the London Clay and the Tarkhanian of Kerch). The richest insect-bearing marine deposits were accumulated in offshore palaeoenvironments, where sediment disturbance by wave action was negligible but at a relatively short distance (probably up to 50–100 km) from the nearest land.

The importance of bottom anoxia is perhaps overemphasised in many taphonomical papers; even the black *Posidonia* Shales could have been deposited under moderately oxygenated conditions (KAUFFMAN 1979). Actualistic observations confirm that insect remains occur not only in coastal seas and at the oceanic shelf but also at the continental slope (e.g., in the Pacific Ocean at the depth of 600 m near the Californian coast: PIERCE 1965). In some layers of Tarkhanian in Kerch a fish species with well-developed light organs, *Vinciguerria merklini* Danilchenko (Photichthyidae) (Fig. 48), occurs together with insects; interestingly, non-marine plant macrofossils are virtually absent suggesting that the distance between this site and the land was considerable. On the other hand, the intertidal zone is unfavourable for insect burial because of continuous sediment reworking. However, it seems that scattered remains only occur in areas distant from land. It should be noted in this connection that small islands are probably unimportant as sources of insects for burial because their fauna is scarce and usually composed mainly or exclusively by flightless species. FOSBERG (1969) found that even the plant pollen from these islands, where the land area is less than 4 square miles and maximum elevation is no greater than 20 feet is practically missing in modern sediment samples.

Lagoon deposits are more or less intermediate between marine and lacustrine ones from the taphonomical viewpoint. Lagoons separated from the sea by natural barriers are generally rather good taphotopes

Fig. 48 An aphodiine dung beetle buried in association with deep-water marine photichthyid fish *Vinciguerria merklini* Danilchenko; Early Miocene (Tarkhanian) of Kerch in Ukraine (photo by D.E. Shcherbakov); elytron 3.7 mm long

for insects because they provide a low energy environment in close proximity to land. Also the invertebrate activity of lagoons is often low because of their brackish or hypersaline conditions.

There are a number of important insect Lagerstätten of lagoon origin, e.g. the Iva-Gora Beds of northern European Russia (Late Permian, see Soyana in Chapter 4.1), the saliniferous deposits of Alsace (latest Eocene–Early Oligocene), Oligocene deposits of southern France (Aix-en-Provence, Camoins-les-Bains, Dauphin, Puy-Saint-Jean, etc.) and the so-called *Hydrobia* Limestone (Hydrobienkalk) of the Mainz Basin, Germany (Miocene). Lagoon palaeoenvironments are connected by rather gradual transitions to shallow sea facies, on the one hand, and to deltaic, estuarine, marshy and other coastal environments, on the other, such that the borderlines are somewhat arbitrary. In particular, the interpretation of many Palaeozoic insect localities is uncertain in this respect (Chapter 1.4.3.2.2). The Solnhofen lithographic limestone was accumulated within a large quiet water area (about 3000 square kilometres) separated by barrier reefs (K. BARTHEL 1970) which was not a true lagoon but a back-barrier basin. In the Late Miocene sequence at Monte Castellaro, Italy, the bituminous shales with terrestrial winged insects deposited in a sea bay extend upwards into the lagoon marls of the Colombacci Formation with dragonfly nymphs (GENTILINI 1984, 1989). For the Early Cretaceous Crato Formation of Brazil a chemically stratified lagoon model has been proposed (MARTILL 1993) but the diverse aquatic insect fauna there suggests a rather freshwater lake environment, and the marine influence is only evident in the overlaying Santana Formation. Also some Oligocene insect-bearing deposits of southern France (e.g. Ceréste: LUTZ 1984a) have originated in coastal saline lakes rather than in lagoons.

Unlike open sea deposits, lagoon oryctocoenoses can contain not only winged but also flightless terrestrial arthropods, sometimes in considerable numbers (e.g., dozens of undescribed spiders from Aix-en-Provence are deposited in the National Natural History Museum, Paris). Some insects buried in brackish-water lagoons are probably authochthonous (e.g., libellulid dragonfly nymphs and culicid pupae in the Oligocene and Miocene deposits of southern France and Italy).

Consequently, some trace fossils can also be autochthonous. The insects are represented mainly by compression fossils, but the preservation quality is often high; articulated remains are common. The size selectivity of burial is probably weak. Phosphatisation of three-dimensional fossils occurs occasionally (e.g., in the Late Cretaceous of Central Asia: NESSOV 1995, 1997); a subfossil phosphatised spider was recorded from the lagoon phosphorites of Nauru, a raised atoll (GILL 1955).

Fossil insect **containers** are represented in both marine and lagoon deposits by fossil resins as well as by damaged plant fossils. Pieces of resin occasionally occur in considerable numbers, but are usually restricted to relatively few layers. Sometimes they are accompanied with abundant marine fossils (e.g., in the Late Cretaceous of Shavarshavan, Armenia, in the Eocene of Corbières, France, and in the London Clay: ALIEV 1977, BRETON et al. 1999, JARZEMBOWSKI 1999a). However, more commonly the marine fossil assemblages are impoverished suggesting somewhat abnormal environments, probably of a low salinity. Some deposits are interpreted as lagoon (e.g., in the Cretaceous of Asturia, Spain, and supposedly in the Dominican Republic: ARBIZU et al. 1999; CARIDAD 1999). In other cases resin deposits are associated with peripheral areas of river deltas and should be regarded as fluvial rather than marine (Chapter 1.4.3.2.3). When a resin fauna contains few aquatic taxa, river transport is improbable. Some resins are re-deposited in marine strata from unknown primary deposits (e.g., the burmite: ZHERIKHIN & ROSS 2000). Other types of containers are rare in marine deposits. Perhaps in the London Clay some pyritised insects were buried primarily in wood containers that were subsequently destroyed.

Insect finds in marine deposits are important first of all from the stratigraphic viewpoint because of the greater possibility of exact dating of accompanying fossils in accordance with the marine biostratigraphic scale. It should be noted that marine deposits often reflect the fauna of islands which might be impoverished and dysharmonic in comparison with continental biota.

1.4.3.2. Non-marine subaquatic palaeoenvironments

1.4.3.2.1. Lacustrine deposits

Lakes are short-lived in comparison with the sea, and the mean sedimentation rate is much higher; consequently, the time interval documented by a lake record is usually relatively short, rarely more than 1 million years. The lacustrine sequences are also generally less continuous than marine ones, however, they are the main source of palaeoentomological material, and the majority of the most important insect Lagerstätten (except for the fossil resins) are connected with lake sediments. Among them are the localities and deposits (mostly listed in Appendix, Chapter 4.1, q.v. for references and details), including the Late Carboniferous Commentry deposits in France, the Triassic Protopivka Formation (Ukraine, see Garazhovka), Madygen (Kyrgyzstan) and Tologoy Formations (Kazakhstan, see Kenderlyk), the Newark Supergroup (eastern North America), the Jurassic Dzhil and Sogul Formations (Central Asia, see under Sogyuty and Kyzyl-Kiya), the Late Jurassic Karabastau Formation (Kazakhstan, see Karatau), almost all highly productive insect-bearing Jurassic and Early Cretaceous formations of Siberia, Mongolia and China, the Early Cretaceous Montsec and Las Hoyas deposits (Spain), the Emanra Formation (Russian Far East, see Khetana), the Crato Formation (Brazil, see Santana), the Koonwarra Fossil Beds

(Australia), the Late Cretaceous deposits of Russian Far East (e.g., Obeshchayushchiy), Orapa in Botswana, the Palaeocene Menat deposits (France), the Tadushi Formation (Russian Far East), the Eocene Messel and Eckfeld deposits (Germany), the Green River Series (USA), the latest Eocene or Early Oligocene Bembridge Marl of England, the Oligocene Enspel and Sieblos deposits (Germany), Bol'shaya Svetlovodnaya (Russian Far East), the Florissant Fossil Lake, Creed and Clarkia deposits, numerous lacustrine sites in Montana and British Columbia (North America), the Miocene deposits of Spain (e.g., Izarra, Bellver de Cerdanya, Ribesalbes, Rubielos de Mora), Rott, Randeck and Oeningen deposits (Germany), the Shanwang Formation (China), numerous Miocene sites in Japan, the Barstow Formation (California, USA), the Late Miocene–Early Pliocene deposits of Central Massif in France, the Pliocene Willershausen deposits (Germany), and the Pleistocene Shiobara deposits (Japan) as well as a great number of less rich (or less investigated) localities.

Many important insect occurrences in fluvial (including deltaic and floodplain) and coastal plain sedimentary complexes are in fact also connected with lenses of lake and pond sediments. The insect distribution within fluvial sequences has been analysed in detail for the Triassic Molteno Formation, South Africa (J. ANDERSON et al. 1998) and the Jurassic deposits of southern Siberia, northern Kazakhstan and western Mongolia (ZHERIKHIN 1985, ZHERIKHIN & KALUGINA 1985). In both cases the most diverse and well-preserved fossils are restricted to oxbow lake sediments.

Directly buried remains predominate strongly in lacustrine palaeoenvironments while the role of containers is low; in particular, no important resin deposits are known. Trace fossils are rather common and diverse including autochthonous locomotion traces and burrows as well as diverse traces on allochthonous plant remains. Autochthonous aquatic taxa (including immature stages and moulted

Fig. 49 Antennal segment of the parasitic wasp *Tanychora petiolata* Townes (Ichneumonidae) with well preserved structures associated with sensilla; Early Cretaceous of Baissa in Siberia (holotype, SEM by H.H. Basibuyuk)

skins) are well represented, often constituting the bulk of fossils. Allochthonous terrestrial insects are common as well, but represented mostly or exclusively by winged stages (except for the hypoautochthonous inhabitants of the lake shore). The role of allochthonous aquatic insects transported from tributaries is probably negligible.

Both intact and disarticulated remains are common. Compression fossils usually dominate but in many sites more or less flattened three-dimensional specimens, and occasionally even practically undeformed three-dimensional cavities are common (e.g., in Izarra and the

Fig. 50 Antennal cleaning device (fore leg basitarsus and tibial spur) of *Cleistogaster buriatica* Rasnitsyn (Hymenoptera: Megalyridae) from the Early or Middle Jurassic of Novospasskoye in Siberia (from QUICKE 1997)

Bembridge Marls). Mineralised three-dimensional inclusions occur in some exceptional sites (e.g., in the Crato and Barstow Formations as well as in the Late Oligocene–Early Miocene of Riversleigh, Australia). The preservation of morphological details is often fine, allowing even electronic microscopy studies of most delicate structures (Figs. 49, 50). The best quality of ultrastructural and chemical cuticle preservation was also observed in a lacustrine site (Enspel).

The abundance of insect remains varies strongly, from scattered specimens up to mass burial, the latter not necessary being connected with any mass mortality events but often resulting from simple accumulation. In general, abundant well-preserved remains occur in finely laminated micritic rocks of different composition, especially in marls, tuffites and diatomites. Usually there are many insect-bearing layers within a lacustrine sequence but in volcanoclastic deposits the insects are often restricted to few thin ash layers. The composition of insect assemblages varies strongly, both within and between sections. Generally, the oryctocoenoses of small (e.g., oxbow) lakes contain many aquatic and relatively few terrestrial insects, mainly medium- to large-sized, and the same is true for shallow-water near-shore deposits of larger lakes. In the offshore area of large lakes, benthic immatures become less abundant or even disappear while the share of small-sized aerial insects is increasing. The differences may be quite dramatic, making the oryctocoenoses nearly incomparable. The area inhabited by benthic insects becomes reduced in warmer regions; there are also some differences in this respect between the deposits of different ages (Chapter 3.3).

1.4.3.2.2. Swamp, marsh and other wetland deposits

Heavily vegetated shallow-water palaeoenvironments are well represented in the geological record, mostly by diverse lignite- and coal-bearing deposits. However, their palaeoentomological importance is rather limited. Insect remains are not rare in coal-bearing strata and in some types of coals themselves (e.g., PIERCE 1961) but their preservation state is usually rather poor. The oryctocoenoses are strongly dominated by the most resistant parts (beetle elytra, cockroach forewings, etc.) and contain few if any articulated fossils. The remains are more or less compressed and usually coalified; sculpture and other fine details are usually obscured. Aerial insects are rare, indicating a low trapping potential of the thanatotope, and the oryctocoenoses are usually strongly dominated by local hypoautochthonous elements lived directly on swamp vegetation. In extreme cases even a rather large collection may contain representatives of only a few species (e.g., in the Permian of Kerbo, Siberia). Insect trace fossils are generally rare except for damaged wood and other plant remains. The Late Cretaceous deposits of the Arkagala Coal Basin (Russian Far East) and the Eocene deposits of Geiseltal (Germany) provide good examples of numerically rich swamp oryctocoenoses. The Geiseltal case is particularly significant because this locality is a well-known vertebrate Lagerstätte, famous for the exceptional preservation state (sometimes even including soft tissues), but its insect fauna seems to be of limited interest. Also the famous permineralised Carboniferous coalballs that contain exceptionally well-preserved plant material are practically devoid of insect remains. When more diverse and better preserved insect assemblages do occur in coal-bearing strata, they are usually restricted to coal-less layers that originated in more open-water lacustrine palaeoenvironments (e.g., in the majority of Palaeozoic localities and in the Jurassic Bayan Teg deposits in Mongolia). Modern studies of Palaeozoic coals suggest that major coal deposits could have originated in flooded rheotrophic mires rather than in ombrotrophic bogs or fens (DiMICHELE & PHILLIPS 1994).

The situation is different in the Pliocene and Quaternary where peat deposits constitute the most important source of palaeoentomological material. The peat assemblages are also dominated by beetle fragments, but their preservation state is much better than in older deposits and the fauna includes mainly or (in the Quaternary) exclusively modern species which can be identified with certainty on the basis of available fragments (see ELIAS 1994 for further detail).

A rare and special case of preservation is represented by a three-dimensional chafer's remains in bog iron ore (Fig. 34) from Protopopovo, Kemerovo Region, Siberia. The age of the specimen is uncertain (ZALESSKY 1961) but should be very young, possibly Holocene. Nevertheless, the remains are fossilised illustrating the role of iron bacteria in a rapid fossilisation.

The most important kinds of containers occurring in wetland deposits are resins and wood. Both contain mainly or exclusively hypoautochthonous material from wetland forests. Fossil resins are not rare in coals but this type of burial is rather unfortunate from the technical point of view because resin extraction from rather hard coals is difficult. Also the share of internal resin bodies lacking any insect inclusions may be high in many cases. Nevertheless, there are some important insect localities including the primary Sakhalinian amber deposits in the Naiba River Basin (Palaeocene), the Fushun amber in China (Eocene), and the resins from the Merit Pila Coalfield, Sarawak, Malaysia (Miocene). The resin fauna is impoverished because of relatively uniform habitats, and in particular, the diversity of aquatic taxa is low.

The only important insect site for wood containers, Piilonsuo in Finland (Holocene) (Chapter 1.4.2.2.2), is connected with peatbog deposits; no significant pre-Quaternary finds are known.

1.4.3.2.3. Fluvial deposits

The fluvial sedimentary sequences are highly complex and include, besides the channel deposits, sediments accumulated in abandoned channels, oxbow lakes, floodplain marshes, levees, crevasse splays, etc., not to mention deltaic and estuarine deposits. Insect remains are not uncommon and sometimes occur in numbers, but are usually fragmentary because of more or less distant transport and repeating sediment disturbance in the taphotope. Unlike swamp deposits, the remains within the same assemblage are almost uniform in size due to hydraulic sorting. Relatively well-preserved remains are restricted to the most low-energy environments such as backwaters, abandoned channels, oxbow lakes and especially deltas. The most important insect Lagerstätten connected with deltaic complexes are Mazon Creek (Late Carboniferous, Illinois, USA), Tshekarda (Early Permian, the Urals, Russia) and the *Voltzia* Sandstone (Mid-Triassic, Vosges, France) as well as the principal Purbeck and Wealden (Early Cretaceous) sites in England (see Chapter 4.1 for references), and even here articulated remains are scarce. As pointed above, when a detailed analysis of insect distribution has been made, it shows that the best state of preservation occurs in fact in lacustrine (oxbow lakes) and not in the true fluvial palaeoenvironments. However, the diversity of terrestrial insects increases with increasing river influence because the assemblages become more polytopic (e.g., in the Jurassic Cheremkhovo Formation, Siberia: ZHERIKHIN & KALUGINA 1985). In Mazon Creek well-preserved insects occur mainly in siderite nodules that originated in swampy areas of a delta (BAIRD et al. 1985a,b). Trace fossils may be common, especially in floodplain palaeoenvironments.

Hence the importance of fluvial deposits from the viewpoint of direct insect burial is rather limited. On the contrary, the richest resin deposits are connected with palaeodeltas where tree logs transported from different parts of river basin became accumulated (e.g., all principal Siberian Cretaceous sites, the New Jersey Late Cretaceous resins and the main deposits of the Baltic amber at the Samland Peninsula, Kaliningrad Region, Russia). The fauna is enriched with aquatic taxa including not only diverse aquatic dipterans but also mayflies, caddisflies, etc., especially in the retinite-type resins buried primarily in those allochthonous deposits.

1.4.3.2.4. Spring deposits

Spring deposits are sporadic and local in space but of a great interest in non-marine palaeontology because the frequently mineralised, and occasionally, hot water favours a high fossilisation rate and exceptional preservation of organic remains. Hence, even when insect remains are uncommon, spring deposits often can be considered as noticeable conservation Lagerstätten. The fossils are calcified (e.g., in the travertine or 'calcareous tuff') or, more rarely, silicified (e.g., in the chert). The best known insect localities of this type are Böttingen in Germany (ZEUNER 1931) and Rusinga Island in Kenya (LEAKEY 1952), both of the Miocene age. Other examples include the Rhynie Chert, Scotland (Devonian; KÜHNE & SCHLÜTER 1985), the so-called "Halbopal" of Lužice, Czech Republic (Oligocene; BEIER 1952), the Kymi deposits, Greece (Early Miocene; BACHMAYER et al. 1971), the cave flowstone of Przeworno, Poland (Miocene; GALEWSKI & GŁAZEK 1973, 1977), the onyx marble of Arizona, USA and Mexico (PIERCE 1950, 1951) of uncertain (probably Tertiary) age, and the travertine deposits of the Hobart area, Tasmania (Miocene; R. JOHNSTON 1888), Weimar, Germany (Pleistocene; JOOST 1984), Tsetsen-Ula, Mongolia (supposedly Neogene or Quaternary, unpublished), and Kobi, Georgia (probably Holocene; ZALESSKY 1961).

The assemblages are mainly allochthonous, with few (if any) aquatic insects (except for some travertine deposits, e.g., Przeworno and Weimar), and can reflect the fauna of habitats poorly represented in other deposits; unusual fossils, in particular flightless terrestrial and soft-bodied insects, are not very rare. The preservation is usually three-dimensional. Trace and container fossils are virtually absent.

1.4.3.3. Subaerial palaeoenvironments

Subaerial taphocoenoses are highly diverse but generally unstable in time, with a low probability of fossilisation and long-term conservation. Most of them are restricted to the Quaternary or nearly so, and only a few types are represented (rarely) in the more ancient record. However, they can provide valuable information on habitats undocumented by other fossils.

Those deposits that originated in **caves and karst crevices** are usually represented by organic debris, clastics and/or chemoclastics accumulated mainly subaerially though there may be some standing and running waters as well. Actualistic data indicate that karst holes can accumulate more than 90% of insects transported by water during occasional flooding (DÉCAMPS & LAVILLE 1975). Subfossil insects occur in organic deposits including but guano, human faeces, ground sloth dung and organic debris (PARMALEE 1967, CALLEN 1970, HEIZER 1970, NELSON 1972, NISSEN 1973, WAAGE 1976, WORTHY 1984, S.A. ELIAS et al. 1992b). Such assemblages are dominated by remains of consumed insects and the puparia of coprophilous dipterans; no specialised troglobionts have been recorded. An exceptional site in a siliceous flowstone deposited by thermal waters in a cave in a marmorised limestone is described from the Late Miocene of Przeworno, Poland (GALEWSKI & GŁAZEK 1973, 1977) where several diving beetle species were found represented by perfectly preserved three-dimensional remains. This locality is unusual both in respect of the preservation state and the presence of aquatic insects in cave deposits. There are also occasional records of insect trace fossils from caves (e.g., a Quaternary bee nest from a cave in Malaysia: STAUFFER 1979). It seems that more attention should be focused on insect remains in palaeontological and archaeological cave studies.

Karst holes and crevices can operate as traps for terrestrial vertebrates; there are several important tetrapod concentration Lagerstätten of such origin (BEHRENSMEYER & HOOK 1992). An unusually high concentration of bones affects chemical diagenesis; in particular, non-marine phosphorite deposits may occur. The most famous karst phosphorites of Quercy, France (Eocene–Oligocene) yield an interesting insect fauna, as described by HANDSCHIN (1947). Rather large amounts of material (several hundred specimens) were collected in Quercy from several different sites. Insects accompany vertebrate fossils also in the karst phosphorites of Ronheim, Germany (Oligocene; HELLMUND & HELLMUND 1996a) and the Dunsinane Site at Riversleigh, Australia ("mid-Tertiary": DUNCAN et al. 1998). Perhaps, insects constitute a common, if not universal, component of the karst phosphorite fauna, but they are often neglected by vertebrate palaeontologists. The phosphorite faunas are of great interest due to both their good preservation state and taxonomic and ecological content. In all cases the insects are phosphatised and preserved three-dimensionally; however, the remains are often incomplete, and the appendages are almost always lost. The assemblages are dominated by flightless terrestrial insects including numerous larvae and pupae; other terrestrial arthropods (e.g., millipedes) are found as well. Almost all insects are either necrophages or predators and were surely attracted by vertebrate carrion; fly puparia being especially abundant. HANDSCHIN (1947) described a braconid wasp pupa found inside a phorid fly puparium; this insect find in an insect container being most unusual. No other container fossils are recorded but perhaps some insects could be found inside vertebrate bones.

Fragmentary beetle remains from Late Triassic fillings in karst crevices in South Wales, UK, are described by GARDINER (1961). The Triassic karst of this area is rather well known and yields an interesting vertebrate fauna (HALSTEAD & NICOLI 1971, BENTON 1989, SIMMS 1990) but insect remains seem to be very rare here. The sediments are not phosphatised, and insect preservation is rather poor. The tetrapod fauna was tentatively interpreted as originated from an upland area but the finds of marine plankton in crevice filling indicate a low hypsometry and proximity to a seacoast (J. MARSHALL & WHITESIDE 1980). Interpretation of Bernissart, the famous Early Cretaceous *Iguanodon* dinosaur locality in Belgium where a few insects were also found (LAMÉERE & SEVERIN 1897), as a collapsed karst terrain is probably incorrect (BEHRENSMEYER & HOOK 1992). In general, ancient karst deposits deserve much attention in palaeoentomology.

Aeolian palaeoenvironments are practically devoid of insect remains. The absence of subfossil insect remains indicates their extremely low preservation potential. On the contrary, insect trace fossils are widespread and may be abundant. Insect burrows, and in particular termite tunnels assigned mainly to the genera *Psammotermes* and *Hodotermes*, are common in the Pleistocene (R. SMITH et al. 1993) and Neogene (SENUT et al. 1994, SEELY & MITCHELL 1986) aeolian sands in Namibia. The Quaternary calcareous sand dunes of the Canary Islands contain abundant solitary bee cells (ELLIS & ELLIS-ADAM 1993), and rather diverse insect burrows are recorded from similar Quaternary deposits of the Bahamas (CURRAN & WHITE 1991). AHLBRANDT et al. (1978) and BRUSSARD & RUNIA (1984) describe dung beetle burrows in Quaternary aeolian sands. Beetle pupation chambers are occasionally common in both interdune and dune deposits in the Quaternary of Australia (TILLEY et al. 1997) and Late Cretaceous of Mongolia and North China (P. JOHNSTON et al. 1996, TILLEY et al. 1997); the latter dune area was situated inland. Numerically rich and diverse insect ichnocoenoses have been discovered in Jurassic aeolian deposits of USA (FAUL & ROBERTS 1951, EKDALE & PICARD 1985) and Brazil (FERNANDES et al. 1990).

Though there is some uncertainty about the definition of **palaeosols** (CATT 1987) this term is widely accepted mainly for buried (or fossil) soils of different ages. Palaeosols occur rather frequently in terrestrial sections of different ages since the Precambrian but like other subaerial deposits become commoner and better preserved in younger, and especially in Quaternary, sequences. For general information on palaeosols see YAALON (1971), RETALLACK (1981a, 1986), WRIGHT (1986), and CATT (1987). Hydromorph soils originating in temporarily flooded environments are best represented in the palaeosol record, but thick weathering crusts formed at ancient watershed areas are also widespread. Well-preserved organic remains in palaeosols are rare and represented mainly by vertebrate bones and silicified plant fragments (BEHRENSMEYER & HOOK 1992). The general preservation potential of insect remains in soils is surely low. Subfossil insect remains are recorded from Holocene soils (NAZAROV & KARASYOV 1992); occasionally insect fragments are preserved in subterranean nests (e.g., termite mandibles in termite tunnels: MACHADO 1982, cited after NEL & PAICHELER 1993). Perhaps, fusainised remains from burned litter layers have a higher preservation potential but their distribution and taphonomy is poorly investigated and their relations to palaeosols are unclear. Plant litter preserved *in situ* is described in some fossil forests buried by mudflows and volcanic ash-falls (e.g., TAGGART & CROSS 1980, YURETICH 1984, BASINGER 1991) and sometimes demonstrates an exceptional state of preservation. Theoretically, such fossilised litter horizons can contain some insect remains as well, and thus merit special attention of palaeoentomologists, but no published records are available. In contrast, trace fossils are widespread and often common in palaeosols, including insect burrows, pupation chambers, subterranean nests and stored food (e.g. dung balls and plant remains) (see Chapter 2.3 for further detail and references). GENISE (2000) stresses that the preservation probability of nests with walls covered by secretions, excrements, plant remains, etc., increases significantly in comparison with simple tunnels. Hence palaeosols provide a highly valuable source of information for insect ichnology. In deeply weathered sequences the state of ichnofossil preservation is usually poor and they may be hardly recognisable.

As discussed above, present-day soil profiles can contain autochthonously buried subfossil resins (copals) with enclosed insects. Resin lumps occur occasionally in palaeosols as well (e.g., in the Eocene organic palaeosols of the Eureka Sound Group at Axel Heiberg Island, Arctic Canada: TARNOCAI & SMITH 1991). Though there are no published records of insects, the palaeosol resins can be potentially fossiliferous and their fauna should be of a great palaeoecological interest.

Various organic remains including food remains, parasites, occasional visitors, etc., can be accumulated in **vertebrate burrows** which may protect their contents providing a valuable source of palaeontological data over a relatively short time scale (i.e. thousands of years) (DINESMAN 1979, 1992). Vertebrate burrows occur in more ancient palaeosols as well (BEHRENSMEYER & HOOK 1992), but no pre-Quaternary insect remains have been recorded from them. In the subfossil assemblages, insects are occasionally abundant but little studied, except for the burrows of packrats (*Neotoma* spp.) in the southern USA and adjacent regions of Mexico (age from about 200 up to more than 43,000 yr BP; ASHWORTH 1976, ELIAS 1987, 1990, 1994, W. HALL et al. 1988, 1989, 1990, MACKAY & ELIAS 1992, VAN DEVENDER & HALL 1994). The content of burrows is cemented with packrat urine forming a mass of midden resistant to both decay and erosion, and insect remains often show an excellent state of preservation except for their original colour which is lost. The assemblages are strongly dominated by the taxa with the most durable exoskeletons, like beetles and ants; the bugs, dipterans, and orthopterans are underrepresented in the subfossil midden in comparison with the contents of present-day packrat burrows. Burial is selective also from the ecological point of view, with the facultative packrat nest inquilines (mainly predators and scavengers) predominating. Packrat parasites (e.g. fleas) are surprisingly rare. The insect remains in packrat faecal pellets are fragmentary and unidentifiable.

Information on subfossil insects from other vertebrate burrows is scarce. Some insects have been recorded from the Late Quaternary badger burrows at Serebryany Bor, Moscow, Russia (PANFILOV 1965, POPOV 1968a, L. MEDVEDEV 1976). The assemblage is dominated by beetles. Unlike the packrat faeces, the badger coprolites contain identifiable remains of beetles and bugs which must have lived outside of the burrows. RATNIKOV (1988) mentioned abundant insects (mainly beetles) from the Late Quaternary bird burrows at Zmeevka, Belgorod Region, Russia, but without any more detailed taxonomic information. He supposes that the entomofauna is represented mostly by remnants of insects preyed upon by the birds. A unique find of well-preserved lice placed to the extinct species *Neohaematopinus relictus* Dubinin has been described from burrows of the ground squirrel (*Citellus*) in the Pleistocene of the Indigirka River Basin, North Siberia, Russia (DUBININ 1948). Both lice and corpses of their hosts have been conserved by low temperature and drying in permafrost.

The importance of **mud flows and landslides** in insect palaeontology seems to be very low. The conservation of insect remains in those sedimentary contexts is improbable. However, some types of insect trace fossils can be preserved (e.g. burrowed wood in fossil forests buried *in situ* by mud flows).

There are some records of insects and other terrestrial arthropods from **volcanogenic deposits** but their taphonomy is unclear; possibly, they were in fact preserved in small hot spring deposits in active volcanic areas. ZALESSKY (1961) mentioned three body segments of an arthropod (tentatively identified as a springtail) found in a Quaternary volcanic breccia in Iran. GILL (1955) illustrates three-dimensionally preserved caterpillars from the Quaternary lava fields in Australia, and PIERCE (1945b) has described a mineralised millipede enclosed in a volcanic rock.

Tar seeps operate as highly effective insect traps, but their deposits are unstable in time and restricted to the Quaternary record. However, in the Quaternary they are known from different regions including California (the famous La Brea and McKittrick asphalt deposits: S. MILLER 1983), the Carpathians (the ozocerite deposits of Boryslaw and Starunia: ŁOMNICKI 1894, SCHILLE 1916), Azerbaidzhan (the Binagady asphalt deposits: BOGACHEV 1939), Trinidad (K. BLAIR 1927) and Peru (the Talara tar seeps: CHURCHER 1966). The assemblages are dominated by beetles but other insects (e.g, orthopterans, bugs, dipterans) are represented as well. Flightless taxa occur occasionally. The preservation state is often perfect, and cuticle can be partially preserved chemically (STANKIEWICZ *et al.* 1998a). Aquatic taxa can occasionally also be common (e.g. aquatic beetles in Binagady), but are represented exclusively by winged adults. Necrophagous fly puparia are found in bone containers in the La Brea asphalt (PIERCE 1945a, GAGNÉ 1982).

While large and relatively old ice shields are restricted to those areas where living insects are absent or nearly so and any significant transport of insects from outside is highly improbable, the smaller and relatively short-lived mountain **glaciers** occasionally contain subfossil insects. Several glaciers in the Rocky Mountains are known as unique localities of abundant grasshopper remains dated from 200 to about 800 years old (LOCKWOOD *et al.* 1988, 1990, 1991, 1992a,b). The remains occur in large numbers at several levels within the ice and become exposed in the course of glacier melting. The assemblage includes several species, both alpine and lowland. Interestingly, the dominant *Melanoplus spretus* (Walsh) is a migrating lowland species which is now extinct. Fragmentary remains predominate even in materials from undisturbed layers indicating that the dead insects have degraded either at the snow surface thanatotope before their burial or in the taphotope due to action of melt-water percolating through the ice layer. Some intact specimens occur as well but none of them retain their original colour. Virtually no other insects have been observed. The glacier grasshopper taphocoenoses are believed to originate as a result of mass mortality of migrating individuals on mountain glaciers. Actualistic observations show that during outbreaks some species of North American grasshoppers (including *M. sanguinipes* (F.) very close to *M. spretus*) occur in large numbers on the surface of mountain snow (G. ALEXANDER 1964, L. PRICE 1985). Recently older insects (about 5,000 years old) have been recorded from an ice core in the Bolivian Andes (L. THOMPSON *et al.* 1998) showing that mountain glaciers in different parts of the world deserve more attention in this respect. No container fossils are known from there.

Glaciers can also conserve large areas covered by ice with its natural vegetation. At Ellesmere Land, Arctic Canada, well-preserved tundra vegetation entombed in the XV–XVII centuries has been observed along the edge of recessing glaciers (BERGSMA *et al.* 1984). No insects were recorded but their remains may occur in the soil while any insects from the soil surface might be displaced by melt water and either destroyed or re-deposited at some distance in recent sediments. The latter possibility should be taken into account in interpretation of Quaternary insect assemblages buried during retreating of glaciers.

Insect remains from **archaeological materials** have been recorded occasionally since XIX century by different authors (e.g. SIEBER 1820, PATTERSON 1835, DOMAISON 1887, F. BLAIR 1908, ALLUAUD 1910, LESNE 1930, ALFIERI 1932, MONTE 1956; see also AMSDEN & BOON 1975) but systematic studies in this field began in the 1970s when archaeoentomology originated as a special scientific discipline (see BUCKLAND 1976, KENWARD 1978, ELIAS 1994, MORET 1996, for reviews). The insect burial environments at ancient human settlements are diverse. Some of them do not differ principally from the non-anthropogenic burial contexts (e.g. in pond or cave deposits) while others are more specific (e.g. the cultural layers formed by anthropogenic debris both indoor and outdoor, the tombs, the contents of storage sites, etc.).

A number of man-made microsites can operate as places of accumulation of insect remains, including wood-paved roads (ULLRICH

Fig. 51 Subfossil remains of the cossonine weevil *Hexarthrum exiguum* Boheman (Coleoptera: Curculionidae) preserved in tunnels in a fragment of a wood construction from the archaeological excavations at Erebuni in Armenia; VIII century B.C. (photo by D.E. Shcherbakov); area shown 12 mm wide

1972), ancient trackways crossing bogs (BUCKLAND 1979), wells (ALVEY 1968, COOPE & OSBORNE 1968), Pueblo Indians cliff caves (S. GRAHAM 1965), cellars (O'CONNOR 1979), latrines (P. MOORE 1981, OSBORNE 1981), pits (BUCKLAND *et al.* 1976), etc. Both natural and man-made containers are also diverse. Wood constructions can be bored by insects and occasionally contain their well-preserved remains (BUCKLAND *et al.* 1974) (Fig. 51). Highly specific assemblages including various consumers of dried flesh (dermestids, clerid beetles of the genus *Necrobia*, tineids) as well as anobiids and other insects that fed on dried plant matter have been discovered in Egyptian mummies (see HUCHET 1995, for a review). Insects have also been sampled from mummified corpses in ancient tombs in other areas including Greenland (J. HANSEN 1989, BRESCIANI *et al.* 1983) and Alaska (HORNE 1979) as well from human corpses occasionally preserved in bog peat (GIRLING 1986). Insects are occasionally buried on human corpses and can also be discovered in ancient graves together with human bones (ROTHSCHILD 1973). The presence of artefacts made from artificial materials (e.g. metals) can modify local chemistry and affect insect preservation around them (BEIER 1955). Occasionally large quantities of insects were accumulated specially by man, perhaps as amulets or ornamentals (SCHLÜTER & DREYER 1985, LEVINSON & LEVINSON 1996). A curious archaeological evidence of a falsified commercial product is the find of *Cetonia* elytra instead of the Spanish fly, *Lytta vesicatoria* (L.), in the cargo of a Dutch ship (HAKBIJL 1987). This is also a rare example of an insect find made during marine archaeological studies; insect infested grain cargo has also been discovered at a Roman ship (PALS & HAKBIJL 1992).

The best state of preservation occurs in anaerobic or highly arid environments as well as in permafrost. Usually beetles and/or fly puparia predominate as the most durable remains. The assemblages are enriched in synanthropic species that are rare or lacking in other types of taphocoenoses. Because of anthropogenic environments even the taphocoenoses buried in rather ordinary conditions may look unusually in comparison with natural palaeoenvironments. For instance, the assemblages discovered from the bottom sediments of ancient wells in England (OSBORNE 1969), North Caucasus, Russia (ANTIPINA *et al.* 1991), and Spain (MORET & MARTÍN-CANTARINO 1996) contain abundant dung beetles suggesting that the wells were used probably for watering cattle. No similar assemblages are known from Quaternary deposits formed in natural environments. The finds of stored product pests (reviewed by BUCKLAND 1981, 1991, KISLEV 1991), technical pests (WARNER & SMITH 1968) and human parasites (e.g. EWING 1933, ROTHSCHILD 1973, FRY 1976, BRESCIANI et al. 1983, BUCKLAND & SADLER 1989, SADLER 1990) are not rare indicating that archaeoentomological observations can elucidate the history of their expansion in considerable detail. Some archaeoentomological data were used for testing of biogeographical hypotheses on the dispersal history of synanthropic species (e.g. TESKEY & TURNBULL 1979, J. ALEXANDER et al. 1991) or for the reconstruction of regional epidemiological history of insect-borne diseases (SUMMERS 1967) but in general the rapidly growing body of archaeoentomological information too often remains unknown to most entomologists.

Ancient insect drawings and sculptures (e.g. LATREILLE 1819, O. KELLER 1913, SCHIMETSCHEK 1978, WARD 1978, LANGER 1982, PARENT 1987) are also of some archaeoentomological interest as well as is evidence of ancient insect cultivation (e.g. E. CRANE & GRAHAM 1985a,b). Such illustrations are usually quite recent but there is a record of a picture of a cave orthopteran (*Troglophilus* sp.) scratched on a bone among Late Palaeolithic archaeological material in France (CHOPARD 1928, cited after ZEUNER 1939).

1.4.3.4. Selectivity of the rock record

It can be seen from the above review that the diversity of palaeoenvironments of insect burial is maximal in the youngest deposits and decreases continuously with increasing age. This pattern reflects the selectivity of the rock record. This aspect was emphasised for the first time by EFREMOV (1950) who proposed a special discipline, litholeumonomy, which should examine this aspect of selectivity. Unlike the word taphonomy, this term has not found acceptance in consequent scientific literature but the phenomenon itself is well established and considered by many authors.

In general, any sediments or rocks situated above the base level of erosion are subject to erosion processes. The rate of erosion varies widely depending upon local and regional physiography, climate, vegetation cover and substrate resistance. Other things being equal, the highest points are eroding more rapidly, and products of their ruination cover local depressions so delaying their erosion for some time. In mountainous areas with large altitude gradients between the valleys and their surroundings the time lag may be rather long. In volcanic areas long term protection against erosion can be assured by overlaying lava flows or thick volcanic ash deposits. A high porosity of ash deposits and certain types of lava are especially favourable, minimising the erosional effects of waterflows due to increasing water filtration (SOKOLOV 1973). However, the only mechanism guaranteeing against erosion is downwarding leading the deposit out its action; and sometimes even the downlifted rocks will be uplifted back to the erosion zone. On the other hand, ocean floor sediments are also disappearing with time because of subduction of the oceanic crust.

This has two most important consequences on sediment preservation. Firstly, the older sediments are, the less the relative volume of them that has been preserved up to present, and the greater the area from which they have been completely eroded. Consequently, the geographical incompleteness of the fossil record increases with time. Secondly, this process is not uniform but selective, and the diversity of palaeoenvironments documented by the sedimentary record decreases

in older deposits. That is why the Late Cainozoic record is most complete and includes a number of palaeoenvironments rarely or never represented in older record. Epicontinental marine deposits have the largest conservation potential, and in fact in the Palaeozoic record any environments other than shallow seas and their coastal areas are almost completely lost. The oldest deposits are most strongly biased from the viewpoint of the habitats they represent. This pattern surely affects our concept of the evolution of biological diversity, but is difficult to say to what degree the general increasing trend in diversity from the Palaeozoic to recent time has been exaggerated due to taphonomy.

1.4.3.5. Insect microfossils and chemofossils

Insect **microfossils** are certainly common and widespread but they are poorly studied. There is no doubt that any deposits with abundant insect remains should also contain large number of diverse microfossils which, however, are rarely noticed. KALUGINA (in RASNITSYN 1986b) mentioned abundant pharyngeal sphincters of chaoborid larvae of the genus *Astrocorethra* Kalugina occurring *en mass* in the same slabs with adults and pupae in the Early Cretaceous lacustrine deposits in western Mongolia. BATTEN (1998) illustrated some insect microfossils from the Wealden of England. Isolated lepidopteran scales are recorded from the Early Cretaceous Lebanese (WHALLEY 1978) and Late Cretaceous French amber-like resins (SCHLÜTER 1978), as well as from the Palaeocene Sakhalin amber (ZHERIKHIN 1978).

There are also some insect microfossil records from deposits that apparently lack any other insect remains. VORONOVA & VORONOVA (1982) and VORONOVA (1984, 1985) mention insect and oribatid mite microfossils (including lepidopteran scales) discovered in palynological samples from the Early Cretaceous of Ukraine. Hepialid moth scales have been found in the Waikato Coal Measures (Late Eocene) in New Zealand (W. EVANS 1931). SCHULZ (1927) recorded lepidopteran scales from Cainozoic tuffite deposits of Germany. Diverse insect microfossils (setae, scales, small cuticle fragments, etc.) are recorded from the lacustrine Eocene of Maslin Bay, South Australia (SOUTHCOTT & LANGE 1971), saliniferous Palaeocene lagoon deposits of Salt Range, Punjab (GEORGE 1952) and organic filling of crevices in Triassic limestone in southern Poland (LIPIARSKI 1971). Three-dimensionally preserved insect eggs and fusainised insect microfossils are known from Late Cretaceous and Eocene organic debris deposits (L. GALL & TIFFNEY 1983, SELLICK 1994, GRIMALDI et al. 2000). One of the oldest insect remains known is a faceted eye fragment from the Late Devonian deltaic deposits of Gilboa, New York State, USA (SHEAR et al. 1984). Insect coprolites which should be regarded as trace microfossils are recorded occasionally both directly from sedimentary rocks and from resin and wood containers (see Chapter 2.3.6 for references).

In general, because published microfossil records are mainly anecdotal, their importance for palaeoentomology is at present limited. Nevertheless, some of them provide valuable information. In particular, dispersed lepidopteran scales may elucidate the distribution and abundance of moths and butterflies better than their scarce body and wing fossil record. Indeed, if dead lepidopterans sink more slowly in water than the majority of other insects, and if they can escape resin traps due to scale autotomy, the scales should be more indicative than any other lepidopteran fossils. The Eocene stick-insect eggs described by SELLICK (1994) make an important addition to our extremely scarce data on the Cainozoic members of their order, to the extent that egg morphology in this group is often diagnostic. Even when the taxonomic placement of microfossils is uncertain, they can be useful for stratigraphy and palaeoecology. Hence further development of insect microfossil studies seems to be desirable. Its potential importance is encouraged by the impressive results of modern works on subfossil insect remains from lake deposits. This field was unexplored until recently but now a number of papers are available, mainly on larval chironomid head capsules from different parts of the world, with important palaeoecological and palaeo-climatological implications (D. STARK 1976, PATERSON & WALKER 1974, WIEDERHOLM & ERIKSON 1979, WARWICK 1980a,b, BINFORD 1982, DÉVAI & MOLDOVÁN 1983, GÜNTHER 1983, HOFMANN 1983, 1986, 1990, 1993, I. WALKER & PATERSON 1983, 1985, SCHAKAU & FRANK 1984, LÖFFLER 1986, SCHAKAU 1986, I. WALKER 1987, 1995, I. WALKER & MATHEWES 1987, 1988, 1989a–c, UUTALA 1990, I. WALKER et al. 1991a,b, MASAFERRO et al. 1993, REYNOLDS & HAMILTON 1993, S. BROOKS et al. 1997, SADLER & JONES 1997, QUINLAN et al. 1998, LOTTER et al. 1999, BRODERSEN & ANDERSON 2000).

Insect **chemofossils** represent a practically unexplored field, though insect cuticular waxes have a high preservation potential and should occur in fossil organic matter. KENIG et al. (1994) demonstrated the presence of long-chain linear alkanes supposedly originated partly from insect cuticles in the Holocene sabkha deposits of Abu Dhabi and speculated on the possible role of insects as the source of hydrocarbons in some more ancient sediments. STANKIEWICZ et al. (1998d) also stress that the contribution of arthropod cuticular material in the formation of kerogene in sediments is not negligible. Diagenetic alteration of arthropod cuticular matter is still insufficiently studied. As discussed above this is surely not a simple process of degradation but a rather complicated chemical process including secondary polymerisation and other chemical changes (STANKIEWICZ et al. 1998d, BRIGGS 1999).

Chemical trace fossils should be also of some interest. In this respect, the contribution of insect coprolites in the formation of non-marine oil shale (e.g. in numerous Early Cretaceous basins in Siberia, Mongolia and China as well as in the Eocene Green River Basin in USA) may be significant (L. BRADLEY & BEARD 1969, IOVINO & BRADLEY 1969, BRADLEY 1970, KALUGINA 1980b). ENGSTRAND (1967) mentioned the presence of bee wax from a Holocene peat in Sweden. KAY et al. (1997) interpret the esters extracted from holes in a wood from the Late Triassic Chinle Formation (Arizona, USA) as probable traces of Dufour's gland secretions, suggesting that the holes could have been constructed by bees. However, any probability of the presence of bees in the Late Triassic should be excluded if other data on the hymenopteran fossil record are taken into account. Under the present-day state of our knowledge of the organic geochemistry of insect-derived products, the interpretation of chemofossils and chemical trace fossils has to be taken with much caution.

In general, any fossil insect assemblages are evidently biased, but the biases vary significantly from one palaeoenvironment to another, and even the deposits poor in insect remains but representing an unusual taphonomical situation may be of great interest.

1.4.4. INSECTS AS CONTAMINANTS IN FOSSIL ASSEMBLAGES

Many insects can penetrate into older sediments either actively, by digging or crawling along crevices, or passively, with percolating water. Thus, at least theoretically, in fossil assemblages some insect specimens may be younger contaminants. With the exception of some subfossil assemblages, modern contaminants can be recognised rather easily on the basis of their different state of preservation. This may be illustrated by the modern calliphorid fly remains recovered in Late Cretaceous sandstone at Nizhnyaya Agapa River, North Siberia, together with amber-like resins (Fig. 52) as well as a modern mayfly nymphal skin found during splitting of Late Permian insect- and plant-bearing mudstone exposed at water level on the bank of the Taymura River near Kerbo, Evenk Autonomous District, Russia (Fig. 53). A modern lathridiid beetle has been found in a sample of

Fig. 52 Remains of modern calliphorid fly *Protophormia terranovae* R.-D. as contaminants in an exposed Late Cretaceous sandstone from Nizhnyaya Agapa River in North Siberia (photo by D.E. Shcherbakov); wing 7.2 mm long

Fig. 53 A moulting nymphal skin (5 mm long as preserved) of a living mayfly found as contaminant in a fissure of a Permian rock of Kerbo in Siberia (photo by A.P. Rasnitsyn)

Palaeozoic saliniferous rocks from the Urals (A.G. Ponomarenko, pers. comm.). Insect and mite fragments discovered in the Palaeocene bauxite from Angara River Basin in Siberia seem to be too fresh and may also represent recent contaminants (ZHERIKHIN 1978). D. SILVA & PEREIRA DE ARRUDA (1976) described a dismembered ant from a drilling hole in north-eastern Brazil as the Early Cretaceous fossil but the remains clearly belong to a present-day insect, probably pumped incidentally with a clay mortar.

More problems can arise if fossilised remains are younger than the fossiliferous layer containing them. Such instances might be rare but they can occur when the fossiliferous deposit had been partly eroded some time in the geological past and then overlain by younger sediments. A find of a supposed thrips nymph in the Early Devonian Rhynie Chert in Scotland may indicate a younger (probably Tertiary) contamination, even though the remains are silicified like other fossils in the chert (CROWSON 1985). However, this case needs further investigation to prove the Crowson's hypothesis (Chapter 3.2.2.1).

Some modern traces of insect activities in exposed rocks has been misinterpreted as fossil remains of other organisms as, for instance, present-day bee brood cells constructed in Precambrian rocks (SANDO 1972), and BARBER (1930) noted that supposed fossil vertebrate eggs from Myanmar were subsequently proved to be the calcareous pupal chambers of a recent longhorn beetle. Such structures are sometimes called "pseudofossils" but it is more correct to restrict this term to inorganic structures imitating fossils (MONROE & DIETRICH 1990).

1.4.5. INSECT ACTIVITIES AS A TAPHONOMIC FACTOR

The taphonomic effect of insects is largely negative like that of other consumers. This is especially true for the diverse and abundant destroyers of dead organic matter both in terrestrial and fresh-water environments, where a large amount of plant and animal remains disappear due to insect feeding. The rate of decomposition of organic debris increases sufficiently as a result of insect and other arthropod activity. The role of phytophagous and predaceous insects as taphonomically negative ecological agents is significant as well, and sediment bioturbation by benthic aquatic insects is significant in many freshwater ecosystems.

DINDO et al. (1992) describe an unusual case of recent insect damage to fossils in which members of the carpenter bees genus *Xylocopa* occasionally construct their nests in mummified Tertiary tree trunks, and in doing so have been classed as "pests" of a fossilised forest monument at Dunarobba, Italy. Some other insects, especially termites, also occasionally damage archaeological sites, either disturbing their original stratigraphy (MCBREARTY 1990) or covering rock painting with their galleries and even damaging human bones at burial sites (WYLIE et al. 1987).

Positive taxonomic roles of insects connected with transferring organic materials to conservation places is limited. Well preserved plant pollen grains, spores and plant cuticles (Figs. 54, 55) have been discovered in the gut contents of fossil insects (KRASSILOV & RASNITSYN 1982, 1997, 1999, A. SCOTT & TAYLOR 1983, CALDAS et al. 1989, SHEAR & KUKALOVÁ-PECK 1990, RASNITSYN & KRASSILOV 1996a,b, 2000, KRASSILOV et al. 1997a,b, 1999) and in insect coprolites. Pollen grains have also been detected on the body surface of some insects, both on compression fossils (LUTZ 1993a, RAYNER & WATERS 1991) and in amber inclusions (WILLEMSTEIN 1980, GRIMALDI et al. 1994b). Plant and fungal spores, fungal hyphae and other recognisable food remains occur occasionally in insect coprolites (A. SCOTT 1977, BAXENDALE 1979, KAMPMANN 1983, T. TAYLOR & SCOTT 1983, REX & GALTIER 1986), and fungal spores have been discovered in specialised mycetangial cavities of an ambrosia beetle preserved in

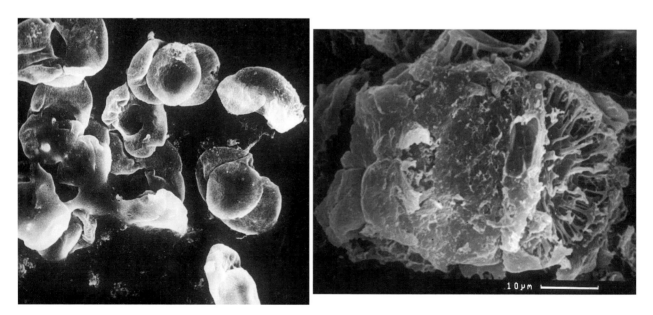

Fig. 54 Best preserved gymnosperm pollen extracted from the intestine of fossil insects: left – *Pinuspollenites entomophilus* Krassilov from the sawfly *Spathoxyela pinicola* (Xyelidae) from the Early Cretaceous of Baissa in Siberia, displaying grains at different maturation stages; right – *Lunatisporites* sp. from the booklouse *Parapsocidium uralicum* G. Zalessky from the Early Permian of Tshekarda in Urals, with internal structures partly exposes due to activity of the insect digesting enzymes (SEM photo by V.A. Krassilov)

Fig. 55 Best preserved leaf fragments of a gnetalean, *Brachyphyllum* or *Pagiophyllum* sp., from the gut of the stick insect *Phasmomimoides minutus* Gorochov (Susumaniidae) from the Late Jurassic of Karatau in Kazakhstan (SEM photo by V.A. Krassilov)

Dominican amber (GRIMALDI *et al.* 1994b). The presence of pollen in fossilised gut contents sometimes demonstrates an unusual state of preservation (e.g. due to partial digestion, Fig. 54, right), or mixture of the grains at different maturation stages (Fig. 54, left) providing valuable palaeobotanical information unavailable in other ways (KRASSILOV & RASNITSYN 1982, 1999, KRASSILOV *et al.* 1999). Studies on such fossils became possible only recently, with the development of new techniques, and surely more cases will be discovered in the future. Thus far, chemofossils in fossilised gut contents have not been investigated.

Phoretic (e.g. mites and pseudoscorpions) and parasitic (e.g. nematodes and hairworms) invertebrates occur occasionally in association with insect inclusions in fossil resins (LARSSON 1978, SCHAWALLER 1981, POINAR 1977, 1984a,b, 1992a, 1999b, POINAR & GRIMALDI 1990, POINAR *et al.* 1991, 1994a,b, 1998, POINAR & POINAR 1999) as do entomopathogenic fungi (GOEPPERT & BERENDT 1845, POINAR & THOMAS

1982, 1984). Similar fossil finds have been recorded from sedimentary rock oryctocoenoses but this is exceptionally rare (VOIGT 1938, 1957). Some unusual plant fossils, including the anthoecia of grasses which are extremely rarely found in a fossilised state, have nevertheless been found in insect burrows in palaeosols (THOMASSON 1982). ALIDOU *et al.* (1977) suppose that plant remains may be fossilised in subterranean termite nests as well, but this fossilisation model needs confirmation. Insect tunnels in fossil and subfossil wood may contain recognisable fungal remains (e.g. MÜLLER-STOLL 1936) and arthropods other than insects (KOPONEN & NUORTEVA 1973), but their contents rarely attracts much attention.

There is a record of viable bacterial spores from the gut contents of a stingless bee enclosed in the Dominican amber (CANO & BORUCKI 1995); however, like other purported records of viable microorganisms from geologically old deposits it needs confirmation and before this, any such record should be considered with much caution.

The taphonomical effects of ant activity are worthy of special mention. Some harvester ants of the genera *Messor* and *Pogonomyrmex* collect small fossils and artefacts and concentrate them near the entrances into their subterranean nests; such ant-collected remains are of some interest for both palaeontologists and archaeologists. Some authors believed that these objects have been simply removed by ants from their burrows during the digging (J. CLARK *et al.* 1967), but experimental observations indicate that surface collecting is important (REYNOLDS 1991). An accumulation of skeletal elements of modern small vertebrates by *Messor barbarus* L. was observed as well, and is believed to be a minor mechanism in the formation of potential fossil assemblages (SHIPMAN & WALKER 1980).

To complete this brief review of the taphonomical role of insects, a curious hypothesis on the origin of the some resin deposits proposed by STUART (1923) should also be mentioned. Stuart drew attention to the presence of the modern stingless bee, *Trigona laeviceps* Smith, in a Burmese resin (probably subfossil). Because *T. laeviceps*, as well as some other stingless bee species, collects natural plant resins and uses them in the construction of its nests, it was supposed that the resin (containing a number of other inclusions as well) may represent not a natural flux but a part of a bee nest. This extravagant hypothesis is absolutely improbable for any extensive resin deposits. Theoretically a resin from a fossil or subfossil bee nest can be found but certainly as an extreme rarity: it could hardly contain any undisturbed inclusions except for bees themselves or their nest cohabitants.

2
CLASS INSECTA Linné, 1758. THE INSECTS (=Scarabaeoda Laicharting, 1781)

A.P. RASNITSYN

(a) **Introductory remarks**. The concept of the class has been modified since ROHDENDORF & RASNITSYN (1980), and it still differs considerably from both classical (e.g. ROHDENDORF 1962, H. ROSS 1965, MACKERRAS 1970), cladistic (e.g. HENNIG 1980, KRISTENSEN 1975, 1981, BOUDREAUX 1979) and radical (KUKALOVÁ-PECK 1991) interpretations. Some ideas and inferences, however, have been borrowed from all of these sources. More deeply the concept used here has been affected by the ideas by MELNIKOV, TIKHOMIROVA and SHCHERBAKOV (MELNIKOV 1970, 1971, 1974a,b, RASNITSYN & MELNIKOV 1984, TIKHOMIROVA 1991, SHCHERBAKOV, submitted).

(b) **Definition**. Size varies from ca. 0.2 mm (some beetles) to 0.7 m (wing span of a Permian dragonfly). In MELNIKOV's scheme (above), body composed of acron, 6 fore larval segments, 11 postlarval segments (produced by growth zone initially separating fore and hind larval segments), and 7 hind larval segments (Fig. 56). Acron plesiomorphically bearing 3 external ocelli, the anterior one of which has a paired origin, and internally, a central, unpaired part of the brain (*pars intercerebralis*). All segments except the 2nd and the two penultimate (lower and fore proctodeal) plesiomorphically bearing variously modified paired limbs, all with a (sub)ventral position, except for the more dorsally situated antenna, wings, cerci, and paracercus. Segments 8 to 17 (i.e. mesothoracic through 8th abdominal) with spiracles framing the entrance into anastomosing tracheae and located between tergum and pleuron. Segments 1–17 each with a pair of ganglia (often variously fused and displaced).

Anterior larval segments form the head tagma (rarely united with thorax or, in larval higher flies, secondarily disintegrated). 1st head segment producing fore (fronto-clypeal) part of head capsule, appendages fused into labrum, and ganglia forming protocerebrum (fused with deutero- and tritocerebrum, ganglia of 2nd and 3rd segments).

Second head segment bearing antenna, compound eyes (composed of numerous ommatidia), though sometimes lost or, particularly in immature stages, disintegrated into separate stemmata, and a deutocerebrum (here, the segmental position of the eyes and mouth opening is taken after SHCHERBAKOV rather than MELNIKOV). Antenna composed of two basal, muscularised segments, called scape and pedicel respectively, and non-muscularised, multisegmented flagellum, the segments of which originate plesiomorphically from a growth zone surmounting the 1st

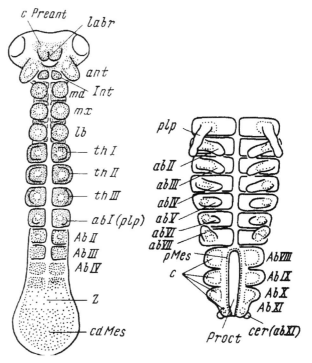

Fig. 56 Insect segmentation as exemplified by two embryonic stages of the termite *Anacanthotermes ahngerianus* Jacobs. (Hodotermitidae) (from MELNIKOV 1970). Fore larval segments: preantennal (cPreant – preantennal coelom, labr – labrum); antennal (ant); intercalary (int); mandibular (md); maxillary (mx); labial (lb). Postlarval segments: thoracic I–III (thI–thIII); abdominal I–VII (abI–abVII; plp – limb rudiment of 1st abdominal segment modified as pleuropodium). Hind larval segments: abdominal VIII–XI (abdVIII–XI; c – coelom, cer – cercus, Proct – proctodeum, pMes – proctodeal mesoderm); three proctodeal segments (abdominal XII–XIV) not shown (not seen in dorsoventral view); cdMes – caudal (hind larval) mesoderm. Z – growth zone (starts between fore and hind larval segments, stops working when between abdominal segments VII and VIII)

flagellomere. Antenna often reduced in number of segments, sometimes to just a single one or to none (in immatures), or otherwise modified.

Third segment bearing tritocerebrum as well as primary mouth opening transformed into pharyngeal entrance within secondary mouth cavity formed by appendages and sterna of gnathal (4th to 6th) segments; its appendages (homologues of crustacean 2nd antennae) suppressed.

Fourth to 6th head segments bearing gnathal appendages, viz. mandibles, maxillae and labium (fused 2nd maxillae) respectively. Their terga forming hind part of head capsule, sterna reduced and modified as hypopharynx, and neuromeres fused to form suboesophageal ganglion. Mandible one-segmented (possibly two fused segments), with fixed rotation axis (that is, functionally dicondylous: KLUGE 1996), reduced or lost in many aphagous insects as well as in adult higher flies. Maxilla plesiomorphically with two basal segments, cardo (proximal) and stipes (distal), and mounted by medial lacinia, sublateral galea and lateral, segmented palpus. In the groundplan, the labium is similarly constructed except for fusion of basal segments to form a proximal postmentum (sometimes subdivided further in submentum and mentum) and distal praementum. Distal labial lobes, medial glossae and lateral paraglossae, often participating in fusion.

Head ectodermal endoskeleton with three pairs of arms, anterior, posterior, and dorsal ones, posterior arms often fused dorsally forming corpotentorium. Anterior arms issued by clypeofrontal sulcus, posterior arms by postoccipital one.

Head plesiomorphically (sub)prognathous (with mouthparts directed obliquely forward and downwards), highly moveable, with mouth cavity and occipital foramen contiguous (separated neither by gular sclerite nor by mesal extensions of head capsule), secondarily often pro- and hypognathous. Adult head with antenna preoral and preocular.

Postlarval segments forming thoracic tagma (7th through 9th) and fore part of abdominal tagma (abdominal segments 1–7), widely varying in scale of differences, and often heteronomous within each tagma. Thoracic segment consists of dorsal tergum (embryologically fused of two halves and often retaining medial longitudinal line of fusion used also as moulting break; same often concerns also fore abdominal terga), ventral sternum, and lateral pleura (originally flattened basal limb segments, subcoxa and possibly epicoxa, KUKALOVÁ 1983). Tergum plesiomorphically extending into free side lobes (paranota) incorporating pleural (limb) material. Pleuron bearing apodeme above coxal articulation, sternum with pair of furcal apodemes and, posteromedially, spinal one (some or all these apodemes can be lost). Thoracic segment normally bearing legs and, meso- and metathoracic ones, often also wings.

Typical insect leg subdivided into coxa, trochanter, femur, tibia, tarsus, and pretarsus. Most mobile articulations occurring between coxa and trochanter, femur and tibia, and, except Machilida and immatures with entire tarsus, between tibia and tarsus. Knee segment (patella) and prefemur cannot be excluded (KUKALOVÁ 1983, but see RASNITSYN & NOVOKSHONOV 1997). Tarsus normally subsegmented except many immatures and paedomorphic adults. Pretarsus variable (claw-like, bladder-like, forming 2 free claws and 1–3 flaps called arolium, pulvillae and so on).

Abdominal segments similar to thoracic ones in being generally composed externally of tergal, sternal, and often separate pleural plates of limb base origin. Abdominal limb, when present, with segmentation reduced compared to thoracic ones, and lacking claws (a possible exception being an aberrant fossil described as "Dasyleptus" sp. by KUKALOVÁ-PECK 1985, 1987a, 1990, but see RASNITSYN 2000b).

Abdominal segments 1–7 usually essentially homonomous (1st one more often modified), sometimes with limb rudiments represented by styli, eversible sacs, gills, pleuropodia, etc. (for their homology see below). Abdominal segments 8–9 in female, 9 in male bearing gonopods, with coxite, style and gonapophysis commonly involved into their formation. Gonopore primarily and usually simple (not paired), opening between sterna 7 and 8 in female, enclosed by penis belonging to segment 9 in male (secondarily sometimes displaced or becoming paired, usually due to morphogenetic arrest).

Abdominal segments 10 and 11 reduced in size and somewhat simplified. Limbs of 11th, the cerci, originally transformed into long annulated threads (except possibly in "Dasyleptus" sp., see above), but often style-like or further reduced. Segments 12–14 invaginated to meet midgut, thus forming ectodermal proctodeum, a unique multisegmented organ claimed to represent a tagma of its own (MELNIKOV 1974b). Segment 14 plesiomorphically retaining limb(s) modified (by either fusion of both or loss of one) into median annulated thread, the paracercus. Alternatively, the 14th abdominal segment may be considered as a telson bearing annulated caudal extension (see below).

Egg normally and plesiomorphically rich in yolk, with superficial cleavage. Blastoderm segregated into embryonic and extraembryonic fractions, with the latter forming embryonic envelopes (primarily during embryo sinking into yolk but commonly by fold formation). Metatrochophoral stage (that with all and only larval segments present) with 4th, 5th, and 6th segments enlarged. Growth zone appearing between 6th and 7th somites, yielding 11 postlarval segments. Development epimorphic (all body segments appearing in egg). Organogenesis ordered and rather standard within class though often obscured due to heterochronies and shift of egg hatching and, in Scarabaeiformes, pupation and eclosion in respect to morphogenetic stages (for details see TIKHOMIROVA 1991). Growth proceeding during and shortly after periodic moulting process which is governed by neuroendocrine system. Timing of development and its tuning to environmental changes regulated mostly by photoperiodic reactions (with modifying effect of other factors) again through neuroendocrine system. Feeding and other bionomic and ecological features highly diverse.

(c) **Synapomorphies** may be outlined only within the framework of a particular hypothesis of insect ancestry. That provided by SHCHERBAKOV (1999 and in preparation), even if appearing rather extreme, appears to fit best with the existing evidence. Its most relevant statements are reproduced below, with proposed insect synapomorphies indicated in bold type.

An old hypothesis of a close insect-crustacean relationship [H. HANSEN 1893, CRAMPTON 1938, SHAROV 1966a] was confirmed by detailed similarities in the structure of compound eyes [NILSSON & OSORIO 1998] and ontogeny of nervous system [WHITTINGTON & BACON 1998]. 18S rRNA data show that Crustacea are paraphyletic with respect to Insecta, linking the latter to Malacostraca and Maxillopoda more closely than to Branchiopoda [GAREY et al. 1996]. Among crustaceans, Eumalacostraca (since Devonian), namely Syncarida [TILLYARD 1930], and of them especially Palaeocaridacea (Carboniferous-Permian, possibly since Devonian), are most similar to insects, namely, Archaeognatha (including Monura; since Devonian) in combining: trunk of 14 segments plus (pleo)telson (see below); carapace undeveloped; mandibles massive; hypopharynx strikingly similar, as well as 1st maxillae; 1st thoracomere free; 1st sometimes also 2nd and 3rd thoracopods specialised in contrast to 4–8th ones; thoracopods with short exopod; elongate abdomen with twisted rope musculature serving strong ventral flexion (for backstroke or leap) [MATSUDA 1957]; pleopods simpler and smaller than thoracopods, with endopod sometimes reduced and in 1st–2nd pairs produced to form male organ; uropods (their endopods homologous to insect cerci) and telson sometimes spike-like; gonopore in male two segments more posterior than in female; embryonic dorsal organ of same structure and position.

The insects are neotenic derivatives of syncarids and retain their embryonic characters (**uniramous 1st antennae, sessile eyes and uniramous uropods**) in adult. Insect thorax corresponds to anterior, maxillipedal part of subdivided malacostracan thorax, **five segments of posterior thorax being homeotically repatterned after abdominal ones**. Caridoid escape reaction gave rise to powerful leap of machilid-like initial insects, which presumably inhabited breakers zone on the seashore and jumped from waves like littoral talitrid amphipods; some modern machilids still live on coastal cliffs and can jump from the water surface [TSHERNYSHEV 1997]. Otherwise they could leap between plants protruding above water in deltaic and proluvial environments, escaping from (mainly chelicerate) predators [KUKALOVÁ-PECK 1987a]. In these first hexapod Atelocerata the abdomen, already adapted for escape reaction, became enlarged at the expense of the thorax, because (1) additional muscle energy is required to overcome gravitation when leaping outside water, and (2) hexapody being optimal for faster gait. Their 1st antennae and uropods turned uniramous with loss of swimming function in younger and older stages, respectively. Their **2nd antennae and mandibular palps (telopodites) had been lost** with loss of swimming larva [BOUDREAUX 1979].

Being so similar in many respects, including total segment number, these groups nevertheless differ in gonopore position: at 6th (female) or 8th (male) trunk segment (i.e. in posterior thorax) in Malacostraca; beyond 10th (female) or 12th (male) trunk segment (i.e. in posterior abdomen) in Thysanura and primitive Pterygota. Along with repatterning of posterior thorax into anterior abdomen in insects, **the set of modified gonopods** (and expressure domain of underlying homeotic gene *Abdominal-B*; AVEROF & AKAM 1995] was **homeotically transferred from the male on to female and shifted caudally to adjoin the transformed tail fan**, obviously due to invention of oviposition habit necessary in open-living terrestrial forms. Male and female gonads of a machilid consist of eight serial elements in the metathorax and abdomen (BIRKETH-SMITH 1974); those of a syncarid at hatching are eight segmental rudiments in the thorax (HICKMAN 1937).

A complexity of homeotic gene system makes body heteronomy much more difficult to achieve *de novo* than to modify once invented (shifting the boundaries between tagmata) or even to lose by turning off the genetic mechanism (derepressing limb development in abdominal segments); hence in Crustacea and Atelocerata an independent appearance of secondarily homonomous forms [Remipedia and Myriapoda, respectively], rather than parallel acquisition of similar heteronomy, is supposed [AVEROV & AKAM 1993]. The loss of heteronomy becomes possible when locomotory abdomen turns inadaptive, e.g. in cryptobiotic forms.

Like Crustacea Remipedia from Maxillopoda, the terrestrial myriapods originated from hexapod ancestors, rather than *vice versa*. Open living machilid leapers retain malacostracan features: tagmosis with rather smooth transition from one tagma to another; escape reaction; tail fan (transformed into cerci + paracercus); exopods (as coxal and abdominal styli); abdominal endopods (as eversible vesicles); leg position (coxo-trochanteral joints close to midventral line); head structure (naupliar eye transformed into ocelli, 1st antennae with flagellum annular, solid sub-ectognathous mandible with incisor process and posterodorsal adductor muscle, 1st and 2nd maxillae alike with palps well developed). Having an anterior articulation incipient, and posterior one hidden under paranotal fold of mandibular segment, machilid mandibles are functionally dicondylous (like in Zygentoma) and represent the type initial for both ectognathous (fully dicondylous) and entognathous (monocondylous or suspended) mandibles [BITSCH 1994].

With the shift to creeping locomotion in coastal debris and litter, trunk tagmosis [OSORIO et al. 1995] and ocelli were suppressed, and eyes reduced in myriapods. Due to abdomen repatterning after thorax, their trunk heteronomy again turned covert, and segment pairing overt (up to diplosegments), a partial reversal to scale worm level. Entognathous hexapods should be considered 'pre-myriapods' rather than 'Parainsecta'. Various Entognatha and Myriapoda had malacostraco-thysanuran features modified or lost: head tagma contrasted to homonomous trunk; paranota reduced; leg bases widely separated; cerci lost; eyes reduced; ocelli suppressed; 2nd maxillae leg-like or lost; palps reduced; sperm

immotile. Mandibles turned entognathous in Entognatha, as well as in Chilopoda and Pauropoda; the condition in Symphyla and Diplopoda is also far from true ectognathy, because their mandibles remain suspended anteriorly on movable levers (futurae). Secondarily disintegrated myriapod mandibles with anterodorsal adductor muscle became similar to maxillae (homeotically repatterned after them?). Myocerata (= Entognatha + Myriapoda) have the flagellum of 1st antennae truly segmented, i.e. with segments having their own intrinsic musculature (homeotically repatterned after palps or legs?). Various reductions in myocerates could have arisen through neoteny, as well as the enlarged limb rudiments at 1st abdominal segment in Protura, Collembola, and some Diplura (from embryonic pleuropodia of Thysanura).

Diplura are related to lepismatids, and Collembola descended from near Diplura [DALLAI 1980]. Ellipura (= Collembola + Protura) were mentioned as a possible sister-group of Myriapoda [ZRZAVY & ŠTYS 1994]. The structure of their abdomen with subterminal gonopore is by no means primitive, even in Protura which anamorphically develop 3 subterminal segments (one more than other insects), possibly mere subdivisions of true 8th segment. The myriapods most similar to hexapods (namely, to Diplura) are Symphyla [TILLYARD 1930]. In both Entognatha and Myriapoda the proto- and deutocerebrum are strongly tilted backwards lying over the stomodaeum, and postantennal organs are developed (Fig. 57). Embryological similarities of Diplura and Collembola with Myriapoda are impressive [ZACHVATKIN 1975]: ovary not subdivided into ovarioles; plagiaxony; ventral bend of germ band; no amnion and serosa; extraembryonic blastoderm transformed into provisional body wall and embryonic dorsal organ (of same type in Diplura, Collembola and Symphyla).

The trunk is not perfectly homonomous in most myriapods, demonstrating usually the first three trunk segments (=insect thorax), and often also the 4th–8th ones (=non-maxillipedal malacostracan thorax converted into anterior abdomen in insects) as subtagmata still distinct from the following segments (=malacostracan abdomen). The difference mostly lies in the change of segment pairing pattern.

Both Remipedia and Myriapoda originated from the most oligopodous representatives of mother lineages, probably because for these latter the loss of trunk heteronomy was an easier way to restore multiple limb-bearing segments (useful for cryptobiotic forms) than the repatterning of abdominal segments one by one. This homeotic transformation is modelled by total deletion of the Homeotic complex in *Drosophila*, resulting in repatterning of all trunk and, moreover, both maxillary segments after prothoracic one [RAFF & KAUFMANN 1983]. A reversal of the 2nd maxillae to leg-like condition in Chilopoda, or their reduction to eversible vesicles (characteristic of trunk segments) in Pauropoda and then complete suppressure in Diplopoda, seem to be mere by-products of the genetic mechanism governing 'myriapodisation'.

Contrary to common opinion, Remipedia and Myriapoda are highly modified rather than primitive subtaxa of Crustacea and Atelocerata respectively, as well as their legless vertebrate analogues: Serpentes and Amphisbaenia within Reptilia, Gymnophiona and Palaeozoic Aistopoda within Amphibia. Like myriapods, snake-like terrestrial vertebrates secondarily acquired superficially homonomous trunk (for creeping or digging locomotion in litter or soil) containing more numerous metameres (vertebrae), and have the head more specialised than in primitive heteronomous forms. Like snakes, myriapods are primarily terrestrial, and one has no reasons to imply that the first of them were aquatic. Both lancelets and snakes are legless, but the latter body plan is highly derived, at least because of the head tagma including differently modified metameres (branchial arches), sometimes bearing poison teeth, and sharply separated from the trunk by neck region. Likewise both trilobites and millipedes (or centipedes) are multi-legged, but the latter is in no way primitive due to limbs of head region being variously specialised (up to poison fangs) and (as well as trunk legs) far from tetraramous trilobite condition, and due to limbless column. Paired limbs are absent in ground-plan vertebrate but present in ground-plan (thoracic) arthropod segment, so both snake leglessness and myriapod total leginess are (partial) reversals to an initial state. To derive Thysanura from Myriapoda is no more logical than to derive lizards from snakes, the more so that the transformation series from trilobites *via* crustaceans to hexapods is nearly as smooth as that from fish to reptiles.

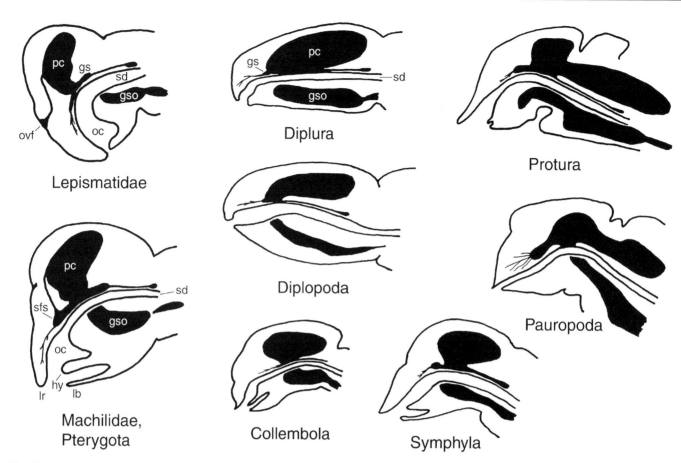

Fig. 57 Schematic section of head of Entognatha and Myriapoda showing position of the protocerebrum (original by O.A. Melnikov, based on MELNIKOV & RASNITSYN 1984): gs–stomodeal ganglion, gso–suboesophagal ganglion, hy–hypopharynx, lb–labium, lr–labrum, oc–oral cavity, ovf–ventral frontal organ, pc–protocerebrum, sd–stomodeum, sfs–frontostomodeal synganglion

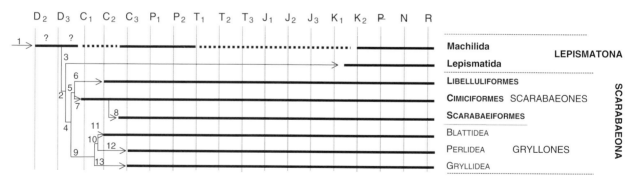

Fig. 58 Phylogeny and system of the insects (Class Insecta)

Time scale intervals are abbreviated as follows: D_2, D_3 – Middle and Late (Upper) Devonian, C_1, C_2, C_3 – Early (Lower), Middle and Late (Upper) Carboniferous, P_1, P_2 – Early (Lower) and Late (Upper) Permian, T_1, T_2, T_3 – Early (Lower), Middle and Late (Upper) Triassic, J_1, J_2, J_3 – Early (Lower), Middle and Late (Upper) Jurassic, K_1, K_2 – Early (Lower) and Late (Upper) Cretaceous, P – Palaeogene (Early Tertiary, viz. Palaeocene + Oligocene + Eocene), N – Neogene (Late Tertiary, viz. Miocene and Pliocene), R – present time (Holocene). Right are names of subclasses (all caps, bold), infraclasses (all caps, light), cohors (small caps, bold), superorders (small caps, light) and, in Lepismatona, orders (lower case, bold). Arrows show ancestry, thick bar – known duration time of taxa (questioned in case of debatable records, dashed for long gaps in the fossil record), figures to synapomorphies of the subtended clades, as follows:

1 – insect synapomorphies (Chapter 2c).

2 – cranio-mandibular articulation double; anterior tentorial arms fused into anterior tentorial bridge; tracheal trunks interconnected; coxal styli lost; ovipositor intromittent, with gonangulum articulated with 1st gonapophysis; cerci long, subequal to paracercus; possibly also habits cryptic, body depressed, and leaping ability lost.

3 – coxa large, flat; tarsal segmentation similar in immature and adult; abdominal musculature impoverished.

(d) Range. Worldwide, with certainty since the latest Early Carboniferous (Namurian A, Arnsbergian, Chapter 2.2.1.2.3.1f). The Early and Middle Devonian records (HIRST & MAULIK 1926, a dicondylous mandible, SHEAR *et al.* 1984, and LABANDEIRA *et al.* 1988, fragments of supposed Machilida) are important but fragmentary and require confirmation.

(e) System and phylogeny are shown at Fig. 58.

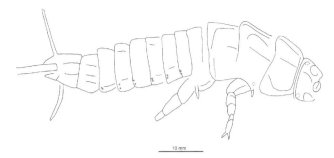

Fig. 59 *Ramsdelepidion shusteri* Kukalová-Peck, a giant enigmatic insect, either a relative of the silverfish or, perhaps the immature of a different, possibly winged insect from the Middle Carboniferous of Mazon Creek in USA (original based on holotype)

2.1
SUBCLASS LEPISMATONA Latreille, 1804. THE WINGLESS INSECTS (=Thysanura Latreille 1796, *s. l.*)

A.P. RASNITSYN

(a) Introductory remarks. The concept of the subclass is basically the same as in ROHDENDORF & RASNITSYN (1980), corrected in details according to KRISTENSEN (1975, 1981) and BOUDREAUX (1979).

'*Dasyleptus*' sp. and *Ramsdelepidion shusteri* Kukalová-Peck from Mazon Creek (later Middle Carboniferous of Illinois, USA, KUKALOVÁ-PECK 1987a) and *Carbotriplura* Kluge from the earlier Middle Carboniferous of Bohemia (Czech Rep.) (KUKALOVÁ-PECK 1985, KLUGE 1996) are excluded from consideration here. The first fossil is not properly described, and its interpretation is very dubious (RASNITSYN 1998a). *Ramsdelepidion* and *Carbotriplura* (Figs. 59, 60) have no specific features of the apterygote insects and can equally

represent immature or neotenic winged insects. This is particularly true for one more enigmatic insect (Fig. 61, see also J. GALL *et al.* 1996, plate IX, Fig. 1) which shows significant similarity to *Carbotriplura*, despite being much younger (of mid-Triassic age). At the same time, the Triassic fossil shows structures, most probably representing the wing pads characteristic of nymphs of the winged insects.

(b) Definition. Small to medium-sized insects lacking wings, with elongated bodies often covered by scales. Head prognathous or hypognathous, with ocelli when present situated before the eyes, head capsule sometimes retaining traces of boundaries of several posterior segments, internally with anterior and dorsal tentorial arms not joining in corpotentorium. Mouthparts chewing, mandibles with dorsal articulation either absent or weak and well distant from ventral one. Postcephalic terga more or less uniform, with paranota present but not delimited from the remaining tergum. Thoracic pleura and sterna composed of free subsclerites. Abdominal venter with coxites,

4 – head hypognathous, with all tentorial arms joining corponentorium; ventral frontal organ fused with stomodeal ganglion to form complex frontal ganglion; lateral cervicalia present; meso- and metathoraces modified into pterothorax, with wings capabale to maintain directed flight and to fold roof-like at rest, with respective changes in structure of sclerites and muscles; leg with tarsus 5-segmented, with pretarsal dactylus modified into arolium in adults; pregenital abdominal segments lacking delimited coxites and eversible sacs; embryo completely enclosed in amniotic cavity by each continuous amnion and serosa; true copulation present, with direct sperm transfer from male to female gonopore; development archemetabolous; tree-dwellers at all developmental stages, with eggs laying into plant tissue cut by female gonapophyses, and with further stages feeding on content of gymnosperm tree sporangia, with both wings and immature winglets using in attitude control during jumps (for further details see Chapter 3c.]).

5 – male gonostyli forming forceps to clasp female during copulation; unknown underlying apomorphy in female genitalia resulting in trend for gonocoxae to acquire role of ovipositor sheath, with only rare and secondary their modification into working (intromittent) part of ovipositor, while gonostylus persisting free unless lost.

6 – Adult antenna short; wings spread at rest; wing surface regularly fluting due to (i) MA being sharply convex and RS and MP concave in addition to the archetypically convex veins C, R, CuA, and A, and concave SC and CuP, and (ii) intercalary veins appearing concave between convex main veins, and convex between concave ones.

7 – pterothoracic sternum invaginated; paracercus lost (secondarily re-gained in some holometabolan larvae); ovarium polytrophic (secondarily panoistic in Thripida: ŠTYS & BILINSKI 1990).

8 – nymphal stages condensed and "embryonicised" as pupal stage; imaginal moulting lost (unless inherited from ancestral Caloneuridea); middle coxae with inner articulation in addition to proximal and distal ones (considering similar character state in Ephemerida as homoplasy).

9 – wings folded with considerable overlapping at rest, with anal region expanded in hind wing and bending down at folding along line anterad 2A (usually anterad 1A), and with common base of anal veins rocking upside down at folding; female gonostylus along with gonocoxa closely associated with gonapophyses resulting in their modification into working (intromittent) part of ovipositor in most subtaxa; paracercus lost.

10 – wide paranota present.

11 – paranota circular, concealing much of head at rest; fore wing with R and RS weakly separated, anal area (clavus) wide, lanceolate, with veins gently curved according its fore margin, and weakly branching, if at all; see Chapter 2.2.2.1c for more details.

12 – hind wing with CuA stock angled at junction with M$_5$; M forming 2 well separated main branches, MA and MP, with MP desclerotised and somewhat depressed sub-basally.

13 – pronotal sides bent downward to form lateral pronotal lobes; fore wing with precostal space large, bearing numerous veins, with MP separated and running in parallel to M$_5$ + CuA$_1$, and with 1A simple; cercal segmentation lost.

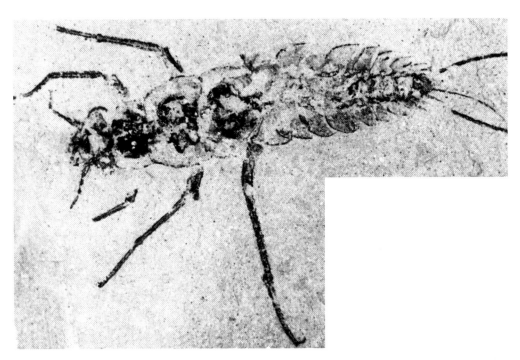

Fig. 60 *Carbotriplura kukalovae* Kluge from the Middle Carboniferous from Boscovice in Czech, first described as an immature giant mayfly *Bojophlebia* Kukalová-Peck but possibly representing either a giant silverfish-relative or the immature stage of an unknown winged insect (from KUKALOVÁ-PECK 1985); body length 102 mm

delimited or not, each bearing a stylus (sometimes subsegmented) and a pair of eversible sacs, modified at genital segments, often lost at one or more segments, always lost at postgenital ones, modified as cerci at segment 11 (sometimes reduced), and as medial paracercus at segment 14. Amniotic cavity open or closed by a plug and not by continuous amnion and serosa. Development gradual, with early postembryonic stage sometimes retaining entire tarsus and large claw-like pretarsus, and lacking cerci despite presence of paracercus. Growth and moulting not arrested upon sexual maturation. Sperm transmission extrasomatic (no copulation). Habits mostly cryptic. Feeding micro- and sapro-phagous, sometimes phytophagous.

(c) Synapomorphies absent, the subclass being paraphyletic (ancestral for Scarabaeona).

(d) Range as in Insecta (Chapter 2d).

(e) System and phylogeny are shown at Fig. 58. In spite of the probable amphibiotic habits of the immediate apterygotan ancestor (Chapter 2.1c), this evolutionary stage was most probably very short for the taphonomic reasons (Chapter 3.2.2.1), the long gap in the insect fossil record (Devonian + Early Carboniferous) being in sharp contradiction to the suggested amphibiotic habits which is taphonomi-cally favourable. Indeed, Devonian records of Lepismatona are extremely rare and debatable, Early Carboniferous absent, and Late Carboniferous ones uncommon and represented either by specialised (neotenic) (semi)aquatic forms (Dasyleptidae) or highly debatable body and trace fossils (Chapters 2.1.1f, 2.1.2e and 2.3.2). Here we accept the old hypothesis that wingless insects were basically soil and/or litter dwellers (Chapter 3.2.2.1) and that they probably managed to enter there having been adapted beforehand for dwelling in tidal flotsam.

2.1.1. ORDER MACHILIDA Grassi, 1888.
THE BRISTLETAILS (=Archaeognatha Börner, 1904, =Microcoriphia Verhoeff, 1904, +Monura Sharov, 1957)

(a) Introductory remarks. Small order of wingless insects with some 350 living species (WATSON & SMITH 1991).

The following is adapted from ROHDENDORF & RASNITSYN (1980). Additional information on fossils, unless stated otherwise, is taken from DURDEN (1978), ZHERIKHIN (1978), SPAHR (1990), ROWLAND (1997), STURM (1997) and RASNITSYN (2000b). "*Dasyleptus*" sp. from Mazon Creek (later Middle Carboniferous of Illinois, USA, KUKALOVÁ-PECK 1987a) is not considered until re-description is made.

(b) Definition. Body not depressed. Head with ocelli present, with anterior tentorial arms free, and with ventral frontal organ fused with sto-modeal ganglion to form complex frontal ganglion (for details see MELNIKOV & RASNITSYN 1984: 33–52). Mandible with single (ventral) morphologically formed articulation but with fixed rotation axis (KLUGE 1996). Maxillary palp 7-segmented, leg-like. Pleura hidden under paraterga and reduced (unknown for Dasyleptidae). Meso- and metatho-racic coxae with styli in Machilidae (unknown for Dasyleptidae, lost in Meunertellidae). Pretarsal dactylus lost except for juvenilised Dasyleptidae. Abdominal spiracles 2–8 only, placed at paranotal apices, tracheae not anastomosing (unknown for Dasyleptidae). Cerci much shorter than paracercus (absent in Dasyleptidae and in some immatures of living groups), paracercus thick basally (streamlined with abdominal apex). Ovipositor not penetrating substrate, only guiding eggs. Leaping ability developed by abdomen that bent quickly, assisted by legs pushing in unison (possibly except Dasyleptidae).

Fig. 61 An undescribed enigmatic fossil from the Middle Triassic of Vosges in France, similar in many respects (three caudal threads, legs with paired claws, abdominal terga with long paranotal outgrowths) to *Carbotriplura* and possibly related to it. However, the pterothoracic lateral structures probably represent wing pads rather than just paranota thus implying that the fossil represents a nymphal winged and not a wingless insect. Many fossils have been collected suggesting an aquatic or near-shore habits. Illustrated are specimens 8036/233 (near complete) and 9209/147 (legs) kept in the Louis Pasteur University (Strasbourg, France; courtesy Dr L. Grauvogel-Stamm)

(c) Synapomorphies. None supposing the group to be ancestral for other insects, and none being positively known for the morphogenetic level of the neotenic Dasyleptidae (RASNITSYN 2000b), with apomorphies of extant groups (posterior head concealed under pronotum, frontoclypeal sulcus lost, eyes hypertrophied, complex frontal ganglion present, pretarsal dactylus and 1st abdominal spiracle lost) being evolved supposedly either later in evolution, or at a later (adult) morphogenetic stage unknown in the Palaeozoic Dasyleptidae.

(d) Range. Worldwide, since the Late or, possibly, later Middle Carboniferous, unless since Devonian (see below), until the present.

(e) System and phylogeny. The circumstances of the origin of Machilida are hypothesised to be the same as for all the insects in general (Chapter 2c). Neotenic, shore-dwelling unless aquatic, Dasyleptidae could perhaps be a model of a stage in that transition,

though hardly a very close one because of the long time interval and the respective environmental differences.

Three families are known in the order, the Palaeozoic Dasyleptidae and living Machilidae and Meunertellidae. The first family (Fig. 62) is usually treated as distinct, primitive order Monura (SHAROV 1957). The reason to reduce its rank (summarised by RASNITSYN 2000b) is that Dasyleptidae are simply a neotenic bristletail, with the majority of their striking characters occurring and then disappearing during development of living Machilida (Fig. 63). These are short antennae, delimited clypeus and terga of gnathal segments, entire tarsus, simple, claw-like pretarsus, free postocular head, similarly developed postcephalic terga (seemingly absent thoracic tagma), fleshy, subsegmented styli, suppressed cercus.

The living families Machilidae and Meunertellidae are synapomorphic in respect to Dasyleptidae in their eyes being large and contiguous, and possibly also in loss of the 1st abdominal spiracle (unknown in Dasyleptidae). Judging from REMINGTON (1954), Machilidae show no undoubted autapomorphy and are possibly ancestral for Meunertellidae which are autapomorphic in small abdominal sterna, reduced number of eversible sacs (not more than one pair per segment), and somewhat reduced scale cover. Machilidae are confined to the northern hemisphere, while Meunertellidae are worldwide though predominantly southern (WYGODZINSKY 1967).

(f) History. The oldest though fragmentary and thus problematic records of unnamed supposed bristletails are reported from the Early Devonian (Early Emsian) at Gaspé in Québec, Canada (LABANDEIRA et al. 1988), and Middle Devonian (Givetian) at Gilboa in New York, USA (SHEAR et al. 1984). The Middle Carboniferous (Westfalian D) record from Mazon Creek in Illinois, USA (KUKALOVÁ-PECK 1987a) badly needs reconsideration (RASNITSYN 2000b). Undoubted Machilida belonging to Dasyleptidae are described as 4 species of Dasyleptus Brongniart from the Late Carboniferous of Commentry,

France, Early Permian (Artinskian) of Elmo in Kansas, USA and earlier Late Permian of Kaltan in Kuznetsk Basin (southern West Siberia) (RASNITSYN 2000b). Commentry and Kaltan have yielded tens of Dasyleptus specimens: for wingless insects this suggests that they populated either the shore or the lake which accumulated deposits with buried fossils. Absence of evident aquatic adaptations does not rule out aquatic habits: many living water dwellers, particularly among beetles and wasps, are superficially indistinguishable from their terrestrial relatives.

The supposed Triassic bristletail Triassomachilis proved to be mayfly nymph (Chapter 2.2.1.1.1.3), so the next oldest are the records of Meinertellidae in the later Early Cretaceous (possibly Aptian) Lebanon amber (STURM & POINAR 1998), and unidentified bristletail from the Early Cretaceous amber from Álava in Spain (Fig. 64). Machilidae (possibly including Meinertellidae) also found in the Late Cretaceous (Santonian) fossil resins of Yantardakh in Taimyr, North Siberia. Further records of Meinertellidae are from the Early Miocene

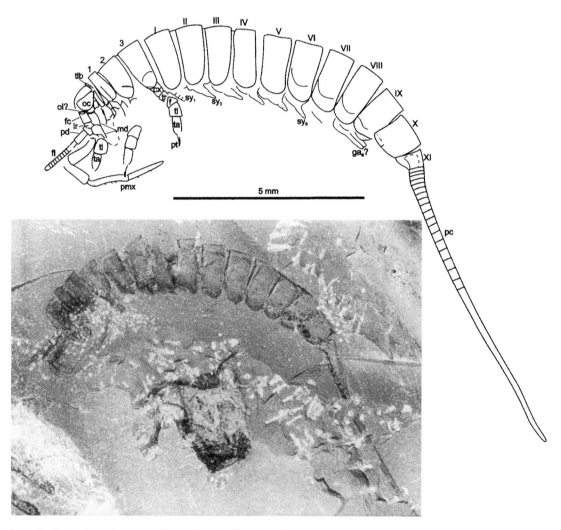

Fig. 62 Neotenic bristletail, *Dasyleptus brongniarti* Sharov (Dasyleptidae), from the Late Permian of Kaltan in SW Siberia: general view of the holotype (photo by D.E. Shcherbakov) and morphological reconstruction based on the type series (from RASNITSYN 2000b); cx–coxa, f–femur, fc–frontoclypeus, fl–antennal flagellum, ga_8–gonapophysis of abdominal segment 8, lr–labrum, md–mandible, oc–eye, ol–ocellus, pc–paracercus, pd–pedicel, pmx–maxillary palp, pt–pretarsus (claw), sc–scape, sy–stylus, sx–subcoxa, ta–tarsus, ti–tibia, tlb–tergum of labial segment, tr–trochanter; Arabian numerals indicate thoracic terga, Roman numerals–abdominal ones

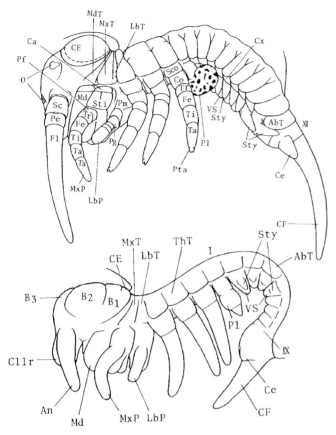

Fig. 63 Late embryonic stages 9 (below) and 10 of the bristletail *Pedetontus unimaculatus* Machida (Machilidae), showing characters also displayed by the adult or late postembryonic *Dasyleptus* (Fig. 62), e.g., suppressed cerci and paired claws, fleshy abdominal prolegs (styli), delimited mandibular, maxillary and labial terga, and so on. AbT – abdominal tergum, An – antenna, B_1, B_2, B_3 – protocephalic bulges 1–3, CE – compound eye anlage, Ce – cercus, CF – caudal filament (paracercus), Cllr – clypeolabrum, Cp – coxopodite, LbP – labial palp, Md – mandible, MxP – maxillary palp, MxT maxillary tergum, Pl – pleuropodium (modified abdominal leg 1), Sc – scape (1st antennal segment), Sty – stylus, ThT – thoracic tergum, Tp – telopodite, VS – ventral (eversible) sac, X, XI – 10th and 11th abdominal segments (from MACHIDA 1981 ©Alan R. Liss, Inc.; reprinted by permission of Wiley–Liss, Inc., a subsidiary of John Wiley & Sons, Inc.)

Dominican amber, from the Bitterfeld amber in Germany (supposed as Early Miocene in age), and also from the Pleistocene copal in Mizunami, Japan. Living genera of Machilidae in a narrow sense are recorded from the Eocene Baltic amber, of Meinertellidae – in the Late Oligocene or Early Miocene Dominican and Mexican amber. Additionally, a possibly Late Tertiary, doubtful bristletail is described in onyx marble from Bonner Quarry in Arizona, USA (PIERCE 1951).

2.1.2. ORDER LEPISMATIDA Latreille, 1804.
THE SILVERFISH (=Thysanura Latreille 1796, *s. str.* =Zygentoma Börner, 1904)

(a) Introductory remarks. The small group of some 370 living species (G. SMITH & WATSON 1991) is treated here after ROHDENDORF & RASNITSYN (1980), with additional information used from REMINGTON

Fig. 64 Unidentified bristletail (order Machilida) from the Early Cretaceous amber from Álava, Spain (from ALONSO *et al.* 2000); body length 3 mm

(1954), WYGODZINSKY (1961), BOUDREAUX (1979) and, for fossils, also from ZHERIKHIN (1978), SPAHR (1990), KLUGE (1996), and STURM (1997, 1998).

(b) Definition. Size medium to small. Body depressed, scaled or not. Head free, with eyes small, ocelli rarely present. Maxillary palp 5-segmented, not particularly leg-like. Ventral frontal organ and stomodeal ganglion free, not forming complex frontal ganglion (MELNIKOV & RASNITSYN 1984). Anterior tentorial arms fused into anterior tentorial bridge. Pleuron external, composed of upper and lower sclerites (anapleurite and katapleurite, respectively), the latter caudally with delimited trochantine. Thoracic coxa large, flatted, coxal styli lost. Tarsus with 2–5 segments. Pretarsus with 2 claws and small median dactylus. All 10 pairs of spiracles present, their respective tracheal trunks interconnected. Abdominal musculature impoverished. Coxites free or fused with sternum, 1st coxite lacking styli and eversible sacs, others with at most one pair of sacs per segment. Ovipositor intromittent, often digging, with gonangulum. Male gonopore double at entire penis. Cerci and paracercus thin basally, subequal in length.

(c) Synapomorphies. Coxa large, flat; tarsal segmentation similar in immature and adult; abdominal musculature impoverished; possibly (unless acquired earlier, before divergence with winged insects): habits cryptic, with body depressed and leaping ability lost. Groundplan number of tarsal segments unclear: either 5 as in Lepidotrichidae, or 3 as in other families.

(d) Range. Worldwide, Early Cretaceous until the present (see below for the alleged Palaeozoic fossils).

(e) System and phylogeny. The silverfish are most probably monophyletic with the winged insects because of numerous synapomorphies (anterior tentorial arms fused into tentorial bridge, interconnected tracheal trunks, intromittent ovipositor, elongated cerci, thin paracercus, lost leaping ability by abdominal stroke; possibly also in cryptic habits, and depressed body). They are plesiomorphic compared to both bristletails and winged insects in having the ventral frontal organ neither shifted ventrocaudally nor fused with the stomatogastric ganglion (MELNIKOV & RASNITSYN 1984): this may indicate that bristletail and winged insects, like some crustaceans mentioned by MELNIKOV & RASNITSYN (1984) have acquired the complex frontal ganglion (fused ventral frontal organ and stomatogastric ganglion) independently.

Ecologically origin of the order is supposed to be connected with a shift to more confined environments, with the body depression, loss of leaping ability, reduction of the eye size, further suppression of ocelli, and possibly also with the hyperthrophy of cerci.

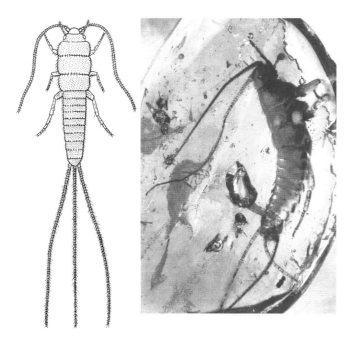

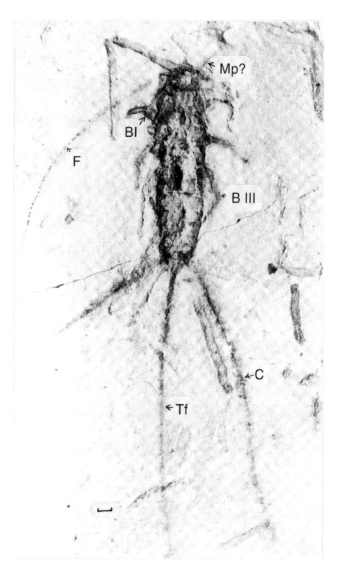

Fig. 65 *Lepidotrix pilifera* Menge (Lepidotrichidae) from the Late Eocene of the Baltic amber (left from HANDLIRSCH 1920, right from WEITSCHAT & WICHARD 1998); length of body 8 mm

Five families are currently appreciated in the order, viz. Lepidotrichidae, Nicoletiidae, Ateluridae, Lepismatidae, and Maindroniidae (for the position of *Carbotriplura* and *Ramsdelepidion* see Chapter 2.1a). Generally the most primitive are Lepidotrichidae (Fig. 65) which retain ocelli, 5-segmented tarsus (unless gained independently of the winged insects), and large abdominal sterna, not fused with coxites. However, some seemingly autplesiomorphic characters of Lepidotrichidae should be rather qualified as reversals due to paedomorphy. These are pubescent instead of scaled body with cuticular pigmentation, and hypognathous head with delimited occipital (maxillary) tergum.

In spite of its primitiveness, Lepidotrichidae are hypothesised by WYGODZINSKY (1961) to be a sister group of Nicoletiidae (+Ateluridae) because of synapomorphies in sensory cones present basiventrally on male paracercus, and in aggregated spermatozoa. The two (or three) families combined form a sister group of Lepismatidae (+Maindroniidae), whose synapomorphies WYGODZINSKY (l.c.) lists as loss of ocelli and eversible sacs, free sternum, acquisition of proventriculus, etc. Within the first of the above two pairs, Nicoletiidae are specialised for the subterranean habits while Ateluridae are adapted to termito- and myrmicophyly. Maindroniidae are also subterranean while Lepidotrichidae and most Lepismatidae are free-living. Lepidotrichidae is now represented by a sole species in the south-west of North America, Maindroniidae – by three species in south-west Asia and on the Pacific coast of South America, the other three families are widespread.

(f) History. Besides the debatable records of the Carboniferous *Ramsdelepidion* and *Carbotriplura* (see above) and possibly the Late Tertiary *Onycholepisma arizonae* Pierce enclosed in onyx marble from Bonner Quarry in Arizona, USA (PIERCE 1951), all ancient silverfish are found in fossil resins. The oldest known is the described and

Fig. 66 Undescribed silverfish (Lepismatidae) from the Early Cretaceous of Santana in Brazil (from STURM 1998); BI, BIII – fore and hind leg, respectively, C – cercus, F – antenna, Mp? – maxillary palp, Tf – paracercus

figured but not named member of Lepismatidae from the mid-Early Cretaceous of Santana in Brazil (Fig. 66). The next oldest is an undescribed Lepidotrichidae found in later Late Cretaceous (Santonian) of Yantardakh in Taimyr (North Siberia); a further member of that family, also generically distinct from the only living one, is *Lepidotrix pilifera* Menge from the Eocene Baltic amber (see Fig. 65). Lepismatidae is represented also in the Baltic amber and in imprecisely dated (Late Cretaceous or earlier Tertiary) Burmese amber. The former fossil is *Allacrotelsa dubia* (Menge) belonging to the living genus (STACH 1972), the latter *A. burmitica* has been described as tentatively congeneric with the former but with the reservation that it probably belongs to an undescribed genus (COCKERELL 1917). Lepismatidae, Lepidotrichidae, Ateluridae, Nicoletiidae are reported by STURM (1997, STURM & MENDES 1998) from the Dominican amber (Early Miocene). Fossil Maindroniidae are not yet known.

2.2
SUBCLASS SCARABAEONA Laicharting, 1781. THE WINGED INSECTS (=Pterygota Lang, 1888)

A.P. RASNITSYN

(a) Introductory remarks. The concept of higher level structure and phylogeny of the subclass is essentially new, though based in part on that in ROHDENDORF & RASNITSYN (1980). The list of synapomorphies relies additionally on data from MATSUDA (1965, 1970, 1976), KRISTENSEN (1975, 1981, 1998), BOUDREAUX (1979), HENNIG (1981), GRODNITSKY (1999).

(b) Definition. Very small (ca. 0.3 mm body length) to very large (0.7 m wing span) insects, winged or secondarily wingless, of extremely diverse appearance and habits, thus with characters difficult to be outlined. When wingless and with well developed paracercus, differing from both bristletails (except the Palaeozoic neotenic *Dasyleptus*, Chapter 2.1), and silverfish by being aquatic. The Carboniferous *Ramsdelepidion* KUKALOVÁ-PECK and *Carbotriplura*

Kluge are insufficiently known and could be either silverfish or immature winged insects (Chapter 2.1a).

(c) Synapomorphies. Head is hypognathous, with mouthparts directed downward and with mandibles moving in plane subparallel to that of occipital foramen (a flight adaptation aiming to bring eyes forward and to save mouthparts in collisions of flying insect with obstacles). Anterior and dorsal tentorial arms are interconnected and join the corpotentorium which is formed by interconnected posterior arms. Lateral cervicalia present. Meso- and metathoraces modified, each with a pair of wings of pleural (limb) origin (Chapter 1c), ontogenetically developing from paranota (seemingly lateral extension of tergum). Wing articulating apparatus permits a rocking movement during flight and a folding movement toward roof-like position at rest. Groundplan flight is functionally four-winged, in-phase, anteromotoric.

Wing membrane is supported by simple or branching veins forming several main vein systems of characteristic structure and position, as follows (Fig. 67).

Costal vein (C) lines wing margin (sometimes, most probably secondarily, runs submarginally, anterobasally, and bears short veinlets there), simple (not branching), and convex (running atop of membrane elevation and stronger on upper wing surface compared to lower one). Subcostal vein (SC) is concave (running along a depression in the membrane and is stronger on lower wing surface), with fore and apical veinlets, apical ones merging with both C and R well before

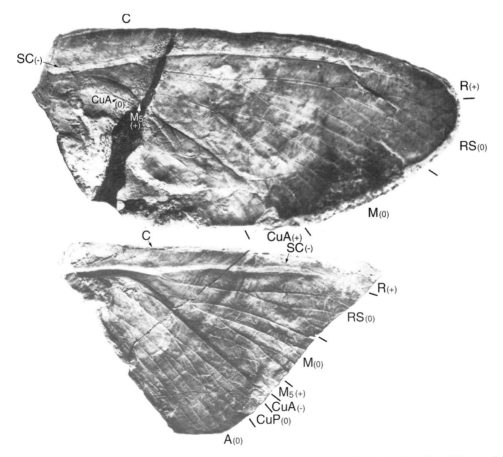

Fig. 67 Wing venation nomenclature as exemplified on plesiomorphic subimaginal wings of *Evenka archaica* Rasnitsyn (Evenkidae, possibly stem Scarabaeones or stem Caloneuridea) from the Late Carboniferous of Chunya in Siberia (from RASNITSYN 1977a; fore wing 58 mm long). Convex veins (elevated above the wing plan) are marked as (+), concave ones (depressed against the wing surface) as (−), and neutral ones as (0)

wing apex. Radial vein (R) is strong, convex, forming main longitudinal axis of wing, branching both forward (toward costal wing margin) and backward, with fore branches subapical, not numerous, and with strong, single, neutral or, rather, weakly concave rear branch called radial sector (RS), which starts submedially in fore wing, sub-basally in hind one (apicad of RS base, main radial stock is labelled either R, or R_1; main RS branches are sometimes re-labelled as R_2 through R_5). Median vein (M) is equally abundantly branching and neutral or weakly concave, with main branches often called M_{1-4} and with sub-basal, convex, weakly branching M_5: this causes M appearance of inverted R (apomorphically M_5 usually merges with CuA and so makes it convex). Cubital vein (Cu) dichotomised sub-basally into anterior CuA and posterior CuP, both concave (unless and until CuA merges with M_5). Anal veins (A) convex, moderately or weakly branching, originating independently or, in part, jointly from common sclerotised base, except free 1A which is sometimes renamed as postcubital (PCu) or empusal (E) vein, with respective re-labelling of following veins as 1A, 2A, etc. (this is justified morphologically but causes taxonomic confusion and therefore is not followed here). Anal area is slightly enlarged in hind wing but plesiomorphically is not rocking down there at wing folding except for small, veinless, posterobasal portion. Groundplan wing bears two concave folds of aerodynamic importance, associated with M and CuP, and the convex posterobasal one which is involved in wing folding at rest.

Groundplan articulating apparatus of the wing can be restored only tentatively as a synapomorphy of living Scarabaeones and Gryllones, being unknown in the most primitive Carboniferous groups. Dorsal apparatus consists basally of three sclerites ascending from lowered tergum margin toward wing, viz. anteriormost tegula which is loosely connected with wing fore margin structures, 1st axillary (1Ax) with wide basal articulation and two distal heads, and 3rd axillary (3Ax) with pointed basal articulation to hind notal process. Articulating zone placed distally of 1Ax varies in construction indicating primary appearance of articulatory sclerites as weakly individualised thickening in unevenly sclerotised region. Coupled with the lack of necessary palaeontological data, this makes it safer to propose only a preliminary hypothesis of multiple independent individualisation of articulating sclerites (basisubcostal, basiradial, median ones, etc.) of varying pattern and shape. Usual pattern is as follows: 1Ax fore head contacts (directly or indirectly) with bases of anterior veins (SC and/or R). Medial part of 1Ax is articulated with 2Ax which in turn connects to medial plates and further to M and Cu. Hind part of 1Ax is articulated with 3Ax which in turn is connected with base of anal veins. Well muscularised 3Ax is responsible for wing folding and for its movement in horizontal plane.

Of the above, only 2Ax is present on the ventral surface: its lower layer rests on the head of the pleural wing process which forms the main wing support at flight. Other ventral articulating sclerites are represented by the anterior basalare (of pleural origin) and posterior subalare (of tergal and particularly postnotal origin), both muscularised and working jointly as wing depressors, and individually as wing pronator and supinator, respectively. Tergum acts as the wing depressor and levator by lifting wing base when it buckles by contraction of dorsal medial longitudinal muscles, and sinks it when being depressed by tergo-sternal, tergo-pedal and some tergo-pleural muscles.

The alternative hypothesis by KUKALOVÁ-PECK (1978, 1983, 1991) of intra-wing metamery claims that the wing and its thoracic tergum are formed by 8 uniform metameres each composed of two branched veins, the convex anterior and the concave posterior ones, with the respective

wing membrane area, tergal subsclerite, and four articulating sclerites in-between. The hypothesis is not accepted here because, firstly, no developmental data support the internal metamery of any insect segment (TRUEMAN's (1990) hypothesis is of no support claiming that the wing consists of only two and not 8 metameres, and not concerning tergum). Secondly, the implied plesiomorphy of the regular wing fluting contradicts both palaeontological and morphological observations (RASNITSYN 1981, 1998a, and below).

Groundplan structure of the pterothoracic segment can be reconstructed as synapomorphy of more advanced Scarabaeones and Gryllones, again because of insufficient knowledge of the Carboniferous fossils. Tergum consists externally of large anterior notum and narrow posterior postnotum, both extending internally as phragmata. Each phragma belongs to both neighbour segments and bears attachments of respective dorsal longitudinal muscles (indirect wing depressors). Groundplan subdivision of notum is debatable (BRODSKY 1991). The most common structures (Fig. 68) are two anterolateral and one postero-medial convexities, the latter is commonly but not always correctly described as scutellum (unless to abandon the notion of scutellum as defined by the presence of intrascutellar muscle t13 and absence of tergo-laterophragmal one, t12). Other common, possibly groundplan notal structures are notauli or parapsidal sutures (although commonly used in non-hymenopterous insects, the term parapsidal sutures has been proposed originally for a different structure of aculeate wasps (e.g. GIBSON 1985) and has gained wide acceptance there, so the term parapsidal line or suture is worth to be abandoned in favour of the notaulus) which delimit the anterolateral convexities medially, V-shaped sulcus delimiting scutum before it and scutellum behind it, and medial longitudinal sulcus (derivable from moulting break line) which divide scutum (not scutellum) into right and left halves. The oldest fossils, especially among Dictyoneuridea, even seemingly adult, often show a notum similar to immatures in the lack of clear sutures except for the moulting line. Unless these fossils were really subimago rather than imago, or unless this is due to their paedomorphy, the difference could imply independent gain of the above structure of the notum in Gryllones and various Scarabaeones).

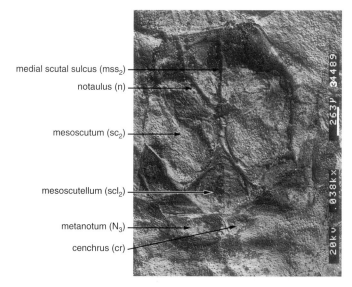

medial scutal sulcus (mss₂)
notaulus (n)
mesoscutum (sc₂)
mesoscutellum (scl₂)
metanotum (N₃)
cenchrus (cr)

Fig. 68 Structure and terminology of the insect pterothoracic tergum as exemplified by the primitive hymenopteran *Angaridyela vitimica* Rasnitsyn (Xyelidae), Early Cretaceous of Baissa in Siberia (based on SEM photo by H.H. Basibuyuk)

The archetypic pleuron is firmly connected with the sternum below, and consists of two subcircular sclerites, anapleuron and katapleuron, which are fused along the paracoxal suture (Fig. 69). Both pleural subsclerites are intersected by subvertical pleural sulcus bearing lateral coxal articulation below, wing articulation above, and pleural apodeme at its crossing with paracoxal suture. Pleural parts separated by pleural sulcus and paracoxal sutures called anepisternum (fore upper quarter), katepisternum (fore lower one), an- and katepimeron (hind upper and lower quarters, respectively). Anepisternum is further intersected by anapleural cleft into anepisternum in the narrow sense (above) and postepisternum (below). Posterodorsal part of anepisternum *s. str.* is delimited as basalar sclerite. Triangular trochantine is separated from katepisternum hind margin to provide anterior coxal articulation.

Groundplan sternum (Fig. 69) consists of fore basisternum (preceded by variously separated and not always easily identifiable presternal sclerites). It is followed by the furcasternum which bears two lateral furcal apodemes, and further by spinisternum with single medial apodeme.

Groundplan adult leg lacks styli on basal segments and has tarsus 5-segmented with pretarsal dactylus lost (probably modified into arolium) and paired claws present. Evidence of styli and genual segments (KUKALOVÁ-PECK 1983, 1991 etc.) needs reconsideration (RASNITSYN & NOVOKSHONOV 1997).

Pregenital abdominal segments lack delimited coxites and eversible sacs, with simple or annulated limb rudiments persisting in many immatures, very rarely in adults (RASNITSYN & NOVOKSHONOV 1997). Female genital segments (8th and 9th, Fig. 70) with gonapophyses working as ovipositor blades, coxites modified and called valvifer 1st and 2nd, respectively, and 8th segment lacking stylus (gonangulum retained). Male genitalia are plesiomorphic (segment 9th with penis, gonapophyses and styli retained, groundplan stylus not modified). Adult paracercus is lost (supposing it has been secondarily retained by some adult mayflies from nymphal stage).

True copulation present, with direct sperm transfer from male to female gonopore. Embryo with amniotic cavity closed and each amnion and serosa continuous (pore lost). Development plesiomorphically archemetabolous, i.e. gradual, with growth and associated moulting continuing in flying insect, and with early immature larva-like, with tarsus entire, possibly even fused with tibia, and with rudiments of imaginal structures (wings, genitalia) appearing as external buds during later (nymphal) section of development (Fig. 71, see also Figs 387, 388).

Groundplan bionomy is supposed as tree-dwelling at all developmental stages, with eggs laid into plant tissue cut by female gonapophyses, and with active stages feeding on sporangia content of gymnosperm trees. Jumps from branch to branch were practised to shorten the distance from one terminal group of sporangia to another, and also to escape predatory chelicerates and myriapods. Both wings and immature winglets were used to make jumps longer and more precise due to body attitude control in the air. For further details see RASNITSYN (1976, 1980, 1981, 1998a) who discuss alternative scenarios as well. The recent skimming hypothesis of insect flight origin is considered below (Chapter 2.2e).

Additional synapomorphies of winged insect are presented by BOUDREAUX (1979).

(d) Range. Latest Early Carboniferous (Namurian A) until the present, worldwide.

(e) System and phylogeny. The higher level structure of the subclass of winged insect is still an area of acute debates (cf. KRISTENSEN 1998, KUKALOVÁ-PECK 1998, RASNITSYN 1998a). The present account starts from the following observations which are considered to be relatively well grounded. The present day winged insects (and this is essentially true for the Mesozoic and Cainozoic Eras as well) form four clear-cut and most probably monophyletic groups. These are the insects with complete metamorphosis (Holometabola, Endopterygota, or Oligoneoptera), psocopteroid-hemipteroid assemblage (Paraneoptera), orthopteroids in the widest sense (Polyneoptera), and mayflies plus dragonflies (Hydropalaeoptera or Subulicornes). They appear below under the typified names Scarabaeiformes, Cimiciformes, Gryllones, and Libelluliformes, respectively. Their synapomorphies are listed in Chapters 2.2.1.3c, 2.2.1.2c, 2.2.2c, and 2.2.1.1c. In short, Scarabaeiformes can be characterised by their complete metamorphosis, Cimiciformes – by the mouthparts with detached, rod- or stylet-like lacinia, Gryllones – by the wings folded flat on their abdomen at rest and with a wider hind wing area, rocked down when in the resting position along the line running anterior of 2A. Characteristic of Libelluliformes are the wings permanently held in near flight position (either outstretched or raised over abdomen), with RS starting free from near wing base, and with highly pronounced fluting of the wing blade, so as the convex and concave veins alternate within each RS, MA, MP, and CuA vein systems. In fact, many cladistically oriented students tend to consider mayflies as the sister group of all other winged insects including dragonflies. However, the similarity of mayfly and dragonfly wings, and particularly of their most plesiomorphic Carboniferous representatives, Syntonopterida (see Figs. 85, 86) and Eogeropteridae (see Fig. 103), is so deep, and their putative synapomorphies are so unique, that the monophyly of Libelluliformes appears to be the best grounded hypothesis.

The picture changes considerably when we go further into the Palaeozoic. The only new, large and more or less clear-cut group found there is the palaeodictyopteroid assemblage (Protorhynchota: here Dictyoneuridea), which is characterised by the piercing beak and the wing blade fluted in the way different from that in Libelluliformes. The entire vein systems RS, MA, MP, and CuA alternate there in being either convex or concave (Chapter 2.2.1.2.3).

Besides the palaeodictyopteroids, in the Palaeozoic we can see a wealth of extinct groups of different rank. Most of them run smoothly to various well known orders. Some others merit creating orders of their own, and nevertheless their affinities appear more or less apparent. Palaeomanteida (miomopterans) probably represent the stem group of holometabolous insects (Chapter 2.2.1.3.1), Jurinida (glosselytrodeans) are a neuropteroid side branch (Chapter 2.2.1.3.3.4), Hypoperlida occupy a position in the roots of both psocidean-rhynchotan and palaeodictyopteran assemblages (Chapter 2.2.1.2.2), Caloneurida and Blattinipseida occur still more basally in the roots of the whole clade of Cimiciformes + Scarabaeiformes (Chapter 2.2.1.2.1). Numerous are the groups of obscure affinities as well. This particularly concerns the Carboniferous insect assemblage, in part because that time the basic insect subclades were not clear-cut yet (cf. the concept of taxa maturation, RASNITSYN 1996). The main reason, however, is that the Carboniferous stage is still the least known in the insect history, because the majority of descriptions were made in the 19th and early 20th centuries and have never been revised since that. The present attempt to re-describe this material at least in part (Rasnitsyn in preparation) has made it possible to reach some new results in terms of our understanding of the gross pterygote system and phylogeny.

As mentioned above, the five main groups of the pterygotes become three while deepening into Palaeozoic, because at the level of the less advanced orders Palaeomanteida, Hypoperlida, Diaphanopterida,

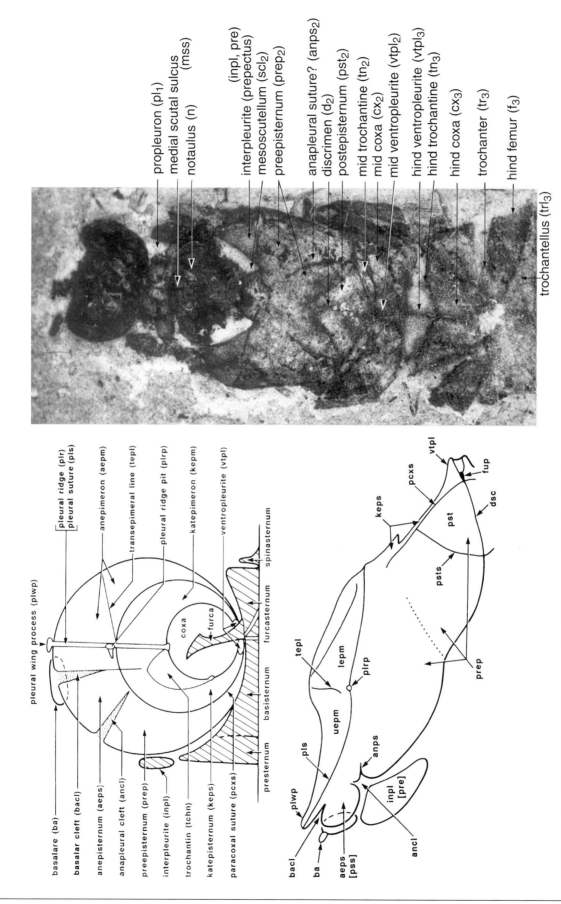

Fig. 69 Structure and terminology of the insect pterothoracic pleuron: groundplan (above left, from GIBSON 1993, based on MATSUDA 1970), same as applied to primitive hymenopteran, *Xyela* sp. (below left, after SHCHERBAKOV 1980 modified by GIBSON 1993), and as exemplified by the holotype of *Eoxyela atra* Rasnitsyn (Xyelidae) from the Late Jurassic of Karatau in Kazakhstan (right, based on photo by D.E. Shcherbakov)

Fig. 70 Posterior of abdomen of female *Permuralia maculata* Kukalová-Peck & Sinitshenkova (Parelmoidae, formerly referred to as *Uralia maculata* Sharov nomen nudum) from the Late Permian of Tshekarda in Urals (PIN 1700/493): cr – cercus, s_8 – abdominal sternum 8, sy_7, sy_9 – styli of 7 and 9 abdominal segments, t_7–t_{10} – abdominal terga, V_1, V_2 – ventral and dorsal ovipositor valves (gonapophyses of segments 8 and 9, respectively), Vr_2 – 2nd valvifer (modified from RASNITSYN & NOVOKSHONOV 1997, with kind permission from *Insect Systematics and Evolution*, formerly *Entomologica Scandinavica*)

Blattinopseida, Caloneurida, the holometabolous, paraneopteran ('psoco-rhynchotan') and palaeodictyopteran clades become very close to each other. This large clade is probably synapomorphic in their cryptosterny: their pterothoracic sterna are characteristically invaginated along the midventral line called discrimen (see Fig. 69), with the furcal arms mounting a common base elevated inside the thorax. Hence, the three main pterygote groups are now Libelluliformes (may- and dragonflies), Gryllones (orthopteroids *s. l.*), and the cryptosternous assemblage. These are our main reference points in the following considerations.

There is a recent alternative hypothesis claiming holometabolans to be monophyletic with Gryllones rather than with Cimiciformes (SHCHERBAKOV 1999). This is based in part on a superficial venational similarity of some miomopterans and the aberrant roach family Nocticolidae. Otherwise the hypothesis relies on similarity in characters that are clearly autapomorphic within at least one of the taxa under comparison. E.g. incipient cryptosterny is observed in some grylloblattidans which is similar to the hypothetical unrecorded ancestral stage passed by some remote cimiciform and holometabolan ancestor. Another example is the long, thin, flexible ovipositor of snakeflies and the grylloblattidan family Sojanoraphidiidae which is

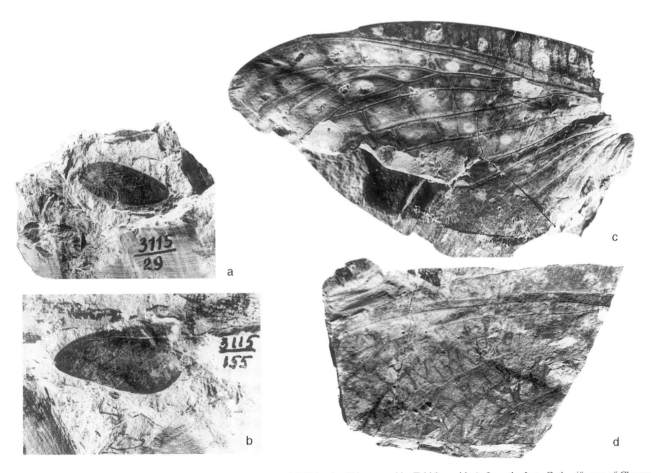

Fig. 71 Archemetaboly in a palaeodictyopteran, *Tchirkovaea guttata* M. Zalessky (Dictyoneurida, Tchirkovaeidae), from the Late Carboniferous of Chunya in Siberia: a–d – wings of half-grown and older nymphs, subimago and imago, respectively (PIN 3115/29, 155, 63, 66, photo by D.E. Shcherbakov) about in a scale (wing or wing fragment length 7, 18.3, 60 and 43 mm, respectively). Note change in size and form of winglet/wing and formation of veins

highly simplified morphologically at least in Raphidiida. There are arguments based on a dissimilarity, with no evidence proposed against the possibility of direct transition between the relevant character states. This concerns devices locking the fore wings in their resting position that occur on the metascutum in holometabolans and on mesopostnotum in hemipterans. There are further similarities and dissimilarities used by Shcherbakov to support his hypothesis. Some of them (including absence of 3rd axillar sclerite from the mayfly wing, embryological similarity between Gryllones and Scarabaeiformes, and some others) have already been considered and rejected (RASNITSYN 1969, 1976, 1980, 1998a). Others look unlikely but need careful examination (e.g. the possibility that cryptosterny is groundplan character state for insect body segment). In general, the present hypothesis of insect wing origin and of insect phylogeny is not considered to be seriously shaken as yet.

Through the course of studying Carboniferous fossils, it was noticed that three of the above groups (Gryllones, Libelluliformes and the cryptosternous assemblage) differ appreciably in the relative space taken by different vein systems over their wing blades. RS was observed to be dominating (occupying a greater wing area compared with M and Cu) in at least the more basal taxa of holometabolans (Palaeomanteida and Hymenoptera, possibly also in the least advanced neuropterans), of the cimiciform assemblage (Diaphanopterida, Blattinopseida and Caloneurida), and in mayflies and dragonflies. Other taxa in the holometabolan and cimiciform assemblages often have M dominating (psocideans, rhynchotans, mecopteroids), or the main vein systems are comparable in their development (many Dictyoneurida), or else they are highly variable in that respect (Hypoperlida). Anyway, Cu is very rarely dominating there, and apparently only in case of the generally modified venation (some Anthracoptilidae, Hypoperlida). In contrast, Cu is often dominant in the gryllonean orders, or Cu and M are subequal in their development, while RS dominates rarely and only in the advanced forms. This is observed in stoneflies but not in their ancestral grylloblattideans (except the advanced or aberrant Tillyardembiidae, Gorokhoviidae, some Sylvaphlebiidae and Blattogryllidae, Fig. 407), and in the specialised Carboniferous *Eucaenus* Scudder (Eucaenidae, Fig. 363).

The diagnostic character proposed here does not look very impressive evidently being subject to homoplasy. On the other hand, this character state can be identified in the majority of fossils, unlike the more important characters of body structure, mode of the wing folding, etc., which are rarely available for the Carboniferous fossils because of the prevailing preservation state (isolated wings and poorly preserved bodies). That is why the careful use of this character, in addition to all others available, is considered justified.

RS is observed to be dominating, besides in the above mentioned forms, also in the following Carboniferous taxa of debatable affinities: *Heterologus* Carpenter (see Fig. 81), *Heterologellus* Schmidt, *Kelleropteron* Brauckmann et Hahn, *Limburgina* Laurentiaux, *Propachytylopsis* Laurentiaux-Vieira et Laurentiaux, *Anthraconeura* Laurentiaux-Vieira et Laurentiaux, *Evenka* Rasnitsyn (Fig. 67), *Klebsiella* Meunier, *Anthracotremma* Scudder (Fig. 79), *Megalometer* Handlirsch (Fig. 80), *Sypharoptera* Handlirsch, *Emphyloptera* Pruvost, *Pruvostiella* Handlirsch, *Boltonaloneura subtilis* Bolton (Fig. 117), *Sthenarocera* Brongniart, *Protokollaria* Brongniart, *Hapaloptera* Handlirsch, *Herdina* Carpenter et Richardson (Fig. 113), *Metropator* Handlirsch (Fig. 114), *Paoliola* Handlirsch (Fig. 115), *Cymenophlebia* Pruvost, *Endoiasmus* Handlirsch, *Geroneura* Matthew (Fig. 116). It is noteworthy that some of these fossils display

character states absent in Gryllones. The most important is the roof-like wing position demonstrated by *Anthracotremma*, *Megalometer*, *Sthenarocera* and *Sypharoptera*. Indicative though not decisive is the similarity of *Sypharoptera*, *Emphyloptera*, *Pruvostiella*, *Sthenarocera*, *Geroneura*, and *Boltonaloneura* to the typical Caloneurida in having elongate wings with narrow costal space and simple, straight CuA and CuP. Approaching to this combination of character states, except for the wider wing, are *Metropator*, *Paoliola* and, possibly, brachypterous *Herdina*. These taxa are probably related to the undoubted Caloneurida (Chapter 2.2.1.2.1.2).

Anthracotremma and *Megalometer* are of particular importance in that they display a combination of roof-like wing position and distant pterothoracic coxae that indicates an absence of cryptosterny (Fig. 79). Therefore they should be attributed, at least preliminary, to the hypothetical stem group of both Libelluliformes and the cryptosternous assemblage (Cimici- and Scarabaeiformes). Still more reasons exist to assign *Heterologus*, *Heterologellus*, *Evenka* and maybe also *Stygne* Handlirsch (Fig. 82) to this group because of their highly plesiomorphic wings, while the remaining taxa, that is, *Limburgina Protopachytylopsis*, *Anthraconeura*, *Protokollaria*, *Hapaloptera*, *Cymenophlebia*, *Endoiasmus* might belong anywhere ranging from the above stem group to the vicinity of Caloneurida (Chapter 2.2.1.2.1.2) and Hypoperlida (Chapter 2.2.1.2.2).

Cu-dominated wings are characteristic of other Carboniferous fossils: Daldubidae (Grylloblattida, Fig. 397), some cockroaches (Fig. 367), some Paoliidae and their possible relatives (*Holasicia vetula* Kukalová (Fig. 77), *Prototettix* Giebel, *Schuchertiella* Handlirsch, *Merlebachia* Waterlot), and the eoblattid-spanioderid-cacurgid assemblage (Figs. 353–362). In that assemblage, the order Eoblattida (Chapter 2.2.2.0.1) was proposed by ROHDENDORF & RASNITSYN (1980) as the stem group of the dictyopteran assemblage (superorder Blattidea, Chapter 2.2.2.1), based on the characteristic form of the fore wing anal area (clavus), and on the gryllonean mode of the wing folding evident in *Protophasma* Brongniart (Fig. 359). However, *Protophasma* has not really large clavus, while *Eoblatta* Brongniart (Fig. 353) is very similar venationally to *Stenoneura* Brongniart (Fig. 354), *Eoblattina* Bolton (Fig. 355), *Ischnoneura* Brongniart, *Ctenoptilus* Laméere, *Ischnoneurilla* Handlirsch, which differ considerably from typical blattoids in general appearance as well (have subquadrate to elongate pronotum lacking wide paranota and often longer legs) and sometimes lack the typical blattoid clavus as well (*Ischnoneura*, *Ischnoneurilla*). Additionally, these fossils form a rather smooth transition to Spanioderidae (via *Ctenoptilus* and *Cacurgus* Handlirsch) and Ischnoneuridae Handlirsch, 1906 (=Aetophlebiidae Handlirsch, 1906, =Narkeminidae Storozhenko, 1996, synn. nov., Fig. 356, probably including *Protodiamphipnoa* Brongniart, Fig. 357, but not *Cnemidolestes* Brongniart which is an isolated wing of unknown affinity). All of them lack blattoid clavus as well, and otherwise they show little similarity to cockroaches. Evidently, the wide, lancet-like clavus is not a synapomorphy of Eoblattida and Blattidea.

The hind wing of Spanioderidae is described as lacking a large anal lobe (BURNHAM 1986, F. CARPENTER 1992a): this would make them unlikely to belong to true Gryllones. However, BURNHAM (1983) has overlooked this lobe in the gerarid hind wing (Fig. 360); the same might be true for Spanioderidae as well. The typically gryllonean anal lobe is definitely developed in the possible eoblattid-spanioderid relatives Ischnoneuridae (STOROZHENKO 1998 as Narkeminidae). Ischnoneuridae (as Narkemidae or Narkeminidae) are usually attributed to Grylloblattida

(ROHDENDORF & RASNITSYN 1980; STOROZHENKO 1998), but their venational similarity to the eoblattid-spanioderid assemblage is much deeper than to Grylloblattida (see above). Still more similar to Eoblattidae and Spanioderidae are *Cacurgus* and its relatives. This makes us able to consider all the eoblattid-spanioderid-cacurgid-narkeminid assemblage as the stem group of true Gryllones termed the order Eoblattida (Chapter 2.2.2.0.1).

Besides the above main assemblages with prevailing RS and Cu, there are many Carboniferous fossils that either have M dominating (e.g. many Geraridae, Fig. 360, some Paoliidae, Fig. 78) or with no evidently dominating system (*Kochopteron* Brauckmann, *Pseudofouquea* Handlirsch, *Ampeliptera* Pruvost, Fig. 122, *Heterologopsis* Brauckmann, Fig. 134, *Protoprosbole* Laurentiaux, *Aenigmatodes* Handlirsch and many others). Many of them can probably be related to one or another of the above main assemblages. Geraridae are undoubted Gryllones, *Ampeliptera*, *Protoprosbole* and *Aenigmatodes* most probably belong to Hypoperlidae, *Heterologopsis* possibly to stem Dictyoneuridea. This permits us to keep the winged insects segregated into two large groups, Scarabaeones with the typically (plesiomorphically?) large RS, and Gryllones with the typically (plesiomorphically?) large Cu.

Unlike the earlier version of the pterygote system (RASNITSYN 1980, ROHDENDORF & RASNITSYN 1980), we cannot include Paoliidae and their relatives (the former order Paoliida or Protoptera) into Scarabaeones. In the above publications Paoliida were hypothesised to be ancestral to all other winged insects, but technically they were placed within Scarabaeones because of the absence of the gryllonean anal lobe. The present approach makes this problematic, because now we have an additional diagnostic feature if not the alternative synapomorphies of each Scarabaeones and Gryllones, that is, the different vein systems prevailed. In this character, some paoliids are close to Gryllones in having large Cu, but no one can be related to Scarabaeones because of really large RS. At the same time, paoliids, like plesiomorphic Scarabaeones, keep their wings roof-like at rest (Figs. 74–76), have no foldable anal lobe in the hind wing (Fig. 78), and in some cases (*Zdenekia* Kukalová-Peck, Fig. 78) closely remind one of some plesiomorphic Scarabaeones (*Ampeliptera*, Fig. 122, *Limburgina*, *Propachytylopsis*, *Heterologopsis*, Fig. 134, etc.) in the form of their fore wing CuA. In contrast, the general body form of the few paoliids known in that respect (BRAUCKMANN 1991 and unpublished, Figs. 74–76) is similar to that of *Protodiamphipnoa*, *Protophasma*, *Gerarus* (Figs. 357, 359, 360) and some Spanioderidae (*Palaeocarria* Cockerell, Fig. 362): all of them are somewhat stick-insect-like in being large and long-legged insects with subquadrate to elongate pronotum without wide paranota. To an extent, however, this can also be true for some ancient Scarabaeones (*Megalometer*, *Heterologopsis*, the Early Permian *Strephoneura* Martynov, Fig. 131).

The general appearance of paoliids as large and heavy, long-legged insects, probably slow both on wings and feet, is in agreement with our earlier hypothesis of the groundplan pterygote habits, that is, feeding on sporangia located on branchlets of the Carboniferous gymnosperm plants (RASNITSYN 1980, ROHDENDORF & RASNITSYN 1980), and so it well might be the groundplan character of the entire subclass Scarabaeona (winged insects). Large body size also fits this hypothesis, because it makes short (comparing the body size) primordial winglets aerodynamically effective in attitude control when jumping from one branch to another (WOOTTON & ELLINGTON 1991). However, long legs might be obstacles for the effective 'pro-flight' of ancestral pterygotes and so they could be acquired later in the evolution, after

the insects have got some flight skill. Long legs and/or pronotum, that is, a kind of life form of giraffe and megatherium, are characteristic of many archaic winged insects of the Palaeozoic, both of the scarabaeonean and gryllonean affinities (Figs. 74–76, 80, 357, 359, 362, 363). At least one of them, short-legged but long-necked *Eucaenus ovalis* Scudder from the Westphalian D of Mazon Creek, Illinois (Fig. 363), is really found to have several lycopod microspores in its guts (A. SCOTT & TAYLOR 1983), though their quantity cannot exclude the possibility that the spores were occasionally and not intentionally ingested by the insects. This implies that these adaptations, and particularly the long-leggedness, might be synapomorphic for all known winged insects.

Taking the above considerations into account, the origin of the winged insects is hypothesised here to start from the bristletail-like habits (that is, the less cryptic one than that in silverfish) and to proceed further through feeding on spores as well as on the gymnosperm plant pollen and ovules, both disperse and enclosed in sporangia, either fallen on the ground or attached to the mother plant. The next stage (Fig. 72) is supposed to be feeding on the sporangia content of gymnosperm trees, with incipient using of movable winglets of the paranotal origin in jumps from one group of terminal sporangia to another, and to escape predators, as mentioned above. At both these stages, the insects are supposed to have large size, relatively short legs and long cerci and paracercus. These assumptions are based, firstly, on the results by WOOTTON & ELLINGTON (1991) that in an insect-like object of larger size (6–10 cm long), even small winglets can be aerodynamically effective in jumping/gliding pre-flight if minimally movable (adjustable against air stream), particularly if assisted by long cerci (and paracercus). The second reason is that a creature with permanently outstretched long winglets (not to say normal wings) would be awkward on its feet, both on the ground and among vegetation, and hardly able to escape extinction unless superior flying ability is gained (RASNITSYN 1976, 1980, 1981, 1998a). And the third reason is a number of indications that paranota are in part of limb origin and so they could retain in their morphogenetic repertoire the relevant mechanisms of limb motility invisible but ready to be re-activated using the customary morphogenetic re-arrangement (TIKHOMIROVA 1991).

At the third stage, when real flight had appeared and wings grew large, hind margins of wings came into contact while at rest. To make possible using narrow spaces as a refuge from predators, wind, rain, etc., it was important to diminish the transverse size of the insect. To this end, the best way is to lay wings one over the other. This creates a problem, however, for being ready for flight, because of the time required to return the wings into the flight position. Therefore the overlapping (flat) wing position at rest is not very likely to be developed until the wing articulation apparatus became reasonably sophisticated. Prior to this, when wings grew wider and met each other along their rear margins while in rest position, it is more likely that the line of wing contact was being pushed upward, and the wings took an oblique (roof-like) rest position. This stage of evolution, like the previous ones, is purely hypothetical.

The next stage, the elongation of legs that permitted insects to behave easily among thick branches, is supposedly represented by diverse Carboniferous fossils (see Figs. 74–76, 80, 357, 359, 362–363). This was probably the key point in the pterygote phylogeny, the starting position for Scarabaeones to begin improving the flight ability, and for Gryllones to enhance adaptations for more cryptic existence.

There are numerous alternative hypotheses about the pterygote origins. Earlier ones are considered elsewhere (RASNITSYN 1980 and

Fig. 72 Hypothesised ancestor of the winged insects restored by A.G. Ponomarenko

references therein). Worth mentioning here is the recent hypothesis that insect flight originated in an amphibiotic ancestor through the surface skimming on feet using pre-wings as an air propeller (proposed by MARDEN & KRAMER 1994, MARDEN 1995, reviewed and further developed by SHCHERBAKOV 1999). This hypothesis does not look appropriate because, firstly, it contradicts all basic taphonomical observations (Chapter 3.2.2.1), and secondly, to be efficient, the airforced locomotory system must rely on a far advanced engine and propelling agent. So it can develop **from** an already evolved, flight system rather than *vice versa*. It does not look occasional that the man designed air-propelled boats similarly appeared using aeroplane motor and propellers, and not as an aeroplane precursor. Equally this makes inappropriate Shcherbakov's (*op. cit.*) hypothesis of the ancestral wings modelled after mayflies, that is permanently outstretched and used by an ephemeral adult simply to mate and disperse (even if one ignores the fact that an ephemeral and hence morphologically simplified adult is not the best starting point for the known variety of adult winged insects).

Taxonomically, we consider the subclass of winged insects as composed of the two infraclasses, Scarabaeones and Gryllones, dealt with below (Chapters 2.2.1 and 2.2.2). According to the basic taxonomic principle of continuum (Chapter 1.1c) we try and follow the supposed relatedness and similarity and not the particular characters as such. That is why, the paradigmatic Gryllones (that is, superorders Blattidea, Perlidea and Gryllidea) are complemented here with the fossils displaying either the posterobasal hind wing area tucking down along the line before 2A (Geraridae, Eucaenidae, Ischnoneuridae,

Protophasmatidae), or those with prevailing CuA and without evident affinity to Scarabaeones (the eoblattid-spanioderid assemblage).

Those with the evident scarabaeonean connections, or with RS prevailing and without the hind wing anal area foldable anterior to 2A (*Anthracotremma, Megalometer, Heterologus, Heterologopsis, Heterologellus, Kelleropteron, Ampeliptera, Protoprosbole, Aenigmatodes, Limburgina, Propachytylopsis, Anthraconeura, Protokollaria, Hapaloptera, Cymenophlebia, Endoiasmus, Sypharoptera, Emphyloptera, Pruvostiella, Sthenarocera, Geroneura, Boltonaloneura, Metropator, Paoliola, Herdina, Evenka*) are referred to as Scarabaeones. Paoliidae cannot be assigned to any of above because of the combination of the roof-like wing resting position and supposedly plesiomorphic Cu prevalence. At the same time, it does not seem wise to create one more infraclass for just a small family, so the order Paoliida is retained for it, unplaced within the subclass Scarabaeona– until a better understanding of insect evolution is gained.

Herbstiala, Schuchertiella, Merlebachia, Hadentomum Handlirsch, *Klebsiella, Homoeodictyon* Martynov (Fig. 73), and many other obscure genera are considered here as Scarabaeona *incertae sedis* (winged insects of obscure taxonomic position). Of them *Klebsiella* and *Homoeodictyon* are of interest because of their convex MA. However the remaining venation has little in common with Dictyoneuridea, so their position in the stem of that group is not very likely though it cannot be ruled out entirely.

The system and supposed phylogeny of the winged insects are shown at Fig. 58.

Fig. 73 *Homoeodictyon elongatum* Martynov, a winged insect of obscure relationships, with convex MA as found in mayflies, dragonflies and dictyoneuridean orders, from the Late Permian of Kargala in South Urals (from RASNITSYN 1980); fore wing 68 mm long as preserved

2.2.0.1. ORDER PAOLIIDA Handlirsch, 1906.
(=Protoptera Sharov, 1966)

(a) Introductory remarks. The order embraces the least advanced winged insects (Figs. 74–76). The chapter is based on data in RASNITSYN (1980), ROHDENDORF & RASNITSYN (1980), and particularly on the results by BRAUCKMANN (1984, 1991), BRAUCKMANN *et al.* (1985, and unpublished), as well as on the above considerations (Chapter 2.2e).

(b) Definition. Large insects with long but otherwise apparently unmodified legs and long (longer than body), setiform antennae. Pronotum lacking wide paranota, other body structures unknown. Wing rest position low roof-like. Wings moderately wide, with fore margin weakly convex or near straight, and with apex often rather narrow. Costal space moderately narrow, with a series of rather irregular, simple or branching, oblique veinlets. Main veins are weakly concave except convex R, M_5 (+CuA) and A in fore wing, only R in hind wing, and distinctly concave SC and, in fore wing, CuP. SC meeting R near apical quarter of wing. SC, R and often also CuP and anal veins lacking long branches. RS, M and CuA irregularly branching, with M and/or CuA having prevalence over RS; fore wing CuA often with series of rather uniform hind branches in its middle part. RS starting near basal quarter, sometimes basal third of wing length. M_5 forming short, oblique, strong cross-vein between M and CuA (not very distinct in *Pseudofouquea* Handlirsch). Cross-veins simple or, mainly, abundantly branching and often forming coarse archedictyon. Hind wing subtriangular, with anal area more or less widened, not tucking down in folded wing.

(c) Synapomorphies. Absent.

(d) Range. Middle Carboniferous (Namurian B – Westphalian A) of Central and West Europe and NE USA.

(e) System and phylogeny. Single family Paoliidae with 10 genera and 12 species which are not yet well characterized enough to justify phylogenetic interpretation.

(f) History. Paoliids are rare insects, and the known history of their order is extremely short. It takes only several million years (since the late Namurian B till late Westphalian A) at the very beginning of the recorded history of the winged insect, in full accord with the supposed stem nature of the group. Most finds are made in Central and West Europe, in the Namurian (two species of *Holasicia* Kukalová, Figs. 74, 77 and one of each *Zdenekia* Kukalová, Fig. 78, *Olinka* Kukalová and *Kemperala* Brauckmann, Fig. 75) and Westphalian A (five species of *Paolia*, *Holasicia*, *Zdenekia*, *Pseudofouquea* and *Sustaia* Kukalová). Two other species of *Paolia* Smith and *Paoliola* Handlirsch are known

Fig. 74 *Holasicia rasnitsyni* Brauckmann (Paoliida: Paoliidae), undescribed specimen from the earliest Middle Carboniferous of Hagen-Vorhalle in Germany (N1041, coll. W. Sippel, deposited in Westfalisches Museum fur Naturkunde–Planetarium in Munster, Westphalia, Germany; courtesy W. Sippel and C. Brauckmann); average length of wings 43–45 mm

from the Namurian of the NE USA (KUKALOVÁ 1958, BRAUCKMANN 1991). No real pattern is discernible in their distribution either in space and time.

Fig. 75 *Kemperala hagenensis* Brauckmann (Paoliida: Paoliidae) (no registration number: origin and source as for *Holasicia rasnitsyni* (see Fig. 74); average length of wings 60–62 mm

Fig. 76 Undescribed paoliid, same origin and source as Figs. 74, 75 (N1006); average wing length 65 mm

2.2.1. INFRACLASS SCARABAEONES Laicharting, 1781

A.P. RASNITSYN

(a) **Introductory remarks**. The chapter is based on data in RASNITSYN (1980) and ROHDENDORF & RASNITSYN (1980), as well as on the above considerations (Chapter 2.2e).

(b) **Definition**. Winged or secondarily wingless insects with basiproximal area of hind wing, if expanded, bending down at rest along the line running behind 2A and not before it. Wings often in roof-like position at rest, with RS commonly and Cu rarely dominating. Female gonocoxa 9 forming ovipositor sheath (or reduced, but only secondarily and rarely used in some holometabolans to penetrate substrate during oviposition). Male gonostyli often working as forceps at copulation.

(c) **Synapomorphies**. Possibly: RS dominating over M and Cu, male gonostyli working as forceps at copulation.

(d) **Range**. Latest Early Carboniferous (Namurian A) up to now, worldwide.

(e) **System and phylogeny**. The origin of Scarabaeones is not hypothesised here to depend on particular ecological changes except further improvement of the flight ability. The stem group of the infraclass may include the Carboniferous Anthracotremmatidae, Heterologidae, Evenkidae and possibly Stygneidae (Chapter 2.2). They are considered unplaced stem taxa (Scarabaeones *incertae sedis*) thus far.

The two main subclades of Scarabaeones both demonstrate further evolution toward the strong flight ability, but do it differently. The cohors Libelluliformes (may- and dragonflies) began with perfection of

Fig. 77 Fore wing of *Holasicia vetila* Kukalová (Paoliida: Paoliidae) from the earliest Middle Carboniferus of Horní Suchà in Czechia (courtesy J. Kukalová); wing 42 mm long as preserved

the wing blades itself (sophisticated fluting of the wing blade), while the clade comprised of the cohorts Cimiciformes and Scarabaeiformes has started improving the thoracic wing engine (cryptosterny, that is invagination of the pterothoracic sterna and their transformation into the thoracic endoskeleton, resulted in appearing of wide and reliable support for the wing musculature). The two contrasting directions of the flight evolution agrees perfectly with, respectively, the inadaptive and (eu)adaptive modes of evolution (KOWALEVSKY 1874, RASNITSYN 1987).

Fig. 78 *Zdenekia grandis* Kukalová (Paoliidae) from the earlier Middle Carboniferous of Horní Suchà in Czechia (courtesy J. Kukalová); fore wing 6.1 mm long as preserved

As already discussed (Chapter 2.2), there are also Carboniferous fossils which cannot be attributed with any reasonable certainty to any of two above clades and may well represent, at least in part, the stem Scarabaeones. These are, first of all, *Anthracotremma* and *Megalometer* (Figs. 79, 80) that explicitly combine wings with large RS (diagnostic of Scarabaeones), a roof-like wing resting position (negatively diagnostic of may- and dragonflies) and pterothoracic coxae widely distant implying sternum exposed (negatively diagnostic of the other Scarabaeones). Only wings are known for *Heterologus* (Fig. 81), *Heterologellus*, *Evenka* (see Fig. 67) and *Stygne* Handlirsch (Fig. 82): they are highly plesiomorphic and show large RS and so can be considered stem Scarabaeones until contrary evidence appears. The case of *Evenka* is somewhat contradictory: its wings are highly plesiomorphic except for the considerably reduced CuA, a feature indicative of Caloneurida *s. l.* (Chapter 2.2.1.2.1.2). It cannot be excluded that *Evenka* takes a position in the stem Caloneurida or nearby. Still less precise can be indicated the position of a larger group of Carboniferous fossils known from wings and, sometimes, poorly preserved bodies. This group (*Limburgina*, *Protopachytylopsis*, *Anthraconeura*, *Protokollaria*, *Hapaloptera*, *Cymenophlebia*, *Endoiasmus*) has been already mentioned as possibly belonging to stem Scarabaeona or to Caloneurida or Hypoperlida (Chapter 2.2).

The proposed system and phylogeny of the infraclass are shown in Fig. 58.

2.2.1.1. COHORS LIBELLULIFORMES Laicharting, 1781 (=Subulicornes Latreille, 1807, =Hydropalaeoptera Rohdendorf, 1968)

A.P. RASNITSYN

(a) Introductory remarks. The cohors is essentially the same as ROHDENDORF's (1968a, 1969) Hydropalaeoptera, that is Palaeoptera minus the palaeodictyopteroid orders.

Fig. 79 *Anthracotremma robusta* Handlirsch, a stem Scarabaeones from the Middle Carboniferous of Mazon Creek in USA (holotype)

(b) Definition. Size, moderate to very large, rarely small. Head essentially plesiomorphic except the antenna are short in adults (confirmed for *Lithoneura* Carpenter by WILLMANN 1999) and in nymphal dragonflies, maxillary galea lost (unknown in Palaeozoic taxa), and mouthparts rudimentary in post-Palaeozoic mayflies. Prothorax never large (small in post-Palaeozoic adults), only exceptionally with enlarged paranota (WOOTTON & KUKALOVÁ-PECK 2000). Pterothorax and wings near homonomous except in post-Palaeozoic mayflies with functionally (sometimes even morphologically) dipterous flight due to better developed mesothorax and its wings. Pterothoracic structure archetypically plesiomorphic, with sterna exposed, but otherwise often modified, particularly so in dragonflies due to their aberrant flight with much enhanced control of movement of individual wing, with all skeletal morphology, musculature, and wing articulation being deeply

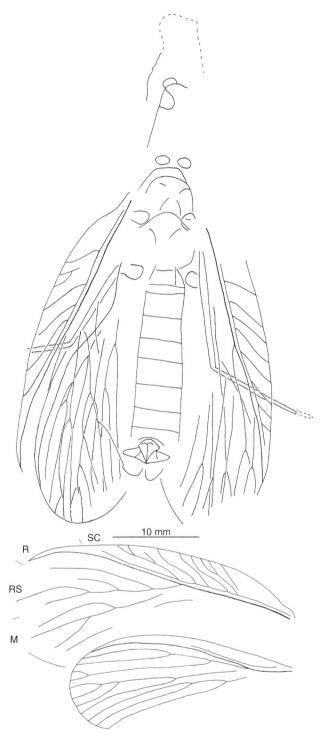

between two convex branches of convex main vein, and convex intercalary between two concave branches of concave main vein. RS always with one basal triad and another on its posterior branch. Costal space narrow, SC simple (lacking branches other than cross-vein-like), straight or nearly so, meeting C. Cross-veins regular, rarely (in some Carboniferous forms) replaced by irregular net. Legs of average form or variously specialised, often with rudiment of patello-tibial suture, particularly in immatures. Abdomen 10-segmented, lacking paranota, with cerci and paracercus variable (long multisegmented to short, simple, or paracercus lost). Male gonocoxites and gonostyli modified into forceps used in copulation (unknown for Carboniferous mayflies, much reduced and replaced by secondary copulatory organ at abdominal base in dragonflies, possibly except Erasipteridae). Ovipositor with 2 pairs of stylets and pair of sheaths (often shorter than stylets) bearing gonostyli, or variously reduced (in many dragonflies) or lost (in post-Palaeozoic mayflies, unknown for Palaeozoic ones). Nymphs aquatic when known (not known with any certainty for Palaeozoic, except protereismatid mayflies, and could well have been terrestrial), often with tracheal gills of various origin, legs with 1–5 tarsomeres and 1–2 claws, dragonfly nymph with labium modified into grasping "mask". Development gradual, with wings, genitalia and tarsal segmentation appearing and progressing after egg hatching, with no pupal stage; in mayflies with moulting flying instar called subimago.

(c) **Synapomorphies**. Adult antenna short; wing rest position spread; wing surface regularly fluted due to (i) MA being sharply convex and RS and MP concave in addition to archetypically convex veins C, R, CuA, and A, and concave SC and CuP, and (ii) triad vein branching, with concave intercalary veins inserted between two convex branches of convex main vein, and convex intercalary between two concave branches of concave main vein, with RS forming basal and posterior triads.

(d) **Range**. Earliest Middle Carboniferous (Namurian B) until the present, worldwide.

(e) **System and phylogeny**. As seen in Fig. 58, the cohors represents a sister group of all the other Scarabaeones. The six orders included are attributed to two superorders: the mayflies and dragonflies, the former being supposedly paraphyletic in respect to the latter. This is because of the hypothesised sister group relationship between Triplosobida and the other orders, as it is supported by a number of synapomorphies of the dragonflies and the other mayflies. The synapomorphies are RS separated from R, connected to MA by short cross-vein sub-basally (usually directly fused with it for a distance), and 1A similarly connected to CuP.

2.2.1.1.1. SUPERORDER EPHEMERIDEA Latreille, 1810. THE MAYFLIES (=Panephemeroptera Crampton, 1928)

(a) **Introductory remarks**. A comparatively small insect group often considered as forming the single order of mayflies but split here into 3 orders. This becomes difficult to avoid after Bojophlebiidae and Syntonopteridae have been incorporated into the group (KUKALOVÁ-PECK 1985) despite of their phenetic distinctness of full ordinal rank, and additionally because of the supposed sister group relationship between Triplosoba Handlirsch and all the dragonflies and the other Ephemeridea (Chapter 2.2.1.1e).

(b) **Definition**. Size small to very large. Head essentially plesiomorphic except antenna short and, in post-Palaeozoic groups, mouthparts rudimentary. Prothorax moderately or, in living groups, fairly small,

Fig. 80 *Megalometer lata* Handlirsch, a stem Scarabaeones from the Middle Carboniferous of Mazon Creek in USA (holotype)

affected. Wing resting position is spread, or contralateral wings approaching each other with their upper surfaces, with respective changes in wing hinge. Wing surface regularly fluted due to (i) MA being sharply convex and RS and MP concave in addition to archetypically convex veins C, R, CuA, and A, and concave SC and CuP, and (ii) triad vein branching, with concave intercalary veins inserted

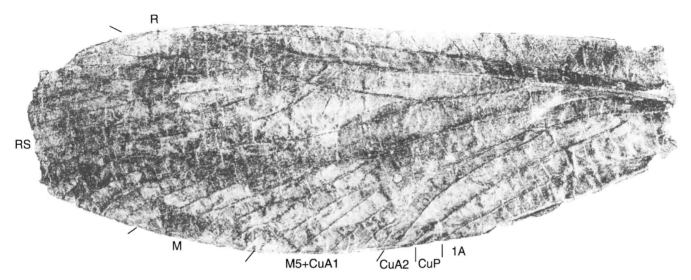

Fig. 81 *Heterologus langfordorum* Carpenter, possibly stem Scarabaeones, from the Middle Carboniferous of Mazon Creek in USA (from CARPENTER 1943a); fore wing 70 mm long

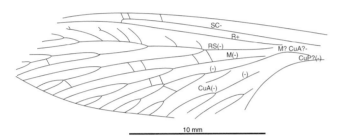

Fig. 82 *Stygne roemeri* Handlirsch, possible stem Scarabaeones from the earliest Middle Carboniferous of Alfred Mine in Poland (holotype lost, drawing is original, based on the photograph in SCHWARZBACH 1939)

lacking expanded paranota except rudimentary ones in *Protereisma* (KUKALOVÁ-PECK 1985). Pterothorax and wings homonomous in Palaeozoic, or with mesothorax and fore wings much better developed (fore and hind wings coupled at flight) in Meso- and Cainozoic members, up to rudimentary or lacking hind wing in some genera. Pterothoracic structure rather archaic, with sternum exposed; living groups with integument having large membranous areas (probably of paedomorphic origin) and with midcoxa-furcosternal articulation (probably gained independently of that characteristic of the holometabolans). Wing rest position spread or upraised, though wing articulating structures responsible for a roof-like folding being hardly modified in living forms (BRODSKY 1974). Wing surface regularly fluted, otherwise venation rather average, varying from moderately rich to poor (in small or otherwise specialised forms), sometimes with only a few remaining veins organised in close, widely spaced pairs. Cross-veins regular, sometimes rare or absent, replaced by a more or less irregular net in some Carboniferous forms. Meso- and Cainozoic forms (possibly except some Triassic ones) with flight functionally (sometimes even morphologically) two-winged, in-phase, anteromotoric (*sensu* GRODNITSKY 1999), in the Palaeozoic functionally four-winged, probably anteromotoric. Legs of average form, more slender in living groups (male fore pair particularly elongated), with simple trochanter and, with few exceptions, 5 tarsomeres (Carboniferous Bojophlebiidae probably with single claw unlike post-Carboniferous

mayflies). Abdomen 10-segmented, lacking paranota, bearing long cerci and paracercus (unknown for Syntonopterida; paracercus sometimes short in post-Palaeozoic groups), and with male gonocoxites and gonostyli modified into forceps which are used in copulation (unknown for Triplosobida and Syntonopterida). Living mayflies with paired male and female genital openings (probably a neotenic reversal). Subimago (known for Protereismatina in addition to living mayfly suborders) differing only slightly compared with adults (wings heavy with less sharply delimited veins and microtrichia present, male genitalia less differentiated and fore leg less elongate). Living forms with malpighian tubes numerous, ovaries panoistic. Nymphs aquatic (known for Protereismatina and living suborders; concerning suggested Carboniferous nymphs see Chapter 2.1), developing fairly gradually, with the general body form of immatures depressed and more robust than in alates. Antenna less reduced than in adult, mouthparts chewing, essentially plesiomorphic. Wing pads in normal position (with upper margin directed laterad and somewhat ventrad). Legs with two claws and 4–5 tarsomeres in Protereismatina, 1 claw and 1 tarsomere in living suborders. Abdomen 10-segmented, segments 1–9 (maximum 1–8 in living suborders) with lateral tergaliae (gill plates *auctorum*) serially homologue to thoracic wings, often modified in form (KLUGE 1989), apical segment with long cerci and paracercus.

(c) **Synapomorphies**. None in respect to Libellulidea.

(d) **Range**. Nearly earliest Middle Carboniferous (Namurian C) until the present, worldwide.

(e) **System and phylogeny** are shown at Fig. 83.

2.2.1.1.1.1. ORDER TRIPLOSOBIDA Handlirsch, 1906 (=Protephemerida Handlirsch, 1906)

(a) **Introductory remarks**. The group known from only one specimen has been variously treated by students in terms of its rank (ranging from family Triplosobidae to order Protephemerida), while its phylogenetic meaning as a member of the mayfly stem group has been generally accepted. The present concept shifts the origin of the group still more basad in considering it as the sister group of all other mayflies combined with the dragonflies (Fig. 84). Interpretation of the fossil is based on F. CARPENTER (1963a).

(b) Definition. General appearance (Fig. 84) of *Protereisma*-like mayfly differing in RS originating from R and in lacking anastomoses between main veins (not certain concerning CuP–A connection). Venation not rich: RS with 3 and MP with 2 branches, other preanal veins simple (intercalars not counted). Cross-veins regular, not particularly numerous. Appendages (including genitalia) unknown except for 3 caudal filaments.

(c) Synapomorphies. Possibly: paranota lost (unless just not preserved in the sole known specimen), venation reduced (unless this is the character of the genus or family and not of all the order). These reservations imply that future findings can show the order to be paraphyletic in respect to other Libelluliformes.

(d) Range. Late Carboniferous of Europe.

(e) System and phylogeny as well as **(f) History** cannot be outlined, for the holotype of *Triplosoba pulchella* (Brongniart) collected in the Stephanian (Late Carboniferous) deposits of France (Commentry) is the only specimen of the order known as yet.

2.2.1.1.1.2. ORDER SYNTONOPTERIDA Handlirsch, 1911

(a) Introductory remarks. The reason to create a new order for the two constituent families is their considerable phenetic difference from the true mayflies as well as their probable paraphyletic (ancestral) position in respect to both true mayflies and dragonflies. The present

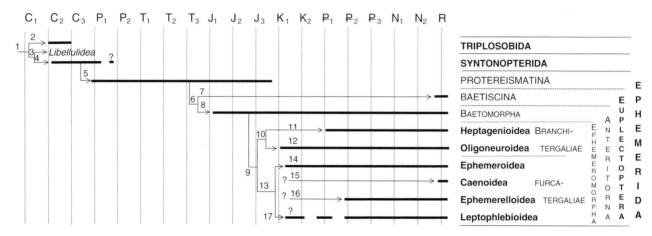

Fig. 83 Phylogeny and system of the superorder Ephemeridea

Time periods are abbreviated as follows: C_1, C_2, C_3 – Early (Lower), Middle and Late (Upper) Carboniferous, P_1, P_2 – Early (Lower) and Late (Upper) Permian, T – Triassic, J – Jurassic, K – Cretaceous, P_1 – Palaeocene, P_2 – Eocene, P_3 – Oligocene, N_1 – Miocene, N_2 – Pliocene, R – present time (Holocene). Right columns are names of taxa. Arrows show ancestry, thick bars are known durations of taxa (dashed in cases with long gaps in the fossil record), figures refer to synapomorphies of the subtended clades, as follows:

1 – wings spread at rest; wing surface regularly fluting due to MA being sharply convex and RS and MP concave in addition to archetypically convex veins C, R, CuA, and A, and concave SC and CuP, and additionally due to intercalary veins inserted concave between convex main veins, and convex between concave ones.

2 – MA and CuA simple.

3 – RS separated from R and connected to MA sub-basally.

4 – costal brace present (possibly, unless gained later, at node 5): ovipositor lost (unless gained at nodes 5 or 6).

5 – nymph aquatic with abdominal tergaliae specialised as tracheal gills, with caudaliae (cerci and paracercus) specialised as horizontal caudal flipper.

6 – hind wing half as along as fore wing, coupled with it at flight; in nymph (unknown for majority of Protereismatina): abdominal tergaliae of pairs VIII and IX lost, tarsi one-segmented with single claw.

7 – fore wing with tornal angle basad of 1A; all trunk nerve ganglia fused; nymph with pronotum, mesonotum and fore wing buds fused in carapax covering metanotum and anterior abdominal segments, abdominal segment VI enlarged and modified; nymphal tergaliae highly differentiated.

8 – fore wing with tornal angle distad to CuP; nymphal maxilla with number of dentisetae reduced to 3.

9 – nymphal maxilla with number of dentisetae reduced to 2.

10 – posterior arms of mesothoracic prealar bridge reduced in adults; eggs with anchors of peculiar structure; nymphal maxillae with ventral longitudinal row of setae; nymphal labial palp with segments 2 and 3 fused; nymphal tergaliae with gill filaments.

11 – nymph with frontal shield above mouth parts (nymphal head and body usually more or less flattened); subimago with peculiar shape of mesonotal sclerotisation; in adults (subimago and imago) anterior paracoxal suture incomplete and turned backward; cubital field of fore wing with 4 (two pairs) long intercalaries.

12 – nymph with fore legs and mouthparts with peculiar filtering specialisation and with gills at maxillae bases.

13 – first tarsal segment of adults shortened; nymphal tergaliae bifurcate.

14 – nymphal mandible with tusk (secondarily lost in Behningiidae); nymphal hind leg with dense stout pointed pectinate setae on inner side of femur and tibia (lost in Behningiidae); nymphal tergaliae of pairs II–VII bifurcate and pinnate; adult mesothorax with peculiar shape of sclerotisation; fore wing with characteristic sub-basal curvature of MP_2 and CuA (possibly all these characters are synapomorphies with Caenoidea being partly lost in Caenoidea).

15 – nymphal tergaliae of pairs II transformed into rounded-quadrangular gill-covers, tergaliae of pairs III–VI semicircular with numerous branching processes; nymphal fore wing buds fused with notum.

16 – in nymph: maxillary palp without muscles (sometimes reduced or lost); paraglossae fused with mentum; second segment of labial palp without muscle; pronotum with anterior V-shaped impression; fore wing buds fused behind notum; tergaliae of pairs II–VI consist of dorsal and a ventral lamellae, ventral lamella being bifurcate with each half bearing processes; in adults: fore wing with constant sub-basal cross-vein between CuP and 1A; gonostylus with second apical segment lost.

17 – nymphal maxilla specialised for filtering, with apical margin widened, truncate, bearing dense long setae and a row of pectinate setae; both costal and anal ribs on nymphal tergaliae lost; on fore wing CuP basally sharply divergent from CuA.

account is based on data from BOLTON (1921–22), F. CARPENTER (1988), KUKALOVÁ-PECK (1985) and WILLMANN (1999). Flight features are as hypothesised by WOOTTON & KUKALOVÁ-PECK (2000).

(b) Definition. Size large to very large. Head structure not sufficiently known. Thorax and abdomen rather slender, only slightly differing in width. Prothorax small, without winglet-like paranota (known in Syntonopteridae). Wing pairs similar in length but not so in width (hind pair wider basally) as well as in venation pattern, particularly comparing Palaeozoic true mayflies, and much less regular, with cross-veins irregular, in part forming meshwork. Costal brace comparatively small. MA sometimes (in *Bojophlebia*) not fused with RS but only connected to it by a short cross-vein. Legs with single strong claw (known in *Bojophlebia* only). Nymph unknown (those described by KUKALOVÁ-PECK 1985 as nymphs of *Bojophlebia* and *Lithoneura* are possibly terrestrial creatures similar to mayfly nymphs only in having three caudal filaments; they equally could be nymphs of a terrestrial winged insect or a giant lepismatid relative (Chapter 2.1a)).

(c) Synapomorphies in relation to the true mayflies and the dragonflies are not definitely known, for the one-clawed tarsus is possibly a synapomorphy of the genus *Bojophlebia* or family Bojophlebiidae, and not for the order as a whole.

(d) Range. Earlier Middle Carboniferous till Early Permian of Europe and North America.

(e) System and phylogeny. The order is monophyletic with Ephemerida and Libellulidea being synapomorphic with them in anastomoses between RS and MA, MP and CuA, and CuA and A_1, and possibly was their ancestor. It embraces 2 families Syntonopteridae and Bojophlebiidae, the latter supposedly apomorphic in its giant size and, possibly, in the single tarsal claw. In contrast, Syntonopteridae probably originated directly from the latter. This inference in based on the syntonopterid autapomorphies in the direct amalgamation of RS and MA sub-basally.

(f) History. Possibly the oldest fossil is *Aedoeophasma anglica* Scudder from the Middle Coal Measure (Westphalian B) of South Lancashire, England. *Bojophlebia procopi* KUKALOVÁ-PECK (Fig. 85), the only member of Bojophlebiidae, comes from the Westphalian C deposits at Vrapice in Central Bohemia (Czech Rep.). Members of Syntonopteridae are found mostly in Mazon Creek (Westphalian D of Illinois, USA) belonging to 3 species of *Syntonoptera* Handlirsch and *Lithoneura* Carpenter (Fig. 86), and in the early Artinskian (Early Permian) deposits in Obora (Czech Rep.; unnamed in KUKALOVÁ-PECK 1985: 940). The enigmatic *Miracopteron mirabile* Novokshonov from the later Early Permian of Tshekarda in the Urals (NOVOKSHONOV 1993a) is possibly the latest member of the order.

2.2.1.1.1.3. ORDER EPHEMERIDA Latreille, 1810.
THE TRUE MAYFLIES (=Ephemeroptera Hyatt et Arms, 1891 (s. l.); =Euephemeroptera Kluge, 2000)

N. Yu KLUGE AND N.D. SINITSHENKOVA

(a) Introductory remarks. An archaic and somewhat aberrant group of amphibiotic insects, the only one among the living insects which has retained a moulting winged stage (subimago), with some 2,000 living species. The present account is essentially original; general system and phylogeny of the order are discussed in KLUGE (1998, 2000), McCAFFERTY (2000).

(b) Definition. Size small to large (2–40 mm). Eyes and three ocelli always developed; at least in Euplectoptera eyes of male with enlarged dorsal portion (sometimes this dorsal portion secondarily reduced). Antennae with varying number of segments, in Euplectoptera very small, often with all segments fused. Mouthparts in Protereismatina chewing, in Euplectoptera imaginal mouth parts lost (only nonfunctional vestiges of nymphal mouth parts can be retained in subimago and imago). Pterothorax and wings homonomous in Protereismatina, or with mesothorax and fore wings much better developed (fore and hind wings coupled at flight) in Euplectoptera, to rudimentary or lacking hind wings in some taxa. Pterothorax structure described by KLUGE (1994). Wing rest position spread or raised-up and wing surface regularly fluted (see Chapter 2.1b). SC and R reaching or nearly reaching wing apex; RS separated from R and connected to MA (as in Libellulidea, in contrast to Triplosobida); RS forming three triads [see **(c)** below]; MA and MP forming one triad each. Cross-veins usually numerous and undetermined in number, sometimes few (in small or otherwise specialised forms). Legs slender,

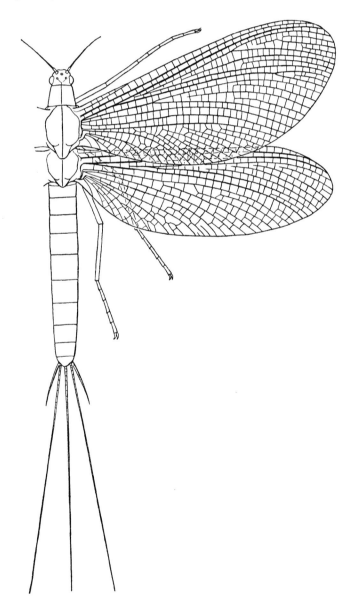

Fig. 84 *Triplosoba pulchella* (Brongniart) from the Late Carboniferous of Commentry in France, restored by N. Yu. Kluge (orig., based on BRONGNIART 1893 and CARPENTER 1963a): wing 25 mm long

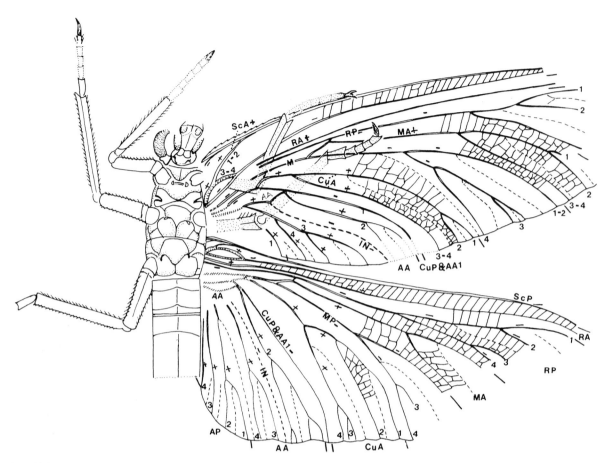

Fig. 85 *Bojophlebia procopi* Kukalová-Peck (Bojophlebiidae) from the Middle Carboniferous of Vrapice in Czechia from KUKALOVÁ-PECK 1985, including venational symbols); fore wing 180 mm long as preserved

usually with 5 tarsomeres [in Tetramerotarsata (or Baetidae *s. l.*) and some others with a lower number of tarsomeres]; at least in Euplectoptera first tarsomere fused with tibia [but in Pentamerotarsata (or Heptageniidae *s. l.*) secondarily separated]. At least in Euplectoptera claws initially unequal: one normal and pointed, another blunt [in some taxa both claws normal and pointed as a result of reversion]. Male fore legs usually particularly elongate, with tarsus able to turn backwards and couple with wing base of female. Abdomen 10-segmented, bearing long cerci and paracercus (paracercus sometimes short in post-Palaeozoic groups), and with male gonostyli modified into secondarily segmented forceps used in copulation. Living mayflies with paired male penes. Ovaries of living forms panoistic. Nymphs aquatic [see **(c)** below], strongly differing from adult. Antenna not shortened. Mouthparts chewing, essentially plesiomorphic, with mandible archaic and superlinguae well developed (in contrast to all other living Pterygota); maxilla with a single biting lobe (see Chapter 2.2.1.1b), maxillary and labial palps 3-segmented (sometimes with less number of segments) (mouth parts structure of Protereismatina unknown). Legs with two claws and 4–5 tarsomeres in Protereismatina, 1 claw and 1 tarsomere in Euplectoptera. Abdomen 10-segmented, segments 1–9 (maximum 1–7 in living suborders) with lateral tergaliae (see Chapter 2.2.1.1.1b) often modified as tracheal gills, apical segment with long multi-segmented cerci and often with long multisegmented paracercus. Malpighian tubes in living forms numerous, with peculiar coiled distal portions and very narrow ducts. Relict moult

from subimago (see Chapter 2.2.1.1.1b) to imago always retained, usually takes place in both sexes (in some short-living species with non-functional legs only male subimagoes moult to imagoes, while female subimagoes do not moult).

(c) Synapomorphies. Wings with costal brace of peculiar shape. Posterior branch of second triad of RS (see Chapter 2.2.1.1) forms a third triad. Ovipositor lost (unknown to other Ephemeridea). Nymph (unknown for other Ephemeridea) aquatic, with closed tracheal system, with abdominal tergaliae (relic serial homologues of thoracic wings) specialised as tracheal gills, initially with peculiar swimming specialisation: legs able to be pressed to body being stretched posteriorly; abdomen capable of undulating dorsoventral swimming movements; cerci and paracercus much shorter than in imago, with swimming setae – i.e. each cercus with a row of setae on inner side only, and paracercus with a pair of such rows of setae on lateral sides (primary swimming, or siphlonuroid setation); when swimming, cerci and paracercus with their setation functioning as a horizontal caudal flipper (this specialisation is initially present at least in Euephemeroptera – in Protereismatina and in many Mesozoic and recent Ephemerina, being lost or substituted by other specialisations in some Mesozoic and recent taxa).

(d) Range. Early Permian until the present, worldwide.

(e) System and phylogeny as shown at Fig. 83, with the following details not evident from the cladogram. The paraphyletic suborder Protereismatina Laméere, 1917 (or Permoplectoptera Tillyard, 1932) includes all true mayflies with homonomous wings: the Permian

Protereismatidae Laméere, 1917 (=Kukalovidae Demoulin, 1970) known for well preserved adults (Fig. 87) and nymphs (KUKALOVÁ 1968; Fig. 88); only adults Misthodotidae Tillyard, 1932 (=Eudoteridae Demoulin, 1954) are known for certain (Fig. 89) (systematic position of nymphs attributed to this family by TSHERNOVA 1965 is doubtful, cf. F. CARPENTER 1979, possibly they belong to *Phtharthus* Handlirsch); Palingeniopsidae Martynov (a single wing is known, Fig. 90), Oboriphlebiidae Hubbard et KUKALOVÁ-Peck and Jarmilidae Demoulin (both known as nymphs and have no distinct difference from Protereismatidae); Jurassic Mesephemeridae Laméere (only adults with

indistinct venation are known, their difference from other families is unclear). The Permian nymphal *Phtharthus netshaevi* Handlirsch, 1904 (=*Ph. rossicus* Handlirsch, 1904, syn. nov.) must also belong to Protereismatina; they have been described and figured by HANDLIRSCH (1906) but not quite correctly; actually these large (16–24 mm) nymphs have no wing pads (but mesonotum and metanotum equally developed, with a relief as in adult winged insects); leg structure is unknown; tergaliae I–VII lamellate and rugose, have normal attachment to posterior-lateral tergal angles (were incorrectly described as pinnate with ventral attachment), tergaliae VIII–IX uncertainly present; cerci and

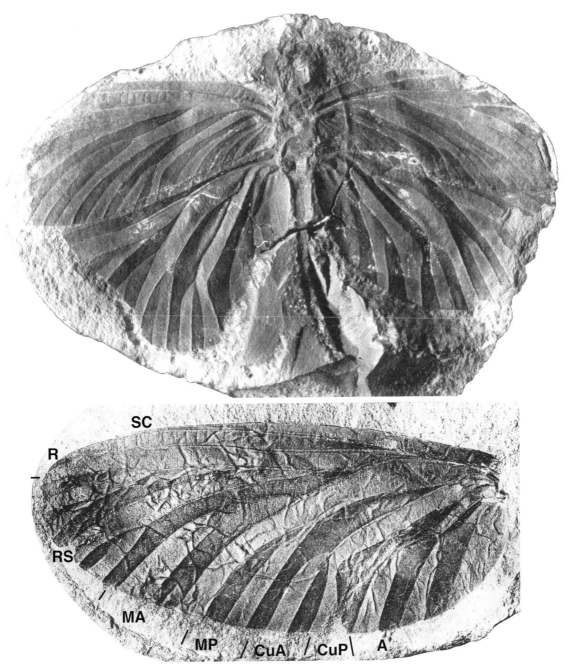

Fig. 86 *Lithoneura lameeri* Carpenter (wing span 64 mm as preserved from WILLMANN 1999, courtesy of *Paläontologische Zeitschrift*) and hind wing (85 mm long) of *L. mirifica* Carpenter (Syntonopteridae), both from the Middle Carboniferous of Mazon Creek in USA (from CARPENTER 1943a)

paracercus with primary swimming setation only [see **(c)** above]. The Triassic Mesoplectopteridae Demoulin (known from a single nymph ascribed to Ametropodidae by DEMOULIN 1955) can be also placed to Protereismatina. This nymph has hind wing pads which are reduced, so its structure is intermediate between typical Protereismatina and Euplectoptera. TSHERNOVA (1980) placed an undescribed Permian adult to this family, but it belongs more certainly to Protereismatidae.

All living mayflies and the majority of Cainozoic and Mesozoic ones form the taxon Euplectoptera Tillyard, 1932. It is characterised by heteronomous wings (hind wings not exceeding half of the fore wing length and coupled with them during flight unless rudimentary or lost) and by some nymphal characters which are less reliable because the permoplectopteran nymphs are insufficiently known.

Euplectoptera includes two suborders, Baetiscina and Ephemerina, which differ in the position of the tornus (the posterior angle of fore wing is present only in Euplectoptera). In Baetiscina Banks, 1900 (or Posteritorna Kluge *et al.* 1995), tornus is posteriad 1A, while in Ephemerina Latreille, 1810 (or Anteriotorna Kluge, 1993) it is always between CuA and CuP, independent of wing shape and of hind wing size. When the hind wing is rudimentary or absent, the tornus moves basad or disappears, but not basad of CuP. The above difference in tornus position may indicate its independent acquisition by basal Baetiscina and

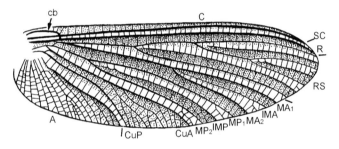

Fig. 87 Hind wing of the primitive mayfly, *Protereisma latum* Sellards (Protereismatidae), from the Early Permian of Elmo in USA (original by N.Yu. Kluge basing on CARPENTER 1933)

Ephemerina, which in turn might suggest a double origin of wing heteronomy in mayflies. The suborder Baetiscina is additionally characterised by numerous unique autapomorphies (KLUGE 1998) and unites the small living families Baetiscidae Banks and Prosopistomatidae Laméere. The suborder Ephemerina includes the majority of known mayflies and is divided into Baetomorpha and Bidentiseta.

The paraphyletic infraorder Baetomorpha Leach, 1815 (or Tridentiseta Kluge *et al.* 1995) includes superfamilies Baetoidea (or Tetramerotarsata) and Siphlonuroidea (including *Ametropus* Albarda). Siphlonuroidea Ulmer, 1920 (1888) includes several families (KLUGE *et al.* 1995) characterised only by plesiomorphies: wings are not modified, anteritornal, nymphs usually with the same swimming specialisation groundplan as in Ephemerida [see **(c)** above]. Isonychiidae and Coloburiscidae are transferred to Bidentiseta Eusetisura (see below). The fine structure of many extinct siphlonurid-like mayflies (Figs. 91, 92) being unknown, we consider them unplaced Siphlonuroidea.

The Jurassic and Cretaceous families Hexagenitidae Laméere, 1917 (=Paedephemeridae Laméere, 1917, Stenodicranidae Demoulin, 1954, Ephemeropsidae Cockerell, 1924; imagoes and nymphs are known: Fig. 93), Epeoromimidae Tshernova and Mesonetidae Tshernova (both known only as nymphs, Figs. 94, 95), whose number of dentisetae and other important infraordinal characters (see below) are unknown, are classified as unplaced Baetomorpha. Epeoromimidae were regarded ancestral to Heptageniidae (TSHERNOVA 1980), but no synapomorphies of these families are known. Mesonetidae were united with Ephemerellidae (TSHERNOVA 1962) or with Leptophlebiidae (TSHERNOVA 1980) but a recent study has failed to confirm their close relation (KLUGE 1989). The Late Cretaceous *Cretoneta* Tshernova, originally placed to Leptophlebiidae (TSHERNOVA 1971), actually belongs to Siphlonuroidea (KLUGE 1993).

Baetoidea Leach, 1815 (or Baetidae Leach, 1815 *s. l.*, or Tetramerotarsata Kluge, 1997) is characterised by four-segmented tarsi of adults and other autapomorphies, and is divided into *Siphlaenigma* Penniket and Liberevenata. Baetidae Leach, 1815 *s. l.* (or Liberevenata Kluge, 1997) is characterised by a transformation of fore wing vein MA$_2$ into an intercalary vein, reduction of penis and some other

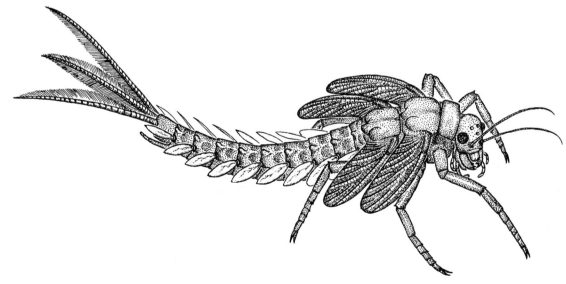

Fig. 88 Primitive mayfly nymph *Kukalová americana* Demoulin (Protereismatidae), from the Early Permian of Wellington Formation in Oklahoma, USA (original by N. Kluge based on KUKALOVÁ-PECK 1978); length of wing pad 5.3 mm

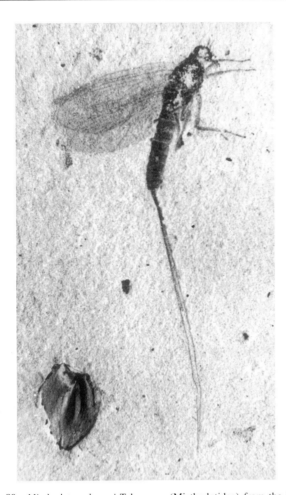

Fig. 89 *Misthodotes sharovi* Tshernova (Misthodotidae) from the Early Permian of Tshekarda in Urals (PIN 1700/388, photo by D.E. Shcherbakov); length of body 10 mm

Fig. 90 *Palingeniopsis praecox* Martynov (Palingeniopsidae) from the Late Permian of Soyana in NW Russia (photo by D.E. Shcherbakov); fore wing 30 mm long

Fig. 91 Nymphal *Stackelbergisca sibirica* Tshernova (Siphlonuridae) from the Late Jurassic of Uda in Siberia (PIN 3053/33, photo by D.E. Shcherbakov); body 18 mm long

autapomorphies. It includes the Late Cretaceous *Palaeocloeon* Kluge and Baetidae Leach, 1815 *s. str.* (or Turbanoculata Kluge, 1997), which is characterised by peculiar turban eyes in the male, and by other autapomorphies (KLUGE 1997b).

In Bidentiseta Kluge, 1993 the number of dentisetae (specialised biting setae in inner-dorsal row of nymphal maxilla) is reduced to 2, in contrast to 3 in the groundplan of Baetomorpha and still more in

Baetiscomorpha (KLUGE 1998). Bidentiseta covers two infraorders, Heptageniomorpha and Ephemeromorpha.

The infraorder Heptageniomorpha Needham, 1901 (or Branchitergaliae Kluge, 1998) is divided into Heptagennota and Eusetisura. The superfamily Heptagenioidea Needham, 1901 (or Heptagennota Kluge, 2000, or Heptageniidae Needham, 1901 *s. l.*) embraces *Pseudiron* McDunnough and Pentamerotarsata. Characteristic of Heptageniidae *s. l.* (or Pentamerotarsata Kluge, 2000) are the restored articulation between the adult tibia and first tarsal segment and some other apomorphies. The taxon is further divided into *Arthroplea* Bengtsson and Heptageniidae *s. str.* (or Radulapalpata Kluge, 2000) which is characterised by a peculiar scraping specialisation of the larval mouthparts (KLUGE 1998, 2000). The Oligoneurioidea Ulmer, 1914 (or Eusetisura Kluge, 1998) is composed of Coloburiscidae Edmunds, Isonychiidae Burks and Oligoneuriidae Ulmer, the latter

Fig. 92 Nymphal *Proameletus caudatus* Sinitshenkova (Siphlonuridae) from the Late Jurassic or Early Cretaceous of Glushkovo Formation in Transbaikalia (PIN 3015/430), photo by D.E. Shcherbakov; body 14 mm long

adaptations for burrowing: fore coxae with additional ventral articulations, fore and middle femora curved anteriad, frons with a projection (in Euthyplociidae a similar projection is present on the clypeus), etc. Fossoriae can be divided into Ichthybotidae Demoulin, Ephemeridae Latreille (characterised by the bases of the tergaliae VII shifted anteriorly), Behningiidae Motas et Bacesco and a holophyletic taxon Cryptoprosternata Kluge, 2000. Characteristic of Cryptoprosternata is a further burrowing specialisation of the nymphal fore legs: additional coxal articulations brought together, trochanter turned to anterior side of femur, femur shape peculiar with proximal convexity, etc. Cryptoprosternata comprises Palingeniidae Albarda, *s. l.* (incl. *Pentagenia* Walsh) and Polymitarcyidae Banks, *s. l.* (incl. Campsurinae). Short-lived imago is gained independently in both families (*Pentagenia* retains plesiomorphic Ephemeridae-like structure of long-lived adult).

The superfamily Caenoidea Newman, 1853 (or Caenotergaliae Kluge, 2000) is characterised by a unique specialisation of the nymphal tergaliae. Caenoidea includes Neoephemeridae Traver which retains the wing venation and some other features as in Ephemeroidea, and Caenidae Newman (or Caenoptera Kluge, 2000) with reduced hind wings, fan-like fore wings and unique autapomorphies in the pterothorax structure (KLUGE 1992).

The superfamily Ephemerelloidea Klapalek, 1909 includes Ephemerellidae Klapalek, *s. str.* with tergaliae II lost, and Tricorythidae Lestage, *s. l.* characterised by a row of setae that crosses the anterior surface of the larval fore femur and continuous on its outer margin. Some nymphal Tricorythidae have tergaliae II operculate, a convergence to Caenoidea. Adult Tricorythidae sometimes have a pterothorax and wing structure strikingly similar to Caenidae, a resemblance that hardly resulted from convergence. Analysis of known characters does not allow us to build a non-conflicting cladogram of Ephemeroidea, Caenoidea and Ephemerelloidea (KLUGE 1992, 1997a); so these superfamilies are accepted as tentative.

The superfamily Leptophlebioidea Banks, 1900, comprising the single family Leptophlebiidae, was previously regarded as very ancient, because various unrelated Mesozoic mayflies were attributed there. Imaginal features of Leptophlebioidea useful for palaeontology are listed by KLUGE (1993).

The Mesozoic family Torephemeridae Sinitshenkova, 1989 has been ascribed to Ephemeroidea (SINITSHENKOVA 1989), but the absence of good synapomorphies makes its position rather unclear: Kluge even doubts if it belongs to mayflies. Systematic position of some other mayfly taxa is unclear as well. This concerns the Jurassic Aenigmephemeridae Tschernova and Triassic Litophlebiidae Hubbard et Riek, 1978 (=Xenophlebiidae Riek, 1976) described from separate wings (TSHERNOVA 1968, RIEK 1976a). Among the nymphal fossils, the Triassic *Triassomachilis* Sharov, 1948 described as Thysanura and known from several nymphs of poor preservation state (SHAROV 1948), Jurassic *Turphanella* Demoulin described in Ephemerellidae (DEMOULIN 1954), *Mesogenesia* Tschernova (as Palingeniidae, TSHERNOVA 1977), *Clephemera* Lin, 1986 (as Ephemerellidae, LIN 1986), and Cretaceous *Leptoneta* Sinitshenkova (as Leptophlebiidae, SINITSHENKOVA 1989) should all be considered unplaced within the Euplectoptera. Systematic position of *Blasturophlebia* Demoulin, *Brevitibia* Demoulin, *Cronicus* Eaton, *Siphloplecton* Demoulin, *Xenophlebia* Demoulin from the Baltic amber (DEMOULIN 1968), and *Philolimnias* Hong from the Eocene Fushun amber (HONG 1979) is not clear and their re-examination is necessary. The Mesozoic *Huizhougenia* Lin originally placed in the Siphlonuridae (LIN 1980), most probably belongs to the dragonfly family Aeschnidiidae and not to mayflies at all.

being further divided into *Chromarcys* Navas and Oligoneuriinae *s. str.*, short-lived mayflies with strongly reduced wing venation.

The infraorder Ephemeromorpha Latreille, 1810 (or Furcatergaliae Kluge, 1998) is characterised by a number of apomorphies (KLUGE 1997a, 1998, 2000) and composed of the superfamilies Ephemeroidea, Caenoidea, Ephemerelloidea and Leptophlebioidea.

Ephemeroidea Latreille, 1810 (or Pinnatitergaliae Kluge, 2000) possesses a number of apomorphies, but it is quite possible that these are synapomorphies with Caenoidea partly retained in Neoephemeridae. Ephemeroidea are subdivided into holophyletic families Potamanthidae Albarda, Euthyplociidae Lestage, and Fossoriae Kluge, 2000. Additionally, there are the Cretaceous genera *Australiphemera* McCafferty, *Microphemera* McCafferty and *Pristiplocia* McCafferty whose diagnoses does not permit them to be separated from other Ephemeroidea, and the Cretaceous family Palaeoanthidae Kluge, characterised by a unique combination of non-unique characters (KLUGE 1993). Autapomorphies of Fossoriae are unique larval

Fig. 93 Giant mayfly, *Ephemeropsis melanurus* Cockerell (Hexagenitidae), from the Late Cretaceous of Baissa in Siberia: adult (restored as in life by A.G. Ponomarenko and Kira Ulanova; photos of adult and nymphal fossils by D.E. Shcherbakov (PIN 1668/216, 217, adult fore wing 40 mm long) and of a nymph (PIN 3064/3313, body 37 mm long)

(f) History (unless cited, the source references can be found in HUBBARD 1987). The oldest, Permian true mayflies, as well as the majority of Triassic ones, belonged to the suborder Protereismatina. The oldest findings from the early Artinskian (Early Permian) of Obora in Moravia (Czechia) all represent nymphs and belong to several species of *Jarmila* Demoulin and *Oboriphlebia* Hubbard et Kukalová-Peck, each attributed to a family of its own by HUBBARD & KUKALOVÁ-PECK (1980). Later in the Permian a number of protereismatid genera and species are found, both as adults and nymphs. The majority of these fossils come from the later Artinskian at Elmo in Kansas, USA, and other North American locality of similar age. Others are recorded later in the Early Permian (Kungurian of Tshekarda in Urals) and in the Late Permian. These are found in the Kazanian of Tikhiye Gory in the Urals and of the northern Rhine in Germany (as Zechstein 1, Kupferschiefer), and in the Tatarian of Kargala in South Urals and of Middle Beaufort Series in South Africa. Some of the above deposits have also yielded members of another family Misthodotidae, represented by 7 species of the only genus in the Artinskian strata in Kansas and Oklahoma and in the Kungurian in Tshekarda, Urals. The third Palaeozoic family Palingeniopsidae

is only known from a single fossil from the Kazanian stage of Soyana in NW Russia.

The Triassic mayflies are the least known (SINITSHENKOVA 2000a). No Early Triassic mayfly has as yet been found, while the Middle and Late Triassic mayfly fauna presents a mixture of Permian and Mesozoic forms. A rich mayfly fauna containing 7 species has been discovered in the Middle Triassic (Anisian) *Voltzia* Sandstone in Vosges, France (MARCHAL-PAPIER 1998), though the only described ones are the nymphs of *Mesoplectopteron* Handlirsch (Mesoplectopteridae). The most abundant nymphs belong to the burrowing type, several nymphs have similarities with the Permian Protereismatidae, and the isolated wings belong to the Permian Misthodotidae. A similar mayfly assemblage is found in the Middle Triassic Bundsanstein deposits of Mallorca I., Spain (COLOM 1988). *Triassomachilis* Sharov from the Ladinian Bukobay Formation of Nakaz in South Urals, described as a machilid (SHAROV 1948) is a mayfly similar to *Mesoneta* Brauer, Redtenbacher et Ganglbauer 1889 (SINITSHENKOVA 2000a). *Turingopteryx* Kuhn was described from the Middle Triassic of Germany as a dragonfly (KUHN 1937) but its wing venation is characteristic of mayflies. A small fragment of a mayfly wing apex was found in the Middle Triassic of

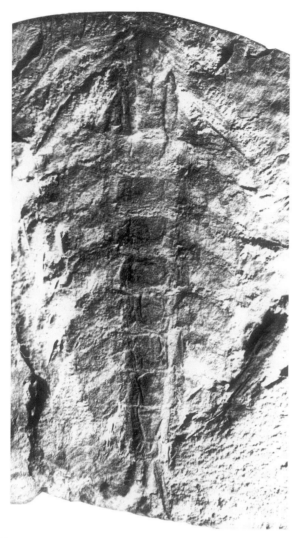

Fig. 94 Nymphal *Epeoromimus kazlauskasi* Tshernova (Epeoromimidae) from the Early Jurassic of Chernyi Etap in Siberia (2245/1, photo by D.E. Shcherbakov), body length 13 mm

Fig. 95 Nymphal *Mesoneta antiqua* Brauer, Redtenbacher et Ganglbauer (Mesonetidae) from the Early Jurassic of Ust-Baley in Siberia (PIN 1873/1, photo by D.E. Shcherbakov); body 8 mm long

Cheshire in England (THOMPSON 1965). A fore wing fragment described as *Montralia* Via from the Late Ladinian of Spain and also ascribed to dragonflies, was subsequently transferred to the mayfly family Mesephemeridae (VÍA & CALZADA 1987). A very interesting mayfly with long and narrow wings and very long legs is discovered in the Middle–Late Triassic deposits of Meride, Ticino canton, Switzerland (KRZEMIŃKI & LOMBARDO 2001). The Late Triassic mayfly *Litophlebia* Hubbard et Riek (Litophlebiidae = Xenophlebiidae, Fig. 96) has been recorded from the Molteno Formation in South Africa (RIEK 1976b). The mayflies from the earlier Middle Triassic of Siberia belong to the Mesozoic families Mesonetidae and Torephemeridae; the Late Triassic mayflies from Garazhovka (Ukraine) represent the Mesozoic genera *Mesoneta* Brauer, Redtenbacher et Ganglbauer and *Mesobaetis* Brauer, Redtenbacher et Ganglbauer (SINITSHENKOVA 2000a).

The post-Triassic mayfly fauna is composed solely by Euplectoptera, with the only exception being the protereismatinan family Mesephemeridae. The latter is known from two species of the type genus that are not uncommon in the famous Late Jurassic lithographic limestone

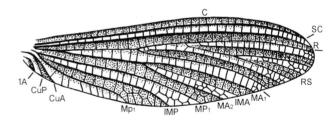

Fig. 96 Fore wing of *Litophlebia optata* (Riek) (Litophlebiidae) from the Late Triassic Molteno Formation in South Africa (original by N. Kluge)

of Solnhofen in Bavaria, Germany, and the monotypic genus *Palinephemera* Lin from the Jiaoshang Formation (Early Cretaceous of Hunan Province, South China; LIN 1986). Otherwise post-Triassic mayflies belong almost exclusively to the infraorder Baetomorpha.

In the Jurassic the extinct families were obviously dominant. The most characteristic of the Jurassic are the extinct Epeoromimidae and

Mesonetidae which were widely frequent in Siberia and Mongolia. Toward the Early–Middle Jurassic boundary another extinct family Hexagenitidae has appeared in Transbaikalia and Mongolia (*Siberiogenites* Sinitshenkova), to become common in Solnhofen (*Hexagenites* Scudder) and particularly in the Early Cretaceous, where it was widespread and often forming mass burials. The earliest Siphlonuroidea are described from South Siberia and Mongolia (*Mesobaetis* Brauer, Redtenbacher et Ganglbauer, *Mogzonurella* Sinitshenkova, *Mogzonurus* Sinitshenkova, Sinitshenkova 1985a). The burrowing mayflies are represented in the Late Jurassic by Torephemeridae (*Archaeobehningia* Tshernova). *Geisfeldiella* Kuhn from the Early Liassic of Nordfranken, Germany, originally described as a member of Lepidoptera, is in fact a mayfly.

In the Early Cretaceous Epeoromimidae and Mesonetidae become rare while Hexagenitidae were usually the only dominant family in mayfly assemblages. They are recorded in the Ukraine and Algeria (*Hexameropsis* Tshernova et Sinitshenkova), Transbaikalia (*Ephemeropsis* Eichwald), Mongolia (*Ephemeropsis* and *Mongologenites* Sinitshenkova), China (*Ephemeropsis*), and Brazil (*Protoligoneuria* Demoulin). The burrowing mayflies occur in Mongolia (Torephemeridae), Spain (Pothamantidae and undescribed Euthyplociidae and Leptophlebiidae; Peñalver *et al.* 1999), and Brazil (*Australiphemera* McCafferty, *Microphemera* McCafferty and *Pristiplocia* McCafferty). One more extant family that appeared in the Early Cretaceous is the Oligoneuriidae (*Colocrurus* McCafferty in the Crato Formation of Brazil, McCafferty 1991). Palaeoanthidae are known from the Late Cretaceous where they are found together with Baetidae (*Palaeocloeon* Kluge), unplaced Siphlonuroidea (*Cretoneta* Tshernova) and an undescribed species of Hexagenitidae in the Taimyr amber (Kluge 1993, 1997). Unplaced Siphlonuroidea (*Promirara* Jell et Duncan, *Australurus* Jell et Duncan, *Dulcimanna* Jell et Duncan) are abundant in the Early Cretaceous Koonwarra bed in Australia (Jell & Duncan 1986). The earliest Leptophlebiidae (*Conovirillus* McCafferty) and Baetidae are found in the Early Cretaceous Lebanese amber (McCafferty 1997). The Late Cretaceous (Turonian) amber of New Jersey (North America) has yielded Polymitarcyidae, Australiphemeridae, and the earliest records of Heptageniidae and Ametropodidae (Sinitshenkova 2000b) as well as Leptophlebiidae (Peters & Peters 2000). Thus Cretaceous mayfly assemblages retain some extinct families but the share of extant ones becomes considerable.

In the Eocene, mayflies are found in Europe, North America and Myanmar (former Burma). Heptageniidae and Neoephemeridae (*Neoephemera* McDunn) are recorded in the Middle Eocene of Republic in western USA (S. Lewis & Wehr 1993, Sinitshenkova 1999a). The Baltic amber mayflies are common and represented by several extant families: Heptageniidae with extant species *Heptagenia fuscogrisea* (Retzius) and several extinct species, Siphlonuroidea with several genera including extant *Siphlonurus* Eaton, Leptophlebiidae with several species including two of the extinct subgenus *Leptophlebia* (*Paraleptophlebia*), and Ephemerellidae. The only fossil representative of the family Prosopistomatidae, *Myanmarella rossi* Sinitshenkova is from the Burmese amber of debatable Late Cretaceous or Palaeogene age (Sinitshenkova 2000c). However, Kluge (unpublished) regards this species as belonging to Anteritorna and probably Liberevenata, rather than to Prosopistomatidae and to Posteritorna in general. Mayfly nymphs have been reported from the Middle Eocene of Messel in Germany (Richter & Krebs 1999).

Ephemeridae is the most abundant mayfly family in the Oligocene. It has been recorded from the terminal Eocene or the earliest Oligocene of Ruby River in Montana, USA, along with the mayflies tentatively attributed to Neoephemeridae and Isonychiidae (S. Lewis 1977, 1978). Six species of *Ephemera* dominate in the basal Oligocene of Florissant in Colorado, USA; additionally found there are *Siphlurites* Cockerell (presumably unplaced Siphlonuroidea) and *Lepismophlebia* Demoulin (attributed to Leptophlebiidae). One more ephemerid species, *Litobrancha palaearctica* McCafferty et Sinitshenkova, is described from Bol'shaya Svetlovodnaya in the Russian Far East (McCafferty & Sinitshenkova 1983).

Mayflies are rare in the Miocene. The nymph of *Miocenogenia* Tshernova (Heptageniidae) has been found in Belyi Yar in West Siberia. From the Early Miocene of Japan an unidentified ?Ephemeridae and a nymph of *Caenis* sp. (Caenidae) are pointed out (Fujiyama 1985, Fujiyama & Nomura 1986). Several specimens of mayflies have been identified in the Early Miocene Dominican amber (Poinar 1992a). Middle Miocene Heptageniidae (*Heptagenia* and *Ecdyonurus*) and Leptophlebiidae are recorded in the Middle Miocene Shanwang Formation of Shandong, China (Zhang 1989). Another heptageniid nymph *Pseudokageronia* Masselot et Nel is described from the Late Miocene of Murat in France (Masselot & Nel 1999). The burrowing *Asthenopodichnium* Thenius from Austria is regarded to have been made by nymphal Asthenopodini (Polymitarcyidae), though otherwise this unusual tropical mayfly group is not recorded in Europe. A wing fragment described from California (North America) as *Aphelophlebodes* Pierce is tentatively attributed to the Ephemeroptera (Demoulin 1962). Identification of nymphs *Cloeon* Leach (Baetidae) from the Pliocene of Australia is considered doubtful (Kluge 1997).

2.2.1.1.2. SUPERORDER LIBELLULIDEA Laicharting, 1781. ORDER ODONATA Fabricius, 1792. THE DRAGONFLIES

A.P. RASNITSYN AND L.N. PRITYKINA

(a) **Introductory remarks**. Comparatively small group of conspicuous and popular insects (Figs. 97, 98). Phylogeny of the order is based mainly on the compendium by Bechly (1998, 1999). However, his strictly cladistic taxonomy is not followed here [Chapter 1.1(c,d)]. History of the order follows in part the ideas by Pritykina (1980, 1981, 1989, and unpublished), with additional records taken from Bechly (1999) and references therein.

(b) **Definition**. Body size moderate to giant, usually large. Mouthparts chewing, eyes large, antennae short setiform. Synthorax oblique, with pleural sutures directed posterodorsally, with left and right mesepisterna fused widely above pronotum, except Namurotypidae, Erasipteridae and Eogeropteridae (the latter family is unique in having pronotal winglets: Wootton *et al.* 1998, Wootton & Kukalová-Peck 2000). Pleural and paracoxal sutures usually much deepened in Meganeurina, with dorsal sections adjacent and subparallel, separated only by a convex fold (Fig. 99); in Libellulina (former Anisoptera + Anisozygoptera) and Calopterygina (former Zygoptera) pleural and paracoxal sutures levelled (not running within deep folds), meeting at wide angle at about lower 1/4 of the former. Pterothoracic nota small, dorsal longitudinal muscle reduced in size and function of indirect wing depressor (wing motor mainly direct). Femora and tibiae with predatory spines, tarsus 4–5-segmented in Meganeurina, otherwise 3-segmented. Wings more or less homonomous, long and narrow, permanently spread or folded over

Fig. 97 *Arctotypus* sp. (Meganeuridae) from the Late Permian of Soyana in NW Russia restored as in life by A.G. Ponomarenko and Kira Ulanova

abdomen at rest, with upper surfaces inside, hind wing slightly widened basally except in (sub)petiolate wings. Wing articulation apparatus simplified, with most sclerites fused into two large ones (already present in Erasipteridae, personal observation on the type of *Erasipteroides valentini* Brauckmann). Wing surface regularly fluted due to alternating convex and concave veins, with dense archedictyon or abundant cross-veins. Venation rather regular, moderately rich to rather poor but covering all wing surface, complicated by kinks and (temporary) fusion between main vein stocks (MA with RS, M stock with CuA, CuP with 1A and then with CuA) resulted in many problems in homologisation that existed until plesiomorphic eogeropterid wings were discovered (RIEK & KUKALOVÁ-PECK 1984). Flight functionally four-winged, in-phase, posteromotoric, at least in living forms can to be changed into anti-phase at the insect's will, in order to reduce body vertical oscillation and so to permit taking precise aim when hunting (GRODNITSKY 1999). Abdomen long, usually cylindrical (wide and probably depressed in *Erasipterodes* Brauckmann, secondarily depressed in some advanced Libellulidae), 10-segmented. Except for at least Namurotypidae (BRAUCKMANN & ZESSIN 1989, BRAUCKMANN 1991), primary male genitalia lost and replaced by secondary ones at segments 2 and 3 (for Meganeurina found on segment 2 in ditaxineuroids, Fig. 100). Additionally, in non-meganeurinan males with non-petiolate wings, abdominal segment 2 bearing lateral auriculi correlated with excised hind wing base. Unless lost, ovipositor with free, cutting gonapophyses 8–9 and simple, sheath-like coxite 9 (2nd valvifer), bearing stylus. Cercus simple (segmented in *Namurotypus*,

BRAUCKMANN & ZESSIN 1989). Development probably terrestrial in Meganeurina (judging from extreme rarity of immature fossils, with those supposedly attributed to immature meganeurinans showing no aquatic adaptations, viz. *Titanophasma fayoli* Brongniart and the fossil figured by KUKALOVÁ-PECK 1983), otherwise aquatic or secondarily subterrestrial. Non-meganeurinan immatures generally similar to adults except wingless, often more stout, with labium modified into elongate grasping organ (mask), wing pads fore margin turned up, and with various aquatic adaptations, including rectal branchial chamber, external gills, anal swimming flaps, etc. Development gradual. Both adult and immature (nymph) predatory, adult aerial, hardly capable of walking locomotion.

(c) **Synapomorphies**. Pterothoracic terga and, hence, indirect wing depressor muscles much reduced; wing articulation with two large composite sclerites (anterior costal plate and posterior radio-anal plate); correlated with a double pleural joint (fulcrum), caused by double pleural sulcus; wings long, narrow, costalised (with C, SC, R straight, near and parallel to each other), with fore margin straight or slightly concave; MP not branched (reversed in Triadophlebioidea); 1A with Z-like kink before fusion to CuP; abdomen elongate, relatively slender (the list is evidently incomplete due to incomplete knowledge of structures other than wings in the oldest and least advanced fossils).

(d) **Range**. Earliest Middle Carboniferous (Namurian B) till now, worldwide (including Antarctica, at least in Jurassic).

(e) **System and phylogeny**. The dragonfly cladogram by BECHLY (1999) is accepted here unlike the system which employs paraphyletic

Fig. 98 *Hemeroscopus baissicus* Pritykina from the Early Cretaceous of Baissa in Siberia, restored as in life by A.G. Ponomarenko and Kira Ulanova

Fig. 99 *Arctotypus* sp. (Meganeuridae) from the Late Permian (Ufimian) Upper Solikamsk Subformation of Mogilnikovo in Middle Urals, Russia (PIN 3473/2, photo by D.E. Shcherbakov); impression of lateral meso- and metathoraces: pleural and paracoxal sutures (ps and pcs respectively)

taxa and different, more traditional rankings (Fig. 101). That cladogram is taken here as the most recent and comprehensive, in spite of some evident errors in it. I particularly mean the case of Hemeroscopidae–Sonidae–Proterogomphidae (BECHLY *et al.* 1998, see also p.11), when two fossil species with well associated and comparatively well described adults and nymphs, *Hemeroscopus baissicus* Pritykina (see Fig. 98) and *Sona nectes* Pritykina were attributed to three unrelated families: adult *H. baissicus* to Hemeroscopidae (basal in libelluloid clade), adult *S. nectes* to Proterogomphidae (gomphoid clade), and both immature *S. nectes* and *H. baissicus* to Sonidae (basal in advanced Libellulina, supposedly synonym of Aeschnidiidae). This construction is unacceptable for several reasons. Firstly, association of the adults and immatures described as *H. baissicus* is very well supported, and evidence in favour of conspecifity of *S. nectes* is also convincing (Chapter 1.2c). Secondly, the libelluloid spoon-like mask is often seen in *H. baissicus* nymphs (Fig. 102); being rather soft, the mask is often a subject of deformation and sometimes looks almost flat: this might explain a claim in the above publication that this material represents a flat gomphoid mask. These observations definitely falsify the hypothesis of BECHLY *et al.* (1998) about the basal phylogenetic position of the nymphs in question. An additional problem arises with the hypothesis of the aeschnidiid nature of both nymphs: Aeschnidiidae are characteristically a seashore group which is rare in deeply intracontinental environments (PRITYKINA 1993), so it would be strange to encounter their nymphs in great numbers in lakes thousands of kilometres from the nearest sea, and in the absence (in West Mongolia) or extreme rarity (in Transbaikalia) of aeschnidiid adults.

NEL *et al.* (1999a) described the monotypic family Lapeyriidae from the Late Permian of Lodève, France, as the sister group of Nodialata (all dragonflies except Meganeuroidea on Fig. 101). However, Campylopteridae take the same position and their relationships are not

Fig. 100 Undescribed Ditaxineuroidea from the Early Permian of Tshekarda in Urals (PIN 1700/325, photo by A.P. Rasnitsyn) showing secondary male genitalia (arrow); left hind wing 27 mm long

discussed, thus preventing us from introducing the new family into the cladogram.

For an alternative view on the earliest stages of dragonfly evolution based on wing functional morphology see PFAU (2000).

(f) History. The Carboniferous dragonflies all belonged to Meganeurina. Being the largest aerial predators of the Palaeozoic, they needed a large open space for hunting and might rely, among other prey, on the numerous (during the Carboniferous) dictyoneuridans.

The oldest, Namurian Meganeurina are classified in Erasipteridae and Namurotypidae (from the Namurian B of Hagen-Vorhalle in Germany and, only Erasipteridae, from the Namurian C of Horní Suchà in Czechia). Slightly younger but less advanced Eogeropteridae (Fig. 103) come from the Westphalian(?) deposits of Malanzán Formation in Argentina. Eogeropteridae and Namurotypidae have not appeared until later, while Erasipteridae survived in the Westphalian of Europe, being far subordinate to Meganeuridae there, and possibly (based on wing fragments) also in the Late Carboniferous of Siberia. In the Westphalian, Meganeuridae was also dominant in North America where endemic Paralogidae are found as well. In the Late Carboniferous (Stephanian) nothing is known except for the unique, enigmatic Campylopteridae, a few meganeurid genera in Europe, and problematic Erasipteridae in Siberia. Because of the scarcity of extra-European records it can be only preliminarily inferred that Meganeuridae preferred the wetland tropical forests, at least in Euramerica.

Since the Permian, the general appearance of the dragonfly assemblages has changed radically. It was not until after the Artinskian

time when the ditaxineuroids (former Protanisoptera) and protomyrmeleontoids (former Archizygoptera, Fig. 104) appeared and took the dominant position in the fauna. Ecological diversity of the dragonfly assemblages has also increased. The complete picture of development of the order is difficult to attain, because richer data exist only for the Artinskian fauna of North America (mostly in Kansas) and for the faunal succession from the Kungurian until Tatarian in the west of Angara continent.

Early Permian meganeuroids are represented by Meganeuridae and, only in North America, by comparatively rare and not diverse Paralogidae. Meganeuridae became several times more diverse than in the Late Carboniferous, and taxonomically, at the generic level, they had practically nothing in common with that fauna. The largest known insect (the North American genus *Meganeuropsis* Carpenter, with its wingspan up to 71 cm) was among these insects, but otherwise the Early Permian Meganeuridae had nothing special comparing the Late Carboniferous ones, at least concerning their wing venation. The Late Permian meganeuroids are known from several meganeurids in the Kazanian of north European Russia (Figs. 97, 105) and Kargalotypidae in the Tatarian of South Urals (though, Pritykina places *Kargalotypus* Rohdendorf to Triadotypidae).

Judging from their wing structure, ditaxineuroid dragonflies were better fliers compared to the Meganeurina. They were probably able to escape from them successfully on the wing, and also to populate more diverse biotopes. As adults, these insects represented analogues of the Meso- and Cainozoic dragonflies (as opposed to the damselflies). During the Early Permian, the ditaxineuroids were represented by the sole family Ditaxineuridae which also persisted into the Late Permian, at that time accompanied by the other ditaxineurinean families Permaeschnidae, Polytaxineuridae, Callimokaltaniidae, and Kaltanoneuridae. Late Permian ditaxineuroids were widespread over the Earth, including Australia and South America.

The protomyrmeleontoids, like true damselflies, probably appeared first as inhabitants of plant thickets, where their petiolate wings should be adaptively superior, the character appearing many times independently both within and beyond Libellulidea. Leg structure indicates that they grasped their weakly mobile prey from their substrate. For the Early Permian, only a few genera of Kennedyidae are known from the Artinskian deposits in North America. It was the Late Permian time when the group was flourishing, with Permolestidae (see Fig. 104), Permepallagidae, and Permagrionidae appearing in addition to Kennedyidae, and with geographic range extending from North Russia (Arkhangelsk Region) to the Falkland Islands.

Both ditaxineuroids and kennedyioids did not show significant specialisation during the Permian, and their Artinskian representatives do not seem primitive compared to the Tatarian (latest Permian) ones.

In general, even in the Palaeozoic it appears that odonatans had managed to enter most of the adaptive zones that are occupied by modern dragonflies and damselflies, and they had also developed similar flight mechanics earlier on (WOOTTON & KUKALOVÁ-PECK 2000).

The Triassic dragonflies are found only in the Middle and Late Triassic deposits, mostly in Madygen in Central Asian where they are represented mainly by Meganeurina: few Kennedyidae and diverse Triadophlebioidea (Triadophlebiidae, Paurophlebiidae, Mitophlebiidae, Zygophlebiidae, Xamenophlebiidae, and Triadotypidae) and Protomyrmeleontidae (Batkeniidae). Three species of Triadotypidae and Batkeniidae are found in the earlier Middle Triassic of Vosges in France, and two species of Protomyrmeleontidae and Triadophlebioidea *inc. sed.* from the Australian Late Triassic. Suborder Libellulina is known

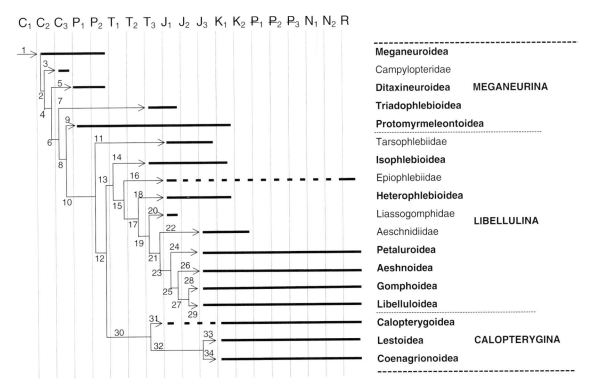

Fig. 101 Phylogeny and system of dragonflies and damselflies (Odonata)

Time scale intervals are abbreviated as follows: C_1, C_2, C_3 – Early (Lower), Middle and Late (Upper) Carboniferous, P_1, P_2 – Early (Lower) and Late (Upper) Permian, T_1, T_2, T_3 – Early, Middle and Late Triassic, J_1, J_2, J_3 – Early, Middle and Late Jurassic, K_1, K_2 – Early and Late Cretaceous, P_1 – Palaeocene, P_2 – Eocene, P_3 – Oligocene, N_1 – Miocene, N_2 – Pliocene, R – present time (Holocene). Right columns are names of suborders (capitals), superfamilies (bold lower case) and families (light lower case). Thin lines show ancestry, thick bars are known durations of taxa (dashed when not confirmed palaeontologically), figures refer to synapomorphies of the subtended clades, as described by BECHLY (1998) for his respective taxa, as follows:

1 – Odonatoptera (our Odonata, see Chapter 2.2.1.1.2.3e).
2 – Odonatoclada.
3 – Campylopterodea.
4 – Nodialata.
5 – Protanisoptera.
6 – Discoidalia.
7 – Triadophlebiomorpha.
8 – Stigmoptera.
9 – Archizygoptera.
10 – Panodonata.
11 – Tarsophlebiidae.
12 – Odonata.
13 – Epiproctophora.
14 – Isophlebioptera + Steleopteridae.
15 – Euepiproctophora.
16 – Epiophlebiidae.
17 – Anisopteromorpha.

18 – Heterophlebioptera + Erichschmidtiidae + Stenophlebioptera.
19 – Pananisoptera.
20 – Liassogomphidae.
21 – Neoanisoptera.
22 – Aeschnidiidae.
23 – Anisoptera.
24 – Petalurida.
25 – Euanisoptera.
26 – Aeshnoptera.
27 – Exophytica.
28 – Gomphidae.
29 – Cavilabiata.
30 – Zygoptera.
31 – Caloptera + Eosagrionidae.
32 – Euzygoptera.
33 – Lestomorpha.
34 – Coenagrionomopha.

from Central Asia, Japan, Australia, South Africa and Argentina, and is represented almost exclusively by Triassolestidae, with a few Cyclothemistidae in Japan. The characteristic nymphs of isophlebioid type (*Samarura* spp.) appear in the Triassic as well (ROZEFELDS 1985a).

Jurassic dragonflies are collected in number and are found in many fossil sites, albeit only in Eurasia, except for a single find in Antarctica. Besides the relict meganeurinean taxa (one species of Piroutetiidae from the earliest Jurassic of France, and more diverse and widespread Protomyrmeleontidae), all Jurassic Odonata belonged to Libellulina and, except for the more advanced, rare Liassogomphidae, to the numerous archaic families (Heterophlebioidea, Isophlebioidea and allies) formerly classified as Anisozygoptera. In the warm and humid climate of the Early Jurassic heterophlebioids and isophlebioids were particularly flourishing and highly morphologically diverse as adults. As far as it is known, their nymphs were rather uniform (usually classified as *Samarura* Brauer, Redtenbacher et Ganglbauer, Fig. 106), reminiscent of large damselfly nymphs with rounded anal flaps, but

Fig. 102 Nymphal *Hemeroscopus baissicus* Pritykina (Odonata: Hemeroscopidae) from the Early Cretaceous of Baissa in Siberia showing mask (arrowheads) and legs with swimming hairs (above, PIN 3064/2796 and 1668/1344, body 49 mm long as preserved; below, 1668/1405–1408, slab height, as shown, 55 mm, photos by A.P. Rasnitsyn and D.E. Shcherbakov)

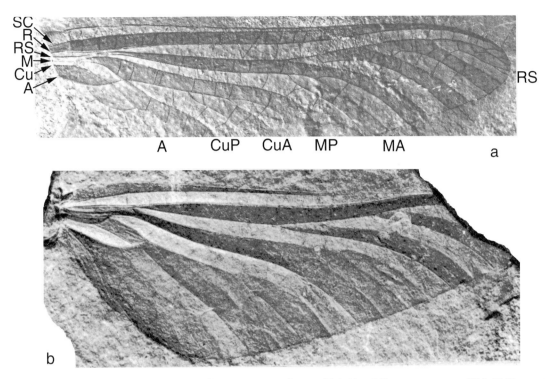

Fig. 103 Hind wings of two of the most archaic known dragonflies, *Eogeropteron lunatum* Riek (a) and *Geropteron arcuatum* Riek (b) (Eogeropteridae) from the Middle Carboniferous of Malanzán Formation in Argentina (length as preserved 35 and 45 mm; respectively; from RIEK & KUKALOVÁ-PECK 1984)

retaining the rectal branchial chamber (PRITYKINA 1985). The Central Asian and especially Siberian fauna was impoverished compared to the European one (possibly because of the less warm climate) and was dominated by Campterophlebiidae or, eastward of Lake Baikal, by Isophlebiidae. The sole Antarctic dragonfly *Caraphlebia antarctica* Carpenter, 1969 from South Victoria Land is placed by BECHLY (1999) to Selenothemistidae (Isophlebioidea).

Little is known about Middle Jurassic dragonflies. The Late Jurassic fauna, known primarily from Solnhofen on Bavaria, Germany, and Karatau Range in South Kazakhstan, has changed considerably compared to the Early Jurassic and had become more diverse. It was still dominated by isophlebioids and heterophlebioids, but the participation of the living superfamilies of Libellulina – Petaluroidea (Protolindeniidae, Aktassiidae), Aeshnoidea (Mesuropetalidae, Cymatophlebiidae, Eumorbaeschnidae), Gomphoidea (Proterogomphidae) and Libelluloidea (Nannogomphidae, Juracorduliidae) becomes really important, particularly in the European assemblages. The former claim that by the Late Jurassic the suborder Calopterygina and several living families of Libellulina have entered the fossil record (PRITYKINA 1989, F. CARPENTER 1992a) has been invalidated by BECHLY (1999) who has placed all the Jurassic libellulinans into extinct families, in part newly formed, and transferred Steleopteridae from the former Zygoptera to the stem of Libellulina. Like in the Early Jurassic, the Late Jurassic assemblages of the East Siberia and Mongolia were impoverished and dominated by Isophlebiidae, with a single find of Campterophlebiidae in Amur Region (ZHERIKHIN 1985).

The Early Cretaceous dragonfly records are widespread (Spain, England, Siberia, Mongolia, China, Brazil, and some minor assemblages in other places). They are almost invariably dominated by the living superfamilies of Libellulina represented mostly by extinct families. However, the stem Libellulina (Tarsophlebiidae, Campterophlebiidae, Aeschnidiidae, etc.) are also found. The other two suborders are present as well, viz., the latest Meganeurina (Protomyrmeleontidae) and the first Calopterygina. The latter suborder from the very beginning is found in both Gondwanaland (Brazil) and Laurasia (England) and represented by all three of its superfamilies: Calopterygoidea (Thaumatoneuridae), Lestoidea (Cretacoenagrionidae, Hemiphlebiidae), and Coenagrionoidea (Protoneuridae). The estimated age of Calopterygina is roughly the same as in Libellulina (since the Triassic), so its late but massive start calls for a biogeographic explanation, e.g. cryptic development in Gondwanaland. This explanation, however, should not obscure the fact that this hypothesis is based solely on the absence of positive data.

The East Asian assemblages were unusual in that the East Siberian, Mongolian and possibly Chinese lake communities were often poor in benthic but rich in nectic insects including, unlike contemporary biota, active swimmers even among the dragonfly nymphs (Hemeroscopidae and Sonidae, see Chapters 1.2c and 2.2.1.1.2e for discussion about these families). The Chinese assemblage described by ZHANG (1992) from the Laiyang Formation in Shandong Prov. is exceptional for all the Mesozoic in being composed only of damselflies (suborder Calopterygina), including a representative of Thaumatoneuridae, subfamily Dysagrioninae (attributed by BECHLY 1999): Dysagrioninae are known otherwise from the North American Eocene and Oligocene, Thaumatoneurinae also from the Aptian of Brazil). REN *et al.* (1995) describe Hemeroscopidae and Aeschnidiidae from the Laiyang Formation, like in many other East Asian assemblages of the earlier Early Cretaceous.

The uncommon Late Cretaceous dragonflies, mostly various Libellulina, are found in a few Cenomanian and Turonian localities of

Fig. 104 Undescribed permolestid from the Early Permian of Tshekarda in Urals (PIN 1700/3247, photo by A.N. Mazin); right hind wing 40 mm long as preserved

Eurasia (in Crimea, Kazakhstan, Krasnoyarsk Province and Magadan Region of Russia) and South Africa (Orapa). Noteworthy is the presence of Aeschnidiidae, one of the latest records of extinct dragonfly families, usually found near to seashores.

The Palaeogene dragonflies are known in number in numerous fossil sites in Eurasia and North America, the majority being the diverse damselflies. The vast majority of the fossils are attributed to extinct genera, and, a rare event for the Cainozoic insects, the extinct families Zacallitidae and Sieblosiidae are present there. In contrast, the Neogene assemblages are usually nearly even in representation of the damsel- and dragonflies (Fig. 107), and rich in living genera of both.

Nevertheless, the extinct family Sieblosiidae (Fig. 108) survived at least till the Late Miocene (in France and Spain, NEL et al. 1996).

2.2.1.2. COHORS CIMICIFORMES Laicharting, 1781

A.P. RASNITSYN

(a) Introductory remarks. The concept of the cohors is basically retained from RASNITSYN (1980) and ROHDENDORF & RASNITSYN (1980), modified according observations mentioned below.

(b) Definition. Very large to very small insects highly diverse in morphology. Pterothorax, or at least pterothoracic segment with leading wing pair, cryptosternous (with sternum invaginated along medial longitudinal line termed discrimen), except in case when flight ability is secondarily weakened or lost, and in some paedomorphic or dorsoventrally depressed insects (in beetles, at least in lower ones, mesothorax is cryprosternous though discrimen lost, cf. PONOMARENKO 1969). Ovipositor with gonocoxa usually forming sheath for gonapophyses (stylets), sometimes shorter than the latter, very rarely intromittent at oviposition. Development from archemetabolous (with larval, nymphal and flying subimaginal stages) to moderately embryonicised (i.e. much of larval development passed before eclosion from egg), with wing rudiments developing gradually, externally, usually with one or several tarsomeres less at eclosion compared with adult stage, and without subimago; rarely with one or two resting postembryonic stages (pupa).

(c) Synapomorphies. No known synapomorphies found, thus making the cohort paraphyletic. Synapomorphies of the entire clade comprising

Fig. 105 *Arctotypus* sp. (Meganeuridae) from the Late Permian of Mogilnikovo in North NW Russia (PIN 3353/87, photo by D.E. Shcherbakov); fore wing 120 mm long along costal margin

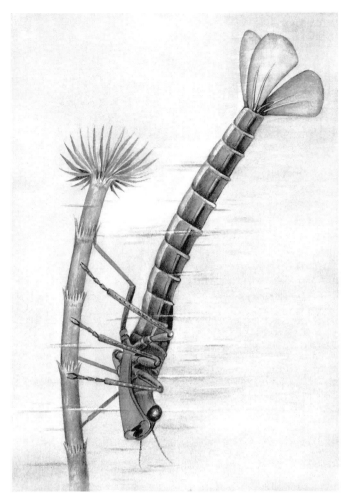

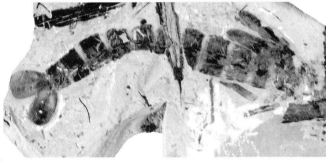

Fig. 106 Dragonfly nymph, *Samarura gigantea* Brauer, Redtenbacher & Ganglbauer (Campterophlebiidae), from the Early Jurassic of Ust-Baley in Siberia, restored as in life by A.G. Ponomarenko and Kira Ulanova, and a fossil (PIN 2375/7, 58 mm long as preserved, photo by A.N. Mazin)

Cimiciformes and their descendant cohort Scarabaeiformes are as follows: pterothorax cryptosternous; possibly (unless inherited from the Scarabaeones stem group) fore wing posterobasal margin lined with vein lying at rest at shallow sublateral impressions of metanotum; paracercus lost (re-gained secondarily in few holometabolan larvae); possibly also: prothoracic paranota narrow or lost (supposing wide circular or winglet-like paranota re-gained secondarily in Dictyoneuridea).

(d) Range. Since Carboniferous (Namurian A – earliest winged insects) until the present, worldwide.

Fig. 107 Two nymphal libellulilids from the Late Miocene of Randecker Maar in Germany (PIN 4391/6, photo by A.N. Mazin); nymph with mask 26 mm long

(e) System and phylogeny are depicted in Fig. 109. No ecological deviation from the ancestral Scarabaeones (as well as from ancestral Scarabaeona) except that a somewhat advanced flight ability can be hypothesised at present to explain the origin of the cimiciformans from their *Anthracotremma*-like ancestor (Chapter 2.2). Both Hemiptera and Psocidea probably originate within Hypoperlida, in or near Hypoperlidae similar to *Martynopsocus* Martynov, but with independent roots (Chapter 2.2.1.2.5e). The hypoperlid suborder Strephocladina is characterised by its head structure (see Figs. 121, 131, 132) and probably function (Chapter 2.2.1.2.2) which would be almost ideal for a precursor of the sucking beak of Dictyoneuridea. The strephocladinan wing venation is often specialised compared to the dictyoneuridean *Heterologopsis*-like (see Fig. 134) ancestor. Nevertheless we consider Strephocladina as ancestral for the dictyoneurideans and expect discovery of a Namurian fossil with the strephocladine-like head and *Heterologopsis*- (or *Ampeliptera*-) like wing.

2.2.1.2.1. SUPERORDER CALONEURIDEA Handlirsch, 1906

(a) Introductory remarks. Description of the superorders is modified from RASNITSYN (1980) and ROHDENDORF & RASNITSYN (1980) basing on the materials discussed above (Chapter 2.2e).

(b) Definition. Medium size and large, rarely small insects with essentially unmodified, chewing mouthparts, flat clypeus, and variously modified wings (undetected stem group probably with plesiomorphic wings). Flight functionally four-winged, in-phase, anteromotoric (GRODNITSKY 1999). Pronotum moderately large, with paranota narrow or lost. Vegetation dwellers supposedly feeding mostly on plant generative organs, or (living zorapterans) cryptobiotic and mostly mycophagous.

(c) Synapomorphies. No because of supposed paraphyly in respect at least to holometabolans unless to, additionally, all the other Cimiciformes. Prothoracic paranota narrow or lost is the probable synapomorphy of the whole clade Cimiciformes + Scarabaeiformes (Chapter 2.2.1.2c), supposed the widened (circular or, usually,

Fig. 108 Undescribed damselfly of the family Sieblosiidae from the Middle Miocene of Stavropol in N. Caucasus (PIN 254/3192, photo by A.N. Mazin); wing 41 mm long

winglet-like) paranota in Dictyoneurida, Mischopterida and rare Diaphanopterida and Hypoperlida are secondarily so.

(d) Range. Middle Carboniferous (Westphalian) till now (with a long gap since Permian till mid-Tertiary); worldwide but predominantly Laurasian during Palaeo- and Mesozoic, mostly circumtropical and subtropical now.

(e) System and phylogeny (Fig. 109). The superorder is considered to form a stem group of Scarabaeiformes unless the whole clade including the remaining of the Cimiciformes and all Scarabaeiformes. Three orders are included: Caloneurida, Blattinopseida, and Zorotypida, the latter representing a recent addition (RASNITSYN 1998b).

2.2.1.2.1.1. ORDER BLATTINOPSEIDA Bolton, 1925

(a) Introductory remarks. The order concept is retained from RASNITSYN (1980) and ROHDENDORF & RASNITSYN (1980), further information is available from KUKALOVÁ (1959, 1965), F. CARPENTER (1966, 1992a) A. MÜLLER (1977) and BURNHAM (1981). For the latest review see HÖRNESCHEMEYER & STAPF (2001).

(b) Definition. Medium-sized to large robust insects with plesiomorphic head (clypeus not much convex, mouthparts chewing) and characteristic wing venation (Fig. 110). Wings wide, convex, sometimes tegminised more or less, roof-like when folded at rest, fore wing with arching impressed nodal line and hind margin indented at CuP. Subcostal area wide, RS pectinate, sometimes irregularly so, issuing part of M branches (with anastomosing M branch persisting in hind wing only), RS originating much more basad in hind wing than in fore one, M with few free branches comparing RS, M_5 anastomosing with CuA, with free portion forming oblique cross-vein, Cu area narrow, wedge-shaped, CuA distally with irregular oblique comb behind, hind wing CuA concave, fore wing anal area oval, similar to that of the roaches but narrower, hind wing with posterobasal area (supported by 3A) enlarged and bending down at rest. Legs apparently not specialised, ovipositor well developed but short, not specialised, cerci segmented, short. Habits unknown except that essentially unmodified ovipositor indicating eggs to be inserted into plant tissues.

(c) Synapomorphies. Nodal line conspicuous, subcostal area wide, M branches partly transferred on RS, M_5 anastomosing with CuA, Cu area narrow, wedge-shaped, fore wing clavus roach-like, hind wing enlarged posterobasally and tucking down behind 2A at rest, cerci short.

(d) Range. Middle Carboniferous to Late Permian, Eurasia and North America.

(e) System and phylogeny. Relationships of the order are outlined above (Chapter 2.2.1.2.1e). The single family Blattinopseidae includes two genera, *Blattinopsis* with near 30 species from Carboniferous and earlier Early Permian, and *Glaphyrophlebia* with 7 species from Middle and Late Carboniferous and Early and Late Permian. There are further fossils of possible blattinopseid affinities whose true position still needs clarification. *Cymbopsis* Kukalová could be an aberrant Blattinopseida with M stock hardly discernible and RS branching disorganised. Among others that are important are *Stephanopsis* Kukalová, which differs from Blattinopseidae mostly in having SC close to R, and *Glaphyrokoris* Richardson, which shares this character but further differs in more rich M and CuA. At least partly these insects formed the blattinopseid stem group which might have given rise to the clade embracing holometabolans and possibly also Hypoperlidea, Dictyoneuridea, Psocidea and Hemiptera (Chapter 2.2.1.2.1).

(f) History. The order is uncommon in Middle Carboniferous (2 species from Mazon Creek, Illinois, USA), but much more common in the Late Carboniferous of both West Europe (France, Germany and Portugal) and North America and in the Permian (Early Permian of Germany, Moravia in Czechia and the Urals in Russia, and Late Permian of Northern and Eastern European Russia), with the Late Permian *Glaphyrophlebia subcostalis* Martynov collected in tens of specimens. Of interest is the inverse correlation that existed between abundance of *Glaphyrophlebia* (in the Middle Carboniferous, earliest and latest Early Permian) and *Blattinopsis* (Late Carboniferous and mid-Early Permian), and exceptional abundance of this, otherwise rather uncommon order, in the earliest Permian Niedermoschel assemblage in Germany (HÖRNSCHEMEIER 1999).

2.2.1.2.1.2. ORDER CALONEURIDA Handlirsch, 1906 (=Caloneurodea Martynov, 1938)

(a) Introductory remarks. Concept of the order is modified since RASNITSYN (1980) and ROHDENDORF & RASNITSYN (1980), mainly in order to accommodate the recent results of type revision of some Carboniferous insects (Rasnitsyn, in preparation). Otherwise the information below is from MARTYNOV (1938c), F. CARPENTER (1943b, 1961, 1970, 1976, 1980, 1992a), SHAROV (1966b), BURNHAM (1984) and other sources cited when appropriate.

(b) Definition (for typical, or traditional Caloneurida). Medium-sized to large insects, elongate and at least sometimes walking-stick alike (Fig. 111), with probably similar habits except feeding on generative rather than vegetative plant tissue. Antenna and legs sometimes very long. Wings moderately to very long and narrow, folding roof-like (Caloneuridae) or horizontal (Paleuthygrammatidae, Permobiellidae) at rest, fore and hind wings hardly differing venationally, hind ones without large anal fan bending down at rest. Venation (Fig. 112) varying little, with RS usually with 3 or more branches, regularly pectinate, rarely simple, M 1–2-branched (sometimes variable), strongly convex M_5 ("CuA" of authors) free, often running close to concave simple CuA (CuA and CuP still more close in Euthygrammatidae), 1–3 simple anal veins. Veins separated by regular cross-veins, rarely by up to 2 rows of irregular cells. Male genitalia when known with long hypopygium and forceps; female with ovipositor somewhat reduced but external, with large eggs, sclerotised, few in number (Fig. 111). Cerci lost.

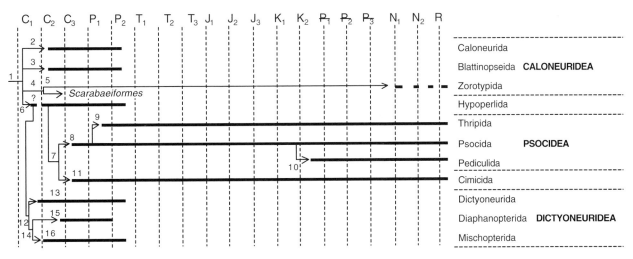

Fig. 109   Phylogeny and system of the cohorts Cimiciformes

Time periods are abbreviated as follows: C_1, C_2, C_3 – Early (Lower), Middle and Late (Upper) Carboniferous, P_1, P_2 – Early (Lower) and Late (Upper) Permian, T_1, T_2, T_3 – Early (Lower), Middle and Late (Upper) Triassic, J_1, J_2, J_3 – Early (Lower), Middle and Late (Upper) Jurassic, K_1, K_2 – Early (Lower) and Late (Upper) Cretaceous, P_1 – Palaeocene, P_2 – Eocene, P_3 – Oligocene, N_1 – Miocene, N_2 – Pliocene, R – present time (Holocene). Two righthand columns are names of orders and superorders, respectively. Arrows show ancestry, thick bars are known durations of taxa (dashed in cases with long gaps in the fossil record, questioned in case of hypothesised range). Figures refer to synapomorphies of the subtended clades, as follows:

1 – pterothorax cryptosternous; pterothoracic paranota lost, paracercus lost.

2 – venation essentially similar in fore and hind wings; CuA, CuP and anals simple and straight or gently curved.

3 – nodal line conspicuous; subcostal area wide; M branches partly transferred on RS; fore wing clavus roach-like, cerci short, hind wing with enlarged posterobasal area bending down at rest.

4 – midcoxa with incipient medial (postepisternal) articulation.

5 – size small; morphology much reduced (antenna 9-segmented, thoracic sides and venter desclerotised, venation reduced with no anal area retained, tarsus 2- and cercus 1-segmented, styli lost, genitalia simplified, etc.); alate and neotenic morphs present in each sex; wings with RS and M fused for a distance and M originating from Cu, fore wing with pterostigma; habits gregarious.

6 – maxilla with lacinia rod- or stylet-like.

7 – cerci lost; habits more sedentary; wings and body short; venation impoverished: RS and CuA each with only 2 branches and M with 4 (supposing higher number of M branches in some Archescytinidae of larger size as well as in various other Cimicida to be secondarily so); CuP simple; 2 simple anal veins; chromosomes holokinetic.

8 – pronotum moderately small; lacinia free from rest of maxilla; hypopharynx with sitophore and associated structures; tarsi 4-segmented.

9 – size small; mouthparts forming short mouth cone with stylet-like mandible; wings unequal, with venation reduced (in fore wing RS, M and CuA at most 2-branched, other veins simple); 2 tarsomeres; ovipositor without sheath (gonocoxae short).

10 – wings lost; head flat, ocelli lost, antenna 5-segmented, lacinia small; parasitising on homoiothermic vertebrates.

11 – mandibles and maxillae transformed into long piercing stylets ensheathed with labium; palpi lost; lateral ocelli adjacent to inner orbits, median one close to postclypeus; SC adjoining R; tarsi 3-segmented; nymph oval, dorsoventrally flattened, with short legs.

12 – mouthparts forming sucking beak sheathed by maxillary (or maxillary and labial) palps; prothoracic paranota winglet-like elongated; wings with M system margined by single convex vein preceded and followed by both concave RS and rest of M, respectively.

13 – head prognathous; wings spread at rest, wide, with fore wing overlapping much of hind one even in flight; M_5 lost; flight mainly gliding.

14 – venation reduced, with archedictyon hardly developed and cross-veins not particularly numerous (but M_5 rarely lost), and with stabilised cross-vein joining bases of RS and MA and bringing them close together.

15 – fore and hind wings almost identical in form and venation; MA simple; tarsi 3-segmented (unless acquired at node 14); pronotal paranota small, abdominal paranota lost; male shedding cerci.

16 – head prognathous; wings spread at rest; possibly: male cercus sinuate basally.

Additionally included in Caloneurida are fossils with RS dominating and CuA and CuP simple or near so, with M more rich in branches than CuA, M_5 fused with CuA and wing sometimes small and not much elongate (see below).

(c) Synapomorphies. Venation essentially similar in fore and hind wings, reduced: SC with cross-vein-like veinlets, CuA, CuP and anal veins simple and straight or gently curved; cerci lost (not proven for Carboniferous families).

(d) Range. Middle Carboniferous to Late Permian; Eurasia and North America.

(e) System and phylogeny. Caloneurida occupy the sister-group position to the Blattinopseida, as confirmed by synapomorphies of the superorder as well as of each order. The traditional system with its 8 families (Caloneuridae, Anomalogrammatidae, Apsidoneuridae, Euthygrammatidae (=Gelastopteridae Carpenter, 1976), Paleuthygrammatidae, Permobiellidae, Amboneuridae, Plesiogrammatidae; F. CARPENTER 1992a) seems formal and overly split now that we tentatively add several more genera, 9 of which being types of their respective families (*Sypharoptera* Handlirsch, *Hapaloptera* Handlirsch, *Herdina* Handlirsch, Fig. 113, and *Metropator* Handlirsch, Fig. 114, from the

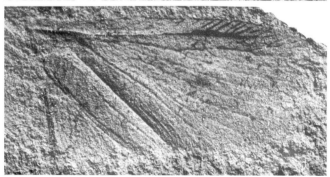

Fig. 110 *Glaphyrophlebia uralensis* (Martynov) from the Early Permian of Tshekarda in Urals (PIN 1700/3361, length of fossil 16.7 mm), and wings of *Glaphyrophlebia subcostalis* (Martynov) from the Late Permian of Soyana in NW Russia (PIN 3353/203, 199, fore and hind wings 9.5 and 8.9 mm long, respectively). Note the hind wing anal area divided by a fold posterior to the second anal vein (photo by A.P. Rasnitsyn)

Middle Carboniferous of USA, *Protokollaria* Brongniart, *Sthenarocera* Brongniart and *Emphyloptera* Pruvost from the Late Carboniferous of France, *Eohymen* Martynov and *Tshecalculus* Novokshonov from the Late Permian of S. Urals). Other genera added are *Paoliola* (Fig. 115) Handlirsch and *Geroneura* Matthew (Fig. 116) from the Middle Carboniferous of Canada, *Pruvostiella* Handlirsch from the Middle Carboniferous of France, and *Boltonaloneura* Rasnitsyn, gen. nov. (*Caloneura subtilis* Bolton, 1925, type species, from the Late Carboniferous of Commentry, France, Fig. 117, generally similar to Caloneuridae but differing from all Caloneurida in having MA fused

with RS next to its base). There are more Carboniferous fossils similar to Caloneurida but incompletely fitting their diagnosis, mainly in having CuA less deeply reduced or not straight enough. *Cymenophlebia* Pruvost, *Kliveria* Handlirsch and *Endoiasmus* Handlirsch from the Middle Carboniferous of France and (*Endoiasmus*) of USA are possible members of Caloneurida and, until we can be more certain, are considered as satellite genera of Caloneurida. More efforts are necessary to propose a balanced family system in the order than is possible now.

Cretaceous monobasic family Mesogrammatidae (HONG 1984) does not belong to Caloneurida: its affinities are unknown. Another Cretaceous fossil claimed to belong to the order and particularly to the family Plesiogrammatidae, *Sinogramma reticularis* Hong (HONG & WANG 1976) most probably represent a roach clavus.

(f) History. Caloneurida were uncommon insects and are known mostly from holotypes, with few species (usually one in any given fossil assemblage) represented by more than 5 specimens.

The Middle Carboniferous caloneurids are found in France (*Pruvostiella*) and in USA in the Allegheny Series of Pennsylvania (Amboneuridae, Permobiellidae, *Metropator* and undescribed), in Mazon Creek, the Carbondale Formation of Illinois (*Sypharoptera*, *Hapaloptera*, *Herdina*), and in Canada in the Lancaster Formation of New Brunswick (*Geroneura*). The Late Carboniferous Caloneurida are known from several localities in Madera Formation in New Mexico, USA, including *Pseudobiella* Carpenter (Permobiellidae), "possible brodiid" by SHEAR *et al.* (1992) and undescribed by ROWLAND (1997). In France, the Late Carboniferous caloneuridans are found in Commentry (*Homaloptila* Carpenter, Apsidoneuridae, *Caloneura* Brongniart, Caloneuridae, *Protokollaria*, *Sthenarocera* and *Boltonaloneura*), in Pas-de-Calais (*Emphyloptera*) and in Montceau-Les-Mines (*Apsidoneura* Carpenter, Apsidoneuridae). Incomplete specimen of uncertain position has also been collected in Lower Alykaeva Formation of Southwest Siberia (Zavyalovo).

The Early Permian fossils come from two principal strata – Artinskian deposits at Elmo, Kansas, USA, and Kungurian ones at Tshekarda and neighbour localities in the Urals, Russia. Seven species have been collected at Elmo, belong to Permobiellidae, Apsidoneuridae, Paleuthygrammatidae, Plesiogrammatidae (2 species), and Anomalogrammatidae and, supposedly, Euthygrammatidae (*Gelastopteron gracile* Carpenter). *Paleuthygramma acutum* Carpenter was fairly common, with 35 specimens having been collected. One species of Plesiogrammatidae (NOVOKSHONOV 1998b) and two of Paleuthygrammatidae are found at Tshekarda and its vicinity, one of which (*Paleuthygramma tenuicorne* Martynov) was also collected in tens of specimens.

Late Permian Caloneurida are known mostly from the Russian North (Early Kazanian deposits at Soyana River, Arkhangelsk Region), where some 80 specimens of *Euthygramma parallelum* Martynov (Euthygrammatidae) have been collected along with two rare species from the same family as well as from Paleuthygrammatidae. Deposits of roughly the same age at Kama River (Tikhiye Gory) yielded one more specimen and species of the above genus, while two undescribed fossils that can be tentatively assigned to the Euthygrammatidae are found in a younger, Late Kazanian Belebey Formation in Kityak, Kirov Region. The earliest Late Permian deposits of Kuznetsk Formation of south-western Siberian (Kaltan locality) produced the unique Eurasian representative of Plesiogrammatidae. Several undescribed specimens belonging, at least in part, to the Apsidoneuridae, have been collected in the Late Permian of Bor-Tologoy in Southern Mongolia.

Fig. 111 *Paleuthygramma tenuicornis* Martynov (Paleuthygrammatidae) from the Early Permian of Tshekarda (Urals): male with clearly visible forceps-like genitalia (PIN 1700/1424, photo by A.G. Sharov; specimen not found at present), and female abdomen with large, sclerotised eggs preserved inside (from RASNITSYN 1980)

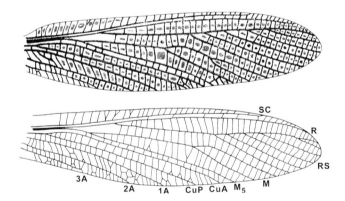

Fig. 112 *Caloneura dowsoni* Brongniart, 1885 (Caloneuridae); Late Carboniferous of Commentry in France (from CARPENTER 1961)

2.2.1.2.1.3. ORDER ZOROTYPIDA Silvestri, 1913 (=Zoraptera Silvestri, 1913)

(a) Introductory remarks. The smallest and the second latest established of the living insect orders (in 1913, that is only 2 years before the most recent one, Grylloblattida), containing only 30 species in a single genus. The present account is based on RASNITSYN (1998b).

(b) Definition (Fig. 118). Size 2–3 mm. Alate and wingless morphs known for both sexes in the same species. Head hypognathous, subtriangular, with mouth cavity and occipital foramen contiguous, in winged morph with 3 ocelli present and with eyes displaced caudad, (wingless with no sight organs). Antenna moniliform, 9-segmented (8-segmented in early instars). Mouthparts chewing, not much modified. Dorsocervicalia fused, ventrocervicalia free, laterocervicalia separated into 2 sclerites each. Pronotum moderately large, lacking paranota. Alate morph with wings (easily breaking off sub-basally, but unlike in termites not using a well-defined line of weakness) and pterothoracic segments heteronomous, venation much reduced and modified, superficially similar to that of more modified bark-lice, pterothoracic morphology somewhat simplified, with sternum invaginated along midventral furrow (discrimen) in both alate and wingless morphs. Wingless morph with thoracic morphology nymph-like, except in some specimens with wing-pad-like projections on meso- and metanotum. Legs moderately short, with coxa subconical, mid coxa may have support on postepisternum process but lacking permanent mesal articulation. Femur and tibia each with row of spines ventrally, tarsus 2-segmented, basitarsus small. Abdomen 10-segmented, with cerci 1-segmented (rudimentary second segment present in Miocene *Zorotypus goeleti* Engel & Grimaldi) and genitalia rather simplified, particularly in female, and not easily comparable with that of other insects. Habits cryptic (in termite tunnels, under bark, etc.), gregarious but not socially advanced (no labour sharing has been recorded within the group, and yet isolated specimen being destined to rapid death). Feeding on fungi (particularly yeast) and dead animal matter.

(c) Synapomorphies. Size small; morphology much reduced (antenna 9-segmented, thoracic sides and venter desclerotised, venation reduced with no anal area retained, tarsus and cercus 2-segmented, styli lost, genitalia simplified, etc.); alate and neotenic morphs present in each sex; wings with RS and M fused for a distance and M originating from Cu, fore wing with pterostigma; habits gregarious.

(d) Range. Worldwide but patchy in warm climates including oceanic islands (Galapagos, Samoa, Hawaii, Fiji, Mauritius, etc.), extending into temperate climates in North America and Tibet. As fossil, recorded only from mid-Tertiary.

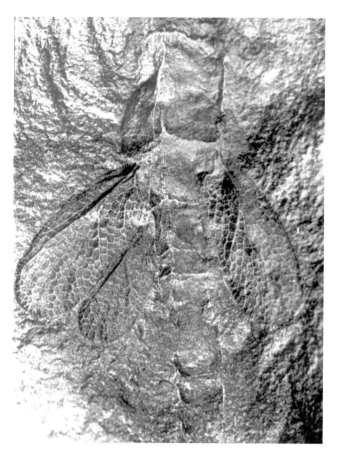

Fig. 113 *Herdina mirificus* Carpenter & Richardson (Herdinidae), a brachypterous insect from the Middle Carboniferous of Mazon Creek in USA (from CARPENTER & RICHARDSON 1971); fossil 16mm long as preserved

(e) System and phylogeny. The only genus, *Zorotypus* Silvestri, is probably synapomorphic with Cimiciformes and Scarabaeiformes in having invaginated pterothoracic sterna and, except many Dictyoneuridea, in lacking prothoracic paranota, and with Scarabaeiformes in incipient medial midcoxal articulation. It is plesiomorphic in respect of the latter group in incomplete metamorphosis and in medial midcoxal articulation not permanent. As a result, *Zorotypus* is phenetically more similar to moderately basal Cimiciformes like Psocida than to holometabolans. ENGEL & GRIMALDI (2000) advocate embiid relationships based on eight synapomorphies all prone to homoplastic development (loss of [detached] gonostyli and reduced cerci are characteristic of most insects; enlarged metafemur of similar form is typical of many cryptic insects, e.g. bethylid wasps; both wing pairs narrow and paddle-shaped, with a reduced anal area are equally trivial, *e.g.* in homopterans, parasitic wasps *etc.*; tarsomere reduction is still more trivial, particularly up to three segments, as in webspinners, unlike the less common 2-segmented state as found in *Zorotypus*; gregarious (maternal) behaviour is widespread in various insect groups which often also possess apterous morphs; dehiscent wings are more characteristic of female ants than of webspinners). The case clearly needs further study.

KUKALOVÁ-PECK & PECK (1993) have created 6 genera additional to *Zorotypus* to house 6 New World species with known wing morphology. Fourteen Old World species, of which the wing venation is known for one, are tentatively lumped under *Zorotypus*. New World species with unknown wings were left without any generic name, other than uncertain attribution to *Zorotypus*, pending discovery of winged forms. This taxonomic rearrangement seems premature considering the present state of our knowledge of the group.

(f) History. The only fossil zorapterans are *Zorotypus paleus* Poinar and *Z. goeleti* Engel & Grimaldi from the Early Miocene Dominican amber (Fig. 119).

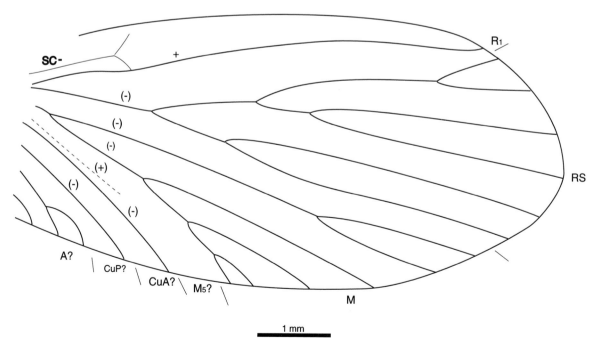

Fig. 114 *Metropator pusillus* Handlirsch from the Middle Carboniferous Allegheny Series of USA (holotype)

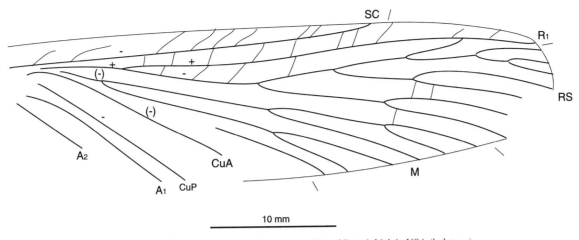

Fig. 115 *Paoliola gourlei* Handlirsch from the Middle Carboniferous (Middle Pottsville) of French Lick in USA (holotype)

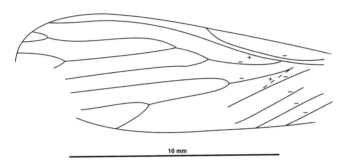

Fig. 116 *Geroneura wilsoni* Matthew from the Middle Carboniferous of Lancaster Formation in Canada (original, based on COPELAND 1957 Pl. IX, Fig. 5)

2.2.1.2.2. SUPERORDER HYPOPERLIDEA Martynov, 1928. ORDER HYPOPERLIDA Martynov, 1928

(a) Introductory remarks. Small extinct group of insects of diverse external appearance (Figs. 120, 121). The chapter is modified after RASNITSYN (1980) and ROHDENDORF & RASNITSYN (1980) including new findings and interpretations (cited whenever appropriate).

(b) Definition. Moderately small to large insects with mouth parts chewing though often beak-like elongate, with lacinia rod- or stylet-like, clypeus convex indicating strong cibarial muscles, or, if flat, mandibles and laciniae long, jointly forming short beak. Pronotum often elongate, with prothoracic paranota inconspicuous (secondarily widened in some Anthracoptilidae). Wings highly variable from near insect groundplan up to venationally polymerised (in *Hypermegethes* Handlirsch, *Rhynomaloptila* Rasnitsyn), much impoverished (in *Boreopsocus* Shcherbakov), or else subelytrised (in *Perielytron* Sharov), folded either roof-like or horizontally with considerable overlapping at rest, hind wing without enlarged anal area, bending down at rest. Flight most probably functionally four-winged, in-phase, antero-motoric (*sensu* GRODNITSKY 1999). Ovipositor with gonocoxite forming sheath. Cerci segmented, usually short. Vegetation dwellers feeding probably on micro- and macrospores in sporangia (for pollinivory see RASNITSYN & KRASSILOV 1996a,b, for feeding on the gymnosperm ovules RASNITSYN 1980).

(c) Synapomorphies. None because of paraphyly in respect to Dictyoneuridea, Psocidea and Cimicidea. Synapomorphy in common with these taxa is maxilla with lacinia detached and rod-like (secondarily stylet-like).

(d) Range. Earliest Middle Carboniferous to Late Permian of Eurasia, North America and Australia.

(e) System and phylogeny. Hypoperlida is considered embracing three suborders: slender, short-headed, more or less oligoneurous Hypoperlina with the roof-like wing rest position; large and heavy, long-headed, often polyneurous Strephocladina with either roof-like or horizontal wing rest position; and rare, monobasic Protelytrina looking like small Hypoperlina with wide, subelytrised fore wing.

SHCHERBAKOV (1995) has proposed to distinguish two families in Hypoperlina, Ampelipteridae and Hypoperlidae. He has defined Ampelipteridae after the branched CuP and has listed five genera there, viz. *Ampeliptera* Pruvost (Fig. 122, Late Namurian A, earliest Middle Carboniferous in The Netherlands; BRAUCKMANN *et al.* 1996), *Protoprosbole* Laurentiaux (Late Namurian B, earliest Middle Carboniferous in Belgium, *op. cit.*), *Fatjanoptera* O. Martynova (Fig. 123, Early Permian of Fatyanikha, East Siberia), *Tshunicola* Rasnitsyn (Late Carboniferous of Chunya, East Siberia), *Tshekardobia* Rasnitsyn (see Fig. 120, later Early Permian of Tshekarda, Urals). Additionally included here in the family are *Limburgina* Laurentiaux (Late Namurian B, earliest Middle Carboniferous in Netherlands), *Aenigmatodes* Handlirsch and *Gyrophlebia* Handlirsch (Westphalian D, Middle Carboniferous of Mazon Creek, Illinois), *Protopachytylopsis* Laurentiaux et Laurentiaux-Vieira (Westphalian A, Middle Carboniferous of Belgium), *Anthraconeura* Laurentiaux et Laurentiaux-Vieira (Westphalian A, Middle Carboniferous of Belgium) and, with some doubts, insufficiently known *Mixotermes* Sterzel (Westphalian, Middle Carboniferous of Germany), as well as *Pruvostia* Bolton and *Boltonocosta* Carpenter (Westphalian, Middle Carboniferous of Middle Coal Measure, England) which may equally belong to Strephocladina. This would result in synonymisation of Protoprosbolidae Laurentiaux, Fatjanopteridae O. Martynova, Aenigmatodidae Handlirsch, Anthraconeuridae Laurentiaux et Laurentiaux-Vieira and Mixotermitidae Handlirsch, under Ampelipteridae Haupt or, rather, all of them under either Aenigmatodidae or Mixotermitidae. However, *Aenigmatodes* is too incomplete to serve well the function

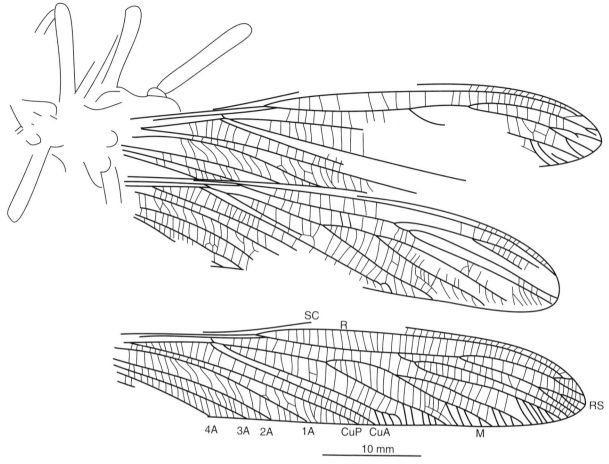

SC
R
RS
4A 3A 2A 1A CuP CuA M
10 mm

Fig. 117 *Boltonaloneura subtilis* (Bolton) from the Late Carboniferous of Commentry in France (holotype)

of the family type, and *Mixotermes* is insufficiently known. That is why they are considered here as 'satellite' genera and not as full members of Ampelipteridae, and only Anthraconeuridae Laurentiaux et Laurentiaux-Vieira, 1986 are formally synonymised here under Ampelipteridae in addition to Protoprosbolidae and Fatjanopteridae synonymised by SHCHERBAKOV (1995). Venationally Ampelipteridae are plesiomorphic enough to be ancestral to the rest of the order. Unfortunately, this hypothesis cannot be tested now because the body morphology is practically unknown for the Carboniferous genera.

Hypoperlidae are delimited by SHCHERBAKOV (1995) to embrace only four genera, *Hypoperla* Martynov (Fig. 124), *Idelopsocus* Zalessky (Fig. 125), *Kaltanelmoa* Rohdendorf and *Boreopsocus* Shcherbakov (later Early Permian of Tshekarda and middle Late Permian of Soyana, European Russia) which have CuP simple. Undescribed Middle Carboniferous fossil from Mazon Creek in Illinois, USA (Fig. 126), probably belongs to an undescribed genus of Hypoperlidae. Two more families are added by NOVOKSHONOV (1997e, 1998b), the monobasic Asiuropidae from Tshekarda with the sinuate, 5-branched CuA (Fig. 127), and the very slender and long-legged Letopalopteridae that were originally described as snakeflies but now found to be dissimilar to them and including two genera, *Letopaloptera* O. Martynova from Soyana and *Permindigena* Novokshonov (Fig. 128) from Tshekarda. *Xenoneura* Scudder from the Middle Carboniferous of Canada also can be related to Hypoperlidae.

Structure of the suborder Strephocladina is more complicated. The most distinct is the monobasic Synomaloptilidae from the Early Permian of Tshekarda, Urals (Figs. 121, 129), with its impoverished venation and horizontal wing resting position. The other forms are more polyneurous and usually with CuA more or less pectinate forward (possibly the suborder synapomorphy). They can be roughly segregated into those with SC meeting R and issuing branchlets which are slightly or moderately oblique and mostly simple, and those with SC branches long, slanting, irregularly branching, or disorganised into irregular rows of cells, and often with SC meeting C. The distinction is not absolute, however, for *Opistocladus* Carpenter is intermediate, but can be maintained for the present. Three families fit the first group: Tococladidae (Fig. 130), Heteroptilidae and Nugonioneuridae, each with the type genus only (except *Opistocladus* whose position in Tococladidae is not certain because of its different form of SC) and all come from the middle Early Permian of Elmo, Kansas, except for one of two species of *Tococladus* which is described from Tshekarda. It seems possible to lump all of them under the name Tococladidae Carpenter, 1966 (=Heteroptilidae Carpenter, 1976, =Nugonioneuridae Carpenter, 1976, synn. nov.).

Several family names have been proposed for those forms with sub-horizontally branching SC: Hypermegethidae, Anthracoptilidae Handlirsch, 1922 (=Strephocladidae Martynov, 1938, syn. nov.) and Permarrhaphidae Martynov, 1931 (=Strephoneuridae Martynov,

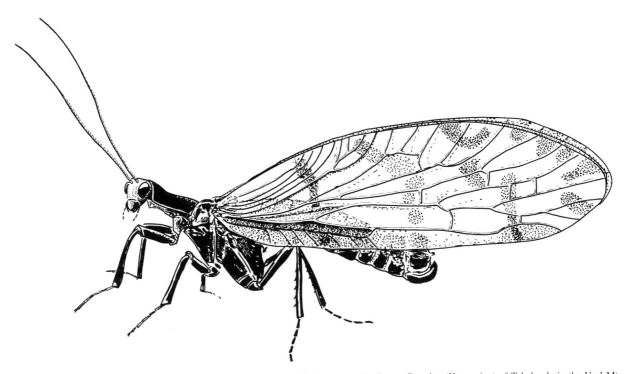

Fig. 119 *Zorotypus palaeus* Poinar, Zorotypidae, body length 2.9 mm, in the Miocene Dominican amber (from POINAR 1988)

Fig. 118 *Zorotypus hubbardi* Caudell, Zorotypidae, extant from above and below (from RASNITSYN 1998b)

Fig. 120 *Tshekardobia osmylina* Rasnitsyn (Hypoperlidae), restored by A.G. Ponomarenko; Lower Permian (Kungurian) of Tshekarda in the Ural Mts. (from ROHDENDORF & RASNITSYN 1980)

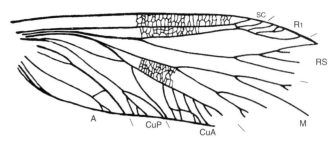

Fig. 122 *Ampeliptera limburgica* Pruvost (Ampelipteridae) from the Middle Carboniferous of Gulpen in the Netherlands (from CARPENTER 1992a); fore wing 10 mm long as preserved.

Fig. 123 *4. Fatjanoptera mnemonica* O. Martynova (Hypoperlidae) from the Early Permian of Fatianikha in East Siberia (holotype, photo by A.G. Sharov); fore wing 11 mm long

Fig. 121 *Synomaloptila longipennis* (Synomaloptilidae) restored as in life by A.G. Ponomarenko (from ROHDENDORF & RASNITSYN 1980)

Fig. 124 *Hypoperla elegans* Martynov (Hypoperlidae), fore and hind wings from the Late Permian of Soyana in NW Russia (PIN 117/2285 and 117/968, 14.4 and 12.2 mm long, respectively, photo by A.G. Sharov)

1940, syn. nov.). Hypermegethidae are known from incomplete giant wings with SC branches disorganised, often forming up to several rows of cells in the costal space; three species of *Hypermegethes* Handlirsch are described from the Middle and Late Carboniferous in USA and England (F. CARPENTER 1992b). Characteristic of Permarrhaphidae is M intercepting branches of CuA, and the roof-like wing rest position. The family includes *Pseudoedischia berthaudi* Handlirsch from the Late Carboniferous of Commentry, France, *Strephoneura robusta* Martynov (Fig. 131) from the Early Permian of Tshekarda and *Permarrhaphus venosus* Martynov from the Late Permian of Tikhiye Gory, eastern European Russia. Anthracoptilidae retain distinct SC

branches as well as forward branches of CuA but the wing rest position data are contradictory: horizontal in *Mycteroptila* Rasnitsyn (Fig. 132) but probably roof-like in *Mesoptilus fayolianus* Handlirsch (inconclusive in other, even bodily preserved fossils). The family is rich in subordinate taxa, including *Strephocladus subtilis* Kliver from the Ottweiler Stufe, earliest Late Carboniferous of the Saar Basin in Germany, *Anthracoptilus perrieri* Meunier, *Mesoptilus dolloi* Laméere, *M. sellardsi* Laméere, *M. fayolianus* and *Ischnoneurona delicatula* (Brongniart) from Commentry, Late Carboniferous of France, with

Fig. 125 *Idelopsocus splendens* (G. Zalessky) (Hypoperlidae) from the Early Permian of Tshekarda in Urals (from ROHDENDORF & RASNITSYN 1980); length of the fossil 14.5 mm

5 genera and 8 species from the Early Permian of Obora in Czechia, Elmo in Kansas, and Tshekarda in the Urals (KUKALOVÁ 1965, F. CARPENTER 1976, RASNITSYN 1977a, NOVOKSHONOV 1998a). Undescribed Anthracoptilidae have also been found in the Late Permian of Belmont in Australia [kept in the Natural History Museum, London, nos. In. 45620, In. 45627, In. 45929 = In. 45933 (part and counterpart)].

Some more Palaeozoic taxa may belong to Hypoperlida, though their positions there are tentative at best. These are *Rhipidioptera* Brongniart from the Late Carboniferous of Commentry, France, *Psoroptera* Carpenter from the Early Permian of Elmo, Kansas, and *Homoeodictyon* Martynov from the Late Permian of Kargala, Urals.

The last to be mentioned is *Perielytron mirabile* G. Zalessky from the Early Permian of Tshekarda, Urals (Fig. 133), which is insufficiently known but is so distinct because of its wide, subelytrised wings that a suborder of its own, the Protelytrina, has been proposed (RASNITSYN 1980).

As outlined, the order is considered a daughter group of the superorder Caloneuridea ancestral to dictyoneuroids, psocopteroids and hemipterans. The present imperfect knowledge of Hypoperlida makes premature attempt of its more formal phylogenetic analysis.

(f) History. This section is based on the sources cited above in this chapter and additionally on NOVOKSHONOV (1995, 2001a) and NOVOKSHONOVA (1998). The oldest, Namurian Hypoperlida (*Ampeliptera, Limburgina* and *Protoprosbole*) belonged to Ampelipteridae and was found in Europe. In the Westphalian several other Ampelipteridae appeared both in Europe and North America, accompanied by rare Hypoperlidae (undescribed taxa and possibly *Xenoneura*) in North

America and Hypermegethidae in North America and Europe. In the Stephanian (Late Carboniferous) Anthracoptilidae appeared as a diverse group in Europe accompanied with rare Strephoneuridae, while rare Ampelipteridae are found in North America and East Siberia and Hypermegethidae in North America. The pattern kept changing in the Permian. The only earliest Permian fossil from the East Siberia is an ampelipterid. The mid-Early Permian hypoperlidans in Kansas and Czechia belonged to Anthracoptilidae and, in Kansas, to Tococladidae. The latest Early Permian assemblage of Tshekarda (Urals) is the most diverse. It includes representatives of all families except Hypermegethidae, with the dominant Hypoperlidae. The latter family is dominant during the Late Permian as well, accompanied only by rare Letopalopteridae in Soyana (Russian North) and Anthracoptilidae in Tikhiye Gory (Tatarian Rep.). Soyana assemblage is unique in having Hypoperlidae fairly common, with *Hypoperla elegans* Martynov and *Boreopsocus danksae* Shcherbakov collected in tens of specimens. The latest Permian (Tatarian) insect assemblages are free of hypoperlidans.

2.2.1.2.3. SUPERORDER DICTYONEURIDEA Handlirsch, 1906 (=Palaeodictyopteroidea)

N.D. SINITSHENKOVA

(a) Introductory remarks. Concept of the superorder is original differing much from previous accounts (cf. ROHDENDORF 1969, ROHDENDORF & RASNITSYN 1980, KUKALOVÁ-PECK 1991). The reason is the accumulation of groups combining characters of different orders and even former superorders and thus obscuring their boundaries (those assigned here to Psychroptilidae and Frankenholziina in Dictyoneurida and to Eubleptina in Mischopterida). This made re-arranging the group as a whole inescapable.

(b) Definition. Body size large to very large, sometimes medium (20–400 mm wingspan). Head with sucking beak formed by mandibular, maxillary, and hypopharyngeal stylets, and sheathed by labrum (dorsobasally), maxillary palps, and sometimes by labium. Prothorax and abdominal segments usually with lateral paranota (pronotal paranota can be circular), sometimes bearing marginal spines, or lost. Wings costalised (with C, SC, and R thick and running relatively close each other), with single convex M vein (either MA or MA stock + MA_1) forming fore margin of M system (unlike mayflies and dragonflies with entire MA convex with concave intercalaries). Costal margin straight to gently convex, costal space narrow, RS predominantly pectinate, MA poor in branching. Hind wing with anal area at most moderately enlarged. Flight variable: fast but not manoeuvrable, often of biplane type (with widely overlapping fore and hind wings) in the largest dictyoneuridans; more manoeuvrable and versatile in respect of flight speed in moderately large forms with wide wing bases; slow and highly manoeuvrable, approaching hovering ability in the case of those with a narrow wing base (WOOTTON & KUKALOVÁ-PECK 2000). Legs hardly much used in walking, often with clinging adaptations, particularly in fore pair (coxae shifted or inclined forward, tarsi sometimes 2-segmented with single thick claw, etc.). Ovipositor short to moderately long, cutting (probably using to place eggs into plant tissues), with sheath shorter than stylets, bearing long stylus. Cerci long, multisegmented. Development gradual, terrestrial (phytophilous), with imaginal moults. Adult and most probably immature feeding by sucking ovule content of gymnosperm plants.

(c) Synapomorphies. Wings costalised, with fore M vein (either simple MA or MA stock + MA_1) convex; possibly: prothoracic paranota

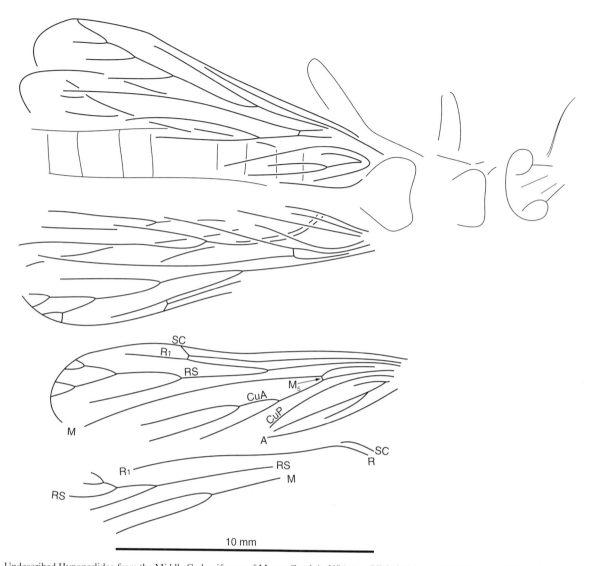

Fig. 126 Undescribed Hypoperlidae from the Middle Carboniferous of Mazon Creek in USA (no. PE 21815 in the Field Museum, Chicago)

winglet-like (unless inherited from ancestral Hypoperlida *Strephocladina* and supposing circular paranota in *Homoioptera* Brongniart a reversal).

(d) Range. Latest Early Carboniferous (early Namurian A) through the Late Permian; widespread but common only in Euramerica and, in Early Permian, in westernmost Angaraland (the Urals).

(e) System and phylogeny. The superorder is hypothesised (see Fig. 109) as having descended from *Heterologopsis*-like forms (Fig. 134) which may also deserve the rank of an order *per se*, but because of insufficient knowledge are considered as unplaced Dictyoneuridea. Otherwise the superorder is taken to be composed of three orders, with Diaphanopterida and Mischopterida forming a sister group in respect to Dictyoneurida. Hence it is supposed that a permanently outstretched wing position had been acquired independently by Dictyoneurida and Mischopterida.

Homoeodiction Martynov (see Fig. 73) from the Late Permian of Kargala in the Urals and *Klebsiella* Meunier from the Late Carboniferous of Commentry in France are similar to dictyoneurideans in the convex MA, but otherwise differ in their venation and so are considered Scarabaeones *incertae sedis* (Chapter 2.2).

2.2.1.2.3.1. ORDER DICTYONEURIDA Handlirsch, 1906 (=Palaeodictyoptera Goldenberg, 1854)

(a) Introductory remarks. One of the most popular groups of Palaeozoic insects, spectacular because of their large size, long and wide, often brightly patterned wings, and long cerci (Figs. 135, 136). The concept of the order is essentially modified compared to ROHDENDORF & RASNITSYN (1980), particularly concerning its taxonomy. Primary data are reviewed and catalogued by HANDLIRSCH (1906–1908, 1922), ROHDENDORF (1962), F. CARPENTER (1967a), KUKALOVÁ (1969a,b, 1970), BRAUCKMANN (1991), and completed by SHAROV & SINITSHENKOVA (1977), SINITSHENKOVA (1979, 1980a,b, 1981, 1992a), F. CARPENTER (1983, 1992a), PINTO (1994a), BRAUCKMANN & GRÖNING (1998), BRAUCKMANN *et al.* (1996).

(b) Definition. Body stout, large to very large (up to 50 cm wingspread or more). Integument rather thick, finely granulated. Head with sucking beak moderately (Spilapteridae) to very long (Eugereonidae), directed obliquely forward. Prothoracic paranota either circular (Fig. 137) or, usually, forming winglets. Wings spread permanently (not folding over abdomen at rest). Both wing pairs

Fig. 127 *Asiuropa uralensis* Novokshonov (Asiuropidae) from the Early Permian of Tshekarda in Urals (holotype, photo by V.G. Novokshonov); length 16 mm

similar except minor venational details and wider base of hind wing, with either rich archedictyon or numerous cross-veins (exceptionally cross-veins less common, forming transverse rows), sometimes with evident nygmata (in both adult and nymphal wings of Homoiopteridae, Lithomanteidae, Breyeriidae), often brightly patterned (with dark stripes in Spilapteridae, Lamproptilidae, Fouqueidae). Longitudinal veins straight or gently curved, rarely with CuP and A more strongly curved (Calvertiellidae, Breyeriidae, Eugereonidae); occasionally merging for a distance and then more sharply bent. C often running submarginally, leaving narrow precostal space. Because of the permanently spread wings, adults only able to fly and to cling on to a plant for feeding, oviposition and rest. Otherwise as in superorder (Chapter 2.2.1.2.3b).

(c) Synapomorphies. Head prognathous; wings permanently spread; M_5 lost.

(d) Range. Latest Early Carboniferous (Namurian A) through Late Permian (Kazanian), with highest abundance during Middle and especially Late Carboniferous; widespread in north hemisphere but common only in Eurameria, solitary records also in Argentina, Tasmania, East Siberia, and Mongolia. Most probably the distribution results from the dictyoneurid thermophily.

(e) System and phylogeny of the order (Fig. 138) are based in part on RIEK (1976c), otherwise are original. Unfortunately the group is possible to be classified almost entirely on the basis of venational characters.

Two suborder and 8 superfamilies are proposed here. Dictyoneurina Handlirsch, 1906, stat. nov. includes 6 superfamilies of the typical dictyoneuridans with wings rich in venation, wide including basally so as fore wing broadly overlapping the hind one, and best is fitted for soaring flight. Eugereonoidea Handlirsch, 1906, stat. nov. (Dictyoptiloidea sensu RIEK 1976c, but Dictyoptilidae have been synonymised with Eugereonidae by F. CARPENTER 1964, and the synonymisation has not been challenged yet) is the largest one with its 11 constituent families (Eugereonidae, Fig. 139, Archaemegaptilidae, Graphiptilidae, Jongmansiidae, Lithomanteidae, Lycocercidae, Megaptilidae, Polycreagridae, Protagriidae, Synarmogidae, Tchirkovaeidae). Characteristic of the group are wing length 2.5 times width, SC reaching C near wing apex, CuA and MA with at most a short apical fork, branching MP and CuP, and the archedictyon replaced by numerous cross-veins or sparse between more spaced veins.

Dictyoneuroidea Handlirsch, 1906, stat. nov. with 3 families [Dictyoneuridae (including *Palaeoneura giligonensis* Hong, described in Neuburgiidae, a synonym of Spilapteridae: HONG 1985b), Peromapteridae, Saarlandiidae] similar to Eugereonoidea but differing in wing length 3.5 times its width, dense archedictyon and often narrow wing apex.

Homoiopteroidea Handlirsch, 1906, stat. nov. (consisting of Homoiopteridae, Figs. 137, 140, and Heolidae) and the closely related superfamily Spilapteroidea Brongniart, 1893, stat. nov. with its 6 families (Spilapteridae, Figs. 135, 136, 141, Aenigmatidiidae, Fouqueidae, Homothetidae, Lamproptilidae, Mecynostomatidae) differ from Eugereonoidea in multibranched MA. They differ from each other in having the veins arching gently at the very wing base of Homoiopteroidea *vs.* not being bent in Spilapteroidea.

Unlike the four above groups, Breyerioidea Handlirsch, 1904, stat. nov. and Calvertielloidea Martynov, 1932, stat. nov., have SC short (scarcely longer than half wing length) and meeting R (except Cryptoveniidae). The former superfamily includes 3 families (Breyeriidae, Stobsiidae, Cryptoveniidae) and only the latter two (Calvertiellidae and Mongolodictyidae which is unusual in having R seemingly falling into SC). Calvertielloidea differ in the strikingly arching longitudinal veins, in simple CuP, and in CuA lost at base and thus seemingly originating from M.

There are also several insufficiently known groups that can only be classified as Dictyoneurina *incertae sedis* (Archaeoptilidae, Lithoptilidae, '*Mecynoptera*' *tuberculata* Bolton).

Suborder Frankenholziina Guthörl, 1962, stat. nov. is less diversified, covering only 10 families in 2 superfamilies. Unlike Dictyoneurina and similarly to Mischopterida its wings are narrow, especially basally, indicating a more active and manoeuvrable flight. Frankenholzioidea Guthörl, 1962, stat. nov. (including Frankenholziidae, Dictyoneurellidae, Caulopteridae, Ancopteridae, Fig. 142, and Psychroptilidae) is less advanced and shows only incipient modifications in the above direction. These changes are better manifested in Arcioneuroidea Kukalová, 1975, stat. nov. (Arcioneuridae, Elmoboriidae, Hanidae, Fig. 143, Eubrodiidae fam. nov. [*Eubrodia* Carpenter, 1967, type genus; the family differs from all other Frankenholziina in having the unique combination of simple MA and CuA, and regular archedictyon covering the hole wing], Eukulojidae).

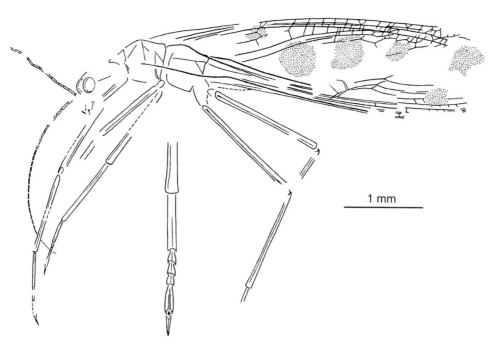

Fig. 128 *Permindigena lientericus* Novokshonov (Letopalopteridae) from the Early Permian of Tshekarda in Urals (from NOVOKSHONOV 1998b)

Fig. 129 *Synomaloptila longipennis* Martynov (Synomaloptilidae) from the Early Permian of Tshekarda in Urals (holotype, photo by A.P. Rasnitsyn); wing length 28 mm

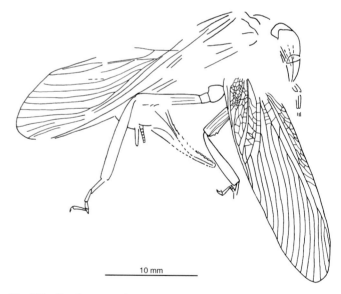

Fig. 131 *Strephoneura robusta* Martynov (Permarrhaphidae) from the Early Permian of Tshekarda in Urals (from RASNITSYN 1980)

Excluded from the order are *Eohymen* Martynov (Caloneurida, Chapter 2.2.1.2.1.2), *Boltonocosta* Carpenter, and *Hypermegethes* Handlirsch (Hypoperlida, Chapter 2.2.1.2.2), *Merlebachia* Waterlot (Scarabaeones *incertae sedis*, probably related to Paoliidae, Chapter 2.2), and *Bardapteron* G. Zalessky (*incertae sedis*).

(f) History. The oldest (Namurian A, Arnsbergian) *Delitzshala bitterfeldensis* (Spilapteridae) is from Germany and is the only Early Carboniferous winged insect (see Fig. 136). Namurian (the latest Early

Fig. 130 *Tococladus similis* Novokshonov (Tococladidae) from the Early Permian of Tshekarda in Urals (holotype, photo by A.G. Sharov); wing length 24 mm

Fig. 132 *Mycteroptila armipotens* Novokshonov (Anthracoptilidae) from the Early Permian of Tshekarda in Urals (holotype, photo by V.G. Novokshonov); length of the fossil 33 mm

Fig. 133 *Perielytron mirabile* G. Zalessky (Perielytridae), Early Permian of Tshekarda in Ural Mts. (PIN 1700/1863); body with wings 10 mm long (from ROHDENDORF 1962)

Fig. 134 *Heterologopsis ruhrensis* Brauckmann (Cacurgidae); Lower Namurian B (lowermost Middle Carboniferous) of Hagen-Vorhalle in Germany (from BRAUCKMANN *et al.* 1985, curtesy of the author)

and earliest Middle Carboniferous) dictyoneurids are found in West Europe and are represented by 11 species belonging to Breyeriidae, Dictyoneuridae, Graphiptilidae, Homoiopteridae (see Fig. 137), Lithomanteidae, and Spilapteridae. The post-Namurian (Westphalian) Middle Carboniferous representatives were much more numerous and diverse, being represented by ca. 140 species of 21 families in the fossil sites of West Europe and North America, with leading families being the Dictyoneuridae and Spilapteridae. Still more diverse and abundant than they were in the Late Carboniferous (Stephanian), this time predominantly in West Europe (20 families, dominated by Dictyoneuridae and Spilapteridae, but also in North America (12 families, dominated by Dictyoneuridae), Siberia (3 families, with most abundant Tchirkovaeidae, see Fig. 71), China, Argentina and Tasmania, each with a single family. Most of the Namurian families persisted in the Westphalian, and all the Westphalian ones survived until the Stephanian. This distribution pattern, and particularly the low diversity of Siberian assemblage of Chunya strongly dominated by the sole species *Tchirkovaea guttata* M. Zalessky indicates that the group was thermophilous and rather uniform and stable over the equatorial belt in the Carboniferous.

Much rarer were the Early Permian dictyoneurids. 18 species are found there, mostly belonging to those families inherited from the Late Carboniferous. These are Dictyoneuridae, Spilapteridae, Calvertiellidae and Homoiopteridae represented in Europe (Obora in Czechia and Tshekarda at the Urals, see Fig. 141), in North America (Elmo in Kansas, see Fig. 135) and Siberia (Fatyanikha at Tunguska River). Several more families have appeared there (all in Europe) for the first time. These families (Ancopteridae, Arcioneuridae, Caulopteridae, and Hanidae, all from Obora in Czechia, see Figs. 142,

Fig. 136 The most ancient winged insect, *Delitzschala bitterfeldensis* Brauckmann & Schneider (Spilapteridae), from the latest Early Carboniferous of E. Germany (combined from BRAUCKMANN & SCHNEIDER 1996, curtesy of the Neue Jahrbuch für Geologie und Paläontologie); hind wing 11 mm long

Fig. 135 *Dunbaria* sp. (Spilapteridae) from the Early Permian of Elmo in USA, restored as in life by A.G. Ponomarenko (from ROHDENDORF & RASNITSYN 1980)

143) represented not only taxonomic, but also morphological and bionomic innovation. They had narrow wings, often with festooned hind margin and surely practised different type of flight comparing normal Dictyoneurida.

Particularly poor was the Late Permian dictyoneurid fauna. The Ufimian beds of Inta Formation in Pechora Basin (North Russia) have yielded 2 species of Spilapteridae (Fig. 144). Another North Russian locality, Soyana (Arkhangelsk Region) of Kazanian age, has given 1 species of Calvertiellidae (Fig. 145) and 4 more species of the highly unusual family Eukulojidae: dipterous, with wide, poorly veined wings (Fig. 146). One species of Mongolodictyidae is known from Bor-Tologoy (Kazanian or Tatarian of Central Mongolia).

It is possible to attribute the Permian decline of the order in part to the decreasing abundance of its food plants (SHAROV 1973), and in part to increasing pressure from flying predators, i.e. dragonflies.

2.2.1.2.3.2. ORDER MISCHOPTERIDA Handlirsch, 1906 (=Megasecoptera Brongniart, 1893 + Archodonata Martynov, 1932)

(a) Introductory remarks. Specialised extinct group of a distinctive general appearance (Fig. 147). The concept of the order is revised here for the reason explained above (Chapter 2.2.1.2.3a). The relevant literature is observed in ROHDENDORF & RASNITSYN (1980) and completed by F. CARPENTER (1963b, 1983, 1992a), BRAUCKMANN (1988b), SINITSHENKOVA (1993), PINTO (1986, 1994b) and PINTO & ORNELLAS (1978).

(b) Definition. Size medium to large (wing length 10–100 mm). Body slender (more robust in Eubleptina). Head small, prognathous, with large convex eyes and long sucking beak. Antenna about as long as body. Thoracic segments of similar size, pronotal paranota large in Eubleptina, reduced to several thorn-like projections or lost in others. Wings spread permanently, fore and hind wings alike, long, often petiolate, with fore margin straight (relatively short and wide in small, dipterous Permothemistidae). Cross-veins not numerous, often forming regular rows. RS and MA bases connected by stabilised cross-vein and usually angular there (unless cross-vein replaced by temporal fusion of both RS and MA for a distance, or cross-vein slanted and thus forming false base of either). Fore leg short, middle, and hind legs poorly known. Thorax and abdomen sometimes with cuticular projections, long or short, simple or branched, and arranged in regular rows.

Fig. 137 *Homoioptera vorhallensis* Brauckmann (Homoiopteridae) from the earliest Middle Carboniferous of Hagen-Vorhalle in Germany (coll. W. Sippel, specimen 3), head and thorax: note the circular pronotal paranota (courtesy C. Brauckmann)

Male genitalia with 2–3-segmented claspers, male cercus sinuate basally (known for Aspidothoracina and Permothemistina). Ovipositor short, wide (known for Protohymenoptera and Scytohymenidae). Nymph (known for Mischopterina, Fig. 148) similar to adult but with beak short, wing pads bent backward when large, pterothoracic segments with lateral spines near wing pad base, abdominal terga with hind margin denticulate. Imaginal moults not proven. Adults supposedly frequented tree crowns while nymphs probably populated lower vegetation, for their spiny armament could be an adaptation against vertebrate predators.

(c) Synapomorphies. Head prognathous; wings spread permanently; possibly also male cercus sinuate basally (unless acquired by a mischopteridan subclade).

(d) Range. Middle and Late Carboniferous, Early and, very rarely, Late Permian; worldwide though rare except Euramery and westernmost Angara.

(e) System and phylogeny. The order considered here comprises 27 families, 55 genera, and ca. 100 species arranged in 4 suborders. Families Ancopteridae, Arcioneuridae, Caulopteridae, Hanidae, Dictyoneurellidae, Frankenholziidae, and Psychroptilidae are transferred to the order Dictyoneurida (Chapter 2.2.1.2.3.1e).

The suborder Eubleptina stat. nov. (=order Eubleptoidea Laurentiaux, 1953) is proposed here to embrace families Eubleptidae (Late Carboniferous of North America, Fig. 149), Namurodiaphidae (earliest Middle Carboniferous of Germany, Fig. 150, Anchineuridae (Late Carboniferous of Spain, F. CARPENTER 1963b), Engisopteridae (Early Permian of Moravia, KUKALOVÁ-PECK 1975), Sphecorydaloididae and Xenopteridae Pinto 1986 (non Riek, 1955) (Late Carboniferous of Argentina, including *"Philiasptilon" hueneckeni* Pinto et Ornellas, which does not belong to *Philiasptilon*; PINTO & ORNELLAS 1978, PINTO 1986, 1994). Unlike other mischopteridans, these forms retained many plesiomorphies also characteristic for the diaphanopteridans, viz. the robust body, prothoracic paranota, wide costal space, wide wing base with 3 independent anal veins, and generally rich venation.

The suborder Mischopterina is essentially the same as that previously called Eumegasecoptera (ROHDENDORF 1962). It includes Allectoneuridae and Moravohymenidae (Early Permian of Moravia), Aykhalidae (Late Carboniferous of Yakutia, Fig. 151, Carbonopteridae (Late Carboniferous of Germany), Corydaloididae, Foririidae, Sphecopteridae and Ichnoptilidae (Late Carboniferous of France), Mischopteridae (Late Carboniferous of Europe and North America, see Fig. 148), Raphidiopseidae, Parabrodiidae (Middle Carboniferous of North America), Vorkutiidae (Late Carboniferous and Late Permian of Kuznetsk and Pechora Basins, respectively, Fig. 152). Mischopterinans are synapomorphic in having the elongate body lacking paranota, the wings are well costalised, homonomous, elongate (rather triangular in Mischopteridae) and usually petiolate, usually with a single pectinate A, cross-veins arranged in transverse rows, and plesiomorphic in having well developed costal space and percurrent SC.

The suborder Aspidothoracina agrees well with the former Protohymenoptera and includes Aspidothoracidae and Aspidohymenidae (Late Carboniferous of France), Brodiidae (Late Carboniferous of England and North America), Bardohymenidae (Middle Carboniferous through Early Permian of Europe, the Urals, and China, Figs. 147, 153), Brodiopteridae (Middle Carboniferous of North America), Protohymenidae (Late Carboniferous through Late Permian of north-east Europe, the Urals, China, and North America), and Scytohymenidae (Early and Late Permian of the Urals and South Africa). Aspidothoracinans are similar to Mischopterina in their wings being rather elongate, costalised, and petiolate, but synapomorphic in the wings being further costalised, (costal space practically lost, and C, SC, and R almost or well touching each other, or else SC lost).

The suborder Permothemistina (the former order Permothemistida= Archodonata) comprises families Permothemistidae (Fig. 154) and Diathemidae (Fig. 155; Early and Late Permian of North and East European Russia, respectively), Ogassidae (Late Carboniferous of Spain, SINITSHENKOVA, MARTÍNEZ-DELCLÒS, in press) and also *Kansasia pulchra* Tillyard (Early Permian of North America) of

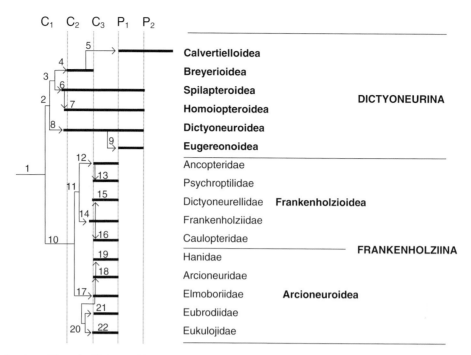

Fig. 138 Phylogeny and system of the order Dictyoneurida

Time periods are abbreviated as follows: C_1, C_2, C_3–Early (Lower), Middle, and Late (Upper) Carboniferous, P_1, P_2–Early (Lower) and Late (Upper) Permian. Three righthand columns are names of families, superfamilies, and suborders, respectively. Arrows show ancestry, thick bars are known durations time of taxa, figures refer to synapomorphies of the subtended clades, as follows:

1 – head prognathous; wings permanently spread; M_5 lost.
2 – wings wide basally, with enriched venation; prothoracic winglets long and wide.
3 – main veins arching basally.
4 – SC scarcely extending beyond wing midlength.
5 – CuP simple, main veins arching throughout.
6 – CuA branching.
7 – archedictyon lost; veins polymerised.
8 – wings long, narrow, particularly apically.
9 – archedictyon lost; cross-veins abundant.
10 – wing base narrow.
11 – wings wide apically.

12 – wing hind margin festooned.
13 – cross-veins lost.
14 – wing narrow apically.
15 – cross-veins irregular, forming intercalary sectors.
16 – wing hind margin festooned.
17 – venation impoverished.
18 – wing base forming long, narrow petiole; wing hind margin festooned.
19 – wing length 9–12.5 times width.
20 – wing strongly costalised; hind wing lost.
21 – narrow meshwork (secondary archedictyon) present throughout wing.
22 – SC fused with C; RS 2-branched; CuA and CuP simple; cross-veins lost.

obscure position. The group differs markedly from all Dictyoneuridans in being of small size (23–36 mm wingspan), having hind wings either reduced either to small scales or completely lost, antennae and cerci exceedingly long, paranota lost, and fore wing wide, with well developed pterostigma and with primary cross-vein single (rs-m), 1A anastomosing with CuP, and with 4–6 anal stalks alternating in being either convex or concave. Doteridae, Rectineuridae, and Permoneuridae are excluded from the suborder and consider *Insecta incertae sedis* for the following reasons. Doteridae differ from Permothemistina in the early origin of RS and in the absence of a pterostigma and characteristic cross-vein between RS and MA bases. Rectineuridae are rich in cross-veins turning here and there into archedictyon, while Permoneuridae have completely different venation with strong concave CuA–a character of hind wing in many insect orders (Palaeomanteida, Grylloblattida, etc.) but not in Dictyoneuridea.

Eubleptina shows no autapomorphy compared to the other three suborders and hence is considered to be their stem group. Aspidothoracina, Mischopterina, and Permothemistina are synapomorphic in having paranota lost, the body slender, and wing venation slightly impoverished. Aspidothoracina follows in the evolutionary direction displayed by Mischopterina beyond the point reached by the latter and shows no autplesiomorphy, so they are considered to be mischopterinan descendants. The same seems possible for Permothemistina, despite their higher number of anal veins compared to the Mischopterina: their fluted anal region suggests that it has experienced considerable re-structuring, evidently in connection with a transition to diptery which generally correlates with an enlarged anal area. Relationships of the families within the suborders is not clear enough to justify constructing a cladogram.

(f) History. The order enters the fossil record as early as the very beginning of the Middle Carboniferous, along with other most ancient insect groups (BRAUCKMANN *et al.* 1996). Two suborders are represented there in Namurian B of Germany, the ancestral one (Eubleptina: *Namurodiapha sippelorum* Kukalová-Peck et Brauckmann), and the highly advanced Aspidothoracina (*Sylvohymen peckae* Brauckmann). Later in the Carboniferous the Mischopterina becomes the dominant group, with Eubleptina and Aspidothoracina taking a subordinate position. Most findings are recorded in West Europe and there have been rare recordings in the Urals, China, and North and South America.

Only three more advanced mischopterid families (Vorkutiidae, Bardohymenidae, Protohymenidae) have crossed the Permo-Carboniferous boundary. The Early Permian mischopteridans, although less common than the Carboniferous ones, have retained the West European centre of diversity and the dominant position of Mischopterina. In contrast, Eubleptina have been replaced there by Permothemistina. Extra-West European records concern North America, Urals, and China, with the American and Ural findings being rather numerous and dominated by Aspidothoracina. The rare Late Permian representatives come from the Kazanian stage of North Russia (Permothemistidae: 6 species of *Permothemis* Martynov and *Ideliella* G. Zalessky) and *Karoohymen delicatulus* Riek (Scytohymenidae) from the Middle Beaufort Series of South Africa.

2.2.1.2.3.3. ORDER DIAPHANOPTERIDA Handlirsch, 1906 (=Diaphanopterodea Handlirsch, 1919)

(a) **Introductory remarks**. The concept of the order is revised here for the reason explained above (Chapter 2.2.1.2.4a). The relevant literature is covered in Rohdendorf & Rasnitsyn (1980) and should be completed by Kukalová-Peck & Sinitshenkova (1992), Kukalová-Peck (1992), F. Carpenter (1993c), Rasnitsyn & Novokshonov (1997).

(b) **Definition**. Robust or slender insects of medium to large size (20–80 mm wingspread) (Fig. 156). Head small with convex lateral eyes and a short to medium length sucking beak directed below (Fig. 157). Prothoracic paranota lateral, small. Wings similar, usually almost identical in form and venation of both pairs, lacking archedictyon, moderately or well costalised (with C, SC, and R thick and running relatively close each other, though costal space sometimes rather wide), with alternating convex (C, R, MA, CuA, A) and concave veins (RS, MP, CuP), with costal margin straight to moderately convex, MA simple, tightened sub-basally by cross-vein to or fused for a distance with RS, M_5 usually retained, cross-vein-like or forming false CuA base. Legs with tarsi 3-segmented, otherwise not evidently specialised. Abdomen segments lacking paranota. Ovipositor cutting, probably used to insert eggs into plant tissues, with sheath shorter than stylets, bearing long stylus (see Fig. 70). Male copulatory apparatus complex but not fully homologised yet. Cerci very long, very rarely found in males. Adult and most probably immature feeding by sucking ovule contents of gymnosperm plants.

(c) **Synapomorphies**. Fore tarsus 3-segmented; archedictyon lost; MA simple, tightened sub-basally by cross-vein with RS; male shedding cerci.

Fig. 139 *Eugereon boeckingi* Dohrn (Eugereonidae) from the Early Permian of the Lebachian Shales in Saar Basin in Germany (from Guthörl 1934); head with beak 38 mm long

Fig. 140 *Parathesoneura carpenteri* Sinitshenkova (Homoiopteridae) from the Late Carboniferous of Chunya in Siberia (holotype, photo by D.E. Shcherbakov); fore wing length 60 mm

Fig. 141 *Paradunbaria pectinata* Sinitshenkova (Spilapteridae) from the Early Permian of Tshekarda in Urals (holotype); body 26 mm long (from Rohdendorf & Rasnitsyn 1980)

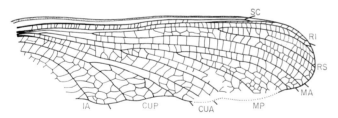

Fig. 142 *Ancoptera permiana* Kukalová-Peck (Ancopteridae) from the Early Permian of Obora in Czechia (from Kukalová-Peck 1975)

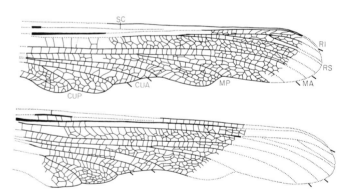

Fig. 143 *Hana filia* Kukalová-Peck (Hanidae) from the Early Permian of Obora in Czechia (from Kukalová-Peck 1975)

(d) Range. Earliest Middle Carboniferous (Namurian B) till Early Permian North Eurasia and North America.

(e) System and phylogeny. 8 families have been established in the order covering 25 genera and about 50 species. Rohdendorf (1962) has proposed segregating them into three suborders, but the following accumulation of data has not confirmed this grouping. Relationships between diaphanopterid families are not clear enough to permit construction of a cladogram.

(f) History. The Carboniferous Diaphanopterida were rare, despite that they were among the oldest winged insects. The Middle Carboniferous species belonging to Prochoropteridae is found in the Westphalian of North America, and a few more (6 species) are known in the Late Carboniferous of France (Diaphanopteridae = Diaphanopteritidae) and Siberia (Kuznetsk Basin, Diaphanopteridae). Permian (Early Permian) Diaphanopterida were much more common and diverse. They are known by 20 genera and 38 species of 6 families in East Europe (Parelmoidae), the Urals (Asthenohymenidae, Biarmohymenidae, Paruraliidae, Parelmoidae, Figs. 156–160), and in Kansas, North America (Asthenohymenidae, Elmoidae, Martynoviidae, Parelmoidae). No Late Permian representatives are known thus far.

2.2.1.2.4. SUPERORDER PSOCIDEA Leach, 1815

A.P. RASNITSYN

(a) Introductory remarks. An insect group of minor size, both in respect to body size and number of taxa covered, but of considerable phylogenetic and outstanding medical importance. The present understanding of the superorder's taxonomy and phylogeny is based primarily on data by KRISTENSEN (1975, 1981, 1995), KIM & LUDWIG (1978a,b, 1982), BOUDREAUX (1979), LYAL (1985), ŠTYS & BILINSKI (1990). The main deviation from the traditional approach, besides the typified names of higher taxa, is, firstly, that the chewing and sucking lice (former Mallophaga and Anoplura) are united in the single order Pediculida because of their deep similarity (except for their mouthpart structure) and close relationship. The second deviation is that thrips are included in this superorder and not afforded a superorder of their own: this is because the extinct family Lophioneuridae practically fills the gap between the Permian Psocida and typical thrips. Results of cladistic studies on the extant psocideans are controversial though sometimes thrips do appear there as a sister group of the Psocida + Pediculida (e.g. WHITING *et al.* 1997: fig. 10).

(b) Definition. Size small, rarely medium. General appearance highly variable depending on free living or parasitic habit and mode of parasitism. Head with clypeus strongly convex to house cibarial muscles (secondarily less developed in some parasitic forms). Mandible chewing (reverted and thus tearing instead of biting in elephant louse, reduced in sucking lice, piercing and asymmetric in thrips except Lophioneuridae where mandibles are symmetrical). Lacinia free of remaining maxilla, rod-like (not changed into stylet in

sucking parasites). Hypopharynx with 2 ovoid sclerites. Pronotum small. Wings often lost, when present, roof-like at rest (horizontal in advanced thrips), rather poor in venation, coupled in flight except in some Palaeozoic psocidans with homonomous wings, and secondarily so in advanced thrips. Legs cursorial or, in parasites, clinging, with tarsus 1–3-segmented (4-segmented in some extinct forms). 1st abdominal sternum lost. Ovipositor present or lost. Aedeagus lost in living Psocidea except in the thrips (cf. LYAL 1986). Cercus lost. Extant forms with abdominal ganglia fused in single ganglionic mass and with 4 or

Fig. 144 Hind wing of *Vorkutoneura variabilis* Sinitshenkova (Spilapteridae) from the Late Permian of Vorkuta in NE European Russia (holotype, photo by D.E. Shcherbakov); wing length, as preserved, 40 mm

Fig. 145 *Sharovia sojanica* Sinitshenkova (Calvertiellidae) from the Late Permian of Soyana in NW Russia, fore and hind wings (holotype and PIN 3353/2, photo by D.E. Shcherbakov) 45 and 53 mm long, respectively

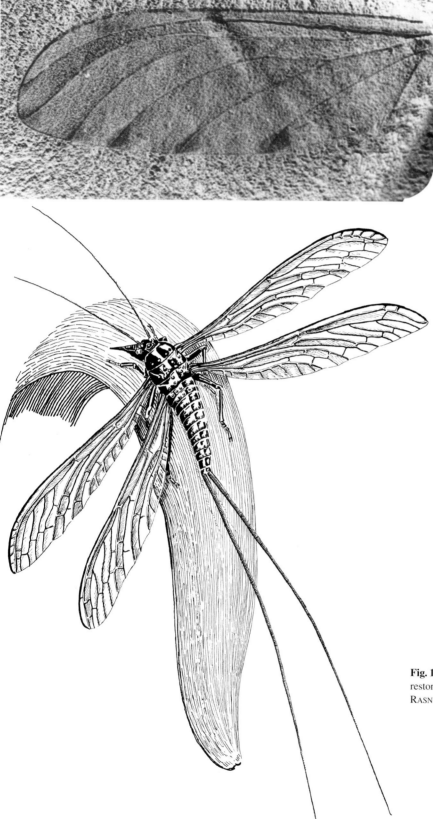

Fig. 146 *Eukuloja cubitalis* (Martynov) (Eukulojudae) from the Late Permian of Soyana in NW Russia (3353/10, photo by D.E. Shcherbakov); fore wing 25 mm long

Fig. 147 *Sylvohymen sibiricus* Kukalová. (Bardohymenidae) restored as in life by A.G. Ponomarenko (from ROHDENDORF & RASNITSYN 1980)

less malpighian tubes. Ovaries polytrophic (secondarily panoistic in Thripida: Štys & Bilinski 1990). Chromosomes holokinetic.

(c) **Synapomorphies**. Pronotum moderately small; hypopharynx with sitophore and associated structures; tarsi 4-segmented; wings with few cross-veins and with the number of branches reduced to 2 in RS and CuA and to 4 in M; CuP simple; 2 simple anal veins (lacinia free of maxilla is most probably inherited from the hypoperlid ancestor).

(d) **Range**. Early Permian till now, worldwide.

(e) **System and phylogeny** (see Fig. 109). The superorder is hypothesised as rooting among Hypoperlidae-like hypoperlideans, close to, but possibly independent of, the basal hemipterans (Chapter 2.2.1.2.5e). Within the superorder, Psocida is most probably the stem group because of absence of any sound autapomorphy. Pediculida and

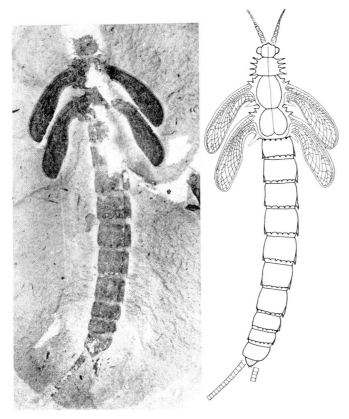

Fig. 148 Nymph of *Mischoptera douglassi* Carpenter et Richardson (Mischopteridae) from the Middle Carboniferous of Mazon Creek in USA (from Carpenter & Richardson 1968); fore wing 13.5 mm long

Fig. 149 *Eubleptus danielsi* Carpenter (Eubleptidae) from the Middle Carboniferous of Mazon Creek in USA (from Carpenter 1983); fore wing 13 mm long

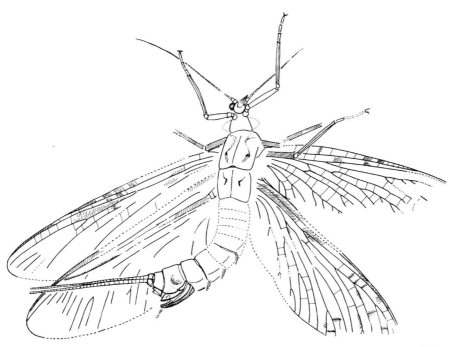

Fig. 150 *Namurodiapha sippelorum* Kukalová-Peck et Brauckmann (Namurodiaphidae) from the earliest Middle Carboniferous of Hagen-Vorhalle in Germany (from Kukalová-Peck & Brauckmann 1990); fore wing 37 mm long

Thripida most probably have independent roots within Psocida, the former starting from near the nidicolous bark lice like Liposcelididae, the latter from a more plesiomorphic ancestor within Permopsocina. The enigmatic Tshekarcephalidae with long wedge- or sickle-shaped mandibles (Fig. 161) from the Early Permian of Tshekarda (Urals Mt.) are tentatively considered here as unplaced psocideans (NOVOKSHONOV & RASNITSYN 2000).

2.2.1.2.4.1. ORDER PSOCIDA Leach, 1815. THE BOOKLICE (=Psocoptera Shipley, 1904 =Copeognatha Enderlein, 1903)

(a) Introductory remarks. Rather small insect order (over 3,000 described and 6,000 hypothesised species according to SMITHERS 1991 and NEW 1987, respectively), inconspicuous even when living free on vegetation, the more so those with the cryptic habits in crevices of leaf

Fig. 151 *Aykhal helenae* Sinitshenkova (Aykhalidae) from near Perm/Carboniferous boundary of Aykhal in Yakutia, Siberia (holotype, photo by N.D. Sinitshenkova); fore wing 22 mm long

litter, under loose bark, stones, in animal nests, and so on. The present account is based mostly on the respective chapter in ROHDENDORF & RASNITSYN (1980), with other sources cited when appropriate. The exception is the section concerning the taxonomy and phylogeny of the suborders Trogiina and Psocina [see **(e)** below].

(b) Definition. Size moderately small to small. General appearance of two main types depending on habits: winged, with body subcylindrical and head hypognathous when free living on vegetation, and depressed, prognathous and mostly wingless when more cryptic in habit. Head freely movable, with inflated clypeus, 11–50-segmented setiform antennae, chewing mandibles, and with eyes and ocelli well developed at least in winged forms. Maxillary and labial palps 4- and 1–2-segmented, respectively. Pronotum short and narrow. Wings when present roof-like in resting position, membranous, rarely sclerotised more or less, almost always heteronomous, with impoverished venation. SC at most with single apical fork, usually incomplete, RS with 2 branches, M with 3, rarely 4 branches, except for some extinct groups lacking its own base and originating from CuA, CuA with characteristic apical fork (hind wing with M and CuA usually simple), CuP weak, simple, 1 or, rarely, 2 anals, cross-vein number well below 10. Flight functionally four-winged, in-phase, anteromotoric (*sensu* GRODNITSKY 1999) in Permopsocina, functionally two-winged in others. Legs cursorial, rarely short with thick femora, with trochanter fused with femur, with tarsus 2–3-segmented (4-segmented in some fossils), 2-segmented in nymphs. Pretarsus with 2 claws. Hind coxae bearing sound producing apparatus mesally (different modes of sound production are known as well, e.g. by striking substrate with abdomen). Abdomen 9–10-segmented. Ovipositor well developed or variously reduced. Male genitalia variable. Spermatozoa enclosed in

Fig. 152 Female *Vorkutia dimina* Novokshonov (Vorkutiidae) from the Early Permian of Tshekarda in Urals (holotype, photo by D.E. Shcherbakov); body 40 mm long

Fig. 153 Male *Sylvohymen robustus* Martynov (Bardohymenidae) from the Early Permian of Tshekarda in Urals (PIN 1700/356, photo by D.E. Shcherbakov); body 33 mm long

spermatophore. Chromosomes 2n=4+1–28+1, usually 16+1 (sex determination system X0) (NEW 1987). Development gradual, with 3–8, usually 6 nymphal instars. Habits solitary or gregarious, sometimes within web nests, not uncommon in association with birds and mammals in their nests. Feeding on small particles of plant and dead animal matter, free living mostly on micro-epiphytes (BROADHEAD & WOLDA 1985).

(c) Synapomorphies. No in respect to lice and thrips.

(d) Range. Since Early Permian till now, worldwide with particular diversity in tropics.

(e) System and phylogeny of the order have been in a state of permanent reshaping for several decades. The version by SMITHERS (1972) is still the only one from which a comparatively full cladogram of the group can be constructed. The resulting cladogram is not displayed here being much too outdated, and additionally somewhat unbalanced, with primitive groups over-split and the advanced ones overly-lumped. SMITHERS (1991) himself has abandoned it in favour of the system by BADONNEL (1951). Partial rearrangement has been proposed by other students as well (e.g. MOCKFORD & GARCIA ALDRETE 1976, MOCKFORD 1978). Comparatively radical is the version by VISHNIAKOVA (1980a). In particular the suborder Trogiina is composed of superfamilies Trogioidea and Psillipsocoidea with 4 and 2 families, respectively. Suborder Psocina consists of 2 infraorders, Amphientomomorpha and Psocomorpha. Amphientomomorpha includes 3 superfamilies, Electrentomoidea (5 families), Liposcelidoidea (3 families), and Amphientomoidea (1 family). Infraorder Psocomorpha embraces 6 superfamilies: Epipsocoidea (6 families), Calopsocoidea (4 families), Caecilioidea (5 families), Elipsocoidea (new superfamily covering Philotarsidae, Elipsocidae, and Mesopsocidae), Archipsocoidea (new monotypic superfamily), and Psocoidea (6 families). Unfortunately, the respective cladogram cannot be constructed at present, for Vishniakova has not recorded synapomorphies of the taxa involved in a consistent fashion. In the next section **(f)** Vishniakova's system is used for the Palaeo- and Mesozoic, with Smithers' taxa added in square brackets when necessary.

(f) History (data from F. CARPENTER 1932, 1933, 1939, TILLYARD 1935a, PIERCE & GIBRON 1962, MOCKFORD 1969, 1986, TSUTSUMI 1974, VISHNIAKOVA 1975, 1976, ROHDENDORF & RASNITSYN 1980, ZHERIKHIN 1980a, GÜNTER 1989, POINAR 1992a, 1993, and original).

Lophioneuridae, Zoropsocidae, and their relatives are excluded from consideration being transferred to the thrips (Chapter 2.2.1.2.6).

The oldest booklice are known from the early Artinskian (Early Permian) deposits of Obora in Moravia (Czech Republic) and in later Artinskian ones at Elmo in Kansas, North America. They belong to the most archaic families Permopsocidae and Psocidiidae (=Dichentomidae), both of the suborder Permopsocina, and were rather common in past communities (ca. 6% at Elmo assemblage and not uncommon in Obora though the precise figure is unknown), with the overwhelming majority being Psocidiidae. The latter family is equally abundant also in the Kungurian (latest Early Permian) of Tshekarda at the Urals (Fig. 162), again taking more than 5% of all the insects collected, and well represented though less abundant in the Late Permian, taking some 2% of the insect assemblage in the Kazanian of Soyana in North European Russia, and possibly 1–3% in the Tatarian of Belmont in New South Wales, Australia. No other certain psocopterans are known from the Permian, for the Archipsyllidae were recorded (VISHNIAKOVA 1976) erroneously (*Eopsylla* Vishniakova belongs in fact to Psocidiidae because it possesses a complete SC unlike the Mesozoic Archipsyllidae), and the identity of *Surijokopsocus* Becker-Migdisova is obscure.

The only Triassic book-louse found among 15 thousand insects from the Late (or possibly later Middle) Triassic deposits of Madygen Formation in Central Asia, represents an aberrant Psocidiidae (undescribed as yet).

The Jurassic and Cretaceous Psocida are also very rare in collections, found in few localities, and constitute, when known, well below 0.5% of non-aquatic insect impression fossils. The German Lias has given a single species of Archipsyllidae (Permopsocina). The earlier (Oxfordian or Kimmeridgian) Late Jurassic assemblage of Karatau in South Kazakhstan is comparatively large (34 specimens found, Fig. 163). It is dominated by Electrentomoidea [Amphientomidae: Electrentominae], with subordinate Archipsyllidae and Elipsocoidea [Psocidae: Elipsocini], all undescribed except Archipsyllidae. The later Late Jurassic deposits of Shar Teg in West Mongolia yielded 4 specimens of Archipsyllidae only. As to the Early or Middle Jurassic *Pseudopsocus* Hong and *Parapsocus* Hong from Haifanggou Formation, Beipiao Basin in North China (HONG 1983a), they do not belong to book-lice. *Parapsocus* is most probably a dipteran, and the relationship of *Pseudopsocus* is obscure.

The earlier Early Cretaceous deposits of Baissa in Transbaikalia, however rich in insects, have yielded only 10 booklice. Half of them belong to Archipsyllidae, among others dominating are Epipsocoidea. In another rich Early Cretaceous (possibly Aptian) insect locality, Bon-Tsagan in Central Mongolia, a single representative of Archipsyllidae is found. 2 specimens possibly belonging to aberrant Permopsocina other than Archipsyllidae are found in Semyon near Chita in Transbaikalia and in Khurilt in Central Mongolia, in deposits supposedly of roughly the same age of Bon-Tsagan. The monotypic family Archaeatropidae closely related to Empheriidae (Atropetae in the traditional system, Trogioidea in Vishniakova's one) has been described recently from the later Early Cretaceous Álava amber in Spain (BAZ & ORTUÑO 2000). The slightly older Lebanese amber has yielded book-lice that compose some 5% of total insect crop (P.E.S. Whalley, unpublished) which are unfortunately undescribed yet. As to the Australian Edgariekiidae (JELL & DUNCAN 1986), it is a synonym of Lophioneuridae, a family of thrips (Chapter 2.2.1.2.6).

The Late Cretaceous and Cainozoic fossils are known entirely or (Cainozoic) predominantly as inclusions in fossil resins. The earliest,

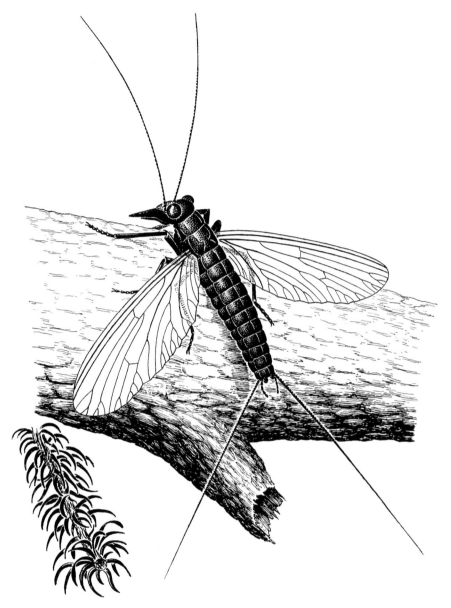

Fig. 154 *Permothemis* sp. (Permothemistidae) from the Late Permian of Soyana in NW Russia restored as in life by A.G. Ponomarenko (from ROHDENDORF & RASNITSYN 1980)

Cenomanian record in Agapa (North Siberia) concerns Trogiidae (Trogioidea) and possibly Sphaeropsocidae (Liposcelidoidea, Fig. 164), while the later, Santonian assemblage from Yantardakh (also in North Siberia) is far richer and includes members of Trogiidae (Trogioidea), Psyllipsocidae (Psyllipsocoidea), Amphientomidae (Amphientomoidea), Lachesillidae (Calopsocoidea) [Psocidae: Psocini], Elipsocidae (Elipsocoidea) [Psocidae: Elipsocini], and several fossils of obscure affinities. Pachytroctidae (Liposcelidoidea) and Psyllipsocidae are found in the Burmese amber dated imprecisely as Late Cretaceous to Eocene. The Palaeocene fossils come also from Salt Range in Pakistan (Liposcelididae: Liposcelidoidea) and in Sakhalin amber (Epipsocidae: Epipsocoidea).

The richest fossil assemblage from the Late Eocene of the Baltic amber includes Trogiidae (Fig. 165), Empheriidae, Amphientomidae, Liposcelidae (these also in the Bitterfeld amber), Sphaeropsocidae, Epipsocidae, Caeciliidae (Fig. 166), and Psocidae (Fig. 167). Next younger comparing the Baltic amber are the Early Oligocene impression fossils of Psocidae from the Bembridge in Isle of White, UK, and a book-louse of uncertain position from Florissant in Colorado, USA.

The next richest assemblages come from the Mexican amber (latest Oligocene or earliest Miocene; recorded are Psyllipsocidae, Amphientomidae, Liposcelidae, Epipsocidae, Caeciliidae, and Psocidae) and the Dominican amber (Early Miocene, mentioned are Lepidopsocidae, Psoquillidae, Psyllipsocidae, Liposcelidae, Troctopsocidae, Amphientomidae, Epipsocidae, Cladiopsocidae, Ptiloneuridae, Polypsocidae, Pseudocaeciliidae, Caeciliidae, Archipsocidae, Philotarsidae, Psocidae, Dolabellopsocidae). Trogiidae are mentioned also for the Middle Miocene Barstow Formation in California.

Fig. 155 *Diathema tenerum* Sinitshenkova (Diathemidae) from the Early Permian of Tshekarda in Urals (holotype, photo by D.E. Shcherbakov); body 13.5 mm long

The youngest, Quaternary fossils are known from the East African (Zanzibar) copal (Lepidopsocidae and Liposcelidae) and 'amber' of Mizunami in Japan (Lachesillidae).

In general, since the Permian till at least the Early Jurassic, book-lice were represented solely by the most archaic suborder Permopsocina, and even in the Late Jurassic and Early Cretaceous the permopsocinan booklice were common, often even dominating among the booklice. In the Late Jurassic, Psocina has appeared with both its infraorders Amphientomomorpha (dominant) and Psocomorpha (rare). Yet it was not before the Late Cretaceous when the latter, now the commonest group in the order, reached a dominant position in assemblages (taking 7 out of 11 fossils identifiable up to infraorder in the Santonian of Yantardakh), though its present-day scale of dominance (80% of total species number) has not been reached even in the mid-Tertiary Chiapas amber assemblage (Zherikhin, unpublished calculation). It is possible to conclude that booklice history reveals only two main stages, the Palaeozoic one dominated by Permopsocina, and the Cainozoic one dominated by Psocina. These are separated by a transitional time covering the Late Jurassic and Early Cretaceous when both suborders played important role in assemblages.

2.2.1.2.4.2. ORDER PEDICULIDA Leach, 1815. THE LICE (=Anoplura Leach, 1815, =Phthiriaptera Haeckel, 1896, =Siphunculata Latreille, 1825, + Nyrmida Leach, 1815 =Mallophaga Nitzsch, 1818)

(a) Introductory remarks. Moderately small (some 2,500 species of chewing and 500 of sucking lice, M. PRICE & GRAHAM 1997) group of permanent ectoparasites of birds and mammals, with morphology profoundly modified due to their parasitic habits. The present review is based on data from DUBININ (1948), VOIGT (1952), KÉLER (1969), EMERSON & PRICE (1985), KIM (1985), LYAL (1985, 1986), CALABY & MURRAY (1991), M. PRICE & GRAHAM (1997) and RASNITSYN & ZHERIKHIN (2000).

(b) Definition. Body size usually small, varying from medium to minute (0.5–20 mm). Body depressed (Fig. 168), hairy (except in Cretaceous *Saurodectes* Rasnitsyn et Zherikhin, Fig. 169). Head weakly movable (except possibly in *Saurodectes*), in chewing lice usually large, of variable form (long rostrate in elephant louse *Haematomyzus* Piaget), in sucking lice narrow, often small, with antennae 3–5-segmented, eyes rudimentary (with 2, 1, or no ommatidia) except might be better developed if not large in *Saurodectes*, ocelli lost. Mouthparts chewing or sucking: in Philopterina chewing, directed ventrad or, in elephant lice (*Haematomyzus*), mounting long beak and directed forward, with mandibles working inward in horizontal plane (ischnocerans, Philopteroidea) or in vertical plane (amblycerans, Ricinoidea), or else (in elephant lice) working outward in horizontal plane using inverted cutting edges to make access to blood vessel in depth of skin; lacinia small and maxillary palp 2–4-segmented or lost. In Pediculina mouthparts sucking: mandibles rudimentary, lacinia and labial palp lost; instead, 3 impair (one left from original pair) stylets (hypopharyngeal, salivary, and labial) present hidden in internal sack in repose. Thorax wingless, with segments often rather small and fused each other to a varying extent. Legs short, with coxae distant of midline and tarsus 1–2-segmented; except for Cretaceous *Saurodectes*, legs thick (less so in *Haematomyzus*), clinging, with coxae short and joining trochanter laterad (in *Saurodectes*, legs thin, coxae elongate, with trochanter joined behind and somewhat medially). Abdomen 8–10-segmented, lacking

Fig. 156 Female *Permuralia maculata* Kukalová-Peck et Sinitshenkova (former *Uralia maculata* Sharov, nomen nudum) (Parelmoidae) from the Early Permian of Tshekarda in Urals, restored as in life by A.G. Ponomarenko (from ROHDENDORF & RASNITSYN 1980)

ovipositor (rudiment of gonapophysis VIII often retained), with rather complex male genitalia comparable to that of Psocida. Nymph essentially adult-like in structure and habits. Eggs relatively large, glued firmly to host's hairs and feathers, bearing operculum to permit easier hatching. All stages inhabiting mammal and bird host permanently and often displaying considerable host specificity (up to extent useful in phylogenetic inferences); *Saurodectes* presumably living on pterosaur wing membrane. Active stages feeding on feathers, hairs, and/or blood.

(c) **Synapomorphies**. Permanently ectoparasitic on homoiotherm vertebrates, ocelli lost; antenna 5-segmented; lacinia small; legs short, coxae distant of thorax midline; egg with operculum, glued to substrate using vaginal secretion (for more complete list and discussion on living forms see LYAL 1985).

(d) **Range**. Since Early Cretaceous, now worldwide.

(e) **System and phylogeny**. The order supposedly descends from within the Psocida and in particular may be monophyletic with Liposcelididae, the inference based mostly on the apomorphies connected with the nest dwelling adaptations of Liposcelididae

(phylogenetic inferences are mostly from LYAL 1985 here). Within the order, Philopterina is probably a paraphyletic group, for the sucking lice are synapomorphic with all known Philopteroidea (=Ischnocera) in the absence of maxillary palps, with Saurodectidae, Haematomyzydae and Trichodectidae in the one-clawed pretarsus, and with elephant lice (*Haematomyzus*) in obligate sucking haematophagy. Eleven families are recognised currently in Philopterina segregated into 3 suborders here considered as superfamilies: Boopiidae, Trimenoponidae, Abrocomophagidae, Gyropidae, Laemobothriidae, Menoponidae, and Ricinidae in Ricinoidea (=Amblycera); Philopteridae, Trichophilopteridae, Heptapsogastridae, and Trichodectidae in Philopteroidea (=Ischnocera), and Haematomyzidae in Haematomyzoidea (=Rhynchophthiraptera). Most of them are bird parasites, with the exceptions being Boopiidae parasitising mostly Australian and Papuan marsupials (but also cassowary), Abrocomophagidae, Gyropidae and Trimenoponidae living on South American caviomorph rodents (the latter also on the marsupials), Trichophilopteridae living on lemurs, Trichodectidae infesting various mammals worldwide, and

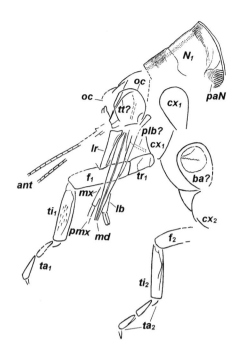

Fig. 157 Head, leg and anterior thorax of *Permuralia maculata* Kukalová-Peck et Sinitshenkova (Parelmoidae) from the Early Permian of Tshekarda in Urals: morphological interpretation (based on specimen PIN 1700/494), and photo of specimen PIN 1700/493; ant–antenna, ba–possibly basalar sclerite, cx–coxa, f–femur, lb–labium, lr–labrum, md–mandible, N–notum, oc–eye, paN–paranotum, plb?–possible labial palp, pmx–maxillary palp, oc–eye, ta$_1$, ta$_2$–fore and mid tarsus, respectively, ti$_1$, ti$_2$–fore and mid tibia, tt?–possible tentorium (head endoskeleton) (from RASNITSYN & NOVOKSHONOV 1997 with kind permission of *Insect Systematics and Evolution* formerly *Entomological Scandinavica*)

Haematomyzidae on elephants and wart-hogs in Africa and Asia. The sucking lice are synapomorphic in sucking mouthparts with rudimentary mandibles and in lacking tentorium. KIM (1985) lists 15 families not organised into superfamilies and parasitising Insectivora, Rodentia and Lagomorpha (Enderleinellidae, Hoplopleuridae, Neolinognathidae [only Insectivora], and Polyplacidae [these infesting primates as well]),

Primates (Pedicinidae, Pediculidae, Phthiridae), Dermoptera (Hamophthiridae), Tubulidentata (Hybophthiridae), Carnivora and Pinnipedia (Echinophthiridae), Artiodactyla (Microthoraciidae, Pecaroecidae, Linognathidae [also on Carnivora and Hyracoidea], and Haematopinidae [these also on Perissodactyla]), and Perissodactyla (Ratemiidae). The system is evidently over-split, with 10 of these 15 families comprising only 1 species (M. PRICE & GRAHAM 1997). For phylogenetic inferences see KIM & LUDWIG (1978a,b, 1982), LYAL (1985).

(f) History is virtually unknown, except for the giant Early Cretaceous chewing louse *Saurodectes* from Baissa in Transbaikalia (Fig. 169), tentatively associated with pterosaurs. Evidence for this comes from the supposed lice eggs found on hairs in the Late Eocene Baltic amber, and the sucking lice *Neohaematopinus relictus* Dubinin (Polyplacidae) found on the frozen corpses of rodent *Citellus* in the Pleistocene of Siberia.

2.2.1.2.4.3. ORDER THRIPIDA Fallen, 1914. (=Thysanoptera Haliday, 1836) THE THRIPS

†*V.V. ZHERIKHIN*

(a) Introductory remarks. Moderately small (some 4,500 living species: MOUND & HEMING 1991) group of small, often highly modified insects. The order is treated here in a broad sense after ZHERIKHIN (1980a) and VISHNIAKOVA (1981) including, besides the traditional Thysanoptera, also the extinct lophioneurids that combine characters of true thrips and the barklice. For additional, more general information on the order see PRIESNER (1949, 1964a,b, 1968), STANNARD (1957, 1968), DYADECHKO (1964), ANANTHAKRISHNAN (1969, 1979), MOUND & O'NEIL (1974), SCHLIEPHAKE & KLIMT (1979), MOUND & WALKER (1982a,b, 1986); for morphology and internal anatomy JORDAN (1888), PETERSON (1915), KLOCKE (1926), REYNE (1927), SHARGA (1933), MELIS (1935), DOEKSEN (1941), LANGE & RAZVYAZKINA (1953), T. JONES (1954), PRIESNER (1957), RISLER (1957), R. DAVIES (1958, 1961, 1969), MICKOLEIT (1961, 1963), MOUND (1971), W. BODE (1975), HEMING (1978), MORITZ (1982 a–d, 1989a,b), CHISHOLM & LEWIS (1992); for metamorphosis TAKAHASHI (1921), K. MÜLLER (1927), DERBENEVA (1967), VANCE (1974), KOCH (1981); and for biology SHULL (1914a,b), ANANTHAKRISHNAN (1973, 1978), T. LEWIS (1973), and BOURNIER (1983).

(b) Definition. Biologically diverse terrestrial insects living either freely (on vegetation and fungi or in litter) or in plant galls. Body usually dark pigmented, distinctly elongate, often sublinear, more or less depressed, small-sized (0.5 to 2 mm, rarely up to 12–14 mm). Head hypo- or opisthognathous. Eyes lateral in position, normally large, prominent, with large compact facets, sometimes (especially in apterous) reduced. Ocelli 3 in number in alates, absent in apterous. Antennae 4- to 11-segmented, slender, filiform or moniliform, inserted at front between eyes. Clypeus undivided. Mouthparts of piercing-and-sucking type. Labium, maxillary stipites and enlarged labrum jointly form mouthcone ensheathing other mouthparts. Mandibles and laciniae transformed into moveable stylets, right mandible often strongly reduced or missing (degenerates during embryogenesis: KIRK 1985), maxillary stylets paired and jointly forming tube-like sucking structure. Maxillary palps 2- to 4-segmented, labial palps 1- to 4-segmented. Prothorax freely movable, short. Meso- and metathoraces more or less fused forming integrated pterothorax. Two pairs of thoracic spiracles (at meso- and metathorax). Legs usually of walking type, often short. Tarsus 2- or, sometimes, 1-segmented, with characteristic large eversible arolium and more or less reduced claws. Flight

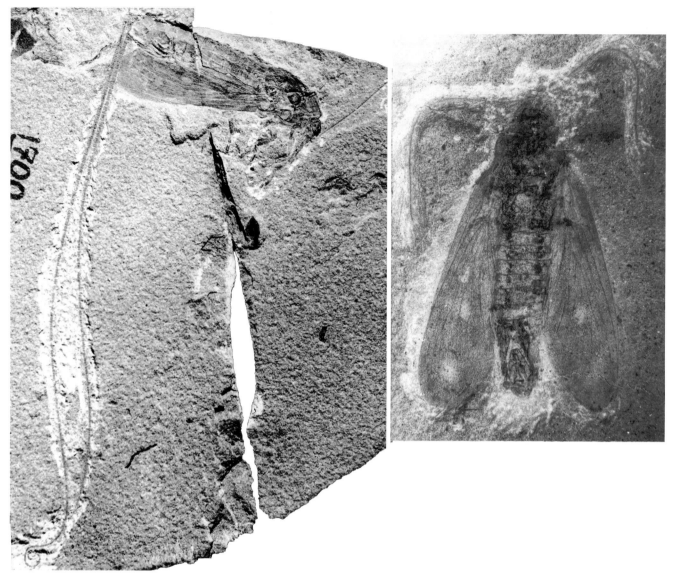

Fig. 158 Female (with long cerci) and male *Permuralia maculata* Kukalová-Peck et Sinitshenkova (Parelmoidae) from the Early Permian of Tshekarda in Urals (PIN 1700/493 and holotype, respectively, photo by A.G. Sharov); fore wing 16 mm long in female, 19 mm in male

probably functionally four-winged, in-phase, anteromotoric (*sensu* GRODNITSKY 1999). Wings slightly heteronomous or homonomous, narrow, membranous, lacking pterostigma, at rest either roof-shaped or lying flatly on abdomen, often with marginal fringe of long hairs. Wing coupling by specialised setae on fore- and hind wing (hooked in the latter); often reduced or absent in one or both sexes and commonly with intraspecific polymorphism in wing development. Wing venation more or less reduced or even lost, veins often setose, cross-veins absent. Fore wings veins weakly flexuose or straight, SC if present freely reaching wing margin, R and M with common stem; R with at most 3 branches; M fused with R basally, then anastomosing with CuA, at most with 2 branches; CuA long, simple; CuP proximally fused with CuA. Hind wing veins straight, SC and A absent. Abdomen long, with 10 visible segments (11th rudimentary), tergum 1 more or less reduced. Spiracles on segments 1 and 8 only. Cerci absent. Gonopore lying between sterna 9 and 10 in male, 8 and 9 in female. Ovipositor lacking

outer valves, reduced in Phloeothripidae with their 10th segment strongly elongate and tube-shaped in both sexes. Male genitalia simplified, symmetrical. Cibarial pump present. Intestine with 4 malpighian tubules and 4 rectal papillae of peculiar structure (W. BODE 1977). Nervous system highly concentrated, with abdominal ganglia fused. Male haploid, with one pair of fusiform testes and 1 or 2 pairs of large accessory glands. Spermatogenesis aberrant, spermatozoa of unique type (W. BODE 1983). Ovaries panoistic. Oviposition into plant tissues or (in Phloeothripidae) free. Metamorphosis of modified hemimetabolous type, with 2 larval and 2 or 3 resting non-feeding nymphal ("pupal") instars. Larvae differ in having lighter body colour, fewer antennal segments, smaller eyes and no ocelli. Nymphs sometimes in silky cocoon. Feeding habit diverse including predation, mycophagy and herbivory but in most cases individual sucking of small objects (insect eggs, mites, plant cells, pollen grains, fungal hyphae, algae), except for some Phloeothripidae ingesting fungal spores. Parthenogenesis is

Fig. 159 *Paruralia rohdendorfi* Kukalová-Peck et Sinitshenkova (Paruraliidae) from the Early Permian of Tshekarda in Urals (holotype, photo by D.E. Shcherbakov); fore wing 19 mm long

Fig. 160 Male *Asthenohymen uralicum* G. Zalessky (Asthenohymenidae) from the Early Permian of Tshekarda in Urals (PU 2/147, photo by D.E. Shcherbakov); length of body 8 mm. Note long cerci which are rarely so well preserved, and male diaphanopteridans may normally have shed them soon after eclosion

widespread, usually arrhenotokous (haplodiploid), rarely thelytokous; few species ovoviviparous. Many species gregarious or perhaps subsocial, true sociality (with defensive "soldier" cast like in some plantlice) being demonstrated in some bark-dwelling (KIESTER & STRATES 1984) and gall-inducing (CRESPI 1992) species. There are some important cross-pollinators of different angiosperms (T. LEWIS 1973).

(c) **Synapomorphies.** Clypeus undivided; labrum, labium and stipites fused forming mouth-cone; mandibles piercing; laciniae forming sucking tube; antennae with 11 segments; pterostigma absent; fore wing M fused with R basally, with 2 branches; CuA simple; CuP fused with CuA basally; hind wings with hooked setae on C; with A absent; tibiae lacking apical spurs; tarsi 2-segmented; claws reduced; arolium strongly enlarged, eversible; ovipositor lacking outer valves. Synapomorphies known in extant thrips only (such as the aberrant spermiogenesis, male haploidy and peculiar metamorphosis) should not be ascribed to the order's groundplan, because the suborder Lophioneurina differs much in many external characters from the living Thripina, and this may well be true for the synapomorphies in question. The most obvious synapomorphies in the mouthpart structures are clearly connected with the unique feeding specialisation (an individual sucking of small objects) which might be ancestral for Thripida and has determined main trends of their subsequent evolution.

(d) **Range**. Since Early Permian until the present, worldwide.

(e) **System and phylogeny**. The order includes two suborders – the extinct Lophioneurina with the single family Lophioneuridae (Early Permian–?Early Tertiary) and the Thripina that corresponds to the traditional Thysanoptera.

Lophioneurina are characterised by a number of plesiomorphies (the body relatively large-sized, with unspecialised chaetome; head hypognathous, with distinct median and frontal sutures; mouthcone symmetrical; right mandible probably well developed; antennae 7- to 11-segmented; wings broad, heteronomous, lacking marginal fringe, roof-like to nearly flat in rest position, with membrane bare, fore wings basically with SC, 1A and 2A, hind wings C with long row of hooked setae, legs long, slender, claws distinct, paired). The suborder is almost certainly paraphyletic being ancestral for other lineages. Two subfamilies are recognised (VISHNIAKOVA 1981). Zoropsocinae have wings possessing short marginal setae, veins setose, fore wings with portion of M between R+M common stem and anastomosis with CuA oblique and relatively long, common stem of CuA and CuP long. In Lophioneurinae, the wing margin and veins lack any setae, fore wings

Fig. 161 Enigmatic long-mandibulate insect of possible psocidean affinities *Tshekarcephalus bigladipotens* Novokshonov & Rasnitsyn (Tshekarcephalidae) from the Early Permian of Tshekarda in Urals (holotype); body length 5.5 mm. (from Novokshonov & Rasnitsyn 2000)

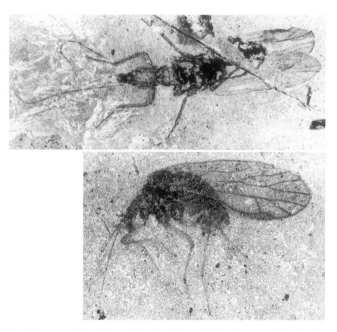

Fig. 162 Archaic booklice of the family Psocidiidae from the Early Permian of Tshekarda in Urals: *Dichentomum* sp. (upper view, PIN 1700/3272, body 7.5 mm long) and *Parapsocidium uralicum* G. Zalessky (side view, PIN 1700/1565, body 3.5 mm long; both photos by D.E. Shcherbakov)

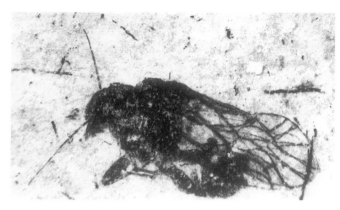

Fig. 163 Undescribed booklouse of the family Mesopsocidae from the Late Jurassic of Karatau in Kazakhstan (PIN 2784/2339, photo by V.N. Vishniakova)

Fig. 164 Undescribed booklouse with characteristically elytrised fore wings, family Sphaeropsocidae, from the Late Cretaceous fossil resin of Agapa in Siberia (PIN 3426/209, photo by V.N. Vishniakova)

M beyond separation from R+M represented by very short transverse cross-vein r+m-cua, common stem of CuA and CuP short. Relations between subfamilies are not very clear. The state of M in Lophioneurinae seems to be apomorphic, suggesting monophyly of the lineage. Polarity of other characters is not obvious. Perhaps Lohioneurinae and Zoropsocinae are sister-groups and could be treated as separate families but the possibility that Zoropsocinae represent a plesiomorphic stem-group cannot be rejected before more detailed morphological studies. Four genera of Zoropsocinae and ten of Lophioneurinae are recognised; the positions of little studied genera *Vitriala* B.-M., *Psococicadellopsis* B.-M. and *Surijokocypha* B.-M. are unclear (see Vishniakova 1981, for a review and keys, Ansorge 1996,

Fig. 165 Wingless booklouse of the family Trogiidae (undescribed) from the Late Eocene of the Baltic amber (PIN 964/165, photo by V.N. Vishniakova); body 0.65 mm long

Fig. 166 Caeciliid booklouse from the Baltic amber (PIN 964/170, photo by V.N. Vishniakova)

for synonymy, and ZHERIKHIN 2000, for an additional genus). Presence of a short mouthcone indicates that Lophioneurina had a piercing-and-sucking feeding habit, perhaps being at least facultatively pollinivorous. However, they could hardly pollinate any plants because of the absence of long setae on the body which are important for pollen transport by thrips (ANANTHAKRISHNAN 1979).

Thripina are probably monophyletic. The suborder is synapomorphic in having a specialised body chaetome, head (Fig. 170) opisthognathous, mouthparts asymmetric (in some cases only atrophy of the right mandible is obvious), head sutures obliterated, antennae with no more than 9 segments, claws forming a plate integrated with arolium, wings very narrow, fringed, nearly homonomous, at rest lying flatly on abdomen; wing venation reduced, basically lacking SC and anal veins in fore wings. Many of those synapomorphies seem to be of a relatively little importance because similar trends occur in different insect taxa and are simply connected with small body size. However, at least the unique asymmetrical mouthparts and the highly derived structure of claws support strongly monophyly of Thripina. Because of setose veins and fringed wings Thripina can hardly be connected with Lophioneurinae and may originate rather from Zoropsocinae; however, none of zoropsocine genera described up to now seem to be closely related with the most primitive Thripina known.

Thripina are divided here into two infraorders (corresponding to the suborders of traditional classification) following ZHERIKHIN (1980a): Thripomorpha (=Terebrantia) and Phloeothripomorpha (=Tubulifera).

The traditional division was criticised by SCHLIEPHAKE (1975) who treated Terebrantia as a paraphyletic group. ANANTHAKRISHNAN (1979) retained both divisions but also believed that the Terebrantia are paraphyletic and Tubulifera have originated from an ancestor allied to the living thripine subfamily Panchaetothripinae. However, later authors (especially MOUND et al. 1980, and MORITZ 1989b) argued that both lineages may be monophyletic sister-groups. BHATTI (1989a) even considered Terebrantia and Tubulifera as two separate orders. Unfortunately, their analysis is based mainly on characters that are rarely or never observable in fossils, for example the possible monophyly of the Thripomorpha which is based mostly on the structure of their sperm and on embryological features. The absence of any pre-Tertiary fossils of Phloeothripomorpha suggests that the group is relatively young, but the Mesozoic Thripina are little studied and their placement to Thripomorpha is based mainly on plesiomorphies and therefore, may be incorrect. The Late Jurassic family Liassothripidae which is now under study by F. Suchalkin is especially interesting in this respect and may be related to the phloeothripomorph lineage

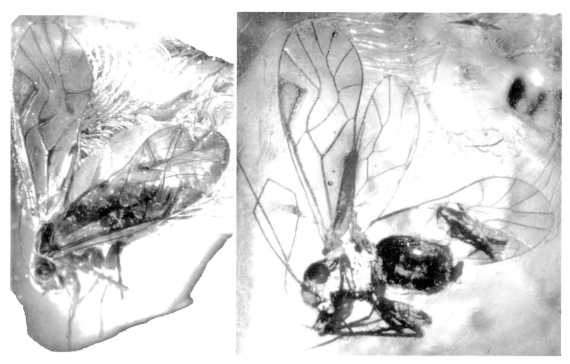

Fig. 167 Undescribed booklice of the family Psocidae from Baltic amber (left, PIN 964/179, right, PIN 964/180, photo by V.N. Vishniakova); fore wing 2.3 mm long

(F. Suchalkin, pers. comm.). Due to the absence of more detailed studies, this family is regarded here as Thripina *incertae sedis*. For a brief historical review of classification at the family level see ANANTHAKRISHNAN (1979).

The important external characters of adult Thripomorpha are as follows: all tarsi 2-segmented; wing fringe setae socketed; wings not overlapping at rest; wing microtrichia present; female 10th abdominal segment ventrally invaginated forming part of ovipositor sheath; ovipositor external, serrate, adapted to oviposition into plant tissues. Some other characteristics include presence of hatching spines in the late embryo, eggs oval, with operculum, two quiescent nymphal stages, wing pads appearing in instar 3 (nymphal 1), larval eyes with 4 ommatidia.

The number of families and superfamilies included to Thripomorpha varies in different classifications. Two superfamilies are accepted here, Aeolothripoidea and Thripoidea. The five families included in Aeolothripoidea are Karataothripidae (extinct, Late Jurassic), Uzelothripidae, Merothripidae, Aeolothripidae, and Stenurothripidae. The latter was originally established only for fossils (BAGNALL 1923) but according to BHATTI (1979, 1989b) it includes also the living genera *Oligothrips* Moulton, *Holarthrothrips* Bagn. and *Adiheterothrips* Ram. (placed by MOUND *et al.* 1980, to Thripoidea as the separate family Adiheterothripidae Shumsher). The Early Cretaceous families created by STRASSEN (1973) for fossils from the Lebanese 'amber' are synonymised by BHATTI (1979, 1989b) with Stenurothripidae, except for Jezzinothripidae which was placed into synonymy with Merothripidae. The Early Tertiary Palaeothripidae are included to Aeolothripidae following BHATTI (1979, 1989b). Karataothripidae (Fig. 171) are known from the single find of *Karataothrips jurassicus* Sharov (SHAROV 1972), and demonstrate the most plesiomorphic wing venation among all known Thripina as well as plesiomorphic, relatively broad wings with unusually short marginal setae, and may

belong to the stem group of the infraorder. Unfortunately, the most important body structures including mouthparts, tarsi and ovipositor cannot be seen clearly in the only specimen available. The oldest known Triassic Thripina are still waiting for detailed study and description; their wing shape, setation and venation seem to be more advanced than in *Karataothrips* resembling rather aeolothripids and merothripids (see FRASER *et al.* 1996: Fig. 2e, and Fig. 171). If Thripina are paraphyletic, Uzelothripidae are likely the sister-group to all other modern thrips families (MOUND *et al.* 1980). However, this aberrant family is still unrecorded in the fossil state. Aeolothripidae with about 20 living genera is the largest modern aeolothripoid family, also well represented in the fossil record; like many ancient insect families they are now nearly restricted to temperate regions of both northern and southern hemispheres. Clearly relict groups are Merothripidae with 3 living litter-dwelling genera mainly in the Neotropics and especially Stenurothripidae with 3 living flower-dwelling genera in Mediterranean area, India and western Nearctic. They are much more common and diverse in fossil assemblages than in the present-day fauna. Carnivory and pollinivory are common in Aeolothripoidea.

The superfamily Thripoidea includes three families, Hemithripidae, Heterothripidae, and Thripidae. In accordance with BHATTI (1979, 1989b) the living family Fauriellidae is considered as a synonym of Hemithripidae which was originally proposed by BAGNALL (1923, 1924) for fossils from the Baltic amber and Rott. As mentioned above, position of Stenurothripidae is disputable between Aeolothripoidea and Thripoidea. Hemithripidae is a small relict family with 3 living genera in West Palaearctic and South Africa including numerous fossils. Heterothripidae are treated by recent authors (MOUND *et al.* 1980, BHATTI 1979, 1989b) as a small family restricted to Neotropics and southern Nearctic; the taxonomic position of the Baltic amber genera *Protothrips* Priesn., *Telothrips* Priesn. and *Archaeothrips* Priesn. originally placed to Heterothripidae (PRIESNER 1924, 1929), needs to be

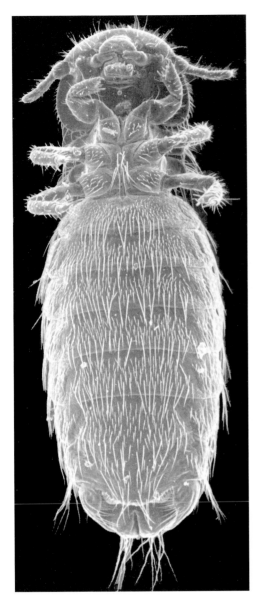

Fig. 168 Angora goat chewing louse, *Bovicola crassipes* (Rudow) (Trichodectidae), extant, ventral view, length 4.5 mm (from M. PRICE & GRAHAM 1997)

confirmed. Thripidae is the largest modern thripomorph family with about 1500 species in 230 genera. A number of suprageneric taxa was proposed, partly treated by some authors (BAGNALL 1912, KARNY 1921, 1922) as separate families. MOUND *et al.* (1980) recognise only two subfamilies, Panchaetothripinae and Thripinae. Biology of Thripoidea is diverse but herbivory and pollinivory predominate.

The infraorder Phloeothripomorpha includes the only large family Phloeothripidae. It is characterised by fore tarsi 1-segmented; wing fringe setae non-socketed, immovable, wings overlapping at rest, lacking microtrichia, venation completely reduced, 10th abdominal segment tubular in both sexes; ovipositor strongly reduced, eversible, chute-like, oviposition freely on surface, embryo lacking hatching spines, eggs reniform, lacking operculum. The phloeothripids have 3 quiescent nymphal stages; wing pads appear in the instar 4

(nymphal 2), larval eyes with 3 ommatidia. Phloeothripidae is the largest living family of the order, with nearly 3000 species known. Older authors (BAGNALL 1912; KARNY 1921, 1922) divided it into a number of families (up to 5) but in modern classifications usually only two subfamilies, Phloeothripinae and Idolothripinae, are accepted following STANNARD (1957, 1968). Many phloeothripomorphs are more or less cryptic, inhabiting litter, bark crevices, galls and other shelters. Their feeding habits are diverse, with carnivory and mycophagy relatively widespread.

The Permian genus *Permothrips* Mart. has been transferred from Thripida to the homopteran family Archescytinidae (SHAROV 1972).

(f) History. Fossil thrips are not rare, both as compression fossils and resin inclusions, but still remain insufficiently studied. They are usually hardly detectable in sedimentary rocks and often underrepresented in collections. The very narrow isolated wings of Thripina are particularly inconspicuous on rock surfaces and difficult to detect during routine palaeontological collecting unlike broader and thus more noticeable lophioneurid wings. On the other hand, the distribution pattern of fossil thrips (including lophioneurids) in sedimentary deposits (mostly in sediments of large lakes and lagoons) suggests that the main source of thrip burials is wind transport which is known to be favourable to thrips (even apterous ones) as a significant component of the air plankton (T. LEWIS 1973; KLIMT 1978).

The conclusion about presence of thrips in the Devonian (KÜHNE & SCHLÜTER 1985) is certainly wrong. It is based on a record of a supposed phloeothripomorph nymph in the Early Devonian Rhynie chert in Scotland which is clearly a result either of much younger (probably Tertiary) contamination (CROWSON 1985) or, perhaps, of a misidentification (GREENSLADE 1988). The oldest thrip fossils are known from the Early Permian of Urals (*Tschekardus* Vishn.; Fig. 172) and North America (*Cyphoneura* Carp., *Cyphoneurodes* B.-M.). All of them belong to Lophioneuridae. *Tschekardus* is a member of Zoropsocinae while both American genera represent Lophioneurinae. The Late Permian lophioneurids are described from European Russia (Soyana), Siberia (the Kuznetsk Basin), Kazakhstan (Karaungir) and Australia, and an undescribed specimen has been collected in Mongolia (Bor-Tologoy). None of the Late Permian genera is common with the Early Permian faunas. Zoropsocine genus *Zoropsocus* Till. is well represented in Siberia, Kazakhstan and Australia, and the Australian lophioneurine genus *Lophiocypha* Till. has been recorded with some reservation from Soyana (VISHNIAKOVA 1981). The lophioneurines *Lophioneura* Till. and *Austrocypha* Till. are restricted to Australia, and zoropsocine *Zoropsocoides* Vishn. as well as somewhat doubtful genera *Vitriala* B.-M. and *Surijokocypha* B.-M. to Siberia.

Very few Triassic finds are known and none of them were investigated in detail. However, the Triassic was surely a very important time from the point of view of thrips evolution because of the appearance of Thripina. Their oldest remains are found in the Late Triassic of North America (Cow Branch Formation, Carnian: FRASER *et al.* 1996) and Kazakhstan (Kenderlyk, Carnian-Norian: Fig. 173). As discussed above, they seem to represent the aeolothripoid lineage resembling living Aeolothripidae and Merothripidae, at least in their wing venation. Surely, more primitive *Karataothrips*-like aeolothripoids should exist undiscovered in the Triassic. In Kenderlyk a poorly preserved lophioneurid has also been found.

In the Jurassic Lophioneurina are still more common and widespread than Thripina which may be partially explained by taphonomical factors as discussed above. A number of species of both Zoropsocinae and Lophioneurinae are described from Germany, Siberia, Kazakhstan, and

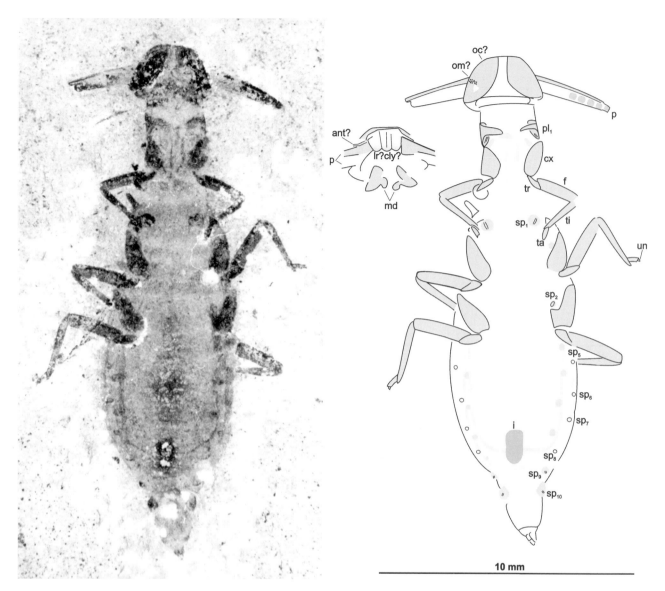

Fig. 169 Supposed pterosaur chewing louse *Saurodectes vrsanskyi* Rasnitsyn & Zherikhin (Saurodectidae) from the Early Cretaceous of Baissa in Siberia (holotype), general view and morphological interpretation (left is head from below); ant?–possibly a part of antenna, cx–coxa, f–femur, i–dark, apparently structureless intestine contens, lr?cly?–anterior boundary of mouth cavity made of either labrum or clypeus (or both), md–mandible, oc?–eye-like sclerotisation, om?–possible ommatidia, p–lateral head process (trabecula?), pl$_1$–propleuron, sp$_1$, sp$_2$–meso- and metathoracic spiracles, respectively, sp$_5$–sp$_{10}$–spiracles of 3rd to 8th abdominal segments, respectively, ta–tarsus, ti–tibia, un–claw (from Rasnitsyn & Zherikhin 1999)

China (Vishniakova 1981; Hong 1983a, 1992a; Ansorge 1996) and many undescribed fossils are represented in collections from the Early Jurassic of Germany (Ansorge 1996, and pers. comm.), Early (Krasnoyarsk) and Middle Jurassic (Kubekovo) of Siberia (Zherikhin 1985), Early (Sogyuty) and Middle or Late Jurassic (Say Sagul) of Central Asia, and Late Jurassic of Mongolia (Bakhar, Shar Teg). *Zoropsocus* Till. (Fig. 174) persisted there since the Late Permian being perhaps the only Late Permian insect genus to have survived up to the Late Jurassic (Uda and Karatau). Two extinct families of Thripina, Karataothripidae and Liassothripidae, are recorded from the Late Jurassic (Sharov 1972). Liassothripidae dominate over Lophioneuridae numerically in the Late Jurassic lacustrine deposits of Karatau in Kazakhstan where insect body fossils are common. A poorly preserved specimen of a member of Thripina has also been found also in the Late Jurassic of Mongolia (Khoutiyn-Khotgor). At the present state of knowledge no evident phylogenetic progress in the Jurassic can be documented in comparison with the Late Triassic. No members of living families are known to appear there.

A decline of Lophioneurina is evident since the beginning of the Early Cretaceous when aeolothripoids become common and diverse. Nevertheless, in some Early Cretaceous localities only lophioneurids have been discovered, though always in a small number. This is particularly true for the Glushkovo, Ukurey and Godymboy Formations in Transbaikalia (Vishniakova 1981, and undescribed materials), all of disputable (Jurassic or Early Cretaceous) age (Chapter 4.1). Equally no Thripina is represented in a small thrip collection from Khotont in

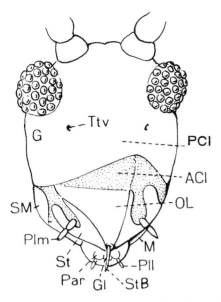

Fig. 170 Head morphology of a thripomorph thrip (from WEBER 1933): Acl – anteclypeus, G – cheek, M – mentum, OL – labrum, Par – paraglossa, PCl – postclypeus, Pll – labial palp, Plm – maxillary palp, SM – submentum, St – stipes, StB – stylet, Ttv – fore tentorial pit

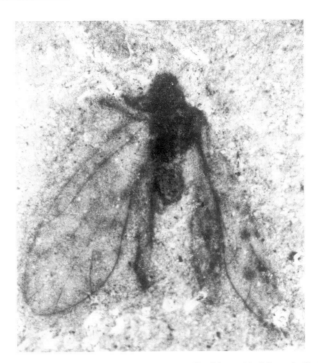

Fig. 172 *Tschekardus hispidus* Vishniakova (Lophioneuridae) from the Early Permian of Tshekarda in the Urals, Russia (holotype, photo by D.E. Shcherbakov); fore wing length 4.2 mm

Fig. 171 *Karataothrips jurassicus* Sharov (Karataothripidae) from the Late Jurassic of Karatau in Kazakhstan (from Sharov 1972); body 2.7 mm long

Mongolia (earliest Cretaceous) which including 3 lophioneurid specimens. One lophioneurid species has been found in Koonwarra (Australia) described as a new barklice genus (*Edgariekia* Jell et Duncan) and family (JELL & DUNCAN 1986), but ANSORGE (1996) has synonymised it with *Undacypha* Vishn. (Lophioneuridae). Finally, a lophioneurid specimen is illustrated by JARZEMBOWSKI (1984) from the Wealden of England where no Thripina were found. Both in Khotont

and in the English Wealden, the absence of Thripina may be explained by taphonomical biases because insect body fossils are generally rare there. However, it is not the case in the above-mentioned formations in Transbaikalia and the Koonwarra Fossil Bed. Interestingly in both cases, climate at the time of deposition of insect-bearing sediments were probably rather cool. It is noteworthy that all the known Cretaceous lophioneurids, including undescribed ones in collections of the Palaeontological Institute RAS in Moscow, and the Wealden specimen figured, seem to belong to Lophioneurinae. The genus *Undacypha* widespread in the Early Cretaceous persisted since the Jurassic faunas (ANSORGE 1996).

Unlike the above, Thripina predominate numerically in large collections from Khutel-Khara in Mongolia (Fig. 173) and from Baissa in Transbaikalia where lophioneurids constitute less than 15% of total thrips finds. Undescribed Thripina are found also in the Early Cretaceous of Western (Gurvan-Ereniy-Nuru, figured in RASNITSYN 1986b) and Central Mongolia (Bon-Tsagan) and in Eastern Transbaikalia (Turga). Thripina are also found in the Early Cretaceous fossil resins of Lebanon (STRASSEN 1973) and Álava in Spain (ORTUÑO 1998, ALONSO et al. 2000). No Early Cretaceous Thripina were described up to now except for several monobasic genera (*Jezzinothrips* Strassen, *Exitelothrips* Strassen, *Scudderothrips* Strassen, *Neocomothrips* Strassen, *Progonothrips* Strassen, *Rhetinothrips* Strassen, *Scaphothrips* Strassen) from the Lebanese amber. As mentioned above all of them but *Jezzinothrips* are now placed into Stenurothripidae while the latter genus is considered a merothripid. A genus similar if not identical to *Scaphothrips* was found in the amber-like resin from the Begichev Formation (Late Albian or Early Cenomanian) of Taimyr Peninsula, Northern Siberia (ZHERIKHIN 1978).

Preliminary observations on undescribed materials show that Aeolothripoidea is the dominant if not the only superfamily of

Fig. 173 Undescribed thrips of the superfamily Aeolothripoidea from the earliest Cretaceous of Khutel Khara in Mongolia (top, PIN 3965/159, body length 1.8 mm), and from the Late Triassic of Kenderlyk in Kazakhstan (below, PIN 2947/297, body length up to the top of ovipositor 1.7 mm; both photos by D.E. Shcherbakov)

Fig. 175 *Jantardachus perfectus* Vishniakova (Lophioneuridae) from the Late Cretaceous of Yantardakh in Siberia (holotype, photo by D.E. Shcherbakov); fore wing length 1.0 mm

Fig. 174 *Zoropsocus itschetuensis* Vishniakova (Lophioneuridae) from the Early or Middle Jurassic of Novospasskoe in Siberia (holotype, photo by D.E. Shcherbakov); fore wing length 2.5 mm

Thripina in all the Early Cretaceous assemblages and that the Early Cretaceous aeolothripoids are of a relatively modern appearance probably representing living families. No Karataothripidae are found. In the Late Cretaceous thrips are found mainly in fossil resins in USA (GRIMALDI & SHEDRINSKI 2000), Canada (PIKE 1994), Siberia and Azerbaidzhan (ZHERIKHIN 1978). Few compression fossils have been discovered in Obeshchayushchiy (Russian Far East). Lophioneurids are very rare in the Late Cretaceous and discovered only in fossils

resins of Yantardakh in Northern Siberia (VISHNIAKOVA 1981) where two species of the lophioneurine genus *Jantardachus* Vishn. are found (Fig. 175). None of the Late Cretaceous Thripina were described but Stenurothripidae (as Scaphothripidae and Scudderothripidae), Aeolothripidae, Thripidae and alleged Heterothripidae have been preliminary reported from Azerbaidzhan and Siberian resins (ZHERIKHIN 1978). The record of a thripid from Yantardakh is the oldest known. In general, the Cretaceous thrip faunas seem to be strongly dominated by aeolothripoids; lophioneurids are rare, in particular in the Late Cretaceous, and thripoids appear in small numbers no later than in the Santonian.

In the Early Tertiary, thrips are found in many localities, both as compression fossils and resin inclusions. The fauna of the Baltic amber is the best studied (see SCHLIEPHAKE 2001 and references herein). A number of taxa have been described also from the Eocene Green River Series of USA (SCUDDER 1890b), the latest Eocene or earliest Oligocene of the Isle of Wight in England (COCKERELL 1917a), the saliniferous Early Oligocene lagoon deposits of Alsace (PRIESNER & QUIÈVREUX 1935), and the Late Oligocene deposits of Aix-en-Provence (HEER 1856; OUSTALET 1873). There are also records from other localities, occasionally with preliminary identifications up to family level (Palaeocene Sakhalinian amber; ZHERIKHIN 1978) or illustrations (Oligocene Sicilian amber: KOHRING & SCHLÜTER 1989, and Late Oligocene of Ceréste, France: LUTZ 1984a). SCHLIEPHAKE

(2001 and references therein) has described numerous new species and several new genera from the Saxonian amber. Since the faunas of other insects in the Saxonian and Baltic amber show very little difference (mainly quantitative), the results by Schliephake are unexpected and need to be confirmed. *Burmacypha*, a new and highly aberrant lophioneurine genus (Fig. 176) was described recently from the Burmese amber of a disputable (Late Cretaceous or Early Tertiary) age (ZHERIKHIN 2000), where also about a dozen of specimens of unidentified Thripina were found (RASNITSYN & ROSS 2000).

In general, the Early Tertiary thrips assemblages seem to be rather uniform in their taxonomic composition. If the Burmese amber is Early Tertiary, *Burmacypha* is the only Cainozoic lophioneurid known. It should be noted that the genus seems to be rather isolated within the Lophioneurinae. However, several wings with a similar aberrant venation were discovered recently in the Early Jurassic of Germany (J. Ansorge, pers. comm.). Aeolothripoids are common and diverse; in particular, Stenurothripidae seem to be more abundant than now. The numerically dominant and most diverse family is, however, Thripidae. Among other thripoids, Hemithripidae are not rare. The oldest known Phloeothripomorpha are described from the Baltic amber; their role in the Early Tertiary assemblages is relatively low. The Baltic amber species of Aeolothripidae, Merothripidae, Thripidae, and Phloeothripidae in part belong to living genera; on the other hand, even in the youngest Oligocene fauna of Rott, many genera are extinct.

Late Tertiary thrips are poorly known. A few taxa were described and recorded from the Miocene of Germany (Oeningen, HEER 1864–1865; the taxa need in revision; and Rott, SCHLECHTENDAL 1887; BAGNALL 1924c) and California (PALMER *et al.* 1957). Recently PEÑALVER (1998) has described and figured diverse thrips from the Early Miocene of Rubielos de Mora (Spain). Only Aeolothripidae, Thripidae, Phloeothripidae and, in the earliest Miocene of Rott, Hemithripidae are recorded. Though the published data are scarce they probably indicate a decline of Stenurothripidae and Hemithripidae in comparison with the Early Tertiary while Aeolothripidae are still very common and Phloeothripidae relatively rare. This may be at least partially a taphonomic bias because modern aeolothripids are often highly active fliers while many phloeothripids are shelter-dwellers. If so, it may be predicted that the former family should be better represented in the Late Tertiary resins. Unfortunately, though thrips are recorded from the Miocene Mexican and Dominican amber (POINAR 1992a), there are no data available on their taxonomic composition.

Although there is a number of modern gall-inducing thrips species no fossil galls or any other ichnofossils have been attributed to thrips.

2.2.1.2.5. SUPERORDER CIMICIDEA Laicharting, 1781 ORDER HEMIPTERA Linné, 1758. The Bugs, Cicadas, Plantlice, Scale Insects, etc.
(=Cimicida Laicharting, 1781, =Homoptera Leach, 1815 + Heteroptera Latreille, 1810)

D.E. SHCHERBAKOV AND YU.A. POPOV

(a) Introductory remarks. One of the leading insect orders, with their 82,000 described species (ARNETT 1985) being the most diverse and economically important group among those that lack complete metamorphosis (although it is the hemipterans and particularly the whiteflies and scale insects which display examples of complete metamorphosis gained independently of holometabolans, with the only other instance being in the thrips). What is particularly worth mentioning is that this

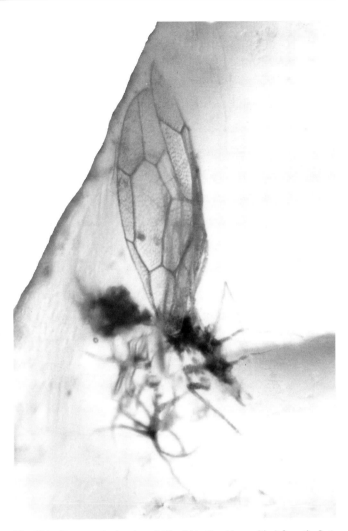

Fig. 176 *Burmacypha longicornis* Zherikhin (Lophioneuridae) from the Late Cretaceous or Early Tertiary Burmese amber (from Ross & York 2000); fore wing length 1.25 mm (courtesy of The Natural History Museum, London)

indisputable evolutionary success has been achieved in spite of significant evolutionary limitations put on the group, due to loss of chewing mouthparts which were replaced by the piercing-and-sucking ones.

The present account is based, besides on original data, on the information gained primarily from J. EVANS (1956), BECKER-MIGDISOVA (1962, 1985), POPOV (1971, 1981), POPOV & WOOTTON (1977), KEILBACH (1982), HEIE (1987, 1999), KOTEJA (1989a), POPOV & SHCHERBAKOV (1996), SHCHERBAKOV (1996, 2000a,b); other sources are cited when necessary.

(b) Definition. Size and appearance various (Figs. 177, 178), size usually small. Head from hypognathous and nearly immobile (most homopterans) to prognathous and freely movable (some predatory bugs). Antennae moderately long to short, usually of fewer than 10 segments. Mouthparts transformed into a sucking proboscis, ensheathed with 1–4-segmented labium and consisting of stylet-like mandibles and maxillae (with bases invaginated into cranium) forming bundle and leaving food and salivary canals medially. Palps lost. Cibarial dilator muscles originating from enlarged postclypeus. Lateral ocelli plesiomorphically close to eyes. Wings homonomous and uncoupled in flight (in Archescytinina only) or heteronomous and

Fig. 177 The most primitive hemipteran, Archescytinidae, restored sucking a strobile of an Early Permian gymposperm plant (by A.G. Ponomarenko, from ROHDENDORF & RASNITSYN 1980))

Fig. 178 The Late Jurassic giant water bug, *Mesobelostomum* sp. (Belostomatidae), restored as in life by A.G.Ponomarenko and V.I.Dorofeev (from ROHDENDORF & RASNITSYN 1980)

coupled at flight by means of (sub)marginal devices, usually with both pairs well developed subdipterous in males of Coccomorpha, (sub)brachypterous with hind wings usually lost in some leafhoppers and bugs, apterous in most aphid and all coccid females). Accordingly, flight functionally four-winged, probably anteromotoric in Archescytinina, functionally (sometimes even morphologically) two-winged, in-phase, anteromotoric in others (GRODNITSKY 1999). Except in Aphidina and Archescytinina, fore wing more sclerotised than hind one and fixed on the thorax in repose, with at least its posterior margin inserted into lateral mesoscutellar groove. Hind wing either diminished or with anal area expanded. SC closely associated with or fused to R, its apical portion simulating a branch of the latter. Venation basically simple with 2 anal veins and 2 cross-veins, secondarily either prolific or reduced up to simple (SC+)R in fore wings of some coccid males. Except in Aphidina, hind leg plesiomorphically modified and armed for jumping, secondarily often similar to middle one. Tarsi 3–1-segmented. Ovipositor laciniate (with cutting blades), or variously modified, or reduced. Male genitalia variable, always including a penis and (except for scale insects) a pair of parameres. Cerci lost. 10th and 11th abdominal segments forming small anal tube, modified or reduced in most Aphidina and Psyllina. Ovarioles telotrophic. Chromosomes holokinetic. Water-shunting filter chambers of various kinds and symbiotic bacteria in mycetomes present in most homopterans and some bugs.

Nymphs dorsoventrally flattened, oval, cryptic, unable to jump, secondarily often more adult-like; whiteflies and male coccids with last instar(s) transformed into non-feeding sedentary metamorphic (pupal) stage(s).

Feeding on contents of phloem, xylem, cambium or parenchyma cells of vascular plants, on seeds, rarely on mosses, fungi or algae (homopterans and some bugs), forming galls (some Psylloidea, Aphidoidea, and Coccoidea), scavengers and predators (majority of bug families), rarely hematophages and/or ectoparasites (some Cimicomorpha and Lygaeidae). In aphids, complicated life cycles occurring with parthenogenetic and sexual generations and usually with host alternation. Some treehoppers and bugs showing parental care. Sound (usually in form of vibration signals transmitted through solid media) commonly produced by either stridulation (in psyllids, bugs, and extinct hopper families Dysmorphoptilidae and Ipsviciidae) or by means of tymbals (cicadas, hoppers, and some Pentatomoidea).

(c) Synapomorphies. Mandibles and maxillae transformed into long piercing stylets ensheathed with labium. Palps lost. Lateral ocelli adjacent to inner orbits, median one close to postclypeus. SC adjoining R. Tarsi 3-segmented. Cerci lost. Nymph oval, dorsoventrally flattened, with short legs. Possibly also: hind legs jumping with apical pectens of setae-bearing teeth on tibia and on enlarged two proximal tarsomeres (either not acquired yet or lost by unknown archescytinoids ancestral to Aphidina).

(d) Range. Since Early Permian until the present, worldwide.

(e) System and phylogeny. The superorder is traditionally divided into two orders, Homoptera and Heteroptera. Unlike this, we consider it comprising 5 major clades and an ancestral group, all treated as suborders within a single order herein (Fig. 179). The system accepted follows subdivisions of Homoptera first proposed by BÖRNER (1904), with his superfamilies turning suborders: Aphidina (including Coccomorpha), Psyllina (including Aleyrodomorpha; both suborders combined are known as Sternorrhyncha), and Cicadina (=Auchenorrhyncha). The splitting of Sternorrhyncha agrees well with both morphological and palaeontological data (SCHLEE 1969a,b; SHCHERBAKOV

1990). The Permian family Archescytinidae, ancestral to both Psyllina and Cicadina (and probably to Aphidina as well), represents a fourth suborder, Archescytinina (=Paleorrhyncha). Cicadina comprise only two infraorders, Fulgoromorpha and Cicadomorpha, the subdivision being thoroughly substantiated by H. HANSEN (1890). Peloridiina (=Coleorrhyncha) and Cimicina (=Heteroptera), usually considered sister-groups, evolved from generalised Cicadomorpha as independent stocks, acquired wing coupling of the same type and dorsoventrally flattened habitus with fore wing overlap in parallel, and therefore are separated at subordinal level (POPOV & SHCHERBAKOV 1991). Heteropteran infraorders correspond rather to superfamilies of Cicadina in degree of their morphological divergence; in contrast to homopterans, only few extinct families and no extinct superfamilies of true bugs are recorded.

Archescytinina, the basal group of Hemiptera, was believed to descend from the early Permopsocina (Psocida; ROHDENDORF & RASNITSYN 1980). However, the only synapomorphy of both orders revealed up to now is the complete loss of cerci, whereas some apomorphic modifications in the wing and body structure of Permopsocina and Archescytinina are different (cf. Figs. 162, 177, 180). The hypothetical common ancestor of both orders (unless the cerci have been lost independently in the two groups) could be similar to some Hypoperlidae in all salient features, and thus easily confused with them in the fossil state, for Hypoperlidae had the cerci already short one-segmented and therefore inconspicuous. Undescribed intermediate forms between Hypoperlidae and Archescytinidae are known to occur in the Early Permian of the Urals (Tshekarda locality).

Some important changes in family-group and higher taxa are proposed in the recent papers or herein. The superfamily Canadaphidoidea was created by HEIE (1981) for Canadaphididae and Palaeoaphididae, the former being transferred to Aphidoidea (HEIE & PIKE 1996), so that the superfamily Palaeoaphidoidea stat. nov. is proposed herein for Palaeoaphididae, Genaphididae, Shaposhnikoviidae, Creaphididae, Triassoaphididae (the family erected by HEIE 1999), and possibly Tajmyraphididae. Boreoscytidae and Pincombeidae are united as Pincombeomorpha and considered the most generalised Aphidina (SHCHERBAKOV 1990); the four-winged aphid-like precoccids constitute a superfamily Naibioidea included in Coccomorpha (SHCHERBAKOV in press). Psylloidea are accepted in the broad sense, including Liadopsyllidae and Malmopsyllidae (BECKER-MIGDISOVA 1985); Neopsylloididae are synonymised under Malmopsyllidae (KLIMASZEWSKI & WOJCIECHOWSKI 1992); Protopsylliidiidae are removed from Pincombeoidea and singled out as a superfamily, according to VONDRÁČEK (1957).

Surijokocixiidae (SHCHERBAKOV 2000b) comprise the Permian and Triassic Fulgoroidea more primitive than Fulgoridiidae. Coleoscytoidea probably represent the earliest Fulgoromorpha; Permian Cicadomorpha are classified into 4 extinct superfamilies (SHCHERBAKOV 1984). Within Cicadomorpha, a group of Mesozoic Clypeata ancestral to all 3 modern superfamilies (Cicadoidea, Cercopoidea, and Membracoidea) is treated as the superfamily Hylicelloidea (SHCHERBAKOV 1996). Membracoidea sensu HAMILTON (1983) embrace Karajassidae (basal for the superfamily) and Cicadellidae s. l. (=Jascopidae?); Paracarsonini are transferred to Ledrinae, and Archijassinae are included in Hylicellidae s. l. (SHCHERBAKOV 1992).

Relic modern Peloridiidae with a far-southern distribution, living cryptic on moss, are the only extant representatives of the suborder Peloridiina and have numerous fossil relatives. Earliest members of this lineage, Progonocimicidae (Progonocimicoidea), were well-jumping

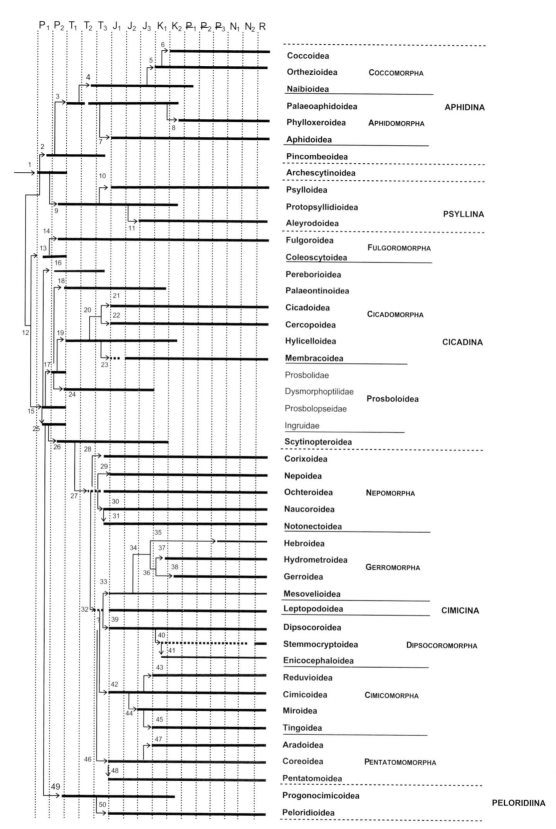

Fig. 179 Phylogeny and system of the order Cimicida

Time periods are abbreviated as follows: P_1, P_2 – Early (Lower) and Late (Upper) Permian, T_1, T_2, T_3 – Early, Middle and Late Triassic, J_1, J_2, J_3 – Early, Middle and Late Jurassic, K_1, K_2 – Early and Late Cretaceous, P_1 – Palaeocene, P_2 – Eocene, P_3 – Oligocene, N_1 – Miocene, N_2 – Pliocene, R – present time (Holocene).

Right columns are names of suborders (all caps), infraorders (small caps), superfamilies (bold lower case) and families (light lower case). Arrows show ancestry, thick bars are known durations of taxa (dashed when hypothetical), figures refer to synapomorphies of the subtended clades, as follows.

1 – mandibles and maxillae transformed into long piercing stylets ensheathed by labium; palpi lost; lateral ocelli adjacent to inner orbits, median one close to postclypeus; SC adjoining R; M_{3+4} basically simple; tarsi 3-segmented; hind legs jumping with apical pectens of setae-bearing teeth on tibia and on enlarged two proximal tarsomeres (either not acquired yet or lost by unknown archescytinoids ancestral to Aphidina); cerci lost; nymph oval, dorsoventrally flattened, with short legs.

2 – fore wing subtriangular with costal space, pterostigma and clavus narrow (R and R_1 in straight line); claval margins of both fore wings widely separated in repose; fore- and hind wing coupled in flight by means of marginal folds; in fore wing 2A fused to 1A at least distally, and the fold extending beyond clavus supported by CuA continued with CuA_2; hind wing rather small with venation simplified; coxae widely separated mesally.

3 – in fore wing cross-vein r-m lost, M concave and fold-like proximally, CuP much reduced, 1A+2A marginal, and RS tending to originate distally; hind wing lacking RS, with hamuli at anterior margin; possibly also (unless acquired at node 2): antenna with processus terminalis; eye with ocular tubercle; C atracheate in fore wing; mesepisternum with lateropleurite; tarsi 2-segmented (1st segment short), pretarsal lobes lacking; ovipositor much reduced.

4 – head flattened dorsoventrally; ommatidia tend to be widely spaced and not numerous; pro- and metanotum considerably reduced; in fore wing SC and R_1 extending distally, M simple, CuA composing of both convex and concave components; membrane tuberculate to corrugate; possibly also (unless acquired later, at node 5): penis proximally enclosed in sheath, parameres lost.

5 – rhinaria and processus terminalis lost; true ocelli lost; hind wing reduced to hamulohaltere, resulting in basal displacement of coupling fold and lack of CuA_2 in fore wing; transverse ventral suture of mesothorax lost; tarsal claw single; ovipositor lost; female apterous, male lacking rostrum.

6 – abdominal spiracles lost (after DANZIG 1980); in female, rostrum 3-segmented, antenna of less than 11 segments, narrow anal lobes developed; in male, compound eyes dissociated, neck region developed, and 3rd instar sedentary.

7 – siphuncular pores developed; anal tube transformed into cauda.

8 – antennal flagellum of 3 to 1 segments, with 4 – 1 enlarged rhinaria; M simple in fore wing; 1 or none oblique veins in hind wing; anal tube lost.

9 – fore wing tough with thick veins, SC reduced (its trachea displaced forwards), M fused with CuA for short distance beyond basal cell, with m-cu lost; fore- and hind wing coupled in flight; fore wing with 2A fused to 1A distally, and with coupling fold restricted to claval margin; hind wing with hamuli at costal margin, M_{1+2} simple and m-cu lost; tarsi 2-segmented.

10 – fore wing with CuA base lost and 1A+2A simple; ovipositor (except for apex) concealed between 7th sternum and 9th tergum; anal tube lost in nymph and female.

11 – fore wing with RS concave, M reduced, both CuA and 1A+2A simple; fore- and hind wing uncoupled in flight; metathorax enlarged; tibiae with setal rows used for wax distribution; abdomen petiolate, desclerotised, with spiracles reduced up to 2 pairs; ovipositor lacking sheath; anal tube transformed into operculum bearing lingula.

12 – (partly after EMELJANOV 1987; true also for nodes 15, 22–25) – antennae short; pronotum produced on to mesoscutum; fore wing sclerotised, with C bicarinate and middle portion of SC merged by R; mesoscutellum triangularly produced on to metascutum; claval margins fixed in scutellar grooves in repose; fore- and hind wing coupled in flight by means of marginal folds; fore wing with 2A fused to 1A distally, and with marginal fold restricted to claval margin; hind wing with marginal fold short (lobe-like), and anal area enlarged.

13 – mesonotum with longitudinal carinae; tegula concealing fore wing base; basicubital triangle short in fore wing; coxae separated mesally.

14 – mandibular stylets broadly contiguous ventrad of maxillary ones; pronotum carinate with paranota appressed to pleura; fore wing with short basal cell and long 1A+2A entering commissural margin before claval apex; metanotal posterior wing process lost; metacoxa immobilised, lacking trochantine; hind leg with trochantero-femoral articulation monocondylar.

15 – pronotum concealing fore wing bases; fore wing punctate at least basally, short 1A+2A entering claval apex; hind wing with anal fold posterior of 2A.

16 – fore wing with venation prolific and apical portion of SC bearing longitudinal branches.

17 – fore wing with commissural space deflected beneath 2A distally (2A seems entering margin).

18 – median ocellus lost; fore wing with nodal flexion line hypertrophied, m-cu bent and elongate; jumping ability lost, legs clinging.

19 – postclypeus enlarged (presumably xylem-feeders); ocelli close each other; fore wing with radial and medial cells always closed, and commissural space reduced distally (2A secondarily free); nymphs probably cryptobiotic (in soil etc.).

20 – maxillary bundle subquadrate in cross-section, rotated at 90° relative to mandibles; metanotal posterior wing process articulated; arolium with central seta; metacoxal meron separated by deep groove; aedeagal endosoma eversible; nymph not flattened.

21 – fore wing with nodal flexion line hypertrophied; arolium reduced; subgenital plates reduced.

22 – fore wing punctate all over, with SC separated from R+M; metepisternal apodeme reduced in adult.

23 – hind coxae transversely enlarged, lacking meracanthae; hind femur elongated, gliding over straight proximal anterior margin of fore wing; fore wing narrow, with M_{3+4} and CuA_1 tending to form anastomosis; hind tibia with longitudinal rows of macrosetae, subsequently replacing apical pectens; unguitractor extended distally; proximal portion of phallobase separated as connective; notopleural suture lost in nymphal prothorax.

24 – fore wing entirely punctate with stridulitrum on underside of costal space (plectrum presumably at hind femur).

25 – fore wing with coupling fold shifted to 1A+2A and shortened, whereas that of hind wing elongated.

26 – fore wing punctate all over, with costal fracture at apex of basal cell (substituting nodal flexion line), hypocostal pit mounting mesepimeral knob in repose (Druckknopfsystem), and SC humped toward C.

27 – median ocellus lost; gula developed; tentorium reduced; maxillary sensory pit lost; fore wings overlapping apically in repose (right one uppermost, or racemic distribution), with overlapping portion separated as membrane, CuA_2 short, transverse, appendix broad, and coupling fold very short; cubital furrow forked in hind wing; scent glands developed; zoophagy.

28 – labium broad, nonsegmented; hind legs natatory, fringed with setae.

29 – fore wing venation reticulate; 8th abdominal tergum modified into a pair of respiratory processes.

30 – ocelli lost.

31 – dorsum tectiform.

32 – antennae elongated.

33 (33–38 after ANDERSEN 1982) – mandibular lever quadrangular; head with 3–4 pairs of trichobothria in deep pits; fore wings not forming claval commissure.

(Continued)

34 – mesoscutellum more or less reduced in winged form; metathoracic spiracle situated latero-dorsally; coxae remote from midventral line.

35 – ventral head lobes hiding basal labial segments.

36 – pronotum produced backwards and mesoscutellum absent in winged form; metasternal scent gland reservoir not lined with glandular cells.

37 – ventral head lobes hiding basal labial segments; metepisternal process present.

38 – egg-burster unpaired; intercalary sclerite between 3 and 4th labial segments; gynatrial sac glandular; only 4 ovarioles in each ovarium.

39 – antennae flagelliform; hind wing venation reduced.

40 – fore wings not forming claval commissure, with medial fracture in preradial position; mesopleurosternum inflated between spaced coxae.

41 – head constricted transversely, ocelli posterior to eyes; fore tarsus opposable to apex of flattened fore tibia.

42 – (42–45 after KERZHNER 1981) – fore wing membrane cells diminished; spermatheca transformed into vermiform gland.

43 – parameres directed backwards; ovipositor plate-like.

44 – only 2 cells on fore wing membrane.

45 – no head trichobothria; tarsi 2-segmented; costal fracture reduced; spermatheca non-functional.

46 – (46–48 mainly after HENRY 1997) – basitarsi of all leg pairs elongate; pulvilli lamellate; spermatheca tubular, apically bulbous; egg without true operculum; embryonic egg burster in nymphs; accessory salivary gland tubular.

47 – body strongly flattened dorsoventrally; extremely long stylets coiled within head; no abdominal trichobothria.

48 – antennae 5-segmented; only 2 lateral trichobothria per abdominal segment; genital aperture directed caudally; egg barrel-shaped with circular eclosion rent.

49 – suprantennal ledge continuous below median ocellus; fore wings overlapping in repose (left one uppermost); fore wing with enlarged posterior apical cells, short transverse CuA_2, and broad appendix; coupling fold very short in fore wing; fore and middle tarsi 2-segmented; pygophore barrel-shaped, parameres protruding and elbowed.

50 – head wide, with pair of areolae at anterior margin; antennae 3-segmented, incrassate; fore wing with SC secondarily free, short, basal cell enlarged, arculus longitudinal, and M_{3+4} fused to CuA_1; hind wing lacking closed cells.

---

but already somewhat flattened; their descendants, earliest Peloridioidea (Karabasiinae) became less vagile and acquired wing polymorphism characteristic of the superfamily (POPOV & SHCHERBAKOV 1991, 1996).

Despite modern ideas that Gerromorpha (COBBEN 1978) or Enicocephaloidea (R. SCHUH 1979, WHEELER *et al.* 1993) are the most primitive true bugs, the basal group of Cimicina could be found among generalised Nepomorpha (not unlike Ochteridae), in accordance with HANDLIRSCH (1906–1908) and REUTER (1912). Scytinopteroidea (Cicadomorpha) are synapomorphic with Cimicina in two unique fore wing characters and considered ancestral to the latter (POPOV & SHCHERBAKOV 1991, SHCHERBAKOV 1996). When scytinopteroids (along with other cicadomorphans) are used as an outgroup, many crucial characters change their polarity to opposite relative to that suggested e.g. by COBBEN (1978, 1981), indicating that neoteny and structural simplification played a greater role in the heteropteran origin than 'anagenetic' differentiation. Dorsoventrally depressed body and elongate antennae of few segments could be nymphal characters retained at the adult stage in true bugs. Nepomorpha (taken as a whole) retain more features of scytinopteroid ancestors than any other infraorder: short tapering antennae without intersegments, hemelytron with broad costal area fixed on mesepimeral knob in repose, costal fracture in proximal position, hind legs with enlarged coxae and elongate basitarsi, tarsus 3-segmented in adult and 2-segmented in first instar, multifaceted eyes without trichobothria in first instar, etc. (tripartite food pump of most Nepomorpha is derivable from bipartite one of Cicadomorpha, not from undivided food pump of Nepoidea which is secondarily simplified, contrary to PARSONS 1966).

Transformation of hopper-like scytinopteroid habitus into the flattened heteropteran one could be explained through emigration from three-dimensional habitat (vegetation) on to the level surface. The first bugs are believed to be scavengers and/or passive predators which used their long 'probing' rostrum to feed on soil microfauna of littoral zone (like Ochteridae, Saldidae and probably Velocipedidae; RIEGER 1976, KERZHNER 1981), or inhabiting floating plant carpets like Mesoveliidae, Veliidae and Hydrometridae (ANDERSEN 1982). Some

primitive members of Dipsocoridae and Enicocephalidae live in moist habitats as well (e.g. *Cryptostemma* Herrich-Schaeffer and *Boreostolus* Wygodzinsky et Štys under stones along streams; R. SCHUH & SLATER 1995). The oldest Diptera and Coleoptera Carabina (= Adephaga) are recorded from the same Triassic deposits as the first Cimicina, and the origin of these three groups may be connected with their expansion into continental waters (KALUGINA 1980b), possibly due to the appearance of floating macrophytes (PONOMARENKO 1996). Both Carabina (PONOMARENKO in ARNOLDI *et al.* 1977) and Cimicina adopted zoophagy at the earliest stages of their evolution.

Infraorders of Cimicina are given after ŠTYS & KERZHNER (1975), except for Enicocephalomorpha reunited here to Dipsocoromorpha *s. l.* (as in MIYAMOTO 1961). The change is based on the intermediate position of Stemmocryptidae (ŠTYS 1983) which demonstrates important synapomorphies with Enicocephaloidea (besides postocular ocelli and anteriorly directed labium mentioned by Štys, these are medial fracture in preradial position, and mesopleurosternum inflated between spaced coxae, the latter implying swarming habits like in enicocephalids) and deserves superfamily rank (Stemmocryptoidea stat. nov.). The change is further supported by the absence of enicocephaloids from the Jurassic in contrast to minimum Early Jurassic age of remaining infraorders (only Nepomorpha are known since the mid-Triassic beyond doubt).

Infraordinal position is identified for all extinct heteropteran families except Early Jurassic Ceresopseidae stat. nov. known by hemelytra only and possibly representing a group ancestral to Gerromorpha. Superfamilies are given chiefly after ŠTYS & KERZHNER (1975), except that those of Gerromorpha follow ANDERSEN (1982). Mesozoic Archegocimicidae (transferred to Leptopodoidea by POPOV 1985) are directly ancestral to Saldidae; Enicocorinae Popov, 1986, stat. nov. (=Mesolygaeidae Zhang, 1991) should be included in the latter family as the most generalised subfamily. Early Jurassic Darniopseinae are here transferred to Velocipedidae as the least advanced subfamily. Tingoidea are accepted including Thaumastocoridae (as in R. SCHUH & ŠTYS 1991) and even Joppeicidae. Coreoidea are treated in the broadest sense, embracing both 'Pyrrhocoroidea' and 'Lygaeoidea', the

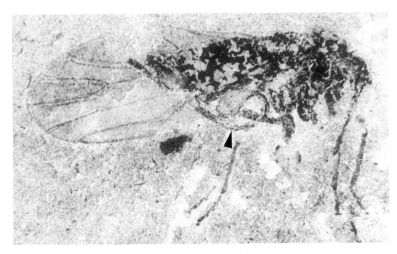

Fig. 180 Unidentified female archescytinid from the Early Permian of Tshekarda in Urals (PIN 1700/60, photo by D.E. Shcherbakov) showing the inner valvulae of ovipositor coiled under the abdomen (arrowhead); body 6.7 mm long

latter division covering Idiostolidae as well as Piesmatidae (according to SCHAEFER 1975). Mesozoic Pachymeridiidae are considered ancestral Coreoidea *s. l.*; they retain the costal fracture (see Fig. 188) similarly to Mesopentacoridae (possible Pentatomoidea, probably synonymous to Early Jurassic Protocoridae) and in contrast to the other, more advanced Pentatomomorpha. The Early Cretaceous cydnid subfamily Clavicorinae (POPOV 1986) along with two families described from the Early Cretaceous (Pricecoridae, Latiscutellidae: PINTO & ORNELLAS 1974) are all synonymised under Amnestinae (POPOV & PINTO 2000). Thaumastellinae are included into Cydnidae (after DOLLING 1981).

(f) History. The oldest Hemiptera, Archescytinidae, are first recorded in the early Artinskian (Early Permian) of Moravia (Obora, undescribed) and very rare in that time, becoming abundant already by the late Artinskian (Kansas, Elmo). In Kungurian faunas of the Urals (Tshekarda and neighbour localities) these insects became still more numerous and diverse, especially in the structure of head and ovipositor. Their different positions of the rostrum base are comparable to those of cicadas and (in *Maueria* G. Zalessky) of true bugs, an example of archaic diversity within primitive group. In some genera the ovipositor was protruding caudally (*Permoscytina* Tillyard), in some others the extremely long needle-like inner valvulae formed a coiled loop under the abdomen in repose (BECKER-MIGDISOVA 1960a,b; Fig. 180), the modifications corresponding to parasitic Hymenoptera rather than Hemiptera. It was hypothesised that these forms laid eggs inside gymnosperm strobiles, and the nymphs dwelt there until ripe strobile would dehisce (BECKER-MIGDISOVA 1972). Smaller forms with much shorter but nevertheless well-developed laciniate ovipositors, like *Permothrips* Martynov, may be compared to xyelid sawflies and hypothesised as developing in staminate cones as well. The origin of Archescytinina (and therefore of the Hemiptera as a whole) seems to be connected to the expansion of Mesophytic gymnosperm groups (such as Peltaspermales or some Coniferales), and flattened nymphal habitus (dissimilar to adult) was possibly acquired for living between cone scales. If so, this suborder evolved somewhat in parallel to oligoneopterans and was overcompeted by them about the end of Permian.

The earliest members of Aphidina (Boreoscytidae; SHCHERBAKOV in press) and both Cicadomorpha (Prosboloidea: Ingruidae and Prosbolopseidae: Fig. 181) and Fulgoromorpha (undescribed

Fig. 181 Unidentified female prosbolopseid (Ivaiinae) from the Early Permian of Tshekarda in the Urals, Russia (PIN 1700/4222, photo by D.E. Shcherbakov); body 4.5 mm long

Coleoscytoidea) have appeared in the same Kungurian beds as minor components of the fauna. The large and clumsily built Early Permian boreoscytids with clinging legs possibly fed on large ovules. Small, usually dorsoventrally depressed hoppers and their flattened cryptic nymphs, both with comparatively small postclypeus and long rostrum, might have fed on phloem of rather thick stems.

Archescytinids survived until the end of the Permian, but never remain so numerous as in the Kungurian (at most being subdominant in the Kazanian faunas). Since the Ufimian–Early Kazanian (Kaltan in S. Siberia, Soyana in NW. Russia), Coleoscytidae, Pereboriidae, Prosbolidae, and Scytinopteridae are known (all nominative families of the respective superfamilies). Boreoscytidae are last recorded and Psyllina (Protopsyllidiidae: somewhat doubtful *Permaphidopsis* Becker-Migdisova) first recorded in the Kazanian. By that time Cicadomorpha were already diversified significantly in the body size and degree of vein polymerisation, the extremities represented with tiny, about 3 mm long species of *Kaltanospes* Becker-Migdisova (Ingruidae; the family presumably ancestral to Peloridiina), and large polyneurous pereboriids and especially their descendants, Ignotalidae (Tatarian–?Early Triassic) reaching 105 mm in fore wing length. In the Tatarian (latest Permian; possibly also in the latest Kazanian), especially in the terminal Permian faunas of Natal, South Africa (Middle Beaufort Series) and New South Wales, Australia (Belmont), the first members of some Mesozoic families have appeared: Pincombeidae (Aphidina; Fig. 182), undoubted Protopsyllidiidae, Surijokocixiidae (the most primitive members of the oldest extant hemipteran superfamily, Fulgoroidea), Dysmorphoptilidae (Prosboloidea), Dunstaniidae (Palaeontinoidea), Serpentivenidae, Stenoviciidae and Paraknightiidae (all Scytinopteroidea; Fig. 183), and Progonocimicidae (the oldest Peloridiina).

The stridulator of dysmorphoptilid hoppers is formed by the strigil on the fore wing underside and a plectrum at the hind knee (J. EVANS 1961; Fig. 184; similar femoro-hemelytral devices are found in some Triassic Ipsviciidae and various heteropterans) is the earliest evidence of insect acoustical communication (which has not been demonstrated yet in Palaeozoic orthopterans). The alarm signal was possibly emitted during an escape leap in these cryptic creatures with bizarrely shaped fore wings. Scytinopteroids, very numerous in Late Permian and Triassic localities (their remains seem to be more abundant in near-shore than off-shore facies), had their coriaceous fore wings securely fixed on the thorax in repose and, like their descendants, Cimicina Nepomorpha, might have been capable of subelytral air storage. If so, these hoppers would be the only amphibiotic homopterans known, feeding possibly on emergent water plants (helophytes). Some families show definite patterns of geographic distribution according to the pronounced Permian climatic zonation; in Angaraland lower palaeolatitude faunas were richer and dominated by Prosbolidae and/or Scytinopteridae, whereas subpolar ones were poorer in species (up to monodominant ones) and composed of the more derived families (SHCHERBAKOV 2000b).

Passing over virtually unknown Early Triassic fauna, one finds the Middle and Late Triassic faunas differing strikingly from the Permian ones. Of the families known since the early Kazanian, only Protopsyllidiidae and Scytinopteridae survived till the Triassic. Creaphididae and Triassoaphididae (Palaeoaphidoidea), Curvicubitidae (Pereborioidea) and Mesogereonidae (Palaeontinoidea) are the Triassic endemics, and Ipsviciidae (Scytinopteroidea) are subendemics (recorded elsewhere in the single Early Cretaceous locality). The earliest members of Aphidomorpha, Clypeata (Hylicellidae) and Cimicina (possibly belonging to Nepomorpha) have their first appearance in the Anisian of Vosges, France (undescribed) and those of Coccomorpha (Naibiidae) in the Madygen Formation of Fergana, Kyrgyzstan. Most Middle and/or Late Triassic faunas (Madygen Formation, Molteno Formation of Cape, Mount Crosby Formation of Queensland) were dominated by hylicellids, scytinopteroids and/or dunstaniids and had dysmorphoptilids and fulgoroids subdominant. Triassocoridae (probably allied to Naucoridae) and possible Ochteroidea (*Heterochterus* Evans) have constituted a minor component of the Ipswich fauna. Some other entomofaunas of that age were dominated either by Triassocoridae (Kenderlyk, E. Kazakhstan) or by the extant water bug families, such as Belostomatidae (Nepoidea), Naucoridae (Naucoroidea) and Notonectidae (Notonectoidea) (Cow Branch Formation of N. Carolina

Fig. 182 *Pincombea* sp. (Pincombeidae) from the Late Permian of Belmont in Australia (NHM In. 45314, photo by D.E. Shcherbakov); fore wing 4 mm long

and Virginia; OLSEN *et al.* 1978, FRASER *et al.* 1996). Most of the Triassic bugs belonged to Nepomorpha (see Fig. 28), i.e. were either true water or shore dwellers; other groups representing littoral (Leptopodoidea), semiaquatic (Mesovelioidea) and terrestrial bugs (Cimicoidea, Coreoidea, Pentatomoidea, probable Dipsocoroidea) have appeared later, about Triassic/Jurassic boundary.

The Clypeata, the only surviving stock of Cicadomorpha, were the first Hemiptera having changed over to xylem-feeding, as indicated by their huge, swollen postclypeus with muscle striations evident already in Hylicellidae (see Fig. 198). The nymphs of the latter family were presumably soil-dwelling (like present-day Cicadoidea and Cercopidae) or otherwise cryptobiotic, so far as in hylicellid *Vietocycla* Shcherbakov the adult fore legs were modified for digging, although less so than in Cicadidae (SHCHERBAKOV 1988b). The phloem- or mesophyll-feeding, and free-living, adult-like nymphs were secondarily acquired by derived members of Membracoidea.

From the Early Jurassic, numerous extant superfamilies and even families enter the fossil record: Aphidoidea (ANSORGE 1996; Oviparosiphidae, still retaining an ovipositor and having cauda and siphuncular pores already developed), Psylloidea *s. l.* (Liadopsyllidae), Cercopoidea (Procercopidae), Cicadoidea (represented with extant Tettigarctidae: Fig. 185; technically the family first appears in the latest Rhaetian, that is the latest Triassic, of England, but this insect assemblage is similar to that from the overlaying Early Jurassic beds; WHALLEY 1983), Membracoidea *s. l.* (Karajassidae recorded since the Middle Jurassic, plus somewhat doubtful undescribed find in the Early Jurassic of Germany), Peloridioidea (Karabasiidae; Fig. 186), Ochteroidea (represented with endemic subfamily Propreocorinae of the extant Ochteridae: POPOV, DOLLING & WHALLEY 1994), Corixoidea (represented with extant Corixidae and Jurassic Shurabellidae), Leptopodoidea (Archegocimicidae; Fig. 187), Mesovelioidea (*Engynabis* Bode and *Sphongophoriella* Becker-Migdisova, probable Mesoveliidae with complete venation), Cimicoidea (represented with endemic subfamily Darniopseinae of the extant Velocipedidae), Coreoidea *s. l.* (Pachymeridiidae, presumably basal for the superfamily; Fig. 188; the family like the above-mentioned Tettigarctidae is first recorded from the terminal Triassic: POPOV, DOLLING & WHALLEY 1994), Pentatomoidea (undescribed Cydnidae and possibly also Protocoridae =Mesopentacoridae?), and probable Dipsocoroidea (endemic Cuneocoridae). Pterocimicidae (Cimicomorpha inc. superfam.; POPOV, DOLLING & WHALLEY 1994) are endemic of the Early Jurassic.

By the Middle Jurassic, Shaposhnikoviidae (Fig. 189) have appeared, and by the Late Jurassic, Genaphididae (both Palaeoaphidoidea), Malmopsyllidae (Psylloidea), Scaphocoridae (Notonectoidea) (three latter groups were Jurassic endemics or subendemics), possibly Drepanosiphidae (extant aphidoid family represented with doubtful *Jurocallis* Shaposhnikov), and extant (super)families Aleyrodoidea (extinct Bernaeinae included in Aleyrodidae *s. l.*: SHCHERBAKOV 2000a; Fig. 190), Cixiidae (the oldest extant fulgoroid family; Fig. 191), Nepidae (Nepoidea), typical Mesoveliidae (see Fig. 488), Miroidea (represented with extant Miridae), Anthocoridae (Cimicoidea; Fig. 192), Coreidae and Alydidae (both Coreoidea;

Fig. 183 Unidentified stenoviciid (Scytinopteroidea) from the Late Permian of Kerbo in Siberia (PIN 3298/4, photo by D.E. Shcherbakov) showing hypocostal pit of 'Druckknopfsystem' and costal fracture (arrowheads) combined with auchenorrhynchous head structure; total length 9 mm

Fig. 184 Unidentified dysmorphoptilid fore wing from the Middle or Late Triassic Madygen Formation of Kyrgyzstan (PIN 2785/3521, photo by D.E. Shcherbakov); incomplete fore wing 15 mm long, and close up of the strigil (arching row of fine ribs) on fore wing underside

POPOV 1968b, ŠTYS & ŘÍHA 1974; Figs. 20, 193) became known. The oldest extant hemipteran genus, *Laccotrephes* Stål (Nepidae), is recorded since the terminal Jurassic (POPOV 1971). Among Triassic families, only Hylicellidae *s. l.* were still abundant in Jurassic, while the dominant position among larger hoppers has been occupied by younger groups (Fulgoridiidae in the Early Jurassic of Europe, Procercopidae in the Jurassic of Central Asia). As to smaller hoppers, scytinopteroids were replaced by Progonocimicidae, a fairly abundant group in some faunas through the Jurassic (Fig. 194). The niche of tiny plantlice has been populated mostly by Protopsyllidiidae (Fig. 195) and/or Liadopsyllidae; Palaeontinidae prevailed among large cicadas,

Corixoidea and sometimes also Naucoridae, Belostomatidae or Archegocimicidae among aquatic and littoral bugs, and Pachymeridiidae and/or Mesopentacoridae (plus Miridae in the Late Jurassic) among land bugs.

Paleontinids abundant in the Jurassic and Early Cretaceous were robust, moth-like, hairy (like relic modern cicadoid *Tettigarcta* White), and probably crepuscular as well, with clinging legs, an extremely long rostrum reaching the end of abdomen, and often (especially in the Jurassic) with disruptive, cryptic wing pattern of dark bands (Fig. 196). They have nearly the same distribution as, and probably fed on, Ginkgoales (BECKER-MIGDISOVA 1960a). A very long rostrum was commonplace in various Mesozoic Cicadina (e.g. Fulgoridiidae) indicating that sucking from trunks and thick branches was more widespread than nowadays. A distractive colour pattern (such as dark 'false eye' spots near the apex of the fore wing) that attracted a predator's attention to the rear end of prey instead of its head seems to have been more common in Jurassic than in recent Fulgoroidea. Another antipredator colour pattern, a pair of 'bird eyes' on the fore wings, is found in one of Triassic Dunstaniidae (undescribed).

Faunistic changes at the Jurassic/Cretaceous boundary were the most conspicuous ones since the beginning of Jurassic (if not since the beginning of Mesozoic) and were probably caused by expansion of proangiosperms (PONOMARENKO 1998a). Cretaceous faunas usually differ from those in Jurassic by dominance of aphids among plantlice, Cicadellidae among smaller hoppers, Notonectidae (see Fig. 492) or Naucoridae among water bugs and Cydnidae or Pachymeridiidae among land bugs. Procercopidae, Corixidae (see Fig. 495) or Leptopodoidea retain dominance in some Early Cretaceous faunas. The dominant leptopodoids belong to the extinct subfamily Enicocorinae, similar to living Chiloxanthinae but peculiar in longer tarsi (ZHANG 1991), and possibly more hydrophilic than present-day shore bugs. Genaphididae, Fulgoridiidae, Dysmorphoptilidae, and Dunstaniidae, so characteristic of the Early Mesozoic, have not survived into the Cretaceous. Near the boundary some extant groups appeared as well: true scale insects (Orthezioidea represented with extant Matsucoccidae and Xylococcidae; KOTEJA 1988, 1989b),

Fig. 185 *Turutanovia karatavica* Becker-Migdisova (Tettigarctidae) from the Late Jurassic of Karatau in Kazakhstan (PIN 2784/1944, photo by D.E. Shcherbakov) showing long rostrum; fore wing 12 mm long

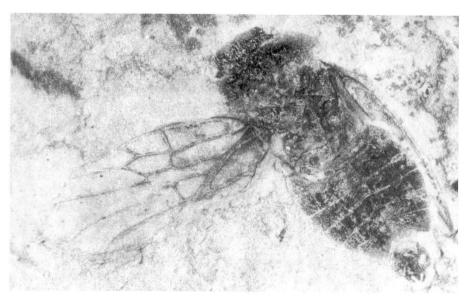

Fig. 186 Male *Karabasia evansi* Popov & Shcherbakov (Karabasiidae: Karabasiinae) from the Late Jurassic or Early Cretaceous Glushkovo Formation in Siberia (holotype, photo by D.E. Shcherbakov); body 3.6 mm long

mycetophagous Achilidae (Fulgoroidea; HAMILTON 1990), Cicadellidae (Membracoidea), Saldidae Enicocorinae (Leptopodoidea), Tingoidea (represented with extant Tingidae Cantacaderinae: *Sinaldocader* Popov, 1989; GOLUB & POPOV 1999), Reduvioidea (represented with extant Reduviidae), Aradoidea (endemic Kobdocoridae; POPOV 1986), and Cydnidae determinable up to subfamily level (Amnestinae). Extant pentatomoid groups other than Cydnidae are absent from the record till the Eocene and might be descendants of that family, especially since some intermediate forms occur in the recent fauna (SCHAEFER, DOLLING & TACHIKAWA 1988).

Protopsyllidiidae and Psylloidea are practically absent from the Cretaceous record, except for the moderately diverse assemblage of the Purbeck in England (CORAM, JARZEMBOWSKI & ROSS 1995, JARZEMBOWSKI & CORAM 1998) and a few specimens from other localities (SHCHERBAKOV 1988a; psyllids were still very rare in the Palaeogene, even in Baltic amber: KLIMASZEWSKI 1997). As for

Fig. 188 Unidentified pachymeridiid hemelytron (Coreoidea *s. l.*) from the Middle Jurassic of Kubekovo in Siberia (PIN 1255/1601, photo by A.P. Rasnitsyn) showing lygaeid-like venation and costal fracture (arrow); length 4.7 mm

Fig. 187 Male shore bug, *Saldonia sibirica* Popov (Archegocimicidae), from the Middle or Late Jurassic of Uda in Siberia (holotype, photo by D.E. Shcherbakov); body 6.8 mm long

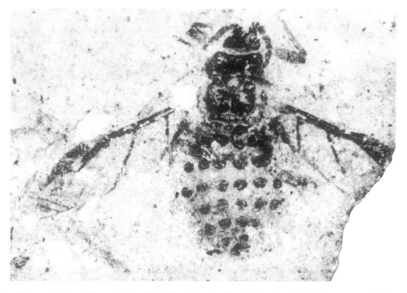

Fig. 189 *Szelegiewiczia maculata* Shaposhnikov (Shaposhnikoviidae Szelegiewicziinae) from the Early Cretaceous of Baissa in Siberia (holotype, photo by D.E. Shcherbakov) showing rows of rounded sclerites (with glands opening in the centre) on the abdominal segments; body 3.0 mm long

Fig. 190 Unidentified bernaeine whitefly (Aleyrodidae) from the Late Jurassic of Karatau in Kazakhstan (PIN 2239/532, photo by A.P. Rasnitsyn); body 2.3 mm long

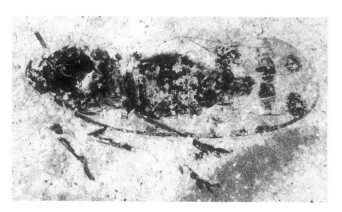

Fig. 191 Unidentified cixiid from the Early Cretaceous of Baissa in Siberia (PIN 3064/4028, photo by D.E. Shcherbakov); total length 12.6 mm

Fig. 193 Unidentified Coreoidea *s. l.* from the Early Cretaceous of Bon-Tsagan in Mongolia (PIN 3559/7676, photo by D.E. Shcherbakov); fore wing 8.3 mm long

Fig. 192 *Eoanthocoris cretaceus* Popov (Anthocoridae: Lyctocorinae *s. l.*) from the Early Cretaceous of Turga in Siberia (paratype PIN 1742/85, photo by D.E. Shcherbakov); body 6.3 mm long

protopsyllidiids, they possibly declined due to the extinction of their host plants, most probably Peltaspermales (KLIMASZEWSKI 1995).

The later Early Cretaceous of Mongolia has yielded the oldest known extant genus of land bugs, *Aradus* Fabricius (POPOV 1989).

Early Cretaceous Karabasiidae Hoploridiinae have lost their jumping ability, show some similarities to aradid and phloeid bugs, and were probably likewise cryptic bark dwellers (POPOV & SHCHERBAKOV 1991; Fig. 197). Extant Hydrometroidea are first recorded from Santana and represented with living Hydrometridae (NEL & POPOV 2000). The oldest insect-bearing fossil resins (late Early Cretaceous Lebanese amber) contain the first representatives of Enicocephaloidea (belonging to the extant family Enicocephalidae; GRIMALDI *et al.* 1993, AZAR *et al.* 1999a) and Thaumastellinae (Cydnidae; POINAR 1992a). Palaeoaphididae and Canadaphididae are Cretaceous endemics.

By the mid-Cretaceous, the remaining Mesozoic families became extinct (except for Naibiidae, also recorded in Palaeogene Sakhalin amber), namely Oviparosiphidae, Protopsyllidiidae, Liadopsyllidae, Palaeontinidae, Hylicellidae (Fig. 198), Procercopidae, Karajassidae, Progonocimicidae, Karabasiidae, Pachymeridiidae, and Mesopentacoridae. Extant Aphrophoridae and Cercopidae (Cercopoidea), Gerroidea (represented with extant Gerridae), Lygaeidae (Coreoidea), and possibly Mindaridae (Aphidoidea) have appeared that time. The first Coccoidea (endemic Labiococcidae and extant Eriococcidae) and the oldest members of extant families Steingeliidae (Orthezioidea; KOTEJA 2000), Vianaididae and Thaumastocoridae (both Tingoidea; GOLUB & POPOV 2000) are found in Turonian New Jersey amber.

Taimyr amber (Yantardakh, Santonian, roughly mid-Late Cretaceous) contains the last Shaposhnikoviidae known, the first Pemphigidae, Aphididae, and undoubted Drepanosiphidae (all Aphidoidea), the first Dictyopharidae (EMELJANOV 1983), and the youngest extinct hemipteran family Elektraphididae (most primitive Phylloxeroidea, known until the Eocene or possibly even Pliocene). Tajmyraphididae (?Palaeoaphidoidea), Cretamyzidae (Aphidoidea), Mesozoicaphididae (Phylloxeroidea), Jersicoccidae, Grimaldiellidae, Electrococcidae (all

three Orthezioidea), Inkaidae (Coccoidea) and Mesotrephidae (Notonectoidea) are known only from the Late Cretaceous. In overall appearance, the Late Cretaceous fauna is similar to the Cainozoic one.

Cicadellidae and Dictyopharidae are the oldest families of Cicadina with free-living, adult-like nymphs whose hind legs became elongate for jumping (such leafhopper nymphs as well as the first planthopper nymphs are known since the Early Cretaceous); the earlier groups of Cicadina presumably had the nymphs either cryptobiotic

Fig. 194 *Mesoscytina* sp. (Progonocimicidae: Cicadocorinae) from the Middle Jurassic of Kubekovo in Siberia (PIN 1255/458, photo by A.P. Rasnitsyn); body without head and prothorax 4.9 mm long

Fig. 195 *Cicadellopsis* sp. (Protopsyllidiidae) from the Middle Jurassic of Kubekovo in Siberia (PIN 1255/1622, photo by A.P. Rasnitsyn); length 4.7 mm

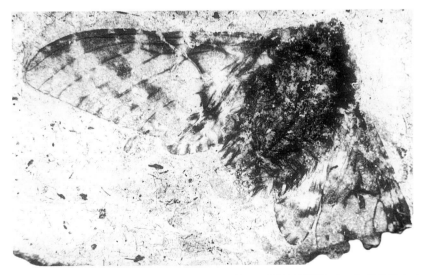

Fig. 196 *Pseudocossus* sp. (Palaeontinidae) from the Late Jurassic of Karatau in Kazakhstan (PIN 2904/445, photo by D.E. Shcherbakov) showing setose body and very long rostrum; fore wing 28 mm long

Fig. 197 Female *Hoploridium dollingi* Popov & Shcherbakov (Karabasiidae: Hoploridiinae) from the Early Cretaceous of Baissa in Siberia (paratype PIN 3064/4552, photo by D.E. Shcherbakov); body 6.7 mm long

(soil-dwelling etc.) or flattened and clinging tightly to the host plant surface. Until the mid-Cretaceous nearly all fulgoroids were cixiid-like, their nymphs probably living in the soil or rotten wood as well.

During the Cainozoic, the number of the still living (super)families further increased. Ricaniidae (Fulgoroidea; FENNAH 1968) and Cicadidae (Cicadoidea; COOPER 1941) are known since the Palaeocene; Fulgoridae (LUTZ 1988, Abb. 103; attributable to modern genus *Dichoptera* Spinola), Issidae (HAUPT 1956), Flatidae (SCUDDER 1890), Delphacidae (COCKERELL 1924; all Fulgoroidea), Pentatomidae (Pentatomoidea; HENRIKSEN 1922; Fig. 199) since the Eocene; Carsidaridae, Psyllidae (both Psylloidea), Clastopteridae (Cercopoidea; SCUDDER 1890) and Colobathristidae (represented with primitive subfamily Dayakiellinae; Coreoidea) since the Oligocene; Greenideidae, Lachnidae (Aphidoidea), Phylloxeridae (Phylloxeroidea; HEIE & PEÑALVER 1999), Triozidae (Psylloidea), Urostylidae (ZHANG 1989), Acanthosomatidae (FUJIYAMA 1987) and Scutelleridae (FUJIYAMA 1967; all Pentatomoidea) since the Miocene; Adelgidae (Phylloxeroidea) and Aphelocheiridae (Naucoroidea; e.g. *Coreoidus latus* Jordan erroneously assigned to Coreidae: JORDAN 1967) since the Pliocene. Piesmatidae first appeared in the Early Eocene Paris amber (NEL *et al.* 1998a: 25, erroneously identified as Tingidae); Anoeciidae, Thelaxidae, Hormaphididae (all Aphidoidea), Ortheziidae, Monophlebidae (both Orthezioidea), Pseudococcidae, Coccidae, Diaspididae (all Coccoidea), oldest Psylloidea *s. str.* (Aphalaridae), Derbidae (Fulgoroidea, EMELJANOV 1994), typical Saldidae, undoubted Dipsocoroidea (represented with extant Schizopteridae, Ceratocombidae and Hypsipterygidae; POPOV & HERCZEK 1993, BECHLY & WITTMANN 2000; Fig. 200), Veliidae (Gerroidea; ANDERSEN 1982), Nabidae (Cimicoidea; KERZHNER 1981), Miridae of the primitive subfamily Isometopinae, and Berytidae (Coreoidea; BACHOFEN-ECHT 1949) in the Late Eocene

Fig. 198 Unidentified hylicellid from the earliest Cretaceous of Khutel Khara in Mongolia (PIN 3965/443, photo by A.P. Rasnitsyn and D.E. Shcherbakov) showing enlarged postclypeus with muscle striations; total length 8.8 mm

Baltic amber; Nogodinidae (Fulgoroidea; STROINSKI & SZWEDO 2000), Leptosaldinae (variously placed within Leptopodoidea), Hebroidea (represented with extant Hebridae; ANDERSEN 1982) and Termitaphididae (Aradoidea; POINAR & DOYEN 1992) in the Mexican amber (Late Oligocene–Early Miocene); Rhinopsyllidae, Rhinocolidae, Ciriacremidae (all Psylloidea; KLIMASZEWSKI 1998), Kinnaridae (EMELJANOV & SHCHERBAKOV 2000), Tropiduchidae (both Fulgoroidea; SZWEDO 2000), Aetalionidae (SZWEDO & WEBB 1999) and Membracidae (both Membracoidea; SHCHERBAKOV 1992) in the Dominican amber (Early Miocene). The oldest living hemipteran species is *Cicadella viridis* known since the Middle Miocene (BECKER-MIGDISOVA 1967); an extant aphid species *Longistigma caryae* (Harris) has been reported from the Late Miocene–Early Pliocene (HEIE & FRIEDRICH 1971).

The Cainozoic fossil assemblages known from impressions in sedimentary rocks are usually dominated by either fulgoroids (often Ricaniidae) or cercopoids, with the former supposedly originating from the more warm and the latter from cooler climate. Aphids were often abundant while psyllids only rarely so; in the Middle Miocene of Stavropol the psyllids were gradually replaced with aphids during the shift from savannah to subtropical forest (BECKER-MIGDISOVA 1967). The dominant water bugs were notonectids or corixids, and

Fig. 199 Unidentified male stink bug (Pentatomidae) from the Late Eocene or Early Oligocene of Bolshaya Svetlovodnaya in Russian Far East (PIN 3429/1525, photo by D.E. Shcherbakov); body 13.3 mm long

Fig. 200 Unidentified schizopterid from the Late Eocene Baltic amber (Zoological Museum of the University of Copenhagen 19-6, photo by Yu.A. Popov); total length 1.1 mm

pentatomoids or coreoids prevailed among the land bugs. In Sakhalin and Baltic amber, the aphids were dominant, while leafhoppers and mirids (mainly Cylapinae and Isometopinae: POPOV & HERCZEK 1993) were subdominant; in Dominican and Mexican amber presumably formed in tropics the planthoppers were most abundant followed by leafhoppers, male coccids and emesine reduviids, whereas aphids were virtually absent. Sea skaters (Gerridae Halobatinae), the only insects to have successfully colonised the open ocean, are recorded since the Early–Middle Eocene (ANDERSEN et al. 1994).

2.2.1.3. COHORS SCARABAEIFORMES Laicharting, 1781. THE HOLOMETABOLANS (=Holometabola Burmeister, 1835, =Endopterygota Sharp, 1899, =Oligoneoptera Martynov, 1938)

A.P. RASNITSYN

(a) Introductory remarks. The concept of the group is modified from RASNITSYN (1980) and ROHDENDORF & RASNITSYN (1980), with ontogenetic features treated according to TICHOMIROVA (1991), other publications used are cited when necessary.

The alleged holometabolan larva from the Westphalian of Mazon Creek (Illinois, USA; KUKALOVÁ-PECK 1991) is excluded from consideration because the personal study of the only known specimen (while on loan to the National Museum of Natural History, Washington, DC; courtesy of Conrad C. Labandeira) has shown some

features strange for insects, e.g. legless segments alternating with the legged ones, and similar structure of more than three anterior leg pairs.

(b) Definition. Insects of variable appearance, morphology and habits, winged or wingless. Hind wing with common anal base not rocking upside down, and with 1A and 2A not bending down in folded wings. Mesothoracic coxa with medial (furcasternal, apparently postepisternal) articulation. Development holometabolous (with complete metamorphosis concentrated in the pupal stage, Fig. 201), larva (Fig. 202) lacking external rudiments of wings and genitalia, experiencing very limited morphogenetic changes except in prepupa (part of last larval stage after feeding stopped; usually concealed in a shelter or cocoon), with postembryonic morphogenesis concentrated at prepupal and pupal stages (larval morphogenesis secondarily is more pronounced in some parasitic or highly social holometabolans; contrary to common belief, morphogenesis is not necessarily coupled with extensive histolysis: TICHOMIROVA 1991). Postembryonic morphogenesis proceeding in prepupal and pupal stages in ordered form, being open to retrieval upon evolutionary demands to appear externally at larval or imaginal stages and thus being responsible for considerable part of evolutionary changes (for detail see TICHOMIROVA 1991). Larval ocelli lost (except mid one in bittacid scorpionflies suggested by KRISTENSEN 1999a to be a case of reversal), eyes, unless lost, replaced by single or few isolated stemmata (except for some Panorpida), antenna at most 7-segmented (except secondarily in a few beetles), thoracic tagma hardly differentiated, legs, unless lost, with tarsus at most one-segmented (often fused with tibia), abdomen often

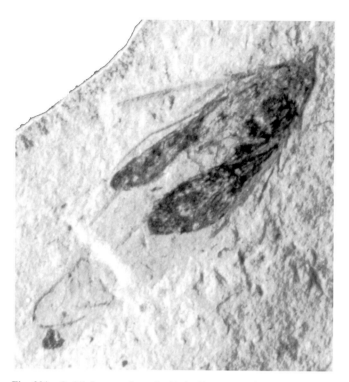

Fig. 201 Caddisfly pupa from the Early Cretaceous of Baissa in Central Siberia (PIN 4210/2471, photo by A.N. Mazin); fossil 19 mm long as preserved

with prolegs (secondarily activated embryonic limb buds). No adult moulting.

(c) Synapomorphies. Holometaboly and medial midcoxal articulation, as described above, both known to originate homoplastically elsewhere, and probably also embryonic abdominal limb buds 11 which are activated in larva as pygopods (unless it is that they appear homoplastically in most groups), and possibly also trend of de-repression of other abdominal limb buds (hardly realised in beetles). Contrary to KRISTENSEN (1999a), ground-plan morphology of both larval and pupal stages is not a holometabolan innovation but inherited from, respectively, larvular and nymphal stages of ancestral archemetabolic development (Chapter 2.2c). The larval stage is apomorphic in that its larvular morphology is stabilised and its morphogenetic functions are lost. The pupal stage is apomorphic in the sense that it allows a number of nymphal moults to be lost, morphogenetic functions are concentrated and hypertrophied at the expense of most other life functions (RASNITSYN 1965, TICHOMIROVA 1991).

Complete metamorphosis has evolved also in some Hemiptera (e.g. in male Coccoidea and in Aleurodoidea) and in thrips, though in a different form, that is, it starts from a later morphogenetic stage after external wing rudiments have appeared. A contrasting example is the external position of the leg and wing rudiments in strepsipteran males which is most probably secondary (KRISTENSEN 1999a), and probably represents a developmental simplification of the sort better studied in some endoparasitic wasps (IVANOVA-KAZAS 1961). Midcoxal articulation has appeared independently in mayflies, while the symplesiomorphic

Fig. 202 Two paddle-legged, predatory larvae of *Coptoclava longipoda* Ping (Coptoclavidae) from the Early Cretaceous of Baissa in Siberia (PIN 4210/404, photo by A.N. Mazin); right larva, body 20 mm long

nature of the similarity seems less likely because that would imply very numerous homoplastic cases of loss of the articulation.

Holometaboly in described form implies the bionomic synapomorphy: preimaginal development in environments unfavourable for the external wing rudiments, that is in narrow shelters. Coupled with the evidence of retained phytophyly (e.g. sheathed cutting ovipositor of Palaeomanteida and Hymenoptera) and phytembryophagy (palinophagy which is observed in both larves and adults of stem hymenopterans, Chapter 2.2.1.3.5f), and has been suggested also for the stem Scarabaeona as well as for the holometabolan ancestors among Cimiciformes), this indicates synapomorphy of Scarabaeiformes in development within rape or about to rape staminate cones of Gymnosperm plants (for details see RASNITSYN 1980). This further implies the eruciform larva as more probably representing the groundplan character state of the cohorts comparing the campodeiform one (PONOMARENKO 1991). The campodeiform larva fits better to predatory habits which is difficult to suppose as a part of conditions provoking the temporary arrest of the postembryonic morphogenesis at its early stage (TICHOMIROVA 1991). Indirectly this suggests also the groundplan character state of the 5-segmented larval leg with a single claw and of the entire tibio-tarsus, though this inference seems less sound.

One more possible synapomorphy is the temporal fusion of frons and clypeus in the pre- or postlarval morphogenesis described by TICHOMIROVA (1991) in beetles and hymenopterans, probable in mecopteroids (Papilionidea) and neuropterans with the entire larval clypeolabrum, and not studied yet in other neuropteroids (Myrmeleontidea).

(d) Range. Middle Carboniferous till now, worldwide.

(f) System and phylogeny. There are four main groups generally appreciated among living holometabolans which are treated as superorders here (Fig. 203). It is their phylogenetic relationship, and to a lesser extent, their ordinal composition, which have been subject of long lasting controversy (BOUDREAUX 1979, HENNIG 1981, KRISTENSEN 1981, 1991, 1995, 1999a). The order Stylopida (Strepsiptera) is the exception: traditionally it is either related to or included into the beetle order as a separate superfamily or just a family basing mostly on their posteromotorism, similar appearance of their wings with that of some parasitic beetles with reduced elytra, and other rather subtle characters (CROWSON 1955, LAWRENCE & NEWTON 1982, KINZELBACH & LUTZ 1985, KINZELBACH 1990, KUKALOVÁ-PECK & LAWRENCE 1993, etc.). However, the unexpected recent claim by WHITING et al. (1997; also see SIDDALL & WHITING 1999 and references therein) of the stylopidan monophyly with Diptera needs close consideration. Both morphological and DNA sequence data are presented. Four morphological synapomorphies of strepsipterans, scorpionflies, flies and fleas are proposed, viz. mandible dagger-like with anterior articulation reduced (characters 117); prelabium without endite lobes/ligula and associated muscles (118); labial palp with segments two or none (119); male abdominal segment IX ring-like (164). However, the scorpionfly mandible has both articulations present and often toothed (MATSUDA 1965, HEPBURN 1969), and this character is inapplicable in fleas whose adults lack mandibles. Other proposed morphological synapomorphies are questioned or discarded by KRISTENSEN (1999a).

The molecular data look more serious. The strepsipterans appear equally monophyletic with Diptera in both the 18-S and 28-S cladograms, despite quite different positions of this clade in the holometabolan tree. The particular scenario proposed by WHITING et al. (1997), the homoeotic re-pattern of the stylopidan meso- and metathoraces after the dipteran meta- and mesothorax, respectively, is hardly

realistic. This morphogenetic mechanism, however real and important, in this particular case would be required to affect too sophisticated and strictly selectively controlled system to undergo an overnight complete re-patterning to permit the species to survive. It seems safer to infer that the same morphogenetic mechanisms which are responsible for the gradual transformation of the stem mecopteroid homonomous pterothorax into the highly heteronomous dipteran one (really including the probable homoeotic processes: SHCHERBAKOV et al. 1995, and 2.2.1.3.4.4f below), were also involved into a similar transformation of the alternative segments of the strepsipteran ancestor. This ancestor might well be related to the dipteran ancestor (or, alternatively, to some different insect group). For the present, this issue is far from being resolved, so the order is treated as *incertae sedis* below.

Concerning the main holometabolan subgroups, a close relationship between beetles and neuropteroids is rather widely accepted because of their basically similar larvae and some other characters (Fig. 203 node 4). Alternative proposals are not uncommon (e.g. Boudreaux 1979), for the respective synapomorphies do not look very convincing (Kristensen 1995, 1999a). Equally widespread but weak is the claim that the hymenopterans form a monophyletic group with the mecopteroids. Kristensen (1999a) stresses two putative larval synapomorphies (single-clawed pretarsus and labial silk production) and one adult one (fully sclerotised floor of sucking pump forming sitophore plate). However, the distribution of larval labial silk production is too whimsical to support monophyly of hymenopterans and mecopteroids: apart from being found in caterpillars and caddis larvae, it is known in Boreidae but not in other Panorpida, in Xyelidae, Tenthredinoidea, Pamphilioidea and Ichneumonoidea + Aculeata but not among the rest of the hymenopterans, and also occurs in book-lice. A single-clawed pretarsus repeatedly appears whenever a particular larval morphogenetic level is stabilised in insect development (Tichomirova 1991), and if eruciform larvae represent the endopterygote groundplan (Chapter 2.2.1.3c), the same holds true for the single-clawed pretarsus. The sitophore is also well known in the psoco-hemipteroid realm as well as in the majority of beetles and so might even enter the groundplan of Cimiciformes + Scarabaeiformes.

In summary, these alleged synapomorphies look weaker than the alternative set shared by the beetles, neuropteroids and mecopteroids (Fig. 203, node 3), and the alternative fits better the geological record. Indeed, monophyly of the hymenopterans with the mecopteroids implies their simultaneous origin which contradicts their basically diachronous record (Fig. 203). Additionally, this hypothesis makes it difficult to explain the deep venational similarity of the least advanced neuropteroids and mecopteroids, as well as the presence of fossils difficult to assign to a superorder because of their intermediate structure (e.g. NOVOKSHONOV 1997a: 17–23). This gives support to the hypothesis of the independent and diachronic divergence of the hymenopterans and the clade embracing the beetles, neuropteroids and mecopteroids from the common ancestor order Palaeomanteida (RASNITSYN 1980, ROHDENDORF & RASNITSYN 1980).

2.2.1.3.0.1. ORDER STYLOPIDA Stephens, 1829 (=Strepsiptera Kirby, 1813)

(a) Introductory remarks. The small (532 living species; KATHIRITHAMBY 1991) order of strepsipterans is a subject of sharp discussion concerning its taxonomic state and phylogenetic position, and yet we are far from any solution of the problem (Chapter 2.2.1.3e). The present state of knowledge of the group is outlined by KINZELBACH (1990), the publication which is used here as the main source of data.

(b) Definition. Adult males of small size, with fore wings reduced into pseudohalteres, hind wings fan-like wide, folding longwise, with veins not branching; flight morphologically two-winged, posteromotoric. Head, all thoracic segments, and abdomen movable. Head more or less hypognathous, lacking gula. Eye ommatidia with central chrystalline cone. Antenna 4–7-segmented, pectinate. Mandible often desclerotised, maxillary palp 2–5-segmented, labial one lost except in *Mengea* Grote. Fore and mid legs with trochanter not separated, hind coxa incorporated into metathorax, trochanter free. Tarsi 2–5-segmented, with or without claws. Abdominal spiracle VIII non-functional. Genitalia lacking paramere. Female larva-like, in suborder Mengeina

(=Mengeiformia) with short legs, leaving host body later in adult life, in Stylopina legless, placed within last larval exuvium, never leaving host, lacking any external genitalia. Development hypermetamorphous, with first instar larva (triunguline) free living, actively seeking for host insect, further instars (up to 4 in number) endoparasitic, sac-like, legless (except Mengeidae). Several male preimaginal stages with external wing and leg buds. Parasitising silverfish, roaches, praying mantids, orthopterans, hemipterans, aculeate hymenopterans, and dipterans.

(c) Synapomorphies. Most of above characters are apomorphic and evidently connected with parasitism. Eye structure is of particular interest, with its apomorphic nature is not proved yet.

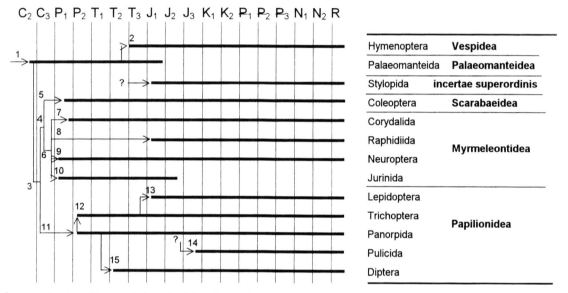

Fig. 203 Phylogeny and system of holometabolans, the Cohors Scarabaeiformes

Time periods are abbreviated as follows: C_2, C_3 – Middle and Late (Upper) Carboniferous, P_1, P_2 – Early (Lower) and Late (Upper) Permian, T_1, T_2, T_3 – Early (Lower), Middle and Late (Upper) Triassic, J_1, J_2, J_3 – Early, Middle and Late Jurassic, K_1, K_2 – Early (Lower) and Late (Upper) – Cretaceous, P_1 – Palaeocene, P_2 – Oligocene, P_3 – Eocene, N_1 – Miocene, N_2 – Pliocene, R – present time (Holocene). Right columns are names of orders (light) and superorders (bold). Arrows show ancestry, thick bars are known durations of taxa, figures refer to synapomorphies of the subtended clades, as follows.

1 – complete metamorphosis, medial midcoxal articulation, and larval pygopodia present; preimaginal development within ripe or near ripe staminate cone of gymnosperm plants.

2 – hymenopteran synapomorphies (Chapter 2.2.1.3.5c).

3 – 1st branch of M (MA) fused with pectinate RS to make eventual "RS" branching dichotomously, with MA base (connecting M and RS) lost (or tending strongly to be lost) in fore wing; pterothoracic coxae with meron; ovipositor with valve 1 much reduced, valve 2 lost, valve 3 forming main intromittent device (or lost); preimaginal development within soil or rotten wood.

4 – M component of RS+M poor in branching resulting in branching of entire RS+M near regularly pectinate; eye disintegrated in larva in 6 stemmata arranged subcircularly (in homoplasy with Trichoptera+Lepidoptera, unless the character state is either acquired at node 3 and reverted within Panorpida, or is due to an underlying synapomorphy acquired at that node).

5 – beetle synapomorphies (Chapter 2.2.1.3.2c).

6 – larva predaceous; ovipositor with 3rd valvulae fused dorsally, muscularised internally.

7 – larva aquatic, with gills, and with both mandible and maxillary stipes elongate.

8 – head elongate behind eyes, with genae contacting mesally, and with hind tentorial pits close each other; pronotum widest anteriorly, inflected widely, ventromesally; ovipositor long, incorporating apical outgrowth of 8th sternum.

9 – male and female gonostylus lost; larva with mandible and maxilla united and modified into piercing stylet with internal canal continuing directly into stomodeum (mouth opening lost), maxillary palp lost, metathoracic spiracle and midgut closed, malpighian tubes silk-producing.

10 – Antenna short; SC short; RS originating far basally; posterior RS and anterior M branches forming straight double vein; M and CuA fused for a distance sub-basally; two hind wing A veins amalgamated apically; ovipositor short if developed.

11 – hind wing with MA base short, displaced far basad; hind wing CuP fused with 1A sub-basally for a distance.

12 – caddisfly synapomorphies (Chapter 2.2.1.3.4.2c).

13 – lepidopteran synapomorphies (Chapter 2.2.1.3.4.3c).

14 – flea synapomorphies (Chapter 2.2.1.3.4.4c).

15 – mandibles piercing; labium modified into proboscis; pro- and metathoraces small, immovably fused with mesothorax; fore wing with venation simplified, with convex vein CuA(M_5?) followed by concave iCu(CuA?); hind wing modified into halter; male lacking 8th abdominal spiracle; larva legless.

(d) Range. Eocene till now, worldwide.

(e) System and phylogeny. The proposed system (KINZELBACH 1990) with 2 suborders and 3 infraorders seems over-split for the small group of ca. 500 species worldwide, especially taking into consideration that some students reduce state of all the order up to a family (CROWSON 1981). The phylogenetic scheme of the order proposed by KINZELBACH (1990) is not grounded except in that less advanced groups parasitise hosts which are also less advanced (silverfish, orthopterans, roaches, hemipterans), while the most advanced develop at the expense of wasps.

(f) History. Because of small size most fossil strepsipterans are found in fossil resins, except a supposed triunguline in the gut contents of a Middle Eocene elaterid beetle from Geiseltal in Germany. This the oldest fossil has been assigned to one of the most advanced family Myrmecolacidae (KINZELBACH & LUTZ 1985) though the very identification of the order seems debatable. The doubt is additionally supported by the fact that the next oldest and indisputable strepsipterans from the Late Eocene Baltic amber for the most part belong to the archaic genus *Mengea* (Fig. 204), while the oldest assemblage of advanced appearance (in fact, very advanced in most of its species, Figs. 205, 206) is known from the Miocene Dominican amber (KINZELBACH 1979). The fossil record (KINZELBACH & POHL 1994) does not support the hypothesis by KINZELBACH (1990) of the Early Mesozoic, unless a Permian origin of the order is considered.

2.2.1.3.1. SUPERORDER PALAEOMANTEIDEA Handlirsch, 1906. ORDER PALAEOMANTEIDA Handlirsch, 1906 (=Miomoptera Martynov, 1927)

(a) Introductory remarks. Small group of extinct insects of ordinary general appearance (Fig. 207) but not at all average phylogenetic importance. The order outline is modified here comparing RASNITSYN (1980) and ROHDENDORF & RASNITSYN (1980) taking into account later

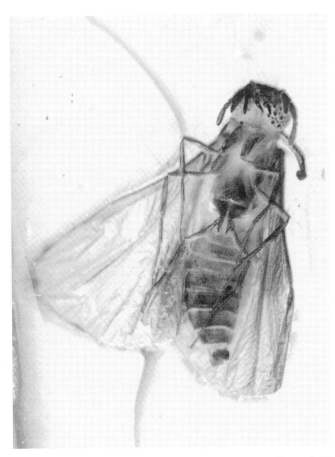

Fig. 204 Male *Mengea tertiaria* Menge (Mengeidae) from the Eocene Baltic amber, the most primitive stylopidan (from WEITSCHAT & WICHARD 1998)

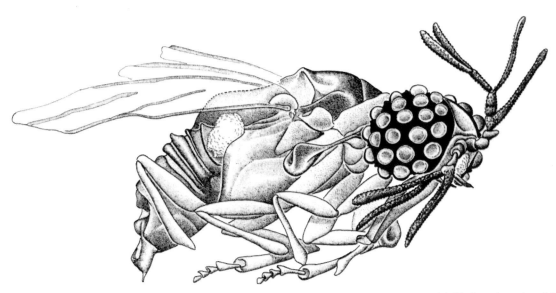

Fig. 205 *Bohartilla megalognatha* Kinzelbach (Bohartillidae), male from the Miocene Dominican amber indistinguishable from the males of this species now occurring in the mainland Central America (Panama and Honduras), thus persisted without apparent change during some 20 million years – uncommon but not unique case in insects (a handful of such are known from the Baltic amber which is about twice as old) (from KATHIRITHAMBY & GRIMALDI 1993, with kind permission of *Insect Systematics and Evolution* formerly *Entomologica Scandinavica*)

Fig. 206 *Caenocholax dominicensis* Kathirithamby & Grimaldi (Myrmecolacidae); congeners are widespread from southern USA to Argentina, parasitising very disparate hosts: males are reared from ants, females from orthopterans and praying mantids (from KATHIRITHAMBY & GRIMALDI 1993, with kind permission of *Insect Systematics and Evolution* formerly *Entomologica Scandinavica*)

publications (F. CARPENTER 1992a, STOROZHENKO & NOVOKSHONOV 1999, NOVOKSHONOV 2000).

(b) Definition. Body size medium. Head orthognathous, with chewing mouthparts, lacking striking specialisation (mandibles somewhat elongate in *Palaeomantina* Rasnitsyn). Antenna long, thin, setiform. Pronotum of medium size (ca. 2–3 times shorter and a little narrower than subequal meso- and metathoraces). Wings folded roof-like at rest (except widely overlapping each other over abdomen in *Permonka* Riek), with regular, sometimes rare cross-veins (lacking archedictyon). Costal space moderately wide to narrow, crossed by numerous or rare oblique branches of SC and R. RS regularly pectinate backward, leaving R near wing base in hind wing, beyond basal 0.3 wing length in fore one. M neutral, forking, sometimes with fore branch fused with last RS branch, sub-basally connected to CuA with short, oblique or subvertical M_5 or more often, particularly in fore wing, directly fused with CuA for a distance. CuA forking, in hind wing at least stem CuA usually clearly concave. CuP simple. Anal veins meeting wing margin separately, sometimes almost looped (1A and 2A approaching each other subapically being joined with short cross-vein here). Hind wing anal area only slightly enlarged, probably not bent under in folded wings. Legs not specialised, or hind pair saltatory, tarsi often 4-segmented. Ovipositor short to moderately long, cutting, sheathed by 3rd valves. Male genitalia weakly known, small, with gonostylus short, simple, or long and curved. Cerci short to medium long, with 3 or more segments. Habits supposedly phytophilous, with eggs placed into plant tissues and larva eating pollens or ovules within gymposperm sporangia.

(c) Synapomorphies. None because of paraphyletic state of the group in respect to hymenopterans and most probably also to all holometabolans (supposing venation and particularly M as secondarily polymerised in Papilionidea + Myrmeleontidea).

(d) Range. Middle Carboniferous until the Early Jurassic; North Eurasia and North America, also rarely in Australia and South Africa.

(e) System and phylogeny of the order is poorly known, and the taxonomy badly needs reconsideration. The fossils currently appreciated forming the order are usually ascribed to three families Palaeomanteidae, Palaeomantiscidae and Permosialidae (Archemiopteridae, Metropatoridae, and Permembiidae are excluded from the consideration, with only insufficiently known *Archemioptera* Guthörl and *Saaromioptera* Guthörl possibly belonging to the order, while *Permembia* Tillyard, *Tychtodelopterum* O. Martynova, and *Metropator* Handlirsch being transferred elsewhere, cf. Chapters 2.2.2.2.1e, 2.2.1.3.4.1e and 2.2.1.2.3e, respectively). With the accumulation of more material, the system has become obsolete and needs reconsideration. The following is a preliminary attempt of this sort.

The most distinct genus in the family Permosialidae is *Permonka* Riek (Fig. 208) with horizontal (not roof-like) wing resting position, with M and CuA free (only connected with M_5), and with characteristic oblique cross-veins between the hind wing CuA and CuP. All other miomopterans have the roof-like wing rest position (for *Permosialis* Martynov confirmed by NOVOKSHONOV & RASNITSYN 2001). Of them, *Permosialis* (Fig. 209) and its closely related, probably synonymous genera *Tologoptera* Storozhenko and *Sarbalopterodes* Storozhenko are similar to *Permonka* in having M usually free (or, when fused with CuA, with free base of the latter and not of M which becomes weak or lost). Characteristic of *Permosialis s. l.* are wide wings with RS always 3-branches.

Other miomopteran genera have M and CuA fused for a distance, with the base of M and not of CuA which is usually weak or lost. In this assemblage, *Palaeomantina* Rasnitsyn (see Fig. 207) and *Sellardsiopsis* G. Zalessky (=*Palaeomantisca* Martynov, cf. NOVOKSHONOV 1998c) (Fig. 210) are unique in having SC ending on R rather than on C. Additionally, *Sellardsiopsis* is unique in the saltatory modification of the hind femur, *Palaeomantina* is unique in having 5-segmented tarsus and convex hind wing CuA. *Delopterum* Sellards (=*Stephanomioptera* Guthörl, =*Miomatoneurella* O. Martynova, synn. nov.; for further synonymy see MARTYNOVA 1962a) (Fig. 211) differs from other genera in having SC short and lacking fore branches except for the apical one. *Epimastax* Martynov (Fig. 212) differs from other genera in having short M+CuA and numerous (more than 3) anal veins and often numerous (up to 7) RS branches.

Most other described genera can be, for the present at least, lumped under *Palaeomantis* Handlirsch, 1904 (=*Miomatoneura* Martynov, 1927, =*Permodelopterum* Kukalová, 1963, =*Perunopterum* Kukalová, 1963, =*Permonia* Kukalová, 1963, =*Permonikia* Kukalová, 1963, synn. nov.) (Fig. 213), which can be diagnosed by the elongate wings, with the wing resting position roof-like, fore wing with the fore margin weakly or not at all convex, SC long and usually with at least several fore veinlets which are often weak and not always seen in fossils. Indeed, SC branchlets are seen clearly, besides *Permonia* and *Permonikia*, in the type series of the type miomopteran *Palaeomantis schmidti* Handlirsch and traceable even on photographs of some other *Palaeomantis* in the present sense, including members of "*Delopterum*" sensu KUKALOVÁ (1963: pl. 5, fig. 2), *Perunopterum* (KUKALOVÁ 1963, pl. 6, fig. 2), *Miomatoneura* (KUKALOVÁ 1963, pl. 11, fig. 3). Identity of *Perunopterum* was based solely on the elongate wing form which is evidently due to the postsedimentation matrix deformation in the Obora locality (cf. proportions of different wings of one and the same fossil: KUKALOVÁ 1963, pl. 7, figs. 1, 2). *Miomatoneura* and *Permonia*, as well as *Miomatoneurella* (under *Delopterum* above) all based solely

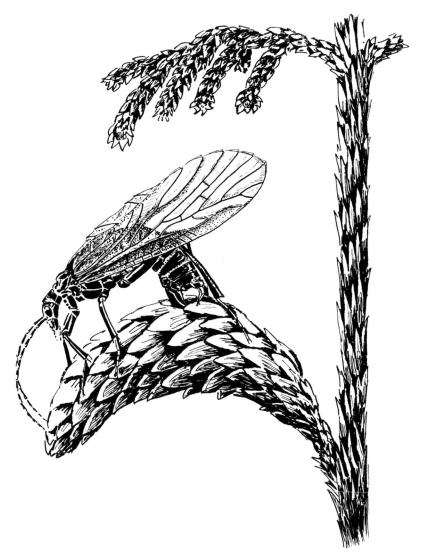

Fig. 207 Primitive palaeomanteid, *Palaeomantina pentamera* Rasnitsyn (Palaeomantiscidae) from the Early Permian of Tshekarda in Urals, restored as in life by A.G. Ponomarenko as ovipositing into a male cone of a Permian conifer (from ROHDENDORF & RASNITSYN 1980)

on the fusion of the adjacent branches of RS and M – a character often appearing independently in different genera (including *Permonka*: RASNITSYN 1977a, fig. 8) and so hardly deserving generic status. *Permodelopterum*, the last genus presently synonymised, was described as different from *Delopterum sensu* KUKALOVÁ (1963) (here *Palaeomantis*) in M fused with R basally. However, in the material available this character appears irregularly as a false fusing (M base hidden under RS base) depending possibly on the subtle deformations of the wing base.

The above considerations make it possible to organise the above genera, at least preliminary, into three families: Permosialidae with *Permosialis* and *Permonka*, Palaeomantiscidae with *Sellardsiopsis* and *Palaeomantina*, and Palaeomanteidae with *Palaeomantis*, *Delopterum* and *Epimastax*.

(f) History. The oldest, Middle Carboniferous miomopteran from Mazon Creek, Illinois, is still undescribed (Fig. 214). Late Carboniferous palaeomanteidans are known from several fossils from Germany, France and SW USA (F. CARPENTER 1967b, OUDARD 1980,

BURNHAM 1984, ROWLAND 1997, Rasnitsyn, personal observations) and belong to *Delopterum* and *Palaeomantis*, mostly undescribed. Worth mentioning is that in the French and USA localities they probably constitute the highest share of the insect assemblage throughout the whole time of the order's existence (e.g. more than 10% in Carrizo Arroyo, New Mexico).

In the Early Permian, exceptional is the earliest Permian Niedermoschel assemblage in Germany with high abundance of a single species of *Palaeomantis* in (HÖRNSCHEMEIER 1999). The Czechian assemblage from Obora includes many species of *Palaeomantis* (KUKALOVÁ 1963), Elmo in Kansas, USA, only *Delopterum* (F. CARPENTER 1933, 1939), while that of Tshekarda (the Urals) is highly diverse including *Permosialis*, *Palaeomantis*, *Delopterum*, *Sellardsiopsis* and *Palaeomantina* (NOVOKSHONOV 1998c, 2000 and references therein, STOROZHENKO & NOVOKSHONOV 1999, Novokshonov and Rasnitsyn, submitted). The only finding in Siberia (Fatyanikha in Tunguska Basin, undescribed) belongs to *Delopterum*.

The Late Permian fossils have been collected in number in North and Eastern Europe (Arkhangelsk and Perm Regions, Tatarian Rep., Pechora basin), Southwest Siberia (Kuznetsk Basin), East Kazakhstan (Saur Range), Australia (Belmont), and South Africa (Middle Beaufort Series in Karoo basin). These represent mainly *Palaeomantis*,

Fig. 208 Male *Permonka* sp. (Permosialidae) from the Late Triassic of Kenderlyk in Kazakhstan (PIN 2497/280); body without head 7.5 mm long

Delopterum, Epimastax and *Permosialis,* except Africa where only *Permonka* is found. The latter genus is otherwise known only from the Mesozoic: in the Middle or Late Triassic of Middle Asia (Madygen Formation in Fergana), Late Triassic of East Kazakhstan (Kenderlyk in Saur Range), and in the Early Jurassic of Middle Asia (Shurab in Fergana and Sogyuty at Issyk-Kul Lake). Another genus crossing the Mesozoic boundary is *Permosialis* with a few specimens collected in the Madygen deposits in Fergana, Middle Asia.

2.2.1.3.2. SUPERORDER SCARABAEIDEA Laicharting, 1781. ORDER COLEOPTERA Linné, 1758. THE BEETLES

A.G. PONOMARENKO

(a) Introductory remarks. The beetles (Fig. 215) form one of the largest orders of living organisms, with more than 350,000 described living species and hence many millions more that have ever existed (the calculation is based on the maximum average species duration that hardly exceeds 5 myr, and the present level of the beetle diversity which persists probably some 100 myr since the later Mesozoic). Ca. 200 beetle families are described thus far, and the figure still tends to grow, making taxonomic and phylogenetic study of the group very difficult. It is enough to say that correct identification of the taxonomic place of a living beetle often depends on information on the structure of the larva and on the internal organs of the adult. More difficult is the task when it comes to fossil beetles, especially their impression fossils, even complete ones which represent a sandwich of several well sclerotised layers of integument. The type of organisation characterising beetles developed during the Permian, and from that time the most common beetle fossils, detached elytra, have been extremely informative concerning their taxonomic position. After that time even fossil elytra seemingly identical to those of, say, Cupedidae, Byrridae, or Artematopidae, and could be easily turn out to belong elsewhere. That is why only a few groups of Mesozoic beetles can be appreciated with

Fig. 209 Fore wing of *Permosialis asiatica* O. Martynova (Permosialidae) from the Late Permian of Kaltan in Siberia (from ROHDENDORF *et al.* 1961); fore wing 19.5 mm long

reasonable confidence, while others have to be classified in artificial systems (cf. Chapter 1.1d and RASNITSYN 1986a, 1996).

Unlike in the Mesozoic, Cainozoic beetle diversity is essentially similar to that of the present day, permitting us to rely solely on the general similarity between fossil and living forms. Thus the morphology of fossils (including that of detached elytra) becomes again taxonomically informative, as it was for the Permian beetles. This is particularly true for the Quaternary beetles, whose study needs, however, especially deep knowledge of the morphology and taxonomy of living relatives of the fossils under examination.

These circumstances should be kept in mind when the reader assesses critically the following considerations which are based on the results of the author's investigations (PONOMARENKO 1969, 1971, 1972, 1973, 1995a, and in ARNOLDI et al. 1977, ROHDENDORF & RASNITSYN 1980) coupled with those by CROWSON (1955, 1960, 1975,

1981), HENNIG (1981), BOUDREAUX (1979), KRISTENSEN (1981), LAWRENCE & NEWTON (1982), BEUTEL (1995, 1997). Additional sources of information are cited when appropriate.

(b) Definition. Size very small to large (body length 0.2–150 mm). Adult body sclerotised, with membranous parts invaginated or otherwise protected by sclerotised ones, except paedomorphic forms. Head capsule closed by gula between occipital foramen and oral cavity, or, secondarily, gular sutures fused or lost. Eyes usually present, ocelli rarely and secondarily so (as reversion), never all 3 at once. Antenna primarily 13- though predominantly 11-segmented, secondarily with number of segment further reduced or, rarely, increased. Mouthparts chewing, very rarely piercing (in some Eucinetidae). Prothorax large, movable. Pronotum bent down and inside laterally. Coxae levelled more or less with ventral body surface, except in paedomorphic forms. Meso- and metathoraces fused immovably, desclerotised dorsally except mesoscutellum. Metathorax increased in size, much more than mesothorax, with wing articulation shifted far cephalad. Fore wings modified into elytra abutting and not overlapping mesally, and with epipleura laterally, with veins hypertrophied and tending to fuse each other, resulting in cells diminished into small windows (in Cupedomorpha), then into pits, and eventually into solid internal columellae supporting two layers of elytral deck, with or without external signs of columellae. Costal and subcostal space lost. Hind wing membranous, folded both lengthwise and crosswise (unless additionally rolled apically), rarely only lengthwise, and hidden under elytra at rest (except in some beetles with short elytra), with radial system diminished and anal one enlarged. Flight functionally or morphologically two-winged, posteromotoric. Legs almost always rather short, with tarsomere number variable. Abdominal venter well sclerotised unlike dorsum (except tergum VIII often sclerotised), with sternum I lost, II reduced, VIII and following internal along with genitalia. Male genitalia rarely complete, usually reduced up to basal piece, paired parameres, and median lobe. Ovipositor usually

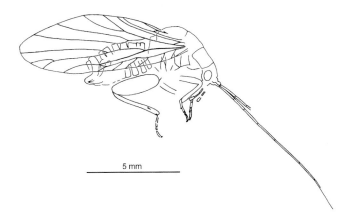

Fig. 210 Female *Sellardsiopsis conspicua* G. Zalessky (=*Palaeomantisca lata* Martynov, Palaeomantiscidae) from the Early Permian of Tshekarda in Urals (From RASNITSYN 1977a)

Fig. 211 Male *Delopterum rasnitsyni* Novokshonov (Palaeomanteidae) from the Early Permian of Tshekarda in Urals (holotype, photo by A.G. Sharov); fore wing 4.9 mm long

Fig. 212 Fore and hind wings of *Epimastax soyanensis* Rasnitsyn (Palaeomanteidae) from the Late Permian of Soyana in NW Russia (PIN 3353/270 and 279); fore wing 12 mm long

desclerotised, often with delimited styli. Cerci lost. Mesothoracic spiracle shifted under pronotum and 8 pairs of abdominal ones usually present. Ovaries poly- or, in polyphagan beetles, telotrophic. Sperm with characteristic accessory bodies.

Larvae variable, campodei- and eruciform, highly active and quiescent, exo- and endobiotic, predaceous, sapro-, xylo-, phyllo-, palino-, spermophagous, omnivorous, ecto- and endoparasitic, and so on. Head capsule well sclerotised, usually with Y-like moulting line ("epicranial suture"). Antenna with 4 or less number of segments, with apical one sometimes further subdivided. Mouthparts chewing or, sometimes, piercing. Leg 1–6-segmented, with single or (in 6-segmented only) double claw. Abdomen normally with 10, sometimes less number of segments, tergum IX often with paired urogomphs (their origin discussed by TICHOMIROVA 1991), X segment with paired or bilobed pygopode(s) of leg origin (*op. cit.*).

Pupa adecticous, rarely obtecta. Male heterochromosome, rarely haploid.

(c) Synapomorphies. Body depressed; ocelli lost; antenna 13-segmented; meso- and metathoraces fused immovably (unlike pro- and mesothorax); flight motor metathoracic; tegula lost; fore wing sclerotised due to vein hypertrophy, forming elytron to protect hind wing and abdominal dorsum; hind wing folded lengthwise at rest, with radial system diminished and anal one enlarged; abdominal terga weakly sclerotised compared to sterna; abdominal sterna beyond VIII and genitalia invaginated.

(d) Range. Since Early Permian (Artinskian) until the present, worldwide.

(e) System and phylogeny of the beetles is an obscure area full of old and new contradicting proposals and problems far from being resolved, for the reasons mentioned above [see **(b)** above]. The version proposed (Fig. 216) is based on opinions of various authors (CROWSON 1955, 1960, 1975, 1981, LAWRENCE & NEWTON 1982, KAVANOUGH 1986, LAWRENCE 1988, BEUTEL 1995, 1997) and yet essentially original. Because of the space limitation only few explanations are included in the present section, and the full discussion on the subject will be published elsewhere. I think that the myxophagan families must be put into the Archostemata if *Micromalthus* is an archostematan beetle. In the adephagan stem, the first branch is the haliploid one based on structure of the abdominal base, wing folding mechanism and venation. Phytophaga and Rynchophora are not descendants of cucujuform beetles but form an independent clade with rynchophorans as ancestral forms.

Names of higher beetle taxa are modified here according to the principles outlined above (Chapter 1.1d), as follows: superorder Coleopteroidea =Scarabaeidea; suborders Polyphaga =Scarabaeina, Adephaga =Carabina, Archostemata + Myxophaga =Cupedina (for the state of Myxophaga see PONOMARENKO 1973); infraorders Dascilliformia (or Elateriformia) =Scarabaeomorpha (because of priority of the latter name based on Scarabaeides Laicharting, 1781), Chrysomeliformia (Phytophaga + Rhynchophora) =Chrysomelomorpha, Cucujiformia =Cucujomorpha, Staphyliniformia =Staphylinomorpha.

Fig. 213 *Palaeomantis aestivus* Novokshonov (Palaeomanteidae) from the Early Permian of Tshekarda in Urals: female (left, PIN 1700/1356; body with wings 19 mm long), and male (right, PIN 1700/1381, fore wing 11 mm long, photos by A.G. Sharov)

(f) History. The oldest known beetles come from the Early Permian (Artinskian) deposits in Obora (Czechia) (KUKALOVÁ 1969c, KUKALOVÁ-PECK & WILLMANN 1990); the claimed older (Asselian) finding (HAUPT 1952) does not belong to a beetle. The beetles from Obora, as well as those from the later Early Permian (Kungurian) strata in Ural region (Tshekarda and neighbouring localities, PONOMARENKO 1969) were very rare in insect assemblages (ca. 1:1,000 unless accumulated by running water, as in Obora), and all belonged to Tshekardocoleidae, true beetles according the metathoracic structures (not possible to prove for Oborocoleidae) with the supposedly xylomycetophagous larvae and aphagous adults looking like under-bark dwellers (Figs. 217, 218). Thus the Early Permian beetles were probably a small rare group distributed at the eastern Laurentia and westernmost Angara.

The Late Permian beetles are best known in the Kuznetsk Basin of the southern West Siberia (PONOMARENKO 1969). Early in the epoch, during the Kuznetsk time (Ufimian Stage) they were still uncommon in the insect assemblages (ca. 1%) and belonged to only 4 families: Permocupedidae (majority of findings, Fig. 219),

Asiocoleidae, Rhombocoleidae, and Schizocoleidae. The first three of the families belonged each to a superfamily of its own, while the fourth one is a form groups known solely from detached elytra (for the concept of form taxa see Chapter 1.1d). Most of the families seemed to live under loose bark though some showed a "schiza", an aquatic adaptation (cf. Fig. 216, node 12). In the next later Ilyinskian (Kazanian) assemblage the beetles were more common (3%), with the dominant family being Taldycupedidae (Permocupedoidea, Fig. 220). Permocupedidae are not found, while a majority of the Schizophoroidea had a "schiza". Still younger in the Late Permian, Erunakovian (earlier Tatarian) beetles have reached the twice higher participation in the insect assemblages. Detached elytra of advanced types (classified as the form groups Permosynidae and Schizocoleidae) were dominant indicating that the majority of beetles lived in or near water, though detritophagous and predaceous forms existed as well (predaceous Late Permian schizophoroid beetle in China has been described by LIN 1982). Unexpectedly the relic Permocupedidae have been found there. From the other side, elytra of advanced types could belong in

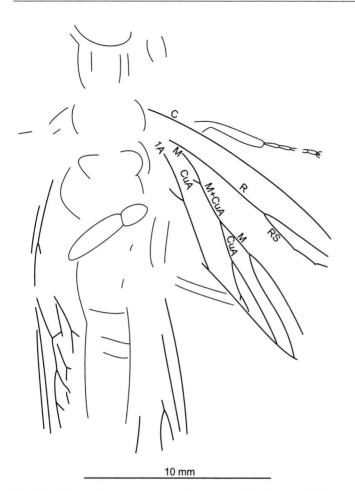

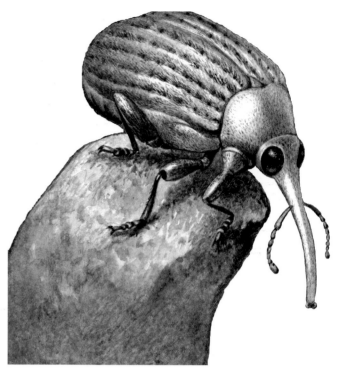

Fig. 215 *Cretonanophyes longirostris* Zherichin (Eccoptarthridae = Caridae) from the Early Cretaceous of Baissa in Siberia, restored as in life by A.G. Ponomarenko and Kira Ulanova. The family was common in the Cretaceous, but now survives as rare relicts developing on members of the conifer family Cupressaceae in Australia and temperate South America; biology of the extinct taxa unknown

Fig. 214 The most ancient, undescribed palaeomanteidan from the Middle Carboniferous of Mazon Creek in USA (Field Museum, Chicago, PE 293590)

part to archaic polyphagan and adephagan beetles, though their correct position cannot be identified with certainty until discovery of better preserved complete fossils.

Other Late Permian insect localities, either from Laurentia, or Gondwana, or from other sites at Angara, have yielded similar though less rich diverse beetle assemblages with very few genera not found in the Kuznetsk Basin (none in Gondwana!). Exceptional is the recently discovered Bor-Tologoy locality in Mongolia where more than 200 beetles have been collected among 885 insect fossils (nearly a quarter of all the insects). The assemblage shows all the families known from the Permian assemblages in the Kuznetsk Basin, and it is generally close taxonomically to the oldest (Kuznetskian) one, except for being dominated by the smooth-elytred Schizophoroidea. Early Kazanian beetles from Northeast European Russia are very rare (about 1%) and dominated by Permocupedidae, with one elytra of a tschecardocoleid genus *Uralocoleus* found as well. No beetle remains, except for isolated hind wings (PONOMARENKO 1973), have been found in numerous Tatarian localities of European Russia. Interesting also are the latest Permian (Late Tatarian) beetles from Karaungir (East Kazakhstan) belonging mostly to more advanced groups and yet including also a few elytra indistinguishable from those found only in the Kuznetskian (early Late Permian) deposits in the Kuznetsk Basin. This observation has been ascribed to the case of appearance of the relics during the major biotic

perturbations at the transition from the Palaeozoic to Mesozoic (PONOMARENKO & ZHERIKHIN 1980).

Mesozoic beetles were much more common and diverse than Palaeozoic, and yet our knowledge of them is far less complete because the Mesozoic fossils are less informative compared to the Palaeozoic ones, and this is particularly true concerning the commonest kind of fossils, the detached beetle elytron (see (**a**) above). There are two main stages of the beetle evolution observed in the Mesozoic, one of which occupying the Triassic and the first half of Jurassic, and another covering the remaining Jurassic and the Early Cretaceous, while the Late Cretaceous belonged in fact to the Cainozoic stage (see below).

More than 50 known Triassic localities yielding beetles are located on all continents except South America and Antarctica, and attributed to all stages of the Period. 250 species are described in 20 families, with the taxonomic position known with sufficient certainty for about 170. This rich material exists only for the later half of the Triassic, however. Early Triassic beetles, the only known in Eurasia had elytra of the schizophoroid and permosynoid types, all of them can be aquatic. Middle Triassic beetles were much more diverse, and Tricoleidae were common among them, but easily identifiable Cupedidae were found only in the Anisian of Vosges, France. That locality yielded numerous and diverse but undescribed beetles remains.

The largest Triassic collection comes from the Madygen Formation in South Fergana (Middle Asia). It is dated as most probably Carnian (earlier Late Triassic). Out of some 15,000 fossil insects 3,500 are beetles, predominantly detached elytra. 65 species are described belonging to 8 families which represents only a fraction of the real

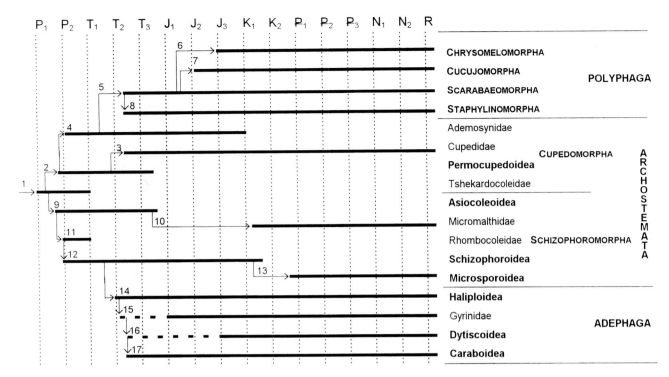

Fig. 216 Phylogeny and system of the beetles (order Coleoptera)

Time periods are abbreviated as follows: P_1, P_2 – Early (Lower) and Late (Upper) Permian, T_1, T_2, T_3 – Early, Middle and Late Triassic, J_1, J_2, J_3 – Early, Middle and Late Jurassic, K_1, K_2 – Early and Late Cretaceous, P_1 – Palaeocene, P_2 – Eocene, P_3 – Oligocene, N_1 – Miocene, N_2 – Pliocene, R – present time (Holocene). Right columns are names of suborders (all caps), infraorders (small caps), superfamilies (bold lower case) and families in monotypic superfamilies (light lower case). Arrows show ancestry, thick bars are known duration times of taxa (dashed for inferred occurrence), figures refer to synapomorphies of the subtended clades, as follows.

1 – order synapomorphies (Chapter 2.2.1.3.2.1c).

2 – elytron not exceeding beyond abdominal apex at rest, with veins aligned along its longitudinal axis, separated mostly by regular double rows of cells, with two veins reduced except basally (traceable in widened interveinal space at shoulder); wing folded crosswise at rest.

3 – fore coxa at hind prothoracic margin; elytron lacking short veins at shoulder.

4 – elytron with rows of cells reduced into punctate striae.

5 – procryptopleury (prothoracic pleuron invaginated).

6 – tarsi pseudotetramerous (with third segment bilobate, fourth reduced, inconspicuous between lobes).

7 – abdominal spiracle VIII lost; larva campodeiform.

8 – larva predatory, campodeiform, with integument sclerotised, with mandibular mola and prostheca lost, with urogomphs segmented.

9 – elytron not extending beyond abdominal apex at rest, with veins aligned along its longitudinal axis, at least in part separated by 3 regular rows of cells; wing folded crosswise at rest.

10 – complex polycyclic development with several larval types, paedomorphically simplified haploid male and larva-like female.

11 – elytral cells reduced into pits comparable in size with vein width.

12 – elytron smooth above, internally with process fixing on abdominal margin at rest (to prevent elytra to uplift with air bubble when under watter)

13 – size very small; adult paedomorphic.

14 – hind coxa bearing large femoral plate, at same level as abdomen (except basal (2nd) abdominal sternum hidden beneath) and thus intervening between metathorax and abdomen.

15 – hind coxa completely immobilised; 2nd abdominal sternum external laterad of coxae.

16 – larva with segments IX and following lost to make easy breathing by air through spiracle VIII.

17 – habits terrestrial.

diversity. The leading family Cupedidae comprises 20–40% of total fossils at different sites and has yielded as many as 29 described species, more than now live in the world. Among the other families (unlike Cupedidae these are identified only for the more complete fossils) most abundant are Schizophoridae (16 described species, Fig. 221). Subordinate positions are taken by Catiniidae (Schizophoroidea), Ademosynidae (Ademosynoidea), Tricoleidae (Asiocoleoidea), Triaplidae (Haliploidea), Trachypachidae (Caraboidea) and, among Scarabaeina (=Polyphaga), members or relatives of Hydrophilidae, Artematopidae, Elateridae (Scarabaeomorpha), and Obrieniidae

(Curculionoidea, Chrysomelomorpha, ZHERIKHIN & GRATCHEV 1993). As in earlier periods, most beetles were detrito- and xylomycetophagous or predatory, while members of the Obrieniidae (Fig. 222) probably developed phytoparasitically in gymnosperm strobiles. Both terrestrial and aquatic (but not actively swimming) forms existed there.

Post-Madygenian (Norian) Triassic assemblages are generally of similar familial composition except for a reduced cupedid abundance and the first appearance of Gyrinoidea in Garazhovka (East Ukraine) and Colymbothetidae (Dytiscoidea, both adults and larvae displaying synapomorphy 17, see Fig. 216) in Kenderlyk (East Kazakhstan)

Fig. 217 *Tshekardocoleus magnus* Rohdendorf (Tshekardocoleidae) from the Early Permian of Tshekarda in Urals (PU 2/323, photo by D.E. Shcherbakov); length of elytron 15.0 mm

Fig. 218 *Sylvocoleus richteri* Ponomarenko (Tshekardocoleidae) from the Early Permian of Tshekarda in Urals (holotype, photo by D.E. Shcherbakov); elytron 8 mm long

(PONOMARENKO 1993). Some East Asian Late Triassic assemblages are characteristic due to the complete absence of Cupedidae. The first appearance of the extant cupedid genus, *Omma* Newman is from the latest Triassic of England.

About a hundred Jurassic beetle localities are known, found almost exclusively in Europe and North Asia, with only two located elsewhere, viz. in India and Antarctica. Around 600 Jurassic species have been described to date and these are ascribed to some 35 families, although the taxonomic position of at most only half of them seems to be satisfactorily based.

The Europe assemblages of the earliest Jurassic (mainly Sinemurian, Early Liassic) beetles were still essentially Triassic in composition being dominated by Cupedina and particularly Cupedidae, including the living genera *Omma* Newman and *Tetraphalerus* Waterhouse. Schizophoridae were less abundant than Cupedidae. Trachypachidae (Eodromeinae) were abundant as well. Of interest is the usual presence of Elateridae. The later Liassic assemblages and particularly those from the Late Toarcian marine black shales in Germany and Luxembourg (A. BODE 1953 and references therein) have been partly

restudied and revealed an absence of Cupedidae, rarity of Schizophoridae and polyphagan beetles (Scarabaeina) represented by Hydrophilidae and lower Scarabaeomorpha. Adephagan beetles (Carabina) and, uniquely for all the Mesozoic, Caraboidea (only Trachypachidae) were dominant in the fauna. The aquatic Carabina, viz. Gyrinidae, Coptoclavidae (gyrinid-like dytiscoids) and lower Dytiscoidea (PONOMARENKO 1992a) were also common.

In Siberia the Jurassic (later Early to Late Jurassic) beetle assemblages were rather uniform in being dominated by various aquatic groups (Parahygrobiidae, Liadytidae and Coptoclavidae in Dytiscoidea, Gyrinidae and Hydrophilidae, Figs. 223–227), for most of which (except Gyrinidae) fossil larvae are also found, all with metapneustic breathing. Larval Coptoclavidae were nectic with paddle-like mid and hind legs (see Fig. 202), other aquatic larvae had walking legs. All of them came from lakes lacking submersed macrophytes (except charophytes) and with uninhabited bottoms, as their undisturbed microlayered deposits witness. Supposedly the lake dwellers with walking legs frequented floating mats formed by water moss, lycopsids and bennettites and covered by green and blue-green algae (PONOMARENKO 1996). Terrestrial beetles (e.g. Figs. 228, 229) are rare in the deposits.

Fig. 219 *Permocupes sojanensis* Ponomarenko (Permocupedidae) from the Late Permian of Soyana in NW Russia (holotype, photo by A.P. Rasnitsyn); length 6 mm

Fig. 220 *Taldycupes reticulatus* Ponomarenko (Taldycupedidae) from the Late Permian of Karaungir in Kazakhstan (holotype, photo by A.P. Rasnitsyn); elytron 5 mm long

In the Late Jurassic, aquatic beetles had hardly increased their diversity unlike the terrestrial ones. There are 2 classical rich localities in Bavaria (Solnhofen) and South Kazakhstan (Karatau), and a third (Shar Teg) is discovered recently in Mongolia. The richest is Karatau with some 1,000 beetle fossils collected and 228 species described. Archostematan beetles were rather abundant taking some 10% of fossils and yielding 30 described species representing mostly Cupedidae, but also Schizophoridae, Catiniidae, and Ademosynidae. Adephagan beetles belonged to Gyrinidae (Fig. 230), Dytiscidae, Coptoclavidae, Carabidae (Fig. 231), and Trachypachidae (Fig. 232). Diversity of the last two families was subequal. Coptoclavidae were represented by several genera, as is usual during the Jurassic unlike Cretaceous. More than 80% fossils belonged to polyphagans many of which are difficult to incorporate into the existing system because of unusual combination of their characters. The list of identified families includes Eucinetidae, Byrridae, Elateridae (Fig. 233), Buprestidae, Lucanidae, Scarabaeidae (Scarabaeomorpha), Staphylinidae, Hydraenidae, Hydrophilidae (Hydrophilomorpha), Peltidae, Trogossitidae, Cleroidea *inc. sed.*, Scraptiidae, Mordellidae, Alleculidae (Cucujomorpha, Fig. 234), Chrysomelidae (Fig. 235), Nemonychidae (Fig. 236) and Curculionoidea *inc. sed.* (Chrysomelomorpha). Worth mentioning are the unusually low diversity of Hydrophilidae and the exceedingly high diversity of Elateridae (107 species) and Curculionoidea (more than 40 species). Karatau lake was hardly inhabited by aquatic insects, with only corixid bugs abundant, and aquatic beetles represented only by adults and dominated by epineuston (Gyrinidae and Coptoclavidae)

and hyponeuston forms (Hydrophilidae). Unlike them, the supposed shore dwellers (most belonging to the Staphylinidae and Carabidae) were quite common. Judging from habits of their living relatives, Chrysomelidae lived at pachycaulous trunks of Cycadophyta, and Curculionoidea could live in gymnosperm strobiles. Most other beetles were probably connected with dead wood that was more or less modified by fungi (PONOMARENKO 1977a).

Two other rich Jurassic assemblages seem alike ecologically and taxonomically despite different genesis of the respective deposits, the Solnhofen fossils coming from the sea bottom while Shar Teg locality is a former lake rich in aquatic insects. It can be mentioned nevertheless that the Solnhofen assemblage was comparatively rich in Cupedidae and Coptoclavidae while in Shar Teg both families were relatively rare, and the same was true for Elateridae as well. One of the most common families was Eucinetidae.

As it has been already mentioned, the Early Cretaceous beetles were more similar to the Jurassic than to the Late Cretaceous ones which were essentially Cainozoic in their taxonomic composition, despite some Mesozoic groups (e.g. *Notocupes* Ponomarenko, Cupedidae) persisted in the Late Cretaceous and not in Tertiary. Number of the Early Cretaceous beetle localities far exceeds a hundred. They are found in almost all continents though predominantly located in East Asia and, to a lesser extent, in Europe. The taxonomic composition of assemblages is close but not identical to that in the Jurassic, however. Archostematan beetles (Cupedina) became comparatively rare and represented practically only by Cupedidae (Fig. 237). Each particular locality yields only 1–2 albeit often very abundant species of Coptoclavidae (see Figs. 202, 491, 493). Carabidae were more

Fig. 221 *Hadeocoleus calus* Ponomarenko (Schizophoridae) from the Late Triassic of Ketmen' in Kazakhstan, ventral view (holotype 8 mm long) and dorsal view of elytra (part and counterpart 7 mm long, PIN 1361/159, photos by D.E. Shcherbakov)

numerous than Trachypachidae, polyphagan beetles became much more diverse when the share of extant groups increased. First entering the fossil record were Scydmenidae, Leiodidae (Staphylinomorpha), Histeridae, Cerophytidae (Scarabaeomorpha), Nitidulidae, Scraptiidae (Cucujomorpha), Attelabidae, Curculionidae (Chrysomelomorpha). The Early Cretaceous resins from Lebanon have given additionally Micromalthidae (Micromalthoidea), Lathridiidae, and Colydiidae (Cucujomorpha). Much increased was the abundance of Scarabaeidae (Fig. 238), while the dynamics of Staphylinidae was reverse. The most important locality in the chronological succession are Montsec in Spain (the earliest Cretaceous), Baissa in Transbaikalian and Las Hoyas in Spain (both mid-Neocomian in my opinion, although there is different view on Baissa and Montsec: see Appendix), Bon-Tsagan in Mongolia (Late Neocomian or Aptian), and Koonwarra in Australia and Santana in Brazil, both Aptian, and Khetana in north-east Asia (Albian). There are numerous Cretaceous localities in China but the stratigraphic position of them and the real systematic position of beetles are vague. The beetles from the oldest Lebanese amber are very interesting but only one of them was described.

The Late Cretaceous is the start of the still lasting Cainozoic stage of the beetle evolution. Since then the archostematan share has fallen to under 1% and taken solely by Cupedidae. No extinct adephagan family has reached the Late Cretaceous. Polyphagan beetles became highly dominant, with the leading role being taken by weevils (Curculionoidea). Cucujoidea, Buprestidae, and Chrysomelidae appeared much more commonly, while less common were Elateridae and Scarabaeidae, as well as aquatic beetles, particularly those other than Gyrinidae.

There are 44 Late Cretaceous beetle localities known, 24 of which are located in Asia. Rich assemblages of better preserved impression fossils are very rare, unlike the beetle inclusions in fossil resins which become less exotic, particularly in North Siberia and North America. Assemblages from both sources are quite different in composition and should be discussed separately.

The oldest Late Cretaceous (Cenomanian) assemblages of the impression fossils are rather similar in general taxonomic composition, though differ in the dominance order. One of two larger assemblages, Orapa (dated as Cenomanian-Coniacian, Chapter 4.1) in Botswana, South Africa, is rich in advanced Carabidae, while at

Fig. 222 The oldest known weevil, *Obrienia kuscheli* Zherikhin & Gratshev (Obrieniidae), from the Middle or Late Triassic of Madygen in Kirghizia (holotype, photo by D.E. Shcherbakov); length including rostrum 1.9 mm

Fig. 223 Aquatic larva of *Parahygrobia natans* Ponomarenko (Parahygrobiidae) from the Late Jurassic of Uda in Siberia (from ROHDENDORF & RASNITSYN 1980); body 6 mm long

another one, Obeshchayushchiy near Magadan, Northeast Siberia, Carabidae are not found, and 20% fossil beetles belong to Staphylinidae. One more Cenomanian locality in East Siberia, Obluchye, has yielded remarkably numerous Nitidulidae. Not a single member of Cupedidae has been found in the Cenomanian as yet, though later, in the Turonian time, they were a regular though rare component, whilst the leadership belonged to Curculionidae and Buprestidae, usual dominant groups at the Late Cretaceous–Cainozoic stage of history of the beetles. Later in the Late Cretaceous representative assemblages of beetle impression fossils are not found.

Unlike that, the time distribution of the beetle inclusions in fossil resin is more even in the Late Cretaceous. There are 19 families found there, and many of them (Callirhypidae, Scirtidae, Ptiliidae, Acanthocnemidae, Corylophidae, Cryptophagidae, Coccinellidae, Endomychidae, Rhypiphoridae, Melandriidae) for the first time. However, only 7 species have been described from the Late Cretaceous resins thus far. It seems probable that the majority of the living beetle families originated during the Late Cretaceous.

The oldest Tertiary, Palaeocene fossil beetle localities are scanty, less then ten being found as yet, mostly in Eurasia and North America. The dominant members were usually Curculionidae, but sometimes Buprestidae or Chrysomelidae. Cerambycidae and Dytiscidae were not uncommon as well. In total only 19 Palaeocene families are recorded up to now, suggesting poor knowledge rather than a poor source fauna.

In contrast, the Eocene beetle fauna is the best known among all ancient ones. 101 family and numerous species are recorded from more than 30 localities, though unfortunately many description are not satisfactory. Their main source is the Baltic amber, which has yielded 88 beetle families. Many oldest family records refer to the Baltic amber, because of its state of knowledge and not of their real origin that time. The Eocene impression fossils are also abundant but for the most part only superficially studied. This makes it difficult to assess the dominance order in the usual way. Counting the number of localities with particular families recorded is of help, however, indicating that Curculionidae was the most common one followed by Buprestidae and Elateridae, and then by Carabidae. More rare were Chrysomelidae, Cerambycidae, and Dytiscidae. These observations show Buprestidae and Elateridae playing a more important role in the Eocene, compared with the present time.

Fig. 224 Diving beetle, *Liadytes longus* (Liadytidae), from the Late Jurassic or Early Cretaceous Glushkovo Formation (Unda) in Siberia (holotype, photo by D.E. Shcherbakov); length 9 mm

Due to different taphonomy, the amber beetles show a quite dissimilar family level dominance order, headed by Scirtidae with the next highest levels occupied by Elateridae, Anobiidae, and Staphylinidae. Of them Elateridae and Staphylinidae were probably most numerous, while the abundance of Scirtidae and Anobiidae is most probably due to their connection to the resin source tree *Pinus succinifera* Goeppert. Scirtidae were most probably abundant due to coincidence regarding the time of maximum resin production and beetle aggregation on the pine tree staminate cones. Most ecological groups of beetles are known from the amber. For details concerning the Baltic amber beetles see LARSSON (1978).

Oligocene beetles follow the Eocene ones in their state of knowledge. There have been 73 families mentioned and several hundred species described there, though the majority of descriptions badly need revision. The list of families known from impression fossils and ranked in the same way as described for the Eocene are close to that for living families ranked after the number of their species: Curculionidae are followed by Carabidae, then by Chrysomelidae, Cerambycidae, and Staphylinidae. Worth mentioning is the relative scarcity of the aquatic

beetles which could indicate their real rarity, for the taphonomic factor tends to exaggerate the abundance of the insects which populated the lakes producing insectiferous deposits. During the Oligocene, participation of the aquatic beetles was growing resulting in their dominant position in the Miocene impression fossil assemblages of beetles. This probably reflects their growing abundance supposedly caused by the expansion of submersed macrophytes in lakes.

The Oligocene fossil resins with beetle inclusions are known in Europe and America. Most important seems the Mexican amber produced by a tropical tree of the fabacean genus *Hymenaea* L., with buried beetles showing more close relation to the subtropical and temperate fauna than to tropical. In contrast, the Dominican amber beetles of same origin and similar age (Early Miocene, ITURRALDE & MACPHEE 1996) had essentially tropical appearance.

Miocene beetles are less known. There are 61 localities explored in Eurasia and North America yielding almost exclusively impression fossils. The list of families ranked as above is headed by Carabidae and Chrysomelidae and followed by Curculionidae, Scarabaeidae, Dytiscidae and Hydrophilidae. As it was mentioned above, the increased participation of aquatic beetles is most probably exaggerated, though reflecting a very real trend. The level of exaggeration can be appreciated by taking into consideration that since the Miocene the participation stopped growing. This suggests that the present day level of abundance of aquatic beetles was reached in the Miocene. The observation implies that pre-Miocene faunas were much impoverished in that respect compared to what we can see now. Indeed, the maximum number of dytiscid species collected in any particular locality has not exceeded two in the Palaeocene, Eocene and Oligocene with the exception of the unbelievably rich collection from the Early Oligocene of Florissant in Colorado, USA, and another from the earliest Miocene (Aquitanian) locality Rott in Germany, both with 8 dytiscid species. Far more modest Middle Miocene locality at Stavropol in the North Caucasus has yielded 9 species, the Late Miocene one, Oeningen in Germany – 11. All larger Miocene collections have given more than two dytiscid species each, except Shanwang in Shandong, China. Diversity of Hydrophilidae was also growing at that time though to a lesser extant when compared with Dytiscidae.

Miocene beetle faunas showed apparent latitudinal zonation. As mentioned above, beetles from the Dominican amber had tropical relatives, while Meigen Is. in Arctic Canada has yielded a boreal beetle assemblage. There were many exotic groups now having southern distribution among European Miocene beetles. The overwhelming majority of the Miocene beetles belonged to living genera and some even to living species. Particularly the amber from Bitterfeld in Germany has been dated as from the Miocene despite the general similarity of its insects to that in the Baltic amber, and yet it includes a higher percentage of beetles belonging to extant species.

It would be of great interest to study thoroughly the assemblages of beetles and other insects from the Messinian (latest Miocene) deposits formed at the bottom of the dried-up Mediterranean sea (e.g. from Gabbro in Italy): their cursory examination has surprisingly shown no unusual features.

Pliocene beetles are still less well known than the Miocene ones. There are only 36 families mentioned from 20 localities in Europe, Asia, and North America; insectiferous copals are found in Australia. As in the Miocene, the leading position has been taken there by Carabidae and Chrysomelidae, followed by Curculionidae and Scarabaeidae. The best studied beetles assemblage from Willershausen in North Germany includes 150 fossils, with almost every one belonging to different

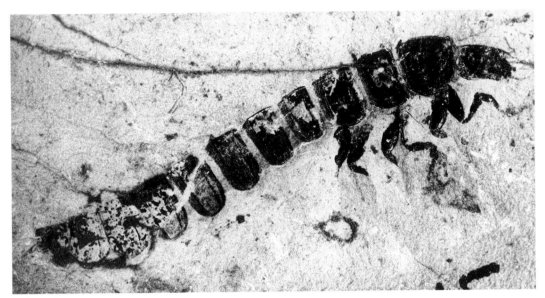

Fig. 225 Larval diving beetle, *Angaragabus jurassicus* Ponomarenko (Liadytidae), from the Early Jurassic of Ust-Baley in Siberia (PIN 2375/205, photo by D.E. Shcherbakov); length 14 mm

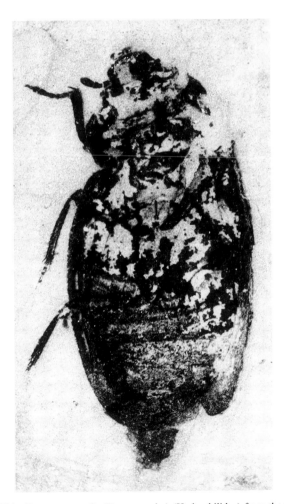

Fig. 226 *Zetemnes tarsalis* (Ponomarenko) (Hydrophilidae) from the Late Jurassic or Early Cretaceous Glushkovo Formation (Unda) in Siberia (PIN 3063/116, photo by D.E. Shcherbakov); length 4 mm

species: 90 specimens have been identified to species level, and a total of 84 species have been found. Eighteen families and 35 genera have been counted, with only 6 extinct genera appreciated. Similarly, out of each 10 identified species 8 are considered extant. Among living taxa, several genera and only two species are not frequent in Europe now (GERSDORF 1969, 1970, 1976).

Impression fossils of the Pleistocene age are rare. Inclusions in subfossil resins (copals) as well in bitumen or ozokerit are not so exceptional. Unlike Tertiary, the usual taphonomic types of the Pleistocene insect localities are moderately or little modified remnants, dismembered or, rarely, intact, of insects with harder integument (predominantly beetles) buried in alluvium, slope sediments, and peat. These remnants are not especially long lived, however, unless protected by permafrost or humic acids. That is why insects older than the last glacial stage are rarely found in territories lacking permafrost.

The Pleistocene fossils (reviewed by ELIAS 1994), though disarticulated, are often of the best preservation and when they belong to a better known group they often permit sound identification. The Pleistocene beetles, even the most ancient of them, have been found to belong, with very few exceptions, to living species (COOPE 1979, MATTHEWS 1980, NAZAROV 1984), in spite of all the environmental and population perturbations they have been experiencing, including fluctuations of quantity and very distant migrations. These migrations, whose scale can be appreciated from, e.g., discovery of a dung beetle from Tibet in England (COOPE 1973) and of a weevil from Yakutia in Belorussia (ZHERIKHIN & NAZAROV 1990), did erase almost all traces of the pre-Pleistocene beetle distribution in the North Hemisphere. This effect was moderated in North America with its mountain ranges directed meridianally and thus permitting the biota to move to and fro following the climatic perturbations. This was not the case in Europe where the fluctuating climate forced animals and plants to migrate far in a longitudinal direction, as the above examples confirm. In general, during a stadial (shortly before and after the territory is covered by glaciers), species from tundra and cold Asian steppe were establishing. During an interstadial they are being replaced by those now living at

Fig. 227 Larval aquatic beetle, *Angarolarva aquatica* Ponomarenko (Hydrophilidae), from the Early Jurassic of Ust-Baley in Siberia (holotype, photo by D.E. Shcherbakov); length 14 mm

that particular territory or still farther southward, up to Mediterranean species and even those from trans-Saharan Africa, as in Britain during the last interstadial. How rapid were these faunistic perturbations can be seen from the case of interstadial Windermere, when South European beetles reached Britain before any forest could established there. These observations make groundless any attempts to restore European pre-Pleistocene zoogeography of beetles based solely on their present distribution. In view of this evidence it seems inescapable to join COOPE (1994) in that only palaeontological study of the Pleistocene beetles will permit us to reconstruct the history of the European beetle fauna.

Different was the Pleistocene history of beetles in Siberia and northeastern North America. It was already in the Late Miocene when the taiga and tundra beetles appeared there, and in the Pliocene the cryophilous fauna took the dominant place. Besides tundra species, it also included cryoxerophilous ones, now characteristic of the cold Mongolian steppe and steppe-like patches in North Siberia. This indicates the origin of the tundrosteppe, a characteristic Pleistocene landscape which was very cold and dry, with almost snowless winter, and with local patches of bogs and thin forest. Accumulation of water in European glaciers made the climate so dry that Siberia had only mountain glaciers. Because of that the environmental differences between stadials and interstadials were less pronounced there, resulting in better continuity of succeeding faunas. The faunistic fluctuations appeared as increasing participation of xerophilous forms when Europe suffered from glaciation, while during the European interstadials the Siberian fauna was enriched with forest and bog beetles. In Eastern Europe the processes took on an intermediate character (NAZAROV 1984).

2.2.1.3.3. SUPERORDER MYRMELEONTIDEA
Latreille, 1802 (=Neuropteroidea Handlirsch, 1903)

A.G. PONOMARENKO

(a) Introductory remarks. Comparatively small but rather clear-cut and popular (mainly for some large and pretty neuropterans) insect group of not fully resolved affinities and obscure internal relationships (see below).

(b) Definition. Size moderate, rarely small or large. Head prognathous, with gular sclerite intervening between mouth cavity and occipital foramen. Antenna long, multisegmented, mostly thread- or bristleshaped. Ocelli plesiomorphically present. Mouthparts chewing, maxilla with galea and lacinia reduced. Body integument not much sclerotised. Prothorax movable, lateral and ventral cervicalia fused ensheathing neck region ventrally and laterally (secondarily desclerotised ventrally in Neuroptera and Sialidae). Meso- and metathoraces similarly shaped. Wing position at rest usually roof-like, rarely horizontal and widely overlapping (in Corydalida other than Sialidae). Fore and hind wing pairs uncoupled in flight, with venation variable, usually rich, with costal space wide and RS more or less regularly pectinate. Flight functionally fourwinged, in-phase, anteromotoric, rarely morphologically two-winged (GRODNITSKY 1999). Legs walking (fore pair sometimes raptorial), with coxae free, long, and tarsi 5-segmented. Male genitalia lacking basal ring, with gonostylus and volsella, or variously reduced. Ovipositor lacking gonapophysis IX (2nd valvula), with at most rudimentary gonapophysis VIII (1st valvula), and with 3rd valvulae fused dorsally, mounted apically with gonostylus, and muscularised internally, or lost. Stomodeum with dorsal diverticulum. Aphagous, palinophagous, or predaceous, often short-living. Male heterochromosomous. Habits terrestrial.

Larva terrestrial or aquatic, predaceous, eruci- or campodeiform. Head prognathous, with 6 stemmata and 4-segmented antenna. Legs two-clawed, with free tibia and tarsus. Pupa free.

(c) Synapomorphies. Larva predaceous; 3rd valvulae fused dorsally, muscularised internally.

(d) Range. Since Early Permian (Artinskian) until the present, Early Permian only Laurasian, since Late Permian also in Gondwana. Now worldwide.

(e) System and phylogeny are not well established. The available fossil material is only fragmentary, and morphological study of the living representatives gives contradictory results concerning both the number and relationships of the constituent orders (ACHTELIG &

Fig. 228 Ground beetle, *Unda cursoria* Ponomarenko (Trachypachidae), Late Jurassic or Early Cretaceous Glushkovo Formation (Daya) in Siberia (holotype, photo by D.E. Shcherbakov); length 7 mm

Fig. 229 Rove beetle, *Anicula inferna* Ryvkin (Staphylinidae), from the Early or Middle Jurassic of Novospasskoye in Siberia (holotype, photo by D.E. Shcherbakov); length 5 mm

KRISTENSEN 1973, KRISTENSEN 1975, 1981, MEINANDER 1975, ŠTYS & BILINSKI 1990).

The superorder is accepted here embracing three living orders Neuroptera (=Myrmeleontida Plannipenia), Corydalida (=Megaloptera), and Raphidiida (=Raphidioptera), and an extinct one Jurinida (=Glosselytrodea). Although Corydalida and Raphidiida are traditionally considered closely related, their similarities are entirely symplesiomorphic. BOUDREAUX (1979) has proposed synapomorphies for Megaloptera + Plannipenia which has been rejected by KRISTENSEN (1981). ŠTYS & BILINSKI (1990) claim that Raphidiida and Sialidae (Corydalida) are synapomorphic in having telotrophic ovaries, and their hypothesis implies that the next more inclusive clade is Corydalida + Raphidiida, as supported by synapomorphy in aquatic development (supposing its secondary loss in Raphidiida). None of the above hypotheses look convincing, and for the present, the relations of the three orders are considered as unresolved.

The position of Jurinida is quite obscure as yet.

2.2.1.3.3.1. ORDER RAPHIDIIDA Latreille, 1810 (=Raphidioptera Navas 1918)
THE SNAKEFLIES

(a) Introductory remarks. Small group (150 living species; ASPÖCK & ASPÖCK 1991) of infrequently seen insects of rather standard general appearance (Fig. 239).

(b) Definition. Size medium to moderately small. Body elongate, with characteristically elongate head and, often, prothorax. Head large, long, most wide before midlength, with eyes shifted toward anterior end, with posterior face closed anteriorly by gular sclerite and by genae meeting posteriorly to form genapont, with hind tentorial pits placed close to each other near intergenal suture. Mouthparts chewing, maxillary palp 4-segmented, labial palp 3-segmented. Pronotum most wide anteriorly, bent downward and mesad to cover cervicalia and almost to meet contralateral margin below. Prosternum and fore coxae shifted toward hind end of prothorax. Meso- and metathoraces similarly developed. Legs walking, tarsi 5-segmented, with 3rd segment

Fig. 230 Whirligig beetle, *Mesogyrus antiquus* Ponomarenko (Gyrinidae), from the Late Jurassic of Karatau in Kazakhstan (holotype, photo by D.E. Shcherbakov); length 9 mm

Fig. 231 Ground beetle, *Protorabus planus* Ponomarenko (Carabidae), from the Late Jurassic of Karatau in Kazakhstan (PIN 2904/928, photo by D.E. Shcherbakov); length 16 mm

bilobed and 4th one small, hidden in excision. Hind wing slightly shorter and narrower than fore one, with narrow anal area. Venation (Fig. 240) forming network of quadrangular or elongate rounded cells, with only few veins ending in terminal forks. Costal space wide, expanding forward in fore wing, with numerous cross-veins. SC meeting C. Pterostigma sclerotised, margined by R posteriorly and apically, crossed by few cross-veins, if any. R, M, and Cu running close each other basally (seemingly forming one vein), diverging either from one point, or M and Cu leaving each other shortly after R. Hind wing with free MA base only rarely leaving M, usually starting from R and then joining RS. Abdomen 10-segmented, in male with gonostylus and volsella present or lost, in female with long ovipositor formed dorsolaterally by fused 3rd valves mounted by gonostyli, and ventrally by ventral valvula (apparently elongated 8th sternum). 8 malpighian tubes. Ovaries telotrophic. Terrestrial, predaceous insects, observed mostly on tree bark, sometimes also feeding on flowers.

Larva campodeiform, holopneustic. Antenna rather long, 4-segmented. Mouthparts chewing, with maxilla rather short, maxillary palp 4-segmented, labial one 3-segmented. Legs with free tibia and tarsus, with pretarsus 2-clawed, bearing empodium. Abdomen with anal pseudopod bilobed apically. Habits terrestrial, predaceous, cryptic (most often under loose tree bark).

(c) Synapomorphies. Head elongate behind eyes, with genae contacting mesally, and with hind tentorial pits close each other; pronotum widest anteriorly, inflected widely, ventromesally; ovipositor long, incorporating apical outgrowth of 8th sternum.

(d) Range. Early Jurassic till now, after Early Cretaceous not found on southern continents.

(e) System and phylogeny. Since all the Palaeozoic families previously ascribed to the order have been found to belong elsewhere (*Letopaloptera* to Hypoperlida, Chapter 2.2.1.2.2, *Sojanoraphidia* to Grylloblattida, Chapter 2.2.2.2.1), 5 snakefly families are acknowledged here, 2 living and 3 extinct. As to two more extinct families from the Mesozoic of China described by HONG (1982a, HONG & CHANG 1989), they are inadequately described and most probably deserve synonymisation with early established one(s). WILLMANN (1994) proved that the Early Jurassic forms, *Priscaenigma* and *Hondelagia*, did not belong to any described families, but did not describe any new. These taxa possibly do not belong to snake-flies because they have SC meeting R instead of C, R is running near the wing margin almost up to its apex, and the characteristic anal cell is absent.

Venational features are highly variable in snakeflies and thus are of little use in the taxonomy of living forms, genitalic characters being most involved there instead. However fossils are represented mostly by wings, or at least the wings are usually best preserved and thus form the bulk of available information on past snakeflies. This does lower the discriminating ability of the palaeontological method. Although quite a few Mesozoic fossilised bodies have now been found, only a few are sufficiently well-preserved to give much information, and even the presence of ocelli (the main trait for discriminating between the two living families) could only be confirmed in a handful of cases.

Judging from the available characters, neither known family could be assessed as ancestral to the other with reasonable certainty. Particularly, the Jurassic Mesoraphidiidae (see Fig. 239) were autplesiomorphic in having short pronotum, and apomorphic due to the very long, weakly sclerotised pterostigma, and the hind wing with MA base transferred to R (probably except *Metaraphidia confusa* Whalley which is, however, insufficiently described; see below). Baissopteridae (Figs. 241, 242) have retained another plesiomorphy: long, sinuate

Fig. 232 Ground beetle, *Karatoma agilis* Ponomarenko (Trachypachidae), from the Late Jurassic of Karatau in Kazakhstan, restored as in life by A.G. Ponomarenko (from ROHDENDORF & RASNITSYN 1980)

MA base starting from M, and yet it has its venation polymerised, and prothorax long at least since the Cretaceous (unknown for the Jurassic fossils). Both cladistic analyses of the whole group, including extinct forms (REN & HONG 1994, WILLMANN 1994) are unsatisfactory. As a result, the taxonomic structure of the order is must remain unresolved.

(f) **History**. After the Palaeozoic forms were excluded from the snake-flies, those described by WHALLEY (1985) from the English Lias (earlier Early Jurassic) took the place of the oldest representatives (re-described WILLMANN 1994). One of these is a member of Mesoraphidiidae. The next youngest snakefly comes from the German Lias (see Fig. 240); at the same time or a little later, snakeflies appeared in Central Asia (Sagul in South Fergana, undescribed). Several snakeflies have also been collected in the Jurassic Kotá Formation, India (undescribed too), these being the oldest Gondwanan record of snakeflies.

Late Jurassic snakeflies are only known from Asia, being found in Mongolia (PONOMARENKO 1988b), China (HONG 1982a), and in particularly large numbers in the Karatau Range of South Kazakhstan (MARTYNOV 1925b, MARTYNOVA 1947a). The latter source has yielded some 600 fossils which take ca. 3% of all the insect material and makes the group the 6th most abundant at ordinal level. All Karatau

fossils seem to belong to several species of *Mesoraphidia* (see Fig. 239), while other assemblages include additionally Baissopteridae with the long "common stock" of R, M, and CuA.

More regularly, albeit less abundantly, snakeflies are encountered in the Cretaceous deposits. They are known there for the Early and the first half of the Late Cretaceous, with Baissopteridae (Figs. 241–243) only, recorded from the Early Cretaceous while Mesoraphidiidae and Alloraphidiidae are reported from both epochs. Early Cretaceous snakeflies are found in England (JARZEMBOWSKI 1984), Spain (GÓMEZ PALLEROLA 1986), Transbaikalia (MARTYNOVA 1961b), Mongolia (PONOMARENKO 1988b), China (HONG & CHANG 1989, REN 1997a), and Brazil (MARTINS-NETO & VULKANO 1990a, OSWALD 1990). The last finding is of particular interest, for the living snakeflies are absent from the South Hemisphere. As to the Late Cretaceous representatives, they come from Middle and East Asia (unpublished in Cenomanian Obeshchayushchiy, Turonian Kzyl-Zhar and Arkagala) and in North America (Alb-Cenomanian Redmond Formation of Labrador: F. CARPENTER 1967c, and Turonian amber of New Jersey, GRIMALDI 2000a). Some Alloraphidiidae even surpassed living snakeflies in their head and pronotum length.

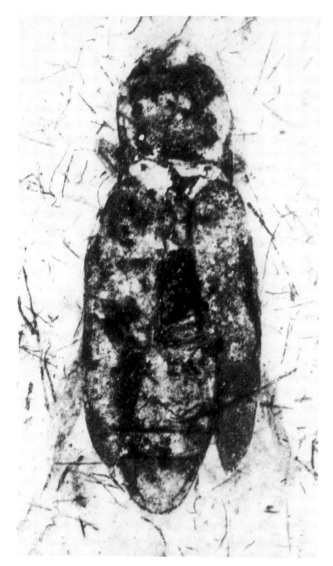

Fig. 234 *Jurallecula grossa* L. Medvedev (Alleculidae) from the Late Jurassic of Karatau in Kazakhstan (holotype, photo by D.E. Shcherbakov); length 19 mm

Fig. 233 Click beetle, *Codemus teres* Dolin (Elateridae), from the Late Jurassic of Karatau in Kazakhstan (PIN 2066/2703, photo by D.E. Shcherbakov); length 9 mm

It is of interest that Mesozoic snakeflies have been collected predominantly in localities that originated from warmer climates, while the living families prefer more temperate ones (ASPÖCK 1998).

The Cainozoic fossils all come from the Palaeogene where they were about as common as in the Late Cretaceous, though being represented here only by living families (with the possible exception of the Palaeocene fossil which reminds one of a member of Alloraphidiidae). The Palaeogene snakeflies are found in Europe, Asia, and North America (CARPENTER 1936). Worth mentioning is that in the North American Early Oligocene of Florissant (Colorado) the genera *Raphidia* L. and *Inocella* Schneid. are recorded which now live outside that continent.

2.2.1.3.3.2. ORDER CORYDALIDA Leach, 1815 (=Megaloptera Latreille, 1802) THE ALDERFLIES AND DOBSONFLIES

(a) Introductory remarks. A small (of about 300 living species, THEISCHINGER 1991a) group of primitive amphibiotic insects

Fig. 235 Leaf beetle, *Pseudomegamerus grandis* L. Medvedev (Chrysomelidae), from the Late Jurassic of Karatau in Kazakhstan (holotype, photo by D.E. Shcherbakov); length 16 mm

(Fig. 244) with long fossil history and low diversity (but not always abundance) both at present and past times.

Fig. 236 Weevil, *Brenthorrhinus mirabilis* L. Arnoldi (Nemonychidae), from the Late Jurassic of Karatau in Kazakhstan (holotype, photo by D.E. Shcherbakov); length including rostrum 8.4 mm

Fig. 237 *Tetraphalerus verrucosus* Ponomarenko (Cupedidae) from the Early Cretaceous of Baissa in Siberia holotype, photo by D.E. Shcherbakov); length 6 mm

(b) Definition. Size moderate to large. Head large, prognathous, with bulging eyes, closed with gular plate between mouth cavity and occipital foramen. All three ocelli either present or lost. Antenna shorter than fore wing length, setiform (rarely moniliform or pectinate in male). Mouthparts chewing, male mandible sometimes much enlarged. Maxillary palp with 5, labial one with 3 segments. Prothorax movable, enlarged more or less, meso- and metathoraces of subequal size and structure. Legs moderately short, tarsi 5-segmented, lacking both pulvilli and empodium. Wing position at rest either roof-like or flat over abdomen. Nygmata present. Costal space crossed with numerous cross-veins. SC joining R. RS pectinate. Hind wing with MA base (abscissa connecting M base and RS) long S-like, short, or absent, anal area wide. Fore wing sometimes with jugum used in wing coupling at flight. Abdomen 10-segmented, with ovipositor rudimentary or lost. Male genitalia lacking gonostylus. 8, rarely 6 malpighian tubes. Ovaries neopanoistic or telotrophic.

Larva aquatic, elongate, somewhat depressed. Head prognathous, closed with gular plate between mouth cavity and occipital foramen, with 6 ommatidia and 4-segmented antennae. Mandible elongate,

teethed toward apex. Maxillary stipes elongate, palp 5-segmented, labial palp 3-segmented. Thoracic terga and prosternum sclerotised. Legs long, tarsi 2-clawed. Sialid abdomen bearing 7 long, segmented limb-derived gills on anterior segments, corydalid one with 8 less long and not segmented gills as well as 7 bundles of short gills. Spiracle of 8th segment mounting long projection. Apical segment bearing unpaired segmented filament in Sialidae, pair of prolegs bearing claw and gill in Corydalidae. Pupa active, decticous, developing out of water in earth cell. Development long, sometimes taking more than a year, with more than 4 larval instars. Adult aphagous, crepuscular, occurring near larval water bodies; larva predaceous.

(c) Synapomorphies. Larva aquatic, with gills, and with both mandible and maxillary stipes elongate.

(d) Range. Late Permian until the present, worldwide; fossils only in Eurasia, except for the sole South African Triassic find.

(e) System and phylogeny. The order consists of two extinct and two extant families, two more fossil ones have now been found to belong to other groups (see Chapters 2.2.1.3.1 for Permosialidae and 2.2.1.3.4.1 for Tychtodelopteridae). Relationships within the order are not clear yet.

Fig. 238 Scarab beetle, *Holcorobeus vittatus* Nikritin (Scarabaeidae), from the Early Cretaceous of Baissa in Siberia (holotype, photo by D.E. Shcherbakov); length 9 mm

Fig. 239 Snakefly, *Mesoraphidia pterostigmalis* O. Martynova (Mesoraphidiidae), from the Late Jurassic of Karatau in Kazakhstan (holotype, photo by D.E. Shcherbakov); fore wing length 13 mm

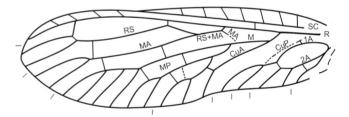

Fig. 240 Fore wing of *Metaraphidia vahldieki* Whalley (Metaraphidiidae) from the German Lias (Early Jurassic), showing the most plesiomorphic venation known within the order because of distal position of the base of vein MA (from WILLMANN 1994, with vein lettering modified)

The polyneurous Corydalidae traditionally have been considered most primitive. The hypothesis has been falsified due to the oligoneury of all more ancient fossils (Parasialidae and Euchauliodidae), this observation implying secondary polymerisation of the corydalid wings. At the same time, Corydalidae are autplesiomorphic in retaining M and Cu independent basally, while Sialidae are autapomorphic in some larval structures (caudal filament and segmented lateral abdominal appendages). Moreover, Sialidae are claimed to be synapomorphic with snake-flies in the telotrophic ovaries, and additionally in the closely approximated bases of R, M, and Cu. Hypothesis of the corydalid relationships of Euchauliodidae (PONOMARENKO 1977b) is not correct because of M originated from CuA.

The intraordinal phylogeny is not clear at present depending first of all on whether the base of M is primarily free in Corydalidae, or secondarily so, e.g., retained from the pupal wing, and whether the telotrophy of corydalidan and snake-fly ovaries has been acquired jointly or independently.

(f) History. The oldest Corydalida are known from the latest Early Permian Kungurian locality Tshekarda in Urals and from the Kazanian

deposits of the Late Permian age at Soyana River in North Russia. All of them belonged to Parasialidae (PONOMARENKO 1976b, NOVOKSHONOV 1993a, Fig. 245), as well as those found in the Mongolian locality Bor-Tologoy possibly of the next, Tatarian stage of the Late Permian (PONOMARENKO 2000). Also a larva of the corydalid type (Fig. 246) is found in the Late Permian (Tatarian) deposits of Kargala (South Urals, SHAROV 1953).

Larval fossils are also found in the East European (Ukrainian) Late Triassic locality Garazhovka (undescribed). The South African Late Triassic (Molteno Series) has yielded the only known member of Euchauliodidae. The Triassic larval fossils, *Mormolucoides* originally assigned to the order (HITCHKOK 1858) do not belong there judging

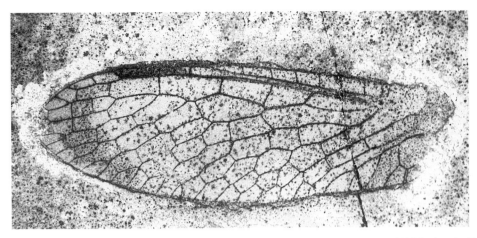

Fig. 241 *Baissoptera sibirica* Ponomarenko (Baissopteridae) from the Early Cretaceous of Baissa in Siberia (PIN 1989/26, photo by D.E. Shcherbakov); wing 18 mm long

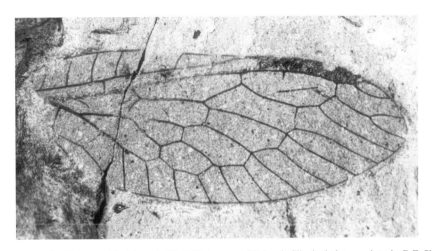

Fig. 242 *Siboptera medialis* Ponomarenko (Baissopteridae) from the Early Cretaceous of Baissa in Siberia (holotype, photo by D.E. Shcherbakov); wing 13 mm long

from the material available courtesy Frank M. Carpenter (Cambridge, USA); the material does not cover the type series, however.

Larvae of corydalid type are found in the Late Jurassic of Shar Teg in West Mongolia (undescribed), though the earliest indisputable Corydalidae come from the lacustrine deposits of the Early Cretaceous (despite the strictly lotic habitats of the living members of the family). An incomplete larva has been found in the early Early Cretaceous deposits of Polovaya in Transbaikalia, and the somewhat younger (but see Chapter 4.1) Baissa locality at Vitim River (also in Transbaikalia) has produced numerous larvae and their moulting casts, as well as adults (PONOMARENKO 1976b; Figs. 244, 247). This type of the corydalid larva was still persisting in the Cenomanian (earliest Late Cretaceous) of the Northeast Siberia (Obeshchayushchiy locality near Magadan, undescribed). Still younger Santonian fossil resin from the Taimyr Peninsula (the northernmost Siberia, Yantardakh locality) has given a young larva structurally intermediate between Corydalidae and Sialidae (PONOMARENKO 1976b). The Late Cretaceous deposits of both France and USA has yielded large egg clutches ascribed to Corydalidae (SCUDDER 1890b).

The Cainozoic Sialida belonged in majority to Sialidae which are known there in the Eocene Baltic amber (WEIDNER 1958) and in the Pliocene locality Willershausen in North Germany (ILLIES 1967). Most of them are identified as *Sialis* L. Of interest is the record of *Indosialis* Lestage from the Miocene of Turkey (NEL 1988). Only one corydalid species is described, albeit unsatisfactorily, from the Baltic amber (PICTET 1854). The Cainozoic fossils are not numerous, except for mandibles of *Sialis* that are abundant in the sapropel of some Quaternary lakes.

2.2.1.3.3.3. ORDER NEUROPTERA Linné, 1758 (=Myrmeleontida Latreille, 1802) THE LACEWINGS and ANTLIONS

(a) Introductory remarks. An old, moderately diverse (about 5,000 living species; NEW 1991) and common group of insects with rather primitive but sometimes large and butterfly-like splendidly coloured adults (Figs. 248, 249), and highly specialised, mostly cryptic larvae. Most fossil forms need re-description (see LAMBKIN 1988).

(b) Definition. Size and structure much as defined for superorder (Chapter 2.2.1.3.3b). Ocelli rarely present. Antenna sometimes pectinate or clavate. Maxillary palp almost always with 5, labial one with 3 (rarely 2) segments. Prothorax sometimes elongate. Wings folded roof-like at rest. Venation usually rich, with end-twigging, often with nygmata (small, rounded, sensory patches) and with gradate series of cross-veins, smallest forms with venation much reduced. Wing margins sometimes with trichosors (thickenings bearing several hairs).

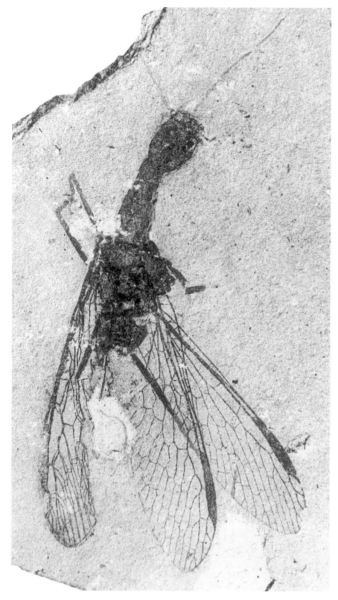

Fig. 243 *Cretoraphidia certa* Ponomarenko (Baissopteridae) from the Early Cretaceous of Romanovka, Vitim River in Siberia (holotype, photo by D.E. Shcherbakov); fore wing length 14 mm

Hind wing sometimes moderately enlarged posterobasally, or reverse, distinctly narrow comparing fore one, even ribbon-like in Nemopteridae. Abdomen with 10, rarely 9 segments. Ovipositor rudimentary, rarely (in Dilaridae) long, lacking gonostylus. Usually 8 malpighian tubes.

Larva predaceous or, rarely, parasitic. Mandible and maxilla united and modified into piercing stylet with internal canal continuing directly into stomodeum (mouth opening lost). Maxillary palp lost, labium reduced but retaining long, sometimes 5-segmented palp. Metathoracic spiracle lost. Midgut closed behind. Malpighian tubes producing silk used for pupal cocoon. Habits terrestrial, rarely aquatic (Sysiridae), semiaquatic (some Osmylidae), or soil-dwelling (Ithonidae, Dilaridae).

(c) Synapomorphies. Male and female gonostylus lost; larva with above mouthpart modifications, closed metathoracic spiracle and midgut, and silk-producing malpighian tubes.

(d) Range. Since Early Permian until the present, worldwide.

(e) System and phylogeny. There is no generally accepted system for the order thus far, and Myrmeleontoidea is the only non-monotypic taxon above family rank considered as indisputably monophyletic. Their monophyly is well founded by a sound set of both adult and larval synapomorphies (rounded body form and sickle-shaped, mesally dentate mandibles in larva, damsel-fly general appearance with long, narrow wings and body, and wings with diminished anal area, characteristic triangle in cubital area, and lost nygmata and trichosors).

Other lacewing families display a diverse array of supposed apo- and plesiomorphies making it difficult to organise them in a convincing cladogram. The present state can be illustrated by the latest published dendrogram by SCHLÜTER (1986), with the 8 main branches leading toward 34 families (including 14 extinct ones) from a common base – an interrogation mark. Far from satisfactory is the state of groups shown as monophyletic: what, for example, do Hemerobiidae have in common with, say, Mantispidae, and Mesithonidae with Promegalomidae?

It seems possible to agree with WITHYCOMBE (1924), who has been supported recently by MEINANDER (1972), that among living lacewings Ithonidae have retained the highest number of plesiomorphies. These are particularly the hind wing with wide base, relatively large anal area, and long, S-like free MA base, and with the wing surface more fluted than in other lacewings (as the very name Planipennia should imply), further the eruciform larva with the short mouthparts and the highest number of instars (5). All these features are rather characteristic of Corydalida. Plesiomorphic though not unique is presence of the nygmata on the wings. The oldest Permian family Permithonidae has the same set of imaginal features. Autapomorphic for Ithonidae are the thick, robust body, lost ocelli, and C-like soil larva. And yet there is not a single fossil lacewing known with the comparably primitive wing form and form of the MA base. No Ithonidae are described as fossils with a certainty, although this is not excluded for *Mesomantispa sibirica* Makarkin from the Early Cretaceous of the Transbaikalia. The plump body is found also in the Jurassic Prohemerobiidae which lack other ithonid characters, however. As a result, the most archaic living lacewing group is impossible to relate to either fossil one.

The next most isolated is the Coniopterygidae. Most of their diagnostic characters form a complex related to their small body size. These are reduced venation and particularly cross-veins (especially in the costal space) and terminal forking. There are autapomorphies not evidently related to the small body size as well, viz. the wax producing glands in adults (also found in *Osmylus*), and the enlarged second segment of the maxillary palp. At the same time in respect of its basal width the coniopterygid hind wing is inferior only to Ithonidae and Permithonidae, and their larva is autplesiomorphic in having the protruding labrum, short, straight mandibles, and 6 malpighian tubes. In spite of their small size, Coniopterygidae are found rather regularly in the fossil state, both as impression fossils and as inclusions in fossil resins (MEINANDER 1975), and yet even the oldest, Jurassic fossils are typical Coniopterygidae and not missing links relating the group to any other living or fossil neuropteran.

The last isolated group is the superfamily Myrmeleontoidea covering living families Myrmeleontidae, Nymphidae, Ascalaphidae, Nemopteridae and the extinct Babinskaiidae. These are almost always insects with long, narrow wings and body, wings having narrow base, reduced anal area, and characteristic triangle in the cubital area. Still more advanced is the larva which has a wide, depressed body, large

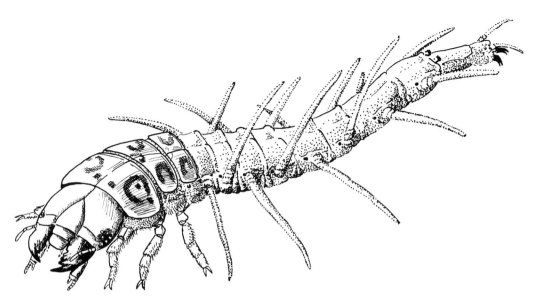

Fig. 244 Larval dobsonfly, *Cretochaulus lacustris* Ponomarenko (Chauliodidae), from the Early Cretaceous of Baissa in Siberia, restored as in life by A.G. Ponomarenko (from ROHDENDORF & RASNITSYN 1980)

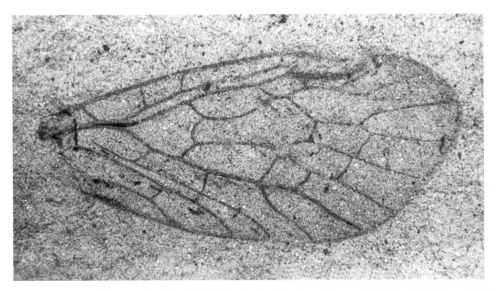

Fig. 245 Fore wing of *Parasialis latipennis* Ponomarenko (Parasialidae) from the Late Permian of Soyana in NW. Russia (holotype, photo by D.E. Shcherbakov); wing 8.7 mm long

head with sickle-shaped mandibles bearing strong teeth mesally, and with gular plate covered by ventral extensions of genae which meet mid-ventrally. Until recently even in the fossil record has not been able to break isolation of the superfamily. However, description of numerous Mesozoic Osmyloidea (particularly by PANFILOV 1968b) and Myrmeleontoidea (MARTINS-NETO & VOLCANO 1989b) makes it possible to derive the latter from the former with reasonable confidence. Re-study of the Jurassic and Early Cretaceous Nymphitidae and adjacent forms could provide dates for connection of the myrmeleontoids with osmyloid neuropterans.

The remaining diverse array of living and fossil families seem impossible to organise into well based superfamilies. The least modified of them are Osmylidae which retain ocelli, trichosors and nygmata, the first plesiomorphy being unique for all the order. As to nygmata, they are known for few Palaeozoic and Mesozoic lacewings, including those typically osmyliod in their wing venation. Possibly many fossils have not been studied with sufficient attention in that respect, or for obscure reason the nygmata are only rarely preserved in fossils. Particularly, the nygmata are found preserved in *Sibithone dichotoma* Ponom. from Jurassic of Transbaikalia, and yet I have failed to find them on any wing from the Jurassic locality Karatau (South Kazakhstan), including the close relatives of *Sibithone*. Trichosors have been seen more often on fossil wings but many authors miss them in their descriptions.

The key synapomorphies of the osmyloid families are probably the narrow hind wing base, and recurrent humeral vein, though many more fossils should be studied for those characters before the synapomorphies will be accepted as well substantiated. Especially important

185

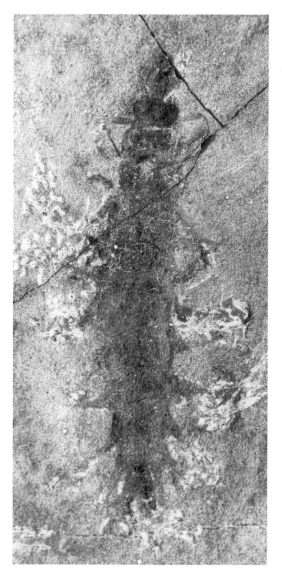

Fig. 246 Dobsonfly-like larva from the Late Permian of Kargala in Urals (PIN 199/213, photo by A.P. Rasnitsyn), described as larval *Permosialis* sp. (SHAROV 1953); body 24 mm long. *Permosialis* is now a member of the extinct order Palaeomanteida, and its larva is supposed to have developed in plants, not in water (Chapter 2.2.1.3.1)

would be descriptions of the rich materials accumulated in the Palaeontological Institute, RAS, in Moscow, which can discover chains of forms connecting extinct and living groups, making it possible to draw more reliable cladogram than is possible for the present. Previously described forms must be re-studied as LAMBKIN (1988) has done. Until then, I propose that it is better to consider the above group of families as a single superfamily, the Hemerobioidea, this being the fourth one in addition to Ithonoidea, Coniopterygoidea, and Myrmeleontoidea.

(f) History. The oldest lacewings are mentioned but not described yet from the earlier Artinskian age (Early Permian) of Obora in Moravia (Czechia; KUKALOVÁ 1964). The oldest described fossil, a hind wing from the later Artinskian deposits of Elmo in Kansas, USA (F. CARPENTER 1976) looks rather strange, with three veins (called MA

and two basal RS branches), which pectinate forward in the medial part of the wing. This form of branching is unknown for the entire order, as well as MA independent of RS in the hind wing. Further Early Permian lacewings are known from the Kungurian age at Tshekarda in the Urals, all representing slender insects. They are described as seven endemic genera of the family Permithonidae (VILESOV 1995) but some of them may be synonyms. Some of them display additionally the double SC bearing strongly oblique cross-veins (Fig. 250).

The Late Permian lacewings are fairly numerous and diverse. They are found in Europe, Asia, Australia, and South America. Their share in collections can reach several percent, as in the Cainozoic oryctocoenoses, thus indicating their abundance at that time as being comparable with that of the present day. 5 families are described thus far from NW European Russia (MARTYNOVA 1962b), Australia (RIEK 1968), South Africa (RIEK 1976b) and Kazakhstan (VILESOV & NOVOKSHONOV 1994), though their more realistic number was probably 1 (Permithonidae) or 2 (NOVOKSHONOV 1996). The wing venation was often polymerised, with SC usually reaching RS, sometimes much basally so. The hind wing was rather wide basally, with the well developed anal area (Fig. 251). The wings bore an irregular spot pattern of a cryptic nature (Fig. 252).

The Triassic lacewings followed the trends displayed by the Late Permian ones. They were roughly as diverse and found on the same continents as before except for South America. All known localities are dated to the Middle and/or Late Triassic. Majority of the Triassic forms had much polymerised venation, often even more so than living Psychopsidae do. There are also fossils which possibly belong to Osmylidae or Polystoechotidae. Several endemic families have been established as well (RIEK 1953, 1955), though their identity and phylogenetic connections do not seem as well based (LAMBKIN 1988). It is possible to state, however, that the Triassic lacewings have reached the diversity level of their wing structure comparable to that of the living groups, except for the long-winged forms.

Features of the Triassic fauna were still valid early in the Jurassic. Unfortunately the Jurassic lacewings are known only in Eurasia. Abundance of the order was rising insufficiently during the Jurassic. Majority of the Early Jurassic forms have been described as Prohemerobiidae with free SC apex (HANDLIRSCH 1906–1908, 1939, A. BODE 1953). Partial revision of the respective fossils has shown these character states as present in only a part of them, predominantly in those buried in the marine deposits of West Europe (WHALLEY 1988b, PONOMARENKO 1995b). The wide-winged forms of the Triassic type were still existing, with the most part of the wing surface taken by densely packed veins. There were also wide-winged fossils with less dense venation leaving small cells separated by cross-veins. They are reminiscent of the Late Jurassic Kalligrammatidae but differ from them in structure of the wing apex and cubital sector. There were long-winged forms appearing that time which mimicked antlions though had quite different venation. Some fossils have been acknowledged as belonging to the living family Osmylidae (LAMBKIN 1988), while others have been attributed to Coniopterygidae (MEINANDER 1975). The characteristic Permian and Triassic wings with SC reaching R well before the wing apex are not found in the Jurassic.

In the Late Jurassic the lacewings became still more numerous, sometimes constituting more than 5% of the total number of the insects, which is much higher than in the Cainozoic. The only locality Karatau in S. Kazakhstan has yielded some 600 fossils. Another important locality is Solnhofen but fossils from that locality must be revised. The Late Jurassic and Early Cretaceous were probably the time when the order

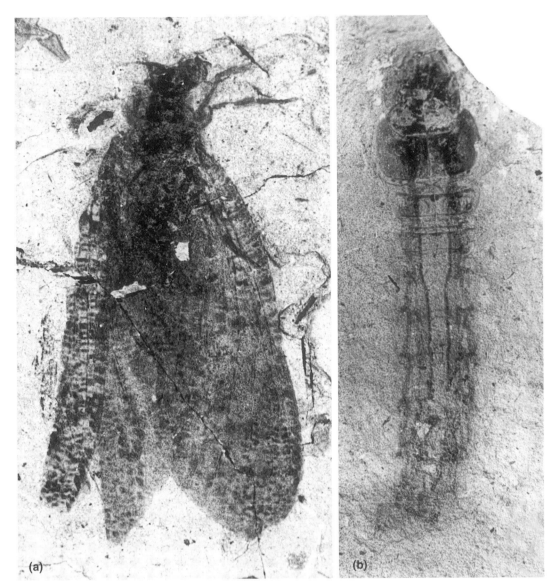

Fig. 247 Dobsonfly (a) adult and (b) larva, *Cretochaulus lacustris* Ponomarenko (Chauliodidae), from the Early Cretaceous of Baissa in Siberia (adult, PIN 4210/7028, length 37 mm, and larval, PIN 1989/2910, length 29 mm, photos by D.E. Shcherbakov)

played the most important role in all its history. As many as 18 families are known from the late Jurassic (HANDLIRSCH 1906–1908, PANFILOV 1980), though their taxonomic relationships remain unclear. Majority of the groups were continuations of the older phyletic lines, and only few of them (Osmylidae, Mantispidae, Nymphidae, Coniopterygidae, Chrysopidae) belonged to living families.

Most common among the Late Jurassic fossils are the broad-winged, abundantly veined forms, usually related to the living Psychopsidae but never showing their characteristic *venae triplicae* and hence representing rather the specific Mesozoic families (Brongniartellidae, Osmylopsychopsidae). Another broad-winged Late Jurassic group was Kalligrammatidae, large insects with darkened wings bearing eye spots (Figs. 253, 254). They were thought to be closely related to the above families (MARTYNOVA 1962b) despite the fact that they have hypertrophied MP instead of RS+MA. One more broad-winged family Panfilovidae combines characters of the two above ones.

The next most abundant are rather large long-winged, rich-veined fossils, predominantly osmylid- or antlion-like (Figs. 255–257). They were described partly in the living families Osmylidae, Polystoechotidae, Nymphidae, or as extinct families Solenoptilidae, Nymphitidae, Osmylitidae, Mesopolystoechotidae, Allopteridae. In the recent revision by LAMBKIN (1988) almost all better preserved fossils of that type are assigned to the mentioned living families, though the majority of Mesozoic forms seem to form a rather natural group descending from Osmylidae and ancestral for Polystoechotidae and Myrmeleontoidea.

One more common Late Jurassic group includes Chrysopidae and their relatives. Among the species described as *Mesypochrysa* Martynov (Fig. 258) there are those certainly belonging to Chrysopidae (*M. intermedia* Panfilov). At the same time, there are fossils forming a succession joining them via those close to *Chrysoleonites* Martynov to the above group of families, thus making it possible to seek the ancestry of Chrysopidae among Hemerobioidea.

Fig. 248 *Calopsychops extinctus* Panfilov (Psychopsidae) from the Late Jurassic of Karatau in Kazakhstan, restored as in life by A.G. Ponomarenko and Kira Ulanova

The other, relatively uncommon Late Jurassic lacewing groups are represented by the aberrant Prohemerobiidae, Mesithonidae (possibly related to Berothidae), Promegalomidae (perhaps close to Hemerobiidae), Coniopterygidae, and the oldest Mantispidae (PANFILOV 1980).

The Early Cretaceous lacewings do not differ much from the Late Jurassic ones, and represent essentially the same types. For instance, the assemblage from the Transbaikalian locality Baissa is by half represented by the typically Mesozoic forms with wings broad and extremely rich venationally. Less abundant are Kalligrammatidae and Hemerobioidea of the Mesozoic type. More common than in the Jurassic are Chrysopidae with the Nothochrysinae-like wings. Rare, aberrant Hemerobiidae and Osmylidae are found which are more similar to the living representatives of their families (MAKARKIN 1990).

Kalligrammatidae and Berothidae are much more abundant in Chinese localities (REN & GUO 1996).

Most characteristic of the epoch (and in particular of its second half) is the diversification of the Myrmeleontoidea exemplified by the Santana assemblage in Brazil (MARTINS-NETO 2000), which is usually referred to as the Late Aptian or Early Albian (the second half of the Early Cretaceous). Found there are antlion-like forms attributed to Araraipeneurinae, a subfamily of Myrmeleontidae that possibly merits the full family state; representatives of the distinct family Babinskaiidae; uncommon fossils attributed to both Nymphidae and Ascalaphidae but apparently better fit to the former family; and, most unexpectedly, numerous and diverse Nemopteridae with the hind wing reduced (otherwise the family is known solely from the Old World,

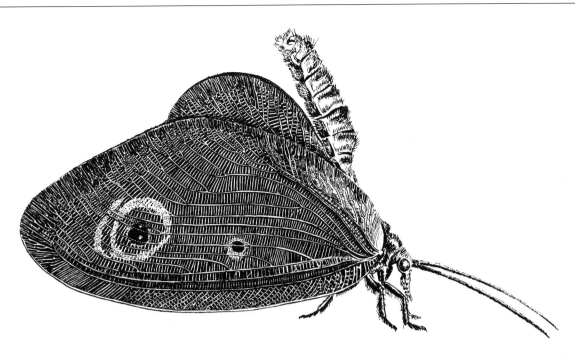

Fig. 249 Female giant butterfly-like lace-wing, *Kalligramma* sp. (Kalligrammatidae), from the Late Jurassic of Karatau in Kazakhstan, restored displaying (male attracting) by A.G. Ponomarenko (from ROHDENDORF & RASNITSYN 1980)

except for a unique find in the Oligocene of Florissant in Colorado, N. America). One more abundant group is Chrysopidae. Other families are rare (MARTINS-NETO & VULCANO 1988, 1989a–c, 1990b, MARTINS-NETO 1990d, 2000). Unlike that, myrmeleontoids take a minority in the somewhat older assemblages from China, Transbaikalia and Mongolia, including the above mentioned Baissa. Nymphidae have existed there that time, including *Mesonymphes* and some other distinct forms. Also found are Araraipeneurinae and Babinskaiidae. Some Asian myrmeleontoids had a fore or hind Banksian line similar to certain living ones (PONOMARENKO 1992b).

Since the mid or later Early Cretaceous (later Neocomian or Aptian age), lacewings are also known from inclusions in fossil resins. Particularly, Coniopterygidae, Berothidae, and myrmeleontoid wing fragments are described from Lebanese amber (WHALLEY 1980).

The Late Cretaceous lacewings are only superficially known. Two broad-winged fossils from West Kazakhstan and Middle Siberia have been described in the living family Psychopsidae (ZALESSKY 1953, MARTYNOVA 1954; Fig. 259), though the attribution does not seem beyond doubt. Several more species have been recently described from various Asian localities (MAKARKIN 1990). *Metahemerobius* Makarkin from the Maastrichtian or Danian of Siberia, *Paleoleon* Rice from the Late Albian or Early Cenomanian of Labrador, North America (RICE 1969) and *Samsonileon* Ponomarenko from Cenomanian of Israel (DOBRUSKINA *et al.* 1997), belongs to primitive myrmeleontoid subfamily (possible family) Paleoleontinae. Further described from the Cenomanian of Northeast Asia are the oldest representative of Myodactylinae (Nymphidae), a member of Mantispidae assigned to the living South American genus *Gerstaeckerella* Makarkin, as well as representatives of the Mesozoic groups (*Paleogetes* Makarkin, Prohemerobiidae). Berothidae and Coniopterygidae are described from fossil resins, the former family in those of Cenomanian age in France (SCHLÜTER & STÜRMER 1984), the latter in the Santonian

resins of Taimyr in North Siberia (MEINANDER 1975). The latter locality has also given undescribed members of Berothidae and Sysiridae. Most Late Cretaceous lacewings belonged to living families.

For the Cainozoic, all the described fossils belong to living families and in the most cases to living genera as well, except the Mesozoic genus *Archegetes* Handlirsch, previously mentioned existing in the Palaeogene of the Russian Far East (MAKARKIN 1991). However, the fossils attributed to Psychopsidae in the same publication may in fact equally belong to the superficially similar broad-winged Mesozoic Prohemerobiidae. Most abundant in the Cainozoic deposits are remains of Hemerobiidae and Chrysopidae mostly belonging to Nothochrysinae. The latter group is rare now, and yet it has embraced majority of the Palaeogene and Neogene (including Pliocene) Chrysopidae from Europe, North America, and Far East (ADAMS 1967, SCHLÜTER 1982). As to the now most common nominotypic subfamily, its fossils are dominated in the Early Miocene of Caucasus (MAKARKIN 1991) and were found in some South France localities. In general, the Cainozoic lacewing zoogeography had little in common with that of the present day, for Nymphidae, now endemic of Australia, have been found in Europe, and the Old World endemic Nemopteridae and Psychopsidae in North America. One more unexpected feature of the Cainozoic neuropterans is the scarcity of Myrmeleontidae and Ascalaphidae, characteristic both for the existing lacewing fauna and for Cretaceous localities.

2.2.1.3.3.4. ORDER JURINIDA M. Zalessky, 1928 (=Glosselytrodea Martynov, 1938)

A.P. RASNITSYN

(a) Introductory remarks. Small, mostly Palaeozoic group with unusual wing form and venation (Fig. 260), with only 22 described species, first established as leafhoppers (TILLYARD 1922b), then commonly treated as orthopteroids due to their "precostal area", and later

Fig. 250 One of the oldest lacewings, *Sylvasenex lacrimabundus* Vilesov (Permithonidae), with double SC, from the Early Permian of Tshekarda in Urals (holotype, photo by A.P. Rasnitsyn); fossil 8 mm long

Fig. 251 *Permithonopsis enormis* O. Martynova (Permithonidae) from the Late Permian of Soyana in NW Russia (PIN 117/2528, photo by D.E. Shcherbakov); hind wing 15 mm long

attributed to holometabolans by SHAROV (1966b). The present account is based on both personal observations and data from MARTYNOV (1938d), F. CARPENTER (1943b), MARTYNOVA (1952, 1961a, 1962c), RIEK (1953), SHAROV (1966b), ROHDENDORF & RASNITSYN (1980), PONOMARENKO (1988b, 2000), VILESOV & NOVOKSHONOV (1994), NOVOKSHONOV (1998a).

(b) Definition. Size medium to rather small (wing length 5–20 mm). Head small, hypognathous (fossilised usually even as opistognathous). Antenna short, moniliform. Prothorax movable, pronotum not particularly small. Meso- and metathoraces of subequal size and structure,

with long, slanting pleura. Legs with coxae long, tarsi 5-segmented. Wing rest position roof-like. Fore wing membranous (strengthened by vein thickening in some Polycytellidae). SC short, rarely reaching wing midlength, meeting R in hind wing (usually very short there), Fore wing C usually bearing individualised branches in secondary "precostal area". RS originating and forming first fork near wing base (distal RS base described in *Glossopterum martynovae* Sharov is a false one formed by a cross-vein, an individual aberration absent in all known specimens but the holotype), pectinate backwards, fore branches being rather irregular (as weak as cross-veins and running in

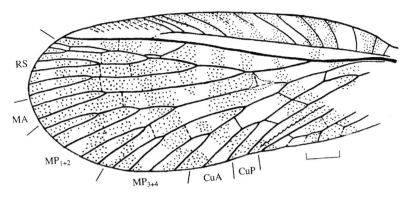

Fig. 252 Fore wing of *Permosisyra paurovenosa* O. Martynova (Permithonidae) from the Late Permian of Soyana in NW Russia, showing cryptic colour pattern (from NOVOKSHONOV 1997a); wing length 11 mm

Fig. 253 *Meioneurites* sp. (Kalligrammatidae) from the Late Jurassic of Karatau in Kazakhstan (PIN 2784/1069, photo by A.P. Rasnitsyn); wing 73 mm long

zigzag manner between them), and last one being straight and running close and parallel to fore M branch (the parallel veins are fused in Polycytellidae and, possibly, running secondarily apart in *Uskatelytrum* Martynova, unless absence of the double vein is due to imperfect preservation). M pectinate forward, in fore wing fused with CuA sub-basally, with free base lost except in Permoberothidae; hind wing M seemingly originating from R, retaining short, oblique M_5. CuA and CuP simple or CuA with short apical branch(es). Fore wing anal veins (all or at least their hind pair) merging with each other apically; hind wing anals free (in some fossils anal area tucking down

before 1A – probably occasionally so, not as normal rest position). Most typical fore wing (in Jurinidae) of very characteristic external appearance: main part long triangular, with symmetric apex, divided submedially, longitudinally, by pair of straight, approaching, parallel veins (posterior RS and anterior M) hardly if at all bearing cross-veins in-between (otherwise cross-veins numerous, transversal), and with similar double veins (two SC branches and, respectively, CuP & A_1) separated it from similarly looking "precostal" and anal areas; main wing part further being almost surrounded (completely so in Polycytellidae) by convex RS and CuA veins, with both "precostal" and anal areas reproducing similar submarginal convex veins. Abdomen narrow, short compared to wings, with terga and sterna similarly sclerotised, with no exserted ovipositor preserved, with cercus four- (in Permoberothidae) to one-segmented. Male genitalia rather small, symmetrical, with ovoid gonocoxa and small articulated gonostylus. Internal structures and preimaginal stages unknown.

(c) Synapomorphies. Antenna short; SC short; RS originating far basally; posterior RS and anterior M branches forming straight double vein; M and CuA fused for a distance sub-basally; two hind A veins amalgamated apically; ovipositor short if developed.

(d) Range. Early Permian (Artinskian) till Middle Jurassic, worldwide, with Jurassic fossils found only in Asia.

(e) System and phylogeny. The number of jurinidan families is reduced here from 7 to 3, with Permoberothidae, Jurinidae, and Polycytellidae considered as valid and the others (Archoglossopteridae, Glosselytridae, Glossopteridae and Uskatelytridae) lumped under Jurinidae (Figs. 261–263). Jurinidae display the next step after Permoberothidae in the trend described in more detail above (submarginal convex vein in "precostal area" secondarily lost due to the general venational impoverishment in *Archoglossopterum* Martynova; the vein is present, contrary to the original description, albeit narrow in *Glosselytron* Martynov). Polycytellidae (Fig. 264) are further modified in two subaxial parallel veins being fused into one and RS and CuA amalgamated apically in an entire submarginal vein. An additional unusual feature of many Polycytellidae, and particularly those described as *Glossopteron* spp. by PONOMARENKO (2000) from the Late Permian of Mongolia, but see also Fig. 264, is the fore wing elytrisation which progressed in a beetle-like manner, that is by thickening of the veins and not the membrane, so as the cells became small and rounded, either in the wing base or over most of the wing.

(f) History. The oldest jurinidan fossil from the earlier Artinskian (mid-Early Permian) sediments of Obora in Czechia is undescribed yet. The oldest described one (*Permoberotha villosa* Tillyard,

Fig. 254 *Kalligramma turutanovae* O. Martynova from the Late Jurassic of Karatau in Kazakhstan (PIN 2784/1070, photo by A.P. Rasnitsyn); fore wing 100 mm long

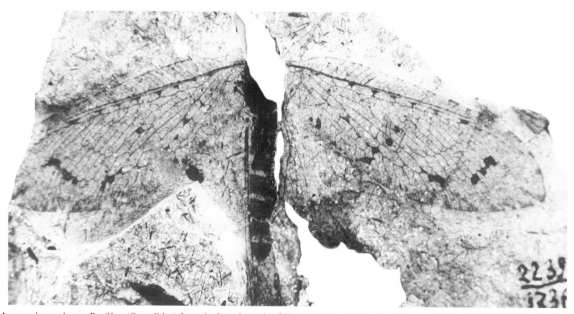

Fig. 255 *Mesosmylus atalantus* Panfilov (Osmylidae) from the Late Jurassic of Karatau in Kazakhstan (holotype, photo by D.E. Shcherbakov); wing 9 mm long

Permoberothidae) comes from Elmo in Kansas, USA. Four specimens belonging to 3 species (one undescribed) of Jurinidae are found in the latest Early Permian of Tshekarda in the Urals (Figs. 260, 261). The Late Permian was the only time when the group was comparatively abundant and diverse, comprising up to a few per cent in some fossil assemblages, which is as much as the snake-flies and lacewings at the time of their highest abundance. The numerous Late Permian fossils are found in Europe, Asia, and Australia, all belonging to Jurinidae (Figs. 262, 263) except for Polycytellidae in the latest Late Permian of Karaungir in the East Kazakhstan (Fig. 264) and in the Kazanian or Tatarian of Bor-Tologoy in Mongolia. The next younger are Jurinidae and Polycytellidae in the Middle or Late Triassic Madygen Formation in Fergana, Central

Asia, Polycytellidae in the Early Jurassic there (Sogyuty at Issyk-Kul Lake and Shurab in South Fergana), and Jurinidae in the Middle Jurassic in Mongolia (Bakhar). Jurinidae (as Uskatelytridae) are also mentioned from the Middle or earlier Late Triassic of the Los Rastros Formation in Argentina (MARTINS-NETO & GALLEGO 2000).

2.2.1.3.4. SUPERORDER PAPILIONIDEA Laicharting, 1781 (=Mecopteroidea Martynov, 1938)

A.P. RASNITSYN

(a) Introductory remarks. The largest superorder which includes two of the four most diverse insect orders, Lepidoptera (moths and

lacking, adult mandible with chewing function subordinate, resulting in solid food particles uncommon in gut. Wings, when present, membranous, homonomous or heteronomous, with venation varying from extremely rich (secondarily) to almost lost, without free base of MA (M branch intercepted by RS) except in hind wing of lowermost scorpion- and caddisflies with MA base short, displaced far basad, usually with RS (+ MA) dichotomously branching (often with two facing combs when more rich), hind wing sometimes with expanded anal region, in some groups one or both wing pairs reduced and lacking aerodynamic function, or lost. Flight plesiomorphically functionally four-winged, in-phase, anteromotoric (*sensu* GRODNITSKY 1999). Advanced flight also anteromotoric but either functionally two-winged (in majority of caddisflies, moths and butterflies, possibly also in some scorpionflies, viz. Aneuretopsychidae with enlarged mesothorax and widened hind wing, and some Permotanyderidae and Liassophilidae with diminished hind wings), or, in dipterans, morphologically two-winged. Primary ovipositor lost as device for positing eggs within substrate (often replaced by telescopically arranged apical abdominal segments). Larva variable, often eruciform, with legs 5-segmented, single-clawed (double-clawed in *Nannochorista*), often with abdominal prolegs of variable shape; some groups with larva simplified, legless and even headless. Pupa variable. Adult and immature habits extremely variable.

(c) **Synapomorphies.** Fore wing lacking free base of MA; hind wing with MA base short, displaced far basad; hind wing CuP fused with 1A sub-basally for a distance. Further synapomorphies cited by BOUDREAUX (1979), HENNIG (1981), KRISTENSEN (1981), WEAVER (1984), are characteristic of more advanced papilionideans (those with fore branch of CuA lost or simulating M branch) but did not necessarily occur in the first Papilionidea.

(d) **Range.** Early Permian till now, worldwide.

(e) **System and phylogeny.** General features of the system and phylogeny of the superorder seem comparatively well established (see Fig. 203) except for the taxonomic state of the paraphyletic Kaltanidae (as a part of Panorpida or as a plesion excluded from the system), relations of caddisflies and moths (whether they are sister groups, or the ancestor and descendant, respectively), and the taxonomic position of strepsipterans (*q.v.* Chapter 2.2.1.3f) and the fleas.

The first problem is purely ideological, and according with our non-cladistic ideology of taxonomy (Chapter 1.1c), the Kaltanidae are retained here in Panorpida. The second problem arose due to the relative youth of moths which only entered the fossil record as late as in the Jurassic and not in the Permian as caddisflies did. This indicates an ancestor-descendant relationship, while the plesiomorphically terrestrial habits of moths, as opposed to the amphibiotic one of the caddisflies, suggests a sister-group relation. The contradiction is apparently possible to resolve taking into consideration all available information from the fossil record and the structure, habits and developmental pattern of living moths and caddisflies. See Chapters 2.2.1.3.4.2f and 2.2.1.3.4.3f for details.

The typical (extant) fleas (order Pulicida) give controversial indications of their ancestry (cf. BOUDREAUX 1979, HENNIG 1981, and KRISTENSEN 1981, 1999a), though it is their monophyly with Panorpida (or even with the family Boreidae) that is usually argued for nowadays (cf. also molecularly based cladograms in WHEELER 1998, SIDDALL & WHITING 1999). The Mesozoic fossils, albeit of rather extravagant appearance and not always easily appreciated as being flea relatives (PONOMARENKO 1976a, 1988c, RASNITSYN 1992b), add further controversy to the debate. They highlight some known flea

Fig. 256 *Kazakhstania fasciata* Panfilov (Osmylidae) from the Late Jurassic of Karatau in Kazakhstan (holotype, photo by D.E. Shcherbakov); wing 33 mm long

butterflies) and Diptera (flies and allies). Additionally, the insects embraced are the most diverse in their structure and habits, ranging from the butterfly and house fly to the flea. As a result, the superorder is the most debatable and contradictory in respect of its phylogeny and taxonomic structure.

This chapter is modified after ROHDENDORF & RASNITSYN (1980) to incorporate some more sound results by BOUDREAUX (1979), HENNIG (1981), KRISTENSEN (1981), as well as other publications cited when appropriate.

(b) **Definition.** Large to very small insects, winged or wingless, highly diverse morphologically and bionomically both in adult and larva. Adult with mouthparts chewing, piercing or licking, rarely

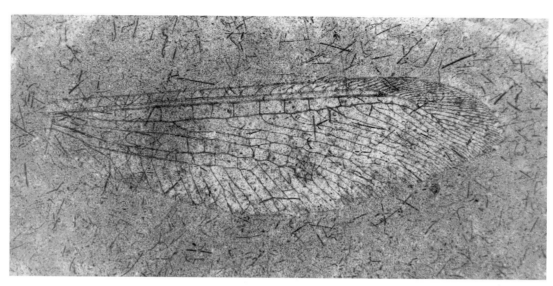

Fig. 257 *Mesonymphes rohdendorfi* Panfilov (Nymphidae) from the Late Jurassic of Karatau in Kazakhstan (holotype, photo by D.E. Shcherbakov); wing length 24 mm

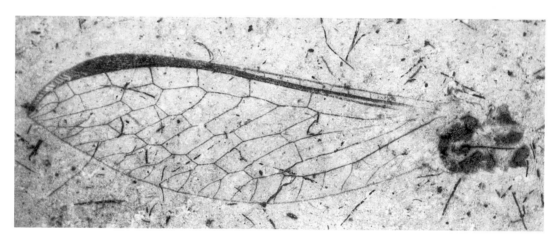

Fig. 258 *Mesypochrysa reducta* Panfilov (Chrysopidae) from the Late Jurassic of Karatau in Kazakhstan (holotype, photo by D.E. Shcherbakov); wing 10 mm long

plesiomorphies (e.g. absence of any indication of the meso- and metathoracic heteronomy so much expectable in case of the dipteran ancestry), and display new ones seemingly absent in any other Papilionidea: male genitalia with volsella and large aedeagus (RASNITSYN 1992b). The free volsella is particularly unexpected being otherwise characteristic of hymenopterans and some archaic beetles (for details see Chapter 2.2.1.3.4.5). This infers a very early (Permian) flea ancestry, as opposed to the Late Jurassic age of even their disputable fossils. This contradiction is not impossible to explain, because a poor fossil record is very characteristic of non-flying ectoparasites of tetrapod vertebrates (cf. chewing and sucking lice, Chapter 2.2.1.2.4.2). So a long cryptic existence of the flea stem lineage parasitising some Permian and Triassic reptiles cannot be ruled out. However, an alternative hypothesis can be proposed, that the seemingly plesiomorphic flea characters might be reversals, paedomorphically stabilised ancestral structures having been extensively recapitulated in the insect's morphogenesis (TICHOMIROVA 1991). That is why it seems possible to root the flea clade near to, or within, the mecopteran clade.

2.2.1.3.4.1. ORDER PANORPIDA Latreille, 1802. THE SCORPIONFLIES (=Mecaptera Packard, 1886, =Mecoptera Comstock et Comstock, 1895, +Neomecoptera Hinton, 1958, +Paratrichoptera Tillyard, 1919, +Paramecoptera Tillyard, 1919)

V.G. NOVOKSHONOV

(a) Introductory remarks. The scorpionflies (Fig. 265) now form a small (480 described species: ARNETT 1985) relic group which has represented one of major elements of the insect world since the Late Permian till Jurassic. The present concept of the order is essentially original, with main source information reviewed in ROHDENDORF (1962), KALTENBACH (1978), WILLMANN (1978, 1989), ROHDENDORF & RASNITSYN (1980) and NOVOKSHONOV (1997a). The fossil history is described based on NOVOKSHONOV (1997a) (for the Permian); ZALESSKY (1933), RIEK (1950, 1955, 1956, 1976d), PONOMARENKO & RASNITSYN (1974), WILLMANN (1978, 1989), UEDA (1991), PAPIER *et al.* (1996a), NOVOKSHONOV (1997a,b,c, 2001b and unpublished), NOVOKSHONOV & SUKATSHEVA (2001) (for Triassic); MARTYNOVA

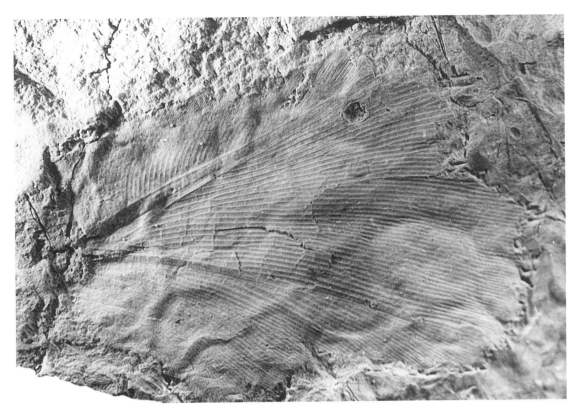

Fig. 259 *Kagapsychops continentalis* Makarkin (Psychopsidae) from the Late Cretaceous of Kzyl-Zhar in Kazakhstan (holotype, photo by D.E. Shcherbakov); wing 32 mm long

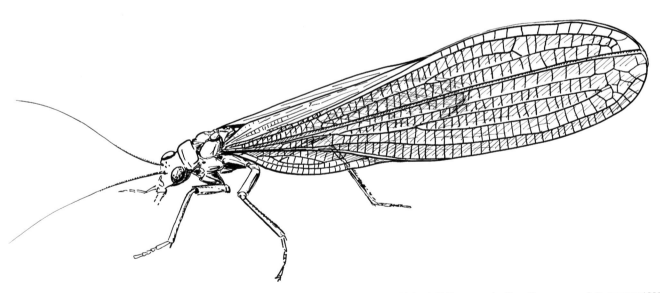

Fig. 260 *Glossopterum* sp. (Jurinidae) from the Early Permian of Tshekarda in Urals restored as in life by A.G. Ponomarenko (from ROHDENDORF & RASNITSYN 1980)

(1948, 1962d), WILLMANN (1978, 1989), LIN (1980, 1985a, 1986), HONG (1984), SUKATSHEVA (1985a, 1990a), JELL & DUNCAN (1986), RASNITSYN & KOZLOV (1990), F. CARPENTER (1992a), SUKATSHEVA & RASNITSYN (1992), NOVOKSCHONOV (1993b, 1997a,c,d), REN (1993, 1997b), ZHANG (1993), REN *et al.* (1995), ANSORGE (1996), SUKATSHEVA & NOVOKSHONOV (1998), WILLMANN & NOVOKSCHONOV 1998a,b) and Novokshonov & Ross (in preparation) (for Jurassic and Cretaceous); WILLMANN (1978), NOVOKSHONOV (1993b) and PETRULEVICIUS 1999a (for Cainozoic).

(b) Definition. Size small to moderately large (2–35 mm), usually medium. Head capsule often with elongate rostrum (see Figs. 270, 274), in living groups closed between occipital foramen and oral cavity, with clypeus and labrum fused borderless, and with anterior mandibular articulations weak. Eyes, ocelli, and antennae rarely

Fig. 261 *Glosselytron martynovae* Sharov (Jurinidae) from the Early Permian of Tshekarda in Urals (holotype, photo by D.E. Shcherbakov); fore wing 10 mm long

Fig. 262 Fore wing of *Eoglosselytrum perfectum* O. Martynova (Jurinidae) from the Late Permian of Kaltan in Siberia (holotype, photo by D.E. Shcherbakov); fore wing 6.2 mm long

modified (ocelli sometimes absent). Mouthparts chewing, except in Aneuretopsychidae and Pseudopolycentropodidae with long, thin, gently curved sucking proboscis of unclear structure (see Figs. 275, 276). Mandible weak, elongate, labium lacking glossa and paraglossa. Thorax of primitive structure, with pronotum usually small. Legs cursory (raptorial in Bittacidae, mid and hind pairs saltatory in Boreidae), slender (femora short, stout in Aneuretopsychidae), sometimes markedly spinose. Wings usually nearly homonomous (hind pair reduced in size in few Permotanyderidae, Liassophilidae and some other forms), shortened in some Panorpodidae, reduced in Boreidae and *Anomalobittacus* (Bittacidae), lost in Apteropanorpidae and

Apterobittacus (Bittacidae). Wing position at rest roof-like (less evident when narrow) or, in broad-winged Eomeropidae, horizontal and widely overlapping. Unless secondarily polymerised (in Englathaumatidae, Meropeidae, Eomeropidae, and especially in Thaumatomeropidae), wing venation lacking pronounced convex veins between R and (M_5+)CuA, veins rarely bending sharply and forming specialised cells (see Figs. 268–274). Pterostigma sometimes present as diffuse membrane sclerotisation near SC apex. SC simple or bearing one to many fore veinlets, usually short, slanting. RS (+MA) and M (=MP) branching dichotomously, rarely with fewer than 4 branches. CuA plesiomorphically forked apically, with fore branch usually lost, in hind

Fig. 263   Undescribed Jurinidae from the Late Permian of Soyana in NW Russia (PIN 3353/881, photo by D.E. Shcherbakov); length of fossil 10 mm

Fig. 264   Fore wings of undescribed Polycytellidae from the Late Triassic of Kenderlyk in Kazakhstan (PIN 2497/286); wing span 12 mm

wing of less advanced groups concave, with free convex M_5 running closely before it (often in form of convex fold of membrane). CuP simple, free in fore wing, anastomosing with 1A sub-basally in hind one. 2–3 anal veins (more in some cases of polymerised venation) reaching wing margin separately. Hind wing with anal area enlarged and tucking down at rest only in Aneuretopsychidae. Abdomen plesiomorphically not much specialised (except reduced, internalised ovipositor and often with ring-like fused male tergum and sternum 9, with 3-segmented cercus), often variously modified (apical segments elongate giving scorpion-like appearance, especially in male, with outgrowths using to fix female wing at copulation, etc., see Fig. 274). Heterochromosome sex male. Ovary polytrophic (panoistic in Boreidae). Sperm tail with $9+2$ axial filaments. Larva eruciform, C-like in Boreidae and Panorpodidae, narrow campodeiform in Nannochoristidae. Head with compound eyes (degenerating and then re-appearing at metamorphosis), with number of stemmata reduced up to 7 in Bittacidae, 3 in Boreidae. Pronotum with sclerotised plate. Legs 4-segmented with claw-like pretarsus (3-segmented in Boreidae, 5-segmented with 2 claws in Nannochoristidae). Abdomen usually with weakly differentiated prolegs, apically with pygopode (replaced by two hooks in Nannochoristidae). Pupa free, decticous. Feeding usually saprophagous or mixed, adult Bittacidae predaceous, larval Boreidae and adult *Brachypanorpa* (Panorpodidae)

phytophagous. Most alate adults common on vegetation, Eomeropidae (*Notiothauma*) cryptic, nocturnal. Larva living in leaf litter, in upper soil, in moss (Boreidae, Apteropanorpidae), at bottom of swamps and creeks (Nannochoristidae). Pupation terrestrial, in cell with walls usually strengthened by saliva (not in silken cocoon, except Boreidae).

(c) **Synapomorphies**. None because of paraphyletic nature of the group with respect to other Papilionidea.

(d) **Range**. From Early Permean until present, worldwide.

(e) **System and phylogeny** as shown at Fig. 266, based on MARTYNOVA (1948, 1959, 1962d), RIEK (1953), PONOMARENKO & RASNITSYN (1974), PENNY (1975), HENNIG (1981), WILLMANN (1981, 1983, 1987, 1989), SUKATSHEVA (1985a), NOVOKSHONOV (1994a,b, 1997a, 1998d and unpublished data) and NOVOKSHONOV & SUKATSHEVA (2001). The following taxonomic structure of the order is proposed here (extinct families asterisked):

Superfamily BOREIDOIDEA Latreille, 1816
(1) *Kaltanidae O. Martynova, 1958 (=Cycloristidae O. Martynova, 1958; =Cyclopteridae O. Martynova, 1958; =Tomiochoristidae O. Martynova, 1958)
(2) *Permopanorpidae Tillyard, 1926 (=Lithopanorpidae Carpenter, 1930; =Martynopanorpidae Willmann, 1989)

197

Fig. 265 Female *Agetopanorpa permiana* O. Martynova (Permopanorpidae) from the Early Permian of Tshekarda in Urals (from Novokshonov 1997a); wing length 7 mm

(3) *Permocentropidae Martynov, 1933

(4) *Belmontiidae Tillyard, 1919 (=Parabelmontiidae Tillyard, 1922)

(5) *Liassophilidae Tillyard, 1933 (=Pseudodipteridae Martynova in Kolosnitsyna et Martynova, 1961; =Laurentipteridae Martynova et Willmann in Willmann, 1978)

(6) *Permotanyderidae Riek, 1953

(7) *Permotipulidae Tillyard, 1929 (=Permilidae Willmann, 1989)

(8) *Robinjohniidae O. Martynova, 1948

(9) Nannochoristidae Tillyard, 1917

(10) *Thaumatomeropidae Willmann, 1978

(11) Boreidae Latreille, 1816

(12) *Permochoristidae Tillyard, 1917 (=Mesopanorpodidae Tillyard, 1918; =Mesochoristidae Tillyard, 1926; =Idelopanorpidae M. Zalessky, 1929; =Agetopanorpidae Carpenter, 1930; =Caenoptilonidae G. Zalessky, 1933; =Protopanorpidae Handlirsch, 1937; =Petromantidae Handlirsch, 1937; =Protochoristidae Handlirsch, 1937; =Eosetidae Tindale, 1945; =Xenochoristidae Riek, 1953; =Tychtopsychidae Martynova, 1958; =Tychtodelopteridae Ponomarenko, 1958; =Petrochoristidae Willmann, 1989; =Choristopsychidae Martynov, 1937)

(13) *Sibiriothaumatidae Sukatsheva et Novokshonov, 1998

Superfamily PSEUDOPOLYCENTROPODOIDEA Handlirsch, 1906

(14) *Mesopsychidae Tillyard, 1917

(15) *Aneuretopsychidae Rasnitsyn et Kozlov, 1990

(16) *Pseudopolycentropodidae Handlirsch, 1906

(17) Meropeidae Tillyard, 1919

Superfamily PANORPOIDEA Latreille, 1805

(18) *Parachoristidae Handlirsch, 1937 (=Neoparachoristidae Willmann, 1978; =Triassochoristidae Willmann, 1989; =Choristopanorpidae Willmann, 1989)

(19) Bittacidae Handlirsch, 1906 (=Neorthophlebiidae Handlirsch, 1925; =Cimbrophlebiidae Willmann, 1977)

(20) *Orthophlebiidae Handlirsch, 1906 (? =Holcorpidae Zherichin, 1969; ? =Austropanorpidae Willmann, 1977)

(21) *Muchoriidae Willmann, 1989

(22) *Dinopanorpidae Carpenter, 1972

(23) *Englathaumatidae Novokshonov et Ross (in preparation)

(24) Eomeropidae Cockerell, 1909 (=Notiothaumatidae Esben-Petersen, 1921)

(25) Apteropanorpidae Byers, 1965

(26) Choristidae Esben-Petersen, 1915

(27) Panorpidae Latreille, 1805

(28) Panorpodidae Byers, 1965

(f) History. Scorpionflies were common and diverse since their first appearance in the Early Permian until the Cretaceous when they shown their participation much decreasing to become rare as fossil since the Palaeogene. During the Palaeogene the fauna was gradually approaching the contemporary family composition.

Only two or three families are known in the Early Permian: Permopanorpidae and Permochoristidae, each with two genera, at Elmo in the Artinskian deposits in Kansas, USA. The Kungurian assemblage of Tshekarda in the Urals includes 9 genera of

Permopanorpidae (Figs. 265, 267). *Moravochorista carolina* Kukalová-Peck et Willmann, 1989 from the early Artinskian of Obora in Czechia may belong to Kaltanidae, though its attribution even to Panorpida is not certain (NOVOKSCHONOV 1998c).

The Late Permian scorpionflies, in contrast to the more ancient ones, are known from all continents (except Antarctica and North America), being most common and diverse in East Europe, Asia and Australia. In the South Africa and South America (Brazil), they are found only in two localities and represented, respectively, by three and two widespread genera of Permochoristidae. In the East Europe, Kazakhstan, Siberia and Mongolia Permochoristidae (Fig. 268) and Kaltanidae (Fig. 269) are much prevailing, with more than 20 and 8 genera, respectively. However, their participation is not alike: Permochoristidae are found in almost every locality, while no Kaltanidae is found in the East Europe (the later appearance of Kaltanidae in the palaeontological chronicle can be explained supposing that they were spread over territories poor in localities of Early Permian insects). In contrast, Permopanorpidae and Permocentropidae are found only in the East Europe (Soyana, one genus per family), Permotipulidae (one species) in Kuznetsk Basin in South Siberia, and Robinjohniidae (also 1 species) in Krasnoyarsk Province, East Siberia. The Late Permian assemblage of Australia (Belmont) consists mostly of Permochoristidae (8 genera), other families being rare (one species is known for each of Belmontiidae, Robinjohniidae, Permotipulidae, while Permotanyderidae and Parachoristidae are represented each by two species of one and two genera, respectively). The whole Permian fauna includes 9 families, two of which (Permochoristidae and Parachoristidae) survived in the Triassic.

The only Early Triassic scorpionfly from Babiy Kamen' in Kuznetsk Basin (S. Siberia) is fragmentary, it may belong to Permochoristidae. The next oldest in the Triassic are rare Permochoristidae, Pseudopolycentropodidae and the first Liassophilidae from the *Voltzia* sandstone (Vosges, France). The Middle and Late Triassic findings are widespread in Europe, Central Asia, Kazakhstan, Australia and South Africa. The most rich assemblages come from Kirghizstan (Madygen Formation), and Australia (Mount Crosby Formation in Queensland and Wianamatta Shales, New South Wales). 5 families, 15 genera and 40 species are known from Kirghizstan, with Permochoristidae (6 genera, 15 species) and Parachoristidae (5, 13; Fig. 270) dominating; other families found are Thaumatomeropidae (3, 6), Mesopsychidae (1, 5; Fig. 271), and Pseudopolycentropodidae (1, 1). Australian scorpionflies were less diverse: those found include Permochoristidae (3, 4?), Parachoristidae (3, 5?), Mesopsychidae (1, 1), and the unique species of the oldest living family Bittacidae (*Archebittacus exilis* Riek). A low diversity of Parachoristidae, Mesopsychidae and Liassophilidae are found in the Late Triassic of Garazhovka in Ukraine, Permochoristidae and Mesopsychidae – in Molteno (South African Republic), Parachoristidae–in Okuhata (Japan); more diverse are Parachoristidae in the Tologoy Formation of Kenderlyk (East Kazakhstan).

In toto, the Triassic scorpionflies belonged to seven families, five of which (Liassophilidae, Permochoristidae, Mesopsychidae, Pseudopolycentropodidae and Bittacidae) entered the Jurassic. It cannot be excluded, however, that the oldest Orthophlebiidae may join these five, for this family, very common in the Jurassic, is not easily identifiable from isolated wings. This particularly concerns one species from Madygen Formation (NOVOKSHONOV 2001b).

Jurassic scorpionflies are found only in Eurasia (West Europe, Kazakhstan, Central Asia, Siberia, Mongolia and China). Since the Early Jurassic, the Permochoristidae lost their former diversity and abundance (except Sogyuty in Kirghizstan with the sole dominant permochoristid species *Liassochorista asiatica* O. Mart.) being replaced by numerous Bittacidae and particularly Orthophlebiidae. Mesopsychidae increased their diversity in Central Asia, the first Nannochoristidae appeared in Siberia, rare Liassophilidae and Pseudopolycentropodidae encountered in West Europe. Muchoriidae entered the fossil record near the Early/Middle Jurassic boundary in Siberia. In the Middle Jurassic Orthophlebiidae and Bittacidae further raised their diversity, Nannochoristidae became very abundant in Siberia (Fig. 273) and China, the oldest Meropeidae are found in Siberia, rare Permochoristidae still persisted in Mongolia and Liassophilidae in Siberia (Fig. 272). Muchoriidae are not found anymore. The Late Jurassic assemblages were still dominated by Orthophlebiidae (Fig. 274), with Bittacidae diverse (about 10 genera) and Nannochoristidae participating regularly. Mesopsychidae are found in Mongolia, the last Pseudopolycentropodidae (Fig. 275) and Permochoristidae – in Kazakhstan and China, respectively. The first Aneuretopsychidae are recorded in Kazakhstan (Fig. 276), Sibiriothaumatidae in Siberia, and one more living family Boreidae – in Mongolia (Fig. 277). In total, 12 families are recorded from the Jurassic, eight of which (Nannochoristidae, Boreidae, Sibiriothaumatidae?, Mesopsychidae, Aneuretopsychidae, Meropeidae, Bittacidae, Orthophlebiidae) survived into the Cretaceous. The oldest living genera are the Jurassic *Nannochorista* Tillyard (Nannochoristidae) and *Orobittacus* Villegas et Byers (Bittacidae).

In the Early Cretaceous, Orthophlebiidae and probably Nannochoristidae decreased in diversity (found in South Siberia, Mongolia, Australia and, only Orthophlebiidae, in China and England). Unlike them, Bittacidae remained diverse, at least in South Siberia and China. Occurring were Boreidae (South Siberia) and the last Mesopsychidae and Aneuretopsychidae (all from S. Siberia) and Englathaumatidae known from nowhere but the English Wealden. Early Cretaceous families are 7 (8?), with Nannochoristidae, Boreidae, Bittacidae and possibly Orthophlebiidae (see below) surviving in the Cainozoic. The list should be completed by Meropeidae known since Jurassic till now, but not yet in the Cretaceous. The Late Cretaceous scorpionflies are unknown.

The Palaeogene scorpionflies are found in many localities in Europe (Bittacidae, Panorpidae, Panorpodidae), Russian Far East (Eomeropidae, Bittacidae, Dinopanorpidae), South America (Bittacidae), North America (Eomeropidae, Bittacidae, Panorpidae, possibly Orthophlebiidae: *Holcorpa* Scudder) and Australia (Choristidae and possibly Orthophlebiidae: *Austropanorpa* Riek). The Neogene scorpionflies (Bittacidae) are found in Eastern Europe (Croatia).

2.2.1.3.4.2. ORDER TRICHOPTERA Kirby, 1813. THE CADDISFLIES (=Phryganeida Latreille, 1810)

V.D. IVANOV AND I.D. SUKATSHEVA

(a) **Introductory remarks**. A moderately small (11,185 extant and 642 fossil species: MORSE 1999) group of insects with moth-like adults and aquatic, rarely semiaquatic or terrestrial larvae that usually use a variety of retreats built of silk and other materials. The chapter is based primarily on the palaeontological data reviewed by MARTYNOVA (1962e), SUKATSHEVA (1982, 1985b, 1990b, 1994), IVANOV (1988, 1992), KULICKA & SUKATSHEVA (1990), KUKALOVÁ-PECK & WILLMANN (1990) and ZHERIKHIN & SUKATSHEVA (1990), and on the phylogenetic considerations by H. ROSS (1967), MALICKY (1973), SUKATSHEVA (1980, 1982), WEAVER (1984), WEAVER & MORSE (1986), WIGGINS & WICHARD (1989), NOVOKSHONOV & SUKATSHEVA (1993), MORSE

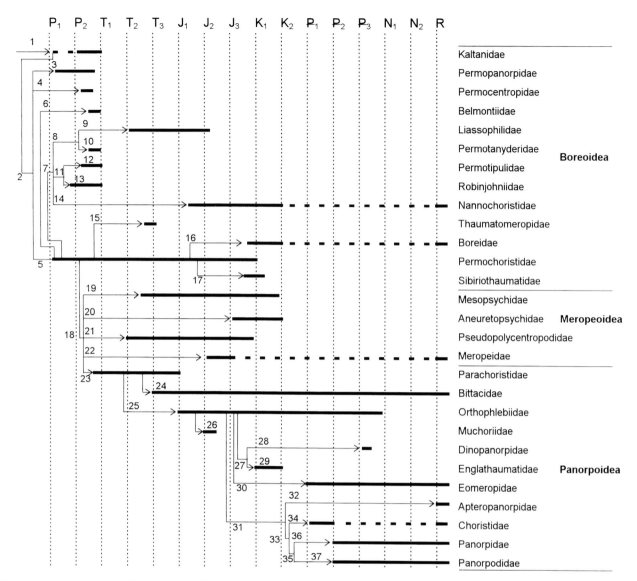

Fig. 266 Phylogeny and system of the order Panorpida

Time periods are abbreviated as follows: P_1, P_2–Early (Lower) and Late (Upper) Permian, T_1, T_2, T_3–Early, Middle and Late Triassic, J_1, J_2, J_3–Early, Middle and Late Jurassic, K_1, K_2–Early and Late Cretaceous, P_1,–Palaeocene, P_2–Eocene, P_3–Oligocene, N_1–Miocene, N_2–Pliocene, R–present time (Holocene). Right columns are names of families and (bold) superfamilies. Thin lines show ancestry, thick bars are known durations of taxa (dashed for long gaps in the fossil record), figures refer to synapomorphies of the subtended clades, as follows:

1 – synapomorphies of the Superorder Papilionidea (Chapter 2.2.1.3.4c).

2 – CuA not forking, fore wing SC with 3 branches.

3 – both wing pairs narrow, fore wing SC short, with single anterior branch.

4 – fore wing anterior margin concave, SC with single fore branch.

5 – MP 6-branched (MP_2 and MP_4 forked); SC tending to have only 1 or 2 fore branches; possibly: gonostyles with stylar organs.

6 – Size large; venation destabilised; fore wing with thyridium shifted apicad (on MP_{1+2} stalk); MP 5-branched.

7 – fore wing RS + MA and MP both 4-branched.

8 – fore wing costal space narrow; outer and posterior wing margins meet at distinct angle; free M_5 tending to become stronger; hind wing shortened, simplified venationally.

9 – fore wing MA fork shorter than RS fork.

10 – fore wing apex subacute; size small.

11 – fore wing SC short; CuA smooth (not angular) at junction with M_5; both wing pairs narrow basally.

12 – RS simple; SC lacking fore branches; possibly: hind wing CuA convex.

13 – fore wing with long fusion of MP and CuA.

14 – fore wing with long fusion of MP and CuA; cross-vein cua-cup shifted basad at level of MP + CuA fork; possibly: basistyles forming entire bubble-like capsule; antennae long; meso- and metascutellum short; larva predaceous, aquatic.

(1997), FRANIA & WIGGINS (1997) and IVANOV (1997); other sources are cited below when necessary. The family Prosepididontidae is synonymised with Geinitziidae (Grylloblattida) (ANSORGE & RASNITSYN 2000) and therefore is not considered here.

(b) Definition. General appearance of adult similar to moths. Body size small to moderately large. Head hypognathous, with eye large, antenna thread-like, mandible weak or lost in Hydropsychina and Phryganeina, better developed in Protomeropina, maxilla and labium lacking inner lobes, hypopharynx well developed, used in licking liquids. Pronotum short; meso- and metathoraces subequal in length and structure. Legs cursorial, tibiae basically with apical and praeapical spurs, tarsi 5-segmented. Wings almost always well developed, covered with setae and numerous short chaetoids among them, rarely with patches of scales, with venation similar to that of scorpion-flies and particularly to that of lower moths, though sometimes specialised. SC sometimes with additional branches, R with apical fork. In most cases RS and M with 2 apical forks (dichotomously 4-branched), CuA with 1 apical fork. Fore wing with anal veins looped. Cross-veins present, numerous in ancient groups. Pterostigma sometimes present. Hind wing often with venation somewhat simplified, sometimes with anal region enlarged, often pleating in folded wings. Male with segment 9 forming entire ring, genitalia generally include aedeagus, gonopods, and modified segment 10. Female primary ovipositor lost; sometimes abdominal segments 8 and 9 elongate to form secondary, often retractable, ovipositor. Immatures aquatic or semiaquatic, without spiracles, using gill or skin breathing. Larva campodeiform, legs well developed, abdominal appendages lacking except claw-bearing anal prolegs attached to segment 9; in Phryganeina (Integripalpia) often with enlarged abdomen ("suberuciform"); unknown in Protomeropina (Permotrichoptera). Mouthparts form maxillolabium with well developed silk-weaving spinneret. Larval leg 6-segmented (seemingly 7-segmented in most groups due to 2-segmented trochanter) with claw-like pretarsus provided with large trichoid sensilla. Pupa with powerful mandibles, leaving retreat to swim actively toward water surface for eclosion. Adult drinking water, honeydew or other liquid matters, sometimes aphagous, ovipositing in or close to water. Larva predatory, algophagous or omnivorous filter-feeder in stationary retreats, web nets and nests in Hydropsychina; mostly phyto- or mixophagous and living in specially constructed portable, usually tube-like case in Phryganeina; preferring running and cool water.

(c) Synapomorphies. Apneustic tracheal system in larvae, transformed mouthparts of adults, and other characters rarely or never visible in fossils. Since status of fossil remnants usually restricted to features of wing venation, we adopt combination of anal loop on fore wing and absence of scales as delimiting characters for fossil Trichoptera. The same criteria are used by taxonomists in the basic classification of extant trichopterans; in most species the ordinal synapomorphies were not observed because immatures are unknown and special anatomical investigations are lacking. We understand that this wide treatment of Trichoptera can contradict current cladistic thinking and tradition of separation of ancestors as taxa "*incertae sedis*". This is generally a matter of agreement, because the Palaeozoic suborder Protomeropina is undoubtedly the common ancestor for both Trichoptera and Lepidoptera. We prefer to include this group in Trichoptera rather that to claim it to be a separate order with no synapomorphies and a very dim diagnosis. This is not an "*incertae sedis*" taxon in the common sense

15 – venation polymerised; hind wing with strong, convex fold in front of CuA.

16 – female apterous, with short, thick abdomen; male wings hook-like reduced, used to fix female at copulation; ocelli widely separated, tending towards loss; possibly: insemination by means of spermatophore.

17 – fore wing SC with 5–6 branches, MP 6-branched (MP_1 and MP_4 forked); wing apex subacute.

18 – leg vestiture forming distinct transverse rings; possibly: female gonopore shifted apicad due to elongation of gonocoxosternum 8; basicoxostyles fused both dorsally and ventrally, thus forming distinct cranial foramen.

19 – CuA base reclivous cross-vein-like; RS + MA and MP both 4-branched; thyridium shifted basad on MP stalk.

20 – body stout; long, thin, flexible, sucking proboscis present; fore wing long, narrow, hind wing with wide jugal lobe tucking down at rest; RS + MA and MP both 4-branched.

21 – body very slender; short, straight (sucking?) proboscis present; fore wing with SC short, RS + MA with 4 branches, MP with 5 (MP_4 with fork lost); hind wing short, with simplified venation.

22 – wings heavy, with tubercular membrane, thick veins and numerous cross-veins; costal vein densely annular; SC multibranched; possibly: stridulatory flap present at posterior proximal margin of fore wing; rostrum present.

23 – RS pectinate; rostrum present.

24 – wings narrow, petiolate; fore wing CuA fused with MP for a distance; MA_2 distinctly convex; MA + MA_1 stalk with characteristic flexure; hind wing CuA convex; possibly: pterothoracic pleura not slanting; antenna short; hind leg raptorial.

25 – male tergum and sternum 8 (possibly 7 as well) fused leaving trace; possibly: MP 5-branched (MP_4 forked); MP_4 stalk and fork base desclerotised; base of M_5 desclerotised; basal segments of female cerci joined with sclerotised bridge.

26 – SC short; R angular at pterostigmal base; RS 2-branched; MA fork short; MP 4-branched.

27 – SC with additional fore branches; MP and CuA connected with long, oblique cross-vein; possibly: RS tending to be polymerised.

28 – possibly: hind wing with apex subacute and R long.

29 – venation polymerised; base of fore wing RS + MA shifted basad.

30 – venation polymerised, reticulate; R, RS + MA and MR running closely adjacent basally, then diverging simultaneously forming trident; pterostigma diffuse; legs spinose.

31 – possibly: female with basal cercal segments fused; basistyles with cranial foramen narrow.

32 – both sexes with wings lost.

33 – stylar organs lost; hind wing CuA convex.

34 – Fore wing costal space wide; SC short; RS 2-branched; CuA base near vertical, joining MP in a point; hind wing cross-vein 1a-2a at level of CuP + 1A.

35 – MP 4-branched; possibly: female cercus with basal segments fused almost completely.

36 – male tergum and sternum 6 almost completely fused; female medigynium with 2 lateral processes.

37 – rostrum short; fore wing MP_4 smooth (not angular) at junction with cross-vein mp-cua; hind wing cross-vein 1a-2a at level of CuP + 1A.

Fig. 267 Female *Seniorita gratiosa* Novokshonov (Permopanorpidae) from the Early Permian of Tshekarda in Urals (from NOVOKSHONOV 1997a); wing length 10 mm

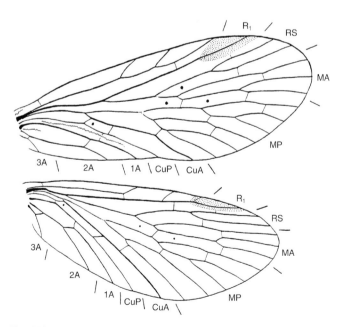

Fig. 268 Wing venation of *Agetopanorpa* sp. (Permochoristidae) from the Late Permian of Soyana in NW Russia (from NOVOKSHONOV 1997a)

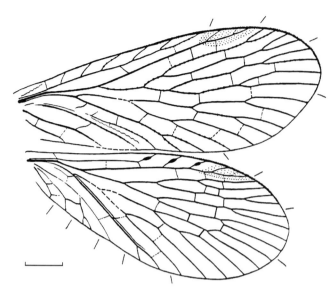

Fig. 269 Wings of male *Altajopanorpa pilosa* (O. Mart.) (Kaltanidae) from the Late Permian of Kaltan in Siberia (from NOVOKSHONOV 1997a); scale bar: 1 mm. Note Carpenter's organs, thickenings on hind wing SC veinlets, also found in males of the oldest caddisflies

because its taxonomic position and limits are easily traceable. The Mesozoic fossils are readily distinguishable from Lepidoptera by their macrotrichia, the long stalk of the anal loop (fore wings), patterns of setose warts of adult bodies, and long legs and cases of immatures.

One of the most striking apomorphies within the Trichoptera is the extremely reduced larval antennae which are situated very close to the mouthparts: apparently the immatures dwelled primarily in habitats poor in olfactory stimuli or where they were of little use. The most appropriate environments that could explain this reduction are soil retreats. Then, later in the evolution of the order, a nomadic life in portable retreats led to a secondary elongation of the antennae in the higher Phryganeina. Hydropsychina retained the ancestral mode of life in bottom retreats so their antennae are still much reduced. Phryganeina originated as phytophagous scrapers living outside of the bottom sediments, and they acquired portable shelters and longer antennae when they invaded the photic zone of stony streams. Pupation in the Trichoptera is associated with the construction of a double shelter: the outer retreat is usually made of sand or small pebbles, and the inner one is made of silk. The semipermeable cocoon has been claimed to be an apomorphy of the Trichoptera (WIGGINS & WICHARD 1989). However, dark silken cocoons are common among lower Lepidoptera and Hymenoptera, so the cocoon itself is not so specific; though in all other orders the cocoon is dry inside and the walls are waterproof. The cocoon of lowermost Trichoptera was filled with a watery solution of unknown chemical composition, perhaps including the remains of the exuvial fluid. The function of the cocoon is thought to be to provide osmotic protection for the developing pupa, though direct experiments on pupal osmoregulation and on the adaptive role of cocoon are lacking. It is possible that a cocoon filled with fluid under osmotic pressure serves like a protective "waterbag" preventing damage of the developing pupa by water currents coming through the holes in the outer stony shelter. Anyway, the fully developed pupa has to pass through the water column before eclosion and hence should have an ability

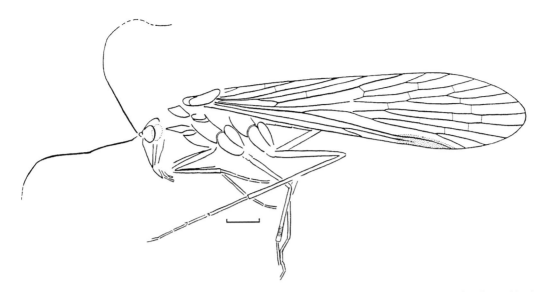

Fig. 270 *Parachorista comica* Novokshonov (Parachoristidae) from the Middle or Late Triassic Madygen Formation Kirghizia (distorted by the matrix deformation; from NOVOKSHONOV 1997a). Scale bar: 1 mm

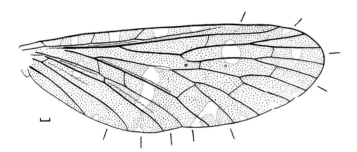

Fig. 271 Fore wing of *Mesopsyche dobrokhotovae* Novokshonov (Mesopsychidae) from the Late Triassic of Garazhovka in Ukraine (from NOVOKSHONOV 1997a). Scale bar: 1 mm

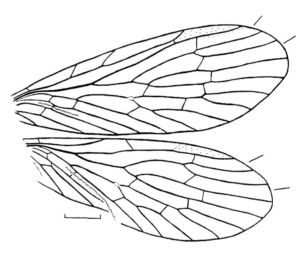

Fig. 273 Wings of *Itaphlebia multa* Novokshonov (Nannochoristidae) from the Middle Jurassic of Kubekovo in Siberia (from NOVOKSHONOV 1997a) Scale bar: 1 mm

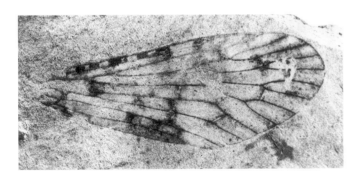

Fig. 272 *Ijapsyche sibirica* Kolosnitsyna et O. Martynova (Liassophilidae) from the early Middle Jurassic of Iya (Ija) in Siberia (holotype, photo by D.E. Shcherbakov); wing length 19.2 mm

to osmoregulate itself. More advanced Trichoptera have developed mechanisms that permit ventilatory movements of the pupa and have elaborate protective mechanisms against accidental damage.

The most characteristic adult autapomorphy of the Hydropsychina + Phryganeina is the specialised mouth with licking haustellar complex represented by the fused hypopharynx, maxillae, and labium with the terminal soft bulb formed by tip of the hypopharynx. This is an

essential adaptation for collecting food from the substrate: these could be wet or dry droplets of honeydew from Homoptera, sugar produced by some epiphytic fungi, or substances secreted by vegetation. Homopteran honeydew would have been very common in biocoenoses since the Early Permian; other sugar sources are probably more ancient. Experiments have shown that Trichoptera can eat dry sugar, and it has also been shown that feeding is essential for mating at least in some species (HOFFMAN 1997). There are no autapomorphic characters in the wing venation of Lepidoptera compared to Trichoptera, thus the ancestor of the Lepidoptera is impossible to find among the fossil Trichoptera wing imprints unless scales are preserved.

Fixed larval retreats characteristic of the suborder Hydropsychina are modifications of the silk tunnel plesiomorphic for Trichoptera, and the major evolutionary trend is acquisition of filtering based on the transformation of this tunnel. Portable cases are produced by larvae of the suborder Phryganeina. There are two larval construction types

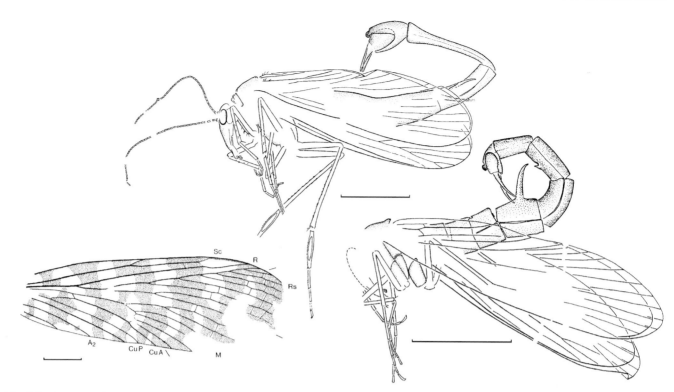

Fig. 274 Structure of Orthophlebiidae from the Late Jurassic of Karatau in Kazakhstan: fore wing of *Orthophlebia grandis* Martynov, the whole male of *O. longicauda* Willmann et Novokshonov, and incomplete male of *O. heidemariae* Willmann et Novokshonov (from WILLMANN & NOVOKSHONOV 1998b, curtesy of *Paläontologische Zeitschrift*). Scale bar 5 mm

within the suborder: the bivalve cases of Glossosomatidae and Hydroptilidae and the tubular cases of other Phryganeina which constitute the monophyletic Integripalpia group. The construction behaviour for bivalvous larval cases is supposed to be transferred from the pupal stage when mature larva makes a dome-shaped outer pupal retreat (another trichopteran apomorphy): the larva manages to add a floor to the pupal dome so turning it into a portable retreat. Tube cases are supposed to have evolved along a different route, probably from fixed tube-like larval constructions (in some Phryganeoidea they can be many times longer than the larva). Observations on larval behaviour (LEPNEVA 1964) show the cylinder to be an initial stage of the dome-shaped construction. The provisional larval shelters in Glossosomatidae (MAJECKI *et al.* 1997) and Hydroptilidae are always tubes; only later on does the larva reconstruct them to form bivalve cases. Some Hydroptilidae (*Neotrichia* Morton, *Mayatrichia* Mosely) built cylindrical cases that probably originated from the provisional larval case. These and many other observations show the cylinder to be an initial type of construction within the Trichoptera and so imply that there were no gaps in the evolution of behaviour within Phryganeina.

(d) Range. From Early Permian until present, worldwide except areas covered with ice or devoid of fresh waters.

(e) System and phylogeny are shown in Fig. 278; for further details see H. ROSS (1967), WEAVER (1984), K. SCOTT & MOORE (1993), FRANIA & WIGGINS (1997), IVANOV (1997) and MORSE (1997). For suborders, the alternative names are Permotrichoptera for Protomeropina, Annulipalpia for Hydropsychina, and Integripalpia for Phryganeina. We use the modified names to stress the shifted limits of traditional Annulipalpia and Integripalpia, after Hydroptiloidea and Glossosomatoidea have been transferred to the Integripalpia. The

synonymy of Uraloptysmatidae with the Microptysmatidae proposed by NOVOKSHONOV (1993c) is not accepted here because re-examination of the Microptysmatidae showed their diagnostic character, viz. the absence of SC branches in the fore wing, to be stable regardless of their state of preservation. The Mesozoic families are defined only by the fore wings and hence are not compliant to the extant ones. Consequently their taxonomic positions are very tentative. Necrotauliidae in the traditional sense is an heterogeneous family comprising basal Lepidoptera and Trichoptera together with specialised offshoots of the primitive Amphiesmenoptera (see below). *Liadotaulius* belongs to the caddisflies, and some *Necrotaulius* are generalised Amphiesmenoptera. Vitimotauliidae are also heterogeneous and should also be considered as *incertae sedis* because they potentially include members of both Leptoceroid and Limnephiloid branches of Integripalpia. Future revisions perhaps will split these groups into well substantiated families. Wing venation of the Prorhyacophilidae is very similar to that of Rhyacophilidae except for the shorter anal loop. Until the body structure is better known, we retain the former as distinct from the latter because of the conservative wing venation and the stable, though undeniably subtle, difference in length of the anal loop.

(f) History. The oldest known caddisflies come from the Artinskian deposits (mid-Early Permian) of Central Europe (Obora in Moravia) and North America (Elmo in Kansas). They were represented by the families Protomeropidae (in both above areas) and Microptysmatidae (in Moravia only). These families persisted up to the latest Late Permian and are found in North and Eastern Europe (Tshekarda in the Urals and Soyana in Arkhangelsk Region) and East Kazakhstan (Karaungir in Saur Range), Protomeropidae also in Australia (Belmont), and Microptysmatidae in Siberia (Sarbala in Kuznetsk

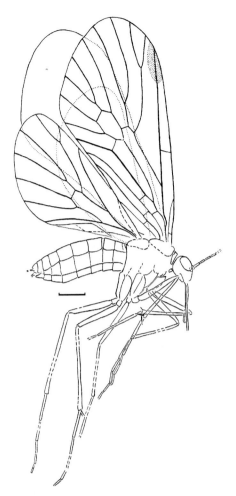

Fig. 275 Female *Pseudopolycentropus latipennis* Mart. (Pseudopolycentropodidae) from the Late Jurassic of Karatau in Kazakhstan (from NOVOKSHONOV 1997c); scale bar 1 mm. Note long proboscis and reduced hind wing

Basin) and South Mongolia (Bor-Tologoy). Uraloptysmatidae, a small family of microptysmatid-like Amphiesmenoptera, is known from Tshekarda. The fourth family of Protomeropina, the Cladochoristidae, appeared in the Late Permian of Australia and survived there till the Late Triassic. Only adults are known for Protomeropina.

Protomeropidae (Fig. 279) were medium-sized insects strikingly resembling mecopterans. Their wings are structurally similar to those of Mecoptera, with numerous cross-veins. However, the arrangement of the hindmost wing veins forming an anal loop is characteristic of Amphiesmenoptera: "2A" and "3A" (vannal veins, V_1 and V_2 of Snodgrass) seemingly meeting each other and "1A" (PCu of Snodgrass) instead of the wing margin (in fact inter-anal cross-veins are aligned so as to simulate bent anal veins parallel to the wing margin, whilst the true apices of the anals are weak, cross-vein-like, or, usually, lost; vein PCu is often designated as 1A with subsequent changes for V_1 to 2A and V_2 to 3A). This construction invoked the hypothesis of an aquatic origin of the Amphiesmenoptera (MARTYNOVA 1959), as an adaptation for taking a store of air beneath the overlapping wings when diving to locate egg-laying places as some extant caddisflies do. Since direct observations on diving Trichoptera show that the film of air actually covers the outer surface of the insect, the closure of the space between the wings is useless for this purpose. Moreover, the absence of immatures in Palaeozoic deposits indicates a terrestrial or soil-dwelling mode of life for the Protomeropina. The functional reason for development of the anal loop is probably the overlapping of fore and hind wing during the downstroke without any firm mechanical coupling (BRODSKY & IVANOV 1983, IVANOV 1985) to provide a continuos airfoil for generation of aerodynamic forces (IVANOV 1990). They probably had the same wing kinematics as Antliophora and lower Amphiesmenoptera with synchronous down-stroke and asynchronous upstroke. Fore and hind wings were not coupled together in flight. In common with the most basal mecopterans (Kaltanidae), CuA had an apical fork. Rarely the anal loop also appears in hind wings, possibly because of a genetic link between the developmental patterns of fore and hind wings (NOVOKSHONOV 1992, NOVOKSHONOV & SUKATSHEVA 1993). Contrary

Fig. 276 *Aneuretopsyche rostrata* Rasnitsyn & Kozlov (Aneuretopsychidae) from the Late Jurassic of Karatau in Kazakhstan (from RASNITSYN & KOZLOV 1990); length of fossil 28 mm. Note moth- or cicada-like form and long proboscis

Fig. 277 Female *Palaeoboreus zherichini* Sukatsheva et Rasnitsyn (Boreidae) from the Late Jurassic of Khoutiyn-Khotgor in Mongolia (from NOVOKSHONOV 1997a); body 3 mm long

to extant Amphiesmenoptera, the base of vein CuA was very long and undulating; and the basal cross-vein connecting M-Cu (M_5) was long and orientated longitudinally. As in the lowermost mecopteroids (Kaltanidae, see Fig. 269), some male hind wings indicate the presence of Carpenter's organs (short protruding spirals) on the subcostal vein branches (Fig. 279d): this putative synapomorphy indicates a sister-group relationship between Protomeropidae and Kaltanidae. The protomeropid head was elongated with a short rostrum present beneath eyes; ocelli grouped on the vertex, and the elongate mandibles had a distal set of teeth indicating that they were carnivorous (or fed on dead insects as some living mecopterans do). Coxae were elongated indicating development of the coxal wing depressors and long tergotrochanteral leg extensors. It is highly probable that these insects jumped with middle and hind legs. The female apodemes were not found so a retractable ovipositor seem to have been absent. The male genitalia display a striking resemblance to those of the mecopterans. It is likely that the heads of mates were oriented unidirectionally during copulation, and the male abdomen was bent and approached the female genitalia from beneath. This mode of copulation is characteristic of mecopterans: Trichoptera normally demonstrate a postcopulatory turn, so as the male genitalia, with the gonopods oriented posterodorsad and clasping inwards (Fig. 279, gcx and gst, respectively), turn so as to direct their gonopods upwards and clasp forwards (So *in cop.* males and females end up facing opposite directions). Segment X short and provided with small one-segmented cerci.

Protomeropidae were widespread and abundant in most Permian localities. The number of specimens in taphocoenoses is high, so they were probably common close to the places of fossilisation. There are a considerable number of bodies fossilised *in situ*, on their sides. Adult adaptations (triangular rostrate head with large eyes, elongated dentate mandibles, larger size) and azonality suggests a predatory habit (omnivory) or feeding on the carcasses of other insects. The well developed chaetome could provide protection against contact and subsequent adhesion to water in wet places, though the immatures probably lived out of water (in soil or in the wet debris).

Microptysmatidae and Uraloptysmatidae (Fig. 280) were less similar to mecopterans than Protomeropidae, both in structure and habits. They were much smaller, with a short head, lacking a rostrum and with short palps and mandibles (studied in material other than that used by IVANOV 1988; see also NOVOKSHONOV 1997a). The thorax outline generally resembles that of Protomeropidae suggesting the same weak type of locomotion. Their legs were long and slender, adapted for quick movements; they possessed apical and subapical spurs suitable for take-off from an unstable surface (especially for jumps from mud, water, and vegetation; similar spurs are characteristic of the jumping legs of Homoptera). The wings of Microptysmatidae and Uraloptysmatidae have venation features (loss of most cross-veins and apical forks) suggesting that they could have been derived from a very tiny ancestor (wing reduction starts usually from its apex whereas the middle and basal parts are more conservative; in the extant short-winged caddisflies the apical wing portions are the most reduced). The anal loop was better shaped though rather short; hence the wings only overlapped in flight at their bases similar to the lowermost moths. The anal loop sometimes appears in the hind wings (Fig. 280f); in these cases it is better shaped than in the hind wings of Protomeropidae (some primitive Lepidoptera and Trichoptera, e.g. Philopotamidae, also feature an anal loop in their hind wing). Unlike the Microptysmatidae, Uraloptysmatidae retained the subcostal cross-veins thus sharing this character with Cladochoristidae, the most advanced protomeropoids. The female abdomen was acute at the apex, with inner apodemes providing movement for the telescopic ovipositor. There were multi-jointed cerci situated at the very apex of the female abdomen; their bases touched each other the same way as in mecopterans. Male genitalia had the well developed 10th segment and the gonopods directed backwards suggesting mating with postcopulatory rotation (the heads of the mating pairs pointing in opposite directions as is usual in the extant Amphiesmenoptera).

Microptysmatidae were abundant close to the water and were a constant component of the Permian taphocoenoses; sometimes they were very numerous (in Tshekarda and Soyana). It seems that the living insects tried to escape from their trap with the wing flaps, since there is a number of bodies with wings outstretched. The female probably put eggs in hidden places, in cracks in soil and in crevices or under debris. Larvae could dwell in very moist soil or litter. The reason for small size could be that their larvae inhabited particularly moist places with reduced levels of available oxygen. This would make high surface area/volume ratio of small soft-bodied larvae adaptive for skin breathing.

Cladochoristidae are known only from their fore wings (Fig. 281), so little can be said about this family. The wing venation is similar to the postulated ground-plan of the extant Trichoptera and Lepidoptera, so this family is apparently the closest one to their common ancestor among the Protomeropina. It is known from few fossils found in the Late Permian of Australia and Late Triassic of Eurasia, and this rarity may be because of a taphonomically unfavourable mode of life far from sedimentation areas, so only separated wings transported downstream could have entered into taphocoenoces. Far southern Permian finds make it possible to hypothesise that the cladochoristid-like common ancestor of Lepidoptera and advanced Trichoptera (Hydropsychina + Phryganeina) was initially cool-adapted, and this facilitated its subsequent transition to semi-aquatic or aquatic habitats. The major problem of an aquatic life is the limited supply of oxygen, particularly in warm waters, and so the frequenting of cool waters might have facilitated its origin. Another problem is a deficiency in certain ions and ion loss in fresh waters, and this was probably already overcome during the soil-dwelling period of larval evolution, because the rectal papillae ("anal gills") which serve as ion absorbers in moist habitats, are well developed in the larvae of mecopterans, Diptera, and Trichoptera. It is unlikely that immature caddisflies initially occupied water torrents of mountain streams, because their extant inhabitants are highly modified, and this could also have been the case in the past. More probably, the order originated in small cool brooks or springs. It should be stressed that the larvae of extant caddisflies usually dwell either within the stream bottom (supposedly the ancestral mode of life) or, if above it, in retreats built using labial silk reinforced by various particles and so imitating in-bottom environments.

Perhaps most of the Triassic Amphiesmenoptera were local in habitats high above the erosion level and far from sedimentation areas. These could have been springs (for the caddisfly ancestor) or moist habitats covered by mosses and liverworts (for the ancestor of Lepidoptera); the ancestral habitat could be the same for both orders. The larval apomorphies of Trichoptera would probably have been formed at these times.

No caddisflies are known from the first half of the Triassic, a generally obscure section of the insect fossil record. Later in the Triassic, caddisflies were still rare compared with the Permian time, but they were widespread. Fossil material comes from Fergana in Central Asia (Cladochoristidae, see Fig. 281, Prorhyacophilidae, Fig. 282, and Necrotauliidae) and from Queensland in Australia (Cladochoristidae and Prorhyacophilidae). They were also taxonomically diverse and distinct, being represented by 4 families with only one (Cladochoristidae) inherited from the Palaeozoic and belonging to Permomeropina.

The Prorhyacophilidae (Fig. 282) is endemic to the Triassic and in its fore wing venation it resembles the postulated ground-plan of Trichoptera and Lepidoptera, except that it has no r and m cross-veins and so fits the venation of more advanced *Rhyacophila* Pict. This family is apparently closer to caddisflies than to moths because they have a long anal loop stalk (ending of PCu) which is not characteristic of the primitive Lepidoptera. Necrotauliidae are known from both hemispheres and have shorter anal veins. The reports on the family Philopotamidae present in the Late Triassic of Middle Asia

(SUKATSHEVA 1982; 1992) are rather tentative: there are no decisive characters indicating the family attribution of the respective wings. The only characteristic feature of the philopotamid wings is an incomplete anal loop (apex of the basalmost anal vein retained in fore wing of some genera): this character never occurs in the Mesozoic material. Further, the suggested philopotamid fossils are difficult to distinguish from either Prorhyacophilidae or Necrotauliidae. Cladochoristidae resemble other Triassic families except they possess the multibranched SC. Wing venation of the Triassic caddisflies looks like a continuum and is difficult to segregate into families.

The most famous Mesozoic family of Amphiesmenoptera, the Necrotauliidae (Fig. 283), was of Triassic origin, though more abundant material originates from the Early Jurassic beds of the West Europe (Lias of Germany and England). The family previously included most of the Early Jurassic Amphiesmenoptera that had been described mainly by HANDLIRSCH (1906, 1939) with minor additions by other writers. Recent revision of this material by ANSORGE (2001) showed that most of Necrotauliidae from the Early Toarcian of Germany described by Handlirsch, except for *Liadotaulius major* (Handlirsch, 1906) and probably *Necrotaulius parvulus* (Geinitz, 1884) and species of *Mesotrichopteridium* Handl., belong to Lepidoptera or Diptera. Necrotauliidae were previously formally included into the suborder Hydropsychina in spite of the fact that they rarely show the fore wing CuA base shifted distad, a diagnostic character of the suborder (except Rhyacophiloidea). Comparison of wings described as Necrotauliidae shows the wide diversity of venation. Some species are primitive enough and could be representatives of the amphiesmenopteran stem-group. Others have venational specialisations, e.g. pterostigma with numerous cross-veins in *Necrotaulius stigmaticus* Till. from Early Lias of Great Britain. Some Necrotauliidae (e.g. *Necrotaulius proximus* Suk., *N. marginatus* Bode) have fore wing with a faint base of CuA and the additional CuA-CuP cross-vein serving as a suspensor for the cubital system. Further development of this trait will give rise to the "shifted base of the cubitus" (strong cross-vein combined with complete reduction of CuA base), a good venational character of higher Hydropsychina (Annulipalpia). *Cretotaulius ultimus* Suk. (Fig. 283a) has CuA separated from CuP (CuA base lost) without development of additional cross veins in the CuA field. This species has very short discoidal and slightly longer medial cells looking like those of the ecnomoid Hydropsychina. In some Necrotauliidae the postcostal field was wider than in Protomeropina suggesting a more prominent wing overlap during flight; in other species it was short and narrow as in its ancestors. Several species, e.g. *Necrotaulius westwoodi* Till., were completely devoid of the fore wing cross-veins. Some members previously included there are not Amphiesmenoptera (*Liassophila* Till. mecopteran) or, as in the genera *Trichopteridium* Geinitz or *Paratrichopteridium* Handl., known only from hind wings, cannot be correctly placed into the system. It is clear that the family is extremely heterogeneous and deserves revision based on the type material (see also SUKATSHEVA 1982). In its present extent, Necrotauliidae include the Lepidoptera ancestor among other materials.

Most of Necrotauliidae are represented by fore wings or wing parts; their bodies are rare, indicating that the family did not inhabit those lentic waters where the fossilisation occurred. The only exception is *Necrotaulius tener* Suk. (Fig. 283b–d) from the Late Jurassic or Early Cretaceous of Siberia (SUKATSHEVA 1990b). This is a female specimen fossilised with remarkably perfect completeness. It demonstrates some characters of the lepidopteran ground-plan and may represent the group ancestral to Lepidoptera. *N. tener* had its head warts, if present, hardly conspicuous. Maxillary palps are very peculiar in having the

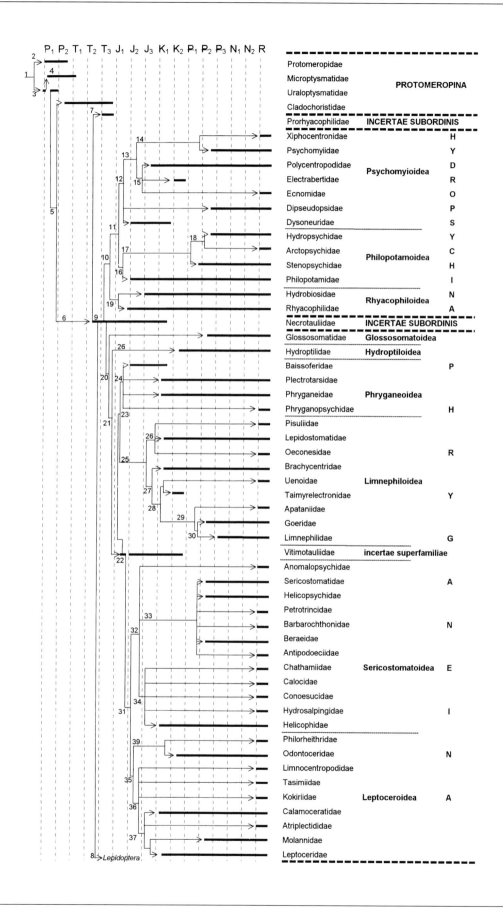

Fig. 278 Phylogeny and system of caddisflies (order Trichoptera) (modified from WEAVER 1984, WEAVER & MORSE 1986, SCOTT & MOOR 1993, FRANIA & WIGGINS 1997, IVANOV 1997, SCHMID 1998)

Time periods are abbreviated as follows: P_1, P_2 – Early (Lower) and Late (Upper) Permian, T_1, T_2, T_3 – Early, Middle and Late Triassic, J_1, J_2, J_3 – Early, Middle and Late Jurassic, K_1, K_2 – Early (Lower) and Late (Upper) Cretaceous, P_1 – Palaeocene, P_2 – Eocene, P_3 – Oligocene, N_1 – Miocene, N_2 – Pliocene, R – present time (Holocene). Three righthand columns are names of families, superfamilies and suborders, respectively. Arrows show ancestry, thick bars are known durations of taxa, figures refer to synapomorphies of the subtended clades, as follows (for synapomorphies of terminal taxa see also MALICKY 1973, SCOTT & MOOR 1993, FRANIA & WIGGINS 1997, SCHMID 1998).

1 – fore wing with longitudinal fold along posterior margin, anal veins with apices displaced and vertical to wing margin, posterior cross-veins aligned parallel to it (adaptations to the wing overlapping during downstroke in flight); size medium, head slightly elongated to form a short beak (adult carnivorous?), pronotum and head with indistinct warts; male gonopods bending upwards (mating with body axes parallel and heads directed to the same side); female abdomen blunt, without telescopic ovipositor.

2 – wings long and narrow, SC joins R apically in both wings.

3 – wing venation simplified, most cross-veins lost, fore wing SC with less than 6 branches, RS and M each with fewer than 7 branches; head short, with a set of 4 medial warts, with maxillary palp rather long; pronotum with 2 pairs of dorsal warts; subapical tibial spurs present (adaptation to jumping, especially useful on the water surface); male genitalia with gonopods directed backwards (mating with antiparallel body axes: heads of the mates directed in opposite sides) and bending inwards, segment 10 well developed, segment 9 ring-like; female abdomen bearing secondary telescopic ovipositor, segments 8 and 9 with apodems. Larval habitat subanoxic (presumably soil filled with water, mud, or semiaquatic); size small (adaptation to partly anoxic larval habitats); larva probably prognathous with reduced tentorium, anal prolegs modified for better recreation in closed spaces; larval labial glands producing silk used in construction tough, water proof cocoon.

4 – SC branches suppressed.

5 – fore wing R and M each dichotomously 4-branched, with cross-veins closing fields in basal forks ("radial and medial cells"); anal loop stalk (apex of PCU) as long as or longer than anal loop (suggesting wider wing overlapping during downstroke); postcostal field wide.

6 – most of fore wing SC branches lost (apparently regained in some Hydropsychidae, Odontoceridae, and Philorheithridae); further synapomorphies of extant caddisflies and moths should be inferred here (see BOUDREAUX 1979, KRISTENSEN 1984, etc., for details) which normally cannot be seen in fossils (no wing coupling at flight inferred here, contrary to BOUDREAUX 1979 and some other authors).

7 – no cross-veins in R and M forks (R and M cells open as in Rhyacophilidae, but unlike it anal loop stalk short).

8 – lepidopteran apomorphies (Chapter 2.2.1.4.4.3c).

9 – (synapomorphies supposed to be acquired by an unknown necrotauliid species ancestral to genuine Trichoptera and the sister group of the Lepidoptera, in contrast to node 1 subtending the stem Amphiesmenoptera and node 6 subtending the crown-group Amphiesmenoptera) – larvae hiding in silken retreats in stream bottom (in aerobic conditions), with tracheal system closed, antenna greatly reduced (two-segmented), bearing a few sensory units; anal prolegs well developed; pupation underwater in special dome-shaped pupal retreat (made of sand and attached to the solid ground), in osmotically active semipermeable cocoon filled with special liquid ensuring ion-friendly environment and protecting from mechanical stress; pupa active, disrupting both cocoon and pupal retreat with mandibles, swimming using middle legs only to stream bank and walking away using both fore and middle legs to eclose in a safe place out of water; adult with specialised sucking mouthparts (including hypopharynx with terminal eversible sack and elongated labrum) for drinking water and other liquids; maxillary and labial palps well developed, with delimited subterminal sensory complex; other mouthparts vestigial; mesonotum with distinct warts along lateral sutures.

10 – larval propleural ridge oblique, mes- and metepisternites each with anteriorly directed apodeme, anal prolegs very prominent; adult imaginal vertex with oblique ridges, subalar wart of mesonotum reduced; larva dependant on immovable bottom retreats (in case of Rhyacophiloidea spaces between stones and other bottom objects making entire huge "retreat").

11 – adult maxillary and labial palps with last segment elongate, bearing network of loose cuticle, flexible (secondary modified in some Dipseudopsidae); base of anal proleg elongate; male parameres lost; fore wing CuA base shifted distad; larval segment 10 reduced, antenna with 3 trichoid sensillae.

12 – larval pronotal hind angles joining sternum behind coxa, prelabio-hypopharynx sclerotised dorsally, prementum not distinguishable, spinneret tending to be long; larva flat, larval cuticle hydrophobous; antenna with one basiconic sensilla; adult scutal warts rounded.

13 – adult head warts enlarged, scutal warts small.

14 – larval labial palps greatly reduced, spinneret elongate and narrow, protrochantine large, ends of pupal mandibles narrow; fore wing thyridial cell small (larval retreats primitively long, larvae detritophagous or algophagous, no filtering habits, body wide).

15 – larva with secondary setae, bottom predator, constructing short retreats; female terminal segments modified.

16 – adult scutal warts suppressed; larva with prosternite sclerotised behind coxae, antenna with 2 basiconic and 3 trichoid sensillae: 1 long and 2 shorter.

17 – larval retreat with sampling screen fastened to frame, base of fore wing 3rd axillary plate with anterior incision, last segment of palps very long; adult with swarming behaviour.

18 – larva with secondary thoracic notal sclerites, secondary chetome, branching abdominal gills, anal proleg with terminal hair brush; sampling screen with regular cells.

19 – larval antenna with two basiconic sensillae and only one trichoid sensilla between them; larva mostly carnivorous, using bottom crevices as retreat, silk producing only before pupation.

20 – larva with head secondarily hypognathous, spinneret opening with wide dorsal lobe, silk thread flat, anal prolegs short and stout at least in last instar, anal claw articulation lateral (recapitulation of ancestral dorsal articulation in some Hydroptilidae; larva making portable tubes as provisional (Glossosomatoidea and Hydroptiloidea) or permanent retreats; pupation in shelter usually transformed from larval retreat or (plesiomorphically) in new retreat structurally similar to larval construction; hind wing cross-vein PCu-1A thin, vertical (plesiomorphically, in Protomeropidae, this vein is reclivous).

21 – larval antenna elongated, meso- and metathoraces with lateral notal sclerites having more than 1 setae, medial notal sclerites usually with more than 2 setae, pleurites large, abdomen with lateral humps.

22 – adult haustellum with well developed parallel channels covered with asymmetrically branching microtrichia; female with ovipositor reduced, apodemes lost, cerci lacking; larval antenna with modified apical sensilla, sterna bearing paired glands, abdomen with lateral projections (humps) on 1st segment (secondarily reduced in a few families), lateral fringe, and forked lamellae, abdominal prolegs short in all instars; eggs laid enclosed in proteinous mass; fore wing M_4 tending to be reduced.

apical segment narrow, acute and short (not annular, contrary to NOVOKSHONOV 1992 and NOVOKSHONOV & SUKATSHEVA 1993), in contrast to two robust and long subapical segments (provided that the palps had 5 segments, the 3rd and 4th ones were greatly enlarged). The palp apex was attenuate and probably had the large apical sensory complex characteristic of extant Rhyacophilidae, Glossosomatidae, Philopotamidae and Stenopsychidae (Trichoptera) and with some modification present in Micropterigidae and Eriocraniidae (Lepidoptera). Tentorium was H-shaped and fitted the ground plan of Trichoptera (FRANIA & WIGGINS 1997). Pronotum was short. There were 4 pronotal setose warts, with medial and lateral warts nearly coalescing into each other (similar to the extant Glossosomatidae). There were large scale-like lepidopteran tegulae on the fore wing bases; other lepidopteran features of wing articulations were the large lateral mesonotal pits and long anterior notal processes. Similarly to Micropterigidae, the 1st abdominal tergum was enlarged and highly modified, with a wide oval desclerotised window covering the most part of it, and the bent hind margin of the segment protruded caudad. The abdominal apex had papillae similar to those in females of Polycentropodidae (Trichoptera). There were no traces of the retractable ovipositor. At least the hind legs had apical and preapical spurs. The fore tibia bore well developed epiphysis and no apical spurs: both characters are autapomorphies of the Lepidoptera. In contrast to all these lepidopteran apomorphies, no sign of scales are found in this fossil: there were only short setae on the legs and wing membrane, and longer setae covered the wing margins and longitudinal veins. The venation pattern is typical of Necrotauliidae of "annulipalpian" type, with the fore wing showing closed short discoidal and perhaps also medial cells; CuA base is not visible. The anal loop is poorly preserved but was certainly shorter than 1/3 of the fore wing length.

The morphology of N. tener indicates that it visited either Mesozoic "flowers" or areas of honeydew production. The tibial epiphysis is an antennal cleaning device that would work in the case of the apical spur being short or absent. Extant caddisflies clean their antennae by pressing them against the substrate and pulling them out between the solid surface and the tarsus. This method is not applicable in "dirty places"

covered with sweat and sticky substances: the substrate-independent device like that present in N. tener (and similar to that developed in hymenopterans and many beetles) is more efficient here. Further, the thick and stout segments of the maxillary palp contain muscles for head support. The extant caddisflies usually spread their maxillary palps over the substrate so that the palp touches the substrate primarily with the lowerside of 2nd segment and the apical sensory complex on the 5th one. Again, this position is hardly possible if the surface is covered with honeydew or sticky pollen. Instead, the "walking" appearance of maxillary palps in lower Lepidoptera and N. tener is functional and serves to unglue the insect from any sticky substrate in the case of undesired contact and adhesion of the head.

The caddisflies were numerous and diverse in the Jurassic. Besides Necrotauliidae and Philopotamidae which had survived since Triassic, there were other families (Hydrobiosidae, Rhyacophilidae, Dysoneuridae, Fig. 284, Baissoferidae, Fig. 285, and Vitimotauliidae, Fig. 286) as adult (mainly wing) fossils, and also larval tubular retreats of Phryganeina one of which could represent Hydroptilidae and the others are incertae sedis (see below). These distinct Trichoptera families appeared in the Middle (cases) or Late Jurassic (adults) for the first time. It seems that the two suborders of advanced Trichoptera, the Hydropsychina and Phryganeina, had evolved before the Middle Jurassic. Most Necrotauliidae have been collected in West Europe, much less so in Siberia and Kazakhstan. Other families are known each from only one or just a few specimens from South Kazakhstan (Dysoneuridae), Siberia (Philopotamidae, Dysoneuridae, and Baissoferidae in Transbaikalia, Rhyacophilidae near rasnoyarsk), and Mongolia (Philopotamidae). Caddis cases similar to those of Hydroptilidae (Fig. 287) are found in Transbaikalia indicating probable origin of archaic Hydroptiloidea before the Middle Jurassic. The tubular shelters are Mongolian in origin. This spatial diversity of findings indicates that the possible faunal differences between Europe and Siberia had appeared by the middle of Jurassic.

Adult Rhyacophilidae are supposed to be the most archaic among the extant Trichoptera: their Jurassic origin fits well this classic opinion. Adults are rather uniform in this family (with rare exceptions like

---

Fig. 278 *Continued*

23 – larval prothorax with Gilson gland opening on sternal projection (prosternal horn); male maxillary palp turned upwards and usually with fewer than 5 segments; adult dorsal tentorial arms secondarily developed.

24 – anal loop with curved hind margin; larva without forked lamellae.

25 – larva with abdomen branched lamellae delayed appearing up to 3 instar; larva eruciform, with thick cylindrical abdomen.

26 – larva with antenna close to eye, with lateral abdominal humps on 8 segment.

27 – larval mesonotum with posterolateral border dark, posterolateral angle bearing sclerotised depression (sometimes with gland opening).

28 – female anal opening sclerotised, used for mate orientation and fastening during copulation; male segment 10 elongate, inserting into female anus at copulation (in more advanced taxa forming claspers with preanal appendages for holding female anal margins).

29 – female genital opening with vulvar lobe.

30 – larval pronotum with tergal ridge obsolete dorsally, posterior external ridge obsolete ventrally.

31 – larva with forked lamellae only on 8 segment; Gilson glands reduced on prosternum, hind legs adopting sensory function and turned upwards.

32 – larval pronotum desclerotised caudally; abdominal proleg seta 7 displaced dorsad to claw base; abdominal tergum 9 membranous.

33 – larval abdominal pleura 2–8 with serrate lamellae; adult scutal warts round, fore wing 3rd axillary sclerite x-shaped.

34 – adult scutal and head warts reduced.

35 – fore wing anal loop with 2A base separated by jugal fold.

36 – adult with anterior part of lateral scutal suture bifurcated and provided with additional anterior scutal warts (rudimentary in Calamoceratidae and absent in more advanced leptoceroids).

37 – adult medial mesoscutal suture reduced, mesoscutellum short.

38 – adult mesoscutellum rectangular, mesoscutum with lateral sutures only; fore wing with narrow postcostal field modified for wing coupling by hooks on midlength of hind wing C.

39 – fore wing anal loop very small, wing coupling by serrate fields; SC sometimes with multiple branches (reversion).

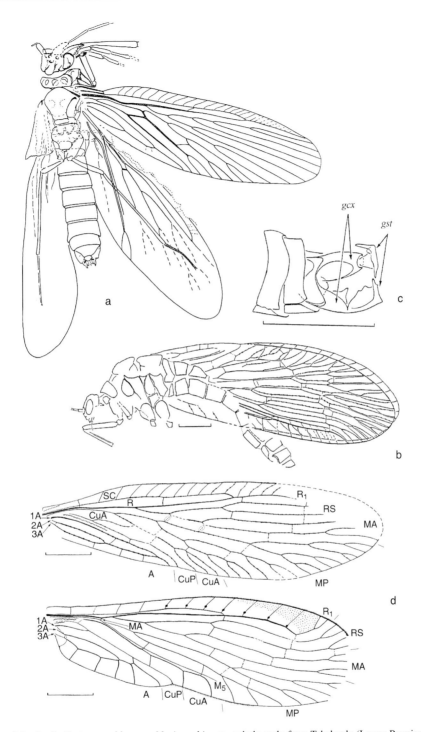

Fig. 279 Archaic caddisflies of the family Protomeropidae: a – *Marimerobius* sp., whole male from Tshekarda (Lower Permian of Urals), based on combined drawing of positive and negative imprints of specimen PIN 1700/1736; b, c – *Permomerope* sp. from the Late Permian of Karaungir in Kazakhstan (b – female specimen PIN 2718/328, c – male terminalia in laterodorsal view, PIN 2718/331; gst – gonostylus, gcx – gonocoxa); d – wings of *Permomerope ramosa* Sukatsheva (holotype); scale bar 1 mm (b–d – from NOVOKSHONOV 1997a)

additional bifurcation of RS and M in *Rhyacophila kaltatica* Levanidova et Schmid), in contrast to their rheophilous larvae which showed a wide range of specialisation.

Hydroptilidae are referred to as microcaddisflies: this is a very speciose and widespread family of small, mostly algophagous Trichoptera. Since they are positioned as predecessors of higher Phryganeina, their occurrence (in the form of a putative larval case) in the Jurassic is not surprising. The most primitive Hydroptilidae are lotic or even hygropetric in habitats, so rarity of their fossils is understandable. There are also poorly preserved small fragments of tiny

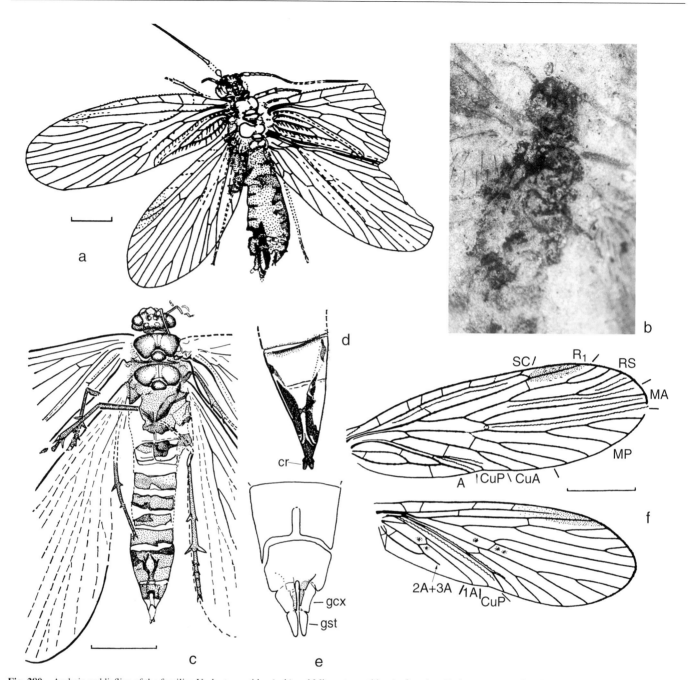

Fig. 280 Archaic caddisflies of the families Uraloptysmatidae (a–b) and Microptysmatidae (c–f): a–b – *Uraloptysma maculata* Ivanov from the Early Permian of Tshekarda in Urals, line drawing (from IVANOV 1992) and photograph (by D.E. Shcherbakov) of head and thorax of holotype from above (fore wing 5.3 mm long); c–e – *Kamopanorpa rotundipennis* Martynov from the Late Permian of Sogana in NW Russia (c – body, PIN 2455/87, d – female terminalia, PIN 3353/1031, e – male terminalia, PIN 3353/1022; from IVANOV 1988); f – wings of *K. pritykinae* Sukatsheva (PIN 2781/261, 265, respectively; from NOVOKSHONOV 1997a).

caddisflies from the Late Jurassic of Karatau which suggest a probable presence of Glossosomatidae.

Jurassic Dysoneuridae are known only from Asia and are represented mostly by isolated fore wings (see Fig. 284). The family was originally assigned to the Phryganeina because of its somewhat reduced venation. However, the Late Jurassic dysoneurid from Mongolia, *Oncovena sharategensis* Ivanov and Novokshonov, a male caddisfly with the body preserved, demonstrates some distinct annulipalpian affinities (Fig. 288) such as oblique head ridges and rounded mesothoracic warts

similar to those of Psychomyioidea. Genitalia are not present, but the 5th segment of the abdomen probably had a triangular sternal "hammer" for vibrational communication which is widespread among the lower caddisflies. The dysoneurid fore wing bulla is actually a flattened projection on the vein, not a groove, and is thus dissimilar to the radial fold of Conoesucidae. It seems plausible to assign all "bullate" wings to males and less specialised dysoneurid wings to females and to include Dysoneuridae tentatively into Hydropsychina along with another male wing, *Bullvenia grandis* Novokshonov and Sukatsheva.

Fig. 281 *Cladochorista multivenosa* Sukatsheva (Cladochoristidae) from the Middle of Late Triassic Madygen Formation in Kirghizia (holotype, photo by D.E. Shcherbakov); wing length 5 mm

Fig. 282 *Prorhyacophila furcata* Sukatsheva (Prorhyacophilidae) from the Middle or Late Triassic of Madygen in Kirghizia (holotype, photo by D.E. Shcherbakov); wing length, as preserved, 7 mm

This fossil shows a bulla at the middle of R similar to that of *O. sharategensis*, and was originally recognised as Hydrobiosidae. Dysoneuridae might be a specialised (with reduced and sometimes distorted fore wing venation) offshoot of Psychomyioidea close to its origin.

The family Baissoferidae, which survived till the Cretaceous, is represented only by fore wings. Baissoferidae (see Fig. 285) resemble the extant family Phryganeidae in their wing venation, especially in the undulated hind margin of the anal loop. These families are probably related as members of the superfamily Phryganeoidea. The cases of "*Folindusia*", some of which might be made by phryganeid larvae, occur as fossils since the Cretaceous. The larval cases of Phryganopsychidae, a primitive extant family of Phryganeoidea, are impossible to appreciate as fossils: they are just heaps of mud, loosely fastened by silk. If Baissoferidae had similar larval cases, the family is not traceable as fossilised larval work. Baissoferidae certainly do not belong to Phryganopsychidae because the wing of the latter has M and Cu connected basally by very strong and elevated cross-vein not present in Baissoferidae.

Vitimotauliidae is mainly a Cretaceous family, though the first representatives occurred in the Late Jurassic. The bulk of the findings originate from Siberia. They do not show any clear affinities to any other families and probably include the roots of several different integripalpian stocks. Certainly this family does not belong to the Phryganeoidea, and we place it tentatively into the leptoceroid assemblage because in Early Cretaceous beds, the wings of Vitimotauliidae occur together with leptoceroid larvae (see below). The family should be revised when more material has been accumulated. Some members of the genus *Multimodus* Sukatsheva are aberrant enough to be placed elsewhere. For example, in a wing from Bon-Tsagan (Barremian or Aptian of Mongolia) the basalmost portion of Cu is greatly widened and the veins constituting the outer edges of anal loop are thick: further development of these features in another specimen from the same locality clearly indicate the wing base of Sericostomatidae.

It can be assumed that only one superfamily of Phryganeina, the Phryganeoidea, was present in the Jurassic; others were not yet evolved though their common ancestor should also have been present.

213

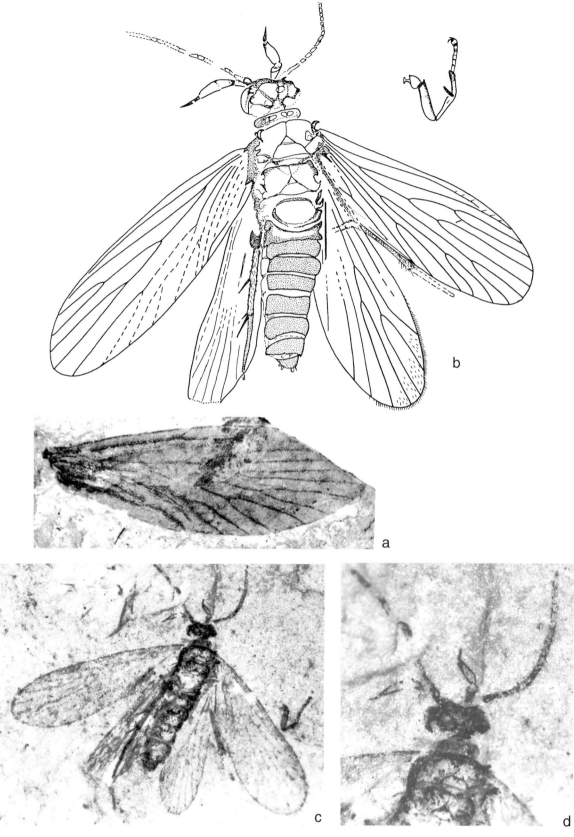

Fig. 283 Key caddisfly family Necrotauliidae: a – *Cretotaulius ultimus* Sukatsheva, wing length 6 mm; b–d – *Necrotaulius tener* Sukatsheva, general view and close up of anterior of body (holotype), wing 3.7 mm long (holotypes from the Late Jurassic of Unda in Siberia, photos by D.E. Shcherbakov)

Fig. 284 *Utanica remissa* Sukatsheva (Dysoneuridae) from the Jurassic or Early Cretaceous Glushkovo Formation in Siberia (holotype, photo by D.E. Shcherbakov); wing length 9 mm

Fig. 285 *Baissoferus latus* Sukatsheva (Baissoferidae) from the Early Cretaceous of Baissa in Siberia (holotype, photo by D.E. Shcherbakov); wing length 12.7 mm

For taphonomic reasons, at least the most abundant groups among the impression fossils could have lived close to, or in, the lentic waters that served as deposition sites. The inference is particularly actual for the caddisfly cases. Inclusions in fossil resins have a different mode of preservation and can be lotic in origin, but such inclusions are not known before the Cretaceous (Chapter 1.4.2.2.1).

The Early Cretaceous caddisflies have all been collected in either Transbaikalia and Mongolia. During the Early Cretaceous the Polycentropodidae and Phryganeidae entered the fossil record. Vitimotauliidae became more abundant in contrast to the Necrotauliidae which decreased considerably in numbers. Philopotamidae, Dysoneuridae and Baissoferidae kept their positions as minor components of fossil assemblages, while the Rhyacophilidae are absent in that section of the fossil record. Also found were members of the Hydrobiosidae, which is now confined to the Southern hemisphere, so its present-time distribution might be refugial. Some poorly preserved remnants could belong to the Psychomyiidae: these are very hairy, tiny and slender wings from the Glushkovo Formation of debatable (Late Jurassic or Early Cretaceous) age in Transbaikalia. Most of the findings were made in numerous fossil lakes of Siberia and Mongolia. Extremely abundant in these were the case-making Phryganeina; sometimes the fossilised cases filled the beds and became the principal component of the bottom sediments. These cases are considered separately below because no bodies were found inside them, except for a few rare instances, making it difficult to attribute than to particular families. A few adult fossils have been found accompanying the cases. Wings of Cretaceous Vitimotauliidae were very hairy and usually with dark pattern made by pigmented pilosity. There were no apomorphic characters in these wings, so the Vitimotauliidae of the Early Cretaceous might have been ancestors of some families of Phryganeina except for the Phryganeoidea which existed at that time. One locality, Baissa in Siberia, remarkable for its richness, provides numerous cases, larval bodies, and a few imaginal remnants classified into Vitimotauliidae, Baissoferidae, and Phryganeidae.

The larvae from Baissa have not yet been described; they are a few body fragments and bodies within cases, and two probably allochthonous caseless bodies. The most numerous were immatures resembling Leptoceridae, with very long hind legs and leg claws, sclerotised pro- and mesonota, and characteristic stick-like metanotal sclerites. They constructed tubular cases of pellets occurring roundabout in the same layer: the cases had the pellets packed into inclined rows with some accuracy as in the extant leptocerid genus *Erotesis* McLachlan. One larger caseless body is highly specialised; this was a filtering predator in streams resembling the extant genus *Oligoplectrum* McLachlan, and so this larva was most probably transported from a nearby stream to the Baissa fossil lake as a drift component. Another caseless larva

Fig. 286 *Multimodus pedunculatus* Sukatsheva (Vitimotauliidae) from the Early Cretaceous of Baissa in Siberia (holotype, photo by D.E. Shcherbakov); wing length 11.3 mm

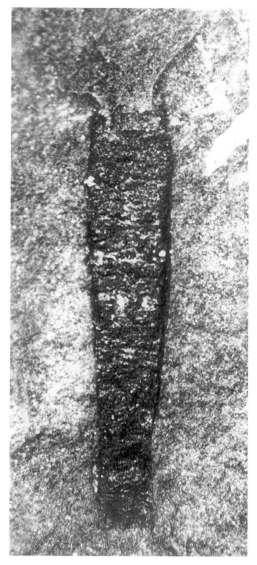

Fig. 287 Case *Scyphindusia hydroptiliformis* Sukatsheva from the Early Jurassic of Bol'shoy Koruy in Transbaikalia, Siberia (holotype, photo by D.E. Shcherbakov), 7 mm long, made of small grains of sand, arranged densely but rather chaotically. Note the attachment bands at the wide (fore) end of the case characteristic of attached pupal retreats

found in Baissa, a very small and young one, probably belongs to the Hydropsychina. The presence of markedly specialised larvae along with the uniform wings of Vitimotauliidae and other lower Phryganeina is puzzling; there are no traces of adult Brachycentridae and Leptoceridae in the Early Cretaceous. The majority of fossil cases in the Early Cretaceous were made of sand grains, not pellets, and hence might belong to other families than the leptocerid larvae found at Baissa.

The Late Cretaceous caddisfly fauna has retained a few older families in its impression fossil assemblages: Philopotamidae in Central Kazakhstan, Phryganeidae in Amur Region, and possibly Vitimotauliidae from imprecisely dated mid-Cretaceous deposits in the west of Khabarovsk Province. The assemblages from fossil resins in Taimyr (North Siberia), Alberta and Manitoba (Canada) and Tennessee (USA) are richer and more diverse. Resin deposits include three families (Leptoceridae, Polycentropodidae and Philopotamidae) persisting from earlier times and 7 more first entering the fossil record. These are either the extinct Late Cretaceous endemic Electralbertidae: Psychomyioidea, and Taimyrelectronidae (Sericostomatoidea or, at Fig. 278, Limnephiloidea), or the extant families Hydrobiosidae, Sericostomatidae, Odontoceridae, and Calamoceratidae. As a result the Late Cretaceous assemblages have a rather modern appearance and, due to the important role of fossil inclusions, better represent the lotic component of the fauna. Thus the wider set of families appeared because of the new type of preservation, i.e. resin inclusions, and should not be

directly compared with the earlier times. Meantime the lake deposits grew poor in caddisflies including fossilised cases (see below). One can assume that at the family level the recent trichopteran fauna was formed mostly in the Late Cretaceous time.

Cainozoic fossils are numerous and diverse having been collected in deposits of various age and provenance. The Baltic amber assemblage described by ULMER (1912) is of primary importance, for it covers all the known Tertiary families except for the Limnephilidae. It is obvious that information on the Eocene Trichoptera would be very defective without the amber fossils so that the faunas in the pre-resin times are surely very incompletely preserved. The amber faunas are very rich and clearly of lotic origin. A few species of Polycentropodidae make the bulk of the remnants (50–80% of specimens). Psychomyiidae and Lepidostomatidae are less abundant; other families are just minor additions. This faunistic composition is typical of large stressed water

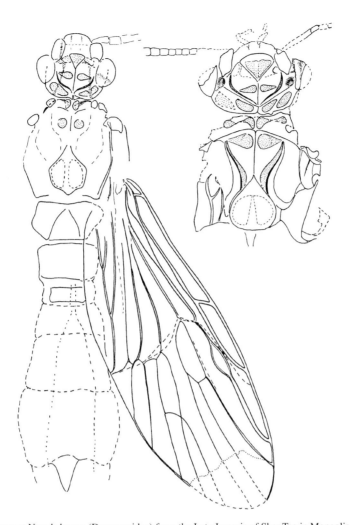

Fig. 288 *Onkovena sharategensis* Ivanov et Novokshonov (Dysoneuridae) from the Late Jurassic of Shar Teg in Mongolia (holotype), part and counterpart

bodies, for example, the large warm rivers of the temperate zone, though in the Quaternary the species dominance changed: the Polycentropodidae lost their dominance in most localities except for some warm river rapids in Europe and West Siberia with an abundance of *Neureclipsis bimaculata* L. and *Polycentropus* spp. Special feature of the European amber fauna, the absence of Limnephilidae, could be explained by the late origin of this family. Diversity at the family level shows that the major lines of caddisfly family-level evolution were well delimited before the Tertiary. Since the impression fossils originated from lakes, and the amber Trichoptera are from the river and its tributaries, the direct comparison of the amber and sediment fossils is not correct. The family Limnephilidae first appears in the Oligocene (or latest Eocene) of Western North America (Ruby River in Montana) where the most primitive limnephilids persisted till now. The present distribution of Limnephilidae shows traces of the past movement from east to west in Eurasia, with more primitive genera occurring in the north Pacific area and the younger ones forming secondary nodes of diversity in more western and southern regions.

The Neogene Trichoptera faunas adopt a contemporary structure, but this process is still obscure. There are various Eurasian Miocene deposits known to contain diverse fossil caddisflies. Since these are generally represented by detached wings which are of low importance for the species-level classification of the extant caddisflies, the possibility of their adequate treatment is limited. For example, the caddisflies in the Miocene fauna of Stavropol (Vishnevaya Balka) in the Caucasus piedmonts consists of numerous well preserved uniform wings of Limnephilidae (Limnephilini). Formally they were described as *Miopsyche kaspievi* O. Mart., though the genus *Miopsyche* Carpenter is now supposed to be a junior synonym of *Limnephilus* Leach (F. CARPENTER 1992a). The low diversity indicates variable conditions of the Miocene environments at that place; perhaps these were warm temporary pools. If so, some Limnephilidae could have become warm-adapted as early as the Miocene.

The most recent major events in the history of caddisflies are the Pleistocene glacial movements. The analysis by MALICKY (1988) shows numerous traces of areal shifts during the Ice Age. The present-day composition of the temperate faunas of both hemispheres should be very young, and the references to the Tertiary climate and palaeogeography made in some studies are not accurate because the early distributions are distorted by the subsequent glaciation and the climatic changes. Since most of the tropical areas experienced drying during the northern glaciations, faunistic changes should have been global. In contrast, evolutionary changes should have been small because little time has passed as yet. However, some Apataniinae

Fig. 290 Case of *Folindusia* (*s. str.*) *undae* Vialov et Sukatsheva from the Late Jurassic or Early Cretaceous Glushkovo Formation in Siberia, made of plant fragments (holotype, photo by A.P. Rasnitsyn), 13.5 mm long as preserved

Fig. 289 Case of *Terrindusia angustata* Vialov & Sukatsheva from the Early Cretaceous of Romanovka (near Baissa) in Transbaikalia, made of diverse mineral particles without strict arrangement (holotype, photo by A.P. Rasnitsyn); length 16 mm

endemic to Lake Baikal appear to be of glacial origin (IVANOV & MENSHUTKINA 1996).

The above short history of caddisflies is based upon their wing and body fossils. However, there is an additional source of evidence, fossil caddis cases, which are the results of the construction behaviours of their larvae. The history of case construction became discussible after an artificial system of the fossil cases had been proposed (VIALOV & SUKATSHEVA 1976). The system is based on the type of building material being used to define genera, and on details of the construction mode used as the species-level characters.

The very first larval constructions occur in the Middle Jurassic of Transbaikalia (East Siberia, SUKATSHEVA 1985b) and Mongolia (SUKATSHEVA 1994). The Transbaikalian findings are remarkable in that they show the only available indication to Mesozoic Hydroptilidae. The small bottle-like case of the Jurassic Hydroptilidae was made of silk with additions of small sand grains (see Fig. 287). The first tubular cases of higher Phryganeina were found in Bakhar locality (Gobi Altai; Togo-Huduk and Ortsag Beds). There were small shallow lakes there, that were rich enough in organic material to produce coal layers within the dominant argillites. No adults were found with these cases, though the younger Jurassic case-bearing layers in Mongolia contain wings of *Multimodus* Suk. (Vitimotauliidae). The earliest

tubular cases are very rough in their construction. The material is not sorted, there are sand grains of different shapes and size, various detritus particles, plant fragments, and some other materials like the shells of Ostracoda. Evidently the larvae selected the materials imprecisely and used the particles of most suitable size to construct slightly conical short tubes. No treatment of material was made by a larva: the material was placed as it was found. These early tubes are classified as *Terrindusia* Vialov (Fig. 289), *Folindusia* Berry (Fig. 290), and *Ostracindusia* Vialov (Fig. 291) depending on the dominant material. However, the selection of material by larva being poor, attribution to these genera might reflect differences in the material abundance, in the age of larvae, or else in the personal preferences of larva, and not necessarily differences in the taxonomic position of the respective caddisworms. Thus it is difficult to decide which of the more than a dozen 'indusi- species' described from the Middle Jurassic actually represent independent zoological species. Most of the cases were empty, though some show 'shadows' within which could indicate presence of larva inside, though the larval morphology is not traceable. Some Middle Jurassic cases had an oblique end, suggesting that this might be a pupal case shaped so as to be fastened to a solid substrate. Among the extant caddisflies an oblique larval case entrance occurs in some more primitive Apataniidae (e.g. Baikalinini). It is unlikely, however, that this family is the most ancient in the Phryganeina, though the similarity of larval cases to the Jurassic ones could indicate its primitiveness.

Fossil cases became more abundant in the Late Jurassic and were particularly numerous in the Early Cretaceous. An interesting feature of this portion of the record is that it directly reflects evolution of the

Fig. 291 Case of *Ostracindusia invisa* Sukatsheva from the Late Jurassic or Early Cretaceous of Onokhovo deposits (same region and probably age as in case of Glushkovo Formation in Siberia), made of shells of ostracods (holotype, photo by D.E. Shcherbakov); length 17 mm

Fig. 292 Case of *Pelindusia conspecta* Vialov & Sukatsheva from the Early Cretaceous of Romanovka (near Baissa) in Transbaikalia, made of shell fragments of bivalve molluscs (holotype, photo by D.E. Shcherbakov); length 15 mm

building abilities of caddisfly larvae (SUKATSHEVA 1980, 1982). The fossil record makes apparent a gradual progression in the construction behaviour of caddis larvae through the Late Jurassic and Early Cretaceous, as reflected in the case structure. The material selection was growing more and more advanced, resulting in cases built of pure sand, or of plant particles, or of conifer needles, or of ostracod shells, or of fragments of bivalve and/or conchostracan shells, and so on (Figs. 292–295). One more important innovation in larval behaviour was the preparation of particles by larva before they were incorporated into the case. The most characteristic is gnawing of the fragments at the sides or cutting the pieces from solid detritus. This preparation provides standard particles for the highly automatic subsequent positioning of the fragment in the case wall: the larvae invested energy in the preliminary treatment to save the time in selection and adjustments of particles in the wall. The progress in the case construction is seen also in the transformation of the case shape. The cross-section sometimes became quadrate or very smoothly and accurately rounded; the packing of particles grew tighter or peculiar in sequence: the particles adopt the spiral disposition or the larger grains were attached to the sides of

the tube. Of importance is that the advanced cases co-occurred with the most unsophisticated case types rather than replacing them, though their relative abundance in the sediments might change.

Another important feature is that the above progression proceeded rapidly and gradually before the mid-Cretaceous (but see below), and was not accompanied with similar evolution of the adults (SUKATSHEVA 1991). In contrast to the rapid progress in the case performance, adult morphology of Vitimotauliidae, most probably the main case builders, remained virtually unchanged since the mid-Cretaceous. However, the almost explosive taxonomic diversification had started at that time (see above), in contrast to the case construction whose all main types have been elaborated up to the mid-Cretaceous. Another aspect of that process is changing diversity of the case taxa. During the Early Cretaceous the cases were abundant and diverse, being represented often by 10–15, sometimes even 20 species in each better explored local assemblage. Unlike that, the Late Cretaceous assemblages feature 1–2, exceptionally up to 5, different species of caddis cases whose structure becomes far less diverse albeit well advanced. These were tubes built either from spirally arranged, uniform, rather large plant fragments, or from equally uniform but small plant fragments that tended to be arranged in transverse rows. The third type represented by the smooth tubes of mineral particles is found only in deposits showing the probable influence of running water. The gradual transition was observed during the Albian (latest Early Cretaceous) time and might reflect the changes in the lake environments caused by the eutrophication from the angiosperm detritus. Cases constructed of plant material might belong either to Phryganeidae or to Leptoceridae; the rheophilous sand cases could hide a variety of families.

During the earlier Tertiary (Palaeocene and Eocene) the cases were only a little more diverse than in the end of Mesozoic, yet retaining the Late Cretaceous types of the case design with few additional ones (e.g. predominantly or entirely secretory, silken tubes). It should be noted that the amber adults have little or no relationship to these lake-dwelling caddis larvae. Only since the Oligocene have the cases became more diverse, with most construction types characteristic of the Early Cretaceous ("winged", frame-like, made from small gastropod shells, etc.) being re-established. The exceptions are those made of the plant seeds and shells of the conchostracan crustaceans; they disappeared completely. These processes were much governed by the ecological evolution of the lake biotas: for more details see Chapter 3.3.

Most of the above inferences have resulted from application of the semi-quantitative analysis of case construction proposed by SUKATSHEVA (1982). She used a system of conventional numbers to assess the degree of 'perfection' of case construction, and used them to calculate the mean number for a particular case sample. It was found that these mean conventional numbers increased rather smoothly from about 10–20 in the early Neocomian (earliest Cretaceous) to 300 in the mid-Cretaceous (Albian and Cenomanian), with hardly any visible further growth. This makes it possible to use the fossil cases as a tool for the correlation and sequencing of the Early Cretaceous non-marine deposits (SUKATSHEVA 1982–1999, ZHERIKHIN & SUKATSHEVA 1990).

Recent palaeontological and stratigraphical approaches and observations have made it necessary to modify the above picture a little (Fig. 296). Firstly, there is the observation that advanced case types can co-exist with earlier and more primitive ones rather than replace them, implying that the highest level of the case performance achieved by a particular geological time level might be particularly informative concerning the behavioural evolution of caddisworms (ZHERIKHIN & SUKATSHEVA 1990). Secondly, it has gradually become more certain that the deposits that have yielded the advanced *Folindusia undae* Vialov & Sukatsheva cases (see Fig. 290) and its close analogue *F. savinensis* Sukatsheva, are from near the Jurassic/Cretaceous boundary rather than from late in the Early Cretaceous. This is particularly apparent in the case of the Khilok Formation in Transbaikalia (RUDNEV & LYAMINA 1990) and Ulugey Formation in Mongolia (PONOMARENKO 1990). This results in the appearance of an unexpected maximum in the left part of both curves (Fig. 296) which can be interpreted as an early and premature (unbalanced, or inadaptive in the sense of RASNITSYN 1987) attempt to prepare a constructively advanced case. The attempt had little effect on the general path of the caddisworm behavioural evolution which continued its gradual development towards the highest constructive ability during all the Late Cretaceous, as it was described above.

2.2.1.3.4.3. ORDER LEPIDOPTERA Linné, 1758. The BUTTERFLIES and MOTHS (=Papilionida Laicharting, 1781)

M.V. KOZLOV, V.D. IVANOV AND A.P. RASNITSYN

(a) Introductory remarks. The second leading numerically (ca. 150,000 living species, KRISTENSEN & SKAILSKI 1999), one of the most important economically and the most attractive aesthetically insect group with comparatively poor fossil records. The chapter is based mostly on morphological data from KRISTENSEN (1984), KRISTENSEN & SKAILSKI (1999) and palaeontological evidence summarised by KOZLOV (1988, 1989).

(b) Definition. Small to large (3–300 mm in wing span) insects with 4 membranous wings covered by scales; sometimes brachypterous or

Fig. 293 Case of *Frugindusia karkenia* Sukatsheva from the Early Cretaceous of Shin-Khuduk in Mongolia, made from seeds of the ginkgoalean plant *Karkenia* sp. (holotype, photo by D.E. Shcherbakov); length 28 mm

even wingless (females only). Head hypognathous, with large eyes and at most 2 ocelli. Mandibles non-functional, reduced or lost except in archaic subfamilies Micropterigoidea, Heterobathmioidea and Agathiphagoidea. In most of more advanced groups maxillary lacinia modified into coilable sucking proboscis. Wings plesiomorphically homonomous, uncoupled during flight and venationally alike plesiomorphic caddisfly (cf. Chapter 2.2.1.3.4.2b); in most of extant taxa heteronomous, coupled and venationally modified. Male genitalia consist of 9th segment forming basal ring (annulus) often divided into articulated dorsal (tegumen) and ventral (vinculum) parts, segment X represented by dorsal lobes (often united into the single sclerotised structure – uncus), ventral articulated lobes (valvae, i.e. undivided gonopode), and aedeagus. Female genitalia lacking primary ovipositor, with one or, in Ditrysia, two orifices (copulatory one on VIII, ovipository opening on IX abdominal segment). Larva (caterpillar) mostly eruciform; in most archaic suborders with prognathous (sometimes retractable) head, thoracic legs weak or lost, typical caterpillar prolegs absent; true caterpillar (with head hypognathous and nonretractable, thoracic legs well developed, apically hooked prolegs developed on abdominal segments IV–VII and X or fewer) appeared

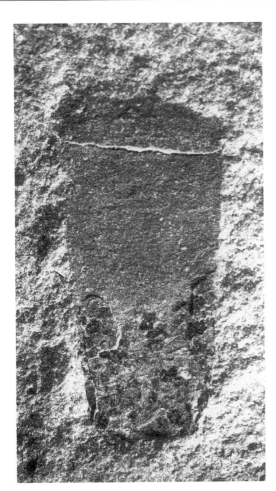

Fig. 294 Silk case of *Secrindusia pacifica* Vialov & Sukatsheva from the Palaeocene of Dal'naya River, Maritime Province in Russian Far East (holotype, photo by D.E. Shcherbakov); length 13 mm

Fig. 295 Case of *Folindusia (Spirindusia) kemaensis* Vialov & Sukatsheva from the Oligocene of Kema, Maritime Province in Russian Far East, made of plant fragments uniformly bitten off and arranged in a spiral (holotype, photo by D.E. Shcherbakov); length 42 mm

probably at the level of Mnesarchaeoidea. Larval antenna 3-segmented. Larval body surface with scarce primary setae, in some advanced taxa also with dense cover of secondary setae. Salivary glands modified to produce silk. Pupae decticous or adecticous, often in silk cocoon; in lower groups (Heterobathmiidae, Eriocraniidae etc.) often with hypertrophied mandibles; pupa plesiomorphically with free appendages but usually obtectate. Some mandibulate adults fed on the content of pollen crushed by mandibles (no hard tissue found in gut); in other groups proboscis is used to suck nectar or other organic liquids (tears, sweat, blood etc.), or water; larva phytophagous or detritophagous, very rarely obligatory zoophagous.

(c) Synapomorphies (abridged from KRISTENSEN 1984, KRISTENSEN & SKAILSKI 1999). Median ocellus lost. Maxillary palp with flexion points at segmental articulation 1/2 and 3/4, with longest segment 4. Labial palp with pit bearing submerged group of specialised sensillae. Laterocervical sclerite with hair plate close to anterior apex. Fore tibia with single apical spur and epiphysis (modified preapical spur). Wing covered densely with broad, flat, widely overlapping scales (secondarily, scales can be narrow and scale cover partially lost). Larva with maxillary palp less than 5-segmented. Possibly (unless acquired by mandibulate moth ancestor after first divergence event): pupal functional mandible retained by adult (apomorphy missed by above authors).

(d) Range. Since Early Jurassic until present, worldwide.

(e) Systematics and phylogeny of the group are still debatable, particularly concerning the suborder Papilionina (HEPPNER 1998, KRISTENSEN 1999b). The version by M.V. Kozlov shown here is based on the wide scope of available information, with special attention being paid to reviews by KUZNETZOV & STEKOLNIKOV (1986) and MINET (1991), and with the superfamily arrangement after CARTER & KRISTENSEN (1999).

At the highest taxonomic level, at least 25 subordinal rank names have been proposed to date (for history see VIETTE 1979, HEPPNER 1998). CARTER & KRISTENSEN (1999) have abandoned all categories above superfamily rank because some of them are paraphyletic. We however consider these names useful for communication and organisation of the higher level lepidopteran diversity. The most widely accepted suborders are currently Micropterigina (Zeugloptera, Laciniata), Agathiphagina (Aglossata), Heterobathmiina, Eriocraniina (Dacnonypha), Neopseustina, Hepialina (Exoporia), and Papilionina (Ditrysia). Other names in use ranked above the superfamily level are the Monotrysia (a paraphyletic taxon covering all suborders other than

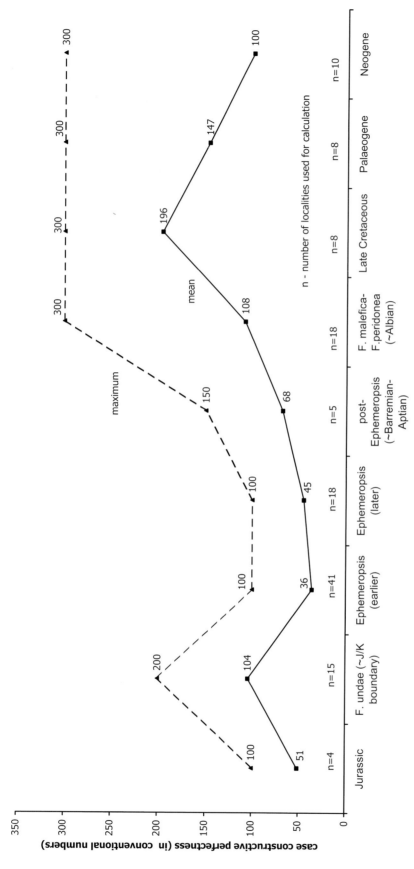

Fig. 296 Perfection of caddis case construction (after SUKATSHEVA 1982) versus time (original by A.P. Rasnitsyn, based on data by A.P. Rasnitsyn and SUKATSHEVA 1982–1999 and unpublished). Upper curve shows the perfection index the best formed cases in the respective assemblage, the lower curve shows the mean of the means calculated for local case samples included into the same assemblage. The assemblages are supposed to succeed one another in time, but because of the often imperfect stratigraphy, some time overlap of neighbouring assemblages cannot be completely excluded

The following are technical details explaining the compositions of the assemblages recognised here. The assemblage called 'Jurassic' includes cases from four localities (Bakhar, Kempendyay, Khoutiyn-Khotgor and Shar Teg, see Chapter 4.1 for details). The next youngest assemblage embraces cases from deposits that either yielded *Folindusia undae* cases (or something very similar to it), or are comparatively well correlated with them (Byankino, Glushkovo, Godymboy, Tergen' and Ukurey Formations in Transbaikalia, Central Siberia, and Ulughey Formation in Mongolia). All of them probably are from somewhere near the Jurassic/Cretaceous boundary. The Earlier *Ephemeropsis* Assemblage is characterised by the presence of giant mayflies, *Ephemeropsis* spp.. combined with the absence of *F. undae*. Also included in this category are cases from deposits that may lack *Ephemeropsis* but are considered to take the stratigraphic position intermediate between those of the *F. undae* and Later *Ephemeropsis* assemblages for various palaeontological or geological reasons. These deposits are, in Transbaikalia, the Bukatshatsha, Daya, Doronino, Ghidari, Kizhinga, Leskovo, Mangut, Murtoy, Narasun, Turga, Ubukun, Ust'-Kara, Utan and Zaza Formations/Beds, and in Mongolia – Dorogot, Gurvan-Eren, Manlay, Mogotuin and Tsagan-Tsab Formations in Transbaikalia, Upper Batylykh Subformation in Yakutia, and Anda-Khuduk, Buylasun and Shin-Khuduk Formations in Mongolia. These deposits equally often produce *Ephemeropsis* mayflies, but they are usually considered somewhat younger than those in the previous assemblage. The Post-*Ephemeropsis* Assemblage consists of cases from the Bainzurkhe Formation in Transbaikalia and the Khurilt Beds in Mongolia. The latter deposits are known to overlay deposits with *Ephemeropsis*, while the former is situated in the upper part of the Transbaikalian Cretaceous section below the Kuti Formation which is possibly of Albian age. The richer assemblage is the *F. peridonea/malefica* one which is defined primarily by the presence of these highly advanced case-species which are collected in reliably dated Albian (latest Early Cretaceous) or Albian-Cenomanian deposits of the Amka and Emanra Formations and, albeit rarely, in the Ottokh Member of the Utshulikan Formation of the Early Albian or Late Aptian age, all in the Okhotsk Sea Region. Other supposedly Albian or Albian-Cenomanian samples are included there as well, viz., the Altshan, Frentsevka and Naliamiaungani Formations in Russian Far East, Ognevka Formation in North Siberia. The same is true for the Kuti and Kholboldzhin Formations in Transbaikalia, also of supposedly Albian age. The Late Cretaceous and Tertiary assemblages in question rely on samples which fit rather unambiguously in the relevant time scale, and so they need no additional comment except that several samples that have been taken to be of Neogene age are in fact recorded as borderline (Oligo-Miocene). This is of lesser importance, however, for transferring them to the Palaeogene would leave the upper curve unaffected, while the mean convention numbers were not modified much (a decrease from 147 to 131 in Palaeogene and an increase from 100 to 102 in Neogene).

Papilionina); the Heteroneura (Nepticulina + Adelina + Papilionina) as opposed to the other moths called the Homoneura; and the Glossata (Haustellata) is referred to all lepidopterans other than Aglossata (Laciniata, the clade above the node 5, Fig. 297). Macro- and Microlepidoptera are less precise terms, the former usually implying taxa from Micropterigoidea up to Copromorphoroidea, and the latter those from Pyraloidea to Bombicoidea on the cladogram. Redefinition of Macrolepidoptera by MINET (1991) on the base of synapomorphy 33 (Fig. 297) is debatable. Further, the taxonomic status of some heteroneuran monotrysian families belonging to the Nepticuloidea, Tischerioidea and Incurvarioidea is still unclear (cf. KRISTENSEN & SKAILSKI 1999). The extinct superfamily Eolepidopterigoidea (including Undopterigidae) can be set off from all other lepidopterans. The relationships between the homoneuran moths are revised in details by KRISTENSEN (1984) and KRISTENSEN & SKAILSKI (1999).

The present cladogram (Fig. 297) shows only taxa that are represented in the fossil record (one extinct and 24 of the 46 extant superfamilies accepted by CARTER & KRISTENSEN 1999). It differs from that of Kristensen particularly concerning those moths with functioning mandibles. While the monophyly of the stem Glossata seems to be indisputable, the non-glossatan superfamilies (Micropterigoidea, Agathiphagoidea, and Heterobathmioidea) are usually considered as a paraphyletic group. Indeed their functional adult mandibles are probably inherited from generalised caddisflies (stem Amphiesmenoptera, Chapter 2.2.1.3.4.2b,f). However, the adults of Micropterigidae and Heterobathmiidae are apomorphic in using their mandibles not for biting but for crushing pollen. The proposed hypothesis (Fig. 297) is based on this character as a synapomorphy for the non-glossatan moths. The mandibles in Agathiphagidae are large and secondarily non-functional. The hypothesised ancestor had unmodified biting mandibles and no filtering structures in the mouth chamber. This ancestor supposedly fed on liquid matter because both sister-orders, Trichoptera and Lepidoptera, have only liquids in their diets. The ancestor gave two independent phyla, the pollen-crushers that might acquire modified, stout asymmetric mandibles and filtering buccal setae in the mouth, and the glossatans which instead drank food with their hypertrophied glossae. Alternatively, there is the currently more popular hypothesis which implies that the pollen crushing mandibles and associated filtering structures were gained by the ancestral moth and then lost independently in *Agathiphaga* Dumbleton and in the common ancestor of the Glossata (node 5 in our cladogram) (KRISTENSEN & SKALSKI 1999, and references therein). Another possible synapomorphy of non-glossatan moths is the loss of long-range pheromone communication, even in groups retaining abdominal sternum V glands (KOZLOV & ZVEREVA 1999).

The system of the suborder Papilionina covering about 99% of recent moths and butterflies (KRISTENSEN & SKAILSKI 1999) is also a subject of broad discussion. It includes at least 70–120 families grouped into 18–33 superfamilies (KUZNETZOV & STEKOLNIKOV 1986, MINET 1986, 1991, HEPPNER 1998, CARTER & KRISTENSEN 1999), subordinated tentatively to 7 infraorders (KOZLOV 1988) not supported by sound synapomorphies.

The recent discussions on ecological aspects of the lepidopteran evolution concentrate on life habit and feeding preferences of the larval stage, with the maximum attention paid to the basal lepidopteran lineages (SINEV 1990, KRISTENSEN 1997), though the presence of their putative plant hosts in the Jurassic and Early Cretaceous is highly questionable and the appropriate fossil material on immatures is lacking. In contrast to these authors, we consider the dense scale cover of both wings and body, evolved on the basis of hair cover of ancestral forms, as the key innovation of the Lepidoptera. This innovation did not demand much genetic change and may have appeared even as a single mutation. At early stages of the lepidopteran evolution, the newly evolved scale cover had relatively low adaptive value, possibly related to the life in habitats that were rich in sticky substances, e.g. the rhynchotan honey dew or plant fructifications bearing sticky pollen and/or nectar (for more details see Chapter 2.2.1.4.4.2f). The scales with metallic heat-reflecting colours in small archaic moths made the additional coverage thus preventing the overheating at daytime and protecting them against the water loss in sunshine. The scale pigmentation stimulated the evolution of bright coloration in certain groups useful in adult protection and signalling. Furthermore, the diurnal imaginal life habit (retained from the ancestor and still kept by many archaic families) might prevent diversification of ancient Lepidoptera because of both competition with other diurnal groups and impact of the diurnal generalist predators.

The sucking proboscis (presumably evolved to maintain the water balance) appeared to be the important preadaptation to exploit the sugar- and nitrogen-rich nectar of angiosperm plants. This energy-rich imaginal food less available for many other insects allowed the development of long rapid flight and stimulated expansion of moths (which was already 'allowed' by larval adaptations) to the environments with limited water resources: exophagous larval feeding on angiosperm plants is known in non-ditrysian moths and is frequent among ditrysian 'micros'. The shift towards the nocturnal life of the imago later on revealed the thermoregulatory function of the scale cover in larger strong flyers. Relatively rapid decrease of global temperatures in Oligocene epoch might have stimulated both diversification and expansion of 'macrolepidopterans', because efficiency of the active thermoregulation increased with the decline in surface/mass ratio. The improvements in pheromone communication (shift towards the long-chain pheromones emitted in very low quantities and growth of the male antennal sensitivity) could be responsible for great species richness of the Ditrysian moths where the species recognition depends mainly on olfactory stimuli, and so the population density can be very low. The inter- and intraspecies competition at low densities are so small, that then allows several species to occupy nearly the same ecological niche; hence species diversity in higher lepidopteran families grows regardless the environment complexity.

(f) History. It is commonly accepted that the moths and advanced (or true) caddisflies (Hydropsychina + Phryganeina) represent the sister groups jointly rooted within Protomeropina (cladistically, the stem Amphiesmenoptera). The majority of the lepidopteran apomorphies concern delicate structures which are difficult (or impossible) to examine in fossils. This naturally hinders identification of the point where moths have diverged from the caddisflies but does not make this impossible. It was found that the oldest among the most important moth synapomorphies appeared already within Necrotauliidae which is supposed to be the stem (though most probably also a catch-all) family of the primordial advanced caddisflies. Noteworthy these synapomorphies appeared one after one, with the characteristically modified maxillary palps and fore tibia first discovered in a scaleless fossil (Chapter 2.2.1.4.4.2f). The result is that we cannot distinguish the earliest moths and caddisflies on the basis of wing morphology.

The Triassic records of the order are erroneous, the fossils concerned being shown to belong to other groups: *Mesochorista proavitella* Tillyard (=*Eoses triassica* Tindale) to Permochoristidae (Panorpida), and *Protorhyacophila* (=*Eocorona*) *iani* (Tindale) to

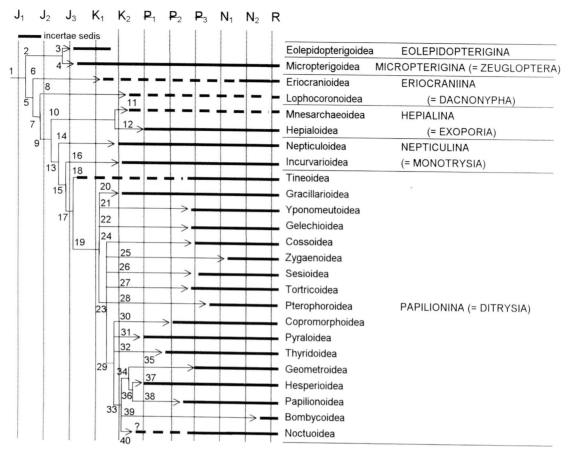

Fig. 297 Phylogeny and system of the butterflies and moths, order Lepidoptera (=Papilionida)

Time periods are abbreviated as follows: J_1, J_2, J_3–Early (Lower), Middle and Late (Upper) Jurassic, K_1, K_2–Early (Lower) and Late (Upper) Cretaceous, P_1– Palaeocene, P_2–Eocene, P_3–Oligocene, N_1–Miocene, N_2–Pliocene, R–present time (Holocene). Righthand column gives names of superfamilies (after CARTER & KRISTENSEN 1999, only including those represented in fossil records; basic arrangement of Glossata [node 5] after KRISTENSEN & SKALSKI 1999). Arrows show ancestry, thick bars are known durations of taxa (dashed ones–supposed/ debatable duration), figures refer to synapomorphies of the subtended clades, as follows:

1 – median ocellus lost; maxillary palp with flexion points at segmental articulation 1/2 and 3/4, with longest segment 4; labial palp apically with recessed group of chemoreceptors; laterocervical sclerite with hair plate close to anterior apex; fore tibia with single subapical spur modified into epiphysis; wing covered densely with broad scales.

2 – functional mandibles re-gained by adult; long-distance pheromones lost.

3 – pronotum sclerotised throughout, at least 0.25 × length of mesonotum.

4 – branched basiconic sensilla present on antenna; further 10 synapomorphies listed by KRISTENSEN (1984).

5 – *clade Glossata*: maxillary galeae forming proboscis; further 5 synapomorphies listed by KRISTENSEN (1984).

6 – tibial spur formula 0-1-4; valvae greatly shortened; phallus with accessory ventral branch; female with complex "vaginal sclerite" in posterior part of genital chamber; ventral diaphragm musculature exceedingly reduced; larval head with dorsal ecdysial line running laterad from antennal base; a single pair of stemmata.

7 – *clade Coelolepida*: scales hollow, with their upper and lower lamellae separated by space.

8 – cranium with weakly developed dorsal condyles at antennal base; epiphysis absent; RS_4 postapical; valve deeply cleft, lower lobe with stout apical sensillum and 2–3 long basal setae; further possible autapomorphies listed by KRISTENSEN (1999).

9 – *clade Neolepidoptera*: pupa adecticous, obtecta; adult mandibles have at most a vestigial musculature; crochet-bearing prolegs present on larval abdominal segments III–VI and X; free profurcal arm displaced laterally; further possible autapomorphies listed by KRISTENSEN (1984, 1999).

10 – female genitalia exoporian, without colleterial glands; further 3 synapomorphies listed by KRISTENSEN (1984).

11 – wings narrow, lanceolate; RS_{1+2} in both wing pairs represented by one vein only; further possible autapomorphies listed by KRISTENSEN (1999).

12 – proboscis reduced, at most as long as the head capsule; further possible autapomorphies listed by KRISTENSEN (1999).

13 – *clade Heteroneura*: heteroneuran wing venation; further 6 synapomorphies listed by KRISTENSEN (1984).

14 – ovipositor greatly shortened, nonpiercing; separate ovipore opening on A_8–A_9.

15 – female pheromone glands at the terminal abdominal segments; female pheromones with long-chain components (LÖFSTEDT & KOZLOV 1997).

16 – ovipositor extensible, piercing; vagina with "guy-wires"; male genitalia with arrow-shaped juxta.

17 – female genitalia of ditrysian type; abdominal sternum II with 2 anterior apodemes; larval head with minute seta G_2 replaced by large O_3.

18 – slender pair of ventral pseudapophyses present within abdominal segment 10 in most females; telescopic "ovipositor" unusually elongated.

19 – maxillary palp short, at most 4-segmented, lacking sharp bend between segments 3 and 4; labial palp without projecting lateral bristles.

20 – early instars mining or boring into plant host; pupa partially extruding from cocoon before eclosion.

21 – male abdominal segment VIII with pleural lobes.

Prorhyacophilidae (Phryganeida) (RIEK 1955, SUKATSHEVA 1982, WILLMANN 1984). Attribution of a mine on a leaf of the peltasperm pteridosperm *Pachypteris* from the Late Jurassic or Early Cretaceous of Queensland, Australia, to Nepticulidae (ROZEFELDS 1988a) is problematic (KRISTENSEN & SKAILSKI 1999). Thus the oldest known moth fossil might be *Archaeolepis mane* Whalley described from the Early Jurassic of Dorset, UK, on the basis of the single, probably hind wing fossil (Fig. 298). The next oldest are a few genera previously described within Necrotauliidae (Trichoptera) by A. Handlirsch and recently transferred to Lepidoptera by ANSORGE (2001). All these small insects were shown to possess scales on the wings and their fore wings have 3 M branches with otherwise primitive venation patterns. The Handlirsch material originates from the Early Toarcian "Green Series" of Germany and are undoubtedly the most ancient Lepidoptera known today. They are represented by the genera *Pseudorthophlebia* Handlirsch, *Nannotrichopteron* Handlirsch, *Pararchitaulius* Handlirsch, *Parataulius* Handlirsch, *Archiptilia* Handlirsch, *Liadoptilia* Handlirsch, *Metarchitaulius* Handlirsch, *Epididontus* Handlirsch. The taxonomic position of all these is however uncertain; poor preservation and lack of suggestive characters make them Lepidoptera *incertae sedis*.

Late Jurassic in age are at least 5 fossils found in South Kazakhstan (Karatau Range) and Transbaikalia (Uda River). They belong to Eolepidopterigidae (2 species, Fig. 299), Micropterigidae (2 species: *Aulipterix mirabilis* Kozlov and possibly *Karataunia lapidaria* Kozlov which was originally assigned to the clade Ditrysia, but evidence for this assignment have proved spurious upon re-examination), and a haustellate moth (*Protolepis cuprealata* Kozlov) possibly of tineoid affinity. There are also undescribed homoneuran fossils (Fig. 300). Further species of Eolepidopterigidae (2 species), Micropterigidae, and Undopterigidae (Figs. 301, 302), come from the deposits of debatable Late Jurassic or Early Cretaceous age in Transbaikalia (Glushkovo Formation) and Mongolia (Khotont). The undoubtedly Early Cretaceous deposits of Baissa in Transbaikalia, Santana in Brazil and of the Lebanese amber have yielded five species of Micropterigidae, two of Undopterigidae and one of Eolepidopterigidae (KOZLOV 1989, MARTINS-NETO & VULCANO 1989c, MARTINS-NETO 1999a). Thus prior to the mid-Cretaceous biocoenotic crisis, the moth fauna consisted almost exclusively of members of three lowermost mandibulate families, two of which failed to cross the mid-Cretaceous boundary, while the third one (Micropterigidae) survived up to now, albeit as a small relict group. However, all of these fossils are considered *incertae sedis* by CARTER & KRISTENSEN (1999).

Although the haustellate moths entered the fossil records in Jurassic, their time came only in the Late Cretaceous when Lophocoronidae have been recorded in Santonian fossil resin of North Siberia, Tineoidea in the Campanian fossil resin of Canada, and mines of Nepticulidae and Bucculatricidae on leaves of various Magnoliophyta in the Cenomanian of USA (LABANDEIRA et al. 1994a) and Czech Rep. and the Turonian of Kazakhstan (see Fig. 475). The finding of a noctuid egg in the Campanian in Eastern USA seems doubtful; other Macrolepidoptera are known since the Eocene, while the oldest Noctuoidea appeared as late as in the Miocene. Because of their large body size and good flight abilities, the late record of the macrolepidopterans can hardly be due to taphonomical distortion. The presence of Noctuoidea in Mesozoic seems to be doubtful; one cannot exclude either misidentification of the mentioned egg or contamination of sediments with younger material.

With all the above reservations, the main grades/clades of the living moths and butterflies certainly originated during the Cretaceous period. Their rapid divergence during the Late Cretaceous was most probably connected with the explosive diversification of the

---

22 – the haustellum has overlapping scales on the dorsal surface from the base extending variously to 1/2 its length.

23 – *clade Apodirtysia*: sternum II with apodemes short, enlarged basally.

24 – none of the previously proposed apomorphies (e.g. larval integument with irregularly distributed spinules; larval pronotum with oblique, dark stripe between setae D_1 and D_2) have been confirmed (EDWARDS et al. 1999).

25 – none of suggested apomorphies (e.g. caterpillar head retractable into prothorax; pupal abdominal sternum II with stigmae covered by pterothecae) have been confirmed (KOZLOV et al. 1998; EDWARDS et al. 1999).

26 – ocular diaphragm more heavily pigmented anteriorly; large patagia extend, ventrad, beyond anterolateral extremities of pronotum; further possible apomorphies discussed by KOZLOV et al. (1998) and EDWARDS et al. (1999).

27 – large flat female ovipositor lobes.

28 – ocelli lost; spinarea (wing-locking device, defined by MINET 1990: 355) lost; hind tibia more than twice as long as femur; larva with proleg crochets arranged in a mesoseries/mesopenellipse (CARTER & KRISTENSEN 1999: fig. 3.2); further possible apomorphies listed by MINET (1991).

29 – *clade Obtectomera*: immobility of pupal abdominal segments I–IV.

30 – fore wing dorsal surface with patches of convex scales; hind wing with basal fringe of piliform scales.

31 – abdominal tympanal organs consisting of paired tympanal chambers on ventral part of segment two, supporting tympanum and conjunctiva, tympanum with sensory scoloparium attached; fore wing RS_2 and RS_3 stalked or fused; hind wing with SC+R and RS approximated or partly fused beyond discal cell.

32 – abdominal tergum I with very large lateroventral lobes immediately behind spiracle; resting posture with mid legs not touching substrate; spinarea absent; hind wing with CuP vestigial; larva with only two L setae occurring on either side of all thoracic segments.

33 – *clade Macrolepidoptera*: CuP completely lost; larval proleg crochets arranged in a mesoseries.

34 – concealment of pupal tibial palps; reduction of the metapostnotal "fenestrae laterales"; internal lamina of secondary metafurcal arms widened, their mesal edges adjacent and parallel, or posteriorly convergent.

35 – spinneret primarily shorter than prementum along its midline; pupa with fore legs projecting far cephalad comparing anterolateral angles of proboscis case, and with abdominal tergum X grooved transversely.

36 – mesothoracic episternum varying from fewer than 1/3 of episternum to virtually absent; second median plate in fore wing base partly or entirely beneath base of vein 1A+2A; further possible apomorphies discussed by ACKERY et al. (1999).

37 – space between antennae is at least two times as wide as scape; eyes with complete marginal ring of reduced ommatidial facets; larvae with "neck"; further apomorphies listed by ACKERY et al. (1999).

38 – anepisternum of mesothorax absent or present as only a tiny sclerite; parepisternal suture running in straight or smoothly curved line from dorsal end to base of sternum; mesophragma with dorsal processes; ventral edge of tegula attached by membrane to the mesonotum; secondary sclerite present behind metascutellum.

39 – mesonotum with prescutal clefts converging dorsally; further 5 possible synapomorphies discussed by MINET (1991).

40 – adult metathorax with tympanal organs; larval metathorax with two MD setae.

Fig. 298 Wing scales of the oldest known moth, *Archaeolepis mane* Whalley (family unknown), from the earliest Jurassic of Dorset in England (holotype), courtesy of P.E.S. Whalley

angiosperm plants (Magnoliophyta) which supplied these insects not only with easily available and digestible food for both adults and caterpillars, but also with a wide array of environments and shelters.

Three fossils of Palaeocene (oldest Tertiary) age are recorded thus far belonging to Hepialidae, Pyraloidea and Hesperioidea (PITON 1940, RUST 1998).

The numerous Eocene moth records are primarily due to the Baltic amber, with only extant families being found there. Most of the inclusions belong to moths whose caterpillars were detritivorous (Psychidae, Tineidae, Oecophoridae), and generally the fauna of Baltic amber seems rather similar to the recent fauna of Scots pine forests. Additionally three butterfly species (two of which belong to Papilionidae and one to Lycaenidae) were found in the Middle Eocene deposits of the Green River Formation in Western USA.

Unlike the Eocene assemblage dominated by microlepidopterans, 24 out of 37 Oligocene fossils and 23 out of 33 Miocene ones (as listed by KOZLOV 1988) belong to Macrolepidoptera (Fig. 303); the Pliocene fossil assemblage is too small for sound comparison. This change results at least partly from taphonomical factors: the Eocene fossils are mostly inclusions in amber where larger insects are rarely encountered, whereas the Oligocene and Miocene assemblages are generally known from impressions where larger insects are often over-represented. It is possible, however, that the above difference is not a mere taphonomical distortion, but an indication of the growing abundance of macrolepidopterans in the source faunas. At least it would be coherent with the general trend which manifests itself in comparison with the Cretaceous moths which are small even among the impression fossils, with the contemporary lepidopteran assemblage that is dominated by forms of intermediate and larger size. However, the available data from the Miocene Dominican amber (SCHLEE 1980, 1990, SKALSKI 1990, POINAR 1992a) indicate its similarity to the Baltic amber in that respect: only 3 out of 12 recorded Dominican families belong to the 'macros'. In any case, it seems obvious that the transition between the fauna of the Mesozoic type, composed of small, usually mandibulate moths with cryptic habits, and the Cainozoic fauna dominated by haustellate forms of larger size and generally of more exposed mode of life, occurred not later than in the Oligocene.

To summarise, the Lepidoptera is one of the youngest insect orders. It is known since the Jurassic, but was rare until the Cretaceous, and

Fig. 299 *Eolepidopterix jurassica* Rasnitsyn (Eolepidopterigidae) from the Late Jurassic of Uda in Siberia (holotype, photo by A.P. Rasnitsyn); body 6.2 mm long

Fig. 300 Undescribed moth, possibly of suborder Eriocraniina, from the Late Jurassic of Karatau in Kazakhstan (PIN 2784/1933, photo by D.E. Shcherbakov); body with wings 5 mm long

Fig. 301 *Undopterix sukatshevae* Skalski (Undopterigidae) from the Jurassic or Early Cretaceous of Unda in Siberia (from ROHDENDORF & RASNITSYN 1980); fore wing 4.6 mm long

Fig. 302 *Daiopterix olgae* Kozlov (Eolepidopterigidae) from the Late Jurassic or Early Cretaceous Glushkovo formation in Siberia (PIN 3063/741, photo by D.E. Shcherbakov); body with wings 6.2 mm long

Fig. 303 *Aglais karaganica* Nekrutenko (Nymphalidae), Middle Miocene of Stavropol in North Caucasian (holotype, photo by D.E. Shcherbakov); hind wing 20 mm long

it is not clear whether the mid-Cretaceous events promptly resulted in considerable diversification of small moths with endophagous caterpillars, or whether this event was postponed (until the Baltic amber time as the latest date). Butterflies (Hesperioidea + Papilionoidea) and particularly Papilionidae retained the diurnal mode of life and became abundant since the Eocene, while the remaining Macrolepidoptera played a visible role in terrestrial insect communities only since the Miocene.

2.2.1.3.4.4. ORDER DIPTERA Linné, 1758. THE TRUE FLIES (=Muscida Laicharting, 1781)

V.A. BLAGODEROV, E.D. LUKASHEVICH AND M.B. MOSTOVSKI

(a) Introductory remarks. The true flies (Fig. 304) are the fourth most diverse, and one of the most important and popular insect orders comprising more than 100,000 described living species. The concept of the order and its history is taken here generally from V. KOVALEV (1984,

common and abundant since the Oligocene. The last inference is weakened, however, by the fact that we have almost no important impression fossil assemblages dated between the Turonian (earlier Late Cretaceous) and Eocene. The mid-Cretaceous biocoenotic crisis and establishment of angiosperm dominated vegetation have probably initiated the early divergence of haustellate groups, as well as diversification of the mining moths. Because of the above mentioned gap in the fossil record,

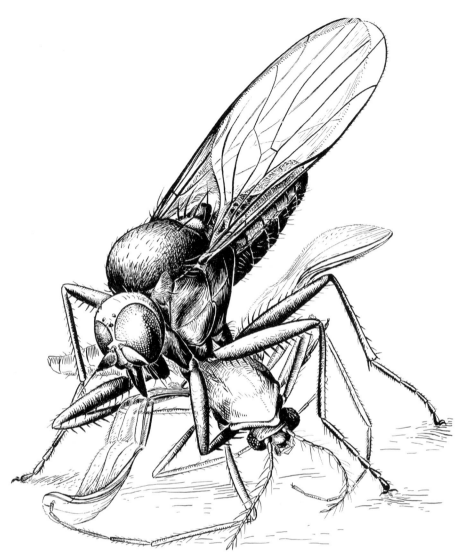

Fig. 304 *Archiplatypalpus cretaceus* Kovalev (Empididae) from the Late Cretaceous of Yantardakh in Siberia, restored as in life by V.G. Kovalev (from Rohdendorf & Rasnitsyn 1980)

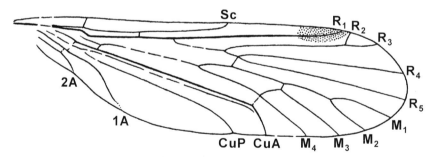

Fig. 305 The most primitive known dipteran wing *Vladiptera kovalevi* Shcherbakov (Vladipteridae) from the Late Triassic of Kenderlyk in Kazakhstan (holotype, after Shcherbakov *et al.* 1995); wing length 3.8 mm

1987), Kalugina & Kovalev (1985) and Shcherbakov *et al.* (1995). Vein nomenclature is given after Wootton & Ennos (1989) with correction according to Shcherbakov *et al.* (1995). Most references of the first records of nematoceran families were given by Shcherbakov *et al.* 1995. Further sources are cited when relevant.

(b) Definition. Minute to medium size, rarely large insects, usually with good flight abilities. Antennae variable, formed by two basal segments and multi-segmented (usually up to 20 flagellomeres) to one-segmented flagellum bearing sensory style or arista in the latter case. Mouthparts licking or biting suctorial. Pro- and metathoraces small, immovably fused with large mesothorax. Mesothoracic wing (Fig. 305), unless rarely lost or reduced, with venation not rich, at most with C, simple SC, simple R, both RS and M dichotomous with 4 branches each, simple CuA, closely followed with concave secondary pseudovein iCu (not homologous to CuP), CuP (A1 auct.), 1A (so-called A2 auct.) (alternative is homology by A. Rasnitsyn, pers. comm., of CuA to M$_5$, and of iCu to CuA, similar to those in hind wings of some other Papilionidea; see Figs. 269, 279a,f). Metathoracic wing reduced to haltere. Male lacking 8th abdominal spiracle and with cerci short, one or two-segmented. Larva legless (sometimes with secondary, not articulated appendages), sometimes headless as well, with mouthparts modified and retracted into thoracic segments, holopneustic to apneustic. Pupa with appendages more or less adherent, sometimes encased into puparium (seed-like, hardened, unhatched larval integument). Adults free-living, rarely parasites, larvae terrestrial or aquatic, sometimes parasitic, concealed in various organic substrate, diet of both highly variable.

(c) **Synapomorphies**. Labium modified into proboscis. Pro- and metathoraces small, immovably fused with mesothorax. Fore wing with venation simplified, with convex vein CuA followed by concave iCu. Hind wing modified into haltere. Male lacking 8th abdominal spiracle. Larva legless.

(d) **Range**. Middle Triassic till now, worldwide.

(e) **System and phylogeny**, as shown at Fig. 306, is followed generally HENNIG (1968, 1973) and V. KOVALEV (1987), with correction according SHCHERBAKOV et al. (1995). System of Cyclorrhapha (=Muscomorpha) is given generally after MCALPINE (1989) with some modifications. Additional apomorphies for subgroups of Brachycera Orthorrhapha are accepted after SINCLAIR (1992) and OVTSHINNIKOVA (1998).

(f) **History**. Diptera originated from some extinct group of Mecoptera. Among scorpionfly families, Permotipulidae and especially Robinjohniidae (Permian; both previously assigned to Paratrichoptera, the suborder discarded as polyphyletic by NOVOKSHONOV & SUKATSHEVA 2001) approach generalised Nematocera most closely in the wing structure. It was suggested that the transformation of the hind wing into a halter took place in a bittacid-like ancestor related to Robinjohniidae, with narrow-based homonomous wings and long cranefly legs, and that several peculiarities of dipteran wing are mecopteran hind wing characters transferred on to the fore wing (SHCHERBAKOV et al. 1995). Alternatively, Diptera are considered descendants from the more generalised scorpionfly family, Permochoristidae (NOVOKSHONOV & SUKATSHEVA 2001). This hypothesis would imply gradual shortening of the hind wing with the compensatory widening of the fore wing, like in some scorpionflies (Permotanyderidae, Liassophilidae etc., Chapter 2.2.1.3.4.1).

There is no reliable record of Diptera until the early Middle Triassic (Anisian), but already in the earliest known assemblage from Vosges, France, various Nematocera co-exist with a single undescribed Brachycera (KRZEMIŃSKI et al. 1994; KRZEMIŃSKI 1998; MARCHAL-PAPIER 1998). Diptera were quite abundant in this locality (uncommon situation for the Triassic, see below) and were estimated as about 280 specimens or 5% of the total number of collected insects, due to numerous aquatic immatures which are not yet described; possibly some of these larvae do not belong to dipterans at all. Up to now a single monobasic family Grauvogeliidae is described from this locality. An undescribed specimen of Diptera was recorded from the Anisian of Mallorca, Spain (COLOM 1988).

In the Middle and Late Triassic the dipterans became widespread and diverse, but not numerous, being found in Central Asia (four

localities in Kyrgyzstan and Kazakhstan; Shcherbakov et al. 1995), Australia (Queensland; V. KOVALEV 1983; BLAGODEROV 1999, LUKASHEVICH & SHCHERBAKOV 1999), North America (Virginia, USA; KRZEMIŃSKI 1992) and Europe (Great Britain; KRZEMIŃSKI & JARZEMBOWSKI 1999). It is impossible to draw any conclusions about dominant groups because of scanty finds: in Triassic assemblages, other groups than Vosges Diptera usually form less than 1% of insect fossils. About 80 specimens are described from the Triassic beds, half a hundred belong to *Grauvogelia* Krzemiński, Krzemińska et Papier, but usually descriptions of new taxa are based on holotypes only.

The true flies never had complicated wing venation pattern with numerous additional (intercalary) veins, or a lot of cross-veins. Descriptions of all such phenomena produced by old authors were based on misinterpretations due to poorly preserved material. In the light of this, the suborder Archidiptera erected by ROHDENDORF (1961) for several taxa should be abolished, because the families included in fact belong to Tipulomorpha and Bibionomorpha. Comparative wing morphology and palaeontology indicate that modern Tipulomorpha belong to the earliest branch of Diptera, and, moreover, that the ancestral dipterans could be assigned to the same infraorder (considered paraphyletic). The most generalised of Triassic Tipulomorpha, Vladipteridae (see Fig. 305), known from Asia (Kenderlyk locality), give us a possibility to reconstruct the groundplan of dipteran wing as already stalked, without alular incision, with sc-r near RS origin and with short oblique R$_2$ reaching wing margin close to R$_1$, and with anal veins free, not yet looped. Such Triassic dipterans as Vladipteridae Psychotipinae (Dzhayloucho locality), combining groundplan characters (free R$_2$), tipulomorphan autplesiomorphies (long convex 1A), and non-tipulomorphan apomorphies (anal loop and alular incision) are difficult to classify, but it is practicable to assign them to Tipulomorpha.

A single large wing in the Triassic (10 mm instead of usual 2–6 mm) was found in Australia (Mt. Crosby). This family Tilliardipteridae, despite of the numerous 'tipuloid' features, should be included in Psychodomorpha sensu Hennig on account of loss of the convex distal 1A reaching wing margin and formation of the anal loop. Similar hypothetical forms with short free R$_2$ (absent in *Tillyardiptera* Lukashevich et Shcherbakov) could represent a group ancestral to all other Psychodomorpha and to Bibionomorpha as well. Discovery of *Tillyardiptera* annectent between the most primitive Tipulomorpha and Psychodomorpha strengthens the hypothesis that separation of the latter from the former was the first divergence in the history of Diptera.

One of the two main phyletic lineages of Psychodomorpha, Psychodoidea s. l. + Blephariceroidea s. l., is still to be recorded from the Triassic. The second, ptychopteroid lineage, is represented by Nadipteridae, Eoptychopteridae (ancestors of the living Ptychopteridae; Triassic find is doubtful) and Hennigmatidae; possibly, Grauvogeliidae belong to it too. The most archaic member of Ptychopteroidea, Nadipteridae, at present seem to be closest to the ancestors of both Culico- and Bibionomorpha.

The main novelty suggested by V. KOVALEV (1987) in his dendrogram, was the independent and heterochronous derivation of both Culicomorpha and Bibionomorpha. Diversity of these infraorders in the Triassic agrees well with the assumption of the later separation of Culicomorpha. Only one chironomid is known from the Late Triassic of Europe (though Culicoidea is still not recorded, but should be represented already too). At the same time bibionomorphan Diptera were rather diverse in the Triassic represented by Procramptonomyiidae (most primitive Bibionomorpha), Alinkidae (described as Brachycera

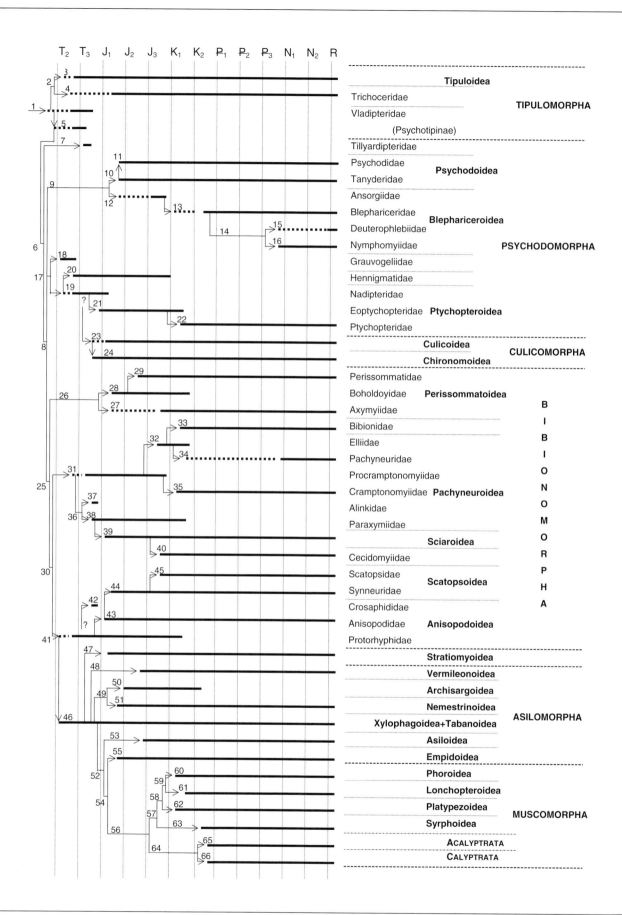

Fig. 306 Phylogeny and system of the Order Diptera

Time periods are abbreviated as follows: T_2, T_3–Middle and Late (Upper) Triassic, J_1, J_2, J_3–Early (Lower), Middle and Late (Upper) Jurassic, K_1, K_2–Early (Lower) and Late (Upper) Cretaceous, P_1,–Palaeocene, P_2–Eocene, P_3–Oligocene, N_1–Miocene, N_2 – Pliocene, R–present time (Holocene). Right three columns are names of families, superfamilies and infraorders, respectively (families are not considered for some superfamilies). Arrows show ancestry, thick bars are known durations of taxa (dashed for hypothesised duration), figures refer to synapomorphies of the subtended clades, as follows (synapomorphies other than in wing structure mainly after WOOD & BORKENT (1989):

1 – Preradial space narrow, R stem kinked, R_{2+3} fork short, R_{4+5} fork very long, R_5 convex, CuA dissociated and curved backwards distally, clavus reduced (2A short, 3A lost), submarginal cross-vein series tends to be reduced, hind wing transformed into haltere, pro- and metathoraces very reduced; (extrapolated from extant groups of Tipulo- and Psychodomorpha) labium modified into proboscis, 8th abdominal spiracle lost in male, pupa with adherent appendages, larva legless with bifold mandibles operating in oblique planes.
2 – R_2 joining R_1, M stem tends to align with M_{3+4}, 2A very short and submarginal, mandibles lost, male genitalia developing partly from pupal ectoderm.
3 – sc-r tending to be terminal, ocelli and free cardo lost; larva metapneustic with head retractile.
4 – 1A shortened (but reaching margin).
5 – Anal loop and alular incision developed
6 – 1A distally reduced (not reaching margin); (unless in Psychotipinae) arolium replaced with empodium.
7 – R_2 lost
8 – R_5 convexity continued basad posterior to RS, CuP entering margin in basal wing half
9 – RS aligned with R_4 (r-m joining R_5 base), im shifted distad of M_{1+2} fork, cu-a replaced with anastomosis, empodium setiform to absent (unless claws vestigial).
10 – sc-r terminal, M_2 convex, ocelli lost, male genitalia inverted.
11 – CuP shortened, anal lobe reduced, 2A forming marginal projection ('wing scale'), im usually absent, body and wing veins densely haired.
12 – SC weak, antennae short, wings permanently extended at rest, legs prehensile with long claws folding against distitarsus, male gonocoxites and 9th sternum fused into small capsule; (possibly also) R_2 reduced, male genitalia dorsoflexed.
13 – C not continued beyond wing tip, M_{1+2} simple, im lost, wings unfolding at emergence; larva apneustic with (at least in 1st instar) paired abdominal prolegs.
14 – Mouthparts and digestive tract atrophied, metasternum partly exposed, femur subdivided, abdominal spiracles vestigial, wings deciduous at least in female
15 – Larva with distal antennal article bifurcate.
16 – Pupal head prognathous, compound eyes holoptic ventrally.
17 – (extrapolated from Ptychopteridae and Culicomorpha) ocelli lost, larva with mandibular comb and invaginated premandibular apodeme.
18 – Broad-based wing; M base subtransverse.
19 – CuP strongly diverging from CuA.
20 – R base weak, RS aligned with long R_{2+3+4}.
21 – R_{2+3} approximated to R_1, R_2 joining R_1, M stem tends to align with M_{1+2}, m-cu joining base of long M_{3+4}, CuP distally strongly convex and curved backwards from claval fold; halter with prehalter.
22 – im lost, mesopleuron lacking transepimeral suture.
23 – Torsion joint at R kink, radial+axillary hair plates, R_2 lost, R_{3+4} stalk formed, MA vein-like and transverse, M_{3+4} simple, im lost, Cu base tends to be captured by cu-a, CuP base angled towards 1A, pedicel enlarged, (post)laterocervicale with ventromedian projection, empodium tends to be setiform; pupa: palpal sheath curved posteromesad, hind leg sheath S-shaped and concealed beneath wing sheath; larva: labral brush complex, torma and premandible invaginated.
24 – R_3 joining R_1, sperm transferred by spermatophore; larva with prothoracic proleg and procerci.
25 – R_2 lost, M base weak to reduced.
26 – M_{3+4} simple, CuP distally reduced, eyes longitudinally subdivided, mandibles lost, palpomeres short, postcervicale lost; larva with antennae reduced and caudal siphon retractile.
27 – R_{3+4} stalk formed, im lost; pupa with abdominal siphon; larva with anal papillae very long and mandibles moving horizontally.
28 – SC distally reduced, pterostigma distad of short R1, ambient vein continued beyond wing tip.
29 – R_{3+4} stalk formed.
30 – Radial+anal hair plates, M_{1+2} fork shifted basad to near im; (unless at 25) chiasmata lost at meiosis in male.
31 – M base oblique, pedicel slightly enlarged, mandibles lost; larval mandibles moving horizontally.
32 – R_{3+4} stalk formed, fore tibiae tend to be fossorial.
33 – R_{3+4} simple; larva with intersegmental fissures interrupted and displaced laterally.
34 – M_3 and im lost.
35 – R_4 fused to R_3 except base.
36 – R_{4+5} fork short.
37 – M stem aligned with M_{1+2}, r-m before R_3 base.
38 – M_3 and im lost.
39 – R_{4+5} simple, meron greatly reduced, 2 spermathecae; pupal leg sheaths coplanar (not superposed); larva: cardo tends to be fused to cranium, metathoracic spiracle lost.
40 – Flagellomeres with encircling setal rows, tibial spurs lost; larval head tiny with stylet-like mandibles.
41 – sc-r basal, MA vein-like and oblique, M base along Cu base, CuP base tends to be associated with 1A.
42 – M_3 and im lost.
43 – RS with 2 branches, mandibles lost, pulvilli (but not empodium) reduced; larval hypostomal bridge lost.
44 – (unless in mycetobiine-like ancestors) M_3 and im lost; M_{1+2} fused to RS for a distance; (in recent members) larval head greatly reduced.
45 – M fork shifted basad, palpus one-segmented.
46 – Veins CuA and CuP with apices approximate, forming a nearly closed cell; antennal flagellomeres reduced to eight; maxillary palps 2-segmented; larval head partly retracted
47 – Larval cuticle encrusted with "warts" of calcium carbonate; pupation within the last larval integument; tibial spurs are absent on front legs; costal vein somewhat abbreviated; anal cell stalked; tendency to costalisation of wing.

but plausible Bibionomorpha close to Procramptonomyiidae), Paraxymyiidae (initial group for Sciaroidea), Protorhyphidae (ancestral to Anisopodidae and Brachycera) and Crosaphididae (in some respects more advanced than some extant members of Bibionomorpha). The only extant families known from the Triassic are Limoniidae (Tipulomorpha) and Chironomidae (Culicomorpha).

The immatures of Triassic Tipulomorpha and Psychodomorpha can be hypothesised as (sub)aquatic (developing in submerged organic material, in detritus or in saturated earth, as living Limoniidae, Tanyderidae and Ptychopteridae do), while Bibionomorpha would be terrestrial (developing within decaying plant matter like the recent Anisopodoidea and Mycetobioidea).

In the Early Jurassic Diptera had become one of the most abundant and diverse insect orders, and retained the position of dominant or subdominant group later. There are a lot of the Jurassic dipteran assemblages in Western Europe, Central Asia and Siberia (e.g. HANDLIRSCH 1906–1908, A. BODE 1953, ROHDENDORF 1962, 1964, HONG 1983a, ANSORGE & KRZEMIŃSKI 1994, 1995, ANSORGE 1996, 1999, BLAGODEROV 1996, LUKASHEVICH 1996a,b, LUKASHEVICH et al. 1998, MOSTOVSKI 1996, 1997a,b, 1998, 1999a). Limoniidae, Nadipteridae, Eoptychopteridae, Hennigmatidae, Chironomidae, Protorhyphidae (Fig. 307), Procramptonomyiidae (Fig. 308) and Paraxymyiidae are passed from Triassic, but only the first family plays a significant role in oryctocoenoses as well as in present. Since that time the living families Trichoceridae (Fig. 309), Tanyderidae (Fig. 310), Psychodidae, Chaoboridae and Dixidae are known.

In the earliest Jurassic (Sogyuty in Kyrgyzstan, probably Sinemurian; German Lias, Toarcian) the records of Culicomorpha are very scanty. But soon culicomorph dipterans began to play an important unless dominant role in many Jurassic freshwater assemblages. In the oligotrophic Jurassic lakes (Chapter 3.3.4) aquatic larvae of these groups became predatory and algophagous (nectic Chaoboridae and benthic Chironomidae, the latter represented with only primitive subfamilies Aenneinae, Ulaiinae, Podonominae and Tanypodinae). Chaoboridae have flourished, and their immatures were often the most numerous fossils among not only the aquatic dipterans but aquatic insects in general. However, chaoborid generic diversity was not great (usually not more than two genera per locality), and the extinct genera were not too exotic (extraordinary characters are not found, only their unusual combinations). For example, two extremities of the chaoborid larval morphotypes still occur now. The first of them is represented by Eucorethra Underwood which lives near water surface and collects its victims from the surface: dark larva, with head capsule broad (Fig. 311), dorsoventrally flattened, antennae widely separated, air sacs not developed, siphon present. The second type is represented by the highly advanced genus Chaoborus Lichtenstein (nectic predator): larva transparent, with head much elongated, antennae approximated, air sacs developed and siphon absent (Figs. 312, 489). Both types, and some genera with the mosaic of above characters, already existed in the Jurassic.

It should be mentioned that as yet no undoubted ancestral forms for any culicomorph family are described (though several extinct, primitive subfamilies are known in Chironomidae, e.g. Aenneinae and Ulaiinae). Architendipedidae from Sogyuty (alleged ancestor of Chironomidae) appeared to be Eoptychopteridae, Protendipedidae from Karatau (Late Jurassic of Kazakhstan) are too poorly preserved to draw any serious conclusions, Rhaetomyidae (Sogyuty and Grimmen) and Dixamimidae (Karatau), supposed ancestors of Chaoboridae and Dixidae, respectively, were shown to be Chaoboridae with standard venation. The single exception among Culicomorpha is the genus Syndixa Lukashevich (Fig. 313), which differs from all dixid genera as well as from all other Culicoidea in having R_3 joining R_1 like in Chironomoidea, and so could be close to chironomoid ancestors.

The diversity of Bibionomorpha was growing dramatically when compared to Culicomorpha. Up to date no one extinct culicomorph family is known, but 15 extinct families of Bibionomorpha are

---

Fig. 306 *Continued*

48 – Attenuate wing base; approximation of M_3 and M_4 ends; distal position of R_{4+5} fork; maxillae and mandibles reduced; slender clavate abdomen in both sexes.

49 – Eyes enlarged; larva parasitic.

50 – R_{4+5} fork somewhat shortened and wide, or R_{4+5} unforked; tendency to loss of *r-m*, and R_{4+5} and M_{1+2} fusion; ovipositor needle-like.

51 – Loss of gonocoxal muscles M33 and pregenital muscles M20; larva hypermetamorphous.

52 – Empodium bristle-like.

53 – Larval posterior spiracle situated in apparent penultimate abdominal segment; presence of three tergosternal muscle pairs $M5^3$; presence of muscles of lateral processes of gonocoxites M38.

54 – Vein M_4 lost or fused with M_3; veins CuA and CuP fused before wing margin; ocellar setae present; maxillary palps one-segmented; 9th abdominal tergum and furca reduced in female; hypandrium with pair of posterior processes; epandrium much reduced, gonocoxites expanded dorsally; 10th abdominal tergum lost in male; larva with 3 instars.

55 – Larva with postcranium modified into a pair of slender metacephalic rods; adult female with a single spermatheca; adult male with paired apodemes of the genital segment attached to the hypandrium.

56 – Adult flagellum composed of first flagellomere and 3 aristomeres; wing with R_{4+5} unbranched and costa distinctly ending at R_{4+5} or M_1; acrostichal and dorsocentral setae differentiated; male terminalia circumverted; pupariation highly evolved.

57 – Larval hypopharyngeal and tentopharyngeal sclerites fused; pupal respiratory horns enlarged.

58 – Frons bristled; cross-vein *sc-r* lost.

59 – Male dichoptic.

60 – SC and R1 fused partially or entirely; tendency to evolve highly costalised wing.

61 – Wings pointed.

62 – Pupal respiratory horns lost.

63 – Puparium globose; apices of R_{4+5} and M_{1+2} approximated.

64 – Ptilinum stabilised, lunule discrete, ptilinal fissure developed.

65 – Pupal respiratory horns reduced; male dichoptic; 2 spermathecal ducts fused.

66 – Dorsal cleft of pedicel, two costal breaks, vibrissae stabilised.

Fig. 308 Unidentified Procramptonomyiidae from the Late Jurassic of Karatau in Kazakhstan (PIN 2239/2104, photo by D.E. Shcherbakov); body 6 mm long

Fig. 307 Unidentified Protorhyphidae from the Late Jurassic of Shar Teg in Mongolia (PIN 4270/2263, photo by D.E. Shcherbakov); body 3 mm long

described for Mesozoic and one can see a number of steps in the development of various lineages. Procramptonomyiidae (see Fig. 308), a stem-group of the whole of the infraorder except for Axymyiiformia, gave several descendants: Elliidae (Late Jurassic–Early Cretaceous, Fig. 314), presumed ancestors of Pachyneuridae and Bibionidae, Cramptonomyiidae (Late Jurassic till now), and Heterorhyphidae (Jurassic). Since the Jurassic, rather scarce Axymyiiformia sensu SHCHERBAKOV *et al.* (1995) are known, i.e. Boholdoyidae (Early Jurassic–Early Cretaceous), Axymyiidae (Late Jurassic till now, Fig. 315), Perissommatidae (Middle Jurassic till now).

Jurassic representatives of the last family were more advanced morphologically then living forms, in particular, Jurassic species of *Palaeoperissomma* Kovalev (Kubekovo, possibly also in the Early Cretaceous of Turga) lacked a discal cell unlike recent *Perissomma* spp. The first Scatopsoidea appeared in the Early Jurassic represented by Protoscatopsidae (Early to Late Jurassic) and Synneuridae (Early Jurassic till now).

The sciaroid lineage founded by Paraxymyiidae, descendants of Procramptonomyiidae, was continued by Protopleciidae (first representative of Sciaroidea, known since the Early Jurassic, when they were diverse and numerous, till the Late Jurassic, and completely disappeared in the Early Cretaceous). The descendants of Protopleciidae were the Pleciofungivoridae (Fig. 316) and Antefungivoridae which

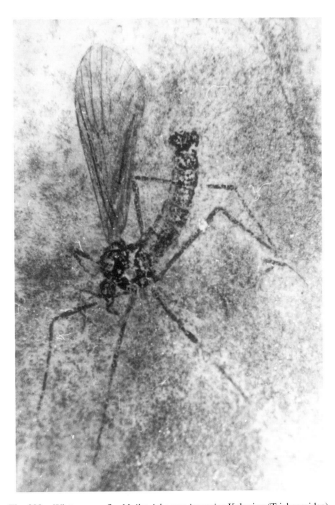

Fig. 309 Winter crane fly, *Mailotrichocera jurassica* Kalugina (Trichoceridae), from the Late Jurassic of Uda in Siberia (holotype, photo by D.E. Shcherbakov); wing length 3.8 mm

233

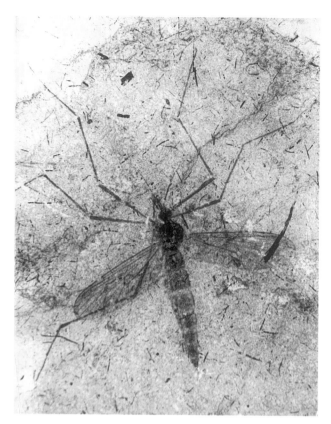

Fig. 310 *Praemacrochile* sp. (Tanyderidae) from the Late Jurassic of Karatau in Kazakhstan (PIN 2554/934, photo by D.E. Shcherbakov); wing length 9.3 mm

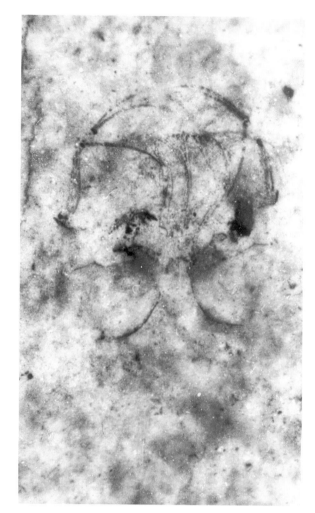

Fig. 311 Larval head of phantom midge, *Hypsocorethra toficola* Kalugina (Chaoboridae), from the Late Jurassic of Uda in Siberia (PIN 3053/1273, photo by D.E. Shcherbakov); head width 1 mm

became more diverse than their ancestors in the Middle and Late Jurassic. The primitive sciaroids were small gnats looking like living Sciaridae and probably also developing in decaying plant matter rich in fungal hyphae, this being probably the commonest diet among Jurassic terrestrial dipterans. Further members of Sciaroidea were: Mesosciophilidae (Middle Jurassic–Early Cretaceous, a sister group of living Mycetophilidae *s. str.*), Eoditomyiidae (Early–Late Jurassic, probable ancestor of Ditomyiidae), and Archizelmiridae (Late Jurassic–Late Cretaceous).

The suggested prevalent biology of Jurassic Bibionomorpha was saprophagy. The larvae of Perissommatidae, Mesosciophilidae and Synneuridae were most probably associated with fungi, the more so that the basidiomycetes entered the fossil record at roughly the same time as three families in question (KRASSILOV 1982).

The main groups of Brachycera Orthorrhapha appeared during the Jurassic as well. The principal sources of our knowledge about these flies for the Jurassic are Laurasian deposits. Exceptionally scarce primitive snipe-flies are found in the Early Jurassic beds of Gondwanaland (India, Kotá Formation). Firstly the brachyceran stock split into the Stratiomyomorpha, represented by Oligophrynidae, and the Asilomorpha. This event occurred not later than the Sinemurian. In fact, brachyceran flies of at least four lineages existed in the Early Jurassic, i.e. Stratiomyoidea (Stratiomyomorpha), Xylophagoidea + Tabanoidea, Nemestrinoidea and Empidoidea (Asilomorpha). Nemestrinoidea represented by the extinct subfamily Archinemestriinae (Nemestrinidae) (Fig. 317) and the extant subfamily Heterostominae (Rhagionemestriidae), and Empidoidea have been found in the Toarcian of Grimmen (German

Lias, J. Ansorge, pers. comm.). One can assume archinemestriins to be parasitic like recent tangle-veined flies, and connected with plants possessing flowers or flower-like organs, as nectar feeders and possible pollinators. Other parasitic brachyceran flies are recorded since either the Middle (first Archisargidae, Fig. 318, and Mythicomyiidae) or Late Jurassic (Eremochaetidae, Acroceridae, Bombyliidae). Aphagy cannot be excluded for the adults of Eremochaetidae. Generally, the Middle and Late Jurassic deposits yield a vast majority of the brachyceran orthorrhaphous flies, namely Stratiomyidae (Beridinae, Fig. 319), Xylomyidae (=Solvidae), Xylophagidae (Coenomyiinae), Kovalevisargidae, Rhagionempididae, Asilidae, Therevidae (Fig. 320), Vermileonidae, Hilarimorphidae (Fig. 321), Scenopinidae, Apystomyiidae, and Empididae. Representatives of more advanced nemestrinids belonging to Hirmoneurinae appeared as well. All the Jurassic dance-flies may be allocated to the extinct subfamily Protempidinae, characterised by a full set of plesiomorphies according to CHVÁLA (1983). Remains of Mesozoic wormlion flies are scarce. Vermileonidae from the Shevia locality in East Transbaikalia (Central Siberia, Ukurey Formation of debatable Late Jurassic or Early Cretaceous age, see Glushkovo Formation in Chapter 4.1 for details) seem to be

Fig. 312 Larval phantom midge, *Chachotosha probatus* Lukashevich (Chaoboridae), from the earliest Cretaceous (or latest Jurassic) of Khotont in Mongolia, showing head and air sacs (PIN 4307/939, photo by D.E. Shcherbakov); air sac 1 mm long

Fig. 313 Meniscus midge, *Syndixa sibirica* Lukashevich (Dixidae), from the Middle Jurassic of Kubekovo in Siberia (PIN 1255/1550, photo by D.E. Shcherbakov); wing length 4.5 mm

plesiomorphic if compared with the recent representatives of the family (B. Stuckenberg, pers. comm.), and may represent ancestral forms of vermileonids. Such an early separation from the rhagionid-like ancestor is supported by analysis of the male genitalia muscles of the recent flies (OVTSHINNIKOVA 1997).

Generally, despite the presence of a number of families persisting until the present, the leading Jurassic terrestrial families (Protopleciidae, Pleciofungivoridae, Antefungivoridae, Protorhyphidae, Rhagionidae, Fig. 322) were more common and taxonomically diverse at that time, compared with later on. As a result the fauna was essentially Mesozoic in appearance. Of interest is the exceptional rarity of dipterans in the later Early and Middle Jurassic insect assemblages of Eastern Siberia westward of Lake Baikal as well as in Northwest Mongolia, and the apparent absolute absence of Culicomorpha

and Bibionomorpha there which were dominating groups in other regions.

V. KOVALEV (1984) distinguished two stages of the Diptera fauna formation during the Jurassic: the Early Liassic (as reflected by the assemblage of Sogyuty in Central Asia) and the rest of the period. He believed the beginning of Liassic to be a separate stage because of the dominance of Protopleciidae, that makes it different from the Triassic as well as from the later Early, Middle and Late Jurassic, when Pleciofungivoridae and Antefungivoridae become dominant, and the first brachyceran flies appeared. Now we know, that brachyceran flies are more ancient than it was believed a few years ago, and the difference in the dominance structure between the earlier Early Jurassic and the other Jurassic dipteran assemblages does not seems so crucial. At the same time, taxonomic composition of the Jurassic assemblages differs radically from that in the Triassic and Cretaceous ones (particularly so from the Late Cretaceous).

The Cretaceous dipteran assemblages are numerous and well-recorded in both Laurasian and Gondwanan deposits (e.g. JARZEMBOWSKI 1984, JELL & DUNCAN 1986, HONG & WANG 1988, 1990, S. V. WATERS 1989a,b, GRIMALDI 1990, V. KOVALEV 1990, KALUGINA 1993, MARTINS-NETO & KUCERA-SANTOS 1994, BLAGODEROV 1995, 1997, 1998a,b, 2000, CORAM et al. 1995, 2000, MOSTOVSKI 1995a,b, 1999a, REN et al. 1995, REN 1998, BLAGODEROV & MARTÍNES-DELCLÒS 2001, MOSTOVSKI & MARTÍNES-DELCLÒS 2000). Another significant source of information about the dipteran assemblages, inclusions in fossil resins, becomes available since the Early Cretaceous (e.g. HENNIG 1971, 1972, POINAR 1992a, BORKENT 1995, 1996, 1997, SZADZIEWSKI 1995, 1996, SZADZIEWSKI & ARILLO 1998, ARILLO & MOSTOVSKI 1999,

Fig. 314 *Ellia colorissima* Krzeminska, Blagoderov et Krzeminski (Elliidae) from the Early Cretaceous of Baissa in Siberia (holotype, photo by D.E. Shcherbakov), wing 7 mm long

Fig. 315 Unidentified Axymyiidae from the Late Jurassic of Karatau in Kazakhstan (PIN 2997/3510, photo by D.E. Shcherbakov); fossil 7 mm long

Fig. 316 The most primitive fungus gnat, *Pleciofungivora* sp. (Pleciofungivoridae); restored as in life (from KOVALEV 1987)

Fig. 317 Tangle-veined fly *Protonemestrius rohdendorfi* Mostovski (Nemestrinidae) from the Late Jurassic of Karatau in Kazakhstan (holotype, photo by D.E. Shcherbakov); body length 10 mm

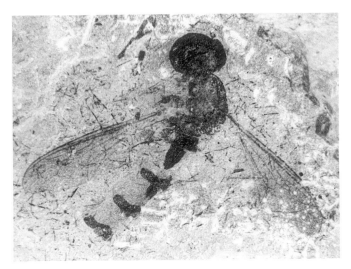

Fig. 318 *Calosargus tatianae* Mostovski (Archisargidae) from the Late Jurassic of Karatau in Kazakhstan (holotype, photo by D.E. Shcherbakov); wing length 6.4 mm

Fig. 319 Unidentified beridine soldier fly (Stratiomyidae) from the Late Cretaceous of Obeshchayushchiy in Russian Far East (PIN 3901/377, photo by D.E. Shcherbakov); body length 7 mm

Fig. 320 *Rhagiophryne bianalis* Rohdendorf (Therevidae) from the Late Jurassic of Karatau in Kazakhstan (PIN 2239/2200, photo by D.E. Shcherbakov); body length 7.5 mm

AZAR *et al.* 1999b, GRIMALDI & CUMMING 1999, MOSTOVSKI 1999b, S. WATERS & ARILLO 1999). Tiny and delicate dipterans are preserved in resins much better than as compression fossils. Moreover, fossil resins yield insect assemblages originating from palaeoenvironments usually differing strikingly from those producing the compression fossils. Fossil resins, especially of the retinite type, were often originated from riverine forests, with the aquatic insects being enriched there by the rheophilous groups, and the terrestrial ones by the tree-trunk dwellers (Chapter 1.4.2.2.1).

Unlike the Jurassic, the Cretaceous fauna consisted mostly of Cainozoic families, with the majority of the Jurassic ones going extinct either near or before the Jurassic-Cretaceous boundary (Nadipteridae, Hennigmatidae, Ansorgiidae, Oligophrynidae, Archisargidae, Kovalevisargidae) or towards the mid-Cretaceous

(Eoptychopteridae, Protorhyphidae, Procramptonomyiidae, Paraxymyiidae, Elliidae, Eoditomyiidae, Boholdoidae, Pleciofungivoridae, Antefungivoridae, Mesosciophilidae, Archizelmiridae, Eremochaetidae). The brachyceran Rhagionempididae are supposed to

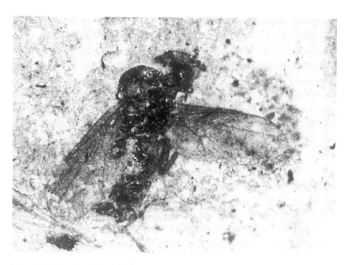

Fig. 321 Unidentified hilarimorphid from the Early Cretaceous of Baissa in Siberia (PIN 4210/5218, photo by D.E. Shcherbakov); body length 1.9 mm

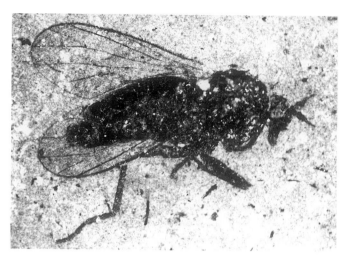

Fig. 323 Unidentified dance fly (Empididae) from the Early Cretaceous of Baissa in Siberia (PIN 3064/9653, photo by D.E. Shcherbakov); body length 3.3 mm

Fig. 322 Snipe fly *Palaeoarthroteles mesozoicus* Kovalev et Mostovski (Rhagionidae) from the Late Jurassic or Early Cretaceous Glushkovo Formation in Siberia (holotype, photo by D.E. Shcherbakov); body length 10 mm

persist till now being represented by genera currently included in the family Apsilocephalidae (NAGATOMI & YANG 1998). As for the endemic Cretaceous families of Nematocera, described only from China, their taxonomic status is too unclear to be discussed here. Re-examination of the type material might result in synonymisation of some of them with recent families (e.g. Gracilitipulidae and Zhangobiidae with Limoniidae, Sinotendipedidae with Blephariceridae).

Despite the extinction of the Jurassic families, the diversity of the order was increasing due to the appearance of Tipulidae, Ptychopteridae, Blephariceridae, Corethrellidae, Culicidae, Ceratopogonidae, Simuliidae, Thaumaleidae, Cramptonomyiidae, Bibionidae (Pleciinae), Bolitophilidae, Diadocidiidae, Keroplatidae, Sciaridae and Cecidomyiidae (although, a Late Jurassic age cannot be excluded for some of these records).

Further asilomorph and cyclorrhaphan families entered the record during the Early Cretaceous, viz. Tabanidae, Athericidae, Bombyliidae, Apioceridae, Dolichopodidae, Opetiidae, although changes of taxonomic composition were more considerable at subfamily level. The eremochaetid subfamily Eremomukhinae with derived wing venation replaced the more archaic Eremochaetinae. Archinemestriinae were disappearing, and different Nemestrininae appeared within the tangle-veined flies. More or less advanced Atelestinae, Oreogetoninae, Hybotinae and Microphorinae substituted Protempidinae in the family Empididae (Fig. 323). Some Early Cretaceous dance flies demonstrate adaptations to a predatory way of life as adults, such as raptorial legs and an elongated proboscis. Representatives of Apioceridae are supposed to be flower-visitors because of their long proboscis. The first cyclorrhaphan flies appeared in the Early Cretaceous represented by Platypezidae, Ironomyiidae (Sinolestinae), and by rarer Lonchopteridae, Sciadoceridae and Phoridae (Prioriphorinae).

The Late Cretaceous time was marked by the appearance of more high-ranking taxa of brachycerans, viz. Parhadrestiinae (Stratiomyidae), Empidinae, Tachydromiinae (see Fig. 304), Trichopezinae (Empididae), Ironomyiinae (Ironomyiidae), Metopininae (Phoridae), Syrphidae, Pipunculidae, Milichiidae, Calliphoridae. The scuttle flies of the subfamily Prioriphorinae became more common in the Late Cretaceous. Findings of Cyclorrhapha increased in number towards the end of the Mesozoic. However, the origin of the Cyclorrhapha remains one of the most controversial issues in dipteran systematics (WIEGMANN *et al.* 1993). There is widespread agreement that the affinities of these flies lie within the orthorrhaphan Heterodactyla, and most probably within the Empidoidea.

In spite of the taxonomic diversification, the adaptive zone of the order only increased weakly. There were only a few cases of origin of new life forms, e.g. those with necrophagous larvae (Sciadoceridae, Calliphoridae and maybe Phoridae: Prioriphorinae); possibly also the first endophytic Cecidomyiidae. The appearance of undoubtedly blood-sucking dipterans (Corethrellidae, Culicidae, Ceratopogonidae, Simuliidae in Culicomorpha, Phlebotominae in Psychodomorpha, and Tabanidae, Athericidae, and *Palaeoarthroteles* (Rhagionidae) in Asilomorpha) was a very important and significant event in the Cretaceous, coinciding with the appearance of other presumed blood-sucking insects (e.g. *Saurophthirus* Ponomarenko, Chapter 2.2.1.3.4.5). It is possible that the sucking of body fluids represents the more ancient diet in lower dipterans, because some recent representatives of relict groups which flourished during the Jurassic (Tanyderidae, Chironomidae Podonominae) have long piercing mouthparts and these are also found in some fossil Tanyderidae (KALUGINA 1991).

The composition of the freshwater dipteran assemblages changed strongly toward the mid-Cretaceous. In the numerous Asian localities of the Early Cretaceous age insect-bearing beds are literally filled with chaoborids, sometimes with the immatures dominating, sometimes the adults (Fig. 324). The density of their remains may be as high as 300–400 specimens per 100 square cm. Towards the end of the Early Cretaceous Chaoboridae reduced their abundance while Chironomidae were assuming their dominant position, which is usually retained till now. The subfamily Chironominae is known since that time, and is now the dominant group of chironomids in eutrophic lakes. Supposedly this was an effect of changes in the trophic regime of water bodies which resulted from the Mid-Cretaceous biocoenotic crisis (Chapter 3.3.6). This was probably the only example of a strong reaction of dipterans to the crisis, and this seems natural. The crisis is hypothesised to have been provoked by changes in plants, or at least it concerned principally those animals closely connected ecologically with plants (Chapter 3.2). Yet the terrestrial Diptera lacked close ecological ties with living plants during the Mesozoic.

Among non-aquatic Diptera dominant roles were taken by Mycetophilidae, various Empididae, and primitive Aschiza since the Early Cretaceous. The fungus gnats of the family Mycetophilidae were diverse and numerous in the beginning of the Early Cretaceous, replacing their ancestors, the Mesosciophilidae. Remarkable is the largely modern taxonomic composition of the Early Cretaceous mycetophilids: at least 6 of their 18 genera known from the Early Cretaceous are extant, an exceptional ratio for all the insects.

The general taxonomic composition of the Palaeogene dipteran fauna was more similar to the Cretaceous than to that of the Neogene. Although the acalyptrate and caliptrate flies, the dominants of the recent dipteran fauna, reached considerable taxonomic diversity during the Palaeogene (e.g. HENNIG 1965; EVENHUIS 1994), their share in assemblages of that time was low (SHATALKIN 2000 removed *Trypaneoides ellipticus* described by HONG 1981 from Lauxaniidae and concluded that it most closely resembles the dolichopodids). The transition from the Late Cretaceous to Palaeogene faunas was gradual, with continual growth of polydominance in the assemblages, and with retained prevalence of phytosaprophagy and mycetophagy as the characteristic diet of the leading terrestrial groups. Another evolutionary trend observable during the Palaeogene was the increasing taxonomic and ecological diversity of the terrestrial dipteran fauna. The complex of bloodsucking dipterans was expanded by brachyceran Carnidae, Hippoboscidae, Glossinidae and, possibly, their relatives, the Eophlebomyiidae. Also increasing was the number of the invertebrate parasites (Conopidae, Cryptochaetidae, Sciomyzidae). It is of much interest that the structure of the Palaeogene fauna differs so much from that of the Neogene and contemporary ones, in spite of the clear dominance of elements of the contemporary fauna. Only two families, Eophlebomyiidae and Proneottiophilidae, became extinct during the Palaeogene. The youngest living nematoceran families with aquatic immatures are known since the Palaeogene – Cylindrotomidae (Mo Clay, Fig. 325) and Nymphomyiidae (Baltic and Bitterfeld amber). The absence of fossil records of Deuterophlebiidae, which also presumably existed in the Palaeogene, can be explained by their connection with mountain streams (Chapter 1.4.2.1.1).

The current stage of dipteran history began at the Palaeogene-Neogene (Oligocene-Miocene) boundary. Characteristic of that stage was a rapid diversification of the cyclorrhaphous flies (Fig. 326), origination of many new families and massive speciation of their members (EVENHUIS 1994). They have radiated into an enormous range of both larval and adult niches, phytophagy, coprophagy, necrophagy, parasitising different invertebrate and vertebrate hosts, a range much broader than that of all other flies. It has been hypothesised that the development of wide open steppe-like territories in the Oligocene and

Fig. 324 Male and (arrowed) pupal phantom midges, *Chironomaptera gregaria* Grabau (Chaoboridae) from the Early Cretaceous of Laiyang in China (PIN 1738/6, photo by D.E. Shcherbakov); Body length of adult 6 mm

Fig. 325 Cranefly, *Cylindrotoma biamoensis* Freiwald et Krzeminski (Cylindrotomidae) from the Late Eocene or Early Oligocene of Bolshaya Svetlovodnaya in Russian Far East (holotype, photo by D.E. Shcherbakov); wing length 16.5 mm

Fig. 326 Hoverfly, *Eristalis miocenicus* (Stackelberg) (Syrphidae) from the Middle Miocene of Stavropol in N. Caucasus (holotype, photo by D.E. Shcherbakov); body length 16 mm

Miocene, with their extensive herds of large vertebrates leaving huge amounts of dung and corpses, was particularly important for that process (Chapter 3.2). The adaptive explosion of Cyclorrhapha had not been reached at the expense of the former dipteran dominants (Empididae, Dolichopodidae, Rhagionidae), for the new groups were mastering new niches and resources. Only a few Cretaceous and Palaeogene families became extinct or rare in the Neogene. As a result, beginning from the Neogene, the dipteran ecological and taxonomic diversity is certain to expand compared with the earlier stages of the order's history. Another important feature of this last period is that it is impossible to single out any particular dominant trophic group.

To summarise, the faunogenesis of the Diptera had several stages in its development. The Triassic fauna consisted nearly completely of extinct families, so it looks very peculiar. The Jurassic one is peculiar as well, due to the dominance of typically Jurassic taxa (Protoplecidae, Pleciofungivoridae, Protorhyphidae, Archisargoidea), despite the appearance of several families persisting till the present, even if some of these (e.g. Limoniidae and Chaoboridae) played a significant role in oryctocoenoces. Through the Cretaceous and Palaeogene the dominance structure was gradually changing towards the modern one. The typical Jurassic families and subfamilies had gone extinct (Eoptychopteridae, Protorhyphidae, Procramptonomiidae, Protoplecidae, Pleciofungivoridae, Eremochaetidae, Prioriphorinae) or relict (Chaoboridae, Perissommatidae, Ironomyiidae, Sciadoceridae), being replaced by extant ones (e.g. Mycetophilidae, Sciaridae, Bibionidae). By the Neogene that process, along with the rapid

radiation of Cyclorrapha, resulted in the appearance of the polydominant fauna of the modern type. Of special interest is the rapid initial radiation of the order, with both suborders appearing in the Triassic, i.e. at the very beginning of its order evolution. The same is true for some subordinate taxa, particularly for brachyceran flies. For example, the orthorrhaphous Heterodactyla, Aschiza and Schizophora seem to appear suddenly, from nowhere, and in a great diversity all at once.

2.2.1.3.4.5. ORDER PULICIDA Billbergh, 1820. THE FLEAS (=Aphaniptera)

A.P. RASNITSYN

(a) Introductory remarks. Small (less than 2000 described species according S. MEDVEDEV 1998) but well studied group of parasitic insects with a poor fossil record and controversial ancestry. There are a few Mesozoic fossils tentatively assigned to the order which could help clarify the question. The present account of the typical fleas is based mainly on data by G. HOLLAND (1964), KRISTENSEN (1975, 1981), ROTHSCHILD (1975), BOUDREAUX (1979), TRAUB (1985); and S. MEDVEDEV (1994, 1998); fossils are described and discussed by PEUS (1968), RIEK (1970a), PONOMARENKO (1976a, 1986b, 1988c), JELL & DUNCAN (1986), RASNITSYN (1992b) and R. LEWIS & GRIMALDI 1997a).

(b) Definition (for typical fleas only: Mesozoic fossils are treated separately [see (**e**) below]. Size small. Body streamline, compressed, apterous, usually with ctenidia (combs of backward directed spines) on thoracic segments, sometimes also on head and some abdominal segments. Head hardly movable, with 2 stemmata only. Antenna 13-segmented with flagellum modified into tight club, hidden in groove, used as additional clasping device in male. Mouthparts piercing, with epipharingeal (labral) and lacinial (maxillary) stylets, with mandible lost, maxillary palp free, and labial one forming beak sheath. Thoracic nota subequal in length (metanotum not reduced), metapleuron the largest due to leading role of hind leg in leaping, legs strong, not particularly long, saltatory, with 5-segmented tarsus and large clinging claws. Abdomen 10-segmented, with 10th bearing patch of specialised sensoria called sensilium, and cerci. Male with sternum 9 modified in clasping organ, with gonocoxa and gonostylus normally retained, volsella lost, and external aedeagus reduced being replaced by large endophallus in its intromittent function. Ovipositor lost. Digestive tract modified to digest blood (salivary pump, small proventriculus furnished with spines called acanthae and used in disintegrating blood cells, large stomach). Four malpighian tubes. Ovaries neopanoistic. Parasitic (most intermittently, sometimes anchoring with mouthparts for days, rarely burrowing into skin as permanent parasite) in fur of mammals (less infesting those either (semi)aquatic, or lacking any kind of nest or lair, or else much involved in grooming, particularly mutual grooming) and, secondarily and less commonly, in bird plumage. Larva worm-like, legless, 13-segmented, with head eyeless, with one-segmented antenna, mandible of chewing type (but not chewing: PILGRIM 1988), and weakly developed maxilla and labium, with 10th abdominal sternum bearing paired unsegmented appendages possibly of cercal nature, living saprophagousely mostly in nest bottom, very rarely parasitic (burrowing host skin). Pupa adecticous, exarate, sometimes with mesothoracic outgrowths sometimes claimed to represent wing buds, enclosed in loose silk cocoon.

(c) Synapomorphies. The above-listed characters of the typical fleas are almost invariably synapomorphies, and a further list of more subtle synapomorphies can be found in the papers referred to in Section (**a**) above. When extended so as to cover the Mesozoic fossils discussed below, the order losses most of the synapomorphies except for the piercing beak and their parasitic habits.

(d) Range. For true fleas, worldwide, since Eocene till now; also several questionable findings in Late Jurassic and Early Cretaceous of East Asia and Australia, discussed below.

(e) System and phylogeny. For the true (Cainozoic) fleas, the most recent taxonomic and phylogenetic review by S. MEDVEDEV (1994, 1998). All the modern flea classifications look over-split, but Medvedev's is especially so. For the group of the size of ordinary family in a larger insect order, he has invented 18 families, 10 superfamilies, and 4 infraorders.

The questionable Mesozoic "pre-fleas" (Figs. 327–329) show many unusual features but all conceivable as pterosauran parasites living permanently on their wing membrane (*Saurophthirus* Ponomarenko and *Strashila* Rasnitsyn) or (*Tarwinia*) both on membrane and in the fur (PONOMARENKO 1976a, 1988c, RASNITSYN 1992b; BAKHURINA & UNWIN 1995 reject SHAROV's 1971 hypothesis of dense fur covering pterosauran body and wing membrane, proposing rather that only the body was furry, nevertheless pterosaurs were probably bat-like enough from the ectoparasite point of view). This inference is based on the general appearance typical of parasites of the bat wing membrane (flies Nycteribiidae, bugs Polyctenidae, mites Spinturnicidae) all being depressed creatures with widely separated coxae, very long legs, and strong claws. The fossils "pre-flies" are similar and supposedly synapomorphic with true fleas in having relatively short (very short in true fleas) moniliform antenna, and also a triangular, hypognathous head mounted with a piercing beak (not seen clearly in *Tarwinia*). Additionally, *Saurophthirus* and *Strashila* are similar in having a weakly sclerotised, extensible abdomen. *Saurophthirus* and true fleas have ctenidia (absent in *Strashila*, unknown in *Tarwinia*). *Tarwinia* and true fleas agree in having a compressed body (weakly so in *Tarwinia*) and sclerotised abdomen. Supposing the above characters as sound phylogenetic arguments, their distribution could be assessed as indicating the flea clade as initially parasitising the pterosauran wing membrane, and then shifting gradually into their or to mammalian fur. Further, due to the absence of ctenidia, *Strashila* may possibly be hypothesised as forming a sister group of the others,

ctenidiate forms, for which the long, almost straight claw with small but distinct basal lobe (claw structure is unknown for *Tarwinia*) may also be synapomorphic. Finally, *Saurophthirus* could form a sister of *Tarwinia* and true fleas combined, a group marked with putative synapomorphies in sclerotised abdomen and a somewhat compressed body. However, until new fossil material and other new evidence are

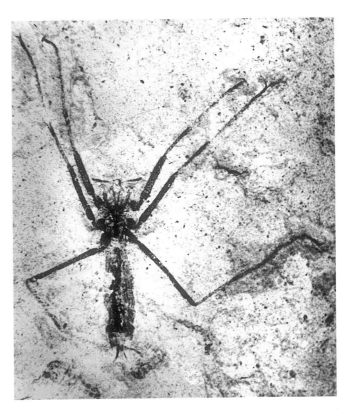

Fig. 327 *Saurophthirus longipes* Ponomarenko from the Early Cretaceous of Baissa in Siberia, supposedly wing parasite of pterosaurs related to fleas (from ROHDENDORF & RASNITSYN 1980); body 12 mm long

Fig. 328 A supposed "pre-flea" parasitising pterosaur wing membrane, *Strashila incredibilis* Rasnitsyn, from the Late Jurassic of Mogzon near Chita in East Siberia (from RASNITSYN 1992b); body 7 mm long

Fig. 330 One of three known flee specimens of *Paleopsylla dissimilis* Peus (Ctenophthalmidae) in the Late Eocene of Baltic amber, holotype (from WEITSCHAT & WICHARD 1998)

Fig. 329 *Tarwinia australis* Jell & Duncan from the Early Cretaceous of Koonwarra in Australia, a possible parasite of pterosaurs (holotype 6.7 mm long, photo courtesy E.F. Riek)

accumulated, the alleged "pre-fleas" should be considered as very tentatively attributed to the order.

(f) History. Mesozoic fossils are found primarily in Transbaikalia (East Siberia). These are *Saurophthirus* (Fig. 327) from the Early Cretaceous deposits in Baissa at Vitim River and, undescribed, from the Semyon locality near Chita, and *Strashila* (Fig. 328) from the Late Jurassic at Mogzon, also near Chita. *Saurophthirodes* Ponomarenko is described from the Early Cretaceous locality Gurvan-Ereniy-Nuru in West Mongolia (PONOMARENKO 1986b). However, this insect is questioned as being parasitic and related to *Saurophthirus* (Chapter 2.2.2). West Australia has yielded the Early Cretaceous *Tarwinia* (Fig. 329) found in Koonwarra, Victoria. As to other alleged fleas from that locality described by JELL & DUNCAN (1986), they do not show sound features justifying the attribution and well may belong elsewhere (PONOMARENKO 1988c). The only true fleas found as fossil are the member of the living genus *Palaeopsylla* Wagner (Ctenophthalmidae) found in the Eocene Baltic amber (Fig. 330), *Pulex (Pulex) larimerius* Lewis et Grimaldi (Pulicidae) and several undescribed Rhopalopsyllidae including *Rhopalopsyllus* sp. from the Miocene Dominican amber (POINAR 1995, R. LEWIS & GRIMALDI 1997).

2.2.1.3.5. SUPERORDER VESPIDEA Laicharting, 1781. ORDER HYMENOPTERA Linné, 1758 (=Vespida Laicharting, 1781)

A.P. RASNITSYN

(a) Introductory remarks. One of the largest (more than 115,000 described living species: A. AUSTIN & DOWTON 2000) and most economically important insect orders which includes such popular forms

as bees, ants, and wasps. The chapter is based mostly on data from RASNITSYN (1969, 1980, 1988c), where the subject is discussed in more detail. Other sources are cited when appropriate.

(b) Definition. Large to very small, usually small insects with 4 unequal membranous wings, sometimes wingless, exceptionally dipterous. Head with occipital and oral orifices either contiguous or separated by hypostoma, postgenae or genae meeting mesally, or else by lower tentorial bridge, but never by gula (independent sclerotisation at or above level of hind tentorial pits). Mandible chewing or cutting (not piercing), maxillae and labium joined in licking or sucking apparatus, palps free unless lost. Thorax variable, often incorporating 1st abdominal tergum (1st sternum lost). Legs generalised except inner (fore) fore tibial spur modified in calcar using in preening of antenna, trochantellus often detached, and tarsus sometimes 3–4-segmented. Wings not roof-like at rest, in flight coupled by hamuli (hook-like setae on hind wing fore margin) linked with a reflected fore wing hind margin. All veins convex except, when present, concave SC and rudimentary CuP. Venation impoverished: veins simple except rarely (in ground-plan) SC branched and fore wing RS forked, CuP rudimentary or lost, anal veins 3 or less, cross-veins 9 or less in fore wing, 7 or less in hind wing, at most 3 cross-veins in longitudinal row. Pterostigmal cell (that bound by diverging and again meeting SC and R just beyond nodal line) thick or lost, exceptionally membranous. M fused with Cu (sub)basally, in fore wing also fused (with few exceptions) with RS near wing midlength (unless respective vein sections lost). Flight functionally (rarely morphologically) two-winged, anteromotoric. Ovipositor rarely reduced, with sheaths only exceptionally penetrating into substrate. Male genitalia integrated (movable as a whole), composed of basal ring (rarely lost) and gonocoxae each bearing apically – gonostylus (movable or fused), ventromedially – volsella (usually mounted with movable digitus and immovable cuspis), and dorsomedially–penial valve (fused to its counterpart at least dorsobasally). Cercus simple or lost. Male sex haploid, female diploid. Malpighian tubes often very numerous. Ovary polytrophic. Adult feeding on nectar (honeydew etc.) and pollen, sometimes licking body fluids of its prey from a puncture made by its ovipositor, rarely using other plant tissues or predatory. Care of progeny usually present, ranging from searching for special condition for ovipositing to highest form of sociality. Larva eruciform, phyto- or zoophagous, ecto- or endobiotic. Head orthognathous (prognathous in some endobiotic larvae). Antenna short, 1–7-segmented, or lost. Eye multiretinous under single lens, or lost. Thoracic and 1–8 abdominal segments uniform except differing in presence of legs, spiracles etc. (if at all). Thoracic legs 1–6-segmented or (in zoophagous forms) lost, claws 1 (rarely 2 with one modified) or none. Abdominal legs (in exobiotic and few endobiotic phytophagous forms) on segments 2–8 or 2–7 and 10, on segment 10 only or, in zoophagous and some phytophagous forms, lost. Cercus short, 1–3-segmented or, usually, lost.

(c) Synapomorphies (those not concerning wings could be inherited from ancestral Palaeomanteida Permosialidae whose body structure is little known). Male haploid. Basal flagellomeres fused in long and thick "3rd" antennal segment. Labial glossae fused in alaglossa. Meta- and especially prothorax reduced in length and simplified in structure, immovably or (pronotum) weakly movably connected to mesothorax; in contrast, propleurosternum highly movable. Metanotum with cenchri fixing wings in rest position. Inner (anterior) fore tibial spur modified into calcar. Wing interlocking apparatus (fore wing hind margin reflected down and hind wing hamuli) developed. Pterostigmal cell thick. Venation impoverished (all veins except SC simple or, fore wing RS, forked). M and Cu fused sub-basally. CuP rudimentary. Anal veins 3 in number, with fore wing 1A bent toward Cu apically, 2A and 3A forming single, sinuate, composite vein which fused with 1A apically; 2A + 3A sinuation with coarse area on lower wing membrane fixing with cenchrus in wing rest position. No vein meeting hind fore wing margin. Hind wing 3A running behind jugal fold, tucking under in folded wing. Cross-veins reduced in number to 3 interradial, 3 radiomedial, 2 mediocubital, 1 or 2 cubitoanal and 2 interanal in fore wing, 3 radiomedial, 1 mediocubital, 1 cubitoanal and 1 interanal in hind wing. 1st abdominal tergum split medially. 1st abdominal sternum lost. Female 8th sternum lost as independent plate. Dorsal ovipositor valves interconnected dorsally, basally. Female gonostylus lost (restored from pupal rudiment in Pamphiliidae). Male genitalia integrated into a unite movable as a whole. Basal ring and gonomacula (membranised gonosylus apex) present. Cercus 1-segmented. Larval eye multiretinous under single lens, cercus short, 3-segmented, labial glands producing silk.

(d) Range. Since Middle or Late Triassic until present, worldwide.

(e) System and phylogeny. Many taxonomic and phylogenetic research papers, both traditional and modern, have been published during the last three decades since RASNITSYN (1969). There are many computerised cladistic and molecular studies among them, considering the order as a whole or its main subdivisions: most of these are either presented or referred to in the Symposium Issue, Phylogeny of Hymenoptera, *Zoologica Scripta* Vol. 28, nos 1–2, 1999. In spite of this wealth of effort, only a very incomplete agreement has been reached thus far, with particularly limited success in the case of the Chalcidoidea. The present version (Fig. 331) follows the consensus achieved, if any, even if my gut feeling is different. In cases of considerable disagreement my own intuitive phylogenetic hypothesis is presented here.

The family structure of the Chalcidoidea is insufficiently understood. No formal cladogram has been proposed for the superfamily, and identification problems of the chalcidoid fossils are hard, so the superfamily appears as a single entity here.

(f) History of the order is comparatively well known, at least at the family level. Of 80 families considered (with Chalcidoidea taken as just one), only four have no fossil record, possibly because they might be of southern origin, and the former Gondwanaland continents are comparatively underexplored palaeoentomologically. Indeed, Pergidae are predominantly and Austrocynipidae entirely southern in distribution, and the same is true for the least advanced subtaxa of Xiphydriidae and Bradynobaenidae (Derecyrtinae and Typhoctinae, respectively). Unless stated otherwise, the review below is based either on RASNITSYN (1980, in part updated in RASNITSYN 1988c and 2000a), or on new unpublished data.

The oldest known hymenopteran fossils come from the Middle or Late Triassic deposits of Central Asia as well as from the Late Triassic of South Africa (SCHLÜTER 2000) and Australia. They show all essential characters of the living family Xyelidae though they differ from younger members of the family clearly enough to be separated as a distinct subfamily, the Archexyelinae. Supposedly they lived in a warm climate and developed in conifer staminate cones as most their living relatives do and as is supposed for their allegedly ancestral palaeomanteids. Adults are also believed to have fed on plant pollen, as it is known for most living xyelid genera as well as for some Early Cretaceous fossil (Fig. 332).

Jurassic hymenopterans are known only from Eurasia. Nevertheless all hymenopteran sub- and infraorders are known upon entering the Jurassic section of the fossil record accompanied by half of the superfamilies, with the majority of the first findings taking place during the first half of the period. It is possible to consider the Early, Middle and Late Jurassic assemblages separately, but the first two, and particularly the Middle Jurassic one are insufficiently known, so the only difference apparent now is the lower taxonomic richness of the older assemblages (see Fig. 481d).

The general appearance of the Jurassic hymenopteran fauna is archaic, with the dominant families being either extinct (Ephialtitidae, Fig. 333, Praeaulacidae, Fig. 334, Mesoserphidae, Fig. 335) or still persisting as undoubted relics (Anaxyelidae, Megalyridae, Fig. 336, Heloridae, Fig. 337), and taking basal positions in their infraorders (see Fig. 331). Among 19 Jurassic families 7 are still living (adding Xyelidae, Pamphiliidae, Siricidae and Roproniidae to the above three families), but again only as relics. The hymenopteran share in non-aquatic assemblages of the Jurassic insects was low, varying from about half a percent (like in Triassic) up to about 2.5% (exceptional cases excluded, for details see RASNITSYN 1980).

The above description concerns only warmer areas of known distribution of the Jurassic hymenopterans (Europe, Central Asia, and to a lesser extent Mongolia). In the more temperate Siberia, Xyelidae still occupy the dominant position (abundance of this family in the Jurassic and Cretaceous, like that of Tenthredinidae in the Cainozoic, reliably indicate a comparatively cool climate). Most regional assemblages, including the Siberian ones (RASNITSYN 1985c), are more or less diverse, except those collected in the West European marine deposits and originated from nearby islands and not from any larger land mass (German Lias: Rasnitsyn & Ansorge, in preparation, and Solnhofen). These assemblages are composed of very few families, even when the number of specimens collected is not very low.

Ecologically the Jurassic hymenopterans occupied a variety of niches. Among these were phytembryophagy (palinophagy: Xyelidae, Xyelotomidae, and perhaps most of the other adult hymenopterans); true (vegetative) phytophagy, probably only or mainly in the form of xylophagy in living (Cephoidea) and dead (Siricoidea) wood; ectoparasitic development on insect larvae in living, dead or decaying

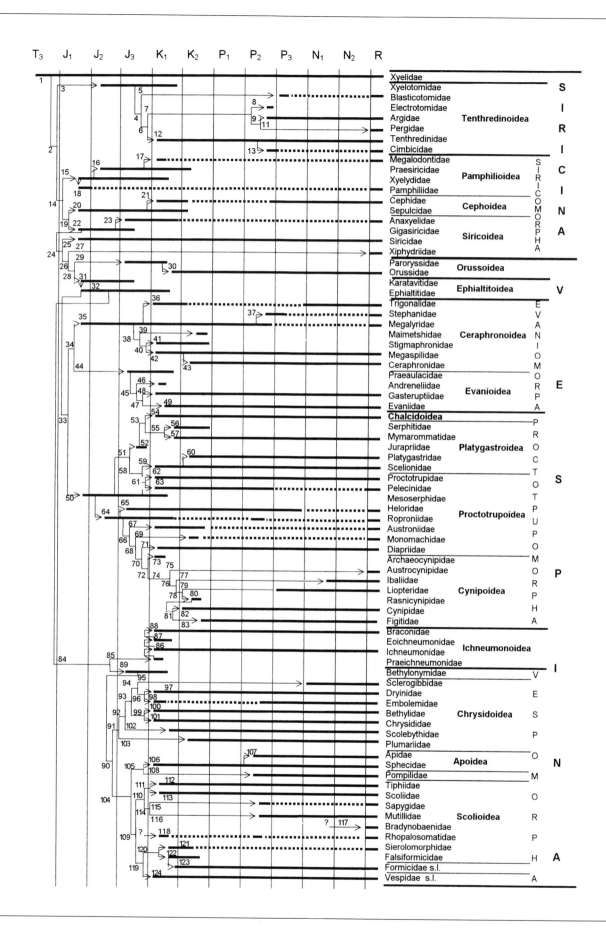

Fig. 331 Proposed phylogeny and system for the Hymenoptera

Time periods abbreviated as follows: T$_3$ – Late (Upper) Triassic, J$_1$, J$_2$ and J$_3$ – Early (Lower), Middle and Late (Upper) Jurassic, K$_1$, K$_2$ – Early and Late Cretaceous, P$_1$ – Palaeocene, P$_2$ – Eocene, P$_3$ – Oligocene, N$_1$ – Miocene, N$_2$ – Pliocene, R – present time (Holocene). Family and superfamily names are in lower case, infraorder and suborder names in CAPITALS; superfamily and suborder names are in **bold**; names of monotypic taxa above family rank are not shown. Arrows show ancestry, thick bars are known durations of taxa (dashed are long gaps in the fossil record); figures refer to synapomorphies of the subtended clades (only more important ones are listed below based on RASNITSYN 1988c, 2000a, BROTHERS 1999, GIBSON 1999, QUICKE et al. 1999, RONQUIST 1999, RONQUIST et al. 1999; unless stated otherwise, venational characters refer to the fore wing):

1 – synapomorphies of the order (Chapter 2.2.1.3.5c).

2 – anapleural cleft intersecting basalar one resulting in detached postspiracular sclerite; maxillary galea simple (unless galea differentiated into endo- and ectogalea is a synapomorphy of Xyelinae + Macroxyelinae as opposed to the Archexyelinae).

3 – flagellar segments thick, reduced in number; fore wing with 2r-rs meeting pterostigma near apex and also M beyond 2r-m, with RS short before RS + M and with 1mcu cell large.

4 – Pronotum short medially; SC lost except as apical branch forming cross-vein in costal space.

5 – Single flagellomere beyond composite 3rd antennal segment.

6 – Spinisternum incorporated into mesothoracic venter anteromesally; male genitalia permanently rotated at 180; larva with eye and antenna separated and with mandible lacking incisive molar flange (for alternative cladogram of Tenthredinoidea *s. str.* see VILHELMSEN 2001).

7 – Larva with subspiracular and suprapedal lobes merged.

8 – Larval legs reduced in segmentation.

9 – Larva free-living with antenna 1-segmented.

10 – Antenna with no flagellomere beyond large, composite 3rd antennal segment; larval leg secondarily two-clawed with posterior claw bubble-like modified.

11 – Antenna with composite segment subequal in size to following segments; hind wing lacking m-cu cross-vein.

12 – Antenna with composite segment subequal in size to following segments; pseudosternal sulci lost; preapical tibial spurs lost; male with abdominal sternum 8 excised apically, hardly seen externally; larva free-living, with prolegs two-segmented.

13 – Antenna clavate; RS lost between submarginal cells; pterostigma long, narrow; larval antenna one-segmented; larval abdominal segments 1–9 with 7 annuli each.

14 – hypostomae meeting between oral and occipital offrices; tentorium issuing upper arms below level of tentorial bridge; propleurae contiguous ventromesally; fore wing RS simple; male gonostylus with gonomacula subapical; larval abdominal longitudinal and oblique sulci lost; larval salivary gland with common envelope and quadrate duct section.

15 – ovipositor small, claw-like.

16 – mesopre-episternum long, reaching fore ventropleuron margin or nearly so; SC lost in both wing pairs.

17 – antenna with 3rd segment subequal to following ones and with flagellum pectinate; R and Cu (including M + Cu) straight from wing base up to pterostigma and 1m-cu, respectively; basal sections of RS and M aligned to form smooth basal vein; 1 + 2A straight, running behind area aspera.

18 – antenna with 3rd segment subequal to following ones at least in width; M + Cu angular (with or without supernumerary cu-a cross-vein at angulation).

19 – outer (hind) foretibial spur reduced; mesofurca with fore arms long, fused in part; larva xylophagous in living plant tissues, with supraanal horn, with appendages reduced in size and segmentation.

20 – SC short, pressed to R, lacking fore branch.

21 – 1r-rs longer than 2r-rs and much longer than RS sections both basad and apicad of 1r-rs; 2 + 3A straight; area aspera and cenchri lost.

22 – costal space moderately narrow; preapical spurs lost (neither of these synapomorphies is particularly sound, so the siricoid stem group might prove to be an ancestor of Cephoidea).

23 – each wing pair with single r-m.

24 – mesoscutum with trans-scutal suture; 3rd antennal segment subequal to following ones at least in width.

25 – head with postgenae contiguous (forming postgenal bridge); mesonotum with lateral sections of trans-scutal suture directed sublongitudinally; basal section of RS subvertical.

26 – mesonotum with incipient adlateral ("parapsidal") lines; SC lost except apical branch forming cross-vein in costal space.

27 – pronotum very short dorsomesally; propleurae neck-like elongate.

28 – postgenae meeting between oral and occipital offrices; tentorial bridge narrow, gate-like bent, issuing corpotendon; incipient lower tentorial bridge present; pronotal spiracular lobes present; postspiracular sclerite lost (fused with basalare?); mesoscutum with adlateral lines well developed; mesopseudosternum wide on fore mesothoracic margin; SC entirely lost; 1r-rs longer than 2r-rs and much longer than RS sections both basad and apicad of 1r-rs; 1st abdominal tergal halves fused into entire propodeum which is abutting with mesopostnotum and well movably hinged with 2nd abdominal segment; male gonostylus lacking gonomacula; larva morphologically simplified, parasitising wood boring larvae.

29 – 1r-rs, 3r-m, 2m-cu and 2 + 3A lost; hind wing lacking all cross-veins.

30 – head with 5 teeth around fore ocellus; female profurcal arms sessile; ovipositor internalised, looped, reaching prothorax rostrally.

31 – wing fixing apparatus (metanotal cenchrus and fore wing area aspera) lost; propodeum sloping backwards.

32 – 1r-rs incomplete toward pterostigma; basal RS section subvertical.

33 – 2 + 3A and 1a–2a lost; metasomal spiracles 1–6 non-functional.

34 – metapostepisternum elongate, articulating mesocoxa well distant of its base.

35 – hind wing with A lost beyond cu-a.

36 – mandibular dentition asymmetrical (left mandible with 3, right with 4 teeth); median mesoscutal sulcus (not line) and respective internal apodeme lost; metasomal sterna hard, convex; ovipositor internalised.

37 – head with 5 teeth around fore ocellus; hind femur dentate ventrally; 2-3r-m and 2m-cu lost; basal section of hind wing RS lost; 1st metasomal segment forming sclerotised tube.

38 – antenna with 16 segments; 1mcu cell small, distant of 2rm; 1st metasomal segment small ring-like.

39 – antennal attachment distant of clypeus.

40 – antenna 11-segmented; tubular veins lost except pterostigma, C, R, 2r-rs and RS apicad of 2r-rs in fore wing and except R in hind wing; 2nd metasomal segment much enlarged; ovipositor internalised.

Fig. 331   *Continued*

41 – hind ocellus almost touching eye; vertex acute; hind coxa disc-like enlarged.

42 – 2 foretibial spurs.

43 – female antenna 10- (male 11-) segmented; single midtibial spur.

44 – propodeum with metasomal articulation far above coxae.

45 – pronotum short medially, immovable attached to mesothorax; mesonotum with median scutal sulcus and internal apodeme lost.

46 – head dorsum (not thoracic dorsum) transversely ridged; fore wing lacking tubular veins posteroapically; hind basitarsus very long; 1st metasomal segment long tube-like.

47 – hind wing with tubular veins lost except R and (M+)Cu.

48 – male antenna 13-, female one 14-segmented; pterostigma wide, semicircular.

49 – thorax short, high, heavily sclerotised; head large, weakly movable; 1st metasomal segment narrow conical (secondarily long tube-like).

50 – medial mesoscutal sulcus lost; 3r cell triangular; 2-3r-m and 2m-cu lost; (M+)Cu straight; hind wing with RS enclosing small cell (or lost), r-m simulating RS, A lost beyond cu-a, replaced by Cu there.

51 – hind wing with no closed cells; possibly: prepectus fused midventrally into symprepectus (reversals occur in various Chalcidoidea, unknown in Jurapriidae, Serphitidae and Mymarommatidae).

52 – antenna 15-segmented; no closed cells except basal, marginal and costal space.

53 – antenna 14-segmented with long scape; M+Cu forming angle with free Cu; 1st metasomal segment tube-like.

54 – flagellar segments with long multiporous plate sensillae elevated and protruding apicad; fore spiracle at mesonotum lateral margin.

55 – 2nd metasomal segment short tube-like.

56 – male antenna with less segment number than female one.

57 – fore- and hind wing with single tubular vein short (about 0.25 fore wing length); hind wing blade hardly surpass short R (based on the least advanced fossil from the Álava amber: No. AA155, Museo de Ciencias Naturales de Alava, Spain).

58 – pronotum short medially, immovable attached to mesothorax; symprepectus fused to pronotum to make it annular; metasomal segments abut and not telescoped (supposing deeply telescoped segments to be secondarily so in Proctotrupidae other than Vanhorniinae); metasomal apex tight at rest; ovipositor internalised.

59 – antenna 14-segmented; no closed cells except basal (1rm), marginal (3r) and costal space; metasomal segments immovably connected (except terminal and internalised ones).

60 – antenna 10-segmented; single tubular vein R distant of wing margin.

61a (if subordinate to node 58) – secondary 'RS$_2$' (probably restored and modifies 2r-m) present.

61b (if subordinate to node 50) – pronotum short medially, immovable attached to mesothorax; symprepectus fused to pronotum to make it annular; secondary 'RS$_2$' (probably restored and modifies 2r-m) present; metasomal apex tight at rest; ovipositor internalised.

62 – metasomal segments moderately elongate in male and extremely so in female, abut instead of telescoped ;

63 – antenna 13-segmented; ovipositor secondarily external, with former sheaths acting as leading intromitting device, and otherwise deeply re-organised.

64 – 1st metasomal segment tube-like.

65 – 1m-cu shortened making 1mcu cell triangular; ovipositor internalised.

66 – antenna distant from clypeus, with scape long, with 15 and 14 segments in male and female, respectively; RS+M lost.

67 – 2cu-a (subdiscoidal) cell receiving 1m-cu far basad of its apex; hind wing lacking closed cells.

68 – hind wing widest apicad of anal area.

69 – antenna with scape elongate; all metasoma and particularly 1st segment much elongate.

70 – 1st (or 1st and 2nd) male flagellar segment(s) modified to have glandular secretion releasing structure; pronotum short medially, immovable attached to mesothorax; no closed cells except basal (1rm), marginal (3r) and costal space; hind wing M+Cu concave.

71 – antenna with scape elongate; metasomal apex tight at rest; ovipositor internalised.

72 – median scutal impression (not true sulcus with internal apodeme) re-gained; RS+M and 2r-rs (and cell 2rm) re-gained; pterostigma faded out apically; propodeum almost lacking dorsal surface.

73 – pterostigma lost apicad of 2r-rs and reduced to short bar basad of it; metasoma short.

74 – C and 1m-cu lost.

75 – antenna naked with last flagellomere short; pronotum projecting over mesopleuron base, covering fore spiracle, median scutal line lost.

76 – pterostigma reduced to vein connecting R apex to wing margin, so as 2r-rs seemingly received by R.

77 – pronotal crest with medial notch; scutellum with 2 rear processes; hind femur short; female 6th metasomal tergum longest.

78 – lower pronotal angle elongate; hind femur thick sub-basally; propodeal spiracle covered by calyptra; metasomal attachment close to that of hind coxa.

79 – occipital carina present (reversal?); male 1st flagellar segment unmodified (reversal); nucha (propodeal articulatory tube) long; hind tibia shorter than femur, with lobe.

80 – mid- and hindcoxae vertical (not directed obliquely backward).

81 – body short, stout, metasoma high and short.

82 – lateral pronotal carina lost; dorsellum narrowed medially; plant feeding (gall inducers and inquilines).

83 – RS+M near to M+Cu fork basally; female 8th metasomal tergum with point of weakness adjacent to articulation between 2nd valvifer and 3rd valvula.

84 – propodeum articulating with metasoma with two tooth-like condyli.

85 – pronotum short medially, immovable attached to mesothorax; medial mesoscutal sulcus (not line) and its internal apodeme lost; fore wing with costal space narrow; hind wing with M straight between M+Cu and r-m; 1st metasomal tergum forming heavily sclerotised plate; 2nd tergum hinged sublaterally with 3rd tergum.

86 – costal space lost (except narrowly, apically); 2r-m (not 3r-m) lost.

87 – 3r-m and 2m-cu with very wide bulla; hind wing r-m received by RS near its base.

88 – 2m-cu lost (supposedly re-gained in Apozyginae).

89 – general appearance bethylid-like; antenna with 13-segments; ovipositor very short externally; female searching for pray within substrate.

90 – ovipositor fully internal, modified into sting.

91 – fore tentorial arm thin, rode-like; profurca proclined; medial mesoscutal sulcus (not line) list; ovipositor with 2nd valvifer intra-articulated.

wood and in the soil (most groups). Endoparasitic development probably originated in the Jurassic Proctotrupoidea, including in less concealed insects, judging from the internal ovipositor in Heloridae (see Fig. 337). Among those hunters of cryptobiotic prey the Bethylonymidae (Fig. 338) are of interest due to their adaptation to pursuing game deep into the substrate, thus opening the way to the primitive habits of the aculeate wasps.

The Jurassic-Cretaceous boundary was a major borderline in the history of the order (technically, this borderline probably lies slightly below the Jurassic-Cretaceous boundary, for the Solnhofen assemblage may already belong to the Cretaceous hymenopteran realm; RASNITSYN et al. 1998). The Late Jurassic extinction was not very impressive, however, while the appearance of new taxa was massive during the Early Cretaceous, when 8 superfamilies and lots of families entered the fossil record for the first time (see Fig. 331). The dominance structure of the hymenopteran assemblages has also changed radically and abruptly, with the exception of the more temperate Siberian assemblages which were still dominated by Xyelidae (Figs. 339–341). In warmer environments, in the earlier section of the Early Cretaceous, the leading position has been taken by archaic Proctotrupidae and Gasteruptiidae (Fig. 342), and later on by early Sphecidae. This was found valid both for Eurasia (including Montsec and Las Hoyas in Spain, Purbeck and Weald Clay in England, Baissa and Turga in Transbaikalia, various localities in Mongolia and probably in China), South America (Santana in Brazil) and possibly Australia (Koonwarra) (RASNITSYN et al. 1998, RASNITSYN & MARTÍNEZ-DELCLÒS 2000). The Koonwarra case might be exceptional because, being free of aculeate wasps as in the early Early Cretaceous assemblages, it is dated as Aptian and so should be rich in Sphecidae. It is of interest that assemblages of the later type often demonstrate lower diversity compared with earlier ones, sometimes in spite of their equally warm climate (RASNITSYN et al. 1998, RASNITSYN & MARTÍNEZ-DELCLÒS 2000).

The ecological diversity of the hymenopterans became enriched in the Early Cretaceous due to the appearance of indisputably phyllophagous Tenthredinidae, angiosperm twig boring Cephidae, nest-building aculeate Sphecidae (Figs. 343, 344) and Vespidae, and oophagous Scelionidae (Fig. 345). Also worth mentioning are the sophisticated

---

Fig. 331 *Continued*

92 – trochantelli lost; RS reclined; 1mcu cell small; hind wing with no closed cells.

93 – prepecti in wide contact midventrally; metapostnotum narrow.

94 – ovipositor with furcula lost.

95 – antenna attached below frontal process, more than 14-segmented; female apterous; embiid parasites.

96 – antenna 10-segmented; homopteran parasites in external cyst made of shed larval moulting casts.

97 – 3r-m lost.

98 – antenna attached well above clypeus; pronotum circular being fused with symprepectus.

99 – ovipositor with 2nd valvifer and 2nd valvula disarticulated; metapostnotum indiscernible externally except laterally.

100 – clypeus with longitudinal carina; intergenal suture long, meeting hypostomal suture at wide angle.

101 – antenna geniculate; apical metasomal segments transformed into telescopic tube.

102 – pronotal collar lost; prosternum large externally; procoxa giving rise to trochanter sub-basally.

103 – male antenna with long erect bristles; female wingless.

104 – 12 antennomeres in female, 13 in male; mandible with cutting edge twisted into moving plane of mandible; hind wing jugal lobe re-established; 1st sting valve with 2-lamellate valve; female cercus lost.

105 – pronotal spiracular lobe inflated; metathoracic interphragmal muscles joining to attach metapostnotum medially; inner (posterior) metatibial spur modified into calcar; female hunting for free-living prey (probably insect nymphs).

106 – pronotal hind margin thick, accommodated into transversal furrow on mesoscutum; metapostnotum long medially (or throughout).

107 – some hairs branched; metatibial calcar lost; hind basitarsus enlarged; larva fed by nectar and pollen.

108 – midcoxal base narrow, tubular; hunting for spiders.

109 – hypopharyngeal pubescence lost; medial mesoscutal sulcus (not line) lost; 1st metasomal sternum thick apically, scarcely overlapping 2nd one.

110 – cells 1mcu and 2rm overlapped; last female sternum depressed toward apex, with lateral flaps delimited by long notch, large, depressed, forming apical offrice transversally slit-like.

111 – mesosubpleuron with lamellae overlaying midcoxal bases; female midtibia short, thick, with outer surface covered with thick spines.

112 – larval metathoracic spiracle lost.

113 – pronotum shortened mesally, immovably attached to mesopleuron; mid and hind coxae level with thoracic venter and widely separated ; cell 2rm starting at (or basad of) pterostigmal base and receiving both 1m-cu and 2m-cu.

114 – last female sternum with apical offrice short slit-like; habits cleptoparasitic on aculeate wasps.

115 – last female sternum long conical with lateral flaps widely overlapping each other.

116 – female apterous.

117 – female apterous; 2nd metasomal tergum with felt lines.

118 – mesosubpleuron with lamellae overlaying midcoxal bases; inner (posterior) metatibial spur modified into calcar; female tarsi very wide; costal space lose except apically.

119 – fore margin of 2nd metasomal sternum with upright lip that closes intersternal gap at their extended position (when sterna are more or less in one plane).

120 – midcoxal base narrow; rotatory movement enhanced between 1st and 2nd metasomal segments; 2nd metasomal sternum with fore lip elongate medially.

121 – RS lost between 1+2r and 2rm cells; 3r-m and 2mcu nebulous (not in Loreisomorpha Rasnitsyn 2000c which should probably be more basal).

122 – female 1st metasomal tergum humped.

123 – antenna geniculate; metapleural glands present; fore wing with 3r-m and 2m-cu lost; both sexes with 1st metasomal segment differentiated as petiole; sociality developed (but not nursing manipulation of progeny).

124 – head behind with subocular furrows (restores occipital sutures); pronotum fixed on mesothorax, short medially, acutely extending above spiracular lobe and tegula; prey air-transported to nest.

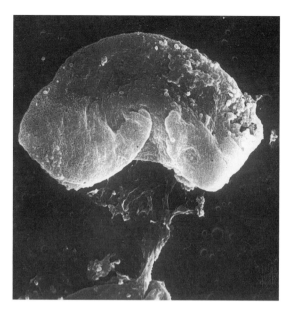

Fig. 332 Perfectly preserved pine-tree pollen (*Pinuspollenites entomophilus* Krassilov) from the gut of *Spathoxyela pinicola* (Xyelidae) from the Early Cretaceous of Baissa in Siberia (from KRASSILOV & RASNITSYN 1982)

Fig. 335 The most primitive proctotrupoid wasp, *Mesoserphus karatavicus* Kozlov (Mesoserphidae), from the Late Jurassic of Karatau in Kazakhstan (from KOZLOV 1968); fore wing 5 mm long

Fig. 333 Archaic parasitic wasp *Stephanogaster magna* Rasnitsyn (Ephialtitidae) from the Late Jurassic of Karatau (from RASNITSYN 1975); body 15 mm long

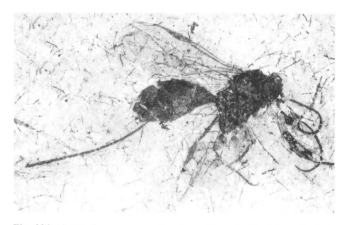

Fig. 334 *Praeaulacus ramosus* Rasnitsyn (Praeaulacidae) from the Late Jurassic of Karatau (holotype); body 9 mm long

Fig. 336 *Cleistogaster buriatica* Rasnitsyn (Megalyridae) from the Early or Middle Jurassic of Novospasskoye in Siberia (from RASNITSYN 1975); body 7.5 mm long

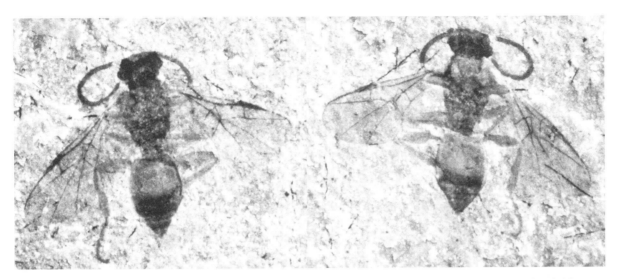

Fig. 337 Part and counterpart of *Protocyrtus* sp. (Heloridae) from the Late Jurassic of Karatau in Kazakhstan (PIN 2784/1220); body length 4.8 mm

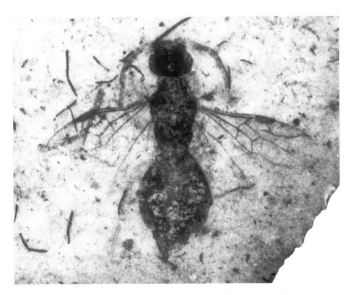

Fig. 338 *Bethylonymellus cervicalis* Rasnitsyn (Bethylonymidae) from the Late Jurassic of Karatau in Kazakhstan (holotype); body length 4.5 mm. Bethylonymidae have probably given rise to all aculeate wasps

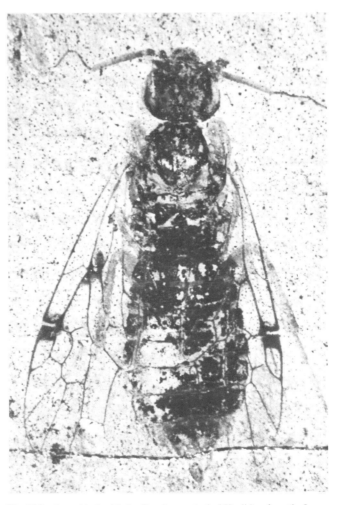

Fig. 339 *Angaridyela vitimica* Rasnitsyn, a typical Xyelidae, from the Lower Cretaceous of Baissa in Siberia (PIN 3064/1919); body 10.5 mm long

habits of Trigonalidae, whose eggs must be eaten by a phytophagous insect larva to hatch and then to continue development in an endoparasitic larva. Also of great note is the first appearance of that most important group of the entomophagous insects, the Ichneumonidae (Figs. 346, 347). Zoogeographically interesting is the occurrence of groups that now have a strictly southern distribution (Fig. 348, for more details on this problem see Chapter 3.4.9).

Towards the end of the Early Cretaceous some assemblages appeared dominated by archaic ants, Armaniidae (Middle Albian of Khetana in Okhotsk Sea region, DLUSSKY 1999a) whose presence is otherwise characteristic of the Late Cretaceous (DLUSSKY 1983; Fig. 349). As to the claim by BRANDÃO *et al.* (1989) that the first ant is found earlier (Myrmeciinae in Santana, Brazil), the fossil in question belongs rather to the sphecid wasp subfamily Ampulicinae. The appearance of Armaniidae indicates that true hymenopteran sociality

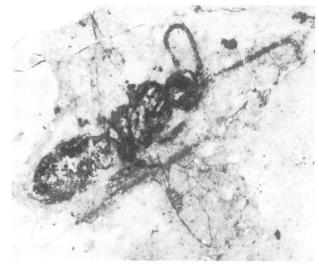

Fig. 340 *Kuengilarva inexpectata* Rasnitsyn, possibly a larva of a member of Xyelidae, from the Jurassic or Early Cretaceous Ukurey Formation in Siberia (holotype, photo by D.E. Shcherbakov); larva 15 mm long

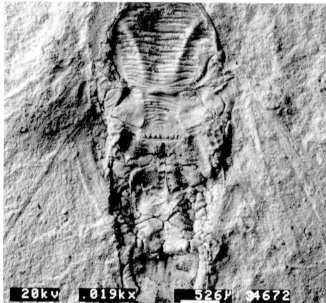

Fig. 342 *Manlaya* spp. (Gasteruptiidae) from the Early Cretaceous of Baissa in Siberia: entire insect (PIN 4210/6434, body 2.8 mm long, photo by D.E. Shcherbakov) and thorax of another, larger species, showing metasomal articulating orifice situated high on propodeum, and transversely-ridged thoracic dorsum that would have enabled the newly emerged insect to chew its way out through wood (PIN 3064/1975, SEM photo by H.H. Basibuyuk)

Fig. 341 Larval Early Cretaceous Xyelidae (Macroxyelinae) restored as in life on an ancient angiosperm by A.G. Ponomarenko and Kira Ulanova

originated in the mid-Cretaceous, only slightly later than termite sociality (Chapter 2.2.2.1.3).

The mid-Albian ants raise another question: when were the hymenopteran assemblages of Early Cretaceous type replaced by those of the Late Cretaceous type (to which the Khetana assemblage belongs)? Like the time borderline between the assemblages of Jurassic and Early Cretaceous type, this one lies lower than the respective stratigraphic boundary, but its precise position is obscure. One reason is that the Khetana assemblage is very imperfectly known (GROMOV *et al.* 1992), while the next youngest Early Cretaceous assemblages of hymenopteran impression fossils (Weald Clay, Montsec, Santana, Bon Tsagan) could all be as old as the Early Aptian

or even Barremian. So the whole of the Aptian and Albian interval may be considered almost as a lacuna in the hymenopteran fossil record.

In fact the above gap is far from complete. Besides the impression fossils, we now have a considerable number of hymenopterans included in both the Late Cretaceous and later Early Cretaceous fossil resins collected in Siberia, Europe, North America, and Lebanon (RASNITSYN & KULICKA 1990, MARTÍNEZ-DELCLÒS *et al.* 1998, Rasnitsyn unpublished), as well as the assemblage from the possibly Cretaceous Burmese amber (RASNITSYN & ROSS 2000). They show appreciable differences in composition depending on the time and place of their origin which vary from near the mid-Early Cretaceous to almost the latest Cretaceous and from the Eurameria and northern

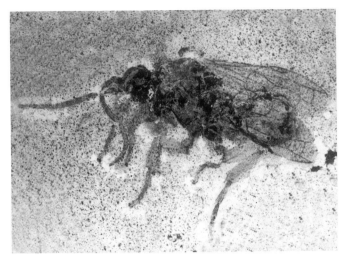

Fig. 343 Ancient digger wasp *Baissodes robustus* Rasnitsyn (Sphecidae) from the Early Cretaceous of Baissa in Siberia (from Rasnitsyn 1975); body with wings 13 mm long

Fig. 345 The oldest, incomplete but otherwise perfectly preserved, scelionid wasp from the Early Cretaceous of Baissa in Siberia (PIN 4210/1190, SEM micrograph by H.H. Basibuyuk); rare in the Early Cretaceous, these wasps were highly abundant in the Late Cretaceous, but not since then

Fig. 344 Another digger wasp *Angarosphex* sp. (Sphecidae) from the Early Cretaceous of Baissa in Siberia (PIN 3064/2062, part and counterpart)

Angara to Gondwana. At the same time, all of them are more similar to each other in their composition compared with the impression fossils, due to taphonomic reasons (Chapter 1.4). In other words, the very major changes in composition of the impression fossil assemblages should take place some time during the Aptian and Albian, and they have no appreciable counterpart in assemblages of the Cretaceous amber inclusions. This might indicate that the above changes occurred near the Barremian-Aptian boundary. However, this hypothesis needs extensive testing.

The transformation of the Mesozoic biota into the Cainozoic one is usually called the mid-Cretaceous one, despite the fact that it implies a long and complicated process (Zherikhin 1978, Dmitriev & Zherikhin 1988, Rasnitsyn 1988b, 1989a). In the case of the hymenopterans, the taphonomic differences between the amber and impression fossil record and the stratigraphic gap in the latter, render the details of the transition quite unclear. If not to specify the precise time of an event within the interval from the Aptian up to the Campanian, it is possible to say that the mid-Cretaceous processes

have resulted in the extinction of most of the Jurassic families (Xyelotomidae, Praesiricidae, Xyelydidae, Sepulcidae, Paroryssidae, Ephialtitidae, Praeaulacidae, Mesoserphidae, Bethylonymidae) and the appearance of numerous new ones (Orussidae, Maimetshidae, Evaniidae, Figitidae, Cynipidae, Serphitidae, Fig. 350, Mymarommatidae, Fig. 351, Plumariidae, Scolebythidae, Sierolomorphidae, Falsiformicidae, Formicidae *s. l.*). Some families that appeared earlier in the Cretaceous or in the Jurassic, changed considerably in their taxonomic composition (Xyelidae, Gasteruptiidae, Sphecidae) or started extensive diversification (Scelionidae, Chalcidoidea, Ichneumonidae), or else lost their most prominent subtaxa (Anaxyelidae, Megalyridae).

The ecological spectrum has changed considerably but not dramatically compared with the earlier Early Cretaceous. First appearing were gall-makers on angiosperms (Cynipidae), highly social ants (Formicidae; Grimaldi *et al.* 1997, Dlussky 1999b) and probably social Vespidae (Wenzel 1990), but not necessarily the bees: the claim by Michener & Grimaldi (1988), even in softened form (Grimaldi 1999), is probably erroneous (Rasnitsyn & Michener 1991). On the other hand, Genise (1999) describes fossil nests of solitary bees from the latest Cretaceous of Uruguay, and the same can be true for the older (Coniacian) nest from Uzbekistan (see Fig. 467); for the Cretaceous and not a Palaeocene age of the Uruguay find see Chapter 3.2.3.2. Among small sized hymenopterans, egg parasites become extremely common: Scelionidae and probably Serphitidae, Mymarommatidae and Tetracampidae.

Late Cretaceous hymenopterans (including the Early Cretaceous amber inclusions, see above) have been collected widely in Eurasia and North America. Incomplete information exists also for the Orapa locality in South Africa (Brothers 1992). No sound and significant differences of a biogeographical nature have been detected thus far, despite the fact that the Lebanese amber and Orapa are Gondwanan in origin unlike other source regions. Highly specific is the Burmese amber assemblage, but it differs equally from all other Cretaceous Gondwanan and Laurasian assemblages, and still more from any Cainozoic one. The reason of this dissimilarity is obscure.

Even taking the problem of a Gondwanan *vs.* Laurasian biogeographic distinction in a wider hymenopteran context, e.g. for all the

Fig. 346 *Tanychora petiolata* Townes, the oldest and least advanced member of the Ichneumonidae from the Early Cretaceous of Zaza near Baissa in Siberia (holotype, fore wing 4.6 mm long): the whole insect (photo from archive of Palaeontological Institute RAS) and antennal segment showing sensillae (SEM micrograph by H.H. Basibuyuk)

Fig. 347 *Tanychorella* sp. (Ichneumonidae) from the Early Cretaceous of Bon-Tsagan in Mongolia (PIN 3559/4492, SEM micrograph by H.H. Basibuyuk), fore wing showing areolet cell (black dot) intermediate in form between those in *Tanychora* and other Ichneumonidae

Mesozoic, the result appears to be the same. We know many groups which are essentially southern now but which are recorded from north continents in the past. These are Megalyridae, Austroniidae, Scolebythidae and Plumariidae. There are also some contrary examples of northern taxa with a southern fossil record (Xyelidae in the Triassic of Australia and South Africa, and Scoliidae Proscoliinae, now strictly Mediterranean, in the Brazilian Early Cretaceous, RASNITSYN & MARTÍNEZ-DELCLÒS 1999). Still more significant is the similarity of particular, roughly contemporaneous, northern and southern assemblages with the same lower ranked taxa. These are the Late Triassic Xyelidae Archexyelinae in S. Africa, Australia and Central Asia, Late Jurassic Sepulcidae Xyelulinae and *Karataus* Rasnitsyn (Ephialtitidae) or similar genus in Karatau in Kazakhstan and in Kotá Formation in India (undescribed fossils kept at the Harvard University, Cambridge, Massachusetts, studied courtesy F.M. Carpenter). The respective earlier Early Cretaceous taxa are *Prosyntexis* Sharkey (Sepulcidae) in Eurasia and Brazil (RASNITSYN & MARTÒNEZ-DELCLÒS 2000), *Westratia* Jell et Duncan (Praeaulacidae) in Australia and Siberia (RASNITSYN 1990a), *Angarosphex* Rasnitsyn in Brazil and widely in Eurasia (RASNITSYN & MARTÒNEZ-DELCLÒS 2000). *Baeomorpha* Brues (Tetracampidae, Chalcidoidea) is common in the Cretaceous amber of the northern continents, the same or a closely related genus is found in the Lebanese amber (unpublished). *Tillywhimia* Rasnitsyn et Jarzembowski (Gasteruptiidae, cf. RASNITSYN *et al.* 1998, family identification is after the Lebanese fossil), *Aposerphites* Kozlov et Rasnitsyn or a closely related genus (Serphitidae) are also reported in both northern and Lebanese Cretaceous amber. The general taxonomic composition of the Lebanese amber assemblage is rather similar to that of northern ones, as described by RASNITSYN & KULICKA (1990), being dominated by Scelionidae (42%) and followed by Megaspilidae and

Fig. 348 Undescribed Austroniidae (the family now rarely caught in Australia) from the Early Cretaceous of Baissa in Siberia (PIN 4210/1685, combined SEM photos by H.H. Basibuyuk)

Bethylidae (11% each) and next by Evaniidae and Falsiformicidae (below 5% each). The vespid genus *Curiosivespa* Rasnitsyn is found both in the Late Cretaceous impression fossil assemblages of Orapa (South Africa, BROTHERS 1992) and in Kazakhstan.

The strict Meso-Cainozoic transition (that is, between the Cretaceous and Tertiary) has been even more gradual than the mid-Cretaceous one (see Fig. 481d). There are many families which went extinct during the Late Cretaceous or first appeared in the Palaeogene

(see Fig. 331). Additionally, near the borderline, Mymarommatidae became very rare and Scelionidae much less common than in the Late Cretaceous. In contrast, Diapriidae, Chalcidoidea, Ichneumonidae and Braconidae became abundant. Nothing indicates, however, that all or the majority of the above events took place close to the Meso-Cainozoic boundary, e.g. during the Maastrichtian and Palaeocene. No important ecological events are noticed around this time, nor significant biogeographic changes, though the latter might be just because

Fig. 349 The most primitive ant *Armania robusta* Dlussky, (Formicidae) from the Late Cretaceous of Obeshchayushchiy in Russian Far East (holotype queen); fossil 13.5 mm as preserved

Fig. 350 *Serphites* sp. (Serphitidae) from the Burmese amber of debatable Late Cretaceous or Early Tertiary age (holotype, body 3.5 mm long). This and related genera are common in the Late Cretaceous and known also in the later Early Cretaceous fossil resins but nowhere else (courtesy of The Natural History Museum, London)

our knowledge of the hymenopteran fossil history is too incomplete for the southern continents.

Later in the Cainozoic, changes in the family composition of the order were modest. This observation is particularly significant because even the incomparably rich assemblages of the Baltic and Dominican amber have yielded only a moderate number of new families. During all the Eocene and Oligocene which are famous for several of the richest insect sites, in addition to the Baltic amber, the first to appear were the Electrotomidae, Blasticotomidae, Argidae, Cimbicidae, Stephanidae, Liopteridae, Pompilidae and Sapygidae, and the Miocene adds Ibaliidae and Sclerogibbidae to the list (for the gall wasps see RONQUIST 1999, the Miocene record of Sclerogibbidae is for Dominican amber and the Middle Eocene record of Pompilidae is for Green River, personal observation at the Paleobiology Department, Smithsonian Institution, Washington). The Electrotomidae was the only hymenopteran family known to have gone extinct during the 65 million years of the Cainozoic.

Of interest are the ecological changes observed in the Cainozoic hymenopteran realm. The most important was the quick diversification and increase of population density of the higher social ants during the Middle and Late Eocene. Their large families practise rapid foraging mobilisation and hence exhibit mass and instant reaction to resource dynamics. As a result, such ant families control effectively, the dynamics of other insects on the controlled territory. This is particularly significant for tropical forests of the modern type, for the high-standing tropical tree depends on the leaf rather than root pump to water its crown. As a result it is very susceptible to defoliation and thus must rely heavily on an

agent able to effectively control the phyllophagous insects, that is on ants (for details see ZHERIKHIN 1978 and Chapter 3.2.4.6. below). The appearance of diverse higher social bees (honeybee close relatives) by the time of formation of the Baltic amber is also noteworthy: being similar to ants in their ability to mass and instantly react to the resource dynamics, they made possible the effective plant pollination under a dense forest canopy where wind is absent and other pollen vectors are scanty.

The above short description of the Cainozoic section of hymenopteran history does not mean its scarcity in further events. Despite the huge amount of accumulated fossil material, the Cainozoic history of the order is unfortunately difficult to appreciate, because the majority of fossils remain unrevised.

2.2.2. INFRACLASS GRYLLONES Laicharting, 1781. THE GRYLLONEANS (=Polyneoptera Martynov, 1938)

A.P. RASNITSYN

(a) Introductory remarks. General scope and concept of the infraclass is modified after ROHDENDORF & RASNITSYN (1980) because of new accumulated evidence.

(b) Definition. Large to medium size, rarely small, insects. Mouthparts chewing. Unless lost, wings resting flat over body (neither roof-like nor spread), with contralateral wings overlapping each other in part or completely, with hind wing anal area enlarged and, when folded, bent down along line running before 2A (unless before 1A), and with common base of anal veins reverted upside down (rarely

Fig. 351 *Palaeomymar agapa* Kozlov et Rasnitsyn (Mymarommatidae) from the Late Cretaceous fossil resin of Agapa in Siberia (from ROHDENDORF & RASNITSYN 1980), body 0.5 mm long. These tiny, probably egg parasitoid, wasps flourished in the Late Cretaceous (rare but widespread now)

enlarged anal area lost in hind wing). Thoracic venter not cryptosternous (except partially so in several extinct Grylloblattida). 3rd valve (gonocoxite 9 with or without its style) forming main working (intromittent) part of ovipositor. Male gonostylus free (not modified into forceps of male genitalia) or, usually, lost. Paracercus lost. Ovaries panoistic (primitive polytrophic in earwigs; ŠTYS & BILINSKI 1990). Development embryonicised, eclosing nymph with compound eyes, ocelli and full number of tarsomeres, and without adult moulting (except Palaeozoic Atactophlebiidae, and in respect of tarsomeres only – Lemmatophoridae).

(c) Synapomorphies. Habits more or less cryptic (in cavities between larger plant debris – trunks, branches, shed lycopsid bark, which formed forest floor in Carboniferous), resulting in tight wing folding over abdomen at rest, with contralateral wings overlapping each other at least partially, and with hind wing anal area enlarged and, when folded, bent down along line running before 2A, and with common base of anals reverted upside down. 3rd valve forming main working part of ovipositor.

(d) Range. Middle Carboniferous (Westphalian) until the present time. Worldwide though less abundant in temperate and especially in cold regions.

(e) System and phylogeny. Taxonomy and relationships of the gryllonean orders are in the state of long lasting debate (cf. HENNIG 1969a, 1980, KRISTENSEN 1975, 1981, 1995, BOUDREAUX 1979, KUKALOVÁ-PECK

1991, WHITING *et al.* 1997, etc.). However, when based on the least advanced taxa and tracing the transitions displayed by the fossil record, their relationships become more apparent. Close relatedness of the lemmatophorinan grylloblattideans and Palaeozoic stoneflies, of roaches, termites (*Masotermes*) and Cretaceous praying mantids, of oedischioid orthopterans, prochresmodoid and aeroplanoid stick insects and titanopterans raises little doubt. For the earwigs, important is the transition from the protoperlinan grylloblattideans to apachelytrid and further, to more advanced earwigs (Chapter 2.2.2.2.3e). The case of the webspinners is less clear. However, *Clothoda* Enderlein (see Fig. 427) shows the characteristically grylloblattidan branching of CuA and, in the hind wing, its angulation at the former junction with M_5, the perlidean synapomorphy. So we follow the three classical superorders Blattidea, Perlidea and Gryllidea, as in ROHDENDORF & RASNITSYN (1980). Besides them, there exists an unplaced residue, the eoblattid-spanioderid-cacurgid assemblage discussed above (Chapter 2.2e). Its members display, both jointly and severally, contradictory evidence of their affinities to better known taxa, and until more knowledge is accumulated, they are treated here as the order Eoblattida, the stem group formally unplaced within Gryllones (Chapter 2.2.2.0.1).

Relationships of the three gryllonean superorders are rather obscure because of absence of really sound synapomorphies. Some cues may follow from new information on the body structure of Paoliida, the most plesiomorphic group of the winged insects (Chapters 2.2e, 2.2.0.1). Their pronotum apparently lacking wide paranota may indicate that the shield-like pronotum with wide paranota is a synapomorphy of the blattideans and perlideans rather than their symplesiomorphy. Therefore they could be considered jointly representing a sister group of the gryllideans. There is an alternative, however, particularly the structure of M. This vein first bifurcates rather early and distantly from its succeeding forking in most Grylloblattida and Orthoptera, with the hind branch desclerotised and concave for a distance sub-basally in grylloblattidans and sometimes in orthopterans as well. None of these characters provide sound evidence of independent development (shield-like pronotum in some palaeodictiopterans, see Fig. 137, individualised MA and MP in many orders). Purely intuitively, a sister group structure (Gryllidea + (Blattidea + Perlidea)) looks slightly more preferable (see Fig. 58).

Tentatively considered as unplaced Gryllones are additionally the Chresmodidae (Fig. 352), a small enigmatic group of the water-strider-like insects known from the mid-Mesozoic (Late and, possibly, Middle Jurassic as well as Early Cretaceous) of Eurasia. Chresmodidae are a subject of long lasting discussion ranging from gerroid bugs to praying mantids, grylloblattidans (as Paraplecoptera) and stick insects (reviewed by MARTÍNEZ-DELCLÒS 1989). They cannot be included into either holometabolous insects, because many of the fossils represent moulting casts of immatures insects, nor into Hemiptera or Psocidea because of the 5-segmented tarsus and well developed cerci (PONOMARENKO 1985a). In general, their position within Cimiciformes is unlikely because the pterothoracic venter shows coxae widely separated and discrimen (midventral suture) apparently absent (Fig. 352). This inference is not obligatory, however, for the discrimen is secondarily lost in a few cimiciform insects including the gerroid bugs which represent the same life form as the Chresmodidae. The bugs being excluded for the above reasons (cerci and tarsi), the cimiciform hypothesis would mean Chresmodidae descending from a long extinct and rather dissimilar group like the Caloneuridea or Hypoperlida, and so looks unlikely. This leaves only Gryllones to which the Chresmodidae may be attributed.

The stick insect hypothesis (based in part on the wing of *Gryllidium oweni* Westwood from the Early Cretaceous Purbeck deposits in England, erroneously ascribed by SHAROV 1968 to *Chresmoda obscura* Germar from the Late Jurassic of Solnhofen in Germany) is refuted by PONOMARENKO (1985a) who failed to find the precostal area in wings of the studied specimens. In turn, this inference is not obligatory either, for some stick insects (e.g. Aerophasmatidae) have a free C lost. However, no serious positive evidence in favour of the phasmatid hypothesis has been presented thus far, while at least the bionomic difference is striking. Stick insects are all plant dwellers and slow on their feet. In contrast, the numerical abundance of both adult and immature *Chresmoda* fossils and their gerroid-like general appearance attest them as an aquatic (sometimes marine) water strider life form (Chapters 3.3.5–6). Hypotheses of a grylloblattidan affinity (PONOMARENKO 1985a)

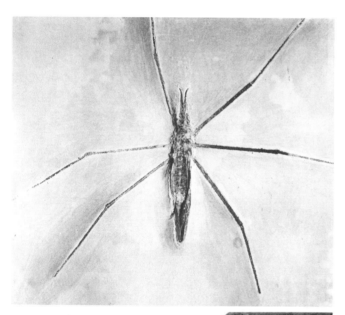

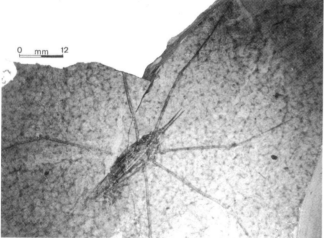

Fig. 352 Chresmodidae, giant water-strider-like gryllonean insects of obscure relationship, *Chresmoda obscura* Germar from the Late Jurassic of Solnhofen in Germany (above; body with antennae and wings 60 mm) from Carpenter 1992a, courtesy of and © 1992, The University of Kansas and Geological Society of America, Inc. and *Ch. aquatica* Martínez-Delclós from the Early Cretaceous of Montsec in Spain (MARTÍNEZ-DELCLÓS 1989, courtesy of Revista Española de Paleontologia)

similarly cannot be tested with the available evidence, and so it is currently better to consider more plausible alternatives.

F. CARPENTER (1992a) has further increased uncertainty by publishing an excellent photograph (Fig. 352a) of a fossil indistinguishable taxonomically from that of the type of *Ch. obscura* (PONOMARENKO 1985a: pl. 3, fig. 2) under the name *Propygolampis giganteus* Germar (unplaced within stick insects). In contrast, *Ch. obscura* appears there unplaced in Orthoptera, with no diagnostic characters mentioned, while the special publication on the subject announced there, was never published (FURTH 1994). Until the puzzle is resolved, it seems safer to consider *Propygolampis* Weyenbergh, 1874 as a junior synonym of *Chresmoda* Germar, 1839, and the family Chresmodidae as unplaced within Gryllones. The family includes, besides *Ch. obscura* from the Late Jurassic of Solnhofen in Germany, the very similar *Ch. aquatica* Martínez-Delclós from the Early Cretaceous of Montsec in Spain (MARTÍNEZ-DELCLÓS 1989; Fig. 352b), the more different *Ch. orientalis* Esaki from the Early Cretaceous of Lingyuan (Lingyen-hsien) in Rehe (Jehol) Province, China (ZHERIKHIN 1978), and an undescribed incomplete fossil from the Middle or Late Jurassic of Bakhar in Mongolia. Incompletely preserved *Saurophthirodes* Ponomarenko from the Early Cretaceous of Gurvan-Ereniy-Nuru in Mongolia (PONOMARENKO 1986b) also might belong here because of a somewhat similar general appearance. The fossils show a moulting slit indicating their nymphal nature which is inconsistent with their alleged relationship to Saurophthiridae (V.V. Zherikhin and D.E. Shcherbakov, pers. comm.) (Chapter 2.2.1.3.4.5).

2.2.2.0.1. ORDER EOBLATTIDA Handlirsch, 1906 (=Cacurgida Handlirsch, 1906, =Protoblattodea Handlirsch, 1906)

(a) Introductory remarks. The present concept of the order is widened and modified since ROHDENDORF & RASNITSYN (1980) because of the new results discussed previously (Chapter 2.2e). In short, the order is a mixture of stem (=ancestral or close to ancestral) and insufficiently known Carboniferous fossils which likely belong to Gryllones and cannot be included, at least for the present, in any of the three gryllonean superorders. Naturally, many of the component taxa have Eoblattida as their temporary home. Others will possibly find their permanent place as stem gryllonean taxa.

(b) Definition of the order is essentially negative: included are fossils free of sound synapomorphies for Scarabaeones (roof-like wing rest position first of all) and showing either gryllonean synapomorphies (hind wing anal area foldable down along the line running anterior of 2A and/or wings laying flat and well overlapping each other at rest), or less sound gryllonean features (e.g. CuA prevailing over RS), or else any significant similarity to a taxon included into Eoblattida for the above reasons. Bionomically the majority of eoblattidans, judging from their characteristic long-legged and/or long-necked appearance, are supposed to retain their ancestral habits of feeding on the plant micro- and macrosporangia (Chapter 2.2e). However, a flat wing resting position indicates that when not foraging, the insects might spend much of their time in various retreats, mostly within dead plant parts. Spiny gerarids could be rather difficult prey for the insectivorous terrestrial vertebrates and possibly for the giant dragonflies as well. Their short, finely notched ovipositor (KUKALOVÁ-PECK & BRAUCKMANN 1992) suggests oviposition into living plant tissues.

(c) Synapomorphies are absent because of supposed stem position of the order in respect to other Gryllones.

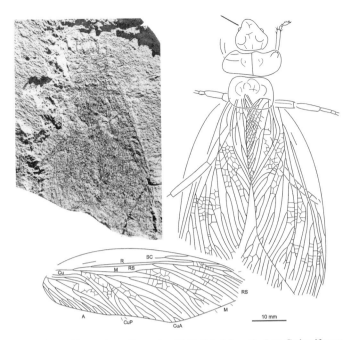

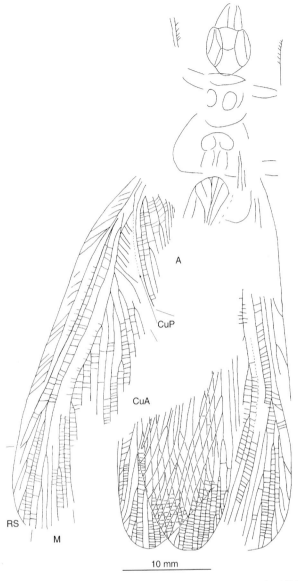

Fig. 353 *Eoblatta robusta* Brongniart (Eoblattidae) from the Late Carboniferous of Commentry in France (holotype): a – general view (from BRONGNIART 1893), b – interpretation of the whole insect and fore wing venation (original)

Fig. 354 *Stenoneura fayoli* Brongniart (Stenoneuridae) from the Late Carboniferous of Commentry in France (holotype); wing span, as preserved, 80 mm (from BRONGNIART 1893)

Fig. 355 *Eoblattina complexa* Bolton (Stenoneuridae) from the Middle Carboniferous of England (holotype); left fore wing moved closer to body in comparison with the original fossil

Fig. 356 *Narkemina angustata* Martynov (Ischnoneuridae) from the Late Carboniferous of Chunya in Siberia; subimaginal and fragment of imaginal fore wing (specimens PIN 3315/250 and 3115/229, respectively, in a scale); subimaginal wing 33 mm, imaginal fragment 14 mm long. Note the difference between the vein structure, thick and diffuse in subimago and thin, clear-cut in imago

Fig. 357 *Protodiamphipnoa* Brongniart (?Ischnoneuridae) from the Late Carboniferous of Commentry in France: a – *P. gaudryi* Brongniart (holotype, original), b – wings of *P. tertrini* Brongniart with eye-spot colour pattern (from CARPENTER 1992a, courtesy of and © 1992. The University of Kansas and Geological Society of America Inc.)

(d) Range. Middle and Late Carboniferous of Europe, Asia (including Siberia), North and South America and South Africa, predominantly in the equatorial belt (BURNHAM 1983). Northern (Siberian) and southern (South American and South African) fossils are not only comparatively rare but also not diverse.

(e) System and phylogeny. No phylogeny can be proposed at present because of our insufficient knowledge of the order. The proposed system is very tentative for the same reason. It seems premature to introduce any taxa above the family level, the more so that the current system looks over-split at both family and genus level and badly needs revision. Descriptions of the taxa below are referred to by BRAUCKMANN (1991) F. CARPENTER (1992a), KUKALOVÁ-PECK & BRAUCKMANN (1992).

Taxonomically, the eoblattid-spanioderid assemblage (Chapter 2.2e) forms a core of the order. It includes Eoblattidae (*Eoblatta* Brongniart, Fig. 353), Stenoneuridae (*Stenoneura* Brongniart, Fig. 354, *Eoblattina* Bolton, Fig. 355, *Anegertus* Handlirsch) and Ischnoneuridae (=Aetophlebiidae, =Narkeminidae; *Ischnoneura* Brongniart, *Ctenoptilus* Lamèere, *Ischnoneurilla* Handlirsch, *Narkemina* Martynov, Fig. 356, *Paranarkemina* Pinto et Ornellas, *Narkeminopsis* Whalley, *?Protodiamphipnoa* Brongniart, Fig. 357). All above genera come from the Late Carboniferous (Stephanian) of Commentry, France, except for *Eoblattina* and *Narkeminopsis* from the Middle Carboniferous (Westphalian B and D, respectively) of England, *Anegertus* from the Middle Carboniferous (Westphalian D) of Mazon

Creek, USA, *Narkemina* from the Late Carboniferous of Siberia, Brazil, Argentina and Madagascar, and *Paranarkemina* from the Late Carboniferous of Argentina. Characteristic of the above families, except (secondarily?) most Ischnoneuridae, are RS that starts near wing base and runs close to R, and except Ischnoneuridae, large lancet-like clavus. Ischnoneuridae retained the supposedly ancestral long-legged general appearance (Chapter 2.2e), and *Protodiamphipnoa* went even further in this direction, while *Eoblatta* displays rather stout habitus with short legs and shield-like pronotum which reminds one of roaches and grylloblattidans. The same is true for *Polyernus* Scudder (Fig. 358) which is different in having RS and the clavus ordinary; it probably deserves family rank and might be only distantly related to *Eoblatta*.

One more family possibly synapomorphic with the above assemblage is Protophasmatidae (*Protophasma* Brongniart, Commentry, Fig. 359) with the clavus somewhat approaching the roach-like form. If this is really a synapomorphy, *Protophasma* may serve as a kind of bridge from the core eoblattidans toward Geraridae, which are famous for their bizarre general appearance (*Gerarus* Scudder from Commentry and Mazon Creek, Fig. 360, *Genentomum* Scudder, *Progenentomum* Handlirsch, *Nacekomia* Richardson, *Anepitedius* Handlirsch, all from Mazon Creek, *Osnagerarus* Kukalová-Peck et Brauckmann from the Late Westphalian of Germany and *Cantabrala* Kukalová-Peck et Brauckmann from the Early Stephanian of Spain) and Homalophlebiidae (*Homalophlebia* Brongniart and

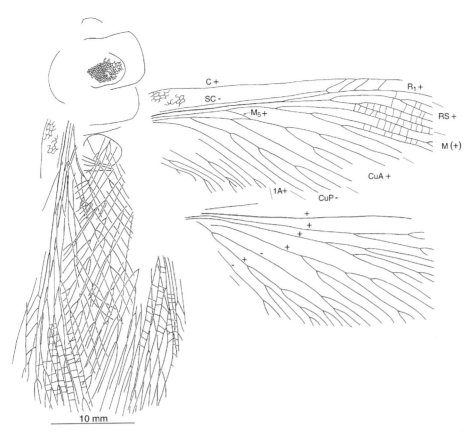

Fig. 358 *Polyernus complanatus* Handlirsch (fam. indet.) from the Middle Carboniferous of Mazon Creek in USA (holotype); general view and interpretation of the wing venation

Parahomalophlebia Handlirsch, Fig. 361, both from Commentry). More important, however, is that their hind wing RS starts from M: this is probably a synapomorphy of Geraridae (Fig. 360), Homaloph-lebiidae (Fig. 361) and possibly Protophasmatidae, whose hind wing is restored with less certainty in that respect (Fig. 359b).

Spanioderidae (*Propteticus* Handlirsch, *Dieconeura* Scudder and possibly *Axiologus* Handlirsch, all from Mazon Creek), Gerapompidae (if distinct of Spanioderidae; *Gerapompus* Scudder, *Cheliphlebia* Scudder and *Palaeocarria* Cockerell (Fig. 362) from Mazon Creek, *?Aenigmatella* Sharov, Late Carboniferous of Zheltyi Yar in Siberia), Apithanidae (*Apithanus* Handlirsch, Mazon Creek) and Cacurgidae (*Cacurgus* Handlirsch and *Spilomastax* Handlirsch from Mazon Creek, *Archimastax* Handlirsch, Namurian C of NE USA, and *Cacurgellus* Pruvost, Westphalian of N. France) are attributed here to Eoblattida, based on the large CuA in conjunction with the absence of alternative evidence of relatedness. This looks like basing a taxon purely on plesiomorphies! It is not illegal for the present approach (Chapter 1.1), but this case may be different, for these taxa usually have CuA so large that it may represent a genuine synapomorphy.

Coseliidae (*Omalia* van Beneden et Coemas, Westphalian A and C of Belgium, and *Coselia* Bolton, Westphalian B of England) are often considered as a subgroup of Cacurgidae with adjacent M and RS branches fused (BRAUCKMANN 1991) and may be real cacurgid relatives. At the same time, Coseliidae share this character with Pachytylopseidae (*Pachytylopsis* De Borre and *Symballophlebia* Handlirsch, Westphalian of Belgium) and additionally they share a rather unusual character, the extra branch of R. The latter character

is found in two other families, Thoronysidae (*Thoronysis* Handlirsch, Westphalian C of Saar Basin in Germany) and Prototettigidae (*Prototettix* Giebel, Westphalian D of Saar Basin in Germany), and may be present in Coseliidae as well (all coseliid wings are incomplete apically). This chain anchors the above families within Eoblattida despite the fact that their CuA is usually only moderately large.

Eucaenidae (*Eucaenus* Scudder, Mazon Creek, Fig. 363) are included into Eoblattidae for a different reason (Chapter 2.2e): these strangely looking insects demonstrate a large RS but show no other affinities to any Scarabaeones. At the same time, it keeps its wings folded horizontally on the abdomen at rest, and supposedly the hind wing anal lobe can be seen bent down under the wing in one fossil ("A" in cited figure).

(f) History of the order is short and uneventful. Most fossils are found in the Westphalian C–D and Stephanian stages of the Middle and Late Carboniferous, respectively. Only *Archimastax americanus* Handlirsch, *Omalia macroptera* van Beneden et Coemas and *Coselia palmiformis* Bolton are recorded from the Namurian C and Westphalian A and B, respectively. With the exception of Spanioderidae and Eucaenidae, all more commonly recorded families (Stenoneuridae, Ischnoneuridae, Geraridae, Cacurgidae), even if over-split, are found in both main Carboniferous localities, Mazon Creek and Commentry, and so both in the Middle and Late Carboniferous as well as in the Old and New World. *Eucaenus*, the only genus of its family, plays the dominant role in the insect assemblage of Mazon Creek. Nowhere else do eoblattidans take so high a position. In spite of this, no Eucaenidae are found elsewhere. Ischnoneuridae are unique in their wide distribution

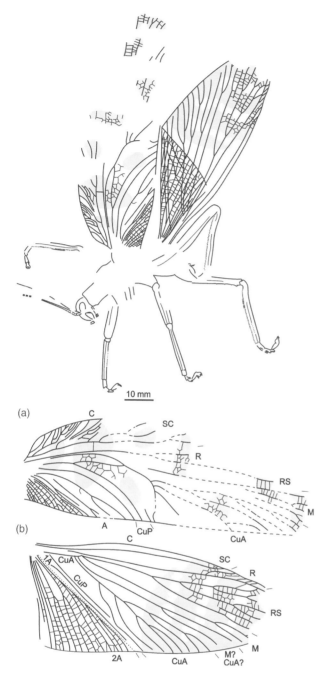

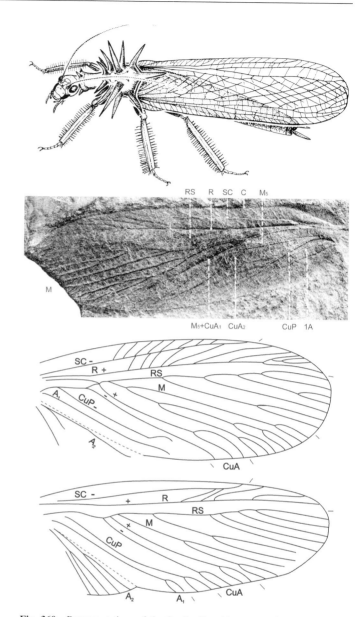

Fig. 359 *Protophasma dumasi* Brongniart (Protophasmatidae) from the Late Carboniferous of Commentry in France (holotype, a – general view, b – wing venation)

Fig. 360 Representatives of the family Geraridae, top to bottom: *Gerarus vetus* Scudder restored as in life (from KUKALOVÁ-PECK & BRAUCKMANN 1992), fore wing of *Gerarus bruesi* Meunier (counterpart with convex veins looking concave, and vice versa, from VIGNON 1929), and hind wings of *G. fischeri* Brongniart (original based on BURNHAM 1983, fig. 17b,c); all from the Late Carboniferous of Commentry in France

covering both the equatorial belt (Europe and N. America) and more temperate territories (Siberia, S. America and S. Africa).

2.2.2.1. SUPERORDER BLATTIDEA Latreille, 1810

(a) Introductory remarks. The present concept of the superorder is essentially new, based on re-interpretation of the order Eoblattida as a part of stem Gryllones rather than as stem blattideans (see Chapters 2.2 and 2.2.2) as well as the new information on the orders included (see below). General information used is from ROHDENDORF & RASNITSYN (1980), KRISTENSEN (1975, 1981, 1995), BOUDREAUX (1979), THORNE & CARPENTER (1992).

(b) Definition. Body size small to large; body shape short and wide, depressed, or elongate, subcylindrical (never compressed). Structure highly diverse. Head capsule with tentorium perforated. Mouthparts chewing, not particularly specialised. Winged, brachypterous or wingless. Fore wing with anal area (clavus), unless reduced, of characteristic structure: wide lanceolate, with veins gently curved according fore or both claval margins, and weakly branching, if at all. Hind wing with CuA stock gently curved (not angled at M_5 junction). Flight functionally four-winged, in-phase, posteromotoric (*sensu* GRODNITSKY 1999; may be different in termites). Cerci usually segmented, mostly not

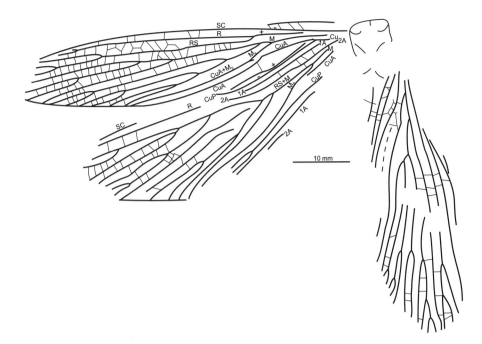

Fig. 361 *Parahomalophlebia courtini* (Homalophlebiidae) from the Late Carboniferous of Commentry in France (holotype)

long. Living forms lacking external ovipositor, with eggs laid in ootheca or, in social forms, naked, in batches or singly. Habits usually cryptic (open but inconspicuous in ambushing mantids), sometimes highly social. Feeding diverse: predatory, phyto- (more often xylo- or mycetophagous), saprophagous, omnivorous, etc.

(c) Synapomorphies. Pronotum shield-like, with paranota circular, much enlarged and concealing head at rest (moderately wide circular paranota can be inherited from a stem gryllonean ancestor: cf. Fig. 358; their appearance in other pterygote taxa including Grylloblattida other than Lemmatophorina as well as in some stem and more advanced Scarabaeones, cf. Figs. 79, 137, is supposedly independent); head opistognathous, fore wing with R and RS weakly individualised, anal area (clavus) wide, lanceolate, with veins gently curved according its fore margin, and weakly branching, if at all (the apomorphic clavus can be inherited from stem Gryllones similar to Eoblattidae, Figs. 353, 355, 358, 359; cercus short. Further synapomorphies (medial ocellus lost, lateral ocelli modified with lens lost, head capsule with tentorium perforated passing circumoesophagal connectives; dorsal longitudinal pterothoracic muscles weak, resulting in postnota and phragmata reduced; proventriculus conical, with ring of teeth; abdominal sternum 1 reduced, female sternum 7 modified into hypopygium) are known for the extant taxa only and may have appeared in advanced and not in stem blattideans.

(d) Range. Since Middle Carboniferous till now; worldwide but much more common and diverse in warmer climates.

(e) System and phylogeny of the superorder is a matter of controversy, despite the fact that the close relatedness of the orders is beyond doubt. The problem is the precise position of the termite and mantid clades on the cladogram of roaches: all possible combinations have been proposed recently with none of the alternative cladograms considered as robust (KRISTENSEN 1995).

Being compared with roaches, the praying mantids display, besides the synapomorphies of the very superorder Blattidea, both the putative synapomorphies with subordinate roach taxa, and their own alleged autplesiomorphies. The latter characters, particularly 3 well defined ocelli, long SC, anal veins joining the hind margin of the fore wing, and 7 abdominal ganglia, seemingly prevent rooting mantids within the roach order, or at least among the post-Carboniferous roaches (HENNIG 1981). At the same time, the putative synapomorphies with higher roaches indicate their much later origin. This concerns the asymmetric three-lobed male genitalia and the "dictyopteran" type of reduced ovipositor (MARKS & LAWSON 1962). The latter feature probably connected functionally with ootheca formation which is different in detail (containing calcium oxalate and hardening slowly to permit attachment to substrate in mantids, while the roach ootheca contain calcium citrate, quickly hardening and portable), and nevertheless indicating synapomorphy in laying the two-row batches of eggs.

Reduction of the external ovipositor is observed in the Jurassic and Early Cretaceous Mesoblattinidae *s. str.* and the Mesozoic families Blattulidae and Umenocoleidae, and is completed in the earliest (Early Cretaceous) representatives of Blattoidea (Blattellidae), Polyphagoidea (Polyphagidae), Mantida and Termitida (Chapters 2.2.2.1.1, 2.2.2.1.2, 2.2.2.1.3).

The fore wing anal veins joining the hind margin (see Fig. 378), and long SC could be re-gained by the mantid ancestor due to elongation of its wings (HENNIG 1981); parallel re-gain of the plesiomorphically long pectinate SC is additionally hypothesised for the Palaeozoic family Archimylacrididae (see Fig. 365). As to the medial ocellus lost in roaches and termites and well developed in mantids, it does represent a serious problem. There are examples, however, of the ocelli lost and re-gained at least in some beetles which are devoid of ocelli except for the medial one in Dermestidae and lateral ones in Derodontidae. If one considers all these character states as secondarily gained by ancestral mantidans, it is possible to derive them directly from the basal polyphagoid family Blattulidae (see Fig. 364).

The termites are also considered here as originating directly and rather lately from a roach ancestor, possibly as a sister group of the polyphagoid lineage (see Fig. 364), for the following reasons. Loss of the

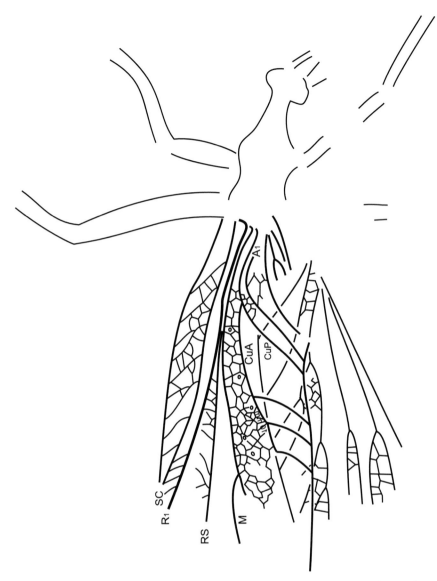

Fig. 362 *Palaeocarria ornata* Cockerell (Spanioderidae) from the Middle Carboniferous of Mazon Creek in USA (holotype); fore wing (left) and hind wing (right) restored from the fragments preserved on both sides of the fossil

fore ocellus and of external ovipositor, lateral ocelli retained but ill-defined externally, and incipient (unless reduced) ootheca indicate termite monophyly with the advanced (non-mylacridoid) roaches, while the retained clypeofrontal suture and not pleated anal fan (in Mastotermitidae) prevent relating them to the higher roaches other than Polyphagoidea. Not without importance is the late (Cretaceous) appearance of termites in the fossil record. Being social insects with mass nuptial flight, termites always had a good chance to be fossilised, and recent numerous and widely distributed findings of the Early Cretaceous but not Jurassic termites make a much older origin of the group unlikely. Additionally an indirect suggestion of a late termite origin is the first Early Cretaceous appearance of two other cases of sociality, in ants (DLUSSKY 1999a) and wasps (WENZEL 1990). As to the proposed synapomorphy of termites with the roach genus

Cryptocercus Scudder (treated variously ranging from a distinct family to a genus within Polyphaginae, e.g., GRANDCOLAS 1997) which share particular intestinal symbionts, this similarity is likely due to direct transfer rather than evolutionary inheritance (ROHDENDORF & RASNITSYN 1980, THORNE 1990, Vidlička *et al.* in preparation). Concerning the early (Triassic) date of the divergence event of termite and polyphagoid lineages inferred from the Middle or Late Triassic record of Blattulidae (see Fig. 364), this contradiction is not fatal: the current study of the Late Mesozoic Mesoblattinidae (Vršanský & Ansorge 2001) seems to indicate a comparatively long existence of an assemblage which is supposedly the stem for both termites and polyphagoids with their descendants. We hypothesise that this assemblage might have given birth to stem termites some time near the Jurassic/Cretaceous boundary.

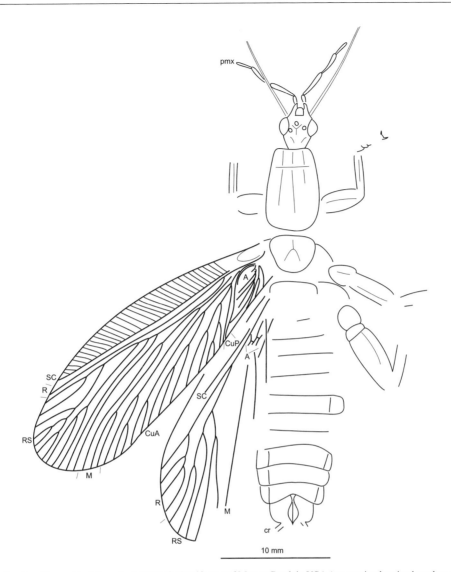

Fig. 363 *Eucaenus ovalis* Scudder (Eucaenidae) from the Middle Carboniferous of Mazon Creek in USA (composite drawing based on the holotype and specimens 38810, PE20790, PE31964, W57, C9242)

2.2.2.1.1. ORDER BLATTIDA Latreille, 1810. THE COCKROACHES (=Blattodea Brunner von Wattenvill, 1882)

PETER VRŠANSKÝ, V.N. VISHNIAKOVA AND A.P. RASNITSYN

(a) Introductory remarks. A comparatively small (about 5000 living and some 1000 extinct species) and ecologically moderately diverse order with mostly cryptic habits and some tendency towards both sociality and carnivory. The cockroach classification system employed here is based on a wide family concept that is more balanced with the available system of extinct taxa.

The main sources used in the chapter are HANDLIRSCH (1906–1908, 1937, 1939), ROHDENDORF *et al.* (1961), RIEK (1962, 1974, 1976b,e), VISHNIAKOVA (1968–1993), LIN (1976–1986), LIN & LIANG (1988), LIN & MOU (1989), SCHNEIDER (1977–1984), SCHNEIDER & WERNEBURG (1993), ZHERIKHIN (1978, 1980a), HONG (1980–1986), PINTO & PURPER (1986), ROTH 1986, DMITRIEV & ZHERIKHIN (1988),

LIN & LIANG (1988), LIN & MOU (1989), THORNE & CARPENTER (1992), SCHNEIDER & WERNEBURG (1993), DOBRUSKINA *et al.* (1997), GRIMALDI (1997a), STOROZHENKO (1997), VRŠANSKÝ (1997–1999), VRŠANSKÝ *et al.* (1999, 2001), VRŠANSKÝ & ANSORGE (2001). Other references are given when appropriate.

(b) Definition. Body wide, depressed, sometimes tuberculate above. Head movable, usually hypognathous and hidden under pronotum, sometimes orthognathous (Umenocoleidae) or possibly prognathous (Raphidiomimidae). Antennae long, multisegmented, setiform, distinctly setose. Eyes well developed, occasionally reduced. Fore ocellus lost, lateral ocelli present (often modified into diffused macula) or reduced. Mouthparts chewing. Pronotum variable in size and form, usually large, ovoid, with wide paranota (except Umenocoleidae), prothoracic sternopleural sclerotisation reduced. Wings normally heteronomous, folded horizontally over abdomen at rest, much reduced or lost in many cases (particularly in female sex). Unless reduced or hard and forming elytron, fore wing wide, coriaceous, often with rich, variable venation, with large, oval anal area called clavus

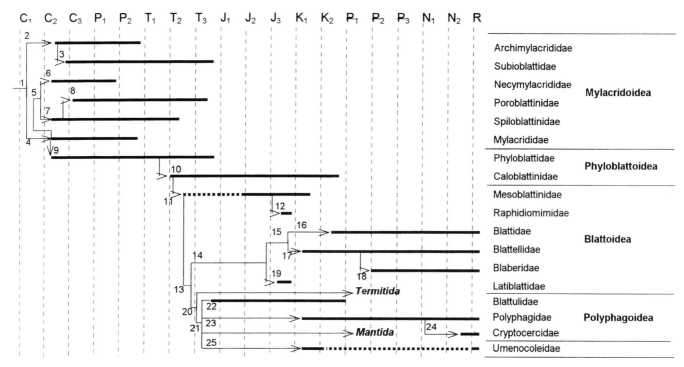

Fig. 364 Phylogeny and system of roaches, the Order Blattida

Time periods are abbreviated as follows: C_2, C_3 – Middle and Late (Upper) Carboniferous, P_1, P_2 – Early (Lower) and Late (Upper) Permian, T_1, T_2, T_3 – Early (Lower), Middle and Late (Upper) Triassic, J_1, J_2, J_3 – Jurassic, K_1, K_2 – Early (Lower) and Late (Upper) Cretaceous, P_1 – Palaeocene, P_2 – Eocene, P_3 – Oligocene, N_1 – Miocene, N_2 – Pliocene, R – present time (Holocene). Two righthand columns are names of families and superfamilies, respectively. Arrows show ancestry, thick bars are known durations time of taxa, dashed lines – hypothesised existence, figures refer to synapomorphies of the subtended clades, as follows:

1 – pronotum shield-like, with paranota much enlarged and concealing head at rest; head opistognathous, fore wing with SC branching mostly sub-basally (subpalmate rather than long pectinate), R and RS hardly separated, anal area wide, lanceolate, with veins gently curved according its margin, and weakly branching, if at all; cercus short; further synapomorphies (e.g., listed by BOUDREAUX 1979) are known for the extant taxa only and may have not appeared until later stages of the roach history.

2 – SC reverted to dense, regular comb of subvertical, weakly branched veinlets.

3 – fore wing small, pigmented, aerodynamically advanced.

4 – costal space very wide, particularly basally, M_5 lost.

5 – basal enlargement of costal space reduced.

6 – fore wings thin, elongate, with R densely branched.

7 – claval boundary particularly distinct, running sub-basally far anterior to wing midwidth.

8 – fore wing with R somewhat reduced, branching regularly; SC short, Cu with numerous terminal branches; size small.

9 – costal space narrow, claval boundary particularly distinct and strongly curved.

10 – Cu branches follow wing posterior margin.

11 – ovipositor very short externally; probably: eggs laid in two-row packages; size small (body length under 20 mm, fore wing about 8–15 mm); fore wing elongate, with fore and hind margins parallel.

12 – head prognathous, long and narrow, with lateral ocelli distinct, wings long, tibia narrow.

13 – fore leg with tibia not longer than femur; fore wing with anal veins giving three or fewer forks.

14 – fore wing venation regular and reduced, with fewer than 55 veinlets meeting wing margin.

15 – external ovipositor lost, ootheca forming within genital chamber.

16 – outer ovipositor valvae fused.

17 – M simple in hind wing.

18 – ootheca rotated and retracted upon formation.

19 – pronotum more than twice as wide as head.

20 – fore wing A lacking apical branchlets; hind wing lacking dense apical branchlets, SC with at most three branches, branches weak, branch bases not fused; 1A branched, concave, arcuate.

21 – venation simplified, with hind wing M fewer than five-branched, CuP simple.

22 – fore wing R with apical branchlets lost

23 – external ovipositor lost, ootheca forming within genital chamber; fore wing more sclerotised, CuP strongly arching.

24 – habits semisocial; wings reduced; intracelular symbiotic bacteria present.

25 – fore wing elytrised, with cup-like structures, head free, orthognathous, pronotum with paranota reduced.

(not clearly expressed in case of highly reduced venation or very narrow wing). Hind wing subtriangular, with ano-jugal space tucking down flat or fan-like pleated, sometimes even rolled at rest, rarely reduced or with apical part folded transversally. Legs cursorial (fore pair raptorial in Raphidiomimidae), coxae large, directed posteromedially, femora slender, bicarinate, spinose particularly below, tibiae long, narrow, abundantly spinose (except Umenocoleidae with legs short, massive). Tarsi 5-segmented, pretarsus with two or rarely one claw, usually with arolium. Abdomen 10-segmented, apical segments modified, often more or less internalised. Cerci multisegmented or with segments fused. Epiproct reduced, paraprocts well defined. Male coxopodites plesiomorphically separated, with styli multisegmented, simple, fused with hypandrium or lost. Male with 2 or 3 asymmetric parameres. Ovipositor external, large or short in most Palaeozoic and Mesozoic roaches, reduced, internalised and used to orient eggs in ootheca in Cainozoic forms, with outer valves formed by coxopodite and occasionally bearing stylus. Breeding biparental (sometimes parthenogenetic), eggs laid inserted into substrate or, in many Cretaceous and all living groups, enclosed into portable ootheca containing two regular rows of eggs; sometimes development viviparous or ovoviviparous. Parental care common, in form of tending ootheca and masking it when laid, or else in incubated immatures in genital chamber and their guarding afterward. Development taking months to years, with 4 to many nymphal instars. Habits cryptic, mostly nocturnal or crepuscular, confined to forest leaf litter and other retreats, thamno-, dendro-, xylo-, or troglobiotic, often myrmico-, termitophilous or synanthropic, rarely aquatic (inhabiting streams, pools and epiphytic bromeliads filled with water; TAKAHASHI 1926). Most groups thermophilous and hygrophilous, selecting humid habitats in xeric environments, heated ones in temperate environments. Diet saprophagous, phytophagous (sometimes of pronounced pollinating activity), rarely predatory, specialised myrmico- and termitobionts often mycophagous, xylobiotic forms with symbiotic bacteria utilising dead wood.

(c) **Synapomorphies**. None because of paraphyly of roaches in respect to mantises and termites.

(d) **Range**. Carboniferous (earliest Westphalian) till now; worldwide, most abundant in warm, both dry and wet climates.

(e) **System and phylogeny** are displayed in Fig. 364 based on the above mentioned sources. Palaeozoic families are after SCHNEIDER (1983), except that Archoblattinidae is synonymised here under Mylacrididae, Compsoblattinidae under Spiloblattinidae, and Diechnoblattinidae under Poroblattinidae. Phyloblattidae are still an heterogeneous group deserving revision. Structure of the Mesoblattinidae-bonded part of the phylogenetic scheme is based on the following hypotheses and observations. Within Mesoblattinidae there are forms (e.g. *Mesoblattina protypa* Handlirsch, *Hispanoblatta sumptuosa* Martínez-Delclòs, *Artitocoblatta asiatica* Vishniakova, etc.) with a short ovipositor and some other characters relating them with the advanced cockroaches (Vršanský & Ansorge in preparation). Due to the revision of the type material of *M. protypa*, all other representatives of Mesoblattinidae are transferred to Caloblattinidae (VRŠANSKÝ 2000). Mesoblattinidae may form a lineage ancestral to the advanced roaches along with the termites and praying mantids, i.e. for all laying egg in two-row packages, with or without true ootheca.

(f) **History**. The oldest roaches representing one of the most archaic families, Archimylacrididae (Figs. 365, 366), come from the oldest Westphalian deposits (Middle Carboniferous) in Germany and Eastern USA. Further in the Carboniferous, the roaches were the dominant insect fossils, at least in the warmer climate of West Europe and North

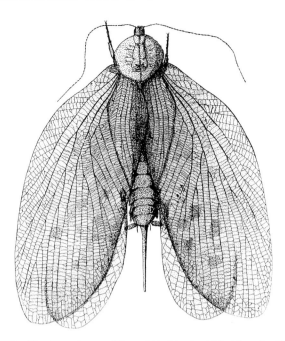

Fig. 365 *Manoblatta bertrandi* Pruvost (Archimylacrididae) from the Middle Carboniferous (Westphalian B) of France, restored by LAURENTIAUX-VIEIRA & LAURENTIAUX (1979); fore wing 44 mm long

America, though since the Late Carboniferous they are known also from Siberia, Tadzhikistan, North Caucasus, China and Brazil. Abundance of the fossil roaches exaggerates their real role in past biocoenoses, however, because the taphonomic factors work in favour of burying of tough, often leaf-like roach fore wings which are better adapted to long distant transport by water currents than most other insect wings.

During the Carboniferous stage of roach evolution two main trends became evident: towards better flight qualities of the relatively narrow wings with more regular venation, and towards better masking function of the wide wing with irregular venation, which was achieved at the expense of the flight quality. The first direction of evolution implied the transition towards a more open and active mode of life, while the second one resulted in the contrary adaptations for cryptic habits. The first direction has been realised independently by Archimylacrididae, Spiloblattinidae, Poroblattinidae and Caloblattinidae, while the second one by Mylacrididae (Fig. 367), whose wings often show striking similarity to pinnules of the widespread Permian and Carboniferous ferns *Neuropteris* Sternberg and *Odontopteris* Brongniart. Dominating the Carboniferous assemblages were the Archimylacrididae, followed in abundance by the Mylacrididae and Spiloblattinidae. Subioblattidae appear in the Late Caboniferous and become rather abundant in the Euroamerican continent. Phyloblattidae appear in the Westphalian B/C. Spiloblattinidae (with fore wings up to 50 mm long) has developed a sexual dimorphism in the wing geometry during that time.

The Permian roach localities are widely distributed, though the majority of fossils are described from West Europe and North America. The Permian fauna differed from the Carboniferous ones mostly in decreased role of roaches in insect assemblages. The Early Permian record lacks local insect assemblages dominated by roaches, and the Late Permian deposits of South Siberia, Australia and South Africa were completely free of roach fossils. Necymylacrididae became

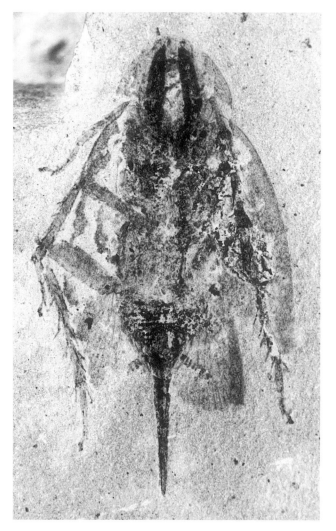

Fig. 366 *Uraloblatta insignis* G. Zalessky (Archimylacrididae) from the Early Permian of Tshekarda in Urals (PU 1/22, photograph by D.E. Shcherbakov); body with ovipositor 25 mm long

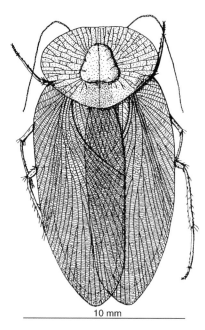

Fig. 367 *Dictyomylacris insignis* Brongniart (Mylacrididae) from the Late Carboniferous of Commentry in France, restored by LAURENTIAUX-VIEIRA & LAURENTIAUX (1981); wing 20 mm long

Subioblattidae (PAPIER *et al.* 1994, 1996b, PAPIER & GRAUVOGEL-STAMM 1994). For the early Middle and Late Triassic, two rich assemblages are known in Middle Asia (Madygen Formation in Fergana) and Australia (Ipswich Series in New South Wales), both dominated by Caloblattinidae and, to a lesser extent, by Spiloblattinidae. Phylogenetically important Blattulidae first appeared there as well. Other assemblages, either of Middle or Late Triassic age, and coming from Siberia, China, Australia, South Africa, and Argentina, are similar to above ones and to each other in being dominated by Caloblattinidae, and as far as known, represent no families absent from Madygen. Phyloblattidae go extinct in the Triassic (Fig. 368). Archimylacrididae and Poroblattinidae probably failed to cross the Perm-Triassic boundary. The Triassic and Jurassic fossils referred to Archimylacrididae agree rather with Phyloblattidae, and the taxonomic position of Mesozoic "Poroblattinidae" is not clear. Roaches are often referred to this family based mainly on their small size, but they look as if they should belong to Mesoblattinidae or to a new family.

Like in other insect groups, the roach fauna was rather uniform throughout the world during at least the second half of the Triassic. Somewhat differing were the Catasian Triassic faunas known in Japan, Maritime Province of Russia, and Vietnam, and characterised by the scarcity of insects other than a few species of Caloblattinidae (several species currently referred to as *Triassoblatta* Tillyard).

Jurassic roach assemblages are known predominantly in Eurasia and can be better referred to the early and late half of the period than to the Early, Middle, and Late Jurassic, because of the absence of rich and reliably dated Middle Jurassic assemblages. The Jurassic assemblages are similar to the Triassic ones in being dominated by Caloblattinidae (Fig. 369). The earlier Jurassic roaches, particularly in Middle Asia, were about as abundant as in the later Triassic. Other earlier Jurassic assemblages, either European or Siberian ones, are less rich in roaches and often dominated by Blattulidae. Mesoblattinidae, presumably ancestral for all extant blattideans including termites and

extinct before the Late Permian. Some peculiar wings resembling that of Umenocoleidae are known from that period. Archimylacrididae and Spiloblattinidae have retained their dominant and subdominant position, respectively, though this is not true for Mylacrididae. Subioblattidae became an advanced group of small size, with protective sclerotised fore wings and enhanced flight abilities. Poroblattinidae with small (only about 5–10 mm long) wings were most abundant in the Euroamerican continent. The reason for the temporary cockroach decline in the Permian was almost certainly the huge diversity explosion of the Permian grylloblattideans which occupied similar, if not the same niches, except that some of them were subcylindrical in form and so better fitted to enter narrow spaces, and that some of the grylloblattideans were pollinivorous (Chapter 2.2.2.2.1).

The Permo-Triassic transition is obscure because of the general paucity of the Lower Triassic insect record. The early Middle Triassic (Late Buntandstein) of Vosges (France) yields the largest amount of material (more than two thousand specimens) representing only 9 cockroach species, dominated by Caloblattinidae, with rare mantises, are known to begin from the Toarcian (later Early Jurassic), although the Triassic records of their alleged descendants Blattulidae implies their earlier origin.

The later Jurassic roaches are known mostly from the rich assemblage in Karatau (South Kazakhstan) and the less rich one in Solnhofen, Germany. They were still dominated by Caloblattinidae, with the second commonest family being Blattulidae (Fig. 370). Mesoblattinidae (Fig. 371) have persisted there as well. Latiblattidae

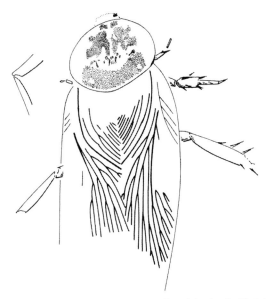

Fig. 368 The latest, undescribed representative of the family Phyloblattidae from the Late Triassic (Newark Supergroup) of Virginia in USA (not registered at AMNH)

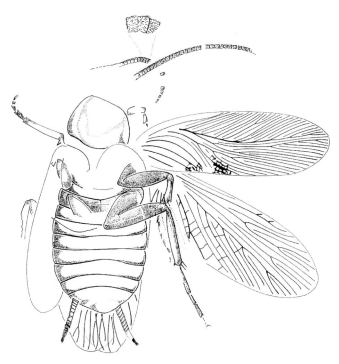

Fig. 370 *Elisama* sp. nov. (Blattulidae) from the Early Cretaceous of Baissa in Siberia (after VRŠANSKÝ 1999b); fore wing about 8 mm long

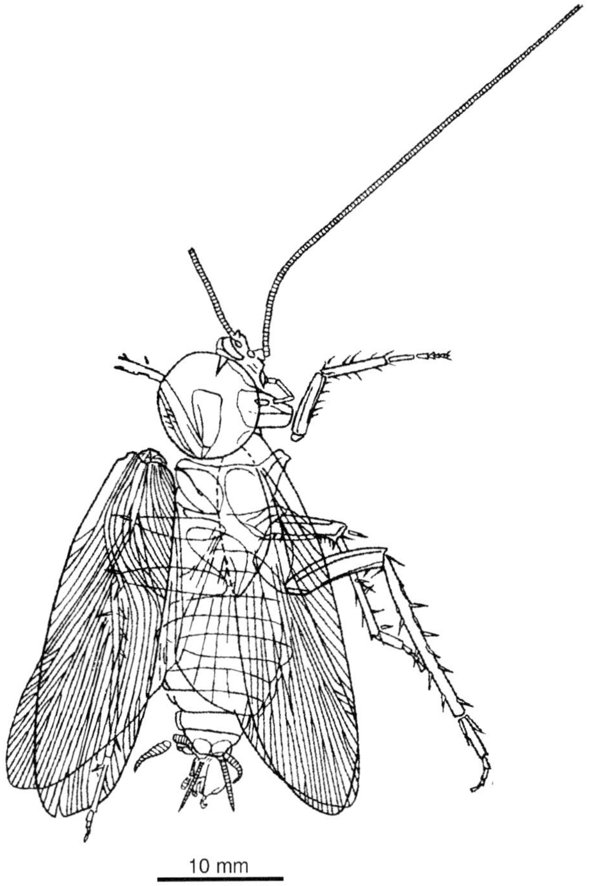

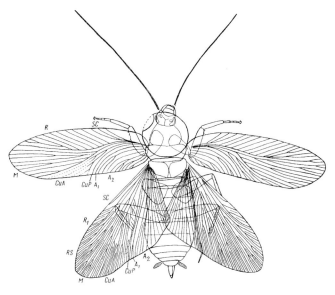

10 mm

Fig. 369 Male *Aktassoblatta pullata* Vishniakova (Caloblattinidae) from the Late Jurassic of Karatau in Kazakhstan (holotype, after VISHNIAKOVA 1971)

Fig. 371 Female *Artitocoblatta asiatica* Vishniakova (Mesoblattinidae) from the Late Jurassic of Karatau in Kazakhstan (from VISHNIAKOVA 1968); fore wing 11.5 mm long

Fig. 372 Predatory cockroach *Raphidiomima cognata* Vishniakova (Raphidiomimidae) from the Late Jurassic of Karatau in Kazakhhstan, general view and close-up of the prognathous head and predatory fore legs (holotype, photo by A.G. Sharov); fore wing 20 mm long

and Raphidiomimidae appeared for the first time in the Late Jurassic of Karatau (South Kazakhstan). These were small groups known only from that locality and thus not changing significantly the general picture of roach fossil history, except that *Raphidiomima* Vishniakova represents the early attempts of roaches to enter the niche of praying mantids (Fig. 372).

The Cretaceous is the most dynamic period in the history of the order. Transition between the Jurassic and Cretaceous is characterised by the change in the dominance order of families, and by appearance of extant families in the fossil record. Caloblattinidae have been replaced by Blattellidae (Fig. 373), Mesoblattinidae (Fig. 374) and, to a lesser extent, by Blattulidae as the dominant families. Early Polyphagidae (Fig. 375) with their small size, sclerotised fore wings and clearly enhanced flight abilities enter the niche earlier occupied by Subioblattidae. Parallel loss of the external ovipositor took place in the common ancestors of the blattellid + blattid clade, in stem termites and independently in polyphagoids and possibly in unenocoleids (see Fig. 364).

Noteworthy is that the advanced blattideans (Blattellidae, Polyphagidae, Mantida, Termitida) appeared as preferring comparatively cool environments, as the composition of two subassemblages of Baissa in Transbaikalia indicates (Chapter 3.2). All the above taxa are found exclusively or predominantly in the comparatively temperate Baissa subassemblage, while Caloblattinidae, Blattulidae and rare Mesoblattinidae fossils concentrate in the warmer subassemblage. It is

of interest that the composition of other Early Cretaceous assemblages accord to one or other of the Baissa subassemblages: those from Purbeck (England), Santana (Brazil), Lebanese amber and Tayasir volcanites (Hauterivian or Barremian of Shomron in Israel) are similar to the cooler subasemblage of Baissa, while those from Bon-Tsagan (Mongolia), Wealden (England) and Montsec in Spain, with their more abundant Mesoblattinidae, resemble the warmer subassemblage. It is evident that this distribution is not only climatically dependent, for many assemblages comparable with the cooler Baissa subassemblage (e.g. Santana) certainly existed in the warm climate. Probably the advanced groups that were less termophilous while entering the fossil record, later occupied warmer environments. All the Gondwanan assemblages (in Brazil, Lebanon and Israel) are dominated by Blattellidae and so resemble the cooler Baissa subassemblage. The Australian assemblage of Koonwarra is known insufficiently to make any conclusion: it includes only a damaged fore wing of Blattulidae (not of Blattidae as JELL & DUNCAN 1986 stated) and several unidentifiable nymphs.

Distribution of the enigmatic Umenocoleidae (Fig. 376) in the Baissa subassemblages is not as selective as in the cases of other families considered. Besides Baissa, they are found in China, Mongolia (Bon Tsagan), Brazil (Santana) and in the Lebanese amber thus far.

The earlier Late Cretaceous (Cenomanian) localities in the East Siberia (Arkagala at Okhotsk Basin, Obluchye at Amur River)

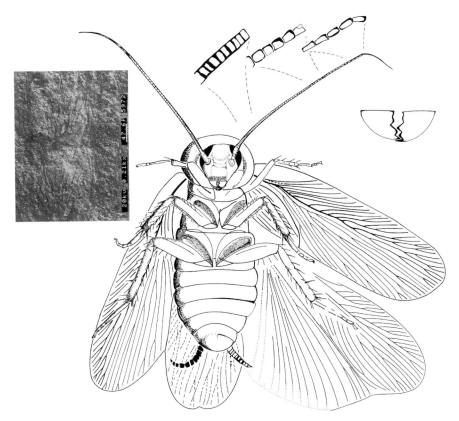

Fig. 373 *Piniblattella vitimica* (Vishniakova) (Blattellidae) from the Early Cretaceous of Baissa in Siberia, general view (modified after Vršanský 1997; fore wing 15 mm long) and close-up of antennal segments (photo by H.H. Basibuyuk & P. Vršanský) to show the preservation state

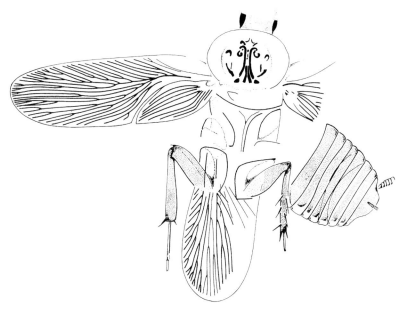

Fig. 374 *Hispanoblatta sumptuosa* Martínez-Delclós (Mesoblattinidae) from the Early Cretaceous of Montsec in Spain (MA1, J. Ansorge private collection, Greifswald; original)

unexpectedly produced the Mesozoic type assemblages dominated by Caloblattinidae. Unlike them, the North American Cenomanian roaches were of modern type (MANSKE & LEWIS 1990). The post-Cenomanian

Late Cretaceous assemblages were essentially Cainozoic in their family composition, including Blattulidae, Blattellidae, and Polyphagidae (Euthyrraphinae) in Turonian of Kzyl-Zhar in Kazakhstan, Blattellidae

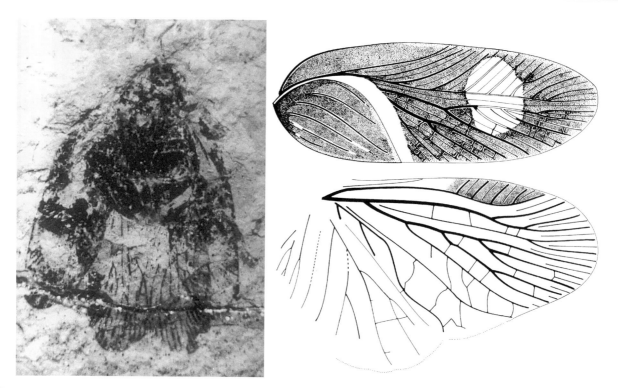

Fig. 375 *Vitisma rasnitsyni* Vršanský (Polyphagidae) from the Early Cretaceous of Baissa in Siberia, general view of holotype and explication of the venation and colour pattern (from VRŠANSKÝ 1999a); wing 6 mm long

in the Turonian of Israel, and Blattidae in China (LIN 1980), although Blattulidae survived till the Santonian.

The Cainozoic roaches look even more modern being represented by living families and predominantly by living genera even in the Late Eocene fauna of the Baltic amber. That assemblage is dominated by Blattellidae, and contains representatives of Polyphagidae (Euthyrrhaphinae, Latindiinae), Blattoidea (Blattidae), and Blaberidae (Perisphaeriinae, Nyctiborinae, and Ectobiinae). Additional taxa found elsewhere in the Cainozoic are Homoeogamiinae (Polyphagoidea) in the Early Oligocene of Colorado, Blaberidae in the latest Oligocene of Germany, Anaplectidae in the Late Oligocene or Early Miocene Mexican amber, and possibly Pycnoscelidiinae in the Late Eocene of England. The impression fossils of the Tertiary period are also dominated by Blattellidae.

2.2.2.1.2. ORDER TERMITIDA Latreille, 1802. THE TERMITES (=Isoptera Brullé, 1832)

N.V. BELAYEVA

(a) Introductory remarks. Rather small (more than 2,300 living species according WATSON & GAY 1991) but an important and common group of social insects. The chapter is based largely on ROHDENDORF & RASNITSYN (1980), with information concerning phylogeny of the order taken from KRISHNA (1970).

(b) Definition. Highly social insects differing from social hymenopterans in having worker and soldier casts represented by immatures of both sexes instead of adult females. Size moderately small to moderately large (6–40 mm long). Only reproductive cast winged (except replacement neotenic reproductives which are wingless), shedding wings after nuptial flight. Mouthparts chewing, mandibles usually asymmetric, in soldiers often large, or soldier nasute (with head nose-like extended forward bearing opening of a frontal gland, Fig. 377). Pronotum flat more or less, moderate size to small, with paranota reduced or lost. Wings long and narrow, folding with full overlap over abdomen, homonomous except Mastotermitidae whose hind wing with anal region bent down at rest without any pleating. Venation varying from impoverished roach-like type in Mastotermitidae (including minute but otherwise typical lancet-like clavus, Fig. 378b) to much reduced in Termitidae (with simple C and RS, weakly branching M and more rich CuA, and without cross-veins), with transverse humeral suture using to break wing off (Fig. 378b). Legs cursorial, not much specialised except for often diminished number of tarsomeres. Abdomen elongate (inflated in elder queen), with 10 segments, sternum 7 enlarged, covering sterna 8 and, partly, 9. Male sternum 9 bearing styli apically. Ovipositor internal, membranous (similar to that in living cockroaches) in Mastotermitidae, lost in other families. Cerci 1–8-segmented. Eggs laid singly or, in Mastotermitidae, in incipient or reduced ootheca: arranged in two rows and covered with film.

Colony (family) consists of dealated male and female, and immatures partly modified into workers and soldiers. In "lower" termites (all except Termitidae) workers are weakly and reversibly modified nymphs of both sexes which carry essentially the same working function as elder unmodified nymphs do, while soldiers are modified irreversibly and more strongly. In Termitidae both workers and soldiers are modified strongly and usually irreversibly, the workers usually being genetically females and soldiers males. Supplementary (replacement) reproductives of both sexes develop neotenically from nymphs of different age (unknown in Termitidae Macrotermitinae). Cast determination depends on various factors, with pheromone regulation being of the most importance.

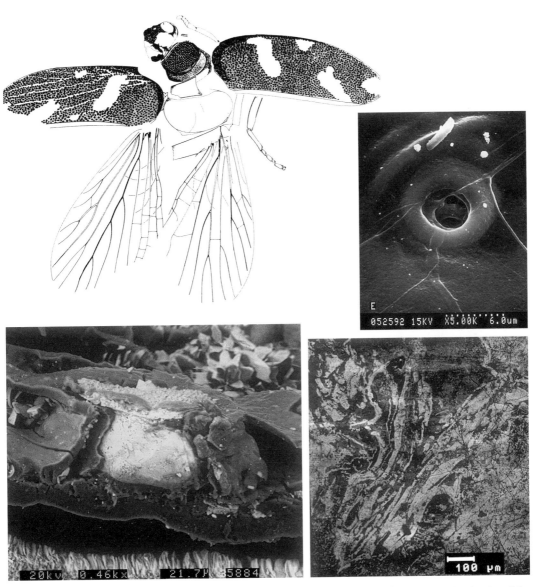

Fig. 376 Beetle-like, diurnal roaches of the family Umenocoleidae, from the Early Cretaceous of Bon Tsagan in Mongolia (line drawing of the whole insect combined from specimens PIN 3559/5783, 5785; wing span 13 mm) and Baissa in Siberia (PIN 3064/7527, SEM photos showing, right, ampulla of the sensilla chaetica and, below left, section of the elytra displaying enigmatic cup-like structures), and *Ponopterix axelrodi* Vršanský et Grimaldi from Santana in Brazil, natural section of the short external ovipositor (below right, from VRŠANSKÝ 1999b)

Termites are feeding mostly on cellulose in various forms (mostly dead wood and other plant material) digested with aid of gut symbionts belonging to the flagellate Polymastigidae and Hypermastigidae in lower termites and to infusoria and bacteria in various Termitidae. The necessity for re-infection with the symbionts that are lost at every moult was probably the main cause of termite sociality. Because of huge size of families and high feeding and digging activity termites play leading role in soil processes in tropic climate.

(c) Synapomorphies (corrected from BOUDREAUX 1979). Sociality in above outlined form; antennae short, moniliform; prothoracic paranota narrow; wings narrow, elongate, both pairs with humeral suture, with venation simplified, with fore branches of R in reduced number but mostly long, with clavus small, hind wing with anal region comparatively small. Loss of the external ovipositor, incipient (or secondarily

reduced) ootheca, and possibly the specialised wood eating combined with flagellate symbionts can be inherited from ancestral roaches (cf. Chapter 2.2.2.1e).

(d) Range. Early Cretaceous till now; warm temperate, subtropic and especially tropic zones (generally within 40° N & S) of all continents.

(e) System and phylogeny, as shown at Fig. 379, is based mostly on data by KRISHNA (1970).

(f) History. The oldest known termites are all from the Early Cretaceous and belonged to the family Hodotermitidae (Figs. 380, 381) (unless stated otherwise, the palaeontological information is from ROHDENDORF & RASNITSYN 1980). They were ancestral for only a part of the order (see Fig. 379), so this is not the very beginning of the order's history. The Early Cretaceous termites comes from England,

Fig. 377 Nasute soldier of *Nasutitermes* sp. (Termitidae) from the Miocene Dominican amber (photo by P. York, courtesy Natural History Museum, London)

Fig. 378 Unidentified mastotermitids from the Middle Miocene of Stavropol in N. Caucasus: (a) winged female with abdomen full of well preserved eggs (PIN 1907/2, body 21 mm long) and (b) incomplete fore wing (PIN 254/2791) 25 mm long as preserved (photos by A.P. Rasnitsyn and D.E. Shcherbakov)

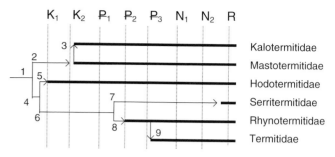

Fig. 379 Phylogeny and system of termites (order Termitida)
Time periods are abbreviated as follows: K_1, K_2 – Early (Lower) and Late (Upper) Cretaceous, P_1 – Palaeocene, P_2 – Eocene, P_3 – Oligocene, N_1 – Miocene, N_2 – Pliocene, R – present time (Holocene). Arrows show ancestry, thick bars are known durations time of taxa, figures refer to synapomorphies of the subtended clades, as follows:

1 – high level of sociality; antennae short, moniliform; prothoracic paranota narrow; wings narrow, elongate, both pairs with humeral suture, with venation simplified, with fore branches of R in reduced number but mostly long, with clavus small, hind wing with anal region comparatively small; intestinal symbionts specialised large flagellates (Hypermastigina, Polymastigina, etc.).

2 – Left mandible with 2 marginal teeth.

3 – Anal lobe lost; tarsi 4-segmented; eggs laid singly.

4 – Anal lobe lost; proventriculus with folds much reduced; ovipositor valves rudimentary; eggs laid singly.

5 – Ocelli lost.

6 – Head with fontanel of frontal gland; tarsi 4-segmented; wing with no SC and with simple R running parallel and close to C.

7 – Mandible with elongated apical tooth widely separated from the only marginal tooth.

8 – Wing with M simple sometimes except apically.

9 – Pronotum and wing scale small; wing without meshwork; intestinal symbionts small protozoans (Sarcodina, Sporozoa, Infusoria) and anaerobic bacteria.

Fig. 380 Unidentified hodotermitid from the Early Cretaceous of Baissa in Siberia (PIN 3064/8583, photo by D.E. Shcherbakov); wing 14 mm long

Spain, Brazil, Mongolia (JARZEMBOWSKI 1981, LACAZA RUIZ & MARTÍNEZ-DELCLÒS 1986, PONOMARENKO 1988b, KRISHNA 1990), and Transbaikalia (Baissa, unpublished; the claim by ROHDENDORF & RASNITSYN 1980 of termites in Transbaikalia refers in fact to mantids:

GRATSHEV & ZHERIKHIN 1993; the alleged termite wing from the Jurassic/Cretaceous boundary in Egypt, SCHLÜTER 1981, is a neuropteran). Hodotermitidae are also found near the Early/Late Cretaceous boundary at Redmond in Labrador, North America,

Fig. 381 *Meiotermes bertrandi* Lacaza Ruiz et Martínez-Delclós (Hodotermitidae) from the Early Cretaceous of Montsec in Spain (from MARTÍNEZ-DELCLÒS 1991, curtesy of the Institut d'Estudis Ilerdencs); fossil length 14 mm

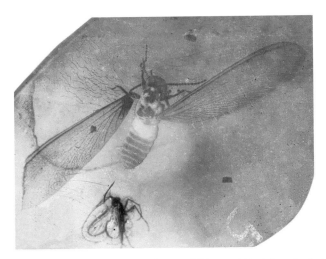

Fig. 382 *Reticulitermes antiquus* Germar (Rhinotermitidae) from the Late Eocene Baltic amber (PIN 964/213, photo by A.P. Rasnitsyn); wing span 14.4 mm

although the undoubtedly Late Cretaceous termites belonged to Mastotermitidae (undescribed specimen from the Cenomanian of Obeshchayushchiy Spring in Magadan Region) and possibly also to Kalotermitidae (this is based on the Burmese amber which has yielded two species of *Kalotermes* but they may be of Early Tertiary age, ZHERIKHIN & ROSS 2000). Further Late Cretaceous (Cenomanian) fossils from French amber were tentatively identified as Kalotermitidae and Mastotermitidae (SCHLÜTER 1978).

In the Palaeogene, termites became more common and diverse, with 9 species of 6 genera coexisting in a single, albeit the best explored, fauna of the Baltic amber; WEIDNER 1955). A new family Rhinotermitidae appeared in addition to those that occurred in the Cretaceous (Fig. 382). And yet the lower termites and particularly Mastotermitidae and Hodotermitidae (Termopsinae) were dominant in all Palaeogene as well as Neogene assemblages. Termitidae, the dominants of contemporary faunas (except those most temperate, populated

mostly by lower termites) first appeared in the Late Oligocene in Aix-en-Provence (NEL 1984b).

The Neogene termite fauna was unexpectedly similar to the Palaeogene one in its taxonomic composition, and different from that which we can see now. As earlier, Mastotermitidae (see Fig. 378) and Termopsinae were the most abundant groups, and even some genera, e.g. *Ulmeriella* Meunier (Hodotermitidae) and *Parastylotermes* Snyder (Rhinotermitidae) persisted for a rather long time, both before and beyond the Oligocene-Miocene boundary (from the Early Oligocene till Late Pliocene and from Late Eocene till Late Miocene, respectively; such a long lifespan is not very common for Tertiary insect genera). Even in the Mexican and Dominican amber, which come from the latest Oligocene and/or Early Miocene age of the equatorial zone, now dominated by Termitidae, that family (see Fig. 377) took a subordinate place (E. WILSON 1971 and A.P. Rasnitsyn, pers. comm.).

2.2.2.1.3. ORDER MANTIDA Latreille, 1802. THE MANTISES (=Mantodea Burmeister, 1838)

†V.V. ZHERIKHIN

(a) Introductory remarks. The mantises comprise a rather small insect group with some 1,800 living species (BALDERSON 1991). They are treated here as a separate order and not as a suborder of Dictyoptera for the reasons explained below. For additional characteristics of the order see BEIER (1968), and a general overview of the fossils was presented by GRATSHEV & ZHERIKHIN (1993).

(b) Definition. Solitary, terrestrial, ambushing, predacious insects living freely either on vegetation or on the ground. Body size moderate to very large (12–130 mm). Body shape usually distinctly elongate, often more or less linear. Head hypognathous, usually extremely moveable, occasionally inserted into prothorax, with large lateral eyes. Antennae slender, filiform, sometimes serrate or pectinate. Ocelli in full number, rarely reduced. Frons with raised plate between fore ocellus and clypeus. Mandibles chewing, strong, toothed. Maxilla with soft galea and toothed lacinia, maxillary palp 5-, labial palp 3-segmented. Prothorax movable, elongate in advanced forms. Pronotal paranota absent, propleura reduced, prosternum sclerotised, with long sternellum. Meso- and metathoraces similar in shape, size and structure, short, with sterna sclerotised, spiracles lost. Fore legs raptorial, with strongly elongate and approximated coxae; with femora more or less swollen, grooved ventrally and with two bordering rows of spines or setae; with tibiae also armed with two rows of spines or rigid setae ventrally (reduced in Amorphoscelidae). Middle and hind legs similar, slender, not specialised; with coxae large and closely approximated; with femora unarmed, with or without small apical spine; with tibiae lacking spines though bearing apical spurs. Tarsi all 5-segmented, very rarely 3- or 4-segmented, lacking arolium and pulvilli. Fore wing narrow, elongate (unless shortened), more or less tegminised (except membranous jugal lobe which can be large and reticulated), usually coloured and occasionally more or less pictured. Precostal area absent or (in some Mantidae, e.g. *Tarachodes* Burm.) very narrow, not wider than C which is strong along fore wing margin. Costal space usually only moderately wide. SC clearly concave, rather long, simple or pectinate, often approximate to R. R most thick compared to other veins, branched or simple; RS usually short, branched or simple. MA approximated more or less to R, branched or simple. M_5 either forming oblique cross-vein or unrecognisable. CuA large, pectinate, with branches simple or partially forked. CuP simple, usually weak. Clavus moderately large to relatively small, flat, not separated by sulcus from remigium. Anal veins 3 to 5, often partially shortened, sometimes interconnected, ending freely in hind wing margin or, in advanced taxa, partially joining CuP. Numerous intercalary folds present between longitudinal veins. Cross-veins numerous, more or less regularly distributed, in part often reticulate. Hind wing broad, especially in advanced taxa, membranous, with wide anal area fan-like folding at rest. C strong along fore wing margin. SC long, simple or pectinate. R branched distally or simple. RS long and usually simple. M and CuA usually with few or no branches. CuP and 1A simple, much approximated and parallel. 2A with first branch (vannal vein according to SMART 1956) usually lost, second branch forked, other anals simple. Wings often shortened and with abnormal venation or even lost either in female or in both sexes. Abdomen flattened more or less, with 8 pairs of spiracles, first being positioned at intersegmental membrane while following ones at ventral margin of respective terga. Tergum 10 forming supraanal plate in both sexes. Sternum 1 indistinguishable. Subgenital plate formed by sternum 7 in female and sternum 9 in male, in latter case bearing paired styli. Cerci multisegmented, long to short. Male genitalia strongly asymmetrical, partially concealed by sternum 9, comprised by basisclerite, paired epiphalli (left with titillator), and hypophallus; hook process (=L3 sclerite after McKITTRICK 1964) missing (KLASS 1997). Ovipositor short, weakened, of dictyopteran type (MARKS & LAWSON 1962): primary external valves reduced and replaced by laminae 1. Alimentary canal relatively straight, gizzard small, with internal teeth, midgut with 8 caeca, salivary glands long and complex. Abdominal ganglia 7. Testes large and very complex; paired vesicula seminalis and a mass of accessory glands which produce spermatophore opening into ejaculatory duct. Female with large accessory glands producing ootheca and with single spermatheca all opening into genital atrium. Ootheca containing calcium oxalate (BOUDREAUX 1979), hardening only after deposition and always attached to substrate. In a few species of Photininae female guarding ootheca and young nymphs (TERRA 1992, 1996). First nymph moulting either within ootheca or while being connected to it by a silk thread ("intermediate moult").

(c) Synapomorphies are partly the predatory adaptations (head and prothorax highly movable; eyes large, lateral; mandibles strongly toothed; fore legs raptorial; gizzard reduced; alimentary canal straight; salivary glands long), and partly those providing the cryptic habits (elongate body and wings, costal space not wide, SC prolonged, R and RS with branching reduced and shifted distad). Further synapomorphies are the raised frontal plate, tarsi lacking arolium and pulvilli, middle and hind legs almost devoid of spines, phallic complex lacking hook-process, and attached ootheca.

(d) Range. Since the Early Cretaceous until the present. Modern mantises are thermophilous (widespread but scarce in the warm temperate and absent in the cold temperate areas).

(e) System and phylogeny. Living mantises are generally classified into 8 to 14 families (BEIER 1968, ROY 1987, TERRA 1995, EHRMANN 1997), and two extinct families have been added by GRATSHEV & ZHERIKHIN (1993). Further undescribed families are recorded from the Cretaceous and Palaeogene [see (**f**) below]. No taxa of intermediate rank between the family and order are currently used, and the system now in use seems artificial (see below).

Cretaceous fossils are unfortunately mostly incomplete, showing either wings or raptorial leg but not both, thus preventing assessment of the heterobathmy level. GRIMALDI *et al.* (2000, fig. 43g) illustrate a completely preserved winged adult from the New Jersey "amber" but this fossil is undescribed and not much detail can be discerned from the photograph. Nevertheless a comparative study of the fossils and less advanced living mantises allows us to hypothesise the minimum array of the groundplan character states of the order. It would include the wing moderately narrow (length ca. three time width) and rather weakly sclerotised, with weakly convex margin; SC moderately long, with numerous, long, S-like bent branches; the subcostal area moderately wide, bearing intercalary vein; R with numerous and partially forked distal branches, with RS unrecognisable among them; M with 3–4 branches; M_5 present as an oblique cross-vein; CuA with the short stem and 6–10 partially forked branches; CuP arched clearly; the clavus large and wide, easily breaking away similarly to that in roaches, with 5 anal veins all reaching the wing hind margin; the jugum indistinct. In hind wing R with several apical branches, 2A with anterior branch, and the anal region moderately large. Front femur only slightly incrassate, with the two rows of the uniform, stout, erect setae along all the lower margin. Front tibia not hooked, with setae similar to that on femur, apical setae being symmetrical, with tarsal articulation terminal. Body relatively wide, pronotum short, body size moderately small, fore wing length about 10–30 mm.

According to P. Vršanský (pers. comm.) an undescribed fossil from the Early Cretaceous of Brazil seems to be very close to the ancestral mantid type in having a short and broad pronotum and abdomen, a broad fore wing with numerous branches on R, M and Cu, and short fore legs with similar and rather thin setae on both the tibiae and femora as well as in its small size (the fore wing about 8 mm long).

In all of the above respects, the Chaeteessidae represents the most plesiotypic group among living mantises. The family is treated here as a paraphyletic complex probably ancestral to other mantid families (except the Brazilian fossil mentioned above). Chaeteessidae comprise, besides the living Neotropical *Chaeteessa* Burm., also a number of extinct Cretaceous and Palaeogene genera known from isolated wings, except for the Late Cretaceous *Chaeteessites minutissimus* Gratshev et Zherikhin which is based on the sole incomplete fossil indistinguishable from living immature *Chaeteessa* (Fig. 383). The Cretaceous *Cretophotina* Gratshev et Zherikhin (Fig. 384) seems to be at least venationally the most plesiotypic genus. More advanced forms occurred in the Early Cretaceous (*Vitimophotina* Gratshev et Zherikhin which is unique in the family in having the fore wing with

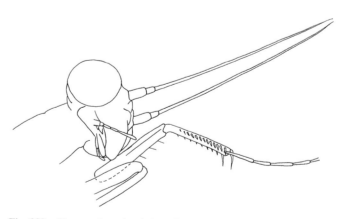

Fig. 383 *Chaeteessites minutissimus* Gratshev & Zherichin (Chaeteessidae) from the Early Cretaceous of Baissa in Siberia (holotype); fore tibia 0.6 mm long

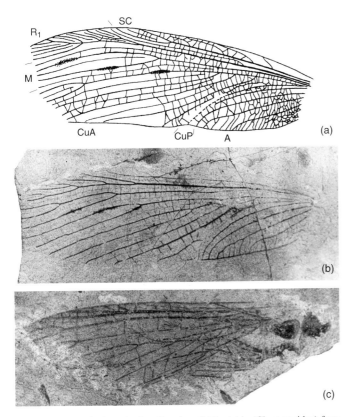

Fig. 384 *Cretophotina tristriata* Gratshev & Zherichin (Chaeteessidae) from the Early Cretaceous of Baissa in Siberia (holotype): (a) line drawing of the holotype (fore wing 30 mm long as preserved), and photos (by D.E. Shcherbakov) of PIN 1989/2487 (fore wing 29 mm long as preserved) and (b) PIN 1989/2489 (hind wings 19 mm long as preserved) tentatively assigned to the same species

Fig. 385 *Cretomantis larvalis* Gratshev & Zherikhin (Cretomantidae) from the Early Cretaceous of Baissa in Siberia (holotype, photo by D.E. Shcherbakov); right hind tibia 2 mm long

SC extremely long and RS with only two branches). The Palaeocene *Arvernineura* Piton may even be synonymous with modern *Chaeteessa* (NEL & ROY 1995). On the other hand, the highly plesiotypic *Lithophotina* Cock. indicates the longer existence of archaic mantises, this time up to the Early Oligocene.

The Early Cretaceous Baissomantidae are clearly apotypic in the fore wing with uniquely long, well individualised, and branched RS, and probably also in hind wing with weak first branch of 1A. In other venational characters Baissomantidae are similar to *Cretophotina* and are likely to constitute an early derived branch of the ancestral chaeteessids.

Jersimantis Grimaldi (GRIMALDI 1997a) from the Late Cretaceous New Jersey amber may represent a family of its own. It combines the plesiomorphic weakly incrassate setose fore femora and setose fore tibiae with apomorphic structure of the fore tibial apex, bearing two strong, articulated, unequal teeth instead of setae, as in Chaeteessidae. The genus is also autapotypic in lacking ocelli.

The Cretaceous Cretomantidae (Fig. 385) are synapomorphic with praying mantids other than Chaeteessidae, *Jersimantis* and possibly Baissomantidae whose leg structure is unknown, in their dentate fore femur and tibia. On the other hand, in the structure of the fore tibial apex cretomantids resemble *Jersimantis*, thus either tooth at tibial apex or spines at femur and tibia should be homoplasies. Unfortunately the wing venation of both *Jersimantis* and Cretomantidae is unknown.

The other, living mantid families are synapomorphic in having hooked fore tibial apex with lateral tarsal articulation. The taxa

grouped here show high level of the heterobathmy which obscures their relationship. In the fore leg structure each Mantoididae, Metallyticidae, Amorphoscelidae, and Eremiaphilidae + Acanthopidae + Mantidae + Empusidae are autapomorphic, seemingly less modified (or secondarily simplified?) in Mantoididae. At the same time, the male genitalia in Mantoididae is of the less advanced type than even in Chaeteessidae: this supports their sister-group relations to all other living mantises (KLASS 1997). Mantoididae, Metallyticidae, Amorphoscelidae, Eremiaphilidae, and some Acanthopidae retain short prothorax. Metallyticidae also keep weakened anterior 2A branch in their hind wing unlike other families (SMART 1956), as well as the plesiomorphic type of the fore wing R and CuA branching along with the arched CuP. Mantoididae, Amorphoscelidae, Acanthopidae, and Mantidae are generally characterised by the simple R and RS, except RS is branched in the mantid subfamilies Choeradodinae and Orthoderinae which are synapomorphic in the pronotum emarginate anteriorly, and R is forked in occasional genera within Acanthopidae and Mantidae (*Helvia* Stål in the former family, *Stagmatoptera* Burm., *Parastagmatoptera* Sauss. in Vatinae, *Coptopteryx* Sauss. in Photininae within Mantidae). An advanced type of the sex determination (male X_1X_2Y, female $X_1X_1X_2X_2$ instead of plesiomorphic male X0, female XX) occurs in Orthoderinae, Choeradodinae, Comsothespinae, Vatinae, some Liturgusinae, Toxoderinae and Mantinae within the Beier's Mantidae, as well as in few Acanthopidae. Such a chaotic distribution of ancestral and derived characters among the "higher mantids" indicates that the system of the group deserves a revision.

(f) History. The mantid fossil record is very fragmentary because of their taphonomically unfavourable terrestrial and highly sedentary habits, combined with too large a body size to be easily trapped into fossil resins. Surprisingly enough, their most ancient Early Cretaceous records seem to be most frequent, thus characterising the oldest mantids as less strictly sedentary than their descendants. Unfortunately, the richest faunas of the Baltic and Dominican amber remain undescribed, and only preliminary data on their taxonomic composition are available (*"Chaeteessa" longialata* Giebel comes in fact from a Quaternary copal and not from the Baltic amber: HENNIG 1966b).

Hypothesising an origin of the order from the Mesozoic Blattulidae (Chapter 2.2.2.1e), we need not follow HENNIG (1966a, 1981), BOUDREAUX (1979) and others in supposing a Palaeozoic origin of praying mantids. The oldest known fossils may well represent the early stage of mantid evolution, and yet their oldest fossil locality Baissa in Transbaikalia has yielded a diverse assemblage including primitive Chaeteessidae and Baissomantidae, more advanced Cretomantidae, and even supposed living Amorphoscelidae (GRATSHEV & ZHERIKHIN 1993). As mentioned above, a somewhat younger undescribed fossil from the Crato Formation, Brazil, possibly represents a family of its own which may be even more primitive than either the Chaeteessidae and Baissomantidae (P. Vršanský, pers. comm.). The numerical dominance of Chaeteessidae and other primitive lineages is likely to have been retained throughout the Cretaceous and possibly even the Palaeogene. Chaeteessidae was probably the commonest family in the Palaeogene deposits occurring in the Palaeocene (Menat, France: NEL & ROY 1995), Eocene (Baltic amber: EHRMANN 1997), and Lower Oligocene (Bol'shaya Svetlovodnaya, Russian Far East, and Florissant, Colorado, USA: SHAROV 1962, GRATSHEV & ZHERIKHIN 1993). Mantoididae are also recorded from the Baltic amber (EHRMANN 1997), and NEL (1998) mentions a primitive mantis probably representing an undescribed family from the Lower Eocene of Paris Basin, France. Several undescribed fossils from the Late Cretaceous or Early Palaeogene Burmese amber possibly represent one more morphologically primitive extinct family (P. Vršanský, pers. comm.). The oldest higher mantid related to living *Choeradodis* Serv. occurs in the Palaeocene of France while the record of Empusidae from the same deposits is erroneous (NEL & ROY 1995). Undescribed Liturgusidae and Mantidae are recorded from the Baltic amber (BEIER 1968, EHRMANN 1999), with the latter family numerically dominating (EHRMANN 1999). An undescribed fossil from the Middle Eocene Green River Formation (Fig. 386) resembles Hymenopodidae in the wing venation, and *Eobrunneria tesselata* Cock. is the only higher mantid described from Florissant. Chaeteessidae and Mantidae occur in the Saxonian amber of disputable age (EHRMANN 1999).

The Neogene fauna is poorly studied but seems to be dominated by advanced families. With the exclusion of the dubious *"Mantis" protogaea* Heer, the only mantids described from the Neogene are *Mantis boettingensis* Zeuner from the Middle Miocene (ZEUNER 1931) and the living species *Mantis religiosa* L. from the Pliocene of Germany (BEIER 1967). Undescribed Chaeteessidae are recorded from the Lower Miocene Dominican amber together with much more abundant higher mantises, including Liturgusidae, Tarachodidae, Mantidae and Vatidae (EHRMANN 1999). Liturgusidae and Tarachodidae are also mentioned from the Quaternary copals of America and Madagascar (EHRMANN 1999).

Fig. 386 A mantis (?Hymenopodidae) from the Middle Eocene of Green River in USA (PIN 4621/347, photo by A.P. Rasnitsyn); fossil 12 mm long as preserved

2.2.2.2. SUPERORDER PERLIDEA Latreille, 1802 (=Plecopteroidea Martynov, 1938)

A.P. RASNITSYN

(a) Introductory remarks. One of the three main groups of the gryllonean insects, which flourished during the Permian and yet retained significant diversity and ecological importance through the Mesozoic and up to the present, in spite of being rather inconspicuous and not particularly familiar to laymen. The present treatment of the superorder is essentially inherited from ROHDENDORF & RASNITSYN (1980), except the order Zoraptera which is transferred to Caloneuridea (Chapter 2.2.1.2.1.3).

(b) Definition. Size small to large. Body more or less depressed or subcylindrical. Head with mouthparts normally directed forward but easily changing their orientation when necessary, or permanently prognathous. Eyes and ocelli usually well developed, sometimes reduced. Antenna variable, rarely short, usually seti- or moniliform. Mouthparts chewing, sometimes reduced more or less, rarely modified. Oral cavity and occipital foramen contiguous or, more often, separated by gular sclerite (or, in webspinners, by postoccipital bridge). Pronotum of large or moderate size, with or without distinct paranota. Pterothoracic segments homonomous more or less, winged or wingless, with wings overlapping each other over abdomen at rest, usually heteronomous, with expanded hind wing anal area, fore wings sometimes heavily sclerotised, long or short, hind wing folding transversely at rest. Rarely wings homonomous due to reduced anal expansion and venational simplification. Wing venation variable, often rich to very rich. Legs variable, usually cursory, with tarsus 3-segmented except usually 5-segmented in stem order Grylloblattida. Abdomen rarely modified pre-genitally except gill formation in some water forms (mostly in immatures). When present (only in most Grylloblattida), ovipositor usually of plesiomorphic structure, long or, less commonly, short. Male genitalia variable, with gonocoxites IX free or fused, with their styli present or lost but not specialised as claspers used in copulation (cerci sometimes forming claspers), and with no true aedeagus. Ovaries panoistic (primitively polytrophic in earwigs). Malpighian tubes numerous. Development gradual, highly embryonicised, with newly hatched nymph differing from adult almost exclusively in absence of wings and genitalia, in the correlated differences in proportions and general appearance, and in less number of antennomeres. Some grylloblattidans (Atactophlebiidae, Lemmatophoridae) are exceptional in having tarsomere number growing postembryonically (Fig. 387) and (Atactophlebiidae) in retaining archemetaboly (winged insects moulting, Fig. 388).

(c) **Synapomorphies.** Hind wing with CuA stock (not CuA fore branch, as in Gryllidea) angled at junction with M_5. Plesiomorphic nature of the character state cannot be ruled out, however, for such bending is quite natural at point of merging of any two veins, and the blattidean state of the character where CuA is straight or gently curved could be secondary so as well. If this is the case, and if additionally the wide circular pronotal paranota represent their symplesio- and not

synapomorphy with the roaches (Chapter 2.2.2e), grylloblattideans might take an ancestral state in respect to Gryllidea and so might have no synapomorphies of their own. Monophyly with the gryllideans can be supported by the possible synapomorphy in M bearing 2 well formed main branches, MA and MP, with MP being (or, rather, having a tendency to be, that is an underlying synapomorphy displaying a tendency to be) desclerotised and somewhat depressed sub-basally (see Figs. 396–398).

(d) **Range.** Late Carboniferous till now, worldwide.

(e) **System and phylogeny.** Affinities of the four orders here included into Perlidea, viz. Grylloblattida, Perlida (stoneflies), Forficulida (earwigs), and Embiida (webspinners), have been the subject of long lasting debate (HANDLIRSCH 1906–1908, MARTYNOV 1938c, ROHDENDORF 1962, 1968a, SHAROV 1968, KRISTENSEN 1975, 1981, 1991, 1995, BOUDRAUX 1979, ROHDENDORF & RASNITSYN 1980). Neither morphological nor molecular researches (including those relying on the modern cladistic computerised methods, cf. for instance various cladograms published by WHITING et al. 1997 and WHEELER 1998) are found to be conclusive. That is why the present proposal is based on the morphological transitions observed between taxa rather than on pure character combinations. These transitions are rather straightforward between the lemmatophorinan grylloblattidans and stoneflies (SINITSHENKOVA 1987, STOROZHENKO 1998), being particularly apparent for their nymphs (compare a grylloblattidan nymph, Figs. 389, 392, a nymph which may belong to either order, Fig. 390, and the undoubtedly stonefly nymphs, Figs. 410, 411, 413–415, 418). Lemmatophorinan grylloblattidans and archaic earwigs (those earlier classified in a distinct order Protelytroptera, now Protelytrina) are also similar enough to infer their close relatedness (STOROZHENKO 1998, Chapter 2.2.2.2.3). Additionally, both stoneflies and the least advanced earwigs (*Apachelytron* Carpenter and Kukalová) display hind

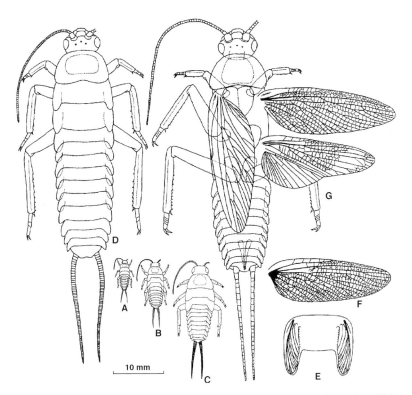

Fig. 387 Postembryonic development of *Gurianovaella silphidoides* G. Zalessky (Atactophlebiidae) from the Early Permian of Tshekarda in Urals (STOROZHENKO 1999)

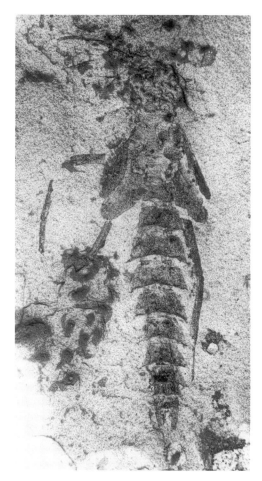

Fig. 388 Mass burial of *Atactophlebia termitoides* Martynov (Grylloblattida: Atactophlebiidae) from the Late Permian of Tikhiye Gory in the eastern European Russia including fore wings of the three succeeding subimaginal stages (younger and older below, intermediate above; part of the material studied in detail by SHAROV 1957, 1966a); the wings, from the youngest to the eldest, 22, 28 and 41 mm long (specimens PIN 2295/45, 84, and 72, respectively, photo by A.P. Rasnitsyn)

Fig. 389 *Liomopterites gracilis* Sharov (Liomopteridae) from the Late Permian of Kaltan in Siberia (holotype), looking like ordinary stonefly nymph except for 5-segmented tarsi and abdominal paranota (from ROHDENDORF *et al.* 1961); body length 9 mm

wing vein CuA angular at junction with M₅, the putative perlidean synapomorphy. Webspinners are much too specialised and have no indisputable pre-Cainozoic fossil record. As a result, their relationship is more obscure, although the characteristic CuA branching and, in hind wing, basal angulation, retained by Clothodidae (see Fig. 427) do indicate grylloblattidean affinities and permits us to root the webspinners preliminary near Sheimiidae and Permembiidae [unless the enigmatic Jurassic Brachyphyllophagidae (Chapter 2.2.2.2.4e) really prove to be webspinner relatives: in that case, the order's relations should be completely reconsidered].

2.2.2.2.1. ORDER GRYLLOBLATTIDA Walker, 1914 (=Notoptera Crampton, 1915, =Grylloblattodea Brues et Melander, 1932, +Protorthoptera Handlirsch, 1906, =Paraplecoptera Martynov, 1925, +Protoperlaria Tillyard, 1928)

S.YU. STOROZHENKO

(a) Introductory remarks. Grylloblattidans are one of the least diverse of today's insect orders, yet they were among the most abundant and diverse insects during the Permian (Figs. 391, 392) and have given birth to all other perlideans (stoneflies, webspinners and earwigs). The present account is based on the recent review by STOROZHENKO (1998).

(b) Definition. Size usually medium but variable, with wings from 2.3 to 91.5 mm long (*Permembia* Tillyard and *Olgaephilus* Storozhenko, respectively). Head prognathous. Mandibles chewing, mouthparts generally not much modified. Antennae multisegmented, usually long (short in Probnidae). Ocelli present in full numbers or (in recent forms) reduced. Pronotum transverse or elongate, with or, rarely, without paranota. Pterothoracic segments homonomous. Legs usually unmodified, with 5-segmented tarsus, but in adult Probnidae modified for burrowing, with 3-segmented tarsus. Tarsal segments increasing in number postembryonicly (known in Atactophlebiidae, Lemmatophoridae and Euryptilonidae). Wings membranous or weakly tegminised with venation variable. Fore wing SC terminating on C, rarely on R. RS arising from R usually before fore wing midlength. M base free or sometimes anastomosed or fused with either R or CuA. M commonly divided into neutral MA and MP which is concave and desclerotised in fore wing near its midlength. M₅ rarely present connecting M stock with CuA, never free. CuA sinuated near wing base, often divided into branched CuA₁ and simple straight CuA₂. CuP

Fig. 390 *Sylvonympha tshekardensis* Novokshonov et Pan'kov, an aquatic nymph from the Early Permian of Tshekarda in Urals (holotype with body 13.2 mm long, photo by D.E. Shcherbakov), which is morphologically intermediate between the grylloblattidan and stonefly nymphs

simple, weak, concave. Anal region of fore wing relatively narrow, usually without rich venation. Hind wing with RS arising more basally than in fore wing and often anastamosing with M. MP base commonly desclerotised. CuA strong, usually concave, straight, angular or sometimes arching basally. Anal area enlarged, tucked down at rest along fold before 2A. Flight functionally four-winged, in-phase, probably mainly anteromotiric in spite that hind wing widened basally (like in stoneflies: GRODNITSKY 1999). Living Grylloblattidae wingless, some Protoperlidae micropterous (see Fig. 412). Abdomen with (in suborder Lemmatophorina) or without (suborders Protoperlina and Grylloblattina) paranota. Ovipositor long or short, sometimes reduced. Styli of 9th segments freely articulated at ovipositor apex in immature Grylloblattidae. Male gonocoxites free, bearing free styli. Cerci usually long, multisegmented, short 2–8-segmented in some Protoperlina, modified in entire forceps in male Chelopteridae. Nymphs terrestrial, except for the semiaquatic Permian Atactophlebiidae and Lemmatophoridae. Feeding zoonecrophagous in living Grylloblattidae, pollinivorous in Permian Ideliidae and Tillyardembiidae, supposedly a wide range of zoo- and phytonecrophagy, pollinivory and predation as general dietary range of the order. Moulting at winged stage known for Palaeozoic Daldubidae and Atactophlebiidae.

(c) **Synapomorphies**. Absent because of paraphyletic state of the order in respect to other perlideans.

(d) **Range**. Late Carboniferous until the present, worldwide in Palaeozoic and Mesozoic, now relict in near-Pacific Holarctic.

(e) **System and phylogeny**. The system and phylogeny of the order (Fig. 393) have been recently considered by STOROZHENKO (1997, 1998). After these publications, the following taxonomic novelties are introduced thus far. Tologopteridae and Perloblattidae are synonymised with Permosialidae (order Palaeomanteida) (STOROZHENKO & NOVOKSHONOV 1999), and new taxa are described by RASNITSYN & KRASSILOV (1996b); NOVOKSHONOV (1997e, 1998b); NOVOKSHONOV & NOVOKSHONOVA (1997); ARISTOV (1998, 2000a,b); STOROZHENKO & ARISTOV (1999). *Prosepididontes calopterix* Handlirsch (Fig. 394) from the German Lias is transferred from Trichoptera to Geinitziidae by ANSORGE & RASNITSYN (2000). According to A. Rasnitsyn (pers. comm.), the follow taxa also belong to Grylloblattida (Protoperlidae): *Protoblattiniella minitissima* Meunier (Fig. 395) and *Stenoneurites maximi* (Brongniart) from Late Carboniferous of France (Protoperlidae), while Narkeminidae is transferred to stem Gryllones (Chapter 2.2.2). Thus 44 families, 176 genera and 331 species of Grylloblattida are considered valid now.

(f) **History**. The Late Carboniferous grylloblattidans are known from both Angaraland and Europe-American Regions. They were not diverse at that time, being represented by only 5 species of Protoperlidae in Commentry (France) (Figs. 395, 396) and 2 species of Daldubidae in Siberia (Fig. 397) (Chunya River, Tunguska Basin).

In contrast, the Permian grylloblattidans were extremely diverse and widespread. The oldest among more representative assemblages is the Artinskian one from the middle Early Permian of Obora (Moravia, Czechia) and Elmo (Kansas, USA). Seven families, 24 genera and 38 species are described from Obora: 16 species of Liomopteridae, 12 Euryptilonidae, 3 Phenopteridae, 3 Havlatiidae, 2 Skaliciidae, 1 Aliculidae and 1 Jabloniidae. There are 9 families, 15 genera and 22 species of Grylloblattida recorded from Elmo, including 8 species of Lemmatophoridae (Fig. 398), 5 Liomopteridae, 3 Euryptilonidae, 1 Permembiidae, 1 Phenopteridae, 1 Probnidae, 1 Protembiidae, 1 Chelopteridae and 1 Demopteridae, with the most abundant (in terms of specimens) Lemmatophoridae and Probnidae. Early Permian grylloblattidans are known also from Germany (one species of Aliculidae and 1 Liomopteridae) and Texas, USA (one species of Megakhosaridae). Younger in the Early Permian is the rich Kungurian assemblage of Tshekarda and minor neighbour fossil sites in Middle Urals. As many as 14 families, 25 genera and 33 species are recorded there, including 5 species of Sylvaphlebiidae (Figs. 391, 399), 4 Sylvardembiidae, 3 Tillyardembiidae (Fig. 400), 3 Ideliidae, 3 Lemmatophoridae, 2 Sojanoraphidiidae, 2 Megakhosaridae, 1 Atactophlebiidae (Figs. 392, 401), 1 Euryptilonidae, 1 Euremiscidae, 1 Permembiidae, 1 Protembiidae, 1 Liomopteridae, 1 Sylvabestiidae and 4 species of Grylloblattida *incertae sedis*. The most abundant are two species: *Tillyardembia antennaeplana* G. Zalessky (Tillyardembiidae, Fig. 400) and *Gurianovella silphidoides* G. Zalessky (Atactophlebiidae, Figs. 387, 392, 401). *T. antennaeplana* is found with the intestine filled with *Cladaitina* pollen produced by the cordait family Rufloriaceae (AFONIN 2000), *Sojanidelia floralis* Rasnitsyn (Ideliidae) – with pollen of *Vittatina* (produced by phylladoderm peltasperms) and *Lunatisporites-Protohaploxypinus* pollen group (produced by *Ulmannia*-like conifers and glossopterids; RASNITSYN & KRASSILOV 1996b).

Seven families, 17 genera and 34 species have been described from the earliest Late Permian, Ufimian assemblage of the Kuznetsk Formation in south-west Siberia (Kaltan, Sarbala and Starokuznetsk localities). 21 species of Liomopteridae (Fig. 402), 4 Ideliidae (Fig. 403), 3 Archiprobnidae, 3 Megakhosaridae, 1 Stegopteridae, 1 Kortshakoliidae

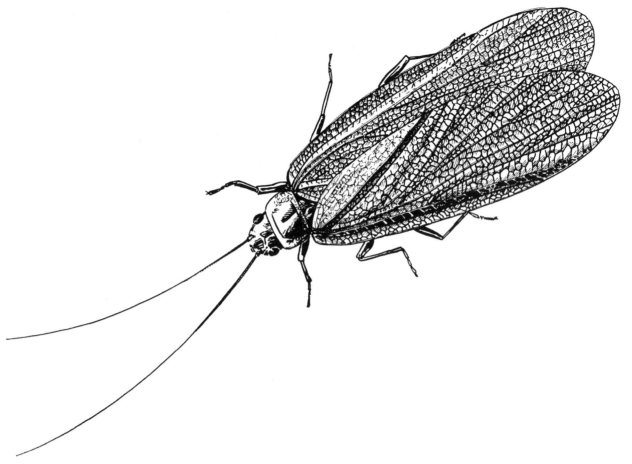

Fig. 391 *Sylvaphlebia tuberculata* Martynov (Sylvaphlebiidae) from the Early Permian of Tshekarda in Urals, restored as in life by A.G. Ponomarenko (from ROHDENDORF & RASNITSYN 1980)

and 1 Euryptilonidae (Fig. 404) are found there. The Lek-Vorkuta Formation of similar age in the Pechora Coal Basin contains two species of Ideliidae, 1 Megakhosaridae and 1 Permotermopsidae.

Later in the Late Permian, the Kazanian assemblage is collected at Sojana River (Asrkhangelsk Region in NW Russia). There are 12 families, 21 genera and 34 species recorded from there, with 11 species of Ideliidae (Fig. 405), 8 Liomopteridae, 4 Megakhosaridae, 2 Atactophlebiidae, 2 Permotermopsidae, 1 Idelinellidae, 1 Sojanoraphidiidae, 1 Sheimiidae (Fig. 406), 1 Phenopteridae, 1 Aliculidae, 1 Sylvardembiidae, 1 Permembiidae. Other Kazanian localities in Russia have yielded 1 species of Ideliidae, 1 Megakhosaridae in Kityak (Kirov Region), and one dominant species of Atactophlebiidae, 1 Camptoneuritidae, 2 Ideliidae, 1 Permotermopsidae, 2 Liomopteridae and genus *Haplopterum* Martynov of uncertain position in Tikhiye Gory (Tatarstan).

The latest, Late Permian, grylloblattidans of the Tatarian stage are known from Kargala in the Orenburg Region (1 species of Ideliidae, 1 Liomopteridae, 1 Megakhosaridae and genus *Kargarella* Martynov of uncertain position), the Eastern Kazakhstan localities Karaungir and Kuryultyu (2 species of Liomopteridae, 1 Euryptilonidae, 1 Aliculidae, 1 Megakhosaridae, 1 Blattogryllidae), Bor-Tologoy in S. Mongolia (4 species of Liomopteridae and Ideliidae, Kazanian age is not excluded, see Chapter 4.1), and Middle Beaufort Series in South Africa (2 species of Liomopteridae, 2 Megakhosaridae and 2 ones of uncertain position). One species of Tunguskapteridae is described from the Late Permian of West Siberia (Krasnoyarsk Region)

The Triassic grylloblattids are most diverse in Asia. For the present, only *Tomia costalis* Martynov (Tomiidae) is known with certainty for the first half of the Triassic of Russia (Babiy Kamen' in Kemerovo Region). The relict *Triasseuryptilon acostai* Marquat (Atactophlebiidae) is found in the Middle Triassic of Argentina. For the second half of period, the very rich assemblage is now available from the Madygen Formation of South Fergana, Central Asia. There are 9 families, 26 genera and 48 species of grylloblattidans recorded from this locality: 14 species of Blattogryllidae, 8 Gorochoviidae, 7 Mesorthopteridae, 7 Ideliidae, 4 Madygenophlebiidae, 3 Megakhosaridae, 3 Geinitziidae, 1 Tunguskapteridae, 1 Mesojabloniidae. The southern hemisphere deposits of similar age have produced 1 species of Mesorthopteridae and 1 Geinitziidae (Molteno Formation in South Africa), and 2 Mesorthopteridae and 1 Geinitziidae in Blackstone Formation (Queensland, Australia). Two species of Mesorthopteridae were described from the latest Late Triassic of Kenderlyk in East Kazakhstan. Two species of Geinitziidae are known from the Late Triassic of each Japan and China.

The Jurassic fauna of grylloblattidans is obviously impoverished. Numerous fossil insect sites of different ages within the Jurassic

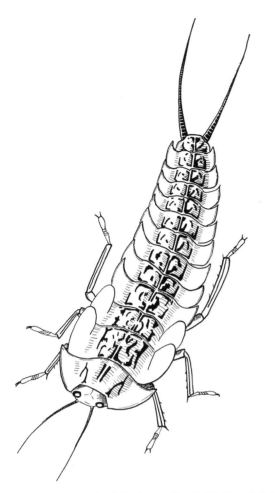

Fig. 392 Late instar nymph of *Gurianovaella silphidoides* G. Zalessky (Atactophlebiidae), which frequented sides of Early Permian lakes and ponds in the vicinity of the present day Tshekarda village in Urals, restored as in life by A.G. Ponomarenko

of Eurasia have only yielded representatives of four families of Grylloblattida. Ten species of *Geinitzia* Handlirsch and one of *Prosepididontes* Handlirsch (Geinitziidae, see Fig. 394), one species of *Dorniella* (Blattogryllidae) and a monotypic genus of Neleidae are recorded from the English and Germany Lias. Six species of *Shurabia* Martynov (Geinitziidae) are known from the Early or Middle Jurassic of Central Asia, Siberia, Mongolia and China, while only a single species of *Geinitzia* is found in Asia (Sagul locality in Kirghizstan). The Jurassic Blattogryllidae are described from Central Asia (1 species in Karatau, Fig. 407, 1 in Sogyuty, 1 in Sagul, and, in Mongolia, 1 from Bahar and 2 from Khoutiyn-Khotgor). The monotypic family Bajanzhargalanidae is endemic of Khoutiyn-Khotgor.

The Cretaceous grylloblattidans are known only from a single species of Blattogryllidae and another of the monotypic family Oecanthoperlidae from the Early Cretaceous Zaza Formation (Baissa in Transbaikalia, East Siberia), and a single fossil of Geinitziidae from the Purbeck of Swanage in England (undescribed in collection of Robert Coram, Swanage, England).

No fossil record of the order exists after the Early Cretaceous. The only living family Grylloblattidae includes four genera and 26 species from Siberia, Northeast China, Korea, Japan, USA and Canada.

2.2.2.2.2. ORDER PERLIDA Latreille, 1810. THE STONEFLIES (=Plecoptera Burmeister, 1839)

N.D. SINITSHENKOVA

(a) Introductory remarks. The order is a moderately small (a little more than 2,000 living species according to THEISCHINGER 1991b) group of amphibiotic insects retaining many plesiomorphic features. The present description is based on ZWICK (1973, 1980, 2000), ROHDENDORF & RASNITSYN (1980), and SINITSHENKOVA (1982, 1987, 1990).

(b) Definition. Body size moderately small to large. Head prognathous, with mouthparts chewing (mandibles sometimes much reduced). Thorax structure little modified, prothorax large, lacking wide paranota, meso- and metathoraces of similar structure. Legs cursory (with swimming hairs in many nymphs), with tarsus 3-segmented. Wings membranous, almost completely overlapping each other over abdomen at rest, with pre-anal venation similar in both pairs, excluding RS which lost its base and originated from M in hind wing (nymphal wing pad with RS base preserved), and CuA which usually (primarily) simple in hind wing; hind wing anal area usually large. Wing venation not especially rich, with main veins branching at or beyond wing midlength, more often forked, RS and CuA up to 4 or 5 branches (in ground plan CuA is similar to M in having 2 branches only). Cross veins not numerous besides 2 characteristic rows aligned CuA, with most constant r-rs and r-m near RS and M forkings (roughly at level of SC apex). RS typically and probably plesiomorphically originating near wing base and running parallel to R. Flight functionally four-winged, in-phase, mainly anteromotoric in spite hind wing widened basally (GRODNITSKY 1999). Abdomen with full number of segments, lacking paranota. Primary ovipositor lost. Cerci always present, usually long, multisegmented, rarely one-segmented in adult. Nymphs aquatic, now populating almost exclusively running or cold water, often bearing external gills of different structure and disposition.

(c) Synapomorphies. Head prognathous; prothoracic paranota narrow; tarsi 3-segmented (possibly inherited from ancestral grylloblattidans similar to *Thaumatophora pronotalis* Riek, *Liomopterites gracilis* Sharov and *Sylvonympha tshekardensis* Novokshonov et Pan'kov (Chapter 2.2.2.2e, see Figs. 389, 390); fore wing with 2 rows cross-veins aligned CuA; male styli lost; nymphs aquatic (inherited from ancestral grylloblattidans).

(d) Range. Since later Early Permian (Kungurian) till now; worldwide though rare in warmer climates.

(e) System and phylogeny. The present system and phylogeny (Fig. 408) is proposed by ZWICK (1973) and further modified by SINITSHENKOVA (1987) to embrace fossil clades. The order is divided into two suborders Perlina and Nemourina. The third suborder Perlopseina proposed for Perlopseidae (now in Nemourina) and Palaeoperlidae (now in Perlina, Perlomorpha) (SINITSHENKOVA 1987) and based mainly on the peculiarities of Perlopseidae, is not expedient. The differences between Perlopseidae and Palaeoperlidae seem to be more important. Perlopseidae (see Fig. 409) resemble nemourines in the second tarsal joint shortened while the first and third are long, the cerci short, the body narrow and slender, the legs long and slender, the wing venation similar (especially with the Permian Palaeonemouridae), particularly in the presence of several rs-m, several cross veins in the costal area, and the 3-branched CuA. This makes it possible to consider the close relationships of Perlopseidae with other representatives of Nemourina, which may originate from as yet unknown less advanced Perlopseidae. The long antennal joints of known Perlopseidae

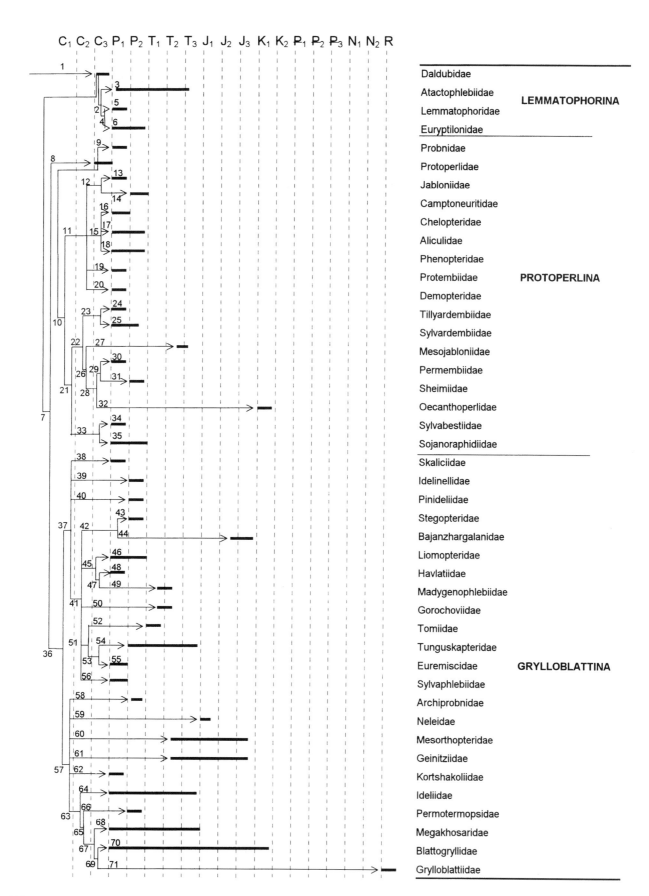

Fig. 393   Phylogeny and system of the Order Grylloblattida

Time periods are abbreviated as follows: C_1, C_2, C_3 – Early (Lower), Middle and Late (Upper) Carboniferous, P_1, P_2 – Early (Lower) and Late (Upper) Permian, T – Triassic, J – Jurassic, K – Cretaceous, P_1 – Palaeocene, P_2 – Eocene, P_3 – Oligocene, N_1 – Miocene, N_2 – Pliocene, R – present time (Holocene). Right two columns are names of families and suborders, respectively. Arrows show ancestry, thick bars are known durations of taxa, figures refer to synapomorphies of the subtended clades, as follows:

1 – synapomorphies of superorder Perlidea (Chapter 2.2.2.2c).

2 – fore wing with cross-veins lost between SC veinlets; M_5 absent in both pairs of wing; hind wing M and CuA anastamosing basally.

3 – fore wing M divided into 3 main branches, viz. MA, MP_1 and MP_2

4 – individual development without subimago; fore wing M and CuA anastamosing basally; fore wing CuA with strong flexure in basal third.

5 – femora and tibia without spines and spurs; fore wing CuA divided into CuA_1 and CuA_2.

6 – fore wing tegminised, coriaceous; fore wing CuA with angular flexure near anastamosis M+CuA.

7 – hemimetabolic metamorphosis with no elements of protometaboly; tergites lacking paranota; hind wing with cubital area wider than that between CuP and 1A.

8 – fore wing CuA and its branches situated in posterior wing half; proximal branch of fore wing CuA forming series of marginal veinlets across posterior wing margin.

9 – legs modified for burrowing; tarsus 3-segmented; fore wing tegminizated, coriaceous, with proximal branch of CuA long, pectinate

10 – hind wing with cubital area considerably wider than that between CuA and 1A.

11 – hind wing with bases of the R, M and CuA forming a single stalk R+M+CuA.

12 – fore wing CuA S-shaped, its flexure strong, so as cubital area very wide basally.

13 – SC reaching only fore wing midlength.

14 – fore wing M branches anastamosing; fore wing tegminised, coriaceous.

15 – fore wing with flexure of CuA strong, so as cubital area very wide basally, SC veinlets H- or Y-shaped, costal area wide.

16 – male cerci modified in forceps; fore wing RS with 2 branches.

17 – fore wing with costal area distinctly emarginate basally, and with RS branches directed to the posterior wing margin.

18 – fore wing with proximal CuA branch long, pectinate.

19 – RS originated from R at fore wing midlength.

20 – fore wing RS with 2 branches; CuA with strong flexure, so as cubital area very wide basally.

21 – pronotal paranota absent.

22 – fore wing CuA almost straight (without basal flexure).

23 – ocelli reduced; RS originated from R at fore wing midlength.

24 – cross-veins forming very narrow regular cells in all areas of fore wing; RS originating from R at fore wing.

25 – in fore wing R, M and CuA forming a single stalk R+M+CuA with branches directed to posterior wing margin.

26 – fore and hind wings with similar shape and venation.

27 – fore wing with at most 1–2 cross-veins in each area; M simple or forked; RS with 1–2 branches.

28 – fore wing R with a few branches directed to anterior wing margin, R and RS anastomosing, CuA_1 S-shaped.

29 – fore wing with A simple or lost.

30 – fore wing with RS+M anastamosis very long.

31 – fore wing with RS originating from R near wing base.

32 – fore wing with M simple.

33 – fore wing M_5 present.

34 – fore wing SC disappearing in distal third of wing; legs stout and long.

35 – ovipositor unusually long; fore wing CuA with flexure angular near M+CuA anastamosis and strong, so as cubital area very wide basally.

36 – hind wing CuA with basal flexure angular; 2A regularly pectinate.

37 – fore wing CuA divided into CuA_1 and CuA_2.

38 – fore wing tegminised, coriaceous, with costal area narrow, RS originating near wing midlength, cross-veins oblique in radial area, MP completely sclerotised.

39 – fore wing with no cross-veins between SC veinlets.

40 – fore wing with costal area narrow, SC with no cross-veins between veinlets, RS and M and, basally, M and CuA anastamosing, CuA without flexure near the base, almost straight.

41 – fore wing CuA_2 simple.

42 – fore wing tegminised, coriaceous, with costal area narrow, SC with no cross-veins between veinlets.

43 – fore wing with RS originating from R near wing midlength, with MP completely sclerotised; hind wing M and CuA fused in common stalk basally.

44 – fore wing CuA without flexure near the base, almost straight.

45 – hind wing M and CuA fused in common stalk basally.

46 – hind wing with R, M and CuA fused in common stalk basally, with RS and MA anastamosing.

47 – fore wing SC S-shaped.

48 – fore wings considerably widened in apical third.

49 – fore wing SC reaching midlength only; hind wing RS anastamosing with MA.

50 – fore wing with RS regularly pectinate, with branches directed toward posterior wing margin, costal area narrow, SC lacking cross-veins between veinlets; hind wing RS ad MA anastamosing.

51 – fore wing with SC lacking cross-veins between veinlets, CuA lacking basal flexure, almost straight.

52 – distal veinlet of fore wing SC angular.

53 – fore wing with costal area narrow; radial area with oblique cross-veins.

54 – fore wing with M having only 1–2 branches, area between M and CuA with simple cross-veins, M_5 present, CuA_1 simple.

55 – fore wing A simple.

56 – ovipositor short; fore wing with CuA branches compressed, costal area narrow; hind wing M and CuA fused in single stalk basally.

57 – fore wing CuA with strong flexure in basal third.

(six times as long as wide) are autapomorphic. The evolution of Nemourina was accompanied by a considerable reduction in wing venation, possibly due to a decrease in the body size.

Palaeoperlidae have the most primitive wing structure, and such features as the convex anterior margin of the fore wing forming a large costal space, the short SC, and the widely rounded wing apex may be common to the stonefly stem group. The rich wing venation and the short first and second tarsal joints of the nymphs allow us to place Palaeoperlidae to Perlina, Perlomorpha.

For more details on the relationships within Nemourina and Perlina see SINITSHENKOVA (1987).

(f) History of the order is incomplete particularly because a significant share of its members were rheophilic and hence escaped entering the fossil record, because they were normally buried in lotic deposits which fell prey to erosion in the geologically short time. Unless stated otherwise, the following information is from ANSORGE (1993) and SINITSHENKOVA (1987, 1990, 1992a,b, 1995, 1999a,b).

The oldest stoneflies come from the Kungurian (latest Early Permian) deposits of the Urals (Tshekarda locality) and are represented there by rare adults of both suborders Nemourina (Perlopseidae, Fig. 409, Palaeonemouridae) and Perlina (Tshekardoperlidae, Fig. 410) as well as by rare nymphs of *Barathronympha victima* Sinitshenkova of obscure taxonomic position (Fig. 411). The latter fossil is of interest because they represent a now extinct morpho-ecological type, the free-living rheophilous nymph.

The Late Permian stoneflies were still uncommon, though they were found in both the Northern and Southern Hemispheres. Palaeonemouridae are known from Northeast Europe (Inta Formation in Pechora Basin and Kityak at Viatka River), south Urals (Kargala), East Kazakhstan (Karaungir in Saur Range), Southwest Siberia (several localities in Kuznetsk Basin), and Antarctica (Messer Ridge). Palaeoperlidae are found in East Kazakhstan, Kuznetsk Basin, and East Siberia (Pelyatka in Tunguska Basin; a nymph of obscure position within the suborder Perlina is found here as well). Grypopterygomorph families are found only in South Africa (Euxenoperlidae in Natal, Middle Beaufort Series) and Australia (Eustheniidae in New South Wales, Belmont, Newcastle Coal Mesures, Fig. 412). Most of the fossils collected represent adults.

The Triassic stoneflies were not common either, and are equally dominated by adult insects in the collections available. They are found in the Late (in some cases possibly Middle) Triassic. Perlomorph fossils are represented by nymphs of obscure position from Ukraine (Garazhovka), East Kazakhstan (Kenderlyk), and South China (Yunnan). Gripopterygomorph families are found both in the Southern Hemisphere (Euxenoperlidae in Molteno Formation in South Africa, Potrerillos Formation, Argentina, and Ipswich Series in Queensland, Australia) and in the Northern Hemisphere (solely nymphs of Siberioperlidae in Garazhovka, Kenderlyk, and in Madygen Formation in Fergana, Central Asia). Nemourina are found only in the Madygen Formation of Fergana and all belong, apart from one species of Mesoleuctridae, to the family Perlariopseidae which is fairly diverse there, however. Undescribed stoneflies are recorded from middle Triassic *Voltzia* Sandstone (Vosges, France) by MARCHAL-PAPIER (1998).

Jurassic stoneflies have been collected, as with most other insect groups, mostly in Eurasia, but unlike many of them, the stoneflies were rare in Europe. There are only three finds from the German Lias, one referred with doubt to Capniidae (*Dobbertinopteryx* Ansorge from Dobbertin; ANSORGE 1993) and two Perlariopseidae from Grimmen, *Dicronemoura furcata* Ansorge and an unidentified wing fragment (ANSORGE 1996). The identification of *Dobbertinopteryx* is based on the isolated wing and hence it is not reliable. If correct, it is the oldest record of the extant family.

In contrast to the Permian and Triassic, the Jurassic stoneflies were numerous and diverse, and their nymphs often took a leading position in insect assemblages. This is particularly true for Mesoleuctridae (Figs. 413, 414) and Platyperlidae (Fig. 415) in the Early and Middle Jurassic assemblages of South Siberia (westward up to Central Kazakhstan, southward to Mongolia). They represented the extinct morpho-ecological type of the free living pelophilous nymphs that populated lakes, particularly the extinct category of the hypotrophic lakes (Chapter 3.3.4). Other stonefly groups fit better to the morpho-ecological classification of the living nymphs as proposed by SINITSHENKOVA (1987). Among them there was the gripopterygomorph family Siberioperlidae (Fig. 416) found in numerous Early and Late Jurassic localities in Transbaikalia and Mongolia and the perlomorph nymphs of obscure position in the Early and Late Jurassic of Cis- and Transbaikalia and West Mongolia. More westward, in the Jurassic deposits of South Kazakhstan and the Middle Asian states of the former USSR, most abundant are adults of the rheophilous Perlariopseidae (Nemourina, Fig. 417), a family not uncommon in the Jurassic of Siberia and Mongolia as well (Fig. 418). Another nemourinan family, Baleyopterygidae, was also widespread in Asia (Fig. 419), but unlike Perlariopseidae it is represented by both adults and nymphs, though adults are more common in collections.

Fig. 393 *Continued*

58 – fore wing with cross-veins forming very narrow regular cells over all wing, SC veinlets lacing cross-veins, MP completely sclerotised.

59 – fore wing with SC desclerotised, R and RS anastamosing, M branches anastamosing, CuA lacking basal flexure (almost straight).

60 – arolium large; fore wing costal area narrow; hind wing RS anastamosed with MA.

61 – fore wing with a few R branches directed to anterior wing margin, RS originating from R near wing midlength, M_5 present.

62 – fore wing RS originating from R near wing midlength.

63 – fore wing cubital area with a series of CuA branches directed to posterior wing margin.

64 – fore wing RS and M anastamosing.

65 – fore wing with CuA branches crossing cubital area terminating on CuP; hind wing RS anastamosing with MA.

66 – fore wing SC with veinlets H- or Y-shaped.

67 – fore wing with intercubital area crossed with S-shaped CuA branches, costal area narrow.

68 – fore wing M_5 present.

69 – head as broad as pronotum; male gonocoxites asymmetrical.

70 – head with dorsal tubercles; claws reduced; arolium large; fore wing with CuA flexure angular near basal anastamosis of M+CuA, SC veinlets H- or Y-shaped.

71 – eyes small or reduced; ocelli lost; cerci with only 7–10 segments; wings absent.

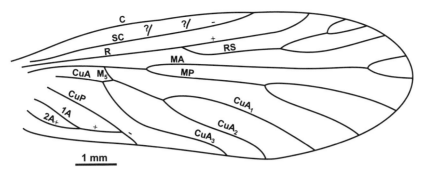

Fig. 394 *Prosepididontes calopterix* Handlirsch (Prosepididontidae) from the German Lias (from ANSORGE & RASNITSYN 2000, with permission of Acta Gèologica Hispanica)

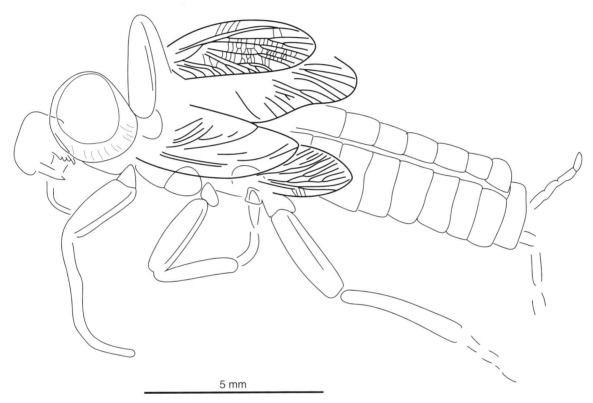

Fig. 395 Brachypterous adult of *Protoblattiniella minutissima* Meunier (Protoperlidae) from the Late Carboniferous of Commentry in France (holotype, original by A.P. Rasnitsyn)

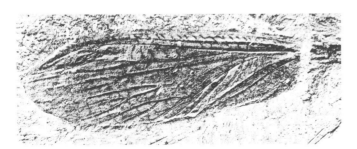

Fig. 396 *Palaeocixius pygmaeus* Meunier (Protoperlidae) from the Late Carboniferous of Commentry in France (from MEUNIER 1912); fore wing 21 mm long

Fig. 397 *Dalduba faticana* Storozhenko (Daldubidae) from the Late Carboniferous of Chunya in Siberia (holotype, photo by D.E. Shcherbakov); fore wing 31 mm long as preserved

Fig. 398 *Lemmatophora typa* (Lemmatophoridae) from the Early Permian of Elmo in USA (from CARPENTER 1935); fore wing 6 mm long

Fig. 399 *Sylvaphlebia tuberculata* Martynov (Sylvophlebiidae) from the Late Permian of Tshekarda of Urals (from ROHDENDORF 1962); fore wing 20 mm long

In the Cretaceous stoneflies became rarer again. They are found only in Australia (Gripopterygidae in the Aptian of Koonwarra, Victoria; JELL & DUNCAN 1986), Siberia, Mongolia and Europe. The nymph

Fig. 400 *Tillyardembia antennaeplana* G. Zalessky (Tillyardembiidae) from the Early Permian of Tshekarda in Urals (PIN 1700/1177, photo by D.E. Shcherbakov); body with ovipositor 22 mm long

Ecdyoperla Sinitshenkova is the only stonefly known from the Early Cretaceous of Europe (England, Ashdown Formation of Wealden; SINITSHENKOVA 1998b). Siberian finds are all from Transbaikalia and Yakutia. They are dated to the Neocomian (earlier Early Cretaceous), except some of them are frequently challenged as being from the Jurassic period or contrarily, post-Neocomian Early Cretaceous. Perlariopseidae, Siberioperlidae, and Baleyopterygidae were persisting there since the Jurassic (or Triassic), and several modern groups joined them as well, particularly Perlodidae and Chloroperlidae (Perlomorpha), as well as Leuctridae, Nemouridae, and Taeniopterygidae (Nemourina). Mongolian localities have produced members of Perlariopseidae, Siberioperlidae, Taeniopterygidae, and Perlomorpha *incertae sedis*. It is of interest that West Mongolian, otherwise rather specific in its Early Cretaceous entomofauna

(RASNITSYN 1986b), was quite average when it comes to the stonefly families represented there (Perlariopseidae, Taeniopterygidae).

Late Cretaceous stoneflies are not yet known. Those of Palaeogene age are rare and, when identified, all belong to living families. They come from near to the Palaeoceae-Eocene boundary (undescribed in Mo Clay, Denmark) and from the Late Eocene (Perlidae, Perlodidae, Taeniopterygidae, Nemouridae, and Leuctridae in the Baltic amber, Leuctridae in the Bitterfeld amber in Germany), Early and Late Oligocene and earliest Miocene (Leuctridae in Bembridge, Isle of White in UK, in Ruby River in Montana, USA, and Rott, Germany, respectively). Neogene fossils are still rarer and mostly undescribed. They are found in the Miocene of Japan (Sadzugawa and Kamiwada, Honshu I.), North America (Leuctridae in Latah Formation in Idaho), Pliocene of Germany (Taeniopterygidae in Willershausen) and Japan (Perlodidae in Iriki-toge, Honshu I.). When identified, all of them are assigned to living genera.

Fig. 401 Elder nymph of *Gurianovaella silphidoides* G. Zalessky (Atactophlebiidae) from the Early Permian of Tshekarda in Urals (PIN 1700/1996, photo by A.G. Sharov); body 30 mm long

Fig. 403 *Archidelia ovata* Sharov (Ideliidae) from the Late Permian of Kaltan in Siberia (from ROHDENDORF *et al.* 1961); wing length 53 mm

Fig. 404 *Euryptilodes cascus* Sharov (Euryptilonidae) from the Late Permian of Kaltan in Siberia, fore wing (holotype, length 16) and nymphal moulting cast (holotype of *E. horridus* Sharov, length of fossil 10.5 mm) (from ROHDENDORF *et al.* 1961)

Fig. 402 *Liomopterites expletus* Sharov (Liomopteridae) from the Late Permian of Kaltan in Siberia (holotype, photo from archive of Palaeontological Institute RAS); wing length 24 mm

Fig. 405 *Sylvidelia latipennis* Martynov (Ideliidae) from the Early Permian of Tshekarda in the Urals (holotype); wing length 42 mm

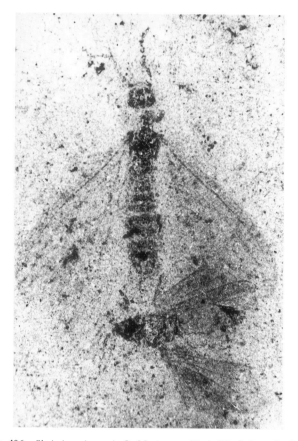

Fig. 406 *Sheimia sojanensis* O. Martynova (Sheimiidae) from the Late Permian of Soyana in NW Russia (holotype, photo by D.E. Shcherbakov); body 3.2 mm long

Fig. 407 *Blattogryllus karatavicus* Rasnitsyn (Blattogryllidae) from the Late Jurassic of Karatau in Kazakhstan (from ROHDENDORF & RASNITSYN 1980); fore wing 24 mm long

2.2.2.2.3. ORDER FORFICULIDA Latreille, 1810. THE EARWIGS AND PROTELYTROPTERANS (=Dermaptera DeGeer, 1773 +Protelytroptera Tillyard, 1931)

D.E. SHCHERBAKOV

(a) Introductory remarks. For the classification of living earwigs, see STEINMANN (1975) and SAKAI (1982), and for the phylogeny, POPHAM (1985) and HAAS (1995; this author introduced many new characters but excluded other important ones from his cladistic analysis). Palaeontology was reviewed by CARPENTER & KUKALOVÁ (1964), KUKALOVÁ (1966), and VISHNIAKOVA (1980b,c). Most fossil records were compiled by F. CARPENTER (1992a) and NEL *et al.* (1994). RENTZ & KEVAN (1991) indicate 1,800 living species.

(b) Definition. Small to medium-sized insects looking like staphylinids or (Permian forms) usual beetles. Head prognathous (rarely hypognathous). Antennae moderately long, with 10–50 segments. Mouthparts chewing. Fore wings transformed into elytra (with venation more or less reduced) covering the entire abdomen or at least most of hind wings in repose, when their overlapping sutural margins are fixed on submedian metascutal setose ridges. Hind wings with large anal fan, longer than fore wings (possibly except for Blattelytridae) and transversely folded in repose (thrice folded in earwigs). Tarsi 3–5-segmented. Ovipositor laciniate to much reduced (concealed by 7th sternum in most earwigs). Male genitalia concealed by 9th sternum, plesiomorphically consist of a pair of distally forked phallomeres. Cerci either short to long with basal segment enlarged, or unsegmented and usually forcipate. Ovarioles of primitive polytrophic type. Chromosomes holokinetic.

Nymphs usually adult-like (four or five instars). However, in Karschiellinae and Diplateinae (Pygidicranidae) nymphal cercus is long and 14–45-segmented, breaking off during penultimate moult beyond enlarged basal segment, the only one developing into short and stout adult forceps (VERHOEFF 1903, BEY-BIENKO 1936). Both phylogenetic and the above ontogenetic sequences show that earwigs' forceps are clearly homologous to other insects' cerci, contrary to MATSUDA (1976).

Most living forms are probably omnivorous, others appear to be rather predators or herbivores; usually cryptobiotic and nocturnal. Females often exhibit maternal care, remaining with the eggs and young nymphs. Two extant families comprise apterous viviparous ectoparasites or rather commensals (NAKATA & MAA 1974): Arixeniidae feed on glandular excretions of bats and (or) on scraps in their roosts; Hemimeridae dwell in the fur and feed on the epidermis of rodents.

(c) Synapomorphies. Fore wings elytrised with sutural margins overlapping in repose. Hind wings transversely folded in repose. Possibly also: basal segment of cercus enlarged; chromosomes holokinetic. Additional synapomorphies are listed by POPHAM (1985).

(d) Range. Since Early Permian until the present, worldwide.

(e) System and phylogeny. A separate order, Protelytroptera, was established for Permian insects presumably ancestral to earwigs by TILLYARD (1931). Protelytropterans originated from Grylloblattida (RASNITSYN 1980), the venation of the most generalised genera (*Apachelytron* Carpenter et Kukalová; Fig. 420) being derivable from near the lemmatophorid one. The discovery of triple hind wing folding of earwig type in Bardacoleidae Zalessky (Fig. 421) and of metascutal setose ridges in Permofulgoridae Tillyard (Fig. 422), both families retaining fully developed elytra, allows us to unite protelytropterans and earwigs under Dermaptera *s. l.* (like normal beetles and Staphylinidae under Coleoptera) as two suborders, Protelytrina and Forficulina. Bardacoleidae are transferred from Coleoptera (ZALESSKY 1947) to Protelytrina (ROHDENDORF 1939; *Uralelytron* Rohdendorf, 1939 = *Bardacoleus* Zalessky, 1947, syn. nov.), Megelytridae Carpenter, 1933 are regarded as a subfamily of Archelytridae Carpenter, 1933 (Archelytrinae = Apachelytridae Carpenter et Kukalová, 1964, syn. nov.), and both Permophilidae Tillyard, 1924 and Labidelytridae Kukalová-Peck, 1987 (=Stenelytridae Kukalová, 1966: KUKALOVÁ-PECK 1987b) as subfamilies of Permofulgoridae (Permofulgorinae = Protocoleidae Tillyard, 1924, syn. nov.). On account of the hind wing structure two superfamilies could be established, Archelytroidea and Protelytroidea (Fig. 423). Blattelytridae Tillyard belonging to the former and Bardacoleidae to the latter. The families with hind wing structure unknown cannot be reliably classified at present, but Artinskian families, except for Protelytridae Tillyard, are tentatively placed in Archelytroidea and all the Late Permian ones into Protelytroidea. Classification of protelytropteran elytra with obliterated venation may sometimes be artificial, as in the case of most isolated beetle elytra (Chapter 2.2.1.3.2.1); e.g. the subordinal allocation of the Dermelytridae Kukalová must be confirmed by hind wing structure which is unfortunately still unknown.

Triassic and most of Jurassic earwigs belong to the infraorder Protodiplateomorpha with a single family Protodiplateidae Martynov, 1925. They differ from living earwigs in the 5-segmented tarsi, segmented adult cerci, external laciniate ovipositor, and also often in their distinct elytral venation, all retained from protelytropteran ancestors, but unlike them Protodiplateidae have their elytra abbreviated (abdomen mostly exposed) and cerci elongated (VISHNIAKOVA 1980c, and unpublished data; Fig. 424). Semenoviolinae Vishniakova are transferred from Pygidicranidae to Protodiplateidae (Protodiplateinae =Dermapterinae Vishniakova, 1980, syn. nov. =Longicerciatidae Zhang, 1994, syn. nov.), all having the same tarsal formula, antennal structure, etc. The parasitic earwig taxa are considered Cainozoic derivatives of free-living Forficulina (GILES 1974) and deserve family (Arixeniidae in Labioidea) or infraordinal rank (Hemimeromorpha: POPHAM 1985; autplesiomorphies listed for the latter are rather restored nymphal characters). Three superfamilies are accepted within Forficulomorpha after POPHAM (ibid.), Forficuloidea descending from ancient Labioidea and the latter from Pygidicranoidea (see Fig. 423).

(f) History. The oldest Protelytrina are recorded from the early Artinskian (Early Permian) of Moravia (Obora); the fauna was already diverse (8 species in 7 genera of at least 4 families: Archelytridae, Blattelytridae, Protelytridae, Elytroneuridae Carpenter, plus doubtful endemic Planelytridae Kukalová, CARPENTER & KUKALOVÁ 1964, KUKALOVÁ 1965). By the late Artinskian (Kansas, Elmo) protelytropterans reach maximal abundance and become about twice as diverse at species level, whereas the generic diversity and family composition of the fauna remain nearly constant (TILLYARD 1931, F. CARPENTER 1933, 1939). The above families are endemic to the Early Permian. In the Kungurian (Tshekarda) and Ufimian (Inta Formation, Vorkuta Coal Basin) faunas of the Urals, only the endemic Bardacoleidae (3 species in 2 genera) are recorded, their hind wings being the most derived of those known for the suborder (hind wings of younger protelytropterans remain undiscovered). Since the beginning of the Kazanian (Soyana River in north European Russia), Permofulgoridae come into existence, represented by *Arctocoleus* Martynov (MARTYNOV 1933). In later Kazanian and Tatarian faunas of South Siberia (Suriyokova and Sokolova localities), small glabrous elytra of unknown family position are found (undescribed). The most diverse fauna of Protelytrina is known from the terminal Permian of Belmont in Australia: 18 species in 11 genera, most belonging to Permofulgoridae *s. l.* and the remaining 4 species to endemic Dermelytridae, are recorded there (TILLYARD 1918, 1924, KUKALOVÁ 1966). Two species of Permofulgoridae are found in the nearly synchronous beds of Natal, South Africa (Middle Beaufort Series; RIEK 1976d).

In the terminal Permian of East Kazakhstan (Saur Mts., Karaungir River), 2 species of the most archaic earwigs are found (undescribed). Their elytra are elongate rather than abbreviated, and their body structure remains unknown; however, their hind wings are twice as long as the fore wings, implying an essentially earwig-like habitus with the abdomen mostly exposed.

Protelytrina were entirely replaced with Forficulina in the Mesozoic insect faunas. Instead, the Cretaceous Umenocoleidae independently acquired elytra similar to those in Protelytrina and thus were previously included in Protelytroptera. Now they are transferred to Blattida (Chapter 2.2.2.1.2).

The findings of Mesozoic Protodiplateidae are scattered over the numerous insect assemblages and remain largely undescribed, so far as they are often represented by isolated veinless elytra. The Triassic finds are from the Madygen Formation (Fergana, Kyrgyzstan) and the

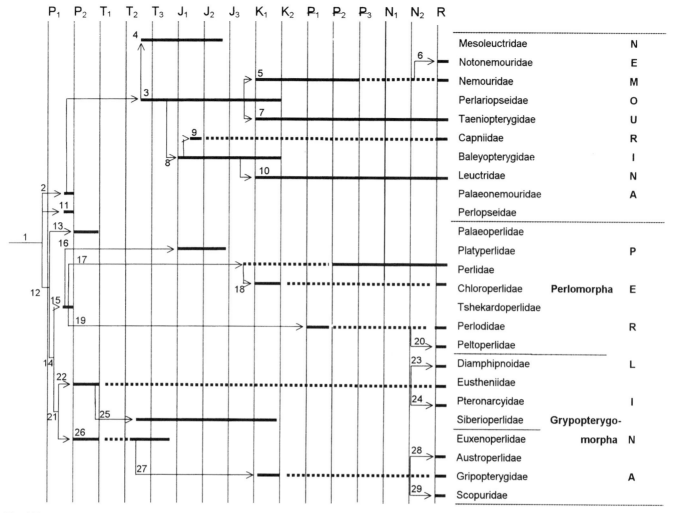

Fig. 408 Phylogeny and system of the stoneflies, Order Perlida

Time periods are abbreviated as follows: C_1, C_2, C_3 – Early (Lower), Middle and Late (Upper) Carboniferous, P_1, P_2 – Early (Lower) and Late (Upper) Permian, T – Triassic, J – Jurassic, K – Cretaceous, P_1 – Palaeocene, P_2 – Eocene, P_3 – Oligocene, N_1 – Miocene, N_2 – Pliocene, R – present time (Holocene). Right three columns are names of families, infraorders (omitted when coincide with suborders) and suborders, respectively. Arrows show ancestry, thick bars are known durations of taxa (dashed for hypothesised duration), figures refer to synapomorphies of the subtended clades, as follows (unless stated otherwise, venational characters refer to fore wing):

1 – head prognathous; prothoracic paranota narrow (usually lost); 2 rows of cross-veins before and behind CuA; 2nd tarsal segment short; male styli lost; possibly (unless inherited from ancestral grylloblattidans): tarsus with 3 segments; nymph aquatic.

2 – in the middle of antennae joint 1,5 times as long as wide.

3 – wing with SC reaching C with 2 (or less) veinlets, with nodal line present.

4 – nymphs with legs long, narrow, abdominal segments long, male subgenital plate narrow, long.

5 – veins around r-rs cross-vein forming X-like figure.

6 – male ductus eiaculatorius thick, well muscularised.

7 – 2nd tarsal segment subequal in length to 1st and 3rd.

8 – adult cercus short, one-segmented in male.

9 – RS fork short; hind wing narrow; male supraanal process well sclerotised, bent under ventral abdominal surface.

10 – wings covering body both dorsally and laterally.

11 – antennal segment length 6 times width.

12 – 1st tarsal segment short.

13 – size small.

14 – fore wing with anterior margin straight; nymphal wing pads wide, diverging.

15 – adult mandible rudimentary; atypical longitudinal musculature well developed (both characters inferred for Tshekardoperlidae and Platyperlidae from living taxa).

16 – nymph with tibiae wide, flat, and abdominal segments long, narrow.

17 – adult with sternal coxal muscles lost and with 6 free abdominal ganglia; male tergum 10 modified into 2 hook-like processes; nymph with thoracic gills.

18 – size small; wing venation impoverished; male tergum 10 with single hook-like process.

19 – wing with meshwork of veins apically; epiproct reduced.

Rhaetian Penarth Group (England; JARZEMBOWSKI 1999b). Jurassic protodiplateids are recorded from the Early Liassic (Late Sinemurian) of Dorset, England (*Brevicula* Whalley; WHALLEY 1985) and of Aargau, Switzerland (*Baseopsis forficulina* Heer, HEER 1864–1865; another species ascribed to the genus by BRAUER, REDTENBACHER & GANGLBAUER 1889 is rather a roach); the Early Jurassic of Sogyuty in Kyrgyzstan; the Middle Jurassic of Kubekovo in South Siberia (VISHNIAKOVA 1985b); and the Middle or Late Jurassic of Bakhar and later Late Jurassic of Shar Teg in Mongolia. The richest fossil earwig fauna occurs in the Late Jurassic (Karabastau Formation) of Karatau, South Kazakhstan: 11 described species in 9 genera of Protodiplateidae (including Semenoviolinae; MARTYNOV 1925b,c, VISHNIAKOVA 1980c; Fig. 424). *Mesoforficula* Ping from the Jurassic of Xinjiang, China (PING 1935) is not a dermapteran, as well as *Forficularia* Weyenbergh from Solnhofen (WEYENBERGH 1869), according to HANDLIRSCH (1906–1908).

The only reliable Early Cretaceous record (and the last find) of Protodiplateidae is from the later Middle Purbeck of Dorset, England (CORAM *et al.* 1995), and 2 species of *Longicerciata* Zhang from the Laiyang Formation of Shandong, China, where they co-occur with the oldest Forficulomorpha assigned to Pygidicranidae Echinosomatinae (*Archaeosoma* Zhang, ZHANG 1994). "*Protodiplatys*" *mongoliensis* Vishniakova (VISHNIAKOVA 1986b) is too poorly preserved to confirm its family placement. Large Pygidicranidae (undescribed; Fig. 425) from Baissa (Zaza Formation, Transbaikalia, possibly Valanginian–Hauterivian but see Chapter 4.1) comprise 0.2% of terrestrial insect numbers (V. Zherikhin, pers. comm.). Undescribed Dermaptera are mentioned from the La Cantera Formation (Neocomian) of San Luis in Argentina (MAZZONI & HÜNICKEN 1984). In Santana (Aptian or Early Albian) of Brazil at least two genera are found: *Cretolabia* Popham (primitive Labiidae; POPHAM in GRIMALDI 1990), *Araripelabia* Martins-Neto and *Caririlabia* Martins-Neto (originally assigned to Pygidicranidae: MARTINS-NETO 1990e; the latter genus is rather similar to and possibly should be synonymised under *Cretolabia*; one more genus, *Caririderma* Martins-Neto, should be transferred to Staphylinidae). The only earwig described from the Late Cretaceous is *Sinolabia longyouensis* Zhou et Chen (Labiidae Spongiphorinae) from Zhejiang in China (ZHOU & CHEN 1983). Earwigs were also reported in Canadian amber (Foremost Formation, Campanian; HEIE & PIKE 1992: 1028).

Cainozoic earwigs (Fig. 426) are usually classified into recent genera (most records being assigned to *Forficula* Linnaeus), but the older descriptions are too imperfect to ascertain their placement. Labiduridae are first recorded in the Palaeogene or Late Cretaceous Burmese amber (COCKERELL 1920a; for the age see ZHERIKHIN & ROSS 2000), Forficulidae (and extant genus *Forficula*) in the Late Palaeocene–Early Eocene of Denmark (WILLMANN 1990), Anisolabididae in the Early Eocene of Argentina (COCKERELL 1925b, age after MARTÍNEZ 1982) and Diplatyinae (and extant genus *Diplatys*) in the Early Miocene of Japan (SAKAI & FUJIYAMA 1989). Extinct genera were described from the Early Oligocene of Florissant, Colorado (*Labiduromma* Scudder, Labiduridae or Forficulidae?; SCUDDER 1890b) and the Middle Miocene of Shandong, China (*Apanechura* Zhang and *Hadanechura* Zhang, Sun et Zhang, Forficulidae; ZHANG 1989, ZHANG, SUN & ZHANG 1994). Earwigs usually constitute a minor component of the Cainozoic faunas (the reported maximum is 0.6% of total insect number; WILLMANN 1990).

Extant species are found in the Late Oligocene – Early Miocene of Nakhitchevan, Caucasus (*Forficula tomis* Kolenati; BOGATCHEV 1940), and the Pliocene of Durham, England (*F. auricularia* Linnaeus; LESNE 1920). *Labidura loveridgei* Zeuner, described as subfossil from Saint Helena (ZEUNER 1962) was synonymised under *L. herculeana* Fabricius, the largest living earwig (endemic of the island and recollected in 1965–67 for the first time since Fabricius' time); however, Zeuner's specimen might be of considerable age, so far as its forceps measured 34 mm in length against 28 in the *L. herculeana* holotype and 24 in the largest of newly collected specimens (BRINDLE 1970).

2.2.2.2.4. ORDER EMBIIDA Burmeister, 1835. THE WEBSPINNERS (=Embioptera Shipley, 1904)

A.P. RASNITSYN

(a) **Introductory remarks**. Small order of inconspicuous and rather uniform insects, with about 200 recorded and 2,000 estimated species (E. ROSS 1991). The present account is based mainly on E. ROSS (1966–2000) as well as on the review by KALTENBACH (1968); data from BOUDREAUX (1979) are used as well.

(b) **Definition**. Size moderate to small. Body elongate. Head large, prognathous, with postoccipital (not gular; RASNITSYN & KRASSILOV 2000) sclerotisation between mouth cavity and occipital foramen, lacking ocelli. Antenna filiform, 12–32-segmented. Mouthparts chewing, male with mandible modified more or less, submentum often shield-like enlarged, sclerotised. Lateral cervicalia divided into 2 pieces. Pronotum lacking paranota. Meso- and metathoraces similarly shaped. Females and many males wingless, nymph-like more or less in thoracic morphology. Wings narrow, homonomous, poor venationally, highly flexible including veins, thus permitting wing to flex over head in backward running in silken tubes. Veins and particularly R surrounded by blood sinus working as vein at flight when filled with blood. Most complete venation with 11 branches only (not

---

20 – wing venation impoverished; nymphal paraproct with gills.

21 – wing with cross-veins abundant; tergal depressor of trochanter lost; nymph with thoracic and abdominal gills.

22 – size large; wings with cross-veins abundant, nodal line present; hind wing with anal area wide, anal veins numerous; nymph with lateral abdominal gills (except Siberioperlidae).

23 – nymphal abdomen with 4 gill pairs basally.

24 – male with ventral vesicle lost; nymphal thoracic sternum with 1 pair of gills between furcal apodemes.

25 – nymph with impair spine mounting tibia apically and with 1st tarsal segment shorter than 2nd one.

26 – wings with cross-veins not numerous; hind wing with reduced number of anal veins.

27 – size rather small; wing venation impoverished; male paraproct membranous, vesiculous in part; nymph with coxa bearing single larger sensorial seta and with anal gills.

28 – nymph with legs very short.

29 – wings lost; adult tarsus with 1st segment long.

Fig. 409 *Perlopsis filicornis* Martynov (Perlopseidae) from the Early Permian of Tshekarda in Urals (from ROHDENDORF 1962); length of body 16.5 mm

Fig. 410 Nymph of *Sylvoperlodes zhiltzova* Sinitshenkova (Tshekardoperlidae) from the Early Permian of Tshekarda in Urals (from SINITSHENKOVA 1987); body length 29.5 mm

counting C) and several weak cross-veins (Fig. 427), still more reduced in other representatives. Flight functionally four-winged, in-phase, anteromotoric (GRODNITSKY 1999). Legs short, femora stout (particularly fore and especially so the hind ones), tarsus 3-segmented, fore basitarsus inflated due to silk gland inside. Abdomen cylindrical or, in male, somewhat depressed, 10-segmented, with cercus 2-segmented, male left cercus usually modified and used in copulation. Male genitalia asymmetrical. Ovipositor lost. All stages living in groups in silk tubes spun using fore basitarsal glands under bark, stones, or, in humid climates, externally on substrate, feeding on plant material, mostly dead matter. Female guarding eggs and younger nymphs.

(c) **Synapomorphies** (practically all resulted from gallery spinning habits). Head permanently prognathous, with ocelli lost, postoccipital bridge present and mouthparts sexually dimorphic; 2 pairs of lateral cervicalia; wings highly homonomous, flexible, with blood sinus, with venation much reduced, RS originating far basally and coalescing with MA for a distance; female wingless due to paedomorphy; legs short, stout, tarsi 3-segmented, with fore basitarsal silk glands used for gallery spinning at all stages but egg; cercus 2-segmented; male genitalia asymmetrical; ovipositor lost.

(d) **Range**. Recorded since Oligocene or, possibly, Late Cretaceous or even Late Jurassic (see below), now world wide in warmer climates.

(e) **System and phylogeny**. Origin of the order is not clear. The Permian grylloblattidan *Sheimia* O. Martynova displays rather close similarity to webspinners in its wing venation (see Fig. 406), and the group it represented could be a promising candidate for the role of the webspinner ancestor, unless there is a huge temporal gap between them. *Sheimia* has even been assigned to the living order (MARTYNOVA 1958), a questionable decision as the fossil shows no adaptation for gallery spinning habits and few other embiid peculiarities (head transversely ovate, female winged, with wings probably heteronomous, judging from absence of any trace of hind pair in fossil and from heteronomous pterothoracic segments, abdomen spindle-shaped rather than cylindrical and with well developed ovipositor; unfortunately fore basitarsus not preserved; cf. also CARPENTER 1976, E. ROSS 2000).

The enigmatic Late Jurassic Brachyphyllophagidae (Fig. 428) are large, phyllophagous, nocturnal insects, winged in females and retaining an external ovipositor, simple fore basitarsus and hind wing with wide anal fan (RASNITSYN & KRASSILOV 2000). In spite of the striking dissimilarity, they share few, but important putative synapomorphies with the webspinners, viz. the cuticular bridge between the occipital and oral orifices and seemingly desclerotised wing veins. Their relationship cannot be ruled out, though further evidence seems necessary to consider the respective phylogenetic scenarios in more detail.

Fig. 411 Nymphal *Barathronympha victima* Sinitshenkova (Nemouromorpha *incertae sedis*) from the Late Permian of Tshekarda in Urals (holotype); body length 5.8 mm (from SINITSHENKOVA 1987)

E. ROSS (1970b) has proposed 4 suborders and 14 families, unnamed in part, for this small and rather uniform group; this seems superfluous, however; KALTENBACH (1968) and E. ROSS (1970a, 1991) used 8 families not grouped into suborders. Clothodidae are unanimously referred to as an ancestral group (E. ROSS 1987). More detailed phylogeny is not worked out yet.

(f) History. Quite a few fossil webspinner taxa are known to date, the oldest being *Burmitembia venosa* Cockerell (assigned to an unnamed monotypic family possibly related to Australembiidae by E. ROSS 1970b) from the imprecisely dated (Late Cretaceous to Eocene, see Chapter 4) Burmese amber. The next oldest is *Electrembia antiqua* (Pictet) from the Baltic amber (Embiidae: E. ROSS 1966), followed by *Lithembia florissantensis* (Cockerell) from the Early Oligocene of Florissant in Colorado, USA, very tentatively assigned to Embiidae (E. ROSS 1984). The youngest Tertiary ones are *Mesembia* sp. (Anisembiidae, Fig. 429) and *Oligembia vetusta* Szumik (Teratembiidae, SZUMIK 1994) in the Early Miocene Dominican amber. Besides these, *Embia* and *Oligotoma* are referred to as found in the Quaternary African copal (HANDLIRSCH 1906–1908). For the possible webspinner relatives from the Late Jurassic see **(e)** above.

2.2.2.3. SUPERORDER GRYLLIDEA Laicharting, 1781 (=Orthopteroidea Handlirsch, 1903)

A.V. GOROCHOV, AND A.P. RASNITSYN

(a) Introductory remarks. The present concept of the superorder is closest, though not identical to that of SHAROV (1968).

(b) Definition. Body size from moderately small to very large. Head plesiomorphically prognathous (with mandibles moving in plane subvertical to that of occipital foramen), but with mouth cavity and occipital foramen contiguous (not separated by gula); sometimes secondary hypognathous, with the two above planes subparallel (adaptation for leaping ability to secure mouthparts). Pronotum lacking free paranota. Fore and middle coxae usually not elongate. Fore wing (see Figs. 432, 454, 462) with precostal space usually well developed (with numerous veins); with MP (usually called MA$_2$) usually well individualised (simple or with few more or less subapical branches), originating mostly close to M$_5$ (usually called MP) and plesiomorphically running in parallels to M$_5$+CuA$_1$; with 1A simple. Hind wing (see Figs. 432, 454, 463) not foldable transversally, wider than fore wing in flying forms, with M$_5$ (MP auctorum) usually meeting CuA$_1$ and not CuA stock which is plesiomorphically straight or gently arching, not angled. Flight functionally four-winged, in-phase, sometimes morphologically two-winged, posteromotoric (GRODNITSKY 1999). Male genitalia and genital plate almost always symmetrical. Ovipositor usually well

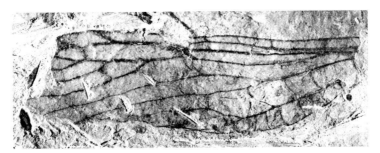

Fig. 412 *Stenoperlidium permianum* Tillyard (Eustheniidae) from the Late Permian of Belmont in Australia (holotype); length of wing fragment 22.5 mm (from TILLYARD 1935b)

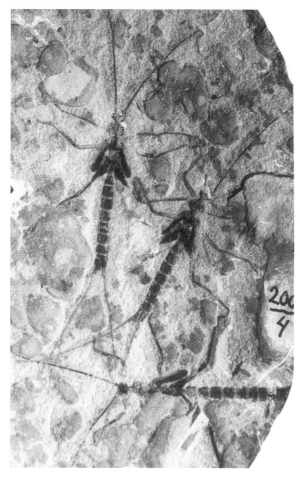

Fig. 413 Nymphal *Mesoleuctra tibialis* Sinitshenkova (Mesoleuctridae) from the Early or Middle Jurassic of Cheremza in Kuznetsk Basin, Siberia (PIN 2990/3–4, photo by D.E. Shcherbakov); body length 18.5 mm

Fig. 414 *Mesoleuctroides saturatus* Sinitshenkova (Mesoleuctridae) from the Early or Middle Jurassic of Novospasskoye in Siberia (PIN 3000/872, photo by D.E. Shcherbakov); body length 12 mm

developed (more or less reduced or almost entirely lost in extant stick insects). Cercus entire. Habits variable, but neither parasitic, nor xylobiotic, nor aquatic. Immatures similar to adults both morphologically and bionomically.

(c) **Synapomorphies**. Pronotal sides bent downward to form side pronotal lobes. Fore wing with precostal space large, bearing numerous veins, with MP individualised and running in parallels to M_5+CuA_1, and with 1A simple. Cercal segmentation lost.

(d) **Range**. Worldwide, since Late Carboniferous until the present.

(e) **System and phylogeny**. Three orders are recognised here to comprise the superorder, viz. Orthoptera (grasshoppers, crickets, etc.), Phasmatida (stick-insects) and Mesotitanida (Fig. 430). The hypothesis of mesotitans as descending from a subgroup among the orthopterans (ROHDENDORF & RASNITSYN 1980) cannot be supported because of their autplesiomorphy in the branched CuP.

2.2.2.3.1. ORDER ORTHOPTERA Olivier, 1789. THE ORTHOPTERANS (=Gryllida Laicharting, 1781)

(a) **Introductory remarks**. A medium-sized order (about 20,000 living species, OTTE 1994–1997), of popular saltatory insects (Fig. 431). The present chapter is based mostly on data by SHAROV (1968) and GOROCHOV (1984–2000), with other references given where appropriate.

(b) **Definition**. Size as described for superorder. Head hypognathous, less often pro- or opistognathous. Antenna long, bristle-shaped, or short, ribbon-like. Prothorax of varying length, meso- and metathoraces usually rather short. Hind legs saltatory, with femur incrassate basally, rarely and secondarily simple. Hind tibia usually and plesiomorphically with two rows of spines dorsally. Tarsi 3–5-segmented, rarely 1–2-segmented. Fore wings (Fig. 432), except in brachypterous forms, bent longitudinally, with fore parts hanging freely, laterally, and hind parts overlapping each other over body dorsum. Fore wing 2A with few basal branches. Both fore and hind wing with CuP simple. Elytral stridulatory apparatus often present, albeit almost exclusively in male, resulting in highly dimorphic

Fig. 415 *Platyperla platypoda* Brauer, Redtenbacher et Ganglbauer (Platyperlidae) from the Early Jurassic of Ust-Baley in Siberia (PIN 2375/5, photo by D.E. Shcherbakov): nymph probably with its moulting cast; body length 18 mm

venation (so many isolated fossil female fore wings cannot be attributed even to a family). Either CuP or a cross-vein modified into stridulatory vein. Hind wing (Fig. 432) with fore branch of 2A lacking distinct comb of branchlets. Ovipositor external (or lost); oviposition into soil, crevices, or in plant tissues. Egg elongate, with thin chorion, lacking operculum. With 4–11 (or more) developmental instars. Nymphal wing cases reverted, with wing fore margins directed dorso-medially. Habits predatory (plesiomorphically), omnivorous, or phytophagous; phytophilous, phytogeophilous, geophilous (sometimes digging), occasionally myrmecophilous, nidicolous, inhabiting caves, hollows of trees, space under loose bark, etc. Few orthopterans displaying incipient form of sociality, parental care, and ability to produce silk used for lining burrows.

(c) Synapomorphies. Head hypognathous; hind legs saltatory; hind tibia with two regular rows of strong spines dorsally; fore wings bent longitudinally, with fore parts hanging freely, laterally; hind wing with fore branch of 2A lacking distinct comb of branchlets, both wings with CuP simple; nymphal wing cases reverted.

(d) Range as in superorder (Chapter 2.2.2.1d).

(e) System and phylogeny is shown at Fig. 430 (for further details see GOROCHOV 1984–2000). Alternative names for suborders are Ensifera for Gryllina and Caelifera for Acridina.

(f) History of the order can be divided into several succeeding stages differing in the composition of their dominant groups, with the boundaries marked by changes in the primary means of defence against predators (GOROCHOV 1989b). This is particularly true for Gryllina, but can also be applied to Acridina. The relevant publications are mostly reviewed by GOROCHOV (1985–2000). Additionally important are HONG (1982b, 1985a, 1986), ZESSIN (1983, 1987, 1988, 1997), and MARTINS-NETO (1990a–c, 1991a,b, 1995a,b).

The first (Palaeozoic) stage has been marked by the abundance of Oedischiodea and practical absence of disruptive coloration. The colour was either uniformly dark (Fig. 434), or the pattern consisted of small, conspicuous dark patches on the light or transparent background (see Fig. 433). Taking into consideration that the sound organs are not found in the Palaeozoic fossils, the above colour pattern could be interpreted as having a role in signalling. The Palaeozoic orthopterans were zoophagous (predaceous or necrophagous) and mostly phytophilous or phytogeophilous with some geophilous features retained (e.g. ovipositing in soil), or possibly sometimes geophilous. The oldest orthopterans come from the Late Carboniferous of France (Commentry) and North China (Shanxi Formation, Shanxi) and all belong to Oedischiinae (Oedischiidae).

The Early Permian fauna was richer. Oedischioidea (Fig. 434) has included both Oedischiidae and Pruvostitidae with 8 subfamilies (retaining Oedischiinae) and found in the Urals (Tshekarda), Kansas (Elmo), China (Yinping in Anhoi, Xiasihezi Formation in Shanxi), Germany (Lebachian Shales) and South Africa (*Afroedischia*, possibly near Mesenoedischiinae, from the Laingsburg Formation). Among Xenopteroidea, a sole member of Adumbratomorphidae is found in Tshekarda. The Early Permian Permoraphidioidea are represented by Permoraphidiidae in Elmo and Pseudelcanidae in Tshekarda. It cannot be excluded that the family Thueringoedischiidae recently described from the Early Permian of Germany (U. Rotliegend of Thüringia) and Czechia (Obora) and venationally intermediate between Permoraphidioidea and Oedischiidae (ZESSIN 1997) also belongs to Permoraphidioidea. Elcanoidea (Permelcanidae) are found in some of the above localities, represented by Permelcaninae (Fig. 435).

The Late Permian Oedischioidea retained considerable diversity (same 2 families with 9 subfamilies in Tataria (Tikhiye Gory), Arkhangelsk (Soyana), Kirov (Kityak), and Orenburg Regions (Kargala) in European Russia, Kuznetsk Basin in Siberia (Kaltan), and in Natal, South Africa). The only other superfamily is Elcanoidea represented in Arkhangelsk and Kirov Regs by the same single subfamily as before.

The second stage embracing the Triassic, and Early and Middle Jurassic, was dominated by Haglidae (Hagloidea). Signal colour patterns have not been encountered at that stage, whereas disruptive patterns had become fairly common (Fig. 436), and stridulatory devices are known to have become established simultaneously in Oedischioidea and Hagloidea (Figs. 437, 439). Otherwise the habits of the majority of orthopterans had not been modified much at the stage, except that obligatory phytophily appeared for the first time in Proparagryllacrididae, and, possibly, phytophagy in some Locustopsoidea and Grylloidea.

Undoubtedly Early Triassic Orthoptera are unknown, except for the sole, oldest, and yet comparatively advanced representative of the suborder Acridina (*Praelocustopsis mirabilis* Sharov, Locustopsidae) from Nizhnyaya Tunguska River in Central Siberia. Three Middle

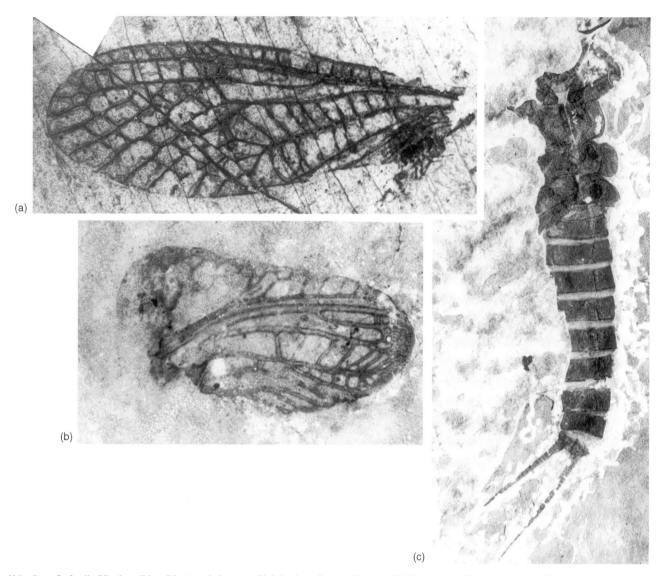

Fig. 416 Stonefly family Siberioperlidae: *Siberioperla lacunosa* Sinitshenkova from the Early or Middle Jurassic of Novospasskoye in Siberia (a, b), and *S. scobloi* Sinitshenkova from the Late Jurassic of Uda in Siberia (c): a–nymph (PIN 3000/871), body length 32 mm, b–fore wing of the brachypterous male 5.6 mm long (PIN 3000/882), c–normal (female) wing 18 mm long (PIN 3053/418) (from SINITSHENKOVA 1987)

Triassic orthopterans have been described recently from the earlier Anisian of Vosges in France including the oldest Gryllavoidea (Gryllavidae) and Hagloidea (Tuphellidae: probably Tuphellinae, and possibly Hagloedischiidae: *Voltziahagla*) (MARCHAL-PAPIER *et al.* 2000). In the Late and possibly Middle Triassic assemblages known from Middle Asia (Madygen Formation in South Fergana) and Australia (Mount Crosby Formation of Queensland), Oedischioidea (Fig. 438) are represented by 3 families (Bintoniellidae, Proparagryllacrididae, and Mesoedischiidae), with 7 subfamilies, none of which are present in the Palaeozoic. Among them the only known (from Fergana) genus of the latter family shows the male elytron modified for stridulatory function (see Fig. 437). Xenopteroidea are found represented only by Xenopteridae, a family endemic to the Middle/Late Triassic of both Madygen and Ipswich. Triassic Elcanoidea were represented (again in Fergana) by a new subfamily of Permelcanidae.

The Hagloidea, and particularly Hagloedischiidae, Tuphellidae, and Haglidae, are found in Fergana (all three families and 5 subfamilies), at Kenderlyk River (East Kazakhstan), Mount Crosby Formation (Australia), and Molteno Formation (South Africa) (only Haglidae). All of them, including the least advanced Hagloedischiidae, had a stridulatory apparatus on the male elytra (see Fig. 439).

Gryllavoidea are known from Fergana (Gryllavidae), and Grylloidea from Molteno (Protogryllidae: Protogryllinae) (Figs. 440, 441). Locustopsoidea are represented (in Fergana and in Ipswich, Denmark Hills, in Australia) by Locustopsidae and (in Fergana only) by both Locustopsidae (as in the Early Triassic) and more primitive Locustavidae. It cannot be excluded that additionally the first Tetrigoidea and Tridactylomorpha are found in the Madygen Formation in Fergana. The pronotum figured by SHAROV (1968: fig. 34d) as belonging to Locustopsidae looks rather like a tetrigid or a

Fig. 417 *Perlariopsis stipitata* Sinitshenkova (Perlariopseidae) from the Early or Middle Jurassic of Novospasskoye in Siberia (from SINITSHENKOVA 1987); length of fossil, as preserved, 11 mm

Fig. 418 Nymphal *Spinoperla spinosa* Sinitshenkova (Perlariopseidae) from the Early or Middle Jurassic of Novospasskoye in Siberia (from SINITSHENKOVA 1987); length (straight from head to abdomen apex) 12 mm

related extinct family retaining fore wings longer than the pronotum. The recently described family Dzhajloutshellidae may represent a separate superfamily of the tridactylomorphs, although this decision would need more information.

In the Early Jurassic, Oedischioidea were still persisting, represented only by the family Bintoniellidae and particularly its nominotypic subfamily known since the Late Triassic. The group is found in the English Lias. Xenopteroidea are not found. In Elcanoidea, a subfamily of Elcanidae, has been recorded from England, Germany, Switzerland, Irkutsk Region in Central Siberia, and again in Fergana (Sagul locality in Kirghizi which might be of the earlier Middle Jurassic age).

There are two hagloid families known in the Early Jurassic of England, Germany, South Fergana and Issyk-Kul Lake in Kirghizia, Kuznetsk Basin and Irkutsk Region in Siberia, and in West Mongolia and China (Shanxi, Inner Mongolia): Haglidae with 5 subfamilies (3 not found earlier), and Prophalangopsidae (unknown for the Triassic). Gryllavoidea are not found. For Grylloidea, only Protogryllidae are known represented by 2 subfamilies in England, Germany, South Fergana, Irkutsk Region and Krasnoyarsk Province (Fig. 442). Locustopsidae, the only Early

Jurassic member of Locustopsoidea, was widespread like Haglidae. Regiatidae (Tridactyloidea) are found in England and possibly in Germany (*Ptotachaeta* Handlirsch in the Upper Lias of Dobbertin).

The Middle Jurassic fauna is poorly explored, with several species of Haglidae, Locustopsidae, and Elcanidae known from Krasnoyarsk Province (Kubekovo) and from Hebei and Liaoning Provinces of China. Hagloidea (Haglidae and Prophalangopsidae) and Grylloidea (Protogryllidae) are found in Mongolia (Bakhar).

The third stage of orthopteran history lasts from the Late Jurassic to the Palaeocene. The interval is marked by the leading position of Prophalangopsidae (Hagloidea), and the decreasing role of the primitive, decamouflaging stridulatory mechanism. In some groups (Stenopelmatoidea, possibly also Phasmomimoidea) it was just lost (Fig. 443), while in others (Tettigonioidea, some Prophalangopsidae) a more cryptic mode of stridulation became established, with the wing bases rather than their apices being most mobile (Figs. 444, 445). Grylloidea were mastering solid substrates with its narrow spaces more actively than before. Obligatory phytophilous Phasmomimidae have replaced their ecological predecessors Proparagryllacrididae. Additional phytophagous groups (Eumastacoidea and some others) have appeared at this stage as well. The orthopterans with old types of habits and diet were still abundant, however.

Fig. 419 Male (larger) and female *Baleyopteryx orthoclada* Sinitshenkova (Baleyopterygidae) from the Early Jurassic of Chernyi Etap in Siberia (PIN 2386/136, photo from archive of Palaeontological Institute RAS), male fore wing 7 mm long

Fig. 420 The most primitive known protelytropteran *Apachelytron transversum* Carpenter et Kukalová (Archelytridae) from the Early Permian of Obora in Moravia (holotype, after CARPENTER & KUKALOVÁ 1964); fore wing 6 mm long

Late Jurassic orthopteran finds are especially abundant albeit concentrated at Eurasia. Elcanoidea (see Fig. 431) are represented in Germany (Solnhofen etc.), South Kazakhstan (Karatau), and possibly in Transbaikalia (Glushkovo Formation whose age can be Early Cretaceous as well) by 2 subfamilies of Elcanidae, one of which appears for the first time. There are 3 families of Hagloidea: the latest Haglidae (single subfamily Cyrtophyllitinae first appearing earlier in Jurassic), Tuphellidae (subfamily Paracyrtophyllitinae unknown before the Late Jurassic), and Prophalangopsidae (see Fig. 444) with 2 subfamilies, one of which is not known before. Hagloids were especially widespread being found in Germany, England, South Kazakhstan, Transbaikalia, West Mongolia, and China (Hebei, Liaoning).

There are two families of Late Jurassic crickets (Grylloidea) known from Germany, South Kazakhstan, and West Mongolia. These are Protogryllidae with 3 subfamilies, known variously since the Late Triassic, Early and Late Jurassic, and Baissogryllidae with 2 subfamilies both known since the Late Jurassic only. The first undoubted Phasmomimoidea (Phasmomimidae, Fig. 446) and Eumastacoidea are also known from South Kazakhstan. Locustopsidae were also persisting (found in Germany and South Kazakhstan, Fig. 447).

The Early Cretaceous Elcanoidea all belonged to the second, younger subfamily of Elcanidae (Fig. 448), found in England, Transbaikalia, West Mongolia, and Brazil. Only Prophalangopsidae have survived in Hagloidea, with 3 subfamilies, either endemic (Termitidiinae) or retained since the Jurassic (found in England, Transbaikalia, Hebei in China, Japan, and Brazil). The oldest Tettigonioidea (Haglotettigoniidae, see Fig. 445) are found in Transbaikalia. Grylloidea are represented (in Transbaikalia, West Mongolia, and Brazil) by 3 families: Baissogryllidae (3 subfamilies, 1 not found in the Jurassic; Fig. 449), Gryllidae (Gryllospeculinae), and Gryllotalpidae. Transbaikalia and West Mongolia have yielded Vitimiidae of unknown taxonomic position within Gryllina. Locustopsidae and the first true Tetrigidae (Tetrigoidea, Fig. 450) and Tridactylidae (Tridactyloidea, Fig. 451) are found in Transbaikalia (Baissa). Locustopsidae (Locustopsinae and Araripelocustinae Martins-Neto, 1995, stat. nov.) and Tridactylidae are additionally found in Brazil, the latter family also in West Mongolia (Fig. 452). Eumastacoidea (Eumastacidae?) are indicated in the Early Cretaceous of China (LIN 1980) and Brazil (MARTINS-NETO 1991b).

Fig. 421 *Uralelytron* spp. (Bardacoleidae) from the Early Permian of Tshekarda in Urals: left *U. insignis* (G. Zalessky) (PIN 1700/508, wingspan 20 mm), right *U. martynovi* Rohdendorf (PIN 1700/509, fore wing 7.0 mm long); arrowhead indicates triple hind wing folding (photo by D.E. Shcherbakov)

Fig. 422 *Arctocoleus ivensis* Martynov (Permofulgoridae) from the Late Permian of Soyana in NW. Russia (PIN 3353/127, photo by D.E. Shcherbakov); wingspan 26 mm

The Late Cretaceous orthopterans are rarely found: Gryllidae (subfamily unknown but not the one known from the Early Cretaceous) and the latest Locustopsidae from Central Kazakhstan (Kzyl-Zhar locality).

The Palaeocene assemblages come from the Pascapoo Formation in Alberta (Canada), Island of Mull (Scotland), Menat (France), Atanikerdluk in Greenland, Redbank Plain Series in Australia, Maíz Gordo in Argentina and possibly Mo Clay in Denmark. These are Prophalangopsidae (possibly same subfamilies as before), earliest Stenopelmatoidea (Anostostomatidae, Fig. 443), Tettigoniidae, the earliest Acridoidea and a probable extant subfamily of Gryllidae. Danish and Argentinian Tettigoniidae may all belong to the extinct subfamily Psedotettigoniinae with the male stridulatory apparatus distinctly less advanced than known in the Baltic amber.

The fourth stage of orthopteran history embraces the Tertiary since the Eocene, and continues up to the present time. It can be characterised through the absence of single or few dominant families. In fact, since the Oligocene, the wide distribution of open herbaceous communities has possibly made Acridoidea the dominant superfamily. The hypothesis is difficult to confirm palaeontologically, however, because inhabitants of such communities are generally not particularly common as fossils. Further, at that stage the orthopterans often gain the highest forms of crypsis, with near perfect imitation of external objects, mimicry, disruptive body form, etc. The fauna becomes similar to the present day one both in habits and diet.

The Eocene assemblages include Tettigonioidea all belonging to Tettigoniidae (Baltic amber and impression fossils from Germany, and Wyoming in USA). The amber fossils belong in part to the living subfamily Tympanophorinae. The only present stenopelmatoid family, Raphidophoridae, is represented by an extinct subfamily in the Baltic amber. Grylloidea are found in the Baltic amber and impression fossils from Italy, Germany, as well as in Wyoming and Colorado (USA). They belong to 5 extant subfamilies of Gryllidae, and to a tribe of Gryllotalpidae. The Baltic amber has given Tetrigidae as well. The latest Eocene or earliest Oligocene Bembridge marls of Isle of Wight (England) have yielded the possibly earliest Acrididae.

In the Oligocene, Tettigoniidae, Gryllidae, Gryllotalpidae, and Acrididae are found in Germany, France, Italy and England, and belong to living subfamilies or in Gryllotalpidae, to tribes. Eumastacidae are mentioned as found in the British Columbia and Montana (S. LEWIS 1976a).

The known Neogene (Miocene and Pliocene) and Anthropogene (Quaternary) assemblages are from Stavropol, from Maritime

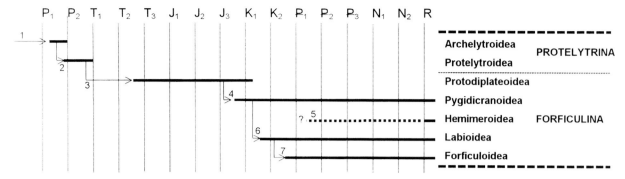

| P₁ | P₂ | T₁ | T₂ | T₃ | J₁ | J₂ | J₃ | K₁ | K₂ | ₽₁ | ₽₂ | ₽₃ | N₁ | N₂ | R |

Archelytroidea

Protelytroidea

PROTELYTRINA

Protodiplateoidea

Pygidicranoidea

Hemimeroidea **FORFICULINA**

Labioidea

Forficuloidea

Fig. 423 Phylogeny and system of earwigs (Order Forficulida).

Time periods are abbreviated as follows: P₁, P₂–Early (Lower) and Late (Upper) Permian, T₁, T₂, T₃–Early, Middle and Late Triassic, J₁, J₂, J₃–Early, Middle and Late Jurassic, K₁, K₂–Early and Late Cretaceous, ₽₁,–Palaeocene, ₽₂–Eocene, ₽₃–Oligocene, N₁–Miocene, N₂–Pliocene, R–present time (Holocene). Right columns are names of suborders (all caps), superfamilies (bold lower case). Arrows show ancestry, thick bars are known durations of taxa (dashed hypothesised range), figures refer to synapomorphies of the subtended clades, as follows.

1 – fore wings elytrised with sutural margins overlapping in repose; hind wings transversely folded in repose; possibly also: basal segment of cercus enlarged; chromosomes holokinetic.

2 – hind wing with proximal remigium (squama) somewhat sclerotised, distal remigium plus anal fan bent under and folded transversely in repose, and fan veins bearing thickenings.

3 – elytra covering at most abdomen base; hind wing with squama fewer than half wing length and with intercalary fan veins; cerci elongate.

4 – tarsi 3-segmented; adult cerci unsegmented; ovipositor much reduced.

5 – eyes lost; thoracic and abdominal segments less sloped; legs gripping with tarsal plantulae; cerci long, not forcipate.

6 – posterior ventral cervical sclerite enlarged; ovipositor lost; one of phallomeres directed caudad in repose.

7 – 2nd tarsomere produced ventrodistally; virga with basal vesicle.

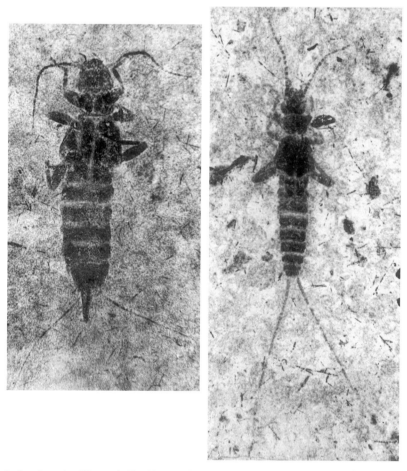

Fig. 424 Protodiplateidae from the Late Jurassic of Karatau in Kazakhstan: left, *Asiodiplatys speciosus* Vishniakova (holotype, body with ovipositor 15.5 mm long), right, *Microdiplatys campodeiformis* Vishniakova (holotype, body excluding cerci 10 mm long) (photo by D.E. Shcherbakov)

(a)   (b)

Fig. 425 Unidentified Pygidicranidae from the Early Cretaceous of Baissa in Siberia, male and female in a scale: a–male (PIN 1989/2491, body including forceps 24 mm long); b–female (PIN 3064/8567, body including forceps 16 mm long)

Province of Russia, from Italy, Croatia, Germany, Colorado (USA) and Dominican amber (Miocene), North Germany (Pliocene), Poland and East Africa (copal, that is subfossil resin) (Pleistocene). They are composed only of members of living subfamilies and represent the majority of existing families.

2.2.2.3.2. ORDER PHASMATIDA Leach, 1915. THE STICK INSECTS
(=Phasmodea Burmeister, 1838)

(a) Introductory remarks. Small (more than 2,500 living species according to KEY 1991) group of inconspicuous, albeit sometimes giant insects, mimicking twigs and leaves (Fig. 453). The present section is based on material described by SHAROV (1968) and GOROCHOV (1992, 1993, 1994a, 2000). For more detail concerning fossils see also MARTINS-NETO (1989a), S. LEWIS (1992), REN (1997c). For living stick insects see KRISTENSEN (1975), J. BRADLEY & GALIL (1977), D. KEVAN (1977), though their classificatory systems are much more split than the present one.

(b) Definition. Body size middle to very large. Body elongate, stick-like, or depressed, leaf-like broad (imitating twigs and leaves, respectively). Head prognathous. Antennae bristle-shaped, long or short. Prothorax short, meso- and metathoraces usually much longer. Hind legs not saltatory, usually with thin femora (in male sometimes incrassate when used to fix female in copulation). Hind tibia with two dorsal rows of spines, if present, never arranged as regular as in Gryllida. Tarsi 3–5-segmented. Fore wings (Fig. 454) not bent or slightly bent longitudinally, often reduced or lost, otherwise overlapping each other over body dorsum, lacking stridulatory apparatus (but there is an exception), with veins parallel more or less, with CuP simple, 2A with at most few basal branches. Hind wing when normally developed with anterior branch of 2A bearing distinct comb of branchlets. Ovipositor external but short in Susumaniidae, reduced and internalised in living forms. Eggs (known for living forms only) usually short, ovoid, with thick, sculptured chorion and operculum, imitating plant seeds, laid on substrate or disseminated freely (rarely laid in soil using abdomen with modified apical sternum instead of ovipositor as digging device). Immatures of up to 6 instars, with wing cases normal (fore margin directed latero-ventrally). Habits phytophilous (obligatory so in living forms), phytophagous, weakly mobile. No sociality or parental care.

(c) Synapomorphies. Body elongate, stick-like; meso- and metathoraces much longer than prothorax; fore wing with veins subparallel; ovipositor short; phytophagy.

(d) Range. Early or Late Permian until the present; fossils uncommon, found in Eurasia, South America and Australia only; living forms common in all tropical regions, rare in warm temperate ones.

(e) System and phylogeny as shown at Fig. 430. The taxonomic position of Xiphopteroidea with its only family Xiphopteridae is not

Fig. 426 Unidentified Forficuloidea nymph from the Late Eocene Baltic amber (PIN 964/827, photo by A.P. Rasnitsyn); body (excluding forceps) 2.8 mm long

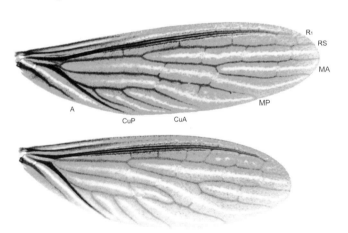

Fig. 427 Wing venation of the primitive living webspinner *Clothoda longicaudata* Ross (Clothodidae) (from E. Ross 1987 courtesy of the California Academy of Sciences)

without uncertainty; its fore wing with the bent apex indicates vertical rather than horizontal rest position. This character state is typical for orthopterans and not known in stick insects, thus indicating that Xiphopteridae could possibly represent an aberrant orthopteran group.

(f) History of the order very poorly known. All known fossils show extinct stick insects as similar to living ones in their habits and general appearance, except for the absence of the highest cryptic forms (those mimicking perfectly, leaves and twigs).

The earliest stick insects are reported from the Bor-Tologoy (Late Permian of Mongolia, Prochresmodoidea: Permophasmatidae) and with some doubts from Tshekarda (Early Permian of Urals). The next oldest are Xiphopteroidea (Xiphopteridae, Fig. 455), Aeroplanoidea (Aeroplanidae, Fig. 456), and Prochresmodoidea (Prochresmodidae, Figs. 453, 457) from the Late (possibly Middle) Triassic of Madygen in Kirghizia (Aeroplanidae also in Ipswich, Australia). Aeroplanoidea (Aerophasmatidae) are known also from the Early Jurassic of Germany and, possibly, England (Chresmodellinae), from the Late Jurassic of Karatau in South Kazakhstan (Aerophasmatinae), and from the Early Cretaceous of Baissa in Transbaikalia (Fig. 458), Purbeck in England (*Blattidium* Westwood, cf. E. CLIFFORD *et al.* 1994) and Santana in Brazil, and the early Late Cretaceous (Turonian) of Central Kazakhstan (Cretophasmatinae). The Middle Jurassic stick insects are unknown. The superfamily Susumanioidea with the only family Susumaniidae (= Hagiphasmatidae: GOROCHOV 2000; Figs. 459, 460) are known beginning from the Late Jurassic of Karatau in Kazakhstan (Phasmomimoidinae), and further from the Early Cretaceous of Baissa in Transbaikalia, Bon Tsagan in Central Mongolia, Laiyang in East China, Khetana near Okhotsk Sea (Susumaniinae), from the Late Cretaceous of Obeshchayushchiy near Magadan and Labrador, East Canada (same subfamily), as well as from the Palaeocene of Alberta (Pascapoo Formation), Early Eocene of Washington (Klondike Mt. Formation) and Late Eocene Baltic amber, possibly again Susumaniinae (Eocene records are based on S. LEWIS 1992 and ZOMPRO 2001: Archipseudophasmatidae may be either a synonym or a subfamily of Susumaniidae). Phytophagy of Phasmomimoidinae is confirmed by intestine contents (leaf fragments of Hirmerellaceae) of a Late Jurassic species (RASNITSYN & KRASSILOV 2000).

Otherwise the only living families, Phasmatidae and Phyllidae, are found in the Cainozoic (in the Early Oligocene of Florissant in Colorado, USA, and in the Early Miocene Dominican amber). They are also possibly present in the Baltic amber: Phyllidae as the areolatae nymph with distinctly heteronomous wing pads (ZOMPRO 2001: Fig. 38), Phasmatidae as the anaorealatae larva (id., Fig. 40). *Electrobaculum* probably belongs to Susumaniidae. Unfortunately the taxonomic position of the respective fossils cannot be further clarified at present.

2.2.2.3.3. ORDER MESOTITANIDA Tillyard, 1925 (=Titanoptera Sharov, 1968)

(a) Introductory remarks. This section is based on data by SHAROV (1968).

(b) Definition. Body size large or very large (Fig. 461). Body moderately elongate and depressed, neither stick- nor leaf-like. Head hypognathous. Antennae long, bristle-shaped. Fore legs strong, raptorial. Hind legs not saltatory, with more or less thin femora. Hind tibiae lacking distinct two dorsal rows of spines. Tarsi 5-segmented. Fore wing (Fig. 462) of general form and rest position as in stick insects, with venation either parallel or not, with 2A bearing numerous branches, with stridulatory apparatus including numerous cross-veins simultaneously. Hind wing (Fig. 463) with anterior branch of CuA bearing comb of branchlets. Both wing pairs with CuP branched. Ovipositor external, oviposition

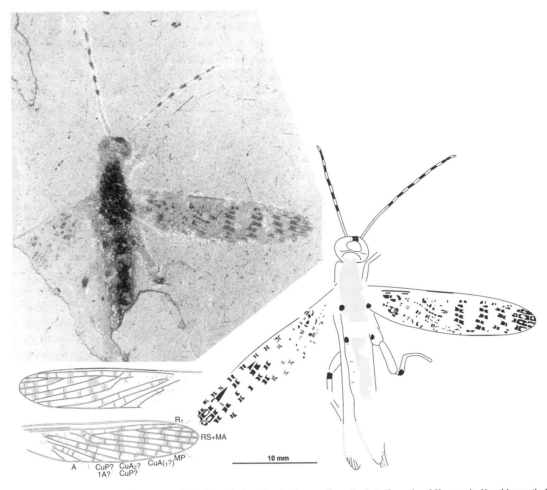

Fig. 428 *Brachyphyllophagus phasma* Rasnitsyn, a possible giant relative of webspinners, from the Late Jurassic of Karatau in Kazakhstan (holotype, photo by D.E. Shcherbakov). Ironically, we know what it had eaten while it was still alive (leaf fragments of a common Mesozoic gymnosperm plant *Brachyphyllum*, now considered related to *Gnetum* among living plants) but not its precise taxonomic position (RASNITSYN & KRASSILOV 2000)

probably into substrate or crevices. Habits most probably phytophilous, predatory, similar to that of larger predatory katydids and mantids.

(**c**) **Synapomorphies**. Head almost hypognathous; fore legs strong, raptorial; elytral stridulatory apparatus including many stridulatory cross-veins.

(**d**) **Range**. Middle and Late Triassic only; Central Asia, Australia, and East Europe (undescribed from Ukraine).

(**e**) **System and phylogeny**. The order includes three families considered in detail by SHAROV (1968). The above synapomorphies were possibly late (Triassic) acquisitions of a group whose earlier representatives could well have been ancestors of both orthopterans and stick insects.

(**f**) **History** of the order cannot be clarified at present because the fossils available all come from deposits of similar age. All three families (Mesotitanidae, Paratitanidae and Gigatitanidae) are known from the Middle or Late Triassic Madygen Formation in Kirghizia, with Mesotitanidae being found also in the Middle Triassic of Brookvale in New South Wales, Australia.

It is of interest that the mesotitanidans acquired their stridulatory apparatus simultaneously with orthopterans though in different and varying ways: always centred near the wing midlength and occupying the space either between MA and MP (Paratitanidae, see Fig. 463), MP

and CuA (Gigatitanidae, see Fig. 461), or all the space between RS and CuA (Mesotitanidae, Fig. 464).

2.3
Insect Trace Fossils

†*V.V. ZHERIKHIN*

2.3.1. INTRODUCTORY REMARKS

Insect trace fossils, particularly in the case where they are described under their own scientific name, represent insect parataxa like the better known category *Insecta incertae sedis* (insects of obscure taxonomic position; for more details on parataxa see RASNITSYN 1996). This is the reason why trace fossils are considered here together with other insect taxa.

Fig. 429 Male *Mesembia* sp. (Anisembiidae) from the Early Miocene Dominican amber (from POINAR 1992a, © The Board of Trustees of the Leland Stanford Jr. University, with the permission of Stanford University Press. www.sup.org)

Besides the fossilised remains of organisms themselves, the fossil record includes also various traces of their activities preserved either on other fossil remnants or as biogenic sedimentary structures. The trace fossils (or *ichnofossils*, from the Greek *ichnos*, a trace) are investigated by a special discipline called *ichnology*. They are widespread and commonly occur even in those palaeoenvironments where few if any organic remains are represented, such as in palaeosols and aeolian sediments. The probability of trace fossilisation depends on the probability of conservation of the trace-bearing matter and not on the fossilisation potential of the trace producer itself. Consequently the history of some organisms that are poorly represented directly in the fossil record (like many actively moving soft-bodied animals including insect larvae), is documented mainly by their trace fossils. Ichnology provides also a unique source of information about the habitat distribution, abundance, biology and behaviour of various organisms in the past.

However, ichnology has its own important constraints which have not to be forgotten when ichnological data are used in a broader biological context. First of all, trace fossils can only rarely be attributed to certain taxa with a good degree of certainty (when they reflect exactly important morphological characters of the trace maker as many vertebrate footprints do, or are highly diagnostic themselves like the galls and mines of some arthropods, or are occasionally accompanied by body fossils). Often an attribution is highly hypothetical or

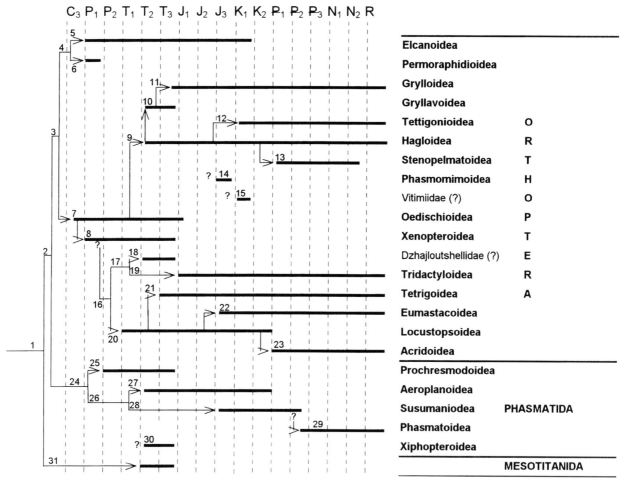

Fig. 430 Phylogeny and system of the Superorder Gryllidea

even impossible, and the traces have to be classified as formal parataxa –*ichnogenera* and ***ichnospecies***. Even when considering morphologically complex fossil traces, it is generally believed that similar ichnofossils may be produced by different organisms (BROMLEY 1990). The probability of misattribution increases with increasing geological age because of increasing differences in the set of taxa. The possibility cannot be rejected that some trace types which are now unique and characteristic for a certain producer could be very similar to traces once produced by very different, extinct taxa. Hence biological and ichnological classifications are largely independent one from another; the ichnotaxa may correspond or not to any taxa of the natural biological system. Accordingly, ichnological data should not be taken uncritically and have to be used with much caution, especially when there is a controversy between them and other data.

There have been several attempts to create a general ichnofossil classification. Modern classifications differ in detail, but are based mainly on the principles proposed by SEILACHER (1953), where the traces are classified in accordance to the kinds of activities of the trace producer (feeding, locomotion, resting, etc.). Originally Seilacher's classification was oriented mainly towards marine invertebrates and does not cover some important types of insect traces. Even its recent more sophisticated version (BROMLEY 1996) is still somewhat unsatisfactory in

this respect. There are also various problems connected with an ethological interpretation when the nature of a trace is unclear or controversial. For instance, insect burrows in a hard substrate (especially in wood) often appear as a result of feeding but are also used as shelters; others (e.g. wood nests of sphecid wasps, bees and ants) are specially constructed by adults as shelters for their brood, while burrows under bark of ambrosia beetles include both the maternal galleries made the by female for oviposition and the larval feeding tunnels. Some authors (e.g. A. MÜLLER 1981) proposed more complex classifications which are, however, not so widely accepted in modern ichnology as Seilacher's original scheme. An alternative (toponomic) trace classification is applied mostly to the biogenic sedimentary structures and based on relations between the trace and sediment (surface, internal, extraneous) (e.g. MARTINSSON 1970). Both ethological and toponomic principles are combined in the classification proposed by VIALOV (1966, 1968).

For a general review on the different aspects of ichnology see ABEL (1935), VIALOV (1966), R. FREY (1975), HÄNTZSCHEL (1975), TEVESZ & MCCALL (1982), EKDALE *et al.* (1984), R. FREY & PEMBERTON (1985), BOTTJER *et al.* (1987), FEDONKIN (1988), BOUCOT (1990), BROMLEY (1990, 1996), and PEMBERTON *et al.* (1990). The arthropod traces are discussed specially by A. MÜLLER (1975),

---

Time periods are abbreviated as follows: C_1–Late (Upper) Carboniferous, P_1, P_2–Early (Lower) and Late (Upper) Permian, T_1, T_2, T_3–Early, Middle and Late Triassic, J_1, J_2, J_3–Early, Middle and Late Jurassic, K_1, K_2–Early and Late Cretaceous, P_1,–Palaeocene, P_2,–Eocene, P_3,–Oligocene, N_1–Miocene, N_2–Pliocene, R–present time (Holocene). Right columns are names of orders (all caps), superfamilies (bold lower case) and families (light lower case). Thin lines show ancestry (question mark denotes questionable or unknown ancestry), thick bars are known durations of taxa, figures refer to synapomorphies of the subtended clades, as follows (Sharov's nomenclature in parentheses).

1 – superorder synapomorphies (Chapter 2.2.2.3c);

2 – fore wing with CuP simple;

3 – orthopteran synapomorphies (Chapter 2.2.2.3.1c);

4 – CuA_1 (= MP + CuA_1 or M_5 + CuA_1) with few branches;

5 – fore wing with lancet area narrow, with composite vein (M&RS) smooth at junction of M and RS;

6 – both wing pairs with M_5 (= base of MP) and basal part of CuA_1 fused with common base of both MA and MP for a distance;

7 – fore wing secondarily slightly elytrised (venation in costal space somewhat disorganised in the most ancient fossils);

8 – fore wing with precostal space decreased; hind wing with 2A branches sinuate and with secondary veinlets;

9 – male elytral stridulatory apparatus with sinuate CuP forming stridulatory vein;

10 – fore wing with CuA_1 area enlarged at expense of space and number of branches of RS;

11 – fore wing with fan-like pleating intercalary triangle: RA, RS and $1MA_1$ fused completely, or at least in part;

12 – both sexes with MP (=MA_2) reduced at least in part; male fore wing with lancet area narrow;

13 – male with elytral stridulatory apparatus lost; both sexes with longitudinal veins parallel more or less;

14 – fore wing with costal space very narrow; possibly: elytral stridulatory apparatus lost;

15 – fore wing with costal space wide, MA_2 ($2MA_1$) single, M_5 cross-vein-like, and possibly with stridulatory apparatus lost;

16 – antenna shortened; ovipositor short, used for digging deepen into substrate along with abdomen while ovipositing;

17 – fore wing elytrised, with costal space wide and false C more or less S-shaped;

18 – fore wing with only two RS branches and RS and MA (= MA_1) anastomosing;

19 – fore wing more or less oval, with longitudinal veins parallel more or less, with characteristic oval area basally;

20 – fore wing with CuP and A_1 almost straight, running parallel and close to each other; both wing pairs with RS branches regular, oblique; midgut with 6 blind diverticles;

21 – pronotum with granular surface and large, acute, posterior process covering wings at least in part;

22 – fore wing with CuA_1 lost; hind wing with one MA (= MA_1) branch lost;

23 – fore wing MA (= MA_1) and with 1 branch; possibly: abdominal spiracles placed on tergum (pleuron sclerotised around spiracle and fused with tergum);

24 – phasmatidan synapomorphies (Chapter 2.2.2.3.2c);

25 – possibly: fore wing with CuA_1 simple;

26 – both wing pairs with RS base shifted basad (hind wing particularly so);

27 – fore wing with longitudinal veins almost perfectly parallel;

28 – both wing pairs with RS branching beginning from their middle part or even more basad;

29 – fore wing distinctly shorter than hind wing;

30 – both wing pairs hook-like bent apically;

31 – titanopteran synapomorphies (Chapter 2.2.2.3.3c).

Fig. 431 A male elcanid from the Upper Jurassic of Karatau in Kazakhstan, restored as in life by A.G. Ponomarenko (from ROHDENDORF & RASNITSYN 1980)

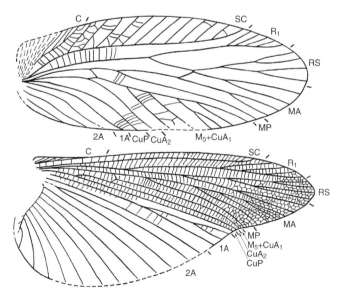

Fig. 432 Wing venation of basal orthopterans: 24 mm long fore wing of *Sylvoedischia crassa* Gorochov (Oedischiidae) and 90 mm long hind wing of *Jubilaeus beybienkoi* Sharov (Tcholmanvissiidae), both from the Early Permian of Tshekarda in Urals (from GOROCHOV 1995a)

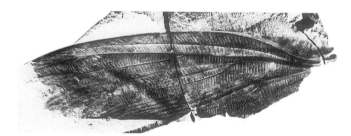

Fig. 433 *Tcholmanvissiella gigantea* Gorochov (Tcholmanvissiidae) from the Late Permian of Soyana, NW Russia (PIN 3353/378, photo by D.E. Shcherbakov); fore wing length 95 mm as preserved

KLUESSENDORF & MIKULIC (1990), HASIOTIS & BOWN (1992), DONOVAN (1994), and BUATOIS *et al.* (1998).

Insect ichnofossils were largely neglected for a long time when only anecdotal records were scattered in the literature and these were not critically reviewed. Considerable advances have been made in this area especially in the last few decades, and further work is necessary in this direction to allow insect ichnology to hold a firm place in palaeoentomology. In particular, formal classification providing the necessary basis for any more detailed studies have only been created for a few types of insect traces.

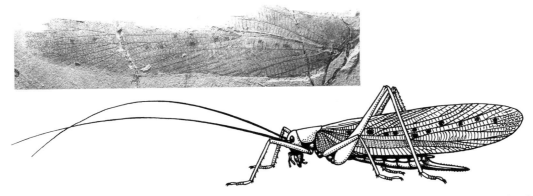

Fig. 434 *Jubilaeus beybienkoi* Sharov (Tcholmanvissiidae) from the Early Permian of Tshekarda in Urals, restored as in life by A.V. Gorochov; holotype fore wing 95 mm long (photo from archive of Palaeontological Institute RAS)

Fig. 435 *Proelcana kukalovae* (Sharov) (Permelcanidae) from the Early Permian of Tshekarda in Urals (from SHAROV 1968); body 12 mm long

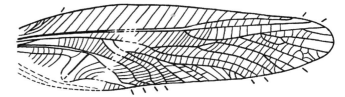

Fig. 437 *Mesoedischia obliqua* Gorochov (Mesoedischiidae) from the Middle or Late Triassic of Madygen in Kirghizia, male fore wing (from GOROCHOV 1995a); wing length 26 mm

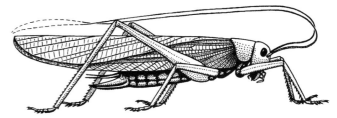

Fig. 438 *Gryllacrimima perfecta* Sharov (Proparagryllacrididae) from Middle or Late Triassic of Madygen in Kirghizia, restored as in life by A.V. Gorochov

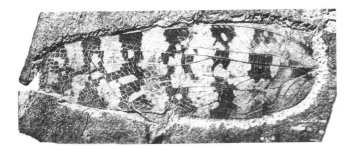

Fig. 436 *Aboilus aulietus* Sharov (Prophalangopsidae) from the Late Jurassic of Karatau in Kazakhstan (from SHAROV 1968); male fore wing 40 mm long

2.3.2. LOCOMOTION AND RESTING TRACES

This type of ichnofossil is represented by tracks and markings produced on surface sediments (Bioexoglyphia *sensu* Vialov, 1968) by body and/or appendages of moving or resting insects. According to the recent version of Seilacher's classification (BROMLEY 1996) they fit to Repichnia (traces of walking and crawling), Cubichnia (traces of resting), Fugichnia (traces of escape), and perhaps partially also to Pascichnia (traces of grazing). Insect locomotion and resting traces occur both in subaquatic (e.g. SCHAIRER & JANICKE 1970, D'ALESSANDRO *et al.* 1987, MORRISON 1987) and subaerial, especially aeolian (e.g. HITCHCOCK 1858, 1865, GILMORE 1927, BRADY 1939, 1947) sediments. They are known in the ichnological record since the Palaeozoic (GÜTHORL 1934, HOLUB & KOZUR 1981, HUNT *et al.* 1993, MÁGNANO *et al.* 1997) and seem to be widespread but insufficiently studied. As a rule, the trace maker cannot be identified; often it is difficult to say even if was an insect or not (for instance, the locomotion traces of insect larvae and worms are hardly distinguishable: METZ 1987a; see Fig. 465 as an example). In particular, many Palaeozoic traces may well belong to non-insect arthropods or other invertebrates. Consequently, the same formal names are commonly used for possible traces of insects and other invertebrates though some have been proposed specially for supposed insect locomotion and resting traces (see HÄNTZSCHEL 1975, for a review; and WALTER 1983, 1985, SKOMPSKI 1991, MÁGNANO *et al.* 1997, for further ichnotaxa established for

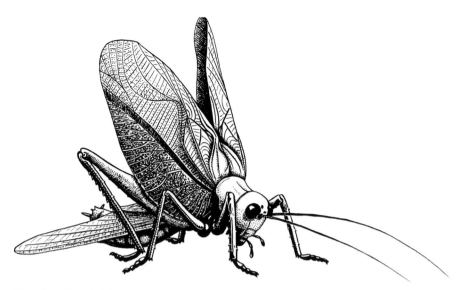

Fig. 439 *Lyrohagla uvarovi* Gorochov (Haglidae) from the Middle or Late Triassic of Madygen in Kirghizia, male restored as if displaying by A.V. Gorochov

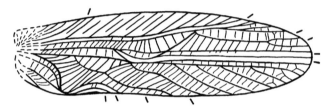

Fig. 440 *Gryllavus madygenicus* Sharov (Gryllavidae) from the Middle or Late Triassic of Kirghizia, male fore wing (from GOROCHOV 1995a); wing 18 mm long

possible insect traces). Important new attempts to improve arthropod trace classification have been proposed recently (WALTER 1984, TREWIN 1995); and the majority of previously described ichnotaxa need revision in accordance with them. Actualistic observations are largely occasional (e.g. BRADY 1939, HATTEN 1976, A. MÜLLER 1977, METZ 1980, 1987a,b, MORRISON 1987).

The unsatisfactory state of classification strongly limits the possibilities for interpreting the traces biologically. Insect walking or crawling trails are useful for recognition of non-marine deposits (BIRON & DUTUIT 1981, BOWLDS 1989). Their frequency and diversity may be used for estimation of insect activity and diversity in some palaeoenvironments. In particular, the traces of benthic larvae are important to distinguish between oxygenated and non-oxygenated near-bottom environments. Some types of insect Repichnia and Cubichnia may elucidate the mode of life of extinct taxa. For instance, *Tonganoxichnus* is attributed to Palaeozoic bristletails of the family Dasyleptidae suggesting their aquatic or semiaquatic habit (MÁGNANO *et al.* 1997; for alternative interpretation see RASNITSYN 2000b). Finally, the non-marine arthropod traces older than Late Carboniferous are of particular interest and should be laboriously investigated because of their potential importance from the point of view of insect ancestry and early evolution.

2.3.3. SHELTERS AND BURROWS

The fossilised shelters constructed by animals belong to Seilacher's group Domichnia. Some of them (e.g. many wood borings)

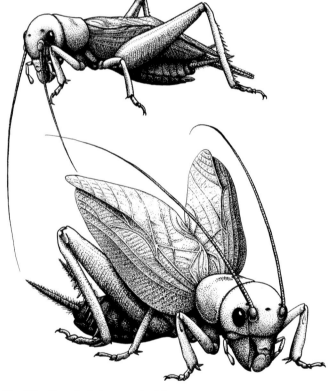

Fig. 441 Hypothetical ancestral crickets (Protogryllidae), male (displaying) and female restored as in life (from GOROCHOV 1995a)

should be classified rather as Fodinichnia (the feeding traces within food substrate) and are discussed separately, though it is not always easy to distinguish between feeding and non-feeding burrows. Some additional divisions have been proposed, e.g. Aedificichnia for structures constructed from materials extraneous to the substrate (BOWN & RATCLIFFE 1988), Calichnia for the shelters constructed specially for larval development (GENISE & BOWN 1993), and Agrichnia for the structures used for gardening or as traps for food (BROMLEY 1996).

The simplest type of insect Domichnia is represented by ***burrows in palaeosols and sediments*** (Fossiglyphia *sensu* VIALOV 1968; Fig. 465). It is often difficult to distinguish between the true Domichnia (i.e., the nids inhabited for some time), and the tunnels simply made during the penetration of a burrowing organism through the substrate which may be regarded as Repichnia. Here all insect tunnels except for more or less obvious feeding traces are conventionally regarded as Domichnia. Like surface locomotion traces they occur widely both in fresh-water and subaerial palaeoenvironments since the Palaeozoic, and are highly palaeoecologically informative (SAUER & SCHREMMER 1969, WEBBY 1970, BELOKRYS 1972, FREY 1973, STANLEY & FAGERSTROM 1974, TOOTS 1975, MEIßNER 1976, RATCLIFFE & FAGERSTROM 1980, AGRAWAL & SINGH 1983, BOWN & KRAUS 1983, MAULIK & CHAUDHURY 1983, BRUSSAARD & RUNIA 1984, M. MILLER 1984, TANDON & NAUG 1984, ARMENTEROS *et al.* 1986, CURRAN & WHITE 1991, DONOVAN 1994, HASIOTIS & DEMKO 1996a,

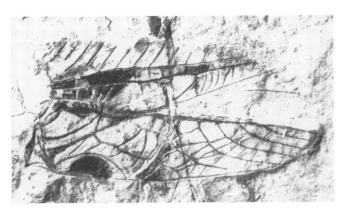

Fig. 442 Male *Angarogryllus angaricus* (Sharov) (Protogryllidae) from the Early Jurassic of Ust-Baley in Siberia (from SHAROV 1968); fore wing 23 mm long

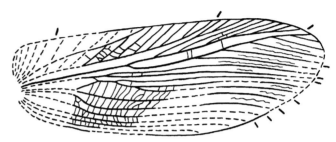

Fig. 443 *Zeuneroptera scotica* (Zeuner) (Anostostomatidae) from the Palaeocene of the Isle of Mull, male (?) fore wing (from GOROCHOV 1995a); wing fragment 27 mm long

Fig. 444 Female *Aboilus* sp. (Prophalangopsidae) from the Late Jurassic of Karatau in Kazakhstan, restored as in life by A.G. Ponomarenko (from ROHDENDORF & RASNITSYN 1980)

BUATOIS *et al.* 1998). However, morphologically simple nids are even less taxonomically indicative than are surface tracks because they demonstrate fewer features reflecting morphology of the producer; they may even be confused with the rhizoliths (pseudomorphs on plant roots). The formal names introduced for such burrows are based mainly on tunnel morphology (for a review see HÄNTZSCHEL 1975), and many of them surely cover the traces of both insects and other invertebrates. Probably many non-marine Fossiglyphia described without indication of their possible producers and partly placed in ichnogenera originally established for marine trace fossils such as *Skolithos*, *Diplocraterion*, *Scoyenia*, etc., represent domiciles of insects (see BUATOIS *et al.* 1998, for a brief review). Actualistic data are discussed in a number of papers (e.g. SILVEY 1936, BRYSON 1939, FIEGE 1961, N. SMITH & HEIN 1971, CHAMBERLAIN 1975, ELDERS 1975, AHLBRANDT *et al.* 1978, JOŇCA 1979, SCHREMMER 1979, RATCLIFFE & FAGERSTROM 1980, HALFFTER *et al.* 1983, M. MILLER 1984, G. CLARK & RATCLIFFE 1989). The most common relatively morphologically simple tunnels in fossil soils and mud are supposedly attributed mainly to various beetles and their larvae (SAUER & SCHREMMER 1969, STANLEY & FAGERSTROM 1974, RATCLIFFE & FAGERSTROM 1980, MAULIK & CHAUDHURI 1983, HASIOTIS & DUBIEL 1993a,b, HASIOTIS & DEMKO 1996a) or, more rarely, to dipteran larvae (e.g. EKDALE & PICARD 1985) but most of them may well have been produced by a number of other insects. BUATOIS *et al.* (1998) pointed out the increasing relative role of burrows in freshwater sediments in the Mesozoic in comparison with the Palaeozoic when crawling traces are strongly dominating; the diversity of nids in terrestrial palaeoenvironments in the Mesozoic increases as well.

Some permanently inhabited burrows are more morphologically characteristic like the **U-shaped nids** in fresh-water sediments attributed to burrowing mayfly nymphs (KRASENKOV 1966) (Fig. 466). Similar tunnels in fossil wood and vertebrate bones (the ichnogenus *Asthenopodichnium* Thenius) are also believed to be mayfly nymph domiciles (THENIUS 1979, 1989, NESSOV 1988). On the other hand, U-shaped biogenic structures are common in marine palaeoenvironments as well and may be produced by various invertebrates.

The **nids with cells** in palaeosols interpreted as being soil nests of digger wasps and solitary bees are mentioned or described in many papers (BUXTON 1932, R. BROWN 1934, 1935, 1941a,b, FRENGUELLI 1938a,b, 1939, 1946, ROSELLI 1938, 1987, FOSSA-MANCINI 1941, SAUER & SCHREMMER 1969, ZEUNER & MANNING 1976, BONINO DE LANGGUTH 1978, ANKETELL & GHELLALI 1984, BÁEZ & BACOLLADO 1984, GAYDUCHENKO 1984, SCHLÜTER 1984, RETALLACK 1984, LAZA 1986a,b, HOUSTON 1987, PETIT-MAIRE *et al.* 1987, RITCHIE 1987, DUCREUX *et al.* 1988, I. FORD 1988, CURRAN & WHITE 1991, ELLIS & ELLIS-ADAM 1993, GENISE & BOWN 1993, 1996, THACKRAY 1994, DOMÍNGUEZ ALONSO & COCO ABIA 1998, ELLIOT & NATIONS 1998, GENISE & HAZELDINE 1998, GENISE 1999, 2000). Some of them were tentatively attributed to present-day subfamilies or even genera. Perhaps, it is possible for some Late Tertiary and Quaternary nids with highly characteristic morphology but most attributions are arbitrary, and a formal classification should be preferred. Several formal names are available [the ichnogenera *Celliforma* Brown, *Uruguay* Roselli (Fig. 467), *Palmiraichnus* Roselli, *Ellipsoideichnus* Roselli, *Rosellichnus* Genise et Bown]. Supposed sphecid (ROSELT & FEUSTEL 1960, DUDICH 1961, KLINGER 1988) and carpenter bee (SCHENK 1937, NEL 1994) nests occur occasionally in fossil woods, but published records are scarce. At present the wood nesting bees and especially sphecid wasps are diverse, and fossil finds show their antiquity. It is possible that their fossil nests have mainly gone unrecognised.

Some morphologically complex nids are interpreted as subterranean ant nests but their records are rare in the literature (LAZA 1982, 1997). The same is true for possible carpenter ant nests in fossil wood (BRUES 1936). Because ants are known in the fossil state since the mid-Cretaceous and are quite common in Tertiary deposits, the rarity of their fossil nests seems to be extremely strange.

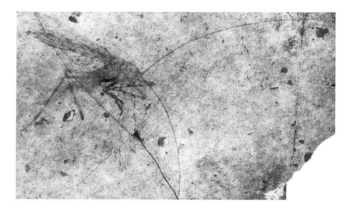

Fig. 446 Undescribed Phasmomimidae from the Late Jurassic of Karatau in Kazakhstan (PIN 2904/1676, photo by D.E. Shcherbakov); length, excluding antennae 36 mm

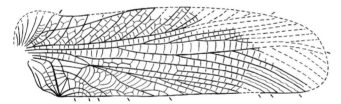

Fig. 445 *Haglotettigonia egregia* Gorochov (Haglotettigoniidae) from the Early Cretaceous of Baissa in Siberia, male fore wing (from GOROKHOV 1995a); wing length 33 mm

Fig. 447 *Locustopsis karatavica* Sharov (Locustopsoidea) from the Late Jurassic of Karatau in Kazakhstan (from SHAROV 1968); fore wing 23 mm long

Complex galleries in palaeosols are largely interpreted as having been constructed by termites (for a review see GRASSÉ 1986, NEL & PAICHELER 1993, BOUCOT 1990; further records are by WATSON 1967, TSEKHOVSKY 1974, R. SMITH *et al.* 1993, GENISE & BOWN 1994, SENUT *et al.* 1994, GENISE 1995, LAZA 1995). Occasionally their attribution is supported by the presence of mandibles or other remains (BARROS MACHADO 1982, cited after NEL & PAICHELER 1993). Some of them, mainly from the Late Tertiary and Quaternary, were tentatively attributed to certain living genera; for the older constructions several formal ichnogeneric names are available (*Termitichnus* Bown, *Syntermesichnus* Bown et Laza, *Krausichnus* Genise et Bown, *Vondrichnus* Genise et Bown, *Fleaglellius* Genise et Bown). Old termite mounds can be preserved and occasionally remain recognisable for up to several thousand years (J. WATSON 1967, CROSSLEY 1984, L. MILLER 1989, J. MOORE & PICKER 1991). Though termites feed on wood, their galleries (ABEL 1933, ROHR *et al.* 1986, SUESS & SCHULTZE-DEWITZ 1987, SCHULTZE-DEWITZ & SÜSS 1988, ROZEFELDS & DE BAAR 1991, NEL 1994, GENISE 1995) may also in this case, be classified as Domichnia, constituting a part of a large and complex nest. The only formal ichnogenus for the alleged termite galleries in wood is *Cycalichnus* Genise. BUATOIS *et al.* (1998) established a peculiar type of ichnofossil assemblages (the *Termitichnus* ichnoguild) as indicative for terrestrial palaeoenvironments.

Besides the nids, Mesozoic and Cainozoic palaeosols and aeolianites also occasionally contain hollow suboval structures interpreted as insect (mostly beetle) *pupal chambers*. The calcretised chambers are rather durable and may occur re-deposited even in marine sediments (P. JOHNSTON *et al.* 1996). They were occasionally cited by different authors (TRUC 1975, FREYTET & PLAZIAT 1982, PICKFORD 1986, NESSOV 1988) but rarely described in detail (LEA 1925, TILLEY *et al.*

1997, P. JOHNSTON *et al.* 1996). Some of them have been misidentified as abiogenic structures or vertebrate eggs, and perhaps a number of misinterpretations still remains unrecognised (P. JOHNSTON *et al.* 1996). The only formal name available is *Fictovichnus* P. Johnston *et al.* 1996 (fossilised dung balls are considered here as the feeding traces and not true Domichnia). Well preserved beetle pupal chambers occur especially in desert soils and are valuable for their diagnostics (VALIAKHMEDOV 1977).

Studies of insect and other invertebrate burrows in palaeosols are in fact still in their early infancy. Available information about their stratigraphic, geographic and facial distribution is too scarce and poorly systematised but their potential scientific importance is great. In particular, future studies in this field could clarify the evolution of pedogenesis – one of the most important and intriguing aspects of the evolution of ecosystems (RETALLACK 1981a). Future studies of complex burrows in palaeosols and in fossil wood may elucidate behavioural evolution pattern especially in aculeate hymenopterans and termites. Some ichnofossils may be indicative from the point of view of time of origin of certain insect taxa. The supposed bee nests are of a particular interest because the fossil record of bees is rather fragmentary. There are strong arguments supporting the interpretation of a number of Late Cretaceous ichnofossils as nests of solitary bees (ELLIOT & NATIONS 1998, GENISE 1999, GENISE & ENGEL 2000), which may indicate a Late Cretaceous origin and a wide distribution of bees even in the absence of any direct fossil evidence. On the other hand, the attribution of some other supposed insect nests which are much older than the oldest known finds of their supposed producers may be seriously doubted. For instance, FEDCHENKO & TATOLI (1983) ascribed tunnels in the non-marine Carboniferous mudstones from Ukraine to beetles on the basis of comparison with the burrows of modern Heteroceridae; however, the existence of any Coleoptera, and in particular of mud-burrowing species, in the Late Carboniferous seems to be implausible because the probability of fossilisation of beetle elytra is high, especially for inhabitants of near-water environments. Other examples are *Pustulichnus* Ekdale et Picard from the Jurassic of Utah (described as possible aculeate hymenopteran nids), alleged Triassic and Jurassic termite, ant and bee tunnels mentioned by different authors (RUSCONI 1948, HASIOTIS & DEMKO 1996a,b, HASIOTIS *et al.* 1995, 1996, KAY *et al.* 1997), and *Archeoentomichnus* Hasiotis et Dubiel (HASIOTIS & DUBIEL 1993a, 1995), the supposed subterranean termite nest from the Triassic of Arizona. The fossilisation potential of termites and ants is too high to suppose their

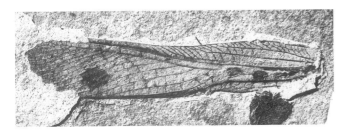

Fig. 448 *Panorpidium sibiricum* (Sharov) (Elcanidae) from the Early Cretaceous of Baissa in Siberia (from SHAROV 1968); fore wing 31 mm long

Fig. 449 Male *Baissogryllus sibiricus* Sharov (Baissogryllidae) from the Early Cretaceous of Baissa in Siberia (from SHAROV 1968); fore wing 23 mm long

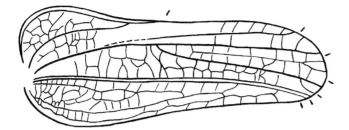

Fig. 450 *Prototetrix reducta* Sharov (Tetrigidae) from the Early Cretaceous of Baissa in Siberia, fore wing (from SHAROV 1968); wing 6.2 mm

Fig. 451 *Monodactylus dolichopterus* Sharov (Tridactylidae) from Early Cretaceous of Baissa in Siberia (from SHAROV 1968); body 14 mm long

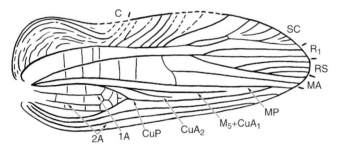

Fig. 452 *Mongoloxya ponomarenkoi* Gorochov (Tridactylidae) from the Early Cretaceous of Bon Tsagan in Mongolia, fore wing (from GOROCHOV 1992); wing length 15 mm

long-time existence undocumented by any direct fossil evidence (Chapter 1.4.2.1.1). The nature of the above-mentioned trace fossils is thus enigmatic, and the possibility of existence of unknown extinct social or subsocial soil-nesting arthropods (not necessarily insects) in the Mesozoic cannot be rejected.

Some fossils are interpreted as **insect nests originally attached to substrate** like the mud cells of some sphecid wasps or paper wasp nests; though in general such constructions have much lower probability of fossilisation than the internal burrows. According to BOWN & RATCLIFFE (1988) they have to be classified as Aedificichnia because their material is different from the enclosing sediments. The ichnogenus *Chubutolithes* Ihering from the Early Tertiary of Argentina discussed by different authors (BOWN & RATCLIFFE 1988, GENISE & BOWN 1990, 1991, B. FREEMAN & DONOVAN 1991) and resembling the

present-day nests of *Sceliphron* probably represents the mud cells constructed by aculeate hymenopterans. Similar ichnofossils from the Late Cretaceous of Central Asia were described in the ichnogenus *Desertiana* Nessov. "*Celliforma*" *favosites* Brown from the Late Cretaceous of USA is probably a fossilised paper wasp comb, very different from other *Celliforma* (WENZEL 1990) and has to be excluded from this ichnogenus (GENISE & BOWN 1993). Another fossil paper wasp nest was described by HANDLIRSCH (1910) and compared with the nests of modern *Polistes*. MARTINS-NETO (1991c) mentions a Quaternary *Stenopolybia* nest from Brazil. Finally, E. WILSON & TAYLOR (1964) have described a completely mineralised nest of the ant *Oecophylla leakeyi* constructed from leaves and containing well-preserved remains of the ant brood in the Miocene of Kenya.

Some insects, both terrestrial and aquatic, create **transportable cases** from various materials, which may be preserved as fossils. This group of trace fossils was called Indusiacea by VIALOV (1973). The caddisworm cases are especially common and widespread, and tubes of chironomid larvae are recorded from the Cainozoic and Mesozoic deposits but are little studied (THIENEMANN 1933, W. EDWARDS 1936, MORETTI 1955, HILTERMANN 1968, KALUGINA 1993). The probability of conservation of cases of terrestrial insects is low and they are recorded mainly from fossil resins, including the cases of microlepidopteran larvae of the families Psychidae and Coleophoridae (summarised by LARSSON 1978 and KOZLOV 1988). The only known lepidopteran (psychid) case from sedimentary deposits is *Psychites pineella* (HEER 1849) from the Miocene of Oeningen, Germany. KOZLOV (1988) introduced *Psychites* as a formal name for fossil psychid cases. WEITSCHAT & WICHARD (1998, Taf. 79) illustrate several lepidopteran cases in the Baltic amber, partly with preserved larvae; it should be noted that one of their figures (pl. 79, fig. d) represents in fact a case-bearing leaf-beetle larva misidentified as a lepidopteran. Two case-bearing leaf beetle larvae from the Dominican amber are figured and briefly described by SANTIAGO-BLAY *et al.* (1996). No isolated chrysomelid cases were recorded. There are also no published records of neuropteran larval cases constructed from food remains.

At present, transportable caddis cases represent the best studied insect ichnofossils and provide an excellent example of their scientific significance. The principles of their formal classification based primarily on the building material have been proposed by VIALOV (1973). VIALOV & SUKATSHEVA (1976) and SUKATSHEVA (1982) revised the previously recorded finds and described a number of new ichnotaxa; further descriptions were published by KRASSILOV & SUKATSHEVA (1979), SUKATSHEVA (1985b, 1989, 1990b, 1991a, 1994), MARTINS-NETO (1989b), GROMOV *et al.* (1993), JARZEMBOWSKI (1995a), HASIOTIS & DEMKO (1996a) and HASIOTIS *et al.* (1997). In total more than 200 ichnospecies are named in the ichnogenera *Terrindusia* Vialov (tubes from mineral particles; with the ichnosubgenera *Mixtindusia* Sukatsheva and *Terrindusia* Vialov), *Secrindusia* Vialov et Sukatsheva (silk tubes), *Pelindusia* Vialov et Sukatsheva (tubes from fragments of bivalve mollusc shells), *Ostracindusia* Vialov (tubes from ostracod shells), *Indusia* Brongniart (=*Boscindusia* Vialov) (tubes from gastropod shells), *Conchindusia* Vialov et Sukatsheva (tubes from conchostracan shells and their fragments), *Folindusia* Berry (tubes from plant particles; with the ichnosubgenera *Profolindusia* Sukatsheva, *Detrindusia* Sukatsheva, *Frugindusia* Sukatsheva, *Acrindusia* Vialov, *Echinindusia* Vialov et Sukatsheva, and *Folindusia* Berry), *Piscindusia* Jarzembowski (tubes from fish scales), *Molindusia* Vialov (cases with broad lateral projections), and *Scyphindusia* Sukatsheva (tubes constricted before narrower end). The ichnogenus *Tektonargus*

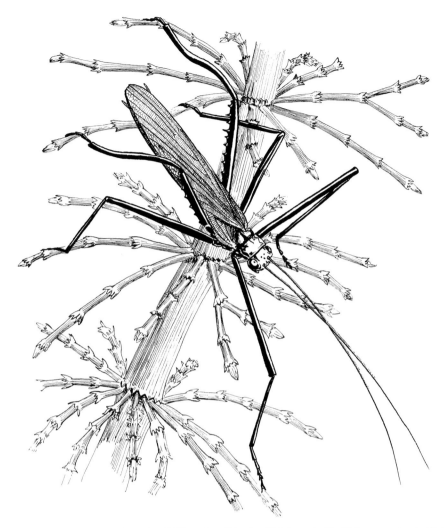

Fig. 453 Male *Triassophasma* sp. (Prochresmodidae) from the Middle or Late Triassic of Madygen in Kirghizia, restored as in life by A.G. Ponomarenko (from ROHDENDORF & RASNITSYN 1980)

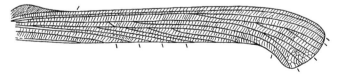

Fig. 455 *Xiphopterum curvatum* Sharov (Xiphopteridae) from the Middle or Late Triassic of Madygen in Kirghizia (from SHAROV 1968); fore wing 63 mm long

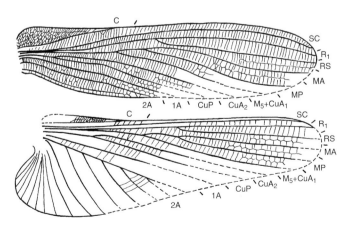

Fig. 454 *Prochresmoda longipoda* Sharov (Prochresmodidae) from the Middle or Late Triassic of Madygen in Kirghizia (from SHAROV 1968); fore and hind wings 38 and 27 mm long, respectively

Fig. 456 Fore wing of *Paraplana affinis* Sharov (Aeroplanidae) from the Middle or Late Triassic of Madygen in Kirghizia (holotype, photo by D.E. Shcherbakov); wing 43 mm long

Fig. 457 Male *Triassophasma* sp. (Prochresmodidae) from the Middle or Late Triassic of Madygen in Kirghizia (from SHAROV 1968); hind femur 20 mm long

Fig. 459 *Phasmomimoides minutus* Gorochov (Susumaniidae) from the Late Jurassic of Karatau in Kazakhstan (holotype, photo by D.E. Shcherbakov), with gut load of leaf fragments of the gnetophytan *Brachyphyllum* or *Pagiophyllum* sp. (from GOROCHOV 1993). Body with wings 44 mm long

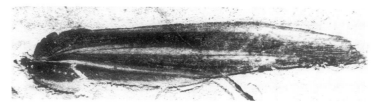

Fig. 458 *Baissophasma pilosa* Gorochov (Aerophasmatidae) from the Early Cretaceous of Baissa in Siberia (from GOROCHOV 1993); fore wing 44 mm long

Fig. 460 Fore wing of *Cretophasmomima baissiensis* (Sharov) (Susumaniidae) from the Early Cretaceous of Baissa in Siberia (from SHAROV 1968); wing length 43 mm

Hasiotis *et al.* 1997, should be probably synonymised with *Terrindusia*. SUKATSHEVA (1980, 1982, 1991b, 1999), ZHERIKHIN & SUKATSHEVA (1990) and ESKOV & SUKATSHEVA (1997) discussed stratigraphic and geographical distribution pattern, and J. JOHNSTON (1999) presented new data on North American Early Tertiary case assemblages.

The oldest transportable cases are found in the Middle Triassic of Argentina and attributed tentatively to chironomid larvae (GENISE *et al.* 2000); however, primitive necratauliid-like caddisflies can not be ruled out as their possible constructors. Other pre-Cretaceous transportable cases are recorded from the Mid- or Late Jurassic of Siberia (SUKATSHEVA 1985), Mongolia (SUKATSHEVA 1994) and USA

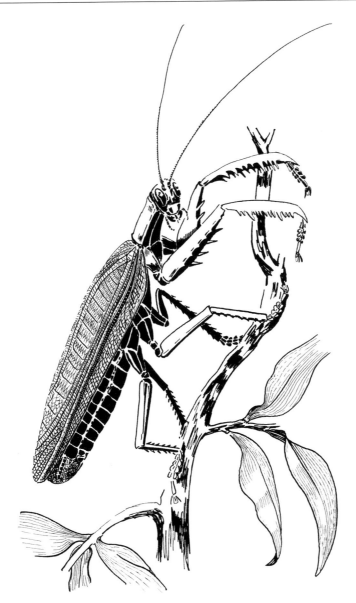

Fig. 461 *Gigatitan* sp. (Gigatitanidae) from the Middle or Late Triassic of Madygen in Kirghizia, restored as in life by A.G. Ponomarenko (from ROHDENDORF & RASNITSYN 1980)

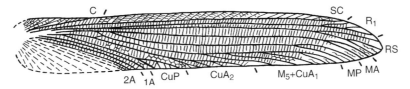

Fig. 462 *Prototitan primitivus* Sharov (Mesotitanidae) from the Middle or Late Triassic of Madygen in Kirghizia (from SHAROV 1968); fore wing 80 mm long

(HASIOTIS & DEMKO 1996a, HASIOTIS *et al.* 1997). Non-transportable shelters attached to substrate are surely older evolutionarily, but absent in the fossil record except for the sand case *Trichopterodomus* Erickson from the Palaeocene of North Dakota (ERICKSON 1988). Since the Early Cretaceous transportable cases become very common in various non-marine deposits. MARTYNOVA (1947b) believed that the ichnofossil *Pectinariopsis* from shallow marine Miocene deposits

of North Caucasus were also constructed by caddisworms but VIALOV (1972) argued that their producers were marine polychaetes.

Portable cases are now constructed only by members of the relatively advanced suborder Phryganeina, and the fossil and ichnofossil data on caddisflies correspond well with each other, because the oldest fossil finds of this suborder and the oldest cases are of the same age. However, the taxonomic position of a caddisfly below the subordinal level can rarely be established on the basis of its case structure, and finds of larvae preserved inside the cases are very rare. Nevertheless, the analysis of ichnofossils demonstrates clearly the general evolutionary trend in larval building behaviour from the most primitive state when particles different in nature, size and shape are arranged nearly chaotically, up to the most sophisticated constructions from highly uniform (sometimes even bitten by larva to standartise their shape and size) and perfectly arranged building materials (SUKATSHEVA 1980, 1982, 1991). Significant and rather rapid evolutionary changes occur mainly in the Early Cretaceous while since the beginning of the Late Cretaceous up to the present day a long-term stasis in the case construction is demonstrated. On the other hand, fossils of adult Phryganeina demonstrate very little evolutionary changes in their morphology before the Late Cretaceous when the radiation of the main modern lineages occurred. This pattern suggests that the evolution of larval behaviour preceded in time the major evolutionary events in the adult stage (SUKATSHEVA 1991a); this conclusion seems to be of a great interest. Another important event is a drastic decrease in case diversity in lacustrine deposits at the very end of the Early Cretaceous, whereas the Late Cretaceous assemblages from alluvial deposits include a more diverse ichnofauna (SUKATSHEVA 1991a, 1999, GROMOV et al. 1993). A semi-quantitative index of the constructive advancement is proposed reflecting the differences in the selection of the building material, its additional reworking, its placement and the general complexity of the construction (SUKATSHEVA 1980). In diverse case assemblages the maximal values of this index are useful for estimation of the minimal age of the assemblage but only before the beginning of the behavioural stasis in the Late Cretaceous (ZHERIKHIN & SUKATSHEVA 1990). The geographic distribution of fossil cases suggests an impoverishment of the lacustrine caddisfly fauna in warmer climatic areas (ESKOV & SUKATSHEVA 1997). Those impressive results clearly demonstrate the potential importance of ichnological studies in palaeoentomology as well as the main prerequisites of their success, namely development of a detailed parataxonomic system and analysis of a representative pool of data. Fossil caddis cases are also discussed in Chapter 2.2.1.3.4.2.

Some types of insect shelters are unrecorded as fossils in spite of their potential ability to be fossilised (for instance, hymenopteran nests in empty plant stems). At the present stage of development of insect ichnology this may simply be due to little interest in them. However, the absence of some types may be significant. In particular, there are no published records of rolled leaf fossils. If leaves rolled by lepidopteran larvae could be misinterpreted as shrivelled by drying (the same may be true for web nests of the yponomeutid moths), the leaf-rolls of attelabid weevils are so characteristic that it is difficult to believe that they have been indeed overlooked by all students dealing with plant fossils. The leaf-rolling attelabids are also virtually unrecorded as fossils, so it could be supposed that they are of late origin (probably not before at least the Late Tertiary) and were initially restricted to some taphonomically unfavourable habitats. Another interesting lacuna in the ichnological record is the absence of insect burrows in fossil resins. At present, coniferous resin fluxes are exploited by a peculiar guild of so-called resinicoles which include larvae of several dipteran families (MAMAEV 1971). Their larvae and pupae are common in modern samples of resin, at least in the temperate forests of Eurasia, but are not found in any known fossil resins (ZHERIKHIN & SUKATSHEVA 1989, 1992). Because virtually no post-Eocene conifer resins have been discovered in the northern hemisphere, it seems possible that the resinicoles have appeared later on, probably only in the Late Tertiary. A possible alternative explanation is that damaged resin has a low preservation potential.

2.3.4. OVIPOSITION TRACES

Traces of insect oviposition should be considered as a peculiar type of shelters made by adults for their progeny. Up to now two types of such shelters are recorded as fossils.

One type is represented by cuttings in plants made by the ovipositor. Endophytic oviposition occur in various present-day insect orders

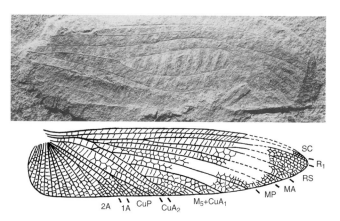

Fig. 463 *Paratitan libelluloides* Sharov (Paratitanidae) from the Middle or Late Triassic of Madygen in Kirghizia: photo of the fore wing 55 mm long (PIN 2555/1498) and hind wing 50 mm long (from SHAROV 1968)

Fig. 464 *Mesotitan scullyi* Tillyard (Mesotitanidae) from the Middle Triassic of Brookvale in Australia (from CARPENTER 1992a); fore wing 125 mm long

Fig. 465 Traces of benthic invertebrates (insects or worms) in an insectiferous lacustrine marl; Early Cretaceous of Baissa in Siberia (PIN 3064/10579, photo by D.E. Shcherbakov); area shown 28 mm high

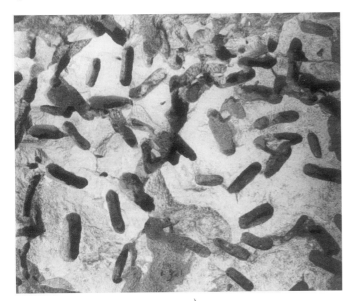

Fig. 466 U-shaped burrows in fresh-water sediments attributed to mayfly nymphs; Quaternary of Moscow Region (not registered, photo by D.E. Shcherbakov); area shown 110 mm wide

Fig. 467 Cretaceous nests of aculeate hymenopterans: above *Desertiana mira* Nessov (tentatively attributed to Vespidae), holotype from the earlier Late Cretaceous (Coniacian) Bissekty Formation of Dzhiraduk in Central Kyzylkum Desert, Uzbekistan (maximum diameter 53 mm, from NESSOV 1997); below vertical section of *Uruguay rivasi* Roselli (tentatively attributed to Apidae) from latest Cretaceous Asensio Formation of Nueva Palmira in Uruguay (scale bar 10 mm, from GENISE 1999; for the Cretaceous age see Chapter 3.2.3.2.)

(odonatans, hemipterans, beetles, hymenopterans, orthopterans, etc.) and could be supposed for some extinct taxa (e.g. for some Palaeozoic Dictyopteridea) on the basis of ovipositor morphology. Oviposition notches are not likely to be rare in the ichnological record (Fig. 468), but for a long time they were either ignored or misinterpreted as normal plant structures or caused by phytopathogenic fungi (see GRAUVOGEL-STAMM & KELBER 1996, and VAN KONIJNENBURG-VAN CITTERT & SCHMEIßNER 1999, for some examples). They were initially recognised by FRIČ (FRIČ & BEYER 1900) on Late Cretaceous leaves but largely ignored by later authors after being regarded with scepticism by HANDLIRSCH (1906–1908). Since the 1980s a number of new records have been published from the Triassic (GRAUVOGEL-STAMM & KELBER 1996, and references herein), Jurassic (VAN KONIJNENBURG-VAN CITTERT & SCHMEIßNER 1999), Cretaceous (HELLMUND & HELLMUND 1996b) and Tertiary (HELLMUND & HELLMUND 1991, 1993, 1996b–d, 1998). Most of them are interpreted as being made by odonatans though some may have been produced by other insects. No formal names have been introduced thus far, though HELLMUND & HELLMUND (1991,

Fig. 469 Lignitised coniferous wood with insect borings from the Kheta Formation, Late Cretaceous of Yantardakh in North Siberia (not registered, photo by D.E. Shcherbakov); wood piece 95 mm long

Fig. 468 Odonatan oviposition notches on plant remains; Early Cretaceous of Gurvan-Ereniy-Nuru in Mongolia (PIN 3792/161, photo by D.E. Shcherbakov); leaf 18 mm long from base to the farthest notch

1996d) recognised several types differing in the placement of notches (the coenagrionid type, the lestid type with double rows, the lestid type with linear rows, the *Sympecma* type, the irregular type). The coenagrionid type is the most common and widespread.

An important type of egg-shelter is represented by secretory coverings produced by special female glands. Their fossils are little studied. J. GALL & GRAUVOGEL (1966) have described several formal genera for egg masses apparently covered with a mucilaginous sheath from the Triassic of Vosges, France (*Monilipartus*, *Clavapartus*, *Furcapartus*) comparing them with egg clusters of modern odonatans, dipterans and caddisflies. If eggs or egg shells are preserved, these remains may be regarded as true fossils rather than ichnofossils. Another type of egg-shelter includes the cockroach and mantid egg capsules (oothecae). Some of their records are erroneous (for example, the supposed Palaeozoic oothecae described and figured by PRUVOST 1919, 1930, and repeatedly cited by later authors as evidence of an early origin of ootheca in roaches are obviously plant remains). Undoubted cockroach oothecae are known from the Early Tertiary (HAUPT 1956, R. BROWN 1957). NESSOV (1995) mentions phosphatised roach oothecae from the Late Cretaceous of Central Asia, and RASNITSYN & ROSS (2000) list supposed mantid oothecae among Late Cretaceous or Early Palaeogene Burmese amber fossils. This is in quite good accordance with the body fossils, indicating that both mantids and advanced roach families appeared in the Early Cretaceous. No formal names have been proposed.

It would be expected that more types of insect oviposition traces will be recognised in the future. In particular, grasshopper egg pods should occur in palaeosols.

2.3.5. FEEDING TRACES

According to Seilacher's classification the tunnels made in a substrate by an animal feeding belong to the group Fodinichnia, and traces of grazing to Pascichnia. Insect feeding traces are diverse, and not all of them are covered satisfactorily by these two main divisions. A more complex classification was introduced by VIALOV (1975) who distinguished the feeding traces in wood and bark (Phagolignichnida), in leaves and stems (Phagophytichnida), in fossilised faeces (Coprinisphaeridea), and in animal remains (Phagozoonichnidea). However, Vialov's classification is not exhaustive and does not

include, for instance, feeding traces in seeds and fructifications. GENISE (1995) has proposed independently Xylichnia as a name for wood borings. No feeding traces documenting insect mycetophagy are known, possibly because fungal fruit bodies are extremely rare as fossils. Insect feeding traces attracted more attention and are better known than most other insect ichnofossils, except for the caddis cases. However, few of them were examined in detail, their large-scale comparative studies have begun only very recently, and their parataxonomy is still poorly developed.

Wood borings (Fig. 469) resemble in many respects burrows in other hard substrates. They demonstrate few features directly reflecting morphology of the hole maker and are often rather simple structurally though sometimes their morphology is characteristic enough. The burrows may have been made both in dead and living wood; the latter case can be recognised on the basis of evidence of tissue wound reaction. Insects are the most diverse and common wood-borers but tunnels in dead wood may be produced by other animals as well (e.g. by worms, oribatid mites and marine borrowing molluscs). Occasionally structures similar to animal borings may appear as a result of a fungal decay (STABBLEFIELD *et al.* 1984) or in the course of wood mineralisation (FISK & FRITZ 1984). Actualistic data on insect damage in wood are very numerous, including regional reviews and keys (e.g. G. BECKER 1950, APEL 1979, 1983, SCHMUTZENHOFER 1985) but are not systematised on a world scale. The oldest wood borings are known from the Carboniferous. Most of them are attributed to organisms other than insects, especially to the oribatid mites (LABANDEIRA *et al.* 1997) but some may have been produced by roaches or other insects feeding on rotten wood (A. SCOTT & TAYLOR 1983). More insect burrows are recorded from younger, especially from the Mesozoic and Cainozoic, strata (see BOUCOT 1990, for a review, and further records in ZHERIKHIN & SUKATSHEVA 1973, PANT & SINGH 1987, MÜLLER-STOLL 1989, SHARMA & HARSH 1989, ZHOU & ZHANG 1989, BENNIKE & BÖCHER 1990, JARZEMBOWSKI 1990, TIDWELL & ASH 1990, FREESS 1991, GUO 1991, NEL 1994, GENISE 1995, GENISE & HAZELDINE 1995, DÒEZ *et al.* 1996, WEAVER *et al.* 1997, PEÑALVER *et al.* 1999). The supposed borers are mostly beetles or, rarely, siricoid hymenopterans; some traces were tentatively assigned to certain present-day genera, but this attribution is usually highly doubtful. Even the most characteristic borings of bark- and ambrosia-beetles could resemble those made by other insects; at least, some borings of this type are found in the Early Cretaceous (JARZEMBOWSKI 1990; see also Fig. 470) and even in the Triassic (M. WALKER 1938) while no fossils of the corresponding beetle taxa are known before the Early Tertiary. It should be noted that no burrows were attributed to wood-boring lepidopterans. The principles of a formal classification have been proposed only very recently (GENISE 1995) and the previously described ichnotaxa need revision. The available ichnogeneric names are *Anobichnium* Linck, *Palaeocerambichnius* Müller-Stoll, *Paleobuprestis* Häntzschel, *Paleoipidus* Häntzschel, *Paleoscolytus* Walker, *Dekosichnus* Genise et Hazeldine, *Ipites* Karpiński, *Scolytolarvariumichnus* Guo, *Stipitichnus* Genise, and *Xylonichnus* Genise. *Phagolignichnus* Vialov, 1975, is a *nomen nudum*.

A peculiar type of feeding trace in fossil wood is represented by characteristic *pitch flecks* interpreted as cambial mines made in living wood and later filled in by irregular parenchymatous cells. GREGUSS (1970) mistreated them as natural wood elements. Similar cambial mines are known to be produced by modern agromyzid flies of the genus *Phytobia* Lioy (SÜSS 1979). Fossil cambial mines are restricted to the Cainozoic (GRAMBAST-FESSARD 1966, SÜSS & MÜLLER-STOLL 1975, 1977, 1979, 1982, SÜSS 1979, 1980, GEISSERT et al. 1981) and may well have been produced by the same group of flies. Two formal ichnogeneric names are available, *Palaeophytobia* Süss et Müller-Stoll (in angiosperm wood) and *Protophytobia* Süss (in coniferous wood). Other pathological patterns in fossil wood may have been caused by insect sucking (e.g. ATTIMS 1969) but they still remain practically uninvestigated.

One more peculiar type of wood structure induced by insect feeding is represented by *false rings*. Unlike the traces discussed above this is not a kind of direct injury but a result of an indirect effect of decreasing tissue growth caused by damage to the photosynthetic tissues. False rings differ from the regular annual rings mainly on the basis of gradual (instead of abrupt) changes in cell size and cell wall thickness; they are not uncommon in both recent and fossil wood and can result from a number of stresses such as droughts, flooding, and pest or pathogen attacks (CREBER & CHALONER 1984). Actualistic observations indicate that insect effect on ring growth may be significant (e.g. KULMAN 1971, MORROW & LaMARCHE 1978, KARBAN 1980, HOLLINGSWORTH et al. 1991). Unfortunately, there are no good criteria for distinction between the insect-induced and other false rings.

The only record of insect *galleries in roots* (rhizoliths) is by HASIOTIS & DEMKO (1996a) from the Late Jurassic of USA. The unnamed galleries were originally interpreted as termite nests but this attribution is extremely doubtful (see above).

Holes in fossil fructifications which should be considered as a separate type of feeding ichnofossil are poorly investigated. In this case the only probable producers are insects; as modern hole producers, beetles, moths, hymenopterans and dipterans should be mentioned. Abundant insect borings in the Early Cretaceous bennettite strobili (REYMANÓWNA 1960, BOSE 1968, DELEVORYAS 1968a, CREPET 1974) show that the cones of those extinct cycadophytes were extensively exploited, and perhaps pollinated, by insects, supposedly by beetles (CROWSON 1991, T. TAYLOR & TAYLOR 1993, LABANDEIRA 1998c). This type of damage deserves much more consideration than has been accorded to it in the past. STOCKEY (1978) mentioned insect borings in female araucarian cones from Argentina, and similar damage occurs in araucarian cones in the Late Cretaceous of Mongolia (V.A. Krassilov, pers. comm.). An insect-damaged pine cone (bored probably by a microlepidopteran caterpillar) was recorded from the Miocene saline deposits of Poland by ZABŁOCKI (1960). No parataxa have been formally named.

Damaged seeds are also little studied. Most of them are eaten from inside (SCHMIDT et al. 1958, BENNIKE & BÖCHER 1990, DECHAMPS et al. 1992, NEL 1994, GENISE 1995). At present diverse insects (beetles, moths, hymenopterans, and dipterans) are the only internal seed parasites, and insect damage can be easily distinguished from external feeding traces produced by vertebrates. The only available name for borings in seeds is *Carporichnus* Genise; damage in larch seeds from the Plio-Pleistocene deposits of northern Greenland were attributed to the living seed-eating wasp genus *Megastigmus* (BENNIKE & BÖCHER 1990). Small round holes occurring on Late Palaeozoic seeds (Fig. 471) represent a distinctive and highly interesting type of external damage which has never been formally named. They are

Fig. 470 Insect borings in wood from the Begichev Formation (latest Early or earliest Late Cretaceous) resembling modern ambrosia beetles tunnels; Zhdanikha in North Siberia (PIN 3308/12, photo by D.E. Shcherbakov); block 72 mm wide along upper side

Fig. 471 A cordaitalean seed, *Samaropsis* sp., damaged by a palaeodictyopteroid insect (supposedly by the diaphanopteridan *Permuralia maculata* Kukalová-Peck et Sinitshenkova) from the Late Carboniferous of Chunya in Siberia (PIN 3115/303, photo by A.G. Sharov); seed 7 mm wide

rather common on *Samaropsis* seeds from the Carboniferous of Siberia; similar damage marks on another Carboniferous seed genus, *Trigonocarpus*, from USA and Britain, are illustrated by JENNINGS (1974, pl. 3, fig. 9) and A. SCOTT & TAYLOR (1983, fig. 7E). No similar trace fossils were recorded from the post-Palaeozoic deposits, and no modern analogues are observed. These holes are interpreted as the feeding traces of Dictyoneurida, an extinct Palaeozoic insect order (SHAROV 1973). Present-day seed-sucking insects (mainly bugs) cause quite different damage (LIVINGSTONE 1978, DUTCHER & TODD 1983, CAMPBELL & SHEA 1990) which could be recognised on anatomically preserved fossils but up to now were not discovered in the ichnological record.

Holes on megaspores recorded from the Carboniferous (DIJKSTRA & PIÉRART 1957, A. SCOTT & TAYLOR 1983) were probably produced by small arthropods other than insects, perhaps by springtails. Though some insects, for instance thrips (GRINFELD 1978, KIRK 1984) suck pollen grains, no record of any feeding traces on fossil pollen are known by me. Most probably, damaged pollen grains are normally ignored by palynologists because they could be deformed and difficult to identify. Future special search for such traces may provide some interesting results concerning the history of insect/plant relationships.

Insect ***feeding traces on leaves and stems*** are common in the fossil record. VIALOV (1975) has divided them into Phagophytichnidae (traces of external feeding by chewing insects), Palaeogallidae (fossil galls) and Palaeominidae (fossil mines). Histopathological structures connected with insect sucking should be added as the fourth group. Three former types are most well-known and used by various authors in discussions of the evolution of plant/insect relationships. However, many illustrations and anecdotal records of damaged leaves scattered in the palaeobotanical literature were never noticed by palaeoentomologists. The available information was briefly summarised in several reviews (e.g. SWAIN 1978, A. SCOTT & PATERSON 1984, CHALONER *et al.* 1991, A. SCOTT 1992, SCOTT *et al.* 1992, STEPHENSON & SCOTT 1992, T. TAYLOR & TAYLOR 1993) but recent progress in this field is so rapid that any review becomes outdated in a few years. In particular, comparative studies of feeding traces on fossil plants, started recently by Labandeira, gave excellent results of much palaeoecological and evolutionary importance (LABANDEIRA 1990, 1991, 1995, 1996, 1997a,b, 1998a–c, LABANDEIRA & BEALL 1990, LABANDEIRA & DILCHER 1993, LABANDEIRA *et al.* 1994a, 1995, A. BECK *et al.* 1996, LABANDEIRA & PHILLIPS 1992, 1996a,b, GREENFAST & LABANDEIRA 1997, A. BECK & LABANDEIRA 1998, WILF & LABANDEIRA 1999).

Leaf-feeding traces comprise four major types, namely margin feeding (Fig. 472), centre feeding (Fig. 473), skeletonisation (Fig. 474), and surface abrasion (LABANDEIRA *et al.* 1994a). They may be produced either by herbivores when the leaf was still living or by detritivores on fallen leaves. Though in principle it is possible to distinguish between the herbivore and detritivore feeding traces on the basis of presence/absence of a wound tissue reaction, this information is missing in most published records thus making their interpretation equivocal in some cases. Besides insects, traces on leaves may be produced by other arthropods (woodlice and millipedes), molluscs and vertebrates. For a useful discussion of the problems connected with the quantitative estimation of leaf consumption on the basis of fossils see A. BECK & LABANDEIRA (1998). The attribution of damage is often more or less arbitrary because their morphology is rarely diagnostic and because some of them may be produced by members of extinct lineages having no modern relatives. Up to now, few chewing traces were formally named (VAN AMEROM 1966, STRAUS 1977, GIVULESCU 1984),

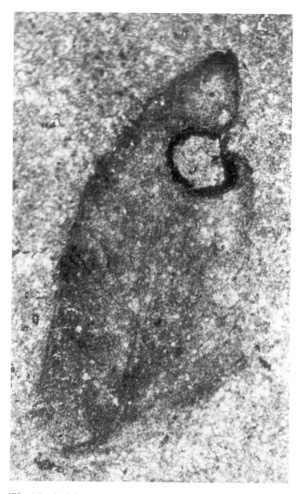

Fig. 472 Marginal-feeding on a fern from the Early Cretaceous of Khetana in Russian Far East (PIN 3800/2062 photo by D.E. Shcherbakov); fern pinnule 6.5 mm long

all in the same ichnogenus *Phagophytichnus* Van Amerom. Some were tentatively attributed to certain families or subfamilial insect taxa (FRIČ & BAYER 1901, COCKERELL 1908b, 1910a, S. LEWIS 1976b, 1992, 1994, STRAUS 1977, KRÜGER 1979) but such attribution is highly doubtful in most cases.

Leaf-chewing traces are known since the Late Palaeozoic (FEISTMANTEL 1881, PLUMSTEAD 1963, VAN AMEROM 1966, VAN AMEROM & BOERSMA 1971, A. SCOTT & TAYLOR 1983, SRIVASTAVA 1987, LABANDEIRA 1990, 1996, 1998a, LABANDEIRA & BEALL 1990, LABANDEIRA & PHILLIPS 1992, A. SCOTT *et al.* 1992, A. BECK *et al.* 1996, M. CASTRO 1997, GREENFAST & LABANDEIRA 1997, A. BECK & LABANDEIRA 1998). ZHERIKHIN (1980a) supposed that in the Palaeozoic, leaf damage could be produced by arthropods other than insects, perhaps by millipedes. However, it is noteworthy that, in spite of the large number of recent publications dealing with evidence of plant/animal interactions in the Palaeozoic, no leaf chewing marks are recorded before the major radiation of the winged insects in the Late Carboniferous. ZHERIKHIN (1980a) and SHEAR (1991) also supposed that Palaeozoic leaf damage could be ascribed to detritivores, but recently, A. BECK & LABANDEIRA (1998) documented in detail unequivocal evidence of herbivory on Early Permian leaves. In the Mesozoic chewed leaves seem to be rare before the Late Cretaceous

Fig. 473 Central feedings on Late Cretaceous leaves from Kzyl-Zhar in Kazakhstan: unregistered (left, leaf 45 mm long) and PIN 4409/64 (right, leaf 67 mm long along central rib) (photo by D.E. Shcherbakov)

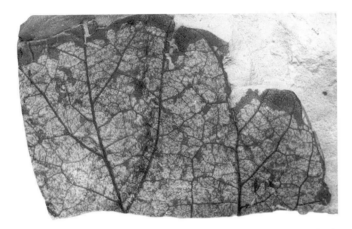

Fig. 474 A skeletonised platanoid leaf from Late Cretaceous of Kzyl-Zhar in Kazakhstan (not registered, photo by D.E. Shcherbakov); leaf 55 mm wide as seen along lower margin

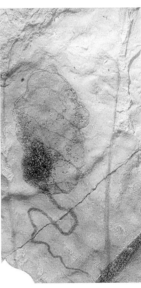

Fig. 475 Mines allegedly of nepticulid moths on leaves of Late Cretaceous (Turonian) angiosperms from Kzyl-Zhar in Kazakhstan: left *Stigmellites tyshchenkoi* Kozlov on "*Platanus*" *latior* Knowlton, leaf fragment 75 mm long; right *S. samsonovi* Kozlov on the leaf of *Trochodendroides arctica* Heer (Berry) (Cercidiphyllaceae), mine length 14.5 mm (holotypes, photo by D.E. Shcherbakov)

(see Fig. 472) (BOCK 1969, GEYER & KELBER 1987, KELBER & GEYER 1989, GRAUVOGEL-STAMM & KELBER 1996, ASH 1997). A peculiar structure at the leaf tip of the Jurassic conifer *Bilsdalea dura* is interpreted as a possible food body which is missing on many specimens because of insect consumption (WCISLO-LURANIEC 1985). There are numerous records of damaged angiosperm leaves (Fig. 473–475) from the Late Cretaceous and Cainozoic (e.g. LESQUEREUX 1892, FRIČ & BAYER 1901, E. BERRY 1924, 1930, 1931, ABLAEV 1974, BAYKOVSKAYA 1974, M. BARTHEL & RÜFFLE 1976, S. LEWIS 1976b, 1992, 1994, RÜFFLE *et al.* 1976, STRAUS 1977, KRÜGER 1979, GIVULESCU 1984, J. JOHNSTON 1993, LABANDEIRA *et al.* 1995, WILF & LABANDEIRA 1999) but they are still insufficiently investigated. More studies are necessary to elucidate evolutionary trends of chewing insect feeding on angiosperms but it is well demonstrated that angiosperm leaves were already eaten extensively very early in the Late Cretaceous (LABANDEIRA *et al.* 1994a). Interestingly, damaged leaves seem to be much more frequent in the Late Cretaceous in warmer areas (southern USA, Kazakhstan) than in temperate Siberia (personal observations). An increase in both leaf attack frequency and diversity of damage traces in Wyoming in the course of the Eocene warming is noted by WILF & LABANDEIRA (1999).

Leaf mines (internal feeding holes) are now produced exclusively by insect larvae. The main present-day miners are microlepidopterans but there are some leaf-mining beetle, dipteran and sawfly larvae as well. Modern mining insects with few exceptions exploit living leaf tissues. The mines may be either narrow and sublinear (often sinuous) or wide, patch-shaped. Actualistic data are summarised by HERING

(1951, 1957a–c). Fossil leaf mines are common (see Fig. 475). Many of them were attributed by different authors to modern insect families and genera with a reasonable degree of certainty because mine morphology is often diagnostic: However, this practice can be doubted from the point of view of the basic principles of ichnotaxonomy. Some Pliocene mines were assigned even to living insect species (STRAUS 1977, KRÜGER 1979) but their identification is not absolutely certain. Several formal ichnogeneric names have been proposed as well, including *Foliofossor* Jarzembowski, *Paleomina* Vialov and *Stigmellites* Kernbach for microlepidopteran mines, *Fenusites* Straus for supposed tenthredinid sawfly mines, *Phytomyzites* Straus, for supposed mines of agromyzid flies and *Cuniculonomus* Straus, 1977, *Loconomus* Straus and *Triassohyponomus* Rozefelds et Sobbe for mines of uncertain provenance. The published records of fossil lepidopteran mines were reviewed by KOZLOV (1988); for additonal records and mines allegedly created by other insects see GÖPPERT (1855), ENGELHARDT (1876), HERING (1930), YAKUBOVSKAYA (1955), STRAUS (1967, 1977), BARTHEL & RÜFFLE (1970), ABLAEV (1974), RÜFFLE & JÄNICHEN (1976), KRÜGER (1979), TIDWELL et al. (1981), GIVULESCU (1984), ROZEFELDS (1985b, 1988a,b), MARTINS NETO (1989, 1991), J. JOHNSTON (1993), DAVIS (1994), LABANDEIRA et al. (1994a), LANG et al. (1995), MUSTOE & GANNAWAY (1977) and WILF & LABANDEIRA (1999).

There are some records of supposed insect mines from the Late Palaeozoic (GRAND'EURY 1890, A. MÜLLER 1982) but these may be doubted (ZHERIKHIN 1980a, A. BECK & LABANDEIRA 1998). Few mine-like structures of unknown provenance are known from the Triassic in the stems of horsetails (R. POTONIÉ 1921) and leaves of conifers (*Heidiphyllum*: ROZEFELDS & SOBBE 1987) and supposed gynkgophytes (?*Glossophyllum*: Fig. 476). ANSORGE (1993) mentioned and figured supposed insect mines in a leaf of *Podozamites*, a Mesozoic conifer, from the Early Cretaceous of Spain, but their producer is unknown. The oldest undoubted insect mine is described by ROZEFELDS (1988b) from the latest Jurassic or earliest Cretaceous of Queensland and attributed to a nepticulid moth. This is the only known alleged lepidopteran mine on an extinct gymnosperm plant (the corystosperm pteridosperm *Pachypteris*). Numerous mines are recorded from the Late Cretaceous where they are locally abundant on angiosperm leaves (CHAMBERS 1882, FRIČ 1882, HAGEN 1882, ABEL 1935, ZHERIKHIN 1978, 1980a, TIDWELL et al. 1981, KOZLOV 1988, LABANDEIRA & DILCHER 1993, LABANDEIRA et al. 1994a, 1995). Interestingly, in the Late Cretaceous they are not only common but also very similar to mines of present-day microlepidopterans and some of them may well be assigned to living genera (e.g. *Stigmella*,

Fig. 476 Supposed insect mine in a leaf of ?*Glossophyllum* from the Middle or Late Triassic of Dzhayloucho in Kyrghyzstan (not registered, 22 mm as preserved, photo by D. E. Shcherbakov)

Ectoedemia, and *Bucculatrix*: KOZLOV 1988, LABANDEIRA et al. 1994a). Like external insect chewing marks, the Late Cretaceous mines seem to be commoner in warmer climatic areas, but this pattern should be proved by quantitative studies. In some cases a long-time constancy of host-miner associations was demonstrated at least at the generic level (OPLER 1973). The Cretaceous and Early Tertiary lepidopteran mines were attributed to only a few of the modern mining moth families (Eriocraniidae, Nepticulidae, Bucculatricidae, Phyllocnistidae and Gracillariidae) while the supposed mines of Incurvariidae, Lyonetiidae, Coleophoridae, and Gelechiidae are recorded only from the Late Tertiary. Fossils of all these families are known at least since the Early Tertiary, so this may indicate that some families maintain the characteristic mine morphology for a longer time than others: this hypothesis needs independent evidence.

Galls are peculiar parasite-induced tumours occurring on different plant organs (leafs, buds, stems, roots). Modern gall producers are very diverse including not only different groups of insects (thrips, aphids, coccids, psyllids, beetles, dipterans, lepidopterans, hymenopterans) but also mites, nematodes, fungi and bacteria. Gall formation is the complex and specific pathogenic process based on intimate physiological and biochemical interactions between the inducer and the plant tissue (SCHÄLLER 1970, PACLT 1972, DREGER-JAUFFRET 1983). Besides the formation of distinctive pathological structures many plant parasites affect developmental processes inducing more or less pronounced teratological deformations of plant organs like shoot shortening, leaf deformation, superfluous branching, etc. Such ***deformed plant organs*** are also considered by many authors as galls in a broad sense, and the borderline between them and true galls is somewhat indefinite (see SLEPYAN 1973, for a review and discussion). Many galls are quite characteristic both morphologically and histologically and are widely used for exact identification of their producers up to the species level. However, attribution of ancient and especially pre-Cainozoic galls is highly tentative. Actualistic data are summarised in a number of papers (e.g. HOUARD 1908, 1913, 1922, 1923, FELT 1940, WEIDNER 1961, SLEPYAN 1962, BUHR 1964, 1965, MANI 1964, MEYER 1987, SHORTHOUSE & ROHFRITSCH 1992, M. WILLIAMS 1994).

Fossil galls have been recorded by many authors (for a review see WITTLAKE 1981, LAREW 1986, 1992, ROSKAM 1992, A. SCOTT et al. 1994, for additional records and figures see IONESCU in POP 1936, PIMENOVA 1954, KRASSILOV 1972a, DIÉGUEZ et al. 1996, GRAUVOGEL-STAMM & KELBER 1996, ASH 1997, ARCHIBALD & MATHEWES, 2000). According to DIÉGUEZ et al. (1996) some structures on fossil leaves described as alleged epiphyllous fungi may in fact represent mite and insect galls as well. Pathologically deformed plant organs should hardly be recognisable in the fossil state and may occasionally be misinterpreted and described under separate names by palaeobotanists (VAN AMEROM 1973). Many fossil galls are unnamed; others were attributed to modern insect genera and even species, sometimes with good reasons. MÖHN (1960) has established a new genus of gall midges (Cecidomyiidae), *Sequoiomyia*, on the basis of galls on a Tertiary redwood from Germany. Several formal ichnogeneric names were proposed for galls of unknown affinities including *Galla* Lesquereux, *Paleogallus* Vialov, *Petiolocecidium* Straus, and *Phyllocecidium* Straus.

The majority of supposed Palaeozoic galls recorded in the literature are rather deformed organs than true complex galls, and their nature is ambiguous (e.g. FLORIN 1938, VAN AMEROM 1973). However, LABANDEIRA & PHILLIPS (1996a) described in histological detail very large galls (up to 24 cm long) on the petiole of the tree-fern *Psaronius* from the Late Carboniferous of Illinois and ascribed it to a holometa-bolan insect. This most interesting find clearly indicates that as early as in the Late Carboniferous, the gall-forming mode of life has been acquired by some phytophagous organisms. However, the attribution of the galls may be doubted. It seems quite possible that they could be produced by a member of one of the numerous Palaeozoic extinct hemimetabolous insect orders or even by an endophytic millipede.

Both deformed organs and supposed true galls are recorded from different Mesozoic gymnosperms (VISHNU-MITTRE 1957, 1959, HARRIS 1969, GRAUVOGEL-STAMM 1978, GRAUVOGEL-STAMM & KELBER 1996, ASH 1997). It is not clear which of these structures, if any, were induced by insects because no direct comparison with any present-day galls is possible. Galls on angiosperm leaves appear in the fossil record as early as in the Albian (later Early Cretaceous; DOYLE & HICKEY 1976, UPCHURCH et al. 1994) demonstrating that even the most complex host-parasite interactions between the flowering plants and their consumers were developed by the time of angiosperm expansion. In the Late Cretaceous, galls (Fig. 477) show the same pattern as other types of feeding traces on angiosperms, being common and diverse since the Early Cenomanian at least in warmer climatic belts (KRASSILOV 1972a, ZHERIKHIN 1978, LABANDEIRA et al. 1994a, 1995). Some Pliocene galls are identical with the present-day galls on the same plant genera in all available significant characters (MARTY 1894, STEINBACH 1967, HEIE 1968, STRAUS 1977, KRÜGER 1979). These observations are of great importance for evolutionary theory because they confirm the long-term (up to 3–5 Myr) stasis in some insect species, operating at the biochemical level in addition to the morphological one. It should be noted that the statement about very poor documentation of the insect-plant relationships in the Quaternary

Fig. 477 An angiosperm leaf from the Late Cretaceous of Kzyl-Zhar in Kazakhstan, showing combined damage (galls and sclerotisation) by different insects (unregistered, photo by D.E. Shcherbakov); leaf fragment 22 mm long along mid rib

(BUATOIS et al. 1998) is incorrect as concerning the galls: there are numerous records of subfossil galls (both from the Pleistocene and Holocene), mostly identified as produced by the present-day insect species (ANDERSSON 1898, BEYLE 1913, 1924, 1926, GERTZ 1914, 1926, JESSEN 1920, HARRISON 1926, OTRUBA 1928, KOLUMBE & BEYLE 1940, LAREW 1987).

The last type of insect damage on leafs and stems are **traces of plant sucking**. The piercing-and-sucking herbivores are diverse (especially among the hemipterans) and geologically ancient, so this type of damage has to be common in the ichnofossil record. However, sucking traces are usually hardly detectable on plant fossils and attract little if any attention. They are briefly mentioned only in some recent papers specially focusing on insect feeding traces (LABANDEIRA et al. 1994a, WILF & LABANDEIRA 1999). The earliest supposed insect sucking traces found on Carboniferous plants are of a particular interest. The only detailed and anatomically-based description of such a mark was published recently by LABANDEIRA & PHILLIPS (1996b). Their attribution of a pin-hole trace in a rhachis of the Late Carboniferous tree-fern *Psaronius* to a dictyoneuriean insect seems to be plausible. In the same paper a review of other possible Palaeozoic sucking traces as well as a brief but useful review of actualistic data can be found. Further studies of sucking traces are highly desirable.

Traces of dung feeding constitute the formal group Coprinisphaeridea (after VIALOV 1975). Presently the insects, first of all, dipterans and beetles are the principal consumers of vertebrate faeces in terrestrial ecosystems. Because the coprolites like most other fossils are preserved mainly in subaquatic environments, the probability of burial of faecal masses burrowed in by insects is extremely low. However, some beetles store dung for their larvae in subterranean nests. Such nests can be constructed either directly under the dung or at some distance. In the latter case dung balls are formed for transportation. Fossil dung beetle nests with preserved food masses are not very rare. The oldest evidence of insect coprophagy was described recently from the Late Cretaceous of North America where burrows filled with dinosaur dung were discovered (CHIN & GILL 1996). This find, documenting insect coprophagy in the Mesozoic for the first time, is of great importance. Numerous dung beetle nests were described and recorded by different authors from South America (see RETALLACK 1990a, and GENISE & BOWN 1993, for a review, and ZEZZA 1974, LAZA 1995, GENISE & LAZA 1998 for further records and discussions). In this region the fossil dung balls probably occur throughout the Cainozoic sequence from Palaeocene up to Pleistocene, some finds being possibly of the latest Cretaceous age. In other regions dung balls are recorded from the Oligocene of USA (RETALLACK 1984) and from the Plio-Pleistocene of South Africa (VRBA 1980). No food remains were found in the supposed dung beetle burrows in the Quaternary of The Netherlands (BRUSSAARD & RUNIA 1984). Some South American finds were tentatively assigned to the present-day dung beetle genera *Megathopa*, *Canthon*, *Onthophagus* and *Phanaeus* (FRENGUELLI 1938a,b) but this attribution may be doubtful (HALFFTER 1959). Several formal ichnogenera were described including *Coprinisphaera* Sauer, *Devincenzichnus* Roselli, 1976 (=*Devincenzia* Roselli, 1938, preoccupied; a junior synonym of *Coprinisphaera* according to GENISE & BOWN 1993), *Fontanai* Roselli, 1938 (=*Fontanaichnus* Roselli, 1987), *Madinaichnus* Roselli, *Martinezichnus* Roselli, *Microicoichnus* Roselli, *Monesichnus* Roselli (a composite ichnofossil made by two different producers according to GENISE & LAZA 1998), *Pallichnus* Retallack, *Rebuffoichnus* Roselli, and perhaps also *Teisseirai* Roselli (though BUATOIS et al. 1998 reject the attribution of this ichnogenus to a dung beetle). *Isociesichnus* Roselli, is probably a synonym of *Teisseirai* (BUATOIS *et al.* 1998).

Palaeoichnological data have been used in some studies on the evolution of dung beetle nesting behaviour (HALFFTER 1959, 1977, HALFFTER & EDMONDS 1982). Their association with steppe-type palaeosols was demonstrated (ANDREIS 1972, RETALLACK 1984, 1990a), and their presence can be used as an indicator of open savannah or steppe palaeolandscapes (ZHERIKHIN 1993a, 1994, BUATOIS *et al.* 1998). If so, future finds would be expected in the Late Tertiary not only within the prairie belt in North America but also in Kazakhstan and Central Asia where steppe palaeosols are widespread (KASIMOV 1988). However, in the Pinturas Formation (Miocene) of Patagonia *Coprinisphaera* occurs in the palaeosols which most probably originated in a subhumid woodland palaeoenvironment (GENISE & BOWN 1993).

Insect *feeding traces on animal remains* (Phagozooichnidea after VIALOV 1975) seem to be uncommon. They include mainly traces of biogenic erosion on vertebrate bones produced supposedly by dermestid beetles. L. MARTIN & WEST (1995) place them among the resting traces (Cubichnia) because the traces are allegedly connected with pupal chambers though they may also be classified into Calichnia (breeding traces) or put into a special ichnotaxon. No formal names are available. Supposed dermestid burrows in dinosaur bones are recorded from the Late Jurassic (LAWS *et al.* 1996, HASIOTIS & FIORILLO 1997) and Late Cretaceous (R. ROGERS 1992) of North America. They occur also in mammal bones in the Late Tertiary and Quaternary of Europe, North America and South Africa (TOBIEN 1965, 1983, ZAPFE 1966, VRBA 1980, L. MARTIN & WEST 1995). They are believed to be taphonomically important as they indicate that vertebrate burial occurred a considerable time after the animal death. They may give some information on the season of death and climatic conditions as well (L. MARTIN & WEST 1995). However, Prof. R.D. Zhantiev (pers. comm.) doubts these kinds of traces, because in natural environments dermestid larvae practically never construct their pupation chambers in bones but pupate in the soil; traces on bones occur only in the laboratory, when more suitable substrate is unavailable. He supposes that the traces on subfossil and fossil bones could be produced by some other kinds of invertebrates, perhaps when a bone is partially macerated in wet soil, and accordingly their taphonomic interpretation based on their attribution to dermestids may be inaccurate.

Finally, some ichnofossils may be interpreted as having been *produced by predators and parasitoids* (Praedichnia after BROMLEY 1996). Their records in literature are anecdotal but there is little doubt that this is simply a result of a lack of special interest in them. GENISE (1995) described *Carporichnus minor* as probable eclosion holes produced by a parasitic wasp in seeds eaten by its host. If this

interpretation is correct, the placement of this ichnofossil to the ichnogenus *Carporichnus*, established for the feeding traces of seed predators, is controversial with respect to the ethological principle of classification. However, Genise's decision seems to be more pragmatic because the interpretation of the holes is purely hypothetical. ELLIS & ELLIS-ADAM (1993) mentioned holes in walls of Quaternary cells of solitary bees from the Canary Islands, which were probably made by meloid beetles or bombyliid flies.

More trace types may be found in the future if more care is taken to search for them. In particular, bird feathers are not rare in the Cretaceous and Cainozoic deposits, so finds of the feeding traces of bird lice would be expected.

2.3.6. COPROLITES

Coprolites in fact constitute a particular type of feeding trace, and in the case of insects they are often associated with them living within burrows in plant tissues (e.g. JURASKY 1932, A. ROGERS 1938, BAXENDALE 1979, CICHAN & TAYLOR 1982, MÜLLER-STOLL 1989, SHARMA & HARSCH 1989, ZHOU & ZHANG 1989, ROZEFELDS & DE BAAR 1991) or in palaeosols (DE 1990). The disperse insect coprolites can rarely be recognised as such (e.g. SCHAARSCHMIDT & WILDE 1986), though sometimes they may be abundant. Morphology of modern insect faeces is little studied (FROST 1928, LADLE & GRIFFITHS 1980), and only a few insect coprolites (e.g. coprolites of termites: LIGHT 1930, A. ROGERS 1938, LANCE 1946, ROZEFELDS & DE BAAR 1991) can be identified. No special names have been proposed except for some finds misinterpreted as other fossils (like "*Lagena*" *samanica* originally described as a foraminiferan: STONE 1950, or "*Microcarpolithes*" *hexagonalis* originally described as a plant seed: HOOKER *et al.* 1995). Sometimes arthropod coprolites contain identifiable food remains, mostly fragments of plant tissues and fungal spores and hyphae (A. SCOTT 1977, BAXENDALE 1979, T. TAYLOR & SCOTT 1983, REX & GALTIER 1986). Insect coprolites containing plant spores from the Early Cretaceous of Germany are of considerable interest; they were tentatively attributed to hymenopterans (KAMPMANN 1983). Aquatic insect coprolites may constitute an important and even the main component of some non-marine sediments enriched with organic matter, as was supposed for the bituminous shales of the Green River Series in USA (W. BRADLEY & BEARD 1969, IOVINO & W. BRADLEY 1969, BRADLEY 1970). The lacustrine bituminous shales widespread in the Early Cretaceous of Siberia, Mongolia and China may be of a similar origin as well, though in this case faeces of nectobenthic chaoborid larvae rather than of benthic chironomids seem to be the probable main source of the organic matter.

3
General Features of Insect History

3.1
Dynamics of Insect Taxonomic Diversity

V.YU. DMITRIEV AND A.G. PONOMARENKO

The study of past dynamics of the taxonomic diversity is a complicated job, threatened by numerous traps and caveats connected mostly with improper or insufficiently representative material and inappropriate calculating methods (ALEKSEEV *et al.* 2001). It is self-evident that the taxa used as operational units in the diversity calculation should be long living enough to have their duration at least roughly recordable in the fossil record available, and yet not too long living to display their appreciable turnover. Accumulated experience (DMITRIEV & ZHERIKHIN 1988, RASNITSYN 1988b, 1989a, JARZEMBOWSKI & ROSS 1993, 1996, DMITRIEV *et al.* 1994a,b, 1995, LABANDEIRA & SEPKOSKI 1993, LABANDEIRA 1994, JARZEMBOWSKI & ROSS 1996) confirms that insect families best satisfy the above conditions given the present state of our knowledge.

To select the appropriate calculation method of the diversity dynamics, a larger discussion is necessary. The first methods coming to mind is just to compare the number of taxa registered at the succeeding stages of the fossil record, for different stratigraphic intervals vary widely in completeness of their palaeontological study. Therefore, the number of taxa recorded at a time interval depends on both their past diversity, their taphonomic properties (chance to enter the fossil record, see Chapter 1.4), and general state of knowledge of respective sections of the fossil record, and these different influences are hardly separable. In addition, the fossils are usually recorded after their stratigraphic position in a particular stratigraphic unit (period, epoch, etc.), rather than by indication of their absolute age in millions of years. The stratigraphic units differ in their absolute time duration, so the longer unit is expected to have the higher number of taxa, an effect deserving to be counterbalanced. The stratigraphic data vary widely in their level of detail and subdivision: one taxon might be recorded just as the Cretaceous, while another as a particular zone within the Hauterivian, one of several stages of the Early Cretaceous. These sources of uncertainty result in problems that are not easy to overcome.

Further, dealing with just the number of taxa recorded results in considerable loss of information compared to the separate use of the first and last appearances of taxa (substitute for their origin and extinction, respectively, events that are hardly possible to be dated precisely). Indeed, the constant number of taxa recorded in two succeeding time intervals may be due to the absence of any change, or may be due to high but equal numbers of taxic origins and extinctions (equal taxic birth and death rates).

Because of these and related problems discussed at length by ALEKSEEV *et al.* (2001), a few relatively safe and informative methods have been selected. One of them is the **momentary** (= instantaneous) **diversity** on the boundary of two adjacent stratigraphic units which is calculated as the number of taxa crossing the boundary (recorded both above and below it). This permits us to avoid or minimise the errors caused by the unequal duration of the stratigraphic units, as well as by irregular distribution of the exceptionally rich fossil sites that fill the fossil record with numerous short lived taxa, a kind of noise in diversity dynamics research.

Instantaneous diversity can be plotted against the time scale (Fig. 478*a*) to display directly the temporal dynamics of taxonomic diversity. Otherwise it can be plotted against the cumulative number of taxa in the group up to the given time (Fig. 478*b*). The advantage of this method is that the resulting chart, though ignoring duration of the stratigraphic units, displays separately both the gain and loss dynamics. The line connecting the succeeding time moments shows the number of gains (first appearances of taxa) when measured along the abscissa, and the difference between the number of gains and losses (first and last appearances of taxa) when measured against the ordinate. As a result, in case of pure losses the line will be directed straight downwards, in case of pure gains – 45° obliquely upwards, in other cases – in intermediate directions.

Other calculation methods used employ numbers of gains and losses plotted separately against the time scale (Figs. 479*a,b*), and the share (per cent) of extinct families in the total insect assemblage (Fig. 480), otherwise known as inverted Lyellian curves (ZHERIKHIN 1978, RASNITSYN 1988b, 1989a). This method displays how a group approaches its contemporary taxonomic composition.

The present calculations are based on the fossil record database in progress at the Arthropoda Laboratory, Palaeontological Institute, Russian Academy of Sciences. 1049 families are considered, less than the 1261 referred to by LABANDEIRA (1994): the extra 212 families are considered synonyms in our list. The stage is taken as the stratigraphic unit of the family duration. The families whose first/last appearance is dated less precisely are not ignored but interpolated in proportion of the precisely dated families. E.g. if an epoch consists of 3 stages with 10, 1, and 40 properly dated families, and 5 more

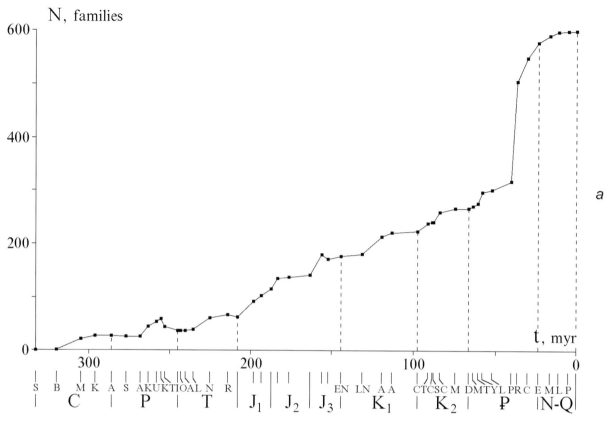

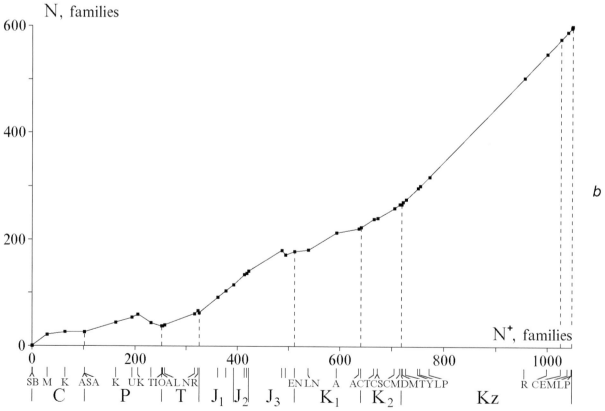

Fig. 478

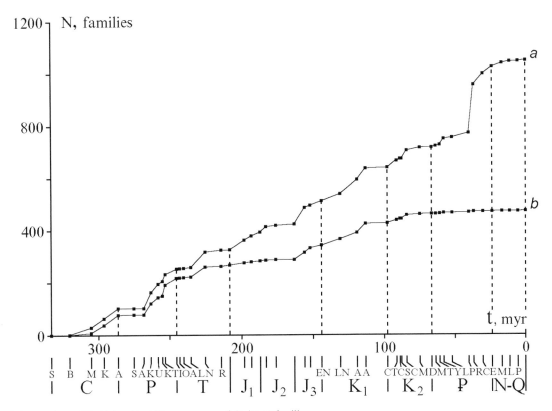

Fig. 479 Cumulative numbers of first (*a*) and last (*b*) appearances of the insect families

families attributed simply to this epoch, the total distribution taken for calculation is 11, 1, 44.

Figure 478 shows the dynamics of the instantaneous number of the insect families plotted against the time (*a*) and the cumulative number of family appearances (*b*). Both curves have a rather simple form. The family number (*a*) is growing slowly in the Carboniferous – Triassic resulting in the near-horizontal pace of the curve. From the Jurassic until the mid-Eocene, the growth is more rapid but almost linear, although the norm of diversity accumulation (*b*) is not constant: it is higher in the Jurassic, but becomes slower near the Jurassic–Cretaceous boundary, and then grows again to reach 1 in the Cainozoic. The incomparably full record of Baltic amber fossils gives a high step in the curve (*a*), then the growth decreases towards the present time, though it is still very high in the Oligocene. In contrast, the Cainozoic growth of (*b*) is linear.

Figure 479 gives further explanation of the above curves. It displays the number of the first (*a*) and last appearances (*b*) of families since the Serpukhovian Stage (latest Early Carboniferous). The slope of the curves depends on the rates of family origin and extinction,

respectively. Long-term deviations in the rate of this sort are clearly apparent, but the short term tendencies are difficult to appreciate correctly because of errors in identification of the duration of the taxonomic and stratigraphic units involved.

The curves in Fig. 479 are of a simple form as well. Interesting is the near the linear part of the first appearance curve from its beginning until the Middle Eocene. In contrast, the last appearance curve (*b*) is essentially convex throughout implying the largely decreasing extinction rates – to nearly 0 in the Cainozoic. Details of the two curves shed additional light to the curves in Fig. 478. Low diversity in the Palaeozoic is connected with extinction rates which are distinctly high compared to those of the Mesozoic. Lower diversity combined with high tempo of origination and extinction means a high rate of the taxonomic turnover resulting in the appearance of numerous short-lived families: these are very characteristic of the Palaeozoic insect fauna.

Since the Triassic, or maybe since the latest Permian, both origination and extinction rates decrease appreciably, and since near the beginning of the Triassic, the family number curves show the start of the

Fig. 478 Insect diversification: a – time versus instantaneous number of families, b – total number of families appearing at a given time versus instantaneous number of families. In figures 478–482, the standard East European stratigraphic scale is used. Dots within the stratigraphic intervals denote, left to right, beginning of the stages (marked along the horizontal axis by the first letter of their names): in the Carboniferous – Serpukhovian, Bashkirian, Moscovian and Kasimovian; in the Permian – Asselian, Sakmarian, Artinskian, Kungurian, Ufimian, Kazanian and Tatarian; in the Triassic – Indian, Olenekian, Anisian, Ladinian + Karnian (because the richest Madygen deposits are not dated within that interval), Norian and Rhaetian. Jurassic material is often dated not precisely enough to permit calculating events on the strict stage boundaries: instead, the figures and respective dots are attributed here to imprecisely defined early, middle and late part of each the Early, Middle, and Late Jurassic. For the Early Cretaceous, dots represent beginnings, left to right, of the lower and upper halves of the Neocomian, Aptian and Albian, for the Late Cretaceous – beginnings of the Cenomanian, Turonian, Coniacian, Santonian, Campanian and Maastrichtian, for the Palaeogene – beginnings of the Danian, Montian, Thanetian, Ypresian, Lutetian, Priabonian, Rupelian and Chattian, in the Neogene – beginnings of the Early, Middle and Late Miocene and Pliocene. The right-most dot denotes the present time. Because of too limited records, dots representing some boundaries may be omitted or, in Fig. (b) coincident

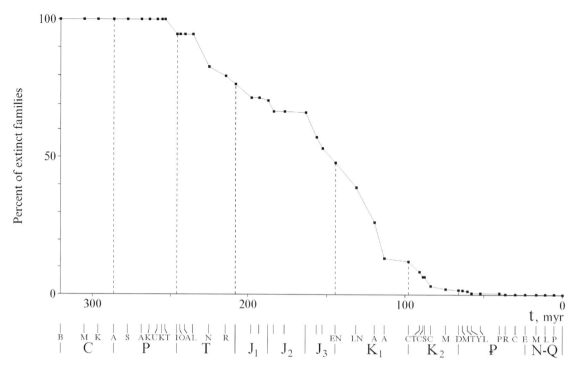

Fig. 480 Share of the extinct insect families (the inverted Lyellian curve)

long-term Meso-Cainozoic growth. Though, towards the end of the Triassic it has not surpassed the Permian value. The near linear growth of family number with time (Fig. 478a), in the scale of the accumulated first appearances (Fig. 478b), breaks down into several characteristic parts. The curves (Fig. 479a,b) reveal the structure of this process. In the Early and Middle Jurassic, extinction rate decreases sharply, and family accumulation is therefore high, and the resulting diversity is growing somewhat more quickly than afterwards. Since the Late Jurassic till the Aptian both the origination and, particularly, extinction rates remain high, so the diversity accumulation is low. However, curve 478a shows no peculiarities there.

During the Late Cretaceous the accumulation norm gradually approaches 1 and remains so in the Cainozoic (Fig. 478b) because of the low extinction rate. The latter effect can be seen because of our high knowledge of living insects (ZHERIKHIN 1978, RAUP 1979), although this does not necessarily exclude really decreasing extinction rates. The Baltic amber insect record is exceptionally complete and can be even approaching that of many contemporary regional faunas. It is separated from them by about 40 million years, and nevertheless very few Baltic amber families went extinct since that time. This implies a really low extinction rate after the Late Eocene. The cause can be the high genus per family ratio of the Cainozoic insects: many extant insect families do include more genera than those in other animal groups, but there is no evidence as yet, that a similar difference exists between the Cainozoic and Palaeo- and Mesozoic insect families. On the other hand, the massive insect diversification is often postulated to be connected with the leading position taken by angiosperm plants, and this event certainly roughly coincides with insect families acquiring their superstability.

In the above general pattern of the insect diversification, one of the noteworthy trends is the low Palaeozoic rate of diversification. Along

with the minimum diversity near the Perm-Triassic boundary, it delimits the distinct Palaeozoic stage, as opposed to the Mesozoic one, of insect taxonomic turnover. The basic pattern in other organisms: in marine animals as a whole and their various subgroups, in freshwater animals (DMITRIEV et al. 1995), in non-marine tetrapods (ALEKSEEV et al. 2001), non-marine plants (BENTON 1993) are all similar.

Calling for a biological interpretation is the near-linear time path of the first appearance curve (Fig. 479a) since the Triassic till at least the Middle Eocene (in a coarser scale and with the Late Eocene step caused by the Baltic amber being mentally dropped, this sublinear interval would extend up to the present time). The origination rate varies around a near constant average value for 200 or maybe even more than 300 million years. 36 families entering the Triassic being compared to 306 entering the Late Eocene suggest more than an 8-fold diversity growth that has resulted in the near zero growth of the origin rate. So the latter is essentially independent of the diversity, and the per family rate is inversely proportional to the diversity. This invokes a diversity–bonded mechanism of divergence control, probably a kind of biocoenotic negative feedback.

In case the divergence probability is independent of biotic interactions (families evolve independently of each other), the instantaneous origination rate is proportional to diversity (to the number of families in existence). If the biotic interactions exist, the instantaneous origination rate will grow overproportional to the degree of diversification. The only case of the first appearance rate, independent of diversity is probably the case of all families tightly interconnected to each other, so as the assemblage is evolving near as a whole. It is not to say that all the insects form a single evolutionary entity: the same effect is expectable in the case of several ecological entities existing which are tightly interconnected within themselves and loosely so in-between. This pattern of biocoenotic organisation appears applicable to living communities. However, we are aware of no direct evidence of this kind of biocoenotic organisation in the remote past.

Biocoenotic stability is not absolute: it is affected by introduction, divergence and extinction events which may cause extremes in the curve. A similar effect may result from the delayed evolutionary reaction of taxa and their assemblages, which depends much on the flexibility of the biocoenotic structure. The subconstant origination rate of insect families suggests that the flexibility is generally neither lost nor varies to too wide an extent.

Analysis of the extinction (last appearance) rate is less straightforward. Figure 479b demonstrates a decreasing extinction rate and, hence, its inversely superproportional dependence on diversity. It is not clear if the above factors are involved here and how they might work. However, the participation of biocoenotic mechanisms does not look dissimilar.

Interpretation of shorter intervals of the above curves, suffers more from our incomplete knowledge than general features of the processes involved. Therefore only a few reliable inferences are given below.

The Carboniferous. The first winged insect, which appears in the Serpukhovian Stage, is *Delitzschala bitterfeldensis* Brauckmann et Schneider from the E2 ammonoid zone of Arnsbergian (Chapter 2.2.1.2.3.1). The main diversification is the Bashkirian resulting in 20 families at its upper boundary. The growth continues in the Moscovian stage, though at the much lower rate, and the diversity stabilises in the Late Carboniferous. Features of the Moscovian – Late Carboniferous interval depend mostly on the steep growth of the extinction rate.

Permian. The apparent zero growth of diversity in the Asselian and Sakmarian is due to the very poor fossil record of this time interval. The norm of diversity accumulation (Fig. 478b) seems to be near constant since the Artinskian till the beginning of the Kazanian, while the origin and extinction rates appear to decrease and the diversification rate only slightly so. The Kazanian and Tatarian stages are characteristic of the quick apparent diversity decrease in the Kazanian and less sharp in the Tatarian. This might result from the comparatively modest Tatarian fossil record compared to the rich Kazanian one (Soyana in the Russian North). However, this pattern finds a parallel in the general Late Permian marine extinction that also has its maximum before the latest Permian. The Permian/Triassic boundary minimum is the most pronounced (or, in a sense, the only well expressed) one in the origin accumulation curve (Fig. 478b). It is less expressed in the diversity curve (Fig. 478a) because of the long subhorizontal section before the Ladinian. The terminal Triassic and Late Jurassic minima are only weakly defined in comparison.

The Triassic interval is insufficiently known in respect of its diversity dynamics. Fig. 478b shows a massive accumulation of the first appearance in the Ladinian-Carnian, while the pre-Ladinian and post-Carnian sections are far reduced. This is because the majority of the Triassic insects come from the Madygen Formation in Central Asia which is dated imprecisely as the Ladinian or Carnian, the interval being shown as a single unit in our charts. The poor record of the remaining Triassic, and particularly its pre-Ladinian section, might exaggerate the Permian/Triassic minimum though it hardly creates it, because this minimum is observed in practically all palaeontological material.

A very slight minimum can be seen at the Triassic/Jurassic boundary as well. This minimum is well expressed in the majority of other groups (ALEKSEEV *et al*. 2001). Together with the supposed coincidence of the Permian/Triassic events, this might indicate that the insects obey, to an extent, the global synchronisation of diversity dynamics.

Jurassic. The low origin and extinction rates observed in the Middle Jurassic may be caused by the relatively poor records of that stage. Being followed by the very rich Karatau fossil site in Kazakhstan, this probably has caused the apparent steep rise in the curves at the beginning of the Late Jurassic: the real pace can be smoother. The nature of the subtle Late Jurassic minimum of family number is unknown: its position is similar but not the same as that of the marine genera at the Jurassic/Cretaceous boundary.

Early Cretaceous. The main insect assemblages here are the Early and Late Neocomian, Aptian and Albian. Before the Albian, the curves follow the general Jurassic direction (excepting the Middle and beginning of Late Jurassic diversions). In the Albian, all the rates of diversification and of origins and extinctions distinctly decrease. The Albian minimum of diversity inferred earlier (ZHERIKHIN 1978, RASNITSYN 1988b, 1989a, DMITRIEV & ZHERIKHIN 1988) is not observed.

Late Cretaceous and Cainozoic. This part of the curve shows no noteworthy events, for the high Late Eocene step, and the comparatively modest Santonian one, most probably reflects the massive records from the Baltic and Yantardakh fossil resin collections, respectively. Unfortunately, the latest Cretaceous (Maastrichtian) record is too scanty to rule out any extinction simultaneous with that described for the marine, freshwater and terrestrial tetrapod assemblages at the Cretaceous/Cainozoic boundary (ALEKSEEV *et al*. 2001).

The process of formation of the contemporary fauna is clearly displayed by the inverse Lyellian curve (Fig. 480). At the family level, the changes can be traced only since the Late Permian until the beginning of the Cainozoic: the global fauna consisted of almost 90% of living families as early as the beginning of the Late Cretaceous and of more than 95% in the Santonian. The subhorizontal intervals in the beginning of the Triassic and Middle Jurassic reflect the poor fossil record rather than any real stasis; the same may be true for the Albian (in recent years the Aptian and earlier Late Cretaceous records were growing quickly unlike the Albian ones, so that the first and last appearances become more evenly distributed and not concentrated in the Albian, as before).

The above generalised picture can be completed with observations on the diversity trends in some particular insect taxonomic and ecological assemblages. The taxonomic variations (Fig. 481) are exemplified by the dictyoneuridean orders (*a*), non-heteropteran hemipterans (*b*), beetles (*c*), dipterans (*d*), hymenopterans (*e*), and moths and butterflies (*f*). Of them, the ecologically most versatile beetles give the curve (Fig. 481c) which most conforms with the general insect pattern, with all the Permian/Triassic and Triassic-Jurassic diversity minima and the Albian plateau and with no distinct event on the Cretaceous/Cainozoic boundary. Similarly, the beetle diversity path is sublinear, except for the Late Eocene step, caused by the exceptional Baltic amber record, and the Early and Late Cretaceous deviations from linearity which are less pronounced but more difficult to explain.

Another group demonstrating diversity dynamics similar to that of insects in general are the hymenopterans (Fig. 481e). The order is represented by only one family before the Jurassic so the diversity starts changing since that period. The decreasing diversification rate is appreciable between the Early and Late Cretaceous and a minimally expressed minimum at the Cretaceous/Cainozoic boundary. The higher diversification rate is registered towards the end of the Early Jurassic and at the Jurassic/Cretaceous boundary. The Late Eocene step is only moderately pronounced, probably because the hymenopteran fossil record is more complete before the Baltic amber time than that of the other orders under comparison.

N, families

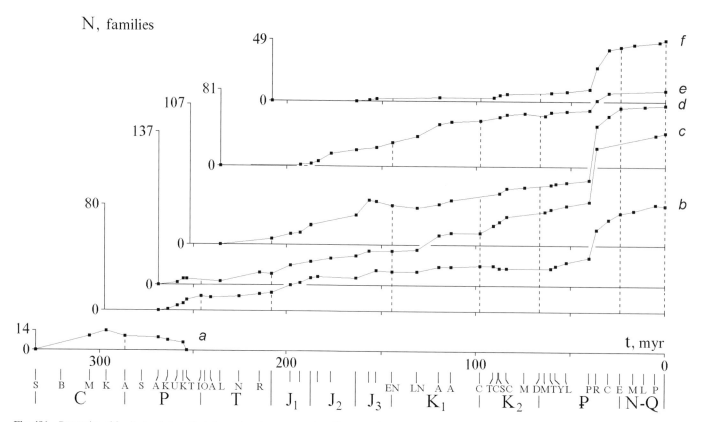

Fig. 481 Dynamics of family level diversity in larger insect taxa: *a* – palaeodictyopteroids (Dictyoneuridea), *b* – homopterans (Cicadina + Aphidina + Psyllina), *c* – beetles (Coleoptera), *d* – flies (Diptera), *e* – Hymenoptera, *f* – butterflies and moths (Lepidoptera). Because of some less complete material within particular groups, some curves lack various points present on previous figures

The homopteran and moth patterns differ more from the average. After their appearance in the Early Jurassic, the moths (Fig. 481*f*) persist as rare fossils for a long time. They become only slightly more diverse in the Late Cretaceous when the angiosperm plants began their spectacular diversification, and moth leaf mines do not appear to be uncommon. However, the really quick growth in moth and butterfly diversity starts only with the global thermal minimum and the appearance of open landscapes before the end of the Palaeogene. Unlike most other insects, at the end of Cainozoic the moth and butterfly diversification rate is still high and even higher than early in the Cainozoic.

The homopterans (Fig. 481*b*) first appear as early as the beetles, in the mid Early Permian, but initially they diversify more intensively. They show no or only slight diversity loss when the general insect curve goes down. In contrast, homopterans show very slight minima in the Early Triassic and earlier and later Late Jurassic. Events near the Cretaceous/Cainozoic boundary are insufficiently known. The group is tightly connected ecologically with plants and they deserve joint study.

The most distinct pattern involves the Diptera (Fig. 481*d*). Like the hymenopterans, they first appear in the Triassic, albeit slightly earlier. Their diversity was growing continuously since then; growth was particularly high in the Late Jurassic, and followed by a prolonged event of diminishing diversity towards the mid-Neocomian. This is followed with a slow but steady growth until the high Late Eocene step, and then, in the Oligocene, with an episode of quickly increasing diversity. No generally appreciated crisis is found to affect the dipteran diversification rate.

The dictyoneuridean diversification pattern (Fig. 481*a*) is limited to a part of the Late Palaeozoic (Serpukhovian through basal Kazanian) and is thus incomparable with the above examples. It forms a simple curve with the single maximum before the end of Carboniferous.

It can be concluded preliminarily that the general insect pattern may be better reproduced by groups of diverse ecology, while the ecologically more uniform taxa often display more specific patterns.

The last case considered is a comparison of the diversity trends in three ecologically distinct groups of phytophagous insects (Fig. 482). These groups are the dictyoneurideans that sucked the liquid content of gymnosperm ovules (*a*), the hemipterans that sucked the plant vascular content (*b*), and the taxonomically diverse assemblage of green mass consumers with chewing (instead of sucking) mouthparts (*c*). Most interesting is their comparison with the diversity of plants. BENTON (1993) describes the plants as diversifying very quickly during the Cretaceous. In contrast, the insect phytophages slowly increased their diversity. Possibly this is because the ecological role of angiosperm plants was growing slowly during the Cretaceous in contrast to their diversity. In the Mesozoic, green mass consumers (*c*) are far less diverse compared to the vascular suckers (*b*), and this gap becomes much more narrow during the Cainozoic. This indicates that the Mesozoic rarity of green mass feeders is probably a real feature of insect evolution and not a mere taphonomic effect. This confirms the earlier suggestion (ZHERIKHIN 1980a) that during the major part of plant history, their green mass was only weakly consumed by the insects.

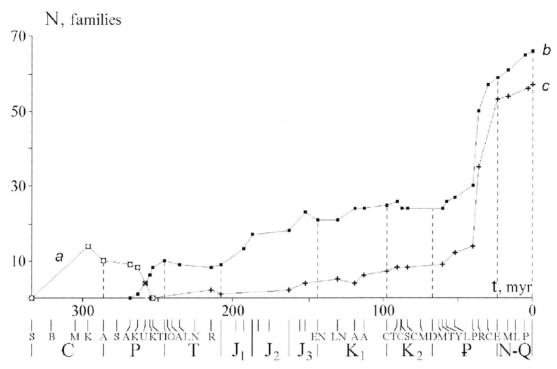

Fig. 482 Dynamics of family level diversity in selected diet-based subassemblages of phytophagous insects: *a* – gymnosperm ovules sucking palaeodictyopteroids (Dictyoneuridea), *b* – vascular content sucking homopterans (Cicadina + Aphidina + Psyllina), *c* – green mass chewing consumers (different orders)

3.2
Ecological History of the Terrestrial Insects

†V.V. ZHERIKHIN

3.2.1. INTRODUCTION

Ecological studies of the geological past are of great importance for evolutionary biology, palaeogeography, palaeoclimatology, etc, as well for ecology itself. Modern biology treats organic evolution as a significantly adaptive, and thus ecological, process. Phylogeny can be reconstructed and formally described on the basis of comparative anatomy but it cannot be interpreted in causal terms without the construction of adaptive, i.e. ecological, scenarios. Often the reconstructed phylogenetic pattern cannot be clarified without recourse to the history of other taxa. In fact, the ecological context of phylogenetic processes is much more substantive for their accurate interpretation than postulated geographical isolation which is the only extrinsic factor routinely (and often, perhaps, inaccurately) analysed in the majority of modern phylogenetic scenarios. Different members of the same taxon may differ considerably in their ecology. In this case their phylogenesis is managed by very different environmental factors, and its resulting adaptive pattern in each of the phylogenetically related but ecologically dissimilar lineages may be similar with those in other taxa, phylogenetically

unrelated but ecologically similar. It is self-evident that in respect to the leading evolutionary challenges and selective forces evolution of predacious aquatic beetles has much less in common with the evolution of terrestrial herbivorous beetles than with the evolution of the odonatans or aquatic bugs.

If an adaptation to the abiotic environment is a one-sided process, biotic interactions result in complicated co-evolutionary phenomena because in this case the adaptation is mutual. Even when the models of highly specific, strictly reciprocal long-term co-evolution in two ecologically connected lineages are often criticised, so-called diffuse co-evolution (JANZEN 1980) involving various multispecific interactions, both direct and indirect, is certainly widespread and important. It should be stressed that in the strongly co-evolutionary interdependent taxa (e.g. in a specialised predator/prey or pollinator/plant system) the co-evolution may be sufficiently inadaptive, with a rapid and very narrow mutual specialisation destroying the previous adaptive compromise system (after RASNITSYN 1987); in this case the extinction of both of the hyperspecialised lineages becomes highly probable. Any causal analysis of phylogeny might take co-evolution into account. An important special case of diffuse co-evolution which is often neglected is the phenomenon of the adaptive value being revised by natural selection (ZHERIKHIN 1978, ZHERIKHIN & RASNITSYN 1980) when an important novelty in one lineage stimulates evolution in a similar direction in many other taxa, mostly through competitive interactions but sometimes in other ways. The members of the same community are always at least somewhat coadapted, and the community is controlling and managing their own evolution in different ways. The palaeontological record shows that biological evolution is remarkably uneven in space and time both in respect of its rate and direction. As a rule, the changes are relatively slow, gradual and channeled towards increasing specialisation within the same principal niches but there are much shorter episodes when they become rapid and drastic (the coherent and incoherent evolution type, respectively, after KRASSILOV 1969). It is hypothesised that coherent evolution is under a strong community control while incoherent processes indicate that this time the coadapted community system has been destroyed (KRASSILOV 1969; ZHERIKHIN 1979, 1987, 1993b).

Palaeoecological considerations are based mainly on indirect evidence and are often started from the ecology of present-day taxa. This direct actualistic approach is the simplest one but it is also rather risky in many cases. Many evolutionary lineages either are completely extinct or include few modern relics which may well be ecologically exceptional, and the functional morphology of insects is not very well-developed leaving adaptive interpretation of too many structures equivocal. As a result, ecological evolutionary scenarios are highly speculative. To test them, NEL (1997) proposed a formal method of probablistic analysis of the distribution of ecological and other unknown characters in a cladogram. He illustrated his method by the example of the biology of the termite family Mastotermitidae (see Fig. 378). The only living species, *Mastotermes darwiniensis* Froggatt, inhabits the woodlands of tropical North Australia, and fossil mastotermitids are often considered as indicators of hot climate (see NEL & PAICHELER 1993). However, cladistic analysis has led Nel to the conclusion that there is not a good basis for such a suggestion even when fossils are assigned to the genus *Mastotermes*. In fact it means that we can say nothing about past climatic preference of mastotermitids.

The palaeontological record itself provides another, and in fact the only reliable possibility of testing palaeoecological hypotheses. Direct evidence such as food remains in fossilised gut contents or coprolites, pollen grains on insect bodies, or plant parasites attached to the host-plant remains, are uncommon. However, there are diverse sources of indirect palaeoecological information including morphological adaptations available on fossil remains, their consistent presence in association with other fossils, lithological and sedimentological features of fossiliferous rocks, pattern of fossil distribution within stratigraphic sequences and on broader geographic scale, etc. Both positive and negative data on fossil distributions are significant. For instance, regarding the above-mentioned example of the mastotermitids, an analysis of the distribution of their fossils shows that at least in the Tertiary they have to have been thermophilous independently of their cladistic affinities with modern *M. darwiniensis* because their finds are restricted to warm (at least subtropical) climatic areas while other termite families (Hodotermitidae, Kalotermitidae, Rhinotermitidae) penetrated much further into temperate regions. In this case the fossil distribution is in agreement with actualistic hypothesis. The snakeflies provide a different example. At present this order is best represented in temperate forest areas and penetrates to subtropical regions mainly along mountains; low temperature (about or below 0°C) is necessary for normal development (ASPÖCK 1998). On the contrary, the Mesozoic snakeflies (see Fig. 239) are much more common and diverse in warmer Euro-Chinese than in the temperate Siberian Realm, suggesting that extinct Mesozoic families differed in their climatic preferences from the modern Raphidiidae and Inocelliidae.

The palaeoecological importance of trace fossils should not be underestimated either; in fact, ichnological information is often crucial for testing palaeoecological hypotheses as discussed in more detail in Chapter 2.3. Taphonomic analysis (Chapter 1.4.2) of reliability and

robustness of palaeontological data is absolutely necessary to avoid serious misinterpretations. In this respect the importance of detailed sedimentological and quantitative data on fossil assemblages has to be stressed; unfortunately, this information has never been published, for too many insect sites including some of the most important Lagerstätten, and the documentation of many museum collections is not detailed enough.

Various ideas about the possible ecology of fossil insects are scattered in the old palaeoentomological literature, and speculative scenarios of ecological evolution became popular in biology after Darwin and Haeckel. However, for a long time the palaeontological data were used in such scenarios only occasionally, at least in entomology. PANFILOV (1967) was the first author who made an attempt to outline a brief general ecological history of insects. His analysis was based nearly entirely on actualistic interpretation of the fossil record of different taxa; many of his conclusions are now considered inconsistent but others have been supported by more recent and diverse sets of evidence. ZHERIKHIN (1980a) gave a more detailed review, but rapid progress in insect palaeoecology in the last decades makes a new synthesis necessary. An important difficulty in this respect is that such a synthesis cannot be limited to palaeoentomological information, but needs the critical analysis of a huge amount of palaeontological and sedimentological data as well. The best and most recent attempt at such an analysis is by BEHRENSMEYER et al. (1992) and that is used here as a general framework for the environmental context of insect evolution. The following is not more than a preliminary outline of the present-day data on the world scale, and more detailed comprehensive analysis is still a matter of future.

3.2.2. PALAEOZOIC

3.2.2.1. Environments of insect origin and early evolution

Early insect history is nearly undocumented. None of the arthropods known from the Palaeozoic could be interpreted as representing a probable ancestral lineage for the insects. A pair of dicondylous mandibles named *Rhyniognatha praecursor* Tillyard has been described from the Early Devonian (Siegenian, 394 to 401 My) Rhynie Chert in Scotland where other arthropods including mites and springtails have been found as well (TILLYARD 1928). This unique specimen may indicate the first insect appearance in the fossil record but it can be doubted for two reasons. Firstly, it may well belong to a different mandibulate arthropod most of which have their mandibles dicondylous (KLUGE 1996). Secondly, the alleged find of a tubuliferous thrips nymph in the same deposit indicates a possibility of much younger (probably Tertiary) contaminants in the chert (CROWSON 1985; but see GREENSLADE 1988). Rhynie is one of the most important Lagerstätten of non-marine Devonian biota, and the absence of unequivocal insect finds here seems to be significant. The Rhynie fossils are preserved in permineralised peat-like matter and represent the likely fauna of semiaqautic vegetation episodically affected by hot spring waters rich in silica (KÜHNE & SCHLÜTER 1985). In the younger Early Devonian (Early Emsian, 390 to 392 My) Battery Point Formation at Gaspé in Quebec, Canada, the head and thorax of a bristletail *Gaspea palaeoentognathae* Labandeira, Beall et Hueber has been found (LABANDEIRA et al. 1988). The remains were found in a clayey layer within a fluvial sequence. Another supposed bristletail fragment was discovered together with diverse other terrestrial arthropods (scorpions, pseudoscorpions, mites, trigonotarbids, spiders, centipedes) in the Late Devonian (Givetian, 376 to 379 My) deltaic Panther Mountain Formation at Gilboa in New York, USA (SHEAR *et al.* 1984). It should be noted that all these Devonian finds occur in some exceptional taphonomical conditions in association with other extremely rare fossils. Another important thing is that all of them are restricted to Atlantic North America and West Europe which at that time would have constituted a single continent located within the equatorial belt.

The genera *Eopteron* Rohdendorf and *Eopteridium* Rohdendorf originally described as primitive winged insects from the Late Devonian of Russia and Ukraine have been found to belong to Crustacea (ROHDENDORF 1972). A number of pre-Carboniferous trace fossils have been attributed to the non-marine arthropods (e.g. H. SCHMIDT 1938, P. KEVAN *et al.* 1975, POLLARD *et al.* 1982, POLLARD & WALKER 1984, ROLFE 1985, SHERWOOD-PIKE et GRAY 1985, HOTTON *et al.* 1996, BUATOIS *et al.* 1998) but none of them are interpreted with any certainty as being made by an insect.

In the Early Carboniferous, insects still remain virtually absent in the fossil record. SCHRAM (1983) mentioned a "*Rochdalia*-like insect" from the shallow-marine Visean (333 to 349 My) deposits of Glencartholm in Scotland without any description or illustrations of the fossil. This record is most probably a misidentification (A. Rasnitsyn, pers. comm.). According to BRAUCKMANN *et al.* (1996), the only unequivocally known Early Carboniferous insect is *Delitzschala bitterfeldensis* Brauckmann et Schneider from the very end of the Early Carboniferous of Germany (early part of Namurian A, Arnsbergian Stage, 320 to 322 My). Morphologically *Delitzschala* is a typical member of the order Dictyopterida (see Fig. 136), and there is a good basis to consider this group as specialised plant-suckers inhabited mainly tree crowns. In the Carboniferous, Bitterfeld was situated within the tropical belt.

Thus there are few points about early insects that are certain at present: that they already existed in the Early Devonian; that they are extremely poorly represented in the fossil record and restricted to exceptional preservation contexts; that they are not found outside of equatorial areas; that the winged insects originated no later than in the latest Early Carboniferous; and that the earliest winged insect known was probably a plant-sucker.

The very scarcity of the early insect record is ecologically significant, especially if the rather high diversity of other Palaeozoic non-marine arthropods is taken into account. It permits us to reject any possibility of primarily marine or lacustrine insect habits because animals with well-developed external skeleton have a high preservation potential in such environments. The oldest insects have to have dwelled in certain taphonomically strongly unfavourable habitats, i.e. strictly terrestrial, running-water, or tidal. This palaeontologically cryptic stage of evolution has to have been rather long, covering not only the early evolution of the insects themselves but also of their immediate ancestors. None of the above three possibilities can be excluded definitively at present; however, the running-water hypothesis seems to be the least, and the terrestrial scenario appears to be the most plausible. Indeed, the watercourse environment is highly specific in many respects, and its inhabitants usually demonstrate a number of peculiar morphological adaptations. The probability of their burial decreases with increasing current speed but also the morphological specialisation increases in fast-running waters. No primitive insects, either extinct or extant, are known to show any features that could be regarded as having been inherited from a specialised rheophilous ancestor. There are few well-documented examples of rheophilous

insects that have secondarily developed a terrestrial mode of life (e.g. some gripopterygid stoneflies). However, in these cases the derived descendants of aquatic lineages retain significant morphological features of their aquatic ancestors; they are in fact semiaquatic rather than truly terrestrial animals. Thus running-water communities seem to be highly conservative and their composition changed very slowly in the geological time; so the probability of complete extinction of a diverse rheophilous lineage without leaving any derived aquatic descendants is low.

A tidal hypothesis is more difficult to test because many inhabitants of the littoral flotsam are morphologically similar to terrestrial forms and some of them probably gave rise to true terrestrial lineages in a way virtually undocumented palaeontologically (e.g. in isopod crustaceans: CLOUDSLEY-THOMPSON 1988). A tidal scenario was discussed recently for the insects by TSHERNYSHEV (1990, 1997). The most reliable way for testing this hypothesis palaeontologically seems to be ichnological studies because diverse trace fossils are not rare in littoral palaeoenvironments. At present, the available information is not enough for any definite conclusion. Tshernyshev supposes that bristletail springing could originally be an adaptation allowing escape from wave action like in the present-day littoral amphipod crustaceans. This specialised mode of locomotion is probably a derived one among the wingless insects, and crawling, like in the silverfish, was most probably plesiomorphic. The terrestrial hypothesis of insect origin seems to be the most plausible. It is in accordance with the palaeontological data and in particular with the presence of other terrestrial organisms in all Devonian fossil assemblages containing insects; it does not suggest rapid extinction of ancestral insect lineages still in the Palaeozoic; and it does not require any hypothetical ancestral specialisation that left no evident traces in the organisation of any known primitive insects. As supposed by GHILYAROV (1949, 1956), the most probable habitat of ancestral insects was wet litter or soil; both are highly unfavourable taphonomically.

Not so long ago it seemed that the Devonian land was still weakly colonised by organisms. However, the large and rapidly growing body of data, accumulated in the last decades indicates that by the time of the first insect appearance in the fossil record, terrestrial life was diverse (for reviews see SHEAR & KUKALOVÁ-PECK 1990, DIMICHELE & HOOK 1992; D. EDWARDS & SELDEN 1993). The following principal milestones of early terrestrial life record are worthy of mention here to outline the ecological context of insect origin and their early evolution.

The oldest tentative evidence of any terrestrial life is the presence of primitive palaeosols supposedly affected by subaerial microbial activity in some Praecambrian sequences with an absolute age up to 3 billion years (RETALLACK *et al.* 1984; RETALLACK 1990b). More developed palaeosols have been discovered in the Late Ordovician, and the burrows in the latest Ordovician (about 440 My) palaeosol of Pennsylvania constructed supposedly by earthworms or myriapods constitute the oldest putative evidence of terrestrial Metazoa (RETALLACK & FEAKES 1987). Another important evidence of progress in terrestrial communities is the appearance of plant spores that are resistant to desiccation in the Late Ordovician and Early Silurian (GRAY 1985, W. TAYLOR 1995). The trackways of myriapod-like creatures have also been discovered in the Late Ordovician (about 440 My) of England (E. JOHNSON *et al.* 1994).

The Mid- to Late Silurian record provides evidence of progress in terrestrial ecosystem organisation. Terrestrial plant radiation is documented by increasing diversity of spore types (GRAY 1985), by plant macrofossils (D. EDWARDS & FANNING 1985; D. EDWARDS & SELDEN

1993), and by better developed palaeosols with alleged rhizome traces (RETALLACK 1990b). The first appearance of undoubted vascular plants is the most important Late Silurian event (P. KEVAN et al. 1975). The tentative ascomycete hyphae and spores of supposedly terrestrial fungi (SHERWOOD-PIKE & GRAY 1985, T. TAYLOR & TAYLOR 1997), and the oldest fossils representing undoubted arachnids and myriapods (D. EDWARDS & SELDEN 1993) have been discovered in the Late Silurian. The vegetation was patchy, with denser cover along stream channels (DIMICHELE & HOOK 1992). It should be stressed that all supposedly terrestrial invertebrates found in the Late Silurian were likely to have been predators; this suggests an existence of an unknown detritivore and/or grazer fauna (DIMICHELE & HOOK 1992) which could include among others also early insects or their ancestors.

In the Early and Middle Devonian the terrestrial plants were rather diverse. Several major groups of vascular plants were appearing, namely the zosterophylls, trimerophytes, horneophytes, barynophytes and lycopsids (MEYEN 1987a), and more ancient plant lineages still survived including the primitive rhynians related to the Silurian genera, and even the hepatic-like producers of the primitive spore types known since the Ordovician (D. EDWARDS et al. 1995). This suggests that the older ecosystems had probably not disappeared but existed perhaps as initial successional stages in more complicated successional systems. The lichens are also found (T. TAYLOR et al. 1997), and the discovery of arbuscules in Aglaophyton documents complex interactions between arbuscular endomycorrhizal fungi and rhynian plants (T. TAYLOR & TAYLOR 1997). However, the local plant diversity was low: even in the polytopic assemblages no more than 10 to 15 plant genera are represented, and normally only 1 to 5 species occur together (DIMICHELE & HOOK 1992). This vegetation still occupied mainly wetlands and areas along streams while watershed areas were probably barren or nearly so. However, RETALLACK (1981a, 1990b) points out that there is less evidence of soil development interruption by clastic sedimentation than in older palaeosols suggesting that the stabilising effect of vegetation has somewhat increased in comparison with the Silurian.

The terrestrial animals were represented there mainly by arachnid and myriapod-like arthropods. Taxonomy and palaeoecology of early myriapods are disputable; in particular, their terrestriality as well as their alleged relations to Diplopoda may be questioned in many cases (ALMOND 1985). Like in the Silurian, the oryctocoenoses were numerically dominated by predators suggesting that the original community structure is inadequately portrayed by fossils (P. KEVAN et al. 1975, SHEAR & KUKALOVÁ-PECK 1990, DIMICHELE & HOOK 1992). Trace fossils support this idea. In Rhynie, plant remains demonstrate occasionally traumatic tissues and other injuries caused rather by biological than by physical agents (P. KEVAN et al. 1975). Some of the pathologies are tentatively interpreted as connected with secondary fungal infection. Similar injuries were discovered on trimerophytes in the Battery Point Formation in Canada (H. BANKS 1981; H. BANKS & COLTHART 1993). The trauma makers are unknown. The mites from Rhynie are more likely to have been predators or fungivores than herbivores (D. EDWARDS & SELDEN 1993), and the springtails belong to Entomobryioidea, a superfamily that does not include plant feeders (GREENSLADE & WHALLEY 1986). P. KEVAN et al. (1975) supposed that trigonotarbids could have been at least facultat sap-suckers or spore-eaters but this hypothesis has been rejected by later authors (ROLFE 1985; SHEAR et al. 1987; SHEAR & KUKALOVÁ-PECK 1990; D. EDWARDS & SELDEN 1993). Rather large (5 to 12 mm long) arthropod coprolites containing plant cuticle were found in the Battery Point

Formation as well; they were tentatively attributed to detritivorous millipedes (HOTTON et al. 1996) but in fact there is no firm basis for any attribution at present. Smaller Early Devonian coprolites from Wales contain spores, sometimes damaged (D. EDWARDS & SELDEN 1993). Some morphological structures of Devonian plants are tentatively interpreted as protective against herbivores (P. KEVAN et al. 1975).

In the Late Devonian and Early Carboniferous there were further important changes in the terrestrial ecosystems. As in the Early and Mid-Devonian, new community types appeared in addition to those previously existed (most probably as later stages of successions) rather than replacing them totally. As a result the structural complexity of terrestrial ecosystems increased considerably without significant extinction of earlier types. The most obvious Late Devonian novelties were the appearance of tall plants forming true forest vegetation (with the trunk diameter reaching 1 meter in the progymnosperm Archaeopteris) and the oldest terrestrial vertebrates (DIMICHELE & HOOK 1992). Forest vegetation was probably restricted to waterlogged soils along streams and on wet floodplains. Both novelties were probably stimulated partly by the progressive development of a terrestrial arthropod fauna and reciprocally affected its evolution. As concerning the origin of trees, P. KEVAN et al. (1975) speculate that besides plant competition for light the spore-feeding arthropods could have stimulated an elevation of sporangia high above the earth surface. The forest communities with increased litter production provided a valuable food source for terrestrial invertebrates, thus allowing a considerable increase in their abundance and biomass. It is well known that living plant tissues, except reproductive parts, are generally poor in nitrogen, and that detritivory on dead plant matter already processed by fungi and bacteria suggests less trophic specialisation than herbivory. Thus the early terrestrial invertebrates should have been predominantly opportunistic detrite-feeders.

The role of terrestrial arthropods and other invertebrates as the key food resource allowing the terrestrialisation of vertebrates is evident, and perhaps an increasing availability of this resource in the Late Devonian forest floor facilitated this process significantly. On the other hand, vertebrate predation on the land surface could have stimulated arthropods moving to tree crowns. The effect was increasing desiccation resistance as well as adaptation to feeding on plant reproductive organs, the only resource of a high nutritive value available for non-predacious animals outside the area of high predation risk. The probability of burial of non-flying crown-dwellers should be very low, even less than for the litter inhabitants in riverine gallery forests. Surely, there were also other protective strategies against vertebrate predation such as cryptic mode of life in the litter and soil, increasing body size (for instance in the Arthropleurida, an extinct myriapod lineage represented by small-sized forms in the Early Devonian but giants in the Carboniferous), development of spines on the body, etc. Myriapods and arachnids which are rather well represented in the Early Carboniferous deposits demonstrate diversity in their presumed protective adaptations. An important piece of evidence for the use of food substrate as a shelter is the presence of rather large tunnels in an enigmatic Devonian organism Prototaxites, perhaps belonging to an extinct lineage of fungi, in the Late Devonian Kettle Point Formation of Canada (HOTTON et al. 1996). The tunnels contain coprolites with fungal remains and were probably constructed in the lifetime of the host. The tunnel maker is unknown and probably represented an extinct mycophagous arthropod lineage.

Other noteworthy Late Devonian and Early Carboniferous events were an expansion of seed plants (the oldest fossil assemblages dominated by seed plants have been discovered in the latest Devonian of

Ireland: DiMichele & Hook 1992); progressive development of heterospory (accompanied probably by an increase in the nutritive value of sporangia: P. Kevan *et al.* 1975); first appearance of the fully terrestrial vertebrates, including the earliest amniotes, in the Visean (the East Kirkton fauna in Scotland: DiMichele & Hook 1992); and progressive geographic differentiation of non-marine biotas. There is no evidence for significant biogeographic differences in land floras before the latest Devonian, but in the Early Carboniferous the differentiation was well established between Euramerica, Angara and Gondwanaland (Meyen 1987a). In particular, the forest vegetation was well developed in Euramerica but not in Angaraland where no plants with large pycnoxylic stumps are known since the Late Famennian up to at least the Late Namurian (Meyen 1987a). The basic ecological differentiation of plant assemblages, with the lepidophyte-dominated ones in the coal-bearing deposits of wetlands (the anthracophilous flora) and the fern- and pterydophyte-dominated assemblages in coal-less deposits that accumulated in drained environments (the anthracophobous flora), is typical of the Middle and Late Carboniferous and has been fully established by the Late Tournaisian (DiMichele & Hook 1992). Meyen (1987a) pointed out that the first appearance of inland coal-bearing deposits in the Tournaisian may indicate an increase in anti-erosional effect of slope vegetation which has reduced the transport of clastic material to the lowlands.

The absence of insect finds in the Early Carboniferous (except its latest part with the first albeit advanced pterygotans) suggests that the pterygotan ancestors continued to evolve in a taphonomically unfavourable environment. The most plausible environment for this most important evolution event over the whole history of insects seems to be tree crowns. The probability of burial of non-flying tree-dwelling arthropods in sedimentary rocks is extremely low even in a near-water forest, but should increase significantly with the acquisition of flight. In the very end of the Early Carboniferous, winged insects are entering to the fossil record and then are continuously present as fossils in diverse sedimentary environments. Shear & Kukalová-Peck (1990) postulate that the main radiation of winged insects must occur well before and that their sudden appearance in the fossil record has to be explained by increasing preservation opportunities in extensive swamps developed in the Late Carboniferous landscapes. However, if this hypothesis is correct, it is difficult to explain why the winged insects have appeared in the fossil record as late as in the Namurian and not in the Tournaisian. Moreover, insect fossils are in fact far from being restricted to swamp palaeoenvironments, and it is difficult to suggest a long-time cryptic existence of winged insects if they did not originally inhabit watershed habitats distant from any water bodies. As discussed below, the frequency of their discoveries is not constant in the Middle and Late Carboniferous when swampy forest environments were constantly widespread but increases continuously since its beginning, and this pattern indicates a significant ecological expansion and diversification of insects from the Namurian to the latest Carboniferous.

It should be stressed also that the recently proposed model of flight origin from aquatic gliding (Marden & Kramer 1994, Marden 1995, Ruffieux *et al.* 1998), theoretically elegant and welcomed by some authorities, is not supported by any palaeontological data; moreover, it is in an obvious controversy with the fact that their origin is undocumented by the palaeontological record (Chapter 2.2e).

Hence the following speculative scenario of early insect evolution looks most plausible in the light of the current palaeontological data (see also Chapter 2.2e). The insects originated in terrestrial environments from a terrestrial ancestor not later than in the Late Silurian or

the earliest Devonian. Up to the Late Devonian they constitute no more than a rare and perhaps weakly radiated group of small-sized myriapods, probably mainly detritivorous, inhabiting low vegetation and litter and restricted to terrestrial habitats. In the Late Devonian they diverged into two main ecological lineages. One retained plesiomorphic feeding (detritivory) and evolved as forest floor inhabitants; there were no ecological prerequisites for development of flight in this lineage. Another line has colonised tree crowns and evolved there as spore-feeders. The sporangia provided valuable (especially after appearance of the heterospory: P. Kevan *et al.* 1975) but sporadically distributed resource. Seeds constituted another important food resource which became more and more important in the course of expansion of seed plants, the more so that plesiomorphic gymnosperm seeds placed on leafless axes and free of any covering fruit structures were easily available for seed predators. A progressive development of hard seed coats and sclerotic tissues in pollen organs during the Early Carboniferous is in accordance with the hypothesis of arthropod consumption (DiMichele & Hook 1992).

The search for food and perhaps also escaping predators (if some predatory arthropods followed the insects to tree crowns) stimulated locomotory evolution from crawling to jumping, with the lateral segmentary flaps acquired to help in attitude control while jumping from one tree branch to another (Rasnitsyn 1976, 1980). Those lateral projections could have become moveable well before the origin of true flight. The absence of similar locomotory adaptations in any recent tree dwellers is not a serious argument against this model because they could be effective only in the absence of any highly active flying predators. It should be stressed that this ecological scenario is compatible with any hypothesis of wing homology. In particular, it did not exclude the serial homology between the wings and the abdominal gills (tergaliae) of the mayfly nymphs as supposed by Kukalová-Peck (1983) and Kluge (1989) but suggests that the gill function has to be secondary. The ancestral pterygote lineage was probably restricted to the equatorial belt of Euramerian land mass.

3.2.2.2. Middle Carboniferous to Permian

3.2.2.2.1. General features of Palaeozoic insects

Though significantly different from each other in their taxonomic composition, the Middle and Late Carboniferous and Permian insect faunas represent the same stage of ecological evolution, with a constant general trend to continuous ecological expansion and adaptive radiation. This trend becomes most evident when the Carboniferous and Permian faunas are considered together.

In the Middle Carboniferous insect fossils become diverse and widespread for the first time in the palaeontological record. As pointed above, the only insect known from the Early Namurian A (Early Carboniferous), *Delitzschala* (see Fig. 136), belongs to the Dictyopterida, an order that is well represented later in the Carboniferous and that is relatively advanced. This indicates (even if the alleged older find of *Rochdalia* in Scotland is not taken into account) that the pterygotan fauna of the latest Early Carboniferous should be in fact of the Middle Carboniferous type and might include some representatives of at least paoliid, caloneurid and hypoperlid lineages as well. Thus the Late Palaeozoic stage of insect evolution started in fact somewhat before the beginning of the Middle Carboniferous.

The Carboniferous insects are known from all present-day continents except Antarctica but mainly from West Europe and eastern North

America which were situated that time in the equatorial belt and had a direct contact with each other. Of course, it may be partly a matter of uneven state of knowledge of fossils in different territories, but anyway the repeated special search for Carboniferous insects in Eastern Europe, Caucasus and Siberia was not very productive suggesting that the extratropical areas really had an impoverished fauna. It cannot be ruled out that in Gondwanaland and especially in Cathaysia (present-day East Asia) the effect of insufficient knowledge is more significant. The Namurian finds are particularly rare. There are seven Namurian insect localities in Europe (in The Netherlands, Belgium, Germany, Poland, and the Czech Republic) as well as one in Utah, USA, and most of them have yielded only one or two specimens each (BRAUCKMANN et al. 1996). Recently insects have also been recorded from the latest Namurian of Northern China (HONG 1998). If their geological age is determined correctly, this is the first Namurian insect locality outside Euramerica; it is noteworthy that the area was situated in the Carboniferous in low latitudes as well. Since the Westphalian the insects occur also in Angaraland (present-day northern Asia) and Gondwanaland (mainly in South America but there are few records from the latest Carboniferous of Africa, Madagascar, Hindustan, and Tasmania as well). However, throughout the Middle and Late Carboniferous West Europe and North America remain to be far more productive in fossil insects than other territories.

The Permian situation is clearly different: the insects are known worldwide including the Antarctic, but East Europe and Asia provide considerably more localities than any other region. In Gondwanaland, Australia is richest in Permian insect finds, followed by South America and Africa; few insects were recorded from Antarctica and Hindustan. Cathaysia still remains the least studied area though a number of localities has been discovered in China during last two decades.

The principal Carboniferous insect Lagerstätten are Hagen-Vorhalle in Germany (Namurian B), Mazon Creek in Illinois, USA (Westphalian D), and Commentry and Montceau-les-Mines in France (Stephanian). They provide very large and taxonomically diverse material but except for Hagen-Vorhalle no full quantitative data have been published, and many taxa badly need revision. The taxonomic situation in roaches is a good illustration. After a revision of Palaeozoic roaches (SCHNEIDER 1977–1984) the number of species and genera has been reduced dramatically, with sometimes as many as 58 synonyms for a single species. It has also been demonstrated that some species are common for different localities in Euramerica, sometimes from both present-day Europe and North America. Nevertheless, in Commentry and Mazon Creek, the number of insect species is clearly more than 100 in each locality which is in a strong contrast with even the richest Namurian locality, Hagen-Vorhalle, with its less than 20 species (ZHERIKHIN 1992b). All localities known in Angaraland are post-Namurian in age, and none of them yield highly diverse faunas. The richest collection from Chunya in the Tunguska Basin includes more than 200 specimens but only about two dozen species, and the dominance pattern is sharply pronounced, with about 50% of specimens belonging to the dictyoneurid *Tchirkovea*, about 25% to the eoblattid *Narkemina*, and about 14% to undescribed but likely rather uniform roaches (ZHERIKHIN 1980a). The assemblages known correspond to different lowland palaeoenvironments but particularly the swampy forests of the equatorial belt.

The most noticeable Permian insect Lagerstätten are Obora in the Czech Republic (Early Sakmarian or Early Artinskian, see Chapter 4.1), Elmo in Kansas, USA (Late Artinskian), Tshekarda in the Urals, Russia (Kungurian), Kaltan (Ufimian) and Suriekovo (Kazanian) in the Kuznetsk Basin, West Siberia, Russia, Soyana in the Arkhangelsk Region, north-eastern European Russia (Kazanian), and Belmont in Australia (Tatarian). There are many other important sites in different parts of the world. Few quantitative data are available for the Permian assemblages. The local diversity of insects can reach several hundreds as in Tshekarda (NOVOKSHONOV 1998c; the evaluation about 1000 species in ZHERIKHIN 1992b, is probably somewhat overestimated). The assemblages known characterise a wide habitat range including inland volcanic areas (e.g. in the Tunguska Basin, Siberia).

Nearly all known Carboniferous and Permian insects are pterygotans. The only apterygotan group which is relatively common in the fossil assemblages is the extinct and supposedly aquatic or at least semiaquatic archaeognathan family Dasyleptidae (see Fig. 62). Regarding the terrestrial fauna two alleged apterygotans should be noted, *Carbotriplura* Kluge (see Fig. 59) from the Westphalian of Czech Republic and *Ramsdelepidion* Kukalová-Peck (see Fig. 60) from Mazon Creek. Both are unusual in having large body size and demonstrate some other characters that question their close relation to any modern apterygotans (Chapter 2.1a). Their affinities with the "propterygotan" lineage of flightless tree-dwellers cannot be ruled out. It is highly probable that this lineage (or, rather, some of its side-branches), which might be strongly underrepresented in the fossil record, did become extinct with the appearance of the true pterygotans but survived for some time in the Middle/Late Carboniferous.

The pterygotans are represented in the Late Carboniferous and Permian by no less than 27 orders nine of which are confined to the Palaeozoic and three more disappeared from the fossil record later, in the Mesozoic (including Grylloblattida which is exceptional in being represented in the modern fauna with a single relict family). Only three orders, Ephemerida, Corydalida and Trichoptera, had their immatures surely or probably aquatic, even if in the case of the Palaeozoic odonatans a terrestrial mode of life is much more probable than an aquatic one. The taxonomic position of many Palaeozoic, and especially Carboniferous, insects is unclear, and the real number of extinct orders may well be more than known. Four Carboniferous orders (Paoliida, Syntonopterida, Triplosobida, and Eoblattida) are not found in the Permian, the two former being unknown even from the Stephanian, and eleven have entered the fossil record in the Permian for the first time. At the family level the system is unsatisfactory in too many cases, and it is hardly possible to estimate the family turnover from the Carboniferous to the Permian accurately. The scarcity of data on the earliest Permian also obscures the extinction pattern. However, total taxonomic richness is increasing in the Early Permian if newly appearing orders are taken into account. Thus there is no basis to postulate any mass extinction event among the insects near the Carboniferous/ Permian boundary.

This obvious increase in the total insect diversity from the beginning of the Middle Carboniferous to the Permian can hardly be explained by taphonomical factors only but should reflect a real diversification process. Another evident trend is a rapid geographical expansion of insects since the Namurian to the Westphalian and Stephanian. This fact seems to be especially remarkable if the general physiographic context is taken into account. The Late Palaeozoic was the time of progressive integration of the lithospheric plates culminating with the formation of the supercontinent Pangea to the end of the Permian when the Cathaysian islands remained the only relatively large land areas lacking any direct connection with it. However, in the Middle and Late Carboniferous this integration was still far from its culmination, and thus the dispersal of the winged insects suggests that since their early times they could cross sea barriers effectively. On the other hand, the vegetation provincialisation increased at the global scale from the Early Carboniferous up to the Late Permian when it has reached a level only comparable with the Late Tertiary (DiMICHELE & HOOK 1992). This provincialisation pattern was a result of an increasing latitudinal climatic gradient which had led to large-scale glaciations in the polar regions and to increasing aridity in low latitudes in the latest Carboniferous and Permian. The intra- and intercontinental insect expansion coincided in time with this increasing in global environmental heterogeneity and thus suggests a large-scale adaptive radiation. In the latest Carboniferous deposits an insect, the palaeodictyopteran *Psychroptilus*, is discovered in the southern polar region in glacial lake sediments of Tasmania. Another late Carboniferous or, perhaps, earliest Permian insect originated from a subpolar area is the Mischopterid *Aykhal* from north-western Yakutia, Siberia (see Fig. 151). Interestingly, both genera belong to families of their own and are not found in lower latitudes while presently the arctic insect faunas are practically devoid of endemic families. The pterygotan expansion from equatorial lowlands up to the most severe periglacial environments has taken about 40 Ma – a time comparable with the average longevity of insect genera in the Tertiary.

The plesiomorphic pterygotan habits can be hypothesised mainly on the basis of comparison between the earliest known representatives of the principal pterygotan lineages. As discussed above, the ancestral pterygotans probably dwelled in tree crowns. If the hypothesis of flight origin accepted here is correct they should be rather large because in this case gliding from branch to branch has to be more effective (Chapter 2.2e). Indeed, one of the most striking features of the Carboniferous insects is their relatively large average size. The small-sized pterygotans (with the wing length 10 mm or less) are quite rare (*Metropator* is one of few examples, see Fig. 114). On the other hand, insect gigantism in the Carboniferous is a widely known but not a very common phenomenon, and there are few truly giant insects (with the wingspan up to several dozen centimetres) mainly restricted to the Stephanian. In the Permian the small-sized insects become more numerous including some homopterans and lophioneurids with the wing length about 2 or 3 mm only. They occur since the Early Permian, and are not rare in the Kungurian of Tshekarda. Thus the general pattern of Palaeozoic evolution in this respect was likely an increasing diversification in body size starting from the moderately large ancestral pterygotans. The decrease in size is profitable in diverse aspects including increasing possibilities to exploit diverse microhabitats for shelters, feeding, oviposition, etc. The main selective advantage of increasing body size is likely the decreasing predation pressure but for the terrestrial arthropods even the maximal possible body size (which is limited both by increasing weight of the exoskeleton and the oxygen diffusion rate in the tracheal system) is still too small for effective protection from larger vertebrates. That is probably why the decrease turned out to be the dominant trend while the increase occurred sporadically in some lineages only and was often reversed.

An intriguing aspect of Palaeozoic arthropod gigantism is the problem of atmospheric chemistry. It was suggested recently that it could indicate hyperoxy, and that the presence of unusually large insects and myriapods in the Late Palaeozoic supports the model of the oxygen concentration pulse with the highest peak (up to 35% in comparison with the present 21%) in the Late Carboniferous and subsequent reduction in the Permian (J. GRAHAM *et al.* 1995, DUDLEY 1998). This model is based on the indirect geochemical estimates and suggests that the oxygen level since the Late Permian up to Cretaceous was lower than now, with the Early Triassic minimum estimated to be about 15% and with only a weak increase by the Late Triassic. However, there is no evidence that the Mesozoic insects were on average smaller than the Cainozoic ones, and in fact some Late Triassic Mesotitanida (see Fig. 461) are well comparable in their body size with the largest Carboniferous pterygotans; the table 1 in DUDLEY (1998) is clearly erroneous in this respect. Thus either the model of the oxygen pulse accepted in the above-mentioned papers is basically inaccurate, or the atmospheric oxygen level was not the principal factor limiting the actual insect body size, and there are other factors which rarely or never allow the insects to reach the maximal body size potentially possible.

Continuous progress in insect flight ability is well documented by fossils. Such easily observable morphological trends as the reduction in the number of longitudinal veins, loss of cross-veins, costalisation, narrowing of the wings and especially of their bases, etc., were developing in different lineages in a parallel way (compare, for instance, Figs. 104, 147 and 156). The reduction of the hind wing pair up to its complete disappearance in Eukulojidae and Permothemistidae (see Figs. 146, 154) is particularly noticeable. But perhaps the most important advance was connected with the evolution of moulting. Archemetaboly (very gradual metamorphosis when the imago is not the only flying stage but one or more subimaginal stages occur as well) is documented in several unrelated Palaeozoic insect taxa including Evenkidae (see Fig. 67), Caloneuridea (e.g. in *Herdina*: SHEAR & KUKALOVÁ-PECK 1990, see Fig. 113), diverse Dictyoneuridea (KUKALOVÁ-PECK 1978, 1983, SINITSHENKOVA 1979, SHEAR & KUKALOVÁ-PECK 1990), Eoblattida (see Fig. 356) and Grylloblattida (see Figs. 387, 388). In the fossil state archemetaboly can be recognised on the basis of the articulated wings in specimens looking like late instar nymphs (see Fig. 71a,b), as well as by the more stout and less sharply delimited veins and less sharp colour pattern on the subimaginal wings (see Figs. 67, 71c, 356), and occasionally directly by finds of moulting specimens. The only living insect order with two flying stages (one subimaginal and one imaginal) is Ephemerida, and thus this state might occur also in the ancestral Libelluliformes. Archemetaboly (with more than one subimago: Fig. 392) is surely plesiomorphic for the pterygotans, and its taxonomic distribution suggests that it has been lost independently in each of their main lineages (Libelluliformes, Cimiciformes, Scarabaeiformes, and Gryllones).

The advantage of the restriction of flight to the single definitive stage is obvious because in order to moult a subimago needs living hypodermic tissue inside the wings; consequently the subimaginal wings are heavy, and their flight is energetically expensive, slow and weakly manoeuvrable. The loss of the subimaginal stages occurred likely in nearly all major pterygotan lineages including some extinct ones. It cannot be excluded that the reduction of subimago has been realised in different ways – in some lineages through its complete loss and in others through its immobilisation; in particular, the homology of the holometabolous pupa with the subimago cannot be ruled out. Even in the Carboniferous, archemetaboly is unequivocally demonstrated for relatively few taxa, and as early as in the Early Permian it has been lost probably in all main orders except for the mayflies, the dictyoneuridean orders and the least advanced grylloblattidan suborder Lemmatophorina. The decreasing frequency of fossil finds of nymphs and isolated nymphal wings also suggests the shift of flight to the later ontogenetic stages in the majority of the Permian insects. This trend is so universal and the evolution in this direction so rapid in many taxa that the selection pressure must have been remarkably high. It hardly can be explained simply by a competitive advantage due to more effective dispersal or something like that. The only selective factor effective enough to induce this pattern seems to be the pressure of the first flying predators, the odonatans (RASNITSYN 1976, ZHERIKHIN 1980a, SHEAR & KUKALOVÁ-PECK 1990, WOOTTON & KUKALOVÁ-PECK 1990). Of course, the loss of the subimaginal stages had to have numerous secondary selective benefits but primarily it was probably an adaptation to escape odonatan predation. The odonatans appeared in the record as early as in the Namurian B, and became rather diverse and common since the Westphalian, at least in the Euramerian Realm. Perhaps, the high frequency of archemetabolous insects (*Evenka*, *Tchirkovaea*, *Narkemina*, Daldubidae) in the Stephanian of Angaraland may be correlated with the rarity of odonatans in cooler areas.

3.2.2.2.2. Herbivores

The evolution of trophic relations is one of the most important aspects of ecology directly affecting the role of the group in the ecosystem turnover. The plesiomorphic pterygotan mouthparts should be of a chewing type more or less similar to those of the present-day generalised Gryllones (e.g. the grylloblattidans, roaches and many orthopterans). As discussed above, early pterygotans probably fed mainly on living plant matter most rich in nitrogen, such as spores, pollen and seeds. Their feeding strategy originally should have been opportunistic and not highly selective in respect to both plant taxonomy or plant organ. Ill or damaged plant tissues could be consumed together with the healthy ones, and perhaps even preferably because of the presence of microorganisms and fungi that might provide additional nitrogen sources; omnivory with facultative predation on smaller arthropods including eggs and the early stages of other insects is also probable.

This generalised opportunistic feeding, which may be supposed for the Paoliida, the ancestral Scarabaeones and perhaps also the ancestral Gryllones, provided a good basis for specialisation in different directions – towards various kinds of specialised herbivory, fungivory, detritivory, and predation; and all these possibilities have been realised in insects. This radiation pattern is partly documented by the mouthpart morphology of fossils (LABANDEIRA 1997a) and partly by other evidence including trace fossils.

In particular, an increase in the herbivorous specialisation is well argued. A brief review of the Late Palaeozoic plant evolution is necessary to represent the general ecological background of this process. The Late Carboniferous and Permian floras were highly diverse and geographically differentiated. The higher plants known include both spore (Pteridophyta) and gymnosperm seed plants (Pinophyta) as well as poorly known bryophytes. All three major plant divisions existed well before the appearance of the oldest known pterygotans. Arthropod feeding on the Palaeozoic Pteridophyta and Pinophyta is documented by trace fossils and coprolite and gut contents; most probably, the bryophytes were also fed on by herbivores. MEYEN (1987a) recognises four classes in the Late Carboniferous and Permian Pteridophyta, namely Lycopodiopsida (lycopsids), Equisetopsida, Polypodiopsida (ferns), and Progymnospermida, and three in Pinophyta including Ginkgoopsida, Cycadopsida, and Pinopsida. All those classes except Progymnospermida are represented in modern floras as well but the majority of Palaeozoic orders are extinct. Most orders were widespread but few were restricted to certain geographic regions (e.g. Gigantonomiales to Cathaysia, and Arberiales to Gondwanaland).

Evidence of arthropods feeding on living plants is available for the lycopsids, the form-genus *Schizoneura* (Equisetopsida *incertae sedis*), the ferns of the order Marattiales, the ginkgoopsids of the orders Peltaspermales (the peltasperm "seed-ferns" or "pteridosperms") and Arberiales (=Glossopteridales), the cycadopsids of the orders Lagenostomales (the lagenopterid "seed-ferns" or "pteridosperms" of older classifications) and Trigonocarpales (the medullosan "seed-ferns" or "pteridosperms"), and the pinopsids of the orders Cordaitanthales (=Cordaitales) and Pinales (conifers). There is also indirect evidence for an insect connection with host plants belonging to other orders (e.g. the equisetopsid order Calamostachyales, or the calamites). This taxonomic pattern suggests that probably all Late Palaeozoic higher plant orders irrespective of their very different morphology and biology were exploited by herbivorous arthropods. However, there is no evidence indicating that any present-day herbivore/host-plant associations could be traced back to the Palaeozoic. Even in the marattialean ferns resembling in their most important characters living Marattiaceae, the documented Palaeozoic herbivory pattern has nothing in common with that observed presently on marattiacean or any other ferns. Interestingly, though the spore plants were more important in the vegetation cover than they are now and dominated in some widespread communities (e.g. the lycopsids in lowland swampy environments), the gymnosperms probably supported a more diverse herbivorous fauna.

The vegetation was structurally complex, and this complexity should provide an additional ground for herbivore specialisation. There were numerous arboraceous forms including some lycopsids (e.g. Lepidocarpaceae) and equisetopsids (e.g. calamites) which are exclusively herbaceous now. The ground cover was also diverse composed mainly of diverse ferns, the bowmanitalean equisetopsids (sphenopsids), and probably lycopsids, and among the ferns there were some vines and perhaps epiphytes (DIMICHELE & HOOK 1992).

The main changes in the Carboniferous and Permian floras are briefly summarised by MEYEN (1987a) and DIMICHELE & HOOK (1992). A decline in the plant diversity is evident in the Euramerican Realm between the Namurian A and B, just at the very beginning of the pterygotan fossil record, followed by a new increase in the Westphalian. In Euramerica, the Westphalian oryctocoenoses were usually dominated by the lepidocarpacean lycopods (e.g. *Lepidodendron*), marattialean ferns and trigonocarpalean cycadopsids in the coal-bearing strata (the so-called "anthracophilous flora"), while diverse ferns and gymnosperms with fern-like leaves (the so-called 'seed-ferns' or "pteridosperms" which are polyphyletic including both ginkgoopsids and primitive cycadopsids) are most common in the coal-less layers. The oldest finds of conifers are noticeable. In the Stephanian the coal-bearing deposits and accordingly the anthracophilous flora became less widespread, and some of its components were disappearing; this pattern is usually interpreted as a result of decreasing humidity. In particular, the role of conifers was increasing continuously since the Stephanian. The Late Permian floras of Euramerica were impoverished and strongly dominated by conifers. The geographic differentiation of floras increased progressively in the Late Carboniferous with appearance of the characteristic cordaite-dominated assemblages in Angaraland and the peculiar strongly impoverished assemblages with *Botrychiopsis* (perhaps, an archaeopteridalean progymnosperm) in Gondwanaland, followed here by the well-known *Glossopteris-* (arberialean-) floras in the Permian. The Cathaysian floras also became highly distinctive in the latest Carboniferous and especially in the Permian.

The earliest pterygotan lineage to make much progress in herbivory is the Cimiciformes. Here opportunistic spore and pollen feeding was probably plesiomorphic like in other pterygotans. This ancestral feeding habit is probable for the Caloneurida and Blattinopseida though the mouthpart structure in both orders is not very well known and no data about the gut content are available. The Hypoperlida of the suborder Hypoperlina demonstrate somewhat modified mouthparts (see Fig. 126) probably adapted to more specialised pollen feeding as postulated by RASNITSYN (1980) and later supported by studies on gut contents of some Early Permian members of the suborder (RASNITSYN & KRASSILOV 1996a, KRASSILOV & RASNITSYN 1999). Smaller Perielytrina (see Fig. 133) were probably also pollen-eaters but their sclerotised fore wings suggest an adaptation to penetrating more protected strobili than in Hypoperlina. The most advanced hypoperlid suborder Strephocladina (see Figs. 121, 131, 132) had the elongate head, the mouthparts with narrow mandibles and highly specialised long and toothed laciniae, and the modified tarsi (sometimes with the strongly widened tarsomeres adapted likely to fixing on plant surface). For this suborder specialisation to feeding on ovules and seeds was supposed (RASNITSYN 1980, ROHDENDORF & RASNITSYN 1980).

Similar specialisation probably evolved in the most primitive Dictyoneuridea (see Fig. 134) and further in more advanced seed-feeders such as the Dictyoneurida, Mischopterida, and Diaphanopterida with their characteristic beak-shaped mouthparts (see Figs. 135, 137, 139, 147, 154, 156–158). These groups were highly successful as indicated by their common presence in the Carboniferous and Permian strata over the world though there are some obvious temporal changes in the dominance pattern. All dictyopteridean orders coexisted as early as since the Namurian B but Dictyoneurida were most common in the Carboniferous, Diaphanopterida in the Early Permian, and Mischopterida in the Late Permian faunas. The find of Delitzschala in the earlier Namurian A shows that this feeding specialisation has been acquired very early, perhaps because the seeds provided the most valuable and highly predictable resource for the herbivores. The mouthpart structure is basically identical in the dictyoneuridean adults and nymphs suggesting that all stages had the same feeding habit. However, KUKALOVÁ-PECK (1987a, SHEAR & KUKALOVÁ-PECK 1990) illustrated a young diaphanopterid nymph from Mazon Creek supposedly containing spores in its gut, and perhaps the ancestral sporivory was retained in some dictyoneurideans as the larval mode of feeding. No recognisable particles has been discovered in the dark gut content of a well-preserved dictyoneurid Paradunbaria from Tshekarda (V.A. Krassilov, pers. comm.) suggesting that the food was completely fluidised.

It should be stressed that none of the present-day herbivores (including the seed predators) demonstrates the robust and rather short beak-like mouthparts resembling those that were common, in the dictyoneuridean orders. From this point of view the remarkably long stylet-like mouthparts of the dictyoneurid Eugereon from the earliest Permian of Germany (see Fig. 139) look in fact less unusual for a sucking herbivore than the less impressive typical dictyoneuridean feeding apparatus. Very long beaks occur also in some other Dictyoneurida (including the Namurian Homaloneura vorhallensis Brauckmann et Koch from Hagen-Vorhalle, see Fig. 137) and Mischopterida (e.g. Monstropterum moravicum Kukalová-Peck) (see LABANDEIRA & PHILLIPS 1996b for a review). It should be noted that Table 2 in LABANDEIRA & PHILLIPS (1996b) is somewhat misleading because it includes only the taxa with well-studied mouthpart morphology. It is

quite understandable that the unusually elongate mouthparts attracted more attention, and in fact the number of dictyoneuridean taxa known with a short beak unstudied in detail is much higher than indicated in this table (see illustrations to the respective chapters). Hence there were many more seed-suckers than the stem- or leaf-sucking taxa. ZHERIKHIN (1980a) even supposed that all dictyoneurideans fed on plant reproductive organs but it is difficult to imagine any seed-protective structure to which Eugereon with its 3-cm-long beak could be adapted.

LABANDEIRA & PHILLIPS (1996b) described in detail the feeding traces in permineralised rhachises of the marattialean tree-fern Psaronius from the Stephanian of Illinois, USA, documenting sucking of photosynthetic tissues. This most interesting find together with other, less studied but likely similar wounds on different Carboniferous plants (A. SCOTT & TAYLOR 1983, LABANDEIRA & PHILLIPS 1996b) indicates the early exploitation of photosynthetic plant tissues by insects with piercing-and-sucking mouthparts, and the dictyoneurideans are the only probable makers of those injuries. LABANDEIRA & PHILLIPS also pointed some dissimilarities between the damage on Psaronius and the feeding traces of modern hemipterans. The hypothesis of seed consumption by dictyoneurideans is supported by ichnological evidence as well, namely by finds of characteristically damaged cordaitalean (see Fig. 471) and primitive cycadopsid seeds in the Carboniferous of both Euramerica and Angaraland (SHAROV 1973, JENNINGS 1974, A. SCOTT & TAYLOR 1983). In this case an insect should either feed on immature seeds with more or less liquid content or, more probably, develop a kind of external digestion with a saliva injected into the seed. Seed eating was likely plesiomorphic in the dictyoneuridean complex because it coincides with the supposed feeding habit of their ancestors, because it requires a less modified mouthpart structure, and because it does not suggest with a necessity the specialised symbiosis with intestinal micro-organisms important for any consumer of the photosynthetic tissues (but see below about possible early origin of such symbiosis in the Cimiciformes). A narrow host specificity of dictyoneurideans is probable (in particular because of their high diversity in some assemblages) though further detailed studies of damage on different plant taxa are necessary to confirm it. Not only mouthparts but also the legs are sometimes strongly modified and probably specialised for fixing on certain types of cones (see Fig. 137; the extremely short fore legs are described by F. CARPENTER 1951 in Mischoptera). The dictyoneuridean finds in the Palaeozoic temperate and even cool areas suggest that some of them might have their life cycle effectively adapted to strongly seasonal climate when fresh food was available only seasonally; however, there are no positive data on their seasonal biology.

In the psocidean lineage within the Cimiciformes, herbivory was surely less advanced than among the dictyoneurideans. Gymnosperm pollen grains have been discovered in the gut contents of some Early Permian barklice (KRASSILOV & RASNITSYN 1999) which however should be rather opportunistic comparing some highly specialised pollinivores, though some Permian permopsocine barklice genera demonstrate specialised head morphology which may indicate feeding on covered sporangia (see Fig. 162). The modern barklice usually feed on fungal hyphae and lichens rather than on higher plant tissues. The primitive lophioneurid thrips appearing in the Early Permian show elongated mouthparts that were supposedly adapted to sucking-out various individual small-sized objects (cf. Fig. 172). At least facultative pollinivory is probable though lophioneurids could well be omnivorous like the modern aeolothripoids. Their small size should have allowed them to penetrate even into well protected cones. Holes in the cuticle of *Permotheca* sp., a peltasperm male fructification from the Early Permian of the Urals, illustrated by NAUGOLNYKH (1998: pl. XXIII, figs. 1–3) and originally attributed to a dictyoneuridean could have been made either by a permopsocine or even a lophioneurid as indicated by their small size (about 40 μ in the diameter), nor can a small-sized hypoperlid be excluded as their possible producer.

One more cimiciform lineage is represented by the hemipterans – the most successful of all Palaeozoic herbivores which, unlike the dictyoneurideans, has not disappeared or even declined by the end of the Palaeozoic. This group known since the Early Permian was probably represented in the Palaeozoic exclusively by herbivorous taxa. The advanced plant sucking was the primary feeding adaptation in this lineage originated probably from Hypoperlida. The ancestral hemipteran suborder Archescytinina shows some characters (e.g. the long flexible ovipositor and long proboscis, see Fig. 177) suggesting that both immatures and adults could feed on gymnosperm ovules and/or immature seeds in cones (BEKKER-MIGDISOVA 1972), which is in accordance with the postulated plesiomorphic feeding on reproductive plant organs in Cimiciformes in general. Shcherbakov (Chapter 2.2.1.2.5) supposes feeding on ovules also for the stem-group of Aphidina (e.g. for Boreoscytidae).

Like the dictyoneurideans, the hemipterans evolved to shift from reproductive organs to photosynthetic tissues. In Cicadina this shift should still have been realised in the Early Permian, and in the Late Permian an intensive radiation occurred, probably connected at least partially with host specialisation. Head morphology of the Permian Cicadina indicates the development of the cibarial musculature adapted to phloem-feeding (SHCHERBAKOV 1996). There is no basis to suggest a xylem-feeding habit for any of the Palaeozoic hemipterans (2.2.1.2.5). While the ancestral Archescytinidae probably lived on gymnosperms, some Permian Cicadina almost certainly fed on spore plants. In particular, in the Late Permian locality Kerbo (Tunguska Basin, Siberia) the flightless nymphs of an undescribed paraknightiid were collected in numbers from a layer containing virtually no plant remains other than the equisetopsids of the genus *Phyllotheca*. The probability and possible distance of transport for the plant stems and flightless insects should be similar so that those equisetopsids are the only probable hosts of the paraknightiid. The specialised symbiosis with intestinal micro-organisms and excretion of surplus carbohydrates should have been acquired by Cicadina at the early stage of their evolution.

The remarkably rapid parallel development of herbivory on nitrogen-poor tissues in both dictyoneurideans and hemipterans originated independently from hypoperlid-like ancestors suggests that the specialised intestinal microflora could have been acquired even at the hypoperlid stage of evolution. The homopteran hemipterans are the only Palaeozoic insects which could induce the majority of galls (which are usually in fact deformed plant organs rather than true galls) known from the Palaeozoic; however, such pathologies may be caused by mites or even by bacterial or fungal infection as well (Chapter 2.3). Some Permian homopterans had to be highly specialised plant parasites as indicated by the finds of sedentary nymphs like *Permaleurodes* Becker-Migdisova and the unnamed Late Permian nymph from South Africa illustrated by SHEAR & KUKALOVÁ-PECK (1990, fig. 50).

RASNITSYN (1980, ROHDENDORF & RASNITSYN 1980) postulated that the ancestral Scarabaeiformes were also biologically connected with gymnosperm reproductive organs. According to his model, the adult Palaeomanteida were probably pollinovorous, and their larvae developed in the male cones like the larvae of the most primitive hymenopterans, xyelids (Chapter 2.2.1.3.5). This hypothesis may well be true though it still doesn't have any direct palaeontological support. The cones should provide both a good shelter and a valuable food source and seem to be nearly ideal microhabitats for the development of the oldest holometabolans. This supposedly ancestral holometabolan mode of life occurs now not only in the most primitive hymenopterans but also in a number of polyphagan beetle families. In the Carboniferous Palaeomanteida is the only order of Scarabaeiformes, and no immatures or their possible damages to cones are recorded. A striking arthropod from Mazon Creek described and figured by KUKALOVÁ-PECK (1991) as a supposed holometabolan larva was almost surely misinterpreted (Chapter 2.2.1.3a). LABANDEIRA & PHILLIPS (1996a) tentatively assigned the large true galls with an interior cavity discovered on petioles of a tree-fern from the Stephanian A of Illinois to holometabolan larvae. The morphology of coprolites and undigested parenchyma fragments discovered in the gall cavity indicates that the gall maker was a chewing, and not a sucking herbivore. The repeating finds (LABANDEIRA 1998a mentions 35 specimens known) of such galls exclusively on *Stipitopteris* (which is the form-genus for the petioles assigned to the marattialean fern *Psaronius*) suggests a high host specificity of the galler. However, the gall is too large for any Palaeomanteida known, and a non-holometabolous chewing insect or even other arthropod seem to be more probable as the gall inducer than a holometabolan. There are many extinct insect orders in the Palaeozoic, and thus the absence, say, of any gall-making Gryllones in the present-day fauna is not a strong argument against their possible existence in the Carboniferous. The same is true for the millipedes. Though the modern Diplopoda are mainly detritivorous, and the opportunistic herbivory in this group occurs only facultatively (HOPKIN & READ 1992), in the Palaeozoic times a highly diverse millipede fauna existed, and some extinct lineages could well include specialised herbivores.

The holometabolan radiation in the Early Permian was rapid, and since the Artinskian all main evolutionary lineages (mecopteroids, neuropteroids, and beetles) are known. However, the ground for any suggestions about their biology is weak. Several undescribed soft-bodied holometabolan larvae discovered in Tshekarda are poorly preserved, and their biology is unknown but their presence as fossils suggests that they hardly could be endophytic. Records of supposed leaf- and stem-mining on Palaeozoic plants are scarce and equivocal (LABANDEIRA 1998a). The modern mecopterans with few exceptions have either detritivorous or predaceous larvae (though both adults and larvae of boreids are moss eaters: SHORTHOUSE 1979), and the larval herbivory occurs in the mecopteroid lineage mainly in advanced and phylogenetically relatively young taxa (e.g. in some Diptera and many Lepidoptera). However, as the adult feeding the ancestral pollinivory could well be retained in primitive mecopteroids. The same is true for the neuropteroids with their nearly exclusively predacious larvae.

The situation in the beetles is more complicated. All living beetle taxa with larvae developing in gymnosperm cones belong to the polyphagans and most of them are phylogenetically advanced. It is generally believed that this habit in the beetles is secondary, and that the ancestral beetles had xylomycetophagous larvae that developed under bark and in dead wood (PONOMARENKO 1969, 1995a, CROWSON 1975). However, this view is based mainly on the biology of the relict living archostematans. If the Triassic obrieniids (see Fig. 223) are not polyphagans but these convergently weevil-like archostematans dwelled gymnosperm cones (Chapter 3.2.2.4), this may also question the biology of ancestral beetles. Fore wing elytrisation should be an adaptation equally effective for hind wing protection both in wood borers and cone inhabitants. The Early Permian Tshekardocoleidae (see Fig. 217), the oldest beetles known, are moderately dorsoventrally flattened and rather small-sized; their body shape is equally suitable for dwelling bark crevices and cones but their limited body size could be better explained by life in the latter. If so, this plesiomorphic habit might be suggested as well for the common ancestral stock of the neuropteroids and beetles. The wood boring habit in the cupedoid archostematans may be apomorphic though it probably had evolved in the Permian. Some Permian beetles perhaps developed in living wood being herbivores, and not detritivores. WEAVER *et al.* (1997) point out that the supposed beetle tunnels in gymnosperm woods from the Late Permian of Antarctica, South Africa and India often show neither evidence of decay nor clear wound reaction. The borings are localised in the late-wood near growth ring boundaries, and it is suggested that the beetle larvae probably fed in winter when the trees were nearly dormant. In this connection it is noteworthy that beetles other than the most primitive tshekardocoleids are virtually absent in the Permian of tropical Euramerica. Beetle xylophagy could have evolved initially from boring in the cone axes. This way theoretically seems to be the most probable but is still not supported by damage to fossil cones. It should be stressed that these alleged beetle borings in living wood are now the best evidence of a true endophytic mode of life available for any Palaeozoic insects.

The origin of Gryllones is hypothetically connected with the secondary invasion of pterygotans from tree crowns to the forest floor habitat, forced probably by increasing pressure of odonatan predation in the air (RASNITSYN 1976, ROHDENDORF & RASNITSYN 1980). This scenario is also accepted by ZHERIKHIN (1980a) and STOROZHENKO (1997) and is in accordance with the changes in the wing resting position, the common tendency towards fore wing sclerotisation with formation a protective cover for the folded hind wings, and with the biology of living primitive Gryllones such as the cockroaches and grylloblattidans. The mouthparts of Gryllones are basically of a primitive chewing type equally suitable for predation, detritivory, and herbivory, and their morphology usually tells us little about their probable feeding habits. The living cockroaches and grylloblattidans are detritivorous, and similar biology was postulated for the ancestral Gryllones suggesting their shift from living to dead plant matter as a synapomorphy for the lineage with secondary reversions in some taxa (ZHERIKHIN 1980a). However, gymnosperm pollen was recently discovered in the gut contents of two Early Permian grylloblattidans belonging to Ideliidae and Tillyardembiidae (RASNITSYN & KRASSILOV 1996b, KRASSILOV & RASNITSYN 1999, AFONIN 2000). Both families are rather advanced but unrelated to one another (see Fig. 393). Earlier A. SCOTT & TAYLOR (1983) recorded lycopsid spores from the gut contents of *Eucaenus* (Eoblattida) and an unidentified "protorthopteran" (which is most probably also an eoblattid because Grylloblattida are rare in the Carboniferous). *Eucaenus* (see Fig. 363) is a morphologically derived genus representing a family of its own and thus could be biologically abnormal; however, all those finds together suggest that ancestral pterygotan pollen feeding and spores was likely to have been inherited by the early Gryllones and retained at least in the adult stage in some lineages for a long time.

It should be noted that *Eucaenus* is not unique among the Eoblattida in its modified morphology. The highly movable prognathous head and more or less elongate prothorax lacking the paranota, like those in the most primitive and ancient paoliidans (see Figs. 74–76), could be adapted for feeding on protected sporangia and occur, sometimes in even more bizarre form, in other eoblattidan families as well (see Figs. 357, 359, 362, 363). The tillyardembiids (see Fig. 400) demonstrate a similar trend of morphological specialisation among the Grylloblattida. Thus specialised sporivory/pollinivory was perhaps not uncommon in some orders of the Palaeozoic Gryllones. Probably it evolved in the lineages which have not moved to the forest litter but preferred rather bark crevices, hollows, lichen crusts, and other shelters available in the tree crowns. This microhabitat preference may well be plesiomorphic for the Gryllones in general.

One of the most intriguing results of palynological studies of the gut contents of Permian insects is that only the saccate, taeniate pollen is common while other morphological types occur occasionally in small numbers, and this pattern is constant in all insects studied independently from their taxonomic position (Hypoperlida, Psocida, or Grylloblattida) (KRASSILOV & RASNITSYN 1999). The taeniate pollen was produced by several unrelated groups of Palaeozoic gymnosperms and is believed to have originated independently in a number of lineages. The morphology of insect-eaten pollen indicates that different insects visited different plants in different habitats but only those which produced this particular pollen type. This most striking specialisation pattern is difficult to explain. KRASSILOV (1997, KRASSILOV & RASNITSYN 1999) suggests that insects could spread micro-organisms between the plants and in this way favour gene transduction. However, independently of the question about general evolutionary importance of the transduction, it is hard to see why the genes transferred should affect just the pollen morphology instead of other characters less directly related with the parts eaten by the carriers. It seems that much more specific, selective and intimate co-evolutionary interrelations than a random gene transfer with viruses might operate to induce those remarkable parallels in pollen. Because the taeniate type is derived, the simplest explanation is that insect feeding stimulated progress in pollen morphology; however, further studies are necessary to prove whether just this unique direction of pollen evolution and not a radiation of advanced morphological types has been stimulated, and if so, why.

New data on the biology of Palaeozoic Gryllones are causing a rebirth of interest in analysis of their possible feeding on living photosynthetic tissues. It is generally believed that folivory is a secondary and more advanced type of living plant matter feeding in comparison with sporivory/pollinivory (SOUTHWOOD 1973, STRONG *et al.* 1984), and the model of pterygote origin accepted here is in accordance with this concept. Chewing marks probably made by mandibulate arthropodans are observed on the leaves of different plants from the Late Carboniferous (e.g. VAN AMEROM 1966, VAN AMEROM & BOERSMA 1971, A. SCOTT & TAYLOR 1983, M. CASTRO 1997) and Permian (e.g. FEISTMANTEL 1881, PLUMSTEAD 1963, SRIVASTAVA 1987, A. BECK & LABANDEIRA 1998, LABANDEIRA 1998a,b). Many authors believed that they could be produced mainly or even exclusively by detritivores on dead leaves while the living photosynthetic tissues were almost unexploited by chewing herbivores (e.g. ZHERIKHIN 1980a, SHEAR & KUKALOVÁ-PECK 1990, DIMICHELE & HOOK 1992). This conclusion was based on occasional, and rarely specified, records in palaeobotanical literature, and it is quite possible that many bite marks on leaves remained unrecognised (A. SCOTT & TAYLOR 1983). The recent special studies conducted by LABANDEIRA (LABANDEIRA & PHILLIPS 1992, A. BECK *et al.* 1996, GREENFEST & LABANDEIRA 1997, LABANDEIRA 1997a, 1998a–c, A. BECK & LABANDEIRA 1998) have demonstrated that the folivory pattern in the Late Palaeozoic was more complicated than first considered. Often leaves were damaged in their lifetime but the frequency of attacks varied considerably in space and time as well as between different plant taxa. In particular, the Early Permian floras from Texas demonstrate nearly a tenfold increase in the overall folivory rate compared to the Sakmarian to Artinskian assemblages (GREENFAST & LABANDEIRA 1997).

Folivory seems to have been highly selective, with the maximum attack intensity on the gymnosperm family Gigantopteridaceae. A. BECK & LABANDEIRA (1998) estimate the attack frequency on *Gigantopteris* leaves as ranging between 40 and 80%, and the share of the total leaf area consumed as 3.1 to 4.4%. Other plant taxa in the same assemblage exhibit significantly lower if any level of damage. The morphology of damage is variable, and some rarer types are restricted to the non-gigantopterid hosts (e.g. the peltasperms *Callipteris* and *Comia* or the cycadalean *Taeniopteris*). It should be noted that according to MEYEN (1987a) the North American Permian leaves usually classified as *Gigantopteris* belong probably to Peltaspermaceae and not to the true Gigantonomiales which seem to be restricted to Cathaysia. If so, the taxonomic distribution of damage is even more unequal, with the peltasperms much more strongly preferred to any other plants. Unfortunately, there is no published information about insect chewing marks on any Chinese Permian plants. Independently from the taxonomic problems it is noteworthy that both the Cathaysian and North American *Gigantopteris* are remarkably angiosperm-like in their leaf morphology. The available data indicate that the biomass of some plant taxa at least in Euramerica was consumed rather intensively as early as in the Early Permian but this was probably not the case for the majority of Palaeozoic plants.

More data are required for more territories, plant taxa, palaeoenvironments and stratigraphic horizons in order to elucidate the distribution pattern of leaf damage and to compare it with the distribution of different insect taxa. In the absence of comprehensive information only very general speculative conclusions are possible about the taxonomic placement of the Palaeozoic folivores.

The gryllonean insects are the most probable candidates for the role of Palaeozoic chewing folivores (A. BECK & LABANDEIRA 1998). However, no leaf cuticle has been discovered in the gut contents of any Palaeozoic insects studied in this respect. The most diverse gryllonean orders are Eoblattida in the Carboniferous and Grylloblattida in the Permian, and their morphological diversity suggests diverse biology. In the Carboniferous of the Illinois Basin the frequency of leaf damage decreases strongly from the Dismoinesean (Late Westphalian) to the Missourian (Early Stephanian) assemblages suggesting an extinction of dominant folivores (LABANDEIRA & PHILLIPS 1992); however, this pattern needs to be confirmed at a wider geographic scale. If Permophasmatidae from the Late Permian of Mongolia are correctly placed into Phasmatida, this exclusively folivorous order is appearing in the fossil record in the Late Permian. One more problematic orthopteroid folivore is *Praelocustopsis* Sharov from the latest Permian or earliest Triassic Bugarikhta Formation of Tunguska Basin, Siberia. It is assigned to the family Locustopsidae related to modern folivorous Acrididae; however, *Praeolucustopsis* body structure is unknown, and its age is somewhat uncertain. Folivory is not absolutely improbable also for some primitive Cimiciformes, especially for some Caloneurida with their relatively weakly specialised mouthparts. Moreover, the insects are not the only probable folivorous arthropods. Leaf chewing cannot be ruled out for Palaeozoic millipedes which could include some specialised herbivorous taxa as discussed above.

The millipedes in general become rarer in the Permian record in comparison with the Carboniferous (SHEAR & KUKALOVÁ-PECK 1990) whereas the rate of damaged leaves is likely to be increasing; however, this is not a very strong argument against millipede folivory because the probability of fossilisation of tree-inhabiting millipedes might be very low and they may even be totally lost from the fossil record.

To summarise, herbivory in the broad sense as feeding on any living plant tissues (including sporivory/pollinivory) was widespread in Palaeozoic pterygotans, with feeding on reproductive organs and sucking on vegetative tissues the predominant styles while chewing on photosynthetic organs was probably less common and restricted to some plant taxa only. The living plants in the Early Carboniferous provided an important underexploited resource, and the insects evolved rapidly and successfully in this potentially complex but unsaturated niche space; perhaps, specialised herbivory was developing contemporaneously but less successfully in some other arthropods as well. The acquisition of symbiotic intestinal microflora and further co-evolutionary progress in this symbiosis must be extremely important for evolution of herbivory. An early development of intestinal symbiosis in Cimiciformes is probable because this lineage was more successful as herbivores than any other Palaeozoic insects. There should also be various co-evolutionary interrelations between the herbivores and their host plants. The development of chemical defence in Late Palaeozoic plants is probable though rarely documented palaeontologically (e.g. by common presence of the "resin-like" bodies in the fossilised tissues of the medullosean cycadopsids and the finds of an amber-like resin in the Late Carboniferous of Scotland: DIMICHELE & PHILLIPS 1994, JARZEMBOWSKI 1999a).

It is interesting to note in this respect that the present-day ferns, horsetails and especially the lycopods have a poor and distinctive herbivore fauna. As regards the latter group I myself have never seen any insects feeding on modern *Lycopodium* or *Selaginella* and have found only two published records of herbivorous insects, a satyrine butterfly caterpillar in North America (SINGER *et al.* 1971) and an eumastacid orthopteran from Indonesia (ANONYM. 1985). This is virtually nothing even in comparison with the horsetails and ferns which are the hosts of diverse homopterans, beetles, flies, sawflies, etc. Mechanical defensive structures are diverse among Palaeozoic plants, including cupules and more or less complex fructifications protecting ovules and seeds, seed coat and sclerotesta in seeds, thick cuticle and lignified fibers in photosynthetic organs, trichomes on leaf and stem surface, etc., documented in a number of plant taxa, though they should be equally effective against other environmental stresses and thus their primarily anti-herbivore nature is not certain (see A. SCOTT & T. TAYLOR 1983, T. TAYLOR & TAYLOR 1993, LABANDEIRA 1998b and references herein). Pollinivory and seed predation could induce the remarkably parallel trends of morphological evolution of the reproductive organs in different phylogenetic lineages noted by MEYEN (1987a) as well as in pollen evolution as mentioned above.

The development of complex well-protected reproductive organs should be especially co-evolutionarily important. It stimulated the development of diverse adaptations allowing insects to reach the most nutritionally valuable tissues not easily available under those covers. The simplest and the least profitable way was likely an elongation of the head and/or prothorax as in some Eoblattida (see Figs. 362, 363). An alternative direction was an adaptation for penetrating into fructifications with either increasing body flexibility (e.g. in some Grylloblattida such as *Tillyardembia*, see Fig. 400) or decreasing body size (in Thripida, see Fig. 171, some hemipterans such as *Permothrips*, and perhaps in the early beetles). This trend provided a significant additional selective advantage: the insect not only could reach the food but also escaped from surface-dwelling predators so that the well-protected plant reproductive organs turned to be the excellent shelters for their eaters.

Predation pressure on insect eggs and flightless immatures was particularly significant because they were more endangered by surface predation than flying adults. The oviposition into plant tissues with the cutting ovipositor was probably acquired already in the ancestral pterygotan stem, and longer flexible ovipositors evolved independently in different taxa allowing better egg protection from both predation and weather in deep crevices of bark and other substrates. In this case the larvae and nymphs still remained highly vulnerable to predation but the appearance of complicated plant reproductive structures allowed them to feed directly inside the shelters thus minimising the predation risk. This way of adaptation has probably been realised in parallel in different phylogenetic lineages. In particular, a number of Permian Grylloblattida including Tillyardembiidae with their pollen-feeding adults shows a specialised long ovipositor with small apical spines (see Fig. 400), and in some archescytinid hemipterans it was extremely long, slender and coiled in repose (see Fig. 180).

Development inside fructifications led, in larger insects, to increasing biological differences between the immatures and adults. This differentiation provided an additional selective benefit because of decreasing competition between the different ontogenetic stages of the same species for food and space but complicated the metamorphosis creating an important new evolutionary challenge. The most radical adaptive solution has been realised in the holometabolans with the acquisition of the pupa as an additional specialised morphogenetic stage between the larva and the imago; this novelty has allowed them to turn into the most successful insects throughout the Mesozoic and Cainozoic up to the present. Similar trends are evident in the herbivorous Cimiciformes including the thrips and advanced homopterans though in this case the evolution toward decreasing adult body size was widespread. Hence a long chain of successive complex and broadly co-evolutionary interrelationships between the insects and their host plants, also involving predator/prey interactions, can be postulated from the ancestral pterygotan opportunistic sporivory on exposed sporangia to the holometaboly and subsequent holometabolan radiation well outside their original trophic niche.

Another important aspect of insect/plant interrelations is entomophily, i.e. plant fertilisation by phoresis of their gametes by insect visitors. Insect pollination is well known in angiosperms but there are some entomophilous gymnosperm taxa as well (for a review see FAEGRI & VAN DER PIJL 1979). Spore transfer by microarthropods (but mostly other than insects) possibly also occurs in some bryophytes though this phenomenon is still poorly investigated (HARVEY-GIBSON & MILLER-BROWN 1927, MUGGOCH & WALTON 1942, KOPONEN & KOPONEN 1978, GERSON 1982). Entomophily is even known outside the Plant Kingdom in some fungi (INGOLD 1953). Thus in principle entomophily can be very old, and zoophily in general could originate even before both insects and seed plants. Zoophily cannot be directly documented by fossils but various indirect evidence can be recognised in plant morphology which is especially important if combined with documented specialised sporivory/pollinivory in certain animal taxa (T. TAYLOR & TAYLOR 1993). In the Palaeozoic the alleged evidence of zoophily is scarce and highly hypothetical. HUGHES & SMART (1967) supposed entomophily for the botryopteridalean ferns producing the Raistrickia-type spores but such spores were never recorded as found

in direct association with any insect fossils. T. TAYLOR & MILLAY (1979) interpreted the capitate hairs on the cupule of Carboniferous Lagenostoma (according to MEYEN 1987a, Lagenostomales is the most primitive cycadopsid order) as possible evidence of nectar secretion; supposed glandular bodies are described also in Early Permian cycadaleans (MAMAY 1976). However, even if this interpretation is correct, nectaries are not rare in non-zoophilous plants serving in this case rather for the distraction of herbivores from other tissues, than for attraction of potential pollinators. ZHERIKHIN (1980a) pointed out that the location of reproductive organs on trunks (cauliflory) in some Palaeozoic lepidophytes may be explained with fertilisation by flightless arthropods but this hypothesis also lacks any further support. Zoophily has been suggested for the gymnosperm producer of Monoletes pollen (assigned to the trigonocarpalean cycadopsids by MEYEN 1987a) which seem to be too large and heavy for effective wind pollination (T. TAYLOR 1978). Later this pollen type has been discovered on the body of the giant Carboniferous millipede Arthropleura, and some authors suppose that Arthropleura could pollinate the Monoletes-producing plants (A. SCOTT & TAYLOR 1983, T. TAYLOR & TAYLOR 1993) while others pointed out that this pollination model is difficult to believe if the extremely large body size of Arthropleura (more than 1 m long) is taken into account (SHEAR & KUKALOVÁ-PECK 1990). Hence the development of zoophily in Palaeozoic plants is still very doubtful. It is generally accepted that the appearance of bisexual strobili should be an important prerequisite for the origin of entomophily because it is difficult to imagine in which way any special adaptations for pollinator attraction could arise in unisexual cones until a diverse guild of specialised anthophilous animals has been formed on bisexual organs; and bisexuality in the gymnosperms has not been documented unequivocally before the Triassic.

Some insects with opportunistic feeding strategy could also disseminate living plant and fungal spores (SHEAR & KUKALOVÁ-PECK 1990) but this is improbable for the specialised sporivores which must have been (and still are) well adapted to their digestion. The development of more specialised mechanisms of insect-mediated dispersal (entomochory), e.g. with the acquisition of specialised food bodies attractive to insects, is not demonstrated, but it would be interesting to analyse Palaeozoic plant morphology carefully from this point of view.

3.2.2.2.3. Detritivores and fungivores

While herbivory among Palaeozoic insects evolved in a niche space nearly free from competitors other than insects themselves, this was not the case for detritivory. The Palaeozoic detritivory is well documented by tunnels in plant remains and arthropod coprolites with undigested plant fragments (for a review see LABANDEIRA et al. 1997, LABANDEIRA 1998b). LABANDEIRA et al. (1997) have given conclusive evidence of oribatid mite importance as the principal decomposers of dead plant matter in Palaeozoic terrestrial ecosystems since the Late Devonian. Analysis of the size distribution of arthropod faecal pellets from Carboniferous deposits indicates that besides the oribatids there were also diverse larger detritivores including probably both millipeds and insects (BAXENDALE 1979, A. SCOTT & TAYLOR 1983). Lycopod tracheid fragments are recorded from the gut contents of arthropleurid millipedes suggesting that they fed on rotten wood (ROLFE & INGHAM 1967). Millipedes are relatively common as fossils in the Palaeozoic, especially in the Carboniferous, and because the probability of their burial is in general, much lower than for the winged insects their abundance in nature must have been remarkably high. They were represented both by cylindrical forms adapted to burrowing in rotten wood or in soil and by flattened litter dwellers (SHEAR & KUKALOVÁ-PECK 1990). Some scavengers may be represented also among the Palaeozoic chelicerates (e.g. some Anthracomartida: SHEAR & KUKALOVÁ-PECK 1990). Hence the detritivore niche space was rather saturated when the pterygotans entered it in the Late Carboniferous. The millipede decline in the Permian and especially their virtual disappearance in the Mesozoic record suggests that the insects have proved to be more successful than the older larger detritivores but failed to replace the smaller ones such as the oribatids and springtails. Perhaps, terrestrial vertebrate predation on larger arthropods played the most important role in the competitive advantage of the pterygotans.

Insects returned back to the forest floor from the trees much better protected from predation than they had left it due to the flight ability of the adults and rapid and manoeuvrable locomotion that the immatures acquired in the crowns. None of the Carboniferous vertebrates were able successfully to catch those striking flying creatures when they re-colonised the ground surface. The superiority in dispersal ability and in ability to search for suitable habitats provided to the winged insects an additional important benefit in comparison with the flightless myriapods. The importance of flight is stressed by the fact that flightlessness (see Fig. 395) is not very common among the detritivorous insects though their wings need to be specially protected against accidental damage during crawling in narrow tunnels and natural cavities. The development of wing folding and fore wing sclerotisation with formation of a protective cover for the hind pair (which consequently become the main flight organ) suggests the adaptation to cryptic habits common in detritivores (but not exclusive for them). Unfortunately, the morphological mouthpart specialisation in detritivorous insects is normally so weak that it cannot indicate their feeding habit with certainty. Thus comparison with modern relatives is often the only, though certainly not an ideal, way to regard extinct insect taxa as probable detritivores.

It is often difficult to distinguish between detritivorous and mycophagous insects. In fact detritivory is usually a sapromycetophagy or microphagy because of the common presence of fungal and bacterial growth on organic debris, and the same might be true for the Palaeozoic. The presence of fungi and other micro-organisms significantly improve the food quality of dead organic matter which itself is normally poor in nutrients and hardly digestible. That is why detritivores are usually specialised toward exploiting debris at a certain stage of its decay but are rather opportunistic in respect to the kind of debris itself. A number of detritivores should be regarded rather as omnivores because of occasional scavenging, predation and feeding on dispersed pollen and spores. The more specialized food substrates such as carrion, dung, fleshy fruits or dead wood at the early stages of decay support highly specific assemblages which, however, become gradually replaced by more opportunistic and ubiquitous detritivores in the course of the progressive substrate decay. Though large supposedly herbivorous vertebrates are known since the Permian (DiMICHELE & HOOK 1992) there is no evidence for specialised arthropod necrophagy or coprophagy, and the trophic opportunists probably strongly dominated the Palaeozoic detritivore guild; many of them were, however, habitat specialists.

Detritivory is commonly postulated for a number of Palaeozoic Gryllones and some Scarabaeiformes; however, it has not been demonstrated definitely for any of them. New data mentioned above question seriously that this feeding habits was ancestral in the Gryllones. It is generally believed that the roaches were detritivorous since their earliest days (the order is known since the earliest Westphalian) but even this cannot be argued by anything more convincing than an analogy with the modern members of the order. The Palaeozoic roaches (see Figs. 365–367) differed considerably from the modern ones in reproductive biology (as indicated by their well-developed and often long external ovipositor) and were probably more active fliers (in particular the dominant Carboniferous family Archimylacrididae) but their morphology tells nothing about their possible trophic biology. There is no data on their gut contents, and the attribution of some Palaeozoic detritivore coprolites to roaches (A. SCOTT & TAYLOR 1983) is possible but not very well grounded. Thus detritivory in Palaeozoic roaches is probable but not proved. It should be noted that the order demonstrates a very evident and intriguing decline in the Late Permian which still remains unexplained. It is not accompanied by any evidence of a renaissance of the millipedes which were probably the main Palaeozoic group of larger detritivorous arthropods. The hypothesis of competition with other allegedly detritivorous insects such as scorpionflies (ZHERIKHIN 1980a) or Grylloblattida (Chapter 2.2.2.2.1) is purely speculative and needs confirmation; now the only reason is that both orders become diverse and abundant in the Permian. Alternatively, the Permian roach decline can be explained by structural changes in their habitats connected with expansion of plants producing more dense and less structurally complex litter (see below). Both hypotheses are not necessarily mutually exclusive because the advanced grylloblattidans with narrow and flexible body as well as holometabolan larvae should have been better adapted to dwell in such litter than the dominant Palaeozoic roach taxa.

In the Grylloblattida detritivory has probably been developed in both Protoperlina and Grylloblattina but this hypothesis is again not very well grounded. The former suborder is the most probable ancestral group for both omnivorous Forficulida and detritivorous Embiida, and some Permian Protoperlina (e.g. Sheimiidae) resemble detritivorous embiids morphologically and could have had similar biology (Chapters 2.2.2.2.1, 2.2.2.2.3 and 2.2.2.2.4). The only living family of Grylloblattina, Grylloblattidae, is detritivorous and so closely related to Megakhosaridae and Blattogryllidae (see Fig. 407) that at least for these two families similar feeding habits are probable. The most primitive Forficulida of the suborder Protelytrina (see Figs. 387–390) entered the fossil record since the Artinskian and were perhaps similar to modern earwigs in diet. Omnivory cannot be excluded also for the Palaeozoic orthopterans though they could well be predominantly predacious (GOROCHOV 1995a,b, STOROZHENKO 1997). However, recent data on pollinivory in the Mesozoic orthopterans (KRASSILOV et al. 1997a, KRASSILOV & RASNITSYN 1999) suggest that this mode of feeding cannot be ruled out for the Palaeozoic members of the order as well.

In general, the morphology of a number of Palaeozoic Gryllones suggests their progressive specialisation towards permanent dwelling sheltered microhabitats such as bark crevices, mosses, litter, etc. where detritivory is more probable than ancestral herbivory (STOROZHENKO 1997, 1998). In particular, an extreme sclerotisation of the fore wings in Forficulida suggests an adaptation to life in narrow tunnels or borings. Nothing better than this indirect and rather weak evidence can be proposed now to argue the development of detritivory in this lineage. The most probable speculative scenario of shifting to detritivory in the Gryllones is via omnivory in natural shelters available in the tree crowns with subsequent expansion to forest litter. In particular, the Carboniferous roaches are most common in the coal-bearing deposits that originated in swampy environments which are very unfavourable for any permanent litter-dwellers because of flooding.

It should be stressed that the litter in Palaeozoic forests was probably rather different from the litter of modern forests (ZHERIKHIN 1980a, SHEAR & KUKALOVÁ-PECK 1990, DiMICHELE & PHILLIPS 1994). It should have contained abundant large-sized debris such as branches, wood and bark fragments as well as large hollow stems, and resembled rather a brushwood in respect of its high structural complexity. Numerous natural cavities, located both between large plant fragments as well as inside them, were very suitable for rather large-bodied insects that lacked any special adaptations for digging.

The true soil inhabitants could well have been lost from the fossil record, and their role in the Palaeozoic communities is difficult to estimate. It may be postulated that the colonisation of soil environment was the next step after adaptation to the life in dense litter. The Late Palaeozoic trace fossil record of soil dwellers is scarce (BUATOIS et al. 1998). However, the Early Permian grylloblattidan genus *Probnis* with its strongly modified legs was likely a highly specialised digger (STOROZHENKO 1998), and the leg morphology in some Permian orthopterans of the family Pruvostitidae also suggests digging adaptations (GOROCHOV 1995a,b).

The Scarabaeiformes are another insect lineage that probably evolved towards detritivory in the Palaeozoic. Here the shift to detritivory is probable because of both common bacterial and fungal infection of damaged plant tissues inside the larval microhabitats and the necessity to complete development in aborted or fallen cones. In particular, detritivory is possible for the Permian mecopterans and could originate even in some Palaeomanteida though there are no positive palaeontological data supporting this hypothesis. If the Permian cupedoid archostematans fed on dead wood, as commonly believed, they might be the most specialised detritivorous insects known from the Palaeozoic; however, as pointed above, the supposed beetle borings in the Permian woods were possibly made at least partially when the tree was still alive. More studies on wood damage are necessary to elucidate the pattern of its utilisation by insects in the Permian. The larger cavities in Permian woods are often interpreted as produced by fungal decay and rarely contain arthropod coprolites even when their morphology suggests arthropods as possible borers (STUBBLEFIELD & TAYLOR 1985, McLOUGHLIN 1992).

Concerning possible co-evolutionary interrelations, the relationships between detritivores and fungi should be of particular importance. As a rule, detritivores depend strongly on the fungal and bacterial processing of organic matter. On the other hand, they themselves affect the fungal and bacterial floras in different ways, including the fragmentation of the substrate with corresponding increase in its total surface area, selective grazing, dissemination, etc. There are many intriguing problems regarding possible effects of morphological and biochemical plant evolution and influence of other organisms on the micro-organism/detritivore co-evolution. In particular, the input of vertebrate faeces could affect it seriously if the supposed Permian large herbivores were terrestrial and not semiaquatic. An additional important event should be the development of sucking herbivorous guild producing liquid excretes rich in carbohydrates and seriously affected the microbial and fungal growth. The dramatic changes in the composition of the soil microarthropod fauna in presence of leafhoppers are experimentally demonstrated in modern ecosystems (ANDRZEJEWSKA 1993). Palaeontological studies of changes in the decay dynamics in space and time are still at their very early stage, and much progress in this field will be probably made in future. At present this seems to be one of the least explored and most promising research fields in non-marine palaeoecology.

In general, the detritivory in Palaeozoic insects is highly probable but scarcely documented, and the relative role of insects in detritus processing cannot be estimated with certainty at the present state of knowledge. The decline of older detritivorous arthropod groups, and especially in the millipedes, is in fact the only evidence suggesting a significant insect expansion within the detritivore niche space in the Permian. Vertebrate predation and changes in the plant debris structure were perhaps of a crucial importance for their replacement by insects.

Even less can be said about the development of specialised mycetophagy. Though some insect orders known since the Palaeozoic now include a number of more or less specialised mycetophages (e.g. the barklice and beetles) none of their Palaeozoic members show close affinities with the present-day mycetophagous taxa. Insect morphology is usually not indicative in this respect, the ichnofossil record of insect/fungi relationships is virtually absent, and the history of fungi is still very poorly known though recent progress in this field is obvious (TIFFNEY & BARGHOORN 1974, ELSIK 1976, PIROZYNSKI 1976, LOCQUIN 1981, TETEREVNIKOVA-BABAYAN 1981, STUBBLEFIELD & TAYLOR 1988).

3.2.2.2.4. Predators

Like the detritivores, the predacious terrestrial invertebrates were diverse well before the appearance of the pterygotes in the fossil record, and early predacious insects had to compete with them. The terrestrial chelicerates were especially common and widespread in the Carboniferous. Besides diverse scorpions and spiders there are other modern (Uropygida, Amblypygida, Solpugida, Opilionida, Ricinuleida) as well as several extinct orders (Trigonotarbida, Anthracomartida, Haptopodida, Kustarachnida, Phalangiotarbida), and the pseudoscorpions and mites, though not found as fossils, surely existed because they are known from the Devonian. A brief review of probable biology of the Palaeozoic chelicerates is given by SHEAR & KUKALOVÁ-PECK (1990). They probably occupied a wide range of habitats. They were highly variable in size from small mites up to the giant forms like the enigmatic *Megarachne* Hünicken with the leg span exceeding 40 cm. They differed in mouthpart morphology suggesting different prey specialisation; they included both sit-and-wait predators and active nomadic hunters. Like the majority of the modern predacious chelicerates, they were most probably trophic opportunists but their high diversity suggests a significant habitat specialisation. It should be noted that the existence of any spiders constructing their aerial webs on vegetation in the Palaeozoic and particularly in the Carboniferous is highly questionable; perhaps, this mode of hunting did not originate until the small-sized pterygotans become abundant in the Permian. The number of chelicerate remains known from the Carboniferous is surprisingly large in comparison with younger sediments and suggests their extreme abundance in nature, especially if their relatively low probability of burial is taken into account. Besides the chelicerates the chilo-pods occur occasionally in fossil assemblages including some Scolopendromorpha and Scutigeromorpha remarkably similar to the living members of these orders (SHEAR & KUKALOVÁ-PECK 1990). Hence the niche space was rather saturated and perfectly subdivided by the time of the insect expansion, and the potential prey were well adapted to escape predation both on vegetation surfaces and in the litter stratum.

In fact, facultative predation including cannibalism is common in many modern herbivorous and detritivorous insects. Thus a transition to predatory habits does not seem to be a hard adaptive problem, probably because of the high nutritional value of animal tissues, and competition with other predators should constitute a major constraint to development of predatory specialisation in the Palaeozoic insects.

A predatory habit is often difficult to demonstrate on insect fossils because the chewing mouthparts of predators often show weak, if any modifications to the ground plan, and mouthparts modified for sucking occur in both predators and herbivores being sometimes morphologically nearly identical (e.g. in thrips and bugs). The leg specialisation to catching prey is more indicative but only occurs in some taxa. Predatory insects are usually visual-orienting, and thus they often have enlarged eyes as well as the head highly movable in different directions but those characters should be used only as indirect evidence of a predatory habit and with reservation. Insect predation is also poorly documented by trace fossils; it should be noted that prey remains are virtually absent in fossilised insect gut contents and coprolites for unknown reasons.

The odonatans were almost certainly the most important predacious insects in the Palaeozoic. Their predatory habits are supported by mouthparts (e.g. the cutting mandibles and maxillary lobes in *Meganeura*: SHEAR & KUKALOVÁ-PECK 1990) and leg structure (PRITYKINA 1989) as well as by their highly movable heads with the large eyes (see Fig. 97) principally similar to those in the modern dragonflies. The odonatans seem to be primarily adapted to hunting flying insects in the air space. This type of specialised predation became possible only with the appearance of the pterygotans and was unavailable for any other invertebrates. There is also no evidence of a Palaeozoic origin of any flying vertebrates so that the adult odonatans have invaded a competition-free niche space and evolved for a long time without competition from other animals. However, the situation might be different for the nymphs, and nothing is known about the adaptations that ensured their success. The nymphal biology of the Palaeozoic odonatans still remains completely enigmatic. There are no unquestionable records of odonatan nymph fossils from the Palaeozoic so that they hardly could be aquatic, and in any terrestrial habitat they should compete with other predacious arthropods (major differences in diet between the immatures and adults in a hemimetabolan insect order are improbable).

The Palaeozoic odonatans show a significant adaptive radiation as demonstrated by their diverse wing and leg morphologies as well as their widely varied size (with wing length from 12–15 up to more than 300 mm). According to WOOTTON & KUKALOVÁ-PECK (2000) the basic flight adaptations of Palaeozoic odonatans were principally the same as in the present-day forms. The oldest odonatans are known from the Namurian B and C of Europe though the Early Westphalian family Eugeropteridae from South America (see Figs. 104, 105) demonstrates the most plesiomorphic wing venation. The earliest odonatans were moderately large insects with relatively weakly specialised wings. According to PRITYKINA (1980, 1989), the Carboniferous odonatans were mainly gliders hunting on flying insects in semi-open habitats. This mode of flight, together with an abundance of large-sized prey and predation by other odonatans stimulated increase in size as a widespread evolutionary trend that culminated in the Stephanian and the Early Permian when some meganeurids reached 63 (*Meganeura monyi* Brongniart from Commentry) and even 71 cm (*Meganeuropsis permiana* Carpenter from Elmo) in the wingspan. SHEAR & KUKALOVÁ-PECK (1990) believe that there was probably a case of ecological escalation in the form of "arms race" (a co-evolutionary phenomenon postulated by VERMEIJ 1983, 1987, primarily for explanation of parallel development of thicker and more protective shells in marine molluscs and specialised teeth system in fish and marine reptiles during the Mesozoic)

between the Palaeozoic odonatans and their prey leading to the progressive increasing in size in both. It is possible that such an "arms race" had resulted in development of an inadaptive hyperspecialisation in both the meganeuroid and dictyoneuroid giants and their extinction by the latest Permian. WOOTTON & KUKALOVÁ-PECK (2000) point out that some Palaeozoic insects demonstrate unique flight adaptations lacking in any living taxa (e.g. the "insect biplanes" with broadly overlapping fore and hind wings); they may represent such early inadaptive lines never again appearing in insect history.

As pointed out above, there were probably other, and much more important evolutionary consequences of odonatan predation, namely the independent loss of archemetaboly in different pterygotan lineages and common trend to a more cryptic life in shelters which has opened diverse ways of further insect specialisation including the separation of Gryllones and Scarabaeiformes as the most important novelties induced. It should be noted that the odonatans themselves were perhaps the oldest pterygotans to lose archemetaboly; at least, supposed subimaginal stages were never recorded for any members of the order. If so, this is a good example of the adaptive value being revisited by natural selection when the acquisition of a new important adaptive advantage in one lineage stimulated parallel evolution in many other lineages because of its increasing selective value. In the Permian the smaller ditaxineuroid (see Fig. 100) and kennedyoid odonatans evolved much towards progress in manoeuvrability rather than in the speed of flight, suggesting that they dwelled less in open air space, but most probably among dense vegetation. This trend was stimulated likely by the pressure of predation of the meganeuroids on both smaller odonatans and their potential prey. The kennedyoids seem to be especially specialised for hunting in densely vegetated space as indicated by their advanced leg structure secondarily adapted to catching prey from the plant surface (PRITYKINA 1980, 1989).

KUKALOVÁ-PECK (1983, 1985) suggests that aquatic nymphs of the early mayflies were also predacious. If this is correct, a predatory habit could be ancestral for the entire libelluliform lineage, and also for the adult Palaeozoic mayflies which had functional mouthparts and so could feed on other insects. DIMICHELE & HOOK (1992) list the ephemerideans among the predatory groups. However, Kukalová-Peck's argumentation is rather weak in this point and may be doubted because the morphology of nymphal mandibles does not indicate the diet with any certainty, and the Carboniferous *Carbotriplura* is probably not a mayfly nymph and not even a pterygotan at all (Chapter 2.1a). The mouthparts of the winged stages of the Palaeozoic ephemerideans are poorly studied but they seem to be too weak for specialised predation and somewhat beak-like suggesting rather an hypoperlidean or dictyopteridean type of feeding. The leg structure also lacks any features of predatory specialisation. It is difficult to imagine which kind of prey could be caught successfully by a *Triplosoba*- or *Protereisma*-like poor flier with unspecialised wings, long fragile caudal filaments, slender unarmed legs, and metamorphosis of the archemetabolic type. Even if the aquatic nymphs of the Permian mayflies were actually predacious nothing can be said presently about their adult feeding. The oldest, Carboniferous ephemerideans differed strongly from the true mayflies (e.g. they probably developed in terrestrial habitats) and are placed now into two separate orders, Triplosobida and Syntonopterida (Chapters 2.2.1.1.1.1 and 2.2.1.1.1.2). Hence the Palaeozoic ephemerideans should be excluded from the guild of terrestrial (aerial) predators.

In the cimiciform orders predation is rare and occurs sporadically in modern true bugs and thrips only. As pointed above, the oldest

Permian lophioneurid thrips (see Fig. 171) could have been omnivorous and feed facultatively on insect eggs and mites like living aeolothripids. For any other Palaeozoic cimiciform insects a predatory habit seems to be improbable. PRUVOST (1919) suggested blood-sucking on vertebrates for some smaller Permian dictyoneurideans. This extravagant and nearly forgotten hypothesis is absolutely groundless and would not be mentioned here if it had not been referred to recently by DiMICHELE & HOOK (1992) in their excellent review.

In the Scarabaeiformes an early origin of predation is quite possible, at least in the larval stage, because of a high probability of cannibalistic behaviour in their closed larval microhabitats inside plant cones. In the virtual absence of palaeontological data on early holometabolan larvae the only and rather weak basis for speculations on their biology is the biology of modern taxa. Among living holometabolans predatory habits are common and occur in all orders even including those adapted to other modes of feeding (e.g. in some moth caterpillars); however, only in the neuropteroid orders is it obligatory or nearly so. Accordingly, it cannot be ruled out for any major taxa of the Palaeozoic Scarabaeiformes including the extinct orders but most probable for the neuropterans and the related Jurinida. Both appear in the record in the Artinskian and become more diverse in the Late Permian. Nothing is known about their larval habitats in the Palaeozoic. For the adults of any Palaeozoic holometabolans specialised predation is unlikely because of the absence of any evident morphological specialisation in this direction, but a facultative predation on smaller arthropods (such as the mites or springtails) and arthropod eggs cannot be ruled out, especially for the neuropteroids and some mecopterans.

Omnivory is also probable for some Palaeozoic Gryllones (e.g. for the orthopterans) but evidence for specialised predation in this lineage are equivocal. Earlier it was postulated for some extinct taxa (e.g. for the Stephanian *Cnemidolestes* Brongniart; ZHERIKHIN 1980a) but mainly on the basis of the misinterpretation of their morphology (Chapter 2.2e). GOROCHOV (1995a) believes that the most primitive oedischioid orthopterans were probably predacious and points out that in the Early Permian genus *Uraloedischia* Sharov the spines on the fore- and midlegs are specialised suggesting that they could operate as an effective catching apparatus. This hypothesis may well be correct but it should be stressed that predatory feeding was supposed on a similar basis also for the Mesozoic hagloids which were in fact pollinivorous as was recently demonstrated. Unspecialised predation cannot be excluded as well for a number of other Palaeozoic Gryllones (e.g. for some Grylloblattida) but here again no definitive conclusions are possible given the present state of our knowledge.

In summary, it may be said that predation is probable for a rather wide range of Palaeozoic pterygotans but, except for the odonatans, in a purely speculative manner only. Hence the actual distribution of predatory habits among the Palaeozoic pterygotans and their relative role as predators in the ecosystems is unclear. Any specialised modes of zoophagy other than predation (parasitoid, ecto- and endoparasitic on the vertebrates including the blood-sucking) are undocumented and almost surely had not developed in any Palaeozoic insects.

3.2.2.2.5. Some ecosystem-level phenomena

Pterygotans were the only predators in Late Palaeozoic air space, and thus the large odonatans constituted the top members in those particular food chains. In the tree crowns there were probably other predatory arthropods as well but no insectivorous vertebrates. On the ground

surface living insects certainly provided an important and intensively exploited food source for other diverse animals, both invertebrates and vertebrates. The main predatory arthropod groups were recorded above. Their ecological interrelations with the insects were complex, including feeding on insects belonging to different trophic guilds as well as both preying by and competition with the predatory insects. The obvious decline of predacious arthropods other than insects in the Permian indicates that the insects were in general more successful. Several chelicerate orders completely disappeared, and some others have survived after the Carboniferous as rare relicts of minor ecological importance (e.g. the modern members of the Ricinuleida and the primitive spider suborder Mesothelae which dominates in the Carboniferous record) (SHEAR & KUKALOVÁ-PECK 1990). However, it is highly possible that their decline was due to increasing vertebrate predation and/or the above-mentioned changes in the structure of the litter as a habitat rather than to direct competition with predacious insects. As pointed out above, diverse adaptations for the protection of eggs and non-flying immatures against predation should have been extremely important in the Palaeozoic insect adaptive radiation.

The terrestrial vertebrates show intensive morphological and ecological radiation in the Late Palaeozoic which had to have had a marked effect on insect evolution. The arthropods probably provided the most important food source for the earliest terrestrial tetrapods though other invertebrates were surely eaten as well. DiMICHELE & HOOK (1992) summarise the available data on possible diets and habitats of the Late Palaeozoic terrestrial vertebrates and listed a number of supposedly insectivorous taxa of both amphibians and early reptiles. The Middle and Late Carboniferous terrestrial vertebrates are known from relatively few localities; the Westphalian assemblages are more ecologically uniform and more specific compared to the Stephanian faunas which demonstrate more affinities with the Early Permian. The appearance of the amniotes in the Westphalian and their Permian radiation were especially important because this indicates a significant progress in vertebrate terrestrialisation in comparison with the amphibians. The Late Permian faunas are very distinctive and especially rich in mammal-like reptiles.

Though there were surely some important co-evolutionary interactions between the insectivorous vertebrates and their prey they are poorly studied. Since ROHDENDORF's (1970) brief paper there were no attempts to compare the data on morphological evolution of the vertebrates with the fossil insect record though large bodies of new data have been accumulated on both. In the papers on vertebrate palaeontology a number of ideas about possible food specialisation in the Palaeozoic can be found but they are never analysed in the light of the palaeoentomological data. Such comparative analysis is highly desirable. The progressive evolution of sensory organs and brain in the smaller vertebrates could be stimulated by hunting on such highly mobile prey as the insects. Vertebrate/insect co-evolution should be equally important also from the point of view of evolution of the sensory and locomotory organs in insects and, as discussed above, their competitive interrelations with other arthropods (DOWNES 1987). For instance, the acquisition of jumping ability in the Palaeozoic orthopterans (see Figs. 434, 435) was probably stimulated by vertebrate predation. Many Palaeozoic insects (see Figs. 148, 360) as well as other terrestrial arthropods demonstrate spines which are often considered as protective structures (SHEAR & KUKALOVÁ-PECK 1990). Some Palaeozoic insects show remarkably long and probably fragile, projections on their bodies (e.g. *Monstropteron moravicum* Kukalová-Peck) which could operate as autotomous structures. The development of cryptic body shape, colour and behaviour is probable.

However, GOROCHOV (1995b) points out the absence of cryptic body shape and disruptive colour pattern in the Palaeozoic orthopterans and supposes that the Palaeozoic insectivorous vertebrates probably could not detect motionless prey. The existence of advanced mimics of plant leaves in the roaches suggested by A. SCOTT & TAYLOR (1983) has been questioned by JARZEMBOWSKI (1994).

No data are available on possible Palaeozoic insect parasites and pathogens though their existence is highly probable. STØRMER (1963) has described supposedly parasitic nematodes found in an Early Carboniferous scorpion suggesting that nematode parasitism on arthropods originated before the appearance of the pterygotans. However, the nematodes were more probably detritivores that fed on the dead scorpion rather than its parasites (POINAR 1984b).

There are no detailed studies of Palaeozoic insect distribution in correlation with plant assemblages either throughout any representative sections or between them which could elucidate their community preferences though the landscape and community succession pattern was surely complex in the Middle and Late Carboniferous and Permian. This is one more important unexplored field waiting for future research.

In general, the ecological role of insects increased dramatically after the appearance of the pterygotans. An intensive ecological radiation occurred in the Late Carboniferous and Early Permian accompanied by a progressive winged insect expansion from their original niche of the crown dwellers feeding on plant reproductive organs to new trophic niches (feeding on photosynthetic tissues, detritivory, predation and probably fungivory) and habitats (forest floor and perhaps soil as well as water bodies). Consequently, the role of insects in ecosystem turnover was increasing significantly while other arthropods became gradually replaced by insects both as detritivores and predators. The most spectacular specialisation level has been reached by the herbivorous and some predatory insects, and probably the plant/herbivore relationships and the relations between the air predators and their prey were the key co-evolutionary phenomena most important for the origin of the most successful insect taxa.

3.2.2.3. The Palaeozoic/Mesozoic transition

The end-Permian extinction as documented by the marine fossil record is one of the largest in the history of life (RAUP & SEPKOSKI 1982, ERWIN 1990). It seems to be significant though less dramatic in the terrestrial biota as well (WING & SUES 1992). In both marine and non-marine taxa the extinction was probably not sudden but rather gradual (ERWIN 1990). Like any other events in the evolution of life the extinction can be detected without any ecological analysis but it needs to be explained. Thus a comparative ecological analysis of the taxa differentially affected by the extinction is necessary to create and test any causal models.

A general model of biotic crises developed recently (ZHERIKHIN 1978, 1992b, KALANDADZE & RAUTIAN 1993a, RAUTIAN & ZHERIKHIN 1997, ARMAND et al. 1999) suggests that the scale of biotic changes depends more on the state of the biological community than on the magnitude of the environmental influences. According to this model, communities generally evolve from assemblages of weakly co-adapted generalists coexisting in an unsaturated niche space to the concerts of highly co-adapted specialists with the niches densely packed in highly saturated systems. In the course of this evolution the community's control over the evolution of its members is increasing progressively with a corresponding decreasing rate and increasing

channelling of the phylogenetic changes. A community constituted mainly by relatively generalised and moderately co-adapted members can survive a large-scale environmental stress without a serious reorganisation because even rather large environmentally caused extinctions can be easily compensated by rapid radiation. As a result, the faunal turnover rate is increasing without a significant drop in the total diversity. In highly advanced communities the rate of speciation is very low, and any further significant radiation is improbable. This stage is marked in the palaeontological record by the decreasing rate of appearance of new taxa. At the same time the probability of extinction under an environmental stress is high because of the narrow specialisations of a large number of taxa. This makes the community in general much more sensitive to environmental changes so that even a relatively small alteration in climatic or other environmental variables may provoke a disproportionally large community destruction accompanying uncompensated mass extinction. The taxa involved in the most intimate and complex co-adaptive interrelations within the system are particularly vulnerable, including the previous community dominants. Most of them are disappearing or at least declining and replaced by more generalised taxa which played a limited role in the pre-crisis communities. In the unsaturated niche space after the mass extinction community control on phylogenetic changes becomes less effective, and the generalists that survived had an opportunity to realise their high phylogenetic potential. A rapid radiation occurs resulting in the appearing of a number of new taxa including some unsuccessful short-living ones. The most successful lineages form the ground of a new stable community type.

The insect dynamics in the Late Permian and Triassic has been analysed recently by PONOMARENKO & SUKATSHEVA (1998), and the sketch below is based mostly on the data from that paper.

The latest Permian (Tatarian, between 253 and 248 Ma BP) insect faunas are known mainly from East Europe, North Asia and the southern continents (South Africa and Australia), that is from both northernmost and southernmost parts of the Late Permian Pangea while the fauna of lower latitudes is practically unknown. The largest collections (several hundred specimens each) are available from Kargala (Orenburg Region, south-eastern part of the European Russia), Sokolova-II, Suriyokova (both in the Kuznetsk Basin, West Siberia), Karaungir (Eastern Kazakhstan), and Belmont (New South Wales, Australia). The material is described only in part, and a number of remains are identified preliminarily only up to family level. The Kargala fauna is likely the oldest, supposedly Early Tatarian, while the Karaungir assemblage is probably one of the youngest Permian insect assemblages known. The Belmont fauna is rather distinctive due to the presence of endemic families and the absence of some important northern taxa, probably because of its isolated geographical position. Thus the data available are rather limited in comparison with the older, Kazanian and especially the Early Permian faunas.

The Early Triassic (248 to 243 Ma BP) insects are even less well studied. The total number of insect fossils collected world-wide from the Early Triassic deposits does not exceed 200–250 specimens, and the largest collection available now (from the Perebor Formation of the Pay-Khoy Range at the Northern Urals) contains only about 100 remains, mostly fragmentary. Besides the localities mentioned by PONOMARENKO & SUKATSHEVA (1998) several insect fossils have been collected in 1999 from the earliest Triassic deposits at Nedubrovo in the Vologda Region (north-eastern European Russia). No Early Triassic insects have been discovered outside East Europe and Northern Asia except for a 'supposed insect wing' from the vicinity of Hobart,

Tasmania (TASCH 1975) and a homopteran from the Rio do Rastro Formation, Brazil (MARTINS-NETO & ROHN 1996). The record of insects in M. BANKS *et al.* (1978) is probably based on TASCH's paper and not on any new finds. The insect fauna of the Buntsandstein of the Vosges Mountains in France referred to occasionally as Early Triassic is in fact not older than the Early Anisian (the earliest Middle Triassic) (J. GALL 1996). Hence the Early Triassic seems to be the most important gap in the insect fossil record since the beginning of the Middle Carboniferous.

The insects collected in the Tunguska Basin in Central Siberia are worthy of a special consideration. Here a number of insect localities have been discovered though none of them has yielded more than 100 specimens. Unfortunately, the position of the Permian/Triassic boundary in the Tunguska Basin is very disputable. The correlation with the non-marine sections in other regions is difficult because of the very distinctive biota that inhabited an uplifted inland volcanic area. The coal-bearing Permian deposits contain plant assemblages similar to the Late Permian floras of other regions while in the overlying volcanic-sedimentary strata they become very different. The rich and peculiar Korvunchan flora from those strata, though including both "Permian" and "Triassic" elements, resembles the Triassic floras and generally is of the Mesozoic (or Mesophytic, to use the palaeobotanical term) appearance (MEYEN 1987a). It is supposed that the Tunguska sequence is more complete than the stratotypic Late Permian sections in the European Russia and may include a post-Tatarian stratigraphic interval missing in Europe (MEYEN 1987a). In particular, the characteristic lycopsid family Pleuromeiaceae appearing westwards in the Early Triassic is not uncommon here. Diverse ferns and peltasperms predominate in the oldest Korvunchan floras while the younger assemblages are dominated by either the "Permian" conifer genus *Quadrocladus* (Lebachiaceae) or, more rarely, the "Triassic" pleuromeiacean lycopsids that probably inhabited more swampy habitats. The oldest Tunguskan insect faunas known from the Pelyatka and Degali Formations are undoubtedly Late Permian (perhaps, Early Tatarian) while others have been discovered at different stratigraphic levels within the disputable interval with the Korvunchan-type floras. Irrespective of their exact geological age, they dwelled in plant communities resembling those of the Early Mesozoic rather than Late Palaeozoic ones. Usually the term "Korvunchan flora" is restricted to the Tunguskan floras only but MEYEN (1987a) assigned to the similar but more southern flora from the Mal'tsevo Series in the Kuznetsk Basin where in Babiy Kamen' few insects have also been collected.

Analysis of the family dynamics (PONOMARENKO & SUKATSHEVA 1998) demonstrates that the number of insect families is maximal in the Kazanian, decreases sharply in the Tatarian and gradually increases again in the Middle and Late Triassic up to the level comparable with the Kazanian one (Fig. 483). Even more informative is the inverted Lyal's curve representing the changes in the share of the insect families extinct before the Jurassic in assemblages dated from the Carboniferous/Permian boundary up to the end of Triassic (Fig. 484). This share is high (about 90%) up to the Kazanian, then it drops suddenly up to about 40% by the end of the Tatarian and, after minor changes in the Early Triassic and the Anisian, further decreases down to 10% or less during the Ladinian. This pattern is remarkably similar to the family level dynamics observed in the Cretaceous (Chapter 3.2.3.2).

When all these patterns are taken into account, a number of symptoms appear evident which were formulated as common for the palaeontologically documented biotic crises (RASNITSYN 1988b, KALANDADZE & RAUTIAN 1993b, BARSKOV *et al.* 1996, RAUTIAN & ZHERIKHIN 1997, ARMAND *et al.* 1999; it should be noted that the English translation of

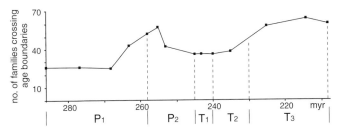

Fig. 483 Dynamics of instantaneous number of the insect families (number of families crossing age boundaries as in Fig. 478a during the Permian and Triassic) (from PONOMARENKO & SUKATSHEVA 1998)

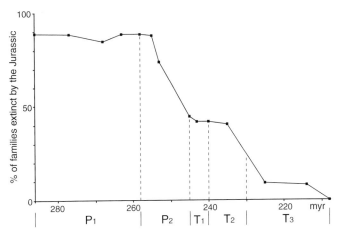

Fig. 484 Share of the families extinct by the Jurassic in the Permian and Triassic insect assemblages (the inverted Lyellian curve; from PONOMARENKO & SUKATSHEVA 1998)

KALANDAZE & RAUTIAN's paper may be misleading because of the inadequate translation of some terms such as "living forms" instead of "life forms"). These symptoms include the remarkable uniformity of the faunas at the world scale before the crisis (in the Kazanian), the episode of somewhat increasing percentage of extinct taxa in the Lyal's curve just before its steeply descending interval, the uncompensated extinction connected first of all with a decreasing rate of appearance of new taxa, the appearance of short-living taxa restricted to the crisis time (e.g. Tomiidae in the Early Triassic, see below), and the increasing rate of radiation of new taxa at the end of the crisis resulting in the restoration of their total number. This complex pattern suggests that the end-Permian event was generally in accordance with the biotic crisis model outlined above.

The extinction is likely to have affected all the major trophic guilds but not equally so. Unfortunately, there are almost no data on the feeding traces and gut contents for the time interval under consideration so that the changes in the trophic guild structure can only be argued by indirect evidence.

The most obvious is the extinction in the guild of ectophytic herbivores feeding on plant reproductive organs. This is equally true for both presumed specialised pollinivores and seed predators. The less advanced and predominantly pollinivorous cimiciform orders Caloneurida, Hypoperlida and Blattinopseida were not uncommon in the Early Kazanian assemblages but none of them occurred after the Late Kazanian. The specialised sucking dictyoneurideans were represented in the Late Permian mainly by Mischopterida but there were also some members of Dictyoneurida. They were common in the Early Kazanian but completely disappeared since the Tatarian. It should be noted that all known Late Permian dictyoneurideans with preserved mouthparts had a short beak and thus probably fed on seeds. The grylloblattidan family Ideliidae, documented as pollinivorous in the Early Permian, has survived the Permian/Triassic boundary. It was only weakly morphologically specialised and so could include genera with a different diet as well. The extinction is less evident in small-sized herbivores with supposedly more endophytic mode of life like the permopsocine barklice and lophioneurid thrips that survived in the Mesozoic. The archescytinid hemipterans occurred up to the Late Tatarian; the latest fossils of this family are discovered in deposits questionably of latest Permian or earliest Triassic age in Vologda Region (Nedubrovo) and Tunguska Basin (Bugarikhta), Russia. The holometabolan Palaeomanteida with supposedly endophytic larvae that dwelled in cones were diverse in the Kazanian and in the Early Tatarian. Unfortunately, the Late Tatarian finds are undescribed, and nothing is known about the Early Triassic miomopterans. Two Permian families (Palaeomantiscidae and Palaeomanteidae) are not recorded from the Triassic while Permosialidae have crossed the Palaeozoic/ Mesozoic boundary as documented by Late Triassic and Jurassic fossils. Moreover, both permosialid genera known from the Triassic are common in the Late Permian though one of them, *Permonka*, is rare in the Permian. Nothing can be said with certainty about other holometabolans because of their unknown larval habits. The most primitive beetle family Tshekardocoleidae which may have been herbivorous disappeared from the fossil record after the Kazanian.

On the contrary, the guild of sucking herbivores that fed on photosynthetic tissues demonstrates a high turnover rate but no decline. In the Late Permian it included only hemipterans but this order was well represented. A number of families were disappearing between the Early Kazanian and Mid-Triassic but this limited extinction was overcompensated by the radiation of new families. Some family-group taxa characteristic of the Mesozoic are known since the Tatarian (e.g. Fulgoroidea and Palaeontinoidea), but the Tunguskan assemblages from those beds with the Korvunchan-type flora are still dominated by Permian taxa. In the earliest Triassic some taxa with strong Late Permian affinities (e.g. Archescytinidae in Nedubrovo: SHCHERBAKOV 2000b) are found. However, in the Middle and Late Triassic assemblages a number of families unknown from the Permian were represented, and the dominance pattern was very different. The phloem-feeders absolutely predominated throughout the Permian and Triassic though first supposed xylem-feeders (Hylicelloidea) appear in the Anisian. Hence there were significant changes at the family and even superfamily level in the taxonomic composition of the guild but no uncompensated extinction is documented suggesting that the radiation rate was constantly high.

The situation with the ectophytic chewing consumers of photosynthetic tissues is unclear. As discussed above, the taxonomic composition of this guild in the Permian is uncertain, and the distribution of chewing marks on the Late Permian and Early Triassic plants was never investigated. An obligate folivory is probable for the stick-insects and the locustopsid grasshoppers. *Permophasma* Gorochov, the oldest supposed stick-insect is discovered in the Late Permian (probably Kazanian) of Mongolia but the only known remains are fragmentary, and no other stick-insects have been discovered in deposits older than the Mid-Triassic. The situation with the locustopsids is similar except for the time of the appearance in the record: the oldest

supposed locustopsid, *Praelocustopsis* Sharov, originated from the lower part of the sequence with the Korvunchan-type flora in the Tunguska Basin and thus is not older than Late Tatarian.

Information about endophytic borers is also extremely scarce. Some borings tentatively attributed to beetles and made perhaps in living trees are recorded from the Late Permian of Gondwanaland but the stratigraphic correlation with the northern hemisphere is strongly disputable. In the Kuznetsk Basin (West Siberia) the Late Permian beetle assemblages of the Kazanian and Tatarian are significantly different, and a decline of cupedoid archostematans is evident. However, taxa related to those found at the different stratigraphic levels in the Kuznetsk Basin are recorded occurring together in the Kazanian of Mongolia and Late Tatarian of East Kazakhstan. So the extinction pattern was more complicated than previously believed. No cupedoid-type beetle elytra occurred in the known Early Triassic assemblages though cupedoids are represented in the younger Triassic faunas. The biology of the Permian cupedoids is uncertain, and most of them could well have been detritivores that developed within dead wood. In any case, there are no cupedoid families in common for the Late Permian and Triassic, and the latest records of the Permian families are from the Late Tatarian of East Kazakhstan and the lowermost part of the Tunguskan Basin deposits with the Korvunchan-type flora (the Limptykon Formation). The Middle and Late Triassic cupedoids belonged mainly to the living xylomycetophagous family Cupedidae *s. l.* though the endemic family Tricoleidae of unknown biology was represented as well.

The changes in the herbivore guilds are not surprising because the plant assemblages were changing significantly during the Late Permian. In particular, in the so-called Subangaran floristic region (after MEYEN 1987a) where the majority of the insect localities were situated, the Late Tatarian flora differed strongly from the Kazanian and Early Tatarian ones. It is known as the *Tatarina* flora named after a characteristic peltasperm form-genus (GOMAN'KOV & MEYEN 1986). However, a comparison of the changes in plants and insects indicates that the Late Permian extinction among the herbivores cannot be easily explained simply by disappearance of their host plants. In particular, the taeniate pollen grains which seem to be the prevailing pollen type in the diet of Early Permian pollinivorous insects not only show no decline but even increase in abundance in the latest Permian, and just this type of pollen was produced by the peltasperms with the *Tatarina*-type leaves (MEYEN 1987a). Also, as pointed out above, different herbivore guilds were unequally affected by the Late Permian extinction, though the endophytic cone dwellers or the phloem-suckers should not be principally less host-specific than the ectophytic pollinivores and seed-suckers. Probably the latter guilds had their niche space over-saturated in niche dimensions other than trophic ones, perhaps in respect of the habitat specialisation. This speculative hypothesis has to be tested by detailed analysis of the palaeoenvironmental distribution pattern of fossils. An overspecialisation with extreme narrowing of niches could arise as a consequence of co-evolution with the odonatan predators. Noteworthy is that the guilds less affected by the extinction are those that were less vulnerable to air predation.

The detritivore/omnivore guild can be outlined only conventionally as including the roaches, grylloblattidans, earwigs, mecopterans and xylomycetophagous beetles. As discussed above, some of them could in fact be herbivores or predators. It is also difficult to subdivide them into habitat guilds though in general the detritivores should be more specialised in respect of microhabitat than diet. All those orders demonstrate important changes in the Late Permian and Triassic in their abundance, diversity and taxonomic composition. In general, the roaches declined in the Late Permian but became more abundant and diverse later. In contrast, the diversity of grylloblattidans, mecopterans and beetles was increasing but then suddenly dropped by the Triassic. Unfortunately, the roaches are insufficiently studied and mostly undescribed; in particular, the Early Triassic fossils are unclassified even to family level. However, since the beginning of the Triassic they evidently became much more abundant as indicated by their finds in nearly all Early Triassic sites and especially by their high abundance in material from the Perebor Formation. The grylloblattidans and mecopterans are the best studied orders though this case the available collections are only partially described as well.

In the alleged detritivore guild the presence of short-lived taxa restricted to the crisis interval is noteworthy. Three mecopteran (Belmontiidae, Robinjohniidae and Permotanyderidae) and one protelytropteran (Dermelytridae) families are confined to the Tatarian of Australia (Belmont), and an undescribed family of the most primitive true earwigs (Forficulina) to the Late Tatarian of Kazakhstan (Karaungir). All these families are rare and it is possible that their stratigraphic distribution is insufficiently known. The case of the grylloblattidan family Tomiidae is more significant. The only described species, *Tomia costalis* Mart., is discovered in the Early Triassic Mal'tsevo Series in the Kuznetsk Basin. Additional undescribed fossils have been collected from the Early Triassic of Yaroslavl' Region (the Rybinsk Formation) and Vologda (the Vokhma Formation) Region in European Russia as well as from the lower part of the Tunguska Basin deposits with the Korvunchan-type flora (the Dvurogino horizon). Thus the family seems to be restricted to the least known stratigraphic interval and completely absent in both older and younger deposits with much better known fauna. This striking distribution pattern suggests that the total diversity of the Early Triassic insects should be remarkably low. Another evidence of low diversity is the remarkable uniformity of the isolated beetle elytra in the same assemblages which, however, can belong to aquatic forms. Unfortunately, they cannot be identified even up to the family level but in any case they give a peculiar appearance to the Early Triassic assemblages differing them from both older and younger faunas. There are also several somewhat longer living families survived in the Middle and earlier Late Triassic but not later, e.g. the roach family Subioblattidae and the grylloblattidan family Tunguskapteridae. On the other hand, few families characteristic of the Mesozoic have appeared in the Latest Permian for the first time (e.g. Blattogryllidae in the Late Tatarian) suggesting that the typically Mesozoic taxa have radiated mainly after the beginning of the Triassic.

The herbivore guild also demonstrates the noteworthy phenomenon of the latest Permian recurrence of some taxa missing from the fossil record for a long time before this. This is documented by the presence of some beetles closely similar to the earliest Late Permian (Ufimian) genera (PONOMARENKO 1995a) and the archaic grylloblattidans (Euryptilonidae: STOROZHENKO 1998) in the Late Tatarian of East Kazakhstan. The euryptilonid *Karaungirella* Storozhenko is particularly interesting because it is the only post-Ufimian member of the family well represented in the Early Permian. The similar pattern was detected in the Cretaceous and explained by a temporary renaissance of some relict groups during the decline of dominants in the course of the community crisis (ZHERIKHIN 1978, 1987, 1993b).

Changes in the guilds of predators have been relatively little studied. Among the aerial predators, a decline of the meganeuroid odonatans is evident, especially after the Early Kazanian, that is more or less synchronously with the disappearance of their supposed main prey, the dictyoneurideans. The only meganeuroid family known from younger

deposits is Triadotypidae appearing in the Early Tatarian (*Kargalotypus* Rohd.) and survived up to the Middle Triassic. The smaller meganeurinan odonatans with the narrowed wing base supposedly adapted to hunting on vegetation were less affected by the extinction though there were some important changes at the family level. The same is true for the orthopterans.

In general, the insect faunas have surely changed significantly around the Permian/Triassic boundary though the available data are too scarce to document those changes in sufficient detail. The extinction of the typically Palaeozoic taxa occurred even in the Late Permian, and many of them were disappearing as early as in the Early Tatarian. The extinction is particularly evident in the guilds of the seed-suckers and the pollinivores. The extinction process was relatively gradual and more or less parallel to the floral changes. Even in the latest Permian the new taxa which could be regarded as Mesozoic ones occurred only sporadically suggesting that the extinction was not compensated for by radiation. The Early Triassic insect fauna was likely significantly impoverished in comparison with both the Late Permian and the younger Triassic.

3.2.3. MESOZOIC

3.2.3.1. Triassic – early Cretaceous

3.2.3.1.1. General features of Mesozoic insects (before mid-Cretaceous)

After the end of the Permian crisis the insect faunas demonstrate rather gradual evolution and a high level of evolutionary continuity for the next approximately 150 Ma up to the mid-Cretaceous extinction event. This period in insect evolution will be referred to here as the Mesozoic stage though it does not coincide exactly with the Mesozoic era time limits. It should be stressed that there is no evidence for any uncompensated insect extinction near the Triassic/Jurassic boundary where some authors postulate a significant extinction event among terrestrial vertebrates (KALANDADZE & RAUTIAN 1983, 1993a, BENTON 1986, 1987). Global insect diversity at the family level in the Late Triassic was still somewhat below the Kazanian numbers, with the pre-crisis level being exceeded only in the Early Jurassic (PONOMARENKO & SUKATSHEVA 1998). This was followed by a continuous increase up to the end of the Early Cretaceous (Chapter 3.1). As discussed below, ecological diversity was increasing as well, with successive development of several new and important niche spaces (**licences** after LEVCHENKO & STAROBOGATOV 1990, that is a 'volume' within the resource space which is considered as existing irrespective of being either occupied, as the **niche** is, or empty) and subsequent intensive radiation within them. On the other hand, local diversity at the species level (estimated on the basis of species numbers in the richest assemblages) changed relatively little since the beginning of the Late Triassic up to the Early Cretaceous ranging between one and several thousand species per locality (ZHERIKHIN 1992b). There are few extinct orders, and the oldest living insect families (e.g., Cupedidae *s. l.* and Xyelidae) appear in the Triassic. In the Early Cretaceous more than 50% of the insect families are still extant. The oldest living genera (the archostematan beetles *Omma* Newman and *Tetraphalerus* Waterhouse) are recorded from the Early Jurassic, and several more are found in younger deposits, though in the Early Cretaceous their total number was still very low. This relatively modern appearance of the fauna makes hypotheses about the Mesozoic insect ecology better grounded in comparison with the Palaeozoic. The dominance pattern differs clearly from the Cainozoic assemblages even at the ordinal level. In particular, the mecopterans, neuropteroids, roaches and orthopterans were much more abundant and diverse there than in the younger faunas. Their high taxonomic diversity suggests that they should also have been biologically diverse, and that some of them probably belonged to extinct life forms. The dipterans and hymenopterans, which were rare in the Triassic, become more and more common during the Jurassic and Cretaceous, and their diversification surely concerned their biology as well. At the ordinal level the most evident difference from younger assemblages is the extreme rarity of the lepidopterans and termites (the latter order appearing only in the Early Cretaceous).

The Mesozoic was the time of progressive fragmentation of the land masses that started from the Triassic Pangea (see Figs. 499–501). The isolation of fragments increased in the Jurassic and had reached its maximum by the Late Cretaceous (see Fig. 501). On the contrary, the climatic differentiation decreased from the strongly provincial Triassic climate to the extremely even Cretaceous one. Even the provincial differences were probably minimal in the Cretaceous because of low effects of continentality on relatively small continents bordered with shallow epicontinental seas. There is no firm evidence of permanent polar glaciations at any time in the Mesozoic, and the high latitudes were constantly occupied by forest vegetation that probably existed in a temperate climate with cool and dark winters. The ice drift deposits occurring in high palaeolatitudes are often cited as possible evidence of glaciations but most probably these originated mostly or even exclusively from seasonal sea ice cover (FRAKES & FRANCIS 1988, 1990, FRANCIS & FRAKES 1993, CHUMAKOV 1995). Supposed crioturbation structures in the Early Cretaceous of South-eastern Australia also may be connected with seasonal freezing and not necessarily with permafrost (CONSTANTINE et al. 1999). Of course, mountain glaciations are quite possible. Information on palaeogeographic and biotic changes during the Mesozoic is summarised in brief by WING & SUES (1992).

Triassic insects are found on all present-day continents except for Antarctica but almost all available material originates from the Middle and Late Triassic. The most representative faunas discovered are in the Buntsandstein of Vosges in France (Anisian; outlined by J. GALL 1996, and J. GALL et al. 1996), the Madygen Formation in Kyrgyzstan (Ladinian-Carnian), the Mid (Brookvale, Ladinian) and Late (Denmark Hill and Mt. Crosby, the Ipswich Series, Carnian) Triassic deposits of Australia, the Molteno Formation (Carnian) of South Africa (outlined by J. ANDERSON et al. 1998), and the Cow Branch Formation (Carnian) of Virginia, USA (outlined by FRASER et al. 1996). There are other important localities (e.g., in Ukraine, Kazakhstan, China, Japan, and Argentina) as well as a number of smaller sites in different parts of the World. The largest insect collections are available from the Madygen Formation where two principal localities, Madygen and Dzhayloucho, have collectively yielded more than 20,000 insect specimens. A brief review of the Triassic insect faunas of Gondwanaland was published by J. ANDERSON & H. ANDERSON (1993); there are no recent reviews for the northern continents. J. ANDERSON et al. (1998) have reconstructed the habitat distribution of insects for the Molteno palaeoenvironments; and this is the only paper analysing the palaeoenvironmental distribution of the Triassic insects in detail.

Jurassic localities are extremely unevenly distributed over the World. There are hundreds of localities of different age within the Jurassic in Europe and the northern half of Asia, and from many of them hundreds or even thousands of specimens are available. In Europe the faunas of the latest Early Jurassic age (Toarcian; so-called German Lias) are particularly rich, especially in Germany (Dobbertin and Grimmen in

Mecklenburg-Vorpommern and a number of sites in the vicinities of Braunschweig in Lower Saxony: A. BODE 1953, ANSORGE 1996, ZESSIN 1996). There are also some important older Early Jurassic (mainly Sinemurian: WHALLEY 1985) faunas in England (English Lias) and a rich Hettangian fauna at Aargau in Switzerland which, however, badly needs a revision. Middle Jurassic finds are quite scarce, and the only rich Late Jurassic fauna originates from the famous Solnhofen lithographic limestone of Bavaria, Germany (Tithonian; see FRICKHINGER 1994, for a review). There are only a few finds in East Europe including European Russia; in Asia numerous localities including a number of important Lagerstätten have been discovered. In Siberia there are several dozens of localities, mostly assigned to the Toarcian-Aalenian but also including several important younger, Middle and Late Jurassic sites (reviewed by ZHERIKHIN 1985). Different stratigraphic levels are also represented in numerous Central Asian (e.g., Sogyuty, Shurab, and Say-Sagul), Mongolian (e.g., Bakhar, Khoutiyn-Khotgor, and Shar Teg), and Chinese localities (e.g., in the Beipiao Basin in Liaoning Province and the Chengde and Luanping Basins in Hebei Province; see LIN 1980, HONG 1998, and references herein). It should be taken into account that the strata usually assigned to the Late Jurassic by Chinese authors contain a fauna of the same type as that found in the Early Cretaceous of Transbaikalia and Mongolia (in particular, the so-called Jehol, or Rehe, biota), and that for many of them an Early Cretaceous age was demonstrated by recent studies (e.g., MAO et al. 1990, KELLY et al. 1994, P. SMITH et al. 1995). In Kazakhstan the famous Karatau Lagerstätte (Oxfordian – Kimmeridgian: see ROHDENDORF 1968b, and DOLUDENKO et al. 1990, for reviews of the fauna and palaeoenvironment) is known as well as several smaller localities. Thus in Eurasia a good succession of insect faunas is documented. On the contrary, only few isolated finds have been made in North America and the southern lands (e.g., India, Madagascar, New Zealand, and the Antarctic) so that their Jurassic insect faunas are quite unknown. The palaeoenvironmental distribution was analysed in detail for the Siberian Jurassic insects (ZHERIKHIN 1985).

Early Cretaceous insects are well represented in both northern and southern hemispheres. The faunas were reviewed by ZHERIKHIN (1978) but since that time the available material has probably almost doubled. The most important new collections have been made in England (both from the Purbeck and Wealden Groups; for general information see JARZEMBOWSKI 1992, 1995b, E. CLIFFORD et al. 1994, RASNITSYN et al. 1998), Spain (for a review see PEÑALVER et al. 1999, ALONSO et al. 2000), Lebanon (AZAR 1997), Siberia, Mongolia, China (see HONG 1998, and references herein), Brazil (see MARTINS-NETO 1999b, and references herein), and Australia (JELL & DUNCAN 1986). There are still virtually no data on the Early Cretaceous insects of North America, Africa south of the Sahara, and Antarctica. The probable habitat distribution of insects was analysed on the basis of material from Western Mongolia (RASNITSYN 1986b), and a synthesis of the palaeoecological information on the Late Jurassic and Cretaceous insects was presented by LABANDEIRA (1998c).

Insect assemblages change dramatically near the Early/Late Cretaceous boundary, and thus the Late Cretaceous is considered separately here.

3.2.3.1.2. Herbivores

The main trophic guilds demonstrate a progressive diversification and increasing structural complexity during the Mesozoic. In particular, the evolution of the herbivore guilds is relatively well documented. In the Mesozoic the insects were surely the main terrestrial small-sized herbivores. Other herbivorous invertebrates (such as nematodes, mites or herbivorous terrestrial molluscs) probably existed as well but their history is obscure, and in any case, their ecological role would probably have been rather limited as in present-day ecosystems.

3.2.3.1.2.1. Anthophily and the evolution of entomophily

Ectophytic chewing pollinivores were probably represented in the Triassic, after the Late Permian extinction of the most specialised taxa, mainly by the opportunistic facultative pollen collectors such as the adult holometabolans (neuropterans, mecopterans, caddisflies, etc.). Perhaps, some more specialised pollinivores had survived among the grylloblattidans and orthopterans. In particular, the grylloblattidan family Ideliidae (in which pollinivory was documented for an Early Permian genus) is represented by several genera in the Madygen Formation. STOROZHENKO (1998) regards the ideliids as the most generalised Grylloblattina and ancestral to a number of other families. He believes that some of their derived Triassic descendants (e.g., the widely distributed family Mesorthopteridae of the Late Triassic) evolved as specialised plant dwellers, though their feeding habit is unknown. In the orthopterans pollinivory was recently demonstrated for the Late Jurassic prophalangopsids of the subfamily Aboilinae which were previously tentatively considered as predators (KRASSILOV et al. 1997a,b). It is quite possible that other hagloids including the families that existed in the Triassic as well as some more primitive taxa (e.g., the oedischioid family Proparagryllacrididae lived probably on plants: GOROCHOV 1995a) fed on pollen. Living orthopterans, and especially some katydids (the group phylogenetically connected with Prophalangopsidae: GOROCHOV 1995a) visit flowers for pollen (e.g., PORSCH 1958, GRINFELD 1962, 1978, SCHUSTER 1974, HABER 1980), and the tettigoniid subfamily Zaprochilinae even includes specialised pollen- and nectar-feeders (RENTZ & CLYNE 1983). It is noteworthy that pollen feeding on both angiosperms and gymnosperms is also recorded for the relict living Prophalangopsidae (MORRIS & GWYNNE 1979, GOROCHOV 1995a).

Further specialisation in pollinivory occurred mostly among the Holometabola. In the Ladinian-Carnian, the oldest hymenopterans present belong to the living sawfly family Xyelidae which has pollinivorous adults. Mesozoic xyelid pollinivory is documented directly by finds of pollen grains in the gut contents of several Early Cretaceous genera from Siberia (KRASSILOV & RASNITSYN 1982, and unpublished data) and Brazil (CALDAS et al. 1989). Adult pollinivory is also highly probable for the majority of the more advanced Mesozoic hymenopterans including both symphytans and apocritans. Among the neuropteroids, and especially in the mecopterans, pollinivorous specialisation in general probably remained comparatively weak, and an opportunistic feeding strategy predominated. However, in the mecopteroid complex more specialised pollinivory has been developed in the caddisflies and especially in the primitive mandibulate lepidopterans which radiated from them in the Jurassic (Chapters 2.2.1.3.4.2–3). Adult pollen feeding is also highly probable for a number of Mesozoic polyphagan beetles (e.g., elaterids, buprestids, some cleroids) and some cupedoid archostematans. Thus by the Late Jurassic and Early Cretaceous the pollinivore guild was probably rich and diverse. As in the Permian, only relatively few types of pollen have been discovered in the guts of the Mesozoic insects studied so far, namely *Classopollis* in Late Jurassic aboiline orthopterans, several peculiar types of the presumably peltasperm and gnetophyte pollen grains in Siberian Early Cretaceous xyelids (KRASSILOV & RASNITSYN 1999), and *Afropollis* in a Brazilian Early Cretaceous

xyelid (CALDAS et al. 1989). *Afropollis* is assigned to the early angiosperms and is supposedly related to the modern family Winteraceae, and the gymnosperm pollen found was probably produced by rather advanced taxa demonstrating some characters in common with the angiosperms (KRASSILOV 1989, 1997). Some insects fed on plant spores as indicated by finds of Early Cretaceous coprolites (tentatively assigned to a hymenopteran) containing fern spores (KAMPMANN 1983).

The sucking pollinivore guild was probably of a minor importance in the Triassic and was represented by the primitive lophioneurid Thripida; later more advanced aeolothripoid (since the Latest Triassic) and thripoid (since the Jurassic) thrips were added. No data on their possible feeding specialisations are available. Modern aeolothripoids are often omnivorous, feeding on both pollen and small invertebrates, and the same habit is probable for many lophioneurids. Living thripoids are usually more stenophagous but only partially pollinivorous, with a number of suckers of photosynthetic plant tissues or fungal hyphae.

A number of Mesozoic insects display mouthpart morphology that is suggestive of fluid feeding but is not of a piercing type. They are conventionally regarded here as constituting the nectarivore guild because nectar (i.e., polysaccharide plant secretions) and honeydew (the hemipteran excretions rich in the carbohydrates) constituted probably their main food sources; however, many of them could feed on different kinds of other exposed fluids of plant, fungal, or animal origin. Fluid feeding is widespread in extant adult holometabolans, in particular in the mecopteroid orders, and also occurs occasionally in orthopterans and other Gryllones. In the Mesozoic the taxonomic composition of the guild was probably similar to that of the present day.

The time of origin of fluid feeding is unclear. No known Triassic fossils demonstrate mouthpart morphologies that indicate highly specialised fluid feeding, but opportunistic fluid consumption is very probable for a number of neuropteroids and mecopteroids, and particularly for the caddisflies and early dipterans. It may be speculated that the first important source of nutritionally valuable liquid food for insects was honeydew. Honeydew production is widespread in different homopteran lineages and probably originated as early as the Permian. In fact, its appearance could have stimulated the development of nectar production in various plants, initially as a mechanism for attracting predacious insects so as to reduce the level of herbivory. With the increasing availability of nectar, more specialised nectarivory could evolve in different insect taxa.

In the Jurassic and Early Cretaceous the radiation and specialisation of nectarivores is documented by a number of fossils with variously derived mouthpart morphologies. The mouthparts of numerous dipterans and hymenopterans show no important differences from their modern relatives and therefore suggest similar feeding habits. Both of these orders appeared in the Triassic but became abundant and diverse from the Jurassic and then their diversity grew continuously up to the Late Early Cretaceous. Several taxa of specialised nectarivorous flies are recorded from the Jurassic and Early Cretaceous (e.g., Nemestrinidae since the Early Jurassic, Mythicomyiidae since the Middle Jurassic, and the Apioceridae and pangoniine Tabanidae since the Early Cretaceous) including some nemestrinids, apiocerids and tabanids (ROHDENDORF 1968c, REN 1998a,b, MOSTOVSKI 1998, MOSTOVSKI & MARTÍNEZ-DELCLÒS 2000) as well as Cratomyiidae, an extinct stratiomyoid family (MAZZAROLO & AMORIM 2000), all having long prosces. It should be noted that the flies described by Ren from the Yixian Formation are Early Cretaceous (P. SMITH et al. 1995) and not Jurassic as stated in his paper, and thus LABANDEIRA's (1998c) conclusion that the Early Cretaceous nectarivory is poorly documented in comparison with the Late Jurassic is invalid.

Besides the dipterans and hymenopterans many Mesozoic mecopterans, caddisflies, neuropterans and beetles were probably at least facultatively opportunistic fluid feeders. Some lineages evolved towards specialised nectarivory in the Late Jurassic and Early Cretaceous as demonstrated by Aneuretopsychidae with strongly modified long proboscis (see Fig. 276), *Necrotaulius tener* Sukatsheva with characteristic tibial cleaning apparatus (see Fig. 283), and Kalligrammatidae with long flexible palps (see Fig. 253). Polycentropodidae may be possibly added to this list but in this case the long proboscis seems to be rather rigid and perhaps of a piercing type (see Fig. 277) so that a different mode of feeding (e.g., blood sucking) cannot be excluded. Strangely, in his review of the palaeontological evidence for plant/pollinator relationships, LABANDEIRA (2000) does not discuss either neuropterans or mecopterans among the most probable Mesozoic pollinators.

The Mesozoic records of glossatan moths are disputable, including *Protolepis* Kozlov from the Late Jurassic (the taxonomic position questioned by KRISTENSEN 1997), the supposed nepticulid mine from the latest Jurassic or earliest Cretaceous of Australia (doubted by KRISTENSEN & SKALSKI 1999), and the scales assigned to an incurvariid from the Lebanese amber (doubted by KOZLOV 1988). Nevertheless, glossatan lepidopterans probably existed in the Late Mesozoic although they were far less common than the pollinivorous mandibulate moths. No evidence of specialised nectarivory is available for the Mesozoic beetles and orthopteroids, although this diet cannot be ruled out there, because many of their living representatives practise it with no conspicuous changes in mouthpart morphology. Fluid-feeding is probable also for some non-orthopteroid Gryllones, e.g., stonefly adults. Modern stoneflies are occasional flower visitors (PORSCH 1958), and SINITSHENKOVA (1986) mentioned the clearly visible gut contents of a specimen of *Gurvanopteryx* from the Early Cretaceous of Mongolia whose intestinal load appears to be amorphous suggesting a liquid food (V. Krassilov, pers. comm.). In general, the diversity and specialisation in the nectarivore guild increased continuously during the Mesozoic, and by the Late Jurassic and Early Cretaceous a high level of specialisation had been reached in a number of unrelated phylogenetic lineages. It should be noted that adult aphagy (non-feeding) seems to have evolved mainly from ancestral fluid feeding and had probably been acquired in the Mesozoic by some dipterans (e.g., Eremochaetidae: V. KOVALEV 1989) and stoneflies (SINITSHENKOVA 1987).

For those endophytic herbivores dwelling in plant reproductive organs, the Mesozoic radiation is less evident. This guild was probably composed mainly or even exclusively of the larvae of several holometabolous orders. The miomopterans had survived up to the later Early Jurassic but are generally uncommon in the Mesozoic. The oldest Triassic hymenopterans belong to the family Xyelidae and probably developed in the male cones of gymnosperms. The modern Xyelinae retain this putatively plesiomorphic habit while the Macroxyelinae have more ectophytic larvae, and both lineages appeared in the Mesozoic (RASNITSYN 1980). The plesiomorphic endophytic habit is probable also for some extinct Xyelotomidae though this family probably included some ectophytic forms as well (RASNITSYN 1990b). Other Mesozoic symphytans evolved toward either ectophytic or stem- and wood-boring larvae and thus outside the original hymenopteran adaptive zone. The stimuli leading to this

evolutionary trend are unclear. It should be noted, however, that endophytic xyelids were common and diverse in the Mesozoic, especially in temperate areas.

The beetles are usually considered the main endophytic order of the Mesozoic, and the damage found on Mesozoic fructifications (REYMANÓWNA 1960, BOSE 1968, DELEVORYAS 1968a, CREPET 1974, STOCKEY 1978) are tentatively attributed to them (SZAFER 1969, CREPET 1974, CROWSON 1981, 1991, T. TAYLOR & TAYLOR 1993, LABANDEIRA 1997a, 1998c, 2000). A number of modern polyphagan beetle larvae (mainly in the cucujomorph but also in some bostrychomorph families such as Anobiidae represented by undescribed fossils in the Early Cretaceous of Transbaikalia) develop in both male and female gymnosperm cones. Some modern beetles (e.g., some genera of Nitidulidae, Boganiidae, Languriidae, and primitive chrysomeloids and curculionoids) infest archaic gymnosperms including Araucariaceae (known in the fossil state since the Jurassic and less certainly from the Triassic: MEYEN 1987a) and Cycadaceae (known in the fossil state undoubtedly since the Late Cretaceous but related to some older Mesozoic taxa), and some of those connections may be very old (CROWSON 1981, 1991, but see Chapter 3.2.3.2). In particular, the weevil families Nemonychidae and Belidae with a number of genera developing in modern araucariacean cones are well represented in the Late Jurassic and Early Cretaceous, especially in warmer areas (Kazakhstan, Spain, Brazil). The majority of Mesozoic genera of those families belong to extinct subfamilies but the extant nemonychid subfamily Rhinorhynchinae living mainly on araucarians is represented in the Early Cretaceous (Santana in Brazil, undescribed). The endophytic guild probably also included several extinct beetle families such as the Jurassic cucujomorph family Parandrexidae (KIREJTSHUK 1994) and Obrieniidae. The latter, known from the Triassic up to the Late Jurassic, had originally been placed to Curculionoidea but with some reservation because of its archostematan-type meso- and metathoraces (ZHERIKHIN & GRATSHEV 1993). Its taxonomic position has been specially discussed during the John Lawrence Celebration Symposium in Canberra in 1999, with a final conclusion that the obrieniids should be regarded rather as an archostematan family convergently similar to the weevils (J.F. Lawrence, V.V. Zherikhin, W. Kuschel, R. Oberprieler, R. Leschen, and J. Muona, in preparation). In particular, an important (though indirect) argument against their curculionoid nature is that otherwise they should be related to highly advanced brentid weevils and not to any of the more primitive families while in the large Mesozoic collections examined, the oldest (and still morphologically primitive) brentids only appear in the Barremian-Aptian (Bon-Tsagan in Mongolia and the Crato Formation in Brazil, undescribed). This 100-million-year-long time gap between the first obrieniids and the first brentids would be difficult to explain if both lineages were closely related. On the other hand, irrespectively of their phylogenetic affinities, the obrieniids might be biologically similar to weevils as indicated by their long rostrum (probably adapted to making a hole in plant tissues for egg laying), clubbed antennae (suggesting a high concentration of sensilla near the apex of the antennae, perhaps for host plant testing), and locking lamellae at the elytral suture (fixing the elytra during crawling in narrow spaces, perhaps between sporophylls in cones). Other obrieniid characters such as the small body size, dorsoventrally flattened body, elongated basal antennomeres and rostral structures supposedly protecting the antennae are in accordance with this hypothesised mode of life. Obrieniids are restricted to localities with a high rate of cycadophyte finds suggesting that they lived either on them or, at least, on some plants that preferred the same plant communities.

One more order that probably took part in the endophytic guild was the Lepidoptera. At least in one living family of primitive non-glossatan moths, Agathiphagidae (still unknown at the fossil state), the larvae are araucariacean seed-feeders. KRISTENSEN (1997) regards this mode of life as autapomorphic and believes that the ancestral lepidopterans had soil-dwelling larvae that fed on mosses, litter and fungal hyphae like modern micropterigids. However, his hypothesis is based on the biology of the recent mecopteran families which may well be apomorphic within the order; and even if it is correct, some Mesozoic moth taxa could have colonised plant cones independently.

Unfortunately, traces of damage on Mesozoic fructifications have not been studied thoroughly. The majority of records are from the Early Cretaceous bennettites (except for a bored araucariacean female cone from the Jurassic of Argentina: STOCKEY 1978). Another interesting point is the absence of bored seed records though recently, insect tunnels have been discovered in seeds of a non-bennettite cycadophyte from the Jurassic of Georgia (V. Krassilov, pers. comm). Perhaps, Triassic and Jurassic holometabolan larvae fed mainly on immature pollen making no easily detectable traces of damage on sporophylls or cone axes, and the pre-dispersal seed consumption in the Mesozoic was limited. However, the available data are too fragmentary and may be misleading; more studies are necessary for any definitive conclusion.

Nor are there any thorough studies of Mesozoic plant reproductive structures for the pathologies caused by insect sucking. The existence of the corresponding insect guild still remains unsupported although it is quite possible that some Mesozoic hemipterans (e.g., lygaeoid and coreoid bugs) and thrips fed on fructifications and seeds.

It is easy to see that the supposed Mesozoic pollinivores and nectarivores belong mainly to the same orders suggesting that the origin of both diet types has been correlated. A number of feeding opportunists combines them now, and the same might have been true for many Mesozoic insects. This pattern suggests that the nectaries in the Mesozoic plants were commonly topographically connected with reproductive structures. The primary function of nectar production is a matter of speculation. For instance, in male strobili it could reduce pollen predation by distracting potential pollinivores, or disorient them by clogging up their sensory organs, or operate as a kind of a sticky trap protecting immature pollen, etc., whereas in the female strobili it could be primarily excreted for pollen trapping as in the living wind pollinated Ginkgo and some conifers. The development of such morphological features as the scaly wing covers in moths and diverse cleaning structures in moths, hymenopterans, beetles, etc., was possibly connected with visiting sticky nectar-bearing structures (Chapter 2.2.1.3.4.2f).

Thus palaeontological data indicate that the superguild of "anthophilous" insects visiting plant reproductive organs for adult feeding and/or oviposition was diverse as early as in the Late Triassic and evolved towards increasing complexity and niche specialisation. The Mesozoic evolution of this superguild strongly affected the evolution of plants in two different (but not mutually exclusive) directions, namely towards more effective anti-herbivore protection of the reproductive organs and towards development of insect/plant mutualism including insect pollination. If, as discussed before, the very existence of entomophily in Palaeozoic plants is still questionable, in the Mesozoic the main point of uncertainty concerns the diversity of the insect-pollinated plant taxa. It should be stressed that even if there were some Palaeozoic insect-pollinated plants their pollination systems were probably completely destroyed in the course of the Late Permian extinction that would have seriously affected nearly all insect taxa that were biologically connected with plant reproductive organs.

In the present-day flora entomophily is widespread first of all in the flowering plants (angiosperms), and it is generally believed that it has been either ancestral for them or acquired very early in their evolution (STEBBINS 1974, A. MEEUSE 1978, DILCHER 1979, FAEGRI & VAN DER PIJL 1979, CREPET & FRIIS 1987, KRASSILOV 1989, 1997). The oldest undoubted fossils of angiosperms are from the Early Cretaceous. Though supposed angiosperm remains were repeatedly reported from older strata all of them were either misidentified or wrongly dated (see MEYEN 1987a, KRASSILOV 1989, 1997); the Chinese Archaefructus described recently by SUN et al. (1998) also originates probably from the Early Cretaceous deposits and not from the Late Jurassic. Angiosperm flowers normally have a low preservation potential because of their delicacy and short life-span. Though the angiosperm flowers discovered in the last decades show that in the Late Cretaceous they were morphologically and biologically diverse (FRIIS & CREPET 1987), the flower morphology of the Early Cretaceous angiosperms still remains poorly known (e.g., KRASSILOV et al. 1983, P. CRANE et al. 1986, 1989a, TAYLOR & HICKEY 1990, SUN 1996, SUN et al. 1998); the data available suggest, however, that at least some of them would have been entomophilous. Pollen morphology (though not always indicative: see WHITEHEAD 1969, HESSE 1981, 1982 for important criteria of entomophily) also supports an Early Cretaceous origin of insect pollination in the flowering plants (HICKEY 1977, CREPET & FRIIS 1987, KRASSILOV 1989, 1997). The only direct evidence of pollinivory on an Early Cretaceous angiosperm is the abovementioned find of Afropollis in the gut of a xyelid sawfly from the Crato Formation. The sculpture of Afropollis indicates that its producer was probably entomophilous though the frequency of this pollen type is high as also in wind-pollinated taxa (DOYLE et al. 1990a,b). Another case of similar controversy was discussed by P. CRANE et al. (1989a): the morphology of Early and Late Cretaceous chloranthoid flowers suggests insect pollination though the corresponding types of dispersed pollen (e.g., Clavatipollenites and Asteropollis) seem to be too abundant. In fact, the differences in pollen production between the anemophilous and entomophilous taxa are not universal (POHL 1937), and a high production occurs when the pollen is the main or the only attractant for pollinators (FAEGRI & VAN DER PIJL 1979). Probably, this was the case of the Afropollis, Clavatipollenites, and Asteropollis producers. When the pollinators are attracted in a different manner, and first of all by nectar, pollen production decreased, and pollen grains should occur in the pollinators gut contents only occasionally and in small numbers. Perhaps, this mode of attraction was widespread in the early angiosperms, and a search of insect body surface for pollen grains would be more informative in respect of their probable role in pollination rather than studies on intestinal contents.

It is often postulated that the angiosperms were primarily cantharophilous (beetle-pollinated). This idea is based on observations on some living morphologically primitive and phylogenetically ancient families (e.g., Magnoliaceae, Degeneriaceae, Winteraceae, Myristicaceae, Annonaceae, Eupomatiaceae, Cyclanthaceae: DIELS 1916, HOTCHKISS 1958, B. MEEUSE 1959, HEISER 1962, GOTTSBERGER 1970, 1974, 1989, G. MEDVEDEV & TER-MINASSIAN 1973, THIEN 1974, 1980, KEIGHERY 1975, ANDOW 1982, BEACH 1982, SAMUELSON 1988, ARMSTRONG & IRVINE 1989, BERGSTRÖM et al. 1991) as well as on the relatively low level of morphological and behavioural specialisation in anthophilous beetles and high pollination costs in cantharophilous flowers (TAKHTAJAN 1969, FAEGRI & VAN DER PIJL 1979). However, the cantharophily model has been criticised for different reasons (e.g., GRINFELD 1975, 1978, A. MEEUSE 1978, ARMSTRONG 1979). In fact the so-called cantharophilous pollination syndrome is not uniform and rarely occurs in the primitive dicots in its "classic" form, with flower visiting for adult feeding only (FAEGRI & VAN DER PIJL 1979). The degree of individual constancy of flower visiting in beetles may be high and certainly comparable with those in other anthophilous insects (e.g., PELLMYR 1985). Not uncommonly, cantharophily in primitive angiosperms is specialised and quite different from its supposedly primitive form (GOTTSBERGER 1970, 1974, ARMSTRONG & IRVINE 1989, BERGSTRÖM et al. 1991). In other cases the pollination system is clearly opportunistic but the pollinator complex includes insects other than beetles, such as dipterans (THIEN 1982, THIEN et al. 1983), moths (THIEN et al. 1985) and perhaps bugs (ANDOW 1982). The morphology of the known Cretaceous angiosperm flowers with a small, free, undifferentiated perianth provides no evidence for any specialised pollination syndrome suggesting rather an opportunistic pollination strategy (KRASSILOV 1989, LABANDEIRA 2000). However, more specialised relations based on attraction of some beetles with herbivorous larvae (e.g., nitidulids and weevils) to breeding sites cannot be excluded. There are no data that definitively support a pre-Late Cretaceous age of any specialised associations between angiosperms and their pollinators though some opportunistic relations may be inherited from the Early Cretaceous (e.g., between some Winteraceae and the micropterigid moths of the *Sabatinca*-group genera as discussed by THIEN et al. 1985).

It should be noted that in modern flowering plants a switch from insect to wind pollination and back again does not seem to be very difficult as illustrated by numerous reversals (LEPPIK 1955, PORSCH 1956, H. CLIFFORD 1962, SODERSTROM & CALDERON 1971, LOROUGNON 1973, LEEREVELD et al. 1976, STELLEMAN & MEEUSE 1976, STELLEMAN 1979, ELVERS 1980, LEEREVELD 1982, PANT et al. 1982). Most probably, both modes of pollination occurred in Late Early Cretaceous flowering plants, and even the relatively rare hydrophily (water-pollination) is suggested for some Aptian and Albian angiosperms (DILCHER 1989). However, wind pollination is ineffective under a low population density. Because palaeontological data indicate that the earliest angiosperms were most probably opportunistic colonists of disturbed habitats (Chapter 3.2.5) forming small, patchy and unstable populations, entomophily might be extremely important for their reproductive success, and a switch to wind pollination became possible only after the appearance of the first angiosperm-dominated plant associations (i.e., hardly before the Aptian).

On the other hand, because of their low population density the early angiosperms could themselves scarcely support a rich anthophilous fauna. This fauna should be initially adopted from other, more common and widespread plants, and thus a diverse complex of anthophilous insects might already exist on certain gymnosperms by the time of the angiosperm origin. Moreover, early angiosperms should inhabit the same (or at least neighbouring) habitats where insect-visited gymnosperms were abundant (LEPPIK 1971, ZHERIKHIN 1978, 1980a). As discussed above, the guilds of pollen-eating, nectarivorous and cone-breeding insects are well represented in the Late Triassic and Jurassic fossil record well before appearance of the angiosperms. This raises the question as to whether some Mesozoic gymnosperms were insect-pollinated.

For modern gymnosperms, entomophily is rather well studied in Cycadaceae and Zamiaceae (Cycadales) (RATTRAY 1913, MARLOTH 1914, NORSTOG & STEVENSON 1980, BRECKON & ORTIZ 1983,

NORSTOG et al. 1986, NORSTOG 1987, TANG 1987, CROWSON 1989, 1991, NORSTOG & FAWCETT 1989, VOVIDES 1991, NORSTOG & VOVIDES 1992, WILSON 1993, OBERPRIELER 1995a,b, DONALDSON 1997, TANG et al. 1999), and in several *Ephedra* species (Ephedraceae) (PORSCH 1910, MEHRA 1950, BINO et al. 1984a,b, A. MEEUSE et al. 1990a,b). Less data are available for *Gnetum* (Gnetaceae) and *Welwitschia* (Welwitschiaceae) (VAN DER PIJL 1953, FAEGRI & VAN DER PIJL 1979, JOLIVET 1986, KATO et al. 1995, LABANDEIRA 2000). Their pollinators are mostly beetles and/or solitary bees but other insects (e.g., the mycetophiloid dipterans) have also been recorded. In the cycads specific associations with cone-parasitising and pollinivorous beetles are documented; in other gymnosperms the pollination system is probably opportunistic and based mainly on nectar attraction. It should be noted that in spite of their strong morphological differences, all living non-cycad entomophilous gymnosperm genera are usually considered either as members of the same order, the Gnetales, or as separate but phylogenetically related orders Ephedrales, Gnetales and Welwitschiales, forming the taxon Gnetopsida; MEYEN's (1984, 1987a) idea about possible ginkgoopsidan affinities of the Ephedrales was not accepted by other authorities.

None of the living entomophilous gymnosperm families are known with certainty before the Late Cretaceous. However, entomophily was postulated for different reasons for a number of Mesozoic gymnosperms (LABANDEIRA 2000). The order Bennettitales is especially often referred to in this respect (e.g., LEPPIK 1971, KRASSILOV 1972a, 1989, 1997, CREPET 1974, 1979a, ZHERIKHIN 1980a, JOLIVET 1986, CREPET & FRIIS 1987, T. TAYLOR & TAYLOR 1993, DILCHER 1995, M. PROCTOR et al. 1996). This purely Mesozoic order was widespread and diverse mainly in warmer climatic belts. The bennettite fructifications are morphologically complex and often bisexual. Indirect evidence of possible bennettite entomophily includes the presence of numerous structures interpreted as nectar glands in the *Weltrichia* male "flowers" (HARRIS 1969, 1973), the differentiation of the "perianth" in the bisexual *Williamsonia* "flowers" with the outer bracteae rich in stomata and probably photosynthetic and the inner ones poor in stomata and probably different in colour (HARRIS 1973), and the above-mentioned insect damage in the bisexual *Cycadaeoidea* fructifications. Those data suggest that the attraction mechanisms were diverse including nectar secretion, visual attractors, and, as in the modern cycads, oviposition sites for visitors. DELEVORYAS (1966, 1968a,b) supposed a self-pollination in *Cycadaeoidea* with presumably permanently closed strobili but insects could penetrate into closed cones for oviposition and thus fertilise the plant. Interestingly, there is no evidence for pollinivory on the bennettites. In fact, the morphologically simple unisulcate bennettite pollen grains are not diagnostic and cannot be separated with certainty from the ginkgoalean and leptostrobalean (czekanowskialean) pollen (KOTOVA 1965, MEYEN 1987a) but no pollen of the ginkgoalean-bennettitalean type has ever been discovered in insect gut contents. A relatively high frequency of fossilised bennettitalean "flowers" suggests that they were mechanically resistant and probably much more long-lived than in the angiosperms and thus the pollination efficiency should be comparatively low (ZHERIKHIN 1978). The most probable *Cycadaeoidea* burrowers were beetles, perhaps primitive belid and nemonychid weevils which were obviously more common in warmer areas where bennettites were abundant. However, both families occur since the Late Jurassic while the damaged *Cycadaeoidea* are recorded from the Early Cretaceous only. The living cycads with similar pollination type are often weevil-pollinated but none of their pollinators (the allocorynine oxycorynids, antliarhinine brentids, and molytine curculionids) are recorded from in the Mesozoic. Other possible pollinators of bennettites were parandrexid and primitive chrysomelid (protoceline) beetles as well as diverse nectar-feeders. In general, the taxonomic and morphological diversity of bennettites suggests that their "anthophilous" fauna was also rich and diverse.

Entomophily is also often suggested for a number of Mesozoic Gnetales and supposedly related gymnosperms on the basis of their reproductive structures and pollen morphology (P. CRANE 1988, 1996, KRASSILOV 1989, 1997, CORNET 1996). In modern entomophilous *Gnetum* and *Ephedra* the reproductive organs are functionally dioecious but morphologically bisexual, with reduced ovuliphores in the male strobili functioning as nectaries. This pattern seems to be characteristic for the Gnetales in a broad sense (KRASSILOV 1989) so that bisexuality can be ancestral in the lineage. Palynological data suggests that the Mesozoic Gnetales were diverse, and many of them had specialised auriculate pollen grains which P. CRANE (1988) regards as evidence of insect pollination. The pollen grains from the gut of *Spathoxyela*, the Early Cretaceous xyelid sawfly from Transbaikalia, is identical with pollen from a gnetalean flower-like organ, *Baisianthus*, found in the same locality (KRASSILOV & RASNITSYN 1999). Besides several other Mesozoic taxa of rather obscure affinities KRASSILOV (1989, 1997) assigned to Gnetales are the Hirmerellaceae (=Cheirolepidiaceae), the family traditionally placed among the conifers. This plant family produced the *Classopollis* pollen which is widespread and often very abundant in the Mesozoic deposits since the Late Triassic up to the Late Cretaceous. Entomophily has been suggested for *Classopollis* by KRASSILOV (1973) because of pollen morphology, and this pollen type was recently discovered in mass in intestinal contents of several Jurassic orthopterans of the genus *Aboilus* Martynov (KRASSILOV *et al.* 1997a,b). The high abundance of *Classopollis* in sediments indicates that if its producer was entomophilous the pollen might be probably the principal attractor for pollinators (KRASSILOV & RASNITSYN 1999). Nectar and pollen were probably principal insect attractors in the Mesozoic Gnetales; no evidence for insect breeding in gnetalean fructifications are available at present.

Entomophily was also supposed for *Sanmiguelia*, a Late Triassic North American gymnosperm of uncertain affinities (CORNET 1989) and for the Caytoniales (KRASSILOV 1973). Peltasperms and cycads are cited as probably entomophilous plants as well (WING & SUES 1992). For a long time it was believed from a botanical point of view that the Mesozoic was "the era of cycads" but recent studies doubt this opinion very much. Not only have the bennettites been found to be only remotely related to modern cycads but also the majority of other Mesozoic plants often classified as Cycadales may belong elsewhere. In particular, the affinities between the Mesozoic Nilssoniaceae and Cycadales are disputable. The former family is regarded by some modern authors as a separate order, Nilssoniales, resembling the true cycads in the male fructifications but differing considerably in the morphology of both female strobili (KRASSILOV 1989) and vegetative parts (SPICER & HERMAN 1996). Thus any direct parallels in pollination biology between them and the living cycads are questionable. However, the Early Cretaceous female strobile *Semionogyna* Krassilov et Bugdaeva was assigned to "Protocycadales" supposedly ancestral to cycads (KRASSILOV & BUGDAEVA 1988).

As discussed above, the probability of origin of entomophily should increase considerably for bisexual plants because in this case no additional attraction mechanisms are necessary to ensure that insects visit female organs after the male ones. Bisexuality is known in the

Bennettitales since the Late Triassic (*Sturiella* Kräusel). Perhaps, the bisexual Triassic bennettites were the first insect-pollinated plants, and the pollinators based on them evolved later on to other gymnosperms. Many bennettites were probably early successional plants (RETALLACK & DILCHER 1981a) for which insect pollination would be especially profitable. However, bennettites were probably not the only bisexual gymnosperms in the Triassic as indicated by the enigmatic genus *Irania* Schweitzer (SCHWEITZER 1977), tentatively assigned to the separate order Iraniales.

The following hypothetical scenario of Mesozoic co-evolution between the anthophilous (in a broad sense) insects and entomophilous plants may be proposed. After the appearance of bisexual strobili in some Triassic gymnosperms (and especially in some bennettites) occasional pollen distribution by visiting insects became highly probable. Insect pollination, which was initially facultative and opportunistic, was supported by natural selection, first of all in early successional plants. As a result additional attraction mechanisms evolved to increase the cross-pollination probability and to minimise pollination costs; their development was accompanied by parallel insect radiation. The early anthophiles were weakly morphologically specialised, and the pollination selectivity would be assured by individual behavioural stereotypes and spatiotemporal patterns of pollen production and pollinator activity. No strong constraint is inferred here for the switch of a potential pollinator from one to another plant within the same or neighbouring associations. The non-entomophilous plants in these associations were also selected for entomophily but the selection pressure in this direction decreased with increasing vegetation uniformity, in particular in late successional stands (dominated by ginkgoaleans and conifers: ZHERIKHIN & KALUGINA 1985). Thus the evolutionary trend towards entomophily was probably widespread in Mesozoic gymnosperms, and parallel evolution of insect attraction mechanisms occurred in a number of phylogenetic lineages as well as the correlated development of protective adaptations minimising the attendant insect damage to the plant. This common selection pattern may explain the similar structural features acquired independently by a number of gymnosperm phylogenetic lineages. KRASSILOV (1974, 1975, 1977, 1989, 1997) has called this process angiospermisation because those features occur in the angiosperms as well. This explanation seems to be more plausible than Krassilov's original idea on gene transduction by viruses; in particular, the transduction hypothesis fails to clarify why the parallel evolution involved mainly the reproductive sphere as was stressed by MEYEN (1987a).

Thus the pattern of diffuse co-evolution was complex including plant adaptations to insect pollination, insect adaptations to search for and to feed on fructifications, insect switches between different plants, plant protective adaptations, plant competition, and expansion of entomophily over a number of plant taxa. All those taxa are regarded by Krassilov as "proangiosperms" which is a polyphyletic grade and not a taxon. Ultimately this process has resulted in the appearance of true angiosperms, their expansion and the large-scale extinction of other entomophilous plants (Chapter 3.2.5). Angiosperm monophyly is still disputable (GREGUSS 1971, KRASSILOV 1989, 1997). It is supported by molecular data (W. MARTIN *et al.* 1989, TROITSKY *et al.* 1991, BOBROVA *et al.* 1995) which, however, suggest at the same time a Palaeozoic age of the angiosperms which is obviously not in agreement the with palaeontological record. The opinions of different authorities vary concerning the phylogenetic relationships between the main gymnosperm lineages and angiosperms but it should be noted that the old hypothesis of the bennettite ancestry of the angiosperms criticised by the majority of modern authors has been revived recently by MEYEN (1984, 1987a, see also HERMAN 1989) who believes that the Gnetales could originate from the bennettite stock as well. If his phylogenetic hypothesis is correct, entomophily in the angiosperms and gnetaleans may be inherited from bennettitalean ancestors. Other authors reject any direct phylogenetic connections between the bennettites and angiosperms but close affinities between the Bennettitales, the Gnetales in a broad sense and the flowering plants are postulated in the majority of the recent phylogenetic schemes (e.g., P. CRANE 1985, DOYLE & DONAGHUE 1986, 1990, DOYLE 1998). In any case, bisexual bennettites were probably at least ecogenetic, if not phylogenetic, precursors of other entomophilous plants. Their appearance would stimulate evolution towards insect pollination in other taxa, and especially in early successional plants which otherwise could not compete with bennettites. This is one more possible case of the adaptive value revisited by natural selection (ZHERIKHIN 1978, 1980a). It seems that not later than in the Jurassic a diverse entomophilous gymnosperm flora has been formed occupying an important place in the vegetation cover.

Some present-day connections between non-pollinating herbivores feeding on fructifications and their host plants have perhaps existed since the Mesozoic (e.g., between the xyeline sawflies or eccoptarthrid weevils and conifers). Some modern associations between archaic angiosperms and their pollinators may go back to the Early Cretaceous as well (e.g., between Winteraceae and micropterigid moths as mentioned above). However, it is improbable that any specific plant/pollinator relationships in the modern gymnosperms are inherited directly since the Mesozoic because both living entomophilous gymnosperm families and their principal pollinators belong to the taxa unknown before the Late Cretaceous.

Highly specialised pollinator/plant relations should originate particularly easily on the basis of attraction for breeding sites because in this case the herbivore/host specialisation is usually a common feature, and additional co-adaptations connected with minimisation of the pollination costs only reinforce this ancestral specialisation. At present this type of interrelations occurs between the cycads and beetles (mainly weevils) as well as between diverse angiosperms and herbivorous beetles (BRUCH 1923, GOTTSBERGER 1970, CROWSON 1988), flies (PELLMYR 1988), moths (BRANTJES 1976a,b, POWELL 1984) and wasps (RAMIREZ 1974, WIEBES 1979, 1986). All those relations are highly specific including some of the most sophisticated mutualistic systems known like the associations between *Yucca* and the prodoxid moths or between *Ficus* and the agaonid wasps. The costs of pollination are variable and may be high not only in primitive but probably also among more recently evolved associations (for example, as between *Melandrium* and hadenine moths: BRANTJES 1976b) and in even more extremely specialised pollination systems as in the case of *Ficus* and its pollinating fig wasps (Agaonidae) (from 25% up to more than 80% of seeds may be sacrificed: JANZEN 1979). In other cases, costs are minimised because insect larvae cause no damage to ovules or seeds, as for example, in Hydnoraceae pollinated by oxycorynid weevils (BRUCH 1923). In the Mesozoic this pollination type is highly probable for *Cycadeoidea*, where the pollination system seems to have been specialised and probably at a low cost to the plant (CREPET 1972). Future studies of damaged fructifications may recover more cases; however, it should be noted that when pollinator larvae are pollinivorous (as in modern meligethine nitidulids) they may not leave any easily detectable traces on the fructifications.

The degree of floral constancy of modern pollinivores and nectarivores varies widely, and the same should probably have been the case for the Mesozoic fauna. The cases of morphologically obvious narrow specialisations are rather rare among those insects in the Mesozoic and thus the selectivity of pollination should be ensured mostly by behavioural and community-level mechanisms (such as spatiotemporal pattern of flowering). Such pollination systems can be effective enough but at the same time their evolutionary plasticity is much higher than in the systems based on breeding specialisation because both morphological and biological constraints of shifting to other plants are generally weak. It should be noted that the most morphologically specialised nectarivores such as the kalligrammatids, pseudopolycentropodids, nemestrinids and perhaps aneuretopsychids were evidently most common and diverse in warmer areas with diverse bennettitaleans.

3.2.3.1.2.2. Other forms of herbivory

Among the consumers of plant vegetative tissues, the sucking herbivores were the most important and diversified from the Triassic. The sucking herbivore guild included mainly diverse hemipterans though some thrips could also belong here. As discussed previously, it was relatively weakly quantitatively affected by the Late Permian extinction though the taxonomic composition of the hemipterans has changed considerably at the family level. Throughout the Mesozoic fossil record the hemipterans represent one of the most frequent terrestrial insect orders in the oryctocoenoses. Their high taxonomic and morphological diversity suggests a high complexity of niche structure and a high level of specialisation within the guild. Niche space division was based on specialisation on different plant tissues, organs and taxa. Since the Triassic, the guild continuously included both phloem- and xylem-feeders, and some mesophyll-feeders probably also existed in younger Mesozoic faunas. The fore leg morphology in some Hylicellidae suggests that in the cicadomorph lineage subterranean mode of life and accordingly root sucking has been acquired rather early in the Mesozoic (SHCHERBAKOV 1988b). According to RETALLACK (1977), vertical burrows in palaeosols of the Triassic Newport Formation in the Sydney Basin, Australia, were most probably constructed by "cicada-like insects". The cydnid bugs appearing in the Jurassic could also feed on plant roots. Although the hemipterans probably fed on all of the main Mesozoic plant taxa or nearly so, the degree of host specialisation should be narrow in many cases. There are few indirect data on possible host plant relations for Mesozoic hemipterans. In the Early or Middle Jurassic deposits of Novospasskoe in Siberia, sedentary protopsyllidiid nymphs are found in association with horsetail remains (ZHERIKHIN 1985). It is highly improbable that they could have originated from any remote habitats. The distribution of Palaeontinidae is correlated with the abundance of Ginkgoales (ZHERIKHIN & KALUGINA 1985, Martínez-Delclòs, pers. comm.). The oldest Cretaceous Tingidae belong to the subfamily Cantacaderinae feeding now mainly of mosses. Moss feeding is probable also for the coleorhynch families Progonocimicidae and Karabasiidae. Some aphids and scale insects occurring in the Cretaceous fossil resins, and particularly those represented by flightless stages should probably have lived on the resin-producing conifers themselves. The scale insect family Matsucoccidae, known since the Early Cretaceous (KOTEJA 1988), is now restricted to coniferous hosts. There are no positive data on possible associations between hemipterans and early angiosperms but the impressive later evolutionary success of herbivorous hemipterans on angiosperm hosts makes their early switch to flowering plants seem quite probable.

Traces of sucking damage on Mesozoic plants are practically unstudied. There are occasional records of supposed puncture holes

(LABANDEIRA 1998c), growth abnormalities and gall-like structures, possibly induced by homopterans (VISHNU-MITTRE 1957, 1959, HARRIS 1969, GRAUVOGEL-STAMM & KELBER 1996), but their attribution is equivocal (LAREW 1992). Morphology of Mesozoic aphids suggests that some of them could be gall-dwellers (SHAPOSHNIKOV 1980), and other specialised, more or less sedentary plant parasites existed in the Mesozoic including psyllomorphs, whiteflies and scale insects. Distribution of the Cretaceous aphids mainly in the northern temperate zone suggests that at least since that time they were well adapted to seasonal climate and probably migrated between primary and secondary hosts. However HEIE (1994) doubts that the alternation of parthenogenetic and sexual generations was the aphid primary adaptation to a seasonal climate. The warm Mesozoic lowlands were probably unsuitable for aphids because the advanced Late Mesozoic families has probably failed to invade the southern temperate zone at least in the Early Cretaceous (they are not found in the rich later Early Cretaceous fauna of Koonwarra in Australia: HAMILTON 1992). At present, tropical lowlands are only colonised by anholocyclic aphids. On the contrary, modern scale insects are much more abundant and diverse in warmer climatic belts, and the same would probably have been true at least for the Cretaceous. This inference is based on their abundance in fossil resins from the Early Cretaceous of Lebanon (AZAR 1997) and earlier Late Cretaceous of New Jersey (GRIMALDI et al. 2000, KOTEJA 2000) and their relative rarity in the Late Cretaceous resins of Siberia (ZHERIKHIN 1978). For unknown reasons, hemipterans, including the scale insects, are surprisingly rare in the Early Cretaceous resins of Spain (ALONSO et al. 2000). Unfortunately, no data are available on the pre-Cretaceous distribution of the scale insects.

The taxonomic composition of the Mesozoic sucking herbivore guild varies considerably in space and time. In particular, psyllomorphs are common in the Triassic and especially in the Jurassic but decline dramatically in the Early Cretaceous. The aphids are uncommon before the Early Cretaceous at the world scale and turn to be extremely abundant and highly diverse in the Early Cretaceous but in the northern temperate zone only. The oldest whiteflies that have been discovered are from the Late Jurassic and the oldest scale insects from the Early Cretaceous. However, their absence in older deposits may well be explained by their general extreme rarity in preservation environments other than fossil resins. The fulgoromorphs seem to be much more common and diverse in warmer regions while the auchenorhynch fauna of more temperate areas was probably dominated by cercopoids and palaeontinids.

There is no firm basis for the quantitative estimation of the rate of consumption of the plant production by sucking herbivores in Mesozoic ecosystems but undoubtedly it was high. Hemipterans belong to the commonest and most widespread terrestrial insect orders in virtually all types of Mesozoic insect oryctocoenoses. Their frequency in many localities is quite impressive; for instance, in the Early Cretaceous deposits of Baissa in Transbaikalia aphids alone constitute about a third of all terrestrial insect fossils. This abundance and ubiquity of sucking herbivores in comparison with other herbivorous functional guilds suggests that they were most probably the dominant primary consumers in all main Mesozoic land ecosystems, ranging from early successional shore communities up to the climax forests. Consequently, their role as prey for diverse secondary consumers has also had to have been very significant. They should also have produced a large amount of honeydew which might have constituted an important food source for other insects and particularly for the fluid-feeders (see ZOEBELEIN 1956a,b, for their trophic role in the modern

ecosystems). At the community level the availability of honeydew could affect the distribution and behaviour of potential pollinators and parasitoids. GRINFELD (1961) and SHAPOSHNIKOV (1980) postulated the Cretaceous age for origin of the mutualistic associations between aphids and ants. However, this dating is rather doubtful because even in the earlier Late Cretaceous the ants are relatively rare and represented by the most primitive taxa only. Observations on recent ecosystems demonstrate that honeydew has an important effect on the soil microflora and fauna as well (OWEN 1977, PETELLE 1980, 1982, CHOUDHURY 1985, ANDRZEJEWSKA 1993). It should also be noticed that nowadays sucking herbivores, and particularly aphids and cicadelloid leafhoppers, are major vectors of many plant pathogens (especially viruses and mycoplasms).

In contrast, the role of the ectophytic chewing folivore guild seems to have been remarkably limited in the Mesozoic before the angiosperm expansion. This guild has changed dramatically in comparison with the Palaeozoic. In the Mesozoic it probably included the stick insects, locustopsid grasshoppers, boreid scorpionflies, larvae of some sawflies and moths (e.g., Micropterigidae), and perhaps some polyphagan beetles, i.e. exclusively the taxa which are either entirely absent in the Palaeozoic or appear firstly in the Late Permian in small numbers only. Leaf fragments have been discovered in the gut contents of Late Jurassic susumaniid stick-insects and in members of the enigmatic gryllonean family Brachyphyllophagidae which is possibly related to the embiids (RASNITSYN & KRASSILOV 2000). Interestingly, in both cases the cuticular structures indicate that the insects fed on leaves of Brachyphyllum or Pagiophyllum (Hirmerellaceae) only. Those leaves were small and scale-like but rich in mesophyll tissue in comparison with conifer needles or fern pinnules. However, there is no correlation in distribution between Hirmerellaceae and Susumaniidae so that the latter should not be restricted to the former as the only or even main host plants. There are no other records of any leaf remains in Mesozoic insect guts or coprolites. In the Permian material studied by A. BECK & LABANDEIRA (1998) the only common plant lacking any traces of folivory was tentatively identified as Brachyphyllum. Most probably, Palaeozoic plants assigned to this form genus were unrelated to the Mesozoic Hirmerellaceae, the more so since the characteristic Classopollis pollen first appeared in the fossil record as late as in the Triassic. Insect damage traces occur occasionally on other Mesozoic plants such as ferns (see Fig. 472) but seem to be remarkably uncommon. No quantitative data on the frequency of damaged leaves in the Mesozoic deposits are available, and taxonomic, geographic and stratigraphic patterns of folivory can be characterised only tentatively on the basis of indirect evidence.

A number of modern sawflies develop on ferns and horsetails, and micropterigid larvae and boreids feed on mosses; those associations may well be ancient and go back to the Late Mesozoic. Adult sawflies including true tenthredinids are not rare in the later Early Cretaceous, especially in the northern temperate belt. Modern sawflies have both endo- and ectophytic larvae, and the presence of the latter in the Mesozoic is confirmed palaeontologically. Two specimens of Kuengilarva inexpectata Rasnitsyn, the ectophytic sawfly larva belonging either to the living xyelid subfamily Macroxyelinae or to the extinct family Xyelotomidae, have been collected in the Late Jurassic or Early Cretaceous Ukurey Formation in Transbaikalia from layers containing practically no plant macrofossils other than horsetails. Any transport of these insects does not seem probably, so that horsetails were the only possible hosts (RASNITSYN 1990b). An additional undescribed larva belonging probably to the same genus is represented in a new collection from the Ichetuy Formation (Early or earliest Middle Jurassic, Transbaikalia) (A. Rasnitsyn, pers. comm.). In those deposits horsetails are the only common plants as well. The distribution of locustopsid grasshoppers in the Jurassic sediments of Siberia suggests that they probably inhabited fern marshes and could have fed on ferns (ZHERIKHIN & KALUGINA 1985). However, they are also common and diverse in the Santana in Brazil (MARTINS-NETO 1990b, 1998a) where the ferns are rare (GRIMALDI 1990). Modern ferns are generally regarded as rarely eaten by insects but nevertheless they do in fact support a rather diverse fauna. For example, the community of herbivorous insects on bracken in Britain (which has a strongly impoverished insect fauna in comparison with continental areas) includes about 30 species belonging to diverse ecto- and endophytic trophic guilds, the occasional visitors and soil-dwelling taxa being excluded (LAWTON & MACGARVIN 1985). Moreover, in some areas with a rich fern flora they are consumed by herbivorous arthropods almost as intensively as angiosperms (BALICK et al. 1978). The stick-insects are common and diverse in the Triassic of Europe (the alleged caloneurid wing from the Middle Triassic of Vosges illustrated by J. GALL et al. 1996: pl. IX, fig. 4, belongs to a stick-insect), Central Asia and Australia and Jurassic of Europe and Kazakhstan. In the Early Cretaceous they also become abundant in temperate Siberia and Mongolia; few remains have been recorded from Gondwanaland (MARTINS-NETO 1989a, 1995b). Consequently their role as folivores might be expected to have varied considerably in space and time within the Mesozoic.

In general, the average rate of consumption of photosynthetic tissue by ectophytic chewing herbivores seems to be rather low in the Triassic and Jurassic in comparison with both the younger and the Permian terrestrial ecosystems. The absence or extreme rarity of such important modern groups as folivorous lepidopterans and leaf-beetles could hardly be compensated by any Mesozoic folivores, and the differences in the abundance of damage traces on fossil plants seem to be quite impressive. The number of finds of the supposed folivores increases in the Early Cretaceous suggesting that folivory was intensifying. Finds of trigonalid wasps which are specialised hyperparasitoids of folivorous insect larvae support this inference. Trigonalid hosts become infested when they swallow eggs laid on leaf surface, and this infestation mechanism hardly can be effective under a low host density. The trigonalids are generally rare in the fossil state but occasionally occur in the Early Cretaceous of Transbaikalia, Mongolia and England (RASNITSYN 1990b, RASNITSYN et al. 1998). The increasing frequency of finds of folivores and their parasitoids does not correlate with the appearance of angiosperms. For example, several genera of both tenthredinid and trigonalid hymenopterans occur in Bon Tsagan in Mongolia where angiosperms are not represented even in palynological samples. On the other hand, the angiosperms almost certainly had been exploited intensively by folivores even in the Early Cretaceous. No Early Cretaceous leaf samples were specially studied in this respect but at the very beginning of the Late Cretaceous different strategies of folivory on flowering plants are documented including centre and margin feeding, skeletonisation, and surface abrasion (LABANDEIRA et al. 1994a).

The Mesozoic history of endophytic herbivory is also insufficiently studied. No relatives of modern endophytic herbivores are known from the Triassic. However, supposed insect mines on leaves of *Heidiphyllum*, a gymnosperm of uncertain taxonomic position, are recorded from the Late Triassic of Australia (ROZEFELDS & SOBBE 1987); linear mine-like markings occur occasionally also on *Glossophyllum* (Peltaspermales) leaves in the Madygen Formation of Kyrgyzstan (see Fig. 476). If those ichnofossils actually represent insect mines, their makers probably belonged to extinct taxa. Supposed Triassic galls could have been induced by sucking herbivores, as pointed out above, or perhaps by bacterial or fungal pathogens. It is quite possible that some Jurassic and Early Cretaceous hymenopterans (e.g., cephoid sawflies) were stem-borers or stem-gallers at the larval stage, and that there were some leaf-miners among moth larvae. However, no representatives of any modern groups of highly specialised gall makers or miners are known with certainty from the Mesozoic. In particular, the Early Cretaceous cynipoid wasps were probably insect parasitoids, the Cretaceous cecidomyiid midges belong to non-herbivorous taxa, and the records of incurvariid moths from the Early Cretaceous need confirmation. It should be noted also that none of the known Mesozoic weevils demonstrate an extremely elongate parallel-sided body shape typical of the modern stem-borers which pupate in narrow tunnels. The oldest known undoubted insect mine was discovered in a *Pachypteris* leaf from the latest Jurassic or earliest Cretaceous of Australia (ROZEFELDS 1988a). If the original attribution of this mine to a nepticulid moths is correct, lepidopteran miners evolved on peltasperms before the appearance of angiosperms; however, this record is still absolutely unique.

Records of supposed insect galls on non-angiosperm hosts are few, and the nature of those galls is as unclear as in the Triassic. In any case, endophytic herbivory on non-angiosperm hosts was probably extremely limited. The oldest galls on angiosperm leaves are found in the Albian of the USA (DOYLE & HICKEY 1976, UPCHURCH et al. 1994), and undescribed mines occur on angiosperm leaves from the Albian of West Kazakhstan. Special ichnological studies of the Early Cenomanian Dakota flora in USA demonstrate that in the earliest Late Cretaceous both miners and gallers developed on angiosperm leaves were common and diverse (LABANDEIRA et al. 1994a). The Dakota materials indicate that niche partitioning was basically the same as now, with different feeding strategies and degrees of host specialisation, and that some mines have no obvious morphological differences from those produced by present-day mining moth genera. There are no accurate comparative data from other regions but in the Late Cretaceous of temperate Siberia and Russian Far East the frequency of damage traces on angiosperm leaves seems to be much lower than in warmer areas (pers. observations; V. Krassilov, pers. comm.).

In general, further studies of damage traces in leaf floras from different stratigraphic levels and regions are necessary to elucidate the Mesozoic history of photosynthetic tissue consumption in more detail, and only quite preliminary conclusions can be made at the moment. It seems that chewing herbivory (ecto- and especially endophytic) on pre-angiosperm vegetation was rather limited, and an explosive evolution of chewing herbivores on angiosperms probably occurred in the Early Cretaceous resulting in a significant increase of the primary production removed by herbivores. This is hardly surprising because of the very high nutritive value of angiosperm tissues: the average nitrogen content in angiosperm leaves is much higher in comparison with both the gymnosperms and spore plants, and nitrogen deficiency is one of the most important problems for the herbivores (STRONG et al. 1984). Effective chemical anti-herbivore defence should be acquired by angiosperms only in later times under increasing herbivore pressure. The tempo of insect invasion into this new licence (see Chapter 3.2.3.1 for difference between licence and niche) and their radiation within it seems to be remarkably high, the more so that any herbivore specialisation on the earliest angiosperms seems to be improbable because of their low population density and should have evolved only when density had increased. Nevertheless, some insect taxa that are supposedly connected biologically with the angiosperms appear in the fossil record when their probable hosts would probably have been rare. This is particularly the case of cephid sawflies now associated exclusively with the flowering plants. The oldest cephid known, *Mesocephus sibiricus* Rasnitsyn, from Baissa in Transbaikalia (perhaps Valanginian – Hauterivian or even older: RASNITSYN *et al.* 1998, ZHERIKHIN *et al.* 1999) was described by RASNITSYN (1968) who predicted on this basis the presence of angiosperms in the locality. An angiosperm leaf described later as *Dicotylophyllum pusillum* Vakhrameev (VAKHRAMEEV & KOTOVA 1977) was collected at Baissa next year after Rasnitsyn's description had been published. This confirmed prediction is quite impressive. It should be noted, however, that an undescribed *Mesocephus* was recorded later, also from Bon Tsagan in Mongolia (RASNITSYN 1980) where no angiosperms are represented even as pollen in spite of the undoubtedly younger age of the locality in comparison with Baissa. Special studies on the oldest leaf floras rich in angiosperms, especially in warmer areas (e.g., the Aptian floras of Portugal) are highly desirable to estimate the rate of herbivore evolution more accurately. PONOMARENKO (1998a) suggests that at least some insect taxa that radiated later on angiosperms could have pre-existed on those Mesozoic gymnosperms with angiosperm-like morphology (the proangiosperms sensu KRASSILOV 1989, 1997). Though in general the angiospermisation process probably affected the vegetative sphere of gymnosperms much less than the reproductive structures, there are indeed some interesting parallels in leaf structure. Thus Ponomarenko's hypothesis merits more attention and is to be tested by special studies on feeding traces on the leaves of such gymnosperm taxa as Peltaspermales and Caytoniales.

The main problem concerning the Mesozoic wood- and bark-borers is to distinguish between herbivores on living plants (bioxylophages) and detritivores restricted to dead tissues (saproxylophages and xylomycetophages). In the modern insect fauna, consumers of living wood are less diverse than are the xylophagous detritivores, but they usually belong to the same families and do not demonstrate any obvious morphological features allowing their feeding habits to be recognised with any certainty. The vast majority of bioxylophages are holometabolans although some termites can also attack living plants. The only probable living wood borers in the Mesozoic are some beetles and siricoid hymenopteran larvae. In the hymenopterans xylophagy has probably evolved from the basis of ancestral herbivory in plant cones (RASNITSYN 1980) and thus feeding on living wood was perhaps primary in comparison with detritivory. The oldest siricoids are known from the Early Jurassic of Central Asia; the superfamily is most diverse in the Late Jurassic faunas (especially in warmer areas) and declines gradually in the Cretaceous (RASNITSYN 1980). The Mesozoic siricoids are not closely related to the living ones and belong mainly to extinct families or subfamilies. Xylophagy on living or dead wood is highly probable for the majority of them except for those taxa with a short ovipositor (e.g., Parapamphiliinae). As discussed above (Chapter 3.2.2.3), a similar evolutionary route from cone dwelling to wood burrowing also cannot be excluded for the beetles though it is generally believed that xylomycetophagy is ancestral in this order. However, even if some Palaeozoic archostematan taxa attacked living trees it is not clear whether any of them had survived into the Early Mesozoic. Among the three main modern xylophagous beetle groups often attacking living plants (Buprestidae, Cerambycidae, and Scolytinae) only the former is known from the Mesozoic with certainty. It appeared in the Middle Jurassic but only became common in

the Early Cretaceous (see ALEXEEV 1999 for a review of the Mesozoic record). Probably, there were also other consumers of living wood among the Mesozoic polyphagan beetles. There are many occasional records of bored wood (e.g., TILLYARD 1922b, JURASKY 1932, WIELAND 1935, M. WALKER 1938, FOSSA-MANCINI 1941, LINCK 1949, SHARMA & HARSCH 1989, ZHOU & ZHANG 1989, JARZEMBOWSKI 1990, TIDWELL & ASH 1990, HASIOTIS & DUBIEL 1993a, GENISE & HAZELDINE 1995, ZHERIKHIN & GRATSHEV 1995) but this type of Mesozoic ichnofossil was never studied thoroughly and the stratigraphic distribution of various types of burrows is insufficiently known. It should be noted, however, that they do not seem to have been rare in the Triassic, well before the oldest known finds both of undoubtedly xylophagous beetles and of wood-boring hymenopterans. Only a few burrows have been described in detail, and it is often unclear whether or not they have been made in living hosts (LABANDEIRA 1998c). Some complex tunnel systems were tentatively assigned to bark-beetles (Scolytidae) (M. WALKER 1938, JARZEMBOWSKI 1990) but this attribution is still unconfirmed by actual insect fossils and is absolutely unbelievable for the Triassic ichnofossils described by Walker. On the other hand, such radiated tunnels in the phloem layer suggests that the larvae constructed them from a central maternal chamber and thus the tunnel maker should indeed have been biologically similar to the present-day bark-beetles. The peculiar tunnels from the Begichev Formation (Albian or Cenomanian) of Taimyr, Siberia (ZHERIKHIN & GRATSHEV 1995; see Fig. 470) are of the same type but they are constructed in the wood like the modern ambrosia beetles do. This interesting find suggests an advanced form of mutualism with ambrosia fungi; however, no fungal remains have been detected in the tunnel contents (V. Krassilov, pers. comm.). Surely, other mutualistic associations between xylophages and fungi might exist earlier in the Mesozoic but virtually no positive data on this type of symbiosis are available, though fungal growth along insect tunnels in conifer wood has been described from the Early Jurassic of Germany (MÜLLER-STOLL 1936).

Large herbivorous tetrapods could be also involved in co-evolutionary interrelations with trees and bioxylophagous insects. In general, their feeding damages plants much more severely than that of smaller vertebrates (WING & SUES 1992). Xylophagous insects normally do not attack healthy trees while branches and trunks beaten or broken by megaherbivores should provide an excellent source for colonisation. Thus evolution of megaherbivory in the tetrapods could affect the frequency of bioxylophagy by insects. There were no true large-sized vertebrate browsers with the feeding level above 1–2 meters until the Late Triassic. In the Jurassic the feeding height has increased perhaps up to 10 meters or more and total megaherbivory had intensified considerably (WING & SUES 1992). The megaherbivores should also stimulate the evolution of protective plant chemicals and mechanical defences, further complicating the resulting co-evolutionary pattern.

A potentially important but largely neglected aspect of the palaeontological study of the history of bioxylophagy is the history of resin secretion. Resins are plant exudates composed mainly of a complex mixture of mono-, di-, sesqui-, and triterpenes (LANGENHEIM 1969, 1990). Primarily accumulation of resin in specialised ducts and cavities could originate simply as a mechanism of deposition of secondary metabolites but later an anti-herbivore effect of resins would be reinforced by natural selection. In a number of recent plants including both conifers and angiosperms, resin content of different plant organs are effective as repellents, feeding deterrents and toxins against a wide spectrum of diverse herbivores including sucking and chewing folivores (SMELYANETS 1968, 1977, RUDNEV *et al.* 1970, GRIMALSKY 1974, WRIGHT *et al.* 1975, MCCLURE 1977, 1984, STUBBLEBINE & LANGENHEIM 1977, HERNANDEZ & RODRIGUEZ 1978, LANGENHEIM *et al.* 1980, GABRID 1982, LANGENHEIM & HALL 1983, B. SCHUH & BENJAMIN 1984, LANGENHEIM 1990), cone-borers and especially wood-borers (POLOZHENTSEV 1947, 1965, BRAUN 1960, R. SMITH 1963, 1966, COUTTS & DELEZAL 1966, ISAEV 1966, VASECHKO & KUZNETSOV 1969, BUIJITENEN & SANTAMOUR 1972, CEREZKE 1972, CHARARAS 1973, HANOVER 1975, ISAEV & GIRS 1975, COYNE & LOTT 1976, ROZHKOV & MASSEL 1982, FERRELL 1983, MESSER *et al.* 1990, NAGNAN & CLEMENT 1990). Intensification of local resin production and development of additional resin ducts is a common protective reaction to an injury. Besides its direct effect on insects the resin also affects bacterial and fungal pathogens including those that are mutualists of the wood-boring insects. Sensitivity to toxic resin components varies strongly between different insect species suggesting complex co-evolutionary interrelations between resin chemistry and insect physiology. In particular, high interspecific and interpopulational variability of many plant resins is probably an evolutionary plant reaction to insect adaptations for their detoxification. These interrelations are further complicated by the common attraction of wood borers to some volatile resin components which signal the presence of damaged hosts potentially suitable for infestation (CHARARAS 1959, 1976, 1977, STARK 1968, RUDNEV *et al.* 1970, SIMPSON 1976, CHARARAS *et al.* 1978, VASECHKO 1978, TOSHIYA *et al.* 1980, ROZHKOV & MASSEL 1982, HAWKESWOOD 1990).

Unlike other natural antiherbivore chemicals, resins have a high fossilisation potential. Unfortunately, the fossil resin record has never been analysed from the point of view of evolutionary plant palaeophysiology and palaeobiochemistry. Resin secretion by diverse gymnosperms is documented since the Late Palaeozoic. However, before the Early Cretaceous the resins occur mainly either as internal inclusions in fossilised plant tissues or as rare and usually quite small isolated pieces, occasionally buried in sediments. Limited internal resin secretion has probably had an antiherbivore function at least since the Jurassic as indicated by the filling of insect burrows in a piece of conifer wood from the Middle Jurassic of China (ZHOU & ZHANG 1989). There are a few records of Jurassic resin deposits which, however, have never been studied in detail (MCLACHLAN & MCMILLAN 1976, BANDEL *et al.* 1997, ZHERIKHIN & ESKOV 1999). In the Early Cretaceous more concentrated resin deposits appear (C. CASTRO *et al.* 1970, DIETRICH 1976, SCHLEE 1984, SAVKEVICH *et al.* 1990, BANDEL *et al.* 1997, JARZEMBOWSKI 1999a, ZHERIKHIN & ESKOV 1999), mainly as numerous small drops in lignitic layers but also occasionally as large pieces (e.g., in the Near East and Brazil). In the Late Cretaceous rich fossil resin deposits become common in many regions over the World. This pattern suggests that resin production has increased significantly in the Cretaceous, in particular in the form of external resin flows. Perhaps, this pattern may be explained by increasing pressure of bioxylophagous insects (ZHERIKHIN & ESKOV 1999). External resin secretion should provide a rather good protection against the wood-boring beetles which usually lay their eggs on the bark or into shallow holes, but hardly could be very effective against the hymenopterans ovipositing rather deep into the wood. Further studies of bioxylophagy trace dynamics in the palaeontological record are necessary to test this hypothesis. Interestingly, some fossil resins have probably been produced by extinct members of extant conifer families which now rarely if ever produce any considerable external resin flows. This is quite possible though still not confirmed for a number of Early Cretaceous

resins, and some Late Cretaceous and Palaeogene resins were probably produced by Taxodiaceae (e.g., the Sakhalin amber and at least partially the New Jersey Late Cretaceous resins: ZHERIKHIN, ESKOV 1999, GRIMALDI *et al.* 2000) and Cupressaceae (e.g., partially the New Jersey resins: GRATSHEV & ZHERIKHIN 2000). These differences between ancient and recent genera of the same families are still difficult to interpret from the point of view of adaptive significance and may be at least partially connected with plant/insect co-evolutionary phenomena.

The evolution of resin chemistry may also be of interest in this respect. The fossil resins are chemically diverse and may be classified into several main types with additional subdivisions (K. ANDERSON *et al.* 1992, K. ANDERSON & BOTTO 1993, K. ANDERSON 1994, 1995, C. BECK 1999). Modern analytical methods allow the recognition of up to several dozens of components in some fossil resins (MILLS *et al.* 1984, GRIMALT *et al.* 1988, K. ANDERSON & WINANS 1991, K. ANDERSON 1995, CZECHOWSKI *et al.* 1996). Chemical studies indicate that in some conifer families the resin composition was changing considerably not only in the Mesozoic but also in the Palaeogene. In particular, many fossil resins demonstrate chemical features resembling modern araucariacean resins but in a number of cases this is in conflict with other data suggesting taxodiacean, cupressacean or pinacean origin of the fossil materials (LARSSON 1978, MILLS *et al.* 1984, C. BECK 1999, GRIMALDI *et al.* 2000). Probably, the "araucarian" resin chemistry is plesiomorphic for a number of conifer lineages though only a few of them have retained this state up to the present day. This chemical evolution could also be stimulated by bioxylophagy. It should be noted that there may be significant differences in the chemical composition between the original fresh resin and products of its fossilisation depending from the age and especially from the diagenetic processes (SAVKEVICH 1970, 1980, K. ANDERSON *et al.* 1992, VÁVRA 1993, KOLLER *et al.* 1997). Unfortunately, in fossil resins mainly the non-volatile diterpenoids are preserved while the volatile components of the original resin, which are most important both as insect repellents and attractants, are largely lost. This makes direct biological interpretation of fossil resin chemistry risky (but see MOSINI *et al.* 1980 for the presence of volatile terpenes in the Baltic amber).

There are very few data on possible Mesozoic chewing root-feeders, either ecto- or endophytic. Though elateroids, scarabaeoids and curculionoids are well represented in the Late Jurassic and Early Cretaceous faunas no modern root-feeding taxa of the click-beetles or weevils are recognised among them and potentially root-feeding chafers of the subfamilies Sericinae and Melolonthinae are quite rare (NIKOLAJEV 1998a). Specialised digging grylloid orthopterans probably related to the mole-crickets occur in the Early Cretaceous of Brazil (MARTINS-NETO 1991a, 1995b) and could be at least facultative root-eaters. HASIOTIS & DEMKO (1996a) describe abundant burrowing traces in rhizoliths from the Late Jurassic Morrison Formation of USA. It is not clear whether those burrows were made before or after plant death. Their original interpretation as termite nests is extremely doubtful. The presence of sucking root-feeders in the Mesozoic has been discussed above.

To summarise the available information on the Mesozoic herbivores, it should be stressed that the guild of the sucking herbivores on photosynthetic tissues (perhaps also wood-borers) has survived the Late Permian extinction more or less successfully (though their taxonomic composition has also changed significantly at the family level). To the contrary, other Palaeozoic herbivorous arthropod guilds have nearly disappeared, and the ecological licences of consumers of plant reproductive organs as well as both ecto- and endophytic chewing herbivores on photosynthetic tissues were filling almost completely *de novo* (see Chapter 3.2.3.1 for difference between licence and niche). As a result, their Mesozoic members belong mostly to the lineages either unknown from the Palaeozoic (hymenopterans, dipterans, polyphagan beetles, advanced mecopteran and neuropteran families) or appeared in a small number in the latest Permian only (stick-insects and short-horned orthopterans). Significant diversification with considerable advance in complexity of the niche space structure occurred continuously in all major guilds of herbivorous insects over the Mesozoic. In the terms of their relative importance for ecosystem turnover, the sucking herbivores strongly dominated at least before the latest Early Cretaceous. In contrast, the chewing herbivores fed mainly on reproductive organs which constitute only a small fraction of total primary production. Thus the removal of plant biomass by insects was highly selective, and their relative importance in consumption of the living hard plant tissues should be much less than now. Those tissues were consumed principally by the vertebrate herbivores, taxonomically and biologically diverse in the Mesozoic, especially since the Jurassic (WEISHAMPEL 1984, NORMAN & WEISHAMPEL 1985, COE *et al.* 1987, WEISHAMPEL & NORMAN 1989). However, because of their relatively slow reproduction and growth rate the vertebrates could hardly support as rapid ecosystem turnover as the chewing insects do now. It seems that generally in the Mesozoic a smaller part of the plant production was consumed by herbivores and a correspondingly larger share was processed in detritic trophic chains in comparison with the present-day situation (ZHERIKHIN 1980a) even if the total production rate of the pre-angiosperm vegetation was relatively low (WING & SUES 1992). However, wide-scale multidisciplinary studies are necessary to provide a better ground for speculations on ecosystem turnover. In any case, it seems that the Mesozoic ecosystems based on the spore plants, gymnosperms, and herbivorous reptiles should differ considerably from the modern grazing ecosystems with grasses and ungulates so that any parallels with the present-day steppes or savannahs are risky. A notable growth of the rate of non-sucking insect herbivory probably occurred in the Early Cretaceous still before the beginning of the angiosperm expansion and was not primarily connected with the angiosperms. Perhaps, this was connected with the angiospermisation effect mentioned above. However, there should be further substantial increase in the later Early Cretaceous directly due to intensive consumption of angiosperm photosynthetic tissues.

3.2.3.1.3. Detriti- and fungivores

Mycetophagy was probably widespread in the Mesozoic and probably played an important role in insect evolution and diversification, but its history is relatively unknown. As noted above (Chapter 3.2.2.3), the palaeontological record of fungi is extremely incomplete and insufficiently studied, and the borderline between mycetophagy and detritivory in the insects is conventional so that the trophic habit in a number of taxa is disputable between both. Mycetophagy is highly probable for diverse larval and adult polyphagan beetles (e.g., Leiodidae, Eucinetidae, Trogossitidae, Nitidulidae, Anthribidae, and almost surely a number of other families still unrecognised among the Mesozoic fossils) and larval dipterans (e.g., Perissommatidae, Mesosciophilidae, Synneuridae, Bolitophilidae, Keroplatidae, Mycetophilidae, Platypezidae and probably a number of extinct families belonging to the sciaroid lineage). No undoubtedly mycetophagous forms are recorded among the Triassic beetles and dipterans but these are insufficiently known and only remotely related to the living taxa so that their biology

is obscure. The oldest probable consumers of fleshy basidiomycete fruit bodies occur in the Jurassic (V. KOVALEV 1984, PERKOVSKY 1999), and some living mycetophilid genera are recorded from the Early Cretaceous (BLAGODEROV 1995, 1997, 1998a,b). It should be noted, however, that BLAGODEROV (1998c) believes that the Mesozoic fungus-gnats inhabited only the basidiomycetes fungi with long-living carpophores (e.g., Polyporaceae) while the taxa connected with ephemeral carpophores only appear in the Cainozoic. According to MATILE (1997), the plesiomorphic mode of larval life in the Sciaroidea was most probably an endobiosis in fungal carpophores but ectobiotic lineages feeding on fruit body tissues and spores have appeared no later than in the Early Cretaceous. The living hemipteran taxa that sucked fungal hyphae (Achilidae and Aradidae) appear in the Early Cretaceous and a similar feeding habit cannot be excluded for some older Mesozoic hemipterans (e.g., some Progonocimicidae) as well as for some Mesozoic thrips. Sporivory on fungal spores is also possible for some Mesozoic insects (e.g., some barklice and polyphagan beetles). Further studies on Mesozoic fungi and mycetophages are necessary to elucidate the history of mycetophagy.

A high production of plant mortmass in Mesozoic communities should have supported a rich and diverse fauna of detritivores, and the observable pattern is in accordance with the hypothesis of a strong predominance of detritic food chains. The abundance and diversity of detritivorous insects in Mesozoic oryctocoenoses are quite impressive. Unlike in the Palaeozoic, large-sized detritivorous arthropods other than insects are nearly absent. Moreover, the guild of larger detritivores in the Mesozoic was probably even more insect-dominated than now; in particular, millipede fossils seem to be rare even in comparison with the Cainozoic. This was not the case of the smaller detritivores which were probably represented largely by oribatid mites and perhaps springtails. Information on the Mesozoic oribatids is scarce but the Jurassic fossils known are remarkably similar to present-day taxa and are partially assigned to modern genera (KRIVOLUTSKY & KRASSILOV 1977).

Roaches and scorpionflies are very common and diverse in Mesozoic deposits from all over the World. They were probably largely opportunistic detritivores specialised more in terms of habitat than diet, though it cannot be excluded that some species were feeding specialists. The Mesozoic roaches demonstrate a high diversity of body size and shape, fore wing sclerotisation, leg and ovipositor structure, etc., indicating that they inhabited a wide variety of habitats in litter, soil, rotten wood, bark crevices, mosses, etc., and varied considerably in locomotion and reproductive biology. The oldest known finds of living families are from the earliest Cretaceous (VRŠANSKÝ 1999b) though the dominant extinct Mesozoic families Caloblattinidae and Blattulidae survived still in the Late Cretaceous. As in the Palaeozoic, the majority of Mesozoic roaches possessed an external ovipositor (though often more or less shortened), and oviposition via oothecae did not evolve before the latest Jurassic. The taxonomic composition varies in space and time but no significant changes in total roach abundance and diversity are observed within the Mesozoic. They also seem to be nearly equally common in temperate and warmer climatic areas and in different palaeoenvironments from the sea coasts up to highlands. Rather common finds of flightless nymphs in some Mesozoic deposits suggest that some roach species probably inhabited abundant organic debris at the lake shores just near the water line.

The dominant Mesozoic scorpionfly families are extinct Orthophlebiidae closely allied to living Panorpidae as well as to several minor extinct families, and the extant Bittacidae. They probably had opportunistic detritivorous or omnivorous larvae inhabited soil and litter. The larval biology of some more distinctive extinct Mesozoic taxa is less certain. Scorpionflies were very abundant in the Triassic and Jurassic throughout the World but evidently declined in the Early Cretaceous for unknown reasons. Other important opportunistic detritivores were probably some orthopterans, grylloblattidans and earwigs. Among the ensiferan orthopterans detritivory is particularly probable among the Grylloidea. There were several cricket families, both extinct and extant, in the Mesozoic, and the oldest finds are known in the Late Triassic. Over the Mesozoic they were probably only common in warmer climates, and some faunas are remarkably rich in crickets (e.g., Santana in Brazil, MARTINS-NETO 1991a, 1995b, 1999b). Modern crickets are often more or less detritivorous or omnivorous. Opportunistic detritivory is also supposed for some other Mesozoic ensiferans (e.g., Elcanoidea: GOROCHOV 1995a). Grylloblattida were common in the Triassic but declined later, leaving the fossil record by the later Early Cretaceous. One of the most widespread Mesozoic families, Blattogryllidae, is closely allied to the modern detritivorous Grylloblattidae. Occasionally earwigs are not uncommon in Jurassic and Early Cretaceous faunas; their morphology suggests that they should have been biologically similar to the present-day forms which are mainly opportunistic detritivores or omnivores. Some Early Cretaceous genera belong to modern families (e.g., Pygidicranidae).

More specialised detritivory is most probable for a number of beetles as well as for the larvae of diverse dipterans and perhaps early lepidopterans. In both beetles (the cupedoid archostematans and many polyphagans) and dipterans (particularly bibionomorphs but probably also some tipulomorphs and psychodomorphs) saproxylophagy was probably widespread. Some siricoid wood wasps could have been saproxylophagous as well. The appearance of the first termites in the Early Cretaceous should have had an important effect on dead wood consumption. As pointed out above, traces of insect feeding in Mesozoic woods are poorly studied. Detailed observations on the pattern of wood degradation are also few, but the available data suggest that the basic modern rot types can be traced back to the Triassic and even to the Late Palaeozoic (STUBBLEFIELD & TAYLOR 1986). The saproxylophagous taxa would probably be specialised in accord with the rotting stages as they are now, and the succession of saproxylophagous invertebrates in decaying wood could be of a substantially modern type, at least since the Jurassic. On the other hand, a number of supposedly saproxylophagous Mesozoic dipterans belongs to extinct families. The dominance pattern in the saproxylophagous beetles also seems to differ considerably from the present-day. In particular, the cupedid archostematans were diverse, and some now relict xylophilous polyphagan taxa were well represented in the Mesozoic (e.g., Cerophytidae were common in the Cretaceous) while some most important living families (e.g., Cerambycidae) are not definitively recorded. The taxonomic composition of saproxylophagous guild varied greatly in space and time. This is particularly well demonstrated for the cupedid archostematans (in a broad sense, including Ommatinae) (PONOMARENKO 1995a). Insect larvae inhabiting rotten wood are not represented in the Mesozoic fossil record (at least until the oldest insect-bearing resins appear) and thus only thorough study of insect activity traces can provide a firmer basis for a thorough analysis of Mesozoic history of saproxylophagy.

The insects would also be expected to have played an important role in the distribution of fungal and bacterial wood decomposers. WHITE & TAYLOR (1989) described fossil fungal remains from silicified peat deposits of the Fremouw Formation (Early to Middle Triassic) of Antarctica. They closely resemble the modern trichomycete order Eccrinales which now includes the intestinal endosymbionts of terrestrial detritivorous arthropods (millipedes and some beetle families). The Triassic fossils are enclosed in the irregular structures interpreted by White and Taylor as arthropod gut cuticle. This most interesting find is still the only palaeontological evidence for a specialised arthropod/fungal symbiosis in the Mesozoic though there is no doubt that diverse fungal and bacterial symbionts were widespread in both detritivorous and herbivorous Mesozoic insects.

Besides the saproxylophages, other specialised detritivore guilds should also have existed in the Mesozoic but the palaeontological information on their possible composition is extremely scarce. The most intriguing in this respect is the problem of exploitation of ephemeral substrates such as fallen fruits, carrion and dung. Those substrates decay rapidly, often in a few days. At present their principal consumers are insects, and first of all the higher (non-asilomorph) brachyceran flies and some beetles highly specialised either in extremely rapid larval development or in food storing. At least among the bibionomorph dipterans, rapid larval development should have been acquired no later then in the Jurassic; however, the living families with rapid larval development (e.g., Sciaridae and Scatopsidae) are remarkably rare in the Mesozoic. Perhaps, fleshy fruits were not widespread in the Mesozoic though the karkeniacean ginkgoaleans had "berries" similar to the fruits of modern Ginkgo, and there was also a thick fleshy outer cover in the seeds of Nilssoniales (WEISHAMPEL 1984). Probably, the decomposition rate of dead plant matter was generally low, and the relatively uncommon, rapidly decomposing parts were utilised by opportunistic detritivores rather than by feeding specialists; if this hypothesis is correct, this slow decomposition should be one of the largest differences between the Mesozoic and modern ecosystems.

Rarity of the substrate is unlikely to be the case with dung and carrion because the large-sized Mesozoic reptiles should have produced a considerable amount of both, at least in some types of landscapes (WING & SUES 1992). Nevertheless, very few insect taxa related to modern specialised coprophages or necrophages are known with certainty from the Mesozoic. The oldest cyclorrhaphan flies appeared in the Early Cretaceous where they were represented only by a few extremely primitive and not very common families (besides the mycetophagous Platypezidae, the Ironomyiidae, Lonchopteridae, Sciadoceridae and prioriphorine Phoridae are also found). Even in the earlier Late Cretaceous faunas there are no other cyclorrhaphan families, and even in the Early Palaeogene their diversity is still limited. The feeding habits of the Mesozoic cyclorrhaphans is obscure. Some of them could have been coprophagous or necrophagous but mycetophagy and opportunistic detritivory is at least equally probable. Lamellicorn beetles appeared in the Jurassic but only become common in the Early Cretaceous. The Early Cretaceous lamellicorn fauna is diverse but dominated by primitive Hybosorinae and Aclopinae (NIKOLAJEV 1998b). Here again the Mesozoic taxa could well be mycetophages, opportunistic detritivores and saproxylophages. No trace fossils documenting specialised dung provisioning are known before the Late Cretaceous. No silphid beetles are recorded from the Mesozoic. Alleged dermestid beetle pupation chambers have been discovered in dinosaur bones from the Late Jurassic (LAWS et al. 1996, HASIOTIS & FIORILLO 1997) but the interpretation of these ichnofossils is subject to doubt (Chapter 2.3.5). Among modern nematocerans, necrophagy and coprophagy are remarkably rare if their high general biological diversity is taken into account.

These data suggest that either the importance of the specialised necro- and coprophagous insects in the Mesozoic was extremely low in comparison with present-day ecosystems, or they were represented by other taxa that were later replaced by the cyclorrhaphan flies and modern beetle lineages. At present, the former hypothesis seems to be more plausible. Perhaps, the principal Mesozoic carrion consumers were vertebrates, as supposed by GRIMALDI & CUMMING (1999), and the dung of herbivorous reptiles, composed mainly of tough undigested fern and gymnosperm tissues, was mostly utilised by opportunistic detritivores together with other plant litter. If so, this is one more signal indicating that the Mesozoic ecosystems with megaherbivores should differ fundamentally from the present-day grazing ecosystems such as steppes and savannahs. The possibility of more specialised coprophagy cannot be ruled out for some Mesozoic roaches, scorpionflies, or nematoceran dipterans, and the two latter orders could also include some extinct necrophagous taxa. In any case, it is improbable that there were specialised guilds of carrion and dung feeders in the Mesozoic comparable with the modern ones in structural complexity and ecological efficiency.

The appearance of nest-building aculeate hymenopterans (e.g., the digger wasps) and the termites in the Early Cretaceous should create a new licence for specialised detritivores but there are no data on their possible commensals except for a record of a supposedly termitophilous scarabaeid beetle from the Early Cretaceous Lebanese amber (CROWSON 1981). There is also no information on the inhabitants of Mesozoic vertebrate holes though the appearance of ectoparasitic mallophagans and of flea-related mecopteroid insects in the Early Cretaceous (see below) suggests that their putative psocopteran and mecopteran ancestors could have been commensals in tetrapods permanent shelters or nests.

3.2.3.1.4. Predators and parasites

The predatory habit was surely widespread in the Mesozoic insects, and predacious insects were abundant in all kinds of terrestrial habitats. All major present-day predatory taxa are represented in the Mesozoic record; some of them (e.g., the mantises) first appeared only in the Early Cretaceous. In some cases (e.g., in the odonatans and neuropterans) the predatory habit was inherited from their Palaeozoic ancestors but in a number of lineages it evolved from taxa with other modes of feeding (e.g., from herbivory in the hemipterans and hymenopterans). There were also some extinct predacious lineages in orders that include no predators in modern faunas. In particular, morphology of the peculiar roach family Raphidiomimidae clearly indicates its predacious habit (Chapter 2.2.2.1.1). The raphidiomimid fore legs are raptorial (though they could also be used for walking, unlike in the mantises: VRŠANSKÝ 1999b), and their free prognathous head is very unusual for the order (see Fig. 372) The family was probably restricted to the warmer parts of continental Asia. It is not uncommon in the Late Jurassic of Karatau in Kazakhstan where probably more species occur than are described. The alleged scorpionfly *Cretaceochorista* from the Early Cretaceous of China (HONG *et al.* 1989) is, judging from the published photograph, almost surely a raphidiomimid. The only extinct insect order restricted to the Mesozoic, the Mesotitanida, was also predacious as indicated in particular by the raptorial fore legs (see Fig. 461). One further extinct group of presumably land predators were the nymphs of the meganerinous odonatans which probably dwelled on land rather than being aquatic (Chapters 3.2.2.2, 3.3.2). Meganeurina were rather common in

the Triassic and survived as a relatively minor faunal component up to the Early Cretaceous where the youngest protomyrmeleontids are found (Chapter 2.2.1.1.2). The high diversity of Mesozoic neuropterans and snakeflies is also noticeable. The modern representatives of both orders are exclusively predacious at least at the larval stage (the adults are often facultative pollinivores), and the mouthparts of all neuropteran larvae are strongly modified for a specialised predatory habit. Modern neuropteran larvae are morphologically diverse and inhabit a wide range of habitats including leaf surface, bark, wood, litter and soil, and though the Mesozoic larvae are unknown, we would expect them to be at least as biologically diverse as the recent ones. Some of the most specialised living families (e.g., Coniopterygidae and Mantispidae) are known since the Late Jurassic. Few living taxa have aquatic larvae but this mode of life is surely derived and was probably not significantly more widespread among the Mesozoic members of the order.

Mesozoic predacious insects were quite diverse in their habitat and hunting specialisations. At least since the Jurassic, the principal predatory guilds were nearly the same as they are now. Diverse predators inhabited vegetation, litter, and soil, and both active hunting and sit-and-wait strategies were probably common. The use of traps is undocumented but possible (e.g., net constructing in the Early Cretaceous keroplatid gnats). A number of predatory insects evolved as specialised dwellers of closed microhabitats such as bark crevices, wood and soil holes, etc. (e.g., the staphilinid beetles known since the Late Triassic and probably diverse asilomorph dipteran larvae in the Jurassic and Cretaceous). Like now, predacious insects fed mainly on various invertebrates though some largest forms (e.g., titanopterans) could also have attacked small vertebrate prey. The morphology of some trachypachid ground-beetles (e.g., the Early Cretaceous genus *Evertus* Ponomarenko: RASNITSYN 1986b) suggests that they could have been specialised predators on terrestrial gastropods. Prey paralysation and storage for larvae in their nests (which is one of the most peculiar types of predatory behaviour known in the insects) probably originated in the aculeate hymenopterans no later than in the Early Cretaceous (Chapter 2.2.1.3.5). Specialised predation on small objects (e.g., mites and insect eggs) is very likely for a number of thrips. The mantispid neuropterans and acrocerid flies known since the Late Jurassic are now highly specialised predators of spider eggs developing in their egg cocoons. The aerial predator guild included, in addition to the odonatans, also bittacid scorpionflies (appearing in the Triassic and particularly common in the Jurassic) and since the Early Cretaceous the asilid and perhaps some empidid flies (though the empidids could take their prey also from vegetation). Interestingly, although many families have disappeared in the Late Permian, the Triassic odonatan faunas were, unlike the majority of other insects, still dominated by groups with Palaeozoic (meganeurine) affinities whereas the most characteristic Mesozoic groups such as Heterophlebioidea, Isophlebioidea and Libellulina become abundant only in the Jurassic. This phenomenon may be partially explained by the appearance and diversification of flying reptiles which would presumably have been both natural enemies and competitors of the odonatans.

Hence the structural complexity and specialisation in the predacious insect guilds increased during the Mesozoic and reached a very high level by the Early Cretaceous. Nevertheless, the total pressure of predacious insects in the Mesozoic terrestrial ecosystems was probably somewhat less than nowadays because of the absence of the ants, one of the most important groups of modern predators. The oldest and very primitive ants only appeared in the latest Early Cretaceous period (see below), and their absence in older times could hardly be compensated for by the abundance of any non-social predatory taxa.

Parasitoids represent a peculiar case of extremely specialised predation and merit special attention. Although there are some parasitoids in other insect orders, this mode of life is widespread only in the hymenopterans and dipterans (QUICKE 1997), and in both it appeared in the Mesozoic. Some aspects of the history of parasitoids are intriguing. Firstly, in both hymenopterans and dipterans the oldest parasitoids appeared synchronously in the fossil record, in the latest Early Jurassic (Toarcian). Secondly, in both cases the most probable original hosts were cryptobiotic (in particular, wood-boring) insect larvae though at present many parasitoids attack ectophytic prey (EGGLETON & BELSHAW, 1992). Both facts are difficult to explain. Interestingly, some of the oldest parasitoid families appearing in the Jurassic are still extant today (e.g., Megalyridae, Roproniidae and Heloridae among the hymenopterans and Mythicomyiidae and Nemestrinidae among the dipterans). Strangely enough, in the hymenopterans more opportunistic predation has evolved only secondarily on the basis of parasitoid specialisation. In this order the parasitoid way of life was acquired only once, in the ancestral apocritans (with several independent secondary shifts to predation and herbivory in some lineages), while in the dipterans it has evolved independently many times (perhaps about 100: FEENER & BROWN 1997) though almost exclusively in the brachycerans. Even in the Mesozoic, parasitic taxa have originated several times in different fly groups (Archisargoidea, Nemestrinoidea). Nevertheless, the hymenopteran parasites are much more diverse than the dipteran ones which presently constitute only about 20% of parasitoid species known (FEENER & BROWN 1997). In the Mesozoic, the share of dipterans among the parasitoids was probably even less than now because there are a number of young Cainozoic parasitoid lineages (including the largest parasitic fly family Tachinidae). The ecological evolution of the parasitic wasps has been discussed by RASNITSYN (1980, 1988c, see also Chapter 2.2.1.3.5), and the history of Mesozoic parasitic flies by MOSTOVSKI (1997a,b, 1998, MOSTOVSKI & MARTÍNEZ-DELCLÒS 2000; see also (Chapter 2.2.1.3.4). In both orders significant taxonomic and biological diversification of parasitoid taxa occurred in the Late Jurassic and Early Cretaceous. At least in the Early Cretaceous the ecological diversity of parasitoids was probably comparable with the present-day situation. They included both ecto- and endoparasitic forms and infested a wide range of hosts, mostly insects (though the above-mentioned predators in spider egg cocoons may be regarded as a special case of parasitoids as well), both ectobiotic and cryptobiotic, at different stages of host life cycle including the egg stage. The mechanisms of prey location and utilisation should also have been quite diverse.

Another mode of feeding that related to predation and appeared in the Mesozoic for the first time is blood sucking (delimited here as piercing-and-sucking on animal prey much larger in size than the sucker). Blood sucking on invertebrate prey is now common in the mites but it is exceptional for the insects (e.g., in some ceratopogonid midges), and this is probably true for the geological past including the Mesozoic. Many modern vertebrate blood-suckers are specialised flightless ectoparasites whose palaeontological record is extremely scarce and incomplete. There are three Mesozoic insect genera that have been interpreted as specialised blood- sucking permanent vertebrate ectoparasites, namely *Strashila* Rasnitsyn from the Late Jurassic of Siberia, *Saurophthirus* Ponomarenko from the Early Cretaceous of Siberia, and *Tarwinia* Jell & Duncan from the Early Cretaceous of Australia. Surprisingly, all of them seem to belong to the

same phylogenetic lineage related to modern fleas. Several other Mesozoic insects have been wrongly assigned here, including *Saurophthirodes* Ponomarenko (which is probably an immature chresmodid, see Chapter 2.2.1.3.4.5), *Niwratia elongata* Jell & Duncan (the illustrations in JELL & DUNCAN 1986, indicate that the unique type specimen is a probably young dragonfly nymph), and two unnamed "pulicid" fossils recorded from Koonwarra (JELL & DUNCAN 1986) which actually appear to be roach nymphs. In general appearance, *Saurophthirus* (see Fig. 327) strongly resembles various arthropods which parasitise modern bats (the polyctenid bugs, nycteribiid flies and spinturnicid mites) and display piercing mouthparts, ctenidia-like structures on the body, weakly sclerotised tensile abdomen, and long flexible legs with tenacious claws, so that PONOMARENKO's original (1976a) hypothesis on its possible ectoparasitism on flying reptiles is still the most plausible. Strangely, *Saurophthirus* is not extremely rare: there are a dozen specimens from Baissa and an additional specimen from another Early Cretaceous Transbaikalian locality, Semyon; conspecificity of specimens from both localities still needs confirmation. *Strashila* (see Fig. 328) is an even more bizarre creature with strongly modified hind legs apparently adapted to grasping, and peculiar lateral projections along the abdomen. Its possible host association is in fact quite obscure. These two genera also demonstrate affinities to *Tarwinia* (see Fig. 329), a remarkably preserved fossil which is intermediate between *Saurophthirus* and modern fleas in a number of characters (Chapter 2.2.1.3.4.5). The host association of *Tarwinia* is unknown. The discovery of this putative ancestral flea lineage related to scorpionflies suggests that blood-sucking could occur in some winged Mesozoic mecopterans as well. Indeed, as discussed above, this mode of feeding might be hypothesised for the Jurassic Pseudopolycentropodidae (see Fig. 277). GRIMALDI et al. (2000) mention *Tarwinia*-like fossils from the early Late Cretaceous resins of New Jersey; I have seen a photograph of one specimen and found that it rather resembles a specialised phoretic first-instar larva (triunguline) of some insect parasite.

Not surprisingly, the history of winged blood-sucking dipterans is better documented palaeontologically. There are some Jurassic finds of the taxa possibly related to modern blood-sucking families but their feeding habit is not confirmed by the mouthpart structure. The blood-sucking dipteran families known from the Early Cretaceous include Ceratopogonidae (Lebanese, Austrian and Spanish resins: SZADZIEWSKI 1996, BORKENT 1997, SZADZIEWSKI & ARILLO 1998, BORKENT 2000), Corethrellidae (SZADZIEWSKI 1995), Simuliidae (JELL & DUNCAN 1986, KALUGINA 1991, and undescribed materials from Siberia), Culicidae (undescribed from England: R. Coram, pers. comm.), phlebotomine Psychodidae (Lebanese amber: AZAR et al. 1999b), and Tabanidae (MARTINS-NETO & KUCERA-SANTOS 1994); some athericid and rhagionid genera were probably also blood-suckers (Chapter 2.2.1.3.4.4). The presence of the modern ceratopogonid genus *Leptoconops* Skuse in the Lebanese amber is worthy of mention. BORKENT (2000) points out that the Early Cretaceous ceratopogonids belong exclusively to lineages feeding on vertebrates whereas those taxa attacking insects only appear in the Late Cretaceous. In many cases a blood-sucking habit is confirmed by mouthpart morphology. Those finds indicate that the guild of flying blood-suckers was diverse and included taxa with different (olfactory and visual) prey location strategies. Interestingly, its appearance in the fossil record coincides with an expansion of the birds (as manifested in particular by numerous finds of feathers in the Early Cretaceous deposits) and probably small mammals. However, homoiothermy is often postulated for some Mesozoic reptiles as well, and feeding on poikilothermous vertebrates has been recorded for modern corethrellids, ceratopogonids, and phlebotomine sandflies.

Recently the first fossil mallophagan, *Saurodectes vrsanskyi*, was described from the Early Cretaceous of Siberia (RASNITSYN & ZHERIKHIN 2000). Its morphology suggests that it probably parasitised a host covered with sparse (or short) hairs, perhaps a pterosaur; interestingly, this unique fossil has been discovered in the same locality as *Saurophthirus*. This is the first evidence for the existence of permanent chewing ectoparasites on the Mesozoic vertebrates.

3.2.3.1.5. Origin of sociality

Perhaps, the most important Mesozoic novelty was the appearance of true social insects. Termite fossils have been discovered in a number of Early Cretaceous localities in Siberia, Mongolia, China, Europe, and Brazil. All Early Cretaceous fossils represent the primitive extant family Hodotermitidae with subterranean nests. The majority of finds are from the Barremian or younger Early Cretaceous strata but undescribed fossils from Baissa are older. The oldest definitive ants have been discovered in the Albian of Russian Far East, and are assigned to the extinct formicoid family Armaniidae (DLUSSKY 1999a). Thus they should be somewhat younger in comparison with the termites. Dlussky (in DLUSSKY & FEDOSEEVA 1988) doubts armaniid sociality, mainly because of their short antennal scape, indicating the absence of the characteristic ant adaptations for brood manipulations in the nest. However, he means the high sociality standard characteristic of even the most primitive living ants. Armaniid remains are very common implying the existence of mass nuptial flights as in true ants. It is most likely that the armaniids were social insects, but their sociality was primitive in comparison with the living ants. No evidence for social wasps or bees are known from the Early Cretaceous or older Mesozoic strata.

There are some records of supposed ant and termite nests from the Jurassic and even from the Triassic (RUSCONI 1948, HASIOTIS & DUBIEL 1993a, 1995, HASIOTIS & DEMKO 1996a,b, HASIOTIS *et al.* 1995, 1996). A long period of existence of social ants or termites without any wing or body fossils is improbable because mass nuptial flight is universal in both groups. Thus the attribution of the above pre-Cretaceous ichnofossils can hardly be accepted. There are several possibilities: perhaps, the tunnels were constructed by some non-social soil invertebrates, or by subsocial invertebrates other than insects (for example, some modern subsocial woodlice live in complex subterranean gallery systems), or by an extinct group (or groups) of social insects. The last possibility is intriguing though at present there is no basis to suppose a social mode of life in any pre-Cretaceous insect taxa.

3.2.3.1.6. Some ecosystem-level phenomena

The insects constituted an important food source for diverse Mesozoic invertebrates and vertebrates but co-evolution of insects and insectivorous animals in the Mesozoic is poorly studied. In particular, the appearance of flying reptiles in the Triassic and the expansion of birds in the Early Cretaceous must have seriousy affected insect evolution. Cryptic body shape is widespread in the Mesozoic insects (e.g., in the stick-insects and orthopterans) suggesting that visually oriented vertebrate predators should have been abundant. GOROCHOV (1995a) stressed that dismembering colour pattern in the orthopterans (and titanopterans)

became common since the Triassic (Chapter 2.2.2.3.1). Some Mesozoic insects (e.g., the kalligrammatid neuropterans) demonstrate large eye-spots on the wings (see Fig. 254) which are commonly interpreted in the modern insects as an anti-predatory adaptation disorienting the enemy. Concerning the invertebrate insect predators, the appearance of the web-building spiders should be noted. An advanced araneoid spider is described from the latest Early Jurassic (ESKOV 1984a) so that ZHERIKHIN's (1980a) statement that the Mesozoic spiders were represented exclusively by hunters is certainly wrong. A special analysis of insectivory in Mesozoic ecosystems is highly desirable.

There is virtually no information on phoretic and parasitic organisms associated with any Mesozoic insects except for records of parasitic mermithid nematodes and mites associated with some inclusions in the Lebanese amber (POINAR *et al.* 1994a,b).

Changes in insect ecology during the Mesozoic must have caused important effects at the community level. However, study of community evolution is a multidisciplinary subject which cannot be discussed in detail here. Some principal consequences of insect ecological evolution for turnover in the terrestrial ecosystem have been discussed above. PANFILOV (1967) and ZHERIKHIN (1980a) speculated on the possible effect of aquatic insect evolution on terrestrial ecosystems. The dispersal of aquatic insects over the watershed areas is poorly studied (KOVATS *et al.* 1996) but probably plays a significant role in nutrient output from water bodies and their return back to terrestrial ecosystem turnover (SAZONOVA 1970, MENZIE 1980, SILINA & GONCHAROV 1985, JACKSON & RESH 1989, ZAIKA 1989, OLECHOWICZ 1990). This transport should partially compensate for the losses due to water migration but it did not exist in the Palaeozoic and earliest Mesozoic. Consequently the appearance of an abundant aquatic insect fauna and especially of aquatic dipterans in the Jurassic has to have changed the general chemical migration pattern at the landscape scale. This problem merits special attention in palaeoecology.

The development of insect pollination should also have been extremely important from the point of view of community evolution. The minimal stable population density for wind-pollinated plants is rather high because of the random pollen transport. Accordingly, the vegetation cover is rather uniform and the potential possibility of narrow habitat specialisation is strongly limited. Entomophilous plants are able to exist at much lower densities than are the anemophiles, and so narrow microhabitat specialisation is easily available to them. Probably, the development of insect pollination should lead to a significant advance in general plant diversity and vegetation complexity, particularly at the substrate-specific, early successional stages. This hypothesis should be tested on palaeobotanical data.

3.2.3.2. The late Cretaceous and the Mesozoic/Cainozoic transition

3.2.3.2.1. General features of late Cretaceous insects

The Late Cretaceous epoch was a time of great biotic changes. The end-Cretaceous extinction attracts much attention as one of the most dramatic in the history of life. The mid-Cretaceous biological events are somewhat less widely known but no less important. Their principal significance is well known in palaeobotany where the Early/Late Cretaceous boundary is regarded for a long time as the borderline between the so-called Mesophytic (with dominance of diverse gymnosperms and ferns) and Cainophytic eras (with angiosperm-dominated floras). This division was criticised by some recent authors (e.g., KRASSILOV 1972b, 1983) as being unnecessary, mainly because the transition was rather gradual, and the zonal watershed vegetation remained conifer-dominated for a long time in the Late Cretaceous. A useful summary of general geological and biological information on the Cretaceous is presented by KRASSILOV (1985).

The Late Cretaceous insects remained nearly unstudied before the 1970s. The available information was summarised and discussed by ZHERIKHIN (1978, 1980b), and important new material has been collected since then. The principal localities of compression insect fossils are Obeshchayushchiy (Cenomanian of NE Siberia), Kzyl-Zhar (Turonian of Kazakhstan), and Orapa (Botswana, the age uncertain between the Cenomanian and mid-Senonian), each yielding about 1000 insect specimens. Insect-bearing resins are widespread; the richest resin faunas have been discovered in the Taimyr peninsula in Northern Siberia (numerous localities from the Late Albian or Early Cenomanian up to the Santonian or even younger Late Cretaceous age, with the richest materials from the Cenomanian and Santonian), Atlantic Coastal Plains, USA (a number of localities with the age ranging from the Turonian up to the Maastrichtian; with the major collections from the Turonian), and in the Campanian of Manitoba and Alberta in Canada (for the locality information see Chapter 4). The Burmese amber should be added in spite of uncertainty concerning its age, because its fauna, even if of the early Palaeogene age, has a clearly Cretaceous appearance (Chapter 4.2). A number of minor or insufficiently studied insect localities are known in Eurasia and North and South America. The most complete sequence of insect assemblages since the Albian up to the Palaeocene has been discovered in Sikhote-Alin Mountains, Russian Far East, but the material from this area is scarce (except for the Palaeocene deposits). The total world collections of Late Cretaceous insects at present hardly outnumber 10,000 specimens, a figure comparable with the material available for each of the known largest Early Cretaceous localities. Additionally, major taphonomical differences between the compression fossil and resin inclusion assemblages cause serious problems in a comparative analysis of the faunas. Nevertheless some important trends are evident enough.

Already the preliminary study of composition of the Late Cretaceous insect assemblages indicated their similarity to the Early Cainozoic rather than with the Early Cretaceous faunas and thus suggested that the insect change pattern should agree with that in the terrestrial plants (ZHERIKHIN & SUKATSHEVA 1973, ROHDENDORF & ZHERIKHIN 1974). Because of its obvious importance, the Cretaceous insect faunal turnover has been analysed in a number of publications, both regarding the general dynamics of the insect taxa (Chapter 3.1) and special analyses (ZHERIKHIN 1978, DMITRIEV & ZHERIKHIN 1988, RASNITSYN 1988b, 1989a, WHALLEY 1988b, A. ROSS et al. 2000). In all cases the analysis concerns the dynamics of insect families. The results differ in many details depending on differences in the methods, databases, taxonomy and stratigraphy used. Particularly important are differences between the papers where the extant taxa disappearing from the fossil record are calculated together with extinct ones (ZHERIKHIN 1978, RASNITSYN 1988b, 1989a) and those where they are treated as surviving. As discussed previously (Chapter 1.4.2.1.1), the former approach is preferable from the taphonomical point of view. The differences are evident from two negative Lyall's curves as shown in Fig. 480, where only the taxa missing in the present-day fauna are taken as extinct, and Fig. 485, where the taxa absent from the Cainozoic record are scored as extinct. The latter drawing additionally displays the curves based on different stratigraphic schemes thus illustrating the

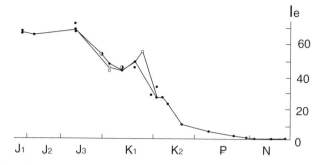

Fig. 485 Share of the extinct families in all insect assemblages of the succeeding time intervals (inverted Lyellian curve) in the Late Mesozoic and Cainozoic. Double lines and differing symbols in the Late Jurassic and Early Cretaceous refer to different hypotheses about the stratigraphic succession of some of the assemblages involved (after RASNITSYN 1988b, q.v. for details)

role of alternative dating of some important localities. Nevertheless, nearly all results agree in two most important points; firstly, there were major changes in the insect faunas between the Albian and Turonian (the mid-Cretaceous event); secondly, they were obviously greater than the changes near the Cretaceous/Palaeogene boundary (the end-Cretaceous event). In other words, the insect faunal pattern agrees better with the Mesophytic/Cainophytic division than with the Mesozoic/Cainozoic. The general model of an ecological crisis, described briefly in Chapter 3.2.2.4, has originally been based on the analysis of the very mid-Cretaceous event (ZHERIKHIN & SUKATSHEVA 1973, ROHDENDORF & ZHERIKHIN 1974, ZHERIKHIN 1987, 1993b) and later applied to other cases with some important specifications and modifications (KALANDADZE & RAUTIAN 1993a,b, BARSKOV et al. 1996, RAUTIAN & ZHERIKHIN 1997, ARMAND et al. 1999).

As discussed before, total taxic and ecological insect diversity both increased continually from the Early Triassic up to the Early Cretaceous. In the course of this process the insect fauna became more and more modern, mainly at the suprageneric level. This faunal modernisation was in fact a resulting pattern of both continuing appearance of extant taxa and disappearance of the extinct ones, but the former process evidently dominated over this time interval. This is reflected in gradual decrease in the share of extinct families as shown at Fig. 3.1.3. During the Late Cretaceous the share of families not represented in the Cainozoic record rapidly drops from about 50% to less than 10%, and the main extinction occurs at the pre-Santonian interval (see Fig. 485). It should be stressed that disappearance of the typical Mesozoic families was in fact even more drastic, because there are numerous apparently short-living taxa restricted to the Late Cretaceous or nearly so. Their stratigraphic distribution pattern is basically similar to the distribution of newly appearing families and differs clearly from the distribution of other extinct families (Fig. 486). The short-living lineages are also responsible for the increase in the extinct family percentage in the later Early Cretaceous (Fig. 486).

The intensive mid-Cretaceous extinction began already in the Albian and was initially uncompensated by the appearance of new taxa. To compare both processes a simple numerical index was used, called the index of coherency (ZHERIKHIN 1978, 1993b):

$$I_k = 100 \frac{(a-b)}{(a+b)}$$

where a and b are the number of taxa of a selected rank (here the families) which are found in the respective time interval for the first (a) and

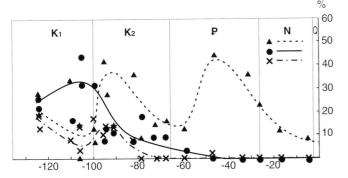

Fig. 486 Composition of the Cretaceous and Cainozoic insect assemblages: proportion of families found in respective time intervals for the first time (triangles) and for the last time in their history (dots), and the share of the families known from nowhere except for the particular time interval (crosses) (after ZHERIKHIN 1978)

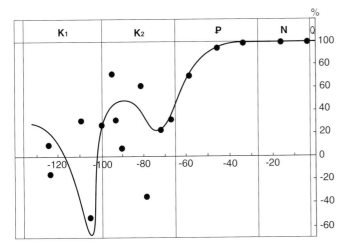

Fig. 487 Variation over time in the coherency index of insect family assemblages during the Cretaceous and Cainozoic (after ZHERIKHIN 1978)

for the last (*b*) time in their entire fossil record. This index value is less than zero when disappearance prevailed over appearance of new taxa. The curve presented in Fig. 487 is drawn considering both the I_k value and the representativeness of the respective assemblages. It indicates that the only time interval with uncompensated extinction was the Albian. An uncompensated extinction pattern suggests that it could not have been caused by competitive interactions within the group considered; and, because the principal insect competitors in the Cretaceous were undoubtedly other insects, it means that the extinction has to be non-competitive at all. As a result, the niche density in ecological space has decreased significantly, provoking a rapid subsequent diversification in unsaturated licences (see Chapter 3.2.3.1 for difference between licence and niche). This diversification is reflected in the palaeontological record by the increasing rate of appearance of new taxa in the Late Cretaceous (Fig. 487). As noted above, some of those taxa were short-lived but most of them have survived until the present. As a result, the Late Cretaceous insect fauna looks like an impoverished Cainozoic one rather than the one forming a transition between the Mesozoic and Cainozoic assemblages.

The mid-Cretaceous extinction seems to have been remarkably non-selective from the ecological point of view. Significant changes occurred in all principal insect guilds, and affect at least some of the most diverse and/or numerically dominant Mesozoic taxa in each of them.

3.2.3.2.2. Herbivores

Among the pollinivorous and nectarivorous guilds, the mid-Cretaceous interval is marked by a lot of families and subfamilies disappearing from the fossil record which were common during the Early Cretaceous. This concerns primarily the hymenopterans, dipterans and especially the neuropterans, but also the mandibulate moths. The latter group remained comparatively common and diverse up to the Aptian (Santana in Brazil and Spanish fossil resins) and later died out (except Micropterigidae which survived but became rare). The only families of the supposed Early Cretaceous nectarivores with strongly specialised mouthparts which survived the mid-Cretaceous event were Nemestrinidae and the pangoniine Tabanidae. A rapid diversification in the Late Cretaceous is more or less evident in the parasitic and aculeate wasps, dipterans and microlepidopterans. It is possible that some macroheteroceran moth lineages were also evolving though the only record (a supposed noctuid egg from the Campanian of USA: L. GALL & TIFFNEY 1983) needs confirmation. The oldest find of an undescribed syrphid fly in a Santonian or, perhaps, younger Late Cretaceous resin from Ugolyak in Taimyr, N. Siberia (ZHERIKHIN & ESKOV 1999) is also noteworthy. A Late Cretaceous origin also seems likely for bees. The only record of a bee fossil from the Late Cretaceous is dubious (RASNITSYN & MICHENER 1991), but a number of ichnofossils interpreted as solitary bee nests have been described or mentioned from the Late Cretaceous, and in particular from the latest Cretaceous Asencio Formation of Uruguay (GENISE 1999). Pollen grains were discovered on the body surface of a staphylinid beetle from Orapa, and on a cranefly from the same deposit demonstrating a specialised elongated proboscis supposedly adapted to nectar-feeding on flowers with a long corolla (RAYNER & WATERS 1991). Most probably, nectarivorous and pollinivorous insects are strongly underrepresented in the Late Cretaceous record because most of them are too large to be enclosed in resins, and the available collections of post-Turonian compression fossils are scarce. Hence the best though indirect evidence of their diversification is the impressive Late Cretaceous radiation of angiosperms, and in particular the recent discoveries of diverse fossil flowers (TIFFNEY 1977, FRIIS & SKARBY 1981, 1982, RETALLACK & DILCHER 1981b, FRIIS 1983, 1984a,b, 1985a,b, 1990, KRASSILOV et al. 1983, DILCHER & CRANE 1984, NISHIDA 1985, DILCHER & KOVACH 1986, FRIIS et al. 1986, 1988, CREPET & FRIIS 1987, FRIIS & CREPET 1987, OHANA & KIMURA 1987, P. CRANE et al. 1989a,b, KRASSILOV 1989, 1997, CREPET et al. 1991, 1992, ENDRESS & FRIIS 1991, 1994, RAYNER & WATERS 1991, HERENDEEN et al. 1992, 1994, NIXON & CREPET 1993, CREPET & NIXON 1994, 1995, 1997, 1998a,b, P. CRANE & HERENDEEN 1996, GANDOLFO et al. 1998a,b). Those finds demonstrate the development of diverse adaptations to advanced insect pollination mechanisms, e.g., specialised nectar glands, spurred anthers, polyads, floral tubes, bilateral symmetry, and a low pollen production, as well as the presence of specialised anemophilous taxa. The Turonian *Palaeoclusia* flower (Clusiaceae) shows the presence of secretory cavities in the receptacle, sepals, and petals suggesting resin production that in a number of modern genera of the family (CREPET & NIXON 1998a). Resin secretion by flowers is a rare and unusual phenomenon associated mainly with bee pollination (ARMBRUSTER 1984). Perhaps, some present-day plant/pollinator associations go back to the Late or even later Early Cretaceous, whereas any older connections are completely lost in the modern flora. As discussed previously (Chapter 3.2.3.1), modern entomophilous gymnosperm taxa appeared in the Late Cretaceous as well. Interestingly, palynological data indicate the gnetalean gymnosperms increased in their diversity in the latest Early and earlier Late Cretaceous (but in low palaeolatitudes only) with subsequent decline (P. CRANE & LIDGARD 1989).

Sucking herbivores have experienced the loss of some of the most important Mesozoic families in the mid-Cretaceous. This particularly concerns the aphids (e.g., Oviparosiphidae) and auchenorrhynchans (e.g., Palaeontinidae, Hylicellidae and Procercopidae). Information on psyllomorphs, whiteflies and scale insects is limited, but almost surely there were also important changes which will be detected by future studies. The feeding habit of the most common Mesozoic terrestrial bug families is often uncertain, but in any case there is an obvious contrast between the Early Cretaceous assemblages, usually dominated by the extinct lygaeoid family Pachymeridiidae (possibly seed predators), and the Late Cretaceous faunas dominated by modern families (e.g., herbivorous or predacious Miridae in Obeshchayushchiy). Undescribed Progonocimicidae (either herbivores or mycetophages) occurred in the Cenomanian of Obeshchayushchiy for the last time in their fossil record. The Late Cretaceous auchenorrhynchans and bugs are insufficiently known (both groups are infrequent in fossil resins). However, the first records of some important extant families (e.g., Aphrophoridae, Cercopidae, Dictyopharidae, Lygaeidae) indicate an important radiation. The Late Cretaceous aphid assemblages are peculiar and differ considerably from both Early Cretaceous and Palaeogene ones. They are dominated by extinct families, partially endemic, and still occasionally appearing in the Early Cretaceous, though some extant families (e.g., Mindaridae, Pemphigidae, and Aphididae) enter the fossil record here for the first time. A similar situation takes place in the scale insects, with the extinct Inkaidae in the Santonian of Taimyr and at least four extinct families in the Late Cretaceous of New Jersey and Canada (KOTEJA 2000). This indicates a rapid evolution of the most specialised plant parasites in the Late Cretaceous (and early Palaeogene) connected with diversification of the angiosperms and Cainophytic conifer lineages. SHAPOSHNIKOV (1980) stressed that the aphid taxa with relatively short antennae and legs, indicating more specialised cryptic habits (including supposed gallers), demonstrate the largest changes during the Cretaceous. KOTEJA (2000) supposes that the absence of finds of females or larvae of the dominant scale insect families Grimaldiellidae and Electrococcidae in Late Cretaceous resins, may also be explained by their gall-dwelling habits. Interestingly, the dominant position of aphids in the northern temperate belt and scale insects in the warmer regions has not been affected.

Significant changes in the chewing herbivore guilds, both ecto- and endophytic, are documented by both insect remains and ichnofossils. The commonest and most diverse Mesozoic siricoid woodwasp families (e.g., Sepulcidae and Anaxyelidae, both still occurring in the Cenomanian) are not found as fossils after the mid-Cretaceous, as well as several important sawfly groups (e.g., Xyelotomidae and Xyelydidae with the latest finds in the Barremian-Aptian). The dominant Mesozoic families of weevils either left the fossil record after the Albian (Nemonychidae and Belidae, both surviving until the present) or decline drastically (Eccoptarthridae) and become replaced by

Brentidae (oldest finds in the Barremian-Aptian, numerically dominant in Orapa: KUSCHEL et al. 1994) and Curculionidae (oldest finds in the Albian, dominate in terms of both specimens and taxa in Kzyl-Zhar, undescribed). There are virtually no data on the Late Cretaceous chrysomeloid beetles, but it is most likely that both Cerambycidae and modern subfamilies of Chrysomelidae s. l. radiated somewhere in the Senonian. Both families are well represented in the Palaeocene, and LABANDEIRA et al. (1995) pointed out that chrysomelid-type feeding traces are common on the leaves from the Meeteetsee Formation (Maastrichtian) of Wyoming, USA. The last finds of the locustopsid grasshoppers are from the Turonian; and all Mesozoic orthopterans (independently of their supposedly herbivorous, omnivorous, detritivorous, or predacious feeding habits) are in fact either lost completely, or at least strongly declining (e.g., Prophalangopsidae). Strikingly, this was not the case of the folivorous stick-insects: the susumaniids not only survived, but became the most common orthopteroids in the Late Cretaceous. They are not rare, even in the Palaeocene, and the latest susumaniid fossil (unnamed) is illustrated from the Eocene of Republic, Washington, USA (S. LEWIS 1992, pl. I, fig. G, as an unidentified orthopteran). The data on lepidopteran taxa with ectophytic larvae are virtually absent, but their radiation in the Late Cretaceous is quite possible.

Wood borings (see Figs. 469, 470) are rather common in the Late Cretaceous (e.g., ZHERIKHIN & SUKATSHEVA 1973, ZHERIKHIN 1978, GENISE 1995) but insufficiently studied, so it is impossible to estimate the frequency of attacks during the host tree lifetime. Their attribution is mostly uncertain (except for the termite tunnels discussed below). The presence of bored seeds in the unspecified Late Cretaceous deposits of Argentina is worth mentioning as they are the oldest definitive evidence of seed parasitism (GENISE 1995; see also 3.2.4.2 on possible seed-beetle presence in the Campanian of Canada). Among the gallers and leaf miners, the mid-Cretaceous diversification, surpassing the extinction, is documented mainly on the basis of trace fossils. Both guilds were probably of minor importance in the Mesozoic, and their increasing role in the Cretaceous is obviously connected with the angiosperm diversification. LABANDEIRA et al. (1994a) hypothesised that the widespread deciduousness of the Cretaceous angiosperms could favour the radiation of leaf miners. However, the deciduous gymnosperms (e.g., ginkgoaleans and czekanowskialeans) common in the Mesozoic temperate areas were nearly unexploited by mining insects. The mid- and Late Cretaceous radiation is documented for leafmining moths (KOZLOV 1988, LABANDEIRA et al. 1994a) and gall wasps (RASNITSYN 1980, O. KOVALEV 1994) but probably occurred in other gall-inducing and mining insects as well (e.g., in sawflies and beetles: LABANDEIRA et al. 1995 mention a coryphoid palm bud bored supposedly by a hispine leaf beetle larva from the Maastrichtian of USA, and WILF et al. 2000 describe presumed hispine damage on ginger leaves of the same age). Observations on mined angiosperm leaves from the Cenomanian and Turonian (KOZLOV 1988, LABANDEIRA et al. 1994a) demonstrate that certain easily recognised mine types are usually restricted to certain leaf morphotypes, suggesting a narrow host specialisation. They are often remarkably similar to the mines produced by present-day moth genera.

It should be noted that larval herbivory was probably remarkably rare among the Late Cretaceous dipterans. In particular, the gall midges seem to be quite rare even in the Campanian resins of Canada (GAGNÉ 1977). Among the brachyceran flies modern herbivorous taxa are scarcely represented even in the Palaeogene. Hence the attribution of some galls and mines on Cenomanian angiosperms to Cecidomyiidae and Agromyzidae respectively (LABANDEIRA & DILCHER 1993) are dubious. The only important dipteran group in the Late Cretaceous which could be partially herbivorous is the Bibionidae. This family probably appeared in the Early Cretaceous but the oldest fossils are extremely rare and somewhat dubious. In the Late Cretaceous bibionids turned out to be common (e.g., in Obeshchayushchiy, undescribed material, and Orapa: RAYNER 1987) as in the Cainozoic deposits where they often numerically dominate over other terrestrial dipterans. Modern Bibionidae include both detritivorous and herbivorous (root-feeding) species.

Although available information is still fragmentary, some important general points should be stressed concerning the Late Cretaceous history of herbivory. The mid-Cretaceous extinction affected all major Mesozoic herbivore guilds. Moreover, it seems that the extinction scale did not correlate strongly with the level of host specialisation. Of course, specialisation of extinct herbivores can be estimated only in a quite hypothetical manner. However, at least among the anthophiles there was a significant extinction of hymenopterans, neuropterans, and snakeflies which should be opportunistic feeders in the adult stage, while the morphologically specialised nectarivorous nemestrinids survived. Also the wood-boring siricoids (which were seriously affected by the extinction) rarely demonstrate a narrow host specialisation nowadays and are unlikely to have been more specialised in this respect in the Mesozoic.

Another important point is a high diversity and a remarkably high level of behavioural and trophic specialisation among the consumers of angiosperm photosynthetic tissues as early as in the Cenomanian and Turonian. However, it should be noted that a quantitative estimation of folivory in a Maastrichtian flora from Wyoming indicated a rather low attack level. The average attack frequency on foliar units is 22%, but the total consumed leaf area constitutes only 6% (LABANDEIRA et al. 1995). Interestingly, the consumed surface area on ferns is even larger than on angiosperms. Though no quantitative comparative data are available, the frequency of all principal damage traces on the Late Cretaceous leaves in temperate Siberia seems to be distinctly lower than in warmer Kazakhstan, Europe and USA. Further special studies on a wide stratigraphic and geographic scale are highly desirable.

The tremendous present-day diversity of herbivorous insects results mainly from their radiation on angiosperm hosts (FARRELL 1998, KRISTENSEN 1999). The drastic mid-Cretaceous vegetation changes had not only to have affected the taxonomic composition of herbivores but also the biology of the Mesozoic survivors. Many plant taxa disappeared completely, and the descendants of their consumers were forced to shift to other hosts. The habitat distribution, abundance and also the physiology and biochemistry of the old plant lineages that survived the crisis had probably changed significantly. Ancient host-plant associations were largely destroyed and very few present-day host relationships, even between obviously ancient insect and plant taxa, may have been inherited directly from the Early Cretaceous or earlier times. In particular, the modern fern fauna is composed mostly by taxa that are of relatively modern origin, including many phylogenetically young hemipterans, brachyceran flies, lepidopterans, etc. (BALICK et al. 1978, COOPER-DRIVER 1978, GERSON 1979), and few if any associations could be treated as having existed since pre-angiosperm times. Almost all present-day cycad insects have colonised their hosts secondarily and probably originate from angiosperm-associated ancestors (OBERPRIELER 1995a,b). I have failed to find any records of truly ancient herbivore taxa from any present-day gnetophytes. The relict

Ginkgo biloba is nearly devoid of herbivores except for occasional attacks from some extremely opportunistic polyphagous species (A. WHEELER 1975). The fauna of araucariacean conifers seems to be more conservative; however, it may be explained, at least partially, by their plesiomorphic protective chemicals that have been lost in other modern conifers, but which were probably present in their ancestors (see chapters 3.2.3.1 and 3.2.4.2). Hence those insects now restricted to Araucariaceae could well have attacked other conifer families in the past. This matter has recently been the subject of attempts to map allegedly old host plant relations on to the results of phylogenetic analyses (e.g., FARRELL 1998, for the cycad consumers and SEQUEIRA et al. 2000, for the araucariacean fauna). However, any hypotheses on pre-mid-Cretaceous age of host associations in modern herbivores also require independent support by palaeontological data (e.g., correlation between the distribution and abundance of fossils, ichnological evidences, etc.) which appears to be absent at present.

3.2.3.2.3. Detriti- and fungivores

It is difficult to analyse the changes in the mycetophagous guilds because of uncertainty in discrimination between them and the detritivores. It may be noted that none of the extinct Mesozoic families of mycetophiloid and anisopodoid dipterans are found in the Late Cretaceous. BLAGODEROV (1998c) believes that the Late Cretaceous Mycetophilidae still only colonised the long-lived bracket fungi as in the Early Cretaceous and not the taxa with ephemeral fleshy carpophores; however, the mycetophilid fauna of the latest Late Cretaceous is unknown.

Detritivores demonstrate significant changes in the middle and Late Cretaceous. In addition to the disappearance of extinct dipteran families, this is evident from the roach fauna. The Cenomanian dipteran faunas are already of the characteristic Late Cretaceous composition. In contrast, rather large roach collections from the Cenomanian of Russian Far East (Arkagala in the Magadan Region and Obluchye in the Hebrew Autonomous Region) are still dominated by Caloblattinidae, while extant roach families only became prevalent since the Turonian (P. Vršanský, pers. comm.). The oviposition mode characteristic of modern roach lineages is confirmed by the finds of mineralised oothecae in the Coniacian deposits in the Kyzylkum Desert, Central Asia (NESSOV 1995). The latest record of an extinct Mesozoic roach family (Blattulidae) is Santonian in age. These data suggest that in the opportunistic detritivores (but not in more specialised guilds such as the saproxylophages), the taxonomic turnover was probably somewhat slower than in the herbivores. Unfortunately, the data on detritivorous beetles are too fragmentary for any analysis.

Termites demonstrate no extinction at the family level in the mid-Cretaceous. However, the characteristic Early Cretaceous subfamily Valditermitinae did not occur in the Late Cretaceous, and the Late Cretaceous radiation of the order is evident. The only Early Cretaceous family was Hodotermitidae, while in the Late Cretaceous all modern families except Termitidae probably existed. It should be stressed that no termitids are found even in the Burmese amber that originated in the tropical belt. Wood borings containing characteristic hexagonal termite coprolites occurred in the Late Cretaceous since the Cenomanian (ROHR et al. 1986, LUDVIGSEN 1993, LABANDEIRA et al. 1994b, GENISE 1995). The termites were widely distributed in the Late Cretaceous reaching subpolar regions (e.g., Taimyr Peninsula). Their evolution should have had an important effect on wood decomposition in the Late Cretaceous. However, other feeding habits occurring now in the advanced Termitidae (e.g., feeding on grass litter and humus) probably did not evolve in the Cretaceous.

The processing of leaf litter in the Late Cretaceous ecosystems is an intriguing problem. SAMYLINA (1988) notes that the leaf cuticle of modern conifer families (Cupressaceae, Taxodiaceae and Pinaceae) and angiosperms in the Cenomanian deposits of Magadan Region is poorly preserved in comparison with gingkgoalean and czekanowskialean leaves from the same sites, and often shows distinct traces of fungal decay. This suggests an increase in decomposition rate of the litter which should provide a valuable food resource for detritivores. However, in contrast to the diverse herbivore fauna exploiting the Cretaceous angiosperms whose existence is well documented, the taxa which could consume decaying leaves seem to be rather rare in the fossil record. Several dipteran families can be tentatively assigned to this guild including Psychodidae, Sciaridae, and Scatopsidae. Psychodidae often develop in aquatic and semiaquatic environments, and their larval biology in the Cretaceous is uncertain. The family is well represented in fossil resins since the Early Cretaceous (e.g., in Lebanon and Spain) but very rare in Siberia. Sciaridae occur in diverse Late Cretaceous resins (e.g., in the Cenomanian and Santonian of Taimyr, in the Turonian of New Jersey, and in the Campanian of Canada) but only in rather small numbers. Their rarity in the Late Cretaceous is in a strong contrast with the Palaeogene situation. Scatopsidae were rare in the Mesozoic but became common and widespread in the Late Cretaceous. However, biology of the Late Cretaceous taxa is uncertain, and they could well be saproxylophagous. Hence the available data suggest that dead leaf consumption evolved later than the folivory on living tissues, but further studies are necessary to test this hypothesis.

The presence of the oldest unquestionable evidences of specialised coprophagy in the Late Cretaceous is noteworthy. CHIN & GILL (1996) describe and illustrate scarabaeid burrows in dinosaur dung composed mainly by undigested conifer remains from the Two Medicine Formation (Campanian) of USA. A number of dung beetle ichnofossils were described from the Asencio Formation of Uruguay (ROSELLI 1938, 1987, GENISE & BOWN 1993, GENISE & LAZA 1998). The ichnofossils are rather diverse suggesting the existence of a well-differentiated guild of coprophages. Unfortunately, in this case no information on the nature of plant materials from the dung provisioned is available. The Late Cretaceous herbivorous reptiles surely exploited angiosperm vegetation (WING & TIFFNEY 1987a), and it is possible that grazing ecosystems based on advanced herbivorous quadrupedal dinosaurs and herbaceous angiosperms other than grasses, were rather similar to modern grassland biomes (ZHERIKHIN 1993a). The age of the Asencio Formation (where the supposed solitary bee nests occur as well) is disputable between the Late Cretaceous and Palaeocene (GENISE 1999). However, the very presence of abundant and diverse dung beetle ichnofossils indicates the latest Cretaceous rather than early Palaeogene age because there were no large herbivorous tetrapods in the Palaeocene after the dinosaur extinction.

The Late Cretaceous record of necrophages is sparse. The find of *Cretophormia* (MCALPINE 1970) in the Edmonton Formation (Maastrichtian) of Alberta, Canada, indicates the appearance of specialised necrophagous flies. The puparia of *Cretophormia* were discovered in a considerable number in association with decomposed organic matter tentatively interpreted as the remains of a decapod crustacean. Originally the genus was placed in the Calliphoridae but this assignment is almost certainly wrong because this family seems to be geologically very young. However, the puparia surely belong to a schizophoran fly

showing that this advanced lineage existed in the latest Cretaceous. More primitive cyclorrhaphans (Sciadoceridae, Ironomyiidae, and Phoridae) occurred rather commonly in the Late Cretaceous but their feeding specialisation is uncertain. The supposed dermestid beetle pupation chambers on dinosaur bones were recorded from the Two Medicine Formation of Campanian age (R. ROGERS 1992). An undescribed dermestid beetle is found in the Turonian resins of New Jersey, but it is closely related to modern detritivorous dermestids living under the bark, and not to the groups feeding on carrion (R. Zhantiev, pers. comm.).

3.2.3.2.4. Predators and parasites

Changes in the taxonomic composition of predacious insects are comparable with those occurring amongst the herbivores. This is particularly evident in the case of the neuropteroids. Both neuropterans and snakeflies, very common and diverse in the Jurassic and Early Cretaceous, declined dramatically from the Aptian to the Cenomanian. Late Cretaceous snakeflies are very rare and insufficiently known; extinct Mesozoic families probably still survived in the Cenomanian, but only as a minor component of the faunas. The neuropterans became uncommon as well, and all extinct Mesozoic families were disappearing in the latest Early Cretaceous except for the Prohemerobiidae which are found in the Cenomanian (MAKARKIN 1990). Other predators, ranging from ground beetles (with the dominant Mesozoic subfamilies Eodromeinae and Protorabinae went extinct) to odonatans, have been seriously affected by the mid-Cretaceous extinction, in spite of being quite different in their habitat preferences and hunting strategies. The Late Cretaceous radiation of predators is still insufficiently known, but undoubtedly occurred in a number of unrelated taxa. In this context the modernisation of predatory beetle fauna should be noted as documented by finds of advanced carabid ground beetles (since the Turonian: ARNOLDI et al. 1977, PONOMARENKO 1995a) and staphylinids (e.g., Steninae and Staphylininae appearing in the Cenomanian: RYVKIN 1988). M. EVANS (1982) emphasises that the morphology of the early ground beetles suggests their adaptation to a cryptic mode of life under stones and/or bark, while since the Late Cretaceous, the taxa with locomotory adaptations to litter dwelling became common. Similarly, RYVKIN (1988) points out that the advanced extant subgenus *Hypostenus* Rey, adapted to active hunting on herbaceous vegetation has appeared as early as in the Cenomanian.

Among the aculeate wasps, the archaic digger wasp subfamily Angarosphecinae (Sphecidae), and to a lesser extent the archaic vespoid wasps (Vespidae Priorvespinae) were common during the later Early Cretaceous (though not recorded from the Albian) but not later on: Angarosphecinae became rare in the Cenomanian and seemingly extinct since that, except for a single wing discovered in the Early Eocene of Canada (PULAWSKI et al. 2000). Priorvespinae are not found after the mid-Early Cretaceous (J. CARPENTER & RASNITSYN 1990). Both subfamilies being replaced by extant ones (Pemphredoninae and Euparagiinae, respectively) in the Late Cretaceous. Characteristic of both families is the behavioural novelty: the prey transport over the earth surface (by foot in Sphecidae, by wings in Vespidae; no prey transport is implied for the basal aculeates). It is understandable that this novel behaviour makes the group less vulnerable to environmental changes. Indeed, the Scoliidae wasps, like their parasitic ancestors, leave the paralysed prey where it is found. The archaic scoliid subfamilies, Archaeoscoliinae and Proscoliinae, crossed the mid-Cretaceous threshold with minor changes only (RASNITSYN & MARTÍNEZ-DELCLÒS 1999). The appearance of social vespids documented by fossilised nest finds in the Bissekty Formation (Coniacian) of Uzbekistan (NESSOV 1988, 1995) and unspecified Late

Cretaceous of Texas, USA (WENZEL 1990) is noteworthy. The appearance and radiation of the ants seems to be of a particular importance. The oldest primitive armaniid ants appeared in the Albian. The Late Cretaceous faunas are dominated by primitive extinct taxa, either armaniids or sphecomyrmine formicids, but revealed also the extant, though primitive, subfamily Ponerinae in Turonian fossil resin of New Jersey (GRIMALDI et al. 1997) and more advanced Dolichoderinae in the Campanian of Canada (DLUSSKY 1999b). It should be noted that in the Late Cretaceous the ants were probably significantly more common and diverse in warmer regions, as indicated by materials from the Turonian of Kazakhstan (DLUSSKY & FEDOSEEVA 1988) and New Jersey (GRIMALDI et al. 1997, D. Grimaldi, pers. comm.). The ants probably inhabited diverse habitats but the ecological structure of the Late Cretaceous ant communities is still unclear. Some taxa demonstrate extremely specialised mandibular morphology suggesting a high level of feeding specialisation (e.g., *Brownimecia* Grimaldi, Agosti & Carpenter from the Turonian of New Jersey, GRIMALDI et al. 1997, and *Haidomyrmex* Dlussky from the Burmese amber, DLUSSKY 1996b). However, because of the absence of highly social species with extremely large families from the Late Cretaceous ecosystems the total pressure of ant predation was probably relatively low in comparison with many present-day communities. According to E. WILSON (1976) the role of ants is remarkably low in modern mountain forests of tropical areas; a strong contrast to lowlands and piedmont belts. Mountain forests are well known as refugia for a number of ancient angiosperm and insect taxa, and the limited ant pressure was probably an important factor of survival of at least some archaic arthropod taxa.

The parasitoid guild demonstrates the same general pattern of changes as in other insects. A number of important Mesozoic taxa disappeared or declined dramatically in the mid-Cretaceous (e.g., Eremochaetidae among the dipterans, Megalyridae, Gasteruptiidae, Mesoserphidae, Heloridae and Eoichneumonidae among the parasitic wasps). The Late Cretaceous radiation is evident in several groups of parasitic wasps (e.g., in Stigmaphronidae, Evaniidae, Scelionidae, Mymarommatidae, Chalcidoidea) but less so in Ichneumonidae and the brachyceran flies. Archaic Braconidae (Protorhyssalinae) became numerous but not diverse (Basibuyuk et al. 1999, D. Quicke, personal communication; presence of the extant genera parasitising mining moths and mentioned by RASNITSYN 1975 needs confirmation).

Interestingly, the changes among the blood-sucking dipterans seem to be minimal. All supposedly blood-sucking Early Cretaceous families occurred in the Late Cretaceous as well, and the differences from the Cainozoic faunas (e.g., the extreme rarity of Culicidae and the absence of any blood-sucking non-asilomorph brachycerans) were slight. The best studied family Ceratopogonidae demonstrates remarkably gradual evolution from the Early Cretaceous (e.g., Lebanese and Spanish resins) up to the present-day fauna. BORKENT (1995, 1996) emphasises that none of the Late Cretaceous ceratopogonids assigned to extinct genera show any autapomorphies indicating that they belonged to extinct lineages lacking any modern descendants. Probably, the radiation of major modern lineages of bird and mammal flightless parasite (at least, fleas, chewing lice and perhaps sucking lice) occurred in the Late Cretaceous but no palaeontological documents are available.

3.2.3.2.5. Pattern and models of the mid-Cretaceous events

The above review clearly shows that the mid- and Late Cretaceous changes were not restricted to herbivores but occurred in a similar manner in nearly all insect trophic and habitat guilds. This becomes even more evident when the large-scale mid-Cretaceous extinction in the aquatic insects is taken into account (Chapter 3.4.7). ZHERIKHIN (1979) stressed that other non-marine animals including the reptiles also demonstrate major changes in the mid-Cretaceous interval. In fact, the only typically Mesozoic group which has retained its major role in the Late Cretaceous, but not in the Cainozoic, was the dinosaurs. However, the Late Cretaceous dinosaur faunas dominated by hadrosaurid ornithopods and ceratopsids were drastically different from the Early Cretaceous ones (WING & SUES 1992). The most probable cause of this revolutionary reorganisation, one of the most impressive in the entire history of non-marine ecosystems, was the expansion of the angiosperms which has deeply modified both trophic webs and habitat characteristics in non-marine ecosystems. The model proposed by ZHERIKHIN (1978, 1979) suggests that plant/insect co-evolution has played a crucial role in this biotic revolution. This model is grounded on RAZUMOVSKY's (1969, 1999) idea about the existence of two fundamentally different plant strategies in relation to plant succession. Razumovsky stressed that in each region there are both successional specialists and opportunists, which he called the coenophilous and coenophobous species, respectively. The coenophiles are integrated into the regional succession system (which implies that the system is constituted by mutually co-adapted species), while the coenophobes are not: they form unstable stochastic associations at disturbed habitats and disappear when the early successional coenophilous species arrive. Unlike true successional pioneers the presence or absence of coenophobous species has no effect on the subsequent succession.

ZHERIKHIN (1978, 1979) supposed that the Early Cretaceous angiosperms were coenophobous (weedy) plants, rapidly colonising disturbed habitats at fresh alluvium, landslides, etc. In those habitats they used the anthophilous insects which existed on entomophilous gymnosperms, as opportunistic pollinators. A rapid evolution occurred in their small population under a strong selection pressure, in particular in the reproductive sphere. By the mid-Cretaceous they acquired highly effective mechanisms of both pollinator attraction and dissemination so that their competitiveness has increased greatly in comparison with the true successional pioneers. As a result, the pioneer stages of the Mesophytic successional systems became replaced by a new vegetation type producing abundant soft tissues. This should alter the pedogenetic processes, so the entire succession system has been destroyed as the subsequent seral stages were not able to colonise those modified habitats any more. Thus angiosperms did not have to directly replace all dominant Mesophytic plants (which is implausible), but only the early successional taxa which constituted a rather limited part of the total flora. This was enough for the indirect extermination of later successional vegetation. By the Cenomanian the Mesophytic vegetation has been almost completely excluded from the most dynamic habitats such as river valleys but still survived at the watershed areas with a low disturbance frequency where the areas occupied by Mesophytic communities declined continuously at least up to the Santonian. New successional systems, adapted to the angiosperm-dominated early stages, evolved gradually in the Late Cretaceous, and this process was accompanied by explosive insect evolution in new and unsaturated biotic environments. Palaeobotanical observations confirm that a mosaic of Mesophytic and Cainophytic communities existed in the middle and earlier Late Cretaceous rather than mixed communities composed by both Mesophytic and Cainophytic elements (SAMYLINA 1974, 1988). The above model suggests that the Mesozoic evolution of biotic interactions (first of all, plant/pollinator co-evolution) has set the stage for the mid-Cretaceous revolution which was in fact a complicated and rather prolonged community-level process which proceeds asynchronously in different landscapes and geographical areas. It also explains why those large-scale biotic changes were not accompanied by any correspondingly great abiotic events but by rather ordinary climatic fluctuations comparable with those which repeatedly occurred, both before and after the mid-Cretaceous interval.

Since this hypothesis was proposed in the 1970's, a large body of new data on the Cretaceous angiosperms has become available which is mainly in accordance with the model. In particular, the hypothesised "weedy" herbaceous habits of the early angiosperms is now supported by various phylogenetic, morphological, palaeontological, and palaeoecological data (D. TAYLOR & HICKEY 1992, DOYLE & DONOGHUE 1993). Very small seed size characteristic of the Cretaceous angiosperms (TIFFNEY 1984) is also typical of coenophobous and early successional plants. New finds of fossil flowers, discussed above, confirm that the Cretaceous angiosperm flowers were ephemeral suggesting high pollination efficiency. The oldest fossil floras rich in angiosperms have been discovered in the Aptian of Portugal. Recent data indicate that in some Portuguese localities more than 100 angiosperm species can be recognised, with about 85% of species belonging to two primitive lineages, the magnoliid dicots and Nymphaeaceae-like monoctos (FRIIS & PEDERSEN 1999).

The effect of introduced *Myrica faya* in the Hawaiian Islands (VITOUSEK *et al.* 1987, APLET 1990, BEARD 1990, MATSON 1990, VITOUSEK 1990, L. WALKER & VITOUSEK 1991) seems to be a good modern example of a successional system destruction, similar to the above mid-Cretaceous scenario. The soil modified by this alien shrub becomes unavailable for native plants and soil-inhabiting invertebrates, so the restoration of natural communities is impossible, even when *Myrica* stands are grubbed out.

There are no detailed data on the insect faunal turnover at the Cretaceous/Tertiary boundary. In some areas, insects are collected from the Maastrichtian-Danian sequences (e.g., from the Tsagayan Formation in Amur Region and from the Bogopol Formation in Sikhote-Alin Mountains in Russian Far East) but the collections are not sufficiently representative. Floristic changes near this boundary seem to have been rapid but not catastrophic (e.g., HICKEY 1981, 1984, ASKIN 1988, SPICER 1989, GOLOVNEVA 1994, HERMAN & SPICER 1995, KRASSILOV 1995, MARKEVICH 1995) though in some areas, an abrupt extinction has been documented (e.g., TSCHUDY *et al.* 1984, LERBEKMO *et al.* 1987, K. JOHNSON & HICKEY 1990). Palaeobotanical data suggest a significant cooling near the boundary though this pattern does not seem to be uniform at the global scale. It may be predicted that the insect extinction pattern should be similar to that of the plants. In any case, very limited if any extinction occurred at the suprageneric level, and hence the Cretaceous/Tertiary event can be at most a second-order boundary in insect history, absolutely incomparable with the mid-Cretaceous one.

The plants and invertebrates together constitute the bulk of the non-marine biological diversity and determine the basic features of terrestrial ecosystem organisation and turnover. From this point of view the non-marine life has acquired essentially Cainozoic general appearance after the end of the mid-Cretaceous extinction, no later than in the Senonian. From this point of view the Late Cretaceous represents the first stage of evolution of the Cainozoic terrestrial biota, and the 'hypnosis' of the dinosaur extinction is the only reason sending the non-marine palaeontologists in search of major changes at the Cretaceous/Palaeogene boundary.

The fundamental importance of the mid-Cretaceous events from the viewpoint of constraints of the actualistic community models has to be emphasised. Even if some present-day taxa probably inherit their ecological connections since the older Mesozoic times, the general organisation of ecosystems has altered so radically after the flowering plant expansion that none of the modern major community types (biomes) could be traced back to the pre-angiosperm times. The possibilities are strongly limited for any direct analogies between the present-day steppes, woodlands, tropical or even conifer forests and Early Cretaceous or older biomes in terms of the basic guild structure and diversity, ecosystem turnover, successional dynamics, etc. On the other hand, some modern warm temperate and subtropical forest ecosystems, especially on isolated islands not colonised by mammals (e.g., New Zealand or New Caledonia), have possibly retained their basic structure since the Late Cretaceous and could be used as really good models in palaeoecological studies.

3.2.4. CAINOZOIC

3.2.4.1. General features of Cainozoic insects

After the mid-Cretaceous there is no evidence for any more uncompensated global insect extinction. Insect evolution in the Late Cretaceous and Cainozoic was gradual and continuous right up to the present, and it was during this period that the present-day insect ecological pattern was formed. No significant ecological insect types which survived the mid-Cretaceous have disappeared subsequently, and the main events of their ecological evolution during the Cainozoic were diversification within some licences (see Chapter 3.2.3.1 for difference between licence and niche), changes in community-level interactions with other organisms, and changes in their geographical pattern.

Very few extinct insect families are known from the Cainozoic, and all of them are rare except for the damselfly family Sieblosiidae. Diversification at the family level generally appeared to be rather weak, but with two important exceptions, namely the lepidopterans and brachyceran dipterans, in both of which many families first enter the fossil record in the Oligocene and Miocene. As early as in the Late Eocene about 50% of insect genera are modern, as exemplified by the well-known fauna of the Baltic amber. However, there are significant differences in the share of modern genera between different orders and families. For instance, the Baltic amber fungus gnats (Mycetophilidae) and biting midges (Ceratopogonidae) belong mainly to recent genera while the bombyliid flies in the somewhat younger Florissant fauna are still represented nearly exclusively by extinct genera, although all these families are known as fossils since the Mesozoic. Records of modern species from the Late Eocene are somewhat questionable (see RÖSCHMANN 1999b, and ZHERIKHIN 1999) but at least in the Oligocene and Early Miocene some species are practically identical with the living ones in their external morphology (e.g., ASKEVOLD 1990, BOGACHEV 1940, MASNER 1969a, MOCKFORD 1972, DOYEN & POINAR 1994). It should be noted that Röschmann, in his discussion, has omitted at least two records of supposedly modern insect species from the Baltic amber, namely the chironomid midge *Buchonomyia thienemanni* Fittkau (MURRAY 1976) and the diapriid wasp *Ambositra famosa* Masner (MASNER 1969b). In the Pliocene the rate of supposedly modern species is about 50% (ZHERIKHIN 1992b), and in the Quaternary very few extinct species have been discovered; in some cases the identity of Quaternary specimens was confirmed by morphology of genitalia (ELIAS 1994). A long duration ecological constancy is supported by finds of plant galls indicating no changes in intimate plant/insect relationships since Neogene times (e.g., MARTY 1894, STEINBACH 1967, STRAUS 1967, 1977, HEIE, 1968, KRÜGER 1979). There is no evidence of increasing controversies between the environmental reconstructions based on different taxa with increasing age, at least during the Quaternary and Late Pliocene which should be expected if some evolution of ecological preferences occurred (ELIAS 1994). This morphological and ecological stasis does not mean that no new species have originated in the Quaternary, but only that the extinction at the species level was negligible except on some isolated islands (e.g., KUSCHEL 1987) and in those ecological guilds that are most closely connected with large mammals (e.g., coprophages, see below).

For a summary of major palaeogeographic, climatic and biotic changes affecting the terrestrial ecosystems during the Cainozoic see WING & SUES (1992) and POTTS & BEHRENSMEYER (1992). In general, climatic shifts were much more pronounced than in the Mesozoic, including the Eocene thermic maximum, the beginning of the Antarctic glaciation near the Eocene/Oligocene boundary with consequent cooling, the Miocene warming, the Pliocene cooling culminating in Quaternary glaciation and large-scale oscillations during the Quaternary. The main trend towards increasing climatic differentiation is obvious and resulted in progressive latitudinal differentiation of the principal biomes. In the Early Palaeogene, forests and woodlands most probably dominated throughout the world, and more open communities were patchily distributed. Later more open landscapes, firstly grasslands and then deserts and tundras, expanded in different parts of the world. This was probably the major mechanism of a significant increase in global biodiversity level from the earliest Palaeogene to now. The Late Cainozoic species constancy is especially remarkable if extreme climatic and other physiographic changes are taken into account; this suggests that there is no simple correlation between the scale of physiographic and evolutionary events.

Cainozoic insects have been discovered on all continents but there are relatively few localities known outside of Eurasia and North America. The stratigraphic distribution of the faunas is also uneven. In particular, the Early Palaeocene (Danian) insects are poorly known. There are some supposedly Danian localities in the Russian Far East, but most of them are insufficiently sampled. The only site where insects were collected specially is Arkhara (the upper member of the Tsagayan Formation) in the Amur Region; several hundred specimens are available, mainly isolated beetle elytra and homopteran fore wings. Information on Mid- and Late Palaeocene and Pliocene insects is limited though several rich local faunas have been discovered; probably, the total world collections of Palaeocene insects can hardly include more than 10,000 specimens, and even less material is available from the Pliocene deposits. To the contrary, there are many Eocene, Oligocene and Miocene localities including a number of important Lagerstätten, some of which yield more insect specimens than the Palaeocene and Pliocene collections combined.

The major European Palaeogene deposits are Menat (Palaeocene, France; PITON 1940), the Fur Formation (=Mo Clay or Moler, the latest Palaeocene or Early Eocene; Denmark and northern Germany; RUST 1999a), Messel (Mid-Eocene, Germany; LUTZ 1990), the Baltic amber (Late Eocene; LARSSON 1978), the Bembridge Marls (England; JARZEMBOWSKI 1980, MCCOBB et al. 1998) and the saliniferous deposits of Alsace (France; FÖRSTER 1891, QUIEVREUX 1935, 1937), both of Late Eocene or earliest Oligocene age; the Oligocene deposits of southern France (numerous localities including Aix-en-Provence, Ceréste, Bois d'Asson, etc.) and Rott (earliest Miocene, Germany; LUTZ 1996). The

fossiliferous amber-like resin recently discovered in the Early Eocene of the Paris Basin, France, is of great interest but its fauna has only been subject to preliminary studies so far (NEL et al. 1998a,b, 1999c). The Baltic amber fauna is the best known one, with about 3,500 named species and probably several times more undescribed ones. A number of other important localities is known. North America is also very rich in Palaeogene insects, especially its western part (see M. WILSON 1978, for a review of faunas). In particular, the Green River Formation (Eocene) and Florissant (latest Eocene or Early Oligocene) are extremely productive but there are dozens of other important deposits. The best North American Palaeocene materials originate from the Paskapoo Formation, Alberta, Canada (MITCHELL & WIGHTON 1979), and the only important North American fossiliferous resin of the Palaeogene age is the so-called Arkansas amber (SAUNDERS et al. 1974).

The Palaeogene insects of Asia are much less well known. Few localities have been discovered outside of the Russian Far East where the Tadushi Formation, the Sakhalinian amber (both Palaeocene; ZHERIKHIN 1978) and Bol'shaya Svetlovodnaya (the latest Eocene or Early Oligocene; ZHERIKHIN 1998) are worthy of mention. Data on Eocene insects of Asia are especially scarce. The only important source of Asiatic Eocene insects is the Fushun amber in China, although published information on the orders other than dipterans is scarce (HONG 1998 and references therein). The age of the Burmese amber is uncertain (not older than the Late Cretaceous and not younger than the Eocene: ZHERIKHIN & ROSS 2000). South America is probably rich in Palaeogene insect fossils but it is insufficiently studied in this respect. Here the Maíz Gordo Formation (Palaeocene) and the Ventana Formation (Eocene or Oligocene) in Argentina (PETRULEVICIUS 1999b) should be mentioned as well as the Tremembé Formation (Oligocene) of Brazil (MARTINS-NETO 1999b). South America provides several important ichnofaunas as well (LAZA 1986a). In Australia the only noticeable Palaeogene insect fauna originates from the Redbank Plain Series, Queensland, which is of Palaeocene age (RIEK 1952, LAMBKIN 1987). In Africa there are some trace fossils from the late Eocene and Oligocene of Egypt (GENISE & BOWN 1994) as well as a few records of undescribed insects from South Africa (REUNING 1931, R. SMITH 1986). Nothing is known on the Palaeogene insects of the Antarctic.

In total, the collections of the Neogene insects are not so huge as those of the Palaeogene but the number of localities is comparable. The European collections are the largest. The most important Lagerstätten are Rubielos de Mora (Early Miocene, Spain; PEÑALVER 1998), Randeck (Early Miocene, Germany; ANSORGE & KOHRING 1995), Stavropol (Mid- to earlier Late Miocene, North Caucasus, Russia), Oeningen (Late Miocene, Germany; HEER 1864–65), the Late Miocene (Messinian) deposits of Italy (Monte Castellaro, Gabbro, etc.; HANDLIRSCH 1906–08, F. BRADLEY & LANDINI 1984, GENTILINI 1989), the latest Miocene–Early Pliocene deposits of the Central Massif in France (PITON & THÉOBALD 1935, BALAZUC 1978, 1989), and Willershausen (Pliocene, Germany; KRÜGER 1979). In America north of Mexico relatively few deposits have been discovered; the richest faunas being known from the Barstow Formation (Middle-Late Miocene, California; PALMER et al. 1957), the Latah Formation (Late Miocene, Pacific Northwest of USA; S. LEWIS 1969), the Beaufort Formation (Late Miocene-Pliocene, Arctic Canada; MATTHEWS 1976a, 1977), the Lava Camp Mine (Pliocene, Alaska: HOPKINS et al. 1971) and the Kap København Formation (Late Pliocene–Early Pleistocene, Greenland; BÖCHER 1989). Southwards there are the most important Neogene fossiliferous resins, the Dominican and Mexican ambers, both of Early Miocene age (HURD et al. 1962, POINAR 1992a,

POINAR & POINAR 1999). The most outstanding Asian Lagerstätte is Shanwang (Middle Miocene, China; ZHANG 1989, ZHANG *et al.* 1994). Other noteworthy Asian deposits are Bes-Konak (Late Oligocene, Turkey; PAICHELER 1977), Chon-Tuz (Middle Miocene, Kyrgyzstan), numerous Miocene and Pliocene localities in Japan (FUJIYAMA 1985), the amber-like resin of the Merit-Pila Coalfield (Early or Middle Miocene, Sarawak, Malaysia; HILLMER *et al.* 1992a,b), and several rich Pliocene sites in the Kolyma River Basin at the north of Russian Far East (KISELYOV 1981). Minor localities are known in other parts of Asia including Siberia, Thailand, Indonesia, and the Philippines. Mainly trace fossils are known in South America (LAZA 1986b). In Africa a rich and exceptionally well-preserved fauna has been discovered in the Early Miocene of the Rusinga and Mfwangano Islands of Kenya (LEAKEY 1952), as well as a rather diverse ichnofaunas (e.g., NEL & PAICHELER 1993, NEL 1994). A supposedly Pliocene propalticid beetle was described from a Kenyan fossil resin (KRINSKY 1985), but the description is not accompanied by any geological information and the possibility that it actually originates from a Quaternary copal cannot be ruled out. Only minor insect occurrences are mentioned from the Neogene of Australia. The recent discovery of insect remains in the Pliocene of Antarctica (the Sirius Group, Transantarcic Mountains) is of great interest (ASHWORTH *et al.* 1997).

Unfortunately, complete lists of taxa and especially quantitative data are not available for the majority of Palaeogene and Neogene assemblages, and in many cases the published identifications need to be confirmed. There have been no modern attempts to carefully reconstruct the landscape distribution of insects for any local faunas, and a detailed ecological analysis of terrestrial insects is only available for the Baltic and Dominican amber (LARSSON 1978, POINAR & POINAR 1999).

Quaternary insects are known from a great number of sites, mainly in Europe, northern Asia and North America; several important faunas have also been discovered in South America. The results of recent extensive studies of Quaternary palaeoentomology in Europe, Siberia and the Americas (excepting trace fossils) are reviewed by ELIAS (1994). Interesting faunas have been discovered in Japan as well (AKUTSU 1964, FUJIYAMA 1968, 1980, UÉNO 1984, FOSSIL INSECT RESEARCH GROUP 1987, YOKOHAMA RESEARCH GROUP 1987, HAYASHI 1995, 1996, 1998, HAYASHI *et al.* 1996), and minor Quaternary assemblages are recorded from Australia (GILL 1955, DE DECKKER 1982a,b), New Zealand (DEEVEY 1955, SCHAKAU 1986, KUSCHEL 1987, KUSCHEL & WORTHY 1996) and some subantarctic islands (COOPE 1963). The Middle Pleistocene fossiliferous resin from Mizunami, Japan, is of great interest (HIURA & MIYATAKE 1974). There are also numerous insect records from subfossil copals of Africa including Madagascar (e.g., MEUNIER 1905, 1906, 1909a, ZEUNER 1943, SCHLÜTER & GNIELINSKI 1987), South Asia (e.g., VIEHMEYER 1913, HELLER 1938), the Caribbean (e.g., CHAMPION 1917) and South America (e.g. COCKERELL 1923, DUBOIS & LAPOLLA 1999), but most of them desperately need revision. The palaeoecological interpretations of Quaternary faunas concentrate primarily on the pattern of climatic oscillations.

The general composition of some trophic and habitat guilds during the Cainozoic demonstrates relatively little change while in others an increase in structural complexity is obvious, at least in some types of ecosystems.

3.2.4.2. Herbivores

The evolution of the herbivore guilds was strongly connected with the evolution of the angiosperms and a global increase in vegetation complexity. This process started in the Cretaceous, with the mid-Cretaceous vegetation changes. All principal modern herbivore insect guilds can be traced back into the Cretaceous, and all of them underwent a more or less apparent diversification in the Cainozoic; some subordinated guilds are perhaps of a Cainozoic origin.

As discussed above (Chapter 3.2.3.2), the major guilds of anthophilous insects on angiosperms, both pollinivores and nectarivores, probably evolved since the very beginning of the angiosperm history, and their main structure was already established in the Cretaceous. In particular, the chewing pollinivore guild probably changed little during the Cainozoic. The principal modern taxa of chewing flower visitors (mainly beetles and hymenopterans) are well represented in the Palaeocene and Eocene. Some families demonstrate a significant radiation in more recent times, but this radiation was probably connected with larval specialisation in other respects rather than with adult pollinivory itself. The pollen-collecting hymenopterans are the only important exception (see below). Nevertheless, the total diversity of pollinivores evidently increased; in particular, the Neogene expansion of treeless communities with diverse and continuously flowering herbaceous plants (e.g., the composites and legumes), would have stimulated their radiation. The Cainozoic history of diverse anthophilous cleroid, clavicornian and heteromeran beetles is incompletely known; some important anthophilous taxa are known since the Palaeocene (e.g. Meloidae) or Eocene (e.g., the alleculine tenebrionids) but their later diversification was probably rather intense. Chrysomelid seed beetles of the subfamily Bruchinae with pollinivorous adults appear in the Eocene fossil record (see below). Some anthophilous scarabaeid beetle taxa are recorded from Florissant (e.g., Glaphyrinae: CARLSON 1980), but one more important and advanced groups of pollinivorous scarabaeids, the Cetoniinae, is not known before the Early Miocene of Kenya (PAULIAN 1976). A rather intensive Cainozoic radiation occurred in diverse families of parasitic and aculeate wasps.

The only important Cainozoic sucking pollinivores are thrips, and in this case the main anthophilous taxa have probably also existed at least since the Eocene. However, some important co-evolutionary interactions between them and entomophilous plants are probably younger. In particular, at present the thrips are very important pollinators of Dipterocarpaceae (APPANAH & CHAN 1981, ANANTHAKRISHNAN 1982), a family poorly represented in the Palaeogene and which probably radiated extensively in the lowland tropics in the Late Oligocene-Miocene (COLLINSON *et al.* 1993). The available palaeontological data on Cainozoic thrips are too fragmentary and do not allow us to elucidate the evolution of anthophily in enough detail.

Changes in the nectarivore guild seem to be more significant, in particular, among Diptera and Lepidoptera. Even if their radiation was primarily stimulated by larval biology, its consequences for the advance of specialised nectarivory were great. Butterflies, moths and flies constitute a very important part of the nectarivore guild in all modern ecosystems. Among the hymenopterans, family-level radiation was less pronounced; however, this order provides evidence of diversification in the most important anthophilous group, the bees. It is noteworthy that the principal groups of specialised nectarivores are represented before the Late Oligocene mainly by extinct genera, suggesting a higher rate of evolution than in the majority of other insects.

Lepidoptera, other than the informal Microlepidoptera group, are extremely rare as fossils before the latest Eocene. The oldest known butterflies have been described from the Middle Eocene (the Green River Formation, DURDEN & ROSE 1978) but there is a record of a supposed hesperiid from the somewhat older Mo Clay deposits (RUST 1998). After the latest Eocene and the Early Oligocene (the Bembridge Marls and Florissant respectively) butterflies occur regularly, though in small numbers (JARZEMBOWSKI 1980, EMMEL et al. 1992). The noctuomorph moths also first appear in the Middle Eocene (DOUGLAS & STOCKEY 1996; a noctuid record from the latest Cretaceous needs confirmation, see Chapter 3.2.3.2) but become more common only from the Miocene onwards. It should be noted that both the supposed Mo Clay hesperiid and the Middle Eocene Canadian fossil assigned to Arctiidae demonstrate a well-developed, long haustellum adapted to nectar feeding. The bombycomorph moths first enter the fossil record as late as in the Miocene. In particular, the oldest known member of the Sphingidae, a highly specialised family well adapted to nectar-feeding on flowers with a long corolla, was discovered in the Middle Miocene of China (ZHANG 1989). The fossil record of large-sized lepidopterans, which rarely occur in resins, is extremely incomplete; nevertheless, the rarity of noctuomorphs and the absence of bombycomorphs in the richest Palaeogene faunas may be significant. Almost all Eocene and Oligocene lepidopteran genera are extinct.

Unlike lepidopterans, the dipterans are well represented as fossils. A number of anthophilous families including Syrphidae have existed since the Late Cretaceous; however, even in the Baltic amber, the brachyceran flies are still represented mainly by asilomorphans. Acalyptratans are relatively rare throughout the Palaeogene, and no anthophilous calyptratan flies are known before the Miocene: both groups include numerous flower visitors. Some important groups of anthophilous asilomorphs also evolved mainly in the Cainozoic. In particular, the first advanced non-mythicomyiine Bombyliidae come from the Green River Formation, and in the Florissant the majority of modern subfamilies are found including specialised flower-visiting Bombyliinae (GRIMALDI 1999). However, almost all Florissant bombyliids still belonged to extinct genera. On the other hand, the Baltic amber Empididae and Syrphidae are of a modern appearance, often belonging to extant genera (GRIMALDI 1999). It seems also that only comparatively minor changes have occurred among the anthophilous nematocerans.

As discussed above (Chapter 3.2.3.2), the existence of bees in the Late Cretaceous is very possible though they are documented only by trace fossils. The appearance of bees manifested an origin of a new anthophilous guild, viz. pollen and nectar gatherers. This guild additionally includes some modern ants and masarine wasps; no palaeontological data are available on the histories of either of these, but both may be of relatively young Cainozoic origin. On the contrary, fossil bees are rather diverse though uncommon. The age of Trigona prisca Michener et Grimaldi 1988, an allegedly Maastrichtian social stingless bee from North America, is dubious (see RASNITSYN & MICHENER 1990, and GRIMALDI 1999, for a discussion), firstly because of the less than certain origin of the type specimen, and secondly, the presence of other advanced insects, not found elsewhere in the Cretaceous, in the same resin piece. Recently, ENGEL (2000) has established a new genus, Cretotrigona, for T. prisca but he confirms that the bee is surprisingly advanced. Only new collections made in situ, or possibly study of the other insects enclosed in the resin with the bee, can resolve this question finally. Before this, the oldest definitively known bee body fossil is the Late Palaeocene Probombus hirsutus Piton from Menat, which is possibly related to the bumblebees (Bombinae). Eocene bees are uncommon but they are diverse, including both solitary and social taxa (ZEUNER & MANNING 1976, GERLACH 1989, LUTZ 1993a, POINAR 1994b, ENGEL 1998). The existence of sociality is strongly supported by a unique find of a dozen bee specimens (unfortunately, unidentified) in a single piece of Baltic amber together with a substance interpreted to be wax or honey (BACHOFEN-ECHT 1949). Those genera found in the Eocene and Early Oligocene are, with few exceptions, extinct. On the other hand, a number of extant genera are recorded from the Late Oligocene and Early Miocene including honey bees and stingless bees (ZEUNER & MANNING 1976, MICHENER & POINAR 1997, NEL et al. 1999b). A surprisingly high relative role of social bee taxa in comparison with solitary ones in the Palaeogene assemblages can be explained by the spatial dominance of forested landscapes where solitary bees are infrequent and patchily distributed. The increasing rate of solitary bee finds in the Neogene might mark an expansion of more open communities (ZHERIKHIN 1978); thus a significant Late Cainozoic radiation apparently occurs even among the phylogenetically old bee lineages.

Hence the total diversity of nectarivores and their specialisation level probably increased continuously since the Late Cretaceous at least up to the Miocene, and in the Palaeogene the taxonomic composition of the nectarivore guild was evidently less modern in comparison with that of the pollinivores.

Though the basic adaptations to insect pollination by angiosperms had already originated in the Cretaceous, the evolution of pollination mechanisms in general and of entomophily in particular, certainly continues into the Cainozoic as documented both by plant and insect fossils. In particular, palaeobotanical data (CREPET & FRIIS 1987, FRIIS & CREPET 1987, T. TAYLOR & TAYLOR 1993) suggest that the relative role of nectar and visual attraction mechanisms was increasing while pollen attraction became of lesser importance. This pattern is in accordance with the above-mentioned palaeoentomological data indicating more progress in nectarivory than in pollinivory during the Cainozoic. CREPET (1979b) points out that fossil flowers from the Middle Eocene (the Claiborn Formation) of the USA demonstrate a high morphological diversity with a number of evident specialisations for entomophily; however, some of the most specialised types (e.g., strongly complex or zygomorphous) are not represented. However, zygomorphous flowers of advanced Fabaceae have been discovered more recently in the earliest Eocene of Tennessee (CREPET & D. TAYLOR 1985) indicating that this morphological type existed rather early in the Palaeogene though it was probably rare. The Palaeogene age of some plant families with highly specialised insect pollination systems is well argued by leaf and/or pollen features even though they are not proved by flowers (e.g., Aristolochiaceae since the Middle Eocene: COLLINSON et al. 1993). Some specialised modern plant/pollinator systems can be traced back to the Palaeogene or at least to the earliest Neogene. A flower of Eoglandulosa, a member of Malpighiaceae, showing the sepals with paired resinous glands as well as characteristic pollen morphology, suggests that it was probably adapted to anthophorine bees as pollinators, like the living species of the family. This was discovered from the Claiborn deposits (D. TAYLOR & CREPET 1987). The Palaeogene age of origin of the most sophisticated of mutualistic pollination systems that is sound in Ficus and based on attraction of wasps for breeding places, can be suggested on the basis of the presence of the agaonid wasp Tetrapus mayri Brues in the Florissant (BRUES 1910), the Ficus-type pollen in the Middle Eocene, and endocarps in the Early Eocene (COLLINSON et al. 1993). Agaonids are also represented in the Dominican amber (POINAR & POINAR 1999). The finds of orchid-specialised euglossine bees in Dominican amber is also noteworthy (ENGEL 1999, POINAR 1999c,d). The fact that both orchids and orchid bees have subsequently disappeared from Hispaniola is an indirect confirmation of their probable mutual dependence since Early Miocene times. On the other hand, some peculiar entomophilous syndromes could hardly appear before the Late Oligocene or even the Neogene. The absence of calyptrate flies before the Miocene suggests that the peculiar phenomenon of pollinator attraction using a carrion-imitating odour should be relatively young. Other supposedly young entomophily types are specialised ant pollination of flowers placed either on trunks (cauliflory) or at the ground (HICKMANN 1974) and hawkmoth pollination of flowers with a strongly elongated corolla. It should be noted that all supposedly young pollination systems are much more widespread in tropical than in temperate areas. The evolution of specialised flower morphology, protecting pollen and nectar from non-target visitors, provoked evolution of new types of insect behaviour (e.g., nectar thieving) and this reciprocally reinforces protective mechanisms in plants (e.g., toxic nectar), but those processes are undocumented palaeontologically.

The entomophilous gymnosperms retained a subordinated position throughout the Cainozoic; no palaeontological data on the history of their specialised pollinators are available. The taxonomic position of *Antliarrhinites* Heer from the Late Miocene of Oeningen (HEER 1864–65) and its relationships with the present-day South African cycad-pollinating weevils of the genus *Antliarhis* Billb. are unclear.

The appearances of bat and bird pollination should affect nectarivorous insects due to increasing competition for nectar (L. CARPENTER 1979). The age of their origin is unclear, but probably it is rather late (post-Eocene).

Among the endophytic herbivores dwelling in plant reproductive organs several important groups appear in the Cainozoic record. Unfortunately, the published information is too fragmentary to trace the history of the guild accurately. At present, different members of the same endophytic tribe, or even genus, often develop in different plant organs so that the taxonomic relations of a fossil should be determined very closely in order to propose any reasonable hypothesis on its biology. Further information could be obtained from trace fossils but the published records of damaged fruit and seed are quite scarce; more studies in this direction are highly desirable. Hence the time of first appearance of principal taxa living in ovules, fruits and seeds (e.g., among beetles, cecidomyiid midges, brachyceran flies, and moths) is largely uncertain. Some taxa have probably been connected with angiosperms since the Cretaceous whilst others seem to be restricted to the Cainozoic, or at least they radiated most intensively in the Palaeogene (e.g., the seed beetles, apionine and curculionine weevils, the tortricid moths) or even in the Neogene (e.g., the tephritid flies). There are few finds of unquestionably specialised fruit parasites in the Palaeocene (e.g., the weevils of the tribe Curculionini: a re-examination of the type specimen of *Curculio elegans* Piton from Menat confirmed its generic placement, and the rather large body size of the species suggests that it developed probably in acorns or nuts). The seed beetles (Chrysomelidae: Bruchinae) is one of very few suprageneric taxa composed exclusively by seed parasites. The oldest fossils has been discovered in the Eocene of British Columbia (ARCHIBALD & MATHEWES 2000) and France (Célas; undescribed). It should be noted, however, that an alleged sagrine beetle from the Late Cretaceous Canadian amber-like resin illustrated by POINAR (1999f) may in fact be a primitive *Rhaebus*-like bruchine. A relatively recent geological age of the subfamily is supported by the fact that only rather primitive genera has been discovered in the earliest Oligocene (Florissant and Bol'shaya Svetlovodnaya; KINGSOLVER 1965, ZHERIKHIN 1989) whereas more advanced tribes appear in the Late Oligocene of France

(undescribed) and earliest Miocene of Germany (Rott: ZHERIKHIN 1989). The host plants of the Eocene and earliest Oligocene genera are not known; some Late Oligocene species belong to genera now restricted to Fabaceae, the plant family supporting the richest fauna of modern Bruchinae. According to KOZLOV (1988), the Baltic amber tortricid *Tortricites skalskii* Kozlov may be related to the modern tribe Lespeyresiini which include many fruit and seed parasites, and at least three species of oecophorid moths from the same deposit represent the subfamily Depressariinae where a number of modern species are seed-eaters; however, the feeding habits of the fossil taxa may be questioned in both of these cases. As regards the Tephritidae, no fossils of this, the largest modern herbivorous fly family, are known before the Miocene (EVENHUIS 1994; it should be noted that in this catalogue the Middle Miocene species *Pseudacidia clotho* Korneyev is erroneously recorded from Kazakhstan instead of North Caucasus, Russia); its complete absence in the very rich Palaeogene dipteran materials seems to be significant.

It should be noted that the main endophytic parasites of fleshy fruits probably appear only later in the Cainozoic. At the present state of our knowledge, no Cretaceous or Palaeocene insects can be assigned to this guild with a high degree of certainty though this mode of life cannot be excluded for some Early Palaeogene rhynchitine weevils (well represented in the Palaeocene as documented by undescribed fossils from Menat) and tortricid moths.

The history of other seed consumers is poorly known. Seed suckers which include mainly lygaeoid, coreoid and pentatomoid bugs probably existed since the Cretaceous but the major radiation of modern subfamilies and tribes (especially among the pentatomoids) probably occurred only in the Cainozoic. Unfortunately, it cannot be proved at present because the majority of Cainozoic species badly need revision. The seed gatherers guild mainly includes some ant groups; it has probably originated in the Cainozoic, no later than in the Miocene, as is documented by plant remains stored in insect burrows in palaeosols (THOMASSON 1982).

The evolution of seed-eating insects (as well as of granivorous birds and rodents) might seriously affect the reproductive success of plants and hence has to have had a pronounced effect on plant evolution. In particular, the seed beetles are a classical object of analysis of herbivore/host interactions (JANZEN 1969, 1971, CENTER & JOHNSON 1974, C.D. JOHNSON 1981), and their further palaeontological study can help to test several co-evolutionary hypotheses.

Consumers of plant vegetative tissues demonstrate the same general evolutionary pattern. All principal trophic (phloem-, xylem-, and mesophyll-feeding) and topic (specialised on different above-ground and underground plant organs) modern guilds of sucking herbivores, with both active and sedentary habits, have existed since the Mesozoic, and they have already colonised the angiosperms and modern conifer taxa by the Late Cretaceous. Their diversity and ecological role might have been high throughout the Cainozoic; however, their taxonomic composition demonstrates important changes. Particularly among the aphids, the differences between the Late Cretaceous and Cainozoic faunas are the largest among the insects, with a number of Cretaceous families not found in the Palaeogene (Palaeoaphididae, Canadaphididae, Creamyzidae, Mesozoicaphididae, Tajmyraphididae; HEIE & WEGIEREK 1998). Two extinct families survived in the Cainozoic, and while one of them, the Naibiidae, has disappeared after the Late Palaeocene, the other, Elektraphididae, occurs in the Baltic amber (several species), Bembridge Marls, Bol'shaya Svetlovodnaya and perhaps even in the Pliocene of Willershausen being the only extinct terrestrial insect family in the Neogene. All Eocene and Oligocene genera, except for *Mindarus* Koch, are extinct; some of them demonstrate a high degree of morphological specialisation (e.g., the pemphigid genus *Germaraphis* Heie with a very long rostrum). Interestingly, the number of *Mindarus* species in the Baltic amber alone (not counting other Palaeogene finds) is comparable with the present-day fauna of the Holarctic Realm as a whole, clearly indicating a decline in this group. The two largest modern families, Aphididae and Lachnidae, are surprisingly poorly represented in the fossil state, and several families enter the fossil record only in the Neogene, either in the Miocene (Lachnidae, Greenideidae, Phylloxeridae; the older records of the latter family based on galls only are questionable: Heie & Peñalver 1999) or even in the Pliocene (Adelgidae). It should be stressed that the fossil history of aphids is rather well studied, and that all Adelgidae and many Lachnidae live on coniferous hosts, so their absence in the Baltic amber and other Palaeogene conifer resins cannot be explained by known taphonomic biases. Moreover, the aphids seem to be consistently poorly represented in equatorial areas, which are weakly investigated palaeontologically, so that it is improbable that either Aphididae or Lachnidae could evolve for a long time somewhere outside Eurasia and North America.

Other Palaeogene herbivorous hemipterans were surely of more modern appearance than the aphids, with no extinct families and a number of living genera being found in the Eocene. Nevertheless, their generic composition and dominance pattern were changing substantially throughout the Cainozoic. The Palaeogene auchenorrhynchan assemblages are usually dominated numerically either by cercopoids (in more temperate areas), or by ricaniids and other fulgoromorphans. The largest modern family, Cicadellidae, occupies a subordinate position, while the second largest one, Membracidae, is extremely rare in the fossil state. A membracoid (not necessary membracid) nymph has been recorded from the Baltic amber (Szwedo & Kulicka 1999), and true membracids have been discovered in Dominican amber (Shcherbakov 1996, Poinar & Poinar 1999) but even here they seem to be rare. This is particularly surprising because the present Neotropical fauna is extremely rich in Membracidae. Cicadellids become dominant from the earliest Miocene (e.g., in Rott: Statz 1950). Similarly, the Palaeogene bug faunas are dominated by Cydnidae and Acanthosomatidae while Miridae and Pentatomidae become common since the Miocene, and Scutelleridae are generally uncommon as fossils. As pointed out above, fossil pentatomoids (as well as mirids) are insufficiently studied and, although a large number of species have been named, most of them need revision. Most probably, the dominance pattern at the subfamily and tribe level in the Palaeogene will be found to differ from that of the present-day, like in another herbivorous bug family, the Tingidae, in which the Eocene fauna is dominated by more primitive Cantacaderinae (which are often moss suckers), while the advanced subfamily Tinginae (connected exclusively with angiosperm hosts and often with advanced ones such as Asteraceae) has been expanding since the Early Oligocene (Golub & Popov 1999). Even more dramatic changes occurred among the Psylloidea. This group is completely missing in the Late Cretaceous and Palaeocogene record and is very rare in the Eocene where, however, several modern families appear for the first time. In the latest Eocene–earliest Oligocene, Psylloidea occur more regularly though they were still uncommon (e.g., in the Bembridge Marls and Florissant), and they became abundant only in the Miocene. The Cainozoic evolution of coccids, and especially the whiteflies, is poorly known but it can be predicted that important changes will be detected here as well.

Surely, the radiation of and changes in sucking herbivores were stimulated by host plant evolution and in turn affected it, but more complex tritrophic interactions were obviously important as well. Mutualism with ants, which is widespread among the present-day homopterans, should have had an especially pronounced effect. There is no firm palaeontological evidence of the time of its origin, however. It is quite possible that some mutualistic relationships, such as primitive guarding, arose in the Early Palaeogene or even in the Cretaceous; however, more complex and highly specialised associations probably evolved in post-Eocene times in parallel with progress in social ant organisation. Perhaps their mutualisms with ants alone could have stimulated the relatively recent radiation of some taxa (e.g., Lachnidae and Membracidae).

Though ectophytic chewing folivory seems to be the simplest herbivore feeding strategy, its rate was probably relatively low in the Mesozoic and increased after the angiosperm expansion in the Cretaceous (Chapter 3.2.3.2). The only accurate quantitative study of damage traces on Cainozoic leaves (Wilf & Labandeira 1999) demonstrates their high diversity in several Palaeocene and especially in the Eocene floras of Wyoming, USA, including various types of external feeding, hole feeding, margin feeding and skeletonisation, some of which are restricted to only one or a few form-genera of angiosperms. The authors explain the increasing rate and diversity of damage traces in the Eocene in terms of climatic warming. Other ichnological observations of folivory in the Cainozoic are too sporadic and non-quantitative. It seems, however, that the frequency of damage varies strongly between different palaeofloras and was controlled by a far more complex set of factors than the mean annual or seasonal temperature. An increase in feeding trace diversity should be forthcoming in the Oligocene and Miocene because of the intensive radiation of folivorous lepidopterans, beetles and orthopterans. Even when the high level of incompleteness of the lepidopteran fossil record is taken into account, it seems that before the Oligocene the share of endophytic and detritivorous moth taxa was disproportionately large in comparison with the present-day situation, and accordingly the diversification of the families with ectophytic caterpillars occurred mainly in the Oligocene and Miocene. The short-horned orthopterans are generally infrequent in the fossil state before the Miocene, and the relative role of the tree- and shrub-dwelling eumastacids is remarkably high in the Palaeogene faunas while the acridid grasshoppers (which are the most important herbivorous insects in modern grasslands and other open communities) become common only since the Miocene (e.g., in Ribesalbes and Rubielos de Mora in Spain, Radoboj in Croatia, Oeningen in Germany and Gabbro in Italy: Zeuner 1941a, 1942, 1944, Peñalver et al. 1996, Peñalver 1998). The taxonomic position of Miocene acridids is often uncertain at the generic or even subfamily level, and very few of them have been assigned to extant genera. According to Stidham & Stidham (2000), *Mioedipoda* Stidh. et Stidh. from the Middle Miocene of Nevada is a primitive genus constituting what is probably the sister lineage to all modern oedipodine genera. The only fossil member of one more acridoid family, the Pyrgomorphidae, *Miopyrgomorpha fischeri* (Heer), has been recorded from Oeningen (D. Kevan 1965). D. Kevan & Wighton (1981) suggest that some North American Early Oligocene taxa assigned to Eumastacidae may represent in fact the extinct Mesozoic family Locustopsidae: if this is correct, the differences between the Palaeogene and Neogene faunas are even greater than stated above. It should be stressed also that when acridids occasionally occur in considerable numbers in Palaeogene deposits their diversity is low.

In particular, in the Bembridge Marls the only catantopine species, *Proschistocerca oligocaenica* Zeuner, is rather common (ZEUNER 1941a), and one can suppose that it was a habitat specialist that inhabited coastal marshes. The data on beetles are less certain but it seems that at least some taxa that are folivorous either as larvae (e.g., the chrysomeline leaf beetles) or at the adult stage (e.g., the broad-nosed weevils and perhaps the blister beetles), were radiating rather intensively in the Oligocene and Miocene times in both forest and open landscapes. As regards the exophytic sawflies (e.g. Tenthredinoidea), their relative role as folivores probably decreased with the diversification of butterflies and heteroceran moths.

As with all herbivores, the evolution of folivores was surely managed first of all by plant radiation and the evolution of anti-herbivore chemicals. The process of co-evolution was certainly complex and many co-evolutionary patterns are difficult to explain at present. Some geologically young folivorous taxa have successfully colonised coniferous hosts (e.g., many heteroceran moths such as Sphingidae, Lasiocampidae, Noctuidae, etc.) while others, and sometimes much older ones, have failed to do so (e.g., the leaf beetles – Chrysomelidae). It should be noted that there is no firm palaeontological evidence for any continuous "arms race" co-evolution models, at least among the folivores. In particular, WILF & LABANDEIRA (1999) note that the total rate of folivore attacks on Palaeocene and Eocene betulacean leaves is high just like in modern members of the family, suggesting that no special protective mechanisms have evolved for more than 50 Myr in spite of intensive insect consumption. Palaeobiochemical studies on Miocene Ulmaceae and Fagaceae indicate only minor differences from present-day species in at least some allelochemicals (e.g., flavonoids and steroids) (NIKLAS & GIANNASI 1978). It is very possible that in other families which have not been investigated in this respect (especially ones with a complex set of toxic compounds, e.g., in Asclepiadaceae, Apiaceae, and Rutaceae: STRONG *et al.* 1984), biochemical evolution was more rapid and fits better with an "arms race" hypothesis. Some modern butterfly taxa living on highly toxic plants are rather old: the nymphalid butterfly subfamily Danainae which is now well adapted to feeding on toxic asclepiadacean and apocynacean hosts has been recorded from the Oligocene of Brazil (BRITO & RIBEIRO 1975), and the papilionid butterfly, *Thaites ruminianus* Scudder, from the Late Oligocene of France, is probably closely allied to extant genera developing on Aristolochiaceae (LEESTMANS 1983). However, their morphological similarity with modern consumers of poisonous plants does not necessarily mean the same degree of feeding adaptations. It seems that other factors, and especially tritrophic interactions, could considerably modify the herbivore/host co-evolution pattern just as in the case of ectophytic herbivores exposed to predators, parasitoids and pathogens as well as to the actions of the weather.

This has probably stimulated the appearance of some minor herbivore guilds that are supposedly restricted to the Cainozoic, namely the leaf rollers and web-nest builders. The former now includes a number of moth genera (e.g., in Oecophoridae, Gelechiidae, Plutellidae, Tortricidae, Thyrididae) and many attelabid weevils (all members of Attelabinae and Apoderinae as well as the rhynchitine tribes Deporaini and Byctiscini). The tortricids, oecophorids and plutellids are among the commonest and most diverse moth families in the Baltic amber, but the modes of life of the Eocene genera are uncertain. Almost all Baltic amber tortricids are tentatively assigned to the subfamily Olethreutinae in which leaf rolling is a common but not the universal mode of life. The majority of amber oecophorids seem to be related to modern detritivores rather than to leaf-rolling genera. All known fossil rhynchitine

attelabids belong to the tribes that include no leaf rollers, and there are only two fossil atelabines, both from the Miocene of Shanwang (ZHANG 1989). It seems that the leaf-rolling attelabids are geologically young, but even in this case their virtual absence in the Neogene record is difficult to explain. The time of appearance of web-building in different moth taxa is unclear. Possibly some Baltic amber Oecophoridae (e.g., Depressariinae) could be assigned to this guild but in a quite tentative manner as no extant genera have as yet been found. The oldest putative web-protected and leaf-rolling sawflies (*Megaxyela* Ashmead, Xyelidae, and *Atocus* Scudder, Pamphiliidae, respectively) are recorded from the Early Oligocene of Florissant in Colorado (RASNITSYN 1980, 1983). Neither rolled leaves nor web nests are recorded as trace fossils but they may simply remain unrecognised among poorly preserved plant remains. Interesting evidence of another highly specialised anti-predator adaptation in an ectophytic herbivore is provided by the discovery of a riodinid caterpillar from the Dominican amber showing the characteristic glandular setae whose secretion would probably have attracted ants that guarding its producer (DE VRIES & POINAR 1997). Protective larval cases of ectophytic caterpillars (e.g., Psychidae and Coleophoridae) are known since the Eocene.

The endophytic folivores (leaf miners and gallers) and stem borers had colonised angiosperms successfully already in the Cretaceous (Chapter 3.2.3.2). Although the practice of attribution of fossil mines to present-day miner genera has been criticised (GRIMALDI 1999), mainly because the mines of too many living taxa still remain undescribed, a constancy of mine morphology indicates that the feeding behaviour was remarkably stable since the Early Late Cretaceous. It should be stressed that in fact this high degree of constancy occurs mostly in mines assigned to lepidopteran larvae while the supposed dipteran, hymenopteran and coleopteran mines are rarely reliably identifiable before the Neogene. Even in the Miocene the taxonomic attribution of non-lepidopteran mines is usually uncertain. One can suppose that the mining habit, found not only among the dipterans but also among the beetles and sawflies, is much younger than in the major families of mining moths; however, this hypothesis does not fit well with other evidence. Several mining moth families (Nepticulidae, Lyonetiidae, Gracillariidae, Elachistidae) have been discovered in the Baltic amber and Bembridge Marls, but the affinities of the fossils to modern genera are uncertain (KOZLOV 1988). Some acalyptratan fly families occurring in the Baltic amber are phytophagous though they include mainly stem- and root-borers rather than leaf miners (LARSSON 1978). The younger acalyptratans are too little studied to be able to date the appearance of leaf-mining taxa with certainty (EVENHUIS 1994). The hispine leaf beetles, which are predominantly leaf miners in the larval stage, are represented by two extinct genera in the Baltic amber (UHMANN 1939). Those weevils that can be tentatively assigned to the leaf-mining curculionine tribe Rhamphini occur occasionally in Palaeogene deposits (the oldest undescribed fossil originates from the Mo Clay). There are no fossil records of leaf-mining buprestid genera (e.g. Trachyini). Modern fern miners (e.g., agromyzid flies) are surely secondary colonisers that rather recently shifted from angiosperm hosts.

The Palaeogene galls are rarely attributed to modern insect genera; in fact, their inducers are rarely identifiable with certainty, even to family level. LARSSON (1978) mentions only two families of supposed gall-makers from the Baltic amber and emphasises that the cynipine gall wasps are extremely rare here and the cecidomyiid gall midges are represented mostly by saproxylophagous genera, with only a few indisputable gall inducers. For the most part, the Palaeogene galls known were probably produced by sucking insects (e.g., aphids), rather than by the holometabolans. Hence the main radiation of modern gall-inducing holometabolan genera, and especially in the taxa connected with herbaceous hosts, most probably occurred in the Oligocene or even in the Neogene.

The history of the stem-boring guild is poorly studied. Here again some modern taxa exist probably since the Early Palaeogene times (e.g., the diopsid flies occurring in the Baltic amber as well as in French and North American Oligocene deposits: EVENHUIS 1994) while others appear in the record later and in relatively small numbers (e.g., the pyralid moths: KOZLOV 1988). The same is true for root-eaters (e.g., the chafers and broad-nosed weevils which probably had root-feeding subterranean larvae) which are found in the Eocene but only become diverse since the Miocene. The oldest definitive anthomyiid fly record is from the Early Miocene (MICHELSEN 1996).

Wood- and bark-borers on living plants (bioxylophages) are well represented in the Cainozoic, including not only ancient xylophagous orders such as the beetles and hymenopterans but also the moths (e.g., some Baltic amber Oecophoridae and Tortricidae were probably bark feeders: LARSSON 1978) and acalyptrate flies. Many bioxylophagous taxa are quite diverse in the Palaeocene fauna of Menat and in the Eocene (e.g., buprestid and cerambycid beetles: PITON 1940, BALAZUC & DESCARPENTRIES 1964, WEIDLICH 1987, HÖRNSCHEMEYER & WEDMANN 1994, WEDMANN & HÖRSCHEMEYER 1994). The bark beetles (Curculionidae: Scolytinae) are diverse in the Baltic amber (SCHEDL 1947) but a number of advanced tribes only appear in the Early Miocene (BRIGHT & POINAR 1994), and the oldest ambrosia beetles (Curculionidae: Platypodinae) occur in the Oligocene Apenninian and Sicilian ambers (SKALSKI & VEGGIANI 1990). It should be stressed that the finds of ambrosia beetles are restricted to the warmest regions (the Oligocene of Italy which was at that time connected with Africa rather than with other European lands, and the Miocene of the Dominican Republic and Mexico). No trace fossils attributable to this group are recorded from any Cainozoic deposits but electron microscopical studies of ambrosia beetle remains from the Dominican amber confirm that the symbiosis with ambrosia fungi had been fully developed, suggesting that the biology of the Early Miocene and modern species might be identical (GRIMALDI et al. 1994b). Some present-day termites attack living trees but this is probably a young phenomenon: except for "Termes" rutoti Meunier, a larva of an uncertain family from an Eocene resin of Belgium (NEL & PAICHELER 1993), all termites found in the Palaeogene resins are alates and were probably trapped during their nuptial flights. Abundant workers appear for the first time in the Dominican amber. The existence of cambium miners is documented since the Eocene by ichnofossils attributed to agromyzid flies (SÜSS & MÜLLER-STOLL 1979, SÜSS 1980). Hence there was an evident post-Eocene diversification at least in some bioxylophagous groups.

The amber faunas provide an interesting opportunity to compare the composition of herbivorous insects on the same plant family, or even genus, in the past with that found nowadays. The rarity or absence of certain modern taxa is of special interest; indeed, if even the species common in an amber did not necessarily inhabit the resin producer (except for the flightless stages), then any of its permanent dwellers (except for the large-sized ones) should occur regularly. At present, only the Baltic amber fauna is known well enough to be compared with the present-day conifer fauna in any detail. The Baltic amber was probably produced by more than one tree species; its main producers

closely resembled the present-day Pinus (subgenus Haploxylon) in leaf morphology and wood anatomy (SCHUBERT 1961) but differed considerably in the resin chemistry (LARSSON 1978, C. BECK 1999), suggesting that the anti-herbivore properties of the resin could have been different. There are some taxa now attacking pines (e.g., the aphid genus Mindarus Koch, the coccid genus Matsucoccus Cock., the cerambycid genus Notorrhina Redtb., the bark beetle genera Hylastes Er. and Hylurgops Lec., the horntail genus Urocerus Fourcroy) but the general taxonomic composition of the herbivores shows a number of important differences from the present-day pine fauna. In particular, the aphid fauna is quite peculiar. The numerically dominant apterae of the genus Germaraphis Heie demonstrate the very long rostrum and strong tarsal claws, suggesting that they were feeding on branches covered with rather thick bark and not on pine needles. Similar adaptations occur in some modern genera living on angiosperms but they are generally uncommon, and no modern conifer aphids represent this morphological type (HEIE 1967). Taxonomically Germaraphis belongs to Pemphigidae, a family connected now mainly with angiosperm hosts and including very few conifer dwellers. A species of Germaraphis has also been recorded from the Oligocene in Rumanian amber, indicating that the genus has survived after the cooling and accompanying significant floristic changes near the Eocene/Oligocene boundary. On the other hand, two aphid groups most diverse on the present-day conifers, the Adelgidae and the cinarine Lachnidae, are completely absent from the Baltic amber. The only aphid group common among the modern conifer fauna is Mindaridae, with Mindarus being the sole living aphid genus found in the Eocene deposits and thus demonstrating a remarkably long-time stasis in both morphology and biology. In addition to Germaraphis, another extinct genus of rather obscure affinities can be tentatively interpreted as a consumer of the amber-producing tree, namely the scale-insect Cancerococcus Koteja, which is represented by several finds of apterous males and females (KOTEJA 2000).

The folivorous guild of Baltic amber insects also seems to be rather unusual in comparison with that of the present-day pines. The wingless nymphs of members of the stick-insect genus Pseudoperla Pictet are so remarkably common that their direct connection with the amber producing tree can hardly be doubted. On the other hand, the extreme rarity of diprionid and pamphiliid sawflies, and especially the absence of heteroceran lepidopterans, is noteworthy. At present, pamphiliids, diprionids and diverse heterocerans (Geometridae, Lymantriidae, Noctuidae, Lasiocampidae, Thaumatopoeidae, Sphingidae) are common and often the most dangerous folivores on pines; even if the heteroceran moths are usually too large to be easily trapped by resin, the complete absence of their young caterpillars can hardly be explained by taphonomical biases alone. Further differences are demonstrated by the bioxylophages. Bark beetles are rather common in the Baltic amber but they are represented by few extant genera, and in particular the advanced Ipini that are quite diverse on modern pines are totally missing. The extreme rarity of Buprestidae, again well represented in the modern pine fauna, is also noteworthy. Although some pine bupresids are too large to be enclosed in resin, others (e.g., the living species of Melanophila Esch., Phaenops Lac. and Anthaxia Esch.) are quite comparable in body size with Notorrhina, a longhorn beetle genus that is not at all rare in the amber. A possible explanation is a wet climate with very low wildfire frequency: at present many buprestid species attack trees damaged by fire and the buprestid fauna on pines seems to be most diverse in the boreal forests and areas with the Mediterranean-type climate where natural wildfires are common.

The most intriguing feature of the Baltic amber fauna is the absence of any relatives of modern male pine-cone dwellers. This guild includes first of all the sawfly genus *Xyela* Dalman (Xyelini) and nemonychid weevils. Both taxa are very ancient and almost surely developed in male conifer cones in the Mesozoic, but no single inclusion is known from the Baltic amber, suggesting that either those groups were absent in the Eocene in the Scandinavian region or that their shift to pines from other conifer hosts occurred in post-Eocene times. Interestingly, in the earliest Miocene several *Xyela* species are found in Rott where pine remains are rare, and in spite of this two of the species identified are found indistinguishable from living European species on the available characters (RASNITSYN 1995). There are also some younger taxa (e.g., some weevil species of the genus *Anthonomus* Germ. and some tortricid moths) with larvae developing in the male cones of recent pines: none of their close relatives are recorded from the amber. The only Baltic amber fossil which could be tentatively interpreted as a male cone dweller is *Electrotoma succini* Rasn., a peculiar endophytic sawfly larva which represents the family on its own (RASNITSYN 1977b). Some modern olethreutine tortricids develop in male pine cones (*Piniphila* Falk.) but no close affinities are demonstrated between them and the Baltic amber species. The composition of the female cone fauna is unclear; the absence of lonchaeid flies, an important group of conifer seed parasites, is noteworthy. In general, although the pine fauna in some subtropical areas (e.g., in Central America and South Asia) is still insufficiently known, it seems that in no areas of their present-day distribution do the pines support a herbivore assemblage similar to that of the Scandinavian Eocene.

It should be forthcoming that the herbivore fauna of the Dominican and Mexican ambers has to resemble the present-day fauna of *Hymenaea* (Fabaceae: Caesalpinioideae). Those resins are much younger than the Baltic amber ones, and their botanical origin is well argued by both botanical and chemical studies (e.g. LANGENHEIM & BECK 1965, 1968, LANGENHEIM 1966, POINAR 1991a) though, interestingly, the producer of the Dominican amber seems to be more similar with modern African than with Neotropical species of the genus (POINAR 1991a). Indeed, POINAR & POINAR (1999) stress a number of similarities between the Dominican amber and modern *Hymenaea* faunas, but unfortunately make no reference to possible points of difference. A special comparison is highly desirable. Interestingly, *Hymaenaea* is exceptional among modern Neotropical legumes in that its seed predator fauna lacks seed beetles (JANZEN 1975) which are scarcely represented in either Dominican or Mexican ambers, and the only described species is clearly related to modern palm seed-beetles and not to those parasitising fabacean hosts (POINAR 1999e).

To summarise the available information on the herbivores, all their major guilds demonstrate a continuous diversification during the Cainozoic but with few principal novelties since the earliest Palaeogene and even latest Cretaceous. The radiation of modern angiosperm families and genera in the Cainozoic is well documented and was surely accompanied with a parallel process in insects. Besides the angiosperm consumers, some important taxonomic changes occurred in the insect fauna of conifers and ferns. In the latter cases, the evolution of hosts was probably generally slower than in angiosperms, with many genera surviving from the Late Cretaceous up until now. Changes in the entomofauna were mainly due to secondary colonisation by taxa that originated primarily on flowering plants. The evolution of herbivorous insects was also affected by the evolution of other herbivorous animals. The relative role of other invertebrate herbivores was probably rather constant and the level of vertebrate

herbivory increased significantly. By the Late Cainozoic their role as primary consumers in some communities becomes comparable with the role of insects, probably for the first time in the history of the terrestrial ecosystem. The diversification of herbivorous mammals was particularly important but development of fruit- and seed-eating habits and of nectarivory among the birds could influence insect evolution as well.

An additional interesting aspect of insect/plant relationships is the appearance of carnivorous plants. All modern carnivorous plants are angiosperms, and no evidence of this peculiar habit is available for any pre-Cainozoic plant taxa. Because the majority of carnivorous plants are restricted to wet habitats, they should be represented in the fossil record, at least by fruits, seeds and pollen with a high preservation potential. JOLIVET (1986) lists six modern carnivorous families. According to COLLINSON *et al.* (1993), three of them are unknown in the fossil state (two are Australian endemics), Droseraceae are represented by pollen since the Early Eocene and by seeds since the Late Eocene (with a questionable record from the Maastrichtian), and the Nepenthaceae and Lentibulariaceae since the Miocene (pollen only).

3.2.4.3. Detriti- and fungivores

The taxonomic composition of the mycetophagous insects was probably more conservative in comparison with that of the herbivores. This trophic guild is well represented by fossils, especially in resins. Almost all living families of fungivorous nematoceran dipterans and beetles are well represented in the Baltic amber. In particular, several dozen species of Mycetophilidae and other mycetophiloids have been described from the Baltic amber, and with very few exceptions, placed into living genera (EVENHUIS 1994). It should be noted that advanced Mycetophilinae and Sciaridae, rare or absent in the Cretaceous, are diverse. There are also numerous records of modern mycetophagous genera of beetles (KLEBS 1910, LARSSON 1978) and, though most of the identifications should be confirmed, the general appearance of the fungivorous beetle fauna is remarkably modern. Even in younger taxa, restricted to the Cainozoic or nearly so, some supposedly mycetophagous species are represented (e.g., in the acalyptratan flies and tineid moths). A significant post-Eocene diversification should be suggested in the calyptratans and in some acalyptratan families, but there are few specialised fungivores in those groups.

Strangely, the opportunistic detritivores which are generally rather conservative, are the only guild showing an important extinction from the Late Cretaceous to Palaeogene, namely the evident decline of the cockroaches. However, it should be stressed that its exact time and rate is in fact not very clear, and it is quite possible that it occurred before the end of Cretaceous. In any case, it could hardly be explained by a cooling near the Cretaceous/Palaeogene boundary, because the faunas of the Eocene thermal maximum are as poor in roaches as the Palaeocene ones. Interestingly, the frequency of juvenile roaches in the Late Cretaceous and Cainozoic resins is nearly the same suggesting that mainly the litter and soil-dwelling taxa were affected by the decline. The Cainozoic roach faunas are numerically dominated by Blattellidae while other families, including Blattidae and Blaberidae, common in the present-day tropics are scarcely represented. The most important Cainozoic opportunistic detritivores are diverse dipterans and beetles feeding mainly on fungi and other micro-organisms. Not surprisingly therefore, many of them belong to the same or related families as the mycetophages and like them demonstrate little change throughout the Cainozoic. It should be noted that the majority of Palaeogene moths probably had detritivorous rather than herbivorous caterpillars. That does not necessarily mean that the relative roles of moths as detritivores was higher than now, but rather that more intensive Late Palaeogene and Neogene diversification occurred in herbivorous lepidopteran taxa.

The saproxylophagous guild includes partly the same beetle and dipteran families as the mycetophages and opportunistic detritivores. At early stages of wood decay the guild is dominated by more specialised beetle taxa related to bioxylophages (e.g., diverse longhorn beetles). The only major saproxylophagous group demonstrating significant changes during the Cainozoic is the termites but their changes had important ecological consequences. Up to the Pliocene this order occurs well outside of its present-day area of distribution. In the Palaeocene and Eocene it is represented mainly by the same relatively primitive families as in the Late Cretaceous. The only family appearing in the Eocene is the Rhinotermitidae but if the scarcity of data on the Cretaceous and Palaeocene termites is taken into account, one could suggest that the rhinotermitids could also be older. In more temperate areas the Hodotermitidae predominate while Kalotermitidae and especially Mastotermitidae are characteristic for warmer regions. The most advanced family, Termitidae, appears in the fossil record only in the Late Oligocene (Aix-en-Provence: NEL & PAICHELER 1993). This group differs from all other termites in having nitrogen-fixing bacterial intestinal symbionts instead of zoomastigophoran protozoans, and hence the termitid production is much less limited by the low nitrogen content in their diet. In the present-day tropics the termitids are the key saproxylohages both in forest and savannah ecosystems, but the family is scarcely represented in subtropical regions or on tropical mountains. The termitids are very important in tropical savannahs as the principal consumers of grass litter, and some genera feed on soil humus. Palaeontological data indicate that in the past the family was restricted to the warmest regions, and, because the Eocene was generally warmer than the Oligocene, its absence in rich Eocene insect deposits both in Europe and North America is noteworthy. It should be noted that no termitids have been found in the Burmese amber, which is not younger than the Eocene. Thus it seems that the family is indeed the youngest one in the order, and could hardly have become common, even in the tropical belt before the Early Oligocene.

Consumers of ephemeral organic substrates, including copro- and necrophages, are the most successful Cainozoic detritivores. They include mainly brachyceran dipterans as well as some nematocerans and beetles, usually more or less related to opportunistic detritivorous taxa and probably arisen from them. All members of this broad superguild are more or less distinctly specialised for rapid larval development and for effectively searching for patchily distributed resources. Many species have a narrow trophic specialisation, being restricted to certain type of organic matter at certain stage of its decay (e.g., rotting fruits and fungal fruit-bodies, fermenting plant exudates, carrion, dung, etc.), while others are feeding generalists, exploiting a rather wide spectrum of food substrates. However, at the generic and suprageneric level trophic specialisation usually varies widely suggesting that the feeding habit is probably phylogenetically plastic. This is very evident in the majority of saprophagous dipterans and some beetle families (e.g., Nitidulidae) but even among the most specialised beetle taxa, shifts from dung to carrion, fruits, fungi, dead leaves, etc., are not uncommon (in the scarabaeids), or from carrion-feeding to coprophagy and even predation and herbivory (in the silphids). As discussed above (Chapters 3.2.3.1 and 3.2.1.2), there is no firm palaeontological evidence of existence of ephemeral substrate detritivores before the latest Cretaceous (although the possibility of this habit cannot be rejected for some

Mesozoic nematoceran dipterans), and the main evolution of recent taxa is expected to have occurred in the Cainozoic.

Indeed, in the Palaeogene this group was still of a limited importance, and its taxonomic composition differed strongly from the present-day one. The history of nitidulids and other biologically similar beetles is insufficiently known but at least in the Eocene they were much less common than other detritivorous beetles connected with more stable substrates. Records of modern genera need confirmation though the more specialised necrophagous silphids are known from the Late Eocene or Early Oligocene of France as well as from Florissant in the USA, and some of them are placed by extant genera. Eocene and Early Oligocene dung beetles are represented mainly by Aphodiinae, the subfamily which includes a number of modern species developing in decayed plant matter and not necessarily in mammal excrement. In particular, their predominance in Florissant is noteworthy. The Late Oligocene and Neogene scarabs are more diverse and represent all principal modern tribes. On the other hand, ichnofossils indicate that dung provisioning by beetles has occurred at least since the Eocene in South America and since the Oligocene in North America (RETALLACK 1990a, GENISE & BOWN 1993). The diversity of ichnofossils indicates the existence of diverse feeding strategies and nesting behaviours. The dung beetles are unusual in respect to the Quaternary extinction pattern. At least three extinct scarab beetle species (including *Copris pristinus* Pierce, the numerically most dominant in the assemblage) have been discovered in the Late Pleistocene of Rancho La Brea, California; two other species in the assemblage being extant and four identified to generic level only (S. MILLER *et al.* 1981). This extinction pattern coincides with the well-known extinction event of large-sized North American mammals and suggests a high degree of beetle specialisation.

Phorid flies are diverse in the Palaeogene, and some modern sapro- and necrophagous genera (e.g., *Aneurina* Lioy, *Spiniphora* Mall., *Phora* Latr.) occur in the Baltic amber (B. BROWN 1999). The biology of Palaeogene acalyptratan flies is uncertain because all Baltic amber genera are extinct (HENNIG 1969b), and the records of living genera from the Oligocene (e.g., MELANDER 1949, STATZ 1940) require confirmation. If the high biological diversity of present-day acalyptratans is taken into account, there are no firm grounds to extrapolate their biology to the Palaeogene species, and this is particularly true in the case of saprophagy, though a rapid larval development in decaying organic matter is quite possible for some Palaeogene taxa (e.g., for the genera of Milichiidae, Dryomyzidae, Heleomyzidae, Sepsidae, Lauxaniidae and Drosophilidae). In any case, the acalyptratans are remarkably uncommon in both the Eocene and earliest Oligocene, and become more abundant only in the Late Oligocene and Miocene (e.g., in the Middle Miocene of Stavropol and Kerch, undescribed). It should be noted that the record of a *Drosophila* species of a modern appearance from the Oligocene of Fontainbleu, France (DESPLAT 1954) is omitted from the catalogue of EVENHUIS (1994).

The fossil record of calyptrate flies is surprisingly scarce. Four families including modern saprophagous, necrophagous and coprophagous genera are recorded from pre-Quaternary deposits, namely Scatophagidae, Anthomyiidae, Muscidae and Calliphoridae They are not represented or at most by very few specimens in the Eocene (the records of scatophagids, anthomyiids and muscids from the Baltic amber and the Green River Formation listed by EVENHUIS 1994 are doubtful). The most interesting find of fly puparia in mammal bones from the Middle Eocene of Geiseltal (WEIGELT 1935) was never described in detail, and their assignment to the Calliphoridae

(EVENHUIS 1994) is not fully argued. *Mecistoneuron* Melander from Florissant is probably not a calyptratan, but a platypezid (MICHELSEN 1996), and the records of Scatophagidae and Muscidae from the same locality as well as of a few calyptratan records from the younger Oligocene of Canada, France and Germany need to be confirmed. Unquestionable Anthomyiidae (MICHELSEN 1996) and Muscidae (PONT & CARVALHO 1997) have been described from the Dominican amber but it should be noted that the latter family is represented by Phaoniinae, a group with predominantly predacious larvae. POINAR & POINAR (1999) also list Calliphoridae from the Dominican amber but no members of this family have been described. In the Miocene of Europe and the North Caucasus the calyptratans seem to be still uncommon. In particular, complete absence of fossil Fanniidae and Sarcophagidae is intriguing because of their wide distribution and abundance in the present-day fauna.

The Late Eocene–Early Oligocene phosphorite deposits in Quercy, France, provide a unique opportunity to investigate rather large fossil assemblage of necrophagous insects (HANDSCHIN 1947). The deposits originated subaerially in carst fissures operating as natural traps for terrestrial vertebrates, and abundant vertebrate remains are accompanied by insects attracted by carrion. Not surprisingly, the assemblage is strongly dominated by fly puparia which, however, all belong to Phoridae identified as belonging to *Megaselia*, *Spiniphora* and an extinct genus. Neither acalyptratans nor calyptratans are recorded. Silphid beetles are represented by a larva and adults of *Ptomascopus aveyronensis* Flach (the commonest species), *Thanatophilus* sp. and *Palaeosilpha fraasi* Flach (the latter species is assigned to an extinct genus); a scarabaeid specimen identified as *Aphodius* sp. was also found. Interestingly, no dermestids are recorded though dried carrion in limestone fissures should represent an excellent food source for their larvae, and there are several pupae and a caterpillar of tineid moths that probably fed on it. It should also be noted that no staphylinid beetles, which are the most common predators of dipteran larvae, are found, and the only predator recorded is an histerid beetle. This most interesting assemblage indicates clearly that the carrion insect fauna was very different from the present-day one, first of all in the complete absence of many important modern taxa.

In general, available data suggest an intensive diversification in all trophic guilds associated with ephemeral substrates, in particular in carrion- and dung-feeders, in the Oligocene and especially in Miocene times.

3.2.4.4. Predators and parasites

No important changes are documented in the predacious insects during the Cainozoic. Probably, there was some radiation in particular taxa (e.g., in the coccinellid beetles which were not very common before the Late Oligocene, and in the praying mantids which, in the Late Eocene–Early Oligocene, were still represented mainly by most primitive chaeteessids) and in some habitat guilds (e.g., in the predators of necro- and coprophagous insects as pointed out above). Relatively few higher taxa are geologically young (e.g., the predacious calyptrate flies). Nevertheless, the total pressure of predacious insects in terrestrial ecosystems probably increased gradually, first of all due to radiation and progress in social organisation in the ants resulting in their increasing density. Advanced ant taxa appear in the fossil record from the Late Eocene, and it should be stressed that they are generally less trophically specialised in comparison with more primitive ones (DLUSSKY & FEDOSEEVA 1988). Other predacious hymenopterans, the

aculeate wasps, are insufficiently known in the Cainozoic. They might be characterised tentatively as rare in the Palaeocene and Eocene (even Sphecidae that were fairly numerous in the later Early Cretaceous) and more abundant and diverse since Oligocene (A. Rasnitsyn, pers. comm.). This holds particularly true for the social wasps which are unknown before the time of the Baltic amber (except for the Late Cretaceous fossil nests ascribed to social Vespidae, Chapter 3.2.3.2) and rather common since the Miocene.

Insect parasitoids demonstrate much more intense and rapid changes in comparison with the predators as documented by important differences between the Palaeogene and modern faunas. Interestingly, the greatest changes occurred probably in the parasitoids of ectophytic and other relatively openly-living (exposed) hosts; it is likely that, they were partially stimulated by co-evolutionary "arms race" effects. Host radiation was surely important as well: a large number of modern parasitoids attack diverse lepidopterans, higher dipterans and short-horned orthopterans (as eggs), the taxa demonstrating the greatest Cainozoic radiation among the insects. Significant evolutionary changes are documented in the parasitic hymenopterans. In particular, the Baltic amber ichneumonoid wasp fauna is peculiar in the taxonomic composition including a number of extinct tribes and subfamilies (Pherhombinae, Townesitinae, Ghilarovitinae, Tobiasitini in Ichneumonidae [incl. Paxylommatidae], Diospilitinae, Acampsoheloninini, Chelonohelconini, Oncometeorini, Prosyntretini in Braconidae: TOBIAS 1987, KASPARYAN 1988a,b, 1994). Some extremely important modern subfamilies (e.g., Ichneumoninae) only enter the fossil record in the earliest Oligocene. The Cainozoic evolution of chalcidoid wasps is less well studied but it was probably also rather intensive (e.g., Chalcididae parasitising mainly lepidopterans and brachyceran dipterans are unknown before the Oligocene). Unlike during the Cretaceous when chalcidoids were always rare compared with the proctotrupoids, their Cainozoic participation is variable but often high (RASNITSYN & KULICKA 1990, A. Rasnitsyn, pers. comm.). Proctotrupoid wasps seem to be more conservative; however, the largest modern family, Diapriidae, is scarcely represented before the latest Eocene–earliest Oligocene, and its radiation in the Oligocene and Neogene was almost surely connected with a radiation of brachyceran dipterans, their main hosts. In contrast, the egg-parasitising family Scelionidae seems to display decreasing diversity (RASNITSYN & KULICKA 1990). A similar pattern occurs in other parasitic insect orders. As mentioned above, among the bombyliid flies significant changes are documented at the generic level, and yet in the Early Oligocene almost all genera are extinct. Pipunculidae and Conopidae appear in the Baltic amber as well as some parasitic acalyptratan families (Odiniidae, Cryptochaetidae). The most important novelty in the parasitoid guild was probably the appearance of parasitic calyptratan flies, and first of all the Tachinidae. This large family is well represented now in different zoogeographical regions and attacks a wide range of hosts, but its radiation is surely geologically young. The Palaeogene finds listed by EVENHUIS (1994) are high questionable, and the oldest reliable fossils are known from the Miocene only. One more important Cainozoic event is the appearance of dipterans infesting non-arthropod hosts (e.g., the acalyptratan family Sciomyzidae rather well represented in the Baltic amber and restricted to molluscan hosts). The strepsipterans and parasitic rhipiphorid beetles first appear in the fossil record in the Eocene, and the meloid beetles in the Palaeocene. Their evolution is poorly studied because of their rather fragmentary fossil record, but their common host associations with the aculeate hymenopterans and acridid orthopterans suggests at least some radiation during the Neogene expansion of solitary bees and acridid grasshoppers.

The rapid evolution of terrestrial vertebrates in the Cainozoic should provoke important evolutionary changes among the vertebrate blood suckers, and especially in their permanent ectoparasites and nest parasites. Unfortunately, their fossil record is extremely scarce. There are few fossil fleas, all from the Baltic and Dominican ambers (see Chapter 2.1.3.4.5 and references herein). The lice are represented by eggs found on mammal hairs in the Baltic amber and identified to only ordinal level (VOIGT 1952). The only fossil flightless parasitic fly known is from the Dominican amber (GRIMALDI 1997b). No fossil parasitic bugs are recorded. There are also few subfossil fleas and lice finds. No palaeontological data on dipterans with blood-sucking larvae developing in bird nests are available. Strangely, all fossil fleas belong to extant genera as well as the Dominican amber carnid fly, which is closely related to modern nest parasites of passerine birds. The only palaeontological evidence of possible rapid changes in the ectoparasite fauna is the find of *Neohaematopinus relictus* Dubinin, an extinct louse species, in the Pleistocene. It should be added that the Cainozoic history of non-blood-sucking vertebrate ectoparasites (e.g., the chewing lice, platypsylline beetles and hemimerid dermapterans) is completely undocumented. Thus, at present, the importance of palaeoentomological data for studies on co-evolution of vertebrates and their ectoparasites is quite limited.

The winged blood-sucking dipterans are much better represented as fossils but rather poorly studied except for ceratopogonid midges. Their connections with the hosts are considerably less than those of the permanent ectoparasites and their evolution should be managed rather by landscape changes and larval adaptations than directly by vertebrate evolution. However, some dipterans (e.g., Tabanidae, Glossinidae and Muscidae) prefer large-sized vertebrates and thus should radiate mainly in the Late Palaeogene and Neogene in parallel with the diversification of large herbivores. The observed palaeontological pattern is only partially in accordance with these predictions. The blood-sucking nematoceran and asilomorphan families are known since the Cretaceous. The best studied family, Ceratopogonidae, demonstrates remarkable stasis at the generic level in the Late Cretaceous and Cainozoic; the ceratopogonids show no preference for large-sized prey. Fossil Simuliidae and Phlebotominae are too poorly studied to say anything about their palaeontological history in the Cainozoic; both groups are known since the Early Cretaceous. Culicidae is the youngest family of blood-sucking nematocerans. The oldest fossil mosquito is discovered in the Late Cretaceous (Campanian) Canadian "amber" (POINAR et al. 2000), and undescribed oldest unquestionable culicids are found in the Mo Clay of Denmark (LARSSON 1975). The family is remarkably rare in the Eocene though extant genera and even subgenera are represented in the Baltic amber (SZADZIEWSKI 1998); however, according to NEL et al. (1999c), culicids are not uncommon in the Early Eocene resin from the Paris Basin. The Oligocene culicid fossils occur more regularly and are occasionally abundant (e.g., in CÉRESTE, France: LUTZ 1984a, 1985a). Interestingly, in the Palaeogene they seem to be more common in the coastal plains palaeoenvironments (e.g., in the BEMBRIDGE MARLS and CÉRESTE) than in inland areas (it should be remembered in this connection that the Mo Clay deposits are of a marine origin). In the Miocene mosquitoes are found at a number of sites and were probably rather common (e.g., in the Dominican amber, in the fossil resin of Merit-Pila Coalfield in Malaysia and in Stavropol), but they are poorly studied. These data suggest increasing culicid abundance in post-Eocene times which,

however, could be easily explained by larval adaptations (e.g., colonisation of small temporary water bodies) as by vertebrate evolution because mosquitoes often attack birds and small mammals. The Cainozoic evolution of Tabanidae is poorly studied. The family is known since the Early Cretaceous but seems to be rather rare even in the Neogene deposits. This scarcity of the fossil record of the horse-flies is somewhat enigmatic if their present-day abundance in near-water environments is anything to go by. The blood-sucking groups of higher brachyceran flies enter the fossil record surprisingly early in the Cainozoic. Tsetse flies (Glossinidae) occur in Florissant, well outside their present-day area of distribution and well before the oldest unquestionable records of the majority of calyptratan families. GRIMALDI (1992) analysed the sister-group relationships between the extinct North American and extant species groups of *Glossina* cladistically and postulated that the Afrotropical lineage should occur in Gondwanaland no later than the American one in the North America. However, a North American origin and subsequent expansion over Eurasia and Africa is demonstrated for a large number of ungulate mammals lineages, and the same zoogeographical scenario also seems to be most plausible for the glossinids. The presence of extinct Eophlebomyiidae which are supposedly related to the Glossinidae, in the Eocene of North America is in accordance with this hypothesis. Hippoboscidae are very rare as fossils and occur in the earliest Miocene of Germany (Rott) and in the Late Miocene of Italy (STATZ 1940, F. BRADLEY & LANDINI 1984). No blood-sucking genera of Muscidae are known from fossils, and their rather recent origin is therefore quite probable.

No definitive fossils of vertebrate endoparasites are known from pre-Quaternary deposits. In insects this habit is restricted to some groups of the calyptratan flies and might be of Cainozoic and most probably of the Neogene origin. The interpretation of abundant fly larvae from the Green River Formation as alleged endoparasites of mammals is fantastic and groundless, and the taxonomic position and biology of *Novoberendtia* Rapp and *Adipterites* Townsend listed by EVENHUIS (1994) under Oestridae are in fact unclear. The extant gasterophilid genus *Cobboldia* Brauer which infests elephants has been discovered in a subfossil state in Siberia in association with mammoths (GRUNIN 1973). Not surprisingly, the Siberian species is extinct and represents a subgenus of its own, *Mamontia* Grunin.

3.2.4.5. Insect sociality and some ecosystem level phenomena

As mentioned above, significant progress in social organisation occurred during the Cainozoic in all major social insect taxa including termites, bees and ants, and had important effects on diverse aspects of ecosystem evolution (turnover of dead organic matter, pollination mechanisms, predation pressure). The social insects are involved in other diverse co-evolutionary interactions as well. Their commensals and specialised predators are uncommon in the fossil record. Paussine ground beetles have been described from the Baltic and Dominican ambers (WASMANN 1929, NAGEL 1987, 1997); their records from Florissant are probably based on misidentifications. In the Baltic amber this group seems to be represented only by relatively primitive genera. The unnamed brentine weevil from Messel illustrated by TRÖSTER (1993) seems to be allied to the *Amorphocephala* group of genera including many myrmeco- and termitophilous species. Four species of termitophilous trichopseniine staphylinids of the genus *Prorhinopsenius* Pasteels et Kistner have been described from the

Mexican and Dominican amber (KISTNER 1998). A highly specialised termitophilous bug of the family Termitaradidae is found in the Dominican amber (POINAR & POINAR 1999). Specialised ant predators (e.g., among the rediviid bugs: WASMANN 1932, POINAR 1991b) and parasitoids (e.g., in the phorid flies: B. BROWN, 2000) occur in the Baltic and Dominican ambers as well, and WUNDERLICH (1988) mentions ant-mimetic spiders from the Dominican and Mexican ambers. Mutualism with ants by the riodinid butterflies (DEVRIES & POINAR 1997) and scale insects (POINAR & POINAR 1999) in the Early Eocene is documented by Dominican amber fossils. The history of aphid-ant mutualism is still obscure; as discussed above, this may be a rather ancient phenomenon but its full development was probably only reached in the Neogene. Nothing is known about the fossil history of specialised ant/plant interactions such as seed dispersal (myrmecochory) or ant attraction by holes in different plant organs that are especially suitable for ant nesting (myrmecophytes). No palaeontological information on the history of social parasites among ants, bees and wasps is available.

The role of insects as a food source for other invertebrates and vertebrates was surely great throughout the Cainozoic but the history of their co-evolutionary interactions with predators is poorly studied. Highly specialised insectivory is well argued for some mammals including bats, pangolins and edentates as early as in the Middle Eocene (STORCH 1978, RICHTER & STORCH 1980, FRANZEN 1984, RICHTER 1988, HABERSETZER & STORCH 1989). Many Palaeogene insects demonstrate the same cryptic body shape and/or colour pattern as their living relatives. Zygaenid moths from the Miocene of Spain (FERNÁNDEZ-RUBIO et al. 1991, FERNÁNDEZ-RUBIO & PEÑALVER 1994) and Germany (REISS 1936, NAUMANN 1987) show the characteristic aposematic coloration quite similar to that of living Zygaenini. The same is true for an undescribed Mylabris species (Coleoptera, Meloidae) from Chon-Tuz in Kyrgyzstan, and typical wasp-like body pattern is very common in diverse fossil syrphid flies as demonstrated by a considerable amount of undescribed material from Stavropol. Thus many anti-predator mechanisms show a stasis for about 20 Ma or more, suggesting that they remained continuously effective during this time interval.

There are occasional records of phoretic and parasitic organisms associated with insects, mainly from fossil resins (see Chapter 1.4.5 and references herein), but the available data provide no firm basis for a detailed analysis of evolutionary history of their interrelations with insects.

3.2.4.6. Origin of principal types of modern terrestrial ecosystems

The above outline of ecological aspects of Cainozoic insect history is cosmopolitan in its scope. However, the pattern of ecological evolution was neither uniform nor synchronous in different regions, and geographical differences in landscapes and ecosystems has increased gradually since the Palaeocene to Quaternary. It is generally believed that this general trend was managed by increasing climatic gradients (both latitudinal and marine/continental), but co-evolutionary effects at the community level were probably important as well, and could lead to the formation of some fundamentally new ecosystem types. The differentiation pattern is rather well-studied in palaeobotany (including palynology) and vertebrate palaeontology but much less so in insects. Only some of the most general aspects of this geographic differentiation and its possible effects on insect evolution can be briefly discussed here.

As discussed previously (Chapter 3.2.3.2), modern wet broad-leaved and sclerophyll forests of temperate and subtropical belts, and especially those of the southern hemisphere, are probably the oldest present-day biomes, retaining their basic organisation and turnover pattern since the Late Cretaceous. Some dryer woodland and sclerophyllous bush ecosystems of the same zones (e.g., in Australia and South Africa) may be of comparable age.

One might find it surprising that lowland tropical rainforests are not mentioned in this context. Indeed, it is generally believed that this ecosystem type is very ancient and has had a much wider range in the geological past than now. This was questioned on the basis of floristic evidence by RAZUMOVSKY (1971, 1999). He pointed out that there are no undoubted records of pre-Eocene fossils of modern true tropical plant taxa if the tropical environments are taken in the strict sense as restricted to the areas with a mean temperature above 18°C in the coolest month (i.e. the equatorial climate). In the recent palaeobotanical literature the Eocene is often recorded as the time of appearance of first fossil assemblages which can be interpreted as produced by tropical rainforests (e.g. NIKLAS et al. 1980, TIFFNEY 1981, UPCHURCH & WOLFE 1987). Insect distribution is strongly controlled by vegetation and hence should demonstrate the same spatio-temporal pattern. ZHERIKHIN (1978) demonstrated that no really ancient (pre-Cainozoic) insect families are restricted to present-day equatorial climate areas. The only exception is Mastotermitidae, but this termite family regularly occurs in the fossil state in association with clearly extratropical taxa, so the biology of the only living species surely cannot be extrapolated to extinct mastotermitids. In the Palaeogene some strictly or predominantly tropical taxa occurred occasionally in Europe and/or North America (e.g., the metrioxenine and brentine weevils, passalid beetles, glossinid flies, termitid termites, diopsid flies and the ant genus Oecophylla F. Smith); some of them (e.g. Metrioxeninae, Diopsidae and Glossinidae) may well be of an extratropical origin while others are very rare and perhaps penetrated from the equatorial belt. The only fauna of a supposedly tropical general appearance is discovered in the Oligocene ambers from Italy (Sicily and Apennines; see KOHRING & SCHLÜTER 1989, and SKALSKI & VEGGIANI 1990, for lists of taxa and geological information). This is the only European fauna resembling the Dominican and Mexican amber faunas in some important features (e.g., in the abundance of the stingless bees constituting about 3% of all insect inclusions and the platypodine ambrosia beetles). DLUSSKY (1981a) points out that the taxonomic composition of ants is similar to the present-day fauna of tropical Asian forests. On the other hand, there are some noticeable differences between the Italian and Neotropical ambers. In particular, KOHRING & SCHLÜTER (1989) mention four aphid specimens in rather limited materials (it should be noted, however, that the specimen figured at fig. 19 in their paper is misidentified and in fact represents a thrip!); the aphids are extremely rare in natural tropical environments, and two specimens only have been discovered up to now in far larger collections of Dominican amber inclusions (HEIE & POINAR 1988, 1999). Another important find is a pine cone mentioned by SKALSKI & VEGGIANI (1991) because no pine trees or any other Pinaceae are represented in modern lowland tropics. Unfortunately, the termites from the Italian ambers still remain undescribed but at least based upon the figures published there are no Termitidae. Data on the insect fossils from tropical latitudes available before the Early Miocene are generally scarce, with one important exception, namely Burmese amber. Its age is uncertain but must be between the Senonian (earlier Late Cretaceous is improbable because of the taxonomic composition of the fauna) and the Eocene (the minimal age of amber-bearing beds) (ZHERIKHIN & ROSS 2000), and throughout this time interval the land of its origin has to have been situated under low latitudes. In the most complete list of its fauna available (RASNITSYN & ROSS 2000) there are several families not found elsewhere but no modern tropical taxa. The Oligocene insects are known from Brazil (MARTINS-NETO 1998b, 1999b, COELHO *et al.* 2000) but it is not clear if they represent a forest fauna or not. The taxonomic composition of those Brazilian faunas seems to be strange in comparison with any present-day tropical areas. In particular, termites are surprisingly rare and represented only by Mastotermitidae; there are no records of any bees or ants, and MARTINS-NETO (1998b) stresses the absence of any orthopterans in the Tremembé Formation from where the largest collection is available. Butterflies are not uncommon and are unusually diverse, and the dipteran fauna is dominated by empidids with very few acalyptratans and no calyptratans. It is likely that some taxa are scarcely represented or missing in the fossil record (e.g., Membracidae, Cetoniinae, Apoderinae, Attelabinae, Calliphoridae, Sarcophagidae) originated somewhere in the equatorial belt and only recently invaded extratropical areas; however, all of them are probably phylogenetically and geologically young. In any case, the oldest insect assemblages closely resembling modern tropical forest faunas has been discovered in the Early Miocene Dominican and Mexican ambers.

The most convincing arguments against a Mesozoic or earliest Cainozoic age for the tropical rain forest biome can be found in the functional structure of its ecosystem where almost all key taxa are geologically young, or at least rather recently radiated (ZHERIKHIN 1978, 1993c). In particular, highly effective pollination systems based mainly on pollinators with advanced behavioural patterns (such as the social bees, ants, birds and bats) are absolutely necessary to support the complex forest stratification and the extreme degree of polydominance so characteristic of modern tropical rain forests. Equally, the uniquely rapid decomposition of dead plant matter, including wood, is primarily due to the activity of Termitidae, the termites with nitrogen-fixing symbionts. The termitids also play an important role in pedogenesis of the peculiar tropical soil profiles which have organic matter dispersed over a thick indefinite profile instead of being concentrated in a well-differentiated surface horizon. This soil type, in turn, is important for the development of deep root systems capable of supporting very tall trees. One more important thing is the extremely high pressure of ant predation reducing the folivory rate. All those features hardly could be developed before the Late Eocene–Early Oligocene, and thus in older times the equatorial regions have to be occupied by ecosystems with significantly lower diversity, simpler spatial structure, slower turnover and basically different soil types. Their structure and history can be elucidated in more detail only in the course of future palaeontological studies in equatorial regions.

The relatively young age of the tropical rain forests (and probably also other modern tropical ecosystems) can be explained in two alternative ways. According to one possible model, intensive Cainozoic evolutionary processes resulted in the replacement of older tropical ecosystems by modern ones. This would predict that certain extinct ecosystem types lacked any close modern analogues existing in the equatorial belt in the Late Cretaceous and Early Palaeogene. An alternative model suggests that not only the modern cool climate (as is commonly accepted) but also the equatorial climate is geologically young, and that the present-day tropical biota have formed under increasing latitudinal climatic gradient since the latest Eocene. In this case, low latitude regions should have been occupied by widely distributed ecosystem types in the Cretaceous and Early Palaeogene. At present, there is no definitive evidence in favour of any of these hypotheses.

The role of insects in open grassland ecosystems is also significant though here the principal key taxa are first of all the grasses and ungulate mammals. The history of those ecosystems is discussed by ZHERIKHIN (1993a, 1994, 1997b). There are some grass-dominated communities in nearly all modern types of successional systems but in undisturbed natural environments they are often restricted to relatively small patches of seral wet and dry meadows within more or less forested landscapes. In some (mainly semi-arid) regions ranging from the temperate steppes to tropical savannahs the grasslands predominate in natural landscapes forming true (intrinsically stable) climaxes stabilised first of all by the activity of grazing megaherbivores. Additional important herbivores are rodents, grasshoppers and occasionally (in some types of tropical savannah) termitid termites. The annual rate of primary production consumption is exceptionally high, reaching 30–45%, and occasionally even 60%. There is a high degree of co-adaptation between the dominant grasses and their principal consumers, including diverse mechanisms of positive regeneration in response to herbivory, and co-evolution between grasses and grazers was of crucial importance for the development of stable grassland ecosystems (PROKHANOV 1965, STEBBINS 1978, OWEN & WIEGERT 1981, MCNAUGHTON 1984, 1986, WING & TIFFNEY 1987a,b). Other fundamentally important guilds are the detritivores and the specialised coprophages, effectively removing grass litter and megaherbivore faeces: in their absence the grazer dung accumulation leads to a rapid ecosystem collapse. It is usually difficult to recognise the difference between the seral meadows and climax grasslands on the basis of insect fossils. Ichnofossils, and in particular the fossilised dung beetle nests, are more informative in this respect. In the earliest Palaeogene there is no firm evidence of the existence of stable grasslands. The finds of specialised grazing mammals, the steppe-type palaeosols and abundant fossilised dung beetle balls in the Middle Eocene (Mastersian) of South America represent the oldest available evidence of the development of stable grasslands. Since the Oligocene, expansion of grasslands is documented (mostly by palaeobotanical and palaeopedological data) in North America where they should originate independently. In the Miocene they appear in Asia and then in Africa under the key influence of North American invaders. The typical open-land grasses with C4 photosynthesis are known since the Miocene (BOCHERENS *et al.* 1994) but their wide expansion is documented in the latest Miocene–earliest Pliocene only (CERLING *et al.* 1997). The available palaeoentomological data are still fragmentary but suggest a significant effect of grassland expansion on insect evolution. Some important examples (e.g. the grasshoppers and solitary bees) are discussed above.

The present-day desert biome is probably even younger than grasslands but its biota should arise from much older dune and beach communities. Its history is poorly documented palaeontologically, and virtually nothing is known about the evolution of desert insect faunas.

Similarly, the present-day boreal forest and tundra biomes should originate from high altitude communities which might have a long but nearly undocumented history. In the Cretaceous and Early Palaeogene the polar regions might be occupied by communities which have no parallel in the present. The structure of this seasonally dark temperate biome in general and the ecological role of insects in particular are still enigmatic. Expansion of boreal and subarctic communities in high latitude areas of the northern hemisphere is documented since the Late Neogene. The Pliocene insect faunas of Alaska correspond well to the modern coniferous forest faunas (HOPKINS *et al.* 1971), and the oldest insect assemblage indicating forest-tundra environments so far discovered is in the Beaufort Formation (Late Miocene-Early Pliocene) at Meighen Island, Canadian Arctic Archipelago (MATTHEWS 1976a, 1977). Still in the latest Pliocene–earliest Pleistocene the boreal insect faunas have been discovered as far north as in northernmost Greenland (BÖCHER 1989). Some mid-latitude glacial assemblages show an unusual combination of modern tundra species with apparently thermophilous elements (e.g., ants). NAZAROV (1984) calls this phenomenon "the latitudinal inversion" and explains it with the longer vegetation season in comparison with modern high-latitude tundra. Frequent and often rapid Pleistocene climatic oscillations provoked large-scale fluctuations in insect distribution. The rate of faunistic change was remarkably high; e.g., the tundra assemblages became replaced by boreal forests, sometimes in 300–400 years (ELIAS 1994). Those rapid distribution shifts occasionally resulted in the appearance of very unusual "mixed" assemblages due to a rapid invasion of species more adequate to new climatic conditions, together with temporary survival of some less adapted populations (COOPE 1969, ELIAS 1994). A high rate of insect invasion is probably responsible for the occasional controversies between palaeobotanical and palaeoentomological data (ELIAS 1994). In Antarctica cold-adapted faunas should appear earlier but no palaeontological data are available. Recent palaeobotanical discoveries in the Pliocene of the Transantarctic Mountains (WEBB & HARWOOD 1993) indicate the existence of dwarf *Nothofagus* heaths resembling present-day vegetation of southernmost South America; insect remains are found as well (ASHWORTH *et al.* 1997) but the available information is too limited to reconstruct the community structure.

Besides the boreal forests and tundra, existence of the peculiar extinct high latitude biome called tundra-steppe is documented in the Late Cainozoic of the northern hemisphere (VELICHKO 1973, MATTHEWS 1976b). Unlike in present-day tundra, the moss cover was strongly reduced and the vegetation was dominated by grasses with much higher annual production which supported a rich mammal fauna resembling the fauna of African savannahs in its basic ecological structure (VERESHCHAGIN & BARYSHNIKOV 1992). The peculiar composition of Quaternary insect assemblages often combining modern cold-adapted and steppe taxa is in accordance with the tundra-steppe model (KISELYOV 1981, ELIAS 1994, BERMAN & ALFIMOV 1998). The oldest faunas demonstrating this "mixed" pattern have been discovered in the Late Pliocene in the north of Russian Far East (KISELYOV 1981). In the glacial stages of the Pleistocene they become extremely widespread reaching Central and West Europe up to the British Isles, but were disappearing in the Holocene. Nevertheless, nearly all tundra-steppe insect species have survived up to now either in the coolest steppe regions of Asia (Kazakhstan, Mongolia, Tibet) or in local grassland habitats in high mountains (e.g., in Altai and the Urals) and boreal areas (Yakutia, Alaska, Arctic Canada). There is no single modern area supporting the entire tundra-steppe insect assemblage. The plants demonstrate the same pattern (YURTSEV 1981) while the mammal mega-herbivores, requiring much larger areas of suitable habitats, have become extinct. As to the climatic conditions, it is generally accepted that the tundra-steppe was significantly dryer than tundra but the temperature evaluations are somewhat controversial. Some authors postulate relatively high summer temperature; e.g. BERMAN & ALFIMOV (1998) estimate the mean July air temperature as 12–13°C and the sum of positive temperatures of the soil surface up to 2000°C on the basis of analysis of present-day distribution of *Stephanocleonus* weevils characteristic members of the tundra-steppe assemblages (see

Fig. 36). VERKHOVSKAYA (1986) also believes that in Northeast Siberia the summer was warmer in the mammoth times than now, and even questions the tundra-steppe model for this region suggesting that there was rather a complex mosaic of larch woodlands with a well-developed grass layer, dwarf pine stands and meadows. However, under dry conditions with high summer insolation the soil temperature may increase significantly even under a low air temperature, especially if the dominant vegetation is of a short-grass type. It should also be noted that the development of a dense grass layer, even in rather sparse larch stand, is difficult to imagine because of a rapid accumulation of needle litter. In connection with the tundra-steppe model, the most intriguing entomological problem is the rarity of coprophages, represented by occasional aphodiine dung beetles and perhaps by some fly puparia. As pointed out above, the grazing ecosystem of a grassland can hardly be stable under a low rate of megaherbivore dung removal by insects.

3.3
Ecological History of the Aquatic Insects

N.D. SINITSHENKOVA

3.3.1. INTRODUCTORY NOTES

As a rule the insect fossils occur in lacustrine deposits. The result is that the fossil record of lentic insects (those living in the still water as contrasts to those living in the running water which are lotic) is much more complete than of the lotic ones, and their role in the lacustrine ecosystems of the past can be reconstructed in much more detail than in the running waters. Our knowledge about the history of aquatic insects was reviewed previously by WOOTTON (1972, 1988) and KALUGINA (1980a). WOOTTON (1988) considers the presence of special adaptations to be the main evidence of the aquatic mode of life of extinct insect taxa. The presence of fossils of "two or more subdivisions of a single group all of whose present-day representatives are aquatic at some stage of their life-cycle, and assumed to be so by descent from an aquatic common ancestor", is considered as less reliable evidence. The least reliable category comprises those situations in which the fossil represents a group "all of whose representatives are aquatic at some stage", as the group may had terrestrial representatives. To the above criteria, evidence of the taphonomical autochthonicity should be added. Worth mentioning here are relatively undamaged fossils of a particular taxon abundantly and/or repeatedly found in lacustrine deposits, especially in association with other fossils whose lentic nature is confirmed independently. Also indicative is the high abundance of the immature stages comparing the adults, particularly when accompanied with the moulting skins.

The ecology of aquatic insects in the past can be hypothetically reconstructed on the basis of diverse evidence such as functional morphology, state of preservation, qualitative and quantitative data on their distribution in different palaeoenvironments, trace fossils, comparison with data on other fossils, etc. However, detailed palaeoecological information is available for relatively few fossil insect sites.

Descriptive papers often do not contain positive data on abundance and distribution of insect remains within the sections and thus can hardly be used for any firm palaeoecological conclusions. The representativeness of museum collections can be doubted all too often, particularly in respect of poorly preserved and fragmentary specimens and ichnofossils. The collectors often neglected those layers containing scarce or poorly preserved fossils, concentrating on the few most productive deposits and thus an important palaeoecological information becomes lost. That is why at present many palaeoecological models are still arbitrary and poorly argued, and more special field observations are necessary to test them.

3.3.2. CARBONIFEROUS

The nonmarine Carboniferous deposits yield numerous crustacean (Conchostraca, Ostracoda, and Decapoda) and mollusc (Bivalvia and Gastropoda) fossils inhabited freshwater and brackish basins. BUATOIS et al. (1998) pointed out that the ichnofossil record indicates the Carboniferous as the time of wide invasion of benthic macroinvertebrates to lacustrine (mainly shallow water) environments, while older freshwater trace fossils are restricted to alluvial and transitional alluvial-lacustrine deposits. The crustaceans (especially Decapoda) were probably the most common and important arthropods in inland waters in the Palaeozoic. There are no Carboniferous insects with undoubted morphological adaptations to aquatic habits.

An aquatic mode of life is commonly supposed for the dragonflies of the suborder Meganeurina (=Palaeodonata) basing on aquatic development of nearly all present-day dragon- and damselflies. However in spite of the common finds of adult Meganeurina in the Late Carboniferous and Permian no their nymphs are known. The earliest undoubtedly aquatic odonatan nymphs occurring in the Late Triassic belong to the Libellulina (=Neodonata). According to PRITYKINA (1980), the Palaeozoic odonatans probably had the terrestrial nymphs which could be litter dwellers. KALUGINA (1980b) stressed that in the wet Carboniferous forests the differences between the terrestrial and aquatic biotopes could not be very sharp. Even the modern rain forests fit some typically aquatic groups like tree leeches, terrestrial ostracods, or the gripopterygid stoneflies living in the wet mosses.

The Carboniferous families Triplosobidae and Bojophlebiidae previously referred to the mayflies are now placed in orders of their own (Chapters 2.2.1.1.1.1 and 2.2.1.1.1.2). The triplosobid immatures are unknown, and the unique Late Carboniferous nymph originally assigned to *Bojophlebia prokopi* Kukalová-Peck (here in Syntonopterida) can be interpreted otherwise (Chapter 2.1a). Thus there is no basis to consider triplosobids or bojophlebiids as aquatic taxa.

Thus no insects closely related to any typically amphibiotic or completely aquatic present-day taxa are found in Carboniferous. There is also no basis to postulate an aquatic mode of life for any extinct Carboniferous winged insects. The only Carboniferous pterygotans rather commonly represented in the fossil record by nymphs are dictyonerids and roaches. However, their nymphs demonstrate no aquatic adaptations, and they are clearly less common as fossils than the adults. They are similar to adults in many respects, including the sucking mouthparts of the alleged dictyoneuridean nymphs which suggest their feeding on higher plants absent in the Palaeozoic waters (CARPENTER & RICHARDSON 1968, 1971; KUKALOVÁ 1970; SHAROV 1971; WOOTTON 1972, 1988; SINITSHENKOVA 1979). Consequently the

assumption of aquatic habit for the immature stages of the dictyoneuridans (HANDLIRSCH 1906–1908, MARTYNOV 1938a) does not seem to be correct. The aquatic mode of life has been suggested for some Permian grylloblattidans (see below), but the nymphs of the Carboniferous families of the order (Daldubidae and Protoperlidae) are unknown. There is no reason to hypothesise the aquatic habits of the Carboniferous (and post-Carboniferous) roaches.

At present the only Carboniferous insects for which an aquatic or semiaquatic habit seems to be probable are Dasyleptidae – a peculiar extinct group allegedly representing the neotenic bristletails in the Carboniferous and Permian (see Chapter 2.1.1 and below). No modern thysanurans are aquatic, and dasyleptids demonstrate no special morphological features which could be treated unequivocally as aquatic adaptations. However, the abundance of dasyleptid remains, including numerous moulting casts, suggests that they should inhabit either aquatic or shore habitats because the probability of burial of terrestrial wingless insects in subaqual deposits is low. It can be supported by the recent finds of *Tonganoxichnus* Mágnano et al., the supposed traces of *Dasyleptus* resting and moving in the Late Carboniferous of Kansas (MÁGNANO et al. 1997), although the evidence is considered inconclusive (RASNITSYN 2000b).The trophic habits of dasyleptids is unknown. Their Carboniferous finds (including the supposed traces) are restricted to the Euramerian Realm situated at that time in the equatorial belt.

In general, the Carboniferous was probably a time of important changes in non-marine aquatic ecosystems, but the role of insects in this process was minimal. The physical and hydrological environments in the Carboniferous inland waters could be highly specific, without any close analogues in the present-day ecosystems (BETEKHTINA 1966, 1974), but unfortunately no detailed and well-grounded reconstruction of the ecosystem structure are available as yet.

KALUGINA (1980b) discussed several models which she believes could explain why insects had hardly colonised fresh water environments in the Carboniferous. One reason might be the supposed instability of the Palaeozoic water bodies because of existence of the primary deserts at the watershed areas, resulting in a highly irregular flowing regime. The reservoirs at the densely forested lowlands could be more stable. However, the habitat was probably unsuitable for insects as well because of a high concentration of humic substances, poorness of nutrients, in particular of nitrogen, and a low content of dissolved oxygen favoured to conservation of plant remains and coal deposition. Nowadays insects do inhabit both temporary and highly humic waters though their fauna is impoverished, but the environment were hardly favourable for the initial aquatic adaptations.

WOOTTON (1972) assumed that aquatic insects were primarily lotic and only secondarily invaded standing waters. As stated above, this point is very difficult to verify because the insect fossil record from running water habitats is quite fragmentary. However, colonisation of running waters suggests more special adaptations for keeping in water stream, certain mechanisms compensating the downstream drift of immatures, and so on. It is more likely that the earliest aquatic insects did not initially develop any specific aquatic adaptations; and that, they inhabited the banks of pools and streams in conditions of constant high humidity and could survive accidental submergence, while looking for food or escaping predators. The assumption (WOOTTON 1972) that the first aquatic insects were mainly predatory seem to be fully logical. Predators could actively penetrate into water while pursuing their preys or searching for new food resources because of serious competition from terrestrial predators.

Thus, Carboniferous aquatic continental ecosystems probably had no recent analogues. Even independently from possibly unique hydrological and hydrochemical parameters of the basins, the very absence of aquatic insects indicates that the ecosystem trophic structure has to be very different. There is no reliable evidence for the presence of aquatic insects there. It is almost certain that they did not inhabit the lentic water-bodies (perhaps except for dasyleptids), and the presence of insects in lotic waters is doubtful. The Late Carboniferous ecological radiation of insects probably affected the inland waters hardly at all.

3.3.3. PERMIAN

Dasyleptidae, probably the only aquatic Carboniferous insects, persisted into the Permian and possibly became more widely distributed (Chapter 2.1.1). They are rather common in the Carboniferous/Permian boundary in New Mexico, USA, as well as in the early Late Permian (Early Kazanian) of West Siberia (Kaltan) and found also in Early Permian (Artinskian) deposits of Kansas (Elmo). Trace fossils that appear to be similar to *Tonganoxichnus* have been discovered in the Early Permian Itararé Subgroup in Brazil (MÁGNANO *et al.* 1997). Even if the latter record is treated as doubtful, Kaltan is situated well outside the equatorial belt. Hence the distribution history of the family is in accordance with the equatorial pump model (Chapter 3.4). The Permian finds add no important data on the possible mode of life of dasyleptids.

The oldest undoubted finds of aquatic immature stages of the winged insects are known from the Permian. They include mayfly and stonefly nymphs and megalopteran larvae; however, finds of aquatic insect larvae and nymphs are still very rare, and adult fossils are numerically predominate. It should be noted that odonatan nymphs are still absent as in the Carboniferous.

The Early Permian mayfly nymphs are known from Oklahoma, USA (Protereismatidae: HUBBARD & KUKALOVÁ-PECK 1980), Obora in Czech Republic (Protereismatidae and Jarmilidae: HUBBARD & KUKALOVÁ-PECK 1980) and Tshekarda in Urals, Russia (Misthodotidae: TSHERNOVA 1965). They demonstrate evident adaptations to aquatic habitat such as fringed caudal filaments and large lateral abdominal tergaliae interpreted as gills. Their feeding habit is unclear. Mayfly adults are known from the same deposits as well. Other nymphs placed to the genus *Phthartus* Handlirsch are recorded from the Late Permian of Russia and South Africa. The Late Permian adult fossils of Misthodotidae are not accompanied by nymphs. In the Late Permian of the Russian North (Arkhangelsk Region) the wing of *Palingeniopsis* Martynov, the oldest known ephemeroid mayfly, was discovered; modern ephemeroids have burrowing nymphs developing mainly in large rivers.

Permian dragonfly diversity was greatly increased compared with the Carboniferous, but the nymphal fossils are absent as well and no other sound evidence is available to consider the Permian dragonflies aquatic.

The supposedly aquatic beetles (schizophoroid Archostemata) are unknown in the Early Permian but they were widely distributed in the Late Permian, occurring in the localities of Europe, Asia, Australia, South America, South Africa (Chapter 2.2.1.3.2). In the well known sequence of the Kuznetsk Basin (South Siberia) they are rare in the Kuznetsk Formation (Ufimian), more numerous in the Iliinskaya one (Kazanian) and particularly in the Erunakovo Formation (Tatarian). The finds are rather numerous, especially toward the end of the Late

Permian. Adult schizophoroid beetles were lacking swimming adaptations, they were rather benthic or inhabited swimming algae; their larvae are unknown. The main reason for supposing an aquatic mode of life is the presence of locking projection on the inner side of the schizophoroid elytra which resemble those of some modern hydradephagans, and which probably prevent the opening of elytra by air retained in the subelytral volume.

The Late Permian megalopteran fossils include their typically aquatic larvae with jointed gills on the abdomen sides (Kargala in South Urals, Chapter 2.2.1.3.3.2f). There is a supposition, however, that this larva resembles that of gyrinid-like beetle (HENNIG 1981). The larva is similar to the extant megalopteran ones and probably was a benthic predator. Both megalopteran wings and larvae are rare in the Permian, and the biotopical preference of these insects is not clear.

The aquatic larval habit cannot be excluded for the numerous and diverse Permian mecopterans. However, the only living family with the aquatic larvae, Nannochoristidae, is not found before the Jurassic, and no other evidence are known on the aquatic habits of the Permian Panorpida (Chapter 2.2.1.3.4.1).

The caddisflies appeared early in the Permian though only adults are known from then. An aquatic habit of their larvae was hypothesised based on the presence of a developing anal loop which was regarded as an adaptation to keep air bubbles under the wings during emergence from the pupa under the water (MARTYNOVA 1959). Currently the anal loop is considered an important structure for the co-ordination of wings movements during the flight (Chapter 2.2.1.3.4.2f). Absence of the palaeontological records of the caddis larvae might be interpreted as indicating development in lotic waters, but on the whole the evidence suggests that water saturated forest litter was their habitat (Chapter 2.2.1.3.4.2).

In the order Grylloblattida, which was flourishing in the Permian, some forms are supposed to have had aquatic nymphs (Chapter 2.2.2.2.1), similar to those in stoneflies (alleged lemmatophorid nymphs from Elmo, several nymphal taxa described from Kaltan), or because the nymphs are abundant (much more numerous than adults) and represent a nearly complete developmental series which is taphonomically possible only for autochthonous fossils (see Fig. 387). The last case is peculiar in that the *Gurianovella* nymph has legs armed with strong spines and spurs which are unusual in aquatic insects and its moulting casts are often shrunken as if they had experienced drying. This could be explained if the nymphs were burrowing in the shores of water-bodies next to the water edge (STOROZHENKO 1998).

By the end of the Early Permian (Kungurian stage of the Urals) several stonefly nymphs of Tshekardoperlidae, and numerous adults of Palaeonemouridae and Perlopseidae appeared. The nymphs are similar to the modern stonefly nymphs and so should be aquatic; most species were probably carnivorous. The nymph *Barathronympha victima* Sinitshenkova (Perlomorpha inc. fam.) apparently adapted to rapid running water environment was also found here (SINITSHENKOVA 1987). In the Late Permian the nymphs and adults of Palaeoperlidae and Palaeonemouridae are found in many sites in the North (Vorkuta) and East (Vyatka) of European Russia, in Kazakhstan, Siberia (Kemerovo Region and Krasnoyarsk Province; SINITSHENKOVA 1987). *Ohionympha* Sinitshenkova (Palaeonemouridae) is found in Messer Ridge in Antarctica (F. CARPENTER 1969). Mooi River in South Africa is a locality of the adult fossils of 3 species of *Euxenoperla* Riek (Euxenoperlidae; RIEK 1973, 1976d). A wing of *Stenoperlidium* Till. (Eustheniidae) has been described from Australia (TILLYARD 1935b).

The nymphs occur more often than in the Early Permian, but the adult fossils numerically still dominate.

The taxonomic position of some Permian stonefly-like nymphs (see Figs. 388, 389) is unclear in terms of stoneflies or grylloblattidans (Chapter 2.2.2.2). However, if the original interpretation of the ventral prothoracic structures observable on the recently described *Sylvonympha* from Tshekarda (NOVOKSHONOV & PAN'KOV 1999) as the coxal gills is correct, it could represent another taphonomically allochthonous (probably transported from the potamal zone) element in the insect assemblage.

Among the numerous extinct Permian insect orders no definite aquatic representatives are known.

No detailed ecosystem reconstruction has been made for any Permian inland waters. The trace fossil record indicates that burrowing benthic animals which were virtually absent in the Carboniferous have appeared both in alluvial and lacustrine environments (M. MILLER 1984; BUATOIS *et al.* 1998). It manifests some significant changes in the ecosystem structure. The find of an ephemeroid wing in the Late Permian suggests that insects could be among those early sediment burrowers; however, the nids of the present-day burrowing mayflies are U-shaped while the Permian non-marine burrows are mainly meniscate. Other possible producers of burrows are *Gurianovella*-like grylloblattidan nymphs which could live in wet bank mud.

The diversity and ecological role of aquatic insects surely increased in the Permian in comparison with the Carboniferous but they were still much less important than later and the crustaceans remain the predominant aquatic arthropod group. Perhaps Permian insects were better represented in running than in standing waters which is in accordance with the Wootton's hypothesis discussed above. However, at least in the Late Permian some insects have probably colonised lacustrine environments as well. In Kaltan and Karaungir stonefly nymphs seem to be taphonomically autochthonous and are accompanied with schizophoroid beetles and (in Kaltan) dasyleptids. All known Permian aquatic insects were probably benthic. Both predators and detritivores are represented, and the presence of algal-feeding forms, at least among the mayflies and beetles, cannot be rejected.

3.3.4. TRIASSIC

The Triassic localities are more widely distributed geographically than those of the Palaeozoic. They occur on all continents, except Antarctica, but their number is considerably less. The Early Triassic insects including the aquatic ones are the least known. The Middle and Late Triassic assemblage of the aquatic insects is more rich, with the nymphal fossils sometimes more numerous than the adults, the case not observed in the Palaeozoic but very typical in the Mesozoic.

Among the Triassic aquatic insects only beetles with schizophoroid and permosynoid type of elytra are found in the Early Triassic deposits in Eurasia. Preliminary dating of insect bearing layers of Super deep hole (СГ-6) in the Tyumen' district as Early Triassic (BOCHKAREV & PURTOVA 1994) later proved to be Middle Triassic (KIRICHKOVA *et al.* 1999). In the Middle Triassic part of this section the aquatic insects are represented by the nymphs of mayflies and stoneflies, the mayflies being identified as *Mesoneta* and *Archaeobehningia*, the genera typical of Mesozoic (SINITSHENKOVA 2000a). So it is possible to discuss features of the Triassic aquatic entomofauna based on the material from the Middle and Late Triassic. The abundant and taxonomically diverse assemblages of undoubtedly aquatic stages of insects occur for the first

time in the palaeontological record in the Anisian deposits of Vosges in France (MARCHAL-PAPIER 1998), in Mallorca (COLOM 1988), Garazhovka (Carnian or Norian of Ukraine), the Newarc Series of Atlantic states of USA (Carnian), Madygen Formations (Ladinian-Carnian of Middle Asia) and Tologoy Formations (Norian-Rhaetian) of Kenderlyk in East Kazakhstan. All these finds are concentrated in the northern hemisphere and mainly in the areas with warmer climates (Euro-Sinian Palaeofloristic Region according to DOBRUSKINA 1980). In the more temperate Siberian region the aquatic stages of the Triassic insects are found only in the South in comparatively young deposits of Kenderlyk. In the southern hemisphere with its numerous Triassic insectiferous deposits in Australia, South Africa and South America, the nymphs are rare. There are no records of aquatic stages of insects in Pacific Asia (Russian Far East, Japan, Vietnam) either.

3.3.4.1. Overview of taxa of aquatic insects

In comparison with older assemblages the Triassic faunas lack the possibly aquatic Machilida of the family Dasyleptidae but contain all other aquatic insect orders known from the Permian and demonstrate their increasing diversity. There are also several important novelties including aquatic bugs and dipterans as well as the oldest undoubtedly aquatic odonatan nymphs.

The Triassic mayfly record was reviewed recently (SINITSHENKOVA 2000a), and the rich Mid-Triassic (Anisian) fauna of the Grés á *Voltzia* deposits of Vosges, France, is under study. The aquatic stages are found in the Mid-Triassic of Vosges, Spain (Mallorca), European Russia (the Bukobay Formation of Bashkortostan) and West Siberia (the Varengayakha Formation of the Tyumen' Region), in the Mid- or Late Triassic of Central Asia (the Madygen Formation of Kyrgyzstan, Ladinian-Carnian) as well as in the Late Triassic of Ukraine (Garazhovka, Carnian-Norian) and the Urals (the Korkino Series of the Chelyabinsk Region, Rhaetian). The supposed bristletails (J. ANDERSON *et al.* 1998) found together with wings of the mayfly *Litophlebia* Hubbard et Riek in the Late Triassic Molteno Formation (South Africa) may well be in fact the nymphs of the same mayfly. Unlike the Permian, the nymphs are clearly more common than remains of the winged stages accompanying them in Vosges and Molteno Formation. In Vosges and probably in Mallorca mayflies occur in deltaic sediments, other deposits are of a lacustrine origin. There are also sporadic finds of winged stages in the Triassic of England, Spain (Tarragona Province), Germany, and Switzerland. In the Madygen Formation very few fragmentary remains were discovered suggesting their taphonomic allochthonicity; other nymphs are probably autochthonously buried.

A bulk of the Triassic mayfly nymphs resemble younger Mesozoic fossils and most of them are assigned to the genera common and widespread in the Jurassic, though some are placed to families restricted to the Triassic (e.g. Mesoplectopteridae). Burrowing ephemeroid nymphs are common in the Grés á *Voltzia* and occur also in Mallorca and in the Tyumen' Region. In Vosges the U-shaped burrows occur in the same deposits and may be tentatively attributed to ephemeroid mayflies. There are also some records of U-shaped nids from coastal lacustrine Triassic deposits of other areas including China and Greenland (BUATOIS *et al.* 1998). At least some of them could have been made by burrowing mayflies though similar burrows may be produced by many other invertebrates. There is some controversy with data on winged stages, most of which differ significantly from younger Mesozoic taxa. The wing from Vosges illustrated by MARCHAL-PAPIER (1998) can be placed tentatively to the Permian family Misthodotidae.

However fragmentary, these data indicate mayfly invasion into lakes and their increasing diversity and abundance during the Triassic. None of the nymphs have their mouth parts preserved well enough to establish their feeding habit with certainty; most probably they were debris collectors or algal-feeders except for burrowing nymphs which could have been filter feeding. The nymphs of *Mesoplectopteron* from Vosges have unusually long and slender legs suggesting that they could represent an enigmatic life-form not represented among younger taxa. Mayflies nymphs from the Bukobay and the Varengayakha Formations assigned to the genus *Mesoneta* are unusually small-sized and so might differ ecologically from their Jurassic congeners.

In addition to those of mayflies, there are about two dozen insect nymphs of uncertain taxonomic position in the collections of the Louis Pasteur University (Strasbourg, France) from the Grés á *Voltzia* (Papier and Sinitshenkova in preparation). As with mayflies they have long cerci and paracercus but differ in having paired tarsal claws. They belong to two different types that are almost equally represented in the collection. One of them (see Fig. 61) has the abdominal segments forming large acuminate lateral projections (immovable paranota and not gill plates) similar to those of *Carbotriplura* Kluge (Chapter 2.1). Their repeated finds could indicate either an aquatic or a shore habitat. The second type of nymph demonstrate branched abdominal gills and were surely aquatic. Both types should be benthic but other aspects of their biology are as enigmatic as their affinities. Perhaps they represent some extinct high-level (of subordinal or even ordinal rank) taxa unknown from the younger strata.

The Triassic odonatans include diverse Meganeurina from Vosges, Central Asia (the Madygen Formation), and Australia and sporadic finds of adults of the suborder Libellulina in Central Asia, Japan, Australia, South Africa and Argentina. As discussed above, Meganeurina could have had terrestrial nymphs. The oldest known odonatan nymphs have been discovered in the Late Triassic of Ukraine (Garazhovka, undescribed, mentioned by KALUGINA 1980b) and Australia (ROZEFELDS 1985a). They resemble closely the Jurassic nymphs of the isophlebioid type (*Samarura* spp.) and could represent the same lineage. Hence odonatans probably invaded lacustrine habitats no later than the Late Triassic, but their abundance and ecological role was probably low. In comparison with the predominant Jurassic forms, the known Triassic nymphs are relatively small-sized and should have fed on smaller prey. Interestingly, no odonatan nymphs are found in the Early Keuper (Ladinian) of Frankonia (Germany) nor in the Lettenkohle Formation of Alsace (France) in spite of the presence of supposed oviposition marks on horsetails (GRAUVOGEL-STAMM & KELBER 1996).

The Anisian records of heteropterans in Vosges, France (MARCHAL-PAPIER 1998) may belong to littoral as well as aquatic bugs (D. Shcherbakov, pers. comm.), so the unique naucoroid fragment from the Madygen Formation of Kyrgyzstan (Ladinian-Carnian, see Fig. 28) is likely to be the oldest known unquestionable aquatic bug. Another undescribed naucoroid has been recorded (POPOV 1980) from Nikolaevka in Ukraine; Carnian after DOBRUSKINA (1980). The naucoroid family Triassocoridae, restricted to the Late Triassic, was described from Australia (Ipswich) and is common also in the Carnian-Norian lacustrine deposits in Kenderlyk in East Kazakhstan (undescribed; pers. comm. by Yu. Popov). Abundant aquatic bugs have been discovered in the Cow Branch Formation (Carnian) in eastern North America (OLSEN *et al.* 1978, FRASER *et al.* 1996), in sediments that allegedly originated in a deep-water zone of large lakes. The bugs belong to the modern families Naucoridae, Belostomatidae and Notonectidae (Yu. Popov, pers. comm.). Like the aquatic beetles, bugs live in water both as larvae and as adults, and winged adults often visit and temporarily inhabit water bodies other than their breeding sites. However, the immature and adult bugs are often hardly distinguishable on the basis of incomplete remains. At least in the Cow Branch Formation some nymphs are found, indicating that the lake was a breeding place. All taxa found in the Triassic are carnivorous. The belostomatids and notonectids are active swimmers while naucoroids should be rather nectobenthic than true nectic forms. Unlike many other aquatic insects bugs use atmospheric oxygen for breathing and so are highly tolerant to the low oxygen concentration.

Among the Triassic aquatic beetles schizophoroids are common and diverse. Three schizophoroid families are known from the Triassic including Schizophoridae and Ademosynidae common with the Late Permian and Catiniidae restricted to the Mesozoic. Completely preserved schizophoroids are known mainly from the Madygen Formation and from the Late Triassic of Australia (PONOMARENKO 1969). The isolated elytra with "schiza" confirm the worldwide distribution of the group being common additionally in a number of localities in Europe, the Urals, Siberia, Kazakhstan, etc., including the Early Triassic where no other aquatic insects have been discovered. The Triassic schizophoroids, like the Permian ones, supposedly inhabited diverse aquatic habitats, both lentic (at least at the adult stage) and lotic. Two more beetle suborders first entered the fossil record in the Triassic, both including some aquatic taxa. Aquatic Adephaga have been discovered in Europe (*Triadogyrus* in Garazhovka; ARNOLDI *et al.* 1977) and Asia (Triaplidae in the Madygen Formation of Kyrgyzstan, and two dytiscoid genera, *Necronectulus* and *Colymbothetis*, in the Tologoy Formation of Kenderlyk in East Kazakhstan). *Colymbothetis* is the most ancient aquatic beetle larva known. It is placed to a family of its own but resembles modern noterid larvae lacking any swimming adaptations and similarly was perhaps a benthic predator (PONOMARENKO 1993). The larvae resembling Colymbothetidae and Coptoclavidae are found also in the Carnian Newark Group in eastern North America (PONOMARENKO 1996). Coptoclavid larvae were highly specialised active swimmers, and are very common in younger Mesozoic deposits. All aquatic Adephaga are predaceous except for the algal-feeding Haliplidae (and perhaps the related extinct family Triaplidae). The Triassic Polyphaga are insufficiently studied but equally include supposedly aquatic forms, e.g. *Peltosyne* from the Madygen Formation which may be closely related to primitive hydrophiloids (PONOMARENKO 1995a). No larvae known in the Triassic could be reliably attributed to the aquatic Polyphaga. Many aquatic beetles, in particular adults, use atmospheric air and do not need dissolved oxygen.

Megalopteran larvae with characteristic abdominal gills occur in Vosges (MARCHAL-PAPIER 1998) and Garazhovka (KALUGINA 1980b); none of them has been described in detail. They were undoubtedly benthic predators like present-day alderfly and dobsonfly larvae. The Vosges larvae probably inhabited a lowland river and those from Garazhovka were probably lacustrine.

None of Triassic neuropterans could be regarded as aquatic with a good reason.

Some Triassic scorpionflies (Liassophilidae) are believed to be related to modern aquatic nannochoristids (Chapter 2.2.1.3.4.1) but there are no evidence of aquatic mode of life for any of them.

The taxonomic composition of caddisflies is changing considerably in the Triassic. The only Permian family survived is Cladochoristidae represented by rare wing fossils in the Madygen Formation in Central

Asia and in the Late Triassic Ipswich Series of Australia. The suborder Hydropsychina also appeared, represented by two families, Prorhyacophilidae (restricted to the Triassic) and Necrotauliidae (common in younger Mesozoic deposits). Both occur together in the Madygen Formation and are found in the southern hemisphere as well (the former in the Ipswich Series, the latter in the latest Triassic of the Mendoza Province, Argentina). Unidentified caddisflies are recorded from the Anisian of Vosges (MARCHAL-PAPIER 1998), and Necrotauliidae occur occasionally in the Late Triassic of West Europe. The specimen from the Newark Series figured as a caddisfly by FRASER *et al.* (1996) is clearly a scorpionfly. As in the Permian, only adults are known suggesting that the order remained exclusively lotic. Their Triassic distribution seems to be highly uneven, with relatively frequent finds in few localities and none in others. This distribution pattern needs ecological interpretation based on analysis of physiographic features of Triassic landscapes. Larval diet of the Triassic caddisfly larvae could be inferred as predatory, though the presence of some algal-feeders cannot be excluded.

One more order of aquatic insects represented in the Triassic is Diptera. They play a great ecological role in the later Mesozoic and Cainozoic aquatic ecosystems. However, Triassic dipterans are poorly studied. Unidentified dipteran larvae and pupae are recorded from the Middle Triassic, from the very beginning of the palaeontological record of the order, and seem to be common enough to treat them as taphonomically autochthonous in deltaic environments (MARCHAL-PAPIER 1998). Adult Limoniidae found in the Newark Series might have aquatic or semiaquatic larvae like many modern members of the family. However, terrestrial Limoniidae are now common as well, so the biology of the Triassic limoniids is unclear.

Triassic Grylloblattida, being rather diverse and abundant as adults, are rarely recorded as nymphs. Probably none of them were aquatic or shore-inhabiting.

The Triassic stoneflies are reviewed by SINITSHENKOVA (1987). Additionally, a few badly preserved nymphs were discovered in the Mid-Triassic of France (Grés á *Voltzia*) and the Tyumen' Region together with the mayflies mentioned above. New finds of stonefly wings were reported from the Molteno Formation as well (J. ANDERSON *et al.* 1998). No families except for the Gondwanan gripopterygomorph Euxenoperlidae, are common with the Permian. Euxenoperlids are represented exclusively by adults and so were most probably lentic. Nymphs of Triassic stoneflies have been discovered only in the northern continents and seem to be rather rare though widespread and occurring in the Mid- and Late Triassic deposits of France, Ukraine (Garazhovka), West Siberia (Varengayakha Formation), Central Asia (Madygen Formation), East Kazakhstan (Kenderlyk), and Southern China (Nalatin Formation, Yunnan Province). Adults are found in small numbers in the same deposits (Garazhovka and the Madygen Formation). Both nymphs and adults belong to those families, and often even to those genera, that are well represented in the Jurassic. Nymphs of the *Siberioperla* Sinitshenkova, *Mesoleuctra* Brauer, Redtenbacher et Ganglbauer and *Trianguliperla* Sinitshenkova from Madygen and Kenderlyk are interpreted as lentic, and the same may be true for nymphs from Garazhovka and Nalatin. The Vosges nymphs probably inhabited the potamal zone of rivers, and adults found in the Madygen Formation in the locality Dzhayloucho probably represent a lotic fauna. *Trianguliperla* Sinitshenkova and supposedly *Triassoperla* Lin and *Berekia* Sinitshenkova were carnivorous while *Siberioperla* and *Mesoleuctra* fed on coarse and fine plant debris respectively. All stonefly nymphs were benthic; unusually long and slender legs of *Mesoleuctra* are interpreted as adapted to crawling on a soft muddy bottom.

3.3.4.2. Aquatic insect assemblages and their typology

In general, the available data, however fragmentary, clearly indicate a considerable radiation of aquatic insects in the Triassic. Some new life-forms appeared at that time including the active nectic carnivores, both chewing (coptoclavid larvae) and sucking (bugs). The role of the latter group was increasing particularly toward the end of Triassic (probably since the Carnian). While apparently very few lacustrine insects are found in the Permian, in the Triassic they became diverse both taxonomically and ecologically. However, the lacustrine fauna was still impoverished in comparison with the lotic one, taking into account unfavourable taphonomic features of the latter. The main trophic guilds of lacustrine insects in the Triassic include benthic and nectic predators, shredders, debris collectors and possibly algal-feeders; presumed filter-feeders (burrowing mayflies) are found in deltaic palaeoenvironments. The Permian relicts are represented by few taxa and are quantitatively scarce while most Triassic families and some genera are common in the Jurassic. Along with the numerical prevalence of aquatic immature stages over respective flying adults in lacustrine oryctocoenoses (a feature unknown in the Permian), this infers the Triassic to be a starting point of a new, Mesozoic evolutionary stage of the lacustrine biocoenoses.

Aquatic insects were never so abundant in the Triassic as in many younger Mesozoic sites where thousands of specimens can be seen on some bedding planes, and other aquatic macroarthropods are relatively common. Some of them are not found in younger sediments like the peculiar extinct arthropod group Euthycarcinoidea (not rare in the Grès a *Voltzia*) and the remarkable crustacean order Kazakharthra. The latter is related to Conchostraca and Notostraca and restricted to the Triassic of Kazakhstan, Central Asia, West China and Mongolia. Those finds stressed a transitional nature of the Triassic aquatic ecosystems where, side by side with the appearance of aquatic assemblages resembling the Jurassic ones, some Palaeozoic characters have been conserved.

Interestingly, all known finds of Triassic lacustrine insect assemblages are confined more or less to northern continents (the Euro-Sinian palaeofloristic region after DOBRUSKINA 1980). Only supposedly lotic taxa are known in the Gondwanaland area, with the possible exception of the supposed mayfly nymphs from the Molteno Formation. At the northern lands the Triassic lacustrine assemblages with autochthonous insects are concentrated mainly in warmer areas (Europe, Central Asia, the Atlantic states of USA). In the more temperate climate of Siberian region they have been discovered only at the southern periphery (West Siberia, East Kazakhstan). The only possible lacustrine insects known from the interior of this region are schizophoroid beetles. There is no record of aquatic stages of any insects in Pacific Asia (Russian Far East, Japan, Vietnam).

KALUGINA (1980b) hypothesised that during the Triassic, the hydrological and hydrochemical regime of inland waters became more stable than previously, mainly because of changes in the watershed vegetation and expansion of the first fresh-water submerged macrophytes (the charophyte algae which firstly appeared in the Permian). The Newark Late Triassic lakes restored as deep chemically stratified (meromictic) basins with a well-oxygenated epilimnion and anoxic hypolimnion (OLSEN *et al.* 1978, FRASER *et al.* 1996), were generally similar to the large Jurassic and especially Early Cretaceous lakes as restored by PONOMARENKO & KALUGINA (1980) and PONOMARENKO (1986a, 1996). However, their trophic structure differed considerably, first of all because of the low number of aquatic dipterans in the Newark lakes. No detailed reconstruction of any Triassic aquatic ecosystems has been made thus far.

The Triassic aquatic insect assemblages were diverse, indicating the existence of the very distinct community types. The following preliminary typology may be proposed:

1. Taxonomically diverse assemblages dominated by various mayflies (including the burrowing ones), with considerable diversity of other aquatic insects and absence of aquatic bugs, recorded from deltaic sediments and probably representing the fauna of slow flowing lowland channels and small ponds with soft bottoms (Vosges and probably Balearic Islands).

2. Assemblages dominated by mayflies and/or stoneflies, often closely related to Jurassic taxa, recorded from lacustrine deposits (probably mainly shallow-water), usually yielding abundant terrestrial plant remains (Garazhovka, Bukobay Formation, Korkino Series, Varengayakha Formation, Madygen).

3. Assemblages dominated by aquatic bugs, but containing also benthic insects including stonefly nymphs and larval and adult beetles; occurred in lacustrine deposits (Kenderlyk).

4. Assemblages strongly dominated by aquatic bugs with few if any immature stages of other aquatic insects; occurred in lacustrine deposits originated in a deep-water anoxic zone (the Newark Group).

5. Assemblages composed of schizophoroid beetles, with any other aquatic autochthonous insects lacking or nearly so; occurred world-wide in various palaeoenvironments.

6. Assemblages including only allochthonous (probably mainly lotic) aquatic insects; also occurred world-wide in various palaeoenvironments.

3.3.5. JURASSIC

The Jurassic insect assemblages vary considerably both in their past environments and taxonomic composition. The North temperate zone and particularly its East Asiatic segment (Siberia, Mongolia and China) is rich in Jurassic localities of the lake genesis. Unlike the Triassic, fossil of immatures usually predominate, and adults of many species have not been found. In contrast, Indo-European Jurassic insect assemblages consist mainly of winged stages, probably because of the predominantly marine genesis of the source deposits. Being allochthonous, they pose problems for ecological analysis. Outside of Eurasia, both in North America and in Gondwanaland, very few Jurassic insect localities have been discovered.

Many extant families and even subfamilies are found in the Jurassic assemblages of the aquatic insects, although the characteristically Mesozoic taxa are common as well. In contrast, the Permian elements are completely absent.

3.3.5.1. Overview of taxa of aquatic insects

Mayflies were very common and diverse in Asia but rare in Europe. Asiatic localities often yield hundreds of specimens, mostly nymphs, while in Europe only winged stages are known. The only European locality that is relatively rich in mayflies is Solnhofen in Germany; outside of it, a single, undescribed specimen from the Early Jurassic of Germany has been recorded (ANSORGE 1996). At this time, epicontinental seas with numerous islands of different sizes, covered much of the present-day territory of Europe. Probably, in the Mesozoic, as well as now, island mayfly faunas were strongly impoverished, and many islands were completely devoid of them.

The most widespread Jurassic mayfly nymphs belong to the morphological type with large tergaliae protruded from the sides of abdomen. They are represented by several genera of Siphlonuridae (*Mesobaetis* Brauer, Redtenbacher et Ganglbauer, *Stackelbergisca* Tshernova, *Mogzonurella* Sinitshenkova), Mesonetidae (*Mesoneta* Brauer, Redtenbacher et Ganglbauer) and Epeoromimidae (*Foliomimus* Sinitshenkova, *Epeoromimus* Tshernova). Both *Mesobaetis* and *Mesoneta* are known from the Triassic as well but in much smaller number. Up to four species may co-occur in the same locality (though in each site only one species dominates numerically) although some assemblages include only one species each. All genera occur in Siberia, *Mesoneta* and *Mesobaetis* are found also in Mongolia, and only *Mesobaetis* has been recorded from China. Unidentified nymphs superficially resembling *Mesobaetis* have also been collected in the late Jurassic of India (A. Rasnitsyn, pers. comm.) being the only Jurassic mayflies found in Gondwanaland. In spite of the high nymph abundance, the winged stages very rarely occur as fossils and are known only for two siphlonurid genera, *Mesobaetis* and *Mogzonurus* Sinitshenkova (the nymphs of the latter are undiscovered). All those nymphs seem to belong to the same broad ecological type of the benthic medium-sized collectors of fine debris inhabited soft muddy bottom of diverse water bodies lacking dense submerged vegetation. They are most abundant and diverse in sediments of either small shallow oxbow lakes or deep and moderately large mountain lakes.

The assemblages of large lowland and piedmont lakes are numerically and taxonomically impoverished (Kubekovo, Khoutiyn-Khotgor, Shar Teg). Undescribed nymphs of the siphlonurid type are abundant also in very fine-grained fluvial deposits (originated probably in a backwater of a lowland river) in Kempendyay in Jakutia (Siberia). The similar but usually larger nymphs of Hexagenitidae (*Siberiogenites* Sinitshenkova) are rare in Siberia and Mongolia and seem to prefer larger lakes, both lowland and mountain, feeding possibly on algal growth. On the contrary, Solnhofen hexagenitids are the most common as adult mayflies; nothing is known about their nymphs. The filter-feeding burrowing nymphs of Torephemeridae (*Archaebehningia* Tshernova) and Palingeniidae (*Mesogenesia* Tshernova) occasionally occur in small number in some Siberian and Mongolian localities, probably as allochthonous lotic element. The mayflies are nearly absent in the Central Asian Jurassic localities (except the only wing of *Aenigmephemera* Tshernova in the Late Jurassic of Karatau in Kazakhstan) despite the presence of insect sites, partly originated in the oxbow lake deposits, which are sedimentologically similar to those in Siberia and Mongolia.

The Jurassic odonatan fauna is characterised by the dominance of diverse Libellulina while the relict Meganeurina are rare. Odonatan nymphs are common in Siberia, Mongolia and perhaps in North China (HONG 1998 mentions odonatan nymphs among the dominant insects in the Haifanggou Formation of Liaoning Province) but absent in Central Asian and European localities. All Jurassic nymphs known are extremely uniform and represent the same peculiar ecological type absent from the present-day fauna. Superficially they resemble the damselfly nymphs (Calopterygina) in having long, slender body and legs and three leaf-shaped caudal flaps. However, unlike in the Calopterygina, these appendages are invariably relatively short and broad, heavily sclerotised and never demonstrate tracheae. They hardly can serve as gills and operated rather as a locomotion organ (caudal fin; PRITYKINA 1985). PRITYKINA (l.c.) has described also the internal rectal gills and stressed that their morphology suggests a high dissolved oxygen content in the habitat.

Classification of these fossil is a matter of confusion because of their extremely uniform morphology. Two genera are established, *Samarura* Brauer, Redtenbacher et Ganglbauer, most probably a collective name for different heterophlebioid nymphs, and *Dinosamarura* Pritykina associated with isophlebiid adults. The mask morphology suggests that their main food was large aquatic insects, most probably mayfly and stonefly nymphs. *Dinosamarura* nymphs were extremely large-sized and should be the top predators in mountain lakes of Transbaikalia where no fish are found. However, in Khoutiyn-Khotgor in Mongolia similar giant isophlebiid nymphs are accompanied by several fish species like somewhat smaller *Samarura*, in numerous Siberian sites. In Siberia *Samarura* is nearly restricted to small oxbow lakes while isophlebiid nymphs occur in deeper and larger lakes, both mountain and lowland. Interestingly, the assemblages with odonatan nymphs are restricted to the temperate Siberian Region while the adult assemblages are much impoverished there compared to the warmer areas like Central Asia and Europe. In Europe the odonatans are much better represented than any other aquatic insects. This is not a puzzle, for being strong fliers and often habitat opportunists, they colonise islands much more easily than mayflies, stoneflies or caddisflies. Several dozen species may co-occur in the same deposits, as in the Late Jurassic Solnhofen lithographic limestone in Germany. This highly diverse fauna should inhabit lotic and/or other taphonomically unfavourable environments.

The Jurassic aquatic bugs were diverse and widespread. They were represented mainly by well-swimming nepomorphs such as the water boatmen Corixidae and related extinct Shurabellidae, and Belostomatidae. The more sedentary nectobenthic Naucoridae and Nepidae were less common. The backswimmers (Notonectidae) were relatively rare. An important ecological novelty was the first appearance of the primitive mesovelioid gerromorphs (water striders) that inhabited the water surface. Both Nepomorpha and Gerromorpha are predacious, but the corixids (and probably shurabellids) can occasionally feed on algae as well. The bugs occur in diverse depositional contexts in the European, Asian and North American Jurassic but their distribution is highly uneven. They are rarely abundant together with other autochthonous aquatic insects in the same layer. One of few examples is Khoutiyn-Khotgor in Mongolia where both nymphs and adults of the corixid *Haenbea badamgaravae* Popov are very common together with diverse aquatic dipterans, mayfly, odonatan and stonefly nymphs, and aquatic beetles. In Siberian oxbow lake deposits rich in aquatic insects few shurabellid, corixid and naucorid specimens (both nymphs and adults) were collected. In contrast, in the mountain lake deposits corixids are occasionally abundant, but only in few layers where no other aquatic insects have been discovered (ZHERIKHIN & KALUGINA 1985).

On the other hand, both nymphs and adults of *Shurabella* Popov are quite abundant in Central Asian floodplain lakes with strongly impoverished insect fauna (POPOV 1971). The same distribution pattern probably occur in China. HONG (1998) mentions the Jurassic assemblages in North China dominated by the corixid *Yanliaocorixa* Hong (the

Jinlongdao assemblage) and by other aquatic insects (the Haifanggou assemblage) as associated with different deposits. In the Early Jurassic of South China the corixid-dominated assemblages with *Lufengonecta* Lin are known (LIN 1977). Some mass finds of aquatic bugs are connected with clearly hydrochemically anomalous palaeoenvironments, where few if any other autochthonous insects are represented, e.g. with the sediments of brackish lagoons or lakes, both at coastal plains (the corixid *Gazimuria* Popov in Transbaikalia: ZHERIKHIN & KALUGINA 1985) and inland (the corixid *Karataviella* Becker-Migdisova and the gerromorph *Karanabis* Becker-Migdisova, Fig. 488 in Kazakhstan; DOLUDENKO *et al.* 1990). Interestingly, rare in the North American assemblages of the Jurassic aquatic insects are often dominated by bugs. In the Todilto Formation in New Mexico mass burial of nepomorph bugs is restricted to the deposits transitional from freshwater to brackish lagoon environments (BRADBURY & KIRKLAND 1966).

Recently aquatic bugs have also been discovered in number in the Late Jurassic Morrison Formation of Sundance (Washington, USA), in association with aquatic beetles, caddis cases and enigmatic insect larvae, and possibly the poorly described *Xiphenax jurassicus* Cockerell (A. Rasnitsyn, pers. comm.). As discussed before, the nepomorph bugs also dominate in the latest Triassic of Atlantic North America. Only adult bugs are found in many oryctocoenoses suggesting that they could either be occasional visitors or temporary inhabitants of the source waterbodies and should develop somewhere outside. It is most obvious for the European marine Jurassic where adult gerromorphs (e.g. *Engynabis tenuis* Bode in the Late Liassic of Germany: Yu. Popov, pers. comm.) and nepomorphs (e.g. *Mesobelostomum deperditum* Germar in Solnhofen: POPOV 1971) are occasionally common. A wide distribution of aquatic bugs in European island faunas suggests their probable opportunistic habitat relations but the direct information about their habitats is nearly absent. In Poland (the Zagaje Formation of the Holy Cross Mountains) the belostomatid *Odrowasicoris polonicus* Popov was found in floodplain deposits together with supposedly aquatic beetles (WEGIEREK & ZHERIKHIN 1997).

Fig. 488 Female water strider, *Karanabis kiritshenkoi* Becker-Migdisova (Mesoveliidae) from the Late Jurassic of Karatau in Kazakhstan (PIN 2554/410, photo by D.E. Shcherbakov); body 5.9 mm long

The aquatic beetles were common and diverse in the Jurassic, including schizophoroids (adults only), hydradephagans (both adults and larvae), and polyphagans (adults only). Together with odonatans and bugs they are well represented not only in continental Asiatic but also in European island faunas. The share of schizophoroids decreases gradually from the Triassic up to the Late Jurassic, though in the Early Jurassic they are still common, at least in Central Asia and in some European assemblages (PONOMARENKO 1995a). Among the hydradephagans the extinct families Coptoclavidae and Liadytidae were most widespread and common both in Europe and Asia since the Early Jurassic. Both families were carnivorous, with nectic (Coptoclavidae) or nectobenthic (Liadytidae) larvae. While the coptoclavid and liadytid adults occurred in diverse palaeoenvironments, the larvae of the former seem to be more common in larger lakes and the latter in smaller ones. Other hydradephagan families (the extinct Parahygrobiidae and the modern Gyrinidae and Dytiscidae) are found in a small number in the Late Jurassic only. Among the polyphagans the hydrophiloid adults were numerous, diverse and widespread, and few Hydraenidae are described from Siberia. An aquatic mode of life cannot be excluded for some other Jurassic polyphagans which have been insufficiently studied, but which are presumably related to modern Scirtoidea, Byrrhoidea and Artematopodidae. No aquatic polyphagan larvae have as yet been recorded except possibly for *Larvula cassa* Ponomarenko whose systematic position as well as ecology are unclear. Probably, Jurassic aquatic polyphagans were ecologically diverse and included predators, algal-feeders and scavengers that inhabited a wide range of habitats.

The megalopterans are rare in the Jurassic (an undescribed larva is found in Shar Teg in Mongolia) and probably existed as a minor, perhaps mainly lotic group of predators. Some neuropterans, e.g. osmyloids, could have lotic larvae as well, but at present there is no positive fossil data supporting this suggestion.

The only modern mecopteran family with aquatic larvae, Nannochoristidae, enters the fossil record in the Early Jurassic (NOVOKSHONOV 1997a). Nannochoristid wings are not too rare in the Jurassic of Siberia in lacustrine (especially in Kubekovo) as well as in fluvial deposits (e.g. in Kempendyay). The larvae very similar to those of present-day nannochoristids have been discovered in the lacustrine deposits in the Ichetuy Formation of Transbaikalia. They are very rare there and so could be transported from running waters. Modern nannochoristid larvae are lotic predators feeding mainly on chironomid larvae. No Jurassic nannochoristids are known from Central Asia and Europe. The liassophilids related to Nannochoristidae survived in the Early Jurassic; their biology is unknown.

The caddisflies are represented in the Jurassic mainly by the family Necrotauliidae, catch-all family supposedly developed, at least in some subtaxa, in streams as algal feeders like the present-day Hydroptilidae. Other families are found as well, including the primitive Phryganeina appearing in the Late Jurassic. The necrotauliids are not rare even in the European marine deposits. The oldest caddis cases have been discovered in the Late Jurassic of Siberia (SUKATSHEVA 1985b), Mongolia (SUKATSHEVA 1994) and North America (HASIOTIS *et al.* 1997, A. Rasnitsyn, pers. comm.). Their appearance indicates the minimal age of the caddisflies colonisation of standing waters, as well as the possible switch of their feeding habit (because the case-building caddisfly larvae are usually algal-feeders or detritivores and not carnivores as most Hydropsychina). However, a lentic mode of life and algal-feeding cannot be excluded for some Triassic necrotauliids, as discussed above. The caddis cases occur in diverse palaeoenvironments, both fluvial

(e.g. in some layers of the Togo-Khuduk Beds in Bakhar, Mongolia) and lacustrine. In particular, they are abundant in some lacustrine black shales (Bol'shoy Koruy in Transbaikalia, and the Ortsag Beds in Bakhar, Mongolia) originated probably from abundant algal materials in eutrophic conditions (ZHERIKHIN & KALUGINA 1985). They often occur together with abundant aquatic bugs (Khoutiyn-Khotgor and the Bayan-Uul Beds in Bakhar, Mongolia, and the Morrison Formation of North America) though their distribution is less wide. Some types of lacustrine deposits (e.g. mountain lake sediments) never contain caddis cases within sequences that are of undoubtedly Jurassic age.

Since the Early Jurassic the aquatic Diptera become extremely common, turning out to be numerical dominants in many oryctocoenoses. Chaoboridae (Fig. 489) and Chironomidae are especially abundant. Their larvae are rare as fossils (or, most probably, their remains are hardly recognisable and mainly overlooked), but pupae are quite common in many lacustrine deposits indicating that both families were well represented in some types of Jurassic lentic ecosystems. The Jurassic chaoborid larvae were nectobentic or, more rarely, nectopleustic predators, mainly planktivorous, and a high chaoborid abundance usually suggests a high density of zooplankton (KALUGINA & KOVALEV 1985, ZHERIKHIN & KALUGINA 1985). The Jurassic Chironomidae belong to the subfamilies Ulaiinae (extinct), Podonominae and Tanypodinae, but only two former are represented by undoubtedly autochthonous lacustrine immature stages (KALUGINA & KOVALEV 1985). Both modern subfamilies prefer well oxygenated environments and have highly mobile benthic larvae diverse in feeding habit (algal-feeders, detritivores, predators, and herbivores including moss-feeders). Ulaiinae were probably similar ecologically with Podonominae. Both chaoborids and chironomids are virtually absent in oxbow lake deposits, and the finds of their immature stages are restricted in the Jurassic to sediments of larger lakes, mainly mountain, of the Siberian Region.

Immatures of other dipterans (Limoniidae, Eoptychopteridae, and Psychodidae) occur as occasional minor component in the Jurassic deposits. The limoniid pupae are found in deposits of some large Jurassic lakes including the brackish Karatau lake where it is likely that they were the only autochthonous lacustrine insects besides the corixid bugs. Pupae of the extinct psychodomorph family Eoptychopteridae have been discovered in small numbers in Sogyuty and Ust-Baley, making them the only dipterans represented by pre-imaginal stages in Jurassic oxbow lake deposits. The only known Jurassic psychodid larva, *Eopericoma zherichini* Kalugina, is found in a drifted assemblage and could be transported from a river (KALUGINA & KOVALEV 1985). Adults of various aquatic nematoceran families occur in diverse lacustrine and marine deposits in Siberia, Mongolia, China, Central Asia, Kazakhstan and Europe; many of them probably belong to lotic species. Among them Limoniidae, Chironomidae, Chaoboridae are most common and widespread and Eoptychopteridae are not uncommon, in particular in the European Jurassic; psychodids are extremely rare, and some additional families (Simuliidae, Dixidae) are found in a small number as well. There is no strong evidence for the existence of any aquatic brachycerans in the Jurassic, although several soldier flies are found in Karatau.

Diverse stonefly nymphs are remarkably abundant in Jurassic (especially Early and possibly Middle Jurassic) lacustrine deposits in Siberia, North Kazakhstan, Mongolia and North China. They often dominate numerically in oryctocoenoses over other aquatic insects, a nearly unique feature for the lacustrine insect assemblages.

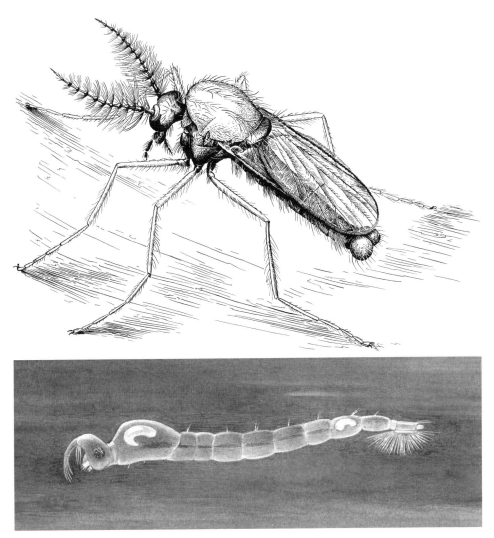

Fig. 489 Phantom midge, *Chironomaptera* sp. (Chaoboridae) characteristic of Jurassic and Early Cretaceous, Asian lakes (imago restored as in life by V.G. Kovalev, from ROHDENDORF & RASNITSYN 1980, and larva by A.G. Ponomarenko and Kira Ulanova, based in part on extant *Chaoborus* spp.)

This particularly concerns the nymphs of Mesoleuctridae (with unusually long and slender legs) and Platyperlidae (with unusually wide and flattened femora and tibia), which are widespread, mainly in sediments of oxbow and mountain lakes. In the latter environments Siberioperlidae are also common. The perlomorph genera *Perlisca* Sinitshenkova, *Perlomimus* Sinitshenkova and *Trianguliperla* Sinitshenkova occur there occasionally. As indicated by mouthparts structure, the platyperlid and other perlomorph nymphs were carnivorous while the mesoleuctrids and siberioperlids fed on plant debris, the siberioperlids being adapted to feeding on very coarse and durable plant tissues, perhaps on horsetail stems rich in silica. The peculiar leg morphology suggests that both mesoleuctrids and platyperlids differed from any modern stonefly nymphs in locomotory adaptations and probably inhabited the soft muddy bottom of the lake. In sediments of large lowland and piedmont lakes, stonefly nymphs are much rarer but represented by the same families, and partly by the same genera. In the Central Asian and European localities only adults (mainly of Perlariopseidae and Baleyopterygidae) occur sporadically in a small number, though some Central Asian sites originated in floodplain palaeoenvironments like Siberian localities rich in mayflies. This distribution pattern suggests that in warmer areas stoneflies could not colonise standing waters, probably due to lower oxygen levels in the water. If so, the stonefly abundance in Siberian lacustrine localities should indicate high oxygen levels, not only in cool-water mountain lakes but also in warmer oxbow lakes.

Finally, the enigmatic family Chresmodidae is worthy of mention. The chresmodids are large insects with narrow body and long slender legs superficially resembling the large-sized water striders; their appearance suggests that they could live on the water surface. Their taxonomic position is unclear as well as the feeding habit; different authors treated them as gerromorph bugs, stick-insects or possible relatives of Grylloblattida (MARTÍNEZ-DELCLÒS 1989). A number of specimens have been collected in Solnhofen and it is possible that chresmodids could colonise seas like the modern marine gerrid bugs. However, an undescribed specimen was also found in the lacustrine Late Jurassic deposits of Bakhar in Mongolia, some distance from any sea.

3.3.5.2. Aquatic insect assemblages and their typology

The following types of Jurassic aquatic insect assemblages can be recognised.

1. Assemblages dominated by the moderately large-sized stoneflies (mesoleuctrids and platyperlids) and/or by mayflies and containing a variety of other insects but nearly devoid of aquatic bugs, dipterans and caddisflies (the *Mesoleuctra-Mesoneta* assemblage by ZHERIKHIN 1985, the assemblage type B by SINITSHENKOVA & ZHERIKHIN 1996). This type is widespread in the temperate Siberian Region and perhaps occurs in some Gondwanaland areas (India) as well, but is absent in warmer areas. ZHERIKHIN & KALUGINA (1985) described it in detail on the basis of Siberian material and proposed a hypothetical reconstruction of the trophic structure of the lake ecosystem. The diagram here (Fig. 490) is based on their data. This assemblage type is restricted to the fine-grained sediments of small oxbow lakes in heavily forested river valleys. Numerous small crustaceans of an uncertain taxonomic position as well as bryozoans and fish accompany the insects. There are also diverse well-preserved terrestrial plant and winged insect fossils. Both taxonomic composition of the aquatic fauna and the

morphology of the odonatan rectal gills suggest high oxygen content and low saprobity, but in more coarse-grained layers, corresponding to an increasing connection with the river flow the aquatic insect fauna becomes impoverished, with the most oxyphilous taxa disappearing. In coal-bearing layers formed at later stages of the lake succession the characteristic insect fauna disappears and is replaced with a community dominated by filter-feeding bivalves and conchostracans.

The presence of a fauna intolerant to oxygen deficiency and high saprobity is unusual for shallow temperate lowland lakes. It indicates a peculiar palaeoenvironment lacking any close present-day analogues. ZHERIKHIN & KALUGINA (1985) refer to this ecosystem type as hypotrophic one and suppose that the oxygen consumption was low because of suppressure of the microbial life by a strong antibiotic effect of the forest leaf fall. They pointed out that the leaves of present-day *Ginkgo* exhibit such effects very clearly. It was also suggested that, unlike the present-day dystrophic peat lakes, the hypotrophic lakes were at most, slightly acid (as indicated by the presence of bivalve molluscs), and that the hypotrophy can be defined as a non-acid dystrophy. PONOMARENKO (1996) has criticised this model: he believes that the lakes were poor in nutrients and that the excellent state of insect preservation suggests oxygen deficiency in the near-bottom water. However, his hypothesis is in controversy with the abundance of small crustaceans and the presence of algal sapropelite coals in the sequences. At present, only the Zherikhin & Kalugina model seems to

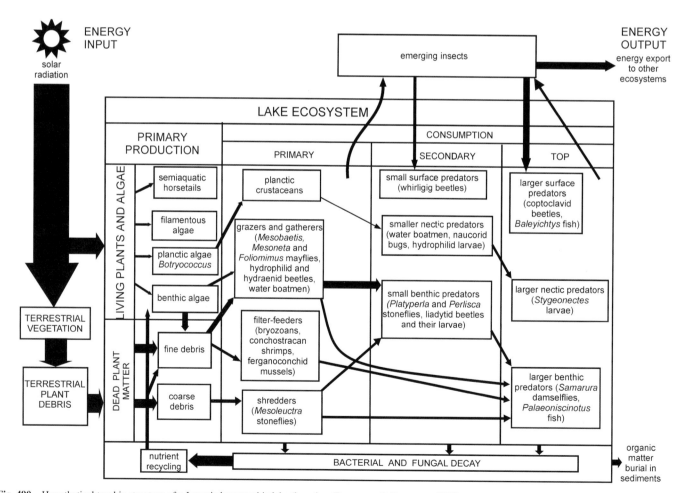

Fig. 490 Hypothetical trophic structure of a Jurassic hypotrophic lake (based on ZHERIKHIN & KALUGINA 1985)

explain the paradoxical decline of the most oxyphilous taxa under increasing input of river flow, which should dilute the lake water reducing the antibiotic concentration in it.

2. Assemblages dominated by aquatic dipterans (Chaoboridae and in a lesser degree Chironomidae) and also contained a variety of other aquatic insects, but almost no aquatic bugs or caddisflies (the assemblage type A after SINITSHENKOVA & ZHERIKHIN 1996). This assemblage type is also described by ZHERIKHIN & KALUGINA (1985) and is restricted in the Jurassic to Transbaikalia where it occurs rather commonly in the volcanic-sedimentary sequences in fine-grained lacustrine deposits with high volcanic ash content. The taxonomic composition differs in different sites. ZHERIKHIN (1985) recognised the two main types, the Early to Middle Jurassic *Mesoleuctroides-Dinosamarura* assemblage and the Middle to Late Jurassic *Stackelbergisca-Siberioperla* assemblage. Both of them, and particularly the latter one, may require further subdivision, particularly to allow for those oryctocoenoses that completely lack particular common species. Nevertheless the principal ecosystem structure seems to be basically similar. In the Ichetuy Formation some mayfly and aquatic beetle species are common with the hypotrophic lakes described above, but other taxa are different. Besides the abundance of dipterans the presence of large-sized siberioperlid stoneflies and the absence of platyperlids is particularly characteristic. The aquatic invertebrates other than insects are uncommon and represented mainly by cladocerans and bryozoans. The ostracods, conchostracans and bivalve molluscs are restricted to few coal-bearing layers where aquatic insect remains are absent. The fish are constantly missing. Plant remains are scarce, except for the horsetails and sometimes mosses; the sapropelite coals are absent. The habitat is interpreted as clear, cool, deep, moderately large oligotrophic dam lakes with a low autochthonous primary production, not very different from present-day oligotrophic mountain lakes. The main difference is that now the most important shredders in such lakes are caddisflies, and not stoneflies. In some sites the peculiar chaoborid genus *Hypsocorethra* Kalugina with the highly specialised larval mouthparts (adapted to feeding on aerial insects fallen on to water surface) is common. Its presence perhaps indicates ultraoligotrophic conditions with extremely low zooplankton density (ZHERIKHIN & KALUGINA 1985).

3. Assemblages strongly dominated by aquatic bugs (mainly by Corixidae or Shurabellidae) with few if any immature stages of other aquatic insects. This assemblage type occurs in diverse sedimentary contexts. In Central Asia and probably in China they replace the stonefly and mayfly dominated assemblages in lacustrine (probably oxbow-lake) lentils within the coal-bearing fluvial sequences. The thin-shelled bivalve molluscs, conchostracans and occasionally unidentified small crustaceans may be abundant. The obvious impoverishment of the fauna in comparison with physiographically similar Siberian habitats may be explained by a lower dissolved oxygen content in warmer climatic conditions. In Eastern Transbaikalia and North America the corixid dominated assemblages occur in coastal plain environments, probably in sediments of brackish pools and lagoons with high algal production. Similar communities probably existed on many European islands but have not yet been discovered here in autochthonous oryctocoenoses. The Karatau lake in Kazakhstan was surely saline (DOLUDENKO *et al.* 1990) but situated inland and lacking any connection with sea. Besides the water boatmen, this large lake was inhabited by craneflies as well as by several species of molluscs, crustaceans and fishes.

4. Assemblages co-dominated by the aquatic bugs and dipterans (mainly Chaoboridae) also containing diverse benthic insect fauna including the mayflies, stoneflies, isophlebiid odonatans, caddis cases

and aquatic beetle larvae and adults. This assemblage type is discovered in Khoutiyn-Khotgor in Mongolia in dark-coloured and extremely finely-laminated deposits (so-called "paper shale") together with conchostracans and several fish species. The stonefly, mayfly and isophlebiid nymphs are relatively uncommon and probably subautochthonous as well as the caddis cases; this hypothesis is supported by rather numerous finds of caddisfly and especially isophlebiid adults. In another Mongolian locality, Bakhar, the sedimentologically similar deposits (Bayan-Uul Beds after SINITZA 1993) contains the same insect assemblage except for lacking both mayflies and stoneflies. In both localities the insect-bearing sediments were accumulated in out-shore areas of large lakes, most probably in anoxic conditions. The dominant corixids and chaoborids were nectic and inhabited water-column. The coptoclavid larvae were also active swimmers but their finds are surprisingly rare. The mayflies, stoneflies, odonatans and caddisflies developed probably in the shallow-water, near-shore areas, inhabited by diverse benthic community. The available data are too scarce to reconstruct it in detail. The considerable abundance of fish and adult isophlebiids indicates a complex trophic structure with numerous top predators and accordingly a high productivity level of the lake ecosystem. Perhaps some features resembled the Early Cretaceous pseudooligotrophic lakes discussed below. However, the corixids are more abundant and chaoborids less so in comparison with that lake type. Other important differences are the presence of stonefly nymphs and rarity of caddis cases. Thus the idea about possible basic similarity between the ecosystems of large Jurassic (at least, Late Jurassic) and Early Cretaceous lakes (ZHERIKHIN & KALUGINA 1985) still needs in proof.

5. Assemblages dominated by caddis cases. This assemblage type occurs very rarely in the Jurassic and seems to be connected with at least two different palaeoenvironments. In Bol'shoy Koruy in Transbaikalia *Scyphindusia* cases are common in the blackish siltstone and accompanied with bivalves, conchostracans and rare remains of adult corixids and aquatic beetles. The lake was probably highly productive, possibly eutrophic, with abundant algal growth (ZHERIKHIN & KALUGINA 1985). In Mongolia, undoubtedly Jurassic assemblages dominated by caddis cases have been discovered in several outcrops and layers in Bakhar (SUKATSHEVA 1994). In the top member of the Ortsag Beds *Terrindusia ochrotrichoides* Sukatsheva is discovered in number in the carbonaceous siltstone lacking any other insect fossils or ichnofossils (site 208, cyclite 5 after SINITZA 1993); the coal material in that siltstone may also be of algal origin. Like in Bol'shoy Koruy the cases closely resemble those of the present-day Hydroptilidae. The modern hydroptilids are algal-feeders as probably were the *Scyphindusia* constructors. Hence the palaeoenvironment was probably similar to the Bol'shoy Koruy Lake though the absence of bivalves and crustaceans is noteworthy. Some layers in the lower part of the Ortsag Beds section contain numerous caddis cases as well, but both the case ichnospecies and the sediments are very different from the top member of the formation. SINITZA (1993) interprets the palaeoenvironment of this part of the section as being alluvial fans and hydrologically unstable lakes, strongly affected by temporarily watercourses; the cases could thus be at least partly allochthonous, transported from running waters. This assemblage also contains some hydroptilid-like cases (*Terrindusia scelerata* Sukatsheva) found together with several other ichnospecies including some *Folindusia*. The latter could be constructed by shredders fed on terrestrial plant debris. The cases from the Brushy Basin Member of the Morrison Formation in Colorado buried in floodplain ponds were probably transported from a river during episodes of flooding (HASIOTIS & DEMKO 1996a).

6. Assemblages dominated by adult aquatic beetles. This type is clearly heterogeneous and, as in the Triassic, occurs in diverse palaeoenvironments. In many cases the oryctocoenoses had probably originated as a result of drift when other insect remains have been destroyed. The Jurassic assemblages differ from the Triassic ones in the relative rarity of schizophoroids, and predominance of hydradephagan or hydrophiloid elytra. ZHERIKHIN (1985) included this type to his *Memptus-Dzeregia* assemblage which is clearly artificial, from both a taxonomic and ecological point of view. PONOMARENKO (1985b) analysed the stratigraphic and geographic distribution pattern of both terrestrial and aquatic beetles in the Jurassic of South Siberia and adjacent territories and demonstrated that it is highly complicated and cannot be interpreted in a simple manner. WEGIEREK & ZHERIKHIN (1997) found that the Early Jurassic aquatic beetle assemblage from the nonmarine deposits of the Holy Cross Mountains in Poland is dominated by *Hydrobiites* Heer (the supposed hydrophilid elytra), as do many Siberian assemblages, but differs from all of them in a high frequency of schizophoroids. Also the hydradephagan (probably liadytid) elytra classified as *Memptus* Handlirsch seem to be less common in Poland than they usually are in Siberia.

7. Assemblages with diverse aquatic insects lacking any well-pronounced dominance pattern. This type is almost surely heterogeneous and includes several assemblages found in sediments of large and medium-sized lowland and piedmont lakes in Siberia (Kubekovo) and Mongolia (Shar Teg, the Togo-Khuduk and Ortsag Beds of Bakhar). The aquatic stages of dipterans and aquatic bugs are also often represented but are never very abundant; mayfly, stonefly and odonatan nymphs and caddis cases can be present or absent, and the share and taxonomic composition of aquatic beetle fauna varies considerably. The aquatic insects are always significantly less abundant than terrestrial ones. In both Shar Teg and Bakhar insect fossils are collected at a number of outcrops from layers corresponding to different stages of physiographic evolution of the basin (SINITZA 1993; GUBIN & SINITZA 1996). Unfortunately, in all cases the collections have only been partly studied taxonomically, and distribution can only be analysed for some species and genera. The common feature of all those assemblages is probably the predominance of subautochthonous insects that dwelled in lake areas (most probably, in shallow waters) other than the area of their burial. In this respect they resemble the above-mentioned Khoutiyn-Khotgor assemblage, but the low abundance or even absence (as in Kubekovo) of nectic bugs and dipterans suggest a very different ecosystem structure. The Bayan-Teg assemblage in Mongolia is not very well fit this type for formal reasons because of the dominance of a stonefly species, *Plutopteryx beata* Sinitshenkova. However, here again burial is almost surely subautochthonous, with extremely abundant adults but relatively rare nymphs. Interestingly, another species of the same genus was discovered in Kubekovo but represented only by few adults. Other aquatic insects in Bayan-Teg seem to be rare, and no nectic taxa have been discovered. It seems doubtful that those assemblages could originate from any of the lake types described above. For both Kubekovo and Bayan-Teg the ecosystem structure has not been reconstructed in detail. ZHERIKHIN & KALUGINA (1985) treated the Kubekovo lake as a "slightly hypotrophied basin" where the influence of the leaf-fall was weak because of a low concentration of the supposed antibiotics in the large water volume. They supposed that the fauna of the shallow-water near-shore areas was perhaps not very different from that in the typically hypotrophic lakes and that it simply could not colonise out-shore areas. This model seems to be highly speculative and requires a thorough analysis of the fossil assemblages.

8. Assemblages co-dominated by mayfly nymphs and caddis cases. This assemblage type is known only from Kempendyay in Yakutia. The age of the fauna is not established with certainty and may be either Late Jurassic or Early Cretaceous. For the present it is placed among the Jurassic assemblages mainly on the basis of the taxonomic composition of stoneflies (SINITSHENKOVA 1992b). The insects were collected from a thin mudstone lentil in a thick cross-bedded fluvial sandstone sequence, most probably formed in a backwater area of a lowland river of the potamal type. The mayflies and caddis cases are numerically abundant but it is likely that each represents a single species. Other aquatic insect stages include very rare stonefly nymphs and probably adult aquatic beetles. Except for stoneflies none of them are studied taxonomically. The aerial adults of aquatic insect taxa are common and diverse including dipterans (Limoniidae, Chironomidae, Eoptychopteridae, etc.), stoneflies, caddisflies and nannochoristid scorpionflies. It is plausible that most of them developed in the river though the possibility cannot be rejected that some species were restricted to smaller tributaries or floodplain ponds.

9. Assemblages including only flying stages of allochthonous (probably mainly lotic) aquatic insects; occur in various palaeoenvironments.

It is difficult to separate the potamal and rhithral species in the allochthonous assemblages. ZHERIKHIN & KALUGINA (1985) made an attempt to do so for the Siberian Jurassic faunas. They postulated that the adult insects found in the oxbow lake deposits of Western and Central Siberia should emerge mainly from large rivers of a potamal type, while those buried in the mountain dam lakes of Transbaikalia inhabited probably the rhithral zone. According to their analysis both potamon and rhithron were taxonomically and ecologically diverse in the Siberian Jurassic, and the rhithral biota was not very different from the present-day one. The same is likely to be true for Central Asia, but the material from this region has been never analysed carefully from this point of view. At present there is no ground for any hypotheses about most probable habitats of diverse European taxa found in the marine Jurassic.

The above typology is clearly a preliminary one. It does not cover a number of insufficiently studied assemblages that can hardly be included to any of the types established. For instance, an assemblage of marine pleistonic insects (now formed mainly by halobatine bugs) may have started in the Jurassic as represented by the enigmatic *Chresmoda* Germar and gerromorph bug *Engynabis* Bode.

In general, the significant taxonomic and ecological diversification of aquatic insects in comparison with the Triassic is obvious, especially in lentic ecosystems. A lot of recent families and even subfamilies have appeared for the first time in the fossil record though the share of the extinct Mesozoic taxa is great. Some Jurassic aquatic ecosystems (first of all, those of cool-water mountain oligotrophic lakes and perhaps also of the rhithral zone) were likely to be only slightly different from the modern ones. Others have no recent analogues and have to be considered extinct at the ecosystem level (SINITSHENKOVA & ZHERIKHIN 1996). PONOMARENKO's (1996) division of the Jurassic lacustrine ecosystems into two main types corresponding more or less to the above types 1 and 4 is oversimplified. In particular, the ecosystems of large lakes were surely diverse and by no means uniform.

3.3.6. CRETACEOUS

The Cretaceous aquatic insects are well represented in the fossil record in both the northern continents and Gondwanaland. The Early Cretaceous assemblages, both taphonomically autochthonous and allochthonous, are diverse and taxonomically rich. In general, the majority of families and some genera are common with the Jurassic faunas so that the age of some assemblages is doubtful. On the contrary, the autochthonous Late Cretaceous aquatic assemblages are strongly impoverished and rather uniform. Since the Cretaceous aquatic insects (mainly their winged stages) are known not only from compression fossils but as resin inclusions as well; those finds provide an important source of information, in particular on the lotic faunas.

3.3.6.1. Overview of taxa of aquatic insects

Among the mayflies, the hexagenitids, siphlonurids, and mesonetids are the most widespread families in the Early Cretaceous lacustrine assemblages; burrowing ephemeroid nymphs are also common in some localities. The data on supposedly lotic fauna are scarce. In comparison with the Jurassic no extinction at the family level is documented but some families appear which are unknown from older deposits. Most of them are modern but the ephemeroid family Australiphemeridae is confined to the Cretaceous. In habitat distribution and feeding habits the Early Cretaceous and Jurassic mayflies seem to be generally similar. *Promirara* Jell et Duncan from the Early Cretaceous of Australia, which have no close relatives in the Jurassic faunas, were probably carnivorous (JELL & DUNCAN 1986). The numerical dominance of Hexagenitidae in many localities is noteworthy when compared with their general rarity in the Jurassic. In the Late Cretaceous deposits mayfly nymphs become extremely rare, though flying stages are well represented in fossil resins. The Australiphemeridae is the only extinct family survived in the Late Cretaceous. None of the Late Cretaceous nymphs is described or even identified up to the family level. Their occasional finds in the lacustrine Cenomanian (Obeshchayushchiy) and floodplain Maastrichtian (Arkhara) deposits of the Russian Far East (undescribed) as well as in the marine Turonian lithographic limestone of the Hvar Island, Croatia (HANDLIRSCH 1937) are probably allochthonous. Several poorly preserved moulting skins were discovered in the Turonian oxbow lake deposits within the fluvial Timmerdyakh Formation in the Vilyui River Basin, Jakutia. Their taphonomic autochthonicity may be doubted, too. About a dozen nymphs were present in a large insect collection from the oxbow lake deposits in Kzyl-Zhar in Southern Kazakhstan. They seem to represent one or two undescribed species and should inhabit the lake, but it is likely that the population density was very low. The *Asthenopodichnium* Thenius burrows in fossil wood known from lagoon and fluvial Late Cretaceous deposits of Central Asia and Russian Far East are assigned to ephemeroid nymphs (SINITSHENKOVA 2000a).

The Early Cretaceous odonatan fauna is much more distinctive. The role of heterophlebioids and isophlebioids decreases considerably in comparison with the Jurassic, the radiation of modern anisopteran lineages of Libellulina is evident, and Calopterygina enter the fossil record for the first time. The Meganeurina represented by rare pseudomyrmeleontids are still surviving. Only adults are known for the majority of taxa. As in the Jurassic, the autochthonous lacustrine assemblages are poor in species. Even when nymphs occur in number they belong to one or a few species while adults are often diverse. It means that for the most part, the species developed outside lakes, perhaps mainly in running waters. High adult diversity in floodplain, deltaic and other non-lacustrine freshwater deposits (e.g. in the Purbeck and Wealden of England) support this hypothesis. Like in the Jurassic, there is an obvious decrease in total odonatan diversity from

the warmer to more temperate areas. In the Early Cretaceous this pattern can be observed not only in Eurasia but also in Gondwanaland (Brazil and South-eastern Australia).

The odonatan nymphs found in the Early Cretaceous belong to the isophlebioids, heterophlebioids, anisopteran Libellulina, and Calopterygina. The isophlebioid nymphs are restricted to the assemblages of disputable Jurassic or Cretaceous age though isophlebiid wings occur occasionally in undoubtedly Early Cretaceous deposits. The *Samarura*-like heterophlebioid nymphs were recorded (as alleged zygopterans) from the Early Cretaceous of Spain (NEL & MARTÍNEZ-DELCLÒS 1993) and also occurs in the Bon-Tsagan Series in Mongolia. True Calopterygina nymphs are known from the Early Cretaceous of Australia (JELL & DUNCAN 1986). The anisopteran nymphs are most widespread. The current state of their taxonomy is unsatisfactory, even familial placement often being uncertain. BECHLY (1998) regards nearly all Early Cretaceous anisopteran nymphs as representing the same lineage and placed them in the Aeschnidiidae. However, he has probably misinterpreted some characters (e.g. the mask structure in *Hemeroscopus* Pritykina: see 2.2.1.2.2) and the polarity of others (e.g. the shape of the paraproct). In any case, the Early Cretaceous anisopteran nymphs surely represent several different ecological types as is clearly evident from their leg structure. The long legs of the nymphal Hemeroscopidae and Sonidae possess a fringe of long hairs interpreted as a swimming adaptation, which is absent in any other known odonatans and so suggests a necto-benthic or even nectic mode of life (PRITYKINA 1977, 1986). Short-legged burrowing gomphoid nymphs are recorded from the Crato Formation of Brazil (MARTILL 1993, BECHLY 1998) and found also in some sites in Asia (e.g. Semyon in Transbaikalia; undescribed) but always in small numbers. The nymphs of *Nothomacromia* Carle from the Crato Formation have long slender legs devoid of the swimming pubescence and represent the third, benthic ecological type. CARLE & WIGHTON (1990) proposed the separate family Pseudomacromiidae (subsequently referred to as Nothomacromiidae) and believed it was lentic, but BECHLY (1998) has subsequently transferred the genus to the Aeschnidiidae and has doubted its lentic mode of life. The taxonomic position and biology of several Chinese genera, also placed to Aeschnidiidae by Bechly, is unclear, and BECHLY (1998) also mentions additional undescribed material from China. Undescribed nymphs from the Early Cretaceous of North Korea demonstrate an elongate ovipositor. This suggests that they probably belong to the Aeschnidiidae, but distant from *Nothomacromia*. Libelluloid- and aeshnid-type nymphs have been recorded from the Early Cretaceous of Spain (MARTÍNEZ-DELCLÒS 1991, NEL & MARTÍNEZ-DELCLÒS 1993) and libelluloid ones from Australia (JELL & DUNCAN 1986). Mask structure has been studied in only a few Early Cretaceous taxa, so the preferred prey type cannot be established for the majority of them.

Relatively few odonatan fossils are known from the Late Cretaceous. All of them appear to represent Cainozoic families except for a single aeschnidiid wing from the marine Cenomanian of the Crimea. No isophlebioid or heterophlebioid adults are found. The Calopterygina seem to be commoner than they were in the Early Cretaceous, at least in the latest Cretaceous (Tsagayan Formation of the Russian Far East, Maastrichtian). Odonatan nymphs are found in small numbers only in Obeshchayushchiy, Kzyl-Zhar and Hvar together with mayflies. All of them are probably allochthonous; in Kzyl-Zhar nymphs of both anisopteran Libellulina and of Calopterygina have been discovered as well as a poorly preserved nymph with short, broad leaf-shaped caudal appendages resembling those of *Samarura*.

Aquatic bugs are common, widespread and diverse in the Early Cretaceous. The only Jurassic family not found in the Early Cretaceous is Shurabellidae. Gerroidea appeared for the first time in the Cretaceous. In general, the ecological types were the same as in the Jurassic, and the abundance of bugs (particularly corixids) and other aquatic insects often remain negatively correlated. However, this was never the case for the Notonectidae which were always accompanied with rich aquatic fauna. It should be noted that the latter family, always rare in the Jurassic, becomes abundant in some lacustrine deposits in Transbaikalia (Zaza Formation), China (Chijiquao, Huihuibao, Xiagou, Yixian, Laiyang, Laoqun, Shouchang, and Zhuji Formations) and Argentina (La Cantera Formation). Bugs occurred sporadically in the Late Cretaceous aquatic and were rarely abundant. They belonged to the Mesotrephidae in Kzyl-Zhar (confined to the Late Cretaceous), Notonectidae (Qujiang Formation of the Zhejiang Province, China), Corixidae (Juezhou Group of Southern China, the Tsagayan Formation of the Amur River Basin), Belostomatidae (Turov Beds of Sakhalin Island), and undescribed large gerroids in the Senonian of the Anadyr' River Basin in the Russian Far East.

The Early Cretaceous aquatic beetle fauna demonstrates only minor differences from the Late Jurassic one both in its taxonomic composition and its ecological spectrum. The beetles occur in various sedimentary contexts, occasionally without any other aquatic insect remains. The schizophoroids become very rare though they still occur occasionally up to the Albian or Early Cenomanian deposits of Labrador in Canada. Aquatic adephagans are represented mainly by the same groups as in the Jurassic (except Parahygrobiidae which are very rare even there). Only Haliplidae appear for the first time in the above-mentioned Labrador fauna. Coptoclavidae (including the highly advanced subfamily Coptoclavinae that have not been found in the Jurassic) are often numerically dominant, at least in Asiatic faunas, and Gyrinidae are widespread and relatively common while Liadytidae seem to be rare. It should be noted that no liadytid larvae are found in the Cretaceous. Dytiscidae are still rare and represented by adults only. The aquatic polyphagans are represented by the same families as in the Jurassic, with Hydrophilidae being the most common and widespread. Not only adult hydrophilids but also their larvae are sometimes abundant. In particular, the genus *Cretotaenia* Ponomarenko originally placed to Adephaga is probably a hydrophilid larva (Chapter 2.2.1.3.2.1). The polyphagan larvae from the Early Cretaceous of Australia were assigned to the Scirtidae (=Helodidae) (JELL & DUNCAN 1986). This family is probably represented among undescribed Jurassic beetle fossils but no larvae have been discovered in the Jurassic.

In the Late Cretaceous the schizophoroid beetles (except the Albian-Cenomanian of Labrador), Coptoclavidae and Liadytidae are completely absent. Some undescribed elytra found in the Maastrichtian Tsagayan Formation are likely to belong to the leaf-beetle subfamily Donaciinae (Chrysomelidae) which feed on aquatic and semiaquatic angiosperms. This is the first appearance of aquatic chrysomelids in the fossil record. Other families are common with the Early Cretaceous and usually occur in small numbers; even hydrophilids seem to be infrequent. No Late Cretaceous aquatic beetle larvae are known.

The megalopterans are represented mainly by corydalid larvae occasionally common in the Early Cretaceous lacustrine deposits (*Cretochaulus* Ponomarenko in the Zaza Formation and in Chernovskie Kopi in Transbaikalia); the same or a closely allied genus is found also in the Cenomanian (Obeshchayushchiy). Adults occur in small number in the Zaza Formation only. One young larva (*Chauliosialis sukatshevae* Ponomarenko) is found in the Santonian amber-like resins of Yantardakh, North Siberia. The record of alleged corydalid eggs from the Campanian of France (SCUDDER 1890b) needs to be confirmed (ZHERIKHIN 1978).

Some Early Cretaceous neuropterans (in particular, osmyloids) could have had lotic larvae but there are no fossil finds of their immature stages. A newly born sisyrid larva has been recorded from Siberian Santonian resins (ZHERIKHIN 1978).

The aquatic mecopterans are known from the Early Cretaceous by rare finds of adult Nannochoristidae in Transbaikalia and their numerous larvae in the lacustrine deposits of Koonwarra in Australia (JELL & DUNCAN 1986). This is the latest find of any aquatic scorpionflies in the fossil record, though several nannochoristid genera are still extant in the temperate zone of the southern hemisphere where they develop in small forest creeks.

The Cretaceous caddisfly fauna is highly distinctive, and rapid evolutionary changes are well documented in this order. Necrotauliidae and other Hydropsychina are relatively rare (except for the Purbeck and Wealden in England where they are numerical subdominants). Only adults are known in the Cretaceous, and the extant families Polycentropodidae and perhaps Psychomyiidae appear for the first time there. Adult Phryganeina are common and diverse in the Early Cretaceous and belong to several families, partly extinct (Dysoneuridae, Baissoferidae, Vitimotauliidae survived from the Jurassic), partly extant and not found earlier (Phryganeidae, Lepidostomatidae, Calamoceratidae, Plectrotarsidae, Helicophidae, and perhaps also Sericostomatidae). The Vitimotauliidae is the numerically dominant family. Common finds of teneral vitimotauliid adults in lacustrine deposits of Transbaikalia and Mongolia demonstrate that some species of the family were surely lentic. As many as 10 to 15 species can be found in the same locality, for example, Baissa in Transbaikalia. However, vitimotauliid wings are also common in the Early Cretaceous floodplain and deltaic deposits of England. The dominant genera are different (*Multimodus* Sukatsheva in Siberia and Mongolia and *Purbimodus* Sukatsheva in England). Dysoneuridae are rare in Siberia and Mongolia but common in England. The phryganeids constitute a minor component in both regions, and Lepidostomatidae, Calamoceratidae, Plectrotarsidae and Helicophidae are known from the English faunas only (SUKATSHEVA 1999). The caddisfly faunas of the Early Cretaceous of Spain and the Crato Formation of Brazil are also diverse but undescribed and their taxonomic composition is unknown.

The caddis cases are very abundant and diverse in different palaeoenvironments in Siberia and Mongolia (probably except the very beginning of the Early Cretaceous where relatively few finds are known), but not recorded from China. Undoubtedly, the case-bearing caddisflies of the suborder Phryganeina become the most important lacustrine shredders and algal grazers in the Early Cretaceous of Northern Asia. However, the caddis cases are very rare in other regions such as Europe, South America and Australia. They have been discovered in numbers only in the English Purbeck, but confined to very few layers there. The taxonomic position of the case makers cannot be identified with certainty, but there is little doubt that the majority of them should belong to Vitimotauliidae. In the localities where both cases and adults are common (Baissa, Bon-Tsagan), their diversity is well comparable suggesting that different case ichnospecies were usually created by different caddisfly species. The cases occur in various non-marine sedimentary environments, both lacustrine and fluvial, and many ichnospecies are recorded from a number of localities. However, with very few exceptions, each species is restricted to certain palaeo-environment indicating strong habitat specialisation.

PONOMARENKO (1996) points out that the sedimentation surfaces show virtually no traces of crawling or dragging associated with cases. He hypothesised that at least in large lakes with fine-grained bottom sediments case-building caddis larvae should therefore have lived mainly on floating algae. However, in this case it is difficult to explain why the distribution of different ichnospecies within the same section differs so clearly, with some of them confined to coarser and others to fine-grained beds, to either coal-less or coal-bearing layers, and so on. Besides this many cases are built entirely of mineral grains, bivalve shell fragments and other materials which could hardly be easily available at floating algal balls or rafts.

Relatively few case types are known in the earliest Cretaceous. Later their diversity increases continuously up to the Albian (SUKATSHEVA 1980, 1982). Some case types (such as the ichnogenus, *Conchindusia* Vialov et Sukatsheva, constructed from conchostracan shells) are restricted to the later Early Cretaceous. In the Late Cretaceous adult caddisflies are nearly disappearing from the assemblages of compression fossils. The latest vitimotauliids (*Multimodus bureensis* Sukatsheva) have been discovered in the latest Early or early Late Cretaceous fluvial deposits of the Kyndal Formation in the Urgal River Basin (Russian Far East), and Philopotamidae and Phryganeidae occur occasionally in floodplain and oxbow lake deposits (SUKATSHEVA 1982). On the contrary, in Late Cretaceous fossil resins adult caddisflies are rather common and diverse. Two families (Electralbertidae and Taimyrelectronidae) are confined to the Late Cretaceous, all others being extant. Several of these are unknown or are doubtfully recorded from older faunas (Hydroptilidae, Hydrobiosidae, Sericostomatidae, Odontoceridae, Calamoceratidae, Leptoceridae).

The Late Cretaceous lacustrine case assemblages in Asia, Europe and North America demonstrate a dramatic decline in diversity. Nearly all Late Cretaceous lacustrine ichnospecies are constructed from either small plant debris or spirally arranged conifer needles, while other types are virtually confined to fluvial deposits (SUKATSHEVA 1991a). Many Early Cretaceous lacustrine assemblages include as many as 10–15 or even 20 ichnospecies each, and sometimes more than ten ichnospecies co-occur in the same layer. In contrast, the maximal number of ichnospecies known in a Late Cretaceous assemblage is 5. According to the data obtained in the Russian Far East, the decline began in the Albian, and the Late Albian assemblages are as poor in ichnospecies as the Late Cretaceous ones (SUKATSHEVA 1991a, SUKATSHEVA in GROMOV *et al*. 1993).

The dominant families of aquatic dipterans were the same in the Early Cretaceous as in the Jurassic. In Siberian, Mongolian and Chinese lacustrine deposits Chaoboridae are even more abundant than in the Jurassic of the same regions, and dozens of specimens can often be seen at the same rock slab. However, their diversity is always low, with only one or, rarely, two or three species found in each locality. Chironomidae were somewhat less abundant but much more diverse. Both chaoborids and chironomids are represented mainly by adults and pupae but this is surely because the larvae are usually poorly preserved and hardly recognisable. The adults and pupae of Limoniidae are also widespread and rather common. Eoptychopterid pupae are found in numbers only in floodplain deposits in Jakutia, though the adults are widespread. Other nematoceran families are of minor importance. In West Europe nematoceran pupae occur in the lacustrine Early

Cretaceous deposits in Spain whereas in the floodplain and deltaic sediments in England yielding diverse adult fossils their pre-imaginal stages are rare. In the lacustrine Early Cretaceous of Koonwarra in Australia a dozen simuliid larvae have been found in addition to adult and immature Chaoboridae, Chironomidae and Limoniidae (JELL & DUNCAN 1986). These finds are rather enigmatic because the simuliids are specialised filter-feeders well adapted to running water and never live in lakes. Perhaps this is one of those very few cases where lotic insects drifted down to a lake and were well-preserved in lacustrine deposits. In the Early Cretaceous resins from England, Spain and Lebanon, chironomids are the most common aquatic Diptera followed by Ceratopogonidae, Psychodidae and Eoptychopteridae (the latter family has been recorded only from Lebanon). These records of Ceratopogonidae, as well as those from Austrian fossil resin, are the oldest. However, these tiny midges are not easily recognisable when buried in sedimentary rocks, so their apparent absence from the Jurassic sediments is not surprising. The only important Early Cretaceous novelty concerning the aquatic dipterans is the appearance of aquatic brachyceran larvae. Numerous larvae of Stratiomyidae and supposed Sciomyzidae are recorded from Montsec in Spain (WHALLEY & JARZEMBOWSKI 1985), though allocation of the latter is highly doubtful because of the total absence of adult Cretaceous acalyptratans. A small number of undescribed asilomorph Diptera larvae have been collected in the Early Cretaceous of Transbaikalia (Baissa, Semyon). At least four asilomorph families with partially aquatic larvae, Stratiomyidae, Tabanidae, Athericidae and Empididae, are represented among the adult brachycerans from the Early Cretaceous, but independent evidence of their aquatic development is absent.

The Late Cretaceous fossils of the immature aquatic dipterans are very rare. In the Cenomanian lacustrine deposits of Obeshchayushchiy in the Russian Far East, few poorly preserved nematoceran pupae and several large asilomorph larvae have been discovered (besides the undescribed material kept in the Palaeontological Institute, Moscow, an additional specimen was observed personally in the collection of the Ten'ka Geological Prospecting Expedition, Ust'-Omchug, Magadan Region, Russia). Among the adult dipterans occurring in Late Cretaceous sedimentary deposits. the only relatively common partially aquatic family is Limoniidae. The only extinct Cretaceous aquatic dipteran family, Eoptychopteridae, is not found in the Late Cretaceous. The Late Cretaceous fossil resins contain numerous adults of Chironomidae and Ceratopogonidae, but not Chironominae which are characteristic for eutrophic environments (KALUGINA 1980b). Psychodidae seem to be common in warmer areas but very rare in North Siberian resins. Limoniidae, Simuliidae and Chaoboridae constitute minor components of the resin faunas. Among other dipteran families recorded from the Late Cretaceous, Stratiomyidae and Empididae now include some species with aquatic larvae.

The stoneflies seem to decline in the Early Cretaceous but it may well be a matter of taphonomy. Indeed, nymphs are found in only a few localities and finds of adults are sporadic, but the total stonefly diversity at the family level increases. A number of living families first appeared there, and none of the Jurassic families disappeared. All Jurassic ecological types of stonefly nymphs have persisted as well. The stonefly decline may be partially explained by competition with the case-building caddisflies of the suborder Phryganeina, as many taxa in both groups are shredders. However, the carnivorous perlomorph stoneflies demonstrate the same pattern, and the detritivorous stoneflies occur together with caddis cases in some localities. No stoneflies are recorded from the Late Cretaceous.

Finally, Chresmodidae still occur in a small number in lacustrine deposits in the Early Cretaceous of Europe (*Chresmoda aquatica* Martínez-Delclòs and *Chresmoda* sp. from Spain) and Asia (*C. orientalis* Esaki from China). *Saurophthirodes* Ponomarenko from Western Mongolia (Gurvan-Ereniy-Nuru) is possibly a chresmodid nymph. Unlike the Jurassic, no chresmodid has been recorded from the marine Early Cretaceous.

Hence the Early Cretaceous aquatic insect taxa, other than odonatans, stoneflies and caddisflies, demonstrate little change in comparison with the Jurassic. Near the Early/Late Cretaceous boundary a large-scale extinction is evident, in particular in the lacustrine insect fauna.

3.3.6.2. Aquatic insect assemblages and their typology

The typology of the Cretaceous aquatic insect assemblages meets various problems. First of all, no quantitative data are available for many important localities, in particular for a number of rich Chinese, European, and South American faunas. This hinders their accurate comparison with the Siberian and Mongolian assemblages. The taxonomic composition of the latter is more variable than in the Jurassic indicating that any comparison based on few taxa can be inaccurate. In particular, some types of faunas established previously on the basis of few characteristic taxa (the *Ephemeropsis-Coptoclava* fauna after ZHERIKHIN 1978) includes in fact several Siberian and Mongolian assemblages with very different ecological structure (SINITSHENKOVA & ZHERIKHIN 1996, SINITSHENKOVA 1999b). The assemblages with *Ephemeropsis* and *Coptoclava* designated as the Jehol fauna by Chinese authors (e.g. HONG 1998) and distributed up to the Anhui and Zhejiang Provinces in Southern China, may not be uniform ecologically, and the published data are not complete enough to permit their satisfactory ecological classification. The age of some important assemblages is disputable even in terms of Cretaceous *vs.* Jurassic. But at least the Late Cretaceous assemblages are very different from the Early Cretaceous ones.

The following types of assemblages can be preliminarily recognised.

1. Assemblages dominated by stoneflies and mayflies and containing diverse other insects but nearly devoid of aquatic bugs and dipterans. This type seems to correspond to the Jurassic Type 1, but differs primarily in the presence of caddis cases and the absence of the heterophlebioid odonatans. ZHERIKHIN & KALUGINA (1985) predicted that assemblages similar to the Jurassic *Mesoleuctra-Mesoneta* faunas should have survived into the Early Cretaceous in small floodplain lakes strongly affected by ginkgoalean and czekanowskialean leaf-fall. Recently this prediction was confirmed with the discovery of the Early Cretaceous insect fauna at the Chernovskie Kopi near Chita, Transbaikalia (the *Siphangarus-Trianguliperla* assemblage after SINITSHENKOVA 1999b). Both stonefly and mayfly nymphs are common and diverse and demonstrate close relations to the Jurassic faunas (SINITSHENKOVA 1998a, 2000d). The hexagenitid mayflies are less abundant than the siphlonurids. Some mayfly genera are common with the Jurassic Type 1 assemblages (*Mesobaetis*, *Mesoneta*, *Siberiogenites*) while others either occur in other types of the Jurassic assemblages (*Stackelbergisca*) or are endemic (*Siphangarus* Sinitshenkova). One specimen is tentatively assigned to the genus *Proameletus*, which is not found in undoubtedly Jurassic deposits. Two filter-feeding species of the genus *Clavineta* Sinitshenkova may be

allochthonous. The mayfly diversity is remarkably high in comparison with any Jurassic assemblages. Except for *Clavineta*, all mayflies were probably plant debris collectors and/or algal-feeders.

The stonefly nymphs include two endemic genera of Siberiopelidae and Mesolectridae and two species of the perlomorph genus *Trianguliperla* Sinitshenkova, widespread since the Late Triassic. Neither siberioperlids nor *Trianguliperla* occur in the Jurassic Type 1 assemblages but the mesolectrids are very characteristic of it. All stoneflies except the predacious *Trianguliperla* were shredders. Stonefly adults are rare and none of them could be associated with any nymphs; no mayfly flying stages were discovered. Other relatively common aquatic insects are carnivorous larvae of a small-sized corydalid and several undescribed caddis case ichnospecies. An incomplete dragonfly larva and an abdomen of a nepomorph bug were found. The aquatic larval stages of dipterans are absent, but a chaoborid wing is present in the collection. No beetle larvae were collected though some adult remains may belong to hydrophilids. Fish scales, small conchostracans and thin-shelled bivalves occur occasionally. No aquatic higher plants are recognised. Though the aquatic fauna is rather of a Jurassic appearance, the terrestrial insects and in particular the susumaniid stick-insects and the homopterans indicate the Early Cretaceous age (V. Zherikhin, pers. comm.; D. Shcherbakov, pers. comm.). According to S. Sinitza (pers. comm.) the material was collected in a tectonically isolated block and the stratigraphic relations of the insect-bearing layers are difficult to establish. The rock is fine-grained and contains abundant plant fossils including numerous ginkgoalean leaves. The available data are not in conflict with the hypothesis that the assemblage represents the fauna of a small hypotrophic lake strongly affected by ginkgoalean leaf-fall. However, the detritivores are surely more diverse than in the Jurassic hypotrophic assemblages while the predator guild is numerically scarce and poor in species. These peculiarities are difficult to interpret yet.

2. Assemblages dominated by mayfly and dragonfly nymphs but lacking both aquatic dipterans and stoneflies. This assemblage type is represented by the Late Cretaceous (Turonian) locality Kzyl-Zhar in Southern Kazakhstan. Aquatic insects are very rare and mostly undescribed. The collection kept at the Palaeontological Institute in Moscow includes a dozen poorly preserved mayfly nymphs belonging to two species (one represented by the single specimen) and about 20 odonatan nymphs (mainly dragonflies but also a specimen of Calopterygina and a *Samarura*-like nymph). The minor components are the endemic aquatic bug *Mesotrephes* Popov, adult beetles (the dytiscid *Cretodytes* Ponomarenko, the gyrinid *Cretotortor* Ponomarenko and an undescribed hydrophilid) and the *Terrindusia* cases. There are also few libellulid dragonfly (*Palaeolibellula* Fleck, Nel et Martínez-Delclòs) and philopotamid caddisfly wings, fragmentary fish and crustacean remains and diverse terrestrial insect fossils. The insects were collected in a mudstone lens within alluvial sandstone and conglomerates containing abundant angiosperm (mainly platanoid) leaves. ZHERIKHIN (1978) misinterpreted the locality as having originated in lagoon palaeoenvironments. The aquatic insects were occasional visitors of an oxbow lake connected with a river rather than its permanent inhabitants; some of them could be carried by flooding. In particular, the *Terrindusia* cases are otherwise restricted to fluvial deposits in the Late Cretaceous. The terrestrial insect remains are often fragmentary and seem to be transported by currents.

In several Early (the Batylykh Formation) and Late (the Timmerdyakh Formation) localities in Jakutia small and incompletely preserved mayfly nymphs were found in small numbers in siltstone and mudstone lentils within fluvial sequences. There are no odonatan nymphs in the small collections available, but the possibility cannot be excluded that the assemblages may belong to the same type.

3. Assemblages dominated by aquatic dipterans (Chaoboridae and to a lesser degree Chironomidae) and diverse mayflies but poor in aquatic bugs. This type is discovered in temperate areas of both Asia and Gondwanaland. It resembles the Jurassic Type 2 in its ecological structure and sedimentary context but differs mainly in the presence of caddis cases (except for the Subtype 3C). The assemblages occur mainly in finely laminated tuffaceous siltstone and mudstone accumulated in still-water areas of rather large lakes situated in mountain or foothill volcanic areas. Terrestrial plant remains are always scarce. The fauna includes a number of benthic oxyphilous taxa suggesting a low saprobity level. The terrestrial fossils often indicate rather cool climatic conditions (except for the Subtype 3C). Other important features of the assemblage are the rarity (at best) of nectic predatory insects except the zooplanktivorous chaoborids, and a low total insect diversity (about 30–50 species). Unlike in the Jurassic Type 2, the insects are often accompanied by abundant and diverse crustacean fauna. The ecosystem seems to be similar to the present-day mountain oligotrophic lakes. This type can be subdivided into several subtypes.

3A. Assemblages with isophlebiid odonatans. This subtype is especially similar to the Jurassic Type 2 not only in ecological structure but also in the taxonomic composition. It is distributed in Eastern Transbaikalia and known from a number of sites, mainly from the Glushkovo (in the Unda-Daya Depression) and Ukurey (in the Olov Depression) Formations (the *Proameletus-Samarura* fauna after ZHERIKHIN 1978, the *Proameletus*-Isophlebiidae assemblage after SINITSHENKOVA 1999b). The insufficiently known fauna of the Ust'-Kara Formation in the Ust'-Kara Depression may belong to the same subtype. Outside Transbaikalia assemblage of the same type is discovered in Khutulyin-Khira in Central Mongolia (SINITZA 1993). The geological age in all cases is disputable between the Late Jurassic and Early Cretaceous, and the data on different insect groups are controversial in this respect (Chapter 4.2, see also RASNITSYN 1988b, 1989a, 1990c, SINITZA 1993). Both aquatic and terrestrial insect fauna is remarkably similar to the Jurassic Transbaikalian faunas, but at the generic and suprageneric levels only, while some species are common with various Early Cretaceous localities. Presence of highly advanced caddis cases types is more characteristic to the mid-Cretaceous (Albian and Cenomanian) faunas (SUKATSHEVA 1982, 1990b, ZHERIKHIN & SUKATSHEVA 1990), but there is an alternative explanation (Chapter 2.2.1.3.4.3f). The presence of several modern stonefly families not found elsewhere in the Mesozoic suggests that the assemblage may be of a post-Neocomian (Aptian or even Albian) age. So the choice between the mid-Cretaceous age of the Glushkovo assemblage and its position near the Jurassic/Cretaceous boundary is a matter of further researches (Chapter 4.1).

The supposedly oxyphilous taxa include the mayflies *Proameletus* Sinitshenkova, the podonomine chironomids and stoneflies (all having the present-day oxyphilous relatives) and perhaps also the mayflies *Epeoromimus* Tshernova, the ulaiine chironomids and the isophlebiid odonatans. The predominant Transbaikalian mayfly genera *Proameletus* and *Furvoneta* Sinitshenkova are almost restricted to this assemblage; *Epeoromimus* occurs in a few sites as only a minor component. In Khutulyin-Khira the only mayfly species is *Mesobaetis mandalensis* Sinitshenkova. All mayflies should be mainly algal grazers because plant debris had to have been scarce. In Transbaikalia the most common aquatic dipterans are several chironomid genera belonging to

Podonominae and Ulaiinae. Not only adults but also pupae and larvae have been discovered in fine-grained deposits. The most common genera *Ulaia* Kalugina and *Oryctochlus* Kalugina occur also in the Jurassic Type 2 but differ at the species level; their larvae were probably algal-feeders. Chaoborids are less abundant and represented by the endemic genera *Mesocorethra* Kalugina (pupae, adults and few larvae found in the Ukurey and Ust'-Kara Formations) and *Baleiomyia* Kalugina (only adults are known in the Glushkovo Formation). *Mesocorethra* was a zooplanktivore, and the same is well possible for *Baleiomyia*. The dipterans from Khutulyin-Khira are unidentified. The shredder guild was probably dominated by caddisworms. Their cases are common though not diverse; *Folindusia undae* Vialov et Sukatsheva, *F. necta* Sukatsheva and *F. savinensis* Sukatsheva dominate numerically in the Glushkovo Formation and *Terrindusia cf. splendida* in the Ust'-Kara Formation. In Khutulyin-Khira both *F. cf. undae* and *Terrindusia* sp. occur together. The most probable adults of the case-builders (Vitimotauliidae and Dasyneuridae) are rare. The large-sized isophlebiid odonatans are common in all Transbaikalian and Siberian sites except Ust'-Kara and represented both by nymphs and adults; in absence of fish they were the top predators in the ecosystem like *Dinosamarura* in the Jurassic Type 2.

Stoneflies are uncommon but diverse in Transbaikalia and include both predators (*Perlitodes* Sinitshenkova, *Dipsoperla* Sinitshenkova, *Savina* Sinitshenkova) and shredders (*Lycoleuctra* Sinitshenkova and probably *Uroperla* Sinitshenkova and *Flexoperla* Sinitshenkova). *Savina* may be taphonomically allochthonous. A peculiar feature of the stonefly fauna is the diversity of nymphs adapted to locomotion in narrow spaces, perhaps between pebbles and boulders (*Flexoperla, Lycoleuctra, Dipsoperla*). Probably they lived mainly in near-shore areas; this may explain the rarity of their finds. *Flexoperla* nymphs are occasionally abundant but only in relatively coarse-grained deposits where other insects (including the caddis cases) are nearly absent. No stoneflies were collected in Khutulyin-Khira. The aquatic beetles in Transbaikalia are uncommon and belong to the predacious Liadytidae (two species of *Liadytes* Ponomarenko) and algal-feeding Hydrophilidae (four genera with four species). However, the single incomplete remains of a beetle larva from the Glushkovo Formation belong to a coptoclavid. In Khutulyin-Khira adult coptoclavids (*Coptoclava* sp.) are not rare. In Transbaikalia some aquatic insects known from adults only probably represent the fauna of rapid mountain creeks (the necrotauliid caddisflies, the limoniid, simuliid and thaumaleid dipterans).

The presence of notostracan (*Prolepidurus* Tshernyshev) and occasionally anostracan (*Prochiracephalus* Trusova) crustaceans is a characteristic feature not observed in the Jurassic. *Prolepidurus* is especially widespread occurring in all localities including Khutulyin-Khira. The conchostracans are moderately abundant in some layers only and poor in species. No fish occur in any localities. The plant fossils are rare and represented mainly by the semiaquatic horsetails. The principal ecosystem structure seems to be the same as in the Jurassic Type 2 assemblages. A strongly impoverished terrestrial insect fauna with abundant Trichoceridae in the Glushkovo Formation suggests cool climatic conditions in the vicinity of the lakes (ZHERIKHIN 1978). The Ukurey lakes were probably situated at a lower altitude, Khutulyin-Khira – at lower latitude: this may explain some differences in their aquatic fauna. The fauna of the locality Polosatik in the Ust'-Kara Depression is somewhat enigmatic, differing not only from the faunas of other regions of Transbaikalia but also from the fauna of the type section of the Ust'-Kara Formation. In Polosatik abundant chaoborid pupae, tentatively assigned to *Mesocorethra*, occur together with copto-clavid larvae identified as *Coptoclava longipoda* Ping (unrecorded from any other localities of this type), a nymph of a *Platyperla* stonefly and abundant caddis cases of the endemic ichnospecies *Terrindusia fulgida* Sukatsheva.

3B. Assemblages with non-isophlebioid odonatans. The assemblage from the Early Cretaceous (probably Aptian) Koonwarra Fossil Beds in Victoria, Australia (JELL & DUNCAN 1986) differs in some important features from the subtype A but can be assigned to the same Type 3. The supposed oxyphilous taxa are the mayflies *Promirara* Jell et Duncan and *Australurus* Jell et Duncan, the stoneflies, and the nannochoristids. The dominant mayfly genus *Australurus* is habitually similar to *Mesobaetis* and *Proameletus* and could be algal grazer and/or faculta-tive debris collector as well as two rarer siphlonurid genera are. Other possible benthic detritivores are relatively uncommon larvae of the scir-tid beetles (not found in any other Cretaceous faunas) and limoniid craneflies. The most important shredders were stoneflies (*Dinotoperla* Jell et Duncan) while caddisworms only occur in small numbers. Both zooplanktivorous chaoborids (*Chironomaptera collessi* Jell et Duncan; the generic placement is somewhat doubtful) and algal-feeding benthic tanypodine chironomids are common. The main difference from Subtype A is the relatively high diversity of the non-planktivorous predatory insects. The guild of benthic predators includes the nymphs of endemic mayflies (*Promirara*) and odonatans (both anisopteran Libellulina and Calopterygina) as well as the nannochoristid larvae. The supposed nectobenthic predators (the dytiscoid and hydrophiloid beetle larvae and rare nepomorph bugs) are less numerous though diverse. Finally, the uncommon pleustic mesoveliid *Duncanovelia* Jell and Duncan occurring in small numbers is the only gerromorph bug recorded from any Type 3 assemblages. If the presence of other preda-tory animals (crustaceans and fishes) is also taken into account it becomes evident that the trophic structure of the ecosystem was more complex than in Subtype A, suggesting a somewhat higher productivity.

3C. Assemblages without any odonatan nymphs. This subtype is discovered in Khutel-Khara in South-eastern Mongolia in the inter-basalt lacustrine deposits assigned to the Tsagan-Tsab Formation. However, the fauna has nothing in common with any other Tsagan-Tsab localities. The supposed oxyphilous taxa are represented by two stonefly genera. An undescribed chaoborid represented by numerous pupae and adults is the numerical dominant; unidentified chironomids also present but seem to be less common. Some larger pupae can belong to limoniid craneflies. The mayfly fauna is diverse and includes 6 species of 5 genera (SINITSHENKOVA 1989). The burrowing *Torephemera* nymph may be allochthonous but other genera probably inhabited the lake. The hexagenitid *Siberiogenites* and the endemic siphlonurid genus *Albisca* are common while *Mesoneta*, *Epeoromimus* and the endemic leptophlebiid *Leptoneta* occur in small numbers. All autochthonous mayfly species were probably algal-feeders and facul-tative debris collectors. The stonefly nymphs are not rare and belong to the genera *Uroperla* (a shredder) and *Trianguliperla* (a predator) (SINITSHENKOVA 1987). No caddis cases are found. The coptoclavid beetles are represented by adults only; the same species *Coptoclavella minor* Ponomarenko is discovered in Transbaikalia in the assemblages 3A and 4 (PONOMARENKO 1990). Both odonatan nymphs and aquatic bugs are completely absent. Conchostracan crustaceans occur occa-sionally, no fish are found. The assemblage is exceptional in extremely impoverished predacious insect fauna. All material has been collected in a single mudstone layer deposited probably in a deep-water area of the lake. Perhaps a more diverse insect community dwelled the

shallow-water zone. The lake probably had rather cold- and clear-water, but none of the terrestrial insects indicate cool climatic conditions comparable with the Subtype 3A.

4. Assemblages numerically dominated by aquatic dipterans (Chaoboridae and in a lesser degree Chironomidae) and *Ephemeropsis* mayflies and containing few if any aquatic bugs (the assemblage type D after SINITSHENKOVA & ZHERIKHIN 1996; the *Ephemeropsis trisetalis–Mesogyrus striatus* assemblage after SINITSHENKOVA 1999b). ZHERIKHIN (1978) included this type to his *Ephemeropsis-Coptoclava* faunas. However, it is completely different from other *Ephemeropsis-Coptoclava* assemblages at the species level (except *Coptoclava longipoda* Ping that may be in fact a species group rather than a single species), and the ecological structure is quite distinct. The type is dis-tributed in the volcanic-sedimentary Early Cretaceous formations of Eastern Transbaikalia (the Ghidari Formation of the Urov River Basin but probably also in the Mangut Formation of the Onon Depression and the Turga Formation in the Turga Depression). Perhaps some Chinese assemblages (e.g., from the Dabeigou Formation of Hebei) belong to this type but this cannot be established with certainty on the basis of published data. The Chinese assemblages differ in the absence of caddis cases and are discussed below. The type seems to be inter-mediate between the Type 3 and Type 5 assemblages both in the taxo-nomic composition of the fauna and in the ecosystem structure. Like the Type 3, the assemblage occurs almost exclusively in fine-grained tuffaceous deposits. The lakes were probably moderately deep and situated in foothill areas at lower altitude than the basins with the 3A Subtype assemblages (SINITSHENKOVA & ZHERIKHIN 1996).

The numerically dominant taxa include the hexagenitid mayfly genus *Ephemeropsis* Eichw., the chaoborid genus *Chironomaptera* Ping (both common with the Type 5), and the chironomid genus *Ulaia* Kalugina (common with the Subtype 3A). Other genera in common with Type 3 are *Liadytes* and *Uroperla*, which never occur in the Type 5 assemblages. The chaoborid density in oryctocoenoses is comparable with that of the Type 5 assemblages reaching several dozens/dm^2 while in the Subtypes 3A, 3B and 3C it never exceeds 10 specimens/dm^2. The supposed oxyphilous taxa include *Ulaia*, *Uroperla*, and the corydalid genus *Cretochaulus* Ponomarenko, with only the former being com-mon. The giant mayfly *Ephemeropsis trisetalis* Eichw. is abundant and represents the principal benthic algal-grazer (perhaps, facultative debris collector) in the ecosystem. Any other mayflies are never pres-ent. *Ulaia* and some caddisworms could be important benthic algal-feeders as well. The shredders are not numerous and include *Uroperla* and diverse caddisworms. The high abundance of the zooplanktivore *Chironomaptera* suggests high zooplankton density. The carnivorous nectic coptoclavid larvae are uncommon; other carnivores include the rare *Cretochaulus* and *Liadytes*. Odonatan nymphs and aquatic bugs are completely absent. The gyrinid beetle *Mesogyrus striatus* Ponomarenko occurs regularly though in small numbers and seems to be characteristic for the assemblage. In general, the predatory insects are very scarcely represented; the main predators in the ecosystem were probably the osteoglossomorph fishes. The filter-feeding crus-taceans are rare to moderately abundant and represented by ostracods and large-sized conchostracans. The high abundance of chaoborids and filter-feeding crustaceans as well as low frequency of the supposedly oxyphilous taxa indicate that the ecosystem should be more productive than in the Type 3, perhaps mesotrophic (SINITSHENKOVA & ZHERIKHIN 1996). On the other hand, the trophic system was rather simple, with very few high level predators, suggesting that the productivity was considerably lower than in the Type 5 assemblages.

5. Assemblages dominated by hexagenitid mayflies and chaoborid midges and containing diverse other insects including abundant aquatic bugs. High abundance of nectic and nectobenthic taxa is the most characteristic feature of this assemblage type. In particular, the density of chaoborid remains is usually much higher than in the Type 3. The diversity of oxyphilous benthic taxa is low. Stoneflies and mayflies other than hexagenitids never occur. The type has no close analogues among the Jurassic faunas discussed above and never occurs in latest Early Cretaceous or younger deposits. ZHERIKHIN & KALUGINA (1985) predicted that similar communities should inhabit large Jurassic lakes as well but they still remain unknown from undoubtedly Jurassic deposits. However, it should be noted that in China most strata with chaoborids and Ephemeropsis were traditionally assigned to the Late Jurassic. Though recent palaeontological (e.g. angiosperm and bird finds) and radiometric data usually confront this view, the result of isotope dating of the Dabeigou Formation in Hebei Province (152.3 Myr) is well before the Jurassic/Cretaceous boundary (ZHANG 1992). The assemblage occurs in lacustrine deposits but the sedimentary environments are rather diverse. Terrestrial plant remains are rarely abundant in the insect-bearing layers. The characteristic large conchostracans of the genus Bairdestheria Jones, diverse ostracods and the osteoglossomorph fish Lycoptera often occur in the same deposits, and occasionally other conchostracan and fish genera accompany the assemblage. This type is only known with certainty from Asia. The abundance and diversity of benthic insects decreases continuously as we move south, and in accordance with this general trend three subtypes can be recognised. Similar assemblages could also exist outside Asia, but the available data are too scarce to be certain. Nymphs of the hexagenitid mayfly Hexameropsis selini Tshernova et Sinitshenkova occur in mass in the Early Cretaceous of Ukraine, and the coptoclavid larva Coptoclava africana Teixeira closely resembling Asiatic C. longipoda Ping is discovered in the Early Cretaceous of Angola.

5A. Assemblages with Cretochaulus and diverse caddis cases (assemblage type C after SINITSHENKOVA & ZHERIKHIN 1996; the Ephemeropsis melanurus–Hemeroscopus baissicus assemblage after SINITSHENKOVA 1999b). The subtype is known from the Early Cretaceous deposits (the Khysekha and Zaza Formations) of Vitim Plateau in Northern Transbaikalia. The Lagerstätte Baissa is the most representative locality known though other smaller and less studied sites in the same region are probably similar (ZHERIKHIN et al. 1999). The richest insect oryctocoenoses are restricted to finely laminated marl layers within rather thick lacustrine sequences; the tuffaceous siltstones constitute a small part of the section, and the bituminous paper shales are widespread but yield virtually no insects, except poorly preserved chaoborids and rare caddis cases. The sediments were accumulated in rather large and deep inter-mountain lakes with stratified limnion (SINITSHENKOVA & ZHERIKHIN 1996). The diversity of aquatic insects is high; their total species number in Baissa can be conservatively estimated between 75 and 110 with no fewer than 50 to 80 of them being lentic autochthones. Zooplanktivorous chaoborids are extremely abundant and represented by no less than two undescribed species indicating a very high density of planktic crustaceans which, however, are only occasionally represented directly in the oryctocoenoses. Nectic predators are abundant and include numerous larvae of the beetle Coptoclava longipoda Ping (Fig. 491) and the notonectid bug Clypostemma xyphiale Popov (Fig. 492). The common corixid bug Diapherinus should also be carnivorous, though facultative algal-feeding cannot be excluded. The nectobenthic Hemeroscopus

(Fig. 493) is the only autochthonous odonatan species and occurs in mass. Other nectic predators (e.g., belostomatid bugs and dytiscid beetles) are rare. Though the feeding strategy of nectic and nectobenthic carnivorous insects as well as of Lycoptera fish was most probably opportunistic, the enormously abundant chaoborids provided the main food resource available for them.

The benthic insect fauna is diverse. The nymphs of Ephemeropsis melanurus Cock. as well as the larvae of common and rather diverse chironomids and caddisflies were likely to be the principal benthic algal-feeders; adult hydrophilids probably fed on both benthic and floating algae. The benthic larvae of the dobsonflies (Cretochaulus) and hydrophilids (Cretotaenia) were carnivorous, and the Hemeroscopus nymphs probably fed partially on benthic organisms. There are also other, rare benthic insects (e.g. naucorid bugs and asilomorph larvae). Adult Coptoclava (Fig. 493) and rare gyrinids and mesovelioid water striders were pleustic, feeding mainly on aerial insects on the lake surface; adult chaoborids probably constitute an important part of their diet. The filter-feeding Bairdestheria and bivalve molluscs as well as rather rare grazing gastropods seem to be restricted to shallow-water coastal areas with abundant algal growth. Ostracods are uncommon. The top predator in the ecosystem was a large sturgeon-like fish Stychopterus occurring in small numbers. The common presence of Cretochaulus larvae and rather high diversity of caddis case ichnospecies in marl layers suggest high oxygen availability in epilimnion. The possibility of cases being transported on floating algal mats should be excluded because in the bituminous shales they are much less numerous, and are represented mainly by a single ichnospecies uncommon in marl layers. There is no evidence of any mass mortality episodes caused by suffocation suggesting that the upper boundary of the anoxic hypolimnion was situated much below the extreme limit of wind and wave action. The ecosystem probably combined high productivity level with low saprobity in a manner quite unusual for the modern lakes (ZHERIKHIN 1978, SINITSHENKOVA & ZHERIKHIN 1996). To explain this SINITSHENKOVA & ZHERIKHIN (1996) postulated a peculiar turnover type called pseudo-oligotrophic. In absence of any submerged higher plants and limited input of terrestrial plant debris, planktic and benthic algae were the only important primary producers in the ecosystem. Their total annual production was high enough to support a complex trophic structure; however the algal standing crop at any time was strongly limited by a very high consumption rate. As a result the ecosystem was of a grazing type, with the detritic food chains playing a minor role only, so that the productivity comparable with modern mesotrophic or slightly eutrophic lakes, was combined with low saprobity level comparable with oligotrophic basins. The hypothetical reconstruction of the trophic structure of the ecosystem proposed by ZHERIKHIN et al. (1999) is illustrated at Fig. 494.

5B. Assemblages with low caddis case diversity and devoid of Cretochaulus (the assemblage type E after SINITSHENKOVA & ZHERIKHIN 1996). This subtype seems to be widely distributed in the lacustrine Early Cretaceous deposits of Mongolia. The depositional environments are rather diverse, varying from relatively small lakes on alluvial fans up to large but probably shallow basins (SINITZA 1980, 1993, SINITZA in RASNITSYN 1986b). Insects occur mainly in laminated siltstones but the rocks are often rather poorly sorted, with admixture of sand grains and occasionally even small pebbles; the finely laminated bituminous mudstones also contain insect remains. The tuffaceous rocks occur only occasionally in some sequences. In Central, Southern and Eastern Mongolia the assemblages are similar to Subtype 5A in the taxonomic composition even at the species level, but

Fig. 491 Hypothetical scene of life in the Early Cretaceous pseudo-oligotrophic lake Baissa in Siberia: specialised nectic beetle larvae *Coptoclava longipoda* Ping preying on the giant mayfly nymph *Ephemeropsis melanurus* Cockerell; restored by A.G. Ponomarenko and V.I. Dorofeev (from ROHDENDORF & RASNITSYN 1980)

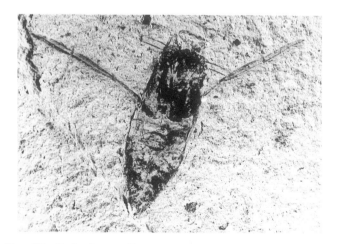

Fig. 492 Backswimmer *Clypostemma xyphiale* Popov (Notonectidae: Clypostemmatinae) from the Early Cretaceous of Baissa in Siberia (PIN 4210/6818, photo by D.E. Shcherbakov); body 19 mm long

less diverse and never include corydalids or notonectids. In Western Mongolian localities the taxonomic composition of the fauna is very different, but the general ecological structure seems to be nearly identical to other Mongolian assemblages. The zooplanktivorous chaoborids are always extremely abundant. They are only partly studied taxonomically so that their diversity is difficult to estimate; at least in some localities several species occur together, demonstrating an uneven distribution between the layers (for instance, two *Chironomaptera* species and *Manlayamyia litorina* Kalugina in Manlay, two *Chironomaptera* species and *Astrocorethra gurvanica* Kalugina in Gurvan-Ereniy-Nuru). Both larvae and adults of *Coptoclava longipoda* are common and widespread except for western faunas. Adults of another coptoclavid genus, *Coptoclavella* Ponomarenko, are common in Western Mongolia and occur as a minor component in Manlay together with *Coptoclava*; their larvae are unknown. The corixid bugs are unevenly distributed, sometimes being much more abundant than in Subtype 5A (e.g. *Diapherinus* Popov in Anda-Khuduk, *Velocorixa* Popov, Fig. 495. *Cristocorixa* Popov, *Bumbacorixa* Popov and *Corixonecta* Popov in different Western

Fig. 493 Hypothetical scene of life in the Early Cretaceous pseudo-oligotrophic lake Baissa in Siberia: specialised nectic predatory beetle *Coptoclava longipoda* Ping preying on the nectic dragonfly nymph *Hemeroscopus baissicus* Pritykina; restored by A.G. Ponomarenko and V.I. Dorofeev (from ROHDENDORF & RASNITSYN 1980)

Mongolian localities) but not found in some localities (e.g. in Manlay). Other aquatic bugs (common naucorid *Mongonecta* Popov and rare *Clypostemma*) are found in Western Mongolia. The nymphs of *Hemeroscopus* occur sporadically in Central Mongolia but are never so frequent as in the previous subtype. Sometimes only adult *Hemeroscopus* were found indicating that the nymphs developed somewhere outside the area of sedimentation; this situation never occurs in the assemblages 5A. In Western Mongolia *Hemeroscopus* is replaced by taxonomically distant but ecologically similar nectobenthic genus *Sona* Pritykina.

The benthic insect fauna seems to be impoverished. Chironomids are uncommon, *Cretochaulus* absent, and *Cretotaenia* very rare, though its possible adult stage, *Hydrophilopsia*, was discovered in several localities in Central Mongolia. The hexagenitids *Mongologenites laqueatus* Sinitshenkova (in Western Mongolia) and *Ephemeropsis*

melanurus Cockerell (in other areas) often occur in mass in shallow-water deposits, but disappear in more deep areas. The total number of caddis case ichnospecies is considerable but relatively few of them occur in each locality; *Terrindusia* cases usually dominate numerically over other ichnogenera. Only rarely up to dozen of ichnospecies occur in the same locality (e.g. in Shin-Khuduk: SUKATSHEVA 1982). On the contrary, the filter-feeding conchostracans, ostracods and bivalves are much more abundant than in insect assemblage 5A. The fish are represented by *Lycoptera* (often very common) and *Stychopterus* (rare). Mass mortality events are documented in some sites and can be explained by episodic suffocation.

PONOMARENKO & KALUGINA (1980) and PONOMARENKO (1986a) reconstructed the ecosystem structure in detail for Lake Manlay in Southern Mongolia and for Western Mongolian basins. Their model suggests that the lakes were stratified and hydrologically unstable, with

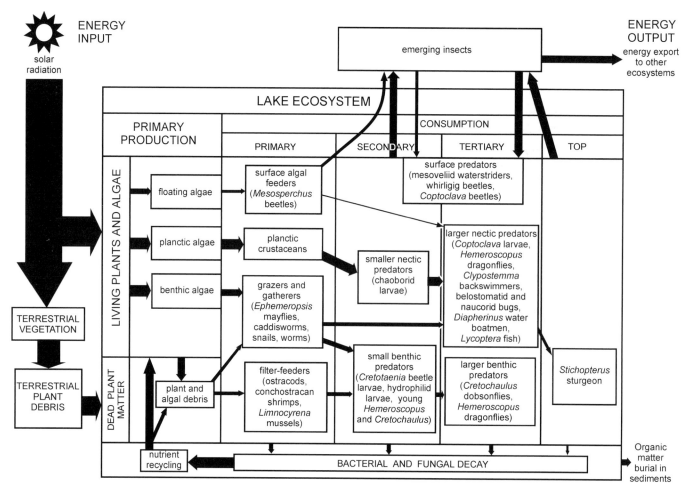

Fig. 494   Hypothetical trophic structure of an Early Cretaceous pseudo-oligotrophic lake (ZHERIKHIN *et al*. 1999)

benthic life strongly suppressed by oxygen deficiency in deep waters, and the frequent drying up of shallow-water areas. Many insects are interpreted as either active or passive (on floating algae) swimmers including even hexagenitid nymphs and caddisworms in the latter. The Western Mongolian lakes are supposedly treated as less productive in comparison with more eastern areas and are probably occasionally salinated. PONOMARENKO (1996) believes that this general model is nearly universal for large Early Cretaceous lakes. At first it seems to be very different from the pseudo-oligotrophic model described above. However, the controversy is in fact not so strong and the two models are not mutually exclusive (SINITSHENKOVA & ZHERIKHIN 1996). Indeed, the key differences between them are in the stability of the hydrological regime and the postulated thickness of the oxygenated epilimnion. Both characters would have been expected to differ between the more humid Transbaikalia and the semiarid Mongolia. In Mongolia, water level fluctuations were probably much more pronounced, indeed, making the shallowest areas less favourable for benthic organisms except for those that are particularly tolerant to periodical drying (as, for example, are conchostracans and ostracods). The different frequencies of archaegocimicid shore bugs that apparently inhabited muddy lake shores, further supports this model. This family is also common in Subtype 5A assemblages, but it never reaches an abundance there that is comparable with that of many Mongolian sites. In accordance with

Ponomarenko's model, the relatively sparse terrestrial vegetation in Mongolia could not have effectively prevented erosion, and the input of terrigenic particles into lakes was therefore probably large. As a result, the water would generally have been turbid, the thickness of photosynthetic zone reduced, and strong winds could have induced suffocation by deep water mixing. None of this was the case in Transbaikalia. Additionally, in Mongolian lakes the level of dissolved oxygen could have decreased to lower than those present in Transbaikalia because of the warmer Mongolian climate.

5C. Assemblages with *Ephemeropsis* and *Coptoclava* but devoid of caddis cases. This subtype is widespread in China and also occurs in some southern Mongolian localities (the Tsagan-Tsab Formation where the sole ichnospecies *Terrindusia sordida* Sukatsheva is found in only one of many Tsagan-Tsab localities, Tsagan-Suburga; SUKATSHEVA 1982). Both *Ephemeropsis* and *Coptoclava* are common in the Early Cretaceous in different parts of China but there are no records of caddis cases in the Chinese literature. Dr. E. Bugdaeva (pers. comm.) confirms that she could not find any caddis cases while collecting fossil plants at several important insect localities in China. Chinese authors often designate this fauna as the Jehol Fauna (e.g. S. WANG 1990, HONG 1992b, 1998, LIN 1994). No quantitative data are available on the main Chinese faunas, the published lists of taxa are probably often incomplete, and also, many taxa need revision. It is most likely that the Chinese faunas

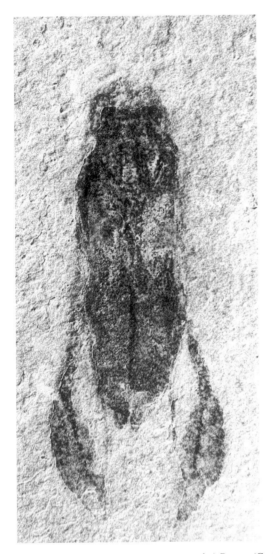

Fig. 495 Water boatman, *Velocorixa ponomarenkoi* Popov (Corixidae: Velocorixinae) from the Early Cretaceous of Myangad in Mongolia (paratype PIN 3152/2494, photo by D.E. Shcherbakov); body 8.6 mm long

under consideration are not uniform but the available information does not provide any firm basis for their further ecological classification. Chinese authors recognise several assemblage types, but mainly in accordance with their supposed stratigraphic distribution and not their ecology (LIN 1983, 1994, ZHANG 1992, HONG 1998). Those types are partly based on terrestrial or near-water insects and help little in the analysis of aquatic ecosystems.

In general, the Chinese assemblages seem to be rich in nectic and nectobenthic insects while benthic insects are scarcely represented. Chaoborids are surely very abundant, though their diversity is difficult to estimate because the generic placement is often doubtful. The aquatic bugs are often common and diverse. For instance, four corixid genera, two species of *Clypostemma* and two species of the endemic naucoroid genus *Schizopteryx* Hong were recorded from the Laiyang Formation of Shandong (ZHANG 1985, 1992, HONG & WANG 1990). Even if some genera and species are in fact synonyms, the total diversity is obviously high. *Schizopteryx* has long swimming legs and seems to be rather nectobenthic than benthic, its placing in the saldoid family

Mesolygaeidae (ZHANG 1991) is erroneous (Yu. Popov, pers. comm.). There are no naucorid records from other regions of China, but both Corixidae (placed to the genera *Jiangxicorixa* Hong, *Linicorixa* Lin, *Ratiticorixa* Lin, *Vulcanicorixa* Lin) and Notonectidae (mainly belonging to *Clypostemma*: Popov, 1989) are widespread. *Coptoclava longipoda* is common in many localities. Several genera of dragonfly nymphs including nectobenthic *Hemeroscopus* were described from different sites but their taxonomy and ecology is not clear enough. Most of them (including aeschnidiid nymphs mentioned by BECHLY 1998) were probably benthic. The mayflies are represented exclusively by *Ephemeropsis* distributed up to Southern China (Zhejiang, Anhui and Fujian Provinces) and generally common. One more important benthic group is Chironomidae. The distribution of both dragonfly nymphs and chironomid pupae seems to be highly uneven and sporadic: they have been never recorded from many strata with chaoborids, *Ephemeropsis* and *Coptoclava*. No other benthic insects are mentioned. Conchostracans, ostracods and bivalves are often no less abundant than in Mongolian faunas. Several fish genera are common including *Lycoptera*, *Stychopterus*, and several Chinese endemics. The archaegocimicid shore bugs of the genus *Mesolygaeus* Ping are extremely abundant in many localities. In general, the Chinese assemblages demonstrate further reduction of the lacustrine benthic insect fauna in comparison with the Subtype 5B. The simplest explanation is a warmer climate resulting in decreasing oxygen availability. If so, many taxa have to survive in running waters only. In particular, it may be predicted that caddis cases will be found in China in fluvial deposits like in Europe (see below, under Type 13). Indeed, rare caddisfly wings including vitimotauliids, were recorded from a number of Chinese localities. Some assemblages may in fact belong to the Type 4 because no aquatic bugs were recorded. However, none of them include caddis cases suggesting that the reduction of benthic fauna was universal in this region. In the Dabeigou Formation chaoborids, *Ephemeropsis* and *Coptoclava* are found together with thin-shelled bivalves and notostracans (YANG & HONG 1980); this combination occurs in no other areas.

6. Assemblages with locally abundant hexagenitid mayflies and diverse caddis cases. This assemblage type is known from several localities in Central Transbaikalia (Semyon near Chita, the Leskovo Beds in the Unda-Daya Depression). *Ephemeropsis trisetalis* is restricted here to very few bedding surfaces where it is highly abundant; sedimentologically these surfaces show no differences from other beds where no hexagenitids are found. Caddis cases are highly diverse; more than 20 ichnospecies can be previously recognised in Semyon and 6 ichnospecies were recorded from rather small collection made in the Leskovo Beds. Nectic taxa are scarcely represented. In Semyon chaoborids never occur in mass; in the Leskovo Beds aquatic dipterans were not found at all. The coptoclavids are absent from large collections from Semyon; in the Leskovo Beds a single larva and an elytron were collected. Aquatic bugs are represented by rare naucorids. Other autochthonous aquatic insects include rare burrowing gomphid nymphs and hydrophilid beetles in Semyon and supposed liadytid and hydrophilid elytra in the Leskovo Beds. The insects are accompanied by abundant conchostracans and ostracods. In Semyon a peculiar fish fauna is discovered, composed of the endemic genus *Irenichthys* Yakovlev and unusually abundant *Stychopterus*, while in Leskovo only fragmentary fish remains were found. Terrestrial plant remains are very abundant in Semyon but scarce in the Leskovo Beds. The assemblage is likely to represent a shallow-water insect community of a lake type different from any discussed previously. A reduction of the nectic

fauna, at least in Semyon, can be explained by high levels of fish predation. *Stychopterus* is much more common than usually also in the Utan Formation of the Olov Depression. The Utan insect assemblage cannot be formally included to this type because the caddis cases are uncommon and represented by two ichnospecies only. Like in Semyon, chaoborid pupae are only moderately abundant. No other aquatic insects were found here, but conchostracans and ostracods are common.

The diversity of assemblages with hexagenitids (types 4–6; cf. also Type 10) is intriguing. Probably either both *Ephemeropsis melanurus* and *E. trisetalis* were extremely opportunistic in the habitat or, more probably, each of them represent rather a species groups than single species. Even in Transbaikalia some localities with *Ephemeropsis* do not fit satisfactorily to any types described above. For instance in the Bukatshatsha Formation *E. melanurus* occurs in coal-bearing deposits together with thin-shelled bivalves and endemic caddis cases constructed from the oogonia of Characeae. Both sedimentary context and associated fossils are unusual for *Ephemeropsis*.

7. Assemblages with non-hexagenitid mayflies, abundant chaoborids and aquatic bugs (the *Shelopuga-Bolbonectes* assemblage after SINITSHENKOVA 1999b). This type resembles the Jurassic Type 4 in combination of non-hexagenitid mayflies and isophlebiids with abundant corixid bugs but differs in its absence of stoneflies. It is restricted to a small area at the extreme south-east of Transbaikalia and occurs only in the Byankino Formation of a disputable age, most probably similar to that of the Glushkovo assemblage (see subtype 3A above). The insects are preserved in finely laminated tuffaceous mudstones and siltstones, and the sedimentary context is similar to the deposits containing the 3A assemblages in other regions of Transbaikalia. On the other hand high abundance of nectic insects resembles rather the Type 5 assemblages. The numerical dominants are corixids (*Shelopuga sinitsae* Popov) and adults and larvae of coptoclavids (*Bolbonectes intermedius* Ponomarenko). Both genera occasionally occur in small numbers in some other Transbaikalian localities but are always very rare; both species are probably endemic. Another adult coptoclavid belonging to the subfamily Necronectinae is also found. Chaoborids are very rare; a single pupa tentatively assigned to the Jurassic genus *Praechaoborus* is found (KALUGINA 1993). Their rarity in combination with high abundance of nectic predators is enigmatic. Perhaps *Shelopuga* fed on benthic prey or even on algae like some modern opportunist-feeding corixids, and in turn provided the main food resource for *Bolbonectes*. Benthic insects are also unusual. The only mayfly species, the rather common *Bolbonyx ludibriosus* Sinitshenkova, belongs to an endemic genus. Its mode of life is unclear but the small gills and legs with unusually long claws suggest an unusual adaptive specialisation, not known in any other Mesozoic mayflies. Another highly unusual and rather common endemic benthic insect is *Bolboia mira* Kalugina, a large worm-like dipteran larva tentatively assigned by KALUGINA (1989) to Eoptychopteridae. A large isophlebiid resembling the Jurassic *Dinosamarura* is common and should be regarded as the top predator species. The hydrophilid beetles, their larvae, podonomine chironomids of the genus *Oryctochlus* Kalugina and *Terrindusia* caddis cases are rare and constitute the minor components of the fauna. Uncommon small conchostracans are the only aquatic animals other than insects. Terrestrial plant remains are quite scarce. The unique taxonomic and ecological pattern of the assemblage is evident but the cause of its specificity is difficult to explain. *Bolbonyx, Bolboia*, chironomids and caddisworms could be benthic algal feeders; both shredders and planktivores seem to be very rare or absent. On the other hand, the ecosystem has to support rather high predator density. Perhaps the lake was situated at a lower altitude than other lakes inhabited by isophlebiids, and the input of nutrients (maybe, with volcanic ash) was significant. This could result in considerable benthic algal growth. It is difficult to say why the plankton was so limited.

8. Assemblages with abundant chaoborids and *Coptoclava* devoid of mayflies. This assemblage type occur in the Early Cretaceous Bon-Tsagan Series (after SINITZA 1993) of Central Mongolia; some Chinese faunas (e.g. of the Dalazi Formation in Jilin Province; the *Coptoclava* assemblage after LIN 1983) may also belong here. Both in Mongolia and China the type is likely restricted to the second half of the Early Cretaceous. Insect fossils are preserved in marls, siltstones and bituminous paper shales. The nectic insects are numerically abundant but poor in species. Chaoborids of the genus *Chironomaptera* Ping and *Coptoclava* larvae are the most common and widespread aquatic insects, though the density of chaoborid remains is generally lower than in many Type 4 and 5 assemblages. *Hemeroscopus* occurs in some layers only. Aquatic bugs are extremely rare in Mongolian sites but *Cristocorixa* was recorded from the Dalazi Formation (LIN 1994). Benthic insects are rather scarcely represented. No mayflies were found except for a damaged siphlonurid nymph from Kholbotu-Gol in Mongolia (SINITSHENKOVA 1989). Caddis cases are found in Mongolia only. In total, 13 ichnospecies are recorded (SUKATSHEVA 1982). However nearly all of them occur in one or few layers only (SINITZA 1993). There are two common and widely distributed ichnospecies, *Folindusia ponomarenkoi* Vialov et Sukatsheva and *Ostracindusia popovi* Vialov et Sukatsheva. Though formally placed to different ichnogenera they are very similar, differing only in the ratio of plant particles and ostracod shells in their material, and may well be constructed by one and the same caddisfly species (VIALOV & SUKATSHEVA 1976). Hence the great majority of caddisfly species should be habitat specialists restricted to certain local microenvironments within the lakes, perhaps mainly in the near-shore areas, and it seems likely that only one opportunistic species was widespread. Cases found in association with algal stromatolites (SINITZA 1993) suggest that at least some species, including the commonest one could be algal feeders. Unidentified caddisfly pupae and vitimotauliid and phryganeid wings are not rare. Chironomid pupae are uncommon. Heterophlebioid odonatan nymphs are the only probable benthic predators; interestingly, they occur mainly in the same layers as *Hemeroscopus* does. The seston-feeding ostracods and thick-shelled bivalves are highly abundant and diverse while conchostracans occur only rarely (SINITZA 1993). *Lycoptera* is common in Mongolia, and other fish genera are found in China. Mass burials of *Lycoptera* in some layers (e.g. in the so-called "fish marl" in Bon-Tsagan) could possibly indicate episodic suffocation. In general the assemblage is strongly impoverished, even in comparison with the low diversity 5B and 5C assemblages. The lakes inhabited by this type of fauna were probably large and hydrologically unstable, with strongly limited oxygenated epilimnion and thick anoxic hypolimnion. The model proposed by Kalugina and Ponomarenko for the Mongolian lakes with the 5B assemblages is applied to this type as well (PONOMARENKO 1996). It is not clear why in Type 8, the benthic insect community is further impoverished including perhaps only one common and widespread species, the producer of *F. ponomarenkoi* and *O. popovi* cases.

9. Assemblages with abundant burrowing mayflies devoid of hexagenitids. This assemblage type is known only from the Early Cretaceous lithographic limestone of Montsec, Spain. For insect and other fossil lists and geological information see MARTÍNEZ-DELCLÒS

(1991, 1995) and PEÑALVER *et al.* (1999). FREGENAL-MARTÍNEZ *et al.* (1992) and MARTÍNEZ-DELCLÒS & MARTINELL (1993) discuss the palaeoenvironments of the lake. Benthic insects clearly predominate in the assemblage. Burrowing mayfly nymphs are common; *Mesopalingea lerida* Whalley and Jarzembowski dominates, but a second species tentatively assigned to Euthyplociidae occurs in a small number. They were probably filtering seston-feeders, and the commonness of filter-feeding insects distinguishes the Montsec assemblage from any Cretaceous assemblages discussed above. Modern burrowing mayflies develop mainly in rivers, but there is a number of facultative or even obligate lentic species. They are moderately sensitive to the dissolved oxygen level because of continuous water current through their nids and occur sometimes in rather deep parts of lakes though disappear under eutrophication. The distribution of burrowing mayflies within the Montsec Lake is unclear. No characteristic U-shaped burrows were recorded (LACASA 1991) suggesting that the nymphs should be subautochthonous (U-shaped nids possibly made by burrowing mayflies are rather common in near-shore sediments in some lacustrine Cretaceous sequences in other regions; BUATOIS *et al.* 1998). On the other hand, their frequency seems to be too high, for burrowing organisms lived far outside the sedimentation place. Some present-day species colonise floating wood but wood fragments are too rare in the Montsec deposits to be a source for any common fossils. The same problem exists with another most common benthic group, the brachyceran dipterans. Their frequency in Montsec is unique for the Early Cretaceous. The larvae of at least two different species are recorded, tentatively identified as a stratiomyid and a sciomyzid though the latter assignment is almost surely wrong. It is difficult to believe that they inhabited a remote lake area but the fine insect-bearing laminites demonstrate no trails or bioturbations that could be produced by fly larvae. Perhaps they burrowed in rather thin algal mats, and not directly in bottom sediments. Algal-feeding commonly occur in modern aquatic flies, and many of them are highly tolerant to low oxygen level. Other, much less common benthic insects are represented by predacious heterophlebioid (of a *Samarura*-type) and anisopteran (*Palaeaeschna*) nymphs, three rare mayfly species including a leptophlebiid (probably algal and debris feeders) and rare nematoceran dipterans, perhaps chironomids. Nectic coptoclavid larvae and bugs are very rare. No caddis cases are found. The pleustic genus *Chresmoda* occurs as a minor component.

Other common aquatic animals are decapod crustaceans (a crawfish and a shrimp, probably predators or scavengers), ostracods, and diverse fishes. More than 20 species recorded numerically dominated by a leptolepid teleostean, *Ascalabos voithi* Muenster, and the actinopterygians *Notagogus ferreri* Wenz and *Propterus vidali* Sauvage. Both *Ascalabos* and *Notagogus* are small-sized, up to 4–5 cm long; *Propterus* reaches 140 mm in length. Large-sized fish are rare. The hybodontid sharks represented by very rare fragmentary remains are the largest predators found. The feeding habit of the Montsec fish was diverse, including probably pleustic-, nektic- and benthic-feeders. Gastropods, bivalves, and frogs are rare, and no conchostracan has been recorded. The algal stromatolites occur in some parts of the sequence. The aquatic macrophytes were likely represented not only by characeans but also by a *Ranunculus*-like angiosperm and an enigmatic genus *Montsechia*. Plant debris and terrestrial plant remains are occasionally abundant but not in the layers with insects. The trophic structure of the ecosystem seem to be complex but the role of insects in comparison with other animals was rather limited; especially carnivorous insects, which are scarcely represented. The fish predation

could be an important factor limiting insect density both in nectic and benthic communities. This may explain the dominance of burrowing benthic insect taxa which would be better protected from fish. It is generally believed that the lake was large, rather deep and permanently stratified, with a thick anoxic hypolimnion unaffected by wind and wave actions. However, bottom anoxia is argued for mainly by the excellent preservation state of fossils, and it should be noted that this evidence is not absolutely reliable.

10. Assemblages dominated by aquatic bugs and containing diverse other aquatic insects but few if any dipterans. This type is represented by the later Early Cretaceous (probably Late Aptian) Crato Formation fauna in Brazil. The insects occur mainly in finely laminated micritic limestones in the lower part of the formation. The published lists of aquatic insects (GRIMALDI 1990, MARTILL 1993) provide no quantitative data for the majority of taxa but BECHLY (1998) stressed that aquatic insects other than bugs are uncommon and represented mainly by flying stages. The water bugs are highly diverse even at the family level and include nectobenthic naucorids, nectic belostomatids, notonectids and corixids, rare benthic or nectobenthic nepids, and pleustic gerromorphs (e.g. wingless Hydrometridae). The most common families are Naucoridae and Belostomatidae. There are no accurate data about their diversity and distribution. No less than 3 different species of burrowing mayfly nymphs and 5 hexagenitid nymph species are recorded (MARTINS-NETO & CALDAS 1990, MARTINS-NETO 1996); the combination of hexagenitids and ephemeroids in the same assemblage is unique. More taxa are known from winged stages only. The mayfly distribution was analysed by MARTINS-NETO (1996) who found that they are abundant in shallow-water deposits only and that different hexagenitid species (*Cratogenites corradiniae* Martins-Neto, *Palaeobaetodes costalimai* Martins-Neto, *P. britoi* Martins-Neto, or *Protoligoneuria limai* Demoulin) dominate in different parts of the basin. *P. costalimai* is the most widespread species. The most diverse mayfly assemblage dominated by *Cratogenites* is restricted to the part of the basin where sedimentological data indicate an increasing river influence; in particular, the burrowing mayfly nymphs and another possible filter-feeder, *Cratoligoneuriella* Martins-Neto, are not represented outside this area. Both *Palaeobaetodes* species are more common in sediments of still-water areas while *Protoligoneuria* seems to be restricted to the shallowest water areas. Though adult odonatans are highly diverse (more than 30 species are listed by BECHLY 1998), their nymphs are uncommon and belong to few taxa (*Nothomacromia* and gomphids). Caddis cases are extremely rare. There are some aquatic dipterans (Tipuloidea, Chaoboridae, and Simuliidae) but they are likely to be represented by adults only as also are the megalopterans. Several aquatic beetle families are mentioned on the basis of their adults; no detailed information about their abundance, diversity and distribution is available. Other aquatic animals are ostracods, very rare decapods and conchostracans, and fish. The fish fauna is of a low diversity unlike the fauna of the overlying Santana Formation in a strict sense (after MARTILL 1993). The dominant fish, *Dastilbe elongatus* Jordan, is small-sized and common; MARTILL (1993) explains mass burial of this species in some layers by mass mortality events. Both oil shales and algal stromatolites occur within the section; the presence of any aquatic macrophytes in the flora is disputable. Terrestrial plant remains are generally uncommon. The presence of salt pseudomorphs at several levels within the limestones is noteworthy.

Two different palaeoenvironment models have been proposed for the Crato Formation. Some authors (e.g. M. SILVA 1986,

CARLE & WIGHTON 1990, MARTINS-NETO 1996) regarded the sedimentation basin as an initially freshwater, later gradually salinated lake, and the aquatic insect fossils as belonging partly to autochthonous lentic fauna disappearing in the course of progressive salination. The alternative model argued by MARTILL (1993) and supported by BECHLY (1998) suggests that the Crato deposits have been accumulated in a stratified saline lagoon affected by input of fresh waters from a river. The aquatic fauna is considered as allochthonous and at best temporarily survived in surface fresh water tongues above the chemocline. Martill adduces the absence of bioturbations and exceptional state of fossil preservation as the main reasons in favour of the chemically stratified lagoon model. In this particular case this model cannot be rejected because of obvious evidence of marine influence in the overlying Santana Formation. If so, the Crato assemblage is nearly unique in respect of the regular presence of diverse allochthonous aquatic insect stages. However, too many Early Cretaceous lacustrine assemblages, discussed above, resemble the Crato assemblage in a number of important features. The water stratification, bottom anoxy, a decline of the benthic fauna in offshore areas, and episodic suffocation is postulated for many of them, especially in warmer regions. Similar sedimentation pattern is common, including the presence of more or less thick deposits rich in calcium carbonate, algal stromatolites and bituminous shales in lacustrine sequences. Moreover, data on the Crato mayfly distribution support the freshwater stratified lake model, because it is difficult to explain how *Palaeobaetodes* nymphs could have been transported distances further from their source than other mayflies. It is more likely that a diverse mayfly community inhabited the lake bottom area around the river mouth, with more oxygenated water and a larger input of allochthonous debris, and became progressively impoverished with a decreasing river influence. *Palaeobaetodes* was likely more eurytopic than other mayflies, less dependent from the aeration and allochthonous debris input (perhaps, an algal-grazer). If the Crato Lake was fresh, other benthic insects have to demonstrate similar distribution pattern, and the nectic species including the immature stages should be more widespread. More observations on the distribution of aquatic insects are necessary for a definite conclusion. The disappearing of aquatic insects in upper part of the formation may indeed indicate the progressive salination of the basin.

11. Assemblages strongly dominated by aquatic bugs with few if any immature stages of other aquatic insects. This type corresponds to the Jurassic Type 3 and occurs in diverse palaeoenvironments both in the Early and Late Cretaceous. In the lacustrine Early Cretaceous the assemblages of this type have been discovered in Europe (Las Hoyas in Spain) and South America (the La Cantera Formation in Argentina). The Las Hoyas assemblage is dominated by the belostomatid *Iberonepa romerali* Martínez-Delclòs, Nel et Popov represented by both nymphs and adults (MARTÍNEZ-DELCLÒS, NEL & POPOV 1995). Two more belostomatid species are undescribed. Other aquatic insects are scarce. In the most recent list of the Las Hoyas insect fauna only two rare species of mayfly nymphs tentatively placed to Leptophlebiidae are mentioned (PEÑALVER *et al.* 1999). In the La Cantera Formation the notonectid *Canteronecta irajae* Mazzoni is represented by numerous nymphs and adults (MAZZONI 1985). According to POPOV (1989a) this species belongs perhaps to the genus *Clypostemma* Popov. MAZZONI (1985) also mentioned undescribed corixids and an alleged belostomatid but no other aquatic insects. Both in Las Hoyas and La Cantera the insects occur in finely laminated micrites accumulated probably in open parts of large lakes. The absence of other nectic insect taxa, and in particular of chaoborids, is noteworthy. Perhaps more diverse insect

communities dwelled in the near-shore areas of the same lakes and belonged to some of the types discussed above.

In the predominantly fluvial and deltaic Early Cretaceous deposits, aquatic bugs are occasionally abundant in the English Purbeck and Wealden (Popov et al. 1998). The Purbeck assemblages are strongly dominated by an unidentified bug combining the characters of Naucoridae, Notonectidae and Belostomatidae, which is likely to be represented by both adults and nymphs. The naucorids and belostomatids are uncommon, and the presence of their nymphal stages is uncertain. No corixids were found. In the Wealden the dominant Purbeck bug had disappeared and the few naucorid and belostomatid remains are fragmentary, suggesting a transport by currents; no nymphal stages have been discovered. There may be some autochthonous aquatic beetles and dipterans as well but their remains seem to be relatively rare except for occasional "swarms" of unidentified dipteran larvae and pupae (POPOV *et al.* 1998). Caddis cases are never found here in mixed assemblages with bugs (see below). In spite of a high adult odonatan diversity their nymphs are completely absent.

In the Late Cretaceous, assemblages of this type have been discovered in China and the Russian Far East. LIN (1980) described two aquatic bug species, *Siculocorixa estria* Lin (Corixidae) and *Clypostemma limna* Lin (Notonectidae) from the Qujiang (or Juezhou) Group of the Zhejiang Province. Later LIN (1983, 1994) has recorded a similar fauna designated as the *Siculocorixa* assemblages also from the Xingning region in the Guandong Province. *Siculocorixa* is the numerical dominant. No aquatic insects other than bugs were recorded. The assemblage occurs in red-coloured deposits assigned to the latest Early and earlier Late Cretaceous and containing also ostracods, conchostracans, molluscs, and charophytes. In the Amur River Basin of the Russian Far East several specimens of *Mesosigara kryshtofovichi* Popov were collected at Belye Kruchi at the Bureya River from the floodplain deposits of the upper part of the Tsagayan Formation (Maastrichtian). Other aquatic insects are found in the same formation in other localities (see below). It is probable that in all cases this assemblage type occurs in palaeoenvironments strongly unfavourable for the majority of aquatic insects because of either oxygen deficiency, or an unstable hydrological and/or hydrochemical regime.

12. Assemblages dominated by caddis cases with few if any other aquatic insects. This type corresponds to the Jurassic Type 5 but is much more widespread in the Cretaceous and occurs in quite diverse palaeoenvironments. Other aquatic insects occasionally represented in a small number are almost exclusively adult beetles and bugs. In the Early Cretaceous this type is very common in Transbaikalia and Mongolia, especially in the coal-bearing deposits where other aquatic insects are nearly always absent and in the coarse-grained sediments where other insect remains could be destroyed mechanically. In the later Early Cretaceous case assemblages appear in northern Siberia (up to Taimyr Peninsula) and the Russian Far East. The Early Cretaceous assemblages are often highly diverse, including 5–10 and occasionally even more than 20 ichnospecies each (SUKATSHEVA 1982, 1991a). Outside those areas the Early Cretaceous cases are found only locally in the English Wealden in fluvial deposits devoid of any other insect remains (JARZEMBOWSKI 1995a). Interestingly, none of the Early Cretaceous assemblages known includes any hydroptilid-type cases occasionally common in the Jurassic. From the Late Cretaceous case assemblages have been discovered in Siberia, the Russian Far East, Central Asia, Europe, and North America but, in spite of the broader geographic area of finds, they are much more uniform than in the Early Cretaceous (SUKATSHEVA 1991a). The Late Cretaceous lacustrine assemblages usually include only 1 or 2

ichnospecies each, and never more than 6. As a rule, only cases constructed from small plant debris and/or from spirally arranged elongate plant particles (mainly conifer needles) are represented. Other case types are nearly restricted to fluvial palaeoenvironments in the Late Cretaceous. The decline in case diversity is evident as early as in the later Early Cretaceous (Albian), and in fact the Late Albian assemblages are essentially similar to the Late Cretaceous ones (SUKATSHEVA 1991a).

13. Assemblages dominated by adult aquatic beetles. This assemblage type is remarkably rare in the Early Cretaceous in comparison with the Jurassic, perhaps because the Cretaceous drifted assemblages are often dominated by caddis cases instead of beetle fragments. It occurs mainly in those regions where caddis cases are rare (e.g. in the English Purbeck and Wealden) or absent (e.g. in Egypt). The assemblage from the Albian or Cenomanian of Labrador is noteworthy because of the high rate of schizophoroid elytra which are generally quite rare in the Cretaceous. Probably, in the Late Cretaceous this type is more common but the Late Cretaceous beetles are insufficiently studied and it is difficult to say which of them could be aquatic. In the floodplain deposits of the Tsagayan Formation (Maastrichtian) of Arkhara (Amur River Basin, Russian Far East), the supposed donacine leaf beetle elytra are not rare, and hydrophilid and gyrinid elytra occasionally occur together with rare naucorid bugs and caddis cases.

14. Assemblages dominated by ptychopteroid dipterans. This assemblage type is known only from the earliest Cretaceous Batylykh Formation of Northern Jakutia, Siberia. Numerous pupae and few wings of *Crenoptychoptera* Kalugina (Eoptychopteroidea) and *Zhiganka* Lukashevich (Ptychopteridae) were collected together from a siltstone lens within fluvial section (LUKASHEVICH 1995). There are also badly damaged fragmentary remains of unidentified mayfly nymphs. The assemblage may represent the fauna of a shallow floodplain pond or a backwater habitat.

15. Compression fossil assemblages including mainly or exclusively flying stages of allochthonous (probably mainly lotic) aquatic insects. The assemblages occur in various palaeoenvironments including marine deposits and are probably polytopic, so it is usually difficult to distinguish between insects which originated from different habitats. It should be noted that in the Late Cretaceous insect oryctocoenoses from lacustrine deposits often contain very few if any autochthonous lentic insects: this situation is unusual in comparison with both the Jurassic and Early Cretaceous. For one such locality, the Orapa diamond pipe in Botswana, a "dead-lake" model was proposed suggesting that the lake was strongly affected by toxic volcanic emissions (MCKAY & RAYNER 1986, RAYNER 1993). However, the same situation occurs in many Russian Late Cretaceous localities (e.g. in the Cenomanian deposits of Obeshchayushchiy in the Magadan Region and Obluch'ye in the Amur River Basin as well as in a number of sites in the Primorye Province ranged from the Cenomanian or Turonian up to the Maastrichtian). Though in Obeshchayushchiy, three fragments of aquatic immature insects were found including a mayfly nymph, a dragonfly nymph and a dobsonfly larva, their remains are unusually scarce and could be allochthonous. All those lakes were situated in volcanic areas. However, it is difficult to believe that all of them were poisonous, especially if the common presence of lentic insects in similar palaeoenvironments in the Jurassic, Early Cretaceous and Tertiary is taken into account. It is more probable, that the absence of lacustrine taxa can be explained by a steep lake profile with very narrow shallow near-shore area inhabited by the benthic fauna.

16. Inclusions in fossil resins. These assemblages appear in the Early Cretaceous and have no analogues in the older deposits (Chapter 1.4.2.2.1). Insect burial occurs in terrestrial environments, the aquatic taxa are also represented there, almost exclusively by the flying stages, and finds of aquatic immatures are exceptional (WICHARD & WEITSHAT 1996). The assemblages are highly polytopic. Insects emerged from different water bodies and mixed together, making their palaeoecological interpretation difficult. The share of aquatic taxa varies widely. Those assemblages most rich in them are believed to have originated when the resin was transported to its burial place on logs, branches and bark by river flow directly from riverine forests (ZHERIKHIN & SUKATSHEVA 1992). In this case lotic insects that developed in large potamal-type rivers should predominate numerically; and this is the case of the best studied Cretaceous resins such as the Late Cretaceous Taimyr and New Jersey ones.

At present no complete data on the aquatic insect taxa are available for any Cretaceous resins. In particular, only preliminary information is published on the Early Cretaceous resin assemblages from the English Wealden, Lebanon and Spain. More taxonomic papers regard the Late Cretaceous resins, in particular from the Taimyr Peninsula, Siberia, and New Jersey, USA. At the present state of knowledge an accurate comparison between the Early and Late Cretaceous assemblages is impossible. It is quite evident, however, that unlike lacustrine assemblages the Late Cretaceous resin fauna does not show any significant impoverishment.

As a rule, resin assemblages are strongly dominated by dipterans, and particularly by diverse Chironomidae. Unfortunately, this family is poorly studied taxonomically, and there are no comparative data on its generic composition. KALUGINA (1974) noted that the Siberian fauna was mainly comprised of orthocladiines, indicating a low saprobity and high water quality. The advanced subfamily Chironominae most tolerant to saprobic environments is extremely rare. This pattern might be proved on a wider geographic scale. Two better-studied orders, the mayflies and caddisflies, demonstrate important differences between the Siberian and New Jersey resin faunas, probably connected with different climatic conditions. In more temperate Siberia the mayflies are more abundant but less diverse, with 32 identifiable specimens representing only 3 families (Siphlonuridae and Baetidae with algal-feeding or detritivorous nymphs and burrowing Australiphemeridae), while 6 specimens from New Jersey belong to 5 families with different nymphal biology (Polymitarcyidae, Australiphemeridae, Heptageniidae, Ametropodidae, and Leptophlebiidae) (SINITSHENKOVA 2000b, PETERS & PETERS 2000). On the contrary, caddisflies are represented in New Jersey mainly by several genera of the algal-feeding Hydroptilidae accompanied by a few predacious Dipseudopsidae and Philopotamidae (BOTOSANEANU et al. 1998, WICHARD & BÖLLING 2000); the dominance of the hydroptilids suggests an abundant algal growth. No integripalpians were found. The identifiable caddisflies from Siberia belong to the predacious Rhyacophilidae, Hydrobiosidae and Polycentropodidae as well as to 3 integripalpian families with grazing or shredding larvae (Taymyrelectronidae, Calamoceratidae and Leptoceridae; BOTOSANEANU & WICHARD 1983) with no hydroptilids found.

In general, there is a striking difference between the Early and Late Cretaceous assemblages and particularly between the lentic ones. The Early Cretaceous faunas are quite diverse and differ from the Jurassic ones in their taxonomic composition rather than their basic ecological structure. In fact, the decline in lentic insect diversity is evident even before the end of the Early Cretaceous, with some Albian assemblages looking like the Late Cretaceous ones (see Type 12 above). Unfortunately, the age of some important sites is debatable, and thus the presence of typically rich Early Cretaceous faunas in the Albian is uncertain. The lotic fauna is insufficiently studied but its changes do not seem to be dramatic.

The Early Cretaceous lakes known can be roughly classified into three main physiographic classes, namely the small oxbow ones (with assemblage types 1, 11, 12, and 14), larger lowland lakes of variable depth (with assemblage types 5B and C, 6, 8, 9, 10, and 11), and larger mountain lakes (with assemblage types 3A–C, 4, 5A, and 7). The Cretaceous oxbow lakes are much less studied than the Jurassic ones, but resemble them in the taxonomically (and usually also quantitatively) scarce nectic fauna. Their abundant benthos is sometimes represented by the same orders and occasionally families as in the Jurassic (assemblage type 1), but usually the caddis cases predominate absolutely. In the larger lakes the nectic insects (and particularly chaoborids and/or aquatic bugs) are usually quite abundant. The benthic fauna is rich in the mountain lakes but reduced at the lowlands, especially in warmer climatic areas. The majority of lacustrine assemblages do not satisfactorily fit any present-day type and probably represent some sorts of extinct ecosystems with no modern analogues, though some mountain lakes like in the Jurassic probably resembled recent oligotrophic clear-water basins (assemblage types 3A and B).

The Late Cretaceous lacustrine deposits often do not contain any obviously autochthonous insect remains (assemblage type 15) irrespective of the lake physiography. This is particularly striking in the case of the fossil lake bodies in the volcanic-sedimentary sequences. In the Jurassic and Early Cretaceous more or less diverse and numerically rich aquatic insect assemblages occur regularly in volcanic ash deposits which accumulated in lacustrine environments, though are usually restricted to a relatively few thin finely laminated layers within each section. The flying insect stages are represented as well but are always less abundant. This pattern is so characteristic for those palaeoenvironments (generally poor in other fossils) that SCOBLO (1968) has established a special facial type called "the insect-bearing shales". The same burial pattern occurs quite frequently in the Early Cretaceous, at least in Siberia, Mongolia, and China but was never observed in sedimentologically similar Late Cretaceous sections where aerial insects are absolutely predominating. When fossils of aquatic stages are more or less frequent in the lacustrine Late Cretaceous, they are represented nearly exclusively either by very few caddis case types or, more rarely, by the corixid bugs (assemblage types 12 and 13, respectively). The virtual absence of aquatic dipteran larvae and pupae is particularly striking in comparison with their frequency in the Jurassic and Early Cretaceous. Sporadic finds of rare immatures in the lake deposits (e.g. a corydalid larva, few fragments of mayfly and dragonfly nymphs, and very rare fly larvae in the Cenomanian of Obeshchayushchiy in Magadan Region) suggest either occasional transport from the lake tributaries or an extreme reduction of the areas suitable for aquatic insects within the lake body. It should be emphasised that, unlike the Early Cretaceous, at least the offshore lake area was nearly devoid of both benthic and nectic insects.

Another noticeable point is that nearly all of today's extinct aquatic families completely disappeared during the latest Early Cretaceous, and this concerns some of those most common still in the Aptian (e.g. Hexagenitidae and Coptoclavidae). Few of them are crossing the mid-Cretaceous boundary (e.g. Australiphemeridae and Aeschnidiidae) but none is represented by autochthonous aquatic stages in the Late Cretaceous oryctocoenoses. The latter is equally true for the families appearing in the fossil record in the Late Cretaceous, both modern and restricted to the Cretaceous.

This general pattern is quite obvious and indicates a large-scale uncompensated extinction of lentic insects probably ending in the Albian. In the Late Cretaceous only the running waters supported a rich aquatic insect fauna, and all major aquatic taxa evolved in lotic environments, both of rhithral and potamal types. This Mid-Cretaceous extinction event can be compared with nothing in the whole history of aquatic insects. It was explained tentatively by a eutrophication caused by a drastic increasing in the nutrient input from the land after the expansion of the angiosperms which produce a large amount of a rapidly decomposing litter (KALUGINA 1974, 1980b, KALUGINA & ZHERIKHIN 1975, ZHERIKHIN 1978, SINITSHENKOVA 1984, SINITSHENKOVA & ZHERIKHIN 1996). The eutrophication hypothesis has been criticised by PONOMARENKO (1996, 1998a) who particularly stressed that the important changes in the lentic ecosystem are detectable well before the time of the angiosperm expansion. He believes also that highly eutrophic lakes were quite widespread at least since the Late Triassic, and that the angiosperm expansion should reduce the land erosion and thus the nutrient transport from the land could even decrease.

The problem is really complicated, and there are many points of uncertainty. As discussed before, the Mesozoic lakes regarded by Ponomarenko as eutrophic should have a special turnover type, which does not exist in the modern lakes. The anti-erosional effects of the pre-angiosperm land vegetation were never estimated accurately on the basis of large-scale comparative sedimentological studies and thus are still the matter of speculation. The age of many faunas, especially in the most productive areas such as Siberia, Mongolia and China, is uncertain; however, the main extinction event should occur after the Aptian which is the maximum supposed age for several sites yielding a rich lentic fauna (e.g. Koonwarra and the Crato Formation) and probable for a number of other typical Early Cretaceous assemblages (e.g. Montsec, Bon-Tsagan, and some Chinese localities).

However, the eutrophication hypothesis does face a number of unresolved problems. Even if the general temporal pattern of extinction is not seriously confronted with the time of the angiosperm expansion in the Albian, there are many local situations which are difficult to explain. In particular, there are a number of Albian and Late Cretaceous sites with the impoverished aquatic fauna typical of the Late Cretaceous. The flora is dominated by conifers while the angiosperm remains are rare and sporadic. This is the case of many sites in the Russian Far East, e.g. the Emanra (Albian) and Arkagala (Cenomanian-Turonian) Formations and the Amka Beds (Cenomanian) with the Type 12 assemblages containing few caddis case ichnospecies, and Obeshchayushchiy (Cenomanian) with the Type 15 assemblage.

The eutrophication hypothesis fails to explain also the differences between the Jurassic and Early Cretaceous faunas. ZHERIKHIN & KALUGINA (1985) postulated that those differences should be mainly an artefact of the physiographic differences between the best-studied areas in the Jurassic and Cretaceous. This idea can be neither rejected nor accepted at the present state of knowledge. The recent discovery of the Type 1 assemblage confirms the prediction of those authors that the characteristic fauna of the shallow oxbow lakes (Jurassic Type 1) was not restricted to the Jurassic but has survived with some minor changes possibly up to the time of the angiosperm expansion. However, the evidence for this assemblage type in the Cretaceous is still too scarce, and another prediction (about the presence of the assemblages similar to the Cretaceous Type 4 and 5 in the Jurassic) remains unconfirmed.

Nevertheless we assess the hypothesis of the Mid-Cretaceous eutrophication as currently the most plausible explaining the drastic extinction of the lacustrine insect fauna. An interesting aspect of the eutrophication hypothesis not discussed previously is the possible decrease in the output of nutrients from the lakes due to flying insects. The production of aquatic dipterans in large Jurassic and Early Cretaceous lakes seems likely to have been very high, and Kalugina (1980b) even suggested that the bituminous shales widespread in the Early Cretaceous of Asia could be formed mainly by the faecal pellets of the chaoborid larvae. If so, billions of emerging midges should bring a considerable amount of organic matter from the lakes, and its removal could be compensated only in part with dead adults having fallen back into the water.

3.3.7. CAINOZOIC

†V.V. ZHERIKHIN AND N.D. SINITSHENKOVA

Though the total number of the Cainozoic aquatic insect localities is great, their present state of knowledge is far from satisfactory. Some of the most important orders including the aquatic beetles and especially dipterans have been poorly investigated taxonomically, with too many taxa either still undescribed or badly in need of revision. Even for the odonatans, which are the best studied Cainozoic aquatic insects, the nymphs are practically unclassified so that the diversity of autochthonous lentic genera and species cannot be estimated. No quantitative data are available for the great majority of assemblages, including some of the richest and most representative ones. Outside of West and Central Europe and North America, the aquatic insects are poorly known, and the information about the Late Cainozoic (Pliocene and Quaternary) is generally scarce. For those reasons the Cainozoic history of aquatic insects can only be outlined in a very general manner though potentially their close affinities with the modern fauna should allow much more detailed and better grounded ecological analysis in comparison with the older assemblages.

Almost all Cainozoic aquatic insects belong to living families and subfamilies, the only exception being the odonatans which include several extinct families (Sieblosiidae, Zacallitidae, Palaeomacromiidae) and subfamilies (Dysagrioninae, Eodichrominae), are restricted to the Cainozoic. Sieblosiidae is the only aquatic extinct insect family known from the Neogene. A number of living genera have been recorded as early as in the Late Eocene, many of them on the basis of excellently preserved amber inclusions, and their identification does not raise any serious doubt. Moreover, two aquatic insects from Late Eocene Baltic amber have been assigned to living species, namely the heptageniid mayfly *Heptagenia fuscogrisea* (Retzius) (KLUGE 1986) and the chironomid midge *Buchonomyia thienemanni* Fittkau (MURRAY 1976). ASKEVOLD (1990) has found that the slightly younger *Donacia primaeva* Wickham, an aquatic leaf-beetle from the earliest Oligocene of Florissant, Colorado, is indistinguishable from the living species *Plateumaris nitida* Say. More living species are recorded (though partly with some reservation) from the Miocene and Pliocene (e.g., ESAKI & ASAHINA 1957, UEDO *et al.* 1960, GALEWSKI & GŁAZEK 1977, KISELYOV 1981, PETERS 1994). In Quaternary deposits, finds of extinct taxa are quite exceptional (e.g., the aquatic leaf-beetle *Donaciella nagaokana* Hayashi: HAYASHI 1998).

3.3.7.1. Overview of taxa of aquatic insects

Mayflies are generally not very common in Cainozoic deposits. Occasionally their nymphs occur in lacustrine assemblages, especially since the Oligocene, but they are never really abundant, and the winged stages are known mainly from fossil resins, especially from Baltic and Dominican amber. The most striking feature of the Cainozoic fossil ephemeropteran record is the virtual absence of Baetidae, the largest living family which has a very wide geographic and ecological distribution. The reasons for its rarity in the recent fossil record are unknown, especially as no obvious differences between the Palaeogene and Neogene faunas are established in the dominance patterns.

Odonatans demonstrate a surprisingly low similarity between the Cainozoic fossil assemblages and the modern fauna in comparison with any other insect order. This is especially true for the Palaeogene where the frequencies of such families as Thaumatoneuridae (including Dysagrioninae), Sieblosiidae and Megapodagrionidae is remarkably high while Lestidae, Coenagrionidae, Corduliidae and Libellulidae are uncommon. The rarity of nymphs in general is also noticeable. The odonatan fauna was changing significantly in the Late Oligocene and Miocene when the libellulid nymphs became very common. The frequency of Lestidae, Coenagrionidae and Libellulidae among the adult remains increases, though the Sieblosiidae and Megapodagrionidae are still not rare even in the Late Miocene of Western Europe (NEL *et al.* 1996). Sieblosiidae are the only extinct insect family that was rather widespread in the Cainozoic; and it only disappeared in the Pliocene.

Aquatic bugs are common and widespread in Cainozoic deposits; usually with either the Notonectidae or the Corixidae being the numerically dominant families. The presence of Electrobatinae, an extinct gerrid subfamily, in the Early Miocene Dominican amber is also worthy of mention as the most interesting finds of the marine gerrids (*Halobates ruffoi* Andersen, Farma, Minelli et Piccoli in the Late Eocene of Monte Bolca, Italy) and veliids (*Halovelia electrodominica* Andersen et Poinar from the Dominican amber) (ANDERSEN 1998). No significant differences between the Palaeogene and Neogene have been demonstrated in their taxonomic composition.

Cainozoic aquatic beetles are rather diverse and represented mostly by living genera. At family level the absence of definitive pre-Quaternary records of Haliplidae, a widely distributed family of algal-feeding adephagans, is difficult to explain. Perhaps the most interesting trend observed in the Cainozoic is a rapid increase in the abundance of Dytiscidae in the Late Oligocene (PONOMARENKO & ZHERIKHIN 1980, KALUGINA 1980b) which more or less coincides in time with an increase in another important group of lentic predators, libellulid dragonfly nymphs. The dytiscid genus *Palaeogyrinus* Schlechtendal from the earliest Miocene of Germany demonstrates adaptations of a surface water swimmer which are unusual for the family. It was originally placed into a family of its own but STATZ (1939) has dropped its taxonomic rank to tribe level, and GALEWSKI & GŁAZEK (1977) place it near the living genus *Laccophilus* Leach. Another large modern family of aquatic beetles, Hydrophilidae, seems to be rather well represented both in the Palaeogene and Neogene. Finds of beetle larvae that are morphologically well adapted to fast flowing water in lacustrine deposits from the Eocene of France (BERTHRAND & LAURENTIAUX 1963) and Germany (LUTZ 1985b, 1990) are intriguing.

The megalopterans and aquatic neuropterans are represented by rare adults belonging mainly to extinct genera. In the Pliocene lake deposits of Willershausen in Germany the larvae of a megalopteran, *Sialis strausi* Illies, were discovered in numbers (ILLIES 1967).

No aquatic mecopterans are recorded in the fossil state from the Cainozoic, and the history of the few aquatic moth taxa is undocumented with fossils.

Caddisflies and their cases are common. All Cainozoic fossils described up to now have been assigned to modern families and many of them belong to living genera (SUKATSHEVA 1982, WICHARD & WEITSCHAT 1996); however, some genera are rather isolated and show no close affinities to present-day taxa (e.g., the phryganeid genus *Amagupsyche* Cockerell from the Oligocene of the Russian Far East: SUKATSHEVA 1982). There are obvious differences in the dominance pattern at the family level between the Palaeogene and Neogene assemblages of both compression fossils and resin inclusions. Prior to the Oligocene, the largest modern families are either scarcely represented (Hydropsychidae, Leptoceridae) or even completely absent (Limnephilidae), and the assemblages are numerically dominated by Polycentropodidae, Psychomyiidae and Lepidostomatidae (SUKATSHEVA 1982); the former family is also remarkable diverse. Limnephilids which currently dominate the Holarctic fauna first appear in the fossil record in the Early Oligocene (S. LEWIS 1973) but only become abundant in the Miocene (SUKATSHEVA 1982). The caddis case assemblages are usually rather low in their diversity, *Folindusia* being the dominant ichnogenus in the majority of localities. Few new case types appear in comparison with the Cretaceous, but include *Indusia* (constructed from small gastropod shells) which is occasionally abundant in the Oligocene of Europe, and undescribed coiled cases resembling those of modern Helicopsychidae found in the Eocene Wilcox Formation of USA (D.L. Dilcher, pers. comm.). It should be noted that PONOMARENKO's (1996) hypothesis that caddisworms inhabited floating algal mats seems to fit some Cainozoic localities better than the Cretaceous situations for which that biology was originally proposed. In the Tadushi Formation of Russian Far East (Palaeocene) there are all intermediate types between the silky *Secrindusia* cases and typical *Folindusia* which are entirely covered by plant particles; the latter type being the rarest. Because the cases are rather uniform in size and shape it is quite possible that they were constructed by the same caddisworm species with opportunistic building behaviour under varying availability of plant materials. A similar situation occurs in the Eocene deposits of Eckfeld in Germany (LUTZ *et al.* 1991). LUTZ (1990, 1991a) has suggested that in another Eocene German locality, Messel, the case builders dwelled among the leaves of floating angiosperms but in this case the algal model seems to be at least equally suitable though in Messel, unlike Tadushi and Eckfeld, only *Terrindusia* cases composed from sand grains are found. In all three sites, the cases are much less abundant and diverse than in the majority of Lower Cretaceous deposits.

Aquatic dipterans are common in the Cainozoic but with a few exceptions, they are still insufficiently studied. The relatively well-known Late Eocene Baltic amber fauna is distinctive for the rarity of Culicidae (SZADZIEWSKI 1998) and the chironomid subfamily Chironominae (KALUGINA 1980a) as well as in the absence of any aquatic higher brachycerans, both acalyptrate and calyptrate. Culicids are, however, recorded from the latest Eocene or earliest Oligocene of England (F. EDWARDS 1923), and their pupae are extremely abundant in the lacustrine Oligocene deposits of Céréste in France (LUTZ 1984a). The fossil history of Chironominae and aquatic higher brachycerans is nearly unknown, and reported records in the literature need to be substantiated. It is noteworthy that brachyceran larvae are occasionally quite abundant unlike in any pre-Cainozoic assemblages. In the Oligocene and Miocene they are represented mainly by several stratiomyid genera (HEIN 1954, DUCREUX & TACHET 1985, BACHMEYER *et al.* 1991, KÜHBANDNER & SCHLEICH 1994). The taxonomic placement of *Lithohypoderma* larvae, very common in some

layers in the Eocene Green River Series of the USA, is unclear (see below). Nymphomyiidae, with their aquatic immatures, are recorded from both Baltic and Bitterfeld amber (T. WAGNER *et al.* 2000). Besides the beetles, dipterans are the only aquatic insect order extensively studied in the Quaternary, but the available data mainly concerns the head capsules of chironomid larvae which are often quite abundant as subfossils.

The stoneflies are scarcely represented in the Cainozoic, mainly as resin inclusions. Most of them are placed into living genera. The nymphs are nearly absent in the fossil state suggesting that the order was restricted to lotic habitats or nearly so.

3.3.7.2. Aquatic insect assemblages and their typology

Because of the incomplete published information only a preliminary typology of the Cainozoic aquatic assemblages can be proposed here.

1. Assemblages with abundant and diverse aquatic beetle larvae. This assemblage type is known only from the Paskapoo Formation (Late Palaeocene) of Alberta, Canada. The preliminary insect list published by MITCHELL & WIGHTON (1979) gives a rather good idea on its composition though many identifications are tentative and precise numerical data are only available for less abundant taxa. The beetles are represented by numerous hydradephagan (probably Noteridae) and polyphagan (probably mainly Scirtidae) larvae as well by adult dytiscids and donaciine chrysomelids. The noterid larvae were probably benthic or nectobenthic predators and the scirtids would probably have been benthic detritivores; donaciine larvae are herbivores on submerged or semiaquatic angiosperms. No nectic dytiscid larvae are recorded. Aquatic dipterans are also common but their larvae are unidentified below the ordinal level, however, there are characteristic pupae of craneflies (Tipuloidea) with either detritivorous or herbivorous larvae, and several cranefly taxa are represented by adults (WIGHTON 1980). Caddis cases are not abundant; their constructors should be either detritivores or algal grazers. Predacious benthic dragonfly nymphs are also present, and mayfly nymphs (unidentified, so their ecology is unclear) and predacious nectic aquatic bugs (belostomatids) are very rare (one specimen each). The fossiliferous sediments were accumulated in abandoned channels, oxbow lakes and/or floodplain ponds episodically affected by floodings (M. WILSON 1996). This is the only diverse and numerically rich aquatic insect assemblage discovered in the Palaeocene. It shows little similarity with any assemblage types found in similar palaeoenvironments in the Mesozoic which are often dominated by mayfly and stonefly nymphs (the Jurassic and Cretaceous Types 1) and only rarely rich in aquatic dipterans (the Cretaceous Type 14). Anisopteran dragonfly nymphs occur in the Mesozoic oxbow lakes and/or only occasionally and in small numbers (in the Cretaceous Types 1 and 2), and polyphagan beetle larvae are lacking or nearly so. Hence the only important point of similarity with the older oxbow lake sites seems to be the extreme rarity of the aquatic bugs.

2. Assemblages with abundant chaoborids and caddis cases. This assemblage was discovered in the Mid-Eocene Tallahatta Formation of Mississippi, USA, and is still unrecorded elsewhere. Aquatic insects are numerically abundant but are remarkably low in diversity and include only dipterans and caddis cases. The dipterans are represented exclusively by numerous pupae of a species of *Chaoborus* Lichtenstein (JOHNSTON & BORKENT 1998); the larvae were probably zooplanktivorous. The caddis cases belong to two different types, the dominant *Folindusia* and a relatively rare ichnospecies of *Terrindusia*

(JOHNSTON 1999). The caddisworm that constructed the *Folindusia* cases was probably a shredder. *Terrindusia* cases are rather small and remarkably narrow resembling some modern hydroptilid cases and could have been made by a member of this algal-grazing family. Their low frequency may indicate a limited algal growth. There are also some remains of adult odonatans but no nymphs were recorded (JOHNSTON 1993) suggesting that they probably developed in another water body. The insect-bearing deposits originated in an oxbow-lake environment (JOHNSTON & BORKENT 1998, JOHNSTON 1999). The assemblage has nothing in common with the preceding one though the physiographic situation should have been similar. It also differs principally from any Mesozoic oxbow lake assemblages known which never contain chaoborid immatures.

3. Assemblages with relatively abundant mayfly nymphs. Though mayflies are never truly abundant in the Cainozoic they regularly occur in some lacustrine deposits being probably the best indicators of well oxygenated habitats among the Cainozoic aquatic insects. This justifies their inclusion in the same general assemblage type though different sites are not very similar in the composition of other insects.

3A. Assemblages with burrowing mayfly nymphs and diverse other insects. This subtype is represented by the famous Florissant locality in Colorado, USA, which is of earliest Oligocene or perhaps the latest Eocene age (WOLFE 1992). The insect fauna of contemporaneous or only slightly younger lacustrine deposits of the neighbouring Chase Basin may belong to the same subtype although the published data are too scarce to draw a definitive conclusion (DURDEN 1966). In both cases the insects occur in sediments of moderately large and deep barrier lakes episodically affected by volcanic ashfall. Interestingly, no aquatic insects are recorded from a Late Oligocene palaeolake in Creed at the same region (CARPENTER *et al.* 1938). The fauna of Bol'shaya Svetlovodnaya is considered separately (see Type 6A) because the autochthonicity of the only burrowing mayfly specimen is doubtful. Subfossil fragmentary remains of hexageniid nymphs are occasionally abundant in the bottom sediments of North American Great Lakes (REYNOLDS & HAMILTON 1993) but there are no records of other aquatic insects except for larval chironomid head capsules. The ichnofossils assigned to the burrowing mayflies from the Neogene of Austria, Germany and North Africa (*Asthenopodichnium* Thenius: THENIUS 1979, 1989) and from the Pliocene and Quaternary of European Russia (KRASENKOV 1966) are not accompanied by other insect remains. The enormous Florissant insect collections have only been studied in part (for a summary of the published records at the family level see M. WILSON 1978), and no quantitative data have been published since SCUDDER (1890b).

The Florissant assemblage is probably numerically dominated by chironomids but their immature stages were never investigated. On the basis of adults MELANDER (1949) has described a number of species assigned mainly to Chironominae. If this pattern is confirmed, the Florissant fauna would be the oldest known one dominated by the chironomines. Before the taxonomic composition of the chironomid midges is studied more carefully nothing can be said about their possible biology because of a high ecological diversity of the chironomids as a whole. Several other dipteran families which could develop in water have been recorded including Tipulidae, Limoniidae, Ptychopteridae, Tabanidae and Stratiomyidae but there are no data on the presence of immature stages of any of them. Mayfly nymphs are probably uncommon but the figures in SCUDDER (1890b) are comparable with other aquatic insects except for chironomids which are referred to as abundant. SCUDDER (1890b) recognised six "*Ephemera*"

species, and *Lepismophlebia platymera* (Scudder) originally described as a silverfish is now regarded to be a leptophlebiid nymph (HUBBARD & SAVAGE 1981). Two "*Ephemera*" species were represented in Scudder's materials by five specimens each. Although their generic placement is unsubstantiated from the point of view of modern mayfly taxonomy, Scudders's illustrations show that at least the majority of his specimens almost certainly belonged to the Ephemeroidea; their burrowing nymphs would be expected to have been filter-feeders. The number of species is probably overestimated, and at least some of them should be synonymised. Their presence suggests that the nutrient level was probably not very high because ephemeroids are rather sensitive to eutrophication. The remains of winged mayfly stages are extremely rare. Unidentified mayflies are also mentioned from the Chase Basin.

Other aquatic orders seem to be nearly as sporadic in their occurrence as are the mayflies. Odonatan adults are diverse, especially the damselflies, but it is likely that most of them belong to species which developed elsewhere. Only very few nymphal specimens were recorded (SCUDDER 1890b, COCKERELL 1910b) those being assigned to two species of uncertain taxonomic position, "*Agrion*" *telluris* Scudder (a damselfly) and "*Aeschna*" *larvata* Scudder (a dragonfly). Both were surely predacious and probably inhabited relatively shallow lake areas. The aquatic bugs are not very common but they are diverse, including at least six species of Nepomorpha (POPOV 1971) and three species of Gerromorpha (ANDERSEN 1998). The omnivorous or algal-feeding nectic corixids are more abundant than the very rare benthic (Nepidae, Naucoridae), nectic (two notonectid species) and water-surface (two gerrids and a veliid) predators; interestingly, no belostomatids are mentioned. The corixids probably belonged to two species, and the commoner one, *Corixia vanduzeei* Scudder, has also been recorded from Chase. The aquatic beetles are poorly studied although a number of species have been described, mainly in Dytiscidae and Hydrophilidae. No data about their relative abundance are available, and it is unknown whether there are any autochthonous beetle larvae. Because the adult beetles may be occasional lake visitors, their role in the ecosystem cannot be estimated. The published record of immature caddisflies is scarce (SUKATSHEVA 1982) and includes a larva of uncertain taxonomic position ("*Phenacopsyche*" *larvalis* Cockerell), a pupa ("*Setodes*" *portionalis* Scudder), and the only rare case ichnospecies, *Ostracindusia cypridis* (Cockerell). On the contrary, the adult caddisflies are diverse; 26 species are named but some of them may well represent allochthonous (including lotic) elements. The numerically dominant family is the Polycentropodidae. The adults belong to families with diverse feeding habits (algal-feeders, predators, filter-feeders, and shredders) but only the latter trophic group seems to be represented among the autochthonous immatures.

In general, it seems that the described materials originate mainly from deep offshore areas populated almost exclusively by chironomines, which are often tolerant to oxygen deficiency. The shallow near-shore areas were probably inhabited by a rather diverse insect community, which is, however, very incompletely represented in the fossil record. It should be noted that even nectic taxa were probably virtually restricted to those lake regions, perhaps because of the scarcity of available food in the central parts of the lake. Similar assemblages occur relatively rarely in the Jurassic and Early Cretaceous where the open water sediments usually contain abundant remains of chaoborids, aquatic bugs and coptoclavids. When nectic fauna did not populate this area, there are no benthic chironomids and thus autochthonous lentic fossils are rare or absent (the Jurassic Type 7 and Cretaceous Type 15 in lake deposits).

3B. Assemblages with non-burrowing mayfly nymphs. This subtype does not include the deposits with very rare and possibly allochthonous mayflies nor the Messel fauna (see Type 13) and occurs sporadically over the World (West Europe, China, North America, Argentina, Australia). No assemblages assigned here are known with certainty before the latest Eocene or even Early Oligocene. In the Mid-Eocene Klondike Mountain Formation of Washington State, USA, mayflies as well as other aquatic insects are very rare and may be taphonomically allochthonous (SINITSHENKOVA 1999a). The age of the Ventana Formation in Argentina is probably Oligocene according to the radiometric data (PETRULEVICIUS 1999b). Hence the oldest definitive occurrence of the subtype is probably from the Ruby River Basin in Montana, USA, where sediments previously assigned to the Oligocene may actually be from the latest Eocene (WOLFE 1992). More finds are known from the Neogene, including the Early Miocene Wada Formation (FUJIYAMA 1983) of Japan, the Mid-Miocene Shanwang Formation of the Shandong Province, China (ZHANG 1989), the latest Miocene-Early Pliocene of Central Massif in France (MASSELOT & NEL 1999), and the Pliocene of Vegetable Creek in Australia (RIEK 1954). The mayflies are represented by different families at different sites. Records of the predominantly lotic family Heptageniidae are intriguing, but the French fossil genus *Pseudokageronia* Masselot et Nel was probably lentic and the supposed heptageniid nymphs from Shanwang may be misidentified (MASSELOT & NEL 1999). In the Early Miocene of Hachiya Formation in Japan, caenid nymphs are recorded; this family is well adapted to muddy bottom areas and may colonise turbid waters. The Ruby River fauna includes several species but only the neoephemerid *Potamanthellus rubiensis* Lewis is represented by a number of specimens (S. LEWIS 1977). Leptophlebiids are recorded from the Ventana Formation and Shanwang, and the systematic position of the Vegetable Creek mayflies is uncertain. *Pseudokageronia* is the only mayfly described from France but other families are represented among the undescribed fossils in the Paris Museum (Zherikhin, personal observation).

The mayflies are accompanied by other aquatic insects in all localities. Complete faunal lists are available only for rather small Japanese collections and include a donaciine beetle and chironomid pupae from Hachiya (FUJIYAMA & NOMURA 1986), an odonatan nymph, several chironomids, adult stoneflies and craneflies from the Wada Formation (FUJIYAMA 1983). Information for other localities is incomplete and includes the records of odonatan nymphs and notonectid bugs from the Ventana Formation (PETRULEVICIUS 1999b), caddisfly cases and wings from the Ruby River Basin (S. LEWIS 1972, 1973), nepid, notonectid, belostomatid and corixid bugs, dytiscid and hydrophilid beetles, adult odonatans, craneflies and chironomids from Shanwang (ZHANG 1989, ZHANG *et al.* 1994), notonectid and naucorid bugs, donaciine leaf beetles, a sialid larva and several adult sialids, a caddis case and a caddisfly, odonatan and chironomid adults from the Central Massif in France (PITON & THÉOBALD 1935, NEL 1991), and aquatic beetle larvae and supposed chironomids from Vegetable Creek (ETHERIDGE & OLLIFF 1890). These data, though incomplete, indicate clearly that mayfly nymphs always occur together with other taxonomically and biologically diverse aquatic insects. This subtype seems to be restricted to diatomite and tuffaceous lacustrine deposits of volcanic regions, suggesting a connection with more or less clear of mountain water and piedmont lakes, probably of an oligotrophic nature. However, for unknown reasons no autochthonous mayflies were discovered in rather large collections from the volcanic-sedimentary deposits of the Russian Far East.

4. Assemblages with abundant sialid larvae and naucorid bugs. This peculiar type is known only from the Pliocene of Willershausen, Germany. The list of insect taxa published by KRÜGER (1979) is evidently uncritical and should be taken with reservation. The most complete taxonomic and numerical data are available on predacious insect taxa. Their taxonomic composition is very unusual. The abundance of sialid larvae (*Sialis strausi* Illies: ILLIES 1967) and naucorids (represented probably by only one common species described under four different names: POPOV 1971) is absolutely unique; both are benthic predators. There are no other aquatic bugs, and GERSDORF (1969, 1970, 1976) does not refer with certainty to any aquatic beetles or their larvae. The odonatans are represented by a few dragonfly adults and gomphoid burrowing nymphs (SCHUMANN 1967) which are generally very rare in lentic habitats. Hence predators are probably represented by only a very few benthic species. Their high abundance suggests, however, a high prey biomass. The most probable prey would probably have been aquatic dipterans. Indeed, KRÜGER (1979) lists craneflies, culicids and chironomids, but without further data on their ontogenetic stages and abundance. He also mentions a caddisfly, again without any further information. The insect fauna suggests that the lake was probably eutrophic, with a taxonomically impoverished, but numerically rich benthic community. The submerged vegetation was probably well developed (WILDE *et al.* 1992). LUTZ (1997) describes the habitat as a rather small but relatively deep meromictic lake in a salt tectonics area, so somewhat anomalous hydrochemical conditions are possible. The insects occur in finely laminated carbonaceous clays, partially bituminous, and were probably deposited in a deep offshore area. Perhaps, a more diverse insect community inhabited the near-shore habitats.

5. Assemblages with abundant lestid nymphs. Damselfly adults are common and diverse in Cainozoic deposits but their nymphs are surprisingly rare suggesting that many taxa were restricted to taphonomically unfavourable (perhaps lotic and/or swampy) environments. The only family whose nymphs are occasionally abundant is the Lestidae. Nymphs assigned to the genus *Lestes* Leach are recorded in considerable numbers from Sieblos (Early Oligocene, Germany) and Bes-Konak (Late Oligocene, Turkey) (MARTINI 1971, PAICHELER 1977, NEL & PAICHELER 1994). Preliminary data on the Sieblos insect fauna are summarised by MARTINI (1971) and WILLMANN (1988). Besides the *Lestes* nymphs the only common aquatic insect species is *"Pissodes" effossus* Heyden, a small weevil probably related to modern aquatic Tanysphyrini or Bagoini (WILLMANN 1988). Its larvae would probably have been herbivorous, most probably leaf or petiole miners on an aquatic monocotyledonous plant. According to MARTINI (1967), the commonest aquatic plant in Sieblos is *Nymphaeites rhoenensis* Kurtz (Nymphaeaceae), and some living Bagoini are biologically connected with water lilies. Other aquatic insect taxa are rare and include a species of predacious notonectid bug (*Notonecta sieblosiensis* Martini) and dipteran larvae assigned to Tipulidae, Culicidae and Dolichopodidae. The insects are virtually restricted to laminated marl layers with abundant diatom remains. The more deep-water bituminous "paper shales" contain mainly ostracod and isopod crustaceans and fish remains but very few insects. The Sieblos lake was probably meromictic and more or less eutrophic; LUTZ (1997) suggests unstable water level and episodic salination events. Information on the Bes-Konak insect fauna is incomplete (PAICHELER 1977, PAICHELER *et al.* 1977). In addition to the lestids, there are rare dragonfly nymphs (though adult dragonflies are commoner than the damselflies), rather common notonectid and rarer nepid and naucorid bugs, and abundant chaoborid pupae suggesting a meromictic stratification. Adult sialids

are recorded as well. NEL (1989) points out that no aquatic beetles have been discovered. It is not clear what both localities with damselfly nymphs had in common from the point view of their ecology which would make them different from other Cainozoic meromictic lakes. Modern lestid nymphs prefer habitats with abundant submerged vegetation but this was probably not the case for the sites under consideration.

6. Assemblages with abundant dipterans and notonectid bugs. The common feature of the assemblages assigned to this type is also a rarity of both odonatan nymphs and caddis cases. Two subtypes can be recognised. The fauna of Kuãelin in the Czech Republic may also belong here. The common Kuãelin notonectid species *Anisops heydeni* (Deichmüller) is closely allied to *A. deichmuelleri* (Schlechtendal) from Rott (the Subtype 6B) (ŠTYS & ŘÍHA 1975). The information on other insects is scarce, and no data on the dipteran fauna are available. The age of the fauna is uncertain (perhaps, Latest Eocene: COLLINSON 1992).

6A. Assemblages lacking corixids and chaoborids. This subtype is known from the Early Oligocene (or perhaps the latest Eocene) of Bol'shaya Svetlovodnaya, Russian Far East (ZHERIKHIN 1998). The only common aquatic insects are an undescribed notonectid bug and unidentified chironomids. There are also few adult aquatic beetles (Gyrinidae and Hydrophilidae), the only specimen of a burrowing mayfly nymph (*Litobrancha palearctica* McCafferty et Sinitshenkova), two undescribed nymphs of a gomphid dragonfly and one specimen of a caddis case (*Folindusia arcuata* Sukatsheva). Adult caddisflies (Phryganeidae, Philopotamidae, Sericostomatidae), craneflies, chironomids and very rare damselflies are also found. The material is collected at a small outcrop of inter-basalt lake deposits represented by whitish laminated mudstones occasionally containing abundant terrestrial plant remains. Other aquatic animals are nearly absent except for a very rare salmonid fish (*Brachymystax bikinensis* Sytchevskaya: SYTCHEVSKAYA 1986). The presence of a salmonid suggests high water quality probably in an oligotrophic mountain lake. The oryctocoenosis probably reflects the fauna of a deep-water area where mainly nectic bugs occurred, perhaps as temporary visitors from near-shore habitats rather than as residents. The fauna of shallow-water lake parts should be much more diverse but is barely represented in the collection. A low productivity level may explain the absence of nectic planktivores (chaoborids), differing the assemblage from the Subtype 6B.

6B. Assemblages with corixids and chaoborids. This subtype is represented by at least two European localities, Rott (earliest Miocene, Germany) and Rubielos de Mora (Early Miocene, Spain); for general information see KOENIGSWALD (1989) and PEÑALVER (1998) respectively. The assemblages are much more diverse than Subtype 6A ones. The dominant benthic group are Chironomidae belonging mainly to the Chironominae; this subfamily is generally tolerant to oxygen deficiency. The presence of the genus *Glyptotendipes* Langton with phytophagous (seed-boring) larvae in Rubielos de Mora is noteworthy. There are also rare cranefly pupae (adults are diverse, at least in Rott) and larvae and pupae of aquatic brachycerans (in Rott only; two genera of Stratiomyidae and a syrphid are recorded, and the former are not rare). The fly larvae probably fed on plant debris. Both lakes were also evidently inhabited by caddisflies but the cases (probably all belonging to the same ichnospecies of *Terrindusia*) are recorded from Rubielos de Mora only, while in Rott the presence of autochthonous caddisflies is documented exclusively by pupae (STATZ 1936). Because the caddisfly pupae actively swim before the emergence of adult they surely

originate from a larval habitat within the lake, which is not represented directly in the fossiliferous section. The caddisflies should be either shredders or algal grazers. The caddisfly adults from Rott are referred to Phryganeidae but the *Terrindusia* cases from Spain should be constructed by a member of a different family. The repeated caddisfly finds suggest that the larvae were probably abundant in their original habitats. The benthic predators are unrecorded from Rubielos de Mora; in Rott they are represented by rare odonatan nymphs (both damselflies and dragonflies) and rather common nectobenthic naucorid bugs. Adult hydrophilid beetles, most of which were probably nectobenthic algal-feeders are too common in Rott to have been occasional visitors, especially a *Berosus* species. This family was not recorded from Rubielos de Mora but may well be represented there among the numerous unidentified beetles. The herbivorous donaciine leaf beetles occur in Rott only occasionally (GOECKE 1960). The nectic insects are numerically abundant in both localities and represented by zooplanktivorous Chaoboridae (at least in Rott by the genus *Chaoborus* Lichtenstein: BORKENT 1978), omnivorous corixid bugs (more abundant in Rubielos de Mora), predacious notonectid bugs (more abundant in Rott) and dytiscids (both larvae and adults). Daphniid crustaceans are recorded from both localities and could constitute an important food source for the chaoborids. Occasional finds of belostomatid bugs and hydrobiid beetles are also mentioned in Rott (the latter family is unrecorded in the fossil state from any other localities). As pointed out above, *Palaeogyrinus* found in Rott represents an unusual ecological dytiscid type with gyrinid- or coptoclavid-like adaptations to surface water swimming. In both localities the insects occur in finely laminated mudstone accumulated in offshore palaeoenvironments. However, the dark bituminous paper shale that originated in the deepest lake area is nearly devoid of insects. This assemblage type resembles the Mesozoic assemblages with abundant nectic taxa; the absence of any strongly oxyphilous taxa is noteworthy.

6C. Assemblages with abundant corixids but lacking chaoborids. This subtype was discovered in the Tremembé Formation (Oligocene) of Brazil (MARTINS-NETO 1998b). The fossils occur in bituminous shale that was probably deposited in a deep-water area of a large lake. Corixids seem to be the only relatively common aquatic insects in the assemblage; three species have been described, each of which has been placed in a different extinct genus; however, all of three of these may in fact be synonymous (Yu. Popov, pers. comm.). Other aquatic taxa recorded are hydrophilid beetles, rare caddis cases composed from plant particles, and insect eggs, assigned for an unknown reason to a hymenopteran, but most probably belonging to an aquatic dipteran. The remains described as a hebrid bug (MARTINS-NETO 1997) also resemble a nematoceran dipteran more closely, perhaps a chironomid midge. There are also adult craneflies and horseflies but no fly larvae were recorded. This assemblage is placed with the other Type 6 assemblages with some reservation because the information on supposedly aquatic dipterans is equivocal. MARTINS-NETO (1998b) stresses that no odonatans are found.

7. Assemblages with abundant dragonfly nymphs. Though remains of anisopteran nymphs occur occasionally at a number of Cainozoic sites they are never common before the Late Oligocene. The oldest finds of this assemblages type known so far have been made in France (Aix-en-Provence, Bois d'Asson). In the Neogene they become widespread in West and Central Europe, including Spain (e.g., Ribesalbes, Early Miocene; Cuenca de Sorbas, Miocene; Libros, Late Miocene: PEÑALVER *et al.* 1999), Italy (e.g., Alba, Gabbro, Monte Castellaro, Late Miocene; San Angelo, Pliocene: HANDLIRSCH 1908,

GENTILINI 1984, 1988, CAVALLO & GALLETTI 1987), Germany (Randecker Maar, Early Miocene; Oeningen, Ries, Late Miocene: HEER 1864–1865, BOLTEN et al. 1976, ANSORGE & KOHRING 1995), Austria (e.g., several Early Miocene localities in Weingraben: BACHMAYER 1952, BACHMAYER et al. 1991), Hungary (e.g., Tállya, Late Miocene: PONGRÁCZ 1923), and Rumania (Sãcel, Late Miocene: BARBU 1939). Some Central Asian (Chon-Tuz in Kyrgyzstan, Miocene) and North American (the Mid-Pliocene Ridge Basin Group of Ridge Basin, California, USA: SQUIRES 1979) localities fit this type as well. A high abundance of predacious dragonflies suggests a high biomass of their prey, probably mainly aquatic dipterans. However, in many cases no aquatic insects other than dragonfly nymphs are recorded; perhaps, they are often simply overlooked. Usually the dominant or the only dragonfly family in this assemblage type is the Libellulidae. This family is biologically diverse and includes a number of extant species that are highly tolerable of generally unfavourable conditions, such as a low oxygen content, high saprobity, episodic or seasonal droughts, and brackish and thermal water environments. It seems that some fossil assemblages of this type correspond to palaeoenvironments with strongly taxonomically impoverished insect faunas, but in other cases it is obviously not so, and several subtypes can be recognised on this basis. At present, the majority of localities cannot be placed into any subtype because of the lack of data on the insects in them other than their dragonflies.

7A. Dragonfly-dominated assemblages with relatively diverse other aquatic insects. This subtype includes the faunas of Ribesalbes, Randeck, Oeningen and Weingraben. At least in Ribesalbes and Oeningen the dragonfly nymphs are represented by both Aeshnidae and Libellulidae (HEER 1864–1865, PEÑALVER et al. 1996); the materials from Randeck and Weingraben are undescribed. There are also other predacious insect taxa, both benthic and nectic, which are never numerically abundant. They include rare damselfly nymphs and nepid bugs in Ribesalbes and Oeningen, uncommon notonectid bugs in Ribesalbes and Weingraben, rare belostomatid and naucorid bugs in Oeningen, and adult dytiscid beetles in Randeck (SCHAWALLER 1986) and Oeningen. Nectic omnivorous corixid bugs from Randeck, Oeningen and Weingraben may be added to this list. In Randeck the corixids are represented by Diacorixa germanica Popov – a species common with the Subtype 7C (see below). The benthic and nectobenthic detritivores and algal-feeders are represented by chironomid and stratiomyid immatures, caddisworms, and adult hydrophilid beetles. In Weingraben all aquatic chironomid stages including the eggs are very abundant (BACHMAYER et al. 1991), and chironomid pupae are recorded from Randeck (KÜHBANDNER & SCHLEICH 1994). Only adult chironomids are mentioned from Ribesalbes and Oeningen although their immatures very probably occur in these localities as well. The taxonomic composition of chironomids is poorly studied, but at least in Weingraben, the Chironomini seem to predominate. Stratiomyid larvae occur regularly, and sometimes in considerable numbers, in Ribesalbes (PEÑALVER 1996), Randeck (KÜHBANDNER & SCHLEICH 1994) and Weingraben (BACHMAYER et al. 1991); there are no records of any brachyceran larvae from Oeningen. Adult craneflies occur in Ribesalbes and Oeningen but there are no records of their immature stages. Terrindusia antiqua (Heer) caddis cases are very rare and recorded only from Oeningen. The hydrophilid adults are known from Randeck (SCHAWALLER 1986) and Oeningen. The herbivorous donaciine leaf beetles are represented by a few specimens in Randeck and Oeningen. The presence of nectic filter-feeding culicids is documented by pupae found in Ribesalbes. All finds of this subtype are connected

with finely laminated deposits of offshore areas of large and deep meromictic lakes, either with marls (in Ribesalbes and Oeningen), or with "paper shales" (in Weingraben), or both (in Randeck). High predator abundance suggests a high productivity, and the taxonomic composition is rather similar with the modern fauna of meso- or slightly eutrophic lakes with abundant submerged vegetation. LUTZ (1997) suggests that the Randeck Lake was probably hydrologically unstable and temporary brackish. This is in accordance with the presence of Diacorixa (see below, 7C).

7B. Taxonomically impoverished assemblages with dragonfly nymphs and chaoborids. This subtype is represented by the fauna of Aix-en-Provence and perhaps also that of Bois d'Asson (HENROTAY 1986), both in France. Besides the common dragonfly nymphs and chironomid pupae, adult aquatic beetles are not rare (NEL 1989 mentions a small-sized hydrophilid as a common species in Aix-en-Provence), and stratiomyid larvae are recorded from Forcalquier near Bois d'Asson (DUCREUX & TACHET 1985). In Aix-en-Provence, the dragonfly nymphs can occasionally be seen side by side with marine fossils (NEL et al. 1987). In both cases the insects occur in sediments of coastal lakes which are probably periodically affected by marine influences (LUTZ 1997). Taking this into consideration, the impoverishment of the insect fauna is not surprising.

7C. Taxonomically impoverished assemblages with dragonfly nymphs and Diacorixa. The distribution of the extinct corixid genus Diacorixa Popov is unusual. It includes two species, D. miocenica Popov from Miocene of Chon-Tuz, Kyrgyzstan (POPOV 1971), and D. germanica Popov from the Late Miocene of Ries, Germany (POPOV 1989b). In both localities Diacorixa occurs en mass together with abundant libellulid nymphs. Other aquatic insect taxa are represented by rare finds of their winged stages. Both localities probably originated in inland brackish lakes (see BOLTEN et al. 1976 for Ries) with numerically abundant but extremely taxonomically impoverished faunas. D. germanica also occurs in the Early Miocene of Randeck; because the Randeck Lake was probably temporarily saline (see Subtype 7A) Diacorixa is probably an indicator of brackish lake environments.

8. Assemblages dominated by brachyceran fly larvae with few if any other aquatic insects. This type occurs in the Middle Eocene Green River Series in the Rocky Mountains Region, USA. This lacustrine formation yields an enormously rich aerial insect fauna including a number of taxa with aquatic larvae (see M. WILSON 1978, for a review of families). However, it is generally surprisingly poor in autochthonous aquatic insect fossils. Nevertheless, some authors have suggested that the bituminous shales constituting a considerable part of the section, in fact originated as a copropel – organic mud composed mainly by insect (supposedly chironomid) faecal pellets (W. BRADLEY & BEARD 1969, IOVINO & BRADLEY 1969, W. BRADLEY 1970). No immature chironomids have been described from the Green River deposits, but a collection of Green River insects in the Palaeontological Institute, Moscow, which was available for examination, contains some small nematoceran pupae probably belonging to this family though they appear to be uncommon. However, there are numerous records of brachyceran larvae assigned to the genus Lithohypoderma Townsend from the Green River Series (SCUDDER 1890b, TOWNSEND 1916, STOKES 1978, GRANDE 1984). The taxonomic position of Lithohypoderma is uncertain. Its interpretation as a parasitic fly larva is not credible; and no adult brachycerans described from Green River are sufficiently large to be associated with Lithohypoderma. It is possible that Lithohypoderma alone produced the Green River copropels. It should inhabit soft bottom areas with highly saprobic conditions.

9. Assemblages dominated by culicid immatures. Culicids are widespread in various present-day water bodies but are surprisingly rare in the fossil state. Adults have been recorded from some Cainozoic deposits but usually in small numbers. The only known locality yielding very abundant culicid larvae and pupae is Ceréste in France (LUTZ 1984a, 1985a); they are assigned to the genus *Culex* L. Other aquatic insects are only represented by flying adults except for a supposed stonefly nymph mentioned by LUTZ (1984a); the photograph in that paper shows the remains of an incomplete insect which seems more likely to be a damselfly nymph. The lake was situated in a coastal area and could have been influenced by fluctuations in sea level. Modern culicid larvae are often highly tolerant to brackish waters as well as of periodic droughts.

10. Assemblages dominated by ceratopogonids. This type is represented by the fauna of the Middle to Late Miocene Barstow Formation in California, USA. The insects occur in lacustrine deposits but the state of preservation is very unusual, with three-dimensional mineralised remains included in calcareous nodules (PALMER *et al.* 1957). The taxonomic composition of the assemblage is also unique, with very abundant ceratopogonid pupae (mainly *Dasyhelea antiqua* Palmer) and small dytiscid larvae (*Schistomerus californicus* Palmer) accompanied by uncommon libellulid dragonfly nymphs and very rare chironomid pupae. The unique preservation suggests an unusual hydrochemical environment, perhaps saline-alkaline (PARK 1995).

11. Assemblages containing mainly or exclusively caddis cases. This type, as with the corresponding Cretaceous Type 12, is surely heterogeneous and occurs in various palaeoenvironments. The localities of Cainozoic caddis cases were reviewed by SUKATSHEVA (1982, 1989). In Europe the cases occur mainly in small numbers in other assemblage types, but in the Late Oligocene–Early Miocene of France and Germany the *Indusia* cases constructed from small gastropod shells constitute the so-called *Indusia*–limestone (BOSC 1805, SERRES 1829, OUSTALET 1870, SCHMIDTGEN 1928, BERTRAND-SARFATI *et al.* 1966). Those mass occurrences of *Indusia* are probably connected with abundant algal growth in lakes. In the marls laterally replacing the limestone, few other insects were found including dytiscid beetles and their larvae. *Folindusia* cases are recorded from the Miocene of Denmark (MATHIESEN 1967), and *Molindusia* cases, constructed very probably by predacious molannid larvae, are occasionally not rare in fluvial Miocene sands of north-eastern Ukraine (REMIZOV 1957). In North American deposits, caddis cases also occasionally occur in number in lacustrine deposits (*Terrindusia calculosa* (Scudder) in the Eocene of Horse Creek, Wyoming: SCUDDER 1878). However, they seem to be not uncommon in floodplain and oxbow lake deposits of the Early Eocene Wilcox Formation in south-eastern USA (E. BERRY 1927, 1930, D.L. Dilcher, pers. comm.) and the Miocene Latah Formation of the Pacific Northeast (S. LEWIS 1970), together with beetle elytra which can partially belong to aquatic beetles. In Siberia, Cainozoic cases (only *Terrindusia* spp.) occur sporadically in both floodplain and lacustrine deposits (Kartashovo, Ust'-Dzhilinda: SUKATSHEVA 1982). In Kartashovo (Early Miocene) they are accompanied by very rare hydrophilid beetles and mayfly nymphs. In the Naran Bulak Formation (Late Palaeocene) in southern Mongolia *Terrindusia* cases are probably not rare (SUKATSHEVA 1982); no other insects are found there.

In the Russian Far East, localities with caddis cases are remarkably numerous at various stratigraphic levels from the Early Palaeocene up to the latest Miocene or perhaps even the Early Pliocene (SUKATSHEVA 1982, 1989, ZHERIKHIN 1998). Undescribed *Folindusia* cases have also been collected from the earliest Palaeocene or, perhaps, Late Maastrichtian deposits at Bogopol' and Ustinovka in the same region. Unlike in any other areas, in the Russian Far East the cases occur mainly in lacustrine deposits, often together with more or less diverse aerial insect faunas. However, other aquatic insects are very scarcely represented and include an undescribed gerrid bug from Amgu (Oligocene), a few chironomid pupae from Amgu and Grossevichi (Late Miocene-Early Pliocene), and perhaps some fragmentary remains of aquatic beetles. The depositional environments are diverse and in some cases uncertain; however, the majority of the Far Eastern finds are connected with deposits of mountain lakes affected by volcanic ash falls. It is difficult to say why they differ so much from those occurring in similar palaeoenvironments elsewhere. The diversity of cases is low, usually with 1–3 ichnospecies at each locality, except for the Voznovo Formation in the Kavalerovo District where as many as seven ichnospecies have been collected from several closely neighbouring sites (SUKATSHEVA 1989). Some ichnospecies are widespread and recorded from different stratigraphic levels within the Oligocene-Miocene interval. *Folindusia* cases dominate numerically while *Terrindusia* occur in small numbers and only at a few localities.

12. Assemblages dominated by adult aquatic beetles. As with older deposits, this assemblage type depends more on taphonomic conditions and thus unfortunately, under-represent those with more fragile insects. Thus they can be characterised more by these factors and so represent a range of different palaeoenvironments. In the Cainozoic this assemblage type is especially widespread in Quaternary peat deposits (ELIAS 1994). Assemblages that are basically similar to the Quaternary ones have also been recorded from the Pliocene and perhaps the Late Miocene of arctic regions (MATTHEWS 1976a, 1977, KISELYOV 1981, BÖCHER 1989, MATTHEWS *et al.* 1990) and the Pliocene of Europe and Japan. In older Cainozoic deposits it occurs only sporadically. The dominant taxa in the pre-Quaternary sites are usually either Dytiscidae or Donaciinae (Chrysomelidae) (mainly in brown-coal deposits accumulated in swamps: GOECKE 1943, UEDO *et al.* 1960, FUJIYAMA 1980). Other aquatic insects occasionally occur together with beetle elytra in transported assemblages. For example, the commonest aquatic insect remains in the latest Oligocene floodplain deposits of the Abrosimovka Formation in West Siberia are dytiscid elytra, mainly of several *Cybister* species (ŘÍHA 1974 and undescribed material). In one of the localities, Kompassky bor, a supposedly allochthonous lotic mayfly nymph (*Miocenogenia gorbunovi* Tshernova) was found (TSHERNOVA 1962). In another site, Ekaterininskoe, a fore leg of a belostomatid bug (*Lethocerus turgaicus* Popov) was discovered (POPOV 1971). An ecological interpretation of transported assemblages is difficult because they can be polytopic. The oldest known assemblage with donaciine leaf beetles is from the brown coals of Geiseltal in Germany accumulated in a taxodiacean swamp (HAUPT 1950, 1956). There are also other aquatic beetles including a dytiscid infested by a gordiid worm (VOIGT 1938) as well as rare fossils of adult damselflies (PONGRÁCZ 1935) and mayflies (DEMOULIN 1957). No immatures of any aquatic insects were recorded, and the record of a caddisfly case is apparently in error (SUKATSHEVA 1982).

13. Assemblages with rare autochthonous insect remains. As in the Late Cretaceous, Palaeogene lacustrine deposits often contain few if any fossils of supposedly autochthonous lentic insects. This is especially evident when large and representative insect collections from different parts of the section are available. In Menat (Late Eocene, France), numerous and diverse terrestrial insects were collected,

mostly from several layers of bituminous lacustrine laminites with abundant plant debris. Aquatic insects are represented by a mayfly nymph, two specimens of adult gyrinid beetles and several caddis cases (NEL & ROY 1996). About 3000 insect specimens have been collected from lacustrine tuffites and clays of the Tadushi Formation (Late Palaeocene) in several outcrops at Zerkalnaya River in the Russian Far East (ZHERIKHIN 1978). Among this material there are a libelluloid dragonfly nymph, five corixid specimens, a naucorid nymph, two adult dytiscids, fewer than 50 small nematoceran pupae (perhaps belonging to Chironomidae) and about 200 caddis cases. Though the latter are more numerous than any other aquatic insects, the Tadushi assemblage is not assigned here to Type 11 because of the relative rarity of caddis case finds. The cases found were assigned to the ichnogenera *Folindusia* Vialov and *Secrindusia* Vialov et Sukatsheva, but differ mainly in the relative tube area covered by plant particles and could be constructed by the one and same caddisfly species.

A quite enormous collection has been built in the last few decades, from the Mid-Eocene lacustrine bituminous laminites in Messel, Germany (LUTZ 1990, 1991a,b). The supposedly autochthonous lentic insects are represented by several dozen unidentified mayfly nymphs (RICHTER & KREBS 1999), a gerrid nymph, very few notonectid nymphs, dytiscid larvae and adults (TRÖSTER 1993), and uncommon caddis cases probably belonging to two unnamed ichnospecies of *Terrindusia*. The presence of psephenid-like beetle larvae is intriguing. LUTZ (1985b) believes that they were transported from a lotic habitat; however, some modern beetles with morphologically similar larvae develop in coastal areas of large lakes (J. DAVIS 1986). From another German Middle Eocene locality, Eckfelder Maar, besides rather rare *Terrindusia* and *Secrindusia* cases only a supposed chironomid pupa has been recorded (LUTZ 1993b, LUTZ et al. 1991). The insects are found in Eckfeld in bituminous laminites deposited in a maar lake. LUTZ et al. (1991) supposed that the morphology of a crater lake made it unsuitable for aquatic insects because of a very steep profile. However, younger sediments which originated from morphologically similar lakes (e.g., in Rott and Randeck) provide evidently more taxonomically and numerically rich lentic insect faunas. No autochthonous insects except for a nepid bug have been reported from the Mid-Eocene lacustrine bituminous laminites and marls in Monteils, France (LUTZ 1997). The Eocene lacustrine deposits of British Columbia, Canada, have been sampled for fossils in a number of sites (M. WILSON 1977, 1978, 1988a, 1991), representing different fossil lakes in a volcanic area. Lentic insects are uncommon there and represented by rare mayfly nymphs, several gerrid species and sporadically distributed caddis cases. This situation is especially surprising because the fish fauna indicates that the water quality in some British Columbia lakes should be high. As discussed above, the Mid-Eocene Green River Series of USA also seem to be very poor in autochthonous lentic insects except for layers with *Lithohypoderma* larvae. No undoubtedly autochthonous insects were recorded from the Klondike Mountain Formation (Mid-Eocene) of Washington State, USA, though some of the rare mayfly nymphs could represent the original lake fauna. Interestingly, this assemblage type, which was so widespread from the Palaeocene up to the Mid Eocene, rarely occurs in younger Cainozoic lacustrine deposits. The Late Oligocene lacustrine assemblage from the bituminous volcanoclastic shales of Enspel in Germany is probably one of few examples. WEDMANN (1998) points out that aquatic insects are generally rare in Enspel and mentions rare odonatan nymphs, belostomatid and gerrid bugs, hydrophilid beetles and caddis cases. More localities with rich allochthonous insect faunas, but few presumably autochthonous fossils (mostly chironomid pupae), are known from the Early Oligocene of Alsace and Baden but in these cases the insect-bearing sediments have probably originated in saline or even hypersaline lakes (LUTZ 1997).

14. Compression fossil assemblages including exclusively flying stages of aquatic insects. This type occurs primarily in marine deposits where the presence of any autochthonous insects is most improbable (e.g., in the Late Palaeocene-Early Eocene Fur Formation of Denmark and Germany and in the Miocene deposits of the Crimea Peninsula, Ukraine, and Stavropol Region, North Caucasus, Russia, rich in allochthonous insect fossils). The records of supposedly autochthonous benthic insects from these deposits (caddis cases from Stavropol and a chironomid larva from Fur Formation) have been based on misinterpreted fossils (see VIALOV 1972, and RUST 1998, respectively). Adult aquatic beetles were not rare in Stavropol (ŘíHA 1974) and gerromorph bugs occurring in the Fur Formation (ANDERSEN 1998) should represent occasional visiting elements of the fauna. *Halobates ruffoi* Andersen, Farma, Minelli et Piccoli from the marine Middle or Late Eocene of Monte Bolca in Italy is probably the only true marine insect discovered in any Cainozoic marine deposits (ANDERSEN 1998). Some freshwater localities should be formally placed to this type too, but most of them are insufficiently investigated and perhaps belong to the previous type (e.g., the Late Palaeocene Redbank Plain Series of Queensland, Australia, the Late Eocene Célas deposits in France or the above-mentioned Oligocene locality Creed in Colorado, USA). However, there are some better studied deposits where the absence of aquatic insect immatures cannot easily be explained by the scarcity of data. This is first of all the case of the latest Eocene insect-bearing horizon at Isle of Wight, England, known as the Bembridge Marls. Diverse terrestrial insects are common in this deposit, and adults of aquatic groups are represented including odonatans, stoneflies, caddisflies and several aquatic dipteran families (JARZEMBOWSKI 1980, McCOBB et al. 1998). The marls contain freshwater gastropod shells and numerous anostracan crustaceans and probably originated in pools at a coastal plain (HOOKER et al. 1995, McCOBB et al. 1998). Perhaps, the Bembridge pools were temporary and somewhat anomalous in hydrochemistry (alkaline) (McCOBB et al. 1998). However, it seems improbable that they were devoid of any aquatic insects including dipteran larvae.

15. Aquatic assemblages of resin inclusions. If the best studied Cretaceous resin faunas of Taimyr and New Jersey reflect mainly the aquatic fauna of relatively large potamal rivers (see the Cretaceous Type 16), this kind of assemblage seems to be rare in the Cainozoic. It may be represented by the recently discovered Early Eocene resin from the Paris Basin in France, which occurs in sedimentary environments very similar to those of the Taimyr resin deposits (Zherikhin, personal observation). NEL et al. (1998a) stressed the presence of several aquatic orders that are generally uncommon in fossil resins, such as mayflies, stoneflies and sialid megalopterans which is in accordance with this idea. The fauna is still undescribed, and thus no detailed data are available. The Palaeocene Sakhalinian amber is originated from coal deposits accumulated at a taxodian swamp and mainly reflects the fauna of the swamp (ZHERIKHIN & ESKOV 1999). The aquatic insect taxa are not very abundant and are only represented by dipterans (Chironomidae and rather rare Ceratopogonidae and Limoniidae), helodid beetles and very rare caddisflies (ZHERIKHIN 1978). This impoverished aquatic assemblage confirms the hypothesis of a swampy palaeoenvironment. The Eocene Fushun amber from China (HONG 1979, 1981) probably contains a similar fauna, and

another possible example is the Miocene resin from Merit Pila Coalfield in Sarawak, deposited in sediments of a tropical lowland dipterocarp swamp (HILLMER *et al.* 1992a,b). This latter resin contains mainly dipterans including culicids which are very rare in other fossil resins. Data on the best known Late Eocene Baltic amber aquatic fauna are summarised by LARSSON (1978) and WICHARD & WEITSCHAT (1996). The aquatic fauna seems to be highly polytopic and includes the taxa now inhabiting diverse lotic environments (both small springs and larger water courses), marshes and ponds. The lotic elements seem to predominate but some important families (e.g., Chironomidae) are poorly studied, and it should be noted that Chironominae are scarcely represented (KALUGINA 1980b). More detailed palaeoecological considerations are possible only on the basis of special studies which might include an analysis of the co-occurrence of different aquatic taxa in the same amber pieces (so-called syninclusions), but this would take a lot of time. Another possible approach is the examination of inclusions in different mineral species from Baltic deposits. SAVKEVICH (1970) suggested that the true amber (succinite) predominating in the Baltic area has been primarily buried for some time in forest soil and then washed out and transported to its definitive burial place while the so-called "rotten amber" (gedanite) was rapidly buried in deltaic sediments without any significant processing in forest soil or litter. If his hypothesis is correct, the inclusions in gedanite might represent only the fauna of an area situated immediately along river bank like in the Taimyr, New Jersey and Paris Basin resins. Unfortunately, comparative studies of the Baltic succinite and gedanite assemblages have never been undertaken. The aquatic fauna of the Early Miocene Dominican and Mexican amber is probably similarly polytopic. Those of the Dominican amber fauna have been analysed by POINAR & POINAR (1999) who stressed the presence of supposedly lotic taxa and a rather rich assemblage of insects that probably inhabited aquatic microhabitats in basal leaves of the epiphytic tank bromeliads; ANDERSEN (1999) has described *Microvelia polhemi*, a fossil water strider closely allied to modern Neotropical species dwelling in such phytotelmatic microhabitats. This peculiar type of aquatic ecosystems is undocumented by any other sources available for palaeontologists. The aquatic fauna of other Cainozoic resins is poorly known.

3.3.7.3. Models of Cretaceous and Cainozoic evolution of aquatic insect assemblages

In general, the Early Cainozoic record of the lentic insects is strikingly sparse and fragmentary in spite of the existence of a number of rich insects localities in general. Their rarity in lacustrine deposits can be explained in each case by *ad hoc* hypotheses of unfavourable local environments, and this is exactly the interpretation used by many authors when they are discussing one or few localities or formations. However, if all data available are viewed together, hypotheses appealing to some strictly local causes become unconvincing. The share of the insect sites with the Type 13 assemblages containing few autochthonous fossils in the Palaeocene and Eocene is strikingly high in comparison with any other post-Palaeozoic time interval except the Late Cretaceous and suggests that the lentic insect fauna was generally highly impoverished on a global scale. The Early Palaeogene lakes with the Type 13 assemblages are often basically similar in physiography and hydrology with both older (Jurassic and Early Cretaceous) and younger (Oligocene and Neogene) fossil lakes yielding much more diverse and numerically rich autochthonous faunas. This is especially true for the mountain lakes of volcanic regions such as the Palaeocene Tadushi Lake or the Eocene

lakes of British Columbia. This pattern confirms a large-scale extinction of lentic taxa in the Cretaceous and suggests that it has not be compensated for, either by lake recolonisation from lotic refugia or by any extensive radiation of survivors *in situ* at least until the latest Eocene.

Hence a thorough analysis of available information supports KALUGINA's (1974, 1980b) and ZHERIKHIN's (1978) hypothesis on a long-time "vacuum" in the lentic insect fauna in the Late Cretaceous and Early Palaeogene. However, their model of a virtually "dead" hypereutrophic lake inhabited by very few species of bugs, brachyceran flies and caddisflies most tolerable to saprobic conditions and oxygen depletion, needs some corrections in the light of recent data. Sporadic but regular presence of other insect taxa, and in particular of mayfly nymphs in Type 13 assemblages suggests that certain habitats supported significantly more diverse insect communities should have existed somewhere within the lakes although their total area has to have been small. It should be stressed that even the distribution of nectic insects breathing with atmospheric air such as corixids, notonectids and dytiscids was remarkably restricted, perhaps because of food deficiency over open-water areas. The same should also be true for the Late Cretaceous. However, the role of those small within-lake refugia in the survival of aquatic insects was probably limited in comparison with lotic habitats. The find of a Type 2 assemblage with abundant chaoborids in Eocene oxbow lake deposits which never contain fossils of this midge family in any older or younger strata supports the hypothesis that lentic insects survived in some refugial habitats connected with running waters. It is also noteworthy that the only taxonomically diverse aquatic assemblage known from the Early Palaeogene, the Type 1 assemblage from the Paskapoo Formation, occurs in sediments of abandoned river channels.

A re-colonisation of offshore lake areas by insects is documented by the appearance of Type 3 and 6 assemblages near the Eocene/Oligocene boundary. It should be stressed that the Florissant and Ruby River faunas are found in association with fossil floras indicating that they preceded the time of the Oligocene cooling (WOLFE 1992), and thus the expansion of aquatic insects over extensive lake areas cannot be explained by increasing oxygen content under decreasing water temperature. The same is possibly true though somewhat less certain, for Kuäelin in Europe (COLLINSON 1992) and Bol'shaya Svetlovodnaya in the Russian Far East (L. Fotyanova, pers. comm.) so that the re-colonisation could have occurred nearly contemporaneously over the mid-latitudes of the northern hemisphere. Information for other regions is extremely scarce. The geological age of the Ventana Formation in Argentina is disputable within the Late Palaeocene-Oligocene time interval (PETRULEVICIUS 1999b), and thus the possibility of an earlier appearance of Type 3 assemblages in the southern hemisphere cannot at present be rejected.

The re-colonisation process is documented by records of both benthic and nectic taxa though it seems to be somewhat less intensive and slower in the former case. It should be stressed that even isolated finds of aquatic insects which cannot be included directly to the above typology are in accordance with the general pattern described above.

The secondary expansion of insects in lacustrine habitats would be expected to be accompanied with an extensive radiation of lentic taxa. Available data on such groups as coenagrionid damselflies, dytiscid beetles and limnephilid caddisflies do not confront this hypothesis. However, it needs further confirmation, in particular by studies on the most diverse but least palaeontologically studied family Chironomidae. At present it seems that the process of lake re-colonisation was more or less completed during the Early Miocene when the fauna of meso- and eutrophic lakes acquired an essentially modern appearance as demonstrated by Type 3, 6 and 7 assemblages (e.g., Shanwang, Rubielos de Mora and Weingraben, respectively). The Late Cainozoic changes in aquatic insect faunas were probably connected mainly, with spatial faunal shifts under changing climatic conditions and were not followed by any further significant radiation at least at the suprageneric level.

It is worth mentioning in this connection that no matter how intensive the mid-Cainozoic lake re-colonisation was, it has most probably failed to surpass insect faunal diversity of the mid-Mesozoic. The above analysis infers that, contrary to PONOMARENKO (1996), the known insect diversity in the larger Jurassic and early Cretaceous lakes was comparable or even superior to that in the mid and later Cainozoic lakes. Taking into consideration those taphonomic processes which tend to reduce representation of past assemblages over geological time (Chapter 1.4.3.4), it is possible to hypothesise that the mid-Mesozoic level of lentic insect diversity has not been surpassed even by now, at least in meso- and eutrophic lakes.

Even if new data support the pattern of Cainozoic lentic insect evolution recognised previously, their contribution to its causal interpretation will still be limited. ZHERIKHIN (1978) and KALUGINA (1980b) associated the Mid-Cainozoic renaissance of the lentic insect fauna with the development of a larger amount of submerged vegetation, so increasing the structural complexity of the habitat and stabilising the oxygen content and nutrient cycles in the lake ecosystem. They believed that although aquatic angiosperms had appeared very early, in the Mid-Cretaceous, they were represented in the Late Cretaceous and Early Palaeogene mainly by semiaquatic and floating forms with limited structural and stabilising effect. More recent data indicate that some submerged angiosperms probably existed in the Late Cretaceous and Palaeocene (e.g., KRASSILOV 1979, COLLINSON 1982, DILCHER 1989, HERENDEEN et al. 1990, KRASSILOV & MAKULBEKOV 1995) but their remains have mainly been discovered in fluvial sequences and most of them were probably restricted to oxbow lakes. According the table 1 in LUTZ (1997), known Palaeocene and Eocene insect localities are indeed nearly free of aquatic angiosperm remains except for Nymphaeaceae-like forms with floating leaves. However, the same is also true for the Oligocene localities, and in fact fossil remains that have been assigned to submerged angiosperms only regularly accompany lacustrine insect fossils since the Miocene. The only Oligocene locality with several genera of submerged angiosperms referred by Lutz is Enspel with a poor, Type 13, aquatic fauna. Hence the submerged vegetation effect could hardly have been very pronounced in pre-Miocene lakes. However, it was probably of some importance for the final Early Miocene radiation of lentic insects rather than for the beginning of their secondary entrance into deep-water areas in the Late Palaeogene. A possible explanation is that highly mobile insects mark the time of the appearing of a densely vegetated subaquatic biotope in large lakes more exactly than plant fossils, perhaps still when their area was rather small. Indeed, the oldest assemblages, Type 3 and Type 6, are dominated by nectic insects which could emigrate to open-water areas from potentially overpopulated near-shore habitats (e.g., in Florissant, Bol'shaya Svetlovodnaya, Rott and Kuãelin). If so, future palaeobotanical studies should confirm the presence of very rare submerged aquatic angiosperm remains in those and similar localities. Other speculative hypotheses (e.g., on a possible increase in the concentration of atmospheric oxygen just before the Oligocene cooling) can also be proposed but equally need to be verified. However, it should be stressed once more that the hypothesis of a cooling effect is not in agreement with the available data as discussed above.

Another phenomenon which still remains to be satisfactorily explained is the supposed Mid-Cainozoic expansion of the taxa tolerant to oxygen deficiency. If the eutrophication model of the Mid-Cretaceous extinction is correct, it is difficult to see why such groups as Chironominae have not invaded deep water areas before the general renaissance of the lentic fauna. This problem can be resolved only by future studies on Cretaceous and Cainozoic history of those taxa. In fact, KALUGINA's (1974, 1980b) statement on a limited chironomine role in the Early Palaeogene is only based on a preliminary examination of resin inclusions and thus may be refuted by studies of lacustrine fossils (e.g., from the Green River Series). The evident Mid-Cainozoic increase in chironomid (and very probably chironomine) abundance in lacustrine deposits may be connected with increasing food availability after the expansion of submerged vegetation over extensive lake areas. It is also possible that in hypereutrophic lakes, such as at Menat or Messel, only the head capsules of chironomid larvae would be well preserved, and there can be detected only with special micropalaeontological techniques.

The extremely uneven chaoborid distribution in the Cainozoic is intriguing. The sedimentary data indicate that the majority of lakes represented in the Cainozoic insect record should be meromictic. Modern chaoborids inhabit rather diverse standing waters but are often abundant in permanently stratified lakes, and their subfossil remains are used in palaeolimnology as indicators of meromictic conditions (STAHL 1959, LÖFFLER 1986, UUTALA 1990). However, for unknown reasons Subtype 6B assemblages with high chaoborid frequency are surprisingly rare in the pre-Quaternary record.

In general, the Cainozoic history of lentic insects is still a puzzle, with only a few fragments probably properly understood. Most probably, it will only be satisfactorily explained when the insect data are analysed together with those for other aquatic organisms, and so the evolution of the entire lake ecosystem will be interpreted.

The Cainozoic lacustrine faunas allow us to test some ecological hypotheses proposed for more ancient faunas. In particular, a strong contrast between total abundance and diversity of benthic insects between the Mesozoic and Cainozoic assemblages is in accordance with the hypothesis of higher and more stable oxygen concentration in large Mesozoic lakes, and the very pronounced differences between the Mesozoic and Cainozoic assemblages support the idea about fundamental differences in their ecosystem structure.

As pointed out above, the lotic insect fauna was probably much less affected by the Mid-Cretaceous extinction than the lentic fauna. Consequently, modern lotic ecosystems should retain a much more ancient basic structure. Their evolution was probably more slow and gradual, especially in the rhithral zone which was less affected by direct angiosperm invasion than the potamal. On the other hand, if the dominant Palaeogene damselfly families were indeed lotic, the gradual changes in their composition continuing in the Neogene should indicate that those gradual biotic changes were more continuous than in the lentic habitats.

For general taphonomic reasons, the completeness of the Cainozoic fossil record should increase in comparison with older times. Indeed, some Cainozoic sites seem to represent peculiar environments undocumented in the previous fossil record. Besides the supposed Type 10 alkaline lake biota some smaller sites should be mentioned in this respect including the thermal spring deposits in the Early Miocene of Böttingen, Germany (ZEUNER 1931) and Late Miocene of Przeworno, Poland (GALEWSKI & GŁAZEK 1977) which contain remains of dipteran larvae and adult hydrophilid and dytiscid beetles.

3.4
Geographical History of Insects

K.YU. ESKOV

3.4.1. INTRODUCTORY REMARKS

Geographical distribution of any taxon in any period of its history has been formed by two main factors: by the actual climatic gradients (equatorial-temperate, humid-arid etc.), and by the previous dispersal routes (sea barriers for the terrestrial groups, continents for the marine ones). The significance of every component in this couple of factors seems to be quite different for various groups. For instance, terrestrial vertebrates (tetrapods) seem relatively autonomous from their environment, and their distribution pattern would reflect mainly the direct isolation in past. In contrast, the vascular plants are able to cross the marine barriers as seeds or spores, and their distribution pattern at every period would be formed mainly by the actual climatic boundaries. In this respect the winged insects seem to be much closer to the plants than to the vertebrates. Due to this reason the geographical history of insects is anticipated to be quite similar to one of the vascular plants, and we should base our reconstructions mainly on the palaeofloristic regions of the World.

DARLINGTON (1957) have supposed that the formation of the new taxa is restricted mainly to the tropical equatorial zone, and proliferated descendants oust the ancestors to the extratropical regions; he called this model the "tropical pump". MEYEN (1987b) presented a similar model, termed "phytospreading", and supported it with a huge amount palaeobotanical data. He demonstrated that during the Late Palaeozoic (i.e. the time of establishment of the main evolutionary lines of vascular plants) almost all new taxa (of familial or higher rank)

appeared in the low latitudes at the more deep stratigraphic levels than at high latitudes. Hence, new taxa, as a rule, really originate in the equatorial zone and later "spread" to the extratropical latitudes. The opposite situation, i.e. the origin of a taxon in boreal zone and its subsequent invasion to the tropics, never have been recorded (MEYEN 1987b).

Later, RASNITSYN (1989b) and ESKOV (1994, 1996) have modified Meyen's model. They supposed that the phenomenon of "phytospreading" is rather ecological than geographical in nature. It means the localisation of macroevolutionary processes in regions of the world where the climatic conditions are most favourable (that is, the abiotic component of natural selection is weakest), followed by anisotropic penetration ("spreading") of the newly formed taxa into regions with harsher abiotic conditions. So, in the epochs of the sharp climatic zonation, such as the recent time or the Late Palaeozoic studied by Meyen, the main area of macroevolution is quite localised in the wet and hot equatorial zone. But at the epochs of effaced climatic zonation, such as the Mesozoic, the restriction of the centres of faunogenesis to the equatorial zone would be less expressed. A "falsifiable prediction" has been made (ESKOV 1994): the number of exceptions to Meyen's scheme would have increased with the weakening gradient in global climatic zonation; this supposition was tested on the distribution patterns of various terrestrial arthropod groups (ESKOV 1996) and seems to be confirmed.

3.4.2. DEVONIAN STAGE OF INSECT EVOLUTION

Very few insect remains are known from the Devonian strata (Fig. 496). All these remains are supposed to belong to Archaeognatha or other wingless forms. No strict geographical conclusions may be based on this scanty material; however, it should be mentioned that all of the three Devonian insect localities lie within 10° from the Devonian equator (LABANDEIRA, BEALL & HUEBER 1988).

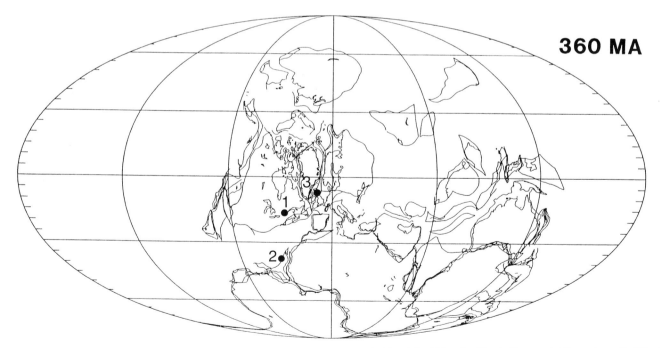

Fig. 496 Devonian insect sites shown on the Earth as it looked 360 million years ago. 1 – Gaspé, 2 – Gilboa, 3 – Rhynie (for more details see Appendix [Chapter 4]). Mobilistic map by SCOTESE (1994)

3.4.3. CARBONIFEROUS STAGE OF INSECT EVOLUTION

This time is known as the period of sharp climatic zonation ("cryo-era"), similar to the modern one. The main palaeofloristic regions were the tropical Euramerican and Cathaysian Realms, north-extratropical Angaran Realm and south-extratropical Gondwanan Realm (Fig. 497). The climate of both Euramerica and Cathaysia was wet and hot, the climate of Angaraland and Gondwanaland varied from cool-temperate to glacial; at the Late Carboniferous several huge continental glaciations developed in Gondwanaland and, according to the modern data, in Angaraland as well (CHUMAKOV 1994). Periglacial plains of Gondwanaland were covered by so-called "Bothrichiopsis tundra"; the landscape of Angaraland, by analogy, is usually called the "Cordaites taiga" (MEYEN 1987b).

All 15 insect orders known from the Carboniferous originated in the tropical zone of the Euramerican region. Two of them (Triplosbida and Syntonopterida) never occurred outside of it, and the remainder appeared in the higher latitudes only after some interval of geological time. Nine (Paoliida, Eoblattida, Blattida, Caloneurida, Grylloblattida, Hypoperlida, Dictyoneurida, Diaphanopterida and Mischopterida) appeared in Angaran and/or Gondwanan realms in the Late Carboniferous, while five (Blattinopseida, Palaeomanteida, Ephemerida, Odonata and Orthoptera) "spread" from tropics only in the Permian. For example, the insect order Hypoperlida, which arose in the tropics, penetrated to the extratropical regions of both Gondwanaland and Angara, and later (in the Early Permian) flourished in the very northern extratropical latitudes – a situation like that noted by MEYEN (1987a) in the history of cordaites and glossopterids. Of the insect families found in the Angaran and Gondwanan faunas, only four (the Psychroptilidae,

Aykhalidae, and Tschirkovaeidae of the Dictyoneurida and the Hypoperlidae of the Hypoperlida) have no confirmed tropical origins.

A similar distribution pattern is demonstrated by other Palaeozoic arthropods. The arachnids were extremely abundant and diverse during the Late Devonian and the Early and Middle Carboniferous in the Euramerican region; all 14 of their orders known in the Palaeozoic originated in this region (Scorpionida, Thelyphonida, Phrynida, Araneida, Kustarachnida, Chernetida, Ricinuleida, Solpugida, Acarida, Haptopodida, Architarbida, Opilionida, Trigonotarbida and Anthracomartida). Of these, only the Scorpionida and Araneida subsequently appeared in the Late Carboniferous of the Angara region, the Trigonotarbida in the Late Carboniferous of Angaraland and Gondwana, and the Opilionida in the Permian of the Subangara region. The Acarida and Chernetida did not appear in extratropical regions until the Jurassic and Cretaceous, respectively, but this lag may be merely due to taphonomic causes. The remaining 8 orders have never been found outside the tropical zone.

The diplopods are represented in the Early and Middle Carboniferous by three orders, which originated in Euramerica: the Amynilyspedida, Spirostreptida, and Spirobolida. The first two of these never occurs outside the tropics, and the last appears only in the Late Permian of Subangara and the Early Triassic of Siberia. Thus, both these classes of arthropods fully correspond to Meyen's model for the Late Palaeozoic. At the beginning of the Permian, however, they practically disappeared from the fossil record, and their Mesozoic-Cainozoic history is known so fragmentarily that no generalisations are possible.

The structure of insect population in the tropical and extratropical regions in Carboniferous is extremely different. Extratropical faunas consists of few species represented by a relatively huge series of specimens. For instance, in Chunya locality (Angaran Realm) the great

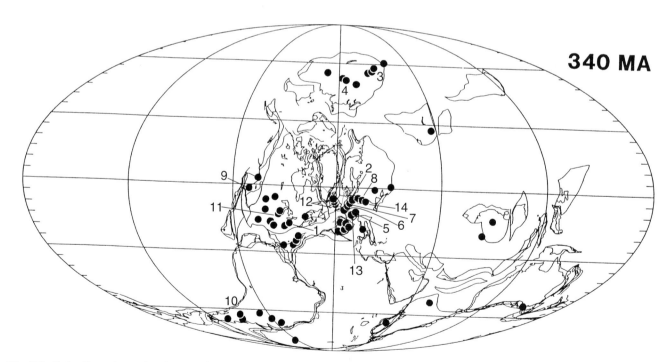

Fig. 497 Carboniferous insect sites shown on the Earth as it looked 340 million years ago. Numbered are sites/site groups catalogued in Appendix (Chapter 4), as follows: 1 – Allegheny Series, 2 – Bitterfeld/Delitz, 3 – Cheremichkino, Zavyalovo, Zheltyi Yar etc., 4 – Chunya, 5 – Commentry, 6 – Gulpen, 7 – Hagen-Vorhalle, 8 – Horní Suchà and Vrapice, 9 – Madera Formation, 10 – Malanzán Formation, 11 – Mazon Creek, 12 – Middle Coal Measure, 13 – Montceau-Les-Mines, 14 – Saar Basin. Mobilistic map by SCOTESE (1994)

majority (87%) of specimens belong to: a single species of dictyoneurid (more than 100 imprints), a single species of eoblattdan (more than 60 imprints), and a series of highly uniform blattid fragments (more than 30 imprints). The situation in the tropical Euroamerican Realm is the opposite. For instance, in Commentry (France) 13 species of dictyoneuridans are found, and only two of them are present by two imprints; 35 insect remains from the Middle Carboniferous of the Saar Basin presents 31 species. So, the species diversity of insects in tropic belt in the Carboniferous was much higher than in the temperate one; in extratropical regions a strong dominance of a few species is observed, known as a character of recent extremal habitats, such as tundra etc. (ZHERIKHIN 1980a).

3.4.4. PERMIAN STAGE OF INSECT EVOLUTION

Like the Late Carboniferous, the Early Permian was a cryo-era: it was a period of sharp climatic zonation accompanied with the continental glaciations, at least in Gondwanaland. At the beginning of the Early Permian the arrangement of the main phytogeographic regions are almost identical to these in the Late Carboniferous (MEYEN 1987a). The type of faunogenesis did not change, and the "equatorial pump" continued to operate steadily without interruption.

All the many higher taxa that appeared in the Early Permian seem to have originated in the tropical equatorial zone of Euramerica; these were the Jurinida, Neuroptera, Coleoptera, Trichoptera, Panorpida, Psocida, Thripida and Hemiptera, as well as the suborders Ephemerina in the mayflies, Grylloblattina in grylloblattidans, and Protelytrina in earwigs. The majority of the above orders and suborders have spread to the extratropical regions only in the latest Early Permian and often

flourished then and in the Late Permian in Angara and possibly in Gondwanaland (the existing records for Australia and South America are not enough for a sure conclusion). These very taxa have formed the basis of Late Palaeozoic–Early Mesozoic global fauna of the insects (though the taxa of the Carboniferous origin, i.e. Odonata, Palaeomanteida, Blattida, Grylloblattida (other than Grylloblattina), Orthopterea also contributed appreciably to that fauna). Of them, only ditaxineurine and kennediine dragonflies and protelytrine earwigs have not survived the Permian-Triassic boundary. Thus, the entire Carboniferous and Early Permian history of the higher taxa of insects shows not a single (!) deviation from Meyen's "phytospreading" scheme.

The global climatic pattern was somewhat changed towards the end of the Early Permian (beginning from the Kungurian). The temperature contrasts were no longer an issue: there was a frostless climate in both Gondwanaland and Angaraland. The southern part of the latter (the Urals, Kazakhstan, Dzhungaria, Mongolia, Manchuria) formed a new palaeofloristic region – the Subangaran one (Fig. 498). Subtropical (or warm-temperate) SubAngaraland is regarded by palaeobotanists as the very place of origin for the majority of taxa, that later composed the bulk of the Mesozoic flora of the Earth (MEYEN 1987a). So the role of "generating forge" had changed hands: from the tropics to the warm-temperate zone. Probably this was because the abiotic conditions in the equatorial zone became quite hard due to global aridisation.

The biogeographic pattern observed in insects seems to be very similar to the plant one. The orders Corydalida, Perlida, and Phasmatida, as well as the earwigs (Forficulina), the hemipteran suborders Cicadina and Psyllina, the grasshoppers (superfamily Tettigonioidea) first appeared in Subangaran region. Stoneflies, in particular, were represented here by both of its main infraorders, the Perlomorpha and the

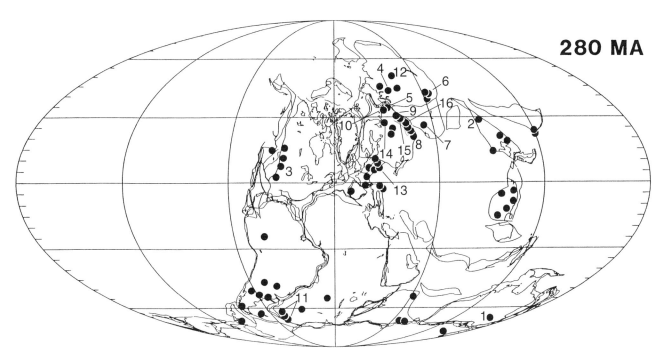

Fig. 498 Permian insect sites shown on the Earth as it looked 280 million years ago. Numbered are sites/site groups catalogued in Appendix (Chapter 4), as follows: 1 – Belmont, 2 – Bor-Tologoy, 3 – Elmo, 4 – Fatyanikha, 5 – Inta Formation, 6 – Kaltan, Erunakovo Formation, Ilyinskaya Formation etc., 7 – Karaungir, 8 – Kargala mines, 9 – Kityak, 10 – Lek-Vorkuta Formation, 11 – Middle Beaufort Series, 12 – Nizhnyaya Tunguska, 13 – Obora, 14 – Soyana, 15 – Tikhiye Gory, 16 – Tshekarda. Mobilistic map by SCOTESE (1994)

Nemouromorpha. The third stone-fly infraorder, the Gripopterygomorpha, appeared during the Late Permian in south-extratropical Gondwanaland (the recent gripopterygomorph stoneflies are endemics of the "Gondwanan" continents, but in the Mesozoic the infraorder was widely spread in Eurasia, see Chapter 2.2.2.2.2f). So, the main sources of new high-rank insect taxa were extratropical (warm-temperate) zones, both northern and southern; for instance, Panorpida flourished in both these zones. On the other hand, in the Late Permian in both Subangaran and Gondwanan regions the cockroaches, the dominant carboniferous group, have almost disappeared.

3.4.5. TRIASSIC STAGE OF INSECT EVOLUTION

In the Triassic all the continents reunited to form the super-continent, Pangea. The climate displayed intermediate contrasts (with a huge arid belt in the equatorial zone), and the super-continent was peculiar by its homogeneous biota. Palaeobotanists emphasise the loss of all obvious floristic boundaries (MEYEN 1987a); the palaeofloristic pattern sketched out by DOBRUSKINA (1982), may be tentatively referred to as a system of diffuse florogenetic centres. Relatively small climatic gradients together with the spatial unity of the Triassic continents account for the very quick dispersion of both insect and plant taxa across the Pangaea; new taxa appearing almost simultaneously over all the present-day continents. Hence, it becomes practically impossible to identify the place of origin for many taxa, and this is also true for all high-level Triassic taxa. The dispersion pattern of Vertebrates is governed by the same regularities.

The Triassic is known as one of the key stages in insect evolution. Numerous taxa arose at that time, and almost all of them appeared synchronously in geographically distant and/or climatically different localities: Diptera in the France (Vosges), North America, the Balearic Islands, and Central Asia, Hymenoptera in Central Asia and Australia, Mesotitanida in Central Asia, Australia, and Ukraine, the advanced dragonflies (suborder Libellulina) and the crickets of the superfamily Grylloidea (Central Asia, Australia, Argentina, South Africa and, only dragonflies, in Japan), the annulipalpian caddis-flies (England, Central Asia and Argentina), the true bugs (Hemiptera: Cimicina) (Vosges, North America, Central Asia and Australia), the aphids (Hemiptera: Aphidomorpha) (Vosges, Central Asia and Australia); the mesoblattid cockroaches quickly spread all over the world, so that in the poor and highly uniform East Asian faunas they became a main, or sometimes even the only component. It should be noted, that the age of Vosges is now appreciated to be more ancient than the other above localities; however, the spread of such taxa as dipterans, bugs, etc., from the equatorial zone may be stated only with reservations, due to absence of contemporary extratropical faunas.

The most intensive faunogenetic processes, however, may be supposed for the palaeofloristic Central Asian Area, which quite strictly corresponds to the Late Permian SubAngaran Area (cf. Figs. 498 and 499) and, probably, has inherited its role of "generating forge" from the latter. Subtropical Central Asia seems to be an area of appearance of the above Triassic high-rank taxa; several insect taxa were in the Late Triassic, the endemics of this region, and spread out sufficiently later: the infraorder Coccomorpha (Hemiptera), the true thrips (suborder Thripina), the curculionoid beetles (the most advanced and proliferated group of coleopterans). On the other hand, there is the first precedent of the appearance of a high-rank taxon in high latitudes and its subsequent spread to the low latitudes, the locusts (suborder Acridina) that first appeared in the Late Triassic of Siberia.

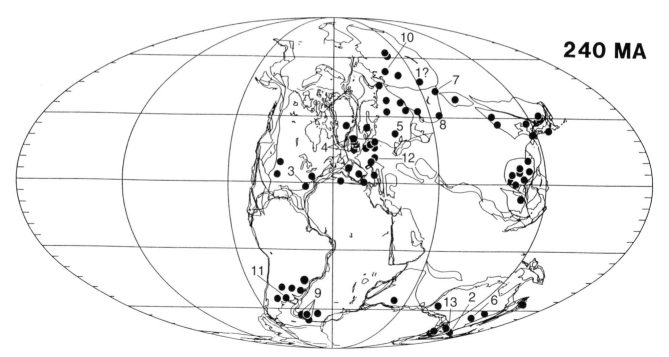

Fig. 499 Triassic insect sites shown on the Earth as it looked 240 million years ago. Numbered are sites/site groups catalogued in Appendix (Chapter 4), as follows: 1 – Babiy Kamen', 2 – Brookvale, 3 – Cow Branch, 4 – English Rhaetian, 5 – Garazhovka, 6 – Ipswich, 7 – Kenderlyk, 8 – Madygen Formation, 9 – Molteno Formation, 10 – Pelyatka, 11 – Potrerillos, 12 – Vosges, 13 – Wianamatta. Mobilistic map by SCOTESE (1994)

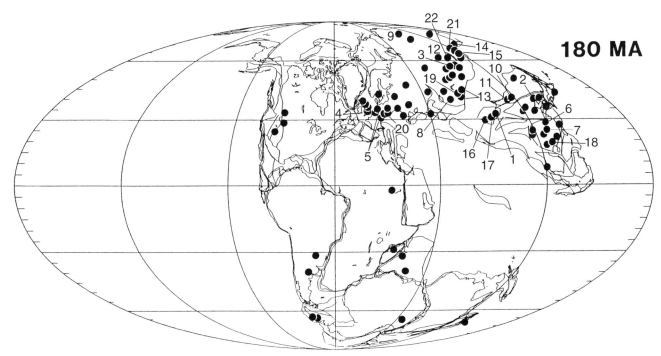

180 MA

Fig. 500 Jurassic insect sites shown on the Earth as it looked 180 million years ago. Numbered are sites/site groups catalogued in Appendix (Chapter 4), as follows: 1 – Bakhar, 2 – Beipiao, 3 – Chernyi Etap etc., 4 – English Lias, 5 – German Lias, 6 – Hansan Formation, 7 – Jiaoshang Formetion, 8 – Karatau, 9 – Kempendyay, 10 – Khotont, 11 – Khoutiyn-Khotgor, 12 – Kubekovo, 13 – Kyzyl-Kiya, Sagul, Shurab, 14 – Mogzon, 15 – Novospasskoye, 16 – Oshin-Boro-Udzur-Ula, Zhargalant, 17 – Shar Teg, 18 – Shiti, 19 – Sogyuty, 20 – Solnhofen etc., 21 – Uda, 22 – Ust-Baley, Iya etc. Mobilistic map by SCOTESE (1994)

3.4.6. JURASSIC STAGE OF INSECT EVOLUTION

The Jurassic was a time of drastic tectonic differentiation of the land masses. The Tethys Ocean separated Laurasia (more or less corresponding modern Eurasia) from the block of southern continents, i.e. Gondwana. The newly formed Indian and Atlantic Oceans had divided the Southern continental block into the Western Gondwanaland (South America, Antarctica, Australia, New Zealand) and Eastern Gondwanaland, which began dividing further into Africa, Hindustan and Madagascar. On the other hand, the Late Triassic, the phytogeographic differentiation of world vegetation became almost the same as in the Jurassic (MEYEN 1987a): the main areas were nothal, equatorial and Siberian (Fig. 500).

The majority of high ranking taxa originated during the Jurassic in the equatorial belt, and subsequently appeared at higher latitudes (in Siberia and Gondwana). There were: moths (England and Kazakhstan), snake-flies (Rhaphidiida) (England and Central Asia), advanced dragonflies (higher Libellulina) (Germany, Kazakhstan), the bark-lice of the suborder Psocina and hemipterans of the infraorder Aleyrodomorpha (Kazakhstan), and the culicomorph dipterans (Central Asia). The parasitic hymenopterans ("Parasitica") present a contrasting history in that they appeared almost simultaneously (though they are very rarely found), in both the equatorial zone (Germany and Central Asia) and in Siberia; later, in the Late Jurassic, they flourished in Kazakhstan.

On the other hand, it was the time when apparent movement occurred in the opposite direction, from high latitudes to low ones. There was only one such instance in the Triassic – the Acridina (see above). During the Early and Middle Jurassic, however, there are a whole series of taxa which had originated in the temperate Siberian region (including Mongolia and northern China) and some time later penetrated into the Equatorial region. These include caddis-flies of the most advanced suborder Integripalpia (Phryganeina), both the leading superfamilies of mayflies (the Ephemeroidea and Siphlonuroidea) and several important hymenopteran superfamilies (Proctotrupoidea, Evaniomorpha, Tenthredinoidea). Due to this reason ZHERIKHIN (1980a) notes that the Siberian entomofaunas of Early Jurassic are very advanced and seem to be more similar with the Late Jurassic ones than to the equatorial (Indo-European) faunas of the Early Jurassic.

Both of the above trends can be regarded as departures from the operating regime of the "equatorial pump"; they can be directly attributed to the establishment of a less contrasting world climate in the geologic time under consideration.

3.4.7. CRETACEOUS STAGE OF INSECT EVOLUTION

The Cretaceous is known both as the time when continental masses were most fragmented, and when climatic gradients were most obliterated: everywhere on the Earth the climate was similar to that of the present-day subtropics to warm-temperate regions ("thermal-era"). The arrangement of the main palaeofloristic regions in the Cretaceous generally correspond to those found in the Jurassic, named Siberio-Canadian, Euro-Sinian, Equatorial and Nothal (Fig. 501).

No results of the operation of the "tropical pump" have been found in the Cretaceous with its almost uniform climate and very weak latitudinal zonation. The only case of the origination of a taxon in relatively low latitudes and its farther propagation into the higher latitudes is represented by the order Termitida. The first termites appeared in the Neocomian of Spain and the Weald of

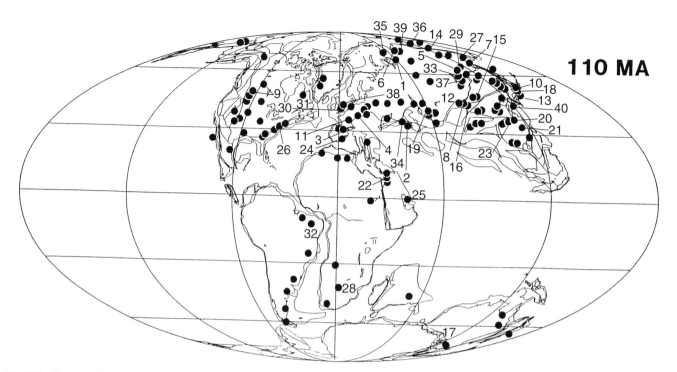

110 MA

Fig. 501 Cretaceous insect sites shown on the Earth as it looked 110 million years ago. Numbered are sites/site groups catalogued in Appendix (Chapter 4), as follows: 1 – Agapa, 2 – Agdzhakend, 3 – Álava, 4 – Austrian amber, 5 – Baissa, 6 – Begichev Formation, 7 – Bolboy etc., 8 – Bon-Tsagan, Khurilt etc., 9 – Canada amber, 10 – Chosi, 11 – French amber, 12 – Gurvan-Eren Formation, 13 – Iwaki, 14 – Khetana, 15 – Khutel Khara, 16 – Khutuliyn, 17 – Koonwarra, 18 – Kuji, 19 – Kzyl-Zhar, 20 – Laiyang Formation, 21 – Laocun, 22 – Lebanese amber, 23 – Lushangfeng Formation, 24 – Montsec, 25 – Near East amber (deep sea sample), 26 – New Jersey amber, 27 – Obluchye, 28 – Orapa, 29 – Polovaya, 30 – Purbeck, 31 – Redmond, 32 – Santana, 33 – Semyon, 34 – Shavarshavan, 35 – Taimyr Lake retinites, 36 – Timmerdyakh, 37 – Turga, 38 – Wealden, 39 – Yantardakh, 40 – Yixan Formation. Mobilistic map by SCOTESE (1994)

England, but it appears in the extratropical territories of eastern Asia (the Neocomian of Transbaikalia and Aptian of Mongolia) so rapidly, that we cannot establish the region of termite origin with certainty; it was most likely the same as in the case of the Triassic termite taxa (see above).

All the other higher-rank taxa originated in extratropical regions of Gondwanaland or Siberio-Canadian Province. Mantida (as well as "pre-fleas" *Saurophthirus* – see below) appeared during the Neocomian in Transbaikalia (Baissa locality), precisely in the layers corresponding to cold climatic episodes. It should be noted that these same layers also contain the first advanced cockroaches (Blattellidae), whereas the "warm" layers are dominated, as before, by archaic groups (Chapter 2.2.1). The most advanced and flourishing group of the higher muscomorph flies – the muscoid group ("Calyptrata") – appeared in the Campanian and the Maastrichtian of Canada. All the infraorders of higher butterflies (Eriocraniina, Incurvariina, Nepticulina, Papilionina) originated in various points of Siberio-Canadian Province. Near the Jurassic-Cretaceous boundary in Mongolia appeared the first stinging hymenopterans (the "Aculeata") and the most advanced and flourishing groups of parasitic hymenopterans – the superfamilies Platygastroidea, Chalcidoidea and Ichneumonoidea. The first ants appear in the Albian of the Okhotsk region (NE Russia), while the true, high social Formicidae are found in the Turonian of New Jersey (USA). The true weevils (Curculionidae – the largest family of the animal kingdom, with about 70,000 Recent species) originated in the Aptian of Mongolia. Siberio-Canadian Province was also the region of proliferation of several key post-Jurassic groups, such as aphids. A remarkable

situation is observed in Pulicida: the first representatives of so-called "pre-fleas" (*Strashila* and *Saurophthirus*) were found in the Late Jurassic and the Neocomian of Siberia, respectively, and, some later, in the Aptian of the high-latitude part of Gondwanaland, in Koonwarra (*Tarwinia*), whereas in the equatorial zone this order has not been recorded up to the appearance of the true fleas (Baltic and Dominican ambers).

The trichopteran fossil cases demonstrate the same pattern as their imagoes. The first cases appeared in the Mid- or Late Jurassic in Laurasia, both extratropical (Transbaikalia and eastern Mongolia) and tropical (western Mongolia), and remained there until the end of the Jurassic. It is worth mentioning that the construction of cases was started by Annulipalpia and Integripalpia practically simultaneously. Reliable data on case construction activities in the Cretaceous are unknown for Annulipalpia, whereas the Integripalpia cases became much more diverse and numerous in the north-eastern Asia. Later in the Early Cretaceous they appeared for the first time in Gondwanaland (Brazil and Australia), although they remain infrequent there. The Early Cretaceous cases had not been discovered in the low latitude of Laurasia until they were encountered quite recently in the Weald deposits in England. Extremely rare are the Low Cretaceous cases in Manchuria, and they are completely absent in the more southern China regions despite their considerable proximity to the Siberian records. In any case, a real extension of the cases construction area, doubtlessly, took place in the Late Cretaceous. Since that time the cases have been reported from a number of sites in Central Asia, Europe, eastern North America.

3.4.8. CAINOZOIC STAGE OF INSECT EVOLUTION

In the Cainozoic the arrangement of the continents was similar to the present, except for Australia (which separated from Antarctica in the Eocene), Hindustan (connected with Eurasia in the Eocene) and South America (isolated from North America up to the Pliocene) (Figs. 502, 503). The Palaeogene was still the "thermo-era", and beginning from the Neogene the "cryo-era" climatic regime (with sharp thermal gradient and polar glaciations) begin to be established. So, it may be expected that the main faunogenetic pattern in the Palaeogene was quite similar to that of the Cretaceous.

ZHERIKHIN (1978, 1980a) supposed that during the Cretaceous and the Palaeogene, the global climate was extremely different from the present and analogues of both modern boreal and equatorial climates were absent. If it is true, tropical ecosystems of the modern type (e.g. hylean rain forests and savannahs) should have appeared only quite recently, simultaneously with the boreal ones (e.g. tundra and taiga). This hypothesis seems to be supported by the composition of a single huge entomofauna from the equatorial zone of the Palaeogene, i.e. Burmese amber (unless its age is actually Late Cretaceous: see Chapter 4.2). RASNITSYN (1996) concluded that "past Burmese forest that has yielded the fossiliferous amber differed fundamentally from the tropical rain forest as we know it".

In fact, a number of groups which played an important role in the functioning of the modern tropical ecosystems originated in the Palaeogene outside the tropical zone. Among these, for example, were the termites of the family Termitidae (Oligocene of Europe), the stingless meliponine bees (Eocene of Europe), and the tsetse flies, Glossinidae (Oligocene of North America).

No taxa of ordinal, subordinal or even superfamilial rank have originated during the Cainozoic (even in the Palaeogene), and it is impossible to test the hypothesis of the "tropical pump" on the modern insect faunas. The situation in the Cainozoic is complicated by the insufficient availability of fossil material from the tropics. It should be noted, however, that the fauna of the Miocene Dominican amber displays several features typical of the modern tropical faunas: it includes some families that are now virtually limited to the tropics, (e.g. brentid beetles), or which are now key elements of the present tropical ecosystems (Termitidae), as well as the almost complete absence of aphids, known to be one of the most important elements of the post-Jurassic temperate entomofaunas.

3.4.9. PROBLEM OF THE SO-CALLED "GONDWANAN" RANGES OF THE RECENT TAXA

ZHERIKHIN (1978, 1980a) have emphasised that numerous "far-southern" taxa, now restricted to the Southern Hemisphere, were, however, present in the Cretaceous and Palaeogene entomofaunas of Eurasia and/or North America. Such, for instance, is the chironomid midge subfamily Aphroteniinae, found as fossils in the Late Cretaceous amber of Yantardakh in North Siberia, while its recent members are restricted to South Africa, Australia and South America. These findings put a new perspective on the problem of the so-called "Gondwanan ranges".

Modern biogeographers prefer to explain the origin of such distribution patterns (as in aphroteniines), in relation to the existence in the Mesozoic of the protocontinent Gondwanaland and to its subsequent

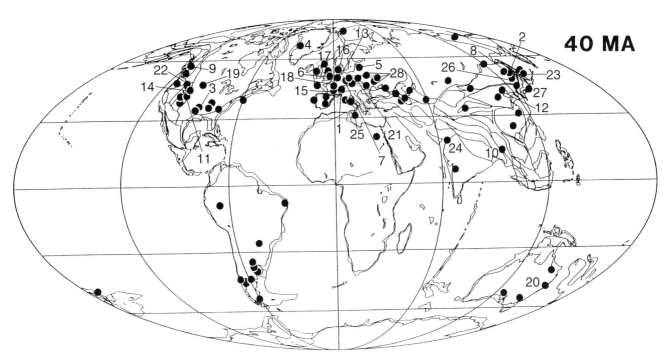

Fig. 502 Palaeogene insect sites shown on the Earth as it looked 40 million years ago. Numbered are sites/site groups catalogued in Appendix (Chapter 4), as follows: 1 – Aix-en-Provence, 2 – Amgu, 3 – Arkansas amber, 4 – Atanikerdluk, 5 – Baltic amber, 6 – Bembridge Marl, 7 – Bitterfeld amber, 8 – Bol'shaya Svetlovodnaya, 9 – British Columbia amber, 10 – Burmese amber, 11 – Florissant, 12 – Fushun, 13 – Geiseltal, 14 – Green River, 15 – Menat, 16 – Mo Clay, 17 – Mull Is., 18 – Oise amber, 19 – Pascapoo Formation, 20 – Redbank Plain, 21 – Romanian amber, 22 – Ruby River, 23 – Sakhalin amber, 24 – Salt Range, 25 – Sicilian amber, 26 – Tadushi, 27 – Ube, 28 – Ukrainian amber. Mobilistic map by SCOTESE (1994)

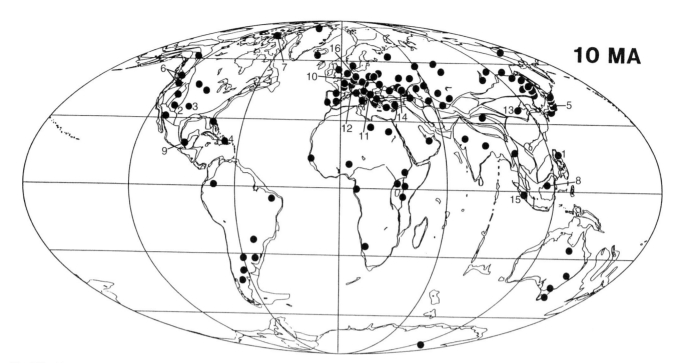

Fig. 503 Neogene insect sites shown on the Earth as it looked 10 million years ago. Numbered are sites/site groups catalogued in Appendix (Chapter 4), as follows: 1 – Bandoc, 2 – Barstow, 3 – Bonner Quarry, 4 – Dominican amber, 5 – Kamiwada, 6 – Latah, 7 – Meigen I., 8 – Merit-Pila, 9 – Mexican amber, 10 – Oeningen, 11 – Radoboj, 12 – Rott, 13 – Shanwang, 14 – Stavropol, 15 – Sumatra resin, 16 – Willershausen.. Mobilistic map by SCOTESE (1994)

fragmentation due to continental drift. The fragments are therefore often believed to have retained representatives of the initial Gondwanan biota even to this day. This approach, recently called "mobilistic biogeography", attempts to fit the cladogram of a studied taxon to the sequence drawn by geophysics for dispersal of the fragments of Pangea and Gondwanaland. This viewpoint, proposed by J. Hooker, now seems to be a basic paradigm of historical biogeography.

Fossil insect data, however, support an alternative hypothesis proposed by A. Walles. His "ousted relics" concept identifies the Southern Hemisphere disjunct ranges as the results of extinction of "intermediate links" at northern continents. Numerous "Gondwanan" insect taxa have been encountered as fossils in Eurasia and North America, and their number is still increasing (see the check-lists in ESKOV 1987, 1992). There are, for instance, such classic objects of the "mobilistic biogeographers", as aphroteniine midges, gripopterygomorph stoneflies, coleorrhynchan bugs, nannochoristid scorpion-flies.

So, it may be supposed that the southern most, so-called "Gondwanan" groups are only relics of a broader, possibly, global distribution. The present-day Southern Hemisphere fauna of many insect groups inherited several elements of the Mesozoic and Palaeogene faunas. Attempts to consider their places of origin as being the same as their present-day distribution seem to be inadequate (ESKOV 1984b, 1987; RASNITSYN 1996).

It should be emphasised that the conservation of Mesozoic and Palaeogene relics is not a monopoly of the Southern continents. Several "Laurasian" taxa, restricted to Eurasia and/or North America, were found as fossils at southern, Gondwanan, continents (e.g. the snake-flies, the most archaic hymenopterans, Xyelidae). These cases seem to be much less common than the "Gondwanan ranges"

(such as Aphrotheninae), but it is possibly an artefact of a relative deficiency of palaeoentomological material from the Southern Continents. In fact, a new study of rich Gondwanan localities (e.g. Santana and Koonwarra) over the past decade has enlarged the check-list of the taxa that are now extinct in Gondwanaland by more than a factor of two (ESKOV 1992), and the number of such patterns is increasing (e.g. the currently Mediterranean Scoliidae-Proscoliinae found as fossils in Santana: RASNITSYN & MARTÍNEZ-DELCLÒS 1999).

The above conclusions don't refute the existence of real Gondwanan groups of insects, which were connected with the Southern Continents during all their history, such as penguins among the vertebrates and *Nothofagus* among the plants. For instance, the modern ranges of the hymenopteran families Pergidae and Monomachidae, as well as of the basal Xiphydriidae (subfamily Derecyrtinae), are restricted to the Southern Continents, and their past ranges are either southern as well (in case of Monomachidae) or totally unknown (Chapter 2.2.1.3.5). But the conclusion about the Gondwanan origin of a taxon should be based on an adequate number of fossils; the value of the modern range appears to be of very limited significance.

3.4.10. SOME GENERAL PHYLOGENETIC PATTERNS

Thus, the "falsifiable prediction" made earlier seems to have been confirmed. The number of exceptions to Meyen's scheme actually increased as the global gradient of climatic zonation weakened, and in the maximally azonal interval – the Cretaceous period – the model ceased to work entirely. This supports the hypothesis that the phenomenon of "phytospreading" is not so much geographic as ecological in nature. In the present context, phytospreading is the localisation of macroevolutionary processes in regions of the world where the climatic conditions are most favourable (that is, the abiotic component of natural selection is the weakest), followed by anisotropic propagation ("spreading") of the newly formed taxa into regions with harsher abiotic conditions. This mechanism can function in different regimes dictated by external conditions, and its operation in the equatorial pump regime is no more than a widespread but partial case. There are three such regimes (determined by the global climatic gradient), and they form a closed three-element cycle.

1. The Late Palaeozoic was characterised by a sharply manifested climatic zonation that included the glaciations of both Gondwanaland and Angaraland. The very powerful equatorial-polar temperature gradient reduced all the other climatic gradients to the status of fluctuations. Phytospreading operated in the equatorial pump regime: all the new taxa of high rank originated in the tropics and then migrated into the higher latitudes, where they sometimes survived even after their disappearance from the regions of their origin. The faunal changes that took place in this interval (including such a radical one as the Carboniferous-Permian) had no effect on the type of faunogenesis.

2. In the Mesozoic and Early Palaeogene, the climatic zonation became increasingly weaker. The lessening and disappearance of the single equatorial-polar temperature super-gradient led to the appearance (and increasing manifestation) of second-order gradients – humid-arid, lowland-highland, etc. As a result, the single tropical-polar phylogenetic gradient also gradually broke down; the equatorial pump began to operate intermittently and then ceased to function altogether. Phytospreading then began to operate in the regime of Dobruskina's diffuse centres of speciation.

3. The Late Palaeogene and Early Neogene was characterised by a transition from an azonal to a zonal epoch. Establishment of the new climatic regime was due to redistribution of the thermal balance resulting from tectonic movements (the separation of Antarctica from South America, leading to the inception of the Antarctic circulation, etc.). The biotic reorganisations (mass extinctions in the circumequatorial and circumpolar regions, concentration of the Mesozoic-Palaeogene relicts in the subtropical and warm-temperate zones) can be characterised by Zherikhin's zonal stratification mechanism.

4. During the interval from the Neogene to the present, the three-stage cycle was completed, and we are now returning to the initial phase of a zonal epoch; phytospreading again operates in the regime of the equatorial pump. Darlington (1957), in fact, proposed this term for the current period of geologic history. The principal macroevolutionary events are thus taking place in the tropics once again. One might even attempt to guess which of the "monsters" that abound in the present tropical faunas will become the initial forms of new orders and suborders to propagate over the earth after the next 10–15 Myr. Unfortunately, of course, it is now obviously impossible to make falsifiable predictions.

On the other hand, the relict survivors from preceding epochs turn out to be mainly localised in the extratropical regions; their future fate will be governed by its own quite rigorous laws. The successive stages in the reduction of the initial amphitropical (bipolar) geographic ranges of such relicts (for example, the "circum-Pacific domino scheme"; ESKOV & GOLOVACH 1986) have been considered in terms of both terrestrial (ESKOV 1987, 1992) and marine (NEWMAN & FOSTER 1986) arthropods.

4
Appendix: Alphabetic List of Selected Insect Fossil Sites

4.1
Impression Fossils

A.P. RASNITSYN AND †V.V. ZHERIKHIN

Aix-en-Provence (see Figs. 5: 1, 502: 1). Famous Oligocene gypsum bearing marls near Aix in Bouche-du-Rhône; France, yielded rich insect and plant material. Correlated to the nanoplankton zone NP 23 and mammalian zone MP 26 (MAI 1995) and hence to the earliest Late Oligocene (basal Chattian: SCHMIDT-KITTLER 1987). Many insects described by THÉOBALD (1937) and other earlier students badly need revision.

Alfred Mine: see **Bitterfeld/Delitz**

Allegheny Series (now Formation) (see Figs. 3: 4, 496: 1) of the later Middle Carboniferous (Westphalian D), of the Appalachian area, Eastern USA (PHILLIPS & PEPPERS 1984).

Amgu (Amagu) (see Figs. 5: 2, 502: 2). The tuffaceous mudstones of the Khutsin Formation at the right side of Kudya River (a tributary of Amgu, or Amagu, or Granatovaya, in Terney District in the Maritime Province of the eastern Russia) have yielded some 500 fossil insects. The deposits were often dated as the Late Oligocene or Early Miocene, but currently are considered older, earlier in the Oligocene or even Late Eocene (ZHERIKHIN 1998).

Arkagala (see Figs. 4: 40). The supposedly Cenomanian (earliest Late Cretaceous) mudstones of the Arkagala Formation are exposed at the Arkagala coal quarry at upper Kolyma basin, East Siberia (ZHERIKHIN 1978: 94). More than 1000 insect remains (mainly undescribed beetles) are available.

Atanikerdluk (see Figs. 5: 4, 502: 4) The insects are collected in the Early Palaeocene (Danian) deposits at the Atanikerdluk Mt. (Nugssuaq Peninsula) and nearby in western Greenland; some insects are thermophilous (ZHERIKHIN 1978: 102).

Babiy Kamen' (see Figs. 4: 1, 499: 1). A few insects, myriapods and other fossils are found in that locality in the Tom' River, Kuznetsk Basin, S. Siberian, in the upper part of the Maltsevo Formation which is of the Early Triassic age (PONOMARENKO & SUKATSHEVA 1998).

Baissa (see Figs. 4: 42, 501: 5). The most important earlier Early Cretaceous insect locality represents marls and limestones of the Zaza Formation outcropping on the left bank (facing downstream) of Vitim River in its upper stream (West Transbaikalia in Central Siberia). More than 10,000 fossil insects, often of excellent preservation state, have been collected there, some of them even displaying well preserved pollen food within their gut (KRASSILOV & RASNITSYN 1982). Another important feature of the locality is the quick climatic change reflected in changes of the insect assemblage composition (e.g. VRŠANSKÝ 1997, ZHERIKHIN et al. 1999 and references therein; earlier these changes were described as cyclic, but the upper part of the section revealing a 'cooler' insect assemblage similar though not identical to that from below in the section has been found to represent the tectonically uplifted block: ZHERIKHIN et al. 1999). Proposed correlation is to the **Purbeck** (RASNITSYN et al. 1998), although in calculations for Chapter 3.1 the age is taken as the Late Neocomian, and there are also alternative proposed datings (see ZHERIKHIN et al. 1999).

Bakhar (see Figs. 4: 14, 500: 1). Insectiferous marls, mudstones, and paper shales of the Togo-Khuduk Member, Bakhar Series, in Central Mongolia (Bayan-Khongor Aymag, 12 km NE of Tsetsen-Ula Mt.) have yielded some 5,000 fossil insects (SINITZA 1993). Usually correlated to the Middle Jurassic, though the Late Jurassic age is not excluded (PONOMARENKO 1998b).

Barstow Formation (see Figs. 5: 32, 503: 2) (later Middle Miocene, EBERTLY & STANLEY 1978) in Mojave Desert, California, USA. Calcareous nodules contain small silicified arthropods (crustaceans, arachnids and insects, mainly immatures biting midges and diving beetles) of excellent preservation state (SEIPLE 1983 and references therein).

Bayan-Teg (see Figs. 4: 15). Some 1,000 insects have been collected in marls at the Bayan Teg Coal Quarry in the Uver-Khangay Aymag, Mongolia, of supposedly Middle Jurassic age.

Beipiao (see Figs. 4: 16, 500: 2). Rich insectiferous deposits of Jiulongshan (or Haifanggou) Formation in Beipiao Co. (Liaoning Prov., China) are usually considered Early or Middle Jurassic (HONG 1983a, REN et al. 1995), and can be correlated with Ust-Baley *q.v.*. Similar deposits yield fossil insects in Luanping Co., Hebei Prov. (HONG 1983a).

Belmont (see Figs. 3: 18, 498: 1). The Australian Late Permian (Tatarian) insects are found in numbers in fine-grained chert of the

Newcastle Coal Measures at Belmont and Warner's Bay in N.S.W. The fauna is remarkably unbalanced being dominated by homopterans and mecopterans, with gryllonean and other exopterygote orders (except Hemiptera and Psocida) virtually absent (RIEK 1970b).

Bembridge Marls (see Figs. 5: 6, 502: 6) (called also Insect Beds, Insect Limestone) are a predominantly argillaceous formation revealed north in Isle of Wight. The age is near Eocene/Oligocene boundary or just above or below that. 3900 rock blocks accumulated in the Natural History Museum (London, UK) bear several times as many insect fossils (JARZEMBOWSKI 1980).

Bitterfeld/Delitz (see Figs. 3: 5, 497: 2). The only indisputable, albeit the latest, Early Carboniferous (earlier Namurian A, Arnsbergian) insect *Delitzschala bitterfeldensis* Brauckmann et Schneider that is found in the area of Bitterfeld/Delitz in East Germany and belongs to the Dictyoneurida (BRAUCKMANN & SCHNEIDER 1996). *Ampeliptera limburgica* Pruvost (Hypoperlida, see Chapters 2.2.1.2.2e and 2.2.1.2.3e) from Gulpen in South Limburg (see Figs. 3: 9, 497: 6), the Netherland, and *Stygne roemeri* Handlirsch (Paoliida, see Chapter 2.2.1) from the Alfred Mine, Upper Silesian Basin, Poland, are often referred to as of the Early Carboniferous age, but in fact they are younger, ranging from the Chokierian or (*Stygne*) Alportian to Kinderscoutian, that is, above the Early/Mid-Carboniferous boundary (BRAUCKMANN *et al.* 1994). The Malanzán Formation at Cuesta de la Herradura (La Rioja, Argentina, see Figs. 3: 9, 497: 6), where three primitive dragonflies and one megasecopteran have been found, their age is the earlier Westphalian rather than the earliest Namurian (BRAUCKMANN *et al.* 1994). A *Rochdalia*-like insect (that is, similar to a nymph of Dictyoneurida) mentioned in the Glencartholm Volcanic Beds, Visean stage, at Eskdale in Scotland (SCHRAM 1983) actually belongs to a non-insect arthropod (Rasnitsyn, personal observation, 1998).

Bol'shaya Svetlovodnaya (former Biamo) (see Figs. 5: 8, 502: 8). The insectiferous lake diatomites of the Late Eocene or Early Oligocene age exposed on the left bank of Barachek Creek 3 km upstream of its mouth at Bol'shaya Svetlovodnaya (former Biamo) River, Pozharsky District, Maritime Province, East Russia. The fossil site has produced more than 1,500 fossil insects of 17 orders. The insect assemblage is characteristic of a woodland in a moderately thermophilous climate and can be possibly correlated to **Florissant** (ZHERIKHIN 1998).

Bolboy (see Figs. 4: 44, 499: 7). Some 300 fossil insects have been collected here in tuffaceous mudstones of Byankino Formation of debatable Late Jurassic or Early Cretaceous age, at the watershed of Bolboy and Urtuy Rivers in Priargunskiy District, Chita Region in Siberia (RASNITSYN 1985a: 11).

Bonner Quarry (see Figs. 5: 42, 503: 12). Various soil arthropods including three insects are described enclosed in the onyx marble of possibly Late Tertiary age from the quarry about 10 miles south-west of Ashfork in Arizona, USA (PIERCE 1951)

Bon-Tsagan (see Figs. 4: 45, 501: 8). The principal insect locality of the later Early Cretaceous in the north hemisphere. Insect fossils are taken from the Khurilt rock unit of supposedly Barremian or Aptian Stage; taken as the Aptian in calculations for the Chapter 3.1. RASNITSYN *et al.* (1998) correlate it with the Weald Clay of **Wealden** (Hauterivian and Barremian, see below). The numerous outcrops of mudstone and marl are widely distributed in Central Mongolia south of the lake Bon-Tsagan-Nur. Ten thousand of fossil insects have been collected there (see also ZHERIKHIN 1978: 44, and for geology SINITZA 1993).

Bor-Tologoy (see Figs. 3: 19, 498: 2). A rich insect locality in deposits of the Tsankhi and Tabun Tologoy (Tavan Tologoy) Formations, supposedly of the Kazanian (PONOMARENKO & SUKATSHEVA 1998) or Tatarian Stage of the Late Permian (SHCHERBAKOV 2000b), in the Central Gobi Aymag, 16 km SE Tsogt-Tsetsiy: almost a thousand insects have been collected but very few have as yet been described. Worth mentioning is that one and the same insect species are recorded in both above formations (A. Ponomarenko, pers. comm.).

Brookvale (see Figs. 4: 2, 499: 2). Usually very well preserved, albeit not particularly numerous, insects are collected in lenticular shale in the Middle Triassic Hawkesbury Sandstone near Sidney, N.S.W., Australia (RIEK 1970a).

Cheremichkino (see Figs. 3: 6, 497: 3). A minor insect locality in deposits of the Lower Alykaeva Formation (Late Carboniferous) in the Kuznetsk Basin, Southwest Siberia (see Zheltyi Yar for age and references).

Chernyi Etap (see Figs. 4: 17, 500: 3). The locality (also called Lyagushye) is on the left bank of Tom' River near the former village Chernyi Etap and mouth of Lyagushye River. Mudstones of the upper Abashevo and lower Osinovskiy Fms yield the extensive insect assemblage sufficiently uniform in both formations and similar to that of Ust-Baley. KIRICHKOVA & TRAVINA (2000) date these deposits as Early Jurassic (plinsbachian and Toarcian, respectively) age. For details see RASNITSYN (1985a,b).

Chunya (see Figs. 3: 7, 497: 4). The only extratropical locality rich in Carboniferous insects is located on the left side of Chunya River (right tributary of Kamennaya Tunguska, Central Siberia) and comes from mudstones of the upper part of Kata Formation (correlated to Gzelian and Kasimovian Stages by ZHURAVLEVA & ILYINA 1988). Unlike more thermophilous assemblages from Europe, North and South America, and even South Siberia, which either display diverse composition or are strongly dominated by roaches, that from Chunya has a lower diversity, with only about half-dozen orders occurring, and roaches are only moderately common. A single species of Dictyoneurida is dominant (Chapter 2.2.1.2.4.1), and less so are members of the Grylloblattida (Chapter 2.2.2.2.1), while roaches take third position being represented by only some 5 species.

Commentry (see Figs. 3: 8, 497: 5). The fine-grained micaceous sandstone of the uppermost Upper Coal Measures of the Commentry Basin in Allier, France, is known as the richest source of the Stephanian (Late Carboniferous) insects (BRONGNIART 1893, MEUNIER 1909b, 1912, BOLTON 1925, etc.).

Cow Branch Formation (see Figs. 4: 3, 499: 3). Abundant insects are reported from the bituminous shales of this upper member of the formation (Carnian Stage of the Late Triassic, N. Carolina and Virginia, USA; OLSEN *et al.* 1978). This is the oldest insect-rich assemblage of undoubtedly lacustrine genesis including both autochthonous aquatic and diverse terrestrial taxa.

Devonian insect sites are far from numerous (only 3 of them are known thus far), and the respective fossils are not well preserved and do not permit unequivocal identification (Chapter 1.2). They are identified either as Archaeognatha (=Machilida), or (in Rhynie chert) just as an insect with the primitive dicondylar mandible. All of them could equally belong to an otherwise unknown group(s) still more primitive than Machilida, or even to a non-insect group independently acquired a dicondylar mandible and/or faceted eye. Additionally, an immature thrips has been recorded from Rhynie chert which was never been described or figured: this finding is contradictory to all that we know about the insect history and, despite the contrary statement by KÜHNE & SCHLÜTER (1985) it should be disregarded as a subsequent contamination (CROWSON 1970, 1985) or even as a misidentification, at least until the voucher specimen has been studied more closely. The Devonian fossil insects are all obtained there either by dissolving of the rock or by searching in thin chert sections. All three known Devonian localities lie within 10° from the Devonian equator (LABANDEIRA *et al.* 1988). Among these localities the oldest is the Rhynie chert in Scotland, UK (see Figs. 3: 3, 496: 3). The insects are found there in the Early Devonian (Sigenian) chert which has also produced well preserved psylophytan flora as well as a variety of small animals including Collembola, a group of hexapodous myriapods traditionally attributed to the insects (WHALLEY & JARZEMBOWSKI 1981 and references therein). Other Devonian insects are found in Gaspé, Quebec, Canada (see Figs. 3: 1, 496: 1), in mudstone of the Battery Point Formation, Emsian stage of the Early Devonian (LABANDEIRA *et al.* 1988), and in the Middle Devonian (Givetian) mudstone at Gilboa (see Figs. 3: 2, 496: 2) (N.Y., USA.; SHEAR *et al.* 1984).

Eckfeld: see **Geiseltal**.

Elmo (see Figs. 3: 20, 498: 3). The famous locality of the fossil insects is located near the town of Elmo in Kansas, USA. Numerous insect remains, often of near perfect preservation, come from the lacustrine Elmo limestone, of the Wellington shales, Artinskian Stage of the Early Permian. The insect environment is supposed to have been "a humid spot in a regional environment of more or less pronounced and long-continued aridity" (C.O. Dunbar in F. CARPENTER 1930b: 72).

English Lias (see Figs. 4: 18, 500: 4). Insect bearing marine deposits of the Lias (Early Jurassic) extend over south part of England from Somerset to Glocestershire, Warwickshire, and Worcestershire. The age varies from the Early (e.g. Late Sinemurian calcareous mudstone at Dorset, WHALLEY 1985) to the latest Lias ("insect limestone" from Dumbleton, Aldertone, etc., roughly equivalent to insectiferous deposits of the Lias epsilon of Dobbertin etc., see German Lias; ZEUNER 1939).

English Rhaetian (see Figs. 4: 4, 499: 4). Insectiferous deposits of the latest Triassic (Westbury Formation, *Pseudomonotis* Beds) are found in southern Britain (GADRINER 1961, STORRS 1993, POPOV *et al.* 1994).

Erunakovo Formation (see Figs. 3: 23, 500: 6). Several insect sites in deposits of this formation (Tatarian Stage of the Late Permian after ZHURAVLEVA & ILYINA 1988) in the Kuznetsk Basin, South Siberia are considered by ROHDENDORF *et al.* (1961).

Fatyanikha (see Figs. 3: 21, 497: 4). A minor insect locality at the Fatyanikha River in Krasnoyarsk Province, East Siberia (Nizhnyaya Burguklya Formation, correlated to Asselian and Artinskian Stages, that is to almost all of the Earlier Permian, by ZHURAVLEVA & ILYINA 1988).

Florissant (see Figs. 5: 12, 502: 11). A famous and one of the world's richest insect localities, Colorado, USA. The insects are collected in lake deposits of the Early Oligocene (Rupelian) Antero Formation and existed in the subtropical or warm temperate climate (MACGINITIE 1953). Modern radiometric data indicate an age near the Eocene/Oligocene boundary, and the fossil flora shows that the time of sedimentation should have preceded the Early Oligocene cooling (WOLFE 1992).

Garazhovka (see Figs. 4: 5, 499: 5). Insectiferous clay of the Protopivka Formation outcropping near Garazhovka Village, vicinity of Izyum, East Ukraine, is dated by DOBRUSKINA (1980, 1982) as Late Carnian or Early Norian (Late Triassic); see also PONOMARENKO & SUKATSHEVA (1998).

Gaspé: see **Devonian** insect sites

Geiseltal, Messel and **Eckfeld** (see Figs. 5: 14, 17, 11, 502: 13). The Middle Eocene (Lutetian) deposits at Geiseltal near Halle, Messel near Darmstadt, and Eckfelder Maar, Eifel, Germany, rich of insects, often of excellent preservation state (KRUMBIEGEL *et al.* 1983, SCHAAL & ZIEGLER 1988, LUTZ 1991a,b). These three localities are of similar age but originated in different palaeoenvironments: Geiseltal in a swampy area, Messel in a large deep lake, and Eckfeld in a crater lake of the maar type. Together they give an excellent picture of the Middle Eocene fauna of Central Europe though many taxa have only been identified preliminary.

"German" Lias (see Figs. 4: 19, 500: 5). The historical name of marine deposits (*Posidonia* Shales and carbonate *elegantulum* concretions) of the later Early Jurassic (early Toarcian) which are widely distributed through Germany, as well as in Switzerland, Luxembourg and neighbour countries. They are well known as a rich source of fossil insects. The principal fossil localities are at Dobbertin (Mecklenburg), Grimmen (Vorpommern) and Schandelah (Lower Saxony) (ZEUNER 1939, ANSORGE 1996). A survey of the insects from Schandelah is presented by BODE (1953), though the taxonomic concept of the book is questionable.

Gilboa: see **Devonian** insect sites

Glushkovo Formation (see Fig. 4: 50). A formation in Central Siberia of strongly disputed age (Late Jurassic vs. later Early Cretaceous, compare SINITZA & STARUKHINA 1986, *vs.* ZHERIKHIN 1978, RASNITSYN 1988b, DMITRIEV & ZHERIKHIN 1988), taken as belonging to unspecified Late Jurassic age in calculations for Chapter 3.1. However, the bulk of evidence now seems to suggest that its position is near the Jurassic/Cretaceous boundary (Chapter 2.2.1.3.4.2f). The insectiferous tuffaceous mudstones are exposed in a number of outcrops in Baley and Shelopogino Districts, Chita Region, Transbaikalia and represent deposits of cold mountain lakes in a volcanic region. Similar insect assemblages also occur in other areas within East Transbaikalia, in particular in the Ukurey Formation in the Olov Depression, and provide similar problems in their dating.

Green River Formation (see Figs. 5: 15, 502: 14) A Middle Eocene deposit widespread across western USA (Utah, Colorado, Wyoming) rich in fossil insects. The insectiferous deposits of past lakes Uinta, Gossuite and Fossil Lake belong to Parashute Member accumulated mostly during the Middle Eocene (W. BRADLEY 1964, FRANCZYK *et al.* 1992). Best collected deposits are from the Piceance Basin (eastern bay of the Uinta Lake) with some 12,000 rock pieces with about 80,000 insect fossils housed in the Paleobiology Department, Smithsonian Institution, Washington DC; further material is kept in a number of other institutions. Dominating the fauna are dipterans (~40%) followed by beetles (~20%) hemipterans (~10%) and hymenopterans (~4%). Some 300 insect species have been named (M. WILSON 1978, GRANDE 1984) but many more are waiting description.

Gulpen: see **Bitterfeld/Delitz**

Gurvan-Eren Formation (see Figs. 4: 50, 501: 12). The earlier Early Cretaceous insectiferous mudstones of this formation are distributed in the West Mongolia near and southward of Khovd. They are considered in detail by RASNITSYN (1986b).

Hagen-Vorhalle (see Figs. 3: 10, 497: 7). The oldest of the rich insectiferous localities in Ruhr Region, Germany (Vorhalle Beds, Middle Marsdenian, later Namurian B, early Middle Carboniferous), which have yielded various Dictyoneurida and Paoliida, and representatives of several other insect orders, as well as Arachnida (BRAUCKMANN 1991).

Hansan Formation (see Figs. 4: 20, 500: 6). Insects found in Hansan deposits in Hansan Co. (Anhui Prov., China) are described mostly by LIN (1985a) and considered Early or Middle Jurassic.

Horní Suchà (see Figs. 3: 11, 497: 8). Latest Namurian C (Yeadonian, earlier Middle Carboniferous) deposits of President Gottwald mine in Central Bohemia, Czech Republic (KUKALOVÁ 1958, BRAUCKMANN *et al.* 1992).

Iljinskoye Formation (see Figs. 3: 23, 498: 6). Insects from sites in Iljinskian deposits (Kazanian Stage of the Late Permian according ZHURAVLEVA & ILYINA 1988) in the Kuznetsk Basin, South Siberia are studied by ROHDENDORF *et al.* (1961).

Inta Formation (see Figs. 3: 22, 498: 5). A rather extensive insect collection from the Ufimian Stage, earliest in the Late Permian, is composed of core samples from various bore-holes made in the Vorkuta and Inta parts of the Pechora Coal Basin, in the north-east European Russia (age after DEDEEV 1990).

Ipswich (see Figs. 4: 6, 499: 6). The famous insect-bearing shales in the Ipswich Coal Measures, Australia, belong to the Mount Crosby Formation, NE New South Wales, and Blackstone Formation, SE Queensland, and are Carnian (Late Triassic) in age (J. ANDERSON & H. ANDERSON 1993).

Isle of Mull (see Figs. 5: 19, 502: 17). The Palaeocene interbasalt Ardtun Beds in the I. of Mull, Scotland, UK, have produced a number of insects (ZEUNER 1941b, JARZEMBOWSKI 1989).

Iya (see Figs. 4: 36, 500: 22). The rich insect site at the left side of Iya River (upper Angara Basin, Irkutsk Region, Siberia, Russia) in the uppermost member of the Cheremkhora Fm. (Aalenian, Middle Jurassic: KIRICHKOVA & TRAVINA 2000); for details see RASNITSYN 1985a,b).

Jiaoshang Formation (see Figs. 4: 21, 500: 7). Insect beds of this formation are known in Yiyang Co. (Hunan Prov., China). Chinese students attribute them to the Early Jurassic (HONG 1986).

Jiulongshan Formation. See **Beipiao**.

Kaltan (see Figs. 3: 23, 498: 6) is the richest among several insect localities of the earliest Late Permian (Ufimian) age found in the Kuznetsk Formation, Kuznetsk Basin in Southwest Siberia (ROHDENDORF *et al.* 1961, dating after ZHURAVLEVA & ILYINA 1988, SHCHERBAKOV 2000b argues for the younger, Kazanian age of the Mitino Horizon which Kaltan deposits belong to).

Kamiwada (see Figs. 5: 35, 503: 4). Nearly 200 Late Miocene insects, mainly nematoceran dipterans and ants, have been collected in the Wada Formation, in Yamagata Pref., NE Honshu, Japan (FUJIYAMA 1983).

Karatau (see Figs. 4: 22, 500: 8). The famous Late Jurassic insect locality or, rather, a group of localities in the southern part of the Karatau Range in South Kazakhstan. The main locality is Aulie (called also Mikhailovka) while that near the village Uspenovka (former Galkino) is subordinate, both yielding ca. 18 thousand insect fossils (DOLUDENKO *et al.* 1990, see also ROHDENDORF 1968b). The age is the Oxfordian or Kimmeridgian (KIRICHKOVA & DOLUDENKO 1996), taken as the Oxfordian in calculations for the Chapter 3.1.

Karaungir (see Figs. 3: 24, 498: 7). The latest Late Permian (Late Tatarian) insect assemblage from which about 700 specimens have been collected in lacustrine mudstone of Maichat and Ak-Kolka Formations at the Karaungir River in Saur Range, East Kazakhstan (SHCHERBAKOV 2000b).

Kargala (see Figs. 3: 25, 498: 8). The insects of Early Tatarian age (Late Permian) come from dumps of several past copper mines in the Belozersky District of Orenburg Region of Russia.

Kempendyay (see Figs. 4: 23, 500: 9). The insects of the Early Cretaceous or, rather, Late Jurassic age are found in mudstones at the right side of Kempendyay River (tributary of Vilyui in Yakutia, E. Siberia) below the mouth of Namdyr River (SINITSHENKOVA 1992b).

Kenderlyk (see Figs. 4: 7, 499: 7). Insectiferous mudstones of the Tologoy Formation, correlated with the Norian and Rhaetian of the Late Triassic (DOBRUSKINA 1980, 1982). The insects are collected in the West Kazakhstan at the Akkolka River in Saikan Mts., Saur Range (PONOMARENKO & SUKATSHEVA 1998).

Khetana (see Figs. 4: 52, 501: 14). The latest Early Cretaceous (Middle Albian) insects are collected at the Khetana River (a system of the Ulya River south of Okhotsk, East Siberia) in tuffaceous mudstones of Emanra Formation (GROMOV *et al.* 1993).

Khotont (see Figs. 4: 24, 500: 10). Almost 1,500 insect fossils collected in rocks of the northern slope of Ukha Mt., Ara-Khangay Aymag (Region) of Mongolia 6 km west of Somon (Village) Khotont. The deposit is supposedly from near the Jurassic/Cretaceous boundary.

Khoutiyn-Khotgor (see Figs. 4: 25, 500: 25). Some 2,000 fossil insects of 16 orders and some 50 families have been collected in the paper shales of the Ulan-Ereg Formation of the latest Late Jurassic in the Dund-Gobi Aymag of Mongolia near Bayan-Zhargalant (PONOMARENKO 1998b).

Khurilt (see Figs. 4: 45, 501: 8). A mid-Early Cretaceous insect locality of the Khurilt rock unit, though minor in comparison to Bon Tsagan (*q.v.* for correlation) in the Bayan Khongor Aymag of Mongolia, 10 km NE Tsetsen Ula Mt. (SINITZA 1993).

Khutel Khara (see Figs. 4: 53, 501: 15). Insectiferous interbasalt mudstones of the lower Tsagan Tsab Formation in 75 km SE Sain Shand (East Gobi Aymag, Mongolia) have yielded more than 3,000 fossil insects that reveal both Jurassic and Cretaceous affinities and so are considered the earliest Cretaceous in age (PONOMARENKO 1990, RASNITSYN *et al.* 1998). The taxonomic composition of the assemblages differs greatly from all other localities assigned to the same formation in SE Mongolia.

Khutuliyn (see Figs. 4: 54, 501: 16). Insectiferous tuffaceous mudstones of the Ulughey Formation (25 km S Mandal Obo, South Gobi Aymag, Mongolia) have yielded some 1,300 fossils of both Jurassic and, to a lesser extent, of Cretaceous affinities, thus indicating the latest Jurassic age (PONOMARENKO 1990, 1998b, SINITZA 1993).

Kityak (see Figs. 3: 26, 498: 9). A few hundred of insects have been collected in the copper-beds of the Belebey Formation (Kazanian stage, Late Permian) at the right side of Kityak River opposite Bolshoy Kityak Village (Kirov Region, north European Russia).

Koonwarra (see Figs. 4: 55, 501: 17). The fine-grained lacustrine mudstone of the Koonwarra Fossil Bed (Korumburra Group, South Gippsland, Victoria, Australia) have produced an extensive collection of insects, as well as plants and various other invertebrates and fishes, of the mid-Early Cretaceous (Aptian) age. For details see JELL & ROBERTS (1986).

Kubekovo (see Figs. 4: 26, 500: 12). The insectiferous mudstones of the Itat Formation (Aalenian or Bathonian, Middle Jurassic) outcrop at the left bank of Yenissey River downstream of Krasnoyarsk, Central Siberia. For details see RASNITSYN (1985a,b).

Kuznetsk Formation: see **Kaltan**

Kyzyl-Kiya (see Figs. 4: 27, 500: 13). The insects are collected in number in earlier Early Jurassic mudstones of a coal-bearing formation near Kyzyl-Kiya railway station, South Fergana, Kirghizstan (MARTYNOV 1937).

Kzyl-Zhar (see Figs. 4: 57, 501: 19). Insects of the Turonian stage (earlier Early Cretaceous) come from mudstone lenses in fluvial deposits at Kzyl-Zhar Hill (NE Karatau Range, Kzyl-Orda Reg. of Kazakhstan; ZHERIKHIN 1978: 77, assignement of the deposits to the Beleuty Formation in ZHERIKHIN 1978 is erroneous).

Laiyang Formation (see Figs. 4: 58, 501: 20). Insectiferous paper shales of this formation in Laiyang Co. (Shandong Prov., East China) are usually considered by Chinese students as Late Jurassic (ZHANG 1992). However, the insect assemblage is similar to and probably nearly contemporaneous with that from Baissa (earlier Early Cretaceous). Taken as the Early Neocomian in calculations for the Chapter 3.1.

Laocun (Laoqun) Formation (see Figs. 4: 59, 501: 21). Several species have been described from the yellow-greenish mudstones and shales in Jiande District, Zhejiang Province, China, as Late Jurassic (LIN 1980), but the insect assemblage is of the same type as in the East Siberian and Mongolian earlier Early Cretaceous.

Latah Formation (see Figs. 5: 36, 502: 5). The Miocene volcanic-sedimentary deposits widespread in north-western USA have produced rich insect fauna collected in a number of sites in eastern Oregon and northern Idaho (S. LEWIS 1969, S. LEWIS et al. 1990).

Lek-Vorkuta Formation (see Figs. 3: 27, 498: 10) in Pechora Basin (NE European Russia) represents an alternation of sand-, silt- and mudstones of marine and brackish-water origin rich in various fossils including insects. It was usually correlated with the Kungurian Stage of the Early Permian but presently considered the Ufimian (earliest Late Permian, GRUNT et al. 1998). The formation is subdivided into the lower Ayachyaga and upper Rudnitskaya Subformations.

Lushangfeng Formation (see Figs. 4: 61, 501: 23). Insectiferous deposits of the formation are recorded near Beijing, China. They overlay those containing *Ephemeropsis* (REN et al. 1995) and thus could preliminary be correlated with Bon-Tsagan and Wealden (mid-Early Cretaceous).

Madera Formation (see Figs. 3: 12, 498: 9) of the Late (possibly latest Late) Carboniferous in New Mexico, USA, is reported to have yielded 5 fossil insects (cockroaches, a palaeodictyopteran, a caloneuridean and a bristletail) of good preservational state (F. CARPENTER 1970).

Madygen Formation (see Figs. 4: 8, 499: 8), Ladinian and/or Carnian according DOBRUSKINA 1980, 1982, that is of either later Middle or earlier Late Triassic age, of south Fergana Valley (an area of junction of Uzbekistan, Kirghizia and Tadzhikistan) includes clay layers very rich in fossil insects, with the principal localities being Dzhayloucho and Madygen (PONOMARENKO & SUKATSHEVA 1998).

Malanzán Formation: see **Bitterfeld/Delitz**

Mazon Creek (see Figs. 3: 14, 497: 11). The assemblage of the Westphalian D age (later Middle Carboniferous) is unique in its general faunal diversity including its insects. The fossils come from siderite concretions in the lower 3–5 m of the Francis Creek Shale Member of the Carbondale Formation, Illinois, USA., near the NE coastal line of the Carboniferous in Illinois (BAIRD 1997). Two faunal types, the marine Essex fauna and the non-marine to minimally brackish Braidwood one are identified, each with equal (0.03 per cent) participation of insects, with no difference mentioned in the insect composition between these faunas (BAIRD & ANDERSON 1997). Mazon Creek insects are reviewed briefly by F. CARPENTER (1997) and KUKALOVÁ-PECK (1997).

Meigen I. (see Figs. 5: 37, 503: 6) (Canada Arctic Archipelago). Insects (mostly beetles closely related to but slightly differing from the living boreal species) are collected in the deposits of the Late Miocene or Early Pliocene deposits of the Beaufort Formation (MATTHEWS 1976a, 1977).

Menat (see Figs. 5: 16, 502: 15). Rich insectiferous deposits at Menat (Puy-de-Dóme, France) are now considered Palaeocene (GAUDANT 1979, OLLIVIEIER 1985). Insects as well as diverse other fossils occur in lacustrine bituminous shales. Many insect descriptions (e.g. PITON 1940) need reconsideration.

Messel: see **Geiseltal**.

Middle Beaufort Series (see Figs. 3: 28, 498: 11) (terminal Permian of Natal, South Africa) is the source of diverse homopterans as well as some other insect fossils (Mischopterida, Palaeomanteida, Grylloblattida, Perlida, Forficulida) (RIEK 1976d).

Middle Coal Measure (see Figs. 3: 15, 497: 12) is widespread in Britain and represents marine (below) to freshwater (above) deposits correlated to Westphalian B – C of the later Middle Carboniferous. Various insects (Blattida, Dictyoneurida etc.) have been described (BOLTON 1921–22, JARZEMBOWSKI 1987, C. PROCTOR & JARZEMBOWSKI 1999).

Mo Clay (see Figs. 5: 18, 502: 16). Many fossil insects have been collected in marine diatomites of the so-called Mo Clay (Fur Formation, analogous to the Moler Formation, Late Palaeocene – Early Eocene) in several localities at Limfjord, Mors Peninsula and Fur Island in Denmark (WILLMANN 1990, RUST 1999a).

Mogzon (see Figs. 4: 28, 500: 14). Insects are found in several sites in the Mogzon Depression westward of Chita (East Siberia, Russia) in tuffaceous mudstones of Zun-Nemetey (or Bada) Formation, a correlative of Uda Formation of the Late Jurassic age. For details see RASNITSYN (1985a,b).

Molteno Formation (see Figs. 4: 9, 499: 9) in the Cape and Natal Provinces, South Africa (ANDERSON 1974) is a well known source of earlier Late Triassic (Carnian) age (RIEK 1974, 1976b, J. ANDERSON et al. 1998).

Montceau-Les-Mines (see Figs. 3: 16, 497: 13). Almost 150 insect fossils have been reported from ironstone nodules and, to a lesser extent, in shales of Stephanian B age (Late Carboniferous) in the Montceau-Les-Mines (Central Massif, France) (BURNHAM 1981). Despite belonging to the same coal basin and stratigraphic position as Commentry, the locality yields a different insect assemblage with unexpectedly high participation of immature insects and smaller size insect fossils, and low participation of cockroaches (op. cit.).

Montsec (see Figs. 4: 62, 501: 24) is an important source of the Early Cretaceous insects and other fossils that are collected at several outcrops of the lithographic limestones located at the Montsec Range in the Lleida Province, eastern Spain (MARTÍNEZ-DELCLÒS 1991). The above author concludes that the stratigraphic position of the limestones is generally accepted to be near the Berriasian-Valanginian transition, that is early in the Early Cretaceous. However, composition of the hymenopteran assemblage is similar to the Wealdean one and thus agrees rather with the

Early Barremian correlation proposed by ANSORGE (1993). Taken as the Late Neocomian in calculations for the Chapter 3.1.

Nizhnyaya Tunguska (see Figs. 3: 29, 498: 12) River near the mouth of the Nizhnyaya Bugarikhta in Central Siberia, lower part of Bugarikhta Formation, the latest Permian or earliest Triassic (for review of the transitional Perm-Triassic insect sites in a wider context see PONOMARENKO & SUKATSHEVA 1998, SHCHERBAKOV 2000b).

Novospasskoye (see Figs. 4: 29, 500: 15). The rich insectiferous mudstone of the lacustrine Ichetuy Formation in the Tugnuy Depression in the Southwest Transbaikalia (East Siberia) are considered as the later Early or earlier Middle Jurassic in age (RASNITSYN 1985a,b). However, the deposits involved may be of somewhat different age (PONOMARENKO 1993).

Obeshchayushchiy (see Fig. 4: 65). The earliest Late Cretaceous (Cenomanian) tuffaceous mudstone of the lacustrine Ola Formation in the Russian Far East (Obeshchayushchiy Creek in the basin of Nil River, a tributary of Arman') has yielded an extensive collection of insects, mainly undescribed (RASNITSYN 1980: 156, SAMYLINA 1988, GROMOV *et al.* 1993).

Obluchye (see Figs. 4: 66, 501: 27). A minor insect locality in the Cenomanian (earliest Late Cretaceous) tuffaceous mudstones near Obluchye Town (Hebrew Autonomous Region in East Siberia; ZHERIKHIN 1978: 90).

Obora (see Figs. 3: 30, 498: 13). Rich Early Permian locality with a wide array of insects collected in the grey lacustrine mudstone of the Middle Bačov Beds in Boskovice Furrow at Obora near Skalice nad Svitavou in Moravia, Czech Republic (KUKALOVÁ 1964). Currently the locality is usually referred to as the earlier Artinskian in age (KUKALOVÁ-PECK & WILLMANN 1990), but the claim is made that it is older (earlier Sakmarian, biozone *Xenacanthus decheni*, ZAJÍC 2000, see also SHCHERBAKOV 2000b).

Oeningen (see Figs. 5: 40, 503: 9). Rich Late Miocene (Sarmatian) insect locality in the freshwater limestones near Schrotzburg (Schienerburg, S. Baden) in Germany. Most descriptions made by HEER (1847–1853, 1867, etc.) badly need revision.

Orapa (see Figs. 4: 67, 501: 28). A potentially important but currently underexplored source of South African Cretaceous insects. The fossils are collected in numbers (ca. 3,000) in Orapa Mine in Botswana, South Africa, and dated as earlier Late Cretaceous (Cenomanian-Coniacean) (RAYNER *et al.* 1998 and references therein).

Oshin-Boro-Udzur-Ula, Zhargalant (see Figs. 4: 30, 500: 16). The insects are found in mudstones of the Zhargalant Formation in the northeastern Mongolia (Mongolian Altai Mts., Dzhargalant Range). One of two insect assemblages recorded here is similar to that of Ust-Baley indicating similar age of deposits (Toarcian in the Early Jurassic afer KIRICHKOVA & TRAVINA 2000); the other has no known counterparts, though the difference is not necessary age dependant (RASNITSYN 1985b, PONOMARENKO 1998b).

Pascapoo Formation (see Figs. 5: 21, 502: 19). Palaeocene (Tiffanian) insects have been collected in several localities (Red Deer, Robb, Edson, Sundance) in Alberta, Canada (MITCHELL & WIGHTON 1979).

Pelyatka River (see Figs. 4: 10, 498: 10) Located 10 km from the river mouth (90 km from Nizhnyaya Tunguska mouth), Central Siberia; Pelyatka Formation, originating from near the Permian/Triassic boundary, with the latest Permian age most probable (PONOMARENKO & SUKATSHEVA 1998, SHCHERBAKOV 2000b).

Polovaya (see Figs. 4: 68, 501: 29). A minor locality in deposits of the Ghidari Formation in East Transbaikalia (Chita Region, Nerchinskiy Zavod District, left side of Urov River Basin, Polovaya Valley) (RASNITSYN 1985a: 11). The age is most probably Early Cretaceous, although the correlation of the Ghidari Formation to the insectiferous deposits of Baissa (ZHERIKHIN 1978: 23) becomes rather problematic since the main indicator, the giant mayfly *Ephemeropsis* Eichwald, was found to belong to different species in East Transbaikalia (SINITSHENKOVA & ZHERIKHIN 1996).

Potrerillos Formation (see Figs. 4: 11, 499: 11) (Cacheuta Group). Lake deposits of the late Middle or early Late Triassic age (Late Ladinian – Early Carnian) in Cacheuta basin in Mendoza Province, NE Argentina (J. ANDERSON & H. ANDERSON 1993).

Purbeck (see Figs. 4: 69, 501: 30). The non-marine Purbeck Limestone Group of southern England (maximum thickness some 120 m in the type area in Dorset) consists of limestones, shales, clays and evaporites, and is subdivided into two formations, the Lulworth Formation below and the Durlston Formation above. The insectiferous deposits are Berriasian (earliest Early Cretaceous) in age (RASNITSYN *et al.* 1998).

Radoboj (see Figs. 5: 41, 503: 10). Large collection of the Early Miocene (Burdigalian) insects from Radoboj in Croatia, was described long ago (e.g. HEER 1847–1853, 1867, etc.) and needs revision.

Redbank Plain (see Figs. 5: 22, 502: 20). Fossil insects have been collected in the Palaeocene deposits of the Redbank Plain Formation in several localities in Southeast Queensland, Australia (e.g. RIEK 1952).

Redmond (see Figs. 4: 70, 501: 31). Insects of the latest Early or earliest Late Cretaceous age are collected in the irony mudstones of the Redmond Formation in the Labrador Peninsula, Eastern North America (DORF 1967, ZHERIKHIN 1978).

Rhynie chert: see **Devonian** insect sites.

Rott (see Figs. 5: 42, 503: 11). A famous locality in the earliest Miocene (Aquitanian) paper coal and polishing shales near Bonn in Germany (KOENIGSWALD 1989). The most important Georg Statz Collection of fossils (STATZ 1936–1950, and references therein) is now kept in the Los Angeles County Museum (SPHON 1973). Taken as the Chattian in calculations for the Chapter 3.1.

Ruby River (see Figs. 5: 24, 502: 22). Insects of 13 orders and 29 families have been collected from deposits of the Passamari Formation in the Ruby River Basin in Montana; only a small part of fossils have thus far been described. The sediments previously assigned to the Oligocene (e.g. M. WILSON 1978) may be the latest Eocene in age (WOLFE 1992).

Saar Basin (see Figs. 3: 17, 497: 14). The Palaeozoic insectiferous deposits of the Saar Basin in the western Germany belong to the coal-bearing Middle Carboniferous (Westphalian C and D, Saarbrückner Layers) and Late Carboniferous (Stephanian, Ottweiler Layers), and the Early Permian (Lower and Upper Rotliegend), coal free. For details see GUTHÖRL (1934).

Sagul (see Figs. 4: 27, 500: 13). Early or Middle Jurassic deposits of the Sogul Formation in South Fergana in the northern part of Osh region of Kyrghyzstan.

Salt Range (see Figs. 5: 26, 502: 24). Small insects and insect fragments have been collected from the Palaeocene Saline Series, Salt Range in Punjab, Pakistan (GEORGE 1952).

Santana (see Figs. 4: 71, 501: 32). Numerous and diverse insects and other fossils of Aptian age (later Early Cretaceous) have been collected in the lacustrine limestone of the Santana Group (Crato Member or

Formation) that is widely distributed over the Araripe Plateau, south of Ceara State, Brazil. This is the principal Early Cretaceous locality of the south hemisphere. For details see GRIMALDI (1990), MARTINS-NETO (1991c), MARTILL 1993 and references therein.

Semyon (see Figs. 4: 72, 501: 33). Insectiferous mudstones of disputable age within the Early Cretaceous are found at Semyon Creek near Elizavetino Village, westward of Chita (Transbaikalia in Central Siberia). ZHERIKHIN (1978: 23) correlated it to **Baissa** but extensive new collections show that the taxonomic composition of the fauna is very different.

Shanwang (see Figs. 5: 43, 503: 13). 400 insect species belonging to 84 families and 12 orders have been described from the Middle Miocene diatomites of the Shanwang Formation in 22 km East of Linqu, Shandong Province, China (ZHANG 1989, ZHANG et al. 1994).

Shar Teg (see Figs. 4: 31, 500: 17). A diverse Late Jurassic insect assemblage (some 3,000 fossils of more than 200 families) collected in mudstone of the Shar Teg Beds at the Shar Teg Ula Mt. SW of Adzh-Bogdo Mt., Gobi-Altai Aymag of Mongolia (GUBIN & SINITZA 1996, PONOMARENKO 1998b).

Shiti (see Figs. 4: 32, 500: 18). Some 20 insect species have been described from the volcanoclastic Shiti Formation (Early Jurassic) in Xiwan, Zhongshan, Guanxi Province, China (LIN 1986).

Shurab (see Figs. 4: 27, 500: 13). Early and Middle Jurassic clayey deposits of the Sogul and Sulyucta Formations in South Fergana near Shurab in Tadzhikistan rich in insect fossils (mainly isolated wings).

Sogyuty (see Figs. 4: 33, 500: 19). The territory known under this name and situated at the south shore of Issyk-Kul Lake near Kadzhi-Say has yielded insectiferous clays of several successive formations. Yet the name is Applied here only to the most productive Early Jurassic Dzhil Formation; taken as the Sinemurian in calculations for the Chapter 3.1.

Solnhofen (Solenhofen; see Figs. 4: 34, 500: 20). Famous Solnhofen lithographic limestone of the Tithonian (latest Late Jurassic) age is of marine origin. It is worked around Solnhofen and Eichstadt in Bavaria, Germany (K. BARTHEL 1971, 1972). The insects described from it are listed by KUHN (1961), see also PONOMARENKO (1985a), FRICKHINGER (1994).

Soyana (see Figs. 3: 31, 498: 14). Iva-Gora limestones available from several outcrops along the left bank of Soyana River (left tributary of Kuloy, Arkhangelsk Region, north European Russia) has yielded the most representative Late Permian insect assemblage of the Early Kazanian age (MARTYNOV 1938b).

Stavropol (see Figs. 5: 44, 503: 14). Numerous and diverse insects are found in marls and calcified mudstone nodules of the Middle Miocene, exposed in Vishnevaya Balka (Valley) next to Senghileevskoye Lake near Stavropol, as well as at Temnolesskaya ear. Stavropol (Central North Caucasus, southern Russia). According GONTSHAROVA (1989: 65), these deposits are Late (latest in Vishnevaya) Tchokrakian (East Paratethian regional stage of the Neogene). Insects indicate savannah-like landscape for the lower part of the section and subtropical forests for the upper one (BECKER-MIGDISOVA 1964, 1967, DLUSSKY 1981b).

Tadushi Formation (see Figs. 5: 28, 502: 26) in Sikhote Alin Range near Suvorovo Town (Maritime Province, Russian Far East) includes tuffaceous mudstones rich in fossil insects of the Palaeocene age (ZHERIKHIN 1978: 119; taken as the Tanetian in calculations for the Chapter 3.1).

Tikhiye Gory (see Figs. 3: 32, 498: 15). A lost (now flooded) locality of Late Permian insects in the *Lingula*-mudstone of the Baitugan

Formation at Kama River (right bank near Naberezhnye Chelny in Tatarian Republic, Russia), essentially of the same age as Soyana (MARTYNOV 1938b).

Tshekarda (see Figs. 3: 33, 498: 16). Very rich insect locality of the Kungurian (latest Early Permian) age in mudstones of the Koshelevka Formation at the left bank of Sylva River, west slopes of the Urals (PONOMARYOVA et al. 1998).

Turga (see Figs. 4: 76, 501: 37). The tuffaceous mudstones of the Turga Formation (earlier Early Cretaceous) outcrop at Turga River near Borzia (East Transbaikalia in Central Siberia). Radiometric data indicate position near the Jurassic/Cretaceous boundary (K-Ar 134 ($\pm$2), Rb-Sr 131 ($\pm$5) for the lower subformation, pers. comm. by A.V. Rublev and Yu.P. Shergina of VSEGEI, St-Petersbourg, Russia). The correlated to **Baissa** (ZHERIKHIN 1978: 23) should be taken as only tentative because in fact the taxonomic composition of the faunas differ significantly.

Uda (see Figs. 4: 35, 500: 21). The tuffaceous insectiferous deposits of the Late Jurassic Uda Formation are of roughly same age as Karatau. They outcrop at the right bank of upper stream of Uda River in West Transbaikalia (Central Siberia). For details see RASNITSYN (1985a,b).

Ukurey Formation: See **Glushkovo Formation**.

Ust-Baley (see Figs. 4: 36, 500: 22). The famous Ust-Baley insectiferous mudstone outcrops at the right bank of Angara River downstream of Ust-Baley Village (South Cisbaikalian) and belong to the lower member of the Upper Cheremkhovo Subformation (Toarcian, Early Jurassic) (KIRICHKOVA & TRAVINA 2000). For details see RASNITSYN (1985a,b).

Vosges (see Figs. 4: 12, 499: 12). Insects are found mostly in the upper, deltaic in origin, part of the *Voltzia* sandstone, Upper Buntsandstein, earlier Anisian Stage of the Middle Triassic (Alsace, France; J. GALL 1971, 1985). This is the most important Mid-Triassic locality but only a minor part of insects therein have been described in detail.

Vrapice (see Figs. 3: 11, 497: 8). Deposits of Whetstone Horizon, Westphalian C (later Middle Carboniferous, Moscovian) at the President Antonín Zapotozký mine near Vrapice, Central Bohemian Coal Basin, Czech Rep. (KUKALOVÁ-PECK 1985)

Wealden (see Figs. 4: 77, 501: 38). The non-marine Wealden Supergroup of southern England reaches a thickness of about 1200 m in the type area of the Weald. It is subdivided into two groups, the Hastings Beds (Valanginian) and the overlying Weald Clay Group. The Hastings Beds consist of sandstones, silts, shales and clays, and is subdivided into the Ashdown, Wadhurst Clay, Lower and Upper Tunbridge Wells Sand, and the Grinstead Clay Formations. The Weald Clay Group consists of silty clays with minor sandstone and limestone beds, and is subdivided into the Lower Weald Clay Formation (Hauterivian) and Upper Weald Clay Formation (Barremian). The rich insect assemblage from several localities correlated to near the boundary between the Lower and Upper Weald Clay Formations is of outstanding importance because it is reliably dated. For details see RASNITSYN et al. (1998).

Wianamatta (see Figs. 4: 13, 499: 13). Coastal floodplain deposits of the Wianamatta Group (Late Ladinian – Early Carnian, later Middle or early Late Triassic) in New South Wales, Australia (RIEK 1970a).

Willershausen-am-Harz (see Figs. 5: 46, 503: 16). Famous Late Pliocene (Ruscinian) fossil site in North Germany near Osterode which lacustrine deposits have yielded many fossil insects besides other animals and plants (GERSDORF 1968 and other papers in that issue).

Yixan Formation (see Figs. 4: 79, 501: 40). Important insectiferous deposits in NE China (near Beipiao in Liaoning Province and Pingqun County, Hebei Province) are now considered Barremian (mid-Early Cretaceous; SWISHER *et al.* 1999) and not Jurassic as Chinese students used to correlate (HONG 1984, REN *et al.* 1995, REN & GUO 1996a,b, REN 1997a,b,c). However, composition of the insect assemblage, and particularly of the hymenopterans described by HONG (1984) and REN *et al.* (1995) is comparable with those from Purbeck and not Weald Clay (see above) being almost free of aculeate wasps. This implies the early Neocomian and not Barremian age (RASNITSYN *et al.* 1998).

Zavyalovo (see Figs. 3: 6, 497: 3). The insects (mostly cockroaches) in the same Late Carboniferous deposits as in Zheltyi Yar are found near Zavyalovo Village, westward of Kemerovo (Kuznetsk Basin in Southwest Siberia, for details see ROHDENDORF *et al.* 1961).

Zhargalant: see **Oshin-Boro-Udzur-Ula**.

Zheltyi Yar (see Figs. 3: 6, 497: 3). The most rich of several localities of the Late Carboniferous insects found in deposits of the Lower Alykaeva Formation at the Tom' River in Kuznetsk Basin, Southwest Siberia (ROHDENDORF *et al.* 1961, dating after ZHURAVLEVA & ILYINA 1988). The majority of the fossils are roach elytra transferred by running water and buried among numerous fern winglets which they were similar to in both general appearance and hydrodynamic properties.

4.2
Fossil Resins

K.YU. ESKOV

Many insect records in fossil resins are catalogued by KEILBACH (1982) and SPAHR (1981–1993). The Mesozoic and Early Tertiary insectiferous fossil resins in the former USSR are reviewed by ZHERIKHIN & ESKOV (1999). Recent reviews of more popular insectiferous fossil resins and insect inclusions are POINAR (1992a), GRIMALDI (1996b), and WEITSCHAT & WICHARD (1998).

Agapa (see Figs. 4: 37, 501: 1). The earlier Late Cretaceous (Late Cenomanian) fossil resins (retinites) are collected from sand of the Dolganian Formation at the Nizhnyaya Agapa River in the West Taimyr Peninsula (North Siberia). Some 700 insect and spider inclusions have been found there thus far (ZHERIKHIN & SUKATSHEVA 1973).

Agdzhakend (see Figs. 4: 38, 501: 2). The resin is found in Yukhary Agdzhakend, the Azerbaidzhanian part of the Minor Caucasus Mountains, in lagoon and deltaic deposits of the Early Cenomanian (earliest Late Cretaceous) age. It superficially resembles some varieties of Lebanese, Spanish, and French Cretaceous "ambers". More than 100 inclusions have been found representing 8 insect orders (dipterans are dominant), as well as spiders and mites (ZHERIKHIN 1978).

Álava (see Figs. 4: 39, 501: 3). This recently discovered amber originates from Basque-Canthabrian Basin near Álava (Northern Spain); its age is estimated as Late Aptian-Middle Albian (later Early Cretaceous; ALONSO *et al.* 2000). To date more than 1500 inclusions have been found. Besides numerous Diptera and Hymenoptera, some rare groups were recorded such as Orthoptera and Crustacea Isopoda (ORTUÑO 1998).

Arkansas amber (see Figs. 5: 3, 502: 3). This amber originates from the hard coals of the Claiborn Formation near Malvern Coalfield in Arkansas (USA). The age of the Claiborn Formation is estimated as the Middle Eocene. Several hundred of insect and spider inclusions have been found, numerous aphids should be noted (see SAUNDERS *et al.* 1974).

Austrian amber (see Figs. 4: 41, 501: 4). Probably the oldest known insectiferous fossil resins are found in Groling, Austria. Their age is estimated as Early Cretaceous, probably Hauterivian. A few dipterans were found in this amber (SCHLEE 1984, BORKENT 1997).

Baltic amber (see Figs. 5: 5, 502: 5). The most famous source of the fossil insects of the best preservation is collected in bulk since antiquity re-deposited along the eastern part of the south beach of Baltic. The least re-deposited is probably the amber dug out from so called blue earth (Prussian Formation) of Late Eocene age. The original source of the resin is supposed to be open mixed pine forests that covered the Scandinavian Shield territory in the Eocene times. The number of fossils collected is innumerable, and that of the insect species described or indicated is near 3,000. For details see LARSSON (1978), ZHERIKHIN (1978, 1998), POINAR (1992a), GRIMALDI (1996), and WEITSCHAT & WICHARD (1998).

Bandoc (see Figs. 5: 31, 503: 1). An "earthy amber" similar to the Sumatra resin is known from the Pliocene deposits of the Bandoc Peninsula at Luzon Island, Philippines; a single dipteran inclusion has been recorded (DURHAM 1956).

Begichev Formation retinite (see Figs. 4: 43, 501: 6). The arthropod inclusions are discovered in four localities in the Khatanga River basin (Eastern Taimyr). The Begichev Formation of the mid-Cretaceous (Albian to the Early Cenomanian) age is represented by the fluvial and deltaic sands with lenses of the lignitised wood containing the retinite. The inclusions are relatively rare. More than 40 arthropods have been found representing 7 insect orders (dipterans are dominant), as well as spiders and mites (ZHERIKHIN & SUKATCHEVA 1973, ZHERIKHIN 1978).

Bitterfeld (Saxonian) amber (see Figs. 5: 7, 502: 7). The amber, similar to the Baltic one but claimed to be the Early Miocene in the age, comes from near Bitterfeld in Saxony, Germany. The fauna is also close to the Baltic one though most species described differ from that from the Baltic amber at least at the species level (BARTEL & HERTZER 1982, SCHUMANN & WENDT 1989, POINAR 1992a, RÖSCHMANN 1999a).

British Columbia amber (see Figs. 5: 9, 502: 9). The amber was found at the Hat Creek coalfield at the British Columbia (Canada). Its age is estimated as Early to Middle Eocene. Several insect orders were recorded, as well as arachnids and nematodes (POINAR *et al.* 1999).

Burmese amber (see Figs. 5: 10, 502: 10). The burmite fossil resin is found in the Middle Eocene deposits in North Myanmar (Burma), though it may be re-deposited from older strata. Composition of the arthropod fossils reveals a mixture of the Late Cretaceous, Tertiary (Palaeogene), and endemic elements of enigmatic origin and date, so the age is disputable within the above limits. A collection comprising some 1200 invertebrate (mostly insect) inclusions is kept at the Natural History Museum, London. For details on the age, origin and composition of the fossil assemblage see A. ROSS (2000).

Canadian amber (see Figs. 4: 46, 501: 9). The Canadian amber (mineralogically called chemavinite) has been collected initially re-deposited at beach of the Cedar Lake in Manitoba, Canada, and later in situ, in the Campanian (later Late Cretaceous) Foremost Formation at the Grassy Lake near Medicine Hat in Alberta, Canada

(F. CARPENTER *et al.* 1937, MCALPINE & MARTIN 1969, ZHERIKHIN 1978: 96, PIKE 1994). The latter collection comprises some 3000 fossils.

Choshi (see Figs. 4: 47, 501: 10). The amber originates from the Choshi Group at Choshi Peninsula, eastern coast of Honshu Island (Japan); its age is estimated as Barremian-Aptian (mid-Early Cretaceous). No information about the general composition of the fauna is available; few Hymenoptera species have been described (FUJIYAMA 1994).

Dominican amber (see Figs. 5: 34, 503: 3). The fossilised resin of a leguminose tree *Hymenaea*, of the Early to Middle Miocene age, is been collecting in various sites of the Dominican Republic (ITURRALDE & MACPHEE 1996). Dominican arthropods are extremely numerous and diverse: the collection kept in the Smithsonian Institution (Washington DC, USA.) comprises some 5,000 pieces, about 3,000 of them being studied more carefully have yielded more then 9,000 fossils (Rasnitsyn unpublished). The general appearance of the fauna is that of the average Central American tropical forest, except for a few exotic elements, and most species and some genera being extinct (POINAR 1992a, and references therein).

French Cretaceous amber (see Figs. 4: 48, 501: 11). Fossil resins (retinites) of the Early Cenomanian (earliest Late Cretaceous) are known from various sites in the Central and Northwest France, with the majority of fossils being found at Bezonnais (Sarthe Dept.). A detailed account of the resins and their inclusions is presented by SCHLÜTER (1978). Additional more recently collected materials are now being studied (A. Nel, pers. comm.).

Fushun (see Figs. 5: 13, 502: 12). This amber originates from the hard coals of the Fushun Coalfield in north-eastern China. The age of the Fushun coals is estimated as the Middle Eocene. No data on the general composition of the fauna are published; some mayflies, roaches, dipterans, hymenopterans, and a spider have described (see HONG 1981, HONG *et al.* 1980).

Iwaki (see Figs. 4: 51, 501: 13). The resin occurs in the coal-bearing fluvial deposits of the Santonian (mid-Late Cretaceous) age in Iwaki, Honshu Island, Japan. Few inclusions have been mentioned (SCHLEE 1990).

Kuji (see Figs. 4: 56, 501: 18). The resin was found at the north-eastern part of Honshu Island (Japan) in the Kuji Series (Santonian-Campanian, Late Cretaceous). Only preliminary data on the fauna have been published, without exact information of its content (SCHLEE 1990).

Lebanese amber (see Figs. 4: 60, 501: 22). One of the oldest known insectiferous fossil resins is found in Jezzin (30 km east of Saida) and numerous other sites in Lebanon. Their age is disputable, with the least dispersion of estimations being from Hauterivian to Aptian (Early Cretaceous). Several thousands of inclusions have been collected there representing 13 insect orders (dipterans and hymenopterans are dominant), as well as 3 arachnid orders: spiders, mites and pseudoscorpions (SCHLEE & DIETRICH 1970, ZHERIKHIN 1978, AZAR 1997).

Merit-Pila (see Figs. 5: 38, 503: 7). This resin was found in Merit-Pila coal-pit in central Sarawak, Malaysia. The coal belongs to the Nylau Formation (Early-Middle Miocene). Seven insect orders (mainly dipterans and hymenopterans), as well as spiders, mites, diplopods and chilopods have been recorded (HILMER *et al.* 1992).

Mexican amber (see Figs. 5: 39, 503: 8). The amber of probably the same botanical origin as the Dominican amber (LANGENHEIM 1966) and with the age of Late Oligocene or Early Miocene. It comes from a number of localities in the Simojovel area in the Mexican state Chiapas (HURD *et al.* 1962). The fossil fauna is rather similar to that of Dominican amber, although identical species are uncommon (POINAR 1992a).

Near East amber. Numerous localities of Early Cretaceous resins are known in Lebanon, Jordan, Israel, and Syria. The resins are of the retinite type (and supposedly of araucariaceous origin) and occur in deltaic deposits of Aptian and Albian age (the estimation of the age by previous authors was quite different: it had varied from the Late Jurassic up to the Albian). The best known is the **Lebanon amber** (*q.v.*). Few insect inclusions, all representing dipterans, have been recorded from the Jordan "amber" (BANDEL *et al.* 1997). In Israel only some plant and fungi have been found (TING & NESSELBAUM 1986). The unidentified arthropod inclusions are mentioned also from an Albian fossil resin from the bottom of Persian Gulf (see Figs. 4: 63, 501: 25) (SAVKEVICH *et al.* 1990).

New Jersey Amber (see Figs. 4: 64, 501: 26). Several important localities of insectiferous fossil resin are known from the Atlantic Coastal Plain (New Jersey and adjacent states of USA); the main locality being Cliffwood. The principal resin occurrences are restricted to the Magothy Formation of Turonian-Coniacian (Early/mid-Late Cretaceous) age, but usually are referred to as Turonian. 19 arthropod orders, 100 families and more than 200 species are recorded, as well as flowers, mushrooms and vertebrate remains (GRIMALDI *et al.* 1989, GRIMALDI 1998, 2000b).

Oise (**Paris Basin Amber**, see Figs. 5: 20, 502: 18). This amber has been discovered only recently in a quarry in the Department Oise (France). It is associated with a rich mammal fauna and is dated as the earliest Eocene. More than 200 new species, in about 40 families and 21 insect orders have been preliminary identified but still await description (NEL *et al.* 1998a,b, 1999).

Romanian amber (see Figs. 5: 23, 502: 21). The Romanian amber (mineralogically called rumanite) is known from several localities in the Romanian part of the Carpathian Mountains; the main site being Buzeu. The age of these deposits is estimated as Oligocene. Few insect species have been described; the general composition of the fauna is still unknown (ZHERIKHIN 1978).

Sakhalinian amber (see Figs. 5: 25, 502: 23). Numerous insects and other fossils are found in the rumanite-type amber collected at the Okhotsk sea beach of Sakhalin Island near Starodubskoye (ZHERIKHIN 1978: 116). Recently the amber has been found *in situ* nearby at the Naiba River embedded in the coal of the Lower Due Formation of the Palaeocene age; amber samples from the coast and from Naiba have been found to be chemically identical (A. Shedrinsky, pers. comm.).

Shavarshavan (see Figs. 4: 73, 501: 34). The resin occurs at the Shavarshavan locality in the Armenian part of the Minor Caucasus Mountains in shallow marine deposits containing abundant marine fossils of Coniacian (mid-Late Cretaceous) age. Few insects and spider inclusions are found (ZHERIKHIN 1978).

Sicilian amber (see Figs. 5: 27, 502: 25). The Sicilian amber (mineralogically called simetite) had initially been collected as a re-deposit on a beach on the Simeto River in Sicily, Italy; resins of the same type containing a similar fauna were later discovered *in situ* in several places in the Northern Apennines (the main locality being Campaolo in Forli Province). The age of simetite, previously estimated as Oligocene-Miocene based on its fauna, has been re-defined as Early Oligocene on the basis of northern Apennine occurrences (SKALSKI & VEGGIANI 1990). Numerous insect inclusions are described, in particular hymenopterans (e.g. SKALSKI 1988), but the general composition of the fauna is still as outlined by KOHRING & SCHLÜTER (1989).

Sumatra resin (see Figs. 5: 45, 503: 15). An "earthy amber" has been collected from several Miocene formations in Central Sumatra, Indonesia, and a few insect inclusions have been recorded including a beetle (DURHAM 1956).

Taimyr Lake retinite (see Figs. 4: 74, 501: 35). The resin occurs at the Baikura-Neru Bay at Taimyr Lake (Central Taimyr) in cross-bedded sands similar with those of Dolgan and Kheta Formations. The data about the age of the deposits are controversial, and the exact position of the locality within the Late Cretaceous is unclear. More than 300 inclusions have been found; they represent 11 insect orders (dipterans and hymenopterans predominate), as well as spiders and mites (RASNITSYN 1980).

Timmerdyakh (see Figs. 4: 75, 501: 36). The arthropod inclusions are known at Timmerdyakh-Khaya at the Vilyui River, Central Yakutia. The fragile transparent resin have been found in fluvial sands of the Timmerdyakh Formation (Cenomanian-Turonian, Late Cretaceous;

ZHERIKHIN 1978). More than 50 inclusions have been found representing 6 insect orders (dipterans dominate).

Ube (see Figs. 5: 29, 502: 27). The resin is found in the Late Eocene coal-bearing deposits of Honshu Island (Japan). A single inclusion, an ant, has been recorded (SCHLEE 1990).

Ukrainian amber (see Figs. 5: 30, 502: 28). This amber is known from several localities in Ukraine, the main of which is Klesovo in Rovno Region. For a long time it was considered to be re-deposited Baltic amber, but recently it was found *in situ* in Late Eocene. The fauna of Ukrainian amber is still poorly investigated (s. SREBRODOLSKY 1980), but at least its ant assemblage differs appreciably from that of the Baltic amber in its taxonomic composition (DLUSSKY in press).

Yantardakh (see Figs. 4: 78, 501: 39). Retinites of Santonian age (RASNITSYN 1980: 157, footnote 1) are gathered in the bulk in sand of the Kheta Formation at the Maimecha River in the East Taimyr Peninsula (North Siberia). Nearly 3,000 fossils have been found there (ZHERIKHIN & SUKATSHEVA 1973). Much smaller collections were made from several other localities in the same deposits on the Kheta River.

References

ANONYMOUS, 1985. Primitive grasshoppers eat primitive plants. *New Sci.* **105**: 22.

ABDERHALDEN, E., and HEYNS, K. 1933. Nachweis von Chitin in Flügelresten von Coleopteren des oberen Mitteleozäns (Fundstelle Geiseltal). *Biochem. Zeitschr.* **259**: 320–321.

ABEL, O. 1933. Ein fossiles Termitennest aus dem Unterpliocän des Wiener Beckens. *Verhandl. zool.-botan. Ges. Wien* **83**: 38–39.

ABEL, O. 1935. *Vorzeitliche Lebensspuren*. G. Fischer Verlag, Wien: 644 p.

ABLAEV, A.G. 1974. *Pozdnemelovaya flora Vostochnogo Sikhote-Alinya i ee znachenie dlya stratigrafii* (Late Cretaceous Flora of Eastern Sikhote-Alin' and Its Stratigraphical Significance). Nauka, Novosibirsk: 180 p. (in Russian).

ACHTELIG, M., and KRISTENSEN, N.P. 1973. A re-examination of the relationships of the Raphidioptera (Insecta). *Zeitschr. Zool. Syst. Evolutions Forsch.* **11**: 268–274.

ACKERY, P.R., DE JONG, R., and WANE-WRIGHT, R.I. The butterflies: Hedyloidea, Hesperioidea and Papilionoidea. In: KRISTENSEN, N.P. (ed.) *Lepidoptera, Moths and Butterflies.* Vol. 1. Handbook of Zoology **4** Arthropoda: Insecta Part 35. Walter de Gruyter, Berlin–New York: 263–300.

ADAMS, P.A. 1967. A review of the Mesochrysinae and Nothochrysinae (Neuroptera: Chrysopidae). *Bull. Mus. Compar. Zool. Harvard Coll.* **135**: 215–238.

ADAMUS, P.R., and GAUFIN, A.R. 1976. A synopsis of Nearctic taxa found in aquatic drift. *Amer. Midl. Natur.* **95**: 198–204.

AFONIN, S.A. 2000. Pollen grains of the genus *Cladaitina* extracted from the gut of the Early Permian insect *Tillyardembia* (Grylloblatida). *Paleontol. Zhurnal* (5): 105–109 (in Russian; English translation: *Paleontol. J.* **34**(5): 575–579).

AGRAWAL, S.C., and SINGH, I.B. 1983. Palaeoenvironment and trace fossils of the Middle Siwalik sediments, Hardwar, Uttar Pradesh. *J. Paleontol. Soc. India* **28**: 50–55.

AHLBRANDT, T.D., ANDREWS, S., and GWYNNE, D.T. 1978. Bioturbation in aeolian deposits. *J. Sediment. Petrol.* **48**: 839–848.

AIZAWA, J. 1991. Fossil insect-bearing strata of the Triassic Mine Group, Yamaguchi Prefecture. *Bull. Kitakyushu Mus. Nat. Hist.* **10**: 91–98 (in Japanese, with English summary).

AKUTSU, J. 1964. The geology and palaeontology of Shiobara and its vicinity, Tochigi Prefecture. *Sci. Rep. Tohoku Univ.* (2), **35**(3): 211–293 (in Japanese, with English summary).

ALEKSEEV, A.S., DMITRIEV, V.Yu., and PONOMARENKO, A.G. 2001. *Evolyutsiya takso-nomicheskogo raznoobraziya* (Evolution of taxonomic diversity). Ecosystem restructures and the evolution of biosphere **5**. Geos, Moscow: 126 p. (in Russian).

ALEXANDER, G. 1964. Occurrence of grasshoppers as accidentals in the Rocky Mountains of Northern Colorado. *Ecology* **45**: 77–86.

ALEXANDER, J.B., NEWTON, J., and CROWE, G.A. 1991. Distribution of Oriental and German cockroaches, *Blatta orientalis* and *Blattella germanica* (Dictyoptera) in the United Kingdom. *Med. Veter. Entomol.* **5**: 395–402.

ALEXEEV, A.V. 1999. A survey of Mesozoic buprestids (Coleoptera) from Eurasian deposits. *Proc. First Palaeontol. Conf., Moscow 1998.* AMBA projects Internat., Bratislava: 5–9.

ALFIERI, A. 1932. Les insectes de la tombe de Toutankhamon. *Bull. Soc. entomol. Egypte* **24**: 188–189.

ALIDOU, S., HANG, J., and HUGAS, G. 1977. Sur un rôle possible joué par les termites dans la fossilisation des restes végétaux. *Sci. géol. Bull.* **30**(3): 203–206.

ALIEV, O.B. 1977. New data about the age of Copal Suite in the Minor Caucasus. *Izvestiya Akad. Nauk Azerbaidzhan. SSR* Ser. nauk o zemle (1): 3–10 (in Russian, with English summary).

ALLISON, P.A. 1986. Soft-bodied animals in the fossil record: the role of decay in fragmentation during transport. *Geology* **14**(12): 979–981.

ALLISON, P.A. 1988. The role of anoxia in the decay and mineralization of proteinaceous macrofossils. *Palaeobiology* **14**: 139–154.

ALLUAUD, C. 1910. Notes sur les coléoptères trouvès dans les momies d'Egypte. *Bull. Soc. entomol. Egypte* (1908) **1**: 29–36.

ALMOND, J.E. 1985. The Silurian-Devonian fossil record of the Myriapoda. *Phil. Trans. R. Soc. London* B **309**: 227–237.

ALONSO, J., ARILLO, I., BARRON E., CORALL, J.C., GRIMALT, J., LOPEZ, R., MARTÍNEZ-DELCLÒS X., ORTUÑO V., PEÑALVER E., and TRICAO, P.R. 2000. A new fossil resin with biological inclusions in Lower Cretaceous deposits from Sierra de Cantabria (Alava, Northern Spain, Basque-Cantabrian Basin). *J. Paleontol.* **74**(1): 158–178.

ALVEY, R.C. 1968. Beetle remains: a Roman well at Bunning, Nottinghamshire. *Trans. Thoronton Soc. Notts.* **71**: 5–10.

AMITROV, O.V. 1989. On the degree of completeness of faunal assemblages. In: SOKOLOV, B.S. (ed.). *Osadochnaya obolochka Zemli v prostranstve i vremeni. Stratigrafiya i paleontologiya.* (Sedimentary Cover of the Earth in Space and Time. Stratigraphy and Palaeontology). Nauka, Moscow: 78–84 (in Russian, with English summary).

AMSDEN, A.F., and BOON, G.C. 1975. C. O. Waterhouse's list of insects from Silchester (with a note on early identifications of insects in archaeological contexts). *J. Archaeol. Sci.* **2**(2): 129–136.

ANANTHAKRISHNAN, T.N. 1969. *Indian Thysanoptera. CSIR Zool. Monogr.* **1**. New Delhi: 171 p.

ANANTHAKRISHNAN, T.N. 1973. *Thrips Biology and Control.* MacMillan, New Delhi: 120 p.

ANANTHAKRISHNAN, T.N. 1978. *Thrips Galls and Gall Thrips. Zool. Surv. India Techn. Monogr.* **1**. New Delhi: 69 p.

ANANTHAKRISHNAN, T.N. 1979. Biosystematics of Thysanoptera. *Annu. Rev. Entomol.* **24**: 159–183.

ANANTHAKRISHNAN, T.N. 1982. Thrips and pollination biology. *Current Sci.* **51**(4): 168–172.

ANDERSEN, N.M. 1982. *The Semiaquatic Bugs (Hemiptera, Gerromorpha): Phylogeny, Adaptations, Biogeography and Classification.* Entomonograph 3. Scandinavian Science Press, Klampenborg: 455 p.

ANDERSEN, N.M. 1998. Water striders from the Paleogene of Denmark with a review of the fossil record and evolution of semiaquatic bugs (Hemiptera, Gerromorpha). *Biol. Skr. Dansk. Vid. Selsk.* **50**: 1–157.

ANDERSEN, N.M. 1999. *Microvelia polhemi*, n.sp. (Heteroptera: Veliidae) from Dominican amber: the first fossil record of a phytotelmic water strider. *J. New York Entomol. Soc.* **107**(2–3): 135–144.

ANDERSEN, N.M., FARMA, A., MINELLI, A., and PICCOLI, G. 1994. A fossil *Halobates* from the Mediterranean and the origin of sea skaters (Hemiptera, Gerridae). *Zool. J. Linn. Soc.* **112**: 479–489.

ANDERSON, J.M., and ANDERSON, H.M. 1993. Terrestrial flora and fauna of Triassic: Part 1. Occurrence. In: LUCAS, S.G., and MORALES, M. (eds.) *The Nonmarine Triassic. Bull. New Mexico Mus. Natur. Hist. and Sci.* **3**: 1–12.

ANDERSON, J.M., ANDERSON, H.M., and CRUICKSHANK, R.I. 1998. Late Triassic ecosystems of the Molteno/Lower Elliot biome of Southern Africa. *Palaeontology* **41**(3): 387–421.

ANDERSON, K.B. 1994. The nature and fate of natural resins in the geosphere. – IV. Middle and Upper Cretaceous amber from the Taimyr Peninsula, Siberia – evidence for a new form of polylabdanoid of resinite and revision of the classification of Class I resinites. *Organ. Geochem.* **21**(2): 209–212.

ANDERSON, K.B. 1995. New evidence concerning the structure, composition and maturation of Class I (polylabdanoid) resinites. In: ANDERSON, K.B., and CRELLING, J.C. (eds.). *Amber, Resinite, and Fossil Resins. AMC Symposium Series 617.* American Chem. Society, Washington, DC: 105–129.

ANDERSON, K., and BOTTO, R.E. 1993. The nature and fate of natural resins in the geosphere – III. Re-evaluation of the structure and composition of *Highgate Copalite* and *Glessite*. *Organ. Geochem.* **20**(7): 1027–1038.

ANDERSON, K.B., and WINANS, R.E. 1991. The nature and fate of natural resins in the geosphere. 1. Evaluation of pyrolysis-gas chromatography/mass spectrometry for the analysis of natural resins and resinites. *Analyt. Chem.* **63**: 2901–2908.

ANDERSON, K.B., WINANS, R.E., and BOTTO, R.E. 1992. The nature and fate of natural resins in the geosphere. – II. Identification, classification and nomenclature of resinites. *Organ. Geochem.* **18**(6): 829–841.

ANDERSON, N.H., and LEHMKUHL, D.M. 1968. Catastrophic drift of insects in a woodland stream. *Ecology* **49**(2): 198–206.

ANDERSSON, G. 1898. Studier öfver Finlands torfmossar och fossila kvartärflora. *Bull. Comm. Geol. Finl.* **8**: 1–210.

ANDOW, D. 1982. Miridae and Coleoptera associated with tulip tree flowers at Ithaca, New York. *J. New York Entomol. Soc.* **90**(2): 119–124.

ANDREIS, R.R., de. 1972. Paleosuelos de la Formación Masters (Eoceno medio), Laguna del Mate, Prov. de Chubut, Rep. Argentina. *Rev. Asoc. Min. Petr. Sedim. Argentina* **3**: 91–98.

ANDRZEJEWSKA, L. 1993. Consequences of plant-soil habitat transformation by Homoptera-Auchenorrhyncha. *Proc. 8th Auchenorrhyncha Congress, Delphi, Greece, 9–13 Aug., 1993*: 51–52.

ANKETELL, J.M., and GHELLALI, S.M. 1984. Nests of a tube-dwelling bee in Quaternary sediments of the Jeffara Plain, S.P.L.A.J. *Libyan Studies* **15**: 137–141.

ANSORGE, J. 1993. Bemerkenswerte Lebensspuren und ?*Cretosphex catalunicus* n. sp. (Insecta; Hymenoptera) aus den unterkretazischen Plattenkalken del Sierra del Montsec (Provinz Lerida, NE-Spanien). *Neues Jahrb. Geol. Paläontol. Monatshefte* **190**: 19–35.

ANSORGE, J. 1996. Insekten aus dem oberen Lias von Grimmen (Vorpommern, Norddeutschland). *Neue Paläontol. Abhandl.* **2**: 1–132.

ANSORGE, J. 1999. *Aenne liasina* gen. et sp. n. – the most primitive non-biting midge (Diptera: Chironomidae: Aenneinae subfam.n.) from the Lower Jurassic of Germany. *Polsk. Pismo Entomol.* **68**: 431–443.

ANSORGE, J. 2001. Revision of the "Trichoptera" described by Geinitz and Handlirsch from the Lower Toarcian of Dobbertin (Germany) basing on new material. In: *Proc. 10th Int. Symp. Trichoptera – Nova Suppl. Ent., Keliern* **15**: 55–74.

ANSORGE, J., and KOHRING, R. 1995. Insekten aus dem Randecker Maar. *Fossilien* **2**: 80–90.

ANSORGE, J., and KRZEMIŃSKI, W. 1994. Oligophrynidae, a Lower Jurassic dipteran family (Diptera, Brachycera). *Acta Zool. Cracov.* **37**(2): 115–119.

ANSORGE, J., and KRZEMIŃSKI, W. 1995. Revision of the Lower Jurassic genera *Mesorhyphus* Handlirsch, *Eoplecia* Handlirsch, *Heterorhyphus* Bode (Diptera: Anisopodomorpha, Bibionomorpha) from the Upper Liassic of Germany. *Paläontol. Zeitschr.* **69**(1/2): 167–172.

ANSORGE, J., and RASNITSYN, A.P. 2000. Identity of *Prosepididontus calopteryx* Handlirsch 1920 (Insecta: Grylloblattida: Geinitziidae). *Acta Geol. Hispanica* **35**(1–2): 19–23.

ANTIPINA, E.E., NAZAROV, V.I., and MASLOV, S.P. 1991. Insects from a well at a wine-making place of Chaika settlement. In: SHCHAPOVA, Yu.L., and YATSENKO, I.V. (eds.). *Pamyatniki zheleznogo veka v okrestnostyakh Evpatorii* (Iron Age Relics in Eupatoria Vicinities). Moscow Univ., Moscow: 155–161 (in Russian).

APEL, K.H. 1979. Erkennung der wichtigsten Stammschädlinge an Kiefer, Fichte und Larche. *Soz. Forstwirt.* **29**(1): 30–31.

APEL, K.H. 1983. Erkennungsmerkmale wichtiger Stammschädlinge an Laubhölzern. *Soz. Forstwirt.* **33**(4): 122–125.

APLET, G.H. 1990. Alteration of earthworm community biomass by the alien *Myrica faya* in Hawai'i. *Oecologia* **82**(3): 414–416.

APPANAH, S., and CHAN, H.T. 1981. Thrips: the pollinators of some dipterocarps. *Malays. Forester* **44**(2–3): 235–252.

ARBIZU, M., BERNÁRDEZ, E., PEÑALVER, E., and PRIETO, M.A. 1999. El ámbar de Asturias (España). *Estud. Mus. Cienc. Natur. Alava* **14**(Núm. Espec. 2): 245–254.

ARCHIBALD, S.B., and MATHEWES, R.W. 2000. Early Eocene insects from Quilchena, British Columbia, and their paleoclimatic interpretations. *Canad. J. Zool.* **78**(8): 1441–1462.

ARILLO, A, and MOSTOVSKI, M.B. 1999. A new genus of Prioriphorinae (Diptera, Phoridae) from the Lower Cretaceous amber of Alava (Spain). *Studia dipterol.* **6**(2): 251–255.

ARISTOV, D.S. 1998. A new species of the Permian genus *Kungurmica* (Grylloblattida: Permembiidae). *Far Eastern Entomol.* **65**: 1–9.

ARISTOV, D.S. 2000a. Grylloblattids of the family Megakhosaridae (Insecta: Grylloblattida) from the Lower Permian of the Middle Urals. *Paleontol. Zhurnal* (2): 69–71 (in Russian, English translation: *Paleontol. J.* **34**(2): 186–188).

ARISTOV, D.S. 2000b. A new family of early Permian grilloblattids (Insecta: Grylloblattida) from Ural mountains. *Far Eastern Entomol.* **85**: 1–4.

ARMAND, A.D., LYURI, D.I., ZHERIKHIN, V.V., RAUTIAN, A.S., KAYDANOVA, O.V., KOZLOVA, E.V., STRELETZKY, V.N., and BUDANOV, V.G. 1999. *Anatomiya krizisov* (Anatomy of Crises). Nauka, Moscow: 238 p. (in Russian).

ARMBRUSTER, W.S. 1984. The role of resin in angiosperm pollination: ecological and chemical considerations. *Amer. J. Botany* **71**: 1149–1160.

ARMENTEROS, I., DABRIO, C., ALONSO, G., JORQUERA, A., and VILLALOBOS, M. 1986. Laminación y bioturbación en carbonatos lagunares: Interpretación genética (Cuenca del Guadiana, Badajoz). *Estud. Geol.* **42**: 271–280.

ARMSTRONG, J.A. 1979. Biotic pollination mechanisms in the Australian flora – a review. *New Zeal. J. Botany* **17**(4): 467–508.

ARMSTRONG, J.A., and IRVINE, A.K. 1989. Floral biology of *Myristica insipida* (Myristicaceae), a distinctive beetle pollination syndrome. *Amer. J. Botany* **76**(1): 86–94.

ARNETT, R.H. 1985. *American Insects. A Handbook of the Insects of North America North of Mexico.* Van Nostrand Reinhold Co., New York: 864 p.

ARNOLDI, L.V., ZHERIKHIN, V.V., NIKRITIN, L.M., and PONOMARENKO, A.G. 1977. *Mezozoyskie zhestkokrylye* (Mesozoic Coleoptera). *Trudy Paleontol. Inst. Akad. Nauk SSSR* **161**. Nauka, Moscow: 204 p. (in Russian, English translation 1991 by Oxonian Press, New Delhi).

ASH, S.R. 1997. Evidence of arthropod-plant interaction in the Upper Triassic of the southwestern United States. *Lethaia* **29**: 237–248.

ASHLOCK, P.D. 1971. Monophyly and associated terms. *Syst. Zool.* **20**: 63–69.

ASHMOLE, N.P., and ASHMOLE, M.J. 1988. Insect dispersal on Tenerife, Canary Islands: High altitude fallout and seaward drift. *Arctic and Alpine Res.* **20**(1): 1–12.

ASHMOLE, N.P., NELSON, J.M., SHAW, M.R., and GARSIDE, A. 1983. Insects and spiders on snowfields in the Cairngorms, Scotland. *J. Natur. Hist.* **17**: 599–613.

ASHWORTH, A.C. 1976. Fossil insects from the Sonoran Desert, Arizona. *Amer. Quatern. Assoc. Abstr.* **1976**: 122.

ASHWORTH, A.C., HARWOOD, D.M., WEBB, P.N., and MABIN, M.G.C. 1997. A weevil from the heart of Antarctica. In: ASHWORTH, A.C., BUCKLAND, P.C., and SADLER, J.P. (eds.). *Studies in Quaternary Entomology. An Inordinate Fondness for Insects.* Quatern. Proc. **5**. John Wiley & Sons, New York: 15–22.

ASKEVOLD, I.S. 1990. Classification of Tertiary fossil Donaciinae of North America and their implications about evolution of Donaciinae (Coleoptera: Chrysomelidae). *Canad. J. Zool.* **68**: 2135–2145.

ASKIN, R.A. 1988. The palynological record across the Cretaceous/Tertiary boundary on Seymour Island, Antarctica. In: FELDMANN, R.M., and WOODBURNE, M.O. (eds.). *Geology and Palaeontology of Seymour Island, Antarctic Peninsula.* Geol. Soc. America Mem. **169**: 155–162.

ASPÖCK, H. 1998. Distribution and biogeography of the order Raphidioptera: updated facts and a new hypotesis. *Acta Zool. Fennica* **209**: 33–44.

ASPÖCK, H., and ASPÖCK, U. 1991. Raphidioptera. In: Naumann, D. (ed.). *The Insects of Australia.* 2nd ed. Melbourne Univ. Press, Carlton: 521–524.

ATTIMS, Y. 1969. Bois du Crétacé inférieur. Étude d'un bois du Crétacé inférieur de Tarfaya: *Protophyllocladoxylon aff. libanoticum* (Edwards) Kraeusel. *Notes Mém. Serv. géol. Maroc* **210**: 51–57.

AUSTIN, A.D., and DOWTON, M. 2000. The Hymenoptera – an introduction. In: AUSTIN, A.D., and DOWTON, M. (eds.) *Hymenoptera: Evolution, Biodiversity and Biological Control.* CSIRO, Collingwood: 3–7.

AUSTIN, J.J., ROSS, A., SMITH, A., FORTEY, R., and THOMAS, R. 1997. Problem of reproducibility – does geologically ancient DNA survive in amber-preserved insects? *Proc. R. Soc. London* B **264**: 467–474.

AVEROF, M., and AKAM, M. 1993. *HOM/Hox* genes of *Artemia*: implications for the origin of insect and crustacean body plans. *Curr. Biol.* **3**: 73–78.

AVEROF, M., and AKAM, M. 1995. *Hox* genes and the diversification of insect and crustacean body plans. *Nature* **376**: 420–423.

AZAR, D. 1997. *L'ambre du Crétacé inférieur du Liban. Analyse paléoenvironmentale. Systématique et phylogénie des Insectes Diptera Psychodoidea et Heteroptera Enicocephalidae.* Diplôme d'Études Approfondies, Univ. Pierre et Marie Curie, Paris: 34 p.

AZAR, D., FLECK, G., NEL, A., and SOLIGNAC, M. 1999a. A new enicocephalid bug, *Enicocephalinus acragrimaldii* gen. nov., sp. nov., from the Lower Cretaceous amber of Lebanon (Insecta, Heteroptera, Enicocephalidae). *Estud. Mus. Cienc. Natur. Alava* **14** (Núm. Espec. 2): 217–230.

AZAR, D., NEL, A., SOLIGNAC, M., PAICHELER, J.-C., and BOUCHET, F. 1999b. New genera and species of psychodid flies from the Lower Cretaceous amber of Lebanon. *Palaeontology* **42**(6): 1101–1136.

BACHMAYER, F. 1952. Fossile Libellenlarven aus miozänen Süsswasserablagerungen. *SB Akad. Wiss. Wien* Abt. 1 **161**: 135–140.

BACHMAYER, F., RÖGL, F., and SCEMANN, R. 1991. Geologie und Sedimentologie der Fundstelle miozäner Insecten in Weingraben (Burgenland, Österreich). In: LOBITZER, H., and CZÁSZÁR, G. (eds.). *Jubiläumsschrift 20 Jahre Geologische Zusammenarbeit Österreich-Ungarn, Wien, Sept. 1991* **1**. Wien: 53–70.

BACHMAYER, F., SYMEONIDIS, N., and THEODOROPOULOS, D. 1971. Einige Insektenreste aus den Jüngtertiären Süsswasserablagerungen von Kumi (Insel Euboea, Griechenland). *Ann. Géol. Pays Hellénique,* 1 sér. **23**: 165–174.

BACHOFEN-ECHT, A. 1949. *Der Bernstein und seine Einschlüsse.* Springer-Verlag, Wien: 204 p. (Re-issued: 1996. Jörg Wunderlich Verlag, Straubenhardt: 230 p.)

BADA, J.L., WANG, S.X., POINAR, H.N., PÄÄBO S., and POINAR, G.O., Jr. 1994. Amino-acid racemization in amber-entombed insects – implications for DNA preservation. *Geochim. Cosmochim. Acta* **58**: 3131–3135.

BADONNEL, A. 1951. Ordre des Psocoptères. In: GRASSÉ P.-P. *Traite de Zoologie.* Masson, Paris **10**: 1301–1340.

BÁEZ, M., and BACOLLADO, J.J. 1984. Los fossiles de Canarias. In: Bacollado, J.J. (ed.). *Fauna (marina y terrestre) del Archipèlago Canario.* Edirea, Las Palmas de Gran Canaria: 343–344.

BAGNALL, R.S. 1912. Some considerations in regard to the classification of the order Thysanoptera. *Ann. Mag. Nat. Hist.* Ser. 8. **10**: 220–222.

BAGNALL, R.S. 1923. Fossil Thysanoptera. 1. *Entomol. Monthly Mag.* **59**: 35–38.

BAGNALL, R.S. 1924. Von Schlechtendal's work on fossil Thysanoptera in the light of recent knowledge. *Ann. Mag. Nat. Hist.* Ser. 9. **14**: 156–161.

BAIRD, G.C. 1997. Geological setting of the Mazon Creek area fossil deposits. In: SHABICA C.W. and HAY, A.A. (eds.). *Richardson's Guide to the Fossil Fauna of Mazon Creek.* Northeastern Illinois Univ., Chicago: 16–20.

BAIRD, G.C., and ANDERSEN, J.L. 1997. Relative abundance of different Mazon Creek organisms. In: SHABICA, C.W., and HAY, A.A. (eds.). *Richardson's Guide to the Fossil Fauna of Mazon Creek.* Northeastern Illinois Univ., Chicago: 21–29.

BAIRD, G.C., SHABICA, C.W., ANDERSEN, J.L., and RICHARDSON, E.S., Jr. 1985a. Biota of a Pennsylvanian muddy coast: habitats within the Mazonian Delta Complex, Northeast Illinois. *J. Paleontol.* **59**: 253–281.

BAIRD, G.C., SROKA, S.D., SHABICA, C.W., and BEARD, T.L. 1985b. Mazon-Creek-type fossil assemblages in the U.S. midcontinent Pennsylvanian: their recurrent character and paleoenvironmental significance. *Philos. Trans. R. Soc. London* **B** 311: 87–99.

BAIRD, G.C., SROKA, S.D., SHABICA, C.W., and KUECHER, G.J. 1986. Taphonomy of Middle Pennsylvanian Mazon Creek area fossil locality δ^{13} C northern Illinois: significance of exceptional fossil preservation in syngenetic concretions. *Palaios* **1**: 271–285.

BAKHURINA, N.N., and UNWIN, D.M. 1995. A survey of pterosaurs from the Jurassic and Cretaceous of the former Soviet Union and Mongolia. *Historical Biol.* **10**: 197–245.

BALAZUC, J. 1978. Les insectes fossiles de la région de Privas (Ardèche). *L'Entomologiste* **34**(4–5): 200–203.

BALAZUC, J. 1989. Quelques insectes fossiles des diatomites de Saint-Bauzile (Ardèche). *Bull. mens. Soc. linn. Lyon* **58**(8): 240–245.

BALAZUC, J., and DESCARPENTRIES, A. 1964. Sur *Lampra gautieri* Bryant et quelques autres Buprestidae fossiles des schistes de Menat (Puy-de-Dôme) (Col.). *Bull. Soc. Entomol. France* **69**(1–2): 47–56.

BALDDERSON, J. 1991. Mantodea. In: NAUMANN, D. (ed.). *The Insects of Australia.* 2nd ed. Melbourne Univ. Press, Carlton: 348–356.

BALICK, M.J., FURTH, D.G., and COOPER-DRIVER, G. 1978. Biochemical and evolutionary aspects of arthropod predation on ferns. *Oecologia* **35**(1): 55–89.

BANDEL, K., SHINAQ, R., and WEITSCHAT, W. 1997. First insect inclusions from the amber of Jordan (Mid Cretaceous). *Mitt. Geol. Paläontol. Inst. Univ. Hamburg* **80**: 213–223.

BANKS, H.P. 1981. Peridermal activity (wound repair) in an Early Devonian trimerophyte from the Gaspé Peninsula, Canada. *Palaeobotanist* **28/29**: 20–25.

BANKS, H.P., and COLTHART, B.J. 1993. Plant-animal-fungal interactions in Early Devonian trimerophytes from Gaspé, Canada. *Amer. J. Botany* **80**: 992–1001.

BANKS, M.R., COSGRIFF, J.W., and KEMP, N.R. 1978. A Tasmanian Triassic stream community. *Austral. Nat. Hist.* **19**(5): 150–157.

BARASH, M.S., KHARIN, G.S., and EMEL'YANOV, E.M. 1970. An asphalt lake at Trinidad Island. *Priroda* (2): 65–69 (in Russian).

BARBER, C.T. 1930. Note on the alleged occurrence of fossil eggs at Yenangyaung, Upper Burma. *Rec. Geol. Surv. India* **62**: 454–455.

BARBU, I.Z. 1939. Insectes fossiles du tertiaire de l'Olténie. *Bull. Soc. Roum. Géol.* **4**: 119–128.

BARONI-URBANI, C., and GRAESER, S. 1987. REM-Analysen an einer pyritisierten Ameise aus Baltischem Bernsteins. *Stuttg. Beitr. Naturk.* B **133**: 1–16.

BARSKOV, I.S., ZHERIKHIN, V.V., and RAUTIAN, A.S. 1996. Problems of evolution of biological diversity. *Zhurnal obshchey biol.* **57**(2): 14–39 (in Russian).

BARTHEL, K.W. 1970. On the deposition of the Solnhofen lithographic limestone (Lower Tithonian, Bavaria, Germany). *Neues Jahrb. Geol. Paläontol. Abhandl.* **135**(1): 1–18.

BARTHEL, K.W. 1972. The genesis of the Solnhofen lithographic limestone (Lower Tithonian, Bavaria, Germany): further data and comments. *Neues Jahrb. Geol. Paläontol. Monatshefte* (3): 133–145.

BARTHEL, K.W., SWINBURNE, N.H.M., and CONWAY MORRIS, S. 1990. *Solnhofen. A Study in Mesozoic Palaeontology.* Cambridge Univ. Press, Cambridge: 236 p.

BARTHEL, M., and HERTZER, H. 1982. Bernstein-Inklusen aus dem Miozän des Bitterfelder Raumes. *Zeitschr. angew. Geol.* **28**: 314–336.

BARTHEL, M., and RÜFFLE, L. 1970. Vegetationsbilder aus dem Alttertiär (Eozän) der Braunkohle des Geiseltales. *Wiss. Zeitschr. Humboldt-Univ. Berlin, mathem.-naturwiss. Reihe* **19**: 273–275.

BARTHEL, M., and RÜFFLE, L. 1976. Ein Massenvorkommen von Symplocaceenblättern als Beispiel einer Variationsstatistik. *Abhandl. Zentr. Geol. Inst., Paläontol. Abhandl.* **26**: 291–305.

BASIBUYUK, H.H., RASNITSYN, A.P., ACHTERBERG, K., von FITTON, M.G., and QUICKE, D.L.J. 1999. A new, putatively primitive Cretaceous fossil braconid subfamily from New Jersey amber (Hymenoptera, Braconidae). *Zool. Scripta* **28**: 211–214.

BASINGER, J.F. 1991. The fossil forests of the Buchanan Lake Formation (Early Tertiary), Axel Heiberg Island, Canadian Arctic Archipelago: preliminary floristics and palaeoclimate. In: CHRISTIE, R.L., and McMILLAN, N.J. (eds.). *Tertiary Fossil Forests of the Geodetic Hills, Axel Heiberg Island, Arctic Archipelago.* Geol. Surv. Canada Bull. **403**: 39–65.

BATTEN, D.J. 1998. Palaeoenvironmental implications of plant, insect and other organic-walled microfossils in the Weald Clay Formation (Lower Cretaceous) of southeastern England. *Cretaceous Res.* **19**(3–4): 279–315.

BAYKOVSKAYA, T.N. 1974. *Verkhnemiotsenovaya flora Yuzhnogo Primor'ya* (Upper Miocene Flora of Southern Primorye Province). Nauka, Leningrad: 143 p. (in Russian).

BAXENDALE, R.W. 1979. Plant-bearing coprolites from North American Pennsylvanian coal balls. *Palaeontology* **22**(3): 537–548.

BAZ, A., and ORTUÑO, V.M. 2000. Archaeatropidae, a new family of Psocoptera from the Cretaceous Amber of Alava, northern Spain. *Ann. Entomol. Soc. Amer.* **93**(3): 367–373.

BEACH, J.A. 1982. Beetle pollination of *Cyclanthus bipartitus* (Cyclanthaceae). *Amer. J. Botany* **69**: 1074–1081.

BEARD, J. 1990. Alien shrub runs riot in Hawaii. *New Sci.* **125**: 30.

BECHLY, G. 1998. *Phylogeny and Systematics of Fossil Dragonflies (Insecta: Odonatoptera), with Special Reference to Some Mesozoic Outcrops. Dissertation zur Erlangung des Grades eines Doctor der Naturwissenschaften.* Eberhard-Karls-Univ. Tübingen: 755 p.

BECHLY, G. 1999. *Phylogenetic Systematics of Odonata.* Internet document available at: http://members.tripod.de/GBechly/phylosys.htm

BECHLY, G., NEL, A., MARTÍNEZ-DELCLÒS, X., and FLEC, G. 1998. Four new dragonfly species from the Upper Jurassic of Germany and the Lower Cretaceous of Mongolia (Anisoptera: Hemeroscopidae, Sonidae and Proterogomphidae fam. nov.). *Odonatologica* **27**(2): 149–187.

BECHLY, G., and WITTMANN, M. 2000. Two new tropical bugs (Insecta: Heteroptera: Thaumastocoridae – Xylastodorinae and Hypsipterygidae) from Baltic amber. *Stuttg. Beitr. Naturk.* B **289**: 1–11.

BECK, A.L., and LABANDEIRA, C.C. 1998. Early Permian insect folivory on a gigantopterid-dominated riparian flora from north-central Texas. *Palaeogeogr., Palaeoclimatol., Palaeoecol.* **142**: 139–173.

BECK, A.L., LABANDEIRA, C.C., and MAMAY, S.H. 1996. Host spectrum and intensity of insect herbivory on a Lower Permian riparian flora: implications for the early sequestring of vascular plant tissues. *Geol. Soc. Amer. Abstracts with Programs* **28**(7): A105.

BECK, C.W. 1999. The chemistry of amber. *Estud. Mus. Cienc. Natur. Alava* **14** (Núm. Espec. 2): 33–48.

BECKER, G. 1950. Bestimmung von Insektenfraßschäden an Nadelholz. *Zeitschr. angew. Entomol.* **31**: 275–303.

BECKER-MIGDISOVA, E.E. 1960a. Paleozoic Homoptera of the USSR and problems of phylogeny of the order. *Paleontol. Zhurnal* (3): 28–42 (in Russian).

BECKER-MIGDISOVA, E.E. 1960b. Die Archescytinidae als vermutliche Vorfahren der Blattläuse. In: *Verhandl. XI Internat. Kongr. Entomol., Wien, Aug. 17–25, 1960* **1**: 298–301.

BECKER-MIGDISOVA, E.E. 1962. Superorder Rhynchota. In: Rohdendorf, B.B. (ed.). *Osnovy paleontologii. Trakheinye i khelitserovye* Nauka, Moscow: 161–226 (in Russian; English translation: ROHDENDORF, B.B. (editor-in-chief). *Fundamentals of Palaeontology.* Vol. 9. Arthropoda, Tracheata, Chelicerata. Amerind Publ. Co. 1991: 216–317).

BECKER-MIGDISOVA, E.E. 1964. *Tretichnye ravnokrylye Stavropol'ya* (Tertiary Homoptera of Stavropol). *Trudy Paleontol. Inst. Akad. Nauk. SSSR* **104**. Nauka, Moscow: 108 p. (in Russian).

BECKER-MIGDISOVA, E.E. 1967. Tertiary Homoptera of Stavropol and a method of reconstruction of continental palaeobiocoenoses. *Palaeontology* **10**: 542–553.

BECKER-MIGDISOVA, E.E. 1972. Phylogeny of Psyllomorpha in view of feeding adaptation to host plants. In: *Sessiya, posvyashchennaya stoletiyu so dnya rozhdeniya akademika A.A. Borisyaka* (Session Dedicated to the Centenary of Academician A.A. Borissiak). Nauka, Moscow: 3–4 (in Russian).

BECKER-MIGDISOVA, E.E. 1985. *Iskopaemye nasekomye psillomorfy* (Fossil Psyllomorphous Insects). *Trudy Paleontol. Inst. Akad. Nauk SSSR* **206**. Nauka, Moscow: 94 p. (in Russian).

BEHRENSMEYER, A.K., DAMUTH, J.D., DiMICHELE, W.A., POTTS, R., SUES, H.-D., and WING, S.L. (eds.). 1992. *Terrestrial Ecosystems through Time. Evolutionary Paleoecology of Terrestrial Plants and Animals.* Univ. Chicago Press, Chicago – London: 568 p.

BEHRENSMEYER, A.K., and HOOK, R.W. 1992. Paleoenvironmental contexts and taphonomic modes. In: BEHRENSMEYER, A.K., DAMUTH, J.D., DiMICHELE, W.A., POTTS, R., SUES, H.-D., and WING, S.L. (eds.). *Terrestrial Ecosystems through Time. Evolutionary Paleoecology of Terrestrial Plants and Animals.* Univ. Chicago Press, Chicago – London: 15–136.

BEIER, M. 1952. Miozäne und Oligozäne Insekten aus Österreich und den unmittelbar angrenzenden Gebieten. *SB. Österr. Akad. Wiss., mathem.- naturwiss. Klasse* Abt. 1, **161**(2–3): 129–134.

BEIER, M. 1955. Insektenreste aus der Halstattzeit. *SB Österr. Akad. Wiss., mathem.-naturwiss. Kl.*, Abt. 1 **164**(9): 747–749.

BEIER, M. 1967. *Mantis religiosa* im Pliozän des Harzvorlandes. *Ber. naturhist. Ges. Hannover* **111**: 63–64.

BEIER, M. 1968. Mantodea (Fangheuschrecken). In: Kukenthal, W. (ed.). *Handbuch der Zooloogie* **4**(2): 2/12. Walter de Gruyter, Berlin: 47 p.

BELOKRYS, L.S. 1972. Insect nids in the Sarmatian deposits of Southern Ukraine and their geological importance as indicators. *Palaeontol. Sbornik* **9**. L'vov: 67–72 (in Russian).

BENNIKE, O., and BÖCHER, J. 1990. Forest-tundra neighbouring the North Pole: plant and insect remains from the Plio-Pleistocene Kap København Formation, North Greenland. *Arctic* **43**(4): 331–338.

BENTON, M.J. 1986. The Late Triassic tetrapod extinction events. In: PADIAN, K. (ed.). *The Beginning of the Age of Dinosaur.* Cambridge Univ. Press, New York: 303–320.

BENTON, M.J. 1987. Mass extinctions among families of non-marine tetrapods: The data. *Mem. Soc. géol. France*, n.s., **150**: 21–32.

BENTON, M.J. 1989. Fossil reptiles from the ancient caves. *Nature* **337**: 309–310.

BENTON, M.J. (ed.). 1993. *The Fossil Record.* 2nd edition. Chapman and Hall, London: 839 p.

BERGER, L.P. 1992. Causes of migrations of Lepidoptera in the Palaearctic. *19 Internat. Congr. Entomol., Beijing, June 28–July 4, 1992. Proc.: Abstracts.* Beijing: 192.

BERGSMA, B.M., SVOBODA, J., and FREEDMAN, B. 1984. Entombed plant communities released by a retreating glacier at central Ellesmere Land, Canada. *Arctic* **37**(1): 49–52.

BERGSTRÖM, G., GROTH, I., PELLMYR, O., ENDRESS, P.K., THIEN, L.B., HÜBENER, A., and FRANCKE, W. 1991. Chemical basis of a highly specific mutualism: chiral esters attract pollinating beetles in Eupomatiaceae. *Phytochemistry* **30**(10): 3225–3231.

BERMAN, D.I., and ALFIMOV, A.V. 1998. The simulation of Late Pleistocene climates of Asiatic and Central Beringia based on entomological data. *Vestnik Dal'nevostochnogo Otdeleniya Ross. Akad. Nauk* (1): 27–34 (in Russian).

BERNER, R.A. 1968. Calcium carbonate concretions formed by decomposition of organic matter. *Science* **159**: 195–197.

BERNER, R.A. 1971. *Principles of Chemical Sedimentology.* McGraw Hill, New York, 240 p.

BERNER, R.A. 1976. The benthic boundary layer from the viewpoint of a geochemist. In: McCAVE, I.N. (ed.). *The Benthic Boundary.* Plenum, New York: 33–55.

BERNER, R.A. 1981. Authigenic mineral formation resulting from organic matter decomposition in modern sediments. *Fortschr. Mineral.* **59**(1): 117–135.

BERRY, E.W. 1924. The Middle and Upper Eocene floras of Southeastern North America. *U.S. Geol. Surv. Prof. Paper* **92**: 1–201.

BERRY, E.W. 1927. A new type of caddis case from the Lower Eocene of Tennessee. *Proc. US Nation. Mus.* **71**(14): 1–5.

BERRY, E.W. 1930. Revision of the Lower Eocene Wilcox Flora of the southeastern states, with descriptions of new species from Tennessee and Kentucky. *U.S. Geol. Surv. Prof. Paper* **156**: 1–196.

BERRY, E.W. 1931. An insect-cut leaf from the Lower Eocene. *Amer. J. Sci.* Ser. 5, **21**(124): 301–303.

BERRY, R.E., and TAYLOR L.R. 1968. High-altitude migration of aphids in marine and continental climates. *J. Anim. Ecol.* **37**(3): 713–722.

BERTRAND-SARFATI, J., FREYTET, P., and PLAZIAT, J.-C. 1966. Les calcaires concrétionnés de la limite Oligocène – Miocène des environs de Saint-Pourçain-sur-Sioule (Limagne d'Allier): rôle des algues dans leur édification, analogie avec les stromatolites et rapports avec la sédimentation. *Bull. Soc. géol. France* Ser. 7. **8**: 652–662.

BETEKHTINA, O.A. 1966. *Verkhnepaleozoiskie nemorskie dvustvorki Sibiri i Kazakhstana* (The Upper Paleozoic Non-marine Bivalvians of Siberia and Kazakhstan). Nauka, Moscow: 220 p. (in Russian).

BETEKHTINA, O.A. 1974. *Biostratigrafiya i korrelyatsiya uglenosnykh otlozheniy pozdnego paleozoya po nemorskim dvustvorkam* (Biostratigraphy and Correlation of the Coal-bearing Deposits of the Late Paleozoic from Non-marine Bivalvians). Nauka, Novosibirsk: 179 p. (in Russian).

BEUTEL, R.G. 1995. The Adephaga (Coleoptera): phylogeny and evolutionary history. In: PAKALUK, J., and SLIPIŃSKI, S.A. (eds). *Biology, Phylogeny, and Classification of Coleoptera: Papers Celebrating the 80th Birthday of Roy A. Crowson.* Muzeum i Inst. Zool. PAN, Warszawa: 173–217.

BEUTEL, R. 1997. Über Phylogenese und Evolution der Coleoptera (Insecta), insbesondere der Adephaga. *Abhandl. Naturwiss. Vereins Hamburg* N.F **31**, 164 ss.

BEY-BIENKO, G.YA. 1936. *Nasekomye kozhistokrylye* (Insecta Dermaptera). *Fauna SSSR* (N. Ser.) **5**. Akad. Nauk SSSR, Moscow – Leningrad: 240 p. (in Russian, with English summary).

BEYLE, M. 1913. Über einige Ablagerungen fossiler Pflanzen der Hamburger Gegend. I. *Jahrb. Hamburg Wiss. Anst.* **30**: 83–99.

BEYLE, M. 1924. Über einige Ablagerungen fossiler Pflanzen der Hamburger Gegend. II. *Mitt. mineral.-geol. Staatsinst. Hamburg* **6**: 1–309.

BEYLE, M. 1926. Über einige Ablagerungen fossiler Pflanzen der Hamburger Gegend. IV. *Mitt. mineral.-geol. Staatsinst. Hamburg* **8**: 111–132.

BHATTI, J.S. 1979. A revised classification of Thysanoptera. *Workshop on Advances in Insect Taxonomy in India and the Orient, Manali 1979:* 46–48.

BHATTI, J.S. 1989a. The orders Terebrantia and Tubulifera of the superorder Thysanoptera (Insecta). A critical appraisal. *Zoologie* **1**: 167–240.

BHATTI, J.S. 1989b. The classification of Thysanoptera into families. *Zoologie* **2**: 1–23.

BIDASHKO, F.G. 1987. A combined procedure of insect remain collecting with use of the kerosene flotation and some recommendations for preparing of insect remains for identification. *Zool. Zhurnal* **66**(7): 1086–1089 (in Russian).

BINFORD, M.W. 1982. Ecological history of Lake Valencia, Venezuela: Interpretation of animal microfossils and some chemical, physical, and geological features. *Ecol. Monogr.* **52**: 307–333.

BINO, R.J., DAFNI, A., and MEEUSE, A.D.J. 1984a. Entomophily in the dioecious gymnosperm *Ephedra aphylla* Forsk. (=*E. alte* C.A. Mey.), with some notes on *E. campylopoda* C.A. Mey. 1. Aspects of the entomophilous syndrome. *Proc. K. Nederl. Akad. Wetensch.* **C 87**: 1–13.

BINO, R.J., DAFNI, A., and MEEUSE, A.D.J. 1984b. Entomophily in the dioecious gymnosperm *Ephedra aphylla* Forsk. (= *E. alte* C.A. Mey.), with some notes on *E. campylopoda* C.A. Mey. 2. Pollination droplets, nectaries and nectarial secretion in *Ephedra*. *Proc. K. Nederl. Akad. Wetensch.* **C 87**: 15–24.

BIRKET-SMITH, S.J.R. 1974. On the abdominal morphology of Thysanura (Archaeognatha and Thysanura *s. str.*). *Entomol. Scand.* Suppl. **6**: 1–67.

BIRON, P.E., and DUTUIT, J.-M. 1981. Figurations sédimentaires et traces d'activité au sol dans le Trias de la formation d'Argana et de l'Ourika (Maroc). *Bull. Mus. nation. hist. natur. Paris* C**3**(4): 399–427.

BITSCH, J. 1994. The morphological groundplan of Hexapoda: critical review of recent concepts. *Ann. Soc. Entomol. France* (N.S.) **30**(1): 103–129.

BLAGODEROV, V.A. 1995. Fungus gnats of the tribe Sciophilini (Diptera, Mycetophilidae) from the Early Cretaceous of Transbaikalia. *Paleontol. Zhurnal* (1): 55–63 (in Russian; English translation: *Paleontol. J.* **29**(1): 72–83).

BLAGODEROV, V.A. 1996. Revision of Nematoceran family Protopleciidae (Insecta: Diptera) from the Early Jurassic Sogyuty locality, Kyrgyzstan. *Paleontol. Zhurnal* (2): 85–90 (in Russian; English translation: *Paleontol. J.* **30**(2): 210–216).

BLAGODEROV, V.A. 1997. Fungus gnats of the tribe Gnoristini (Diptera, Mycetophilidae) from the Lower Cretaceous of Transbaikalia. *Paleontol. Zhurnal* (6): 44–49 (in Russian; English translation: *Paleontol. J.* **31**(6): 609–615).

BLAGODEROV, V.A. 1998a. Fungus gnats of the tribes Gnoristini and Leiini (Diptera, Mycetophilidae) from the Lower Cretaceous of Transbaikalia. *Paleontol. Zhurnal* (1): 58–62 (in Russian; English translation: *Paleontol. J.* **32**(1): 54–59).

BLAGODEROV, V.A. 1998b. Fungus gnats (Diptera, Mycetophilidae) from the Lower Cretaceous of Mongolia. *Paleontol. J.* **32**(6): 598–604.

BLAGODEROV, V.A. 1998c. *Mezozoiskie gribnye komariki (Diptera, Mycetophilidae)* [Mesozoic Fungus Gnats (Diptera, Mycetophilidae)]. *Autoreferat dissert. cand. biol. nauk.* Palaeontol. Institute Ross. Akad. Nauk, Moscow: 24 p. (in Russian).

BLAGODEROV, V.A. 1999. New Bibionomorpha from Triassic of Australia and Jurassic of Central Asia with notes on the family Paraxymiidae Rohdendorf (Insecta, Diptera). *Proc. First Palaeoentomol. Conf., Moscow 1998.* AMBA projects Internat., Bratislava: 11–16.

BLAGODEROV, V.A. 2000. New fungus gnats (Diptera: Mycetophilidae) from the Cretaceous and Paleogene of Asia. *Paleontol. J.* **34**(suppl. 3): S355–S359.

BLAGODEROV, V.A. and MARTÍNES-DELCLÒS, X. 2001. Two fungus gnats (Insecta: Diptera: Mycetophilidae) from the Lower Cretaceous of Spain. *Geos*.

BLAIR, F.E. 1908. Some ancient beetles from Egypt and Mesopotamia. *Proc. R. Entomol. Soc. Lond.* **10**: 19.

BLAIR, K.G. 1927. Insect remains from oil sand in Trinidad. *Trans. R. Entomol. Soc. London* **75**(1): 137–141.

BLOCH, D. 1779. Beytrag zur Naturgeschichte des Kopals. *Beschäft. Ges. Naturf. Fr. Berlin* **2**: 91–196.

BLOM, C.D., and ALBERT, N.R. 1985. Carbonate concretions: an ideal sedimentary host for microfossils. *Geology* **13**(3): 212–215.

BOBROVA, V.K., GOREMYKIN, V.V., TROITSKY, A.V., VALIEJO-ROMAN, K.M., and ANTONOV, A.S. 1995. Molecular studies of the angiosperm plant origin. *Zhurnal obshchey biol.* **56**: 645–660 (in Russian).

BÖCHER, J. 1989. Boreal insects in northernmost Greenland: palaeoentomological evidence from the Kap København Formation (Plio-Pleistocene), Peary Land. *Fauna norveg.* B **36**: 37–43.

BOCHERENS, H., FRIIS, E.M., MARIOTTI, A., and PEDERSEN, K.R. 1994. Carbon isotopic abundances in Mesozoic and Cenozoic fossil plants: palaeoecological implications. *Lethaia* **26**: 347–358.

BOCHKAREV, V.S., and PURTOVA, S.I. 1994. The complete transection of the Triassic of West Siberia. *Problemy Severa. Tezisy dokladov nauchnykh chteniy, posvyashchennykh 100-letiyu so dnya rozhdeniya prof. V.A. Khakhlova, Tomsk, 30 marta – 1 aprelya 1994* (Problems of the North. Abstracts of the Scientific Lectures Devoted to the Prof. V.A. Khakhlov's 100th Anniversary, Tomsk, March, 30–April,1, 1994). Tomsk **1**: 107–108 (in Russian).

BOCK, W. 1969. The American Triassic flora and global distribution. *Geol. Center Research Series (North Wales, Pennsylvania)* **3–4**: 1–406.

BODE, A. 1953. Die Insektenfauna der Ostniedersächsischen oberen Lias. *Palaeontographica* A **103** (1–4): 395 S.

BODE, W. 1975. Die Ovipositor und die weiblichen Geschlechtswege der Thripiden (Thysanoptera, Terebrantia). *Zeitschr. Morphol. Tiere* **81**: 1–53.

BODE, W. 1977. Die Ultrastruktur der Rektalpapillen von Thrips (Thysanoptera, Terebrantia). *Zoomorphologie* **90**: 53–65.

BODE, W. 1983. Spermienstruktur und Spermatohistogenese bei *Thrips validus* Uzel (Insecta, Thysanoptera). *Zool. Jahrb. (Anat.)* **109**: 301–318.

BOGACHEV, V.V. 1939. *Binagady – kladbishche chetvertichnoy fauny na Apsheronskom poluostrove* (Binagady – The Quaternary Fauna Cementary at Apsheron Peninsula). Azerbaidzhan. Filial Akad. Nauk SSSR, Baku: 84 p. (in Russian).

BOGATCHEV, A.V. 1940. Fossil *Forficula* from salt-bearing deposits of Nakhitchevan ASSR. *Izvestiya Azerbaidzhan. Filiala Akad. Nauk. SSSR* (5): 66–68 (in Russian).

BOLTEN, R., GALL, H., and JUNG, W. 1976. Die obermiozäne (sarmatische) Fossil-Lagerstätte Wemding im Nördlinger Ries (Bayern). Ein Beitrag zur Charakterisierung des Riessee-Biotops. *Geol. Blätt. NO-Bayern* **29**(2): 75–94.

BOLTON, H. 1921–22. *A Monograph of the Fossil Insects of the British Coal Measures.* Palaeontographical Society Monograph, London: 156 p.

BOLTON, H. 1925. *Insects from the Coal Measures of Commentry.* British Museum (Natural History), London: 56 p.

BONDE, N. 1974. Palaeoenvironment as indicated by the "Mo-Clay Formation" (Lowermost Eocene of Denmark). *Tertiary Times* **2**: 29–36.

BONESS, M. 1975. Arthropoden im Hochwassergenist von Flüssen. *Bonn. zool. Beitr.* **26**: 383–401.

BONESS, M., and STARÝ, P. 1988. Aphid parasitoids in river flood debris. *Entomol. Gener.* **13**(3/4): 251–254.

BONINO DE LANGGUTH, V. 1978. Nidos de insectos fósiles del Cretácico Superior del Uruguay. *Rev. Mus. Argent. Cienc. Natur. (Palaeontología)* **2**: 69–75.

BORKENT, A. 1978. Upper Oligocene fossil pupae and larvae of *Chaoborus tertiarius* (von Heyden) (Chaoboridae, Diptera) from West Germany. *Quaest. Entomol.* **14**: 491–496.

BORKENT, A. 1995. *Biting midges in the Cretaceous amber of North America (Diptera: Ceratopogonidae).* Backhuys Publ., Leiden: 237 p.

BORKENT, A. 1996. Biting midges from Upper Cretaceous New Jersey amber (Ceratopogonidae: Diptera). *Amer. Mus. Novitates* **3159**: 1–29.

BORKENT, A. 1997. Upper and Lower Cretaceous biting midges (Ceratopogonidae: Diptera) from Hungarian and Austrian amber and the Koonwarra Fossil Bed of Australia. *Stuttg. Beitr. Naturk.* B **249**: 1–10.

BORKENT, A. 2000. Biting midges (Ceratopogonidae: Diptera) from Lower Cretaceous Lebanese amber with a discussion of the diversity and patterns found in other ambers. In: GRIMALDI, D.A. (ed.). *Studies on Fossils in Amber, with Particular Reference to the Cretaceous of New Jersey.* Backhuys Publ., Leiden: 355–451.

BÖRNER, C. 1904. Zur Systematik der Hexapoden. *Zool. Anz.* **27**: 512–533.

BOSC, L. 1805. Note sur un fossil remarquable de la montagne de Saint-Gerand-le-Puy, entre Moulins et Roanne, département d'Allier, appelé l'Indusie tubuleuse. *J. Mines* **17**(1): 397–400.

BOSE, M.N. 1968. A new species of *Williamsonia* from the Rajmahal Hills, India. *J. Linn. Soc. London (Botany)* **61**: 121–127.

BOTOSANEANU, L., JAHNSON, R.O., and DILLON, P.R. 1998. New caddisflies (Insecta: Trichoptera) from Upper Cretaceous amber of New Jersey, USA. *Polsk. Pismo entomol.* **67**(3–4): 219–231.

BOTOSANEANU, L., and WICHARD, W. 1983. Upper Cretaceous Siberian and Canadian amber caddisflies (Insecta: Trichoptera). *Bijdrag. Dierkunde* **53**: 187–217.

BOTTJER, D.J., DROSER, M.L., and SAVDRA, C.E. 1987. *New Concepts in the Use of Biogenic Sedimentary Structures for Paleoenvironmental Interpretation.* Soc. Economic Paleontol. and Mineralogists, Pacif. Section **52**: 65 p.

BOUCOT, A.J. (ed.). 1990. *Palaeobiology of Behaviour and Co-evolution.* Elsevier, Amsterdam: 725 p.

BOUDREAUX, H.B. 1979. *Arthropod Phylogeny with Special References to Insects.* John Wiley & Sons, New York etc.: 320 p.

BOURNAUD, M., and THIBAULT, M. 1973. La dérive des organismes dans les eaux courantes. *Ann. hydrobiol.* **4**(1): 11–49.

BOURNIER, A. 1983. *Les thrips: biologie, importance agronomique.* INRA, Paris: 128 p.

BOWDEN, J., and JOHNSON, C.G. 1976. Migrating and other terrestrial insects at sea. In: CHENG, L. (ed.). *Marine Insects.* North-Holland Publ. Co., Amsterdam: 97–117.

BOWLDS, L.S. 1989. Tracking down the Early Permian. *Geotimes* **34**(5): 12–14.

BOWN, T.M., and KRAUS, M.J. 1983. Ichnofossils of the alluvial Willwood Formation (Lower Eocene), Bighorn Basin, northwest Wyoming. *Palaeogeogr., Palaeoclimatol., Palaeoecol.* **43**(1–2): 95–128.

Bown, T.M., and Ratcliffe, B.C. 1988. The origin of *Chubutolithes* Ihering, ichnofossils from the Eocene and Oligocene of Chubut Province, Argentina. *J. Paleontol.* **62**: 163–167.

Boyer, B.W. 1982. Green River laminites: does the playa-lake model really invalidate the stratified-lake model? *Geology* **10**: 321–324.

Bradbury, J.P., and Kirkland, D.W. 1966. Upper Jurassic aquatic Hemiptera from the Todilto Formation, Northern New Mexico. *Geol. Soc. Amer. Abstracts for 1966, Spec. Paper* **101**: 24.

Bradley, F., and Landini, W. 1984. I fossili del "tripoli" messiniano di Gabbro (Livorno). *Palaeontogr. Ital.* **73**: 5–33.

Bradley, J.C., and Galil, B.S. 1977. The taxonomic arrangement of the Phasmatodea with keys to the subfamilies and tribes. *Proc. Entomol. Soc. Washington* **79**: 176–208.

Bradley, W.H. 1964. Geology of Green River Formation and associated Eocene rocks in southwestern Wyoming and adjacent parts of Colorado and Utah. *U.S. Geol. Surv. Prof. Paper* **496-A**: 1–86.

Bradley, W.H. 1970. Green River oil shale – concept of origin extended, an interdisciplinary problem being attacked from both ends. *Geol. Soc. America Bull.* **81**(4): 985–1000.

Bradley, W.H., and Beard, M.E. 1969. Mud Lake, Florida; its algae and alkaline brown water. *Limnol. Oceanogr.* **14**(6): 889–897.

Brady, L.F. 1939. Tracks in the Coconino sandstone compared with those of small living arthropods. *Plateau* **12**: 32–34.

Brady, L.F. 1947. Invertebrate tracks from the Coconino sandstone of Northern Arizona. *J. Paleontol.* **21**: 466–472.

Brandão, C.R.F., Martins-Neto, R.G., and Volcano, M.A. 1989. The Earliest known fossil ant (first Southern Hemisphere Mesozoic record) (Hymenoptera: Formicidae: Myrmeciinae). *Psyche* **96**: 195–208.

Brantjes, N.B.M. 1976a. Riddles around the pollination of *Melandrium album* (Mill.) Garcke (Caryophyllaceae) during the oviposition by *Hadena bicruris* Hufn. (Noctuidae, Lepidoptera). I. *Proc. Kon. Nederl. Akad. Wetensch.* C **79**(1): 1–12.

Brantjes, N.B.M. 1976b. Riddles around the pollination of *Melandrium album* (Mill.) Garcke (Caryophyllaceae) during the oviposition by *Hadena bicruris* Hufn. (Noctuidae, Lepidoptera). II. *Proc. Kon. Nederl. Akad. Wetensch.* C **79**(2): 127–141.

Brauckmann, C. 1984. Weitere neue Insekten (Palaeodictyoptera; Protorthoptera) aus dem Namurium B von Hagen-Vorhalle. *Übers. naturwiss. Verein Wuppertal* **37**: 108–115.

Brauckmann, C. 1988b. Zwei neue Insekten (Odonata, Megasecoptera) aus dem Namurium von Hagen-Vorhalle (West-Deutschland). *Dortmunder Beitr. Landesk. naturwiss. Mitt.* **22**: 91–101.

Brauckmann, C. 1991. Arachniden und Insekten aus dem Namurium von Vorhalle-Schichten (Ober-Karbon, West-Deutschland). *Veröff Fuhlrott-Mus.* **1**: 1–275.

Brauckmann, C., Brauckmann, B., and Grönung, E. 1996. The stratigraphical position of the oldest known Pterygota (Insecta, Carboniferous, Namurian). *Ann. Soc. géol. Belgique* **117**: 47–56.

Brauckmann, C., and Gröning, E. 1998. A new species of *Homaloneura* [Palaeodictyoptera: Spilapteridae] from the Namurian (Upper Carboniferous) of Hagen-Vorhalle (Germany). *Entomol. Gener.* **23**(1/2): 77–84.

Brauckmann, C., Koch, L., and Kemper, M. 1985. Spinnentiere (Arachnida) und Insekten aus den Vorhalle-Schichten (Namurium B; Ober-Karbon) von Hagen-Vorhalle (West-Deutschland). *Geologie und Paläontologie in Westfalen* **3**. 132 S.

Brauckmann, C., and Schneider, J. 1996. Ein unter-karbonisches Insekt aus dem Raum Bitterfeld/Delitz (Pterygota, Arnsbergium, Deutschland). *Neues Jahrb. Geol. Paläontol. Monatshefte* (1): 17–30.

Brauckman, C., and Zessin, W. 1989. Neue Meganeuridae aus dem Namurium von Hagen-Vorhalle (BRD) und die Phylogenie der Meganisoptera. *Deutsche entomol. Zeitschr.* **36**: 177–215.

Brauer, F., Redtenbacher, J., and Ganglbauer, L. 1889. Fossile Insekten aus der Jura-formation Ost-Sibiriens. *Mem. Acad. Imper. Sci. St. Petersbourg* Ser. 7. **36**(15): 1–22.

Braun, J.H. 1960. Histogene und gummöse Abwehrreaktion nach Borkenkäfer-Infektion bei *Tsuga heterophylla* (Raf.) Sarg. *Allg. Forst- und Jagdzeitung* **131**(5): 189–190.

Breckon, G., and Ortiz, V.N. 1983. Pollination of *Zamia pumila* by fungus gnats. *Amer. J. Botany.* **70**: 106–107.

Bresciani, J., Haarlov, N., Nansen, P., and Moller, G. 1983. Head louse (*Pediculus humanus* subsp. *capitis* de Geer) from mummified corpses of Greenlanders, A.D. 1460 (±50). *Acta Entomol. Fenn.* **42**: 24–27.

Breton, G., Gauther, C., and Vizcaino, D. 1999. Land and freshwater microflora in Sparnacian amber from the Corbière (South France): first observations. *Estud. Mus. Cienc. Natur. Alava* **14** (Núm. Espec. 2): 161–166.

Brezinova, D. 1977. Paläobotanische Zugehörigkeit und Fossilisationsmöglichkeiten versteinerter Hölzer. *Fundgrube* **13**(3–4): 79–95.

Briggs, D.E.G. 1999. Molecular taphonomy of animal and plant cuticles: selective preservation and diagenesis. *Philos. Trans. R. Soc. Lond.* B **354**: 7–17.

Briggs, D.E.G., Evershed, R.P., and Stankiewicz, B.A. 1998a. The molecular preservation of fossil arthropod cuticles. *Ancient Biomolecules* **2**: 135–146.

Briggs, D.E.G., Stankiewicz, B.A., Meischner, D., Bierstedt, A., and Evershed, R.P. 1998b. Taphonomy of arthropod cuticles from Pliocene lake sediments, Willershausen, Germany. *Palaios* **13**: 386–394.

Bright, D.E., and Poinar, G.O., Jr. 1994. Scolytidae and Platypodidae (Coleoptera) from Dominican Republic amber. *Ann. Entomol. Soc. Amer.* **87**: 170–194.

Brindle, A. 1970. Dermaptera. In: *La faune terrestre de l'île de Sainte-Helene*. *Ann. Mus. R. Afrique centr., Zool.* **181**. Tervuren: 213–227.

Brito, I.M., and Ribeiro, F.A.M. 1975. Ocorrência de Lepidoptera nos Folhelos de Tremembé e algumas considerações sobre a bacia geológica do Paraíba, Estado de São Paulo. *An. Acad. Brasil. Ciênc.* **47**(1): 105–111.

Britton, E.B. 1960. Beetles from the London Clay (Eocene) of Bognor Regis, Sussex. *Bull. Brit. Mus. (Natur. Hist.), Geol.* **4**(2): 27–50.

Broadhead, E., and Wolda, H. 1985. The diversity of Psocoptera in two tropical forests in Panama. *J. animal Ecol.* **54**: 539–754.

Brodersen, K.P., and Anderson, N.J. 2000. Subfossil insect remains (Chironomidae) and lake-water temperature inference in the Sisimiut-Kangerlussuaq region, Southern West Greenland. *Geol. Greenland Surv. Bull.* **186**: 78–82.

Brodsky, A.K. 1974. Evolution of the mayfly flight apparatus. *Entomol. Obozrenie* **53**(2): 291–303 (in Russian).

Brodsky, A.K. 1991. Structure, functioning and evolution of tergum in alate insects. I. Generalized model of structure. *Entomol. Obozrenie* **70**(2): 297–315 (in Russian. English translation: *Entomol. Review* 1992 **70**(2): 64–82).

Brodsky, A.K., and Ivanov, V.D. 1983. Functional assessment of wing structure in insects. *Entomol. Rev.* **1**: 35–52.

Bromley, R.G. 1990. *Trace Fossils. Biology and Taphonomy.* Unwin Ltd., London: 280 p.

Bromley, R.G. 1996. *Trace Fossils: Biology, Taphonomy and Applications.* Chapman & Hall, London: 361 p.

Brongniart, Ch. 1893. *Recherches pour servir à l'histoire des insectes fossiles des temps primaires etc.* Théolier et Cie, Saint-Etienne: 493 p.

Brooks, S.J., Lowe, J.J., and Mayle, F.E. 1997. The Late Devensian Lateglacial palaeoenvironmental record from Whitrig Bog, SE Scotland. 2. Chironomidae (Insecta: Diptera). *Boreas* **26**: 297–308.

Brothers, D.J. 1992. The first Mesozoic Vespidae from the Southern hemisphere, Botswana. *J. Hymenopt. Res.* **1**: 119–124.

Brothers, D.J. 1999. Phylogeny and evolution of wasps, ants and bees (Hymenoptera, Chrysidoidea, Vespoidea and Apoidea). *Zool. Scripta* **28**: 233–249.

Brown, B. 1999. Re-evaluation of the fossil Phoridae (Diptera). *J. Natur. Hist.* **33**: 1561–1573.

Brown, B. 2000. Revision of the "*Apocephalus-miricauda* group" of ant-parasitizing flies (Diptera: Phoridae). *Contrib. in Sci.* **482**: 1–62.

Brown, R.W. 1934. *Celliforma spirifer*, the fossil larval chamber of mining bees. *J. Washington Acad. Sci.* **24**: 532–538.

Brown, R.W. 1935. Further notes on fossil larval chambers of mining bees. *J. Washington Acad. Sci.* **25**: 526–528.

Brown, R.W. 1941a. The comb of a wasp nest from the Upper Cretaceous of Utah. *Amer. J. Sci.* **239**: 54–56.

Brown, R.W. 1941b. Concerning the antiquity of social insects. *Psyche* **48**: 105–110.

Brown, R.W. 1957. Cockroach egg case from the Eocene of Wyoming. *J. Washington Acad. Sci.* **47**(10): 340–342.

Bruch, C. 1923. Coleopteros fertilizadores de "*Prosapanche Burmeisteri*" de Bary. *Physis* **7**: 82–88.

Brues, C.T. 1910. The parasitic Hymenoptera of the Tertiary of Florissant, Colorado. *Bull. Mus. Compar. Zool. Harvard Coll.* **54**(1): 1–125.

Brues, C.T. 1933. Progressive changes in the insect population of forest since the Early Tertiary. *Amer. Naturalist* **47**: 385–405.

Brues, C.T. 1936. Evidences of insect activity preserved in fossil wood. *J. Paleontol.* **10** (7): 637–643.

Brussard, L., and Runia, L.T. 1984. Recent and ancient traces of scarab beetle activity in sandy soils of the Netherlands. *Geoderma* **34**(3–4): 229–250.

Bryson, H.R. 1939. The identification of soil insects by their burrow characteristics. *Trans. Kansas Acad. Sci.* **42**: 245–254.

Buatois, L.A., Mángano, M.G., Genise, J.F., and Taylor, T.N. 1998. The ichnologic record of the continental invertebrate invasion: Evolutionary trends in environmental expansion, ecospace utilization, and behavioural complexity. *Palaios* **13**: 217–240.

Buckland, P.C. 1976. The use of insect remains in the interpretation of archaeological environments. In: Davidson, D.A., and Shackley, M.L. (eds.). *Geoarchaeology: Earth Science and the Past.* Westview Press, Boulder, Colorado: 360–396.

Buckland, P.C. 1979. Thorne Moors: A palaeoecological study of a Bronze Age site. *Univ. Birmingham Dept. Geogr. Occas. Publication* **8**: 1–173.

Buckland, P.C. 1981. The early dispersal of insect pests of stored products as indicated by archaeological records. *J. Stored Product Res.* **17**: 1–12.

Buckland, P.C. 1991. Granaries stores and insects. The archaeology of insect synanthropy. *La préparation alimentaire des céréales (PACT 26).* Bruxelles: 69–81.

Buckland, P.C., Greig, J.R.A., and Kenward, H.K. 1974. York: an early medieval site. *Antiquity* **48**: 25–33.

Buckland, P.C., Holdsworth, P., and Monk, M. 1976. The interpretation of a group of Saxon pits in Southampton. *J. Archaeol. Sci.* **3**: 61–69.

Buckland, P.C., and Sadler, J. 1989. A biogeography of the human flea, *Pulex irritans* L. (Siphonaptera: Pulicidae). *J. Biogeogr.* **16**: 115–120.

Buhr, H. 1964. *Bestimmungstabellen der Gallen (Zoo- und Phytocecidien) an Pflanzen Mittel- und Nordeuropas* **1**. Gustav Fischer Verlag, Jena: 762 p.

Buhr, H. 1965. *Bestimmungstabellen der Gallen (Zoo- und Phytocecidien) an Pflanzen Mittel- und Nordeuropas* **2**. Gustav Fischer Verlag, Jena: 763–1572.

Buijtenen, J.P., van, and Santamour, F.S., Jr. 1972. Resin crystallization related to weevil resistance in white pine (*Pinus strobus*). *Canad. Entomol.* **104**(2): 215–219.

Burchak-Abramovich, N.I., and Dzhafarov, R.D. 1955. The Binagady locality of Upper Quaternary fauna and flora at Apsheron Peninsula. *Trudy Estestvenno-Istoricheskogo muzeya imeni G. Zardabi* **10**: 89–146 (in Russian).

BURKE, H.R., AMOS, A.F., and PARKER, R.D. 1991. Beach-drifted insects on Padro Island National Seashore, Texas. *Southwest. Entomol.* **16**(3): 199–203.

BURNHAM, L. 1981. Fossil insect from Montceau-les-Mines (France): a preliminary report. *Bull. Soc. hist. natur. Autun* **100**: 5–12.

BURNHAM, L. 1983. Studies on Upper Carboniferous insects: 1. The Geraridae (Order Protorthoptera). *Psyche* **90**: 1–57.

BURNHAM, L. 1984. Les insectes du Carbonifére Supérieur de Montceau-les-Mines. *Ann. Paléont.* **70**: 167–180.

BURNHAM, L. 1986. *Studies on Upper Carboniferous insects: 1. The Spanioderidae (Order Protorthoptera).* Unpublished Ph.D. thesis, Cornell University, Ithaca.

BUXTON, P.A. 1932. Ancient workings of insects, perhaps bees, from Megiddo, Palestine. *Proc. R. Entomol. Soc. London* **7**(1): 2–4.

CALABY, J.H., and MURRAY, M.D. 1991. Phthiriaptera (Lice). In: NAUMANN, D. (ed.). *The Insects of Australia.* 2nd ed. Melbourne Univ. Press, Carlton: 421–428.

CALDAS, M.B., MARTINS-NETO, R.G., and LIMA-FILHO, F.P. 1989. *Afropollis* sp. (polén) no trato intestinal de vespa (Hymenoptera: Apocrita: Xyelidae) no Cretáceo da Bacia do Araripe. *Atas II Simpos. Nacion. Estud. Tectonico, Soc. Brasil. Geologia*: 195–196.

CALLEN, E.O. 1970. Diet as revealed by coprolites. In: BROTHWELL, D., and HIGGS, E. (eds.). *Science in Archaeology.* Praeger, New York: 235–243.

CAMPBELL, B.C., and SHEA, P.J. 1990. A simple staining technique for assessing feeding damage by *Leptoglossus occidentalis* Heidemann (Hemiptera: Coreidae) on cones. *Canad. Entomol.* **122**(9–10): 963–968.

CANO, R.J., and BORUCKI, M.K. 1995. Revival and identification of bacterial spores in 25 to 40 million year old amber. *Science* **268**: 1060–1064.

CANO, R.J., POINAR, H., and POINAR, G.O., Jr. 1992a. Isolation and partial characterisation of DNA from the bee *Proplebeia dominicana* (Apidae: Hymenoptera) in 25–40 million year old amber. *Med. Sci. Res.* **20**: 249–251.

CANO, R.J., POINAR, H.N., ROUBIK, D.W., and POINAR, G.O., Jr. 1992b. Enzymatic amplification and nucleotide sequencing of portions of the 18S rRNA gene of the bee *Proplebeia dominicana* (Apidae: Hymenoptera) isolated from 25–40 million year old Dominican amber. *Med. Sci. Res.* **20**: 619–622.

CANO, R.J., POINAR, H.N., PIENIAZEK, N.J., ACRA, A., and POINAR, G.O., Jr. 1993. Amplification and sequencing of DNA from a 120–135 million-year-old weevil. *Nature* **363**: 536–538.

CARIDAD, J. 1999. Ámbar dominicano. *Estud. Mus. Cienc. Natur. Alava* **14** (Núm. Espec. 2): 141–147.

CARLE, F.L. and WIGHTON, D.C. 1990. Odonata. In: Grimaldi D. (ed.). *Insects from the Santana Formation, Lower Cretaceous, of Brasil. Bull. Amer. Mus. Natur. Hist.* **195**, New York: 51–68.

CARLSON, D.C. 1980. Taxonomic revision of *Lichnanthe* Burmeister (Coleoptera: Scarabaeidae). *Coleopterists' Bull.* **34**(2): 177–208.

CARPENTER, F.M. 1930a. The fossil ants of North America. *Bull. Mus. Compar. Zool. Harvard Coll.* **70** (1): 1–66.

CARPENTER, F.M. 1930b. The Lower Permian insects of Kansas. Part 1. Introduction and the order Mecoptera. *Bull. Mus. Compar. Zool. Harvard Coll.* **52**: 69–101.

CARPENTER, F.M. 1932. The Lower Permian insects of Kansas. Part 5. Psocoptera, and the additions to Homoptera. *Amer. J. Sci.* **24**: 1–22.

CARPENTER, F.M. 1933. The Lower Permian insects of Kansas. Part 6. Delopteridae, Protelytroptera, Plectoptera, etc. *Proc. Amer. Acad. Arts Sci.* **68**: 411–503.

CARPENTER, F.M. 1935. The Lower Permian insects of Kansas. Part 7. The order Protoperlaria. *Proc. Amer. Acad. Arts Sci.* **70**: 103–146.

CARPENTER, F.M. 1936. Revision of the Nearctic Raphidiodea (recent and fossil). *Proc. Amer. Acad. Arts Sci.* **71**(2): 89–157.

CARPENTER, F.M. 1939. The Lower Permian insects of Kansas. Part 8. Additional Megasecoptera, Protodonata, Homoptera, Psocoptera, Protelytroptera, Plectoptera, and Protoperlaria. *Proc. Amer. Acad. Arts Sci.* **73**: 29–70.

CARPENTER, F.M. 1943a. Carboniferous insects from the vicinity of Mazon Creek, Illinois. *Illinois State Mus. Sci. Papers* **3**(1): 9–20.

CARPENTER, F.M. 1943b. The Lower Permian insects of Kansas. Part 9. The Orders Neuroptera, Raphidioptera, Caloneurodea and Protorthoptera (Probnisidae), with additional Protodonata and Megasecoptera. *Proc. Amer. Acad. Art Sci.* **75**: 55–84.

CARPENTER, F.M. 1951. Studies on Carboniferous insects from Commentry, France; Part 2. The Megasecoptera. *J. Paleontol.* **25**(3): 336–355.

CARPENTER, F.M. 1961. Studies on Carboniferous insects of Commentry, France: Part 3. The Caloneurodea. *Psyche* **68**: 145–153.

CARPENTER, F.M. 1963a. Studies on Carboniferous insects of Commentry, France. Part 4. The genus *Triplosoba*. *Psyche* **70**: 120–128.

CARPENTER, F.M. 1963b. A Megasecopteron from Upper Carboniferous strata in Spain. *Psyche*, **70**: 44–49.

CARPENTER, F.M. 1963c. Studies on Carboniferous insects of Commentry, France. Part 5. The genus *Diaphanoptera* and the order Diaphanopterodea. *Psyche* **70**(4): 240–256.

CARPENTER, F.M. 1964. Studies on Carboniferous insects of Commentry, France. Part 6. The genus *Dictyoptilus* (Palaeodictyoptera). *Psyche* **71**: 104–116.

CARPENTER, F.M. 1966. The Lower Permian insects of Kansas. Part 11. The orders Protorthoptera and Orthoptera. *Psyche* **73**: 46–88.

CARPENTER, F.M. 1967a. Studies on North American Carboniferous insects. 5. Palaeodictyoptera and Megasecoptera from Illinois and Tennessee, with a discussion of the order Sypharopteroidea. *Psyche* **74**: 58–84.

CARPENTER, F.M. 1967b. The structure and relationships of *Stephanomioptera* Guthörl (Miomoptera-Palaeomanteidae). *Psyche* **74**: 224–227.

CARPENTER, F.M. 1967c. Cretaceous insects from Labrador 2. A new family of snake-flies (Neuroptera: Alloraphidiidae). *Psyche* **74**(4): 270–275.

CARPENTER, F.M. 1969. Fossil insects from Antarctica. *Psyche* **76**(4): 418–425.

CARPENTER, F.M. 1970. Fossil insects from New Mexico. *Psyche* **77**: 400–412.

CARPENTER F.M. 1976. The Lower Permian insects of Kansas. Part 12. Protorthoptera (continued), Neuroptera, additional Palaeodictyoptera, and families of uncertain position. *Psyche* **73**: 46–88.

CARPENTER, F.M. 1979. Lower Permian insects from Oklahoma. Part 2. Orders Ephemeroptera and Palaeodictyoptera. *Psyche* **86**(2–3): 261–290

CARPENTER, F.M. 1980. Studies on North American Carboniferous insects. 6. Upper Carboniferous insects from Pennsylvania. *Psyche* **87**: 107–119.

CARPENTER, F.M. 1983. Studies on North American Carboniferous insects. 7. The structure and relationships of *Eubleptus danielsi* (Palaeodictyoptera). *Psyche* **90**: 81–95.

CARPENTER, F.M. 1988. Review of the extinct family Syntonopteridae (order uncertain). *Psyche* **94**: 373–388.

CARPENTER, F.M. 1992a. *Treatise on Invertebrate Palaeontology. Pt. R. Arthropoda 4. Vol. 3. Superclass Hexapoda.* Geol. Society of America, Boulder, Colorado, and Univ. of Kansas, Lawrence, Kansas: 655 p.

CARPENTER, F.M. 1992b. Studies on the North American Carboniferous insects. 8. New Palaeodictyoptera from Kansas, USA. *Psyche* **99**(2–3): 141–146.

CARPENTER, F.M. 1997. Insecta. In: SHABICA, C.W., and HAY, A.A. (eds.) *Richardson's Guide to the Fossil Fauna of Mazon Creek.* Chicago, Northeastern Illinois Univ.: 184–193.

CARPENTER, F.M., FOLSOM, J.W., ESSIG, E.O., KINSEY, A.C., BRUES, C.T., BOESEL, M.W., and EWING, H.E. 1937. Insects and arachnids from Canadian amber. *Univ. Toronto Stud., Geol. Ser.* **40**: 7–62.

CARPENTER, F.M., JAMES, M.T., ALEXANDER, C.P., and HULL, F.M. 1938. Fossil insects from the Creed Formation, Colorado. *Psyche* **45**: 105–118.

CARPENTER, F.M., and KUKALOVÁ, J. 1964. The structure of the Protelytroptera, with description of a new genus from Permian strata of Moravia. *Psyche* **71**: 183–197.

CARPENTER, F.M., and RICHARDSON, E.S., Jr. 1968. Megasecopterous nymphs in Pennsylvanian concretions from Illinois. *Psyche* **75**(4): 295–309.

CARPENTER, F.M., and RICHARDSON, E.S. 1971. Additional insects in Pennsylvanian concretions from Illinois. *Psyche* **78**: 267–295.

CARPENTER, J.M., and RASNITSYN, A.P. 1990. Mesozoic Vespidae. *Psyche* **97**: 1–20.

CARPENTER, L.F. 1979. Competition between hummingbirds and insects for nectar. *Amer. Zool.* **19**(4): 1105–1114.

CARTER, D.J., and KRISTENSEN, N.P. 1999. Classification and key to higher taxa. In: KRISTENSEN, N.P. (ed.). *Lepidoptera, Moths and Butterflies.* Vol. 1. *Handbook of Zoology 4 Arthropoda: Insecta* Part 35. Walter de Gruyter, Berlin – New York: 27–40.

CASTRO, C., de MENOR, E. de A., and ALVES COMPANHA, V. 1970. Discoberta de resinas fósseis na Chapada do Araraipe, Municipio de Pórteira – Ceará. *Univ. Federal Pernambuco Inst. Geociênc., Sér. C: Notas prévicias* **1**(1): 1–11.

CASTRO, M.P. 1997. Huellas de actividad biológica sobre plantas del Estefaniense superior de La Magdalena (León, España). *Rev. Español. Paleontol.* **12**(1): 52–66.

CATT, J.A. 1987. Palaeosols. *Progr. Phys. Geogr.* **11**(4): 487–510.

CAUSSANEL, C., DIA, A.T., NEL, A., OULD BOURAYA, I., and THIBAUD, J.-M. 1997. Prospection préliminaire de la biodiversité des insectes des sables littoraux de Mauritanie. *Bull. Soc. entomol. France* **102**(1): 67–72.

CAVALLO, O., and GALLETTI, P.A. 1987. Studi di Carlo Sturani su Odonati e altri insetti fossili del Messiniano albese (Piemonte) con descrizione di *Oryctodiplax gypsorum* n.gen. n. sp. (Odonata, Libellulidae). *Boll. Soc. paleontol. ital.* **26**: 151–176.

CENTER, T.D., and JOHNSON, C.D. 1974. Co-evolution of some seed beetles (Coleoptera: Bruchidae) and their hosts. *Ecology* **55**: 1096–1103.

CEREZKE, H.F. 1972. Effects of weevil feeding on resin duct density and radial increment in lodgepole pine. *Canad. J. Forest Res.* **2**(1): 11–15.

CERLING, T.E., HARRIS, J.M., MACFADDEN, B.J., LEAKEY, M.G., QUADE, J., ELSENMANN, V., and EHLERINGER, J.R. 1997. Global vegetation change through the Miocene–Pliocene boundary. *Nature* **389**: 153–158.

CHALONER, W.G., SCOTT, A.C., and STEPHENSON, J. 1991. Fossil evidence for plant-arthropod interactions in the Palaeozoic and Mesozoic. *Philos. Trans. R. Soc. London* B **333**: 177–186.

CHAMBERLAIN, C.K. 1975. Recent lebensspuren in nonmarine aquatic environments. In: FREY, R.W. (ed.). *The Study of Trace Fossils.* Springer, New York–Berlin: 431–438.

CHAMBERS, V.T. 1882. Burrowing larvae. *Nature* **25**: 529.

CHAMPION, G.C. 1917. Notes on the Coleoptera recorded from "resin animé" by the rev. F.W. Hope. *Entomol. Monthly Mag.* **53**: 7–8, 244–246.

CHANEY, R.W. 1938. The Deschutes Flora of Eastern Oregon. *Carnegie Inst. Washington Publ.* **476**: 1–105.

CHARARAS, C. 1959. L'attractivité exercée par les coniféres à l'égard des Scolytides et rôle des substances terpeniques extraites des oléoresines. *Rev. pathol. végét. et entomol. agric.* **38**(2): 113–129.

CHARARAS, C. 1973. Faculté d'adaptation d'*Orthotomicus erosus* Woll. à des coniféres autres que ses essences-hôtes habituelles. *C.R. Acad. Sci. Paris* D **276** : 555–558.

CHARARAS, C. 1976. Étude de l'attraction primaire et secondaire chez les *Phloeosinus* (Coléoptères Scolytidae). *C.R. Acad. Sci. Paris* D **282**: 1793–1796.

CHARARAS, C. 1977. Attraction chimique exercée sur certains Scolytidae par les pinacées et les cupressacées. *Colloq. internat. CNRS* **265**: 165–185.

CHARARAS, C., DUCAUSE, C., and REVOLON, C. 1978. Étude comparative du pouvoir attractif de certains coniféres sur divers Scolytidae (Insectes Coléoptères). *C.R. Acad. Sci. Paris*, D **286**: 343–346.

CHENG, L. 1982. Beach and coastal insects. *Encyclopedia of Beach and Coastal Environments.* Hutchinson Ross, Philadelphia: 489–492.

CHENG, L., and BIRCH, M.C. 1977. Terrestrial insects at sea. *J. Marine Biol. Assoc. U.K.* **57**: 995–997.

CHENG, L., and BIRCH, M.C. 1978. Insect flotsam: an unstudied marine resource. *Ecol. Entomol.* **3**(2): 87–97.

CHERNYKH, E.N. 1997. *Kargaly. Zabyty mir.* (Kargaly. The Forgotten World). Nox, Moscow: 175 p. (in Russian).

CHIN, K., and GILL, B.D. 1996. Dinosaurs, dung beetles, and conifers: participants in a Cretaceous food web. *Palaios* **11**: 280–285.

CHISHOLM, I. E., and LEWIS, T. 1984. A new look at thrips (Thysanoptera) mouthparts, their action and effect of feeding on plant tissue. *Bull. Entomol. Res.* **74**: 663–675.

CHOUDHURY, D. 1985. Aphid honeydew: a re-appraisal of Owen and Wiegert's hypothesis. *Oikos* **45**(2): 287–290.

CHUMAKOV, N.M. 1994. Evidence of Late Permian glaciation in the Kolyma River Basin: a repercussion of the Gondwanaland glaciation in North Asia? *Stratigrafiya. Geol. korrelyatsiya* **2**(5): 130–150 (in Russian, English translation: *Stratigraphy and Geol. Correlation* **2**(5): 426–444.

CHUMAKOV, N.M. 1995. The problem of the warm biosphere. *Stratigrafiya. Geol. korrelyatsiya* **3**(3): 3–14 (in Russian, English translation: *Stratigraphy and Geol. Correlation* **3**(3): 205–215).

CHURCHER, C.S. 1966. The insect fauna from the Talara tar-seeps, Peru. *Canad. J. Zool.* **44**: 985–993.

CHVÁLA, M. 1983. The Empidoidea (Diptera) of Fennoscandia and Denmark. II. General part. The families Hybotidae, Atelestidae and Microphoridae. *Fauna Entomol. Scand.* **12**: 279 p.

CICHAN, M.A., and TAYLOR, T.N. 1982. Wood-borings in *Premnoxylon*: plant-animal interactions in the Carboniferous. *Palaeogeogr., Palaeoclimatol., Palaeoecol.* **39**: 123–127.

CLAGG, H.B. 1966. Trapping of air-borne insects in the Atlantic-Antarctic area. *Pacif. Insects* **8**(2): 455–466.

CLARK, G.R., and RATCLIFFE, B.C. 1989. Observations on the tunnel morphology of *Heterocerus brunneus* Melasheimer (Coleoptera: Hetroceridae) and its paleoecological significance. *J. Paleontol.* **63**(2): 228–232.

CLARK, J., BEERROWER, J.R., and KIETZKE, K. 1967. Oligocene sedimentation, stratigraphy and palaeoecology and palaeoclimatology in the Big Badlands of South Dakota. *Field. Geol.*, **5**: 1–58.

CLAYTON, R.B. 1970. The chemistry of non-hormonal interactions: terpenoid compounds in ecology. In: SANDHEIMER, E., and SIMEONE, J.B. (eds.). *Chemical Ecology.* Academic Press, New York: 235–280.

CLEVE-EULER, A., and HESSLAND, I. 1948. Vorläufige Mitteilung über eine neuentdeckte Tertiärablagerung in Süd-Schweden. *Bull. Geol. Inst. Uppsala* **32**: 155–182.

CLIFFORD, E., CORAM, R., JARZEMBOWSKI, E.A., and ROSS, A. 1994. A supplement to the insect fauna from the Purbeck Group of Dorset. *Proc. Dorset Nat. Hist. Archaeol. Soc.* **115**: 143–146.

CLIFFORD, H.T. 1962. Insect pollinators of *Plantago lanceolata* L. *Nature* **193**: 196.

CLOUDSLEY-THOMPSON, J.L. 1988. *Evolution and Adaptation of Terrestrial Arthropods.* Springer-Verlag, Berlin: 141 p.

COBBEN, R.H. 1978. Evolutionary trends in Heteroptera. Part II. Mouthpart-structures and feeding strategies. *Meded. Landbouwhogesch. Wageningen* **78**(5): 1–407.

COBBEN, R.H. 1981. Comments on some cladograms of major groups of Heteroptera. *Rostria* **33** Suppl.: 29–39.

COCKERELL, T.D.A. 1908a. Florissant, a Miocene Pompeii. *Pop. Sci. Monthly* **73**: 112–126.

COCKERELL, T.D.A. 1908b. A fossil leaf-cutting bee. *Canad. Entomol.* **40**(1): 31–32.

COCKERELL, T.D.A. 1910a. A Tertiary leaf-cutting bee. *Nature* **82**: 429.

COCKERELL, T.D.A. 1910b. Fossil insects and a crustacean from Florissant, Colorado. *Bull. Amer. Mus. Natur. Hist.* **28**(25): 275–288.

COCKERELL, T.D.A. 1917a. New Tertiary insects. *Proc. U.S. Nation. Mus.* **52**: 251–384.

COCKERELL, T.D.A. 1917b. Arthropods in Burmese amber. *Amer. J. Sci.* **44**: 360–368.

COCKERELL, T.D.A. 1920a. Fossil arthropods in British Museum. 4. *Ann. Mag. Nat. Hist.* Ser. 9. **6**: 211–214.

COCKERELL, T.D.A. 1923. Insects in amber from South America. *Amer. J. Sci.* Ser. 5, **5**: 331–333.

COCKERELL, T.D.A. 1924. Fossil insects in the United States National Museum. *Proc. U. S. Nation. Mus.* **64**: 1–15.

COCKERELL, T.D.A. 1925a. Tertiary insects from Argentina. *Proc. U.S. Nation. Mus.* **68** (1): 1–5.

COCKERELL, T.D.A. 1925b. Tertiary fossil insects from Argentina. *Nature* **116**: 711–712.

COE, M.J., DILCHER, D.L., FARLOW, J.O., JARZEN, D.M., and RUSSELL, D.A. 1987. Dinosaurs and land plants. In: FRIIS, E.M., CHALONER, W.G., and CRANE, P.R. (eds.). *The Origins of Angiosperms and Their Biological Consequences.* Cambridge Univ. Press, New York: 225–258.

COELHO, R.R., MENDES, M., and MARTINS-NETO, R.G. 2000. The paleoentomofauna from the Fonseca Formation (Fonseca Basin) (Oligocene, Minas Gerais State, Brazil). *I Simpósio Brasileiro de Paleoartropodología, I Simpósio Sudamericano de Paleoartropodologia, I Internat. Meeting on Paleoarthropodology, Ribeirão Preto – SP, Brazil 3–8.9.2000. Abstracts*: 45.

COHEN, A.S. 1984. Effect of zoobenthic standing crop on laminae preservation in tropical lake sediment, Lake Turkana, East Africa. *J. Paleontol.* **58**(2): 499–510.

COLLINS, N. 1980. Population ecology of *Ephydra cinerea* Jones (Diptera: Ephydridae), the only benthic metazoan of the Great Salt Lake, U.S.A. *Hydrobiologia* **68**: 99–112.

COLLINSON, M.E. 1982. A reassessment of fossil Potamogetoneae fruits with description of new material from Saudi Arabia. *Tertiary Res.* **4**: 83–104.

COLLINSON, M.E. 1992. Vegetational and floristic changes around the Eocene/Oligocene boundary in Western and Central Europe. In: PROTHERO, D.R., and BERGGREN, W.A. (eds.). *Eocene-Oligocene Climatic and Biotic Evolution.* Princeton Univ. Press, Princeton: 437–450.

COLLINSON, M.E., BOULTER, M.C., and HOLMES, P.L. 1993. Magnoliophyta ("Angiospermae"). In: BENTON, M.J. (ed.). *The Fossil Record* **2**. Chapman & Hall, London: 809–841.

COLOM, F.C. 1988. *Estratigrafia y sedimentologia de la litofacies Bunsandstein de Mallorca.* Unpublished thesis, Univ. Iles Baleares & Univ. Barselona: 130 p..

CONSTANTINE, E., CHINSAMY, A., VICKERS-RICH, P., and RICH, T.H. 1999. Periglacial environments and polar dinosaurs. *Paleontol. Zhurnal* (2): 59–65 (in Russian).

CONWENTZ, H. 1890. *Monographie der baltischen Bernsteinbäume.* Danzig: 151 p.

COOPE, G.R. 1963. The occurrence of the beetle *Hydromedion sparsutum* (Mull.) in a peat profile from Jason Island, South Georgia. *Bull. Brit. Antarctic Surv.* **1**: 25–26.

COOPE, G.R. 1969. The response of Coleoptera to gross thermal changes during the Mid-Weichselian interstadial. *Mitt. Internat. Verein Limnol.* **17**: 173–183.

COOPE, G.P. 1973. Tibetan species of dung beetle from Late Pleistocene deposits in England. *Nature* **245**, no 5424: 335–336.

COOPE, G.P. 1979. Late Cenozoic fossil Coleoptera: evolution, biogeography and ecology. *Annu. Rev. Ecol. Syst.* **10**: 247–267.

COOPE, G.R. 1986. Coleoptera analysis. In: BERGLUND, B.E. (ed.). *Handbook of Holocene Palaeoecology and Palaeohydrology.* John Wiley & Sons, New York: 703–713.

COOPE, G. R. 1994. The response of insect faunas to glacial-interglacial climatic fluctuations. *Philos. Trans. R. Soc. London* B **344**: 19–26.

COOPE, G.R., and LISTER, A.M. 1987. Late Glacial mammoth skeletons from Condover, Shropshire, England. *Nature* **330**: 472–474.

COOPE, G.R., and OSBORNE, P.J. 1968. Report on the coleopterous fauna of the Roman well at Barnsley Park, Gloucestershire. *Trans. Bristol and Gloucestershire Archaeol. Soc.* **86**: 84–87.

COOPER, K.W. 1941. *Davispia bearcreekensis* Cooper, a new cicada from the Paleocene, with a brief review of the fossil Cicadidae. *Amer. J. Sci.* **239**: 286–304.

COOPER-DRIVER, G. 1978. Insect-fern associations. *Entomol. Exper. Appl.* **24**: 110–116.

COPELAND, M.J. 1957. The Arthropod fauna of the Upper Carboniferous rocks of the Maritime Province. *Geol. Survey Canada Memoir* **286**: 1–110.

CORAM, R., JARZEMBOWSKI, E.A., and MOSTOVSKI, M.B. 2000. Two rare Eremoneuran flies (Diptera: Empididae and Opetiidae) from the Purbeck Limestone Group. *Paleontol. J.* **34**(suppl. 3): S370–S373.

CORAM, R., JARZEMBOWSKI, E.A., and ROSS, A.J. 1995. New record of Purbeck fossil insects. *Proc. Dorset Nat. Hist. & Archaeol. Soc.* **116**: 145–150.

CORNET, B. 1989. The reproductive morphology and biology of *Sanmiguelia lewisii*, and its bearing on angiosperm evolution in the Late Triassic. *Evol. Trends Plants* **3**: 25–51.

CORNET, B. 1996. A new gnetophyte from the Late Carnian (Late Triassic) of Texas and its bearing on the origin of the angiosperm carpel and stamen. In: TAYLOR, D.W., and HICKEY, L.J. (eds.). *Flowering Plant Origin, Evolution and Phylogeny.* Chapman & Hall, New York: 32–67.

COUTTS, M.P., and DELEZAL, J.E. 1966. Polyphenols and resin in the resistance mechanism of *Pinus radiata* attacked by the wood wasp, *Sirex noctilio*, and its associated fungus. *Leafl. Forest. Timber Bureau Austral.* **101**: 1–19.

COYNE, J.F., and LOTT, L.H. 1976. Toxicity of substances in pine oleoresin to southern pine beetles. *J. Ga. Entomol. Soc.* **11**(4): 301–305.

CRAMPTON, G.C. 1938. The interrelationships and lines of descent of living insects. *Psyche* **45**: 165–181.

CRANE, E., and GRAHAM, A.J. 1985a. Bee hives of the Ancient World. 1. *Bee World* **66**(1): 23–41.

CRANE, E., and GRAHAM, A.J. 1985b. Bee hives of the Ancient World. 2. *Bee World* **66**(2): 148–170.

CRANE, P.R. 1985. Phylogenetic analysis of seed plants and the origin of angiosperms. *Ann. Missouri Botan. Garden* **72**(4): 716–793.

CRANE, P.R. 1988. Major clades and relationships in the "higher" gymnosperms. In: BECK, C.B. (ed.). *Origin and Evolution of Gymnosperms.* Columbia Univ. Press, New York: 218–272.

CRANE, P.R. 1996. The fossil history of the Gnetales. *Indian J. Plant Sci.* **157**(suppl.): S50–S57.

CRANE, P.R., DRINNAN A., FRIIS, E.M., PEDERSEN, K.R., and LIDGARD, S. 1989a. Paleobotanical evidence of the early radiation of non-magnoliid (higher) dicotyledons. *Amer. J. Botany* **76**(6)(suppl.): 160–161.

CRANE, P.R., FRIIS, E.M., and PEDERSEN, K.R. 1986. Cretaceous angiosperm flowers: fossil evidence on early radiation of dicotyledons. *Science* **232**: 852–854.

CRANE, P.R., FRIIS, E.M., and PEDERSEN, K.R. 1989b. Reproductive structure and function in Cretaceous Chloranthaceae. *Plant Syst. Evol.* **165**(3–4): 211–266.

CRANE, P.R., and HERENDEEN, P.S. 1996. Cretaceous floras containing angiosperm flowers and fruits from eastern North America. *Rev. Palaeobot. Palynol.* **90**: 319–337.

CRANE, P.R., and LIDGARD, S. 1989. Angiosperm diversification and paleolatitudinal gradients in Cretaceous floristic diversity. *Science* **246**: 675–678.

CREBER, G.T., and CHALONER, W.G. 1984. Influence of environmental factors on the wood structure of living and fossil trees. *Botan. Rev.* **50**: 357–448.

CREPET, W.L. 1972. Investigations on North American cycadeoids: pollination mechanism in *Cycadaeoidea*. *Amer. J. Botany* **59**: 1048–1056.

CREPET, W.L. 1974. Investigations of North American cycadeoids: the reproductive biology of *Cycadeoidea*. *Palaeontographica* B **148**: 144–169.

CREPET, W.L. 1979a. Insect pollination: a palaeobotanical perspective. *Bioscience* **29**: 102–108.

CREPET, W. 1979b. Some aspects of the pollination biology of Middle Eocene angiosperms. *Rev. Palaeobot. Palynol.* **27**(3–4): 213–238.

CREPET, W.L., and FRIIS, E.M. 1987. The evolution of insect pollination in angiosperms. In: FRIIS E.M., CHALONER, W.G., and CRANE, P.R. (eds.). *The Origins of Angiosperms and their Biological Consequences*. Cambridge Univ. Press, Cambridge: 181–201.

CREPET, W.L., FRIIS, E.M., and NIXON, K.C. 1991. Fossil evidence for the evolution of biotic pollination. *Philos. Trans. R. Soc. London* **B 333**: 187–195.

CREPET, W.L., and NIXON, K.C. 1994. Flowers of Turonian Magnoliidae and their implications. *Plant Syst. Evol.*, Suppl. **8**: 73–91.

CREPET, W.L., and NIXON, K.C. 1995. The fossil history of stamens. In: D'ARCY, W.G., and KEATING, R.C. (eds.). *The Anther: Form, Function, and Phylogeny*. Cambridge Univ. Press, Cambridge: 25–57.

CREPET, W.L., and NIXON, K.C. 1997. The appearance of floral innovations in the Cretaceous and implications regarding angiosperm diversity. *Amer. J. Botany* **84**: 185.

CREPET, W.L., and NIXON, K.C. 1998a. Fossil Clusiaceae from the Late Cretaceous (Turonian) of New Jersey and implications regarding the history of bee pollination. *Amer. J. Botany* **85**: 1122–1133.

CREPET, W.L., and NIXON, K.C. 1998b. Two new fossil flowers of magnoliid affinity from the Late Cretaceous of New Jersey. *Amer. J. Botany* **85**: 1273–1288.

CREPET, W.L., NIXON, K.C., FRIIS, E.M., and FREUDENSTEIN, J.V. 1992. Oldest fossil flowers of hamamelidaceous affinity from the Late Cretaceous of New Jersey. *Proc. Nation. Acad. Sci. USA* **89**: 8986–8989.

CREPET, W., and TAYLOR, D.W. 1985. The diversification of the Leguminosae: first fossil evidence of the Mimosoideae and Papilionoideae. *Science* **228**: 1087–1089.

CRESPI, B.J. 1992. Eusociality in Australian gall thrips. *Nature* **359**: 724–726.

CROSSKEY, R.W. 1966. The putative fossil genus *Palexorista* Townsend and its identity with *Prosturmia* Townsend (Diptera: Tachinidae). *Proc. R. Entomol. Soc. London* B **35**: 133–137.

CROSSLEY, R. 1984. Fossil termite mounds associated with stone artefacts in Malawi, Central Africa. *Palaeoecol. Africa Surround. Islands* **16**: 397–401.

CROWSON, R.A. 1955. *The Natural Classification of the Families of Coleoptera*. Lloyd, London: 187 p. (2nd edition: 1967. E.W. Classey Ltd., Middlesex: 214 p.)

CROWSON, R.A. 1960. The phylogeny of Coleoptera. *Annu. Rev. Entomol.* **5**: 111–134.

CROWSON, R.A. 1975. The evolutionary history of Coleoptera, as documented by fossil and comparative evidence. *Atti X Congr. naz. Ital. Ent.*, Sassari: 47–90.

CROWSON, R.A. 1981. *The Biology of the Coleoptera*. Academic Press, London: 802 p.

CROWSON, R.A. 1985. Comments on Insecta of the Rhynie Chert. *Entomol. Gener.* **11**: 97–98.

CROWSON, R.A. 1988. Meligethinae as possible pollinators (Coleoptera: Nitidulidae). *Entomol. Gener.* **14**(1): 61–62.

CROWSON, R.A. 1989. Observations on two species of Curculionidae (Col.) associated with cycad cones in South Africa. *Entomol. Monthly Mag.* **125**: 129–131.

CROWSON, R.A. 1991. The relations of Coleoptera to Cycadales. In: ZUNINO, M., BELLÉS, X., and BLAS, M. (eds.). *Advances in Coleopterology*. European Assoc. Coleopterology, Barcelona: 13–28.

CURRAN, H.A., and WHITE, B. 1991. Trace fossils of shallow subtidal to dunal ichnofacies in Bahamian Quaternary carbonates. *Palaios* **6**(5): 498.

CZECHOWSKI, F., SIMONEIT, B.R.T., SACHANBINSKI, M., CHOJAN, J., and WOLOWIEC, S. 1996. Physicochemical structural characterization of ambers from deposits in Poland. *Applied Geochem.* **11**: 811–834.

CZECZOTT, H. 1961. Skład i wiek flory bursztynów bałtyckich. Częsc pierwsza. *Prace Muz. Ziemi* **4**: 121–145.

D'ALESSANDRO, A., EKDALE, A.A., and PICARD, M.D. 1987. Trace fossils in fluvial deposits of the Duchesne River Formation (Eocene), Uinta Basin, Utah. *Palaeogeogr., Palaeoclimatol., Palaeoecol.* **61**(3–4): 285–301.

DALLAI, R. 1980. Considerations on Apterygota phylogeny. *Boll. Zool.* **47**(suppl.): 35–48.

DANZIG, E.M. 1980. Koktsidy Sovetskogo Dal'nego Vostoka (*Homoptera, Coccinea*) [*Coccids of the Soviet Far East* (Homoptera, Coccinea)]. Nauka, Leningrad: 367 p.

DARLINGTON, P.J., Jr. 1957. *Zoogeography. The Geographical Distribution of Animals.* New York (cited after Russian translation of 1966, Progress, Moscow: 588 p.).

DAVIES, R. 1958. Observations on the morphology of the head and mouthparts in the Thysanoptera. *Proc. R. Entomol. Soc. London* A **33**: 97–106.

DAVIES, R. 1961. The postembryonic development of the female reproductive system in *Limothrips cerealis* Haliday (Thysanoptera: Thripidae). *Proc. zool. Soc. London* **136**: 411–437.

DAVIES, R. 1969. The skeletal musculature and its metamorphosis in *Limothrips cerealium* Haliday. *Trans. R. entomol. Soc. London* **121**: 167–233.

DAVIS, D.R. 1994. Neotropical Microlepidoptera XXV. New leaf-mining moths from Chile, with remarks on the history and composition of Phyllocnistinae. *Tropical Lepidoptera* **5**(1): 65–75.

DAVIS, J.A. 1986. Revision of the Australian Psephenidae (Coleoptera): systematics, phylogeny and historical biogeography. *Austral. J. Zool.* Suppl. Ser. **119**: 1–97.

DAVITASHVILI, L.Sh. 1964. On the classification of assemblages of organisms and organic remains. *Obshchie voprosy evolutsionnoy paleobiologii* (General Problems of Evolutionary Palaeobiology), I. Metsniereba, Tbilisi: 5–18 (in Russian).

DE, C. 1990. Upper Barakar lebensspuren from Hazaribagh, India. *J. Geol. Soc. India* **36**(4): 430–438.

DÉCAMPS, H., and LAVILLE, H. 1975. Invertébrés et matiers organiques entrainés lors des crues a l'entrée et a la sortie du systeme karstique du Baget. *Ann. limnol.* **11**(3): 287–296.

DECHAMPS, R., SENUT, B., and PICKFORD, M. 1992. Fruits fossiles pliocènes et pleistocènes du Rift Occidental ougandais. Signification paléoenvironmentale. *C.R. Acad. Sci. Paris* Sér. 2, **314**(3): 325–331.

DE DECKKER, P. 1982a. Holocene ostracods, other invertebrates and fish remains from cores of four maar lakes in southeastern Australia. *Proc. R. Soc. Victoria* **94**(3–4): 183–220.

DE DECKKER, P. 1982b. Australian aquatic habitats and biota: their suitability for palaeolimnological investigations. *Trans. R. Soc. S. Austral.* **106**(3): 145–153.

DEDEEV, V.A. (ed.). 1990. Uglenosnyie formatsii Pechorskogo Basseyna (*Coal bearing Formations of the Pechora Basin*). Nauka, Leningrad: 176 p. (in Russian).

DEEVEY, E.S. 1955. Palaeolimnology of the upper swamp deposit, Pyramid Valley. *Rec. Canterbury Mus.* **6**: 291–344.

DELETTRE, Y. 1990. Aerial drift of Chironomidae (Diptera) above terrestrial habitats: preliminary results. *2nd Internat. Congr. Dipterology, Bratislava, Aug. 27–Sept. 1, 1990, Abstract Volume*. Bratislava: 48.

DELEVORYAS, T. 1966. Investigations of North American Cycadeoids: microsporangiate structures and phylogenetic implications. *Palaeobotanist* **14**(1–3): 89–99.

DELEVORYAS, T. 1968a. Investigations of North American Cycadeoids: structure, anatomy and phylogenetic considerations of cones of *Cycadeoidea*. *Palaeontographica* B **121**: 122–133.

DELEVORYAS, T. 1968b. Some aspects of Cycadeoid evolution. *Botan. J. Linn. Soc.* **61**: 137–146.

DELL, B., and MCCOMB, A.B. 1978. Plant resins – their formation, secretion and possible functions. *Adv. Botan. Res.* **6**: 275–316.

DEMOULIN, G. 1954. Les Ephéméroptères jurassiques de Sinkiang. *Bull. Ann. Soc. Entomol. Belgique* **90**(11–12): 322–326.

DEMOULIN, G. 1955. Quelques remarques sur les composantes de la famille Ametropodidae (Ephemeroptera). *Bull. Ann. Soc. R. Entomol. Belgique* **91**: 342.

DEMOULIN, G. 1956. Nouvelles recherches sur *Triplosoba pulchella* (Brongniart) (Insectes Éphéméroptères). *Bull. Inst. R. Entomol. Belgique* **32**(14): 1–8.

DEMOULIN, G. 1957. A propos de deux Insectes éocènes. *Bull. Inst. R. Sci. Natur. Belgique* **33**(45): 1–4.

DEMOULIN, G., 1962. A propos de l'ordre des Aphelophlebia (Insecta Palaeoptera). *Bull. Inst. R. Sci. natur. Belgique* **38**(8): 1–4.

DEMOULIN, G. 1968. Deuxième contribution à la connaissance des Éphéméroptères de l'ambre oligocène de la Baltique. *Deutsch. entomol. Zeitschr.* **15**(1–3): 233–276.

DENDY, J.S. 1944. The fate of animals in stream drift when carried into lakes. *Ecol. Monogr.* **14**: 334–357.

DERBENEVA, N.N. 1967. New data on biology and structure of preimaginal phases and stages of the predatory thrips *Aeolothrips intermedius* Bagnall (Thysanoptera, Aeolothripidae). *Entomol. Obozrenie* **46**: 629–644 (in Russian).

DESALLE, R. 1994. Implications of ancient DNA for phylogenetic studies. *Experientia* **50**: 543–550.

DESALLE, R., BARCIA, M., and WRAY, C. 1993. PCR jumping in clones of 30 million-year-old DNA fragments from amber preserved termites (*Mastotermes electrodominicanus*). *Experientia* **49**: 906–909.

DESALLE, R., GATESY, J., WHEELER, W., and GRIMALDI, D. 1992. DNA sequencing from a fossil termite in Oligo-Miocene amber and their phylogenetic implications. *Science* **257**: 1933–1936.

DESPLAT, Y. 1954. Note preliminaire sur des insectes fossiles de l'oligocène parisien. *L'Entomologiste* **10**(5–6): 93.

DÉVAI, G., and MOLDOVÁN, J. 1983. An attempt to trace eutrophication in a shallow lake (Balaton, Hungary) using chironomids. *Hydrobiologia* **103**: 169–175.

DE VRIES, P.J., and POINAR, G.O., Jr. 1997. Ancient butterfly–ant symbiosis: direct evidence from Dominican amber. *Proc. R. Soc. London* B **264**: 1137–1140.

DIÉGUEZ, C., NIEVES-ALDREY, J.L., and BARRÓN, E. 1996. Fossil galls (zoocecids) from the Upper Miocene of La Cerdaña (Lérida, Spain). *Rev. Palaeobot. Palynol.* **94**(3): 329–343.

DIELS, L. 1916. Käferblumen bei den Ranales und ihre Bedeutung für die Phylogenese der Angiospermen. *Ber. Deutsch. Botan. Ges.* **34**: 758–774.

DIETRICH, H.-G. 1976. Zur Entstehung und Erhaltung von Bernstein-Lagerstätten. – 2: Bernstein-Lagerstätten im Libanon. *Neues Jahrb. Geol. Paläontol. Abhandl.* **152**(2): 222–279.

DÍEZ, J.B., CANUDO, J.I., FERRER, J., MUÑOZ-BARRAGÁN, P., RUÍZ-OMEÑACA, J.I., and SORIA, A.R. 1996. Transporte y resedimentación de troncos silicificados en el Albiense (Fm. Utrillas, Castellote, Cordillera Ibérica). *Com. II Reunión Tafonomía y Fosilización, Zaragoza*: 97–102.

DIJKSTRA, S.J., and PIÉRART, P. 1957. Lower Carboniferous megaspores from the Moscow Basin. *Meded. Geol. Stichting.* Nieuw. Ser. **11**: 5–19.

DILCHER, D. 1979. Early angiosperm reproduction: an introductory report. *Rev. Palaeobot. Palynol.* **27**(3–4): 291–328.

DILCHER, D. 1989. The occurrence of fruits with affinities to Ceratophyllaceae in Lower and Mid-Cretaceous sediments. *Amer. J. Botany* **76**(suppl. 6): 162.

DILCHER, D. 1995. Plant reproductive strategies: using the fossil record to unravel current issues in plant reproduction. *Monogr. Syst. Botany Missouri Botan. Garden* **53**: 187–193.

DILCHER, D.L., and CRANE, P.R. 1984. *Archaeanthus*: an early angiosperm from the Cenomanian of the Western Interior of North America. *Ann. Missouri Botan. Garden* **71**: 351–383.

DILCHER, D.L., and KOVACH, W.L. 1986. Early angiosperm reproduction: *Caloda delevoryana* gen. et sp. nov., a new fructification from the Dakota Formation (Cenomanian) of Kansas. *Amer. J. Botany* **73**: 1230–1237.

DiMICHELE, W.A., and HOOK, R.W. 1992. Paleozoic terrestrial ecosystems. In: BEHRENSMEYER, A.K., DAMUTH, J.D., DiMICHELE, W.A., POTTS, R., SUES, H.-D., and WING, S.L. (eds.). *Terrestrial Ecosystems through Time. Evolutionary Paleoecology of Terrestrial Plants and Animals*. Univ. Chicago Press, Chicago – London: 205–325.

DiMICHELE, W.A., and PHILLIPS, T.L. 1994. Paleobotanical and paleoecological constraints of models of peat formation in the Late Carboniferous of Euramerica. *Palaeogeogr., Palaeoclimatol., Palaeoecol.* **106**: 39–90.

DINDO, M.L., CAMPADELLI, G., and GAMBETTA, A. 1992. Note su *Xylocopa violacea* L. e *Xylocopa valga* Gerst. (Hym. Anthophoridae) nidificanti nei tronchi della foresta fossile di Dunarobba (Umbria). *Boll. Ist. Entomol. "G. Grandi" Univ. Bologna* **46**: 153–160.

DINESMAN, L.G. 1979. Studies on history of biocoenoses based on mammal burrows. In: SOKOLOV, V.E., and DINESMAN, L.G. (eds.). *Obshchie metody izucheniya istorii sovremennykh ekosistem* (General Methods of Studies on the History of Modern Ecosystems). Nauka, Moscow: 76–101 (in Russian).

DINESMAN, L.G. 1992. Reconstruction of the history of recent biocoenoses on the base of long-time shelters of mammals and birds. *Vekovaya dinamika biogeotsenozov (The Centuries-old Dynamics of Biogeocoenoses)*. *Doklady na 10 Ezhegodnom chtenii pamyati V.N. Sukacheva.* (Lectures on the 10th Annual Readings in Memory of V.N. Sukachev) Nauka, Moscow: 4–17 (in Russian).

DIOLI, P. 1992. Esame del popolamento degli eterotteri (Insecta, Hetroptera) negli strati bassi dell'atmosfera sul Delta del Po. *Boll. Mus. civ. Stor. natur. Venezia* **41**(1990): 183–205.

DLUSSKY, G.M. 1981a. *Murav'yi pustyn'* (Desert Ants). Nauka, Moscow: 230 p. (in Russian).

DLUSSKY, G.M. 1981b. Miocene ants (Hymenoptera, Formicidae). In: VISHNIAKOVA, V.N., DLUSSKY, G.M., and PRITYKINA, L.N. 1981. *Novye iskopaemye nasekomye s territorii SSSR* (New Fossil Insects from the Territory of USSR). *Trudy Paleont. Inst. Akad. Nauk. SSSR* **183**. Nauka, Moscow: 64–83 (in Russian).

DLUSSKY, G.M. 1983. A new family of Upper Cretaceous Hymenoptera: an "intermediate link" between the ants and the scolioids. *Paleontol. Zhurnal* (3): 65–78 (in Russian, English translation: *Paleontol. J*, **17**(3): 63–76).

DLUSSKY, G.M. 1988. Ants from (Paleocene?) Sakhalin amber. *Paleontol. Zhurnal* (1): 50–61 (in Russian, English translation: *Paleontol. J.* **22**(1): 50–60).

DLUSSKY, G.M. 1996. Ants (Hymenoptera: Formicidae) from Burmese amber. *Paleontol. Zhurnal* (3): 83–89 (in Russian, English translation:*Paleontol. J.* **30**(4): 449–454).

DLUSSKY, G.M. 1999a. The first find of the Formicoidea (Hymenoptera) in the Lower Cretaceous of the Northern Hemisphere. *Paleontol. Zhurnal* (3): 62–66 (in Russian, English translation: *Paleontol. J.* **33**: 274–277).

DLUSSKY, G.M. 1999b. New ants (Hymenoptera: Formicidae) from the Canadian amber. *Paleontol. Zhurnal* (4): 73–76 (in Russian, English translation: *Paleontol. J.* **33**(4): 409–412).

DLUSSKY, G.M., and FEDOSEEVA, E.B. 1988. Origin and early evolution of ants. In: PONOMARENKO, A.G. (ed.). *Melovoy biotsenoticheskiy krizis i evolyutsiya nasekomykh* (Cretaceous Biocoenotic Crisis and Evolution of the Insects). Nauka, Moscow: 70–144 (in Russian).

DMITRIEV, V.Yu., PONOMARENKO, A.G., and RASNITSYN, A.P. 1994a. Dynamics of the taxonomic diversity of nonmarine biota. In: Rozanov, A. Yu., and Semikhatov, M.A. (eds.). *Ekosistemnye perestroiki i evolyutsia biosfery* (Ecosystem Re-organization and Evolution of Biosphere). Nedra, Moscow: 167–174 (in Russian).

DMITRIEV, V.Yu., PONOMARENKO, A.G., and RASNITSYN, A.P. 1994b. Organic diversity in geological past. State of the problem. In: SOKOLOV, V. E. & RESHETNIKOV, Yu.S. (eds). Bioraznoobrazie. Stepen' taksonomicheckoy izuchennosti (Biodiversity. Degree of Taxonomic Knowledge). Nauka, Moscow: 12–19 (in Russian).

DMITRIEV, V.Yu., PONOMARENKO, A.G., and RASNITSYN, A.P. 1995. Dynamics of the taxonomic diversity of nonmarine aquatic biota. *Paleontol. Zhurnal* (4): 3–9 (in Russian, English translation: *Paleontol. J.* 1996 **30**(3): 255–259).

DMITRIEV, V.Yu., and ZHERICHIN, V.V. 1988. Changes in diversity of insect families as revealed by the method of accumulated appearances. In: PONOMARENKO, A.G. (ed.). *Melovoy biotsenoticheskiy krizis i evolyutsiya nasekomykh* (Cretaceous Biocoenotic Crisis and Evolution of the Insects). Nauka, Moscow: 208–215 (in Russian).

DOBERENZ, A.R., MATTER, P., III, and WYCKOFF, R.W.G. 1966. The microcomposition of some fossil insects of Miocene age. *Bull. S. Calif. Acad. Sci.* **65**(4): 229–236.

DOBRUSKINA, I.A. 1980. *Stratigraficheskoye polozhenie triasovykh floronosnykh otlozheniy Evrazii (Stratigraphic Position of Triassic Plant-bearing Beds of Eurasia)*. Trudy Geol. Inst. Akad. Nauk SSSR **346**. Nauka, Moscow: 164 p.

DOBRUSKINA, I.A. 1982. *Triasovye flory Evrazii* (Triassic Floras of Eurasia). *Trudy Inst. Akad. Nauk SSSR* **365**. Nauka, Moscow: 196 p.

DOBRUSKINA, I.A., PONOMARENKO, A.G., and RASNITSYN, A.P. 1997. Fossil insect findings in Israel. *Paleontol. Zhurnal* (5): 91–95 (in Russian, English translation: *Paleontol. J.* **31**(5): 528–533).

DOEKSEN, J. 1941. Bijdrag tot de vergelijkende morphologie der Thysanoptera. *Meded. Landbbouwhoogeschool Wageningen* **45**: 1–114.

DOLUDENKO, M.P., SAKULINA, G.V., and PONOMARENKO, A.G. 1990. *La géologie du gisement unique de la fauna et de flora du Jurassique supérieur d'Aulie (Karatau, Kazakhstan du Sud)*. Geol. Inst. Acad. Sci. URSR, Moscow: 38 p. (in Russian, with French and English summary).

DOLLING, W.R. 1981. A rationalized classification of the burrower bugs (Cydnidae). *Syst. Entomol.* **6**: 61–76.

DOMAISON, L. 1887. [Untitled note]. Ann. Soc. entomol. France, sér. 6, **7**, Bull.: CCIV.

DOMÍNGUEZ ALONSO, P., and COCA ABIA, M. 1998. Nidos de avispas minadoras en el Mioceno de Tegucigalpa (Honduras, América Central). *Coloq. Paleontol.* **49**: 93–114.

DONALDSON, J.S. 1997. Is there a floral parasite mutualism in cycad pollination? The pollination biology of *Encephalartos villosus* (Zamiaceae). *Amer. J. Botany* **84**(10): 1398–1406.

DONOVAN, S.K. 1994. Insects and other arthropods as trace-makers in nonmarine environments and paleoenvironments. In: DONOVAN, S.K. (ed.). *The Palaeobiology of Trace Fossils*. John Hopkins Univ. Press, Baltimore: 200–220.

DORF, E. 1967. Cretaceous insects from Labrador. 1. Geological occurrence. *Psyche* **74**: 267–268.

DOUGLAS, S.D., and STOCKEY, R.A. 1996. Insect fossils in middle Eocene deposits from British Columbia and Washington State: faunal diversity and geological range extensions. *Canad. J. Zool.* **74**: 1140–1157.

DOWNES, W.L., Jr. 1987. The impact of vertebrate predators on early arthropod evolution. *Proc. Entomol. Soc. Washington* **89**(3): 389–406.

DOYEN, J.T., and POINAR, G.O., Jr. 1994. Tenebrionidae from Dominican amber (Coleoptera). *Entomol. Scand.* **25**: 27–51.

DOYLE, J.A. 1998. Phylogeny of vascular plants. *Annu. Rev. Ecol. Syst.* **29**: 567–599.

DOYLE, J.A., and DONAGHUE, M.J. 1986. Seed plant phylogeny and the origin of angiosperms: an experimental cladistic approach. *Botan. Rev.* **52**(4): 321–431.

DOYLE, J.A., and DONOGHUE, M.J. 1993. Phylogenies and angiosperm diversification. *Paleobiology* **20**: 89–92.

DOYLE, J.A., and HICKEY, L.J. 1976. Pollen and leaves from the Mid-Cretaceous Potomac Group and their bearing on early angiosperm evolution. In: BECK, C.B. (ed.). *Origin and Early Evolution of Angiosperms*. Columbia Univ. Press, New York: 139–206.

DOYLE, J.A., HOTTON, C.L., and WARD, J.V. 1990a. Early Cretaceous tetrads, zonasulcate pollen, and Winteraceae. I. Taxonomy, morphology, and ultrastructure. *Amer. J. Botany* **77**(12): 1544–1557.

DOYLE, J.A., HOTTON, C.L., and WARD, J.V. 1990b. Early Cretaceous tetrads, zonasulcate pollen, and Winteraceae. II. Cladistic analysis and implications. *Amer. J. Botany* **77**(12): 1558–1568.

DRAKE, V.A., and FARROW, R.A. 1988. The influence of atmospheric structure and motion on insect migration. *Annu. Rev. Entomol.* **33**: 183–210.

DRAKE, V.A., and FARROW, R.A. 1989. The "aerial plankton" and atmospheric convergence. *Trends Ecol. Evol.* **4**: 381–385.

DRAKE, V.A., and GATEHOUSE, A.G. (eds.). *Insect Migration: Tracking Resources in Space and Time*. Cambridge Univ. Press, Cambridge: 478 p.

DRAKE, V.A., GATEHOUSE, A.G., and FARROW, R.A. 1995. Insect migration: a holistic conceptual model. In: DRAKE, V.A., and GATEHOUSE, A.G. (eds.). *Insect Migration: Tracking Resources in Space and Time*. Cambridge Univ. Press, Cambridge: 427–457.

DREGER-JAUFFRET, F. 1983. Diversity and unity by arthropod galls. An example: the bud galls. *Adaptation to Terrestrial Environments. Internat. Symposium, Khalkidiki, Sept. 26–Oct. 2, 1982*. New York–London: 77–87.

DUBININ, V.B. 1948. Finding of the Pleistocene Anoplura and nematods in corpses of the fossil susliks from Indigirka. *Doklady Akad. Nauk SSSR* **62**: 417–420 (in Russian).

DuBOIS, M.B., and LAPOLLA, J.S. 1999. A preliminary review of Colombian ants (Hymenoptera: Formicidae) preserved in copal. *Entomol. News* **110**(3): 162–172.

DUCREUX, J.-L., BILLAUD, Y., and TRUC, G. 1988. Traces d'insectes dans les paléosols rouges de l'Eocéne supérieur du nord-est du Massif central français: *Celliforma arvernensis* ichnosp. nov. *Bull. Soc. géol. France* **4**(1): 167–175.

DUCREUX, J.-L., and TACHET, H. 1985. Larves de Stratiomyidae (Insecta, Diptera) dans laminites stampiennes du bassin de Forcalquier (Alpes-de-Haute-Provence, France). *Geobios* **18**(4): 517–524.

DUDICH, F. 1961. Rovarlelet a Szentgáli fás barnakoszénbol (An insect find from the Szentgal lignite). *Földtani Közlöny* **91**(1): 20–31 (in Hungarian with German summary).

DUDLEY, R. 1998. Atmospheric oxygen, giant Paleozoic insects and the evolution of aerial locomotor performance. *J. Exp. Biol.* **201**(8): 1043–1050.

DUNCAN, I.J., BRIGGS, D.E.G., and ARCHER, M. 1998. Three-dimensionally mineralized insects and millipeds from the Tertiary of Riversleigh, Queensland, Australia. *Palaeontology* **41**(5): 835–851.

DURDEN, C.J. 1966. Oligocene lake deposits in Central Colorado and a new fossil insect locality. *J. Paleontol.* **40**: 215–219.

DURDEN, C.J. 1978. A dasyleptid from the Permian of Kansas, *Lepidodasypus sharovi* n.gen., n. sp. (Insecta: Thysanura: Monura). *Pearce-Sellards Series*, **30**: 1–9.

DURDEN, C.J., and ROSE, H. 1978. Butterflies from the Middle Eocene: the earliest occurrence of fossil Papilionoidea (Lepidoptera). *Pearce – Sellards Series* **29**: 1–25.

DURHAM, J.W. 1956. Insect bearing amber in Indonesia and Philippine Islands. *Pan-Pacif. Entomol.* **32**(2): 51–53.

DUTCHER, R., and TODD, J.W. 1983. Hemipteran kernel damage of pecan. *Entomol. Soc. America Misc. Publ.* **13**: 1–11.

DUVIARD, D., and ROTH, M. 1973. Utilisation des pièges à eau colorés en milieu tropical. Exemple d'une savanne préforestière de Côte d'Ivoire. *Cahiers ORSTOM, Biol.* **18**: 91–97.

DYADECHKO, N.P. 1964. *Tripsy ili bakhromchatokrylye nasekomye evropeiskoy chasti SSSR* (Thrips, or Fringed-Winged Insects, of the European Part of the USSR). Naukova Dumka, Kiev: 387 p. (in Russian) (English translation: 1977. Amerind Publ., New Delhi: 344 p.).

EBERLY, L.D., and STANLEY, T.B. 1978. Cenozoic stratigraphy and geologic history of southwestern Arizona. *Bull. Geol. Soc. America* **89**: 921–940.

EDWARDS, D., and FANNING, U. 1985. Evolution and environment in the Late Silurian – Early Devonian: the rise of the pteridophytes. *Philos. Trans. R. Soc. London* B **309**: 147–165.

EDWARDS, D., and SELDEN, P.A. 1993. The development of early terrestrial ecosystems. *Botan. J. Scotl.* **46**(2): 337–366.

EDWARDS, D., DUCKETT, J.G., and RICHARDSON, J.B. 1995. Hepatic characters in the earliest land plants. *Nature* **374**: 635–636.

EDWARDS, E.D., GENTILI, P., KRISTENSEN, N.P., HORAK, M., and NIELSEN, E.S. 1999. The Cossoid/Sesioid assemblage. In: KRISTENSEN, N.P. (ed.) 1999. *Lepidoptera, Moths and Butterflies.* Vol. 1. *Handbook of Zoology 4 Arthropoda: Insecta* Part 35. Walter de Gruyter, Berlin–New York: 181–197.

EDWARDS, F.W. 1923. Oligocene mosquitoes in the British Museum, with a summary of our present knowledge concerning fossil Culicidae, *Quart. J. Geol. Soc. London* **79**: 139–155.

EDWARDS, J.S. 1987. Arthropods of alpine aeolian ecosystems. *Annu. Rev. Entomol.* **32**: 163–179.

EDWARDS, J.S., and BANKO, P.C. 1976. Arthropod fallout and nutrient transport: a quantitative study of Alaskan snowpatches. *Arctic and Alpine Res.* **8**(3): 237–245.

EDWARDS, J.S., and SCHWARTZ, L.M. 1981. Mount St. Helene ash: a natural insecticide. *Canad. J. Zool.* **59**(4): 714–715.

EDWARDS, W.N.A. 1936. Pleistocene chironomid-tufa from Crèmieu (Isère). *Proc. Geol. Assoc.* **48**: 197–198.

EFREMOV, I.A. 1950. *Tafonomiya i geologicheskaya letopis'* (Taphonomy and Geological Record). *Trudy Paleontol. Inst. Akad. Nauk SSSR* **24**. Akad. Nauk SSSR, Moscow: 177 p. (in Russian).

EGGLETON, P., and BELSHAW, R. 1992. Insect parasitoids: an evolutionary overview. *Philos. Trans. R. Soc. London* **B 337**: 1–20.

EGLINGTON, G., and LOGAN, G.A. 1991. Molecular preservation. *Philos. Trans. R. Soc. London* **B 333**: 315–328.

EHRLICH, H.L. 1981. *Geomicrobiology.* Dekker, New York: 387 p.

EHRMANN, R. 1997. Systematik der Ordnung Mantoptera (Mantodea) (Insecta: Dictyoptera). *Arthropoda* **5**(2): 6–12.

EHRMANN, R. 1999. Gottesanbeterinnen in Kopal und Bernstein (Insecta: Mantodea). *Arthropoda* **7**(3): 2–8.

EKDALE, A.A., BROMLEY, R.G., and PEMBERTON, S.G. 1984. *Ichnology: The Use of Trace Fossils in Sedimentology and Stratigraphy. Soc. Econ. Palaeontol. Mineral. Short Course* **15**. Tulsa, Okla: 317 p.

EKDALE, A.A., and PICARD, M.D. 1985. Trace fossils in a Jurassic aeolianite, Entrada Sandstone, Utah. *Soc. Econ. Paleontol. Miner. Spec. Publ.* **35**: 3–12.

ELDERS, C.A. 1975. Experimental approaches in neoichnology. In: FREY, R.W. (ed.). *The Study of Trace Fossils.* Springer, New York–Berlin: 513–536.

ELIAS, S.A. 1987. Palaeoenvironmental significance of Late Quaternary insect fossils from packrat middens in south-central New Mexico. *Southwest. Naturalist* **32**(3): 383–390.

ELIAS, S.A. 1990. Observations on the taphonomy of Late Quaternary insect fossil remains in packrat middens of the Chihuahuan Desert. *Palaios* **5**(4): 356–363.

ELIAS, S.A. 1994. *Quaternary Insects and Their Environments.* Smithsonian Inst. Press, Washington–London: 284 p.

ELIAS, S.A., MEAD, J.I., and AGENBROAD, L.D. 1992b. Late Quaternary arthropods from the Colorado Plateau, Arizona and Utah. *Great Basin Natur.* **52**: 59–67.

ELIAS, S.S., SHORT, S.K., and PHILIPS, R.L. 1992a. Palaeoecology of Late Glacial peats from the Bering Land Bridge, Chukchi Sea shelf region, Northwestern Alaska. *Quatern. Res.* **38**: 371–378.

ELIAS, S.S., SHORT, S.K., NELSON, C.H., and BIRKS, H.H. 1996. Life and time of the Bering Land Bridge. *Nature* **382**: 60–63.

ELLIOT, D.K., and NATIONS, J.D. 1998. Bee burrows in the Late Cretaceous (Late Cenomanian) Dakota Formation, northeastern Arizona. *Ichnos* **5**: 243–253.

ELLIS, W.N., and ELLIS-ADAM, A.C. 1993. Fossil brood cells of solitary bees on Fuerteventura and Lanzarote, Canary Islands (Hymenoptera: Apoidea). *Entomol. Berichten* **53**(12): 161–173.

ELSIK, W.C. 1976. Fossil fungal spores. In: WEBER, D.J. and HESS, W.M. (eds.). *The Fungal Spore. Form and Function.* John Wiley & Sons, New York: 849–862.

ELTON, C.S. 1925. The dispersal of insects to Spitzbergen. *Trans. R. Entomol. Soc. London* **73**: 289–299.

ELVERS, I. 1980. Pollen eating *Trichops* flies (Diptera Muscidae) on *Arrhenantherum pubescens* and some other grasses. *Botan. Not.* **133**: 49–52.

EMELJANOV (YEMEL'YANOV), A.F. 1983. A dictyopharid from the Cretaceous of Taimyr (Insecta, Homoptera). *Paleontol. Zhurnal* (3): 79–85 (in Russian, English translation: *Paleontol. J.* **17**(3): 77–82).

EMELJANOV, A.F. 1987. Phylogeny of Cicadina (Homoptera) according the data from comparative morphology. In: TOBIAS, V.I. (ed.) *Morfologicheskie osnovy filogenii nasekomykh* (Morphological Grounds of the Insect Phylogeny). Trudy Vsesoyuznogo Entomol. Obshchestva. **69**. Nauka, Leningrad: 19–109 (in Russian).

EMELJANOV, A.F. 1994. The first find of the fossil Derbidae, and a redescription of Paleogene achilid *Hooleya* Cockerell (Insecta, Homoptera, Fulgoroidea). *Paleontol. Zhurnal* (3): 76–82 (in Russian, English translation: *Paleontol. J.* **28**(3): 92–101).

EMELJANOV, A.F., and SHCHERBAKOV, D.E. 2000. Kinnaridae and Derbidae (Homoptera, Fulgoroidea) from the Dominican amber. *Neues Jahrb. Geol. Paläontol. Monatshefte* (7): 438–448.

EMERSON, K.S., and PRICE, R.D. 1985. Evolution of Mallophaga om mammals. In: KIM K.Ch. (ed.). *Co-evolution of Parasitic Arthropods and Mammals.* New York, etc., John Wiley & Sons: 233–255.

EMMEL, T.C., MINNO, M.C., and DRUMMOND, B.A. 1992. *Florissant Butterflies. A Guide to the Fossil and Present-Day Species of Central Colorado.* Stanford Univ. Press, Stanford: 118 p.

EMSLIE, S.D., and MESSENGER, S.L. 1991. Pellet and bone accumulation at a colony of western gulls (*Larus occidentalis*). *J. Vertebr. Paleontol.* **11**(1): 133–136.

ENDRESS, P.K., and FRIIS, E.M. 1991. *Archamamelis,* hamamelidacean flowers from the Upper Cretaceous of Sweden. *Plant Syst. Evol.* **175**: 101–114.

ENDRESS, P.K., and FRIIS, E.M. (eds.). 1994. *Early Evolution of Flowers.* Plant Syst. Evol. Suppl. 8: 229 p.

ENGEL, M.S. 1998. A new species of the Baltic amber bee genus *Electrapis* (Hymenopera: Apidae). *J. Hymenoptera Res.* **7**(1): 94–101.

ENGEL, M.S. 1999. The first fossil *Euglossa* and phylogeny of the orchid bees (Hymenoptera: Apidae; Euglossini). *Amer. Mus. Novitates* **3272**: 1–14.

ENGEL, M.S. 2000. A new interpretation of the oldest fossil bee (Hymenoptera: Apidae). *Amer. Mus. Novitates* **3296**: 1–11.

ENGEL, M.S. 2001. A monograph of the Baltic amber bees and evolution of the Apoidea (Hymenoptera). *Bull. Amer. Mus. Nat. Hist.* **259**: 192 p.

ENGEL, M.S., and GRIMALDI, D.A. 2000. A winged *Zorotypus* in Miocene amber from Dominican Republic (Zoraptera: Zorotypidae), with discussion on relationships of and within order. *Acta Geol. Hispanica* **35**(1–2): 149–164.

ENGELHARDT, H. 1876. Tertiärpflanzen aus dem Leitmeritzer Mittelgebirge. *Nova Acta Leopold.* **38**(4): 60–107.

ENGSTRAND, L.G. 1967. Stockholm natural radiocarbon measurements. VII. *Radiocarbon* **9**: 387–438.

ERMISCH, K. 1936. Über Versteinerungen im Kali, besonders einen reichen Fossilfund im Tertiärkali des Elsaß. *Zeitschr. Kali* **30**: 31–35.

ERWIN, D.H. 1990. The end-Permian mass extinction. *Annu. Rev. Ecol. Syst.* **21**: 69–91.

ESAKI, T., and ASAHINA, S. 1957. On two Tertiary dragonfly species from the Ôya formation in Kazusa, Nagasaki Prefecture. *Kontyū* **25**: 82–88.

ESKOV, K.Yu. 1984a. A new fossil spider family from the Jurassic of Transbaikalia (Araneae: Chelicerata). *Neues Jahrb. Ceol. Paläontol. Monatshefte* (11): 645–653.

ESKOV, K.Yu. 1984b. Continental drift and the problems of historical biogeography. In: CHERNOV, Yu.I. (ed.). *Faunogenez i filocenogenez* (Faunogenesis and Phylocenogenesis). Nauka, Moscow: 24–92 (in Russian).

ESKOV, K.Yu. 1987. A new archaeid spider (Chelicerata: Araneae: Archaeidae) from the Jurassic of Kazakhstan, with notes on so called "Gondwanan" ranges of recent taxa. *Neues Jahrb. Geol. Paläontol. Abhandl.* **175**(1): 81–106.

ESKOV, K.Yu. 1992. Archaeid spiders from Eocene Baltic amber (Chelicerata: Araneae: Archaeidae), with remarks on so called "Gondwanan" ranges of recent taxa. *Neues Jahrb. Geol. Paläontol. Abhandl.* **185**(3): 311–328.

ESKOV, K.Yu. 1994. On macrobiogeographic patterns of phylogenesis. In: ROZANOV, A.Yu., and SEMIKHATOV, M.A. (eds.). *Ekosistemnye perestroiki i evolyutsia biosfery* (Ecosystem Restructures and the Evolution of Biosphere) **1**. Nedra, Moscow: 199–205 (in Russian).

ESKOV, K.Yu. 1996. A test of the "Phytospreading" and "Zonal stratification" biogeographic models on various groups of arthropods (preliminary results). *Paleontol. J.* (1995) **29**(4): 105–111.

ESKOV, K.Yu., and GOLOVACH, S.I. 1986. On the origin of Trans-Pacific disjunctions. *Zool. Jahrb. Syst.* **113**(2): 265–285.

ESKOV, K.Yu., and SUKATCHEVA, I.D. 1997. Geographical distribution of the Paleozoic and Mesozoic caddisflies (Trichoptera). *Proc. 8th Internat. Symp. Trichoptera.* Ohio Biol. Surv.: 95–98.

ETHERIDGE, R., and OLLIFF, A.S. 1890. The Mesozoic and Tertiary insects of New South Wales. *Mem. Geol. Surv. N. S. Wales* (Palaeontol.) **7**: 1–12.

EUGSTER, H.P., and HARDIE, L.A. 1975. Sedimentation in an ancient playa-lake complex: the Wilkins Peak Member of the Green River Formation of Wyoming. *Geol. Soc. America Bull.* **86**: 319–334.

EUGSTER, H.P., and SURDAM, R.C. 1973. Depositional environment of the Green River Formation of Wyoming: a preliminary report. *Geol. Soc. America Bull.* **84**: 1115–1120.

EVANS, J.W. 1946. Robin John Tillyard (1881–1937). *Proc. Linnean Soc. N. S. Wales* **71**: 252–256.

EVANS, J.W. 1956. Palaeozoic and Mesozoic Hemiptera (Insecta). *Austral. J. Zool.* **4**: 165–258.

EVANS, J.W. 1961. Some Upper Triassic Hemiptera from Queensland. *Mem. Queensl. Mus.* **14**: 13–23.

EVANS, M.E.G. 1982. Early evolution of the Adephaga – some locomotory speculations. *Coleopterists' Bull.* **36**(4): 597–607.

EVANS, W.R. 1931. Traces of a lepidopterous insect from the Middle Waikato Coal Measures. *Trans. Proc. New Zeal. Inst.* **62**: 99–101.

EVENHUIS, N.L. 1994. *Catalogue of the Fossil Flies of the World (Insecta: Diptera).* Backhuys Publ., Leiden: 600 p.

EWING, H.E. 1933. Some peculiar relationships between ectoparasites and their hosts. *Amer. Naturalist* **67**(711): 365–373.

FAEGRI, K., and VAN DER PIJL, L. 1979. *The Principles of Pollination Ecology.* Pergamon Press, Oxford etc. (cited after Russian translation: *Osnovy ekologii opyleniya.* Mir, Moscow, 1982: 379 p.).

FAHN, A. 1979. *Secretory Tissues in Plants.* Academic Press, New York: 302 p.

FARRELL, B.D. 1998. "Inordinate fondness" explained: why are there so many beetles? *Science* **281**: 555–559.

FARROW, R.A. 1982. Aerial dispersal of microinsects. *Proc. 3rd Australasian Conference on Grassland Invertebrate Ecology,* Adelaide: 51–55.

FARROW, R.A. 1984. Interactions between synoptic scale and boundary-layer meteorology on micro-insect migration. In: DANTHANARAYANA, W. (ed.). *Insect Flight: Dispersal and Migration.* Springer, Berlin–Heidelberg–New York: 185–195.

FAUL, H., and ROBERTS, W.A. 1951. New fossil footprints from the Navajo (?) Sandstone of Colorado. *J. Paleontol.* **45**: 781–795.

FEDCHENKO, Y.I., and TATOLI, I.A. 1983. To the time of origin of the insect order Coleoptera. In: VIALOV, O.S. (ed.). *Iskopaemaya fauna i flora Ukrainy* (Fossil Fauna and Flora of Ukraine) *Materialy 3ey Sessii Ukrainskogo palaeontol. obshchestva, Kerch', 13–17 maya 1980 g.* Kiev: 148–151 (in Russian).

FEDONKIN, M.A. 1988. Basic concepts and problems of palaeoichnology. In: MENNER, V.V., and MAKRIDIN, V.P. (eds.). *Sovremennaya palaeontologiya* (Modern Palaeontology) **1**. Nedra, Moscow: 400–415 (in Russian).

FEENER, D.H., Jr., and BROWN, B. V. 1997. Diptera as parasitoids. *Annu. Rev. Entomol.* **42**: 73–97.

FEISTMANTEL, O. 1881. The fossil flora of the Gondwanaland System. II. The flora of the Damuda and Panchet Divisions. *Mem. Geol. Surv. India, Paleontol.* **3**: 59–64.

FELT, E.P. 1940. *Plant Galls and Gall Makers.* Hafner, New York – London: 364 p.

FENNAH, R.G. 1968. A new genus and species of Ricaniidae from Palaeocene deposits in North Dakota. *J. Natur. Hist.* **2**: 143–146.

FERNANDES, A.C.S., CARVALHO, I.S., and NETTO, R.G. 1990. Icnofosseis de invertebrados da Formação Botucatu, São Paulo (Brasil). *An. Acad. Brasil. Ciênc.* **62**: 242–249.

FERNÁNDEZ-RUBIO, F., PEÑALVER, E., and MARTÍNEZ-DELCLÒS, X. 1991. *Zygaena*? *turolensis*, una nueva especie de Lepidoptera Zygaenidae del Mioceno de Rubielos de Mora (Teruel). Descripción y filogenia. *Estud. Mus. Cienc. Natur. Alava* **6**: 77–93.

FERNÁNDEZ-RUBIO, F., and PEÑALVER, E. 1994. Un nuevo ejemplar de *Zygaena*? *turolensis* Fernández-Rubio, Peñalver y Martínez-Delclòs, 1991 (Lepidoptera: Zygaenidae). *Estud. Mus. Cienc. Natur. Alava* **9**: 39–48.

FERRELL, G.T. 1983. Host resistance to the fir engraver, *Scolytus ventralis* (Coleoptera: Scolytidae): frequencies of attacks contacting cortical resin blisters and canals of *Abies concolor. Canad. Entomol.* **115**(10): 1421–1428.

FERSMAN, A.E. 1924. On the character of hypergenic processes in the regions with desert climate. *Doklady Russian Akad. Nauk* A (July–Sept.): 97–98 (in Russian).

FIEGE, K. 1961. Beobachtungen an rezenten Insekten-Fährten und ihre palichnologische Bedeutung. *Meyniana* **2**: 1–7.

FISHER, J.B. 1982. Effects of macrobenthos on the chemical diagenesis of freshwater sediments. In: McCALL, P.L., and TEVESZ, M.J.S. (eds.). *Animal-Sediment Relations: The Biogenic Alteration of Sediments.* Plenum, New York–London: 177–218.

FISK, L.H., and FRITZ, W.J. 1984. Pseudoborings in petrified wood from the Yellowstone "fossil forest". *J. Paleontol.* **58**(1): 58–62.

FLORIN, R. 1938. Die Koniferen des Oberkarbons und des unteren Perms. 1. *Palaeontographica* B **85**: 1–62.

FORD, I. 1988. Conglomerados con nidos de insectos fósiles Formación Palmitas (provisorio) – Terciario inferior (tentativo). *Actas del 6to Panel de Geol. del Litoral y 1era Reunión de Geol. del Uruguay*: 47–49.

FÖRSTER, B. 1891. Die Insekten des "Plattigen Steinmergels" von Brunstatt. *Abhandl. Geol. Spezialkarte Elsaß-Lothringen* **3**(5): 333–594.

FOSBERG, F.R. 1969. Paleobotany of the oceanic Pacific islands. In: LEOPOLD, E.B. (ed.) *Miocene Pollen and Spore Flora of Eniwetok Atoll, Marshall Islands. U.S. Geol. Surv. Prof. Paper* **260–II**: 1140–1163.

FOSSA-MANCINI, E. 1941. Les bosques petrificados de la Argentina según E. Riggs y G. Wieland. *Notas Mus. La Plata, Geol.* **6**: 59–92.

FOSSIL INSECT RESEARCH GROUP FOR NOJIRI-KO EXCAVATION. 1987. Fossil insects obtained from the Nojiri-ko Excavations in the 9th Nojiri-ko Excavation and the 4th Hill Site Excavation. *Monogr. Assoc. Geol. Collabor. Japan* **32**: 117–136 (Japanese with English summary).

FRAKES, L.A., and FRANCIS, J.E. 1988. A guide to Phanerozoic cold polar climates from high-latitude ice-rafting in the Cretaceous. *Nature* **333**: 547–549.

FRAKES, L.A., and FRANCIS, J.E. 1990. Cretaceous palaeoclimates. In: GINSBURG, R.N., and BEAUDOIN, B. (eds.). *Cretaceous Resources, Events and Rhythms.* Kluwer Acad. Publ., Dordrecht: 273–287.

FRANCIS, J.E., and FRAKES, L.A. 1993. Cretaceous climates. *Sedim. Rev.* **1**: 17–30.

FRANCZYK, K.J., FOUCH, T.D., JONHNSON, R.C., MOLENAAR, C.M., and COBBAN, W.A. 1992. Cretaceous and Tertiary palaeogeographic reconstructions for the Uinta-Piceance Basin study area, Colorado and Utah. *U.S. Geol. Surv. Bull.* **1787**: Q1–Q37.

FRANIA, H.E., and WIGGINS, G.B. 1997. Analysis of morphological and behavioural evidence for the phylogeny and higher classification of Trichoptera (Insecta) *R. Ontario Museum Life Science Contrib.* **160**: 1–67.

FRANZ, E. 1942. Ein verkiestes Insekt aus dem Miozän Rumäniens. *Senckenbergiana* **25**: 87–90.

FRANZEN, J.L. 1984. Spuren des Lebens im Magen der Fossilien. 50 Millionen Jahre alte Funde in der Grube Messel. *Mitt. Deutsche Forschungsgemeinsch.* **1**: 6–9.

FRASER, N.C., GRIMALDI, D.A., OLSEN, P.E., and AXSMITH, B. 1996. A Triassic Lagerstatte from eastern North America. *Nature* **380**: 615–619.

FREEMAN, B.E., and DONOVAN, S.K. 1991. A reassessment of the ichnofossil *Chubotolithes gaimanensis* Bown and Ratcliffe. *J. Paleontol.* **65**: 702–704.

FREEMAN, J.A. 1945. Studies in the distribution of insects by aerial currents. The insect population of the air from ground level to 300 feet. *J. Anim. Ecol.* **14**: 128–154.

FREESS, W.B. 1991. Beiträge zur Kenntnis von Fauna und Flora des marinen Mitteloligozän bei Leipzig. *Altenburger naturwiss. Forsch.* **6**: 1–74.

FREGENAL-MARTÍNEZ, M., MARTÍNEZ-DELCLÒS, X., MELÉNDEZ, N., and RUIZ DE LOIZAGA, M.J. 1992. Lower Cretaceous lake deposits from La Sierra del Montsec. La Cabrua Fossil Site, Pyrenees. In: CATALAN, J., and PRETUS, J.Ll. (eds.). *Mid-Congress Excursions. XXV SIL Congress, Barcelona, Aug. 21–27 1992.* Barcelona: 12-1–12-10.

FRENGUELLI, G. 1938a. Nidi fossili di Scarabeidi e Vespidi. *Boll. Soc. geol. ital.* **57**(1): 77–96.

FRENGUELLI, G. 1938b. Bolas de escarabeidos y nidos de véspidos fósiles. *Physis* **12**: 348–352.

FRENGUELLI, J. 1939. Nidos fósiles de insectos en el Terciario de Neuquen y Río Negro. *Notas Mus. La Plata, Paleontol.* **4**(18): 379–402.

FRENGUELLI, J. 1946. Un nido de Esfégido del Cretácico superior del Uruguay. *Notas Mus. La Plata, Paleontol.* **11**(90): 259–267.

FREY, R.W. 1973. Concepts in the study of biogenic sedimentary structures. *J. Sediment. Petrol.* **43**: 6–19.

FREY, R.W. (ed.). 1975. *The Study of Trace Fossils.* Springer, New York–Berlin: 562 p.

FREY, R.W., and PEMBERTON, S.G. 1985. Biogenic structures in outcrops and cores. 1. Approaches to ichnology. *Bull. Canad. Petrol. Geol.* **33**(1): 72–115.

FREYTET, P., and PLAZIAT, J.C. 1982. Continental carbonate sedimentation and pedogenesis. Late Cretaceous and Early Tertiary of Southern France. *Contrib. Sedimentol.* **12**: 1–116.

FRICKHINGER, K.A. 1994. *Die Fossilien von Solnhofen.* Goldschneck, Korb: 336 p.

FRIČ, A. 1882. Fossile Arthropoden aus den Steinkohlen- und Kreideformation Böhmens. *Beitr. Paläontol. Geol. Österr.-Ungarn* **2**: 1–7.

FRIČ, A., and BAYER, E. 1901. Studien im Gebiete der Böhmischen Kreideformation. Palaeontologische Untersuchungen der einzelnen Scihten. Perucer Schichten. *Arch. naturwiss. Landesdurchforsch. Böhmen* (1900) **11**(2): 1–180.

FRIIS, E.M. 1983. Upper Cretaceous (Senonian) floral structures of Juglandalean affinity containing *Normapolles* pollen. *Rev. Palaeobot. Palynol.* **39**: 161–188.

FRIIS, E.M. 1984a. Daekfrøede planter fra Øvre Kridt. *Fauna och flora* **79**(4): 169–176.

FRIIS, E.M. 1984b. Organization og bestovningsformer hos blomster fra Øvre Kridt. *Dansk geol. Foren.* (1983): 1–8.

FRIIS, E.M. 1985a. *Actinocalyx* gen. nov., sympetalous angiosperm flowers from the Upper Cretaceous of southern Sweden. *Rev. Palaeobot. Palynol.* **45**: 171–183.

FRIIS, E.M. 1985b. Structure and function in Late Cretaceous angiosperm flowers. *Kongl. Dansk Vidensk. Selskab. Biol. Skr.* **25**: 1–37.

FRIIS, E.M. 1990. *Silvianthemum suecicum* gen. et sp. nov., a new saxifragalean flower from the Late Cretaceous of Sweden. *Biol. Skrift.* **36**: 1–35.

FRIIS, E.M., CRANE, P.R., and PEDERSEN, K.R. 1986. Floral evidence for Cretaceous chloranthoid angiosperms. *Nature* **320**: 163–164.

FRIIS, E.M., CRANE, P.R., and PEDERSEN, K.R. 1988. Reproductive structures of Cretaceous Platanaceae. *Biol. Skrift.* **31**: 1–55.

FRIIS, E.M., and CREPET, W.I. 1987. Time of appearance of floral features. In: FRIIS, E.M., CHALONER, W.G., and CRANE, P.R. (eds.). *The Origins of Angiosperms and Their Biological Consequences.* Cambridge Univ. Press, Cambridge: 259–304.

FRIIS, E.M., and PEDERSEN, K.R. 1999. Current perspectives on basal angiosperms and the palaeobotanical record. *14. Sympos. Diversität & Evolutionsbiol.* Jena 1999: 234.

FRIIS, E.M., and SKARBY, A. 1981. Structurally preserved angiosperm flowers from the Upper Cretaceous of southern Sweden. *Nature* **291**: 484–486.

FRIIS, E.M., and SKARBY, A. 1982. *Scandianthus* gen. nov., angiosperm flowers of Saxifragalean affinity from the Upper Cretaceous of southern Sweden. *Ann. Botan.* **50**: 569–583.

FROHLICH, D.R., and PARKER, F.D. 1985. Observations on the nest-building and reproductive behaviour of a resin-gathering bee: *Dianthidium ulkei* (Hymenoptera: Megachilidae). *Ann. Entomol. Soc. Amer.* **78**(6): 804–810.

FROST, S.W. 1928. Insect scatology. *Ann. Entomol. Soc. Amer.* **21**: 36–46.

FRY, G.F. 1976. Analysis of prehistoric coprolites from Utah. *Univ. Utah Anthropol. Pap.* **97**: 1–45.

FUJIYAMA, I. 1967. A fossil scutellerid bug from marine deposit of Tottori, Japan. (Tertiary insect fauna of Japan, 1). *Bull. Nation. Sci. Mus., Tokyo* **10**: 393–402.

FUJIYAMA, I. 1968. A Pleistocene fossil *Papilio* from Shiobara, Japan. *Bull. Nation. Sci. Mus., Tokyo* **11**(1): 85–96.

FUJIYAMA, I. 1980. Late Cenozoic insects from Takai and Mikawa districts, Central Japan. *Mem. Nation. Sci. Mus., Tokyo* **13**: 21–28 (Japanese with English summary).

FUJIYAMA, J. 1983. Neogene termites from Northeastern districts of Japan, with references to the occurrence of fossil insects in the districts. *Mem. Nation. Sci. Mus., Tokyo* **16**: 83–98.

FUJIYAMA, J. 1985. Early Miocene insect fauna of Seki, Sado Island, Japan, with notes on the occurrence of Cenozoic insects from Sado to San-in District. *Mem. Nation. Sci. Mus., Tokyo* **18**(1): 35–55.

FUJIYAMA, I. 1987. Middle Miocene insect fauna of Abura, Hokkaido, Japan, with notes on the occurrence of Cenozoic fossil insects in the Oshima Peninsula, Hokkaido. *Mem. Nation. Sci. Mus., Tokyo* **20**: 37–44.

FUJIYAMA, J. 1994. Two parasitic wasps from Aptian (Lower Cretaceous) Choshi amber. *Natur. Hist. Res.* **3**(1): 1–5.

FUJIYAMA, J., and NOMURA, T. 1986. Early Miocene insect fauna of Hachia Formation, Gifu Pref., Japan. *Bull. Mizunami Fossil Mus.* **13**: 1–14 (in Japanese with English summary).

FURNISS, M.M., and FURNISS, R.L. 1972. Scolytids (Coleoptera) on snow field above timberline in Oregon and Washington. *Canad. Entomol.* **104**: 1471–1478.

FURTH, D.G. 1994. Frank Morton Carpenter (1902–1994): academic biography and list of publications. *Psyche* **101**: 127–144.

FYE, R.E. 1983. Impact of volcanic ash on pear psylla (Homoptera: Psyllidae) and associated predators. *Environ. Entomol.* **12**: 222–226.

GABRID, N.V. 1982. Host specialization of the pine chermes *Pineus pini* L. (Macq.). *Entomologicheskie Issledovaniya v Kirgizii* **15**: 62–76 (in Russian).

GADRINER, B.G. 1961. New Rhaetic and Liassic beetles. *Palaeontology* **4**: 87–89.

GAGNÉ, R.J. 1977. Cecidomyiidae (Diptera) from Canadian amber. *Proc. Entomol. Soc. Washington* **79**: 57–62.

GAGNÉ, R.J. 1982. *Protochrysomyia howardae* from Rancho La Brea, California, Pleistocene, new junior synonym of *Cochliomyia macellaria* (Diptera: Calliphoridae). *Bull. South Calif. Acad. Sci.* **80**(2): 95–96.

GALEWSKI, K., and GŁAZEK, J. 1973. An unusual occurrence of the Dytiscidae (Coleoptera) in the siliceous flowstone of the Upper Miocene cave at Przeworno, Lower Silesia. *Acta Geol. Polon.* **23**(3): 445–461.

GALEWSKI, K., and GŁAZEK, J. 1977. Upper Miocene Dytiscidae (Coleopter) from Przeworno (Lower Silesia) and the problem of Dytiscidae evolution. *Bull. Acad. Polon. Sci.*, sér. sci. biol. Cl. 2, **25**(12): 781–789.

GALL, J.-C. 1971. Faunes et paysages du Grès à *Voltzia* du Nord des Vosges. Essai paléoécologique sur le Buntsandstein supérieur. *Mem. Service Carte Géol. Alsace et de Lorrain* **34**: 1–138.

GALL, J.-C. 1983. *Ancient Sedimentary Environments and the Habitats of Living Organisms. Introduction to Palaeoecology*. Springer-Verlag, Berlin *etc.*: 219 p.

GALL, J.-C. 1985. Fluvial depositional environment evolving into deltaic setting with marine influences in the Buntsandstein of Northern Vosges (France). In: *Lecture Notes in Earth Sciences* **4**: 449–477.

GALL, J.-C. 1990. Les voiles microbiens. Leur contribution à la fossilisation des organismes au corps mou. *Lethaia* **23**(1): 21–28.

GALL, J.-C. (ed.). 1996. *Triassic Insects of Western Europe*. Paleontol. Lombarda *Nuova Ser.* **5**: 60 p.

GALL, J.C., and GRAUVOGEL, L. 1966. Faune du Buntsandstein 1. Pontes d'invertébrés du Buntsandstein supérieur. *Ann. Paléontol., Invertébrés* **52**(2): 155–161.

GALL, J.C., GRAUVOGEL-STAMM, L., NEL, A., and PAPIER, F. 1996. Entomofauna from the Upper Buntsandstein of Vosges (France). *European Science Foundation Workshop "Fossil Insects", Strasbourg, April 12–13, 1996*, unpaginated.

GALL, L.F., and TIFFNEY, B.H. 1983. A fossil noctuid moth egg from the Late Cretaceous of Eastern North America. *Science* **219**: 597–509.

GANDOLFO, M.A., NIXON, K.C., and CREPET, W.L. 1998a. *Taylerianthus crossmanensis* gen. et sp. nov. (aff. Hydrangeaceae) from the Upper Cretaceous of New Jersey. *Amer. J. Botany* **85**: 376–386.

GANDOLFO, M.A., NIXON, K.C., and CREPET, W.L. 1998b. A new fossil flower from the Turonian of New Jersey: *Dressiantha bicarpellata* gen. et sp. nov. (Capparales). *Amer. J. Botany* **85**: 964–974.

GARDINER, B.G. 1961. New Rhaetic and Liassic beetles. *Palaeontology* **4**(1): 87–89.

GAREY, J.R., KROTEC, M., NELSON, D.R., and BROOKS, J. 1996. Molecular analysis supports a tardigrade-arthropod association. *Invertebr. Biol.* **115**: 79–88.

GARYAINOV, V.A. 1974. On insect burial at present-day salt-depositing lakes of arid zone. *Tafonomiya, ee ekologicheskie osnovy. Sledy zhizni i ikh interpretatsiya* (Taphonomy, Its Ecological Backgrounds. Traces of Life and Their Interpretation). *Tezisy dokladov XX sessii Vsesoyuznogo Paleontol. Obshchestva, Febr. 4–9, 1974, Leningrad*. Leningrad: 11–13 (in Russian).

GARYAINOV, V.A. 1978. To the problem of modern insect deposits at salt-depositing lakes of arid zone. *Voprosy stratigrafii i paleontologii* **3**. Saratov: 99–100 (in Russian).

GATEHOUSE, A.G. 1997. Behaviour and ecological genetics of wind-borne migration by insects. *Annual Rev. Entomol.* **42**: 473–502.

GAUDANT, J. 1979. Mise au point sur l'ichthyofaune paléocène de Menat (Puy-de-Dôme). *C.R. Acad. Sci.* **D 288**, No. 19: 1461–1464.

GAUTIER, A. 1975. Fossile vliegenmaden (*Protophormia terraenovae* Robineau-Desvoidy, 1830) in een schedel van de wolharige neushoorn (*Coelodonta antiquatis*) uit het Onder-Würm te Dendermonde (Oost-Vlaanderen, België). *Natuurwetenschapp. Tijdschr.* **56**(1–4) (1974): 76–84.

GAUTIER, A., and SCHUMANN, H. 1973. Puparia of the subarctic or black blowfly *Protophormia terraenovae* (Robineau-Desvoidy, 1830) in a skull of a Late Eemian (?) bison at Zemst, Brabant (Belgium). *Palaeogeogr., Palaeoclimatol., Palaeoecol.* **14**(2): 119–125.

GAYDUCHENKO, L.L. 1984. On a find of a fossil colony of burrowing bees in the Pavlodar Red Beds. *Geologiya i geofizika* (10): 141–142 (in Russian).

GEISSERT, G., NÖZOLD, T., and SÜSS, H. 1981. Pflanzenfossilien und *Palaeophytobia salinaria* Süss, eine fossile Minierfliege (Agromyzidae, Diptera) aus dem Pliozän des Elsaß. *Mitt. bad. Landesver. Naturk. und Naturschutz* **12**: 221–231.

GENISE, J.F. 1995. Upper Cretaceous trace fossils in permineralized plant remains from Patagonia, Argentina. *Ichnos* **3**: 287–299.

GENISE, J.F. 1999. Fossil bee cells from the Asencio Formation (Late Cretaceous-Early Tertiary) of Uruguay, South America. *Proc. First Palaeoentomol. Conf., Moscow 1998*. AMBA projects Internat., Bratislava: 27–32.

GENISE, J.F. 2000. Insect fossil nests. *I Simpósio Brasileiro de Paleoartropodología, I Simpósio Sudamericano de Paleoartropodología, I Internat. Meeting on Paleoarthropodology, Ribeirão Preto – SP, Brazil 3–8.9.2000. Abstracts*: 114–115.

GENISE, J.F., ARCHANGELSKY, M., and CILLA, G. 2000. Possible chironomid cases (Insecta: Diptera) from the Middle Triassic Los Rastros Formation (Ischigualasto – Villa Union Basin), western Argentina. *I Simpósio Brasileiro de Paleoartropodología, I Simpósio Sudamericano de Paleoartropodología, I Internat. Meeting on Paleoarthropodology, Ribeirão Preto – SP, Brazil 3–8.9.2000. Abstracts*: 118–119.

GENISE, J.F., and BOWN, T.H. 1990. The constructor of the ichnofossil *Chubutolithes*. *J. Paleontol.* **64**: 482–483.

GENISE, J.F., and BOWN, T.H. 1991. A reassessment of the ichnofossil *Chubutolithes gaimanensis* Bown and Ratcliffe: reply. *J. Paleontol.* **65**: 705–706.

GENISE, J. F., and BOWN, T.M. 1993. New Miocene scarabaeid and hymenopterous nests and Early Miocene (Santacrucian) paleoenvironment, Patagonian Argentina. *Ichnos* **2**: 1–11.

GENISE, J.F., and BOWN, T.M. 1994. New trace fossils of termites (Insecta: Isoptera) from the Late Eocene – Early Miocene of Egypt, and the reconstruction of ancient isopteran social behaviour. *Ichnos* **3**: 1–29.

GENISE, J.F., and BOWN, T.M. 1996. *Uruguay* Roselli and *Rosellichnus* n. ichnogen., two ichnogenera for cluster of fossil bee cells. *Ichnos* **4**: 199–217.

GENISE, J.F., and ENGEL, M.S. 2000. The evolutionary history of sweat bees (Hymenoptera: Halictidae): integration of paleoentomology, paleoichnology, and phylogeny. *I Simpósio Brasileiro de Paleoartropodología, I Simpósio Sudamericano de Paleoartropodología, I Internat. Meeting on Paleoarthropodology, Ribeirão Preto – SP, Brazil 3–8.9.2000. Abstracts*: 116–117.

GENISE, J.F., and HAZELDINE, P.L. 1995. A new insect trace fossil in Jurassic wood from Patagonia, Argentina. *Ichnos* **4**: 1–5.

GENISE, J.F., and HAZELDINE, P.L. 1998. 3D-reconstruction of insect trace fossils: *Ellipsoideichnus meyeri* Roselli. *Ichnos* **5**: 167–178.

GENISE, J.F., and LAZA, J.H. 1998. *Monesichnus ameghinoi* Roselli: a complex insect trace fossil produced by two distinctive trace makers. *Ichnos* **5**: 213–223.

GENTILINI, G. 1984. Limoniidae and Trichoceridae (Diptera, Nematocera) from the Upper Miocene of Monte Castellaro (Pesaro, Central Italy). *Bol. Mus. Civ. Stor. Natur. Verona* **11**: 171–190.

GENTILINI, G. 1989. The Upper Miocene dragonflies of Monte Castellaro (Marche, Central Italy) (Odonata Libellulidae). *Mem. Soc. entomol. ital.* **67**: 251–271.

GEORGE, P.P. 1952. On some arthropod microfossils from India. *Agra Univ. J. Sci.* **1**: 83–108.

GERLACH, J. 1989. Bienen-Inklusen der Gattungen *Dasypoda*, *Megachile* und *Apis* in Baltischen Bernstein. *Münsteraner Forsch. Geol. Paläontol.* **69**: 251–260.

GERMANPRÉ, M., and LECLERCQ, M. 1994. Les pupes de *Protophormia terraenovae* associées à des mammifères pléistocènes de la Vallée flamande (Belgique). *Bull. Inst. Roy. Sci. Natur. Belgique, Sci. de la Terre* **64**: 265–268.

GERMAR, E.F. 1837. Ueber die versteinten Insecten des Juraschiefers von Solenhofen. *Isis* **4**: 421–424.

GERMAR, E.F. 1839. Die versteinerten Insecten Solenhofens. *Acta Acad. Leopold. Carol.* **19**: 187–222.

GERSDORF, E. 1968. Neues zur Ökologie des Oberpliozäns fon Willershausen. *Beih. Ber. Naturhist. Ges.* **6**: 83–94.

GERSDORF, E. 1969. Käfer (Coleoptera) aus dem Jungtertiär Norddeutschlands. *Geol. Jahrb.* **87**: 295–331.

GERSDORF, E. 1970. Weitere Käfer (Coleoptera) aus dem Jungtertiär Norddeutschlands. *Geol. Jahrb.* **88**: 629–669.

GERSDORF, E. 1976. Dritter Beitrag über Käfer (Coleoptera) aus dem Jungtertiär von Willershausen, Bl. Northeim 4226. *Geol. Jahrb. Reihe A* **36**: 103–145.

GERSON, U. 1979. The associations between pteridophytes and arthropods. *Fern Gazette* **12**: 29–45.

GERSON, U. 1982. Bryophytes and invertebrates. In: SMITH, A.J.E. (ed.). *Bryophyte Ecology*: 291–332.

GERTZ, O. 1914. Fossila zoocecidier å kvartära väztlämningar. *Geol. Fören. Stockholm Förhandl.* **36**: 533–540.

GERTZ, O. 1926. Stratigrafiska och paleontologiska studier över torvmossar i södra Skåne. 1. *Bilaga till årsredogörelse för Lunds högra allmänna läroverk 1925–26*: 1–64.

GEYER, G., and KELBER, K.-P. 1987. Flügelreste und Lebensspuren von Insekten aus dem Unteren Keuper Mainfranckens. *Neues Jahrb. Geol. Paläontol. Abhandl.* **174**(3): 331–355.

GHILYAROV, M.S. 1949. *Osobennosti pochvy kak sredy obitaniya i ee znachenie v evolyutsii nasekomykh* (Soil as Habitat and Its Importance for Insect Evolution). Izdatel'stvo Akad. Nauk SSSR, Moscow – Leningrad: 280 p. (in Russian).

GHILYAROV, M.S. 1956. Soil as environment of the invertebrate transition from the aquatic to the terrestrial life. *VI Congr. Internat. Sci. du Sol* **3**: 307–315.

GIANNO, R., and KOCHUMMEN, K.M. 1981. Notes on some minor forest products. *Malays. Forester* **44**: 566–568.

GIBSON, G.A.P. 1985. Some pro- and mesothoracic structures important for phylogenetic analysis of Hymenoptera, with a review of terms used for the structurers. *Canad. Entomol.* **117**: 1395–1443.

GIBSON, G.A.P. 1993. Groundplan structure and homology of the pleuron in Hymenoptera based on a comparison of the skeletomusculature of Xyelidae (Hymenoptera) and Raphidiidae (Neuroptera). *Mem. Entomol. Soc. Canada* **165**: 165–187.

GIBSON, G.A.P. 1999. Sister-group relationships of the Platygastroidea and Chalcidoidea (Hymenoptera) – an alternative hypothesis to Rasnitsyn 1988. *Zool. Scripta* **28**: 125–138.

GILES, E.T. 1974. The relationship between the Hemimerina and the other Dermaptera: a case for reinstating the Hemimerina within the Dermaptera, based upon a numerical procedure. *Trans. R. Entomol. Soc. London* **126**: 189–206.

GILL, E.D. 1955. Fossil insects, centipede and spider. *Victorian Natur.* **72**: 87–92.

GILMORE, C.W. 1927. Fossil footprints from the Grand Canyon. 2nd contribution. *Smithson. Misc. Coll.* **80** (3): 1–78.

GIRLING, M. 1977. Bird pellets from a Somerset Levels Neolithic trackway. *Naturalist* **102** (941): 49–52.

GIRLING, M.A. 1986. The insects associated with Lindow Man. In: STEAD, I.M., BOURKE, J.B., and BROTHWELL, D. (eds.). *Lindow Man: The Body in the Bog*. British Museum, London: 90–91.

GIVULESCU, R. 1984. Pathological elements on fossil leaves from Chiuzbaia (galls, mines and other insect traces). *Dariseama sedint. Inst. geol. geofiz.: Paleontol.* (1981) **68**: 123–133.

GLICK, P.A. 1939. The distribution of insects, spiders, and mites in the air. *U.S. Dept. Agric. Techn. Bull.* **673**: 1–150.

GLICK, P.A., and NOBLE, L.W. 1961. Airborne movements of the pink bollworm and other arthropods. *U.S. Dept. Agric. Techn. Bull.* **1255**: 1–20.

GOECKE, H. 1960. Donaciinen der oligocänen Ablagerungen von Rott. *Decheniana* **112**: 279–281.

GOEPPERT, H.R., and BERENDT, G.C. 1845. Die Bernstein und die in ihm befindlichen Pflanzenreste der Vorwelt. In: BERENDT, G.C. (ed.) *Die im Bernstein befindlichen organischen Reste der Vorwelt* **1**. Berlin: 61–126.

GOLOVNEVA, L.B. 1994. *Maastrikht-datskie flory Koryakskogo nagorya* (Maastrichtian-Danian Floras of Koryak Upland). *Trudy Botan. Inst. Ross. Akad. Nauk* **13**. St.-Petersburg: 148 p. (in Russian with English summary).

GOLUB, V.B., and POPOV, YU.A. 1999. Composition and evolution of Cretaceous and Cenozoic faunas of bugs of the superfamily Tingoidea (Heteroptera: Cimicomorpha). *Proc. First Palaeoentomol. Conf., Moscow 1998*. AMBA projects Internat., Bratislava: 33–39.

GOLUB, V.B., and POPOV, YU.A. 2000. A remarkable fossil lace bug from Upper Cretaceous New Jersey amber (Heteroptera: Tingoidea, Vianaididae), with some phylogenetic commentary. In: GRIMALDI, D.A. (ed.). *Studies on Fossils in Amber, with Particular Reference to the Cretaceous of New Jersey*. Backhuys Publ., Leiden: 231–239.

GOMAN'KOV, A.V., and MEYEN, S.V. 1986. *Tatarinovaya flora* (The *Tatarina* Flora). Nauka, Moscow: 173 p. (in Russian).

GÓMEZ PALLEROLA, J.E., 1986. Nuevos insectos fósiles de las calizas litográficas del Cretácico Inferior del Montsec (Lérida). *Bol. Geol. Minero* **97**: 717–736.

GONTSHAROVA, I.A. 1989. *Dvustvorchatye mollyuski Tarkhanskogo i Chokrakskogo basseinov* (Bivalve Molluscs of Tarkhanian and Tshokrakian Basins). *Trudy Paleontol. Inst. Akad. Nauk SSSR* **234**. Nauka, Moscow: 200 p. (in Russian).

GOODAY, G.W., POSSER, J.I., and HILLMAN, K. 1990. Degradation of chitin in the Ythan estuary, Scotland. *Abstr. 5th Internat. Congr. Ecology, Yokohama, Aug. 23–30, 1990*. Yokohama: 204.

GÖPPERT, H.R. 1855. *Die tertiäre Flora von Schossnitz in Schlesien*. Görlitz, Königsberg: 52 p.

GOROCHOV, A.V. 1984. On classification of the living Grylloidea (Orthoptera), with description of new taxa. *Zool. Zhurnal* **63**: 1641–1651 (in Russian, with English summary).

GOROCHOV, A.V. 1985. Grylloidea (Orthoptera) from the Mesozoic of Asia. *Paleontol. Zhurnal* (2): 59–68 (in Russian, English translation: *Paleontol. J.* **19**(2): 56–66).

GOROCHOV, A.V. 1986a. Triassic orthopterans of the superfamily Hagloidea (Orthoptera). *Trudy Zool. Inst. Akad. Nauk SSSR* **143**. Zool. Inst. Akad. Nauk SSSR, Leningrad: 65–100 (in Russian).

GOROCHOV, A.V. 1986b. On the system and morphological evolution of the family Gryllidae (Orthoptera), with description of new taxa. Contributions 1 & 2. *Zool. Zhurnal* **65**: 516–527, 851–858 (in Russian, with English summary).

GOROCHOV, A.V. 1987a. New fossil Bintoniellidae, Mesoedischiidae fam. n., and Pseudelcanidae fam. n. *Vestnik zoologii* (1): 18–23 (in Russian, with English summary).

GOROCHOV, A.V. 1987b. Permian orthopterans of the infraorder Oedischiida (Orthoptera, Ensifera). *Paleontol. Zhurnal* (1): 62–75 (in Russian, English translation: *Paleontol. J.* **21**(1), 72–85).

GOROCHOV, A.V. 1987c. New fossil orthopterans of the families Adumbratomorphidae fam. nov., Pruvostitidae, and Proparagryllacrididae (Orthoptera, Ensifera). *Vestnik zoologii* (4): 20–28 (in Russian, with English summary).

GOROCHOV, A.V. 1988a. Taxonomy and phylogeny of katidids (Gryllida = Orthoptera, Tettigonoidea). In: PONOMARENKO, A.G. (ed.). *Melovoy biotsenoticheskiy krizis i evolyutsiya nasekomykh* (Cretaceous Biocoenotic Crisis and Evolution of the Insects). Nauka, Moscow: 145–190 (in Russian).

GOROCHOV, A.V. 1988b. Orthopterans of the superfamily Hagloidea (Orthoptera) from the Lower and Middle Jurassic. *Paleontol. Zhurnal* (2): 54–66 (in Russian, English translation: *Paleontol. J.* **22**(2), 50–61).

GOROCHOV, A.V. 1988c. System and phylogeny of the living orthopterans from the superfamilies Hagloidea and Stenopelmatoidea (Orthoptera), with description of new taxa. Contributions 1 & 2. *Zool. Zhurnal* **67**: 353–366, 518–529 (in Russian, with English summary).

GOROCHOV, A.V. 1988d. On the classification of the fossil Phasmomimoidea, with description of new taxa. *Trudy Zool. Inst. Akad. Nauk SSSR* **178**. Zool. Inst. Akad. Nauk SSSR, Leningrad: 32–44 (in Russian).

GOROCHOV, A.V. 1989a. New taxa of orthopterans of the families Bintoniellidae, Xenopteridae, Permelcanidae, Elcanidae, and Vitimiidae. *Vestnik zoologii* (4): 20–27 (in Russian, with English summary).

GOROCHOV, A.V. 1989b. Main stages of the historical development of the suborder Ensifera (Orthoptera). *Trudy Zool. Inst. Akad. Nauk SSSR* **202**. Zool. Inst. Akad. Nauk SSSR, Leningrad: 211–217 (in Russian).

GOROCHOV, A.V. 1992. New fossil Orthoptera and Phasmoptera from the Mesozoic and Paleozoic of Mongolia. In: Rozanov, A.Yu. (ed.) *Novye vidy iskopaemykh bespozvonochnykh Mongolii* (New Species of Fossil Invertebrates of Mongolia). *Trudy Sovmestnoy Rossiysko-Mongol'skoy Paleontol. Expeditsii* **4**. Nauka, Moscow: 117–121 (in Russian).

GOROCHOV, A.V. 1993. Fossil stick-insects (Phasmatoptera) from the Jurassic and Cretaceous. In: Ponomarenko, A.G. (ed.). *Mezozoiskie nasekomye i ostrakody Azii* (Mesozoic Insects and Ostracods of Asia). *Trudy Paleontol. Inst. Ross. Akad. Nauk* **252**. Nauka, Moscow: 112–116 (in Russian).

GOROCHOV, A.V. 1994a. Permian and Triassic Phasmoptera from Eurasia. *Paleontol. Zhurnal* (4): 64–75 (in Russian, English translation: *Paleontol. J.* **28**(4): 83–98).

GOROCHOV, A.V. 1994b. New data on Triassic Orthoptera from Middle Asia. *Zoosystematica Rossica* **3**(1): 53–54.

GOROCHOV, A.V. 1995a. *Sistema i evolyutsiya pryamokrylykh podotryada Ensifera (Ortrhoptera)* [System and Evolution of the Suborder Ensifera (Orthoptera)]. Pt. 1, 2. *Trudy Zool. Inst. Ross. Akad. Nauk* **260**. Zool. Inst. Ross. Akad. Nauk, St.-Petersburg: 224 +212 p. (in Russian).

GOROCHOV, A.V. 1995b. On the system and evolution of the Order Orthoptera. *Zool. Zhurnal* **74**(10): 39–45 (in Russian, English translation: Contribution to the systematics and elucidation of the evolution of the order Orthoptera. *Entomol. Review* 1996 **75**(6): 156–162).

GOROCHOV, A.V. 2000. Phasmomimidae: are they Orthoptera or Phasmatoptera? *Paleontol. Zhurnal* (3): 67–72 (in Russian, English translation: *Paleontol. J.* **34**(3): 295–300).

GOTTSBERGER, G. 1970. Beiträge zur Biologie von Annonaceen-Blüten. *Österr. botan. Zeitschr.* **118**: 237–279.

GOTTSBERGER, G. 1974. The structure and function of the primitive angiosperm flower. *Acta botan. neerland.* **20**: 461–471.

GOTTSBERGER, G. 1989. Beetle pollination and flowering rhythm of *Annona* sp. (Annonaceae) in Brazil. *Plant Syst. Evol.* **167**(3–4): 165–187.

GRAHAM, J.B., DUDLEY, R., AGUILLAR, N.M., and GANS, C. 1995. Implications of the late Palaeozoic oxygen pulse for physiology and evolution. *Nature* **375**: 117–120.

GRAHAM, S.A. 1965. Entomology: An aid in archaeological studies. *Amer. Antiquity Mem.* **19**: 167–174.

GRAMBAST-FESSARD, N. 1966. 4ᵐᵉ contribution à l'étude des flores tertiaires des régions provencales et alpines: deux bois nouveaux de dicotyledones du Pontien de Castellane. *Mém. Soc. géol. France*, n.s. **105**: 130–146.

GRANDCOLAS, P. 1997. What did the ancestors of the woodroach *Cryptocercus* look like? A phylogenetic study of the origin of subsociality in the subfamily Polyphaginae (Dictyoptera, Blattaria). In: GRANDCOLAS, P. (ed.) *The Origin of Biodiversity in Insects: Tests of Evolutionary Scenarios*. Mém. Mus. nation. Hist. natur. **173**. Paris: 231–252.

GRAND'EURY, F.C. 1890. *Géologie et paléontologie du bassin houiller du Gard*. Saint-Etienne: 354 p.

GRANDE, L. 1984. Palaeontology of the Green River Formation, with a review of the fish fauna. *Wyoming Geol. Surv. Bull.* **63**: 1–333.

GRASSÉ, P. 1986. *Termitologia* **3**. Masson, Paris: xi + 715 p.

GRATSHEV, V.G., and ZHERICHIN, V.V. 1993. New fossil mantids (Insecta, Mantida). *Paleontol. J.* **27**(1A): 148–165.

GRATSHEV, V.G., and ZHERIKHIN, V.V. 2000. The weevils from the late Cretaceous New Jersey amber (Coleoptera: Curculionidae). In: GRIMALDI, D.A. (ed.). *Studies on Fossils in Amber, with Particular Reference to the Cretaceous of New Jersey*. Backhuys Publ., Leiden: 241–254.

GRAUVOGEL-STAMM, L. 1978. La flore du Grès à *Voltzia* (Buntsandstein supérieur) des Vosges du Nord (France). Morphologie, anatomie, interprétation phylogénique et paléogéographique. *Sci. Géol. Mem.* (Strasbourg) **50**: 1–225.

GRAUVOGEL-STAMM, L., and KELBER, K.-P. 1996. Plant-insect interactions and co-evolution during the Triassic in western Europe. In: GALL, J.-C. (ed.). *Triassic Insects of Western Europe*. Paleontol. Lombarda (N.S.) **5**: 5–23.

GRAY, J. 1985. The microfossil record of early land plants: advances in understanding early terrestrialization. *Philos. Trans. R. Soc. London* B **309**: 167–195.

GREENFAST, E.F., and LABANDEIRA, C.C. 1997. Insect folivory on a Lower Permian (Sakmarian) riparian flora from north-central Texas. *Geol. Soc. Amer. Abstracts with Programs* **29**(6): 262.

GREENSLADE, P.J.M. 1988. Reply to R.A. Crowson's "Comments on Insects of the Rhynie Chert" (1985 *Entomol. Gener.* **11**(1/2): 097–098). *Entomol. Gener.* **13**(1–2): 115–117.

GREENSLADE, P., and WHALLEY, P.E.S. 1986. The systematic position of *Rhyniella praecursor* Hirst & Maulik (Collembola), the earliest known hexapod. In: Dallai, R. (ed.). *2nd Internat. Seminar on Apterygota*. Univ. Siena, Siena: 319–323.

GREGUSS, P. 1970. Ein *Callitris*-ähnlichen Holz aus dem Tertiär von Limburg (Niederlande). *Senckenberg. Lethaea* **51**: 265–275.

GREGUSS, P. 1971. Der polyphyletische Ursprung der Angiospermen. *Ber. Deutsch. Ges. geol. Wiss.* A **16**(6): 705–718.

GRESSITT, J.L., COATSWORTH, J., and YOSHIMOTO, C.M. 1962. Airborne insects trapped on "Monsoon Expedition". *Pacif. Insects* **4**: 319–323.

GRESSITT, J.L., LEECH, R.E., LEECH, T.R., SEDLACEK, J., and WISE, K.A.J. 1961. Trapping of air-borne insects in the Antarctic area. Pt. 2. *Pacif. Insects* **3**: 559–562.

GRESSITT, J.L., LEECH, R.E., and O'BRIEN, C.W. 1960. Trapping of air-borne insects in the Antarctic area. *Pacif. Insects* **12**: 303–325.

GRESSITT, J.L., and NAKATA, S. 1958. Trapping of air-borne insects on ships in the Pacific. *Proc. Hawaiian Entomol. Soc.* **16**(3): 363–365.

GRIMALDI, D. (ed.) 1990. Insects from the Santana Formation, Lower Cretaceous, of Brasil. *Bull. Amer. Mus. Nat. Hist.* **195**: 1–191.

GRIMALDI, D. 1992. Vicariance biogeography, geographic extinctions, and the North American tsetse flies. In: NOVACEK, M.J., and WHEELER, Q.D. (eds.). *Extinction and Phylogeny*. Columbia Univ. Press, New York: 178–204.

GRIMALDI, D.A. 1993. The care and study of the fossiliferous amber. *Curator* **36** (1): 31–49.

GRIMALDI, D.A. 1996a. Captured in amber. *Scientific American* (April): 70–77.

GRIMALDI, D.A. 1996b. *Amber: Window to the Past*. Harry N. Adbrams, New York: 216 p.

GRIMALDI, D.A. 1997a. A fossil mantis (Insecta, Mantodea) in Cretaceous amber of New Jersey, with comments on the early history of the Dictyoptera. *Amer. Mus. Novitates* **3204**: 11 p.

GRIMALDI, D. 1997b. The bird flies, genus *Carnus*: species revision, generic relationships, and a fossil *Meconeura* in amber (Diptera: Carnidae). *Amer. Mus. Novitates* **3190**: 1–30.

GRIMALDI, D. 1998. North American ambers and an exceptionally diverse deposit from the Cretaceous of New Jersey. *World Congress on Amber Inclusions, Museo de Ciencias Naturales de Alava*: 81.

GRIMALDI, D. 1999. The co-radiation of pollinating insects and angiosperms in the Cretaceous. *Ann. Missouri Botan. Garden* **86**(2): 373–406.

GRIMALDI, D.A. 2000a. A diverse fauna of Neuropteroidea in amber from Cretaceous of New Jersey. In: GRIMALDI, D.A. (ed.). *Studies on Fossils in Amber, with Particular Reference to the Cretaceous of New Jersey.* Backhuys Publ., Leiden: 259–303.

GRIMALDI, D.A. (ed.) 2000b. *Studies on Fossils in Amber, with Particular Reference to the Cretaceous of New Jersey.* Backhuys Publ., Leiden: 498 p.

GRIMALDI, D.A., and CUMMING, J. 1999. Brachyceran Diptera in Cretaceous ambers and Mesozoic diversification of the Eremoneura. *Bull. Amer. Mus. Nat. Hist.* **239**: 124 p.

GRIMALDI, D.A., and MAISEY, J. 1990. Introduction. In: GRIMALDI, D.A. (ed.). *Insects from the Santana Formation, Lower Cretaceous of Brazil. Bull. Amer. Mus. Nat. Hist.* **195**: 5–14.

GRIMALDI, D., BECK, C.W., and BOON, J.J. 1989. Occurrence, chemical characteristics, and palaeontology of the fossil resins from New Jersey. *Amer. Mus. Novitates* **2948**: 1–28.

GRIMALDI, D., MICHALSKI, C., and SCHMIDT, K. 1993. Amber fossil Enicocephalidae (Heteroptera) from the Lower Cretaceous of Lebanon and Oligo-Miocene of the Dominican Republic, with biogeographic analysis of *Enicocephalus. Amer. Mus. Novitates* **3071**: 1–30.

GRIMALDI, D., SHEDRINSKY, A., ROSS, A., and BAER, N.S. 1994a. Forgeries of fossils in "amber": history, identification and case studies. *Curator* **37**(4): 251–274.

GRIMALDI, D.A., AGOSTI, D., and CARPENTER, J.M. 1997. New and rediscovered primitive ants (Hymenoptera: Formicidae) in Cretaceous amber from New Jersey, and their phylogenetic relationships. *Amer. Mus. Novitates* **3208**: 1–43.

GRIMALDI, D.A., BONWICH, E., DELLANOY, M., and DOBERSTEIN, W. 1994b. Electron microscopic studies of mummified tissues in amber fossils. *Amer. Mus. Novitates* **3097**: 1–31.

GRIMALDI, D., SHEDRINSKY, A. and WAMPLER, T.A. 2000. A remarkable deposit of fossiliferous amber from the Upper Cretaceous (Turonian) of New Jersey. In: GRIMALDI, D.A. (ed.). *Studies on Fossils in Amber, with Particular Reference to the Cretaceous of New Jersey.* Backhuys Publ., Leiden: 1–76.

GRIMALSKY, V.I. 1974. Tree stands resistance to needle- and leaf-eating pests in connection with trophic theory of insect abundance dynamics. *Zool. Zhurnal*, **53**(2): 189–198 (in Russian).

GRIMALT, J.O., SIMONEIT, B.R.T., HATCHER, P.G., and NISSENBAUM, A. 1988. The molecular composition of ambers. *Organ. Geochem.* **13**: 677–690.

GRINFELD, E.K. 1961. Origin of symbiosis in ants and aphids. *Vestnik Leningrad. Gosudarstv. Univ.* **15**: 73–84 (in Russian).

GRINFELD, E.K. 1962. *Proiskhozhdenie antofilii u nasekomykh* (Origin of Anthophily in Insects). Leningrad Univ., Leningrad: 186 p. (in Russian).

GRINFELD, E.K. 1975. Anthophily of beetles (Coleoptera) and critics of the cantharophily hypothesis. *Entomol. Obozrenie* **54**(3): 507–514 (in Russian).

GRINFELD, E.K. 1978. *Proiskhozhdenie i razvitie antofilii u nasekomykh* (Origin and Evolution of Anthophily in Insects). Leningrad Univ., Leningrad: 204 p. (in Russian).

GRODNITSKY, D.L. 1999. *Form and Function of Insect Wing. The Evolution of Biological Structures.* John Hopkins Univ. Press, Baltimore–London: 261 p.

GROMOV, V.V., DMITRIEV, V. YU., ZHERIKHIN, V.V., PONOMARENKO, A.G., RASNITSYN, A.G., and SUKATSHEVA, I.D. 1993. Cretaceous insects from Ul'ya River basin (Western Okhotsk region). In: PONOMARENKO, A.G. (ed.). *Mezozoiskie nasekomye i ostrakody Azii* (Mesozoic Insects and Ostracods from Asia). *Trudy Paleontol. Inst. Ross. Akad. Nauk* **252**. Nauka, Moscow: 5–60 (in Russian).

GRUNIN, K.Y. 1973. The first discovery of the mammoth bot-fly *Cobboldia* (*Mamontia*, subgen. n.) *russanovi* sp. n. (Diptera, Gasterophilidae). *Entomol. Obozrenie* **52**(1): 228–233 (in Russian).

GRUNT, T.A., ESAULOVA, N.K., and KANEV, G.P. (eds.). 1998. *Biota vostoka Evropeiskoy Rossii na rubezhe ranney i pozdney permi* (Biota of East European Russia at the Early/Late Permian Boundary). Geos, Moscow: 355 pp. (in Russian).

GUBIN, YU.M., and SINITZA, S.M. 1996. Shar Teg: a unique Mesozoic locality of Asia. In: MORALES, M. (ed.). *The Continental Jurassic. Mus. South Arizona Bull.* **60**: 311–318.

GUILMETTE, J.E., Jr., HOLZAPFEL, E.P., and TSUDA, D.M. 1970. Trapping of air-borne insects on ships in the Pacific. Part 8. *Pacif. Insects* **3**: 559–562.

GÜNTER, K.K. 1989. Development of Grossensee (Holstein, Germany): variations in trophic status from the analysis of subfossil microfauna. *Hydrobiologia* **103**: 231–234.

GÜNTHER, J. 1983. *Embidopsocus saxonicus* sp. n., eine neue fossile Psocoptera-Art aus Sächsischem Bernstein des Bitterfelder Raumes (Insecta, Psocoptera: Liposcelidae). *Mitt. Zool. Mus. Berlin* **65**: 321–325.

GUO, Sh.-x., 1991. A Miocene trace fossil of insect from Shanwang Formation in Linqu, Shandong. *Acta Palaeontol. Sinica* **30**(6): 739–742.

GUSEV, O. 1971. Baikal Lake – a giant insect trap. *Nauka i zhizn'* (3): 135–137 (in Russian).

GUST, S. 1992. Taphonomy and chronology at Rancho La Brea. *J. Vertebr. Paleontol.* **12**(3), Suppl.: 32.

GUTHÖRL, P. 1934. Die Arthropoden aus dem Carbon und Perm des Saar-Nahe-Pfalz-Gebietes. *Abhandl. Preuss. Geol. Landesanst.* N.F. **164**: 1–220S.

HAAS, F. 1995. The phylogeny of the Forficulina, a suborder of the Dermaptera. *Syst. Entomol.* **20**: 85–98.

HABER, E. 1980. *Aster* florets in the diet of a broad-winged bush katydid. *Canad. Field-Natur.* **94**(2): 194–195.

HABERSETZER, J., and STORCH, G. 1989. Ecology and echolocation of the Eocene Messel bats. *Proc. 4th European Bat Research Symp., Prague, Aug. 18–23, 1987.* Praha: 213–233.

HAESELER, V. 1974. Aculeate Hymenopteren über Nord- und Ostsee nach Untersuchungen auf Feuerschiffen. *Entomol. Scand.* **5**(2): 123–136.

HAGEN, H.A. 1882. Fossil insects of the Dakota Group. *Nature* **25**: 265–266.

HAKBIJL 1987. Insect remains: Unadulterated Cantharidum and tobacco from the West Indies. In: GAWRONSKI, J.H.G. (ed.). *Amsterdam Project: Annual Rep. of the VOC Ship "Amsterdam" Foundation 1986.* Amsterdam: 93–94.

HALFFTER, G. 1959. Etología y paleontología de Scarabaeinae (Coleoptera, Scarabaeidae). *Ciencia* **19**: 165–178.

HALFFTER, G. 1977. Evolution of nidification in the Scarabaeinae (Coleoptera: Scarabaeidae). *Quaest. Entomol.* **13**: 231–253.

HALFFTER, G., ANDUAGA, S., and HUERTA, C. 1983. Nidification des *Nicrophorus* (Col. Silphidae). *Bull. Soc. entomol. France* **88**(7–8): 648–666.

HALFFTER, G., and EDMONDS, W.D. 1982. The nesting behaviour of dung beetles. An ecological and evolutive approach. *Publ. Inst. Ecol. México* **10**: 1–176.

HALL, H.J. 1977. A paleoscatological study of diet and disease at Dirty Shame Rock-shelter, southeast Oregon. *Tebiwa: Miscell. Papers Idaho State Univ. Mus. Natur. Hist.* **8**: 1–14.

HALL, W.E., OLSON, C.A., and VAN DEVENDER, T.R. 1989. Late Quaternary and modern arthropods from the Ajo Mountains of southwestern Arizona. *Pan-Pacif. Entomol.* **65**: 322–347.

HALL, W.E., VAN DEVENDER, T.R., and OLSON, C.A. 1988. Late Quaternary arthropod remains from Sonoran Desert packrat middens, southwestern Arizona and northwestern Sonora. *Quatern. Res.* **29**: 277–293.

HALL, W.E., VAN DEVENDER, T.R., and OLSON, C.A. 1990. Late Quaternary and modern arthropods from the Puerto Blanco Mountains, Organ Pipe Cactus National Monument, southwestern Arizona. In: BETANCOURT, J.L., VAN DEVENDER, T.R., and MARTIN, P.S. (eds.). 1990. *Packrat Middens: Late Quaternary Environments of the Arid West.* Univ. Arizona Press, Tucson: 363–379.

HALSTEAD, L.B., and NICOLI, P.G. 1971. Fossilized caves of Mendip. *Studies Speleol.*, **2**: 93–102.

HAMILTON, K.G.A. 1983. Classification, morphology and phylogeny of the family Cicadellidae (Rhynchota: Homoptera). *Proc. 1st Internat. Workshop on Leafhoppers and Planthoppers of Econ. Importance, London, 1982*: 15–37.

HAMILTON, K.G.A. 1990. Homoptera. In: GRIMALDI, D.A. (ed.). *Insects from the Santana Formation, Lower Cretaceous, of Brazil. Bull. Amer. Mus. Nat. Hist.* **195**: 82–122.

HAMILTON, K.G.A. 1992. Lower Cretaceous Homoptera from the Koonwarra Fossil Bed in Australia, with a new superfamily and synopsis of Mesozoic Homoptera. *Ann. Entomol. Soc. Amer.* **85**(4): 423–430.

HANDLIRSCH, A. 1906–1908. *Die fossilen Insekten und die Phylogenie der rezenten Formen.* Wilhelm Engelmann, Leipzig: 1430S.

HANDLIRSCH, A. 1910. Fossile Wespennester. *Ber. Senckenb. Ges.* **41**: 265–266.

HANDLIRSCH, A. 1920. Paläontologie. In: SCHRÖDER, C. (ed.). *Handbuch der Entomologie.* **3**. Jena: 117–306.

HANDLIRSCH, A. 1922. *Fossilium catalogus.* I: *Animalia.* Pars 16: Insecta Palaeozoica. W. Junk, Berlin: 230 p.

HANDLIRSCH, A. 1937. Neue Untersuchungen über die fossilen Insekten mit Ergänzungen und Nachträgen sowie Ausblicken auf phylogenetische, palaeogeographische und allgemein biologische Probleme. I. Theil. *Ann. Naturhist. Mus. Wien* **48**. S1–140.

HANDLIRSCH, A. 1939. Neue Untersuchungen über die fossilen Insekten mit Ergänzungen und Nachträgen sowie Ausblicken auf phylogenetische, palaeogeographische und allgemein biologische Probleme. II. Theil. *Ann. Naturhist. Mus. Wien* **48**. S1–240.

HANDSCHIN, E. 1947. Insekten aus den Phosphoriten des Quercy. *Schweiz. paläontol. Abhandl.*, **64**(4) (1944): 1–23.

HANDSHIN, E. 1958. Die systematische Stellung der Collembolen. *Proc. 10th Internat. Congr. Entomol.* **1**. Montreal: 499–508.

HANOVER, J.W. 1975. Physiology of tree resistance to insects. *Annu. Rev. Entomol.* **20**: 75–95.

HANSEN, H.J. 1890. Gamle og nye hofvedmomenter til Cicadariernes morphologi og systematik. *Entomol. Tidskr.* II: 19–76 (English translation: Hansen, H.J. 1900–1903. On the morphology and classification of the auchenorrhynchous Homoptera. *Entomologist* **33**: 116–120, 169–172, 334–337; **34**: 149–154; **35**: 214–217, 234–236, 260–263; **36**: 42–44, 64–67, 93–94).

HANSEN, H.J. 1893. Zur Morphologie der Gliedmassen und Mundtheile bei Crustaceen und Insekten. *Zool. Anz.* **16**: 193–198, 201–212.

HANSEN, J.P.H. 1989. The mummies from Qilakitsoq – paleopathological aspects. *Meddel. Grønl.* **12**: 69–82.

HÄNTZSCHEL, W. 1975. *Trace Fossils and Problematica.* In: Teichert, C. (ed.). *Treatise of Invertebrate Palaeontology.* Part W. Suppl. 1. Geol. Society of America & Kansas Univ. Press, Lawrence, Kansas: 269 p.

HARDING, P.T., and PLANT, R.A. 1978. A second record of *Cerambyx cerdo* L. (Coleoptera: Cerambycidae) from sub-fossil remains in Britain. *Entomol. Gazette* **29**(3): 150–152.

HARDY, A.C., and CHENG, L. 1986. Studies in the distribution of insects by aerial currents. III. Insect drift over the sea. *Ecol. Entomol.* **11**(3): 283–290.

HARRELL, J.C., and HOLZAPFEL, E. 1966. Trapping of air-borne insects on ships in the Pacific. Pt. 6. *Pacif. Insects* **8**(1): 33–42.

HARRELL, J.C., and YOSHIMOTO, C.M. 1964. Trapping of air-borne insects on ships in the Pacific. Pt. 5. *Pacif. Insects* **6**(2): 274–282.

HARRIS, T.M. 1969. The Yorkshire Jurassic flora. 3. Bennettitales. *Brit. Mus. (Nat. Hist.) Publ.* **675**: 1–201.

HARRIS, T.M. 1973. *The Strange Bennettitales. 19th Sir A.C. Seward Memorial Lecture, 1970.* B. Sahni Inst. Paleobotany, Lucknow, 11 p.

HARRISON, J.W.H. 1926. Zoocecidia from a peat-bed near Britley, Co. Durham, with some reference to other insect remains. *Entomologist* **59**: 121–123.

HART, G.F. 1986. Origin and classification of organic matter in clastic sediments. *Palynology* **10**: 1–23.

HARTNOLL, R.G., and BRYANT, A.D. 1990. Size-frequency distributions in decapod Crustacea – the quick, the dead, and the cast-offs. *J. Crustac. Biol.* **10**(1): 14–19.

HARVEY-GIBSON, R.J., and MILLER-BROWN, D. 1927. Fertilization of Bryophyta. *Polytrichum commune* (preliminary note). *Ann. Botany* **41**: 190–191.

HASIOTIS, S.T., and BOWN, T.M. 1992. Invertebrate trace fossils: The backbone of continental ichnology. In: MAPLES, C.G., and WEST, R.R. (eds.). *Trace Fossils. Palaeontol. Soc. Short Courses* **5**: 64–104.

HASIOTIS, S.T., and DEMKO, T.M. 1996a. Terrestrial and freshwater trace fossils, Upper Jurassic Morrison Formation, Colorado Plateau. In: MORALES, M. (ed.). *The Continental Jurassic. Bull. Mus. Northern Arizona* **60**: 355–370.

HASIOTIS, S.T., and DEMKO, T.M. 1996b. Ant (Hymenoptera: Formicidae) nest ichnofossils, Upper Jurassic Morrison Formation, Colorado Plateau: Evolutionary and ecologic implications. *Geol. Soc. Amer. Abstracts with Programs*: A106.

HASIOTIS, S.T., and DUBIEL, R.F. 1993a. Continental trace fossils of the Upper Triassic Chinle Formation, Petrified Forest National Park, Arizona. In: LUCAS, S.G., and MORALES, M. (eds.). *The Nonmarine Triassic. Bull. New Mexico Mus. Natur. Hist. Science* **3**: 175–178.

HASIOTIS, S.T., and DUBIEL, R.F. 1993b. Trace fossil assemblage in Chinle Formation alluvial deposits at the Tepees, Petrified Forest National Park, Arizona. In: LUCAS, S.G., and MORALES, M. (eds.). *The Nonmarine Triassic. Bull. New Mexico Mus. Natur. Hist. Science* **3**: G42–G43.

HASIOTIS, S., and DUBIEL, R.F. 1995. Termite (Insecta: Isoptera) nest ichnofossils from the Upper Triassic Chinle Formation, Petrified Forest National Park, Arizona. *Ichnos* **4**: 119–130.

HASIOTIS, S., DUBIEL, R.F., and DEMKO, T.M. 1995. Triassic hymenopterous nests: Insect eusociality predates angiosperm plants. *47th Annual Meeting Geol. Soc. America, Rocky Mountain Section, Abstracts with Programs*: 13.

HASIOTIS, S., and FIORILLO, A.R. 1997. Dermestid beetle borings in dinosaur bones, Dinosaur National Monument, Utah: Additional keys to bone bed taphonomy. *Geol. Soc. Amer. Abstracts with Programs* **29**: A13.

HASIOTIS, S.T., BOWN, T.M., KAY, P.T., DUBIEL, R.F., and DEMKO, T.M. 1996. The ichnofossil record of hymenopteran nesting behaviour from Mesozoic and Cenozoic pedogenic and xylic substrates: Example of relative stasis. *Sixth North American Paleontol. Convention Abstracts of Papers.* (*Paleontol. Soc. Spec. Publ.* **8**): 165.

HASIOTIS, S.T., KIRKLAND, J.I., WINDSCHEFFEL, G.W., and SAFRIS, C. 1997. Fossil caddisfly cases (Insecta: Trichoptera), Upper Jurassic Morrison Formation, Fruita Palentological area, Colorado. *Modern Geol.* **23**: 1–10..

HATTEN, H.J. 1976. Lebensspuren produced by insect wings. *J. Paleontol.* **50**(5): 833–840.

HAUPT, H.1950. Die Käfer (Coleoptera) aus der eozänen Braunkohle des Geiseltales. *Geologica* **6**: 1–168.

HAUPT, H. 1952. Insektenfunde aus den Goldlauterer-Schichten des Thuringer Waldes. *Hall. Jahrb. mitteldeutsch Erdgeschichte* **1**: 241–258.

HAUPT, H. 1956. Beitrag zur Kenntnis der eozänen Arthropodenfauna des Geiseltales. *Nova Acta Leopold.* **18** Nr 128: 1–90.

HAWKESWOOD, T.J. 1990. Observations on the biology, host plants and immature stages of *Dihammus tincturatus* Pascoe (Coleoptera: Cerambycidae Lamiinae) in Papua New Guinea. Part 1. General biology and host plants. *Giorn. ital. entomol.* **5**(25): 95–101.

HAYASHI, M. 1995. Middle Pleistocene insect fossils obtained from the Nodono Formation in Annaka, Central Japan. *Bull. Nojiri-ko Mus.* **3**: 35–44 (Japanese with English summary).

HAYASHI, M. 1996. Insect fossil assemblage and paleoenvironments from the Early Pleistocene Bushi Formation in Saitama Prefecture, Japan. *Earth Sci.* **50**(3): 223–237 (Japanese with English summary).

HAYASHI, M. 1998. Early Pleistocene ground beetles (Coleoptera: Carabidae) from the Ookui Formation in Nagano Prefecture, central Japan, and their biogeographical and paleoenvironmental significance. *Quatern. Res.* **7**(2): 117–129.

HAYASHI, M., KATO, M., and KOBAYASHI, I. 1996. Insect fossils from the upper part of the Uonuma Formation, Nagaoka City, Niigata Prefecture, Japan. *Bull. Nagaoka Municipal Sci. Mus.* **31**: 109–116 (Japanese with English summary).

HEER, O. 1847. *Die Insektenfauna der Tertiärgbilde von Oeningen und von Radoboj in Croatien. Vol. 1. Käfer.* Engelmann, Leipzig: 222 p.

HEER, O. 1849. *Die Insektenfauna der Tertiärgbilde von Oeningen und von Radoboj in Croatien. Vol. 2. Heuschrecken, Florfliegen, Alderflügen, Schmetterlinge, und Fliegen.* Engelmann, Leipzig: 264 p.

HEER, O. 1853. *Die Insektenfauna der Tertiärgbilde von Oeningen und von Radoboj in Croatien. Vol. 3. Rhynchoten.* Engelmann, Leipzig: 138 p.

HEER, O. 1856. Über die fossilen Insekten von Aix in der Provence. *Viertelj. naturf. Ges. Zürich* **1**: 1–40.

HEER, O. 1864–1865. *Die Urwelt der Schweiz.* F. Schulthess Verlag, Zürich: 628 p.

HEER, O. 1867. Fossile Hymenopteren aus Oeningen und Radoboj. *Neue Denkschr. allgem. schweiz. Ges. gesam. Naturwiss.* **22**: 1–42.

HEIE, O. 1967. Studies on Fossil Aphids (Homoptera: Aphidoidea), Especially in the Copenhagen Collection of Fossils in Baltic Amber. *Spolia Zool. Mus. Haunien.* **26**: 1–274.

HEIE, O. 1968. Pliocene aphids from Willershausen (Homoptera: Aphidoidea). *Beih. Ber. naturhist. Ges. Hannover* **6**: 25–39.

HEIE, O.E. 1981. Morphology and phylogeny of some Mesozoic aphids (Insecta, Hemiptera). *Ent. Scand.* Suppl. **15**: 401–415.

HEIE, O.E. 1987. Palaeontology and phylogeny. In: MINKS, A.K., and HARREWIJN, P. (eds.). *Aphids: Their Biology, Natural Enemies and Control* A. Elsevier Sci. Publ., Amsterdam: 367–391.

HEIE, O. 1994. Why are there so few aphid species in the temperate areas of the southern hemisphere? *Europ. J. Entomol.* **91**: 127–133.

HEIE, O.E. 1999. Aphids of the past (Hemiptera, Sternorrhyncha). *Proc. First Palaeoentomol. Conf., Moscow 1998.* AMBA projects Internat., Bratislava: 49–56.

HEIE, O.E., and FRIEDRICH, W.L. 1971. A fossil specimen of the North American hickory aphid (*Longistigma caryae* Harris), found in Tertiary deposits in Iceland. *Entomol. Scand.* **2**: 74–80.

HEIE, O.E., and PEÑALVER, E. 1999. *Palaeophylloxera* nov. gen., the first fossil specimen of the family Phylloxeridae (Hemiptera, Phylloxeroidea); Lower Miocene of Spain. *Geobios* **32**(4): 593–597.

HEIE, O.E. and PIKE, E.M. 1992. New aphids in Cretaceous amber from Alberta (Insecta, Homoptera). *Canad. Entomol.* **124**: 1027–1053.

HEIE, O.E., and PIKE, E.M. 1996. Reassessment of the taxonomic position of the fossil aphid family Canadaphididae based on two additional specimens of *Canadaphis carpenteri* (Hemiptera, Aphidinea). *Europ. J. Entomol.* **95**(4): 617–622.

HEIE, O., and POINAR, G.O., Jr. 1988. *Mindazerius dominicanus* nov. gen., nov. sp., a fossil aphid (Homoptera, Aphidoidea, Drepanosiphidae) from Dominican amber. *Psyche* **95**: 153–165.

HEIE, O., and POINAR, G.O., Jr. 1999. A fossil aphid (Hemiptera: Sternorrhyncha) in Dominican amber. *Proc. Entomol. Soc. Washington* **101**(4): 816–821.

HEIE, O., and WEGIEREK, P. 1998. A list of fossil aphids (Homoptera: Aphidinea). *Ann. Upper Silesian Mus. Bytom, Entomol.* **8/9**: 159–192.

HEIN, L. 1954. Larve der Waffenfliege *Eulalia* sp. im Dysodil des Randecker Maar. *Geologie* **3**: 644–645.

HEINRICH, A. 1988. Fliegenpuppen aus eiszeitlichen Knochen. *Cranium* **5**: 82–83.

HEISER, C.B. 1962. Some observations on pollination and compatibility in *Magnolia. Proc. Indiana Acad. Sci.* **72**: 259–266.

HEIZER, R.F. 1970. The anthropology of prehistoric Great Basin coprolites. In: BROTHWELL, D., and HIGGS, E. (eds.). *Science in Archaeology.* Praeger, New York: 244–250.

HELLER, K.M. 1938. Indomalaiische, vorwiegend javanische Rüsselkäfer (mit einer Bestimmungstabelle der *Syrotelus*arten). *Entomol. Blätt.* **34**(6): 313–327.

HELLMUND, M., and HELLMUND, W. 1991. Eiablageverhalten fossiler Kleinlibellen (Odonata, Zygoptera) aus dem Oberoligozän von Rott im Siebengebirge. *Stuttg. Beitr. Naturk.* B **177**: 1–17.

HELLMUND, M., and HELLMUND, W. 1993. Neufund fossiler Eilogen (Odonata, Zygoptera, Coenagrionidae) aus dem Oberoligozän von Rott im Siebengebirge. *Decheniana* **146**: 348–351.

HELLMUND, M., and HELLMUND, W. 1996a. Phosphoritisierte Insekten- und Annelidenreste aus der mitteloligozänen Karstspaltenfüllung "Ronheim 1" bei Harburg (Bayern, Süddeutschland). *Stuttg. Beitr. Naturk.* B **241**: 1–21.

HELLMUND, M., and HELLMUND, W. 1996b. Zum Fortpflanzgsmodus fossiler Kleinlibellen (Insecta, Odonata, Zygoptera). *Paläontol. Zeitschr.* **70**(1–2): 153–170.

HELLMUND, M., and HELLMUND, W. 1996c. Fossile Zeugnisse zum Verhalten von Kleinlibellen aus Rott. In: KOENIGSWALD, W., von. (ed.). *Fossillagerstätte Rott bei Hennef im Siebengebirge.* 2. Aufl. Siegburg: 57–60.

HELLMUND, M., and HELLMUND, W. 1996d. Zur endophytischen Eiablage fossiler Kleinlibellen (Insecta, Odonata, Zygoptera), mit Beschreibung eines neuen Gelegetyps. *Mitt. Bayer. Staatssamml. Paläont. histor. Geol.* **36**: 17–25.

HELLMUND, M., and HELLMUND, W. 1998. Eilogen von Zygopteren (Insecta, Odonata, Coenagrionidae) in unteroligozänen Maarsedimenten von Hammerunterwiesenthal (Freistaat Sachsen). *Abhandl. Staatl. Mus. Miner. Geol. Dresden* **43–44**: 281–192.

HEMING, B.S. 1978. Structure and function of the mouthparts in larvae of *Haplothrips verbasci* (Osborn). *J. Morphol.* **156**: 1–37.

HENLE, M. 1992. Der Einfluss von Mikroorganismen auf das Erosionsverhalten von Wattsedimenten. *GKSS Rept.*, **E110**: 1–77.

HENNIG, W. 1965. Die Acalyptratae des Baltischen Bernstein. *Stuttg. Beitr. Naturk.* **145**, 215S.

HENNIG, W. 1966a. *Phylogenetic Systematics.* Illinois Univ. Press, Urbana: 263 p.

HENNIG, W. 1966b. Einige Bemerkungen über die Typen der von Giebel 1862 angeblich aus dem Berstein beschriebenen Insectenarten. *Stuttg. Beitr. Naturk.* **162**: 1–7.

HENNIG, W. 1968. Kritische Bemerkungen über den Bau der Flügelwurzel bei den Dipteren und die Frage nach der Monophylie der Nematocera. *Stuttg. Beitr. Naturk.* **193**: 1–23.

HENNIG, W. 1969a. *Die Stammesgeschchte der Insekten.* W. Kramer, Frankfurf a/M.: 436 S.

HENNIG, W. 1969b. Neue Übersicht über die aus dem Baltischen Bernstein bekannten Acalyptratae (Diptera: Cyclorrhapha). *Stuttg. Beitr. Naturk.* **209**: 1–42.

HENNIG, W. 1971. Insektenfossilien aus der unteren Kreide. III. Empidiformia ("Microphorinae") aus der unteren Kreide und aus dem Baltischen Bernstein; ein Vertreter der Cyclorrhapha aus der unteren Kreide. *Stuttg. Beitr. Naturk.* **232**: 28S.

HENNIG, W. 1972. Insektenfossilien aus der unteren Kreide. IV. Psychodidae (Phlebotominae), mit einer kritischen Übersicht über das phylogenetische System der Familie und die bisher beschriebenen Fossilien (Diptera). *Stuttg. Beitr. Naturk.* **241**: 1–69.

HENNIG, W. 1973. Ordnung Diptera (Zweiflügeler). Kukenthal, W. (ed.). *Handbuch der Zoologie* **4**(2) 2/31. Walter de Gruyter, Berlin: 337 p.

HENNIG, W. 1981. *Insect Phylogeny.* John Wiley & Sons, Chicester etc.: 514 p.

HENRIKSEN, K.L. 1922. Eocene insects from Denmark. *Denmarks Geol. Undersøgelse* **2**: 1–36.

HENROTAY, M. 1986. Découverte d'un nouveau gisement de Lepidoptères, d'autres insectes et d'araignées fossiles à Dauphin (Hte. Provence, France). *Linneana Belg.* **10**(6): 266–279.

HENRY, T.J. 1997. Phylogenetic analysis of family groups within the infraorder Pentatomomorpha (Hemiptera: Heteroptera), with emphasis on the Lygaeoidea. *Ann. Entomol. Soc. Am.* **90**(3): 275–301.

HENWOOD, A. 1992a. Exceptional preservation of dipteran flight muscle and the taphonomy of amber. *Palaios* **7**: 203–212.

HENWOOD, A. 1992b. Soft-part preservation of beetles in Tertiary amber from the Dominican Republic. *Palaeontology* **35**(4): 901–912.

HENWOOD, A. 1993a. Ecology and taphonomy of Dominican Republic amber and its inclusions. *Lethaia* **26**: 237–245.

HENWOOD, A. 1993b. Recent plant resins and the taphonomy of organisms in amber: a review. *Modern Geol.* **19**: 35–59.

HEPBURN, H.R. 1969. The skeleto-muscular system of Mecoptera: the head. *Univ. Kansas Sci. Bull.* **48**: 721–765.

HEPPNER, J.B. 1998. Classification of Lepidoptera. Part 1. Introduction. *Holarctic Lepidoptera* **5**, Suppl. 1: 1–148.

HERENDEEN, P.S., LES, D.H., and DILCHER, D.L. 1990. Fossil *Ceratophyllum* (Ceratophyllaceae) from the Tertiary of North America. *Amer. J. Botany* **78**: 1–12.

HERENDEEN, P.S., CREPET, W.L., and NIXON, K.C. 1992. *Chloranthus*-like stamens from the Upper Cretaceous of New Jersey. *Amer. J. Botany* **80**: 865–871.

HERENDEEN, P.S., CREPET, W.L., and NIXON, K.C. 1994. Fossil flowers and pollen of Lauraceae from the Upper Cretaceous of New Jersey. *Plant Syst. Evol.* **189**: 29–40.

HERING, M. 1930. Eine Agromyziden-Mine aus dem Tertiär (Dipt. Agromyzidae). *Deutsche entomol. Zeitschr.* **1930**: 63–64.

HERING, M. 1951. *Biology of the Leaf Mines.* W. Junk, 's-Gravenhage: 420 p.

HERING, M. 1957a. *Bestimmungstabellen der Blattminen von Europa, einschließlich des Mittelmeerbeckens und der Kanarischen Inseln* **1**. W. Junk, 's-Gravenhage: 648 p.

HERING, M. 1957b. *Bestimmungstabellen der Blattminen von Europa, einschließlich des Mittelmeerbeckens und der Kanarischen Inseln* **2**. W. Junk, 's-Gravenhage: 649–1185.

HERING, M. 1957c. *Bestimmungstabellen der Blattminen von Europa, einschließlich des Mittelmeerbeckens und der Kanarischen Inseln* **3**. W. Junk, 's-Gravenhage: 221 p.

HERMAN, A.B. 1989. The S.V. Meyen's hypothesis on gamoheterotopic origin of angiosperms. In: AKHMETYEV, M.A. (ed.). *Paleofloristika i stratigrafiya mezozoya* (Palaeofloristics and Stratigraphy of the Mesozoic). Geol. Institut, Akad. Nauk SSSR, Moscow: 151–153 (in Russian).

HERMAN, A.B., and SPICER, R.A. 1995. Latest Cretaceous flora of Northeastern Russia and the "terminal Cretaceous event" in the Arctic. *Paleontol. J.* **29**(2A): 22–35.

HERNANDEZ, G.S., and RODRIGUEZ, P.M.A. 1978. Rendimientos de resina en pinares plagados y no plagados por *Zadiprion vallicola* Rohwer (Hymenoptera: Diprionidae) en la Meseta Tarasca, Michoacan. *Chapingo* **13/14**: 3–14.

HESPENHEIDE, H.A. 1977. Dispersion and the size composition of the aerial insect fauna. *Ecol. Entomol.* **2**(2): 139–141.

HESSE, M. 1981. The fine structure of the exine in relation to the stickness of angiosperm pollen. *Rev. Palaeobot. Palynol.* **35**(1): 81–82.

HESSE, M. 1982. Zur Mechanik des Pollentransports durch blütenbesuchende Insekten. *Stapfia* **10**: 99–110.

HICKEY, L.J. 1977. Early Cretaceous fossil evidence for angiosperm evolution. *Botan. Rev.* **43**(1): 90–104.

HICKEY, L.J. 1981. Land plant evidence compatible with gradual, not catastrophic, change at the end of the Cretaceous. *Nature* **292**: 529–531.

HICKEY, L.J. 1984. Changes in the angiosperm flora across the Cretaceous-Tertiary boundary. In: BERGGREN, W.A., and VAN COUVERING, J.A. (eds.). *Catastrophes in Earth History: The New Uniformitarianism.* Princeton Univ. Press, Princeton: 279–313.

HICKMAN, V.V. 1937. The embryology of the syncarid crustacean, *Anaspides tasmaniae.* *Pap. & Proc. R. Soc. Tasm.* **1936**: 1–35.

HICKMAN, J.C. 1974. Pollination by ants: a low-energy system. *Science* **184**: 1290–1292.

HILLMER, G., VOIGT, P.C., and WEITSCHAT, W. 1992a. Bernstein im Regenwald von Borneo. *Fossilien* **6**: 336–340.

HILLMER, G., WEITSCHAT, W., and VÁVRA, N. 1992b. Bernstein aus dem Miozän von Borneo. *Naturwiss. Rundschau* **45**(2): 72–74.

HILTERMANN, H. 1968. Gehäuse von Insekten-Larven, insbesondere von Chironomiden, in quartären Sedimenten. *Mitt. Geol. Inst. Techn. Univ. Hannover* **8**: 34–53.

HILTON, J. 1946. Fossils while you wait. *Desert Magaz.* (Aug.): 12.

HIRST, S., and MAULIK, S. 1926. Contribution à l'étude de *Rhyniella praecursor*, Collembole fossile du Devonien. *Rev. Ecol. Biol. Sol* **4**: 497–505.

HITCHKOK, E. 1858. *Ichnology of New England. A Report on the Sandstone of the Connecticut Valley, Especially its Fossil Footmarks, made to the Government of the Commonwealth of Massachusetts.* W. White, Boston: 220 p.

HITCHCOCK, E. 1865. *Supplement of the Ichnology of New England.* Wright & Porter, Boston: 96 p.

HIURA, I. 1982. Fossil pellets? *Nature Study* **28**(11): 127–130 (in Japanese).

HIURA, I., and MIYATAKE, Y. 1974. Mizunami amber and fossil insects. General remarks on Arthropoda. *Bull. Mizunami Fossil Mus.* **1**: 385–392 (in Japanese, with English summary).

HOFFMANN, A. 1997. Adult feeding and reproduction in *Lasiocephala basalis* (Kol.) females (Trichoptera: Lepidostomatidae). *Proc. 8th Internat. Symp. Trichoptera, Minneapolis/St. Paul and Lake Itasca, Minnesota, 9–15 July 1995.* Ohio Biol. Surv.: 145–149.

HOFMANN, W. 1983. Stratigraphy of Cladocera and Chironomidae in a core from a shallow North German lake. *Hydrobiologia* **103**: 235–239.

HOFMANN, W. 1986. Chironomid analysis. In: BERGLUND, B. (ed.). *Handbook of Holocene Palaeoecology and Palaeohydrology.* John Wiley & Sons, New York: 715–727.

HOFMANN, W. 1990. Weichselian chironomid and cladoceran assemblages from maar lakes. *Hydrobiologia* **214**: 207–212.

HOFMANN, W. 1993. Late Glacial/Holocene changes of the climatic and trophic conditions in three Eifel maar lakes, as indicated by faunal remains. II. Chironomidae (Diptera). In: NEGEDANK, J.F., and ZOLITSCHKA, B. (eds.). *Palaeolimnology of European Maar Lakes. Lecture Notes in Earth Sciences* **49**: 421–433.

HOLLAND, E.F., ZINGMARK, R.G., and DEAN, J.M. 1974. Quantitative evidence concerning the stabilization of sediments by marine benthic diatoms. *Marine Biol.* **27**: 191–196.

HOLLAND, G.P. 1964. Evolution, classification, and host relationships of Siphonaptera. *Annu. Rev. Entomol.* **9**: 123–146.

HOLLINGSWORTH, R.G., BLUM, U., and HAIN, F.P. 1991. The effect of adelgid-altered wood on sapwood conductance of Fraser fir Christmas trees. *IAWA Bull.* **12**(3): 235–239.

HOLLOWAY, R.G. 1990. Analysis of a coprolite from James Creek Shelter. *Anthropol. Pap. Univ. Utah* **115**: 313–315.

HOLTZMANN, R.C. 1979. Maximum likelihood estimation of fossil assemblage composition. *Paleobiology* **5**(2): 77–89.

HOLUB, V., and KOZUR, H. 1981. Arthropodenfährten aus dem Rotliegend der CSSR. *Geol. Paläont. Mitteil.* **11**: 95–148.

HOLZAPFEL, E. 1978. Transoceanic aeroplane sampling for organisms and particles. *Pacif. Insects* **18**(3–4): 169–189.

HOLZAPFEL, E.P., CLAGG, H.B., and LEE, G.M. 1978. Trapping of air-borne insects on ships in the Pacific. Pt. 9. *Pacif. Insects* **19**(1–2): 65–90.

HOLZAPFEL, E.P., and HARRELL, J.C. 1968. Transoceanic dispersal studies of insects. *Pacif. Insects* **10**(1): 115–153.

HOLZAPFEL, E.P., and PERKINS, B.D. 1969. Trapping of air-borne insects on ships in the Pacific. Pt. 7. *Pacif. Insects* **11**: 455–476.

HOLZAPFEL, E.P., TSUDA, D.M., and HARRELL, J.C. 1970. Trapping of air-borne insects in the Antarctic area. Pt. 3. *Pacif. Insects* **12**(1): 133–156.

HONG, Y.-C. 1979. On Eocene *Philolimnias* gen. nov. (Ephemeroptera, Insecta) in amber from Fushun Coalfield, Liaoning Prov. *Scientia Sinica* (1): 67–74 (in Chinese; English translation: **22**(3): 331–339).

HONG, Y.-C. 1980a. New genus and species of Mesoblattinidae (Blattodea, Insecta) in China. *Bull. Chinese Acad. Geol. Sci.* (VI) **1**: 49–60 (in Chinese, with English summary).

HONG, Y.-C. 1980b. Fossil insecta. In: *Mesozoic Stratigraphy and Palaeontology of Shaan-Gian-Ning Basin.* Geol. Publ. House, Beijing: 111–114 (in Chinese).

HONG, Y.-C. 1981. *Eocene Fossil Diptera (Insecta) in Amber of Fushun Coalfield.* Geol. Publ. House, Beijing: 166 p. (in Chinese with English summary).

HONG, Y.-C. 1982a. *Mesozoic Fossil Insects of Jiuquan Basin in Gansu Province.* Geol. Publ. House, Beijing: 188 p. (in Chinese).

HONG, Y.-C. 1982b. Fossil Haglidae (Orthoptera) in China. *Scientia sinica* **25**: 1118–1129.

HONG, Y.-C. 1983a. *Middle Jurassic Fossil Insects in North China.* Geol. Publ. House, Beijing: 223 p. (in Chinese, with English summary).

HONG, Y.-C. 1983b. Fossil Blattods (Insecta) from Linjia Formation of Benxi in Liaoning Province and discussion about their age. *Bull. Shenyang Inst. Geol. Miner. Res.* **8**: 57–61 (in Chinese, with English summary).

HONG, Y.-C. 1984. Insecta. In: *Palaeontological Atlas of North China. II. Mesozoic volume.* Geol. Publ. House, Beijing: 128–185 (in Chinese).

HONG, Y.-C. 1985a. Insecta. In: *Palaeontological atlas of North China. I. Paleozoic volume.* Geol. Publ. House, Beijing: 489–510 (in Chinese).

HONG, Y.-C. 1985b. New fossil genera and species of Shanxi Formation in Xishan of Taiyuan. *Entomotaxonomia* **7**: 85–91 (in Chinese).

HONG, Y.-C. 1986. New fossil insects of Haifanggou Formation, Liaoning Province. *J. Changchun Coll. Geol.* **4**: 10–16 (in Chinese, with English summary).

HONG, Y.-C. 1992a. On the homonyms of four generic names for fossil insects. *Mem. Beijing Natur. Hist. Mus.* **3**: 51–53 (in Chinese with English summary).

HONG, Y.-C. 1992b. The origin, development, flourish and disappearance of the Late Mesozoic Jehol biota in Eastern Asian paleocontinent. *19 Internat. Congr. Entomol., Beijing, June 28–July 4, 1992. Proc.: Abstr.* Beijing: 30.

HONG, Y.-C. 1998. Establishment of fossil entomofaunas and their evolutionary succession in North China. *Entomol. Sinica* **5**(4): 283–300.

HONG, Y.-C., and CHANG, J. 1989. A new family Jilinoraphidiidae of Raphidioptera (Insecta). *Geoscience* **3**(3): 290–296.

HONG, Y.-C., and WANG, W. 1990. Insecta. In: *Palaeontological atlas of North China. Inner Mongolia 2. Mesozoic and Cenozoic.* Science Press, Beijing,: 81–87 (in Chinese).

HONG, Y.-C., and WANG, W. 1988. Sinolestinae, an Early Cretaceous new subfamily of Platypezidae (Insecta, Diptera) from Laiyang basin, Shandong province. *Geoscience* **2**(3): 386–392.

HONG, Y.-C., and WANG, W. 1990. Fossil insects from the Laiyang basin, Shandong province. *Stratigraphy and Palaeontology of Laiyang Basin, Shandong.* Geol. Publ. House, Beijing: 44–189 (in Chinese, with English summary).

HONG, Y.-C., YANG, T.-C., WANG, S.-T., SUN, H.-C., Tu, N.-C., and SUN, M.-R. 1980. *The Study of Stratigraphy and Faunas of Fushun Coalfield, Liaoning Province.* Science Press, Beijing: 98 p. (in Chinese with English summary).

HONG, Y.-C., YAN, D., and WANG, D. 1989. Discovery of Early Cretaceous *Cretacochorista* gen. nov. Insecta: Mecoptera from Jiuquan Basin, Gansu Province. *Mem. Beijing Nat. Hist. Mus.* **44**(9): 1–9 (in Chinese, with English summary).

462

HOOD, M.A., and MYERS, S.P. 1974. The biology of aquatic chitinoclastic bacteria and their chitinolytic activity. *Bull. Soc. franco-japan. d'océanogr.* **11**: 213–229.

HOOKER, J.J., COLLINSON, M.E., VAN BERGEN, P.F., SINGER, R.L., DE LEEUW, J.W., and JONES, T.P. 1995. Reconstruction of land and freshwater palaeoenvironments near the Eocene-Oligocene boundary, southern England. *J. Geol. Soc.* (London) **152**: 449–468.

HOPKIN, S.P., and READ, H.J. 1992. *The Biology of Millipeds*. Oxford Univ. Press, Oxford: 233 p.

HOPKINS, P.M., MATTHEWS, J.V., Jr., WOLFE, J.A., and SILBERMAN, M.L. 1971. A Pliocene flora and insect fauna from the Bering Strait region. *Palaeogeogr. Palaeoclimatol., Palaeoecol.* **9**: 211–231.

HORNE, P. 1979. Head lice from an Aleutian mummy. *Palaeopathol. Newsletter* **25**: 7–8.

HÖRNSCHEMEIER, Th. 1999. Fossil insects from the Lower Permian of Niedermoschel (Germany). *Proc. First Palaeoentomol. Conf., Moscow 1998*. AMBA projects Internat., Bratislava: 57–59.

HÖRNSCHEMEIER, Th., and STAPF, H. 2001. Review of Blattinopsidae (Protorthoptera) with description of new species from the Lower Permian of Niedermoschel (Germany). *Neues Jahrb. Geol. Paläontol. Abhandl.* **221**: 81–109.

HÖRNSCHEMEYER, T., and WEDMANN, S. 1994. Fossile Prachtkäfer (Coleoptera: Buprestidae: Buprestinae) aus dem Mitteleozän der Grube Messel bei Darmstadt, Teil 1. Willi Ziegler-Festschrift III. *Courier Forsch.-Inst. Senckenberg* **170**: 85–136.

HOTCHKISS, A.T. 1958. Pollen and pollination in the Eupomatiaceae. *Proc. Linn. Soc. N.S.Wales* **83**: 86–91.

HOTTON, C.L., HUEBER, F.M., and LABANDEIRA, C.C. 1996. Plant-arthropod interactions from early terrestrial ecosystems: two Devonian examples. In: REPETSKI, J.E. (ed.). *Sixth North American Paleontol. Convention Abstr. of Papers, Smithsonian Institution, Washington, D.C., June 9–12, 1996*. Paleontol. Soc. Spec. Publ. **8**: 181.

HOUARD, C. 1908. *Les zoocecidies des plantes d'Europe et du bassin de la Méditerranée* **1–2**. Paris: 1248 p.

HOUARD, C. 1913. *Les zoocecidies des plantes d'Europe et du bassin de la Méditerranée* **3**. Paris: 1560 p.

HOUARD, C. 1922. *Les zoocecidies des plantes d'Afrique, d'Asie et d'Océanie* **1**. Paris: 499 p.

HOUARD, C. 1923. *Les zoocecidies des plantes d'Afrique, d'Asie et d'Océanie* **2**. Paris: 500–1057.

HOUSTON, T.F. 1987. Fossil brood cells of stenotritid bees (Hymenoptera, Apoidea) from the Pleistocene of South Australia. *Trans. R. Soc. S. Australia* **111**(1–2): 93–97.

HOWARD, J.J. 1985. Observations on resin collecting by six interacting species of stingless bees (Apidae: Meliponinae). *J. Kansas Entomol. Soc.* **58**(2): 337–345.

HOWDEN, H.F. 1977. Beetles, beach drift, and island biogeography. *Biotropica* **9**(1): 53–57.

HOWES, F.N. 1949. Vegetable gums and resins. *Chronica Botan.* **22**: 1–188.

HOWLAND, D.E., and HEWITT, G.M. 1994. DNA analysis of extant and fossil beetles. *Biomolecular Palaeontology: Lyall Meeting Volume. NERC Earth Sciences Directorates Spec. Publ.* **94/1**: 49–51.

HSU, K.J., MENTADERT, L., BERNOUILLI, D., CITA, M.B., ERICKSON, A., GARRISON, R.E., KIDD, R.B., MELIERES, F., MULLER, C., and WRIGHT, R. 1977. History of the Mediterranean salinity crisis. *Nature* **267**: 399–403.

HUBBARD, M.D. 1987. Ephemeroptera. In: WESTPHAL, F. (ed.). *Fossilium Catalogus* **1**: *Animalia*. Pars 129. Kugler Publ., Amsterdam: 99 p.

HUBBARD, M.D., and KUKALOVÁ-PECK, J. 1980. Permian mayfly nymphs: new taxa and systematic characters. In: FLANNAGAN, J.F., and MARSHALL, K.E. (eds.). *Advances in Ephemeroptera Biology (Proc. 3rd Internat. Conf. Ephemeroptera)*. Plenum Publ. Corp.: 19–31.

HUBBARD, M.D., and SAVAGE, H.M. 1981. The fossil Leptophlebiidae (Ephemeroptera): a systematic and phylogenetic review. *J. Paleontol.* **55**: 810–813.

HUCHET, J.-B. 1995. Insectes et momies égyptiennes. *Bull. Soc. linn. Bordeaux* **24**(3): 29–39.

HUCKE, K., and VOIGT, E. 1967. *Einführung in die Geschiebeforschung*. Nederlandse Geol. Vereinigung, Oldenzaal: 132 p.

HUDON, C. 1994. Biological events during ice breakup in the Great Whale River (Hudson Bay). *Canad. J. Fish. Aquat. Sci.* **51**: 2467–2481.

HUGHES, N.F., and SMART, J. 1967. Plant-insect relationships in Paleozoic and later time. In: Harland, W.B., Holland, C.H., House, M.R., Hughes, N.F., Reynolds, A.B., Rudwick, M.J.S., Satterthwaite, G.E., Tarlo, A.B.H., and Willey, E.C. (eds.). *The Fossil Record*. Geol. Soc., London: 107–117.

HUNT, A.P., LOCKLEY, M.G., LUCAS, S.G., MacDONALD, J.P., HATTON, N., and KRAMER, J. 1993. Early Permian tracksites in the Robledo Mountains, South-Central New Mexico. *Bull. New Mexico Mus. Natur. Hist. Science* **2**: 23–31.

HURD, P.D., SMITH, R.F., and DURHAM, J.W. 1962. The fossiliferous amber of Chiapas, Mexico. *Ciencia* **21**: 107–118.

HURST, G.W. 1969. Insect migrations to the British Isles. *Quart. J. R. Meteorol. Soc.* **95**(404): 435–439.

HUTCHESON, J. 1992. Observations on the effects of volcanic activity on insects and their habitat on White Island. *New Zeal. Entomol.* **15**: 72–76.

HYNES, H.B.N. 1970. *The Ecology of Running Waters*. Univ. Toronto Press, Toronto: 555 p.

HYNES, J. 1975. Downstream drift of invertebrates in a river in southern Ghana. *Freshwater Biol.* **5**(6): 515–531.

ICBN: STAFLEU, F.A. (ed.). 1980. *International Code of Botanical Nomenclature*. Nauka Press, Leningrad: 284 p. (Russian edition has been consulted to).

ICZN 1985: RIDE, W.D.L., SABROSKY, C.W., BERNARDI, G., and MELVILLE, R.V. (eds.) 1985. *International Code of Zoological Nomenclature*. Internat. Trust for Zool. Nomencl. and Univ. of California Press, London, Berkeley and San Francisco: 338 p.

ILLIES, J. 1967. Megaloptera und Plecoptera (Ins.) aus den jungpliocänen Süsswassermergeln fon Willershausen. *Ber. naturhist. Ges. Hannover* **111**: 47–55.

ILYINSKAYA, I.A. 1958. Fossil monotopic and polytopic floras and assemblages. *Doklady Akad. Nauk SSSR* **119**: 797–799 (in Russian).

INGOLD, C.T. 1953. *Dispersal in Fungi*. Oxford Univ. Press, Oxford (cited after Russian translation: *Puti i sposoby rasprostraneniya gribov*. Izdatel'stvo Inostrannoy Literatury, Moscow, 1957: 182 p.).

IOVINO, A.J., and BRADLEY, W.H. 1969. The role of larval Chironomidae in the production of lacustrine copropel in Mud Lake, Marion County, Florida. *Limnol. Oceanogr.* **14**(6): 898–905.

IRVINE, J.R. 1985. Effects of successive flow perturbations on stream invertebrates. *Canad. J. Fish. Aquat. Sci.* **42**: 1922–1927.

IRWIN, M.E., and ISARD, S.A. 1992. The effect of vertical zonation on weakly flying insects on the direction of long distance movement in the planetary boundary layer. *19th Internat. Congr. Entomol., Beijing, June 28–July 4, 1992. Proc.: Abstracts*. Beijing: 152.

ISAEV, A.S. 1966. The importance of resinous substances of conifers in the consortion of tree and xylophagous insects. *Voprosy zoologii: Materialy 3 Soveshchaniya zoologov Sibiri* (Problems of Zoology: Materials 3rd Meet. Siberian Zoologists). Tomsk: 59–61 (in Russian).

ISAEV, A.S., and GIRS, G.I. 1975. *Vzaimodeystvie dereva i nasekomykh-ksilofagov* (Interactions between Tree and Xylophagous Insects). Nauka, Novosibirsk, 344 p. (in Russian).

ITURRALDE, M.A., AND MACPHEE, R.D.E. 1996. Age and paleogeographical origin of Dominican amber. *Science* **273**: 1850–1852.

IVANOV, V.D. 1985. Comparative analysis of wing kinematics in caddisflies. *Entomol. Obozrenie* **64**: 273–284 (in Russian).

IVANOV, V.D. 1988. The structure of the Paleozoic caddis flies of the family Microptysmatidae (Insecta). *Paleontol. Zhurnal* (3): 64–69 (in Russian, English translation: *Paleontol. J.* **22**(3): 63–69).

IVANOV, V.D. 1990. Comparative analysis of flight aerodynamics in caddisflies (Insecta, Trichoptera). *Zool. Zhurnal* **69**(2): 46–60 (in Russian).

Ivanov, V.D. 1992. New family of caddisflies from Paleozoic of the Middle Urals (Insecta, Trichoptera). *Paleontol. Zhurnal* (4): 31–35 (in Russian, English translation: *Paleontol. J.* **26**(4): 36–41).

IVANOV, V.D. 1997. Rhyacophiloidea: a paraphyletic taxon. *Proc. 8th Internat. Symp. Trichoptera. Minneapolis/St. Paul and Lake Itasca, Minnesota, 9–15 July 1995*. Ohio Biol. Surv.: 189–193.

IVANOV, V.D., and MENSHUTKINA, T.V. 1996. Endemic caddisflies of the lake Baikal (Trichoptera, Apataniidae). *Braueria* **23**: 13–28.

IVANOVA-KAZAS, O.M. 1961. *Ocherki po sravnitel'noy embriologii pereponchatokrylykh* (Essays on the Comparative Embryology of Hymenoptera). Leningrad Univ. Press, Leningrad: 266 p. (in Russian).

JACKSON, J.K., and RESH, V.H. 1989. Distribution and abundance of adult aquatic insects in the forest adjacent to a northern California stream. *Environm. Entomol.* **18**: 278–283.

JANZEN, D.H. 1969. Seed-eaters versus seed size, number, toxicity and dispersal. *Evolution* **23**: 1–27.

JANZEN, D.H. 1971. Seed predation by animals. *Annu. Rev. Ecol. Syst.* **2**: 465–492.

JANZEN, D.H. 1975. Behaviour of *Hymenaea courbaril* when its pre-dispersal seed predator is absent. *Science* **189**: 145–147.

JANZEN, D.H. 1979. How many babies do figs pay for babies? *Biotropica* **11**(1): 48–50.

JANZEN, D.H. 1980. When is it co-evolution? *Evolution* **34**: 611–612.

JARZEMBOWSKI, E.A. 1980. Fossil insects from the Bembridge Marls, Palaeogene of the Isle of Wight, Southern England. *Bull. Brit. Mus.* (*Nat. Hist.*) *Geol. Ser.* **33**: 237–293.

JARZEMBOWSKI, E.A. 1981. An early Cretaceous termite from southern England (Isoptera: Hodotermitidae). *Syst. Entomol.* **6**: 91–96.

JARZEMBOWSKI, E.A. 1984. Early Cretaceous insects from Southern England. *Modern Geol.* **9**: 71–93.

JARZEMBOWSKI, E.A. 1987. The occurrence and diversity of Coal Measure insects. *J. Geol. Soc., London* **144**: 507–511.

JARZEMBOWSKI, E.A. 1989. A century plus of fossil insects. *Proc. Geol. Assoc.* **100**(4): 433–449.

JARZEMBOWSKI, E.A. 1990. A boring beetle from the Wealden of the Weald. In: BOUCOT, A.J. (ed.). 1990. *Palaeobiology of Behaviour and Co-evolution*. Elsevier, Amsterdam: 373–376.

JARZEMBOWSKI, E.A. 1992. A provisional checklist of fossil insects from the Purbeck Beds of Dorset. *Proc. Dorset Nat. Hist. & Archaeol. Soc.* **114**: 175–179.

JARZEMBOWSKI, E.A. 1994. Fossil cockroaches or pinnule insects? *Proc. Geol. Assoc.* **105**: 305–311.

JARZEMBOWSKI, E.A. 1995a. Fossil caddisflies (Insecta: Trichoptera) from the Early Cretaceous of southern England. *Cretaceous Res.* **16**: 695–703.

JARZEMBOWSKI, E.A. 1995b. Early Cretaceous insect faunas and paleoenvironment. *Cret. Res.* **16**: 681–693.

JARZEMBOWSKI, E.A. 1999a. British amber: a little-known resource. *Estud. Mus. Cienc. Natur. Alava* **14** (Num. Espec. 2): 133–140.

JARZEMBOWSKI, E.A. 1999b. Arthropods 2: Insects. In: SWIFT, A., and MARTILL, D.M. (eds.). *Fossils of the Rhaetian Penarth Group. Palaeontol. Association Field Guides to Fossils, no. 9*. Palaeontol. Association, London: 149–160.

JARZEMBOWSKI, E.A., and CORAM, R. 1998. New fossil insect records from the Purbeck of Dorset and the Wealden of the Weald. *Proc. Dorset Nat. Hist. & Archaeol. Soc.* (1997) **118**: 119–124.

JARZEMBOWSKI, E.A., and ROSS, A.J. 1993. Time flies: the geological record of insects. *Geology today* (Nov.–Dec.): 218–223.

JARZEMBOWSKI, E.A., and ROSS, A.J. 1996. Insect origination and extinction in the Phanerozoic. In: Hart, M.B. (ed.). *Biotic Recovery from Mass Extinction Events. Geol. Soc. Spec. Publ.* **102**: 65–78.

JELL, P.A., and DUNCAN, P.M. 1986. Invertebrates, mainly insects, from the freshwater, Lower Cretaceous, Koonwarra Fossil Bed (Korumburra Group), South Gippsland, Victoria. In: JELL, P.A., and ROBERTS, J. Plants and Invertebrates from the Lower Cretaceous Koonwarra Fossil Bed, South Gippsland, Victoria. *Mem. Ass. Australas. Palaeontol.* **3**: 111–205.

JELL, P.A., and ROBERTS, J. 1986. *Plants and invertebrates from the Lower Cretaceous Koonwarra Fossil Bed, South Gippsland, Victoria. Mem. Ass. Australas. Palaeontol.* **3**: 205 p.

JENNINGS, J.R. 1974. Lower Pennsylvanian plants of Illinois. I: A flora from the Pounds Sandstone Member of the Casyville Formation. *J. Paleontol.* **48**(3): 469–479.

JESSEN, K. 1920. Moseundersøgelser i det nordøstlige Sjælland. *Danm. Geol. Unders.* R.**2**, **34**: 1–239.

JOHNSON, C.D. 1981. Interactions between bruchid (Coleoptera) feeding guilds and behavioural patterns of pods of the Leguminosae. *Envir. Entomol.* **10**: 249–253.

JOHNSON, C.G. 1957. The distribution of insects in the air and the empirical relation of density to height. *J. Anim. Ecol.* **26**: 479–494.

JOHNSON, C.G. 1963. Physiological factors in insect migration by flight. *Nature* **198**: 423–427.

JOHNSON, C.G. 1967. The progress of research at Rothamstead on aerial migrations of insects. *Biometeorology* **2**, Pt. 2. Pergamon Press, Oxford *etc.*: 570–572.

JOHNSON, C.G. 1969. *Migration and Dispersal of Insects by Flight.* Methuen & Co Ltd., London: 763 p.

JOHNSON, D.E. 1931. Some observations on chitin destroying bacteria. *J. Bacteriol.* **24**: 335.

JOHNSON, E.W., BRIGGS, D.E.G., SUTHREN, R.J., WRIGHT, J.L., and TUNNCLIFF, S.P. 1994. Non-marine arthropod traces from the subaerial Ordovician Borrowdale Volcanic Group, English Lake District. *Geol. Magaz.* **131**: 395–406.

JOHNSON, K.R., and HICKEY, L.J. 1990. Megafloral change across the Cretaceous/Tertiary boundary in the northern Great Plains and Rocky Mountains, U.S.A. *Geol. Soc. America Spec. Paper* **247**: 433–444.

JOHNSON, S.Y. 1992. Insect migration, weather, and climate in the Americas. *19th Internat. Congr. Entomol., Beijing, June 28–July 4, 1992. Proc.: Abstracts.* Beijing: 151.

JOHNSTON, J.E. 1993. Insects, spiders, and plants from the Tallahatta Formation (Middle Eocene) in Benton County, Mississippi. *Mississippi Geol.* **14**(4): 71–82.

JOHNSTON, J.E. 1999. Caddisfly cases from the Middle Eocene (Lower Lutetian) of Mississippi, USA. *Proc. First Palaeoentomol. Conf., Moscow 1998.* AMBA projects Internat., Bratislava: 61–64.

JOHNSTON, J.E., and BORKENT, A. 1998. *Chaoborus* Lichtenstein (Diptera: Chaoboridae) pupae from the middle Eocene of Mississippi. *J. Paleontol.* **72**(3): 491–493.

JOHNSTON, P.A., EBERTH, D.A., and ANDERSON, P.K. 1996. Alleged vertebrate eggs from Upper Cretaceous redbeds, Gobi Desert, are fossil insect (Coleoptera) pupal chambers: *Fictovichnus* new ichnogenus. *Canad. J. Earth Sci.* **33**: 511–525.

JOHNSTON, R.M. 1888. *A Systematic Account on the Geology of Tasmania.* Hobart, Government Printing: 408 p.

JOLIVET, P. 1986. *Insects and Plants. Parallel Evolution and Adaptations. Flora & Fauna Handbook* **2**. E.J. Brill Flora & Fauna Publications, New York: 197 p.

JOŃCA, W. 1979. Współczesne psedomorfozy zoogeniczne na Pogórzu Walbrzyskim. *Czas. geogr.* **50**(1–2): 65–73.

JONES, T. 1954. The external morphology of *Chirothrips hamatus* (Trybom) (Thysanoptera). *Trans. R. Entomol. Soc. London* **105**: 163–187.

JOOST, W. 1984. Fossile Reste aquatischer Insekten aus dem Travertin von Weimar. *Abhandl. Ber. Inst. Quartärpaläontol. Weimar* **5**: 321–324.

JORDAN, K. 1888. Anatomie und Biologie der Physopoda. *Zeitschr. wiss. Zool.* **47**: 541–620.

JORDAN, K.H.C. 1967. Wanzen aus dem Pliozän von Willershausen. *Ber. Naturhist. Ges. Hannover* **111**: 77–90.

JURASKY, K.A. 1932. Frassgänge und Koprolithen eines Nagekäfers in liassischer Steinkohle. *Zeitschr. Deutsche Geol. Ges.* **84**: 656–657.

KAISILA, J. 1952. Insects from arctic moutain snows. *Ann. entomol. fennici* **18**: 8–25.

KALANDADZE, N.N., and RAUTIAN, A.S. 1983. The place of Central Asia in zoogeographical history of the Mesozoic. In: TATARINOV, L.P. (ed.). *Iskopaemye reptilii Mongolii* (Fossil Reptiles of Mongolia). *Trudy Sovmestnoy Sovetsko-Mongol'skoy Paleontol. Expeditsii* **24**. Nauka, Moscow: 6–44 (in Russian with English summary).

KALANDADZE, N.N., and RAUTIAN, A.S. 1993a. The Jurassic ecological crisis and a heuristic model of the attended evolution of a community and its biota. In: ROZANOV, A.Y. (ed.). *Problemy doantropogenovoy evolyutsii biosfery* (The Problems of Pre-anthropogenous Biosphere Evolution). Nauka, Moscow: 60–95 (in Russian).

KALANDADZE, N.N., and RAUTIAN, A.S. 1993b. Symptoms of ecological crises. *Stratigrafiya. Geol. korrelyatsiya* **1**(5): 473–478 (in Russian, English translation: *Stratigraphy and Geol. Correlation* **1**(5): 473–484).

KALTENBACH, A. 1968. *Embiodea (Spinfüsser).* In: Kukenthal, W. (ed.). *Handbuch der Zoologie* **4**(2) 2/8. Walter de Gruyter, Berlin: 29 p.

KALTENBACH, A. 1978. *Mecoptera (Schnabelhafte, Schnabelfliegen).* In: Kukenthal, W. (ed.). *Handbuch der Zoologie* **4**(2) 2/28. Walter de Gruyter, Berlin: 111 p.

KALUGINA, N.S. 1974. Changes in the subfamilian composition of midges (Diptera, Chironomidae) as indicator of probable eutrophication of the Late Mesozoic water bodies. *Byulleten' Moskovskogo Obshchestva Ispytateley Prirody, Biol.* **79**: 45–55 (in Russian).

KALUGINA, N.S. (ed.) 1980a. *Rannemelovoe ozero Manlay* (Early Cretaceous Lake Manlay). *Trudy Sovmestnoy Sovetsko-Mongol'skoy Paleontol. Expeditsii* **13**. Nauka, Moscow: 92 p. (in Russian).

KALUGINA, N.S. 1980b. Insects in fresh water ecosystems of the past. In: ROHDENDORF, B.B., and RASNITSYN, A.P. (eds.). 1980. *Istoricheskoe razvitie klassa nasekomykh* (Historical Development of the Class Insecta). *Trudy Paleontol. Inst. Akad. Nauk SSSR* **175**. Nauka, Moscow: 224–240 (in Russian).

KALUGINA, N.S. 1986. Dipterans. Muscida (=Diptera). Infraorders Tipulomorpha and Culicomorpha. In: RASNITSYN, A.P. (ed.). *Nasekomye v rannemelovykh ekosistemakh Zapadnoy Mongolii* (Insects in the Early Cretaceous Ecosystems of West Mongolia). *Trudy Sovmestnoy Sovetsko-Mongol'skoy Paleontol. Expeditsii* **28**. Nauka, Moscow: 112–125 (in Russian).

KALUGINA, N.S. 1989. New Mesozoic psychodomorph dipteran insects from of Siberia (Diptera: Eoptychopteridae, Ptychopteridae). *Paleontol. Zhurnal* (1): 65–77 (in Russian, English translation: *Paleontol. J.* **23**(1): 62–73).

KALUGINA, N.S. 1991. New Mesozoic Simuliidae and Leptoconopidae and the origin of bloodsucking in the lower dipteran insects. *Paleontol. Zhurnal* (1): 69–80 (in Russian, English translated in *Paleontol. J.* **25**(1): 66–77).

KALUGINA, N.S. 1993. Chaoborids and non-biting midges from the Mesozoic of Eastern Transbaikalia (Diptera: Chaoboridae and Chironomidae). In: PONOMARENKO, A.G. (ed.), *Mezozoiskie nasekomye i ostrakody Azii* (Mesozoic Insects and Ostracods of Asia). *Trudy Paleontol. Inst. Ross. Akad. Nauk* **252**: 117–139 (in Russian).

KALUGINA, N.S., and KOVALEV, V.G. 1985. *Dvukrylye nasekomye yury Sibiri* (Diptera in Jurassic of Siberia). Nauka, Moscow: 199 p. (in Russian).

KALUGINA, N.S., and ZHERIKHIN, V.V. 1975. Changes in insect limnofauna in the Mesozoic and Cainozoic and their ecological interpretation. *Istoriya ozer v mezozoe, paleogene i neogene.* (Lake History in the Mesozoic, Palaeogene and Neogene.) *Tezisy dokladov IV Vsesoyuznogo simpoziuma po istorii ozyor* (Abstracts of IV All-Union Symposium on Lake History) **1**. Leningrad: 55–61 (in Russian).

KAMPMANN, H. 1983. Ein Insekten-Exkrement aus Sporen aus dem unterkretazischen Saurlager von Nehden (Sauerland, Westfalen). *Paläontol. Zeitschr.* **57**: 75–77.

KARBAN, R. 1980. Periodical cicada nymphs impose periodical oak tree wood accumulation. *Nature* **287**: 326–327.

KARNY, H. 1921. Zur Systematik der orthopteroiden Insekten. III. Thysanoptera. *Treubia* **1**: 211–261.

KARNY, H. 1922. Zur Phylogenie der Thysanopteren. *Treubia* **3**: 29–37.

KARTASHEV, N.N., and KRYZHANOVSKY, O.L. 1954. On mass insect burial at saline water bodies at Western Uzboy. *Byulleten' Moskovskogo Obshchestva Ispytateley Prirody, Biol.* **59**(2): 31–32 (in Russian).

KASIMOV, N.S. 1988. *Geokhimiya stepnykh i pustynnykh landshaftov* (Geochemistry of Steppe and Desert Landscapes). Moscow Univ., Moscow: 252 p. (in Russian).

KASPARYAN, D.R. 1988a. A new subfamily and two new genera of ichneumonid wasps (Hymenoptera: Ichneumonidae) from the Baltic amber. *Trudy Zool. Inst. Akad. Nauk SSSR* **175**: 38–43 (in Russian).

KASPARYAN, D.R. 1988b. New taxa of ichneumonoid wasps of the family Paxylommatidae (Hymenoptera, Ichneumonoidea) from the Baltic amber. *Trudy Vsesoyuznogo Entomol. Obshchestva* **70**: 125–131 (in Russian).

KASPARYAN, D.R. 1994. A review of the ichneumonid wasps of the subfamily Townesitinae subfam. nov. (Hymenoptera, Ichneumonidae) from Baltic ambers. *Paleontol. Zhurnal* (4): 86–96 (in Russian, English translation: *Paleontol. J.* **28**(4): 114–126).

KATHIRITHAMBY, J. 1991. Strepsiptera. In: NAUMANN, D. (ed.). *The Insects of Australia.* 2nd ed. Melbourne Univ. Press, Carlton: 684–695.

KATHIRITHAMBY, J., and GRIMALDI, D.A. 1993. Remarkable stasis in some Lower Tertiary parasitoids: descriptions, new records, and review of Strepsiptera in the Oligo-Miocene amber of the Dominican Republic. *Entomol. Scand.* **24**: 31–41.

KATO, M., INOUE, T., and NAGAMITSU, T. 1995. Pollination biology of *Gnetum* (Gnetaceae) in a lowland mixed dipterocarp forest in Sarawak. *Amer. J. Botany* **82**: 862–868.

KAUFFMAN, E.G. 1979. Benthic environments and paleoecology of the Posidonienschiefer (Toarcian). *Neues Jahrb. Geol. Paläontol. Abhandl.* **157**: 18–36.

KAVANNAUGH, D.H. 1986. A systematic review of amphizoid beetles (Amphizoidae: Coleoptera) and their phylogenetic relationships to other Adephaga. *Proc. Calif. Acad. Sci.* **44**(6): 67–109.

KAWALL, J.H. 1876. Organische Einschlüsse im Bergkrystall. *Bull. Soc. Natural. Moscou* **1**(3): 170–173.

KAY, P.T., KING, J.D., and HASIOTIS, S.T. 1997. Petrified Forest Park Upper Triassic trace fossils yield biochemical evidence of phylogenetic link to modern bees (Hymenoptera, Apoidea). *Geol. Soc. Amer. Abstracts, Oct. 20, 1997:* A13168.

KEIGHERY, G.J. 1975. Pollination of *Hibbertia hypericoides* (Dilleniaceae) and its evolutionary significance. *J. Natur. Hist.* **9**(5): 681–684.

KEILBACH, R. 1982. Bibliographie und Liste der Arten tierischer Einschlusse in fossilen Harzen sowie ihrer Aufbewahrungsorte. Teil 1, 2. *Deutsche entomol. Zeitschr. N.F.* **29**: 129–286, 301–391.

KELBER, K.P., and GEYER, G. 1989. Lebensspuren von Insekten an Pflanzen des Unteren Keuper. *Courier Forsch.-Inst. Senckenberg* **109**: 165–174.

KÉLER, S., VON. 1969. *Mallophaga (Federlinge und Haarlinge).* In: Kukenthal, W. (ed.). *Handbuch der Zoologie* **4**(2) 2/17, Walter de Gruyter, Berlin: 72 p.

KELLER, E.A., and SWANSON, F.J. 1979. Effects of large organic debris material on channel form and fluvial processes. *Earth Surface Process* **4**: 361–380.

KELLER, O. 1913. *Die Antike Tierwelt.* Bd.2. Vögel, Reptilien, Fische; Insekten, Spinnentiere, Tausendfüssler, Krebstiere, Würmer, Weichtiere, Stachelhäuter, Schauchtiere. Engelmann, Leipzig: 618 p.

KELLY, S.R.A., WANG, Y.-G., and ZHANG, J. 1994. A revised Cretaceous age for ammonites originally identified as Middle Jurassic from Eastern Heilongjiang, China. *Acta Palaeontol. Sinica* **33**(4): 509–517.

KENIG, F., SINNINGHE DAMSTÉ, J.S., DE LEEUW, J.W., and HAYES, J.M. 1994. Molecular palaeontological evidence for food-web relationships. *Naturwissenschaften* **81**: 128–130.

KENWARD, H.K. 1975. Pitfalls in the environmental interpretation of insect death assemblages. *J. Archaeol. Sci.* **2**: 85–94.

KENWARD, H.K. 1978. The value of insect remains as evidence of ecological conditions on archaeological sites. In: BROTHWELL, D.R., THOMAS, K.D., and CLUTTON-BROCK, J. (eds.). *Research Problems in Zooarchaeology. Inst. Archaeol. Univ. London Occas. Publ.* **3**: 25–38.

KERZHNER, I.M. 1981. *Poluzhestkokrylye semeystva Nabidae* (Heteropterans of the family Nabidae). *Fauna SSSR* (N.S.). Nasekomye poluzhestkokrylye **13**, 2. Nauka, Leningrad: 327 p. (in Russian).

KEVAN, D.K. 1965. *Miopyrgomorpha fischeri* (Heer) – a fossil pyrgomorphid bush-hopper (Orthoptera: Acridoidea). *Mitt. Schweiz. Entomol. Ges.* **38**(1–2): 66–70.

KEVAN, D.K.McE. 1977. Suprafamilial classification of "Orthopteroid" and related insects; a draft scheme for discussion and consideration. *Mem. Lyman Entomol. Mus. Res. Lab.* **4**(1): 1–44.

KEVAN, D.K.McE., and WIGHTON, D.C. 1981. Paleocene orthopteroids from south-central Alberta, Canada. *Canad. J. Earth Sci.* **18**: 1824–1837.

KEVAN, P.G., CHALONER, W.G., and SAVILLE, D.B.O. 1975. Interrelationships of early terrestrial arthropods and plants. *Palaeontology* **18**: 391–417.

KEY, K.H.L. 1991. Phasmatodea. In: NAUMANN, D. (ed.). *The Insects of Australia.* 2nd ed. Melbourne Univ. Press, Carlton: 394–404.

KIESTER, A.R., and STRATES, E. 1984. Social behaviour in a thrips from Panama. *J. natur. Hist.* **18**: 303–314.

KIM, K.C. 1985. Evolution and host associations of Anoplura. In: KIM, K.C. (ed.) *Co-evolution of Parasitic Arthropoda and Mammals.* John Wiley & Sons, New York: 197–231.

KIM, K.C., and LUDWIG, H.W. 1978a. The family classification of the Anoplura. *Syst. Entomol.* **3**: 249–284.

KIM, K.C., and LUDWIG, H.W. 1978b. Phylogenetic relationships of parasitic Psocodea and taxonomic position of Anoplura. *Ann. Entomol. Soc. Amer.* **71**: 911–922.

KIM, K.C., and LUDWIG, H.W. 1982. Parallel evolution, cladistics, and classification of parasitic Psocodea. *Ann. Entomol. Soc. Amer.* **75**: 537–548.

KINGSOLVER, J.M. 1965. A new fossil bruchid genus and its relationships to modern genera (Coleoptera: Bruchidae: Pachymerinae). *Coleopterists' Bull.* **19**: 25–30.

KINZELBACH, R.K. 1979. Das erste neotropische Fossil der Fächerflügler (Stuttgarter Bernsteinsammlung: Insecta: Strepsiptera). *Stuttg. Beitr. Naturk.* **B 52**: 1–14.

KINZELBACH, R.K. 1990. The systematic position of Strepsiptera (Insecta). *Amer. Entomol.* **36**: 292–303.

KINZELBACH, R.K., and LUTZ, H. 1985. Stylopid larva from the Eocene – a spotlight on the phylogeny of the stylopids (Strepsiptera). *Ann. Entomol. Soc. Amer.* **78**(5): 600–602.

KINZELBACH, R., and POHL, H. 1994. The fossil Strepsiptera (Insecta: Strepsiptera). *Ann. Entomol. Soc. Amer.* **87**(1): 59–70.

KIREJTSHUK, A.G. 1994. Parandrexidae fam. nov. – Jurassic beetles of the infraorder Cucujiformia (Coleoptera, Polyphaga). *Paleontol. Zhurnal* (1): 57–64 (in Russian, English translation: *Paleontol. J.* **28**(1): 69–78).

KIRICHKOVA, A.I., and DOLUDENKO, M.P. 1996. New data on the Jurassic phytostratigraphy in Kazakhstan. *Stratigrafiya. Geol. korrelatsiya.* **4**(5): 35–52 (in Russian, English translation: *Stratigraphy and Geol. Correlation* **4**(5): 440–49).

KIRICHKOVA, A.I., KULIKOVA, N.K., OVCHINNIKOVA, L.L., TIMOSHINA, N.A., TRAVINA, T.A., and FEDOROVA, V.A. 1999. Biostratigraphic subdivision of Mesozoic deposits penetrated by the Tyumen superdeep borehole. *Stratigrafiya. Geol. korrelatsiya.* **7**(1): 71–85 (in Russian, English translation: *Stratigraphy and Geol. Correlation* **7**(1): 59–63).

KIRK, W.D.J. 1984. Pollen feeding in thrips (Insecta: Thysanoptera). *J. Zool.* (London) **204**: 107–117.

KIRK, W.D.J. 1985. Egg-hatching in thrips (Insecta: Thysanoptera). *J. Zool.* (London) **207**: 181–190.

KISELYOV, S.V. 1981. *Pozdnekainozoiskie zhestkokrylye Severo-Vostoka Sibiri* (The Late Cainozoic Coleoptera of the Northeast of Siberia). Nauka, Moscow: 116 p. (in Russian).

KISLEV, M.E. 1991. Archaeobotany and storage archaeoentomology. In: RENFREW, J.M. (ed.). *New Light on Early Farming. Recent Developments in Palaeoethnobotany.* Edinburgh Univ. Press, Edinburgh: 121–136.

KISTNER, D.H. 1998. New species of termitophilous Trichopseniinae (Coleoptera: Staphylinidae) found with *Mastotermes darwiniensis* in Australia and in Dominican amber. *Sociobiology* **31**(1): 51–76.

KLASS, K.-D. 1997. The external male genitalia and the phylogeny of Blattaria and Mantodea. *Bonner Zool. Monogr.* **42**: 1–341.

KLAUSNITZER, B. 1989. Marienkäfersammlungen am Ostseestrand (Col., Coccinellidae). *Entomol. Nachr. und Ber.* **33**(5): 189–194.

KLEBS, R. 1910. Über Bernsteineinschlüsse in allgemeinen und die Coleopteren meiner Bernsteinsammlung. *Schrift. phys.-ökon. Ges. Königsberg* **51**: 217–242.

KLIMASZEWSKI, S.M. 1995. Supplement to the knowledge of Protopsyllidiidae (Homoptera, Psyllomorpha). *Acta Biol. Siles.* **27**: 33–43.

KLIMASZEWSKI, S.M. 1997. New psyllids from the Baltic amber (Insecta: Homoptera, Aphalaridae). *Mitt. Geol.-Paläont. Inst. Univ. Hamburg* **80**: 157–171.

KLIMASZEWSKI, S.M. 1998. Further data concerning Dominican amber jumping plant lice (Homoptera, Psylloidea). *Acta Biol. Siles.* **32**: 19–27.

KLIMASZEWSKI, S.M., and WOJCIECHOWSKI, W. 1992. Relationships of recent and fossil groups of Sternorrhyncha as indicated by the structure of their fore wings. *Prace Nauk. Uniw. Śląsk. Katowice* **1318**: 1–50.

KLIMT, K. 1978. Thysanopteren im freien Luftraum. *Hercynia* **15**(1): 10–16.

KLINGER, R. 1988. Insektenfunde in jungpleistozänen Hölzern der nördlichen Oberrheinebene. *Zur Paläoklimatologie der letzten Interglazials Nordteil Oberrheinebene.* Springer, Stuttgart – New York: 173–191.

KLOCKE, F. 1926. Beiträge zur Anatomie und Histologie der Thysanopteren. *Zeitschr. wiss. Zool.* **128**: 1–36.

KLUESSENDORF, J., and MIKULIC, D.G. 1990. Temporal patterns in the arthropod trace-fossil record. In: MIKULIC, D.G. (ed.). *Arthropod Paleobiology. Palaeontol. Soc. Short Courses Notes* No.3: 66–98.

KLUGE, N.Ju. 1986. A recent mayfly species (Ephemeroptera, Heptageniidae) in Baltic amber 2. *Paleontol. Zhurnal* (2): 111–112 (in Russian; English translation: *Palaeontol. J.* **20**(2): 106–107).

KLUGE, N.Ju. 1989. The problem of the homology of the tracheal gills and paranotal processi of the mayfly larvae and wings of the insects with reference to the taxonomy and phylogeny of the order Ephemeroptera. *Chteniya pamyati N.A. Kholodkovskogo, 1988* (Lectures in Memoriam of N.A. Kholodkovsky, 1988). Nauka, Leningrad: 48–77 (in Russian).

KLUGE, N.Ju. 1992. Redescription of *Leptohyphes eximius* Eaton and diagnoses of the genera *Leptohyphes* and *Tricorythodes* based on the structure of the pterothorax (Ephemeroptera: Tricorythidae, Leptohyphinae). *Opuscula zool. fluminensia* **98**: 1–16.

KLUGE, N.Ju.1993. New data on mayflies (Ephemeroptera) from fossil Mesozoic and Cenozoic resin. *Paleontol. J.* **27**(1A): 35–49.

KLUGE, N.Ju. 1994. Pterothorax structure of mayflies (Ephemeroptera) and its use in systematics. *Bull. Soc. entomol. France* **99**(1): 41–61.

KLUGE, N.Ju. 1996. A new suborder of Thysanura for the Carboniferous insect initially described as larva of *Bojophlebia*, with comments on characters of the orders Thysanura and Ephemeroptera (Insecta). *Zoosystematica Rossica* **4**(1), 1995: 71–75.

KLUGE, N.Ju. 1997a. A paradoxical problem in the phylogeny of Furcatergalia (Ephemeroptera). In: LANDOLT, P., and SARTORI, M. (eds.) *Ephemeroptera & Plecoptera. Biology – Ecology – Systematics. Proc. VIII Internat. Conf. on Ephemeroptera and XII Internat. Symposium on Plecoptera, Aug. 1995, Lausanne.* Mauron+Tinguely & Lacht SA, Fribourg/Switzerland: 520–526.

KLUGE, N.Ju. 1997b. Classification and phylogeny of the Baetidae (Ephemeroptera) with description of the new species from the Upper Cretaceous resins of Taimyr. In: LANDOLT, P., and SARTORI, M. (eds.) *Ephemeroptera & Plecoptera. Biology – Ecology – Systematics. Proc. VIII Internat. Conf. on Ephemeroptera and XII Internat. Symposium on Plecoptera, August 1995, Lausanne.* Mauron+Tinguely & Lacht SA, Fribourg/Switzerland: 527–535.

KLUGE, N.Ju. 1998. Phylogeny and higher classification of Ephemeroptera. *Zoosystematica Rossica* **7**(2): 255–269.

KLUGE, N.Ju. 2000. *Sovremennaya sistematika nasekomykh. Printsipy sistematiki zhivykh organizmov i obshchaya sistema nasekomykh s klassifikatsiey pervichnobeskrylykh i drevnekrylykh* (Modern Systematics of Insects. Principles of Systematics of Living Organisms and General System of Insects with Classification of Primary Wingless and Paleopterous Insects). Lan', S.-Peterburg: 336 p. (in Russian).

KLUGE, N.Ju., STUDEMANN, D., LANDOLT, P., and GONSER, T., 1995. A reclassification of Siphlonuroidea (Ephemeroptera). *Bull. Soc. entomol. Suisse* **68**: 103–132.

KOCH, F. 1981. Zur präimaginalen Ontogenese des Fransenflüglers *Hercinothrips femoralis* (O.M. Reuter) (Thysanoptera, Insecta). *Zool. Jahrb. (Anat.)* **106**: 412–419.

KOENIGSWALD, W., VON (ed.) 1989. *Fossillagerstätte Rott bei Hennef am Siebengebirge.* Rheinlandia-Verlag, Siegburg: 82 p.

KOHRING, R., and SCHLÜTER, T. 1989. Historische und paläontologische Bestandaufnahme des Simetits, eines fossilen Harzes mutmaßlich mio/pliozänen Alters aus Sizilien. *Documenta naturae* **56**: 33–58.

KOHRING, R., and SCHLÜTER, T. 1995. Erhaltungsmechanismen känozoischer Insekten in fossilen Harzen und Sedimenten. *Berl. geowiss. Abhandl.* (E) **16**: 457–481.

KOLBE, H.J. 1888. Zur Kenntniss von Insektenbohrgängen in fossilen Hölzern. *Zeitschr. Deutsche Geol. Ges.* **40**: 131–137.

KOLLER, J., BAUMER, B., and BAUMER, U. 1997. Die Untersuchung von Bernstein, Bernsteinölen und Bernsteinlacken. *Neue Erkenntnisse zum Bernstein, Internat. Symposium, Bochum, 1–17 Sept. 1996. Metalla* Sonderheft: 85–102.

KOLUMBE, E., and BEYLE, M. 1940. Neue Interglaziale Flora aus Schleswig-Holstein und Hamburg. *Mitt. geol. Staatsinst. Hamburg* **17**: 59–74.

KOPONEN, A., and KOPONEN, T. 1978. Evidence of entomophily in Splachnaceae (Bryophyta). *Bryophyt. Bibliotheca* **13**: 569–577.

KOPONEN, M., and NUORTEVA, M. 1973. Über subfossile Waldinsekten aus dem Moor Piilonsuo in Südfinnland. *Acta Entomol. Fenn.* **29**: 4–84.

KOPP, F., and MARKIANOVICH, E. 1950. On chitin-destroying bacteria in Black Sea. *Doklady Akad. Nauk SSSR* **75**(6): 859–861 (in Russian).

KORNILOVICH, N. 1903. Is the structure of striated muscles preserved in fossil amber? *SB Naturforsch. Ges. Dorpat* **13**: 198–203 (in Russian).

KOSMOWSKA-CERANOWICZ, B. 1999. Succinite and some other fossil resins in Poland and Europe (deposits, finds, features and differences in IRS). *Estud. Mus. Cienc. Natur. Alava* **14** (Núm. Espec. 2): 73–117.

KOSMOWSKA-CERANOWICZ, B., KOCISZEWSKA-MUSIAL, G., MUSIAL, T., and MÜLLER, C. 1990. Bursztynonosne osady trzeciorzędowe okolic Parczewa. *Prace Muz. Ziemi* **41**: 21–35.

KOSMOWSKA-CERANOWICZ, B., KOVALIUKH, N., and SKRIPKIN, V. 1996. Sulphur content and radiocarbon dating of fossil and sub-fossil resins. *Prace Muz. Ziemi* **44**: 47–50.

KOTEJA, J. 1988. *Eomatsucoccus* gen. n. (Homoptera, Coccinea) from Siberian Lower Cretaceous deposits. *Ann. Zool.* **42**: 141–163.

KOTEJA, J. 1989a. Palaeontology. In: ROSEN, D. (ed.). *Armoured Scale Insects, Their Biology, Natural Enemies and Control.* A. Elsevier Sci. Publ., Amsterdam: 149–163.

KOTEJA, J. 1989b. *Baisococcus victoriae* gen. et sp. n. – a Lower Cretaceous coccid (Homoptera, Coccinea). *Acta Zool. Cracov.* **32**: 93–106.

KOTEJA, J. 1996. Syninclusions. *3. Internat. Fachgespräch fossile Insekten in Friedrichsmoor, 14–16, Juni 1996*: 27.

KOTEJA, J. 2000. Scale insects (Homoptera, Coccinea) from Upper Cretaceous New Jersey amber. In: GRIMALDI, D.A. (ed.) 2000. *Studies on Fossils in Amber, with Particular Reference to the Cretaceous of New Jersey.* Backhuys Publ., Leiden: 147–229.

KOTOVA, I.Z. 1965. Palaeofloristic regions of USSR in Jurassic and Early Cretaceous times and the data of spore-pollen analysis. *Paleontol. Zhurnal* (1): 115–124 (in Russian).

KOVALEV, O.V. 1994. Palaeontological history, phylogeny and systematics of Brachycleistogasteromorpha infraorder n., and Cynipomorpha infraorder n., (Hymenoptera) with descripton of new fossil and recent families, subfamilies and genera. *Entomol. Obozrenie* **73**(2): 385–426 (in Russian, English translation: *Entomol. Review* **74**. 4. 105–147).

KOVALEV, V.G. 1983. New dipteran family from the Triassic of Australia and its presumed descendants (Diptera, Crosaphididae fam.n., Mycetobiidae). *Entomol. Obozrenie* **62**(4): 800–805. (In Russian; English translation: *Entomol. Review* **62**(4): 130–136).

KOVALEV, V.G. 1984. Stages and factors of the historical development of the Dipteran fauna. In: CHERNOV, Yu.I. (ed.) *Faunogenez i filocenogenez* (Faunogenesis and Phylocenogenesis). Nauka, Moscow: 138–154 (in Russian).

KOVALEV, V.G. 1987. Classification of Diptera in the light of palaeontological data. In: NARTSHUK, E.P. (ed.) *Dvukrylye nasekomye: sistematika, morfologiya, ekologiya* (Two-winged insects: systematics, morphology and ecology). Zool. Inst. Akad. Nauk SSSR, Leningrad: 40–48 (in Russian).

KOVALEV, V.G. 1989. Eremochaetidae, the Mesozoic family of Diptera Brachycera. *Paleontol. Zhurnal* (2): 104–108 (in Russian, English translation as "Bremochaetidae [sic]...": *Paleontol. J.* **23**(2): 100–105).

KOVALEV, V.G. 1990. Flies. Muscida. In: RASNITSYN, A.P. (ed.). *Pozdnemezozoiskie nasekomye Vostochnogo Zabaikal'ya* (Late Mesozoic Insects of Eastern Trasbaikalia). *Trudy Paleontol. Inst. Akad.Nauk SSSR* **239**. Nauka, Moscow: 123–177 (in Russian).

KOVATS, Zs.E., CIBOROWSKI, J.J.H., and CORKUM, L.D. 1996. Inland dispersal of adult aquatic insects. *Freshwater Biol.* **36**: 265–276.

KOWALEVSKY, W. 1874. Monographie der Gattung *Anthracotherium* Cuv. und Versuch einer natürlichen Classification der fossilen Huftiere. *Palaeontographica* N.F., II **3** (XXII): 291–346.

KOZLOV, M.A. 1968. Jurassic Proctotrupidea (Hymenoptera). In: ROHDENDORF, B.B. (ed.). *Yurskie nasekomye Karatau* (Jurassic Insects of Karatau). Nauka, Moscow: 237–240 (in Russian).

KOZLOV, M.V. 1988. Palaeontological data and the problems phylogeny of the order Papilionida. In: PONOMARENKO, A.G. (ed.). *Melovoy biotsenoticheskiy krizis i evolyutsiya nasekomykh* (Cretaceous Biocoenotic Crisis and Evolution of the Insects). Nauka, Moscow: 16–69 (in Russian).

KOZLOV, M.V. 1989. New moths (Papilionida) from the Upper Jurassic and Lower Cretaceous. *Paleontol. Zhurnal* (4): 37–42 (in Russian, English translation: *Paleontol. J.* **23**(4), 34–39).

KOZLOV, M. V., KUZNETZOV, V. I., and STEKOLNIKOV, A. A. 1998. Skeleto-muscular morphology of the pterothorax and male genitalia of *Synemon plana* Walker (Castniidae) and *Brachodes appendiculata* (Esper) (Brachodidae), with notes on phylogenetic relationships of tortricoid-grade moth families (Lepidoptera). *Invertebr. Taxonomy* **12**: 245–256.

KOZLOV, M.V. and ZVEREVA, E. L. (1999). A failed attempt to demonstrate pheromone communication in archaic moths of the genus *Sabatinca* Walker (Lepidoptera, Micropterigidae). *Ecology Letters* **2**: 215–218.

KRASENKOV, R.V. 1966. Nids of the larvae of Pliocene and recent mayflies from Voronezh Region. In: *Organizm i sreda v geologicheskom proshlom* (Organism and Environment in the Geological Past). Nauka, Moscow: 214–221 (in Russian).

KRASSILOV, V.A. 1969. Phylogeny and systematics. In: KRASNOV, E.V. (ed.). *Problemy filogenii i sistematiki* (Problems of Phylogeny and Systematics). USSR Acad. Sci. Far East Reserch Centre, Vladivostok: 12–30 (in Russian).

KRASSILOV, V.A. 1972a. *Paleoekologiya nazemnykh rasteniy (osnovnye printsipy i metody)*. (Paleoecology of Terrestrial Plants. Basic Principles and Techniques). Vladivostok: 212 p. (in Russian; English translation: 1975. John Wiley and Sons, New York & Toronto: 283 p.).

KRASSILOV, V.A. 1972b. On the coincidence of the lower borders of Cainozoic and Cainophytic. *Izvestiya Akad. Nauk SSSR* Ser. geol. (3): 9–16 (in Russian).

KRASSILOV, V.A. 1973. On the modes of pollination in ancient plants. In: BOLKHOVITINA, N.A. (ed.). *Morphologiya i sistematika iskopaemykh spor i pyl'tsy* (Morphology and Systematics of Fossil Spores and Pollen). Nauka, Moscow: 6–9 (in Russian).

KRASSILOV, V.A. 1974. Mesozoic plants and the problem of angiosperm origin. *Lethaia* **6**(2): 163–178.

KRASSILOV, V.A. 1975. Ancestors of angiosperms. *Problemy evolutsii (Novosibirsk)* (Problems of Evolution), **4**: 75–106 (in Russian).

KRASSILOV, V.A. 1977. The origin of angiosperms. *Botan. Rev.* **43**(1): 143–176.

KRASSILOV, V.A. 1979. *Melovaya flora Sakhalina* (Cretaceous Flora of Sakhalin). Nauka, Moscow: 183 p. (in Russian).

KRASSILOV, V. 1982. Early Cretaceous flora of Mongolia. *Palaeontographica* B **181**(1–3): 1–43.

KRASSILOV, V.A. 1983. Evolution of the flora in the Cretaceous period: is the Cainophytic Era necessary? *Paleontol. Zhurnal* (3): 93–96 (in Russian; English translation: *Paleontol. J.* **17**(3): 89–93).

KRASSILOV, V.A. 1985. *Melovoy period. Evolyutsiya zemnoy kory i biosfery* (The Cretaceous Period. Evolution of the Earth Crust and Biosphere). Nauka, Moscow: 240 p. (in Russian).

KRASSILOV, V.A. 1989. *Proiskhozhdenie i rannyaya evolyutsiya tsvetkovykh rasteniy* (Origin and Early Evolution of Flowering Plants). Nauka, Moscow, 264 p. (in Russian).

KRASSILOV, V.A. 1995. Models of plant and plant community evolution: Cretaceous-Paleocene transition examples. *Paleontol. J.* **29**(2A): 1–21.

KRASSILOV, V.A. 1997. *Angiosperm Origins: Morphological and Ecological Aspects.* Pensoft, Sofia: 270 p.

KRASSILOV, V.A., and BUGDAEVA, E.V. 1988. Protocycadopsid pteridosperms from the Lower Cretaceous of Transbaikalia and the origin of cycads. *Palaeontographica* B **208**: 27–32.

KRASSILOV, V.A., and MAKULBEKOV, N.M. 1995. Maastrichtian aquatic plants from Mongolia. *Paleontol. J.* **29**(2A): 119–140.

KRASSILOV, V.A., and RASNITSYN, A.P. 1982. A unique find: pollens in the intestine of Early Cretaceous sawflies. *Paleontol. Zhurnal* (4): 83–96 (in Russian, English translation: *Paleontol. J.* **16**(4): 80–94).

KRASSILOV, V.A., and RASNITSYN, A.P. 1997. Pollen in the guts of Permian insects: first evidence of pollinivory and its evolutionary significance. *Lethaia* **29**: 369–372.

KRASSILOV, V.A., and RASNITSYN, A.P. 1999. Plant remains from the guts of fossil insects: evolutionary and paleoecological inferences. *Proc. First Palaeoentomol. Conf., Moscow 1998.* AMBA projects Internat., Bratislava: 65–72.

KRASSILOV, V.A., SHILIN, P.V., and VAKHRAMEEV, V.A. 1983. Cretaceous flowers from Kazakhstan. *Rev. Palaeobot. Palynol.* **40**: 91–113.

KRASSILOV, V.A., and SUKATSHEVA, I.D. 1979. Caddisfly cases built from *Karkenia* seeds (Ginkgoales) in the Lower Cretaceous deposits of Mongolia. *Trudy Biol.-Pochvennogo Inst. Dal'nevostochnogo Nauchnogo Tsentra Akad. Nauk SSSR*, novaya seriya **53**: 119–121 (in Russian).

KRASSILOV, V.A., ZHERIKHIN, V.V., and RASNITSYN, A.P. 1997a. *Classopollis* in the guts of Jurassic insects. *Palaeontology* **40**(4): 1095–1101.

KRASSILOV, V.A., ZHERIKHIN, V.V., and RASNITSYN, A.P. 1997b. Pollen in guts of fossil insects as evidence for co-evolution. *Doklady Akad. Nauk* **354**(1): 135–138 (in Russian, English translation: *Doklady Biol. Sci.* **354**: 239–241).

KRASSILOV, V.A., RASNITSYN, A.P., and AFONIN, S.A. 1999. Pollen morphotypes from the intestine of a Permian booklouse. *Rev. Palaeobot. Palynol.* **106**: 89–96.

KRINSKY, W.L. 1985. A *Propalticus* species in Kenyan amber (Coleoptera: Propalticidae). *Coleopterists' Bull.* **39**(2): 101–102.

KRISHNA, K. 1970. Taxonomy, phylogeny and distribution of termites. In: KRISHNA, K., and WEESNER, F.M. (eds.) *Biology of Termites* **2**. Academic Press, New York–London: 127–152.

KRISHNA, K. 1990. Isoptera. In: GRIMALDI, D.A. (ed.) *Insects from the Santana Formation, Lower Cretaceous, of Brasil. Bull. Amer. Mus. Nat. Hist.* **195**: 76–81.

KRISTENSEN, N.P. 1975. The phylogeny of the hexapod "orders". A critical review of recent accounts. *Zeitschr. Zool. Syst. Evolutions Forsch.* **13**: 1–44.

KRISTENSEN, N.P. 1981. Phylogeny of insect orders. *Annu. Rev. Entomol.* **26**: 135–157.

KRISTENSEN, N.P. 1984. Studies on the morphology and systematics of primitive Lepidoptera (Insecta). *Steenstrupia* **10**: 141–191.

KRISTENSEN, N. P. 1995. Forty years' insect phylogenetic systematics. *Zool. Ber.* N.F. **36**: 83–124.

KRISTENSEN, N. P. 1997. Early evolution of the Lepidoptera + Trichoptera lineage: phylogeny and the ecological scenario. In: GRANDCOLAS, P. (ed.). *The Origin of Biodiversity in Insects: Tests of Evolutionary Scenarios. Mém. Mus. nation. Hist. natur.* **173**. Paris: 253–271.

KRISTENSEN, N. P. 1998. The groundplan and basal diversification of the hexapods. In: FORTEY, R.A., and THOMAS, R.H. (eds.) *Arthropod Relationships. Systematic Assoc. Spec. Volume Series* **55**. Chapman & Hall, London etc.: 281–293.

KRISTENSEN, N.P. 1999a. Phylogeny of endopterygote insects, the most successful lineage of living organisms. *Europ. J. Entomol.* **96**(3): 237–253.

KRISTENSEN, N.P. (ed.). 1999b. *Lepidoptera, Moths and Butterflies* **1**. *Handbook of Zoology* **4** Arthropoda: Insecta Part 35. Walter de Gruyter, Berlin–New York: 487 p.

KRISTENSEN, N.P., and SKALSKI, A.W. 1999. Phylogeny and palaeontology. In: KRISTENSEN, N.P. (ed.). 1999. *Lepidoptera, Moths and Butterflies* **1**. *Handbook of Zoology* **4** Arthropoda: Insecta Part 35. Walter de Gruyter, Berlin–New York: 7–25.

KRIVOLUTZKAYA, G.O., and NECHAEV, V.A. 1963. Volcanic ashfall. *Priroda* (9): 82–85 (in Russian).

KRIVOLUTSKY D.A., and KRASSILOV V.A. 1977. The oribatid mites in the Upper Jurassic deposits of USSR. In: BALASHOV, Yu.S. (ed.). *Morfologiya i diagnostika kleshchey* (Morphology and Diagnostics of Mites). Nauka, Leningrad: 16–24 (in Russian).

KROGER, R.L. 1974. Invertebrate drift in the Snake River, Wyoming. *Hydrobiol.* **44**: 369–380.

KRÜGER, F.J. 1979. Tongrube Willershausen, ein geologisches Naturdenkmal. *Aufschlüss* **30**(11): 389–408.

KRUMBIEGEL, G., and KRUMBIEGEL, B. 1996a. *Bernstein – fossile Harze aus aller Welt. Fossilien.* Sonderband 7. Goldschneck-Verlag, Korb: 112 p.

KRUMBIEGEL, G., and KRUMBIEGEL, B. 1996b. Bernsteinlagerstätten und -vorkommen in aller Welt. In: GANZELEWSKI, M., and SLOTTA, R. (eds.). *Bernstein – Tränen der Götter.* Deutsches Bergbau-Mus., Bochum: 31–46.

KRUMBIEGEL, G., RÜFFLE, L., and HAUBOLD, H. 1983. *Das Eozäne Geiseltal, ein mitteleuropäisches Braunkohlenvorkommen und seiner Pflanzen- und Tierwelt. Die Neue Brehm Bücherei* **237**. A. Ziemsen Verl., Wittenberg Lutterstadt: 227 p.

KRZEMIŃSKI, W. 1992. Triassic and Lower Jurassic stage of Diptera evolution. *Mitt. Schweiz. Entomol. Ges.* **65**: 39–59.

KRZEMIŃSKI, W. 1998. Origin and the first stages of evolution of the Diptera Brachycera. *4th Internat. Congr. Dipterology. 6–13th Sept. 1998. Keble College, Oxford, UK. Abstract Volume.* Oxford: 113–114.

KRZEMIŃSKI, W., and JARZEMBOWSKI, E.A. 1999. *Aenne triassica* sp.n., the oldest representative of the family Chironomidae (Insecta: Diptera). *Polskie Pismo Entomol.* **68**: 445–449.

KRZEMIŃSKI, W., KRZEMIŃSKA, E., and PAPIER, P. 1994. *Grauvogelia arzvilleriana* sp.n. – the oldest Diptera species (Lower/Middle Triassic of France). *Acta Zool. Cracov.* **37**(2): 95–99.

KRZEMIŃSKI, W., and LOMBARDO, C. 2001. New fossil Ephemeroptera and Coleoptera from the Ladinian (Middle Triassic) of Canton Ticino (Switzerland). *Riv. Ital. Paleontol. Strat.* **27**(1): 69–78.

KUGLER, H.G. 1927. Report on a fossil locality for Arthropoda and vertebrates in Trinidad. *Trans.R. Entomol. Soc. London* **75**(1): 141–142.

KÜHBANDNER, M., and SCHLEICH, H.M. 1994. *Odontomyia*-Larven aus dem Randecker Maar (Insecta: Diptera, Stratiomyidae). *Mitt. Bayer. Staatssamml. Paläontol. histor. Geol.* **34**: 163–167.

KUHN, O. 1937. Insekten aus dem Buntsandstein von Thüringen. *Beiträge zur Geol. Thüringen* **4**: 190–193.

KUHN, O. 1961. Die Tier- und Pflanzenwelt der Solnhofener Schiefers, mit vollständigem Arten- und Schriftenverzeichnis. *Geol. Bavar.* **48**: 5–68.

KÜHNE, W.G., and SCHLÜTER, T. 1985. A fair deal for the Devonian Arthropoda fauna of Rhynie. *Entomol. Gener.* **11**(1–2): 91–96.

KUKALOVÁ, J. 1958. Paoliidae Handlirsch (Insecta-Protorthoptera) aus dem Oberschlesischen Steinkohlenbecken. *Geologie* **7**: 935–959.

KUKALOVÁ, J. 1959. On the family Blattinopsidae (Insecta, Protorthoptera). *Rozpravy Československ. Acad. Ved* **69**(1): 32 p.

KUKALOVÁ, J. 1963. Permian insect of Moravia. Part I – Miomoptera. *Sbornik geol. ved* (P) **1**: 7–52.

KUKALOVÁ, J. 1964. Permian insect of Moravia. Part II – Liomopteridea. *Sbornik geol. ved* (P) **3**: 39–118.

KUKALOVÁ, J. 1965. Permian Protelytroptera, Coleoptera and Protorthoptera (Insecta) of Moravia. *Sbornik geol.ved* (P) **6**: 61–98.

KUKALOVÁ, J. 1966. Protelytroptera from the Upper Permian of Australia, with a discussion of the Protocoleoptera and Paracoleoptera. *Psyche* **73**: 89–111.

KUKALOVÁ, J. 1969a. Revisional study of the order Palaeodictyoptera in the Upper Carboniferous shales of Commentry, France. Pt. 1. *Psyche* **76**: 163–215.

KUKALOVÁ, J. 1969b. Revisional study of the order Palaeodictyoptera in the Upper Carboniferous shales of Commentry, France. Pt. 2. *Psyche* **76**: 439–486.

KUKALOVÁ, J. 1969c. On the systematic position of the supposed Permian beetles, Tshecardocoleidae, with a description of a new collection from Moravia. *Sbornik geol. ved. Paleontol.* (P) **11**: 139–159.

KUKALOVÁ, J. 1970. Revisional study of the order Palaeodictyoptera in the Upper Carboniferous shales of Commentry, France. Pt. 3. *Psyche* **77**(1): 1–44.

KUKALOVÁ-PECK, J. 1975. Megasecoptera from the Lower Permian of Moravia. *Psyche* **82**(1): 1–19.

KUKALOVÁ-PECK, J. 1978. Origin and evolution of insect wings and their relation to metamorphosis, as documented by the fossil record. *J. Morphol.* **15**: 53–126.

KUKALOVÁ-PECK, J. 1983. Origin of insect wing and wing articulation from the arthropodan leg. *Canad. J. Zool.* **61**: 1618–1669.

KUKALOVÁ-PECK, J. 1985. Ephemeroid wing venation based upon new gigantic Carboniferous mayflies and basic morphology, phylogeny, and metamorphosis of pterygote insects (Insecta, Ephemerida). *Canad. J. Zool.* **63**: 933–955.

KUKALOVÁ-PECK, J. 1987a. New Carboniferous Diplura, Monura, and Thysanura, the hexapod ground plan, and the role of thoracic side lobes in the origin of wings (Insecta). *Canad. J. Zool.* **65**: 2327–2345.

KUKALOVÁ-PECK, J. 1987b. A substitute name for the extinct genus *Stenelytron* Kukalová (Protelytroptera). *Psyche* **94**: 339.

KUKALOVÁ-PECK, J. 1991. Fossil history and the evolution of hexapod structure. In: NAUMANN, D. (ed.). *The Insects of Australia*. 2nd ed. Melbourne Univ. Press, Carlton: 144–182.

KUKALOVÁ-PECK, J. 1992. The "Uniramia" do not exist: the ground plan of the Pterygota as revealed by Permian Diaphanopterodea from Russia (Insecta: Palaeodictyopteroidea). *Canad. J. Zool.* **70**: 236–255.

KUKALOVÁ-PECK, J. 1997. Mazon Creek insect fossils. In: SHABICA, C.W., and HAY, A.A. (eds.) *Richardson's Guide to the Fossil Fauna of Mazon Creek*. Chicago, Northeastern Illinois Univ.: 194–207.

KUKALOVÁ-PECK, J. 1998. Arthropod phylogeny and 'basal' morphological structures. In: FORTEY, R.A., and THOMAS, R.H. (eds.). *Arthropod Relationships. Systematic Assoc. Spec. Volume Series* **55**. Chapman & Hall, London etc.: 249–268.

KUKALOVÁ-PECK, J., and BRAUCKMANN, C. 1990. Wing folding in pterygote insects, and the oldest Diaphanopterodea from the early Late Carboniferous of West Germany. *Canad. J. Zool.* **68**: 1104–1111.

KUKALOVÁ-PECK, J., and BRAUCKMANN, C. 1992. Most Paleozoic Protorthortera are ancestral hemipteroids: major wing braces as clues to new phylogeny of Neoptera (Insecta). *Canad. J. Zool.* **70**: 2452–2473.

KUKALOVÁ-PECK, J., and LAWRENCE, J.F. 1993. Evolution of the hind wing in Coleoptera. *Canad. Entomol.* **125**: 181–258.

KUKALOVÁ-PECK, J., and PECK, S.B. 1993. Zoraptera wing structures: evidence for new genera and relationship with the blattoid orders (Insecta: Blattoneoptera). *Syst. Entomology* **18**: 333–350.

KUKALOVÁ-PECK, J., and RIEK, E.F. 1984. A new interpretation of dragonfly wing venation based upon early Upper Carboniferous fossils from Argentina (Insecta: Odonatoidea) and basic character states in pterygote wings. *Canad. J. Zool.* **62**: 1150–1166.

KUKALOVÁ-PECK, J., and SINICHENKOVA, N.D. 1992. The wing venation and systematics of Lower Permian Diaphanopterodea from the Ural Mountains, Russia (Insecta: Paleoptera). *Canad. J. Zool.* **70**: 229–235.

KUKALOVÁ-PECK, J., and WILLMANN, R. 1990. Lower Permian "Mecopteroid-like" insects from Central Europe (Insecta, Endopterygota). *Canad. J. Earth Sci.* **27**: 459–468.

KULICKA, R., and SUKATSHEVA, I.D. 1990. Rodziny kopalnych Trichoptera mezozoiku i kenozoiku (The families of fossil Trichoptera of the Mesozoic and Cenozoic Eras). *Prace Muz. Ziemi* **41**: 65–75 .

KULMAN, H.M. 1971. Effect of insect defoliation on growth and mortality of trees. *Annu. Rev. Entomol.* **16**: 289–324.

KUSCHEL, G. 1987. The subfamily Molytinae (Coleoptera: Curculionidae): General notes and descriptions of new taxa from New Zealand and Chile. *New Zeal. Entomol.* **9**: 11–29.

KUSCHEL, G., OBERPRIELER, R.G., and RAYNER, R.J. 1994. Cretaceous weevils from southern Africa, with description of a new genus and species and phylogenetic and zoogeographical comments (Coleoptera: Curculionoidea). *Entomol. Scand.* **25**: 137–149.

KUSCHEL, G., and WORTHY, T.H. 1996. Past distribution of large weevils (Coleoptera: Curculionidae) in the South Island, New Zealand, based on Holocene fossil remains. *New Zeal. Entomol.* **19**: 15–22.

KUSMER, K.D. 1990. Taphonomy of owl pellet deposition. *J. Paleontol.* **64**(4): 629–637.

KUZNETSOV, S.I. 1970. *Mikroflora ozyor i ee geokhimicheskaya deyatel'nost'*. Nauka, Leningrad: 440 p. (in Russian) (English Translation: *The Microflora of Lakes and its Geochemical Activity*. Univ. Texas Press, Austin, 1970).

KUZNETSOV, S.I., IVANOV, M.V., and LYALIKOVA, N.N. 1962. *Vvedenie v geologicheskuyu mikrobiologiyu* (Introduction to Geological Microbiology). Izdatel'stvo AN SSSR, Moscow: 239 p. (in Russian).

KUZNETZOV, V.I., and STEKOLNIKOV, A.A. 1986. The system of higher taxa of Lepidoptera based on the data on the functional morphology of genitalia. In: TOBIAS, V.I. (ed.). *General Entomology. Trudy Vsesoyuznogo Entomol. Obshchestva*. **68**. Leningrad, Nauka: 42–46 (in Russian).

LABANDEIRA, C.C. 1990. Rethinking the diets of Carboniferous terrestrial arthropods: evidence for a nexus of arthropod/vascular plant interactions. *Geol. Soc. Amer. Abstracts with Programs* **22**(7): A265.

LABANDEIRA, C.C. 1991. Evidence for Pennsylvanian stem-mining and the early history of complete metamorphosis in insects. *Geol. Soc. Amer. Abstracts with Programs* **23**(5): A405.

LABANDEIRA, C.C. 1994. A compendium of fossil insect families. *Milwaukee Public Mus. Contrib. Biol. Geol.* **88**: 1–71.

LABANDEIRA, C.C. 1995. The evolution of behavioural diversity through time: how the fossil record of plant/insect interactions is used for testing hypotheses in evolutionary biology. *Geol. Soc. Amer. Abstracts with Programs* **27**(3): A66.

LABANDEIRA, C.C. 1996. The presence of a distinctive insect herbivore fauna during the Late Paleozoic. *Sixth North American Paleontol. Convention Abstracts of Papers. Paleontol. Soc. Spec. Publ.* **8**: 227.

LABANDEIRA, C.C. 1997a. Insect mouthparts: ascertaining the paleobiology of insect feeding strategies. *Annu. Rev. Ecol. Syst.* **28**: 153–193.

LABANDEIRA, C.C. 1997b. Multiply way of documenting the fossil record of insect feeding strategies. *Geol. Soc. Amer. Abstracts with Programs* **29** (6): A461–462.

LABANDEIRA, C.C. 1998a. The Paleozoic origin of diverse and significant insect herbivory: rejection of the expanding resources hypothesis. *First Paleoentomological Conference. 30 Aug.–4 Sept. 1998. Moscow, Russia. Abstracts.* Palaeontological Inst., Russian Acad. Sci., Moscow: 21.

LABANDEIRA, C.C. 1998b. Macroevolutionary study of plant-insect associations. *Geotimes* (Sept.) (not paginated).

LABANDEIRA, C.C. 1998c. The role of insects in Late Jurassic to Middle Cretaceous ecosystems. In: LUCAS, S.G., KIRKLAND, J.I., AND ESTEP, J.W. (eds.). *Lower and Middle Cretaceous Terrestrial Ecosystems. Bull. New Mexico Mus. Natur. Hist. and Science* **14**: 105–124.

LABANDEIRA, C.C. 2000. The paleobiology of pollination and its precursors. In: GASTALDO, R.A., and DiMICHELE, W.A. (eds.). *Phanerozoic Terrestrial Ecosystems.* Paleontol. Soc. Papers **6**: 233–269.

LABANDEIRA, C.C., and BEALL, B.S. 1990. Arthropod terrestriality. In: MIKULIC, D.G. (ed.). *Arthropod Paleobiology. Short Courses in Palaeontology* **3**: 214–256.

LABANDEIRA, C.C., BEALL, B.S., and HUEBER, F.M. 1988. Early insect diversification: Evidence from Lower Devonian bristletail from Québec. *Science* **242**: 913–916.

LABANDEIRA, C.C., and DILCHER, D.L. 1993. Insect functional feeding groups from the Mid-Cretaceous Dakota Formation of Kansas and Nebraska: evidence for early radiation of herbivores on angiosperms. *Geol. Soc. Amer. Abstracts with Programs* **25**(6): A390.

LABANDEIRA, C.C., DILCHER, D.L., DAVIS, D.R., and WAGNER, D.L. 1994a. Ninety-seven million years of angiosperm-insect association: paleobiological insight into the meaning of co-evolution. *Proc. Nation. Acad. Sci.* **91**: 12278–12282.

LABANDEIRA, C.C., NUFIO, C., WING, S., and DAVIS, D. 1995. Insect feeding strategies from the Late Cretaceous Big Cedar Ridge Flora: comparing the diversity and intensity of Mesozoic herbivory with the present. *Geol. Soc. Amer. Abstracts with Programs* **27**(6): A447.

LABANDEIRA, C.C., and PHILLIPS, T.L. 1992. Ecological response of plant consumers to Middle-Upper Pennsylvanian extinction in Illinois Basin coal swamps: evidence from plant/arthropod interactions. *Geol. Soc. Amer. Abstracts with Programs* **24**(7): A120.

LABANDEIRA, C.C., and PHILLIPS, T.L. 1996a. A Carboniferous insect gall: insight into early ecological history of the Holometabola. *Proc. Nation. Acad. Sci. USA* **93**: 8470–8474.

LABANDEIRA, C.C., and PHILLIPS, T.L. 1996b. Insect fluid-feeding on Upper Pennsylvanian tree ferns (Palaeodictyoptera, Marattiales) and the early history of the piercing-and-sucking functional feeding group. *Ann. Entomol. Soc. Amer.* **89**(2): 157–183.

LABANDEIRA, C.C., PHILLIPS, T.L., and NORTON, R.A. 1997. Oribatid mites and the decomposition of plant tissues in Paleozoic coal-swamp forests. *Palaios* **12**(4): 319–353.

LABANDEIRA, C.C., SCOTT, A.C., MAPES, R., and MAPES, G. 1994b. The biologic degradation of wood through time: new insights from Paleozoic mites and Cretaceous termites. *Geol. Soc. Amer. Abstr. with Progr.*, **26**(7): A123.

LABANDEIRA, C.C. and SEPKOSKI, J.J., Jr. 1993. Insect diversity in the fossil record. *Science* **261**: 310–315.

LACASA, A. 1991. Icnología de les calcàries litogràfiques del Montsec. In: MARTÍNEZ-DELCLÒS, X. (ed.). *Les calcàries litogràfiques del Cretaci inferior del Montsec. Deu anys de campanyes paleontologiques*. Inst. d'Estudis Illerdencs, Lleida: 151–152.

LACAZA RUIZ, A., and MARTÍNEZ DELCLÒS, X. 1986. Meiatermes, *nuevo género fósil de insecto isóptero (Hodotermitidae) de las calizas neocomienses del Montsec (Provincia de Lérida, España)*. Publ. Inst. d'Estud. Ilerdencs, Lleida: 65 p.

LADLE, M., and GRIFFITHS, B.S. 1980. A study on the faeces of some chalk stream invertebrates. *Hydrobiol.* **74**(2): 161–171.

LAICHARTING, J.N., VON. 1781. *Verzeichniss und Beschriebung der Tyroler Insecten* **1**. Fuessley, Zurich: xii+248 p.

LAKSHMANAPERUMALSAMY, P. 1983. Preliminary studies on chitinoclastic bacteria in Vellar estuary. Mahasagar. *Bull. Nation. Inst. Oceanogr.* **16**(3): 293–298.

LAMBERT, J.B., FRYE, J.S., and POINAR, G.O., Jr. 1985. Amber from the Dominican Republic: analysis by nuclear magnetic resonance spectroscopy. *Archaeometry* **27**: 43–51.

LAMBERT, J.B., JOHNSON, S.C., POINAR, G.O., Jr., and FRYE, J.S. 1993. Recent and fossil resins from New Zealand and Australia. *Geoarchaeology* **8**(2): 141–155.

LAMBKIN, K.J. 1987. A re-examination of *Euporismites balli* Tillyard from the Palaeocene of Queensland (Neuroptera: Osmylidae: Kempyninae). *Neuroptera Internat.* **4**(4): 295–300.

LAMBKIN, K. J. 1988. A re-examination of *Lithosmylidia* Riek from the Triassic of Queensland with notes on Mesozoic 'Osmylid-like' fossil Neuroptera (Insecta: Neuroptera). *Mem. Queensl. Mus.* **25**(2): 445–458.

LAMÉERE, A., and SEVERIN, G. 1897. Les insectes de Bernissart. *Ann. Soc. entomol. Belg.* **41**: 35–38.

LANCASTER, J., HILDREW, A.G., and GJERLOV, C. 1996. Invertebrate drift and longitudinal transport processes in streams. *Canad. J. Fish. Aquat. Sci.* **53**: 572–582.

LANCE, J.F. 1946. Fossil arthropods of California. 9. Evidence of termites in the Pleistocene asphalt of Carpinteria, California. *Bull. S. Calif. Acad. Sci.* **45**: 21–27.

LANDIN, J., and SOLBRECK, Ch. 1974. Med luftströmmarna kommer insekterna. *Forsk. och framsteg* **6**: 3–10.

LANG, P.J., SCOTT, A.C., and STEPHENSON, J. 1995. Evidence of plant-arthropod interactions from the Eocene Branksome Sand Formation, Bournemouth: Introduction and description of leaf mines. *Tertiary Res.* **15**: 145–174.

LANGE, A.B., and RAZVYAZKINA, G.M. 1953. Morphology and development of the tobacco thrips. *Zool. Zhurnal* **32**(4): 576–593 (in Russian).

LANGENHEIM, J.H. 1966. Botanical source of amber from Chiapas, Mexico. *Ciencia* **24**(5–6): 201–210.

LANGENHEIM, J.H. 1969. Amber: a botanical inquiry. *Science* **163**: 1157–1169.

LANGENHEIM, J.H. 1990. Plant resins. *Amer. Sci.* **78**: 16–24.

LANGENHEIM, J.H., and BECK, C.W. 1965. Infrared spectra as a means of determining botanical sources of amber. *Nature* **149**: 52–55.

LANGENHEIM, J.H., and BECK, C.W. 1968. Catalogue of infrared spectra of fossil resins (ambers). I. North and South America. *Harvard Univ. Botan. Mus. Leafl.* **22**(3): 65–120.

LANGENHEIM, J.H., FOSTER, C.E., and McGINLEY, R.B. 1980. Inhibitory effects of different quantitative compositions of *Hymenaea* leaf resins on a generalist herbivore, *Spodoptera exigua. Biochem. Syst. and Ecol.* **8**(4): 385–396.

LANGENHEIM, J.H., and HALL, G.D. 1983. Sesquiterpene deterrence of a leaf-tying lepidopteran, *Stenoma ferrocanella*, on *Hymenaea stigonocarpa* in central Brazil. *Biochem. Syst. and Ecol.* **11**(1): 29–36.

LANGER, T.W. 1982. Europas aeldste sommerflugelafbildning? *Lepidoptera* **4**(3): 81–84.

LAREW, H.G. 1986. The fossil gall record: a brief summary. *Proc. Entomol. Soc. Washington* **88**: 385–388.

LAREW, H.G. 1987. Two cynipid wasp acorn galls preserved in the La Brea tar pits (Early Holocene). *Proc. Entomol. Soc. Washington* **89**: 831–833.

LAREW, H.G. 1992. Fossil galls. In: SHORTHOUSE, J.D., and ROHFRITSCH, O. (eds.). *Biology of Insect-Induced Galls*. Oxford Univ. Press, New York: 50–59.

LARSSON, S.G. 1975. Paleobiology and mode of burial of the Lower Eocene Mo-clay of Denmark. *Medded. Dansk geol. foren.* **24**: 193–209.

LARSSON, S.G. 1978. *Baltic Amber – A Paleobiological Study (Entomonograph* **1**): 192 p.

LATREILLE, P.A. 1819. Des insectes peintes ou sculptés sur les monuments antiques de l'Egypte. *Mém. Mus. nation. Hist. natur. Paris* **5**: 249–269.

LAURENTIAUX-VIEIRA, F., and LAURENTIAUX, D. 1979. Nouvelle contribution á la connaissance du genre Westphalien *Manoblatta* Pruvost (Blattaires Archimylacridiens). *Ann. Soc. Géol. Nord* **99**: 415–423.

LAURENTIAUX-VIEIRA, F., and LAURENTIAUX, D. 1981. Mise en évidence d'un dimorphisme sexuel chez les Blattes *Dictyomylacris* du Stéphanien de Commentry (Allier). *Ann. Soc. Géol. Nord.* **100**: 175–182.

LAWRENCE, J.F. 1988. Rhinorhipidae, a new beetle family from Australia, with comments on the phylogeny of the Elateriformia. *Invertebr. Taxonomy* **2** (1987): 1–53.

LAWRENCE, J.F., and NEWTON, A.F., Jr. 1982. Evolution and classification of beetles. *Annu. Rev. Ecol. Syst.* **13**: 261–290.

LAWS, R.R., HASIOTIS, S.T., FIORILLO, A.R., CHURE, D.J., BREITHAUPT, B.H., and HORNER, J.R. 1996. The demise of a Jurassic Morrison dinosaur after death – Three cheers for the dermestid beetle. *Geol. Soc. Amer. Abstracts with Programs* **28**: A299.

LAWTON, J.H., and MacGARVIN, M. 1985. Interaction between bracken and its insect herbivores. *Proc. R. Soc. Edinburgh* **B 86**: 125–131.

LAZA, J.H. 1982. Signos de actividad attribuiables á *Atta* (Myrmicidae) en el Mioceno de la Provincia de La Pampa, República Argentina. Significación paleozoogeográfica. *Ameghiniana* **19**: 109–124.

LAZA, J.H. 1986a. Icnofósiles de paleosuelos de Cenozoico mamalífero de Argentina. I. Paleogeno. *Bol. Asoc. Paleontol. Argentina* **15**: 13.

LAZA, J.H. 1986b. Icnofósiles de paleosuelos de Cenozoico mamalífero de Argentina. II. Neogeno. *Bol. Asoc. Paleontol. Argentina* **15**: 19.

LAZA, J.H. 1995. Signos de actividad de insectos. In: ALBERDI, M.T., LEONE, G., and TONNI, E.P. (eds.). *Evolución biológica y climática de la región pampeana durante los ultimos cinco millones de años*. Mus. Nacion. Cienc. Natur. y Consejo Superior de Invest. Científicas, Madrid: 340–361.

LAZA, J.H. 1997. Signos de actividad atribuibles a doa especies de *Acromyrmex* (Myrmicinae, Formicidae, Hymenoptera) del Pleistocene en la Provincia de Buenos Aires, República Argentina. Significado paleoambiental. *Rev. Univ. Guarulhos, Geociências* **2**(6): 56–62.

LEA, A.M. 1925. Notes on some calcareous insect puparia. *Rec. S. Austral. Mus.* **3**: 35–36.

LEAKEY, L.S.B. 1952. Lower Miocene invertebrates from Kenya. *Nature* **169**: 624–625.

LEAR, D.W. 1961. Occurrence and significance of chitinoclastic bacteria in pelagic waters and zooplankton. *Bacterial Proc.* **61**: 13.

LEEREVELD, H. 1982. Anthecological relations between reputedly anemophilous flowers and syrphid flies. 3. Worldwide survey of crop and intestine contents of certain anthophilous syrphid flies. *Tijdschr. Entomol.* **125**(2): 25–35.

LEEREVELD, H., MEEUSE, A.D.J., and STELLEMAN, P. 1976. Anthecological relations between reputedly anemophilous flowers and syrphid flies. II. *Plantago media* L. *Acta Botan. Neerland.* **25**: 205–211.

LEESTMANS, R. 1983. Les Lépidoptères fossiles trouvés en France (Insecta Lepidoptera). *Linnaea Belg.* **9**(1): 64–89.

LEKANDER, B., KOPONEN, M., and NUORTEVA, M. 1975. *Eremotes* larva (Col., Curculionidae) found as subfossil. *Ann. entomol. fenn.* **41**(4): 121–123.

LENGERKEN, H., VON. 1913. Etwas über den Erhaltungszustand von Insekten-Inklusen im Bernstein. *Zool. Anz.* **41**: 284–286.

LENGERKEN, H., VON. 1922. Über fossile Chitinstrukturen. *Verhandl. Deutsche zool. Ges.* **1922**: 73–74.

LEONOVA, T.B., and ROZANOV, A.Yu. (eds.) 2000. *Palaeontological Institute 1930–2000*. Palaeontological Inst. RAS, Moscow: 144 p.

LEPNEVA, S.G. 1964. Lichinki i kukolki podotryada kol'chatoshchupikovykh (Annulipalpia) (Larvae and Pupae of the Suborder Annulipalpia). Fauna SSSR. Nasekomye rucheyniki **2**(1). Nauka, Moscow – Leningrad: 560 p. (in Russian).

LEPPIK, E.E. 1955. *Dichromena ciliata*, a noteworthy entomophilous plant among Cyperaceae. *Amer. J. Botany* **42**: 455–458.

LEPPIK, E.E. 1971. Palaeontological evidence on the morphogenic development of flower types. *Phytomorphology* **2**(2–3): 164–174.

LERBEKMO, J.F., SWEET, A.R., and ST. LOUIS, R.M. 1987. The relationship between the iridium anomaly and palynological floral events at three Cretaceous-Tertiary boundary localities in western Canada. *Geol. Soc. America Bull.* **99**: 325–330.

LESNE, P. 1920. Quelques insectes du Pliocène supérieur du Comté de Durham. *Bull. Mus. nation. Hist. natur. Paris*: 388–394, 484–488.

LESNE, P. 1930. Le dermeste des cadavres (*Dermestes frischi* Kug.) dans les tombes de l'ancienne Égypte. *Bull. Soc. entomol. Egypte* N.S **14**: 21–24.

LEVCHENKO, V.F., and STAROBOGATOV, Ya.I. 1990. Succession changes and evolution of ecosystems (Some problems in evolutionary ecology). *Zhurnal obshchey biol.* **51**(5): 619–631 (in Russian).

LEVINSON, H., and LEVINSON, A. 1996. *Prionotheca coronata* Olivier (Pimeliinae, Tenebrionidae) recognized as a new species of venerated beetles in the funerary cult of pre-dynastic and archaic Egypt. *J. Appl. Entomol.* **120**: 577–585.

LEWIS, R.E., and GRIMALDI, D.A. 1997. A pulicid flea in Miocene amber from the Dominican Republic (Insecta: Siphonaptera: Pulicidae). *Amer. Mus. Novitates* **3205**: 1–9.

LEWIS, S.E. 1969. Fossil insects in the Latah Formation (Miocene) of Eastern Washington and Northern Idaho. *Northwest Sci.* **43**(3): 99–115.

LEWIS, S.E. 1970. Fossil caddisfly (Trichoptera) cases from the Latah Formation (Miocene) of Eastern Washington and Northern Idaho. *Ann. Entomol. Soc. Amer.* **63**(2): 621–622.

LEWIS, S.E. 1972. Fossil caddisfly (Trichoptera) cases from the Ruby River Basin (Oligocene) of southwestern Montana. *Ann. Entomol. Soc. Amer.* **65**(2): 518–519.

LEWIS, S.E. 1973. A new species of fossil caddisfly (Trichoptera: Limnephilidae) from the Ruby River Basin (Oligocene) of southwestern Montana. *Ann. Entomol. Soc. Amer.* **66**(5): 1173–1174.

LEWIS, S.E. 1976a. A new specimen of the fossil grasshopper (Orthoptera: Caelifera) from the Ruby River Basin (Oligocene) of southwestern Montana. *Ann. Entomo. Soc. Amer.* **69**: 120.

LEWIS, S.E. 1976b. Lepidopteran feeding damage of live oak leaf (*Quercus convexa* Lesquereux) from the Ruby River Basin (Oligocene) of southeastern Montana. *J. Paleontol.* **50**(2): 345–346.

LEWIS, S.E. 1977. Two new species of fossil mayflies (Ephemeroptera: Neoephemeridae and Siphlonuridae) from the Ruby River Basin (Oligocene) of Southwestern Montana. *Proc. Entomol. Soc. Washington* **79**: 583–587.

LEWIS, S.E. 1978. An immature fossil Ephemeroptera (Ephemeridae) from the Ruby River Basin (Oligocene) of Southwestern Montana. *Ann.Entomol. Soc. Amer.* **71**(4): 479–480.

LEWIS, S.E. 1992. Insects of the Klondike Mountain Formation, Republic, Washington. *Washington Geol.* **20**(3): 15–19.

LEWIS, S.E. 1994. Evidence of leaf-cutting bee damage from the Republic sites (middle Eocene) of Washington. *J. Paleontol.* **68**(1): 172–173.

LEWIS, S.E., and WEHR, W.C. 1993. Fossil mayflies from Republic, Washington. *Washington Geol.* **21**: 35–37.

LEWIS, S.E., HEIKES, P.M., and LEWIS, K.L. 1990. Entomofauna from Miocene deposits near Juliaetta, Idaho. *Occ. Pap. Paleobiol.* **4**(9): 1–22.

LEWIS, T. 1973. *Thrips, Their Biology, Ecology and Economic Importance.* Academic Press, London–New York: 349 p.

LIGHT, S.F. 1930. Fossil termite pellets from the Seminole Pleistocene. *Univ. Calif. Publs. Geol. Sci.* **19**: 75–77.

LIN, Q.-B. 1976. The Jurassic fossil insects from Western Liaoning. *Acta Paleontol. Sinica* **15**: 97–118 (in Chinese, with English summary).

LIN, Q.-B. 1977. Fossil insects from Yunnan. In: *Mesozoic Fossils from Yunnan, China* **2**. Beijing, Science Press: 373–381 (in Chinese).

LIN, Q.-B. 1978a. On the fossil Blattoidea of China. *Acta Entomol. Sinica* **21**: 335–342 (in Chinese, with English summary).

LIN, Q.-B. 1978b. Upper Permian and Triassic fossil insects of Guizhou. *Acta Paleontol. Sinica* **17**: 313–317 (in Chinese, with English summary).

LIN, Q.-B. 1980. Mesozoic insects from Zhejiang and Anhui. In: *Division and Correlation of Mesozoic Volcano-sedimentary Formation in Zhejiang and Anhui Provinces.* Science Press, Beijing: 211–238 (in Chinese).

LIN, Q.-B. 1982. Insecta. In: *Palaeontological Atlas of East China. 3. Volume of Mesozoic and Cenozoic.* Geol. Publ. House, Beijing: 148–155 (in Chinese).

LIN, Q.-B. 1983. Cretaceous succession of insect assemblages in China. *Zitteliana* **10**: 393–394.

LIN, Q.-B. 1985a. Insect fossils from the Hanshan Formation at Hanshan County, Anhui Province. *Acta Paleontol. Sinica* **24**: 300–310 (in Chinese, with English summary).

LIN, Q.-B. 1985b. A new cockroach from the Upper Shihhotse Formation (Upper Permian) in Yu County, Henan Province. *Acta Paleontol. Sinica* **24**: 122–124 (in Chinese, with English summary).

LIN, Q.-B. 1986. *Early Mesozoic Fossil Insects from South China.* (*Paleontol. Sinica* **170** (Ser. B, no 21)). Sci. Press, Beijing: 112 p. (in Chinese, with English summary).

LIN, Q.-B. 1994. Cretaceous insects of China. *Cretaceous Res.* **15**: 305–316.

LIN, Q.-B., and LIANG, X.-Y. 1988. A Permian cockroach tegmen from Gongxian, Henan, China. *Acta Paleontol. Sinica* **27**: 640–642 (in Chinese, with English summary).

LIN, Q.-B., and MOU, Ch.-J. 1989. On insects from Upper Triassic Xiaoping Formation, Guangzhou, China. *Acta Paleontol. Sinica* **28**: 598–603 (in Chinese with English summary).

LINCK, O. 1949. Fossile Bohrgänge (*Anobichnium simile* n.g. n.sp.) an einem Keuperholz. *Neues Jahrb. Min., Geol. Paläontol.* B **4–6**: 180–185.

LINDAHL, T. 1997. Facts and artifacts of ancient DNA. *Cell* **90**(1): 1–3.

LINDBERG, H. 1949. Zur Kenntniss der Insektenfauna im Brackwasser des Baltichen Meeres. *Comment. Biol.* **10**(9): 1–206.

LINNAEUS, C. 1751. Philosophia botanica in qua explicatur fundamenta botanica cum definitionibus partium, exemplis terminorum, observationibus rariotum, adjectis figuris aeneis. G. Kiesewetter, Stockholm: 362 p.

LIPIARSKI, I. 1971. On fossil remains of arthropods and plants, and organic matter from cavities in the Triassic rocks of the Cracow area. *Bull. Acad. Polon. Sci. Sér. terre* **19**(2): 71–83.

LISTER, A.M. 1993. The Condover mammoth site: excavation and research 1986–93. *Cranium* **10**: 61–67.

LIVINGSTONE, D. 1978. Phytosuccivorous bugs and cecidogenesis. *J. Indian Acad. Wood Sci.* **9**(1): 39–45.

LIVINGSTONE, D., and LIVINGSTONE, C. 1865. *Narrative of an Expedition to the Zambesi and Its Tributaries; and of the Discoveries of the Lakes Shirwa and Nyassa 1858–1964.* John Murray, London (cited after Russian translation: LIVINGSTON, D., and LIVINGSTON, C. 1956. *Puteshestvie po Zambezi s 1858 po 1864 gg.* Gosudarstnnoe Izdatel'stvo geogr. literatury, Moscow: 393 p.).

LOCKE, A., and COREY, S. 1986. Terrestrial and freshwater invertebrates in the neuston of the Bay of Fundy, Canada. *Canad. J. Zool.* **64**(7): 1535–1541.

LOCKWOOD, J.A., NUNAMAKER, R.A., and PFADT, R.E. 1988. Characterization of Grasshopper Glacier. *18 Internat. Congr. Entomol., Vancouver, July 3 – 9, 1988. Proc.: Abstr. & Author Index.* Vancouver: 60.

LOCKWOOD, J.A., CLAPP, D.A., CONINE, J., and ROMBERG, M.H., Jr. 1989. Grasshopper swimming as a function of sexual and taxonomic differences: evolutionary and ecological implications. *J. Entomol. Sci.* **24**(4): 489–495.

LOCKWOOD, J.A., NUNAMAKER, R.A., PFADT, R.E., and DEBREY, L.D. 1990. Grasshopper Glacier: Characterization of a vanishing biological resource. *Amer. Entomol.* **36**: 18–27.

LOCKWOOD, J.A., THOMPSON, C.D., BEBRAY, L.D., LOVE, C.M., NUNAMAKER, R.A., and PFADT, R.E. 1991. Preserved grasshopper fauna of Knife Point Glacier, Fremont County, Wyoming, U.S.A. *Arctic and Alpine Res.* **23**(1): 108–114.

LOCKWOOD, J.A., DEBREY, L.D., and NUNAMAKER, R.A. 1992a. Preserved grasshopper deposits in the glaciers of the Rocky Mountains, U.S.A. *19 Internat. Congr. Entomol., Beijing, June 28–July 4, 1992. Proc.: Abstr.* Beijing: 174.

LOCKWOOD, J.A., SCHELL, S.P., WANGBERG, J.K., DEBREY, L.D., DEBREY, W.G., and BOMAR, C.R. 1992b. Preserved insects and physical condition of Grasshopper Glacier, Carbon County, Montana, U.S.A. *Arctic and Alpine Res.* **24**(3): 229–232.

LOCQUIN, M.V. 1981. Relations chronophénétiques entre les champignons fossiles et actuels. *Cah. micropaléontol.* **1**: 87–89.

LÖFFLER, H. 1986. An early meromictic stage in Lobsigensee (Switzerland) as evidenced by ostracods and *Chaoborus. Hydrobiologia* **143**: 309–314.

LÖFSTEDT, C. and KOZLOV, M. V. 1997. A phylogenetic analysis of pheromone communication in primitive moths. In: CARDÉ, R. T., and MINKS, A. (eds.). *Insect Pheromones: New Directions.* Chapmann & Hall, New York: 473–489.

LOMNICKI, A.M. 1894. Plejstocenskie owady z Boryslawia. *Wydawn. Muzeum im. Dzieduszyckich* **4**: 92–99.

LOROUGNON, G. 1973. Le vecteur pollinique chez les *Napania* et les *Hypolytrum*, cyperacées du sous-bois des forêts tropicales ombrophiles. *Bull. Jard. Botan. Nation. Belg.* **43**: 33–36.

LOTTER, A.F., WALKER, I.R., BROOKS, S.J., and HOFMANN, W. 1999. An intercontinental comparison of chironomid palaeotemperature inference model: Europe vs North America. *Quatern. Sci. Rev.* **18**: 717–735.

LUCAS, J., and PREVOT, L. 1985. The synthesis of apatite by bacterial activity: mechanism. *Mem. Soc. Géol. Strasbourg* **1985**: 83–92.

LUDVIGSEN, R. 1993. A cryptic fossil of importance. *Gulf Island Guardian* 1993 (Spring): 6–8 (cited after LABANDEIRA 1998c).

LUKASHEVICH, E.D. 1995. First pupae of the Eoptychopteridae and Ptychopteridae from the Mesozoic of Siberia (Insecta: Diptera). *Paleontol. J.* **29**(4): 164–171.

LUKASHEVICH, E.D. 1996a. Mesozoic Dixidae (Insecta: Diptera) and the systematic position of *Dixamima* Rohdendorf and *Rhaetomyia* Rohdendorf. *Paleontol. Zh.* 1: 48–53. (In Russian; translated in *Paleontol. J.* **30**(1): 46–51).

LUKASHEVICH, E.D. 1996b. New chaoborids from the Mesozoic of Mongolia (Diptera: Chaoboridae). *Paleontol. Zh.* N 4: 55–60 (In Russian; English translation: *Paleontol. J.* **30**(5): 551–558).

LUKASHEVICH, E.D., ANSORGE, J., KRZEMIŃSKI, W., and KRZEMIŃSKA, E. 1998. Revision of Eoptychopterinae (Diptera: Eoptychopteridae). *Polsk. Pismo Entomol.* **67**: 311–343.

LUKASHEVICH, E.D, and SHCHERBAKOV, D.E. 1999. A new Triassic family of Diptera from Australia. *Proc. First Palaeoentomol. Conf., Moscow 1998.* AMBA projects International, Bratislava: 81–90.

LUTZ, H. 1984a. Beitrag zur Kenntnis der unteroligozänen Insektenfauna von Ceresté <sic> (Süd-Frankreich). *Documenta naturae* **21**: 1–26.

LUTZ, H. 1984b. Parallelophoridae – isolierte Analfelder eozäner Schaben (Insecta: Blattodea). *Paläontol. Zeitschr.* **58**: 145–147.

LUTZ, H. 1985a. Eine fossile Stechmucke aus dem Unter-Oligozan von Ceréste, Frankreich (Diptera: Culicidae). *Paläontol. Zeitschr.* **59**(3/4): 269–275.

LUTZ, H. 1985b. Eine wasserlebende Käferlarve aus dem Mittel-Eozän der Grube Messel bei Darmstadt. *Courier Forsch.-Inst. Senckenberg* **124**: 1–165.

LUTZ, H. 1988. Riesenameisen und andere Raritäten – die Insektenfauna. In: Schaal S. and Ziegler W. (eds.). *Messel – ein Schaufenster in die Geschichte der Erde und des Lebens.* W. Kramer Verlag, Frankfurt a. Main: 55–67.

LUTZ, H. 1990. Systematische und palökologische Untersuchungen an Insekten aus dem Mittel-Eozän der Grube Messel bei Darmstadt. *Courier Forsch.-Inst. Senckenberg* **124**: 1–165.

LUTZ, H. 1991a. Autochthone aquatische Arthropoda aus dem Mittel-Eozän der Fundstätte Messel (Insecta. Heteroptera; Coleoptera; cf. Diptera-Nematocera; Crustacea: Cladocera). *Courier Forsch.-Inst. Senckenberg* **139**: 119–125.

LUTZ, H. 1991b. *Fossilfundstelle Eckfelder Maar.* Landessammlung für Naturk. Reinland-Pfalz, Mainz: 51 p.

LUTZ, H. 1993a. *Eckfeldapis electrapoides* nov.gen. n.sp., eine "Honigbiene" aus dem Mittel-Oligocän des "Eckfelder Maares" bei Manderscheid/Eifel, Deutschland (Hymenoptera: Apidae, Apinae). *Mainzer naturwiss. Archiv* **31**: 177–199.

LUTZ, H. 1993b. Grabungskampagne im "Eckfelder Maar": 1992. *Mitt. Rheinisch. Naturforsch. Ges.* **14**: 53–59.

LUTZ, H. 1996. Die fossile Insektenfauna von Rott. Zusammensetzung und Bedeutung für die Rekonstruktion des ehemaligen Lebensraums. In: KOENIGSWALD, W., VON (ed.). *Fossillagerstätte Rott bei Hennef im Siebengebirge.* 2. Aufl. Rheinlandia, Siegburg: 41–56.

LUTZ, H. 1997. Taphozönosen terrestrischer Insekten in aquatischen Sedimenten – ein Beitrag zur Rekonstruktion des Paläoenvironments. *Neues Jahrb. Geol. Paläontol. Abhandl.* **203**: 173–210.

LYAL, C.H.C. 1985. Phylogeny and classification of the Psocodea, with particular reference to the lice (Psocodea: Phthiraptera). *Syst. Entomol.* **10**: 145–165.

LYAL, C.H.C. 1986. External genitalia of Psocodea, with particular reference to lice (Phthiraptera). *Zool. Jahrb. Anat.* **114**: 277–292.

MACHIDA, R. 1981. External features of embryonic development of a jumping bristletail, *Pedetontus unimaculatus* Machida (Insecta, Thysanura, Machilidae). *J. Morphol.* **168**: 339–355.

MACGINTIE, H.D. 1953. Fossil plants of the Florissant beds, Colorado. *Carnegie Inst. Washington Publ., Contrib. to Palaeontol.* **599**: 198 p.

MACKAY, R.J. 1992. Colonization by lotic macroinvertebrates: a review of processes and patterns. *Canad. J. Fish. Aquat. Sci.* **49**: 617–628.

MACKAY, W.P., and ELIAS, S.A. 1992. Late Quaternary ant fossils from packrat middens (Hymenoptera: Formicidae): implications for climatic change in the Chihuahuan Desert. *Psyche* **99**(2–3): 169–184.

MACKERRAS, I.M. 1970. Superclass Hexapoda. In: CSIRO. *The Insects of Australia.* Melbourne Univ. Press, Melbourne: 152–167.

MÁGNANO, M.G., BUATOIS, L.A., MAPLES, C.G., and LANIER, W.P. 1997. *Tonganoxichnus*, a new insect trace from the Upper Carboniferous of eastern Kansas. *Lethaia* **30**: 113–125.

MAI, D.H. 1995. *Tertiäre Vegetationsgeschichte Europas*. Gustav Fischer Verlag, Jena etc.: 691 p.

MAJECKI, J., HIGLER, L.W.G., VERDONSCHOT, P.F.M., and SCHOT, J. 1997. Influence of sand cover on mortality and behaviour of *Agapetus fuscipes* larvae (Trichoptera: Glossosomatidae). *Proc. 8th Internat. Symp. Trichoptera*. Ohio Biol. Surv.: 283–288.

MAKARKIN, V.N. 1990. New Neuroptera from the Upper Cretaceous of Asia. In: AKIMOV I.A. (ed.). *Novosti faunistiki i sistematiki* (News in Faunistics and Systematics). Naukova Dumka, Kiev: 63–68 (in Russian).

MAKARKIN, V.N. 1991. Miocene insects (Neuroptera) from the northern Caucasus and Sikhote-Alin. *Paleontol. J.* **25**(1): 55–65.

MALICKY, H. 1973. Trichoptera (Köcherfliegen). In: KUKENTHAL, W. (ed.). *Handbuch der Zoologie* **4**(2) 2/29. Walter de Gruyter, Berlin: 114 p.

MALICKY, H. 1988. Spuren der Eiszeit in der Trichopterofauna Europas (Insecta, Trichoptera) *Riv. Idrobiol.* **27**(2–3): 247–297.

MAMAEV, B.M. 1971. Insect use of conifer resins as a habitat (with a review of the resinicolous insects). *Zhurnal obshchey biol.* **32**(4): 501–597 (in Russian).

MAMAY, S.H. 1976. Paleozoic origin of the cycads. *U.S. Geol. Surv. Prof. Paper* **934**: 1–48.

MANI, M.S. 1945. Description of some fossil arthropods from India. *Indian J. Entomol.* **6**: 61–64.

MANI, M.S. 1947. Some fossil arthropods from the Saline Series in the Salt Range of the Punjab. *Proc. Indian Nation. Acad. Sci.* **B 16**(1–4): 43–56.

MANI, M.S. 1964. *The Ecology of Plant Galls. Monogr. Biol.* **12**. W. Junk, The Hague: 434 p.

MANI, M.S. 1968. *Ecology and Biogeography of High Altitude Insects*. W. Junk, The Hague: 527 p.

MANSKE, L.L., and LEWIS, S.E. 1990. Two fossil adult cockroaches (Blattaria) from the Cretaceous of Minnesota. *J. Paleontol.* **64**: 159–161.

MAO, S., YU, J., and LENTIN, J.K. 1990. Palynological interpretation of Early Cretaceous non-marine strata of Northeast China. *Rev. Palaobot. Palynol.* **65**: 115–118.

MAPLETON, C.W. 1879. Great flight of beetles. *Entomol. Mon. Mag.* **16**(80): 18–19.

MARCHAL-PAPIER, F. 1998. *Les insectes du Buntsandstein des Vosges (NE de la France). Biodiversité et contribution aux modalites de la crise biologique du Permo-Trias.* Unpublished PhD thesis. Louis Pasteur Univ, Strasbourg.

MARCHAL-PAPIER, F., NEL, A., and GRAUVOGEL-STAMM, L. 2000. Nouveaux Orthoptéres (Ensifera, Insecta) du Trias des Vosges (France). *Acta Geol. Hispanica* **35**(1–2): 5–18.

MARDEN, J.H. 1995. Flying lessons from a flightless insect. *Natur. Hist.* **104**(2): 4–8.

MARDEN, J.H. and KRAMER, M.G. 1994. Surface-skimming stoneflies: a possible intermediate stage in insect flight evolution. *Science* **266**: 427–430.

MARGOLIES, D.C. 1987. Conditions elucing aerial dispersal behaviour in Banks grass mite, *Oligonychus pratensis* (Acari, Tetranychidae). *Environ. Entomol.* **16**: 928–932.

MARKEVICH, V.S. 1995. *Melovaya palinoflora severa Vostochnoy Azii.* (Cretaceous Palynoflora of the North of Eastern Asia). Dal'nauka, Vladivostok: 200 p. (in Russian).

MARKS, E.P., and LAWSON, F.A. 1962. A comparative study of the dictyopteran ovipositor. *J. Morphol.* **3**: 132–171.

MARLOTH, R. 1914. Note on the entomophilous nature of *Encephalartos*. *Trans. Roy. Soc. S. Africa* **4**: 69–71.

MARSHALL, C.R. 1991. Estimation of taxonomic ranges from the fossil record. In: GNILINSKY, N.L., and SIGNOR, P.W. (eds.). *Analytical Palaeobiology. Short Courses in Palaeontology* **4**: 19–37.

MARSHALL, J.E.A., and WHITESIDE, D.I. 1980. Marine influence in the Triassic "uplands". *Nature* **287**: 627–628.

MARTILL, D.M. 1987. Prokaryote mats replacing soft tissues in Mesozoic marine reptiles. *Modern Geol.* **11**(5): 265–269.

MARTILL, D. 1993. *Fossils of the Santana and Crato Formations, Brazil. (Palaeontol. Assoc. Field Guides to Fossils* **5**). Palaeontol. Association, London: 159 p.

MARTIN, L.D., and WEST, D.L. 1995. The recognition and use of dermestid (Insecta, Coleoptera) pupation chambers in paleoecology. *Palaeogeogr., Palaeoclimatol., Palaeoecol.* **113**: 303–310.

MARTIN, W., GIERL, A., and SAEDLER, H. 1989. Molecular evidence for pre-Cretaceous angiosperm origin. *Nature* **339**: 46–48.

MARTINI, E. 1967. Die oligozäne Fossilfundstätte Sieblos an der Wasserkuppe. *Natur u. Mus.* **97**(1): 1–8.

MARTINI, E. 1971. Neue Insektenfunde aus dem Unter-Oligozän von Sieblos/Rön. *Senckenberg. Lethaea* **52**(4): 359–369.

MARTINELL, J., and MARTÍNEZ-DELCLÒS, X. 1990. Observaciones de laboratorio sobre la flotabilidad de los insectos. *Com. Reunión de Tafonomía y Fossilización (Madrid)*: 201–209.

MARTINEZ, S. 1982. Catalogo sistematico de los insectos fossiles de America del Sur. *Rev. Facultad de Humanidades y Ciencias*. Series Ciencias de la Terra **1**(2): 29–83.

MARTÍNEZ-DELCLÒS, X. 1989. *Chresmoda aquatica* n. sp. insecto Chresmodidae del Cretácico inferior de la Sierra del Montsec (Lleida, España). *Rev. Español. Paleontol.* **4**: 67–74.

MARTÍNEZ-DELCLÒS, X. (ed.) 1991. *Les calcàries litogràfiques del Cretaci inferior del Montsec. Deu anys de campanyes paleontològiqies.* Inst. d'Estudis Ilerdencs, Lleida: 160 p. (Catalonian) + 106 p. (English version).

MARTÍNEZ-DELCLÒS, X. (ed.). 1995. *Montsec and Mont-ral-Alcover, Two Konservat-Lagerstätten, Catalonia, Spain. II Internat. Symposium on Lithographic Limestones. Field Trip Guide Book. July 9–11, 1995.* Inst. d'Estudis Ilerdencs, Lleida: 97 p.

MARTÍNEZ-DELCLÒS, X. 1996. El registro fósil de los insectos. *Bol. Asoc. Español. Entomol.* **20**(1–2): 9–30.

MARTÍNEZ-DELCLÒS, X., and MARTINELL, J. 1993. Insect taphonomy experiments. Their application to the Cretaceous outcrops of lithographic limestones from Spain. *Kaupia* **2**: 133–144.

MARTÍNEZ-DELCLÒS, X., NEL, A., and POPOV, Yu.A. 1995. Systematics and functional morphology of *Iberonepa romerali* n. gen. and sp., Belostomatidae from the Spanish Lower Cretaceous (Insecta, Heteroptera). *J. Paleontol.* **69**(3): 496–508.

MARTÍNEZ-DELCLÒS, X., PEÑALVER-MOLLÁ, E., and RASNITSYN, A. 1998. Los Hymenoptera del ámbar del Cretácico inferior de Álava (País Vasco, España). *World Congress on amber inclusions 20th–23th October 1998* Vittoria-Gasteiz: 119.

MARTINS-NETO, R.G. 1989a. Primero registro de Phasmatodea (Insecta, Orthopteromorpha) na Formação Santana, Bacia do Araripe (Cretáceo Inferior) nordeste do Brasil. *Acta Geol. Leopoldensia* (12): 91–104.

MARTINS-NETO, R.G. 1989b. Novos insetos terciários do Estado de São Paulo. *Rev. Brasil. Geociênc.* **19**(3): 375–386.

MARTINS-NETO, R.G. 1990a. *Systemática dos Ensifera Insecta (Orthopteroidea) da Formação Santana (Cretáceo Inferior do nordeste do Brasil).* Diss. Maestrado. Univ. Sao Paulo, Sao Paulo: 267–275.

MARTINS-NETO, R.G. 1990b. The family Locustopsidae (Insecta, Caelifera) in the Santana Formation (Lower Cretaceous, Northeast Brasil). I – Description of two new species of the genus *Locustopsis* Handlirsch and three new species of the genus *Zessinia* n. gen. *Atas I Sympós. Bacia do Araripe e Bacia Interiores do Nordeste. Crato, 14 a 16 junho 1990*, DNPM: 277–291.

MARTINS-NETO, R.G. 1990c. Um novo gênero e duas novas espécies de Tridactylidae (Insecta, Orthopteridea) na Formação Santana (Cretáceo Inferior do nordeste do Brasil). *An. Acad. brasil. Ciên.* **62**: 51–59.

MARTINS-NETO, R.G. 1990d. Neuropteros (Insecta: Planipennia) da Formação Santana (Cretaceo Inferior) Bacia do Araripe, Nordeste do Brasil. VI – Ensaio filogenetico das especies do genero *Blittersdorffia* Martins-Neto & Vulcano, com descricao de nova especie. *Acta Geol. Leopoldensia* (31) **13**: 3–12.

MARTINS-NETO, R.G. 1990e. Primeiro registro de Dermaptera (Insecta, Orthopteromorpha) na Formação Santana (Cretáceo Inferior), Bacia do Araripe, Nordeste do Brasil. *Rev. brasil. Entomol.* **34**: 775–784.

MARTINS-NETO, R.G. 1991a. Systemática dos Ensifera Insecta (Orthopteroidea) da Formação Santana (Cretáceo Inferior do nordeste do Brasil). *Acta Geol. Leopoldensia* (33): 3–162.

MARTINS-NETO, R.G. 1991b. Primeiro registrado de Eumastacoidea (Insecta, Caelifera) da Formação Santana, Cretàceo Inferior do Nordeste do Brasil. *An. Acad. Brasil. Ciên.* **63**(1): 91–92.

MARTINS-NETO, R.G. 1991c. *Paleoarthropodologia aplicada.* Univ. Guarulhos, Saõ Paolo: 178 p.

MARTINS-NETO, R.G. 1995a. Araripelocustidae, fam.n. Una nova familia de gafanhotos (Insecta, Caelifera) da Formação Santana, Cretàceo Inferior do Nordeste do Brasil. *Rev. brasil. Entomol.* **39**(2): 311–319.

MARTINS-NETO, R.G. 1995b. Complementos ao estudo sobre os Ensifera (Insecta, Orthopteroida) da Formação Santana, Cretáceo Inferior do Nordeste do Brasil. *Rev. brasil. Entomol.* **39**(2): 321–345.

MARTINS-NETO, R.G. 1996. New mayflies (Insecta, Ephemeroptera) from the Santana Formation (Lower Cretaceous, Araripe Basin, Northestern Brazil). *Rev. Español. Paleontol.*, **11**(2): 177–192.

MARTINS-NETO, R.G. 1997. A paleoentomofauna da Formação Tremembé (Bacia de Taubaté) Oligoceno do Estado de São Paulo: descrição de novos hemípteros (Insecta). *Rev. Univ. Guarulhos Sér. Geociências* **2**(6): 66–69.

MARTINS-NETO, R.G. 1998a. A new genus of the family Locustopsidae (Insecta Caelifera) in the Santana Formation (Lower Cretaceous, Northeast Brazil). *Rev. Español. Paleontol.* **13**(2): 133–138.

MARTINS-NETO, R.G. 1998b. A paleoentomofauna da Formação Tremembé (Bacia de Taubaté) Oligoceno do Estado de São Paulo: novos Hemiptera, Auchenorrhyncha, Hymenoptera, Coleoptera e Lepidoptera (Insecta). *Rev. Univ. Guarulhos Sér. Geociências* **3**(6): 58–70.

MARTINS-NETO, R.G. 1999a. New genus and new species of Lepidoptera (Insecta, Eolepidopterigidae) from Santana Formation (Lower Cretaceous, Northeast Brazil). *Boll. 5° Simp. Cretác. Brasil*: 531–553.

MARTINS-NETO, R.G. 1999b. Estado actual del conocimiento de la paleoentomofauna brasileña. *Rev. Soc. Entomol. Argent.* **58**(1–2): 71–85.

MARTINS-NETO, R.G. 2000. Remarks on the neuropterofauna (Insecta, Neuroptera) from the Brazilian Cretaceous, with keys for the identification of the known taxa. *Acta Geol. Hispanica* **35**(1–2): 97–118.

MARTINS-NETO, R.G., and CALDAS, E.B. 1990. Efêmeras escavadoras (Insecta, Ephameroptera, Ephemeroidea) na Formação Santana (Cretáceo inferior), Bacia do Araripe, Nordeste do Brasil: descrição de três novas espécies (ninfas). *Atas do I Simpósio sobre a Bacia do Araripe e Bacias Interiores do Nordeste, Crato, 14 a 16 junho 1990.* DNPM: 265–275.

MARTINS-NETO, R.G., and GALLEGO, O.F. 2000. First record of Glosselytroptera (Insecta) in the Los Rastros Formation (Triassic, Argentina). *I Simpósio Brasileiro de Paleoartropodologia, I Simpósio Sudamericano de Paleoartropodologia, I Internat. Meeting on Paleoarthropodology. Abstracts. Ribeirão Preto, SP, Brazil 3 to 8.9.2000.* Ribeirão Preto, USP/Soc. Brasil. Paleoarthropodol.: 37.

MARTINS-NETO, R.G., and KUCERA-SANTOS, J.C. 1994. Um novo genero e nova especie de mutuca (Insecta, Diptera, Tabanidae) da Formação Santana (Cretaceo Inferior), Bracia do Araripe, Nordeste do Brazil. *Acta Geol. Leopoldensia* **17** (39): 289–297.

MARTINS-NETO, R.G. and ROHN, R. 1996. Primeiro registro de inseto na Formação Rio do Rasto, Bacia do Paraná, com descrição de novo táxon. *Rev. Geociências* **15**(1): 243–251.

MARTINS-NETO, R.G., and VULCANO, M.A. 1988. Neuropteros (Insecta: Planipennia) da Formacão Santana (Cretaceo Inferior), Bacia do Araripe, Nordeste do Brasil. I – Familia Chrysopidae. *An. Acad. Brasil. Ciênc.* **60**(2): 189–201.

MARTINS-NETO, R.G., and VULCANO, M.A. 1989a. Neuropteros (Insecta: Planipennia) da Formacão Santana (Cretaceo Inferior), Bacia do Araripe, Nordeste do Brasil. II – Superfamilia Myrmeleontoidea. *Rev. brasil. Entomol.* **33**(2): 367–402.

MARTINS-NETO, R.G., and VULCANO, M.A. 1989b. Neuropteros (Insecta: Planipennia) da Formacão Santana (Cretaceo Inferior), Bacia do Araripe, Nordeste do Brasil. IV. – Complementos as partes I e II, com descricão de novos taxa. *An. Acad. Brasil. Ciên.* **61**(3): 311–318.

MARTINS-NETO, R.G., and VULCANO, M.A. 1989c. Amphiesmenoptera (Trichoptera + Lepidoptera) da Formacão Santana (Cretaceo Inferior), Bacia do Araripe, Nordeste do Brasil. I – Lepidoptera (Insecta). *An. Acad. Brasil. Ciênc.* **61**: 459–466.

MARTINS-NETO, R.G., and VULCANO, M.A. 1990a. Primeiro registro de Raphidioptera (Neuropteroidea) na Formacão Santana (Cretaceo Inferior), Bacia do Araripe, Nordeste do Brasil. *Rev. brasil. Entomol.* **34**(1): 241–249.

MARTINS-NETO, R.G., and VULCANO, M.A. 1990b. Neuropteros (Insecta, Planipennia) da Formacão Santana (Cretaceo Inferior) Bacia do Araripe, Nordeste do Brasil. III. Superfamilia Mantispoidea. *Rev. brasil. Entomol.* **34**(3): 619–625.

MARTINSSON, A. 1970. Toponomy of trace fossils. In: CRIMES, T.P., and HARPER, J.C. (eds.). *Trace Fossils.* (Geol. J., Special Issue 3) Seel House, Liverpool: 323–330.

MARTY, P. 1894. De l'ancienneté de la *Cecidomyia fagi. Feuille Jeun. Natur.* **24**: 173.

MARTYNOV, A.V. 1925a. Über zwei Grundtypen der Flügel bei den Insekten und ihre Evolution. *Zeitschr. Ökol. Morphol. Tiere* **4**: 465–501.

MARTYNOV, A.V. 1925b. To the knowledge of fossil insects from Jurassic beds in Turkestan. 1. Raphidioptera. *Bull. Acad. Sci. Russie* **1925**: 233–246.

MARTYNOV, A.V. 1925c. On a new interesting fossil beetle from the Jurassic beds in North Turkestan. *Russ. Entomol. Obozrenie* **19**: 73–78 (in Russian with English summary).

MARTYNOV, A.V. 1933. Permian fossil insects from the Arkhangelsk district. Pt. II. Neuroptera, Megaloptera and Coleoptera, with the description of two new beetles from Tikhiye Gory. *Trudy Paleozool. Inst.* **2**. Moscow: 63–96 (in Russian with English summary).

MARTYNOV, A.V. 1937. *Liasovye nasekomye Shuraba i Kizil-Kii (*Liassic Insects from Shurab and Kyzyl-Kiya). *Trudy Paleontol. Inst. Akad. Nauk SSSR* **7**(1). Moscow – Leningrad : 231 p.

MARTYNOV, A.V. 1938a. Fossil insect localities in USSR. *Trudy Paleontol. Inst. Akad. Nauk SSSR* **7** (3). Moscow: 6–28 (in Russian).

MARTYNOV, A.V. 1938b. Permian fossil insects from Arkhangelsk district. Part V. The family Euthygrammatidae and its relatives. *Trudy Paleontol. Inst. Akad. Nauk SSSR* **7**(3). Moscow: 69–80 (in Russian, with English summary).

MARTYNOV, A.V. 1938c. An essay on the geological history and phylogeny of the insect orders (Pterygota). Pt. I. Palaeoptera and Neoptera-Polyneoptera. *Trudy Paleontol. Inst. Akad. Nauk SSSR* **7**(4). Moscow: 148 p. (in Russian, with French summary).

MARTYNOV, A.V. 1938d. On a new Permian order of orthopteroid insects Glosselythrodea. *Izvestiya Akad. Nauk SSSR* Ser. Biol (1): 187–206 (in Russian, with long English summary).

MARTYNOVA, O.M. 1947a. Two new snake flies (Raphidioptera) from the Jurassic shales of Karatau. *Doklady Akad. Nauk SSSR* New Ser. **56**(6): 635–637.

MARTYNOVA, O.M. 1947b. On the nature of the *Pectinariopsis* Andr. tubes (Trichoptera, non Polychaeta). *Entomol. Obozrenie* **29**(3): 152–153 (in Russian).

MARTYNOVA, O.M. 1948. *Materialy po evolyutsii Mecoptera* (Materials on the Evolution of Mecoptera). *Trudy Paleontol. Inst. Akad. Nauk SSSR* **14**(1). Izdatel'stvo Akad. Nauk SSSR, Moscow – Leningrad: 77 p. (in Russian).

MARTYNOVA, O.M. 1952. Order Glosselytrodea in Permian deposits of Kemerovo Region. *Trudy Paleontol. Inst. Akad. Nauk SSSR* **40**. Moscow: 187–196 (in Russian).

MARTYNOVA, O.M. 1953. *Nastavlenie dlya sborov iskopaemykh nasekomykh* (Instruction to Collecting of the Fossil Insects). Iadatel'stvo Akad. Nauk SSSR, Moscow: 16 p.

MARTYNOVA, O.M. 1954. Neuroptera from the Cretaceous deposits of Siberia. *Doklady Akad. Nauk SSSR* New Ser. **94**(6): 1167–1169.

MARTYNOVA, O.M. 1958. New insects from the Permian and Mesozoic deposits of USSR. *Materialy k Osnovam paleontologii* **2**. Paleontol. Inst. Akad. Nauk SSSR, Moscow: 69–94 (in Russian).

MARTYNOVA, O.M. 1959. Phylogenetic relationships of insects in the mecopteroid complex. *Trudy Inst. Morfologii Zhivotnykh im. A.N. Severtzova* **27**: 221–230 (in Russian).

MARTYNOVA, O.M. 1961a. Order Glosselytrodea. In: ROHDENDORF, B.B., BECKER-MIGDISOVA, E.E., MARTYNOVA, O.M., and SHAROV, A.G. (eds.). 1961. *Paleozoiskie nasekomye Kuznetskogo basseina* (Paleozoic Insects of the Kuznetsk Basin). *Trudy Paleontol. Inst. Akad. Nauk SSSR* **85**. Akad. Nauk SSSR: 247–269 (in Russian).

MARTYNOVA, O.M. 1961b. Extant and extinct snake-flies (Insecta, Raphidioptera). *Paleontol. Zhurnal* (3): 73–83 (in Russian).

MARTYNOVA, O.M. 1962a. Order Miomoptera. In: ROHDENDORF, B.B. (ed.) 1962. *Osnovy paleontologii. Tracheinye i khelitserovye.* Akad. Nauk SSSR, Moscow: 140–142 (in Russian; English translation: ROHDENDORF, B.B. (editor-in-chief) *Fundamentals of Palaeontology* **9**. Arthropoda, Tracheata, Chelicerata. Amerind Publ. Co., New Delhi 1991: 183–185).

MARTYNOVA, O.M. 1962b. Order Neuroptera. In: ROHDENDORF, B.B. (ed.) 1962. *Osnovy paleontologii. Tracheinye i khelitserovye.* Akad. Nauk SSSR, Moscow: 157–159 (in Russian; English translation: ROHDENDORF, B.B. (editor-in-chief) *Fundamentals of Palaeontology* **9**. Arthropoda, Tracheata, Chelicerata. Amerind Publ. Co., New Delhi 1991: 209–212).

MARTYNOVA, O.M. 1962c. Order Glosselytrodea. In: ROHDENDORF, B.B. (ed.) 1962. *Osnovy paleontologii. Tracheinye i khelitserovye.* Akad. Nauk SSSR, Moscow.: 157–159 (in Russian; English translation: ROHDENDORF, B.B. (editor-in-chief) *Fundamentals of Palaeontology* **9**. Arthropoda, Tracheata, Chelicerata. Amerind Publ. Co., New Delhi 1991: 209–212).

MARTYNOVA, O.M. 1962d. Ordo Mecoptera. Scorpionflies. In: ROHDENDORF, B.B. (ed.) 1962. *Osnovy paleontologii. Tracheinye i khelitserovye.* Akad. Nauk SSSR, Moscow: 283–294 (in Russian; English translation: ROHDENDORF, B.B. (editor-in-chief) *Fundamentals of Palaeontology* **9**. Arthropoda, Tracheata, Chelicerata. Amerind Publ. Co., New Delhi 1991: 405–424).

MARTYNOVA, O.M. 1962e. Order Trichoptera. In: ROHDENDORF, B.B. (ed.) 1962. *Osnovy paleontologii. Tracheinye i khelitserovye.* Akad. Nauk SSSR, Moscow: 294–302 (in Russian; English translation: ROHDENDORF, B.B. (editor-in-chief) *Fundamentals of Palaeontology* **9**. Arthropoda, Tracheata, Chelicerata. Amerind Publ. Co., New Delhi 1991: 424–437).

MASAFERRO, J., LAMI, A., GUILIZZONI, P., and NIESSEN, F. 1993. Record of changes in the fossil chironomids and other parameters in the volcanic Lake Nemi (central Italy). *Verhandl. Internat. Ver. Limnol.* **25**: 1113–1116.

MASNER, L. 1969a. A scelionid wasp surviving unchanged since Tertiary (Hymenoptera Proctotrupoidea). *Proc. Entomol. Soc. Washington* **71**(3): 397–400.

MASNER, L. 1969b. The geographical distribution of recent and fossil Ambositrinae. *Tagungsber. Deutsche Akad. Landwirtschaftwiss. Berlin* **80**(1): 105–109.

MASSELOT, G., and NEL, A. 1999. *Pseudokageronia thomasi* gen. nov., sp. nov. from the Upper Miocene of Murat (France) [Ephemeroptera: Heptageniidae]. *Ephemera* **1**(1): 61–73.

MATHIESEN, F. J. 1967. Notes on some fossil phryganidean larval tubes from theTertiary of Denmark and Greenland. *Medd. Dansk. Geol. Foren.* **17**(1): 90–94.

MATILE, L. 1997. Phylogeny and evolution of the larval diet in Sciaroidea (Diptera, Bibionomorpha) since the Mesozoic. In: GRANDCOLAS, P. (ed.). *The Origin of Biodiversity in Insects: Phylogenetic Tests of Evolutionary Scenarios. Mém. Mus. nation.Hist. natur. (Paris)* **173**(zool.): 273–303.

MATSON, P. 1990. Plant-soil interactions in primary succession at Hawaii Volcanoes National Park. *Oecologia* **85**(2): 241–246.

MATSUDA, R. 1957. Comparative morphology of the abdomen of a machilid and a rhaphidiid. *Trans. Amer. Entomol. Soc.* **83**: 39–63.

MATSUDA, R. 1965. *Morphology and Evolution of the Insect Head.* Mem. Amer. Entomol. Inst. **4**. Amer. Entomol. Inst., Ann Arbor: 334 p.

MATSUDA, R. 1970. *Morphology and Evolution of the Insect Thorax.* Mem. Entomol. Soc. Canada. **76**. Entomol. Soc. Canada, Ottawa: 431 p.

MATSUDA, R. 1976. *Morphology and Evolution of the Insect Abdomen.* Pergamon Press, Oxford etc.: 534 p.

MATTHEWS, J.V., Jr. 1976a. Insect fossils from the Beaufort Formation: geological and biological significance. *Geol. Surv. Canada Papers* **76** 1B: 217–227.

MATTHEWS, J.V., Jr. 1976b. Arctic steppe – an extinct biome. *AMQUA Abstr.* **4**: 73–77.

MATTHEWS, J.V., Jr. 1977. Tertiary Coleoptera fossils from the North American Arctic. *Coleopterists' Bull.* **31**: 297–308.

MATTHEWS, J.V., Jr. 1980. Tertiary land bridges and their climate: backdrop for development of the present Canadian insect fauna. *Canad. Entomol.* **112**: 1089–1103.

MATTHEWS, J.V., Jr., OVENDEN, L.E., and FYLES, J.G. 1990. Plant and insect fossils from the Late Tertiary Beaufort Formation on Prince Patrick Island, N.W.T. In: HARRINGTON, C.R. (ed.). *Canada's Missing Dimension: Science and History in the Canadian Arctic Islands.* **1**. National Museums of Canada, Ottawa: 105–139.

MAULIK, P.K., and CHAUDHURI, A.K. 1983. Trace fossils from continental Triassic red beds of the Gondwana sequence, Pranhita-Godavari Valley, South India. *Palaeogeogr., Palaeoclimatol., Palaeoecol.* **41**(1–2): 17–34.

MAYHEW, D.F. 1977. Avian predators as accumulators of fossil mammal material. *Boreas* **6**(1): 25–31.

MAZZAROLO, L.A., and AMORIM, D.S. 2000. *Cratomyia macrorrhyncha*, a Lower Cretaceous brachyceran fossil from the Santana Formation, Brazil, representing a new species, genus and family of the Stratiomyomorpha (Diptera). *Insect Syst. Evol.* **31**: 91–102.

MAZZONI, A.F. 1985. Notonectidae (Hemiptera, Heteroptera) de la Formación La Cantera (Cretácico inf.), Provincia de San Luis, Argentina. *Bol. Acad. Nacion. Cien.* **56**(3–4): 759–773.

MAZZONI, A. and HÜNICKEN, M.A. 1984. Ontogenia de los Notonectidos (Insecta, Heteroptera) del Cretácico inferior de San Luis, Argentina. *Mem. III Congr. Latinoamer. Paleontol.*: 388–393.

MCALPINE, J.F. 1970. First record of calypterate flies in the Mesozoic era (Diptera: Calliphoridae). *Canad. Entomol.* **102**: 342–346.

MCALPINE, J.F. 1989. Phylogeny and classification of Muscomorpha. In: MCALPINE, J.F. and WOOD, D.M. (coordinators). *Manual of Nearctic Diptera* **3**. Research Branch Agricult. Canada Monogr. **3**. Ottawa: 1397–1518.

MCALPINE, J.F., and MARTIN, J.E.H. 1969. Canadian amber – a palaeontological treasure chest. *Canad. Entomol.* **101**: 819–838.

MCBREARTY, S. 1990. Consider the humble termite: Termites as agents of postdepositional disturbance at African archaeological sites. *J. Archaeol. Sci.* **17**: 111–143.

MCCAFFERTY, W.P. 1991. Toward a phylogenetic classification of the Ephemeroptera (Insecta): a commentary on systematics. *Ann. Entomol. Soc. Amer.* **84**: 343–360.

MCCAFFERTY, W.P. 1997. Discovery and analysis of the oldest mayflies (Insecta, Ephemeroptera) known from amber. *Bull. Soc. Hist. Natur. Toulouse.* **133**: 77–82.

MCCAFFERTY, W.P. 2000. Phylogenetic systematics of the major lineages of Pannote mayflies (Ephemeroptera: Pannota). *Trans. Amer. Entomol. Soc.* **126**(1): 9–101.

MCCAFFERTY, W.P., and SINITSHENKOVA, N.D. 1983. *Litobrancha* from the Oligocene in Eastern Asia (Ephemeroptera: Ephemeridae). *Ann. Entomol. Soc. Amer.* **76**(2): 205–208.

MCCALL, P.L., and TEVESZ, M.J.S. 1982. The effects of benthos on physical properties of freshwater sediments. In: MCCALL, P.L., and TEVESZ, M.J.S. (eds.). *Animal-Sediment Relations: The Biogenic Alteration of Sediments.* Plenum, New York – London: 105–176.

MCCLURE, M.S. 1977. Population dynamics of the red pine scale, *Matsucoccus resinosae* (Homoptera: Margarodidae): the influence of resinosis. *Environ. Entomol.* **6**(6): 789–795.

MCCLURE, M.S. 1984. Influence of cohabitation and resinosis on site selection and survival of *Pineus boerneri* Annand and *P. coloradensis* (Gillette) (Homoptera: Adelgidae) on red pine. *Environ. Entomol.* **13**(3): 657–663.

MCCOBB, L.M.E., DUNCAN, I.J., JARZEMBOWSKI, E.A., STANKIEWICZ, B.A., WILLS, W.A. and BRIGGS, D.E.G. 1998. Taphonomy of the insects from the Insect Bed (Bembridge Marls), late Eocene, Isle of Wight, England. *Geol. Mag.* **135**: 553–563.

MCKAY, I.J., and RAYNER, R.J. 1986. Cretaceous fossil insects from Orapa, Botswana. *J. Entomol. Soc. S. Africa* **49**: 7–17.

MCKITTRICK, F.A. 1964. Evolutionary studies on cockroaches. *Cornell Univ. Agricult. Exper. Station Mem.* **389**: 1–197.

MCLACHLAN, I.R., and MCMILLAN, I.K. 1976. Review and stratigraphic significance of southern Cape Mesozoic palaeontology. *Trans. Geol. Soc. S. Africa* **79**(2): 197–212.

MCLOUGHLIN, S. 1992. Late Permian plant megafossils from the Bowen Basin, Queensland, Australia. Part 1. *Palaeontographica* B **228**: 105–149.

MCNAUGHTON, S.J. 1984. Grazing lawns: animals in herds, plant form, and co-evolution. *Amer. Naturalist* **124**(6): 863–886.

MCNAUGHTON, S.J. 1986. On plants and herbivores. *Amer. Naturalist* **128**(5): 765–770.

MEDLER, J.T., and GHOSH, A.K. 1968. Apterous aphids in water, wind and suction traps. *J. Econ. Entomol.* **61**: 422–270.

MEDVEDEV, G.S., and TER-MINASSIAN, M.E. 1973. A new nitidulid species of the genus *Haptoncus* Murray (Coleoptera, Nitidulidae) from Fiji Islands. *Entomol. Obozrenie* **52**(1): 151–153 (in Russian).

MEDVEDEV, L.N. 1976. On the composition of the entomoassemblages from the Holocene badger coprolites from the vicinities of Moscow. In: SUKACHEV, V.N. (ed.). *Istoriya biotsenozov SSSR v golotsene* (The History of Biocoenoses of the USSR in the Holocene). Nauka, Moscow: 183–193 (in Russian).

MEDVEDEV, S.G. 1994. Morphological basis of the classification of fleas. *Entomol. Obozrenie* **73**(1): 22–43 (in Russian, English translation: *Entomol. Review* **73**(1): 30–51).

MEDVEDEV, S.G. 1998. Classification of the flea order and its theoretical foundations. *Entomol. Obozrenie* **77** (4): 904–922 (in Russian, English translation: *Entomol. Review* **77**(4): 1080–1093).

MEEUSE, A.D.J. 1978. Nectarial secretion, floral evolution, and the pollination syndrome in early angiosperms. *Proc. K. Nederl. Akad. Wetensch.* C **81**: 300–326.

MEEUSE, A.D.J., DE MEIJER, A.H., MOHR, O.W.P., and WELLINGA, S.M. 1990a. Entomophily in the dioecious gymnosperm *Ephedra aphylla* Forsk. (= *E. alte* C.A. Mey.), with some notes on *E. campylopoda* C.A. Mey.). III. Further anthecological studies and relative importance of entomophily. *Israel J. Botany* **39**: 113–123.

MEEUSE, A.D.J., DE MEIJER, A.H., MOHR, O.W.P., and WELLINGA, S.M. 1990b. Entomophily in the dioecious gymnosperm *Ephedra aphylla* Forsk. (= *E. alte* C.A. Mey.), with some notes on *E. campylopoda* C.A. Mey.). IV. The cohesive and adhesive properties of *E. aphylla* pollen. In: MEEUSE, A.D.J. (ed.). *Flowers and Fossils Supplement*. Delft, Eburon: 1–9.

MEEUSE, B.J.D. 1959. Beetles as pollinators. *Biologist* **42**: 22–32.

MEHRA, P.N. 1950. Occurrence of hermaphrodite flowers and the development of female gametophyte in *Ephedra intermedia* Schrenk. et Mey. *Ann. Botany* n.s. **14**: 164–180.

MEINANDER, M. 1972. A revision of the family Coniopterygidae (Planipennia). *Acta Zool. Fennica* **136**: 1–223.

MEINANDER, M. 1975. Fossil Coniopterygidae (Neuroptera). *Notulae Entomol.* **55**: 53–57.

MEIßNER, B. 1976. Das Neogen von Ost-Samos. Sedimentationsgeschichte und Korrelation. *Neues Jahrb. Geol. Paläontol. Abhandl.* **152**(2): 161–176.

MELANDER, A. 1949. A report on some Miocene Diptera from Florissant, Colorado. *Amer. Mus. Novitates* **1407**: 1–64.

MELIS, A. 1935. Tisanotteri italiani, studio anatomo-morfologico e biologico de Liothripidae dell olive (*Liothrips oleae* Costa). *Redia* **21**: 1–188.

MELLETT, J.S. 1974. Scatological origin of microinvertebrate fossil accumulations. *Science* **185**: 349–350.

MELNIKOV, O.A. 1970. Embryogenesis of *Anacanthotermes ahngerianus* (Isoptera, Hodotermitidae), larval segmentation and nature of labrum. *Zool. Zhurnal* **49**: 838–854 (in Russian, with English summary).

MELNIKOV, O.A. 1971. Primary heteronomy of body in Articulata. *Zhurnal obshchey biol.* **32**: 597–612 (in Russian, with English summary).

MELNIKOV, O.A. 1974a. On the problem of anterior larval segments in arthropods with special reference to their promorphology and morphological evolutionary trends in these animals. *Zhurnal obshchey biol.* **35**: 858–873 (in Russia with English summary).

MELNIKOV, O.A. 1974b. On morphogenesis of proctodeum in insects with reference to tagmosis of their body. *Zool. Zhurnal* **52**: 1786–1797 (in Russian, with English summary).

MELNIKOV, O.A., and RASNITSYN, A.P. 1984. Zur Metamerie des Arthropoden-Kopfes: Das Acron. *Beitr. Entomol.* **34**: 3–90.

MENGE, A. 1856. Lebenszeichen vorweltlicher Thiere im Bernstein eingeschlossener Thiere. *Programm Petrischule Danzig* **1856**: 1–32.

MENZIE, C.A. 1980. Potential significance of insects in the removal of contaminants from aquatic systems. *Water, Air and Soil Pollution* **13**: 473–480.

MESSER, A.C. 1985. Fresh dipterocarp resins gathered by megachilid bees inhibit the growth of pollen-associated fungi. *Biotropica* **17**: 175–176.

MESSER, A., MCCORMICK, K., SUNJAYA, HAGEDORN, H.H., TUMBEL, F., and MEINWALD, J. 1990. Defensive role of tropical tree resins: antitermitic sesquiterpenes from Southeast Asian Dipterocarpaceae. *J. Chem. Ecol.* **16**(12): 3333–3352.

METZ, R. 1980. Control of mud-crack patterns by beetle larvae traces. *J. Sediment. Petrol.* **50**(3): 841–842.

METZ, R. 1987a. Sinusoidal trail formed by a recent biting midge (family Ceratopogonidae): trace fossil implications. *J. Paleontol.* **61**(2): 312–314.

METZ, R. 1987b. Insect traces from nonmarine ephemeral puddles. *Boreas* **16**(2): 189–192.

MEUNIER, F. 1905. Sur quelques Diptères (Cecidomyidae, Tachinidae, Chloropidae, Phoridae) et un Hyménoptère (Chalcididae) d'un copal recent de Madagascar. *Miscell. Entomol.* **13**: 89–94.

MEUNIER, F. 1906. Sur quelques insectes (Diptères, Hyménoptères, Neuroptères, Orthoptères) du copal fossile, subfossile et recent du Zanzibar et du copal recent d'Accra, de Togo et de Madagascar. *Ann. Soc. Sci. Bruxelles* **30**: 211–213.

MEUNIER, F. 1909a. Quelques considérations sur la faune d'insectes du copal fossile de Zanzibar, du copal récent d'Accra, de Zanzibar et de Madagascar. *Ann. Soc. Sci. Bruxelles* **33**: 141–142.

MEUNIER, F. 1909b. Nouvelles recherches sur les insectes du terrain houiller de Commentry (Allier). *Ann. Paléontol.* **4**: 125–152.

MEUNIER, F. 1912. Nouvelle recherches sur quelques insectes du terrain houiller de Commentry (Allier). Pt. 2. *Ann. Paléontol.* **7**: 1–19.

MEYEN, S.V. 1984. Basic features of gymnosperm systematics and phylogeny as shown by the fossil record. *Botan. Rev.* **50**(1): 1–111 (cited after Russian translation: MEYEN, S.V. 1992. *Evolutsiya i sistematika vysshikh rasteniy po dannym paleobotaniki*. Nauka, Moscow: 40–105).

MEYEN, S.V. 1987a. *Fundamentals of Palaeobotany*. Chapman and Hall, London–New York: 432 p.

MEYEN, S.V. 1987b. Geography of macroevolution in Angiospermians. *Zhurnal obschey biol.* **48**(3): 291–309 (in Russian).

MEYER, J. 1987. *Plant Galls and Gall Inducers*. Gebr. Borntraeger, Berlin: 291 p.

MEYN, L. 1876. Der Bernstein der norddeutschen Ebene aus zweiter, dritter, vierter, fünfter und sechster Lagerstätte. *Zeitschr. Deutsche geol. Ges.* **28**: 171–198.

MICHELSEN, V. 1996. First reliable record of a fossil species of Anthomyiidae (Diptera), with comments on the definition of recent and fossil clades in phylogenetic classification. *Biol. J. Linn. Soc.* **58**: 441–451.

MICHENER, C.D., and GRIMALDI, D.A. 1988. A *Trigona* from Late Cretaceous amber of New Jersey (Hymenoptera: Apidae: Meliponinae). *Amer. Mus. Novitates* **2917**: 1–10.

MICHENER, C.D., and POINAR, G.O., Jr. 1997. The known bee fauna of the Dominican amber. *J. Kansas Entomol. Soc.* **69**: 353–361.

MICKOLEIT, E. 1961. Zur Thoraxmorphologie der Thysanoptera. *Zool. Jahrb. (Anat.)* **79**: 1–92.

MICKOLEIT, E. 1963. Untersuchungen zur Kopfmorphologie der Thysanopteren. *Zool. Jahrb. (Anat.)* **81**: 101–150.

MIERZEJEWSKI, P. 1976. On application of scanning electron microscope to the study of organic inclusions from the Baltic amber. *Ann. Geol. Soc. Poland* **46**: 291–295.

MIERZEJEWSKI, P. 1978. Electron microscopy study on the milky impurities covering arthropod inclusions in the Baltic amber. *Prace Muz. Ziemi* **28**: 81–84.

MILLER, L.R. 1989. Sub-fossil termite mounds in the Simpson Desert. *Northern Terr. Naturalist* **11**: 27–30.

MILLER, M.F. 1984. Distribution of biogenic structures in Paleozoic nonmarine and marine-margin sequences: an actualistic model. *J. Paleontol.* **58**(2): 550–570.

MILLER, R.F. 1991. Chitin palaeoecology. *Biochem. Syst. and Ecol.*, **19**(5): 401–411.

MILLER, R.F., VOSS-FOUCART, M.-F., TOUSSAINT, C., and JEUNIAUX, C. 1993. Chitin preservation in Quaternary Coleoptera: preliminary results. *Palaeogeogr., Palaeoclimatol., Palaeoecol.* **103**: 133–140.

MILLER, S.E. 1983. Late Quaternary insects of Rancho La Brea and McKittrick, California. *Quatern. Res.* **20**: 90–104.

MILLER, S.E., GORDON, R.D., and HOWDEN, H.F. 1981. Reevaluation of Pleistocene scarab beetles from Rancho La Brea, California (Coleoptera Scarabaeidae). *Proc. Entomol. Soc. Washington* **83**(4): 625–680.

MILLS, J.S., WHITE, R., and GOUGH, L.J. 1984. The chemical composition of Baltic amber. *Chemical Geol.* **47**: 15–39.

MINET, J. 1986. Ebache d'une classification moderne de l'ordre des Lepidoptères. *Alexanor* **14**: 291–313.

MINET, J. 1990. Nouvelles frontières, géographiques et taxonomiques, pour la famille des Callidulidae (Lepidoptera, Calliduloidea). *Nouv. Rev. Entomol.* (N.S.) **6**: 351–368.

MINET, J. 1991. Tentative reconstruction of ditrysian phylogeny (Lepidoptera, Glossata). *Entomol. Scand.* **22**: 69–95.

MITCHELL, P., and WIGHTON, D. 1979. Larval and adult insects from the Paleocene of Alberta, Canada. *Canad. Entomol.* **111**: 777–782.

MIYAMOTO, S. 1961. Comparative morphology of alimentary organs of Heteroptera with the phylogenetic consideration. *Sieboldia* **2**: 197–259.

MIYAMOTO, Sh., YAMAMOTO, H., and SEKI, H. 1991. Chitin dynamics in the freshwater environment. *Biochem. Syst. and Ecol.* **19**(5): 371–377.

MOCKFORD, E.L. 1969. Fossil insects of the order Psocoptera from Tertiary amber of Chapas, Mexico. *J. Paleontol.* **43**: 1267–1273.

MOCKFORD, E.L. 1972. New species, records and synonymy of Florida *Belaphotroctes* (Psocoptera: Liposcelidae). *Florida Entomol.* **55**: 153–163.

MOCKFORD, E.L. 1978. A generic classification of family Amphipsocidae (Psocoptera: Caeciletae). *Trans. Amer. Entomol. Soc.* **104**: 139–190.

MOCKFORD, E.L. 1986. A preliminary survey of Psocoptera from Tertiary amber of the Dominican Republic. *Entomol. Soc. Amer. Conf. (Reno, Nevada; December 7–11, 1986)*: 112.

MOCKFORD, E.L., and GARCIA ALDRETE, A.N. 1976. A new species and notes on the taxonomic position of *Asiopsocus* Günther (Psocoptera). *Southwest. Nat.* **21**: 335–346.

Möhn, E. 1960. Eine neue Gallmücke aus niederrheinischen Braunkohle, *Sequoiomyia kreuseli* n.g., n.sp. (Diptera, Itonididae). *Senckenberg. Lethaea* **41**: 513–522.

Monetta, A.M., and Pereyra, R.E. 1986. Nuevos datos sobre la morfología alar de *Paranarkemina kurtzi* (Insecta, Paraplecoptera) de la Formación Bajo de Veliz (Carbonífero superior), San Luis, Argentina. *IV Congr. Argent. Paleontol. Bioestratigr., Mendoza*: 139–142.

Monroe, J.S., and Dietrich, R.V. 1990. Pseudofossils. *Rocks and Miner.* **65**(2): 150–158.

Monte, G., del. 1956. La presenza di insetti dei granai in frumento trovato negli scavi di Erculano. *Redia* **41**: 23–28.

Moore, J.M., and Picker, M.D. 1991. Heuweltjies (earth mounds) in the Clanwilliam District, Cape Province, South Africa: 4000-year-old termite nests. *Oecologia* **86**(3): 424–432.

Moore, P.D. 1981. Life seen from a medieval latrine. *Nature* **294**: 614.

Moret, P. 1996. Arqueo-entomología: cuando los insectos fósiles contribuyen al conocimiento de nuestro pasado. *Bol. Soc. Entomol. Arag.* **16**, Volumen Monográfico: PaleoEntomología: 183–188.

Moret, P., and Martín-Cantarino, C. 1996. L'utilisation des Coléoptères subfossiles dans la reconstruction des paléo-environments: l'exemple du port antique de Santa Pola (Espagne). *Bull. Soc. entomol. France* **101**(3): 225–229.

Moretti, G. 1955. Sulla presenza dei foderi dei Tricotteri e dei Ditteri Tanytarsi sui fondi del Lago Maggiore. *Mem. Ist. ital. Idrobiol. Marchi* suppl. **8**: 205–219.

Moritz, G. 1982a. Zur Morphologie des Kopfinnenskeletts (Tentorium) bei den Thysanoptera. *Deutsche entomol. Zeitschr.*, n. F. **29**: 17–26.

Moritz, G. 1982b. Zur Morphologie und Anatomie des Frasenflüglers *Aeolothrips intermedius* Bagnall. 1. Mitteilung: Der Kopf. *Zool. Jahrb. (Anat.)* **107**: 557–608.

Moritz, G. 1982c. Zur Morphologie und Anatomie des Frasenflüglers *Aeolothrips intermedius* Bagnall. 2. Mitteilung: Der Thorax. *Zool. Jahrb. (Anat.)* **108**: 55–106.

Moritz, G. 1982d. Zur Morphologie und Anatomie des Frasenflüglers *Aeolothrips intermedius* Bagnall. 3. Mitteilung: Das Abdomen. *Zool. Jahrb. (Anat.)* **108**: 293–304.

Moritz, G. 1989a. Die Ontogenese der Thysanoptera (Insecta) unter besonderer Berücksichtigung des Frasenflüglers *Hercinothrips femoralis* (O.M. Reuter, 1891) (Thysanoptera, Thripidae, Panchaetothripinae). V. Mitteilung: Imago – Thorax. *Zool. Jahrb. (Anat.)* **118**: 393–429.

Moritz, G. 1989b. Die Ontogenese der Thysanoptera (Insecta) unter besonderer Berücksichtigung des Frasenflüglers *Hercinothrips femoralis* (O.M. Reuter, 1891) (Thysanoptera, Thripidae, Panchaetothripinae). VI. Imago – Abdomen. *Zool. Jahrb. (Anat.)* **119**: 157–217.

Morris, G.K., and Gwynne, D.T. 1979. Geographical distribution and biological observations on *Cyphoderris* (Orthoptera: Haglidae) with a description of a new species. *Psyche* **85**: 147–167.

Morrison, A. 1987. Chironomid larvae trails in proglacial lake sediments. *Boreas* **16**(3): 318–326.

Morrow, P.A., and LaMarche, V.C. 1978. Tree ring evidence for chronic insect suppressure of productivity in subalpine *Eucalyptus*. *Science* **201**: 1244–1246.

Morse, J.C. 1997. Phylogeny of Trichoptera. *Annu. Rev. Entomol.*. **42**: 427–450.

Morse, J.C. (ed.) 1999. *Trichoptera World Checklist*. http://entweb.clemson.edu/database/trichopt/index.htm Effective 22 May 1999.

Mosini, V., Forcellese, M.L., and Nicoletti, R. 1980. Presence and origin of volatile terpenes in succinite. *Phytochemistry* **19**: 679–680.

Mostovski, M.B. 1995a. New representatives of Platypezidae (Diptera) from the Mesozoic, and the main directions in the evolution of the family. *Paleontol. Zhurnal* 2: 106–118 (in Russian, English translation: *Paleontol. J.* **29**. 2: 130–146).

Mostovski, M.B. 1995b. New taxa of ironomyiid flies (Diptera, Phoromorpha, Ironomyiidae) from Cretaceous deposits of Siberia and Mongolia. *Paleontol. Zhurnal* 4: 86–103 (in Russian, English translation: *Paleontol. J.* **30**. 3: 318–331).

Mostovski, M.B. 1996. New species of the genus *Mesosolva* Hong, 1983 (Diptera, Archisargidae) from the Jurassic of Kazakhstan and Mongolia. In: Morales, M. (ed.). *The Continental Jurassic*. Bull. Mus. North. Arizona **60**: 329–332.

Mostovski, M.B. 1997a. To the knowledge of Archisargoidea (Diptera, Brachycera). Families Eremochaetidae and Archisargidae. *Russian Entomol. J.* **5**(1–4): 117–124.

Mostovski, M.B. 1997b. To the knowledge of fossil dipterans of superfamily Archisargoidea (Diptera, Brachycera). *Paleontol. Zhurnal* 1: 72–77 (in Russian, English translation: *Paleontol. J.* **31**(1): 72–78).

Mostovski, M.B. 1998. A revision of the nemestrinid flies (Diptera, Nemestrinidae) described by Rohdendorf, and a description of new taxa of the Nemestrinidae from the Upper Jurassic of Kazakhstan. *Paleontol. Zhurnal* 4: 47–53 (in Russian, English translation: *Paleontol. J.* **32**(4): 369–375).

Mostovski, M.B. 1999a. A brief review of brachyceran flies (Diptera, Brachycera) in the Mesozoic, with descriptions of some curious taxa. *Proc. First Palaeoentomol. Conf., Moscow 1998*. AMBA projects Internat., Bratislava: 103–110.

Mostovski, M.B. 1999b. Curious Phoridae (Insecta, Diptera) found mainly in Cretaceous ambers. *Estud. Mus. Cienc. Natur. Alava*. **14** (Num. espec. 2): 233–245.

Mostovski, M.B., and Martínes-Delclòs, X. 2000. New Nemestrinoidea (Diptera: Brachycera) from the Upper Jurassic – Lower Cretaceous of Eurasia. Taxonomy and palaeobiology. *Entomol. Problems* **31**(2): 137–148.

Mound, L.A. 1971. The feeding apparatus of thrips. *Bull. Entomol. Res.* **60**: 547–548.

Mound, L.A., and Heming, B.S. 1991. Thysanoptera. In: Naumann, D. (ed.). *The Insects of Australia*. 2nd ed. Melbourne Univ. Press, Carlton: 510–515.

Mound, L.A., Heming, B.S., and Palmer, J.M. 1980. Phylogenetic relationships between the families of recent Thysanoptera (Insecta). *Zool. J. Linn. Soc. (London)* **69**: 111–141.

Mound, L.A., and O'Neil, K. 1974. Taxonomy of the Merothripidae with ecological and phylogenetic consideration. *J. Natur. Hist.* **8**: 481–509.

Mound, L.A., and Walker, A.K. 1982. *Terebrantia (Insecta: Thysanoptera). Fauna of New Zealand*. DSIR, Wellington: 113 p.

Mound, L.A., and Walker, A.K. 1986. *Tubulifera (Insecta: Thysanoptera). Fauna of New Zealand*. DSIR, Wellington: 340 p.

Mudd, A., and Corbet, S.A. 1975. Use of pine resin in nests of pemphredonine wasps. *Trans. R. Entomol. Soc. London* **127**(3): 255–257.

Muggoch, H., and Walten, J. 1942. On the dehiscence of the antheridium and the part played by surface tension in the dispersal of spermatocytes in Bryophyta. *Proc. R. Soc. London* B **130**: 448–461.

Müller, A.H. 1975. Zur Ichnologie limnisch-terrestrischer Sedimentationsräume mit Bemerkungen über Ichnia landbewohnender Arthropoden als Gegenwart und geologischer Vergangenheit. *Freiberger Forschungsheft* C **304**: 79–87.

Müller, A.H. 1977. Über fossile und rezente Ichnia vom Typ *Helicorhaphe* und *Belorhaphe* sowie ihre Erzeuger auf Flugsanddunen der Gegenwart. *Freiberger Forschungsheft* **6**: 55–63.

Müller, A.H. 1981. Zur Ichnologie, Taxiologie und Ökologie fossiler Tiere. *Freiberger Forschungsheft* C **151**: 5–49:

Müller, A.H. 1982. Über Hyponome von fossiler und rezenter Insekten, erster Beitrag. *Freiberger Forschungsheft* C **366**: 7–27.

Müller, H. 1977. Zur Entomofauna des Permokarbon: 2. Über einige Blattinopsidae (Protorthoptera) aus dem Unterrotliegenden (Unteres Autun) von Mitteleuropa. *Zeitschr. geol. Wiss.* **5**: 1029–1051.

Müller, K. 1927. Beiträge zur Biologie, Anatomie, Histologie und inneren Metamorphose der Thripslarven. *Zeitschr. wiss. Zool.* **130**: 251–303.

Müller, K.J. 1985. Exceptional preservation in calcareous nodules. *Philos.Trans. R. Soc. London* B **311**: 67–73.

Müller-Stoll, W.R. 1936. Pilzzerstörungen an einem jurassischen Koniferenholz. *Paläont. Zeitschr.* **18**: 202–212.

Müller-Stoll, W.R. 1989. Fraßgänge mit Koprolithen eines holzbewohnenden Käfers miozänen Alters aus dem Lettengraben (Herrenwasser) in der Rhön. *Archaeopteryx* **7**: 51–57.

Mumcuoglu, Y., and Stix, E. 1974. Milben in der Luft. *Rev. suisse zool.* **81**(3): 673–677.

Murray, D.A. 1976. *Buchonomyia thienemanni* Fittkau (Diptera, Chironomidae), a rare and unusual species recorded from Killarney, Ireland. *Entomol. Gazette* **27**(3): 179–180.

Mustoe, G.E., and Gannaway, W.L. 1997. Palaeogeography and palaeontology of the Early Tertiary Chuckanut Formation, northwest Washington. *Washington Geol.* **25**(3): 3–18.

Nagatomi, A., and Yang, D. 1998. A review of extinct Mesozoic genera and families of Brachycera (Insecta, Diptera, Orthorrhapha). *Entomol. Mon. Mag.* **134**: 95–192.

Nagel, P. 1987. Fossil ant nest beetles (Coleoptera, Carbidae, Paussinae). *Entomol. Arb. Mus. G. Frey* **35/36**: 137–170.

Nagel, P. 1997. New fossil paussids from Dominican amber with notes on the phylogenetic systematics of the paussine complex (Coleoptera: Carabidae). *System. Entomol.* **22**: 345–362.

Nagnan, P., and Clement, J.L. 1990. Toxins for subterranean termites of the genus *Reticulitermes* (Isoptera: Rhinotermitidae)? *Biochem. Syst. and Ecol.* **18**(1): 13–16.

Nakata, S., and Maa, T.C. 1974. A review of the parasitic earwigs (Dermaptera: Arixeniina; Hemimerina). *Pacif. Insects* **16**: 307–374.

Naugolnykh, S.V. 1998. *Kungurskaya flora Srednego Priural'ya* (Kungurian Flora of the Middle Cis-Urals). Trudy Geol. Inst. Ross. Akad. Nauk **509**. Geos, Moscow: 200 p. (in Russian).

Naumann, C.F. 1858. Lehrbuch der Geognosie. Bd. 2. W. Engelmann, Leipzig: 960 p.

Naumann, C.M. 1987. On the phylogenetic significance of two Miocene zygaenid moths (Insecta, Lepidoptera). *Paläont. Zeitschr.* **61**(3–4): 299–308.

Nazarov, V.I. 1984. *Rekonstruktsiya landshaftov Belorussii po paleoentomologicheskim dannym (antropogen)* [Landscape Reconstruction of Belorussia Basing on Paleoentomological Data (the Anthropogene)]. Trudy Paleontol. Inst. Akad. Nauk SSSR **205** Nauka, Moscow: 96 p. (in Russian).

Nazarov, V.I., and Karasyov, V.P. 1992. The Holocene entomofauna of buried soils. In: *Fauna i flora kainozoya Belorussii* (Fauna and Flora of the Cainozoic of Belorussia). Navuka i tekhnika, Minsk: 5–11 (in Russian).

Nazarov, V.I., and Kovalyukh, N.N. 1993. Dating of palaeoecological events on insect chitin. *Doklady Akad. Nauk Belarusi* **37**(3): 92–94 (in Russian).

Nel, A. 1984b. Description d'une nouvelle espèce de termite fossile du Stampien d'Aix-en-Provence (Dictyoptera, Termitidae, Termitinae). *Entomol. Gall.* 1(3): 159–160.

Nel, A. 1988. Les Sialidae (Megaloptera) fossiles de diatomites de Murat (Cantal, France) et de Bes-Konak (Anatolie, Turquie). *Neuroptera Internat.* **5**(1): 39–44.

Nel, A. 1989. Les Gyrinidae fossiles de France (Coleoptera). *Ann. Soc. entomol. France* n.s. **25** (3): 321–330.

Nel, A. 1991. Nouveaux insectes neuropteroids fossiles de l'Oligocéne de France. *Bull. Mus. nation. Hist. natur. (Paris)*, 4-e sér. C **12**(3–4): 327–349.

Nel, A. 1994. Traces d'activités d'insectes dans des bois et fruits fossiles de la formation de Nkondo (Mio-Pliocéne du Rift occidental, Ouganda). *Geology and Palaeobiology of the Albertine Rift valley, Uganda – Zaire* 2. Palaeobiology. *CIFEG Occas. Publ.* **29**: 47–57.

Nel, A. 1997. The probabilistic inference of unknown data in phylogenetic analysis. In: Grandcolas, P. (ed.). *The Origin of Biodiversity in Insects: Tests of Evolutionary Scenarios. Mém. Mus. nation. Hist. natur. (Paris)* **173**: 305–327.

Nel, A. 1998. A new fossiliferous amber from the Deprtment of Oise (Paris Basin, France). *World Congress on amber inclusions*. Mus. Cienc. Natur. Alava: 85.

NEL, A., ALBOUY, V., CAUSSANEL, C., and JAMET, C. 1994. Réflexion paléo-entomologique sur la systématique des Dermapteres. Quatre nouveaux forficules fossiles de l'Oligocene de Provence (France) (Dermaptera). *Bull. Soc. entomol. France* **99**: 253–266.

NEL, A., ARILLO, A., and MARTÍNEZ-DELCLÒS, X. 1996. New fossil Odonata (Insecta) from the Upper Miocene of France and Spain (Anisoptera and Zygoptera). *Neues Jahrb. Geol. Paläontol. Abhandl.* **199**(2): 167–219.

NEL, A., DE PLÖEG, G., DEJAX, J., DUTHEIL, D., DE FRANCESCHI, D., GHEERBRANT, F., GODINOT, M., HERVET, S., MENIER, J.J., AUGÉ, M., BIGNOT, G., CAVAGNETTO, C., DUFFAUD, S., GAUDANT, J., HUA, S., JOSSANG, A., BROIN, F.L., POZZI, J.-T., PAICHELER, J.-C., BEUCHET, F., and RAGE, J.-C. 1999c. Un gisement sparnacien exceptionel à plantes, arthropodes et vertébrés (Eocène basal, MP 7): Le Quesnoy (Oise, France). *C.R. Acad. Sci., Paris*, Sci. terre **229**: 65–72.

NEL, A., DE PLÖEG, G., GHEERBRANDT, E., GODINOT, M., MENIER, J.-J., and JOSSANG, A. 1998a. Un gisement a ambre fossilifère dans l'Oise. Création de la première collection nationale d'ambre français. *Minéraux & Fossiles* (264): 25–29.

NEL, A., GAND, G., and GARRIC, J. 1999a. A new family of Odonatoptera from the continental Upper Permian: the Lapeyriidae (Lodève Basin, France). *Geobios* **32**(1): 63–72.

NEL, A., GILL, G.A., and NURY, D. 1987. Découverte d'empreintes attribualbles à des Coelentérés *Siphonophores chondrophorides* dans l'Oligocène de Provence. *C.R. Acad. Sci. Paris Sér.* II **305**: 637–641.

NEL, A., LACAU, S., DE PLÖEG, G., MENIER, J.J., PAICHELER, J.C., and BOUCHET, F. 1998b. A new fossiliferous amber from the department of Oise (Paris Basin, France). *World Congress on Amber Inclusions, 20–23 Oct. 1998, Vitoria-Gasteiz, Alava, Basque Country, Abstr. with Progr.*: 85.

NEL, A., and MARTÍNEZ-DELCLÒS, X. 1993. Nuevos Zygoptera y Anisoptera (Insecta: Odonata) en el Cretacico inferior de España. *Estud. Geol.* **49**(5–6): 351–359.

NEL, A., MARTÍNEZ-DELCLÒS, X., ARILLO, A., and PEÑALVER, E. 1999b. A review of the Eurasian fossil species of the bee *Apis. Palaeontology* **42**(2): 243–285.

NEL, A., and PAICHELER J.-C. 1993. Les Isoptera fossiles. État actuel des connaissances, implications paléoécologiques et paléoclimatologiques (Insecta, Dictyoptera). *Cah. Paléontol.*: 101–179.

NEL, A., and PAICHELER, J.C. 1994. Les Lestoidea fossiles: un inventaire critique (Odonata, Zygoptera). *Ann. Paléontol.* **90**(1): 1–59.

NEL, A., and POPOV, Yu.A. 2000. The oldest known fossil Hydrometridae from the Lower Cretaceous of Brazil (Heteroptera: Gerromorpha). *J. Natur. Hist.* **34**: 2315–2322.

NEL, A., and ROY, R. 1996. Revision of the fossil "mantid" and "ephemerid" species described by Piton from the Palaeocene of Menat (France) (Mantodea: Chaeteessidae, Mantidae; Orthoptera: Tettigonioidea). *Europ. J. Entomol.* **93**: 220–234.

NELSON, B.C. 1972. Fleas from the archaeological site at Lovelock Cave, Nevada. *J. Med. Entomol.* **9**(3): 211–218.

NESSOV, L.A. 1988. Activity traces of organisms of the Late Mesozoic-Palaeocene of Mid-Asia and Kazakhstan as the indicators of palaeoenvironments of vertebrates. In: *Sledy zhiznedeyatel'nosti i dinamika sredy v drevnikh biotopakh* (Trace Fossils and Dynamics of Environments in Ancient Biotopes. *Trudy 30th Sessii Vsesoyuznogo Paleontol. Obshchestva i 7th Sessii Ukrainskogo Paleontol. Obshchestva.* Naukova Dumka, Kiev: 76–90 (in Russian).

NESSOV, L.A. 1995. *Dinozavry Severnoy Evrazii: novye dannye o sostave kompleksov, ekologii i paleobiogeografii* (Dinosaurs of Northern Eurasia: New Data about Assemblages, Ecology and Palaeobiogeography). Univ. St.-Petersburg, Institute Zemnoy Kory, St.-Petersburg: 156 p. (in Russian).

NESSOV, L.A. 1997. *Nemorskiye pozvonochnyie melovogo perioda Severnoy Evrazii* (Non-marine Vertebrates of the North Eurasia). Univ. St.-Petersburg, Institute Zemnoy Kory, St.-Petersburg: 218 p., 60 plates (in Russian).

NEVEU, A. 1980. La dérive des invertébrés aquatiques et terrestres dans un petit fleuve côtier de l'ouest des Pyrénées, la Nivelle. *Acta oecol. Oecol. appl.* **1**(4): 317–339.

NEVEU, A., and ECHAUBARD, M. 1975. La dérive estivale des invertébrés aquatiques et terrestres dans un ruisseau du Massif Central: la Couze-Pavin. *Ann. hydrobiol.* **6**(1): 1–26.

NEW, T.R. 1987. Biology of Psocoptera. *Oriental Insects* **19**: 115–120.

NEW, T.R. 1991. Neuroptera. In: NAUMANN, D. (ed.). *The Insects of Australia.* 2nd ed. Melbourne Univ. Press, Carlton: 525–542.

NEWMAN, W.A., and FOSTER, B.A. 1987. Southern Hemisphere endemism among the barnacles: explained in part by extinction of northern members of amphitropical taxa *Bull. Mar. Sci.* **41**: 361–377.

NICHOLAS, H.J. 1963. The biogenesis of terpenes in plants. In: BERNFIELD, P. (ed.). *Biogenesis of Natural Compounds.* Pergamon Press, New York: 641–691.

NICHOLAS, H.J. 1973. Terpenes. In: MILLER, L.P. (ed.). *Phytochemistry.* Van Nostrand, New York: 254–309.

NIKLAS, K.J., and GIANNASI, D.E. 1978. Angiosperm paleobiochemistry of the Succor Creek Flora (Miocene), Oregon, USA. *Amer. J. Botany* **65**(9): 943–952.

NIKLAS, K.J., TIFFNEY, B.H., and KNOLL, A.H. 1980. Apparent changes in the diversity of fossil plants: a preliminary assessment. In: HECHT, M.K., STEERE, W.C., and WALLACE, B. (eds.). *Evolutionary Biology* **12**. Plenum Press, New York: 1–89.

NIKOLAJEV, G.V. 1998a. Pleurostict lamellicorn beetles (Coleoptera, Scarabaeidae) from the Lower Cretaceous of Transbaikalia. *Paleontol. Zhurnal* (5): 77–84 (in Russian, English translation: *Paleontol. J.* **32**(5): 513–520).

[NIKOLAJEV] NIKOLAEV, G.V. 1998b. Taxonomic composition of the Mesozoic fauna of lamellicorn beetles (Coleoptera, Scarabaeoidea). *First Paleoentomol. Conf., 30 Aug.–4 Sept. 1998, Moscow, Russia. Abstracts*: 30 (in Russian).

NILSSON, D.-E., and OSORIO, D. 1998. Homology and parallelism in arthropod sensory processing. In: FORTEY, R.A., and THOMAS, R.H. (eds.). *Arthropod Relationships.* Chapman & Hall, London: 333–347.

NISHIDA, H. 1985. A structurally preserved magnolialean fructification from the Mid-Cretaceous of Japan. *Nature* **318**: 58–59.

NISSEN, K. 1973. Analysis of human coprolites from Bamert Cave, Amador County, California. In: HEIZER, R.F., and HESTER, T.R. (eds.). *The Archaeology of Bamert Cave, Amador County, California.* Univ. California Archaeol. Facility, Berkeley: 65–71.

NIXON, K.C., and CREPET, W.L. 1993. Late Cretaceous flowers of Ericalean affinity. *Amer. J. Botany* **80**: 616–623.

NORLIN, A. 1964. The occurrence of terrestrial insects on the surface of two lakes in northern Sweden (Ankarvatnet and Blåsjön). *Rep. Inst. Freshwat. Res. Drottingholm* **45**: 196–205.

NORLIN, A. 1967. Terrestrial insects in lake surface. *Rep. Inst. Freshwat. Res. Drottingholm* **47**: 40–55.

NORMAN, D.B., and WEISHAMPEL, D.B. 1985. Ornithopod feeding mechanisms: their bearing on the evolution of herbivory. *Amer. Naturalist* **126**: 151–164.

NORSTOG, K.J. 1987. Cycads and the origin of insect pollination. *Amer. Scientist* **75**: 270–279.

NORSTOG, K.J., and FAWCETT, P.K.S. 1989. Insect-cycad symbiosis and its relation to the pollination of *Zamia furfuracea* (Zamiaceae) by *Rhopalotria mollis* (Curculionidae). *Amer. J. Botany* **76**(9): 1380–1394.

NORSTOG, K.J., and STEVENSON, D.W. 1980. Wind? Or insects? The pollination of cycads. *Fairchild Trop. Gard. Bull.* **35**: 28–30.

NORSTOG, K.J., STEVENSON, D.W., and NIKLAS, K.J. 1986. The role of beetles in pollination of *Zamia furfuracea* L. fil. *Biotropica* **18**: 300–306.

NORSTOG, K.J., and VOVIDES, A.P. 1992. Beetle pollination in two species of *Zamia*: evolutionary and ecological considerations. *Palaeobotanist* **41**: 149–158.

NOVOKSHONOV, V.G. 1993a. New insects (Insecta) from the Lower Permian of Chekarda (Central Urals). *Paleontol. J.* **27**(1A): 172–178.

NOVOKSHONOV, V.G. 1993b. Mückenhafte (Mecoptera Bittacidae) aus dem Jura, Kreide and Paläogen von Eurasien und ihre phylogenetischen Beziehungen. *Russian Entomol. J.* **2**(3–4): 75–86.

NOVOKSHONOV, V.G. 1993c. Caddis flies (Insecta, Trichoptera, Microptysmatidae). *Paleontol. J.* **27**(1A): 90–102.

NOVOKSHONOV, V.G. 1994a. Permian scorpionflies (Insecta, Panorpida) of the families Kaltanidae, Permochoristidae and Robinjohniidae. *Paleontol. Zhurnal* (1): 65–76 (in Russian, English translation: *Paleontol. J.* **28**(1): 79–95).

NOVOKSHONOV, V.G. 1994b. Scoprion flies of the family Permochoristidae, the closest common ancestors of extant scorpion flies. *Zool. Zhurnal* **73**(7–8): 58–70 (in Russian).

NOVOKSHONOV, V.G. 1995. New Hypoperlida (Insecta) from the Lower Permian of Chekarda (Perm Region). *Vestnik Permskogo Univ., Biol.* **1**: 179–183.

NOVOKSHONOV, V.G. 1996. The position of some Upper Permian Neuroptera *Paleontol. Zhurnal* (1): 39–47 (in Russian, English translation: *Paleontol. J.* **30**(1): 38–45).

NOVOKSHONOV, V.G. 1997a. *Rannyaya evolyutsiya skorpionnits* (Early Evolution of Scorpionflies (Insecta: Panorpida). Nauka, Moscow: 140 p. (in Russian).

NOVOKSHONOV, V.G. 1997b. New Triassic scorpionflies (Insecta: Mecoptera). *Paleontol. Zhurnal* no. 6: 63–70 (in Russian, English translation: *Paleontol. J.* **30**(6): 628–635).

NOVOKSHONOV, V.G. 1997c. Some Mesozoic scorpionflies (Insecta: Panorpida = Mecoptera) of the families Mesopsychidae, Pseudopolycentropodidae, Bittacidae, Permochoristidae. *Paleontol. Zhurnal* (1): 65–71 (in Russian, English translation: *Paleontol. J.* **31**(1): 65–71).

NOVOKSHONOV, V.G. 1997d. New and little known Mesozoic Nannochoristidae (Insecta: Mecoptera). *Vestnik Permskogo Univ.* **4**: 126–136 (in Russian).

NOVOKSHONOV, V.G. 1997e. New taxa of fossil insects from the Lower Permian of the Middle Urals. *Paleontol. Zhurnal* (4): 39–44 (in Russian, English translation: *Paleontol. J.* **31**(4): 383–388).

NOVOKSHONOV, V.G. 1998a. New insects (Insecta: Hypoperlida, Mischopterida, Jurinida) from the Lower Permian of Tshekarda (Middle Urals). *Paleontol. Zhurnal* (1): 50–57 (in Russian, English translation: *Paleontol. J.* **32** (1): 46–53).

NOVOKSHONOV, V.G. 1998b. New fossil insects (Insecta: Grylloblattida, Caloneurida, Hypoperlida?, Ordinis Incertis) from the Kungurian Beds of the Middle Urals. *Paleontol. Zhurnal* (4): 41–46 (in Russian, English translation: *Paleontol. J.* **32**(4): 362–368).

NOVOKSHONOV, V.G. 1998c. Fossil insects of Chekarda. In: PONOMARYOVA, G.Yu., NOVOKSHONOV, V.G., and NAUGOLNYKH, S.V. *Chekarda – mestonakhozhdenie permskikh iskopaemykh rasteniy i nasekomykh.* (Chekarda – The Locality of Permian Fossil Plants and Insects). Perm' Univ., Perm': 25–54 (in Russian).

NOVOKSHONOV, V.G. 1998d. Some problems of scorpionfly evolution (Mecoptera). *Zool. Zhurnal* **77**(6): 677–688. (in Russian).

NOVOKSHONOV, V.G. 2000. New Palaeomanteida = Miomoptera from the Lower Permian of Chekarda. *Paleontol. J.* **34** (suppl. 3): S303–S308.

NOVOKSHONOV, V.G. 2001a. New and little-known representatives of the family Hypoperlidae (Insecta: Hypoperlida). *Palaeontol. Zhurnal* (1): 41–44 (in Russian, English translation: *Paleontol. J.* **35**(1): 40–44.

NOVOKSHONOV, V.G. 2001b. New Triassic scorpionflies (Insecta, Mecoptera) from Kyrgyzstan. *Palaeontol. Zhurnal* (3): 57–54 (in Russian, English translation: *Paleontol. J.* **36**(3): 281–288).

NOVOKSHONOV, V.G., and NOVOKSHONOVA, E.A. 1997. *Neraphidia mitis* gen. et sp. nov. (Insecta: Grylloblattida: Protembiidae) from the Lower Permian of Tshekarda (Permian Reg.). *Vestnik Permskogo Univ., Geol.*, **4**: 123–125 (in Russian).

NOVOKSHONOV, V.G., and PAN'KOV, N.N. 1996. A new aquatic larva (Plecopteroidea) from the Lower Permian of the Ural. *Neues Jahrb. Geol. Paläontol. Monatshefte* (4): 193–198.

NOVOKSHONOV, V.G., and RASNITSYN, A.P. 2000. New enigmatic group of insects (Insecta, Psocidea, Tshekarcephalidae) from Tshekarda (Lower Permian of the Middle Urals. *Paleontol. J.* **34**(suppl. 3): S284–S287.

NOVOKSHONOV, V.G., and ROSS, A.J. A new family of the scorpionflies (Insecta; Mecoptera) from the Lower Cretaceous of England (in preparation).

NOVOKSHONOV, V.G., and SUKATSHEVA, I.D. 1993. Early evolution of caddisflies. *Proc. 7th Internat. Symp. Trichoptera, Umeå, Sweden, 3–8 Aug. 1992.* Backhuys Publ., Leiden: 95–100.

NOVOKSHONOV, V.G., and SUKATCHEVA, I.D.2001. The fossil Scorpionflies of "suborder" Paratrichoptera (Insecta: Mecoptera). *Paleontol. Zhurnal* (2): 66–75 (in Russian, English translation: *Paleontol. J.* **35**(2): 173–182).

NOVOKSHONOVA, E.A. 1998. A new species of Hypoperlidae (Insecta, Hypoperlida) from the Lower Permian of the Middle Urals. *Paleontol. Zhurnal* (5): 68–68 (in Russian, English translation: *Paleontol. J.* **32**(5): 503–504).

OBERPRIELER, R. 1995a. The weevils (Coleoptera, Curculionoidea) associated with cycads. 1. Classification, relationships, and biology. *Proc. 3rd Internat. Conf. on Cycad Biology, Pretoria, South Africa, 5–9 July 1993.* Cycad Society of South Africa, Stellenbosch: 295–334.

OBERPRIELER, R. 1995b. The weevils (Coleoptera, Curculionoidea) associated with cycads. 2. Host specificity and implications for cycad taxonomy. *Proc. 3rd Internat. Conf. on Cycad Biology, Pretoria, South Africa, 5–9 July 1993.* Cycad Soc.of South Africa, Stellenbosch: 335–365.

OCHEV, V.G. 1997. On the sources and recent structure of taphonomy. In: OCHEV, V.G., and TVERDOKHLEBOVA, G.I. (eds.). *Tafonomiya nazemnykh organizmov* (Taphonomy of Terrestrial Organisms). Saratov Univ., Saratov: 14–27 (in Russian).

O'CONNOR, J.P. 1979. *Blaps lethifera* Marsham (Coleoptera: Tenebrionidae), a beetle new to Ireland from Viking Dublin. *Entomol. Gazette* **30**: 295–297.

ODIN, G.S. 1994. Geological time scale. *C. R. Acad. Sci., Paris* Ser. II **318**: 59–71.

OHANA, T., and KIMURA, T. 1987. Preliminary notes on the multicarpelous female flower with conduplicate carpels from the Upper Cretaceous of Hokkaido, Japan. *Proc. Japan. Acad.* **B 63**(6): 175–178.

OKAFOR, N. 1966. The ecology of micro-organisms on, and the decomposition of, insect wings in soil. *Plant & Soil* **25**: 211–237.

OLECHOWICZ, E. 1990. Estimation of insect immigration in three Kampinos forest ecosystems, differing in trophy. *Ekol. Polska* **38**(3–4): 399–411.

OLLIVIER, C. 1985. Gisements fossilifères d'Europe occidentale recelant une ichthyofaune cenozoique. *Minér. et fossiles* **11**(118): 15–21.

OLSEN, P.E., REMINGTON, C.L., CORNET, B., and THOMSON, K.S. 1978. Cyclic change in Late Triassic lacustrine communities. *Science* **201**: 729–733.

OPLER, P.A. 1973. Fossil lepidopterous leaf-mines demonstrate the age of some insect-plant relationships. *Science* **179**: 1321–1323.

ORTUÑO, V.M. 1998. Artropodofauna fósil del yacimiento de ámbar Alavés (Cretácico inferior). *World Congress on Amber Inclusions, 20–23 Oct. 1998, Vittoria-Gasteiz, Abstracts*: 115.

OSBORNE, P.J. 1969. An insect fauna of the Late Bronze Age date from Wilsford, Wiltshire. *J. Anim. Ecol.* **38**: 555–566.

OSBORNE, P.J. 1981. Coleopterous fauna from Layer 1. In: GREIG, J.R.A. *The Investigation of a Medieval Barrel Latrine from Worcester.* J. Archaeol. Sci. **8**: 268–271.

OSORIO, D., AVEROF, M., and BACON, J.P. 1995. Arthropod evolution: great brains, beautiful bodies. *Trends Ecol. Evol.* **10**: 449–454.

OSWALD, J.D. 1990. Raphidioptera. In: GRIMALDI, D.A. (ed.). *Insects from the Santana Formation, Lower Cretaceous, of Brazil.* Bull. Amer. Mus. Natur. Hist. **195**: 154–163.

OTRUBA, J. 1928. Príspevek ku poznání quarterní kveteny v okolí Olomouce. *Acta Mus. Morav.* **25**: 237–250.

OTTE, D. 1994–1997. *Orthoptera species file.* Philadelphia. **1**(1994): 120 p., **2**(1994): 162 p., **3**(1994): 241 p.; **4**(1995): 518 p., **5**(1995): 630 p.; **6**(1997): 261 p.;. **7**(1997): 373 p.

OTTO, C., and SJÖSTRÖM, P. 1986. Behaviour of drifting insect larvae. *Hydrobiologia* **131**: 83–88.

OUDARD, J. 1980. Les Insectes des nodules du Stéphanien de Montceau-Les-Mines (France). Le Stratotype de l'Autunien & le Carbonifère du Morvan. Excursion B42 XXVIᵉ Congrès Géol. Interna. *Soc. Hist. Natur. Amis Mus. Autun, Trimestr.* **94**: 37–51.

OUSTALET, E. 1870. Recherches sur les insectes fossiles des terrains tertiaires de la France. *Ann. Sci. Géol.* **2**: 7–178.

OUSTALET, E. 1873. Sur quelques espèces fossiles de l'ordre des Thysanoptères. *Bull. Soc. philomat. Paris*, sér. 6, **10**: 20–27.

OVTSHINNIKOVA, O.G. 1997. Systematic position of the family Vermileonidae within the order Diptera: a comparative analysis of the male genitalia muscles. In: NARCHAK, E.P. (ed.). *Mesto i rol'dvukrylykh nasekomykh v ekosistemakh Diptera (Insecta) in Ecosystems.* Zool. Inst. Ross. Akad. Nauk, St.-Petersburg: 88–89 (in Russian).

OVTSHINNIKOVA, O.G. 1998. A brief review of male genital muscles in Brachycera Orthorrhapha (Insecta, Diptera) with special reference to the phylogenetic relationships of families. *Trudy Zool. Inst. Ross. Akad. Nauk* **276**. St.-Petersburg: 143–147.

OWEN, D. 1977. Are aphids really plant pests? *New Sci.* **76**: 76–77.

OWEN, D.F., and WIEGERT, R.G. 1981. Mutualism between grasses and grazers: an evolutionary hypothesis. *Oikos* **36**(3): 376–378.

PACLT, J. 1972. Zur allgemein-biologischen Deutung der Pflanzengalle. *Beitr. Biol. Pflanz.* **48**(1): 63–77.

PAICHELER, J.C. 1977. Les lacs cretaceés et miocène de la region de Kizilcahamam – Cerkes (Anatolie – Turquie): restes orhaniques, environments et statuts trophiques. In: HORIE, S. (ed.). *Palaeolimnology of Lake Biwa and the Japanese Pleistocene* **5**. W. Junk Publ., Dordrecht – Boston – Lancaster: 288–301.

PAICHELER, J.C., DE BROIN, F., GAUDANT, J., MOURE-CHAUVERÉ, C., RAGE, J.C., and VERGNAULT-GRAZZINI, C. 1977. Le bassin lacustre de Bes-Konak (Anatolie, Turquie). Géologie et introduction à la paléontologie des Vertébrés. *Geobios* **11**: 43–63.

PALM, T. 1949. Ett eksempel påanemohydrochor insektspridning vid Torne Träsk. *Entomol. Tidskr.* **70**: 65–74.

PALMER, A.R., CARVALHO, J.C.M., COOK, D.R., O'NEILL, K., PETRUNKEVITCH, A., and SAILER, R.I. 1957. Miocene arthropods from the Mojave Desert, California. *U.S. Geol. Surv. Prof. Paper* **294**-G: 237–280.

PALS, J.P., and HAKBIJL, T. 1992. Weed and insect infestation of grain cargo in a ship at the Roman fort of Laurium in Werden (Province of Zuid-Holland). *Rev. Palaeobot.* **73**: 287–300.

PAMPALONI, L. 1902. Microflora e microfauna nel disodile di Mellili in Sicilia. *Atti R. Acad. Lincei, Rend.* **11**(9, 2): 248–253.

PANFILOV, D.V. 1965. On subfossil insect remains from Serebryany Bor. *Byulleten' Moskovskogo Obshchestva Ispytateley Prirody, Biol.* **70**(5): 115–116 (in Russian).

PANFILOV, D.V. 1967. On the role of insects in ancient and modern continental biocoenoses. *Zool. Zhurnal* **46**(5): 645–656 (in Russian).

PANFILOV, D.V. 1968a. The ecological-landscape characteristic of the Jurassic insect fauna of Karatau. In: ROHDENDORF, B.B. (ed.). *Yurskie nasekomye Karatau* (Jurassic Insects of Karatau). Nauka, Moscow: 7–22 (in Russian).

PANFILOV, D.V. 1968b. Kalligrammatidae (Neuroptera) from the Jurassic deposits of Karatau. In: ROHDENDORF, B.B. (ed.). *Yurskie nasekomye Karatau* (Jurassic Insects of Karatau). Nauka, Moscow: 166–174 (in Russian).

PANFILOV, D.V. 1980. New representatives of Neuroptera from the Jurassic of Karatau. In: DOLIN, V.G., PANFILOV, D.V., PONOMARENKO, A.G., and PRITYKINA, L.N. *Iskopayemye nasekomye mezozoya* (Fossil Insects of the Mesozoic). Naukova Dumka, Kiev: 88–111 (in Russian).

PANT, D.D., NAUTIYAL, D.D., and CHATURVEDI, S.K. 1982. Insect pollination in some Indian Glumiflorae. *Beitr. Biol. Pflanzen* **57**(2): 229–236.

PANT, D.D., and SINGH, V.K. 1987. Xylotomy of some woods from Raniganj Formation (Permian), Raniganj Coalfield, India. *Palaeontographica* **B 203**: 1–82.

PAPIER, F., and GRAUVOGEL-STAMM, L. 1994. The Triassic Blattoddea: the genus *Voltziablatta* n. gen. from the Upper Bunter of the Vosges Mountains (France). *Palaeontographica* A **235**(4–6): 141–162.

PAPIER, F., GRAUVOGEL-STAMM, L., and NEL, A. 1994. *Subioblatta undulata* n.sp., une nouvelle blatte (Subioblattidae Schneider) du Buntsandstein superieur (Anisien) des Vosges (France). Morphologie, systematique et affinites. *Neues Jahrb. Geol. Paläontol. Monatshefte* (5): 277–290.

PAPIER, F., NEL, A., and GRAUVOGEL-STAMM, L. 1996a. Deux nouveaux insectes Mecopteroidea du Buntsandstein superieur (Trias) des Vosges (France). *Palaeontol. Lombarda* **5**: 37–45.

PAPIER, F., NEL, A., and GRAUVOGEL-STAMM, L. 1996b. New Blattoddea from the Upper Buntsandstein (Triassic) of the Vosges, France. *Palaeontol. Lombarda* **5**: 47–59.

PAPP, R.P. 1978. A nival aeolian ecosystem in California. *Arctic and Alpine Res.* **10**(1): 117–131.

PAPP, R.P., and JOHNSON, J.B. 1979. Origins of psyllid fallout in the central Sierra Nevada of California (Homoptera). *Pan-Pacif. Entomol.* **55**(2): 95–98.

PARENT, G.H. 1987. Les plus anciennes représentations de Lépidoptères (Egypte, Crete, Mycènes) et leur signification. *Linneana Belg.* **11**(1): 19–46.

PARK, L.E. 1995. Geochemical and paleoenvironmental analysis of lacustrine arthropod-bearing concretions of the Barstow Formation, Southern California. *Palaios* **10**: 44–57.

PARMALEE, P.W. 1967. A recent cave bone deposit in southwestern Illinois. *Bull. Nation. Speleol. Soc.* **29**: 119–147.

PARSONS, M.C. 1966. Modifications of the food pumps of Hydrocorisae (Heteroptera). *Canad. J. Zool.* **44**: 585–620.

PASTEELS, J.J. 1977. Une revue comparative de l'éthologie des Anthidiinae nidificateur de l'Ancien Monde (Hymenoptera, Megachilidae). *Ann. Soc. entomol. France* **13**(4): 651–667.

PATERSON, C.G., and WALKER, K.F. 1974. Recent history of *Tanytarsus barbitarsis* Freeman (Diptera: Chironomidae) in the sediments of a shallow, saline lake. *Austral. J. Mar. Freshw. Res.* **25**: 315–325.

PATTERSON, R. 1835. Note relative to the beetles observed in unrolling a mummy at Belfast. *Trans. R. Entomol. Soc. London* **1**: 67.

PAULIAN, R. 1976. Three fossil dung beetles (Coleoptera: Scarabaeidae) from the Kenya Miocene. *J. East Africa Natur. Hist. Soc. Nation. Mus.* **31**(158): 1–4.

PAWŁOWSKI, J., KMIECIAK, D., SZADZIEWSKI, R., and BURKIEWICZ, A. 1996a. Attempted isolation of DNA from insects embedded in Baltic amber. *Inclusion – Wrostek* **22**: 12–13.

PAWŁOWSKI, J., KMIECIAK, D., SZADZIEWSKI, R., and BURKIEWICZ, A. 1996b. Proba izolacji DNA owadów z bursztynu baltyckiego. *Prace Muz. Ziemi* **44**: 45–46.

PECK, S.B. 1994. Aerial dispersal of insects between and to islands in the Galápagos Archipelago, Ecuador. *Ann. Entomol. Soc. Amer.* **87**: 218–224.

PEDGLEY, D.E. 1990. Concentration of flying insects by the wind. *Philos. Trans. R. Soc. London* B **328** (1251): 631–653.

PEDGLEY, D.E., and REYNOLDS, D.R. 1992. Long-range migration in relation to climate and weather. *19 Internat. Congr. Entomol., Beijing, June 28 – July 4, 1992. Proc.: Abstr.* Beijing: 152.

PELLMYR, O. 1985. Flower constancy in individuals of an anthophilous beetle, *Byturus ochraceus* (Scriba) (Coleoptera: Byturidae). *Coleopterists' Bull.* **39**(4): 341–345.

PELLMYR, O. 1988. Seed parasites as pollinators: interactions between *Trollius europaeus* (Ranunculaceae) and *Chiastocheta* (Diptera: Anthomyidae). *Proc. 18th Internat. Congr. Entomol., Vancouver, July 3–9, 1988. Abstracts and Author Index*: 180.

PEMBERTON, S.G., FREY, R.W., and SAUNDERS, T.D.A. 1990. Trace fossils. In: BRIGGS, D.E.G., and CROWTHER, P.R. (eds.). *Palaeobiology, a Synthesis*. Blackwell Scientific Publ., Oxford: 355–362.

PEÑALVER, E. 1996. Tecnicas y metodos de obtención, preparación, conservación y estudio de insectos fósiles. *Bol. Soc. Entomol. Aragon*. **16** (Volumen Monográfico: Paleoentomología): 157–174.

PEÑALVER, E. 1998. *Estudio tafonómico y paleoecológico de los insectos del Mioceno de Rubielos de Mora (Teruel)*. Inst. Estudios Turolenses, Teruel: 179 p.

PEÑALVER, E., NEL, A., and MARTÍNEZ-DELCLÒS, X. 1996. Insectos del Mioceno inferior de Ribesalbes (Castellón, España). Paleoptera y Neoptera poli- y paraneoptera. *Treb. Mus. geol. Barcelona* **5**: 15–95.

PEÑALVER, E., MARTÍNEZ-DELCLÒS, X., and ARILLO, A. 1999. Yacimientos con insectos fósiles en España. *Rev. Español. Paleontol.* **14**(2): 231–245.

PENNY, N.D. 1975. Evolution of the extant Mecoptera. *J. Kansas Entomol. Soc.* **48**: 331–350.

PERKOVSKY, E.E. 1999. Evolutionary development of the specific antennal structure in leiodid beetles and systematic position of Jurassic *Mesecanus communis* and *Polysitum elongatum* (Coleoptera: Staphylinoidea, Leiodidae). *Proc. First Palaeontomol. Conf., Moscow 1998*. AMBA projects Internat., Bratislava: 111–115.

PETELLE, M. 1980. Aphids and melizitose: a test of Owen 1978 hypothesis. *Oikos* **35**(1): 127–128.

PETELLE, M. 1982. More mutualisms between consumers and plants. *Oikos* **38**(1): 125–127.

PETERS, W.L., and PETERS, J.G. 2000. Discovery of a new genus of Leptophlebiidae: Leptophlebiinae (Ephemeroptera) in Cretaceous amber from New Jersey. In: GRIMALDI, D.A. (ed.). *Studies on Fossils in Amber, with Particular Reference to the Cretaceous of New Jersey*. Backhuys Publ., Leiden: 127–131.

PETERSON, A. 1915. Morphological studies of the head and mouthparts of the Thysanoptera. *Ann. Entomol. Soc. America* **8**: 20–66.

PETIT-MAIRE, N., ROSSO, J.C., DELIBRIAS, G., MECO, J., and POMEL, S. 1987. Paléoclimats de l'île de Fuerteventura (Archipel Canarien). *Palaeoecol. Africa Surround. Islands* **18**: 351–356.

PETRULEVICIUS, J.F. 1999a. First hanging fly fossil from South America. *First Paleoentomological Conference. 30 Aug.–4 Sept. 1998. Moscow, Russia. Abstracts*. Palaeontological Inst., Russian Acad. Sci., Moscow: 34.

PETRULEVICIUS, J.F. 1999b. Insectos del Cenozoico de l'Argentina. *Rev. Soc. Entomol. Argentina* **58**(1–2): 95–103.

PETRUNKEVITCH, A. 1935. Striated muscles of an amber insect. *Nature* **135**: 760–761.

PEUS, F. 1968. Über die beiden Bernstein-Flöhe. *Paläontol. Zeitschr.* **42**: 62–72.

PFAU, H.K. 2000. *Erasipteron larischi* Pruvost, 1933, *Eogeropteron lunatum* Riek, 1984 und die Evolution der Verstellpropeller-Flügel der Libellen. *Mitteil. Schweiz. Ent. Ges.* **73**: 223–263.

PHILLIPS, T.L., and PEPPERS, R.A. 1984. Changing patterns of Pennsylvanian coal-swamp vegetation and implications of climatic control on coal occurrence. *Internat. J. Coal Geol.* **3**: 205–255.

PICARD, D.M., and HIGH, L.R., Jr. 1981. Physical stratigraphy of ancient lacustrine deposits. *Soc. Econ. Paleontol. Mineral. Spec. Publ.* **31**: 233–259.

PICKFORD, M. 1986. Cainozoic palaeontological sites of Western Kenya. *Münchner geowiss. Abhandl*. A **8**: 5–151.

PICTET, F.J. 1854. Classe Insectes. In: *Traité de paléontologie, ou Histoire naturelle des animaux fossiles considérés dans leurs rapports zoologiques et géologiques*. Ed. 2. Ateas. **2**: 301–401.

PIERCE, W.D. 1945a. Fossil arthropods of California. 8. A case of Pleistocene myasis from the La Brea pits. *Bull. South Calif. Acad. Sci.* **44**(1): 8–9.

PIERCE, W.D. 1945b. A crystallized milliped from volcanic rock in a well. *Bull. S. Calif. Acad. Sci.* **44**(1): 1–2.

PIERCE, W.D. 1947. Fossil arthropods of California. 10. Exploring the minute world of the California asphalt deposits. *Bull. South Calif. Acad. Sci.* **45**(3): 113–118.

PIERCE, W.D. 1949. A modern asphalt seep tells a story. *Los Angeles County Mus. Quart.* **7**(3): 12–17.

PIERCE, W.D. 1950. Fossil arthropods from onyx-marble. 1–3. *Bull. S. Calif. Acad. Sci.* **49** (2): 101–104.

PIERCE, W.D. 1951. Fossil arthropods from onyx-marble. 4. Hot calcareus water killing insects. *Bull. S. Calif. Acad. Sci.* **50**: 34–49.

PIERCE, W.D. 1961. The growing importance of paleoentomology. *Proc. Entomol. Soc. Washington* **63**(3): 211–217.

PIERCE, W.D. 1965. Fossil arthropods of California. 26. Three new fossil insect sites in California. *Bull. S. Calif. Acad. Sci.* **64**(3): 157–162.

PIERCE, W.D., and GIBRON, J., Sr. 1962. Fossil arthropods from California. 24. Some unusual fossil arthropods from the Calico Mountains nodules. *Bull. S. Calif. Acad. Sci.* **61**: 143–151.

PIGULEVSKY, G.V. (ed.). 1965. *Terpenoidy i kumariny* (Terpenoids and Cumarins*)*. Trudy Botan. Inst. Akad. Nauk SSSR Ser. 5. **12**. Leningrad: 1–198.

PIKE, E.M. 1993. Amber taphonomy and collecting biases. *Palaios* **8**: 411–419.

PIKE, E.M. 1994. Historical changes in insect community structure as indicated by hexapods of Upper Cretaceous Alberta (Grassy Lake) amber. *Canad. Entomol.* **126**: 695–702.

PILGRIM, R.L.C. 1988. Flea larvae – their morphology in relation to taxonomy, biology and phylogeny. *The Results and Perspectives of Further Research of Siphonaptera in Palearct from the Aspect of Their Significance for Practice [sic!]. Symposium, Bratislava 6.6.–11.6.1988*. Bratislava: 107–116.

PIMENOVA, M.V. 1954. *Sarmatskaya flora Amvrosievki* (The Sarmatian Flora of Amvrosievka). Izdatel'stvo Akad. Nauk Ukrain. SSR, Kiev: 96 p. (in Russian).

PING, C. 1935. On four fossil insects from Sinkiang. *Chin. J. Zool.* **1**: 107–115.

PINTO, I.D. 1986. Carboniferous Insects from Argentina. III. Familia Xenopteridae Pinto, nov. Ordo Megasecoptera. *Pesquisas* **18**: 23–29.

PINTO, I.D. 1994a. A new species of Palaeodictyopteran Insecta from Piedra Shotle Formation, Upper Carboniferous, Argentina. *Pesquisas* **21**(2): 107–111.

PINTO, I.D. 1994b. *Sphecorydaloides lucchesei* a new Carboniferous Megasecopteran Insecta from Argentina. *Pesquisas* **21**(2): 85–89.

PINTO, I.D., and ORNELLAS, L.P.de. 1974. New Cretaceous Hemiptera (Insecta) from Codó Formation – Northern Brazil. In: *An. XXVIII Congr. Brasil. Geol.*: 289–304.

PINTO, I.D., and ORNELLAS, L.P., de. 1978. Carboniferous insects (Protorthoptera and Paraplecoptera) from the Gondwanaland (South America, Africa, and Asia). *Pesquisas* **11**: 305–321.

PINTO, I.D., and PURPER, I. 1986. A new Blattoid from the Cretaceous of Brasil. *Pesquisas* **18**: 5–10.

PIROZYNSKI, K.A. 1976. Fossil fungi. *Annu. Rev. Phytopathol.* **14**: 237–246.

PISHCHIKOVA, T.I. 1992. *Srednemiotsenovye ostrakody Ravninnogo Kryma* (The Mid-Miocene ostracods of lowland Crimea). Autoreferat dissert. cand. geol.-mineral. nauk. Moscow Univ, Moscow: 1–24 (in Russian).

PITON, L.E. 1940. *Paléontologie du gisement eocéne de Menat (Puy-de-Dome) (flore et faune)*. Paul Lechevalier, Paris: 303 p.

PITON, L., and THÉOBALD, N. 1935. La faune entomologique des gisements miopliocènes du Massif Central. *Rev. Sci. Natur. Auvergne* **1**: 65–104.

PLATONOFF, S. 1940. Beobachtungen über windgetriebene Insekten im Petsamfjord an der finnischen Eismeerküste. *Notulae Entomol.* **20**: 10–13.

PLOTNICK, R.E. 1986. Taphonomy of a modern shrimp: Implications for the arthropod fossil record. *Palaios* **1**(3): 286–293.

PLUMSTEAD, E.P. 1963. The influence of the plants and environment on the developing animal life of Karroo times. *S. Africa J. Sci.* **59**: 135–145.

POHL, F. 1937. Die Pollenerzeugung der Winterblüter. *Beih. Botan. Centralbl.* **56**A: 365–470.

POINAR, G.O., Jr. 1977. Fossil nematodes from Mexican amber. *Nematologica* **23**(2): 232–238.

POINAR, G.O., Jr. 1984a. First fossil record of parasitism by insect parasitic Tylenchida (Allantonematidae: Nematoda). *J. Parasitol.* **70**: 306–308.

POINAR, G.O., Jr. 1984b. Fossil evidence of nematode parasitism. *Rev. nématol.* **7**(2): 201–203.

POINAR, G.O., Jr. 1988. *Zorotypus palaeus*, new species, a fossil Zoraptera (Insecta) in Dominican amber. *J. New York Entomol. Soc.* **96**: 253–259.

POINAR, G.O., Jr. 1991a. *Hymenaea protera* sp. n. (Leguminosae, Caesalpinioideae) from Dominican amber has African affinities. *Experientia* **47**: 1075–1082.

POINAR, G.O., Jr. 1991b. *Praecoris dominicana* gen. n., sp. n. (Holoptilinae: Reduviidae: Hemiptera) from Dominican amber, with an interpretation of past behaviour based on functional morphology. *Entomol. Scand.* **22**: 193–199.

POINAR, G.O., Jr. 1992a. *Life in Amber*. Stanford Univ. Press, Stanford, CA: 350 p.

POINAR, G.O., Jr. 1992b. Fossil evidence of resin utilization by insects. *Biotropica* **24**(3): 466–468.

POINAR, G.O., Jr. 1993. Insects in amber. *Annu. Rev. Entomol.* **46**: 145–159.

POINAR, G.O., Jr. 1994a. The range of life in amber: significance and implications in DNA studies. *Experientia* **50**: 536–542.

POINAR, G.O., Jr. 1994b. Bees in fossilized resin. *Bee World* **75**(2): 71–77.

POINAR, G.O., Jr. 1995. Fleas (Insecta: Siphonaptera) in Dominican amber. *Med. Sci. Res.* **23**: 789.

POINAR, G.O., Jr. 1999a. Ancient DNA. *Amer. Sci.* **87** (Sept.–Oct.): 446–457.

POINAR, G.O., Jr. 1999b. *Palaeochordodes protus* n.g., n.sp. (Nematomorpha, Chordodidae), parasites of a fossil cockroach, with a critical examination of other fossil hairworms and helminths of extant cockroaches (Insecta: Blattaria). *Invertebr. Biol.* **118**(2): 109–115.

POINAR, G.O., Jr. 1999c. *Paleoeuglossa melissiflora* gen. n., sp. n. (Euglossinae: Apidae), fossil orchid bees in Dominican amber. *J. Kansas Entomol. Soc.* **71**(1): 29–34.

POINAR, G.O., Jr. 1999d. Extinction of tropical insect lineages in Dominican amber from Plio-Pleistocene cooling event. *Russian Entomol. J.* **8**(1): 1–4.

POINAR, G.O., Jr. 1999e. A new species of fossil palm bruchid, *Caryobruchus dominicanus* sp. n. (Pachymerini: Bruchidae) in Dominican amber. *Entomol. Scand.* **30**: 219–224.

POINAR, G.O., Jr. 1999f. Chrysomelidae in fossilized resin: behavioural inferences. In: COX, M.I. (ed.). *Chrysomelidae Biology* **1**. Backhuys, Leiden: 1–16.

POINAR, G.O., Jr., ACRA, A., and ACRA, F. 1994a. Earliest fossil nematode (Mermithidae) in Cretaceous Lebanese amber. *Fundam. Appl. Nematol.* **17**: 475–477.

POINAR, G.O., Jr., ACRA, A., and ACRA, F. 1994b. Animal–animal parasitism in Lebanese amber. *Med. Sci. Res.* **22**: 159.

POINAR, G.O., Jr., ARCHIBALD, B., and BROWN, A. 1999. New amber deposit provides evidence of Early Paleogene extinctions, paleoclimates, and past distributions. *Canad. Entomol.* **131**: 171–177.

POINAR, G.O., Jr., CURČIČ, B.P.M., and COCKENDOLPHER, J.C. 1998. Arthropod phoresy involving pseudoscorpions in the past and present. *Acta Arachnol.* **47**(2): 79–96.

POINAR, G.O., Jr., and DOYEN, J.T. 1992. A fossil termite bug, *Termitaradus protera* sp. n. (Hemiptera: Termitaphididae), from Mexican amber. *Entomol. Scand.* **23**: 89–93.

POINAR, G.O., Jr., and GRIMALDI, D. 1990. Fossil and extant macrochelid mites (Acari: Macrochelidae) phoretic on drosophilid flies (Diptera: Drosophilidae). *J. New York Entomol. Soc.* **98**(1): 88–92.

POINAR, G.O., Jr., and HESS, R. 1982. Ultrastructure of 40-million-year-old insect tissue. *Science* **215**: 1241–1242.

POINAR, G.O., Jr., and HESS, R. 1985. Preservative qualities of recent and fossil resins: electron microscope studies on tissues preserved in Baltic amber. *J. Baltic Stud.* **16**: 222–230.

POINAR, G.O., Jr., and POINAR, R. 1999. *The Amber Forest. A Reconstruction of a Vanished World.* Princeton Univ. Press, Princeton: 239 p.

POINAR, G.O., Jr., and THOMAS, G.M. 1982. An Entomophthorales from Dominican amber. *Mycologia* **74**: 332–334.

POINAR, G.O., Jr., and THOMAS, G.M. 1984. A fossil entomogenous fungus from Dominican amber. *Experientia* **40**: 578–579.

POINAR, G.O., Jr., TREAT, A.E., and SOUTHCOTT, R.V. 1991. Mite parasitism of moths: Examples of paleosymbiosis in Dominican amber. *Experientia* **47**: 210–212.

POINAR, G.O., Jr., WAGGONER, B.J., and BAUER, H.-C. 1993. Terrestrial soft-bodied protists and other micro-organisms in Triassic amber. *Science* **259**: 202–224.

POINAR, G.O., ZAVORTINK, T.J., PIKE, T., and JOHNSTON, P.A. 2000. *Paleoculicis minutus* (Diptera: Culicidae) n.gen., n.sp., from Cretaceous Canadian amber, with a summary of described fossil mosquitoes. *Acta Geol. Hispanica* **35**(1–2): 119–128.

POINAR, H.N., CANO, R.J., and POINAR, G.O., Jr. 1993. DNA from an extinct plant. *Nature* **363**: 677.

POINAR, H.N., HÖSS, M., BADA, J.L., and PÄÄBO, S. 1996. Amino acid racemization and the preservation of ancient DNA. *Science* **272**: 864–866.

POLLARD, J.E., STEEL, R.J., and UNDERSRUD, E. 1982. Facies sequences and trace fossils in lacustrine/fan delta deposits, Hornelen Basin (M. Devonian), western Norway. *Sedimentary Geol.* **32**: 63–82.

POLLARD, J.E., and WALKER, E. 1984. Reassessment of sediments and trace fossils from Old Red Sanstone (Lower Devonian) of Dunure, Scotland, described by John Smith (1909). *Geobios* **17**: 567–576.

POLOZHENTSEV, P.A. 1947. Pine resin and its entomotoxicity. *Trudy Bashkir. Sel'skokhozyaystv. Inst.* **5**: 169–184 (in Russian).

POLOZHENTSEV, P.A. 1965. Anti-insect resistance of trees and influence of their physiological state on pest outbreaks. *Byull. Glavn. Botan. Sada Akad. Nauk SSSR* **59**: 78–83 (in Russian).

PONGRÁCZ, A. 1923. Fossile Insekten aus Ungarn. 1. Tertiärer Odonatenlarven von Tállya. 2. Die fossilen Insekten von Ungarn und ihre Beziechungen zur gegenwärtigen Fauna. *Palaeontol. Hungar.* **1**: 63–76.

PONGRÁCZ, A. 1935. Die eozäne Insektenfauna des Geiseltales. *Nova Acta Leopold.* n.F. **2**: 483–572.

PONOMARENKO, A.G. 1969. *Istoricheskoe razvitie zhestkokrylykh-arkhostemat* (Historical Development of Archostematan Beetles). Trudy Paleontol. Inst. Akad. Nauk SSSR *125*. Nauka, Moscow: 240 p. (in Russian).

PONOMARENKO, A.G. 1971. On the taxonomic position of some beetles from the Solenhofen shales of Bavaria. *Paleontol. Zhurnal* 1: 67–81 (in Russian, English translation: *Paleontol. J.* 1971 **5**(1): 62–75).

PONOMARENKO, A.G. 1972. On nomenclature of the beetle wing venation (Coleoptera). *Entomol. Obozrenie* **51**(4): 768–773 (in Russian).

PONOMARENKO, A.G. 1973. On subdividing of the order Coleoptera into suborders. In: NARCHUK, E.P. (ed.). *Voprosy paleontologii nasekomykh* (Problems of the Insect Palaeontology). *Doklady na 24–m Ezhegodnom chtenii pamyati N.A. Kholodkovskogo, 1971* [Lectures on the XXIV Annual Readings in Memory of N.A. Kholodkovsky (1–2 April, 1971)]. Nauka, Leningrad: 78–89 (in Russian).

PONOMARENKO, A.G. 1976a. A new insect from the Cretaceous of Transbaikalia, a possible parasite of pterosaurians. *Paleontol. Zhurnal* (3): 102–106 (in Russian, English translation: *Paleontol. J.* **10**: 339–343).

PONOMARENKO, A.G. 1976b. Corydalidae (Megaloptera) from the Cretaceous deposits of the North Asia. *Entomol. Obozrenie* **55**(2): 425–433.

PONOMARENKO, A.G. 1977a. Composition and ecological characteristic of Mesozoic Coleoptera. In: ARNOLDI, L.V., ZHERIKHIN, V.V., NIKRITIN, L.M., and PONOMARENKO, A.G. *Mezozoiskie zhestkokrylye.* Trudy Paleontol. Inst. Akad. Nauk SSSR **161**. Nauka, Moscow: 8–16 (in Russian, English translation: ARNOLDI, L.V., ZHERIKHIN, V.V., NIKRITIN, L.M. and PONOMARENKO, A.G. 1991. *Mesozoic Coleoptera.* Oxonian Press, New Delhi. 5–18).

PONOMARENKO, A.G. 1977b. Paleozoic members of Megaloptera (Insecta). *Paleontol. Zhurnal* (1): 78–86 (in Russian, English translation: *Paleontol. J.* **11**: 73–81).

PONOMARENKO, A.G. 1985a. Fossil insects from the Tithonian "Solnhofener Plattenkalke" in the Museum of Natural History, Vienna. *Ann. Naturhist. Mus. Wien* A **87**: 135–144.

PONOMARENKO, A.G. 1985b. Beetles from the Jurassic of Siberia and western Mongolia. In: RASNITSYN, A.P. (ed.) *Yurskie nasekomye Sibiri i Mongolii* (Jurassic Insects of Siberia and Mongolia). *Trudy Paleontol. Inst. Akad. Nauk SSSR* **211**. Nauka, Moscow: 47–87 (in Russian).

PONOMARENKO, A.G. 1986a. Insects in the Early Cretaceous ecosystems of West Mongolia. In: RASNITSYN, A.P. (ed.) *Nasekomye v rannemelovykh ekosistemakh Zapadonoy Mongolii* (Insects in the Early Cretaceous Ecosystems of West Mongolia). Trudy Sovmestnoy Sovetsko-Mongol'skoy Paleontol. Expeditsii **28**. Nauka, Moscow: 183–201 (in Russian).

PONOMARENKO, A.G. 1986b. Scarabaeiformes *incertae sedis.* In: RASNITSYN, A.P. (ed.). *Nasekomye v rannemelovykh ekosistemakh Zapadonoy Mongolii* (Insects in the Early Cretaceous Ecosystems of the West Mongolia). Trudy Sovmestnoy Sovetsko-Mongol'skoy Paleontol. Expeditsii **28**. Nauka, Moscow: 110–112 (in Russian).

PONOMARENKO, A.G. (ed.). 1988a. *Melovoy biotsenoticheskiy krizis i evolyutsia nasekomykh* (Cretaceous Biocoenotic Crisis and Evolution of the Insects). Nauka, Moscow: 230 p. (in Russian).

PONOMARENKO, A.G. 1988b. New Mesozoic insects. In: Rozanov, A.Yu. (ed.). *Novye vidy iskopaemykh bespozvonochnykh Mongolii* (New Species of Fossil Invertebrates of Mongolia). Trudy Sovmestnoy Sovetsko-Mongol'skoy Paleontol. Expeditsii **33**. Nauka, Moscow: 71–80 (in Russian).

PONOMARENKO, A.G. 1988c. The origin of fleas: the evidences from fossil record. *The Results and Perspectives of Further Research of Siphonaptera in Palearct from the Aspect of Their Significance for Practice [sic!]. Symposium: Bratislava 6.6.–11.6.1988.* Bratislava: 3–7.

PONOMARENKO, A.G. 1990. Insects and the Lower Cretaceous stratigraphy of Mongolia. In: KRASSILOV, V.A. (ed.). *Kontinental'ny mel SSSR. Materialy soveshchaniya sovetskoy rabochey gruppy proekta N 245 MPGK. Vladivostok, Oktyabr' 1988* (Nonmarine Cretaceous of the USSR. Materials of Conference of the Soviet working group of the Project 245 IGCP, Vladivostok, Oct. 1988). Vladivostok: 15–18 (in Russian).

PONOMARENKO, A.G. 1991. Main trends in larval evolution and taxonomy of Oligoneoptera. *Verhandl. XII Internat. Sympos. Entomofaunistik Mitteleuropa.* Naukova Dumka, Kiev: 38–47 (in Russian).

PONOMARENKO, A.G. 1992a. Upper Liassic beetles of Lower Saxony, Germany. *Senckenberg. Lethaea* **72**: 179–188.

PONOMARENKO, A.G. 1992b. Neuroptera (Insecta) from the Lower Cretaceous of Transbaikalia. *Paleontol. Zhurnal* (3): 43–50 (in Russian, English translation: *Paleontol. J.* **26**(3): 56–66).

PONOMARENKO, A.G. 1993. Two new species of Mesozoic beetles from Asia. *Paleontol. J.* **27**(1A): 182–191.

PONOMARENKO, A.G. 1995a. The geological history of beetles. In: PAKALUK, J., and SLIPIŃSKI, S.A. (eds.). *Biology, Phylogeny, and Classification of Coleoptera: Papers Celebrating the 80th Birthday of Roy A. Crowson.* Warszawa, Muzeum i Instytut Zoologii PAN: 155–171.

PONOMARENKO, A.G. 1995b. Upper Liassic neuropterans (Insecta) from Lower Saxony, Germany. *Russian Entomol. J.* **4**(1–4): 73–89.

PONOMARENKO, A.G. 1996. Evolution of continental aquatic ecosystems. *Paleontol. J.* **30**(6): 705–709.

PONOMARENKO, A.G. 1997. The peculiarities of taphonomy of organic remains in continental lacustrine and volcanogenic formations. In: OCHEV, V.G., and TVEROKHLEBOVA, G.I. (eds.). *Tafonomiya nazemnykh organizmov* (Taphonomy of Terrestrial Organisms). Izdatel'stvo Saratovskogo Univ., Saratov: 108–116 (in Russian).

PONOMARENKO, A.G. 1998a. Paleobiology of angiospermization. *Paleontol. Zhurnal* (4): 3–10 (in Russian; English translation: *Paleontol. J.* **32**: 325–331).

PONOMARENKO, A.G. 1998b. Paleoentomology of Mongolia. *First Paleoentomological Conference. 30 Aug.–4 Sept. 1998. Moscow, Russia. Abstracts.* Palaeontological Inst., Russian Acad. Sci., Moscow: 36.

PONOMARENKO, A.G. 2000. New alderflies (Megaloptera: Parasialidae) and glosselytrodeans (Glosselytrodea: Glosselytridae) from the Permian of Mongolia. *Paleontol. J.* **34**(suppl. 3): S309–S311.

PONOMARENKO, A.G., and KALUGINA, N.S. 1980. The general characteristic of insects at the Manlay locality. In: KALUGINA, N.S. (ed.) *Rannemelovoe ozero Manlay* (Early Cretaceous Lake Manlay). Trudy Sovmestnoy Sovetsko-Mongol'skoy Paleontol. Expeditsii **13**. Nauka, Moscow: 69–82 (in Russian).

PONOMARENKO, A.G., and POPOV, Yu.A. 1980. Palaeobiocoenoses of Early Cretaceous Mongolian lakes. *Paleontol. Zhurnal* (3): 3–13 (in Russian, English translation: *Paleontol. J.* **14**(3): 1–10).

PONOMARENKO, A.G., and RASNITSYN, A.P. 1974. New Mesozoic and Cenozoic Protomecoptera. *Paleontol. Zhurnal* (4): 59–73 (in Russian, English translation: *Paleontol. J.* **8**: 493–507).

PONOMARENKO, A.G., and SUKATSHEVA, I.D. 1998. Insects. In: LOZOVSKY, V.R., and ESAULOVA, N.K. (eds.). *Granitsa permi i triasa v kontinental'nykh seriyakh Vostochnoy Evropy* (Permian-Triassic Boundary in the Continental Series of the East Europe). Geos, Moscow: 96–106 (in Russian).

PONOMARENKO, A.G., and ZHERIKHIN, V.V. 1980. Superorder Scarabaeidea. In: ROHDENDORF, B.B., and RASNITSYN, A.P. (eds.). 1980. *Istoricheskoe razvitie klassa nasekomykh* (Historical Development of the Class Insecta). Trudy Paleontol. Inst. Akad. Nauk SSSR **175**. Nauka, Moscow: 75–84 (in Russian).

PONOMARYOVA, G.Yu., NOVOKSHONOV, V.G., and NAUGOLNYKH, S.V. 1998. *Chekarda – mestonakhozhdenie permskikh iskopaemykh rasteniy i nasekomykh* (Chekarda – the Locality of Permian Fossil Plants and Insects). Perm' Univ., Perm'. 92 p. (in Russian).

PONT, A.C., and CARVALHO, C.J.B., DE. 1997. Three species of Muscidae (Diptera) from Dominican amber. *Stud. dipterol.* **4**(1): 173–181.

POP, E. 1936. Flora pliocena de la Borsec. *Rev. Univ. Cluj, Fac. Stiinte* **1**: 3–94.

POPHAM, E.J. 1985. The mutual affinities of the major earwig taxa (Insecta, Dermaptera). *Zeitschr. Zool. Syst. EvolutionsForsch.* **23**: 199–214.

POPLIN, C. 1986. Taphocoenoses et restes alimentaires de vertébrés carnivores. *Bull. Mus. nation. Hist. natur. (Paris)* D**8**(2): 257–267.

POPOV, Yu.A. 1968a. The true bugs (Heteroptera) from the Holocenous badger coprolites. In: SUKACHEV, V.N. (ed.) *Istoriya razvitiya rastitel'nogo pokrova tsentral'nykh oblastei evropeiskoi chasti SSSR v antropogene* (The History of Development of the Vegetation Cover of Central Regions of the European Part of the USSR in the Anthropogene). Nauka, Moscow: 129–132 (in Russian).

POPOV, Yu.A. 1968b. Bugs of the Jurassic fauna of Karatau. In: ROHDENDORF, B.B. (ed.) *Yurskie nasekomye Karatau* (Jurassic insects of Karatau). Nauka, Moscow: 99–113 (in Russian).

POPOV, Yu.A. 1971. *Istoricheskoe razvitie poluzhestkokrylykh infraotryada Nepomorpha* (Historical Development of the Heteropterous Infraorder Nepomorpha). Trudy Paleontol. Inst. Akad. Nauk SSSR **129**. Nauka, Moscow: 228 p. (in Russian).

POPOV, Yu.A. 1980. Superorder Cimicidea Laicharting, 1781. In: ROHDENDORF, B.B., and RASNITSYN, A.P. (eds.). *Istoricheskoe razvitie klassa nasekomykh* (Historical Development of the Class Insecta). Trudy Paleontol. Inst. Akad. Nauk SSSR **175**. Nauka, Moscow: 58–69 (in Russian).

POPOV, Yu.A. 1981. Historical development and some questions on the general classification of Hemiptera. *Rostria* **33** Suppl.: 85–99.

POPOV, Yu.A. 1985. Jurassic bugs and Coleorrhyncha of southern Siberia and western Mongolia. In: RASNITSYN, A.P. (ed.). *Yurskie nasekomye Sibiri i Mongolii* (Jurassic insects of Siberia and Mongolia). Trudy Paleontol. Inst. Akad. Nauk SSSR **211**. Nauka, Moscow: 28–47 (in Russian).

POPOV, Yu.A. 1986. Peloridiina (=Coleorrhyncha) et Cimicina (=Heteroptera). In: RASNITSYN, A.P. (ed.). *Nasekomye v rannemelovykh ekosistemakh Zapadnoy Mongolii* (Insects in Early Cretaceous Ecosystems of Western Mongolia). Trudy Sovmestnoy Sovetsko-Mongol'skoy Paleontol. Expeditsii **28**. Nauka, Moscow: 50–83 (in Russian).

POPOV, Yu.A. 1989a. New fossil Hemiptera (Heteroptera + Coleorrhyncha) from the Mesozoic of Mongolia. *Neues Jahrb. Geol. Paläontol. Monatshefte* (3): 166–181.

POPOV, Yu.A. 1989a. On names and taxonomic position of some Mesozoic aquatic bugs (Heteroptera, Nepomorpha). *Paleontol. Zhurnal* (4): 122–124 (in Russian, English translation: *Paleontol. J* **23**(4): 118–121, titled 'The assignment <sic!> and systematic position of certain Mesozoic water bugs').

POPOV, Yu.A., CORAM, R., and JARZEMBOWSKI, E.A. 1998. Fossil heteropteran bugs from the Purbeck Limestone Group of Dorset. *Dorset Proc.* **120**: 73–76.

POPOV, Yu.A., DOLLING, W.R., and WHALLEY, P.E.S. 1994. British Upper Triassic and Lower Jurassic Heteroptera and Coleorrhyncha (Insecta: Hemiptera). *Genus* **5**: 307–347.

POPOV, Yu.A., and HERCZEK, A. 1993. New data on Heteroptera in amber resins. *Ann. Upper Siles. Mus., Entomol.*, Suppl. **1**: 7–12.

POPOV, Yu.A., and SHCHERBAKOV, D.E. 1991. Mesozoic Peloridioidea and their ancestors (Insecta: Hemiptera, Coleorrhyncha). *Geologica et Palaeontologica* **25**: 215–235.

POPOV, Yu.A., and SHCHERBAKOV, D.E. 1996. Origin and evolution of Coleorrhyncha as shown by the fossil record. In: SCHAEFER, C.W. (ed.). *Studies on Hemiptera Phylogeny. Proceedings. Thomas Say Publications in Entomology.* Entomol. Soc. America, Lanham, Maryland: 9–30.

POPOV, Yu.A., and WOOTTON, R.J. 1977. The Upper Liassic Heteroptera of Mecklenburg and Saxony. *Syst. Entomol.* **2**: 333–351.

PORSCH, O. 1910. Ephedra campylopoda C.A. Mey., eine entomophile Gymnosperme. *Ber. Deutsch. Botan. Ges.* **28**: 404–412.

PORSCH, O. 1956. Windpollen und Blumeninsekten. *Österr. Botan. Zeitschr.* **103**: 1–19.

PORSCH, O. 1958. Alte Insektentypen als Blumenaubeuter. *Österr. Botan. Zeitschr.* **104**: 115–164.

POTONIÉ, P. 1910. *Die Entstehung der Steinkohle und der Kaustobiolithe überhaupt.* Berlin: 225 p.

POTONIÉ, R. 1921. Mitteilungen über mazerierte kohlige Pflanzenfossilien. *Zeitschr. Botanik* **13**(2): 79–88.

POTTS, R., and BEHRENSMEYER, A.K. 1992. Late Cenozoic terrestrial ecosystems. In: BEHRENSMEYER, A.K., DAMUTH, J.D., DiMICHELE, W.A., POTTS, R., SUES, H.-D., and WING, S.L. (eds.). *Terrestrial Ecosystems through Time. Evolutionary Paleoecology of Terrestrial Plants and Animals.* Univ. Chicago Press, Chicago–London: 419–541.

POWELL, J.A. 1984. Biological interrelationships of moths and *Yucca schottii*. *Univ. Calif. Publ. Entomol.* **100**: 1–93.

PRICE, L.W. 1985. Grasshoppers on snow in the Wallowa Mountains, Oregon. *Northwest Sci.* **59**(3): 213–220.

PRICE, M.A., and GRAHAM, O.H. 1997. Chewing and sucking lice as parasites of mammals and birds. *U.S. Depr. Agric. Techn. Bull.* **1849**: 1–309.

PRIDHAM, B. (ed.). 1967. *Terpenoids in Plants.* Academic Press, New York: 257 p.

PRIESNER, H. 1924. Bernstein-Thysanopteren. *Entomol. Mitteil.* **13**(4–5): 130–151.

PRIESNER, H. 1929. Bernstein-Thysanopteren II. *Bernstein-Forsch.* **1**: 111–138.

PRIESNER, H. 1949. Genera Thysanopterorum. *Bull. Soc. Fouad Ier Entomol.* **33**: 31–157.

PRIESNER, H. 1957. Zur vergleichenden Morphologie des Endothorax der Thysanopteren. *Zool. Anzeiger*, B **159**(7–8): 159–167.

PRIESNER, H. 1964a. A monograph of the Thysanoptera of the Egyptian deserts. *Publ. Inst. Désert Egypte* **13**: 549 p.

PRIESNER, H. 1964b. *Ordnung Thysanoptera. Bestimmungsbücher zur Bodenfauna Europas* **2**. Akademie–Verlag, Berlin: 242 p.

PRIESNER, H. 1968. Thysanoptera. In: Kükenthal, W. (ed.). *Handbuch der Zoologie.* 4. Walter de Gruyter, Berlin: 32 p.

PRIESNER, H., and QUIÉVREUX, F. 1935. Thysanopteres des couches de potasse du Haut-Rhin. *Bull. Soc. géol. France*, sér. 5, **5**: 471–477.

PRITYKINA, L.N. 1977. New Odonata from the Lower Cretaceous deposits of the Transbaikalia and Mongolia. *Fauna, flora i biostratigrafiya mezozoya i kaynozoya Mongolii* (Fauna, Flora, and Biostratigraphy of the Mesozoic and Cenozoic of Mongolia). Trudy Sovmestnoy Sovetsko-Mongol'skoy Paleontol. Expeditsii **4**. Nauka, Moscow: 81–96 (in Russian).

PRITYKINA, L.N. 1980. The order Libellulida. Dragonflies. In: ROHDENDORF, B.B., and RASNITSYN, A.P. (eds.). *Istoricheskoe razvitie klassa nasekomykh* (Historical Development of the Class Insecta). Trudy Paleontol. Inst. Akad. Nauk SSSR **175**. Nauka, Moscow: 128–134 (in Russian).

PRITYKINA, L.N. 1985. The Jurassic dragonflies (Libellulida = Odonata) from Siberia and West Mongolia. In: RASNITSYN, A.P. (ed.). *Yurskie nasekomye Sibiri i Mongolii* (The Jurassic Insects of Siberia and Mongolia). Trudy Paleontol. Inst. Akad. Nauk SSSR. **211**. Nauka, Moscow: 120–138 (in Russian).

PRITYKINA, L.N. 1986. Two new dragonflies from the Lower Cretaceous deposits of West Mongolia. *Odonatologica* **15**(2): 169–184.

PRITYKINA, L.N. 1989. Palaeontology and evolution of the dragonflies. In: Mordkovich, V.G. (ed.) *Fauna i ekologiya strekoz* (Fauna and Ecology of Dragonflies). Nauka, Novosibirsk: 33–59 (in Russian).

PRITYKINA, L.N. 1993. First dragonflies (Odonata; Aeschnidiidae) from Cenomanian of Crimea. *Paleontol. J.* **27**(1A): 179–181.

PROCTOR, C.J., and JARZEMBOWSKI, E.A. 1999. Habitat reconstructions in the Late Westphalian of southern England. *Proc. First Palaeoentomol. Conf., Moscow 1998.* AMBA projects Internat., Bratislava: 125–129.

PROCTOR, M., YEO, P., and LACK, A. 1996. *The Natural History of Pollination.* Harper Collins, London: 479 p.

PROKHANOV, Ya.I. 1965. The grassland plains and young deserts, their nature and origin. *Trudy Moskovskogo Obshchestva Ispytateley Prirody* **13**: 124–154 (in Russian).

PRUVOST, P. 1919. *Introduction à l'étude du terrain houiller du Nord et du Pas-de-Calais: la faune continentale du terrain du Nord de la France.* Impr. Nationale, Paris: 584 p.

PRUVOST, P. 1930. La faune continentale du terrain huiller de Belgique. *Mem. Mus. R. Hist. natur. Belgique* **44**: 142–166.

PULAWSKI, W.J., RASNITSYN, A.P., BROTHERS, D.J., and ARCHIBALD, S.B. 2000. New genera of Angarosphecinae: *Cretosphecium* from Early Cretaceous of Mongolia and *Eosphecium* from Early Eocene of Canada (Hymenoptera: Sphecidae). *J. Hymenopt. Res.* **9**(1): 34–40.

QUENSTEDT, W. 1927. Beiträge zum Kapitel Fossil und Sediment vor und bei der Einbettung. *Neue Jahrb. Miner., Geol. Paläontol.*, A **58**: 353–432.

QUICKE, D.L.J. 1996. *Principles and Techniques of Contemporary Taxonomy.* Blackie Academic & Professional, London etc.: 311 p.

QUICKE, D.L.J. 1997. *Parasitic Wasps.* Chapman & Hall, London: 470 p.

QUICKE, D.L.J., BASIBUYUK, H.H., FITTON, M.G., and RASNITSYN, A.P. 1999. Morphological, palaeontological and molecular aspects of ichneumonoid phylogeny (Insecta, Hymenoptera). *Zool. Scripta* **28**: 175–202.

QUIEVREUX, F. 1935. Esquisse du mond vivant sur les rives de la lagune potassique. *Bull. Soc. Industr. Mulhouse* **101**: 161–187.

QUIEVREUX, F. 1937. Les fossiles des potasses d'Alsace. *Le Nature* **65**(2): 455–460.

QUINLAN, R., SMOL, J.P., and HALL, R.I. 1998. Quantitative inferences of past hypolimnetic anoxia in south-central Ontario lakes using fossil midges (Diptera: Chironomidae). *Canad. J. Fish. Aquat. Sci.* **55**: 587–596.

RAFF, R.A., and KAUFMANN, T.C. 1983. *Embryos, Genes, and Evolution.* Macmillan, New York–London: 395 p.

RAINEY, R.C. 1978. The evolution and ecology of flight. The "oceanographic" approach. In: DINGLE, H. (ed.). *Evolution of Insect Migration and Diapause.* Springer-Verlag, Berlin–Heidelberg–New York: 33–48.

RAINEY, R.C. 1983. Exploration of wind fields for flying insects. *Bull. OEPP* **13**(2): 121–124.

RAISWELL, R. 1987. Non-steady state microbiological diagenesis and origin of concretions and nodular limestones. In: MARSHALL, J.D. (ed.). *Diagenesis of Sedimentary Sequences.* Geol. Soc. London Spec. Publ. **36**: 41–54.

RAISWELL, R. 1988. Evidence for surface-reaction controlled growth of carbonate concretions in shales. *Sedimentology* **35**: 571–575.

RAMIREZ, B.W. 1974. Co-evolution of *Ficus* and Agaonidae. *Ann. Missouri Botan. Gard.* **61**(3): 770–780.

RASNITSYN, A.P. 1965. Some aspects of interrelations between morphogenesis and growth in the evolution of insect ontogeny. *Entomol. Obozrenie* **44**: 476–485 (in Russian, English translation: *Entomol. Review* **44**: 279–284).

RASNITSYN, A.P. 1968. New Mesozoic sawflies (Hymenoptera, Symphyta). In: ROHDENDORF, B.B. (ed.). *Yurskie nasekomye Karatau* (Jurassic insects of Karatau). Nauka, Moscow: 190–236 (in Russian).

RASNITSYN, A.P. 1969. *Proiskhozhdenie i evolyutsiya nizshikh pereponchatokrylykh.* Trudy Paleontol. Inst. Akad. Nauk SSSR **123**). Nauka, Moscow: 196 p. (in Russian, English translation: *Origin and Evolution of Lower Hymenoptera.* Amerind Co., New Delhi, 1979).

RASNITSYN, A.P. 1975. *Vysshie pereponchatokrylye mezozoya* (Hymenoptera Apocrita of Mesozoic). Trudy Paleontol. Inst. Akad. Nauk SSSR **147**. Nauka, Moscow: 134 p. (in Russian).

RASNITSYN, A.P. 1976. On the early evolution of insect and the origin of Pterygota. *Zhurnal obshchey biol.* **37**: 543–555 (in Russian with English summary).

RASNITSYN, A.P. 1977a. New Paleozoic and Mesozoic insects. *Paleontol. Zhurnal* (1): 64–77 (in Russian, English translation: *Paleontol. J.* 1978 **11**: 60–72).

RASNITSYN, A.P. 1977b. A new sawfly family (Hymenoptera, Tenthredinoidea, Electrotomidae) from the Baltic amber. *Zool. Zhurnal* **56**(9): 1304–1308 (in Russian).

RASNITSYN, A.P. 1980. *Proiskhozhdenie i evolyutsiya pereponchatokrylykh nasekomykh* (Origin and Evolution of Hymenoptera). Trudy Paleontol. Inst. Akad. Nauk SSSR **174**. Nauka, Moscow: 192 p. (in Russian).

RASNITSYN, A.P. 1981. A modified paranotal theory of insect wing origin. *J. Morphol.* **168**: 331–338.

RASNITSYN, A.P. 1982. Proposal to regulate the names of taxa above the family group. Z.N. (S) 2381. *Bull. Zool. Nomencl.* **39**: 200–207.

RASNITSYN, A.P. 1983. Fossil Hymenoptera of the superfamily Pamphilioidea. *Paleontol. Zhurnal* (2): 54–68 (in Russian, English translation: *Paleontol. J.* **17**(2): 56–70).

RASNITSYN, A.P. (ed.) 1985a. *Yurskie nasekomye Sibiri i Mongolii* (Jurassic Insects of Siberia and Mongolia). Trudy Paleontol. Inst. Akad. Nauk SSSR **211**. Nauka, Moscow: 192 p. (in Russian).

RASNITSYN, A.P. (ed.) 1985b. *Yurskie kontinental'nye biotsenozy Yuzhnoy Sibiri i sopredel'nykh territoriy* (Jurassic non-marine biocoenoses of South Siberia and neighbor territories). Trudy Paleontol. Inst. Akad. Nauk SSSR **213**. Nauka, Moscow: 200 p. (in Russian).

RASNITSYN, A.P. 1985c. Hymenopterous insects in Jurassic of the Eastern Siberia. *Byulleten' Moskovskogo Obshchestva Ispytateley Prirody,* Biol. Sect. **58**: 85–94 (in Russian).

RASNITSYN, A.P. 1986a. Parataxon and paranomenclature. *Paleontol. Zhurnal* (3): 11–21 (in Russian, English translation: *Paleontol. J.* **20**(3): 25–33).

RASNITSYN, A.P. (ed.) 1986b. *Nasekomye v rannemelovykh ekosistemakh Zapadnoy Mongolii* (Insects in the Early Cretaceous Ecosystems of the West Mongolia). Trudy Sovmestnoy Sovetsko-Mongol'skoy Paleontol. Expeditsii **28**. Nauka, Moscow: 214 p. (in Russian).

RASNITSYN, A.P. 1987. Tempo of evolution and evolutionary theory (hypothesis of the adaptive compromise). In: TATARINOV, L.P., and RASNITSYN, A.P. (eds.). *Evolyutsiya i biotsenoticheskie krizisy* (Evolution and Biocoenotic Crises). Nauka, Moscow: 46–64 (in Russian).

RASNITSYN, A.P. 1988a. Phylogenetics. In: MENNER, V.V. (ed.). *Sovremennaya paleontologiya* (Modern Palaeontology) **1**. Nedra, Moscow: 480–497 (in Russian).

RASNITSYN, A.P. 1988b. Problem of the global crisis of the non-marine biocoenoses in the mid-Cretaceous. In: PONOMARENKO, A.G. (ed.). *Melovoy biotsenoticheskiy krizis i evolyutsiya nasekomykh* (Cretaceous Biocoenotic Crisis and Evolution of Insects). Nauka, Moscow: 191–207 (in Russian).

RASNITSYN, A.P. 1988c. An outline of evolution of the hymenopterous insects (order Vespida). *Oriental Insects* **22**: 115–145.

RASNITSYN, A.P. 1989a. Dynamics of insect families and a hypothesis of the Cretaceous biocoenotic crisis. In: SOKOLOV, B.S. (ed.). *Osadochnaya obolochka Zemli v prostranstve i vremeni. Stratigrafiya i paleontologiya* (Sedimentary Cover of the Earth in Space and Time. Stratigraphy and Palaeontology). Nauka, Moscow: 35–40 (in Russian, with English summary).

RASNITSYN, A.P. 1989b. Phytospreading from the selectionist's viewpoint. *Zhurnal obshch. biol.* **50**: 581–583 (in Russian, with English summary).

RASNITSYN, A.P. 1990a. New representatives of the hymenopterous family Praeaulacidae from the Early Cretaceous in Buriatia and Mongolia. *Vestnik Zool.* (6): 27–31 (in Russian).

RASNITSYN, A.P. 1990b Vespida. In: RASNITSYN, A.P. (ed.). *Pozdnemezozoyskie nasekomye Vostochnogo Zabaykal'ya* (Late Mesosoic insects of Eastern Transbaikalia). Trudy Paleontol. Inst. Akad. Nauk SSSR **239**. Nauka, Moscow: 177–205 (in Russian).

RASNITSYN, A.P. (ed.) 1990c. *Pozdnemezozoyskie nasekomye Vostochnogo Zabaykal'ya* (Late Mesosoic Insects of Eastern Transbaikalia). Trudy Paleontol. Inst. Akad. Nauk SSSR **239**. Nauka, Moscow: 200 p. (in Russian).

RASNITSYN, A.P. 1992a. Principles of phylogenetics and taxonomy. *Zhurnal obshchey biol.* **53**: 176–185 (in Russian).

RASNITSYN, A.P. 1992b. *Strashila incredibilis*, a new enigmatic mecopteroid insect with possible siphonapteran affinities from the Upper Jurassic of Siberia. *Psyche* **99**: 323–333.

RASNITSYN, A.P. 1995. Tertiary sawflies of the tribe Xyelini (Insecta: Vespida = Hymenoptera, Xyelidae) and their relationships to the Mesozoic and modern faunas. *Contrib. Sci. Natur. Hist. Mus. Los Angeles County* **450**: 1–14.

RASNITSYN, A.P. 1996. Conceptual issues in phylogeny, taxonomy, and nomenclature. *Contrib. to Zoology* **66**: 3–41.

RASNITSYN, A.P. 1998a. Problem of the basal dichotomy of the winged insects. In: FORTEY, R.A., and THOMAS, R.H. (eds.). *Arthropod Relationships. Systematic Assoc. Spec. Volume Series* **55**. Chapman & Hall, London etc.: 87–96.

RASNITSYN, A.P. 1998b. On the taxonomic position of the insect order Zorotypida = Zoraptera. *Zool. Anzeiger* **237**: 185–194.

RASNITSYN, A.P. 2000a. Testing cladograms by fossil record: the ghost range test. *Contrib. to Zoology* **69**(4): 251–258.

RASNITSYN, A.P. 2000b (1999). Taxonomy and morphology of *Dasyleptus* Brongniart, 1885, with description of a new species (Insecta: Machilida, Dasyleptidae). *Russian Entomol. J.* **8**(3): 145–154.

RASNITSYN, A.P. 2000c. An extremely primitive aculeate wasp in the Cretaceous amber from New Jersey. In: GRIMALDI, D.A. (ed.). *Studies on fossils in amber, with particular reference to the Cretaceous of New Jersey.* Backhuys Publisher, Leiden, The Netherlands: 327–332.

RASNITSYN, A.P., and ANSORGE, J. 2000. Two new Lower Cretaceous hymenopterous insects (Insecta: Hymenoptera) from Sierra del Montsec, Spain. *Acta Geol. Hispanica* **35**(1–2): 49–54.

RASNITSYN, A.P., and DLUSSKY, G.M. 1988. Methods and principles of reconstruction of the phylogenesis. In: PONOMARENKO, A.G. (ed.). *Melovoy biotsenoticheskiy krizis i evolyutsiya nasekomykh* (Cretaceous Biocoenotic Crisis and Evolution of the Insects). Nauka, Moscow: 5–15 (in Russian).

RASNITSYN, A.P., JARZEMBOWSKI, E.A., and ROSS, A.J. 1998. Wasps (Insecta: Vespida = Hymenoptera) from the Purbeck and Wealden (Lower Cretaceous) of Southern England and their biostratigraphical and paleoenvironmental significance. *Cretaceous Res.* **19**(3 and 4): 329–391.

RASNITSYN, A.P., and KHOVANOV, G.M. 1972. An improved method of estimation of the local faunal diversity. *Paleontol. Zhurnal* (3): 162–167 (in Russian).

RASNITSYN, A.P., and KOZLOV, M.V. 1990. A new group of the fossil insects: a scorpionfly with adaptations of cicads and moths. *Doklady Akad. Nauk SSSR* **228**: 973–976 (in Russian, English translation: *Doklady Biol. Sci.* **310**(4): 973–976).

RASNITSYN, A.P., and KRASSILOV, V.A. 1996a. First find of pollen grains in the gut of Permian insects. *Paleontol. Zhurnal* (3): 119–124 (in Russian, English translation: *Paleontol. J.* **30**(4): 484–490).

RASNITSYN, A.P., and KRASSILOV, V.A. 1996b. Pollen in the gut contents of fossil insects as evidence of co-evolution. *Paleontol. J.* **30**(6): 716–722.

RASNITSYN, A.P., and KRASSILOV, V.A. 2000. First palaeontologically confirmed phyllophagy of pre-Cretaceous insects: leaf tissues in the guts of the Upper Jurassic insects from Karatau (Kazakhstan). *Paleontol. Zhurnal* (3): 73–81 (in Russian, English translation: *Paleontol. J.* **34**(3): 301–309).

RASNITSYN, A.P., and KULICKA, R. 1990. Hymenopteran insects in Baltic amber with respect to the overall history of the order. *Prace Muz. Ziemi* **41**: 53–64.

RASNITSYN, A.P., and MARTÍNEZ-DELCLÒS, X. 1999. New Cretaceous Scoliidae (Vespida = Hymenoptera) from the Lower Cretaceous of Spain and Brazil. *Cretaceous Res.* **20**: 767–772.

RASNITSYN, A.P., and MARTÍNEZ-DELCLÒS, X. 2000. Wasps (Insecta: Vespida = Hymenoptera) from the Early Cretaceous of Spain. *Acta Geol. Hispanica* **35**(1–2): 55–85.

RASNITSYN, A.P., and MICHENER, C.D. 1991. A Miocene fossil bumble bee from the Soviet Far East with comments on the chronology and distribution of fossil bees (Hymenoptera, Apidae). *Ann. Entomol. Soc. Amer.* **84**: 583–589.

RASNITSYN, A.P., and NOVOKSHONOV, V.G. 1997. On the morphology of *Uralia maculata* (Insecta: Diaphanopterida) from the Early Permian (Kungurian) of Ural (Russia). *Entomol. Scand.* **28**: 27–38.

RASNITSYN, A.P., and PONOMARENKO, A.G. 1967. Contribution to the methods of estimation of tentative diversity of local faunas of the past. *Paleontol. Zhurnal* (3): 98–105 (in Russian).

RASNITSYN, A.P., and ROSS, A.J. 2000. A preliminary list of arthropod families present in the Burmese amber collection at The Natural History Museum, London. In: ROSS, A.J. (ed.). *The History, Geology, Age and Fauna (Mainly Insects) of Burmese Amber, Myanmar.* Bull. Natur. Hist. Mus., Geol. **56**(1): 21–24.

RASNITSYN, A.P., and ZHERIKHIN, V.V. 2000 (1999). First fossil chewing louse from the Lower Cretaceous of Baissa, Transbaikalia (Insecta, Pediculida = Phthiriaptera, Saurodectidae fam. nov.). *Russian Entomol. J.* **8**(4): 253–255.

RATCLIFFE, B.C., and FAGERSTROM, J.A. 1980. Invertebrate lebensspuren of Holocene floodplains: their morphology, origin, and paleoecological significance. *J. Paleontol.* **54**: 614–630.

RATNIKOV, V.Yu. 1988. The Upper Quaternary herpetofaunas of the Belgorod Region. *Paleontol. Zhurnal* (3): 119–122 (in Russian, English translation: *Paleontol. J.* **22**(3): 124–126).

RATTRAY, G. 1913. Notes on the pollination of some South African cycads. *Trans. Roy. Soc. S. Africa* **3**: 259–270.

RAUP, D.M. 1979. Biases in the fossil record of species and genera. *Bull. Carnegie Mus. Natur. Hist.* **13**: 85–91.

RAUP, D.M., and SEPKOSKI, J.J., Jr. 1982. Mass extinctions in the marine fossil record. *Science* **215**: 1501–1504.

RAUTIAN, A.S., and ZHERIKHIN, V.V. 1997. The models of phylocoenogenesis and lessons of the ecological crises in the geological past. *Zhurnal obshchey biol.* **58**(4): 20–47 (in Russian).

RAYNER, R.J. 1987. March flies from an African Cretaceous springtime. *Lethaia* **20**: 123–127.

RAYNER, R.J. 1993. Fossils from a Middle Cretaceous crater lake: biology and geology. *Kaupia* **2**: 5–12.

RAYNER, R.J., BAMFORD, M.K., BROTHERS, D.J., DIPPENAAR-SCHOEMAN, McKAY, I.J., OBERPRIELER, R.G., and WATERS, S.B. 1998. Cretaceous fossils from the Orapa diamond mine. *Paleont. afr.* **33**: 55–65.

RAYNER, R.J., and WATERS, S.B. 1991. Floral sex and the fossil insect. *Naturwissenschaften* **78**: 123–127.

RAZUMOVSKY, S.M. 1969. On the area borders and the floristic lines. *Byull. Glavn. Botan. Sada AN SSSR* **72**: 20–28 (in Russian).

RAZUMOVSKY, S.M. 1971. On the origin and age of tropical and subtropical floras. *Byul. Glavn. Botan. Sada AN SSSR* **82**: 43–51 (in Russian).

RAZUMOVSKY, S.M. 1999. *Izbrannye trudy* (Selected Works). KMK Scientific Press, Moscow: 557 p. (in Russian).

REDFIELD, A.C. 1958. The biological control of chemical factors in the environment. *Amer. Scient.* **46**: 206–226.

REISS, H. 1936. Zygaenenfund aus der Tertiärzeit. *Entomol. Rundsch.* **53**(39): 554–556.

REMINGTON, C.L. 1954. The suprageneric classification of the order Thysanura (Insecta). *Ann. Entomol. Soc. Amer.* **47**: 277–286.

REMINGTON, C.L. 1955. The "Apterygota". In: *A Century of Progress in the Natural Sciences (California Acad. Sci. Centennial Volume)*: 495–505.

REMIZOV, I.I. 1957. A find of fossil caddisflies of the family Molannidae in sands of the Poltava Stage in Ukraine. *Uchenye zapiski Khar'kov. Univ., Geol.* **14**: 269–280 (in Russian).

REN, D. 1993. First discovery of fossil bittacids from China. *Acta Geol. Sinica* **67**(4): 376–381 (in Chinese with English summary).

REN, D. 1997a. Studies on the Late Mesozoic snake-flies of China (Raphidioptera: Baissopteridae, Mesoraphidiidae, Alloraphidiidae). *Acta Zootaxon. Sinica* **22**(2): 172–188 (in Chinese with English summary).

REN, D. 1997b. Studies on Late Jurassic scorpion-flies from Northeast China. *Acta Zootaxon. Sinica* **22**(1): 75–85.

REN, D. 1997c. First record of fossil stick insects from China with analysis of some paleobiological features (Phasmatodea: Hagiphasmatidae fam. nov.). *Acta Zootaxon. Sinica* **22**(3): 268–281.

REN, D. 1998. Late Jurassic Brachycera from northeastern China (Insecta: Diptera). Acta *Zootaxon. Sinica* **23**(1): 65–83.

REN, D., and GUO, Z. 1996a. On the fossil genera and species of Neuroptera (Insecta) from the Late Jurassic of Northeast China. *Acta Zool. Sinica* **21**(4): 461–479.

REN, D., and GUO, Z. 1996b. Three new genera and three new species of dragonflies from the Late Jurassic of Northeast China (Anisoptera: Aeshnidae, Gomphidae, Corduliidae). *Entomol. Sinica* **3**(2): 95–105.

REN, D., and HONG, Y. 1994. A cladistic study on the familial phylogeny of fossil and living Raphidioptera (Insecta). *Bull. Chin. Acad. Geol. Sci.* **29**: 103–118 (in Chinese with English summary).

REN, D., LU, L., GUO, Z., and JI, Sh. 1995. *Faunae and Stratigraphy of Jurassic-Cretaceous in Beijing and the Adjacent Areas.* Seismic Publ. House, Beijing: 222 p. (in Chinese with English summary).

RENTZ, D.C.F., and CLYNE, D. 1983. A new genus and species of pollen- and nectar-feeding katydids from Eastern Australia (Orthoptera: Tettigoniidae: Zaprochilinae). *J. Austral. Entomol. Soc.* **22**(2): 155–160.

RENTZ, D.C.F., and KEVAN, D.K.McE. 1991. Dermaptera. In: NAUMANN, D. (ed.). *The Insects of Australia.* 2nd ed. Melbourne Univ. Press, Carlton: 360–368.

RESH, V.H., BROWN, A.V., COVICH, A.P., GURTZ, M.E., LI, H.W., MINSHALL, G.W., REICE, S.R., SHELDON, A.L., WALLACE, J.B., and WISSMAR, R.C. 1988. The role of disturbance in stream ecology. *J. North Amer. Benthol. Soc.* **7**: 433–455.

RETALLACK, G.J. 1981a. Fossil soils: indicators of ancient terrestrial environments. In: NIKLAS, K.J. (ed.). *Paleobotany, Paleoecology, and Evolution* **1**. Praeger, Ithaca, N.Y.: 55–102.

RETALLACK, G.J. 1981b. Two new approaches for reconstructing fossil vegetation with examples from the Triassic of Eastern Australia. In: GRAY, J., BOUCOT, A.J., and BERRY, W.B.N. (eds.). *Communities of the Past.* Hutchinson Ross, Stroudsburg: 271–295.

RETALLACK, G.J. 1984. Trace fossils of burrowing beetles and bees in an Oligocene paleosol, Badlands National Park, South Dakota. *J. Paleontol.* **58**: 571–592.

RETALLACK, G.J. 1986. Fossil soils as grounds for interpreting long-time controls on ancient rivers. *J. Sedimentary Petrology* **56**: 1–18.

RETALLACK, G.J. 1990a. The work of dung beetles and its fossil record. In: BOUCOT, A.J. (ed.). *Palaeobiology of Behaviour and Co-evolution.* Elsevier, Amsterdam: 214–226.

RETALLACK, G.J. 1990b. *Soils of the Past. An Introduction to Paleopedology.* Unwin Hyman, Boston: 520 p.

RETALLACK, G.J., and DILCHER, D.L. 1981a. A coastal hypothesis for the dispersal and rise to dominance of flowering plants. In: NIKLAS, K.J. (ed.). *Paleobotany, Paleoecology, and Evolution* **2**. Praeger, Ithaca, N.Y.: 27–77.

RETALLACK, G., and DILCHER, D.L. 1981b. Early angiosperm reproduction: *Prisca reynoldsii*, gen. et sp. nov. from Mid-Cretaceous coastal deposits in Kansas, U.S.A. *Palaeontographica* **B 179**: 103–137.

RETALLACK, G.J., and FEAKES, C. 1987. Trace fossil evidence for late Ordovician animals on land. *Science* **235**: 61–63.

RETALLACK, G., GRANDSTAFF, D., and KIMBERLEY, M. 1984. The promise and problems of Precambrian paleosols. *Episodes* **7**(2): 8–12.

REUNING, E. 1931. A contribution to the geology and palaeontology of the western edge of the Bushmanland Plateau. *Trans. R. Soc. S. Africa* **19**: 215–232.

REUTER, O.M. 1912. Bemerkungen über mein neues Heteropteren-system. *Öfv. Finska Vet.-Soc. Forh.* **54A**(6): 1–62.

REX, G.M., and GALTIER, J. 1986. Sur l'évidence d'interactions animal-végétal dans le Carbonifére inférieur français. *C.R. Acad. Sci.* (Paris) sér. 2, **303**(17): 1623–1626.

REYMANÓWNA, M. 1960. A Cycadeoidean stem from the western Carpathians. *Acta Palaeobotan.* (Kraków) **1**(2): 1–28.

REYNE, A. 1927. Untersuchungen über die Mundteile der Thysanopteren. *Zool. Jahrb.* **49**: 391–500.

REYNOLDS, T.D. 1991. Movement of gravel by the "owyhee" harvester ant, *Pogonomyrmex salinus* (Hymenoptera: Formicidae). *Entomol. News* **102**(3): 118–124.

REYNOLDS, T.B., and HAMILTON, A.I. 1993. Historical changes in populations of burrowing mayflies (*Hexagenia limbata*) from Lake Erie based on sediment tusk profiles. *J. Great Lakes Res.* **19**: 250–257.

RHOADS, D.C., YINGST, J.Y., and ULLMAN, W.J. 1978. Seafloor stability in central Long Island Sound. Part 1. Temporal changes in erodability of fine-grained sediments. In: WILEY, M. (ed.). *Estuarine Interactions.* Academic Press, New York: 221–244.

RICE, H.M.A. 1969. An antlion (Neuroptera) and a stonefly (Plecoptera) of Cretaceous age from Labrador, Newfoundland. *Geol. Surv. Canada Paper* **68–65**: 1–12.

RICHTER, G. 1988. Versteinerte Magen/Darm Inhalte, ihre Analyse und Deutung. In: SCHAAL, S., and ZIEGLER, W. (eds.). *Messel – Ein Schaufenster in die Geschichte der Erde und des Lebens.* Senckenberg-Buch **64**(1). W. Kramer, Frankfurt a. M.: 287–289.

RICHTER, G., and KREBS, G. 1999. Larvenstadien von Eintagsfliegen (Insecta: Ephemeroptera) als Sedimenten des eozänen Messelsees. *Natur u. Mus.* **129**(1): 21–28.

RICHTER, G., and STORCH, G. 1980. Beiträge zur Ernährungsbiologie eozäner Fledermäuse aus der Grube Messel. *Natur u. Mus.* **110**(12): 353–367.

RIEGER, C. 1976. Skelett und Muskulatur des Kopfes und Prothorax von *Ochterus marginatus* Latreille. Beitrag zur Klärung der phylogenetischen Vervandtschaftsbeziehungen der Ochteridae (Insecta, Heteroptera). *Zoomorphologie* **83**: 109–191.

RIEK, E.F. 1950. A fossil Mecopteron from the Triassic Beds at Brookvale, N.S.W. *Rec. Austral. Mus.* **22**: 253–256.

RIEK, E.F. 1952. Fossil insects of the Tertiary Redbank Plains Series. Part 1: An outline of the fossil assemblage with description of the fossil insects of the orders Mecoptera and Neuroptera. *Univ. Queensl. Papers, Dept. of Geol.* **4** (new ser.) (1): 3–14.

RIEK, E.F. 1953. Fossil Mecopteroid insects from the Upper Permian of New South Wales. *Rec. Austral. Mus.* **23**: 55–87.

RIEK, E.F. 1954. A re-examination of the Upper Tertiary mayflies described by Etheridge and Olliff from the Vegetable Creek Tin-field. *Rec. Austral. Mus.* **23**: 159–160.

RIEK, E.F. 1955. Fossil insects from the Triassic beds at Mt. Crosby, Queensland. *Austral. J. Zool.* **3**: 654–691.

RIEK, E.F. 1956. A reexamination of the mecopteroid insects and orthopteroid fossils (Insecta) from the Triassic beds at Denmark Hill, Queensland, with descriptions of further specimens. *Austral. J. Zool.* **4**: 98–110.

RIEK, E.F. 1962. Fossil insects from the Triassic at Hobart, Tasmania. *Pap. Proc. R. Soc. Tasmania* **96**: 39–40.

RIEK, E.F. 1968. Undescribed fossil insects from the Upper Permian of Belmont, New South Wales (with an appendix listing the described species). *Rec. Austral. Mus.* **27**: 303–310.

RIEK, E.F. 1970a. Lower Cretaceous fleas. *Nature* **227**: 746–747.

RIEK, E.F. 1970b. Fossil history. In: CSIRO. *The Insects of Australia.* Melbourn Univ. Press, Melbourne: 168–186.

RIEK, E.F. 1973. Fossil insects from the Upper Permian of Natal, South Africa. *Ann. Natal Mus.* **21**: 513–532.

RIEK, E.F. 1974. Upper Triassic insects from the Molteno "Formation", South Africa. *Paleontol. Africana* **17**: 19–31.

RIEK, E.F. 1976a. An unusual mayfly (Insecta: Ephemeroptera) from the Triassic of South Africa. *Palaeontol. Africana* **19**: 149–151.

RIEK, E.F. 1976b. A new collection of insects from the Upper Triassic of South Africa. *Ann. Natal Mus.* **22**: 791–820.

RIEK, E.F. 1976c. Neosecoptera, a new insect suborder based on specimen discovered in the Late Carboniferous of Tasmania. *Alcheringa* **1**(2): 227–234.

RIEK, E.F. 1976d. New Upper Permian insects from Natal, South Africa. *Ann. Natal Mus.* **22**: 755–789.

RIEK, E.F. 1976e. Fossil insects from the Middle Ecca (Lower Permian) of Southern Africa. *Palaeontol. Africana* **19**: 145–148.

RIEK, E.F., and KUKALOVÁ-PECK, J. 1984. A new interpretation of dragonfly wing venation based upon Early Upper Carboniferous fossils from Argentina (Insecta: Odonatoidea) and basic character states in pterygote wings. *Canad. J. Zool.* **62**: 1150–1166.

RIETSCHEL, S. 1983. *Aleurochiton petri* n.sp., eine Mottenschildlaus (Homoptera, Aleyrodina) aus dem Pliozän von Neu-Isenburg, Hessen. *Carolinea* **41**: 97–100.

ŘÍHA, P. 1974. Neue fossile Schwimmkäfer aus dem Tertiär Europas und Westsibiriens (Coleoptera, Dytiscidae). *Acta Entomol. Bohemoslov.* **7**: 398–413.

RISLER, H. 1957. Der Kopf von *Thrips physapus* L. (Thysanoptera, Terebrantia). *Zool. Jahrb.* **76**: 251–302.

RITCHIE, J.M. 1987. Trace fossils of burrowing Hymenoptera from Laetoli. In: LEAKEY, M.D., and HARRIS, J.M. (eds.). *Laetoli, a Pliocene Site in Northern Tanzania.* Oxford Univ. Press, London: 433–438.

RÖDEL, M.-O., and KAUPP, A. 1994. Durch Hochwasser in den Bodensee verdrifte Carabiden (Coleoptera: Carabidae). *Mitt. internat. entomol. Ver.* **19**(1/2): 21–28.

ROGERS, A.F. 1938. Fossil termite pellets in opalized wood from Santa Maria, California. *Amer. J. Sci.* Ser. 5. **35**(215): 389–392.

ROGERS, R.R. 1992. Non-marine borings in dinosaur bones from the Upper Cretaceous Two Medicine Formation, northwestern Montana. *J. Vertebr. Paleontol.* **12**(4): 528–531.

ROHDENDORF, B.B. 1939. A new protelytropteron from the Permian of the Urals. *Doklady Akad. Nauk SSSR* **23**: 506–508.

ROHDENDORF, B.B. 1946. *Evolyutsiya kryla i filogenez dlinnousykh dvukrylykh Oligoneura (Diptera, Nematocera)* [The evolution of the Wing and the Phylogeny of Oligoneura (Diptera, Nematocera)]. Trudy Paleontol. Inst. Akad. Nauk SSSR **13**(2). Akad. Nauk SSSR, Moscow – Leningrad: 108 p. (in Russian, with English summary).

ROHDENDORF, B.B. 1961. The oldest dipteran infraorders from the Triassic of Middle Asia. *Paleontol. Zhurnal* **2**: 90–100 (in Russian).

ROHDENDORF, B.B. (ed.) 1962. Osnovy paleontologii. Chlenistonogie. Trakheinye i khelitserovye. Akad. Nauk SSSR, Moscow: 560 p. (in Russian, English translation: ROHDENDORF, B.B. (ed.) 1991. *Fundamentals of Palaeontology. Vol. 9. Arthropoda, Tracheata, Chelicerata.* Amerind Co., New Delhi: 894 p.).

ROHDENDORF, B.B. 1964. *Istoricheskoe razvitie dvukrylykh nasekomykh* (Historical Development of Diptera). *Trudy Paleontol. Inst. Akad. Nauk SSSR* **100**. Nauka, Moscow: 312 p. (in Russian, English translation: 1974. Alberta Univ. Press, Edmonton).

ROHDENDORF, B.B. 1968a. Insect phylogeny and the fossil record. *Entomol. Obozrenie* **47**: 321–342 (in Russian).

ROHDENDORF, B.B. (ed.) 1968b. *Yurskie nasekomye Karatau* (Jurassic Insects of Karatau). Nauka, Moscow: 252 p. (in Russian).

ROHDENDORF, B.B. 1968c. New Mesozoic nemestrinids (Diptera, Nemestrinidae). In: ROHDENDORF, B.B. (ed.). *Yurskie nasekomye Karatau* (Jurassic Insects of Karatau). Nauka, Moscow: 180–189 (in Russian).

ROHDENDORF, B.B. 1969. Phylogenie. In: KÜKENTHAL, W. (ed.). *Handbuch der Zoologie* IV.2.1.1.4. Walter de Gruyter, Berlin: 28 p.

ROHDENDORF, B.B. 1970. Role of insects in the historical development of terrestrial vertebrates. *Paleontol. Zhurnal* (1): 10–18 (in Russian).

ROHDENDORF, B.B. 1972. The Devonian eopterids are not insects but Crustacea Eumalacostraca. *Entomol. Obozrenie* **51**(1): 96–97 (in Russian).

ROHDENDORF, B.B. 1977. The rationalization of names of higher taxa in zoology. *Paleontol. Zhurnal* (3): 11–21 (in Russian, English translation: *Paleontol. J.* **11**: 149–155).

ROHDENDORF, B.B., BECKER-MIGDISOVA, E.E., MARTYNOVA, O.M., and SHAROV, A.G. 1961. *Paleozoyskie nasekomye Kuznetskogo basseyna* (Paleozoic Insects of the Kuznetsk Basin). Trudy Paleontol. Inst. Akad. Nauk SSSR **85**. Nauka, Moscow: 706 p. (in Russian).

ROHDENDORF, B.B., and RASNITSYN, A.P. (eds.). 1980. *Istoricheskoe razvitie klassa nasekomykh* (Historical Development of the Class Insecta). Trudy Paleontol. Inst. Akad. Nauk SSSR **175**. Nauka, Moscow: 269 p. (in Russian).

ROHDENDORF, B.B., and ZHERIKHIN, V.V. 1974. Palaeontology and nature protection. *Priroda* (5): 82–91 (in Russian).

ROHR, D.M., BOUCOT, A.J., MILLER, J., and ABBOTT, M. 1986. Oldest termite nest from the Upper Cretaceous of West Texas. *Geology* **14**(1): 87–88.

ROLFE, W.D.I. 1985. Early terrestrial arthropods: a fragmentary record. *Philos. Trans. R. Soc. London B* **309**: 207–218.

ROLFE, W.D.I., and INGHAM, J.K. 1967. Limb structure, affinity and diet of the Carboniferous "centipede" *Arthropleura*. *Scott. J. Geol.* **3**: 118–124.

RONQUIST, F. 1999. Phylogeny, classification and evolution of the Cynipoidea. *Zool. Scripta* **28**: 139–164.

RONQUIST, F., RASNITSYN, A.P., ROY, A., ERIKSSON, K., and LINDGREN, M. 1999. Phylogeny of the Hymenoptera: A cladistic reanalysis of Rasnitsyn's (1988) data. *Zool. Scripta* **28**: 13–50.

RÖSCHMANN, F. 1999a. Analysis of the relationship between Baltic and Saxonian amber based on their sciarid and ceratopogonid faunas (Tertiary, Eocene-Miocene). *Proc. First Palaeoentomol. Conf., Moscow 1998*. AMBA projects Internat., Bratislava: 131–134.

RÖSCHMANN, F. 1999b. Revision of the evidence of *Tetracha carolina* (Coleoptera, Cicindelidae) in Baltic amber (Eocene – Oligocene). *Estud. Mus. Cienc. Natur. Alava* **14** (Núm. Espec. 2): 205–209.

ROSELLI, F.L. 1938. Apuntes de geología y paleontología Uruguaya. Sobre insectos del Cretácico del Uruguay o descubrimiento de admirables instintos constructivos de esa época. *Bol. Soc. Amigos Cienc. Natur. "Kraglievich-Fontana"* **1**: 72–102.

ROSELLI, F.L. 1987. Paleoicnología: nidos de la cubiertura mesozoica del Uruguay. *Publ. Mus. Municipal Nueva Palmira* **1**(1): 1–56.

ROSELT, G., and FEUSTEL, H. 1960. Ein Taxodiazeenholz aus der Mitteldeutschen Braunkohle mit Insektenspuren und -resten. *Geologie* **9**: 84–91.

ROSKAM, J.C. 1992. Evolution of the gall-inducing guild. In: SHORTHOUSE, J.D., and ROFFRITSCH, O. (eds.). *Biology of Insect-Induced Galls*. Oxford Univ. Press, Oxford--New York: 34–49.

ROSS, A. 1998. *Amber: The Natural Time Capsule*. The Natural History Mus., London: 73 p.

ROSS, A.J. (ed.). 2000. *The History, Geology, Age and Fauna (Mainly Insects) of Burmese Amber, Myanmar*. Bull. Natur. Hist. Mus., Geol. Series **56**(1). London: 83 p.

ROSS, A.J., JARZEMBOWSKI, E.A., and BROOKS, S.J. 2000. The Cretaceous and Cenozoic record of insects (Hexapoda) with regard to global changes. In: CULVER, S., and RAWSON, P. (eds.). *Biotic Responses to Global Change, the last 145 million years*. Cambridge University Press, Cambridge: 288–302.

ROSS, E.S. 1966. The Embioptera of Europe and the Medditerranean region. *Bull. Brit. Mus. (Natur. Hist.)*, *Entomol.* **17**: 273–326.

ROSS, E.S. 1970a. Embioptera. In: CSIRO. *The Insects of Australia*. Melbourne Univ. Press, Melbourne: 360–366.

ROSS, E.S. 1970b. Biosystematics of the Embioptera. *Annu. Rev. Entomol.* **15**: 157–172.

ROSS, E.S. 1984. A synopsis of the Embiidina of the Unated States. *Proc. Entomol. Soc. Washington* **86**: 82–93.

ROSS, E.S. 1987. Studies in the insect order Embiidina: A revision of the insect family Clothodidae. *Proc. Calif. Acad. Sci.* **45**: 9–34.

ROSS, E.S. 1991. Embioptera. In: NAUMANN, D. (ed.). *The Insects of Australia*. 2nd ed. Melbourn Univ. Press,Carlton: 405–409.

ROSS, E.S. 2000. *Embia. Contribution to the biosystematics of the insect order Embiidina. Part 1. Origin, relationships and integumental anatomy of the insect order Embiidina. Part 2. A review of the biology of Embiidina*. Occas. Pap. Calif. Acad. Sci. **140**: 53 + 36p.

ROSS, H.H. 1965. *A Textbook of Entomology*. John Wiley & Sons, New York etc.: 539 p.

ROSS, H.H. 1967. The evolution and past dispersal of the Trichoptera. *Annu. Rev. Entomol.* **12**: 169–206.

ROTH, L.M. 1986. The genus *Symploce* Hebard. V. Species from mainland Asia (China, India, Iran, Laos, Thailand, South Vietnam, West Malaysia) (Dictyoptera: Blattaria, Blattellidae). *Entomol. Scand.* **16**: 375–397.

ROTHSCHILD, M. 1973. A specimen of *Pulex irritans* found in a Viking pit in Ireland (Note of an exhibit). *Proc. R. Entomol. Soc. London C* **38**(7): 29.

ROTHSCHILD, M. 1975. Recent advances in our knowledge of the order Siphonaptera. *Annu. Rev. Entomol.* **20**: 241–260.

ROWLAND, J.M. 1997. The late Paleozoic insect assemblage at Carrizo Arroyo, New Mexico. *Bull. New Mexico Mus. Natur. Hist. Science* **11**: 1–7.

ROY, R. 1987. General observations on the systematics of Mantodea. In: BACCETTI, B. (ed.). *Evolutionary Biology of Orthopteroid Insects*. Chichester: 488–492.

ROZEFELDS, A.C. 1985a. A fossil zygopteran nymph (Insecta: Odonata) from the Late Triassic Aberdare conglomerate, Southeast Queensland. *Proc. R. Soc. Queensl.* **96**: 25–32.

ROZEFELDS, A.C. 1985b. The first records of fossil leaf mining from Australia. *Hornibrook Symposium, 1985. Extended Abstracts*. New Zeal. Geol. Surv. Rec. **9**: 80–81.

ROZEFELDS, A.C. 1988a. Lepidoptera mines in *Pachypteris* leaves (Corystospermaceae: Pteridospermophyta) from the Upper Jurassic/Lower Cretaceous Battle Camp Formation, North Queensland. *Proc. R. Soc. Queensl.* **99**: 77–81.

ROZEFELDS, A.C. 1988b. Insect leaf mines from the Eocene Anglsea locality, Victoria, Australia. *Alcheringa* **12**: 1–6.

ROZEFELDS, A.C., and DE BAAR, M. 1991. Silicified Kalotermitidae (Isoptera) frass in conifer wood from a mid-Tertiary rainforest in central Queensland, Australia. *Lethaia* **24**: 439–442.

ROZEFELDS, A.C., and SCOBBE, J. 1987. Problematic insect leafmines from the Upper Triassic Ipswich coal measures of southeastern Queensland, Australia. *Alcheringa* **12**: 51–57.

ROZHKOV, A.S., and MASSEL, G.I. 1982. *Smolistye veshchestva khvoinykh i nasekomye-ksilofagi* (Conifer Resins and Xylophagous Insects). Nauka, Novosibirsk, 148 p. (in Russian).

RUDNEV, A.F., and LYAMINA, N.A. 1990. New data on the Cretaceous reference sections of the Khilok-Tchukoyan depression (western Transbaikalia). In: KRASSILOV, V.A. (ed.). *Kontinental'ny mel SSSR. Materialy soveshchaniya sovetskoy rabochey gruppy proekta N 245 MPGK. Vladivostok, Oktyabr' 1988* (Non-marine Cretaceous of the USSR. Materials of Conference of the Soviet working group of the Project 245 IGCP, Vladivostok, Oct. 1988). Vladivostok: 76–84 (in Russian).

RUDNEV, D.F., SMELYANETS, V.P., and VOYTENKO, A.M. 1970. Biological effect of the Norwegian spruce terpenoids on insects and mites. *Visnik sil'skogospodar. nauki* **7**: 71–74 (in Ukrainian).

RUFFIEUX, L., ELOUARD, J.-M., and SARTORI, M. 1998. Flightlessness in mayflies and its relevance to hypotheses on the origin of insect flight. *Proc. R. Soc. London B* **265**: 2135–2140.

RÜFFLE, L., and JÄNICHEN, H. 1976. Die Myrtaceen im Geiseltal und einigen anderen Fundstellen des Eozän. *Abhandl. Zentr. Geol. Inst., Paläontol.* **26**: 307–336.

RÜFFLE, L., MÜLLER-STOLL, W.R., and LIETKE, R. 1976. Weitere Ranales, Fagaceae, Loranthaceae, Apocynaceae. *Abhandl. Zentr. Geol. Inst., Paläontol.* **26**: 199–282.

RUNDLE, A.J., and COOPER, J. 1970. Occurrence of a fossil insect larva from the London Clay of Herne Bay, Kent. *Proc. Geol. Ass.* **82**(5): 293–296.

RUSCONI, C. 1948. Notas sobre fósiles ordovícias y triásicos de Mendoza. VIII. Nidos de hormigas del triásico de Salagasta. *Rev. Mus. Hist. Natur. Mendoza* **2**(4): 254.

RUST, J. 1998. Biostratinomie von Insekten aus der Fur-Formation von Dänemark (Moler, oberes Paleozän/unteres Eozän). *Paläontol. Zeitschr.* **72**(1/2): 41–58.

RUST, J. 1999a. Fossil insects from the Fur and Olst Formations ("Mo-clay") of Denmark (upper Paleocene/lowermost Eocene). *Proc. First Palaeoentomol. Conf., Moscow 1998*. AMBA projects Internat., Bratislava: 135–139.

RUST, J. 1999b. Oldest known pteroplistine cricket and other Grillidae (Orthoptera) from the Palaogene Fur and Olst Formations of Denmark. *Entomol. scand.* **30**: 35–45.

RUST, J. 2000. Fossil record of mass moth migration. *Nature* **405**: 530–531.

RUSTAMOV, A.K. 1947. Some data on faunistic changes at Western Uzboy. *Doklady Akad. Nauk SSSR* **55**(9): 885–888 (in Russian).

RYVKIN, A.B. 1988. New Cretaceous Staphylinidae (Insecta) from the [Russian] Far East. *Paleont. Zhurnal* (4): 103–106 (in Russian, English translation: *Paleont. J.* **22**(4): 100–104).

SADLER, J.P. 1990. Records of ectoparasites on humans and sheep from Viking Age deposits in the former Western Settlement on Greenland. *J. Med. Entomol.* **27**: 628–631.

SADLER, J.P., and JONES, J.C. 1997. Chironomids as indicators of Holocene environmental change in the British Isles. In: ASHWORTH, A.C., BUCKLAND, P.C., and SADLER, J.P. (eds.). *Studies in Quaternary Entomology. An Inordinate Fondness for Insects*. Quatern. Res. **5**. John Wiley & Sons, New York: 219–232.

SAINFELD, P. 1952. Minéralogie de la Tunisie. *Bull. Soc. Sci. natur. Tunisie* **5**(1–4): 115–148.

SAKAI, S. 1982. A new proposed classification of the Dermaptera with special reference to the check list of the Dermaptera of the world. *Bull. Daito Bunka Univ.* **20**: 1–108.

SAKAI, S. and FUJIYAMA, I. 1989. New dermapteran fossil from Sado Islands, Japan with description of a new species (Dermaptera, Diplatyidae). *Special Bull. Daito Bunka Univ.* **38**: 3102–3103.

SAMUELSON, G.A. 1988. Pollen feeding in Alticinae (Chrysomelidae). *Proc. 18th Internat. Congr. Entomol., Vancouver, July 3–9, 1988. Abstracts and Author Index*: 37.

SAMYLINA, V.A. 1974. *Rannemelovye flory Severo-Vostoka SSSR. K probleme stanovleniya flor kainofita* (Early Cretaceous Floras of the Northeast of USSR. Contribution to the Problem of Formation of the Cainophytic Floras). Komarovskie Chteniya (V.I. Komarov's Memorial Readings) **27** (1972). Nauka, Leningrad: 1–55 (in Russian).

SAMYLINA, V.A. 1988. *Arkagalinskaya stratoflora Severo-Vostoka Azii* (The Arkagala Stratoflora of North-East Asia). Nauka, Leningrad: 131 p. (in Russian).

SANDER, P.M., and GEE, C.R. 1990. Fossil charcoal: techniques and applications. *Rev. Palaeobot. Palynol.* **63**: 269–279.

SANDO, W.J. 1972. Bee-nest pseudofossils from Montana, Wyoming, and South-West Africa. *J. Paleontol.* **46**(3): 421–425.

SANTIAGO-BLAY, J.A., POINAR, G.O., Jr., and CRAIG, P.R. 1996. Dominican and Mexican amber chrysomelids, with descriptions of two new species. In: JOLIVET, P.H.A., and COX, M.L. (eds.). *Chrysomelid Biology*. Vol. 1: *The Classification, Phylogeny and Genetics*. SPB Academic Publ., Amsterdam: 413–424.

SAPUNOV, V.B. 1990. Quantitative approach to estimation of the representativeness of palaeontological record. *Izvestiya Akad. Nauk SSSR Ser. biol.* (3): 420–426 (in Russian).

SAUER, E., and SCHREMMER, F. 1969. Fossile Insekten-Bauten aus dem Tertiär der Hegau (S.-Deutschland). *Senckenberg. Lethaea* **50**(1): 1–18.

SAUNDERS, W.B., MAPES, R.H., CARPENTER, F.M., and ELSIK, W.C. 1974. Fossiliferous amber from the Eocene (Claiborne) of the Gulf Coastal Plane. *Geol. Soc. Amer. Bull.* **85**: 979–984.

SAUNDERS, W.W. 1836. Remarks upon innumerable quantities of the dead bodies of *Galeruca tanaceti*, observed at Cleathorpe, in the Lincolnshire. *Trans. R. Entomol. Soc. London* **1**: XXXIII–XXXIV.

SAVKEVICH, S.S. 1970. *Yantar'* (Amber). Nedra, Leningrad, 191 p. (in Russian).

SAVKEVICH, S.S. 1980. New developments in amber and other fossil resins mineralogical studies. *Samotsvety* (Gem Minerals) (*Proc. XI General Meeting of IMA, Novosibirsk, 4–10 Sept. 1978*). Nauka, Leningrad: 17–27 (in Russian).

SAVKEVICH, S.S. 1983. Change processes of amber and other amber-like fossil resins depending on condition of origin and occurrence in nature. *Izvestiya Akad. Nauk SSSR Ser. geol.* (12): 96–106 (in Russian).

SAVKEVICH, S.S., SKALSKI, A.W., and VEGGIANI, A. 1990. Fossil resin in deep deposits of the Persian Gulf. *Prace Muz. Ziemi* 41: 51–52.

SAYLOR, L.W. 1933. Attraction of beetles to tar. *Pan-Pacif. Entomol.* 9: 182.

SAZONOVA, O.N. 1970. Transport of organic matter by the blood-sucking mosquitoes from depressions to interfluves. In: *Sredoobrazuyushchaya deyatel'nost' zhivotnykh* (Environmental Effects of Animal Activities). *Materialy k soveshchaniyu, 17–18 dek. 1979 g.* (Contributions to Meeting, Dec. 17–18, 1979). Moscow Univ., Moscow: 69–71 (in Russian).

SCHAAL, S., and ZIEGLER, W. (eds.) 1988. *Messel – ein Schaufenster in die Geschichte der Erde und die Leben.* Verlag Waldemar Kramer, Frankfurt a/M.: 315 p.

SCHAARSCHMIDT, F., and WILDE, V. 1986. Palmblüten und -blätter aus dem Eozän von Messel. *Courier Forsch.-Inst. Senckenberg* 86: 177–202.

SCHAEFER, C.W. 1975. Heteropteran trichobothria (Hemiptera: Heteroptera). *Internat. J. Insect Morphol. & Embryol.* 4: 193–264.

SCHAEFER, C.W., DOLLING, W.R., and TACHIKAWA, S. 1988. The shieldbug genus *Parastrachia* and its position within the Pentatomoidea (Insecta: Hemiptera). *Zool. J. Linn. Soc.* 93: 283–311.

SCHÄFER, W. 1962. *Aktuo-paläontologie nach Studien in den Nordsee.* W. Kramer Verlag, Frankfurt: 666 p.

SCHAIRER, G., and JANICKE, V. 1970. Sedimentologisch-paläontologische Untersuchungen an den Plattenkalken der Sierra de Montsech (Prov. Lérida, N.-E.-Spanien). *Neues Jahrb. Geol. Paläontol. Abhandl.* 135(2): 171–189.

SCHAKAU, B. 1986. Preliminary study of the development of the subfossil chironomid fauna (Diptera) of Lake Taylor, South Island, New Zealand, during the younger Holocene. *Hydrobiologia* 143: 287–291.

SCHAKAU, B., and FRANK, C. 1984. Die Entwicklung der Chironomiden-Fauna (Diptera) des Tegeler Sees im Spät- und Postglazial. *Verhandl. Ges. Ökol.* 12: 375–382.

SCHÄLLER, G. 1970. Wie entstehen Pflanzengallen? Vorstellungen auf Grund neuer Untersuchungergebnisse mit pflanzensaftsaugenden Insekten. *Biol. Rundsch.* 8(5): 346–350.

SCHAWALLER, W. 1981. Pseudoskorpione (Cheliferidae) phoretisch auf Käfern (Platypodidae) in Dominikanischen Bernstein (Stuttgarter Bernsteinsammlung: Pseudoscorpionidea und Coleoptera). *Stuttg. Beitr. Naturk.* B 71: 1–17.

SCHAWALLER, W. 1986. Fossile Käfer aus miozäne Sedimenten des Randecker Maar in Südwest-Deutschland (Insecta: Coleoptera). *Stuttg. Beitr. Naturk.* B 126: 1–9.

SCHEDL, K.E. 1947. Die Borkenkäfer des baltischen Bernstein. *Zentralbl. Gesamtgeb. Entomol.* 2(1): 12–45.

SCHENK, E. 1937. Insektenfrassgänge oder Bohrlöcher der Pholadiden in Ligniten aus dem Braunkohlenflöz bei Köln. *Neues Jahrb. Miner., Geol. Paläontol.* 77: 392–401.

SCHILLE, F. 1916. Entomologie aus der Mammut- und Rhinoceros-Zeit Galiziens. *Entomol. Zeitschr.* 30: 14–15, 19, 22–23, 26–27, 31–32, 39–40, 42–44, 46–47, 50–51, 55–56, 58–60, 62–63, 67–68, 75, 78–79, 83–84, 87–88, 90–91, 95.

SCHIMETSCHEK, E. 1978. Ein Schmetterlingsidol im Val Camonica aus dem Neolithikum. *Anz. Schädlingsk. Pfl.-Umweltschutz* 51(8): 113–115.

SCHINK, B. 1989. Lebens-gemeinschaften in Gewässersedimenten. *Naturwissenschaften* 76(8): 364–372.

SCHLECHTENDAL, D. 1887. Physopoden aus dem Braunkohlengebirge von Rott im Siebengebirge. *Zeitschr. Naturwiss.* 60: 512–592.

SCHLEE, D. 1969a. Die Verwandtschaftsbeziehungen innerhalb der Sternorrhyncha aufgrund synapomorpher Merkmale. Phylogenetische Studien an Hemiptera II: Aphidiformes (Aphidina + Coccina) als monophyletische Gruppe. *Stuttg. Beitr. Naturk.* 199: 1–19.

SCHLEE, D. 1969b. Der Flügel von *Sphaeraspis* (Coccina), prinzipiell identisch mit Aphidina-Flügeln. Phylogenetische Studien an Hemiptera V: Synapomorphe Flügelmerkmale bei Aphidina und Coccina. *Stuttg. Beitr. Naturk.* 211: 1–11.

SCHLEE, D. 1980. *Bernstein-Raritäten.* Staatliches Mus. für Naturkunde, Stuttgart: 88 p.

SCHLEE, D. 1984. Notizen uber einige Bernstein und Kopal aus alter Welt. *Stuttg. Beitr. Naturk.* C 18: 29–37.

SCHLEE, D. 1990. Das Bernsteinkabinett. *Stuttg. Beitr. Naturk.* C 28: 3–100.

SCHLEE, D., and DIETRICH, H.-G. 1970. Insektenführender Bernstein aus der Unterkreide des Libanon. *Neues Jahrb. Geol. Paläontol. Monatshefte* (1): 40–50.

SCHLEE, D., and PHEN, H.Ch. 1992. Riesenbernsteine in Sarawak, Nord-Borneo. *Lapis* 17(9): 13–23.

SCHLIEPHAKE, G. 1975. Beitrag zur phylogenetischen Systematik bei Thysanoptera (Insecta). *Beitr. Entomol.* 25: 5–13.

SCHLIEPHAKE, G. 2001. Thysanoptera (Insecta) of the Tertiary amber of the Museum of Earth, Warsaw, with keys to the species of the Baltic and Bitterfeld amber. *Prace Muz. Ziemi* 46: 17–39.

SCHLIEPHAKE, G., and KLIMT, K. 1979. *Thysanoptera. Fransenflügler.* In: DAHL, F. (ed.). Die Tierwelt Deutschlands 66. VEB Gustav Fischer Verlag, Jena: 477 p.

SCHLÜTER, T. 1975. Nachweis verschiedener Insekten-Ordines in einem mittelkretazischen Harz Nordwestfrankreichs. *Entomol. German.* 1(2): 151–161.

SCHLÜTER, T. 1978. Zur Systematik und Paläökologie harzkonservierter Arthropoda einer Taphozönose aus dem Cenomanium von NW-Frankreich. *Berl. geowiss. Abhandl.* A 9: 150S.

SCHLÜTER, T. 1981. Fossile Insekten aus dem Jura/Kreide-Grenzbereich Südwest-Ägiptens. *Berl. geowiss. Abhandl.* A 32: 33–61.

SCHLÜTER, T. 1982. *Cimbrochrysa molerensis* n.g., n. sp. und *Hypochrysa hercyniensis* n.sp., zwei fossile Chrysopiden Arten (Insecta: Planipennia) aus dem europaeische Tertiär. *Neues Jahrb. Geol. Paläontol. Monatshefte* (5): 257–264.

SCHLÜTER, T. 1984. Kretazische Lebensspuren von solitären Hymenopteren und ihre Nomenklatur. *Aufschluss* 35: 423–430.

SCHLÜTER, T. 1986. The fossil Planipennia – a review. In: GEPP, J., ASPÖCK, H., and HOLZEL, H. (eds.). *Recent Research in Neuropterology Proc. 2nd Internat. Symposium on Neuropterology.* Hamburg: 103–111.

SCHLÜTER, T. 2000. *Moltenia rieki* n. gen., n. sp. (Hymenoptera: Xyelidae?), a tentative sawfly from the Molteno Formation (Upper Triassic), South Africa. *Paläontol. Zeitschr.* 74(1/2): 75–78.

SCHLÜTER, T., and DREGER, G. 1985. Subfossile Schwarzkäfer-Rests aus dem frühzeitlichen Königsfriedhof von Abydos/Oberägypten (Coleoptera: Tenebrionidae). *Entomol. Gener.* 10(2): 143–148.

SCHLÜTER, T., and GNIELINSKI, F., von. 1987. The East African copal, its geologic, stratigraphic, palaeontologic significance and comparison with fossil resins of similar age. *Nation. Mus. Tanzania Occas. Paper* 8: 1–32.

SCHLÜTER, T., and KÜHNE, W.C. 1974. Die einseitige Trübung von Harzinklusen – ein Indiz gleicher Bildungsumstande. *Entomol. German.* 1(3/4): 308–315.

SCHLÜTER, T., and STÜRMER, W. 1984. Die identifikation einer fossilen Rhachiberothinae-Art (Planipennia: Berothidae oder Mantispidae) aus mittelkretazischen Bernstein NW-Frankreichs mit Hilfe röntgenographischer Methoden. In: GEPP, J., ASPÖCK, H., and HOLZEL, H. (eds.). *Proc. 1st Internat. Symposium on Neuropterology.* Graz: 49–55.

SCHMID, F. 1998. *Genera of the Trichoptera of Canada and Adjoining or Adjacent United States.* The Insects and Arachnids of Canada 7. NRC Research Press, Ottawa: 319 p.

SCHMIDT, H. 1938. Eine Insektenfährte aus dem Mitteldevon des Wuppertals. *Decheniana* 97A: 43–46.

SCHMIDT, W., SCHÜRMANN, M., and TEICHMÜLLER, M. 1958. Biss-Spuren an Früchten des Miozän-Waldes der niederrheinischen Braunkohlen-Formation. *Forstschr. Geol. Rheinl. u. Westfal* 2: 563–572.

SCHMIDT-KITTLER, N. (ed.) 1987. *International Symposium on Mammalian Biostratigraphy and Paleoecology of the European Paleogene – Mainz, February 18th–21st 1987 Münchner Geowiss. Abhandl.* Reihe A Geol. Paläontol. 10: 311 p.

SCHMIDTGEN, O. 1928. Kalk aus Gehäusen von Köcherfliegen-larven. *Natur. und Museum* 58(1–12): 171.

SCHMITZ, M. 1991. Die Koprolithen mitteleozäner Vertebraten aus der Grube Messel bei Darmstadt. *Courier Forsch.-Inst. Senckenberg* 137: 1–143.

SCHMUTZENHOFER, H. 1985. Insektenspuren an berindetem Nadelholz: Eine Anleitung zum Bestimmen von Schädlingsbefall an Nadenholz in Rinde. Agrarverlag, Wien: 166 p.

SCHNEIDER, J. 1977. Zur Variabilität der Flügel paläozoischer Blattodea (Insecta), Teil I. *Freiberg. Forsch.-Heft* C 326: 87–105.

SCHNEIDER, J. 1978a. Zur Variabilität der Flügel paläozoischer Blattodea (Insecta), Teil II. *Freiberg. Forsch.-Heft* C 334: 21–39.

SCHNEIDER, J. 1978b. Zur Taxonomie und Biostratigraphie der Blattodea (Insecta) des Karbon und Perm der DDR. *Freiberg. Forsch.-Heft* C 340: 1–152.

SCHNEIDER, J. 1978c. Revision der Poroblattinidae (Insecta, Blattodea) des europäischen ind nordamericanischen Oberkarbon und Perm. *Freiberg. Forsch.-Heft* C 342: 55–66.

SCHNEIDER, J. 1980a. Zur Taxonomie der jungpaläozoischen Neorthroblattinidae (Insecta, Blattodea). *Freiberg. Forsch.-Heft* C 348: 31–39.

SCHNEIDER, J. 1980b. Zur Entomofauna des Jungpaläozoikums der Boskovicer Furche (CSSR), Teil I: Mylacridae (Insecta, Blattodea). *Freiberg. Forsch.-Heft* C 357: 43–55.

SCHNEIDER, J. 1982. Entwurf einer biostrastigraphischen Zonengliederung mittels der Spiloblattinidae (Blattodea, Insecta) für das kontinentale euramerische Permocarbon. *Freiberg. Forsch.-Heft* C 375: 27–47.

SCHNEIDER, J. 1983a. Die Blattodea (Insecta) des Paleozoicums, Teil I: Systematik, Ökologie und Biostratigraphie. *Freiberg. Forsch.-Heft* C 382: 106–146.

SCHNEIDER, J. 1983b. Taxonomie, Biostratigraphie und Palökologie der Blattodea-Fauna aus dem Stephan von Commentry, Frankreich–Versuch einer Revision. *Freiberg. Forsch.-Heft* C 384: 77–100.

SCHNEIDER, J. 1984. Die Blattodea (Insecta) des Paleozoicums, Teil II: Morphogenese der Flügelstrukturen und Phylogenie. *Freiberg. Forsch.-Heft* C 391: 5–34.

SCHNEIDER, J., and WERNEBURG, R. 1993. Neue Spiloblattinidae (Insecta, Blattodea) aus dem Oberkarbon und Unterperm von Mitteleuropa sowie die Biostratigraphie des Rotliegend. *Veroff. Naturhist. Mus. Schleusingen* 7/8: 31–52.

SCHOTT, C. 1984. *Cerambyx cerdo* dans des bois fossiles. *Bull. Soc. entomol. Mulhouse* (Août–Sept.): 48.

SCHRAM, F.R. 1983. Lower Carboniferous biota of Glencartholm, Eskdale, Dumfriesshire. *Scott. J. Geol.* 19: 1–15.

SCHREMMER, F. 1979. Ethökologische Beobachtungen zum Wohnröhrenbau bei Larven der mitteleuropäischen Sandlaufkäfer-Art *Cicindela silvicola* (Coleoptera: Cicindelidae). *Entomol. Gener.* 5(3): 201–219.

SCHUBERT, K. 1961. Neue Untersuchungen über Bau und Leben der Bernsteinkiefern (*Pinus succinifera* (Conw.) emend.). *Beih. Geol. Jahrb. Hannover* 45: 1–143.

SCHUH, B.A., and BENJAMIN, D.M. 1984. Evaluation of commercial resin as a feeding deterrent against *Neodiprion dubiosus*, *N. lecontei* and *N. rubifrons* (Hymenoptera: Diprionidae). *J. Econ. Entomol.* 77: 802–805.

SCHUH, R.T. 1979. [Review] Evolutionary Trends in Heteroptera. Part II. Mouthpart-structures and Feeding Strategies, by R.H. Cobben. *Syst. Zool.* 28: 653–656.

SCHUH, R.T., and SLATER, J.A. 1995. *True Bugs of the World (Hemiptera: Heteroptera). Classification and Natural History.* Cornell Univ. Press, Ithaca & London: xii + 336 p.

SCHUH, R.T., and ŠTYS, P. 1991. Phylogenetic analysis of cimicomorphan family relationships (Heteroptera). *J. New York Entomol. Soc.* **99**: 298–350.

SCHULTZE-DEWITZ, G., and SÜSS, H. 1988. Fossiler Termitenfraß an Holzresten aus dem Tertiär von Staré Sedlo (CSSR). Ein Beitrag zu den Termiten der Vorwelt. *Zeitschr. geol. Wiss.* **16**(2): 169–173.

SCHUMANN, H. 1967. Fossile Libellen (Odonata) aus dem Oberpliozän am westliches Harzvorland. *Ber. Naturhist. Ges. Hannover* **111**: 31–46.

SCHUMANN, H., and WENDT, H. 1989. Zur Kenntnis der tierischen Inclusen des Sächsischen Bernstein. *Dtsch. entomol. Zeitschr.* **36**: 33–44.

SCHULZ, P. 1927. Diatomeen aus norddeutschen Basalttuffen und -Tuffgeschieben. *Zeitschr. Geschiebekunde u. Flachlandsgeol.* **3**(1): 66–78; (2): 118–126.

SCHUSTER, J.C. 1974. Saltatorial Orthoptera as common visitors to tropical flowers. *Biotropica* **6**(2): 138–140.

SCHWARZ, A. 1931. Insektenbegräbnis im Meer. *Natur u. Mus.* **61**(19): 453–465.

SCHWARZBACH, M. 1939. Die älteste Insektenflügel. Bemerkungen zu einem oberschlesichen Funde. *IX. Jahresber. und Mitteilungen oberrhein. geol. Vereines N.F.* **1939**: 28–30.

SCHWEITZER, H.-J. 1977. Die Rätische Zwitterblüte *Irania hermaphroditica* nov. spec. und ihre Bedutung für die Phylogenie der Angiospermen. *Palaeontographica* **B 161**: 98–145.

SCHWIND, R. 1989. A variety of insects are attracted to water by reflected polarized light. *Naturwissenschaften* **76**(8): 377–378.

SCOBLO, V.M. 1968. Jurassic fossil lakes in volcanic series of the West Transbaikalian Region. In: ZHUZE, A.P., and FLORENSOV, N.A. (eds.). *Mezozoyskie i kainozoyskie ozyora Sibiri* (Mesozoic and Cenozoic Lakes of Siberia). Nauka, Moscow: 9–21 (in Russian).

SCOTESE, C.R. 1994. *Continental drift.* Ed. 6. Paleomap Project. Dept. Geol., Univ. Texas, Arlington (www.scotese.com).

SCOTT, A.C. 1977. Coprolites containing plant material from the Carboniferous of Britain. *Palaeontology* **20**: 59–68.

SCOTT, A.C. 1992. Trace fossils of plant-arthropod interactions. In: MAPLES, C.G., and WEST, R.R. (eds.). *Trace Fossils. Short Courses in Palaeontology* **5**: 197–223.

SCOTT, A.C., and PATERSON, S. 1984.Techniques for the study of plant/arthropod interactions in the fossil record. *Geobios* Mém. spéc. **8**: 449–455.

SCOTT, A.C., and TAYLOR, Th.N. 1983. Plant/animal interaction during the Upper Carboniferous. *Botan. Rev.* **49**: 259–307.

SCOTT, A.C., STEPHENSON, J., and CHALONER, W.G. 1992. Interaction and co-evolution of plants and arthropods during the Palaeozoic and Mesozoic. *Philos. Trans. R. Soc. London* **B 335**: 129–165.

SCOTT, A.C., STEPHENSON, J., and CHALONER, W.G. 1994. The fossil record of leaves with galls. In: WILLIAMS, M.A.J. (ed.). *Plant Galls: Organisms, Interactions, Populations.* Clarendon Press, Oxford: 447–470.

SCOTT, K.M., and MOORE, F.C., de. 1993. Three recently erected Trichoptera families from South Africa, the Hydrosalpingidae, Petrotrinicidae and Barbarochtonidae (Annulipalpia, Sericoistomatoidea). *Ann. Cape Province Mus. (Natur. Hist.)* **18**: 293–354.

SCUDDER, S.H. 1878. An account of some insects of unusual interest from the Tertiary rocks of Colorado and Wyoming. *U.S. Geol. Surv. Bull.* **13**: 176–200.

SCUDDER, S.H. 1890a. A classed and annotated bibliography of fossil insects. *U.S. Geol. Surv. Bull.* **69**: 1–101.

SCUDDER, S.H. 1890b. The Tertiary Insects of North America. *U.S. Geol. Surv. Terr.* **13**: 1–734.

SCUDDER, S.H. 1891. Index to the known fossil insects of the World including myriapods and arachnids. *U.S. Geol. Surv. Bull.* **71**: 1–744 p.

SDZUY, K. 1962. Über das Entzerren von Fossilien (mit Beispielen aus der unterkarbonischen Saukianda-Fauna). *Paläontol. Zeitschr.* **36**(3–4): 275–284.

SEELY, M.K., and MITCHELL, D. 1986. Termite casts in Tsondab Sandstone. *Palaeoecol. Africa Surround. Islands* **17**: 109–112.

SEILACHER, A. 1953. Studien zur Palichnologie. I. Über die Methoden der Palichnologie. *Neues Jahrb. Min., Geol. Paläontol. Abhandl.* **96**: 421–453.

SEILACHER, A., REIF, W.-E., and WESTPHAL, F. 1985. Sedimentological, ecological and temporal patterns of fossil Lagerstätten. In: WHITTINGTON, H.B., and CONWAY MORRIS, S. (eds.). *Extraordinary Fossil Biotas: Their Ecological and Evolutionary Significance.* Philos. Trans. R. Soc. London B **311**: 5–23.

SEIPLE, E. 1983. Miocene insects and arthropods in California. San Bernardino County. *California Geol.* **36**: 246–248.

SEKI, H., MIYAMOTO, Sh., and YAMAMOTO, H. 1990. Chitin dynamics in the freshwater environment. *Abstr. 5th Internat. Congr. Ecology, Yokohama, Aug. 23–30, 1990.* Yokohama: 239.

SEKI, H., and TAGA, N. 1963. Microbiological studies on the decomposition of chitin in marine environments. I. Occurrence of chitinoclastic bacteria in the neritic region. *J. Oceanogr. Soc. Japan* **19**: 101–108.

SEKI, H., and TAGA, N. 1965. Microbiological studies on the decomposition of chitin in marine environments. VIII. Distribution of chitinoclastic bacteria in the pelagic and neritic waters. *J. Oceanogr. Soc. Japan* **21**: 174–187.

SELLICK, J.T.C. 1994. Phasmida (stick insect) eggs from the Eocene of Oregon. *Palaeontology* **37**(4): 913–921.

SENUT, B., PICKFORD, M., and WARD, J. 1994. Biostratigraphie des éolianites néogènes du sud de la Sperrgebiet (Désert de Namib, Namibie). *C.R. Acad. Sci. (Paris)* Sér. 2, **318**: 1001–1007.

SEQUEIRA, A.S., NORMARK, B.B., and FARRELL, B.D. 2000. Evolutionary assembly of the conifer fauna: distinguishing ancient from recent associations in bark beetles. *Proc. R. Soc. Lond.* **B 267**: 2359–2366.

SERRES, P.M., de. 1829. *Géognosie des terrains tertiares ou Tableau des principaux animaux invertébrés des terrains marins tertaires du Midi de la France* **2**. Montpellier: 277 p.

SHAPOSHNIKOV, G.Kh. 1980. Evolution of morphological structures in aphids (Homoptera, Aphidinea) and habits of the recent and Mesozoic representatives of the group. *Entomol. Obozrenie* **59**(1): 39–59 (in Russian; English translation: *Entomol. Review* (1): 29–48).

SHARGA, U.S. 1933. On the internal anatomy of some Thysanoptera. *Trans. R. Entomol. Soc. London* **81**: 185–204.

SHARMA, B.D., and HARSCH, R. 1989. Activities of phytophagous arthropods (wood borers) on extinct plants from the Mesozoic of the Rajmahal Hills, India. *Bionature* **9** (1): 29–34.

SHAROV, A.G. 1948. The Triassic Thysanura from Urals. *Doklady Akad. Nauk SSSR* **61** (3): 517–519 (in Russian).

SHAROV, A.G. 1953. The first find of the Permian megalopteran larva (Megaloptera) from Kargala. *Doklady Akad. Nauk SSSR* **89**: 731–732.

SHAROV, A.G. 1957. The types of insect metamorphose and its relationships (from comparative-ontogenetic and palaeontological data). *Entomol. Obozrenie* **36**(3): 569–576 (in Russian).

SHAROV, A.G. 1962. Redescription of *Lithophotina floccosa* Cock. (Manteodea), with some notes on the manteod wing venation. *Psyche* **69**(3): 102–106.

SHAROV, A.G. 1966a. *Basic Arthropodan Stock.* London, Pergamon Press: 271 p.

SHAROV, A.G. 1966b. On the position of the orders Glosselytrodea and Caloneurodea in the system of Insecta. *Paleontol. Zhurnal* (3): 84–93 (in Russian).

SHAROV, A.G. 1968. *Filogeniya ortopteroidnykh nasekomykh.* Trudy Paleontol. Inst. Akad. Nauk SSSR **118**. Nauka, Moscow: 218 p. (in Russian; English translation: *Phylogeny of the Orthopteroidea.* Keter Press, Jerusalem 1971: 251 p.).

SHAROV, A.G. 1971. New flying reptiles from the Mesozoic of Kazakhstan and Kirghizia. *Trudy Paleontol. Inst. Akad. Nauk SSSR* **130**. Nauka, Moscow: 104–113 (in Russian).

SHAROV, A.G. 1972. On phylogenetic position of the order thrips (Thysanoptera). *Entomol. Obozrenie* **54**: 854–858 (in Russian).

SHAROV, A.G. 1973. Morphological features and habit of palaeodictyopterans. In: NARCHUK, E.P. (ed.) *Voprosy paleontologii nasekomykh. Doklady na 24-m Ezhegodnom chtenii pamyati N.A. Kholodkovskogo, 1971* [Problems of Insect Palaeontology. Lectures on the XXIV Annual Readings in Memory of N.A. Kholodkovsky (1–2 April, 1971)]. Nauka, Leningrad: 49–63 (in Russian).

SHAROV, A.G., and SINICHENKOVA, N.D. 1977. New Palaeodictyoptera from the territory of the USSR. *Paleontol. Zhurnal* (1): 48–63 (in Russian, English translation: *Paleontol. J.* **11**(1): 44–59.

SHATALKIN, A.I. 2000. Keys to the Palaearctic flies of the family Lauxaniidae (Diptera). *Zool. issledovaniya* **5**: 1–101 (in Russian with English summary).

SHCHERBAKOV, D.E. 1980. Morphology of the pterothoracic pleura in Hymenioptera. I. Groundplan. *Zoolog. Zhurnal* **59**: 1644–1652 (in Russian).

SHCHERBAKOV, D.E. 1984. A system and the phylogeny of Permian Cicadomorpha (Cimicida, Cicadina). *Paleontol. Zhurnal* (2): 89–101 (in Russian, English translation: *Paleontol. J.* **18**(2): 87–97).

SHCHERBAKOV, D.E. 1988a. New Mesozoic Homoptera. In: Rozanov, A.Yu. (ed.). *Novye vidy iskopaemykh bespozvonochnykh Mongolii* (New Species of Fossil Invertebrates of Mongolia). Trudy Sovmestnoy Sovetsko-Mongol'skoy Paleontol. Expeditsii **33**. Nauka, Moscow: 60–63 (in Russian).

SHCHERBAKOV, D.E. 1988b. New Cicadina from the Late Mesozoic of Transbaikalia. *Paleontol. Zhurnal* (4): 55–66 (in Russian, English translation: *Paleontol. J.* **22**(4), 52–63.

SHCHERBAKOV, D.E. 1990. Extinct four-winged ancestors of scale insects (Homoptera, Sternorrhyncha). *Proc. 6th Internat. Symposium Scale Insect Studies, Cracow, Aug. 6–12, 1990.* Agricultural Univ. Press, Cracow: 23–29.

SHCHERBAKOV, D.E. 1992. The earliest leafhoppers (Hemiptera: Karajassidae n. fam.) from the Jurassic of Karatau. *Neues Jahrb. Geol. Paläontol. Monatshefte* (1): 39–51.

SHCHERBAKOV, D.E. 1995. A new genus of the Paleozoic order Hypoperlida. *Russian Entomol. J.* (1994) **3**(3/4): 33–36.

SHCHERBAKOV, D.E. 1996. Origin and evolution of the Auchenorrhyncha as shown by the fossil record. In: SCHAEFER, C.W. (ed.). *Studies on Hemiptera Phylogeny. Proceedings. Thomas Say Publications in Entomology.* Entomol. Soc. America, Lanham, Maryland: 31–45.

SHCHERBAKOV, D.E. 1999. Controversions over the insect origin revisited. *Proc. First Palaeoentomol. Conf., Moscow 1998.* AMBA projects Internat., Bratislava: 141–148.

SHCHERBAKOV, D.E. 2000a. The most primitive whiteflies (Hemiptera; Aleyrodidae; Bernaeinae subfam. nov.) from the Mesozoic of Asia and Burmese amber, with an overview of Burmese amber hemipterans. *Bull. Nat. Hist. Mus. Lond. (Geol.)* **56**: 29–37.

SHCHERBAKOV, D.E. 2000b. Permian faunas of Homoptera (Hemiptera) in relation to phytogeography and the Permo-Triassic crisis. *Paleontol. J.* **34**(suppl. 3): S251–S267.

SHCHERBAKOV, D.E., in press. Extinct four-winged precoccids and the ancestry of scale insects and aphids. *Palaeontology.*

SHCHERBAKOV, D.E., LUKASHEVICH, E.D., and BLAGODEROV, V.A. 1995. Triassic Diptera and initial radiation of the order. *Internat. J. Dipterol. Res.* **6**(2): 75–115.

SHEAR, W.A. 1991. The early development of terrestrial ecosystems. *Nature* **351**: 283–289.

SHEAR, W.A., BONAMO, P.M., GRIEDSON, J.D., ROLF, W.D.I., SMITH, E.L., and NORTON, R.A. 1984. Early land animals in North America: Evidence from Devonian age arthropods from Gilboa, New York. *Science* **224**: 492–494.

SHEAR, W.A., HANNIBAL, J.T., and KUKALOVÁ-PECK, J. 1992. Terrestrtial arthropods from Upper Pennsilvanian rocks at the Kinney Brick Quarry, New Mexico. *Bull. New Mexico Bureau Mines & Mineral Resources* **138**: 135–141.

SHEAR, W.A., and KUKALOVÁ-PECK, J. 1990. The ecology of Paleozoic terrestrial arthropods: the fossil evidence. *Canad. J. Zool.* **68**: 1807–1834.

SHEAR, W.A., SELDEN, P.A., RILFE, W.D.I., BONAMO, P.M., and GRIERSON, J.D. 1987. New terrestrial arachnids from the Devonian of Gilboa, New York. *Amer. Mus. Novitates* **2901**: 1–74.

SHERWOOD-PIKE, M.A., and GRAY, J. 1985. Silurian fungal remains: probable records of Ascomycetes. *Lethaia* **18**: 1–20.

SHIPMAN, P., and WALKER, A. 1980. Bone-collecting by harvesting ants. *Paleobiology* **6** (4): 496– 502.

SHORTHOUSE, J.D. 1979. Observations on the snow scorpionfly *Boreus brumatus* Fitch (Boreidae: Mecoptera) in Sudbury, Ontario. *Quaest. Entomol.* **15**(3): 341–344.

SHORTHOUSE, J.D., and ROFFRITSCH, O. (eds.). 1992. *Biology of Insect-Induced Galls.* Oxford Univ. Press, Oxford – New York: 285 p.

SHULL, A.F. 1914a. Biology of the Thysanoptera. I. Factors governing local distribution. *Amer. Natur.* **48**: 161–176.

SHULL, A.F. 1914b. Biology of the Thysanoptera. II. Sex and the life-cycle. *Amer. Natur.* **48**: 236–247.

SIDDAL, M.E., and WHITING, M.F. 1999. Long-branch abstractions. *Cladistics* **15**: 9–24.

SIEBER, F.M. 1820. *Beschreibendes Verzeichniss der in den Creta, Aegypten und Palestina gesammelten Alterhümer und andere Kunst- und Natur-Producte, nebst einer Abhandlung über ägyptische Mumien.* Gräffer, Wien: 86 p.

SIEBURTH, J.McN. 1979. *Sea Microbes.* Oxford Univ. Press, London: 491 p.

SILINA, A.E., and GONCHAROV, M.A. 1985. Chironomids as a factor of organic matter transport from water bodies. *Zhivotny mir Belorusskogo Poles'ya, okhrana i ratsionalnoe ispol'zovanie 4 Oblastnaya itogovaya nauchnaya konferentsiya, Gomel', noyabr' 1985.* Animals of Belorussian Polesye, Protection and Rational Use 4th Regional Sci. Conf., Gomel, Nov. 1985. Gomel': 145 (in Russian).

SILVA, D., DA, and PEREIRA DE ARRUDO, G. 1976. Insetos (Hymenoptera) cretaceos do grupo Araripe – Nordeste do Brasil. *An. Univ. fed. rural Pernambuco* **1**: 45–54.

SILVA, M.A.M., da. 1986. Lower Cretaceous sedimentary environment in the Araripe Basin, northeastern Brazil: a revision. *Rev. brasil. Geociencias* **16**: 311–319.

SILVEY, J.K.G. 1936. An investigation of the burrowing inner-beach insects of some freshwater lakes. *Michigan Acad. Sci., Arts & Letters Pap.* **21**: 655–696.

SIMMS, M.J. 1990. Triassic paleokarst in Britain. *Cave Science* **17**: 93–101.

SIMPSON, R.F. 1976. Bioassay of pine oil components as attractants for *Sirex noctilio* (Hymenoptera: Siricidae) using electroantennogram techniques. *Entomol. exp. et appl.* **19**(1): 11–18.

SINCLAIR, B.J. 1992. A phylogenetic interpretation of the Brachycera (Diptera) based on the larval mandible and associated mouthpart structures. *System. Entomol.* **17**(3): 233–251.

SINEV, S.Yu. 1990. Phylogenesis of Lepidoptera as an adaptive process. In: TOBIAS, V. I., and LVOVSKY, A.L. (eds.) *Advantages of Entomology in USSR: Hymenoptera and Lepidoptera.* Zool. Inst. Akad. Nauk SSSR, Leningrad: 216–218 (in Russian).

SINGER, M.C., EHRLICH, P.R., and GILBERT, L.E. 1971. Butterfly (*Euptichia westwoodi*; Lep., Nymphalidae, Satyrinae) feeding on lycopsid (*Selaginella horizontalis*, Pteridophyta). *Science* **172**: 1341–1342.

SINITSHENKOVA, N.D. 1979. A new family of the Palaeodictyoptera from the Carboniferous of Siberia. *Paleontol. Zhurnal* (2): 74–89 (in Russian, English translation: *Paleontol. J.* **13**(2): 192–205).

SINITSHENKOVA, N.D. 1980a. A revision of the order Permothemistidae (Insecta). *Paleontol. Zhurnal* (4): 91–106 (in Russian, English translation: *Paleontol. J.* **14**(4): 97–112).

SINITSHENKOVA, N.D. 1980b. The order Dictyoneurida. The order Mischopteridae. The order Permothemistida. In: ROHDENDORF, B.B., and RASNITSYN, A.P. (eds.). 1980. *Istoricheskoe razvitie klassa nasekomykh* (Historical Development of the Class Insecta). Trudy Paleontol. Inst. Akad. Nauk SSSR **175**. Nauka, Moscow: 44–49 (in Russian).

SINITSHENKOVA, N.D. 1982. Systematic position of the Jurassic stoneflies *Mesoleuctra gracilis* Br., Redtb., Gangl. and *Platyperla platypoda* Br., Redtb., Gangl., and their stratigraphic distribution. *Byulleten' Moskovskogo Obshchestva Ispytateley Prirody,* Sect. Geol. **57**(4): 112–124 (in Russian).

SINITSHENKOVA, N.D. 1984. The Mesozoic mayflies (Ephemeroptera) with special reference to their ecology. *Proc. 4 Internat. Conf. Ephemeroptera, Bechyne, Sept. 4–10, 1983.* Ceske Budejovice: 61–66.

SINITSHENKOVA, N.D. 1985a. The Jurassic mayflies (Ephemerida = Ephemeroptera) of South Siberia and West Mongolia. In: RASNITSYN, A.P. (ed.). *Yurskie nasekomye Sibiri i Mongolii* (The Jurassic Insects of Siberia and Mongolia). Trudy Paleontol. Inst. Akad. Nauk SSSR **211**. Nauka, Moscow: 11–23 (in Russian).

SINITSHENKOVA, N.D. 1985b. The Jurassic stoneflies of South Siberia and adjoining territories (Perlida = Plecoptera). In: RASNITSYN, A.P. (ed.). *Yurskie nasekomye Sibiri i Mongolii* (The Jurassic Insects of Siberia and Mongolia). Trudy Paleontol. Inst. Akad. Nauk SSSR **211**. Nauka, Moscow: 148–171 (in Russian).

SINITSHENKOVA, N.D. 1986. Stoneflies. Perlida (Plecoptera). In: RASNITSYN, A.P. (ed.). *Nasekomye v rannemelovykh ekosistemakh Zapadonoi Mongolii* (Insects in the Early Cretaceous Ecosystems of West Mongolia). Trudy Sovmestnoy Sovetsko-Mongolskoy Paleontol. Expeditsii **28**. Nauka, Moscow: 169–171 (in Russian).

SINITSHENKOVA, N.D. 1987. *Istoricheskoe razvitie vesnyanok* (Historical Development of the Stoneflies). Trudy Paleontol. Inst. Akad. Nauk SSSR **221**. Nauka, Moscow: 144 p. (in Russian).

SINITSHENKOVA, N.D. 1989. New Mesozoic mayflies (Ephemerida) from Mongolia. *Paleontol. Zhurnal* (3): 30–41 (in Russian, English translation: *Paleontol. J.* **23**(3): 26–37).

SINITSHENKOVA, N.D. 1990. The stoneflies. Perlida. In: RASNITSYN, A.P. (ed.). *Pozdnemezozoyskie nasekomye Vostochnogo Zabaykal'ya* (The Late Mesozoic insects from Eastern Transbaikalia). *Trudy Paleontol. Inst. Akad. Nauk SSSR* **239**. Nauka, Moscow: 207–210 (in Russian).

SINITSHENKOVA, N.D. 1992a. Two new insect species (Insecta: Dictyoneurida = Palaeodictyoptera, Perlida = Plecoptera) from the Late Permian of South Mongolia. In: GRUNT, T.A. (ed.). *Novye taxony iskopaemykh bespozvonochnykh Mongolii* (New Taxa of Fossil Invertebrates from Mongolia). Trudy Sovmestnoy Sovetsko-Mongol'skoy Paleontol. Expeditsii **41**. Nauka, Moscow: 98–100 (in Russian).

SINITSHENKOVA, N.D. 1992b. New stoneflies from the Upper Mesozoic of Yakutia (Insecta: Perlida = Plecoptera). *Paleontol. Zhurnal* (3): 34–42 (in Russian, English translation: *Paleontol. J.* **26**(3): 43–55).

SINITSHENKOVA, N.D. 1993. A new insect family Aykhalidae from the Upper Palaeozoic of Yakutia-Sakha (Insecta: Mischopteridae = Megasecoptera). *Paleontol. J.* **27**(1A): 131–134.

SINITSHENKOVA [SINICHENKOVA], N.D. 1995. New Late Mesozoic stoneflies from Shara-Teeg, Mongolia (Insecta: Perlida = Plecoptera). *Paleontol. J.* **29**(4): 93–104.

SINITSHENKOVA, N.D. 1998a. New Upper Mesozoic stoneflies from Central Transbaikalia (Insecta, Perlida = Plecoptera). *Paleontol. Zhurnal* (2): 64–69 (in Russian, English translation: *Paleontol. J.* **32**(2): 167–173).

SINITSHENKOVA, N.D. 1998b. The first European Cretaceous stonefly (Insecta, Perlida = Plecoptera). *Cretaceous Res.* **19**(3–4): 317–321.

SINITSHENKOVA, N.D. 1999a. A new mayfly species of the extant genus *Neoephemera* from the Eocene of North America (Insecta: Ephemerida = Ephemeroptera). *Paleontol. Zhurnal* (4): 67–69 (in Russian, English translation: *Paleontol. J.* **33**(4): 403–405).

SINITSHENKOVA, N.D. 1999b. The Mesozoic aquatic assemblages of Transbaikalia, Russia. *Proc. First Palaeoentomol. Conf., Moscow 1998.* AMBA projects International, Bratislava: 149–154.

SINITSHENKOVA, N.D. 2000a. A review of Triassic mayflies, with description of new species from Western Siberia and Ukraina (Ephemerida = Ephemeroptera). *Paleontol. J.* **34**(suppl. 3): S275–S283.

SINITSHENKOVA, N.D. 2000b. New Jersey amber mayflies: the first North American Mesozoic members of the order (Insecta: Ephemeroptera). In: GRIMALDI, D.A. (ed.). *Studies on Fossils in Amber, with Particular Reference to the Cretaceous of New Jersey.* Backhuys Publ., Leiden: 111–125.

SINITSHENKOVA, N.D. 2000c. The first fossil prosopistomatid mayfly from Burmese amber (Ephemeroptera: Prosopistomatidae). *Bull. Nat. Hist. Mus. London* (Geol.) **56** (1): 23–26.

SINITSHENKOVA, N.D. 2000d. New mayflies from the Upper Mesozoic locality Chernovskiye Kopi (Insecta: Ephemerida = Ephemeroptera). *Paleontol. Zhurnal* (1): 63–69 (in Russian, English translation: *Paleontol. J.* **34**(1): 68–74).

SINITSHENKOVA, N.D., and KUKALOVÁ-PECK, J. 1992. The wing venation and systematics of Lower Permian Diaphanopterodea from the Urals Mountains, Russia (Insects: Paleoptera). *Canad. J. Zool.* **70**: 229–235.

SINITSHENKOVA, N.D., and MARTÍNEZ-DELCLÒS, X. (in press). A new family of insecta Mischopterida (Dictyoneuridea) from the Upper Carboniferous of Ogassa (Catalonia, Spain). *J. Palaeontol.*

SINITSHENKOVA, N.D., and ZHERIKHIN, V.V. 1996. Mesozoic lacustrine biota: extinction and persistence of communities. *Paleontol. J.* **30**(6): 710–715.

SINITZA, S.M. 1980. The geological description of the Manlay locality. In: KALUGINA, N.S. (ed.). *Rannemelovoe ozero Manlay* (Early Cretaceous Lake Manlay). Trudy Sovmestnoy Sovetsko-Mongol'skoy Paleontol. Expeditsii **13**. Nauka, Moscow: 20–39.

SINITZA, S.M. 1993. Yura i nizhniy mel Tsentral'noy Mongolii (ostrakody, stratigrafiya i paleorekonstruktsiya) (Jurassic and Lower Cretaceous of Central Mongolia). Trudy Sovmestnoy Sovetsko-Mongol'skoy Paleontol. Expeditsii **42**. Nauka, Moscow: 240 p. (in Russian).

SINITZA, S.M., and STARUKHINA, L.P. 1986. New data and problems in stratigraphy and palaeontology of the Upper Mesozoic in East Transbaikalia. In: *Novye dannye po geologii Zabaikalya* (New data on geology Transbaikalia). Ministerstvo Geol. RSFSR, Moscow: 46–51 (in Russian).

SKALSKI, A. 1975. Notes on present status of botanical and zoological studies of ambers. *Studi e recerche sulla problematica dell'ambra* **1**. Consiglio Nazionale della Recerche, Roma: 153–175.

SKALSKI, A.W. 1988. A new fossil trichogrammatid from the Sicilian amber (Hymenoptera, Chalcidoidea, Trichogrammatidae). *Fragm. Ent.* **21**(1): 111–116.

SKALSKI, A.W. 1990. Lepidoptera in Fossil Resins with Emphasis on New Investigations. *Prace Muz. Ziemi* **41**: 163–164.

SKALSKI, A.W. 1992. The possibility of influence of phenological factors on composition of the Lepidoptera in the Baltic and Saxonian amber. *VIII Europ. Congress of Lepidopterology, Helsinki, April 19–23, 1992. Abstracts:* 31.

SKALSKI, A.W., and VEGGIANI, A. 1990. Fossil resin in Sicily and the Northern Apennines; geology and organic content. *Prace Muz. Ziemi* **41**: 37–49.

SKOMPSKI, S. Trace fossils in the deposits of ice-dammed lakes. *Kwart. Geol.* **35**(1): 119–130.

SLEEPER, E.L. 1968. A new fossil weevil from Nevada (Coleoptera: Curculionidae). *Bull. S. Calif. Acad. Sci.* **67**(3): 196–198.

SLEPYAN, E.I. 1962. Nomenclature and classification of arthropod-induced galls and bud teratoses in relation to their place among pathological phenomena. *Botan. Zhurnal* **47**(5): 721–753 (in Russian).

SLEPYAN, E.I. 1973. *Patologicheskie novoobrazovaniya i ikh vozbuditeli u rasteniy. Gallogenez i parazitarnyi teratogenez.* (Pathological Neoformations and Their Makers in Plants. Gallogenesis and Parasitic Teratogenesis). Nauka, Leningrad: 512 p. (in Russian).

SLY, P.G. 1978. Sedimentary processes in lakes. In: LERMAN, A. (ed.). *Lakes – Chemistry, Geology, Physics.* Springer-Verlag, New York: 65–89.

SMART, J. 1956. On the wing venation of *Chaeteessa* and other mantids (Insecta: Mantodea). *Proc. Zool. Soc. Lond.* **127**: 545–553.

SMELYANETS, V.P. 1968. Qualitative resin differences as a factor of resistance of different pine species to insect pests. *Zashchita rasteniy* (8): 101–109 (in Russian).

SMELYANETS, V.P. 1977. Mechanisms of plant resistance in Scotch pine (*Pinus sylvestris*). 4. Influence of food quality on physiological state of pine pests (trophic preferendum). *Zeitschr. angew. Entomol.* **84**(3): 232–241.

SMITH, D.M. 1999. Comparative taphonomy and paleoecology of insects in lacustrine deposits. *Proc. First Palaeoentomol. Conf., Moscow, 1998.* AMBA projects Internat., Bratislava: 155–162.

SMITH, G.B., and WATSON, J.A.L. 1991. Thysanura. In: NAUMANN, D. (ed.). *The Insects of Australia.* 2nd ed. Melbourne Univ. Press, Carlton: 275–278.

SMITH, N.D., and HEIN, F.J. 1971. Biogenic reworking of fluvial sediments by staphylinid beetles. *J. Sediment. Petrol.* **41**: 598–602.

SMITH, P.E., EVENSEN, N.M., YORK, D., CHANG, M.-m., JING, F., LI, J.-l., CUMBAA, S., and RUSSELL, D. 1995. Dates and rates in ancient lakes: 40Ar-39Ar evidence for an Early Cretaceous age for the Jehol Group, Northeast China. *Canad. J. Earth Sci.* **32**: 1426–1431.

SMITH, R.H. 1963. Toxicity of pine resin vapors to three species of *Dendroctonus* bark beetles. *J. Econ. Entomol.* **56**: 827–831.

SMITH, R.H. 1966. Resin quality as a factor in the resistance of pines to bark beetles. In: GERHOLD, H.D., SCHREINER, E.J., McDERMOTT, R.E., and WINIESKI, J.A. (eds.). *Breeding Pest-Resistant Trees.* Pergamon Press, London: 189–196.

SMITH, R.M. 1986. Sedimentation and paleoenvironments of crater-lake deposits in Bushmanland, South Africa. *Sedimentology* **33**: 369–386.

SMITH, R.M.H., MASON, T.R., and WARD, J.D. 1993. Flash-flood sediments and ichnofacies of the Late Pleistocene Homeb Silts, Kuiseb River, Namibia. *Sedim. Geol.* **85**: 579–599.

SMITHERS, C.N. 1972. *The classification and phylogeny of Psocoptera.* The Australian Mus. Mem. **14**. Sidney: 351 p.

SMITHERS, C.N. 1991. Psocoptera. In: NAUMANN, D. (ed.). *The Insects of Australia*, 2nd ed. Melbourne Univ. Press, Carlton: 412–420.

SNEATH, P.H.A., and SOKAL, R.R. 1973. *Numerical Taxonomy. The Principles and Practice of Numerical Classification.* W.H. Freeman, San Francisco: 573 p.

SODERSTROM, T.R., and CALDERON, C.E. 1971. Insect pollination in tropical rain forest grasses. *Biotropica* **3**: 1–16.

SOKOLOV, I.A. 1973. *Vulkanizm i pochvoobrazovanie* (Volcanism and Pedogenesis). Nauka, Moscow: 224 p. (in Russian).

SOUTHCOTT, R.V., and LANGE, R.T. 1971. Acarine and other microfossils from the maslin Eocene, South Australia. *Rec. S. Austral. Mus.* **16**(7): 1–21.

SOUTHWOOD, T.R.E. 1973. The insect/plant relationship – an evolutionary perspective. In: VAN EMDEN, H.F. (ed.). *Insect/Plant Relationships.* Symp. R. Entomol. Soc. London **6**. Blackwell, London: 3–30.

SPAHR, U. 1981a. Bibliographie der Bernstein- und Kopal-Kafer (Coleoptera). *Stuttg. Beitr. Naturk.* B **72**: 1–21.

SPAHR, U. 1981b. Systematischer Katalog der Bernstein- und Kopal-Kafer (Coleoptera). *Stuttg. Beitr. Naturk.* B **80** 1981: 1–107.

SPAHR, U. 1985. Erganzungen und Berichtigungen zu R. Keilbachs Bibliographie und Liste der Bersteinfossilien – Ordnung Diptera. *Stuttg. Beitr. Naturk.* B **111**: 1–146.

SPAHR, U. 1988. Erganzungen und Berichtigungen zu R. Keilbachs Bibliographie und Liste der Bersteinfossilien – Uberordnung Hemipteroidea. *Stuttg. Beitr. Naturk.* B **144**: 1–60.

SPAHR, U. 1987. Erganzungen und Berichtigungen zu R. Keilbachs Biblographie und Liste der Bernsteinfossilien – 'Ordnung Hymenoptera'. *Stuttg. Beitr. Naturk.* B **127**: 1–121.

SPAHR, U. 1989. Erganzungen und Berichtigungen zu R. Keilbachs Bibliographie und Liste der Bersteinfossilien – Uberordnung Mecopteroidea. *Stuttg. Beitr. Naturk.* B **157**: 1–87.

SPAHR, U. 1990. Ergänzungen und Berichtigungen zu R. Keilbachs Biblographie und Liste der Bernsteinfossilien – 'Apterygota'. *Stuttg. Beitr. Naturk.* B **166**: 1–23.

SPAHR, U. 1992. Erganzungen und Berichtigungen zu R. Keilbachs Bibliographie und Liste der Bernsteinfossilien – Klasse Insecta (Ausgenommen: 'Apterygota', Hemipteroidea, Coleoptera, Hymenoptera, Mecopteroidea). *Stuttg. Beitr. Naturk.* B **182**: 1–102.

SPAHR, U. 1993. Ergänzungen und Berichtigungen zu R. Keilbachs bibliographie und Liste der Bernsteinfossilien – Verschiedene Tiergruppen, ausgenommen Insecta und Araneae. *Stuttg. Beitr. Naturk.* B **194**: 1–77.

SPALDING, J.B. 1979. The aeolian ecology of White Mountain Peak, California: windblown insect fauna. *Arctic and Alpine Res.* **11**(1): 83–94.

SPARKS, A.N., JACKSON, R.D., CARPENTER, J.E., and MULLER, R.A. 1986. Insects captured in light traps in the Gulf of Mexico. *Ann. Entomol. Soc. Amer.* **79**(1): 132–139.

SPHON, G.G. 1973. Additional type specimens of fossil Invertebrata in the collections of the Natural History Museum of Los Angeles County. *Contribution in Science, Natural History Museum of Los Angeles County* **250**: 1–75.

SPICER, R.A. 1989. Plants at the Cretaceous-Tertiary boundary. *Philos. Trans. R. Soc. London* B **325**: 291–305.

SPICER, R.A., and HERMAN, A.B. 1996. *Nilssoniocladus* in the Cretaceous Arctic: new species and biological insights. *Rev. Palaeobot. Palynol.* **92**: 229–243.

SQUIRES, R.L. 1979. Middle Pliocene dragonfly nymphs, Ridge Basin, Transverse Ranges, California. *J. Paleontol.* **53**(2): 446–452.

SREBRODOLSKY, B.I. 1980. *Yantar' Ukrainy* (The Amber of Ukraina). Naukova Dumka, Kiev: 123 p. (in Russian).

SRIVASTAVA, A.K. 1987. Lower Barakar flora of Raniganj Coalfield and insect/plant relationship. *Palaeobotanist* **36**: 138–142.

STABBLFIELD, S.P., TAYLOR, T.N., MILLER, C.E., and COLE, G.T. 1984. Studies on Palaeozoic fungi. III. Fungal parasitism in a Pennsylvanian gymnosperm. *Amer. J. Botany* **71**: 1275–1282.

STACH, J. 1972. Owady bezkrydle (Apterygota) z bursztynu baltickiego. *Przegl. zool.* **16**: 416–420.

STAHL, J.B. 1959. The developmental history of the chironomid and *Chaoborus* fauna of Myers Lake. *Invest. Ind. Lakes and Streams* **5**: 47–102.

STANKIEWICZ, B.A., BRIGGS, D.E.G., EVERSHED, R.P., FLANNERY, M.B., and WUTTKE, M. 1997a. Preservation of chitin in 25 million-year-old fossils. *Science* **5318**: 1541–1543.

STANKIEWICZ, B.A., BRIGGS, D.E.G., EVERSHED, R.P., and DUNCAN, I.J. 1997b. Chemical preservation of insect cuticles from the Pleistocene asphalt deposits of California, USA. *Geochim. Cosmochim. Acta* **61**: 2247–2252.

STANKIEWICZ, B.A., BRIGGS, D.E.G., EVERSHED, R.P., MILLER, R.F., and BIERSTEDT, A. 1998a. The fate of chitin in Tertiary and Quaternary strata. In: STANKIEWICZ, B.A., and VAN BERGEN, P.F. (eds.). *Nitrogen-Containing Molecules in the Biosphere and Geosphere.* Amer. Chem. Soc. Symp. Series **707**: 211–225.

STANKIEWICZ, B.A., MASTALERZ, M., HOF, C.H.J., BIERSTEDT, A., FLANNERY, M.B., BRIGGS, D.E.G., and EVERSHED, R.P. 1998b. Biodegradation of the chitin-protein complex in crustacean cuticle. *Organ. Geochem.* **28**: 67–76.

STANKIEWICZ, B.A., POINAR, H.N., BRIGGS, D.E.G., EVERSHED, R.P., and POINAR, G.O., Jr. 1998c. Chemical preservation of plants and insects in natural resins. *Proc. R. Soc. Lond.* B **265**: 641–647.

STANKIEWICZ, B.A., SCOTT, A.C., COLLINSON, M.E., FINCH, P., MÖSLE, B., BRIGGS, D.E.G., and EVERSHED, R.P. 1998d. The molecular taphonomy of arthropod and plant cuticles from the Carboniferous of North America. *J. Geol. Soc., London* **155**: 453–462.

STANLEY, K.O., and FAGERSTROM, J.A. 1974. Miocene invertebrate trace fossils from a braided river environment, western Nebraska, U.S.A. *Palaeogeogr., Palaeoclimatol., Palaeoecol.* **15**(1): 63–82.

STANNARD, L.J. 1957. The phylogeny and classification of the N. American genera of the suborder Tubulifera. *Illinois Biol. Monogr.* **25**: 200 p.

STANNARD, L.J. 1968. The thrips or Thysanoptera of Illinois. *Bull. Illinois State Natur. Hist. Surv.* **29**: 215–552.

STARK, D.M. 1976. Paleolimnology of Elk Lake, Itasca State Park, northwestern Minnesota. *Arch. Hydrobiol.* Suppl. **50**(2–3): 208–274.

STARK, R.W. 1968. Substances attractives chez les Scolytides. *Mitt. Schweiz. entomol. Ges.* **41**(1–4): 245–252.

STAROBOGATOV, Ya.I. 1991. Problems in the nomenclature of higher taxonomic categories. *Bull. Zool. Nom.* **48**: 6–18.

STATZ, G. 1936. Uber neue Funde von Neuropteren, Panorpaten und Trichopteren aus dem Tertiären Schiefern vom Rott an Siebengebirge. *Decheniana* **93**: 208–255.

STATZ, G. 1939. Geradflügler und Wasserkäfer der oligocänen Ablagerungen von Rott. *Decheniana* **99A**: 1–102.

STATZ, G. 1940. Neue Dipteren (Brachycera et Cyclorhapha) aus dem Oberoligocän von Rott. *Palaeontographica* A **91**: 120–174.

STATZ, G. 1944. Neue Dipteren (Nematocera) aus dem Oberoligozän by Rott. *Palaeontographica* A **95**: 123–187.

STATZ, G. 1950. Cicadariae (Zikaden) aus den oberoligocänen Ablagerungen von Rott. *Palaeontographica* A **98**: 1–44.

STATZNER, B., GORE, J.A., and RESH, V.H. 1988. Hydraulic river ecology: observed patterns and potential application. *J. North Amer. Benthol. Soc.* **7**: 307–360.

STAUFFER, P.H. 1979. A fossilized honeybee comb from Late Cenozoic cave deposits at Batu Caves, Malay Peninsula. *J. Paleontol.* **53**: 1416–1421.

STEBBINS, G.L. 1974. *Flowering Plants. Evolution above Species Level.* Harvard Univ. Press, Cambridge (Mass.): 397 p.

STEBBINS, G.L. 1978. Co-evolution of grasses and herbivores. *Ann. Missouri Botan. Gard.* **68**: 75–86.

STEFANESCU, C. 1997. Butterflies and moths (Insecta, Lepidoptera) recorded at sea off Eivissa and Barcelona (Western Mediterranean) in October 1996. *Boll. Soc. Hist. Natur. Balears* **40**: 51–56.

STEINBACH, G. 1967. Zur Hymenopterenfauna des Pliozäns von Willershausen/Westharz. *Ber. naturhist. Ges. Hannover* **111**: 95–104.

STEINMANN, H. 1975. Suprageneric classification of Dermaptera. *Acta Zool. Acad. Sci. Hung.* **21**: 195–220.

STELLEMAN, P. 1979. Insecten en windbloeiers. *Vakbl. Biol.* **59**(16): 264–269.

STELLEMAN, P., and MEEUSE, A.D.J. 1976. Anthecological relations between reputedly anemophilous flowers and syrphid flies. I. The possible role of syrphid flies as pollinators of *Plantago*. *Tijdschr. Entomol.* **119**(2): 15–34.

STEPHENSON, J., and SCOTT, A.C. 1992. The geological history of insect-related plant damage. *Terra Nova* **4**: 542–552.

STIDHAM, T.A., and STIDHAM, J.A. 2000. A new Miocene band-winged grasshopper (Orthoptera: Acrididae) from Nevada. *Ann. Entomol. Soc. Amer.* **93**(3): 405–407.

STIGER, M.A. 1977. *Anasazi Diet: The Coprolite Evidence.* Unpublished Masters Thesis, Dept. of Anthropology, Univ. Colorado, Boulder (cited after ELIAS 1994).

STOCKEY, R.A. 1978. Reproduction biology of Cerro Cuadrado fossil conifers: Ontogeny and reproductive strategies in *Araucaria mirabilis* (Spegazzini) Windhausen. *Palaeontographica* B **166**: 1–15.

STONE, R. 1950. '*Lagena samanica*' Berry. *Micropalaeontology* **4**(2): 17.

STORCH, G. 1978. *Eomanis waldi*, ein Schuppentier aus dem Mittel-Eozän der 'Grube Messel' bei Darmstadt (Mammalia, Pholidata). *Senckenberg. Lethaea* **59**(4–6): 503–529.

STØRMER, L. 1963. *Gigantoscorpio willsi*, a new scorpion from the Lower Carboniferous of Scotland and its associated preying micro-organisms. *Skrift. Norske Vidensk.-Akad. Oslo*, Nj serie, **1**, Naturv. Klasse (8): 1–171.

STOROZHENKO, S.Yu. 1997. Fossil history and phylogeny of orthopteroid insects. In: GANGWERE, S.K., MURALIRANGAN, M.C., and MURALIRANGAN, M. (eds.). *The Bionomics of Grasshoppers, Katydids and Their Kin*. CAB Internat., Oxton & New York: 59–82.

STOROZHENKO, S.Yu. 1998. *Sistematika, filogeniya i evolyutsiya grilloblattidovykh nasekomykh (Insecta: Grylloblattida)* [Systematics, Phylogeny and Evolution of the Grylloblattids (Insecta: Grylloblattida)]. Dal'nauka, Vladivostok: 207 p. (in Russian).

STOROZHENKO, S.Yu., and ARISTOV, D.S. 1999. New genus of the family Liomopteridae (Insecta: Grylloblattida) from Lower Permian of Russia. *Far Eastern Entomologist* **76**: 6–8.

STOROZHENKO, S.Yu., and NOVOKSHONOV, V.G. 1999. To the knowledge of the fossil family Permosialidae (Insecta: Miomoptera). *Far Eastern Entomologist* **76**: 1–5.

STORRS, G.W. 1993. Terrestrial components of the Rhaetian (Uppermost Triassic) Westbury Formation of southwestern Britain. In: LUCAS, S.G., and MORALES, M. (eds.) *The Nonmarine Triassic*. Bull. New Mexico Mus. Natur. Hist. Science **3**: 447–451.

STRASSEN, R., zur. 1973. Fossile Fransenflügler aus mesozoischem Bernstein der Libanon. *Stuttg. Beitr. Naturk.* **A 256**: 1–51.

STRAUS, A. 1967. Zur Paläontologie des Pliozäns von Willershausen. *Ber. naturhist. Ges. Hannover* **111**: 15–24.

STRAUS, A. 1977. Gallen, Minen und anderen Fraßspuren im Pliocän von Willershausen am Harz. *Verhandl. Botan. Ver. Prov. Brandenburg* **113**: 43–80.

STROIŃSKI, A., and SZWEDO, J. 2000. *Tonacatecutlius gibsoni* gen. and sp. nov. from the Oligocene/Miocene Mexican amber (Hemiptera: Fulgoromorpha: Nogodinidae). *Ann. Zool. (Warszawa)* **50**(3): 341–345.

STOKES, W.L. 1978. Transported fossil biota of the Green River Formation, Utah. *Paleogeogr., Paleoclimatol., Paleoecol.* **25**: 353–364.

STRONG, D.R., LAWTON, H., and SOUTHWOOD, R. 1984. *Insects on Plants. Community Pattern and Mechanisms*. Blackwell, Oxford etc.: 313 p.

STUART, M. 1923. Amber and the dammar of living bees. *Nature* **111**: 83–84.

STUBBLEBINE, W.H., and LANGENHEIM, J.H. 1977. Effects of *Hymenaea courbaril* leaf resin on the generalist herbivore *Spodoptera exigua* (beet armyworm). *J. Chem. Ecol.* **3**: 633–647.

STUBBLEFIELD, S.P., and TAYLOR, T.N. 1985. Fossil fungi in Antarctic wood. *Antarct. J. U.S.* **19**(5): 7–8.

STUBBLEFIELD, S.P., and TAYLOR, T.N. 1986. Wood decay in silicified gymnosperms from Antarctica. *Botan. Gazette* **147**(1): 116–125.

STUBBLEFIELD, S.P., and TAYLOR, T.N. 1988. Recent advances in paleomycology. *New Phytol.* **108**(1): 3–25.

STURM, H. 1997. Fossilgeschichte und Taxonomie der Felsenspringer und der fishchenartigen Insecten (Archaeognatha, Zygentoma, "Apterygota", Insecta). *Mitt. Dtsch. Ges. allg. angew. Ent.* **11**: 811–816.

STURM, H. 1998. Erst Nachweis fischerartiger Insekten (Zygentoma, Insecta) für das Mesozoicum (Untere Kreide) Brasilien. *Senckenberg. Lethea* **78**(1/2): 135–140.

STURM, H., and POINAR, G.O., Jr. 1998. *Cretaceomachilis libanensis*, the oldest known bristle-tail of the family Meinertellidae (Machiloidae, Archaeognatha, Insecta) from the Lebanese amber. *Mitt. Mus. Naturk. Berlin, Deutsche entomol. Zeitschr.* **45**(1): 43–48.

STURM, H., and MENDES, L.F. 1998. Two new species of Nicoletiidae (Zygentoma, "Apterygota", Insecta) in Dominican amber. *Amer. Mus. Novitates* **3226**: 1–11.

ŠTYS, P. 1983. A new family of Heteroptera with dipsocoromorphan affinities from Papua New Guinea. *Acta Entomol. Bohemoslov.* **80**: 256–292.

ŠTYS, P., and BILINSKI, S. 1990. Ovariole types and the phylogeny of hexapods. *Biol. Rev.* **65**: 401–429.

ŠTYS, P., and KERZHNER, I. 1975. The rank and nomenclature of higher taxa in recent Heteroptera. *Acta Entomol. Bohemoslov.* **72**: 65–79.

ŠTYS, P., and ŘÍHA, P. 1974. An annotated catalogue of the fossil Alydidae (Heteroptera). *Acta Univ. Carolinae (Prague), Biologica* (1977): 173–188.

ŠTYS, P., and ŘÍHA, P. 1975. Studies on Tertiary Notonectidae from Central Europe (Heteroptera). *Acta Univ. Carolinae (Prague), Biologica* (1973): 163–184.

SUCHÝ, J. 1991. Nouzova potrava sluneček (Col., Coccinellidae) v roce jejich přemnozeni. *Zpr. Muz. Zapadočes. Kraje-Přir.* **41**: 77–80.

SUESS, E. 1979. Mineral phases formed in anoxic sediments by microbial decomposition of organic matter. *Geochim. Cosmochim. Acta* **43**: 339–352.

SUESS, H., and SCHULTZE-DEWITZ, G. 1987. Fossilized feeding pattern of termites on wood remains of the Tertiary carbon formation of Stare Sedlo (CSSR). In: EDER, J., and RENBOLD, H. (eds.) *Chemistry and Biology of Social Insects*: 615–616.

SUKATSHEVA (SUKACHEVA), I.D. 1980. Evolution of the caddisfly (Trichoptera) larval case construction. *Zhurnal obshchey biol.* **41**: 457–469 (in Russian, with English summary).

SUKATSHEVA, I.D. 1982. *Istoricheskoe razvitie otryada rucheinikov* (Historical development of the Caddisflies). Trudy Paleontol. Inst. Akad. Nauk SSSR **197**. Moscow, Nauka: 112 p. (in Russian).

SUKATSHEVA, I.D. 1985a. Jurassic scorpionflies of South Siberia and West Mongolia. In: RASNITSYN, A.P. (ed.). *Yurskie nasekomye Sibiri i Mongolii* (Jurassic Insects of Siberia and Mongolia). Trudy Paleontol. Inst. Akad. Nauk SSSR **211**. Nauka, Moscow: 96–114 (in Russian).

SUKATSHEVA, I.D. 1985b. Jurassic cadisflies of South Siberia. In: Rasnitsyn, A.P. (ed.). *Yurskie nasekomye Sibiri i Mongolii* (Jurassic Insects of Siberia and Mongolia). Trudy Paleontol. Inst. Akad. Nauk SSSR **211**. Nauka, Moscow: 115–119 (in Russian).

SUKATSHEVA, I.D. 1989. Cainozoic caddisflies from the Primor'ye Province. In: KRASSILOV, V.A., and KLIMOVA, R.S. (eds.). *Kaynozoy Dal'nego Vostoka* (The Cainozoic of the Far

East). Biologo-Pochvenny Inst. Dal'nevostochnogo Otdeleniya Akad. Nauk SSSR, Vladivostok: 151–160 (in Russian).

SUKATSHEVA, I.D. 1990a. Scorpionflies. Panorpida. In: RASNITSYN, A.P. (ed.). *Pozdneme-zozoiskie nasekomye Vostochnogo Zabaikal'ya* (Late Mesozoic Insects of Eastern Transbaikalia). Trudy Paleontol. Inst. Akad. Nauk SSSR **239**. Nauka, Moscow: 88–94 (in Russian).

SUKATSHEVA, I.D. 1990b. Caddisflies. Phryganeina. In: Rasnitsyn, A.P. (ed.). *Pozdneme-zozoiskie nasekomye Vostochnogo Zabaikal'ya* (Late Mesozoic insects of Eastern Transbaikalia). Trudy Paleontol. Inst. Akad. Nauk SSSR **239**. Nauka, Moscow: 94–122 (in Russian).

SUKATSHEVA, I.D. 1991a. The Late Cretaceous stage in the history of the caddisfles (Trichoptera). *Acta Hydroentomol. Latvica* **1**: 68–85 (in Russian, with English summary).

[SUKATSHEVA] SUKATCHEVA, I.D. 1991b. Historical development of the order Trichoptera. *Proc. 6th Internat. Symp. Trichoptera, Lódz – Zakopane, Poland, 13–16 Sept. 1989*: 441–445.

SUKATSHEVA, I.D. 1992. New fossil representatives of caddisflies (Phryganeida) from Mongolia. In: GRUNT, T.A. (ed.). *Novye taksony iskopaemykh bespozvonochnykh Mongolii* (New Taxa of Fossil Invertebrates from Mongolia) Trudy Sovmestnoy Sovetsko-Mongol'skoy Paleontol. Expeditsii **41**. Nauka, Moscow: 100–110 (in Russian).

SUKATSHEVA, I.D. 1994. Cases of the Early Jurassic caddisflies (Insecta, Trichoptera) from Mongolia. *Paleontol. Zhurnal* (4): 75–85 (in Russian, English translation: *Paleontol. J.* **28**(4): 99–113).

SUKATSHEVA, I.D. 1999. The Lower Cretaceous caddis case assemblages (Trichoptera). *Proc. First Palaeoentomol. Conf., Moscow 1998*. AMBA projects Internat., Bratislava: 163–166.

SUKATCHEVA, I.D., and NOVOKSHONOV, V.G. 1998. A new family of scorpionflies from the Mesozoic of Yakutia (Insecta: Mecoptera, Sibiriothaumatidae fam. nov.). *Paleontol. Zhurnal* (6): 50–52 (in Russian, English translation: *Paleontol. J.* **32**(6): 596–597).

SUKATCHEVA, I.D., and RASNITSYN, A.P. First member of the family Boreidae (Insecta, Panorpida) from the Upper Jurassic of Mongolia and the Lower Cretaceous of Transbaikalia. *Paleontol. Zhurnal* (1): 126–129 (in Russian, English translation: *Paleontol. J.* **26**(1): 168–172).

SUMMERS, R. 1967. Archaeological distribution and a tentative history of tsetse infestation in Rhodesia and the northern Transvaal. *Arnoldia* **3**(13): 1–18.

SUN, G. 1996. Discovery of the early angiosperms from Lower Cretaceous of Jixi, north-eastern China and its significance for study of angiosperm origin in the World. *Chteniya pamyati Vsevoloda Andreevicha Vakhrameeva. Sbornik tezisov i dokladov. 13–14 noyabrya 1996 g.* (Memorial Conference Dedicated to Vsevolod Andreevich Vakhra-meev. Abstracts and Proceedings. Nov. 13–14, 1996). Geos, Moscow: 68.

SUN, G., DILCHER, D.L., ZHENG, S., and ZHOU, Z. 1998. In search of the first flower: a Jurassic angiosperm *Archaefructus*, from northeast China. *Science* **282**: 1692–1695.

SURDAM, R.C., and WOLFBAUER, C.A. 1975. Green River Formation, Wyoming: a playa-lake complex. *Geol. Soc. America Bull.* **86**: 335–345.

SÜSS, H. 1979. Durch *Protophytobia cupressorum* gen. nov. sp. nov. (Agromyzidae, Diptera) verursachte Markflecke in einem Holz von *Juniperoxylon* aus dem Tertiär von Süd-Limburg (Niederlande) und der Nachweis von Markflecken in einer rezenten *Callitris*-Art. *Feddes Repertorium* **90**(3): 165–172.

SÜSS, H. 1980. Fossile Kambium-Minierer der Familie Agromyzidae (Diptera) in tertiären Laub- und Nadelholzresten. *Zeitschr. geol. Wiss.* **9**: 1217–1225.

SÜSS, H., and MÜLLER-STOLL, W. 1975. Durch *Palaeophytobia platani* n.g. n.sp. (Agromyzidae, Diptera) verursachte Markflecken im Holz fossiler Platanen aus dem ungarischen Miozän. *Wiss. Zeitschr. Humboldt-Univ. Berlin, mathem.-naturwiss. Reihe* **24**(4): 515–519.

SÜSS, H., and MÜLLER-STOLL, W. 1977. Untersuchungen über fossile Platanenholzer. Beiträge zu einer Monographie der Gattung *Platanoxylon* Andreánszky. *Feddes Repertorium* **88**: 1–62.

SÜSS H., and MÜLLER-STOLL, W. 1979. Das fossile Holz *Pruninium gummosum* Platen emend. Süss et Müller-Stoll aus dem Yellowstone Nationalpark und sein Parasit – *Palaeophytobia prunorum* sp. nov. nebst Bemerkungen über Markflecke. In: VENT, W. (ed.). *100 Jahre Arboretum Berlin*. Berlin: 343–364.

SÜSS, H., and MÜLLER-STOLL, W. 1982. Ein Rosaceen-Holz, *Pruninium kraeuseli* (E. Schönfeld) comb. nov. aus dem miozänen Ton von Lauterbach. *Zeitschr. geol. Wiss.* **10**(12): 1553–1563.

SWAIN, T. 1978. Plant-animal co-evolution: a synoptic view of the Paleozoic and Mesozoic. In: HARBORNE, J.B. (ed.). *Biochemical Aspects of Plant and Animal Co-evolution*. Ann. Proc. Phytochem. Soc. Europe **15**: 5–19.

SWISHER, C.C., III, WANG, Y.-Q., WANG, X.-l., XU, X., and WANG, Y. 1999. Crertaceous age for the feathered dinosaurs from Liaoning, China. *Nature* **400**: 58–61.

SYTCHEVSKAYA, E.K. 1986. *Presnovodnaya paleogenovaya ikhtiofauna SSSR i Mongolii* (Palaeogene Freshwater Fish Fauna of the USSR and Mongolia). Trudy Sovmestnoy Sovetsko-Mongol'skoy Paleontol. Expeditsii **29**. Nauka, Moscow: 157 p. (in Russian).

SZADZIEWSKI, R. 1995. The oldest fossil Corethrellidae (Diptera) from Lower Cretaceous Lebanese amber. *Acta Zool. Cracov.* **38**(2): 177–181.

SZADZIEWSKI, R. 1996. Biting midges from Lower Cretaceous amber of Lebanon and Upper Cretaceous Siberian amber of Taimyr (Diptera, Ceratopogonidae). *Studia dipterol.* **3**(1): 23–86.

SZADZIEWSKI, R. 1998. New mosquitoes from Baltic amber (Diptera: Culicidae). *Polskie Pismo entomol.* **67**(3–4): 233–244.

SZADZIEWSKI, R., and ARILLO, A. 1998. Biting midges (Diptera: Ceratopogonidae) from the Lower Cretaceous amber from Alava, Spain. *Polsk. Pismo Entomol.* **67**: 291–298.

SZAFER, W. 1969. *Kwiaty i zwierzęta* (Flowers and Animals). Panstwowe Wydawnictwo Naukowe, Warszawa: 378 p.

SZUMIK, C.A. 1994. *Oligembia vetusta*, a new fossil teratembiid (Embioptera) from Dominican amber. *J. New York Entomol. Soc.* **102**: 67–73.

SZWEDO, J. 2000. First fossil Tropiduchidae with a description of a new tribe Jantaritambiini from Eocene Baltic amber. *Ann. Soc. Entomol. France* (N.S.), **36**(3): 279–286.

SZWEDO, J., and KULICKA, R. 1999. Inclusions of Auchenorrhyncha in Baltic amber (Insecta: Homoptera). *Estud. Mus. Cienc. Natur. Alava* **14** (Núm. Espec. 2): 177–197.

SZWEDO, J., and WEBB, M.D. 1999. First fossil representatives of Aetalionidae (Homoptera, Membracoidea) from Oligocene/Miocene ambers of the New World. *Proc. 10th Internat. Auchenorrhyncha Congress, Cardiff 6–10 Sept. 1999* (unpaginated).

TAGGART, R.E., and CROSS, A.T. 1980. Vegetation change in the Miocene Succor Creek flora of Oregon and Idaho: case study in paleosuccession. In: DILCHER, D.L., and TAYLOR, T.N. (eds.). *Biostratigraphy of Fossil Plants*. Dowden, Hutchinson & Ross, Stroudsburg, Pa.: 185–210.

TAKAHASHI, R. 1921. The metamorphosis of Thysanoptera, with notes on that of Coccidae. *Zool. Mag.* **35**: 80–85.

TAKAHASHI, R. 1926. Observations on the aquatic cockroach, *Opistoplatia maculata*. *Dobutsugaku Zasshi* **38**: 89.

TAKHTAJAN, A.L. 1969. *Flowering Plants: Origin and Dispersal*. Edinburgh (cited after Russian edition: TAKHTAJAN, A.L. 1970. *Proiskhozhdenie i rasselenie tsvetkovykh rasteniy*. Nauka, Leningrad, 145 p.).

TANDON, S.K., and NAUG, B. 1984. Facies – trace fossils relationships in a Plio-Pleistocene fluvial sequence – the Upper Siwalik Subgroup, Punjab, sub-Himalaya. *Palaeogeogr., Palaeoclimatol., Palaeoecol.* **47**(3–4): 277–299.

TANG, W. 1987. Insect pollination in the cycad *Zamia pumila* (Zamiaceae). *Amer. J. Botany* **74**(1): 90–99.

TANG, W., OBERPRIELER, R.G., and YANG, S.-L. 1999. Beetles (Coleoptera) in cones of Asian *Cycas*: diversity, evolutionary patterns, and implications for *Cycas* taxonomy. *Biology and Conservation of Cycads. Proc. 4th Internat. Conference on Cycad Biology, Panzhihua, China.* Internat. Academic Publ., Beijing: 280–297.

TARNOCAI, C., and SMITH, C.A.S. 1991. Paleosols of the fossil forest area, Axel Heiberg Island. In: CHRISTIE, R.L., and McMILLAN, N.J. (eds.). *Tertiary Fossil Forests of the Geodetic Hills, Axel Heiberg Island, Arctic Archipelago*. Geol. Surv. Canada Bull. **403**: 171–183.

TASCH, P. 1975. Non-marine arthropods of the Tasmanian Triassic. *Pap. Proc. R. Soc. Tasmania* **109**: 97–106.

TAYLOR, D.W., and CREPET, W.L. 1987. Fossil floral evidence of Malpighiaceae and an early plant-pollinator relationship. *Amer. J. Botan.* **74**(2): 274–286.

TAYLOR, D.W., and HICKEY, L.J. 1990. An Aptian plant with attached leaves and flowers: implications for angiosperm origin. *Science* **247**: 702–704.

TAYLOR, D.W., and HICKEY, L.J. 1992. Phylogenetic evidence for the herbaceous origin of angiosperms. *Plant Syst. Evol.* **180**: 137–156.

TAYLOR, L.R. 1960. The distribution of insects at low levels in the air. *J. Anim. Ecol.* **29**: 45–63.

TAYLOR, L.R. 1967. The effect of weather on the height of flight of insects. *Biometeorology* **2**(2). Pergamon Press, Oxford etc.: 583–584.

TAYLOR, T.N. 1978. The ultrastructure and reproductive significance of *Monoletes* (Pteridospermales) pollen. *Canad. J. Botany* **56**: 3105–3118.

TAYLOR, T.N., HASS, H., and KERP, H. 1997. A cyanolichen from the Lower Devonian Rhynie chert. *Amer. J. Botany* **84**(7): 992–1004.

TAYLOR, T.N., and MILLAY, M.A. 1979. Pollination biology and reproduction in early seed plants. *Rev. Palaeobot. Palynol.* **27**: 329–355.

TAYLOR, T.N., and SCOTT, A.C. 1983. Interactions of plants and animals during the Carboniferous. *Bioscience* **33**(8): 488–493.

TAYLOR, T.N., and TAYLOR, E.L. 1993. *The Biology and Evolution of Fossil Plants*. Prentice Hall, Englewood Cliffs, New Jersey: 982 p.

TAYLOR, T. N., and TAYLOR, E.L. 1997. The distribution and interactions of some Paleozoic fungi. *Rev. Palaeobot. Palynol.* **95**: 83–94.

TAYLOR, W.A. 1995. Spores in earliest land plants. *Nature* **373**: 391–392.

TEICHMÜLLER, M. 1989. The genesis of coal from the viewpoint of coal petrology. *Internat. J. Coal Geol.* **12**: 1–87.

TERRA, P.S. 1992. Zelo materno em *Cardioptera brachyptera* (Mantodea, Vatidae, Photininae). *Rev. brasil. Entomol.* **36**(3): 493–503.

TERRA, P.S. 1995. Revisão sistemática dos gêneros de louva-a-deus de Região Neotropical (Mantodea). *Rev. brasil. Entomol.* **39**(1): 13–94.

TERRA, P.S. 1992. Zelo materno em *Photina amplipennis* Stål (Mantodea, Vatidae). *Rev. brasil. Entomol.* **40**(1): 9–10.

TESKEY, H.J., and TURNBULL, C. 1979. Diptera puparia from pre-historic graves. *Canad. Entomol.* **111**(4): 527–528.

TETEREVNIKOVA-BABAYAN, D.N. 1981. Advances and perspectives of palaeomycological studies in USSR. *Mikologiya i fitopatologiya* **15**(4): 335–341 (in Russian).

TEVESZ, M.J.S., and McCALL, P.L. 1982. Geological significance of aquatic nonmarine trace fossils. In: McCALL, P.L., and TEVESZ, M.J.S. (eds.). *Animal-Sediment Relations*. Plenum Press, New York: 257–285.

THACKRAY, G.D. 1994. Fossil nest of sweat bees (Halictinae) from a Miocene paleosol, Rusinga Island, western Kenya. *J. Paleontol.* **68**(4): 795–800.

THEISCHINGER, G. 1991a. Megaloptera. In: NAUMANN, D. (ed.). *The Insects of Australia*. 2nd ed. Melbourne Univ. Press, Carlton: 516–520.

THEISCHINGER, G. 1991b. Plecoptera. In: NAUMANN, D. (ed.). *The Insects of Australia*. 2nd ed. Melbourne Univ. Press, Carlton: 311–319.

THENIUS, E. 1979. Lebensspuren von Ephemeropteren-Larven aus dem Jung-Tertiär des Wiener Beckens. *Ann. Naturhist. Mus. Wien* (1977) **82**: 177–188.

THENIUS, E. 1989. Fossile Lebensspuren aquatischen Insekten in Knochen aus dem Jungtertiär Niederösterreichs. *Anz. Österr. Akad. Wiss., mathem.-naturwiss. Kl.* (1988) **125**: 41–45.

THÉOBALD, N. 1937. Les insectes fossiles des terrains oligocenes de France. *Bull. mens. Soc. Sci. Nancy* (N.S.) **2bis**: 1–473.

THIEN, L.B. 1974. Floral biology of *Magnolia*. *Amer. J. Botany* **61**: 1037–1045.

THIEN, L.B. 1980. Patterns of pollination in the primitive angiosperms. *Biotropica* **12**: 1–13.

THIEN, L.B. 1982. Fly pollination in *Drimys* (Winteraceae), a primitive angiosperm. In: GRESSITT, J.L. (ed.). *Biogeography and Ecology of New Guinea* **1**. W. Junk, The Hague etc.: 529–533.

THIEN, L.B., BERNHARDT, P., GIBBS, G.W., PELLMYR, O., BERGSTRÖM, G., GROTH, I., and McPHERSON, G. 1985. The pollination of *Zygogynum* (Winteraceae) by a moth, *Sabatinca* (Micropterigidae): an ancient association? *Science*, **227**: 540–543.

THIEN, L.B., WHITE, D.A., and YATSU, L.Y. 1983. The reproductive biology of a relict – *Illicium floridanum* Ellis. *Amer. J. Botany* **70**: 719–727.

THIENEMANN, A. 1933. Mückenlarven bilder Gestein. *Naturu. Mus.* **63**: 370–378.

THOMAS, B.R. 1969. Kauri resins – modern and fossil. In: EGLINGTON, G., and MURPHY, M.T.J. (eds.). *Organic Geochemistry, Methods and Results*. Longman, London: 599–618.

THOMAS, B.R. 1970. Modern and fossil plant resins. In: HARBORNE, J.B. (ed.). *Phytochemical Phylogeny*. Academic Press, New York: 59–79.

THOMASSON, J.R. 1982. Fossil grass anthoecia and other plant fossils from arthropod burrows in the Miocene of Western Nebraska. *J. Paleontol.*, **56**(4): 1011–1017.

THOMPSON, D.B. 1965. The occurrence of an insect wing and branchiopods (*Euestheria*) in the Lower Keuper marl at Styal, Cheshire. *Mercian Geologist* **1**(3): 237–245.

THOMPSON, L.G., DAVIS, M.E., MOSLEY-THOMPSON, E., SOWERS, T.A., HENDERSON, K.A., ZAGORODNOV, V.S., LIN, P.-N., MIKHALENKO, V.N., CAMPEN, R.K., BOLZAN, J.F., COLE-DAI, J., and FRANCOU, B. 1998. A 25,000–year tropical climate history from Bolivian ice cores. *Science* **282**: 1858–1864.

THORNE, B.L. 1990. A case for ancestral transfere of symbionts between cockroaches and termites. *Proc. R. Soc. Lond.* B **341**: 37–41.

THORNE, B.L., and CARPENTER, J.M. 1992. Phylogeny of the Dictyoptera. *Systematic Entomol.* **17**: 253–268.

THORNTON, I.W.B., NEW, T.R., McLAREN, D.A., SUDARMAN, H.K., and VAUGHAN, P.J. 1988. Air-borne arthropod fall-out on Anak Krakatau and a possible pre-vegetation pioneer community. *Philos. Trans. R. Soc. London* B **322**(1211): 471–479.

TIDWELL, W.D., and ASH, S.R. 1990. On the Upper Jurassic stem *Hermanophyton* and its species from Colorado and Utah, USA. *Palaeontographica* B **218**: 77–92.

TIDWELL, W.D., ASH, S.R., and PARKER, L.R. 1981. Cretaceous and Tertiary floras of the San Juan Basin. In: LUCAS, S.G., RIGBY, J.K., Jr., and KUES, B.S. (eds.). *Advances in San Juan Basin Palaeontology*. Univ. New Mexico Press, Albuquerque: 307–332.

TIFFNEY, B.H. 1977. Dicotyledonous angiosperm flower from the Upper Cretaceous of Martha's Vineyard, Massachusetts. *Nature* **265**: 136–137.

TIFFNEY, B.H. 1981. Diversity and major events in the evolution of land plants. In: NIKLAS, R.J. (ed.). *Paleobotany, Paleoecology, and Evolution* **2**. Praeger, New York: 193–230.

TIFFNEY, B.H. 1984. Seed size, dispersal syndromes, and the rise of the angiosperms: evidence and hypothesis. *Ann. Missouri Botan. Gard.* **71**: 551–576.

TIFFNEY, B.H., and BARGHOORN, E.S. 1974. The fossil record of the fungi. *Occas. Pap. Farlow Herbarium* **7**: 1–42.

TIKHOMIROVA, A.L. 1991. *Perestroyka ontogeneza kak mekhanizm evolutsii nasekomykh* (Ontogeny transformation as a mechanism of the insect evolution). Nauka, Moscow: 169 p. (in Russian).

TILLEY, D.B., BARROWS, T.T., and ZIMMERMAN, E.C. 1997. Bauxitic insect pupal cases from northern Australia. *Alcheringa* **21**: 157–160.

TILLYARD, R.J. 1918. Permian and Triassic insects from New South Wales, in the collection of Mr. John Mitchell. *Proc. Linn. Soc. N.S.Wales* **42**: 720–756.

TILLYARD, R.J. 1922a. An insect wing in a crystal of selenite. *Rec. Geol. Surv. N.S.Wales* **10**: 205–207.

TILLYARD, R.J. 1922b. Mesozoic insects of Queensland. 9. Orthoptera, and additions to the Protorthoptera, Odonata, Hemiptera and Planipennia. *Proc. Linn. Soc. N.S.Wales* **47**: 447–470.

TILLYARD, R.J. 1924. Upper Permian Coleoptera and a new order from the Belmont Beds, New South Wales. *Proc. Linn. Soc. N.S.Wales* **49**: 429–435.

TILLYARD, R.J. 1928. Some remarks on the Devonian fossil insects from the Rhynie Chert Beds, Old Red Sandstone. *Trans. Entomol. Soc. London* **76**: 65–71.

TILLYARD, R.J. 1930. The evolution of the class Insecta. *Pap. & Proc. R. Soc. Tasmania* 1930: 1–89.

TILLYARD, R.J. 1931. Kansas Permian insects. Pt. 13. The new order Protelytroptera, with a discussion of its relationships. *Amer. J. Sci.* **21**(5): 232–266.

TILLYARD, R.J. 1935a. Upper Permian Insects of the New South Wales. III. The order Copeognatha. *Proc. Linn. Soc. N.S.Wales* **60**: 265–279.

TILLYARD, R.J. 1935b. Upper Permian insects of New South Wales. Pt. 5. The order Perlaria or stone-flies. *Proc. Linn. Soc. N.S.Wales* **60**: 385–391.

TING, W.S., and NISSENBAUM, A. 1986. Fungi in Lower Cretaceous amber from Israel. *Spec. Publ. Exploration and Development Res. Center*, Chinese Petroleum Group Mioli (Taiwan): 1–27.

TOBIAS, V.I. 1987. New braconid taxa (Hymenoptera, Braconidae) from the Baltic amber. *Entomol. Obozrenie* 66(4): 845–959 (in Russian, English translation: *Entomol. Review* 67(4): 18–33).

TOBIEN, H. 1965. Insekten-Fraßspuren an Tertiären und Pleistozänen Säugtier-Knochen. *Senckenberg. Lethaea* 46: 441–451.

TOBIEN, H. 1983. Bemerkungen zur Taphonomie der spättertiären Säugtierfauna aus den Dinotheriensanden Rheinhessens (BR Deutschland). *Erwin Rutte-Festschrift*. Weltenburger Akad., Kelheim/Weltenburg: 191–200.

TOMLINSON, A.I. 1973. Meteorological aspects of trans-Tasman insect dispersion. *New Zeal. Entomol.* 5: 253–268.

TOOTS, H. 1975. Distribution of meniscate burrows in non-marine Tertiary sediments of western U.S. *Univ. Wyaoming Contr. Geol.* 14: 9–10.

TOSHIYA, I., NOBUO, E., AKIOMI, Y., KATSUO, O., and TAKAAKI, T. 1980. Attractants for the Japanese pine sawyer, *Monochamus alternatus* Hope (Coleoptera: Cerambycidae). *Appl. Entomol. and Zool.* 15(3): 258–260.

TOWNSEND, C.H.T. 1916. *Lithohypoderma*, a new fossil genus of Oestridae. *Insect. Inscit.* 4: 128–130.

TOWNSEND, J.I. 1994. Carabidae of the Manawatu-Horowhenua lowlands as revealed by collections from coastal flood debris. *New Zeal. Entomol.* 17: 7–13.

TRAUB, R. 1985. Co-evolution of fleas and mammals. In: KIM, K.Ch. (ed.). *Co-evolution of Parasitic Arthropods and Mammals*. John Wiley & Sons, New York, etc.: 295–437.

TREWIN, N.H. 1995. A draft system for the identification and description of arthropod trackways. *Palaeontology* 37(4): 811–823.

TROITSKY, A.V., MELEKHOVETS, Yu.F., RAKHIMOVA, G.M., BOBROVA,V.K., VALIEJO-ROMAN, K.M., and ANTONOV, A.S. 1991. Angiosperms origin and early stages of seed plants evolution deduced from rRNA sequence comparisons. *J. Molec. Evol.* 32: 253–261.

TRÖSTER, G. 1993. Wasserkäfer und andere Raritäten – Neue Coleoptera-Funde aus dem mitteleozänen Tonstein der Grube Messel bei Darmstadt. *Kaupia* 2: 145–154.

TRUC, G. 1975. Sol á profile calcaire différencié et pellicules rubanées dans le Paléogène du Sud-Est de la France. *Colloque "Types de croûtes calcaires et leur répartition régionale", Strasbourg, 1975*: 101–113.

TRUEMAN, J.W.H. 1990. Comment – Evolution of insect wing: a limb exite plus endite model. *Canad. J. Zool.* 68: 1333–1355.

TSCHIRCH, A., and STOCK, E. 1936. *Die Harze. Die botanischen und chemischen Grundlagen unserer Kenntnisse über die Bildung, die Entwicklung und die Zusammensetzung der pflanzlichen Exkrete.* Bd. 2. 2. Hälfte, 2. Teil. Borntraeger, Berlin: i-xv + 1017–1858.

TSCHUDY, R.H., PILLMORE, C.L., ORTH, C.J., GILMORE, J.S., AND KNIGHT, J.D. 1984. Disruption of the terrestrial plant ecosystem at the Cretaceous/Tertiary boundary, Western Interior. *Science* 225: 1030–1032.

TSEKHOVSKY, Y.G. 1974. Activity traces of animals and plants in paleosols of a Late Cretaceous-Palaeocenous savannah of Kazakhstan. In: *Tafonomiya, eyo ekologicheskie osnovy. Sledy zhizni i ikh interpretatsiya.* (Taphonomy, its Ecological Background. Life Traces and their Interpretation). *Tezisy dokladov XX Sessii Vsesoyuznogo paleontol. obshchestva, 4–9 fevr. 1974 g. Leningrad*: 46–47 (in Russian).

TSHERNOVA, O.A. 1962. A mayfly nymph from the Neogene of West Siberia (Ephemeroptera, Heptageniidae). *Zool. Zhurnal* 41(6): 944–945 (in Russian).

TSHERNOVA, O.A. 1965. Some fossil mayflies (Ephemeroptera, Misthodotidae) from Permian beds of the Ural. *Entomol. Obozrenie* 44(2): 353–361 (in Russian, English translation: *Entomol. Review* 44(1): 202–207).

TSHERNOVA, O.A. 1968. A new mayfly from Karatau (Ephemeroptera). In: ROHDENDORF, B.B. (ed.). *Yurskie nasekomye Karatau* (Jurassic Insects of Karatau). Nauka, Moscow: 23–25 (in Russian).

TSHERNOVA, O.A. 1971. A mayfly from the fossil resin of the Cretaceous deposits of Polar Siberia (Ephemeroptera, Leptophlebiidae). *Entomol. Obozrenie* 50(3): 612–618 (in Russian, English translation: *Entomol. Review* 50(3): 346–349).

TSHERNOVA, O.A. 1977. Distinctive new mayfly nymphs (Ephemeroptera: Palingeniidae, Behningiidae) from the Jurassic of Transbaikalia. *Paleontol. Zhurnal* (2): 91–96 (in Russian, English translation: *Paleontol. J.* 11(2): 221–226).

TSHERNOVA, O.A. 1980. The order Ephemerida. Mayflies. In: ROHDENDORF, B.B., and RASNITSYN, A.P. (eds.). *Istoricheskoe razvitie klassa nasekomykh* (Historical Development of the Class Insecta). Trudy Paleontol. Inst. Akad. Nauk SSSR 175. Nauka, Moscow: 31–36 (in Russian).

TSHERNYSHEV, W.B. 1990. Life history of the ancestors of Insecta: A hypothesis. *Dev. Ecol. Perspect. 21st Cent. 5th Internat. Congr. Ecol., Yokohama, Aug. 23–30, 1990: Abstr.* Yokohama: 238.

TSHERNYSHEV, W.B. 1997. The origin of insects and their early evolution from the ecological point of view. *Zhurnal obshchey biol.* 58(3): 5–16.

TSUTSUMI, C. 1974. Mizunami amber and fossil insects: Psocoptera: Lachesillidae. *Bull. Mizunami Fossil Mus.* 1: 415–416.

TURNOCK, W.J., and TURNOCK, R.W. 1979. Aggregations of lady beetles (Coleoptera: Coccinellidae) on the shores of Lake Manitoba. *Manitoba Entomol.* 13: 21–22.

TUROV, S.S. 1950. *Zhizn' ptits* (Life of Birds). Moskovskoe Obshchestvo Ispytateley Prirody, Moscow: 147 p. (in Russian).

TVERDOKHLEBOVA, G.I. 1984. Coprolites of Late Permian tetrapods as possible palaeoenvironmental indicators. *Sledy zhizni i dinamika sredy v drevnikh biotopakh. Tezisy dokladov XX sessii Vsesoyuznogo Paleontol. Obshchestva, Jan. 23–27, 1984, L'vov.* (The Traces of Life and Dynamics of Environments in Ancient Biotopes). L'vov: 69–70 (in Russian).

TYLER, P.A., and BUCKNEY, R.T. 1980. Ferromanganese concretions in Tasmanian lakes. *Austral. J. Mar. Freshwater Res.* 31(4): 525–531.

UEDA, K. A. 1991. Triassic fossil of scorpion fly from Mine, Japan. *Bull. Kitakyushu Mus. Nat. Hist.* N. 10: 99–103.

UEDO, F., IWAI, J., OZAKI, H., and OHAO, M. 1960. On the general geology of Iwagado and adjacent regions, with a note on a Tertiary species of aquatic leaf-beetles. *Bull. Nation. Sci. Mus.* (Tokyo) 5(2): 95–99.

UÉNO, M. 1984. Biogeography of Lake Biwa. In: HORIE, S. (ed.). *Lake Biwa*. W. Junk Publ., Dordrecht – Boston – Lancaster: 625–633.

UHMANN, E. 1939. Hispinen aus baltischem Bernstein. *Bernstein-Forsch.* 4: 18–22.

ULLRICH, W.G. 1972. Untersuchungen über die Käferfauna eines frühgeschichtlichen Bohlweges aus dem Wittmoor bei Duvenstedt. *Faun.-Ökol. Mitteil.* 4(4): 119–126.

ULMER, G. 1912. Die Trichoptera des Baltischen Bernsteins. *Beitr. Naturwiss.Preuss. Schrift. phys.-ökon. Ges. Königsberg* 10: 1–380.

UPCHURCH, G.R., Jr., CRANE, P.R., and DRINNAN, A.N. 1994. The megaflora from the Quantico locality (Upper Albian), Lower Cretaceous Potomac Group of Virginia. *Virginia Mus. Nat. Hist. Bull.* 4: 1–57.

UPCHURCH, G.R., Jr., and WOLFE, J.A. 1987. Mid-Cretaceous to Early Tertiary vegetation and climate: evidence from fossil leaves and woods. In: FRIIS, E.M., CHALONER, W.G., and CRANE, P.R. (eds.). *The Origins of Angiosperms and Their Biological Consequences*. Cambridge Univ. Press, Cambridge: 75–105.

USINGER, R.L. 1958. "Harzwanzen" or "resin bugs" in Thailand. *Pan-Pacif. Entomol.* 34: 52.

UUTALA, A.J. 1990. *Chaoborus* (Diptera: Chaoboridae) mandibles – paleolimnological indicators of the historical status of fish populations in acid-sensitive lakes. *J. Paleolimnol.* 4: 139–151.

VAKHRAMEEV, V.A., and KOTOVA, I.Z. 1977. Ancient angiosperms and associated plants from the Lower Cretaceous deposits of Transbaikalia. *Paleontol. Zhurnal* (4): 101–109 (in Russian).

VALIAKHMEDOV, B. 1977. Pupal chambers of soil invertebrates in serozems of Tadzhikistan and their influence to the formation of the soil profile. *Pochvovedenie* (4): 85–91 (in Russian).

VAN AMEROM, H.W.J. 1966. *Phagophytichnus ekowskii* nov. ichnogen. & ichnosp. nov., eine Missbildung infolge von Insektenfrass, aus dem spanischen Stephanien (Provinz León). *Leidse Geol. Meded.* 38: 181–184.

VAN AMEROM, H.W.J. 1973. Gibt es Cecidien im Karbon bei Calamiten und Asterophylliten? *C.R. 7ème Congr. Internat. Stratigr. et Géol. Carbon., Krefeld, 1971* 2: 63–83.

VAN AMEROM, H.W.J., and BOERSMA, M. 1971. A new find of the ichnofossil *Phagophytichnus ekowskii* Van Amerom. *Geol. en mijnbouw* 50(5): 667–669.

VANCE, T.C. 1974. Larvae of Sericothripini with reference to other larvae of the Terebrantia of Illinois (USA). *Illinois Natur. Hist. Surv. Bull.* 31: 145–208.

VAN DER PIJL, L. 1953. On the flower biology of some plants from Java with general remarks on fly-traps (species of *Annona, Artocarpus, Typhonium, Gnetum, Arisaema* and *Abroma*). *Ann. Bogor.* 1: 77–99.

VAN DEVENDER, T.R., and HALL, W.E. 1994. Holocene arthropods from the Sierra Bacha, Sonora, Mexico, with emphasis on beetles (Coleoptera). *Coleopterists Bull.* 48(1): 30–50.

VAN KONIJNENBURG-VAN CITTERT, J.H.A., and SCHMEIßNER, S. 1999. Fossil insect eggs on Lower Jurassic plant remains from Bavaria (Germany). *Palaeogeogr., Palaeoclimatol., Palaeoecol.* 152: 215–223.

VASECHKO, G.I. 1978. Host selection by some bark beetles. 1. Studies of primary attraction with chemical stimuli. *Zeitschr. angew. Entomol.* 85(1): 66–67.

VASECHKO, G.I., and KUZNETSOV, N.V. 1969. The toxicity of terpenoids for spruce barkbeetles. *Khimiya v Sel'skom Khozyaystve* (Chemistry in Agriculture) 7(12): 33–34 (in Russian).

VÁVRA, N. 1993. Chemical characterization of fossil resins ("amber"). A critical review of methods, problems and possibilities: determination of mineral species, botanical sources and geographical attribution. *Abhandl. Geol.* B 49: 147–157.

VÁVRA, N., and VYCUDILIK, W. 1976. Chemische Untersuchungen an fossilen und subfossilen Harzen. *Beitr. Paläontol. Österr.* 1: 121–135.

VELICHKO, A.A. 1973. *Prirodnyi protsess v pleystotsene* (Natural Process in the Pleistocene). Nauka, Moscow: 256 p. (in Russian).

VENABLES, E.M., and TAYLOR, H.E. 1963. An insect fauna of the London Clay. *Proc. Geol. Assoc.* 73(3): 273–279.

VERESHCHAGIN, N.K., and BARYSHNIKOV, G.F. 1992. The ecological structure of the "Mammoth Fauna" in Eurasia. *Ann. Zool Fenn.* 28(Bjürn Kurten Memorial Vol.): 253–259.

VERHOEFF, K. 1903. Endsegmente des Körpers der Chilopoden, Dermapteren und Japygiden und zur Systematik von *Japyx*. *Abhandl. Kaiserl. Leop.-Carol. Deutsch. Akad. Naturforsch.* 81: 258–297.

VERKHOVSKAYA, N.B. 1986. The environments of the mammoth fauna in the North-East of Siberia (after palaeobotanical data). *Biogeografiya Beringiyskogo sektora Subarktiki. Materialy 10 Vsesoyuznogo Simpoz., Magadan, 1983.* (Biogeography of the Beringen Sector of Subarctics. Materials of the 10th All-Union Simposium, Magadan, 1983). Vladivostok: 194–203 (in Russian).

VERMEIJ, G.J. 1983. Intimate associations and co-evolution in the sea. In: FUTUYMA, D.J., and SLATKIN, M. (eds.). *Co-evolution*. Sinauer Associates Inc., Sunderland, MA: 311–327.

VERMEIJ, G.J. 1987. *Evolution and Escalation*. Princeton Univ. Press, New Haven: 528 p.

VERVOENEN, M. 1991. Pleistocene vleesvliegen-puparia uit hoornpitten van *Bison priscus*. *Cranium* 9(2): 57–58.

VÍA, L. and CALZADA, S. 1987. Artropodos fosiles de Alcover-Montral. 1. Insectos. *Cuadernos Geol. Ibérica* 11: 273–280.

VIALOV, O.S. 1966. *Sledy zhiznedeyatel'nosti organizmov i ikh paleontologicheskoe znachenie* (The Organic Activity Traces and Their Palaeontological Significance). Naukova Dumka, Kiev: 217 p. (in Russian).

VIALOV, O.S. 1968. Materials to classification of trace fossils and traces of activity of organisms. *Paleontol. Sbornik* **5**(1): 125–129 (in Russian).

VIALOV, O.S. 1972. On the worm tubes *Pectinariopsis* Andrusov. *Paleontol. Sbornik* **10**(1): 42–52 (in Russian).

VIALOV, O.S. 1973. Classification of fossil caddis cases. *Dopovidi Akad. Nauk URSR* Ser. B. (7): 585–588 (in Ukrainian).

VIALOV, O.S. 1975. The fossil traces of insect feeding. *Paleontol. Sbornik* **12**(1–2): 147–155 (in Russian).

VIALOV, O.S., and SUKATSHEVA, I.D. 1976. The fossil larval cases of caddisflies (Insecta, Trichoptera) and their stratigraphic significance. In: *Paleontologia i biostratigrafiya Mongolii* (Palaeontology and Biostratigraphy of Mongolia). Trudy Sovmestnoy Sovetsko-Mongol'skoy Paleontol. Expeditsii **3**. Nauka, Moscow: 169–230 (in Russian).

VIEHMEYER, H. 1913. Ameisen aus dem Kopal von Celebes. *Stett. entomol. Zeit.* **74**: 141–155.

VIETTE, P. 1979. Reflexions sur la classifications en sous-ordres de l'ordre des Lepidoptera (Insecta). *Bull. Soc. Entomol. France* **84**: 68–78.

VIGNON, M.-P. 1929. Introduction à de nouvelles recherches de morphologie comparée sur l'aile des insectes. *Arch. Mus.Hist. Natur.* (Paris) Sér. 6. **4**: 89–125.

VILESOV, A.P. 1995. Permian neuropterans (Insecta: Myrmeleontida) from the Chekarda locality in the Urals. *Paleontol. Zhurnal* (2): 95–105 (in Russian, English translation: *Paleontol. J.* **29**(2): 115–124).

VILESOV, A.P., and NOVOKSHONOV, V.G. 1994. New fossil insects (Myrmeleontida, Jurinida) from the Upper Permian in the Eastern Kazakhstan. *Paleontol. Zhurnal* (2): 66–74 (in Russian, English translation: *Paleontol. J.* **28**(2): 81–92).

VILHELMSEN, L. 2001. Phylogeny and classification of the extant basal lineages of the Hymeniotera (Insecta). *Zool. J. Linn. Soc.* **131**: 393–442.

VINOGRADOV, B., and STAL'MAKOVA, V. 1938. On accumulation of animal remains in oil pools. *Priroda* (5): 104–106 (in Russian).

VISHNIAKOVA, V.N. 1968. Mesozoic roaches with the external ovipositor and peculiarity of their reproduction (Blattodea). In: ROHDENDORF, B.B. (ed.). *Yurskie nasekomye Karatau* (Jurassic insects of Karatau). Nauka, Moscow: 55–86 (in Russian).

VISHNIAKOVA, V.N. 1971. Structure of the abdominal appendages of the Mesozoic roaches (Insecta: Blattodea). In: OBRUCHEV, D.V., and SHIMANSKY, V.N. (eds.). *Sovremennye problemy paleontologii* (Current Problems in Palaeontology). Trudy Paleontol. Inst. Akad. Nauk SSSR **130**. Nauka, Moscow: 174–186 (in Russian).

VISHNIAKOVA, V.N. 1973. New roaches (Insecta: Blattodea) from the Upper Jurassic deposits of Karatau Range. In: NARCHUK, E.P. (ed.). *Voprosy paleontologii nasekomykh. Doklady na 24–m Ezhegodnom chtenii pamyati N.A.Kholodkovskogo, 1971* Problems of the Insect Palaeontology Lectures on the XXIV Annual Readings in Memory of N.A. Kholodkovsky (1–2 April, 1971)]. Nauka, Leningrad: 64–77 (in Russian).

VISHNIAKOVA, V.N. 1975. Psocoptera from the fossil Late Cretaceous insect-bearing resins. *Entomol. Obozrenie* **54**: 92–106 (in Russian).

VISHNIAKOVA, V.N. 1976. On the relic Psocoptera (Insecta) of the Mesozoic. *Paleontol. Zhurnal* (2): 76–84 (in Russian, English translation: *Paleontol. J.* **10**(2): 180–188).

VISHNIAKOVA, V.N. 1980a. Order Psocida Leach, 1815. In: ROHDENDORF, B.B., and RASNITSYN, A.P. (eds.). 1980. *Istoricheskoe razvitie klassa nasekomykh* (Historical Development of the Class Insecta). Trudy Paleontol. Inst. Akad. Nauk SSSR **175**. Nauka, Moscow: 53–57 (in Russian).

VISHNIAKOVA, V.N. 1980b. Order Protelytrida Tillyard, 1931. Order Forficulida Latreille, 1810. In: ROHDENDORF, B.B., and RASNITSYN, A.P. (eds.). 1980. *Istoricheskoe razvitie klassa nasekomykh* (Historical Development of the Class Insecta). Trudy Paleontol. Inst. Akad. Nauk SSSR **175**. Nauka, Moscow: 160–164 (in Russian).

VISHNIAKOVA, V.N. 1980c. The earwigs from Upper Jurassic deposits of Karatau Range (Insecta, Forficulida). *Paleontol. Zhurnal* (1): 78–94 (in Russian).

VISHNIAKOVA, V.N. 1981. New Palaeozoic and Mesozoic lophioneurids (Thripida, Lophioneuridae). In: VISHNIAKOVA, V.N., DLUSSKY, G.M., and PRITYKINA, L.N. *Novye iskopaenye nasekomye s territorii SSSR* (New Fossil Insects from the Territory of USSR). Trudy Paleontol. Inst. Akad. Nauk SSSR **183**. Nauka, Moscow: 43–63 (in Russian).

VISHNIAKOVA, V.N. 1982. Jurassic Blattodea of the new family Blattulidae in Siberia. *Paleontol. Zhurnal* (2): 69–79 (in Russian, English translation: *Paleontol. J.* **16**(2): 67–78).

VISHNIAKOVA, V.N. 1983. Jurassic Blattodea of the family Mesoblattinidae in Siberia. *Paleontol. Zhurnal* (1): 79–93 (in Russian, English translation: *Paleontol. J.* **17**(1): 76–89).

VISHNIAKOVA, V.N. 1985a. Roaches (Blattida = Blattodea) in the Jurassic of South Siberia and West Mongolia. In: RASNITSYN, A.P. (ed.). *Yurskie nasekomye Sibiri i Mongolii* (Jurassic Insects of Siberia and Mongolia). Trudy Paleontol. Inst. Akad. Nauk SSSR **211**. Nauka, Moscow: 138–146 (in Russian).

VISHNIAKOVA, V.N. 1985b. A new earwig from the Jurassic of Siberia. In: RASNITSYN, A.P. (ed.). *Yurskie nasekomye Sibiri i Mongolii* (Jurassic Insects of Siberia and Mongolia). Trudy Paleontol. Inst. Akad. Nauk SSSR **211**. Nauka, Moscow: 146–147 (in Russian).

VISHNIAKOVA, V.N. 1986a. Roaches. Blattida (=Blattodea). In: RASNITSYN, A.P. (ed.). *Nasekomye v rannemelovykh ekosistemakh Zapadnoy Mongolii* (Insects in the Early Cretaceous Ecosystems of the West Mongolia). Trudy Sovmestnoy Sovetsko-Mongol'skoy Paleontol. Expeditsii **28**. Nauka, Moscow: 166–169 (in Russian).

VISHNIAKOVA, V.N. 1986b. Forficulida (=Dermaptera). In: RASNITSYN, A.P. (ed.). *Nasekomye v rannemelovykh ekosistemakh Zapadnoy Mongolii* (Insects in the Early Cretaceous ecosystems of West Mongolia). Trudy Sovmestnoy Sovetsko-Mongol'skoy Paleontol. Expeditsii **28**. Nauka, Moscow: 171 (in Russian).

VISHNIAKOVA, V.N. 1993. New Paleozoic Spiloblattinidae from Russia. *Paleontol. J.* **27**(1A): 135–147.

VISHNU-MITTRE. 1957. Fossil galls on some Jurassic conifer leaves. *Current Sci.* **26**(7): 210–211.

VISHNU-MITTRE. 1959. Studies on fossil flora of Nipania (Rajmahal Series), Bihar. Coniferales. *Palaeobotanist* **6**(2): 82–111.

VITOUSEK, P.M. 1990. Biological invasions and ecosystem processes: towards an integration of population biology and ecosystem studies. *Oikos* **57**(1): 7–13.

VITOUSEK, P.M., WALKER, L.R., WHITEAKER, L.D., MUELLER-DOMBOIS, D., and MATSON, P. 1987. Biological invasion by *Myrica faya* alters ecosystem development in Hawaii. *Science* **238**: 802–804.

VOIGT, E. 1937. Der Erhaltungszustand der tierischen Einschlüsse im Bernstein. *Chem. Zeit.* **83**: 122–126.

VOIGT, E. 1938. Ein fossiler Seitenwurm (*Gordius tenuifibrosus* n.sp.) aus der eozänen Braunkohle des Geiseltales. *Nova Acta Leopold.* n.F., **5**(31): 351–360.

VOIGT, E. 1952. Ein Haareinschlüss mit Phthirapteren Eiern in Bernstein. *Mitt. geol. St.-Inst. Hamburg* **21**: 59–74.

VOIGT, E. 1957. Ein parasitischer Nematode in fossiler Coleopteren-Muskulatur aus der eozänen Braunkohle des Geiseltales bei Halle (Saale). *Paläontol. Zeitschr.* **31**(1–2): 35–39.

VORONOVA, M.A. 1984. Some Albian arthropods of Ukraine and their role in the origin of entomophily. *Sledy zhizni i dinamika sredy v drevnikh biotopakh. Tezisy dokladov XX sessii Vsesoyuznogo Paleontol. Obshchestva, Yan. 23–27, 1984, L'vov.* (The Traces of Life and Dynamics of Environments in Ancient Biotopes. Abstracts of the XX Session of the All-Union Palaeontol. Soc. Jan. 23–28, 1984 , L'vov). L'vov: 11–12 (in Russian).

VORONOVA, M.A. 1985. Evidence of oldest entomophily process in the Albian. *Doklady Akad. Nauk Ukrain. SSR* **5**(5): 11–15 (in Russian).

VORONOVA, M.A., and VORONOVA, N.N. 1982. Contribution to the problem of entomophily development in Early Cretaceous plants. In: *Sistematika i evolyutsiya drevnikh rasteniy Ukrainy* (Systematics and Evolution of Ancient Plants of Ukraine). Naukova Dumka, Kiev: 107–111 (in Russian).

VOVIDES, A.P. 1991. Insect symbionts of some Mexican cycads in their natural habitat. *Biotropica* **23**(1): 102–104.

VRBA, E.S. 1980. The Sterkfontein Valley australopithecine succession. *Palaeontol. afr.* **23**: 61–68.

VRŠANSKÝ, P. 1997. *Piniblattella* gen. nov. – the most ancient genus of the family Blattellidae (Blattodea) from the Lower Cretaceous of Siberia. *Entomol. Problems* **28**(1): 67–79.

VRŠANSKÝ, P.1998. Lower Cretaceous Blattodea. *First Paleoentomological Conference. 30 Aug.–4 Sept. 1998. Moscow, Russia. Abstracts.* Palaeontological Inst., Russian Acad. Sci., Moscow: 44.

VRŠANSKÝ, P. 1999a. Two new species of Blattaria (Insecta) from the Lower Cretaceous of Asia, with comments on the origin and phylogenetic position of the families Polyphagidae and Blattulidae. *Entomol. Problems* **31**(2): 85–91

VRŠANSKÝ, P. 1999b. Lower Cretaceous Blattaria. *Proc. First Palaeoentomol. Conf., Moscow 1998.* AMBA projects Internat., Bratislava: 167–176.

VRŠANSKÝ, P. 2000. Decreasing variability – from Carboniferous to the present! (validated on independent lineages of Blattaria). *Paleontol. J.* **34**(suppl. 3): S374–S379.

VRŠANSKÝ, P. and ANSORGE, J. 2001. New Lower Cretaceous polyphagid cockroaches from Spain (Blattaria, Polyphagidae, Vitisminae subfam. nov.). *Cretaceous Research* **22**: 157–162.

VRŠANSKÝ, P., QUICKE, D.L.J., RASNITSYN, A., BASIBUYUK, H., FITTON, M., ROSS, A., and VIDLIČKA, L. 2001. The oldest insect sensillae. In: *AMBA*/B/21.01.3/ABS/D: 8 pp.

VRŠANSKÝ, P., RASNITSYN, A.P., GRIMALDI, D. and BASIBUYUK, H. (in press). Umenocoleidae – an amazing lineage of aberrant insects (Insecta, Blattaria). In: VRŠANSKÝ, P. (ed.) *Contribution to the Living and Fossil Blattaria (Arthropoda, Insecta).* AMBA/AM/CLBF2000/1.00.

WAAGE, J.K. 1976. Insect remains from ground sloth dung. *J. Paleontol.* **50**(5): 991–994.

WAGENSBERG, J., and BRANDÃO, C.R.F. 1998. Criteria for a taphonomy of amber inclusions. *World Congress on Amber Inclusions, 20th–23th Oct. 1998, Vitoria-Gasteiz, Alava, Basque Country* [Abstracts & Program]: 105.

WAGNER, R., HOFFEINS, C., and HOFFEINS, H.W. 2000. A fossil nymphomyiid (Diptera) from the Baltic and Bitterfeld amber. *Syst. Entomol.* **25**: 115–120.

WAGNER, T., NIENHUIS, C., and BARTHLOTT, W. 1996. Wettability and contaminability of insect wings as a function of their surface sculptures. *Acta Zool.* (Stockholm) **77**: 213–225.

WALDEN, K.K., and ROBERTSON, H.M. 1997. Ancient DNA from amber fossil bees? *Molec. Biol. Evol.* **14**: 1075–1077.

WALKER, I.R. 1987. Chironomidae (Diptera) in paleoecology. *Quaternary Sci. Reviews* **6**: 29–40.

WALKER, I.R. 1995. Chironomids as indicators of past environmental changes. In: ARMITAGE, P.D., CRANSTON, P.S., and PINDER, L.C.V. (eds.). *The Chironomidae: Biology and Ecology of Non-Biting Midges.* Chapman & Hall, London: 405–422.

WALKER, I.R., and MATHEWES, R.W. 1987. Chironomidae (Diptera) and postglacial climate at Marion Lake, British Columbia, Canada. *Quatern. Res.* **27**: 89–102.

WALKER, I.R., and MATHEWES, R.W. 1988. Late Quaternary fossil Chironomidae (Diptera) from Hippa Lake, Queen Charlotte Islands, British Columbia, with special reference to *Corynocera* Zett. *Canad. Entomol.* **120**: 739–751.

WALKER, I.R., and MATHEWES, R.W. 1989a. Early postglacial chironomid succession in Southwestern British Columbia, Canada, and its paleoenvironmental significance. *J. Paleolimnol.* **2**: 1–14.

WALKER, I.R., and MATHEWES, R.W. 1989b. Much ado about dead Diptera. *J. Paleolimnol.* **2**: 19–22.

WALKER, I.R., and MATHEWES, R.W. 1989c. Chironomidae (Diptera) remains in surficial lake sediments from the Canadian Cordillera: analysis of the fauna across an altitudinal gradient. *J. Paleolimnol.* **2**: 61–80.

WALKER, I.R., MOTT, R.J., and SMOL, J.P. 1991a. Alleröd – Younger Dryas lake temperatures from midge fossils in Atlantic Canada. *Science* **253**: 1010–1012.

WALKER, I.R., and PATERSON, C.G. 1983. Post-glacial chironomid succession in two small, humic lakes in the New Brunswick – Nova Scotia (Canada) border area. *Freshw. Invert. Biol.* **2**: 61–73.

WALKER, I.R., and PATERSON, C.G. 1985. Efficient separation of subfossil Chironomidae from lake sediments. *Hydrobiologia* **122**(2): 189–192.

WALKER, I.R., SMOL, J.P., ENGSTROM, D.R., and BIRKS, H.J.B. 1991b. An assessement of Chironomidae as quantitative indicators of past climatic change. *Canad. J. Fish. Aquat. Sci.* **48**: 975–987.

WALKER, L.R., and VITOUSEK, P.M. 1991. An invader alters germination and growth of a native dominant tree in Hawai'i. *Ecology* **72**(4): 1449–1455.

WALKER, M.V. 1938. Evidence of Triassic insects in the Petrified Forest National Monument, Arizona. *Proc. U.S Nation. Mus.* **85**(art. 3033): 137–141.

WALTER, H. 1983. Zur Taxonomie, Ökologie und Biostratigraphie der Ichnia limnisch-terrestrischer Arthropoden des mitteleuropäischen Jungpaläozoikums. *Freiberger Forschungsheft* C. **382**: 146–193.

WALTER, H. 1984. Zur Ichnologie der Arthropoda. *Freiberger Forschungsheft* C. **391**: 58–94.

WALTER, H. 1985. Zur Ichnologie des Pleistozäns von Liebegast. *Freiberger Forschungsheft* C. **400**: 101–116.

WANG, S. 1990. Origin, evolution and mechanism of the Jehol fauna. *Acta Geol. Sinica* **64**(4): 350–360 (in Chinese with English summary).

WANG, X.S., POINAR, H.N., POINAR, G.O., Jr., and BADA, J.L. 1995. Amino acids in the amber matrix and in entombed insects. In: ANDERSON, K.B., and CRELLING, J.C. (eds.). *Amber, Resinite and Fossil Resins.* ACS Symposium Series, Washington: 256–262.

WARD, W.A. 1978. *Studies on Scarab Seals.* Vol. I. *Pre-12th Dynasty Scarab Amulets.* Aris & Phillips Ltd., Warminster, Wilts.: 116 p.

WARNES, C.E., and RANDLES C.I. 1980. Succession in a microbial community associated with chitin in Lake Erie sediment and water. *Ohio J. Sci.* **80**(6): 250–255.

WARNER, R.E., and SMITH, C.E., Jr. 1968. Boll weevil found in pre-Columbian cotton from Mexico. *Science* **162**: 911–912.

WARWICK W.F. 1980a. Palaeolimnology of the Bay of Quinte, Lake Ontario: 2800 years of cultural influence. *Canad. Bull. Fish. Aquat. Sci.* **206**: 1–118.

WARWICK, W.F. 1980b. Chironomidae (Diptera) responses to 2800 years of cultural influence: a palaeolimnological study with special reference to sedimentation, eutrophication, and contamination processes. *Canad. Entomol.* **112**: 1193–1230.

WASHBURN, J.O., and WASHBURN, L. 1984. Active aerial dispersal of minute wingless arthropods: exploitation of boundary-layer velocity gradients. *Science* **223**(4640): 1088–1089.

WASMANN, E. 1929. Die Paussiden des baltischen Bernsteins und die Stammesgeschichte der Paussiden. *Bernstein-Forsch.* **1**: 1–110.

WASMANN, E. 1932. Eine ameisenmordende Gastwanze (*Proptilocerus dolosus* n.g., n.sp.) im baltischen Bernstein. *Bernstein-Forsch.* **3**: 1–3.

WASMUND, E. 1926. Biocoenose und Thanatocoenose. Biosoziologische Studie über Lebensgemeinschaften und Totengesellschaften. *Arch. Hydrobiol.* **17**: 1–116.

WATERS, S.B. 1989a. A Cretaceous dance fly (Diptera: Empididae) from Botswana. *Syst. Entomol.* **14**: 233–241.

WATERS, S.B. 1989b. A new hybotine dipteran from the Cretaceous of Botswana. *Palaeontology* **32**: 657–667.

WATERS, S.B. and ARILLO, A. 1999. A new genus of Hybotidae (Diptera, Empidoidea) from Lower Cretaceous amber of Alava (Spain). *Studia dipterol.* **6**(1): 59–66.

WATERS, T.F. 1965. Interpretation of invertebrate drift in streams. *Ecology* **46**: 327–334.

WATSON, J.P. 1967. A termite mound in an Iron age burial ground in Rhodesia. *J. Ecol.* **55**: 663–669.

WATSON, J.A.L., and GAY, F.J. 1991. Isoptera. In: NAUMANN, D. (ed.). *The Insects of Australia.* 2nd ed. Melbourne Univ. Press, Carlton: 330–347.

WATSON, J.A.L., and SMITH, J.B. 1991. Archaeognatha. In: NAUMANN, D. (ed.). *The Insects of Australia.* 2nd ed. Melbourne Univ. Press, Carlton: 272–274.

WCISLO-LURANIEC, E. 1985. New details of leaf structure in *Bilsdalea dura* Harris (Coniferae) from the Jurassic of Kraków, Poland. *Acta palaeobot.* (Kraków) **25**(1–2): 13–20.

WEAVER, J.S., III. 1984. The evolution and classification of Trichoptera, Pt.I: The groundplan of Trichoptera. *Proc. 4th Internat. Symp. Trichoptera* Clemson: 413–419.

WEAVER, J.S., III, and MORSE, J.C. 1986. Evolution of feeding and case-making behaviour in Trichoptera. *J. North Amer. Benthol. Soc.* **5**: 150–158.

WEAVER, L., McLAUGHLIN, S., and DRINNAN, A.N. 1997. Fossil woods from the Upper Permian Bainmedart Coal Measures, Northern Prince Charles Mountains, East Antarctica. *J. Austral. Geol. Geophys.* **16**(5): 655–676.

WEBER, H. 1933 *Lehrbuch der Entomologie.* Gustav Fischer Verlag, Jena: 726 p.

WEBB, P.-N., and HARWOOD, D.M. 1993. Pliocene fossil *Nothofagus* (southern beech) from Antarctica: phytogeography, dispersal startegies, and survival in high latitude glacial-deglacial environments. In: ALDEN, J. *et al.* (eds.). *Forest Development in Cold Climates.* Plenum Press, New York: 135–165.

WEBBY, B.D. 1970. *Brookvalichnus*, a new trace fossil from the Triassic of the Sydney Basin, Australia. In: CRIMES, T.P., and HARPER, J.C. (eds.). *Trace Fossils.* Geol. J. Spec. Issue **3**, Seel House Press, Liverpool: 527–530.

WEDMANN, S. 1998. Insects from the Oligocene deposits of Enspel (Germany). *First Paleoentomol. Conf. 30 Aug.–4 Sept. 1998, Moscow, Russia.* Abstracts: 46.

WEDMANN, S., and HÖRNSCHEMEYER, T. 1994. Fossile Prachtkäfer (Coleoptera: Buprestidae: Buprestinae und Agrilinae) aus dem Mitteleozän der Grube Messel bei Darmstadt, Teil 2. *Willi Ziegler-Festschrift III.* Courier Forsch.-Inst. Senckenberg **170**: 137–187.

WEGIEREK, P., and ZHERIKHIN, V.V. 1977. An Early Jurassic insect fauna in the Holy Cross Mountains. *Acta Palaeontol. Polonica* **42**(4): 539–543.

WEIDLICH, M. 1987. Systematik and Taxonomie der Buprestidae des mitteleozänen Geiseltales (Insecta, Coleoptera). *Hall. Jahrb. Geowiss.* **12**: 29–52.

WEIDNER, H. 1955. Die Bernstein-Termiten der Sammlung des Geologischen Staatsinstituts Hamburg. *Mitt. geol. Staatsinst. Hamburg* **24**: 55–74.

WEIDNER, H. 1958. Einige interessante Insektenlarven aus der Bernsteininklusen-Sammlung des Geologischen Staatsinstitut Hamburg (Odonata, Coleoptera, Megaloptera, Planipennia). *Mitt. geol. Staatsinst. Hamburg* **27**: 50–68.

WEIDNER, H. 1961. Gallen aus Indien und neue Grundsätze für eine Einteilung der Gallen. *Abhandl. Verhandl. naturwiss. Ver. Hamburg*, n. F. **5**: 19–67.

WEIGELT, J. 1935. Was bezwecken die Hallenser Universitäts-Grabungen in der Braunkohle des Geiseltales? *Natur u. Volk* **65**: 347–356.

WEISHAMPEL, D.B. 1984. Interactions between Mesozoic plants and vertebrates: fructifications and seed predation. *Neues Jahrb. Geol. Paläontol. Abhandl.* **167**: 224–250.

WEISHAMPEL, D.B., and NORMAN, D.B. 1989. Vertebrate herbivory in the Mesozoic: jaws, plants, and evolutionary metrics. In: FARLOW, J.O. (ed.). *Paleobology of the Dinosaurs.* Geol. Soc. America Spec. Paper **238**: 87–100.

WEISSMANN, G. 1966. Distribution of terpenoids. In: SWAIN, T. (ed.). *Comparative Phytochemistry.* Academic Press, New York: 97–120.

WEITSCHAT, W. 1997. Bitterfelder Bernstein – ein eozäner Bernstein auf miozäner Lagerstätte. *Metalla* **66**, Sonderheft: 71–84.

WEITSCHAT, W., and WICHARD, W. 1998. *Atlas der Pflanzen und Tiere im Baltischen Bernstein.* Verlag Dr. Friedrich Pfeil, München: 256 p.

WELLINGTON, W.G. 1954. Atmospheric circulation processes and insect ecology. *Canad. Entomol.* **86**: 312–333.

WENINGER, G. 1968. Vergleichende Drift-Untersuchungen an niederösterricheschen Fliessgewässern (Flysch-, Gneiss-, Kalkformaton). *Schweiz. Zeitschr. Hydrol.* **30**: 138–185.

WENZEL, J.W. 1990. A social wasp's nest from the Cretaceous period, Utah, USA, and its biogeographical significance. *Psyche* **97**: 21–29.

WEYENBERGH, H. 1869. Sur les insectes fossiles du calcaire lithographique de la Bavière qui se trouvent au Musée Teyler. *Arch. Mus. Teyler* **2**: 247–294.

WHALLEY, P.E.S. 1978. New taxa of fossil and recent Micropterygidae with a discussion on their evolution and comment on the evolution of Lepidoptera (Insecta). *Ann. Transvaal Mus.* **31**: 71–86.

WHALLEY, P.E.S. 1980. Neuroptera (Insecta) in amber from the Lower Cretaceous of Lebanon. *Bull. Brit. Mus. Nat. Hist. (Geol.)* **33**: 157–164.

WHALLEY, P.E.S. 1983. A survey of recent and fossil cicadas (Insecta, Hemiptera-Homoptera) in Britain. *Bull. Brit. Mus. Nat. Hist. (Geol.)* **37**: 139–147.

WHALLEY, P.E.S. 1985. The systematics and palaeogeography of the Lower Jurassic insects of Dorset, England. *Bull. Brit. Mus. Nat. Hist. (Geol.)* **39**: 107–189.

WHALLEY, P.E.S. 1988b. Insect evolution during the extinction of the Dinosauria. *Entomol. Gener.* **13**(1/2): 119–124.

WHALLEY, P.E.S., and JARZEMBOWSKI, E.A. 1981. A new assessment of *Rhyniella*, the earliest known insect, from the Devonien of Rhynie, Scotland. *Nature* **291** no 5813: 317.

WHALLEY, P.E.S., and JARZEMBOWSKI, E.A. 1985. Fossil insects from the Lithographic Limestone of Montsech (late Jurassic-early Cretaceous), Lérida Province, Spain. *Bull. Brit. Mus. Nat. Hist. (Geol.)* **38**(5): 381–412.

WHEELER, A.G., Jr. 1975. Insect associates of *Ginkgo biloba*. *Entomol. News* **89**(1–2): 37–44.

WHEELER, W.C. 1998. Sampling, groundplans, total evidence and the systematics of arthropods. In: FORTEY, R.A., and THOMAS, R.H. *Arthropod Relationships.* Systematic Assoc. Spec. Volume Series **55**. Chapman & Hall, London etc.: 87–96.

WHEELER, W.C., SCHUH, R.T. and BANG, R. 1993. Cladistic relationships among higher groups of Heteroptera: congruence between morphological and molecular data sets. *Entomol. Scand.* **24**: 121–137.

WHITE, J.F., Jr., and TAYLOR, T.N. 1989. A trichomycete-like fossil from the Triassic of Antarctica. *Mycologia* **81**(4): 643–646.

WHITEHEAD, D.R. 1969. Wind pollination in the angiosperms: evolutionary and environmental considerations. *Evolution* **23**(1): 28–35.

WHITING, M.F., CARPENTER, J.C., WHEELER, QU.D., and WHEELER, W.C. 1997. The Strepsiptera problem: phylogeny of the holometabolous orders inferred from 18S and 28S ribosomal DNA sequences and morphology. *Systematic Biology* **46**: 1–68.

WHITTINGTON, P.M., and BACON, J.P. 1998. The organization and development of the arthropod ventral nerve cord: insights into arthropod relationships. In: FORTEY, R.A., and THOMAS, R.H. (eds.). *Arthropod Relationships.* Systematic Assoc. Spec. Volume Series **55**. Chapman & Hall, London: 349–367.

WHITMORE, T.C. 1980. Utilization, potential and conservation of *Agathis*, a genus of tropical Asian conifers. *Econom. Botan.* **34**: 1–12.

WICHARD, W., and BÖLLING, A.-C. 2000. Recent knowledge of caddisflies (Trichoptera) from Cretaceous amber of New Jersey. In: GRIMALDI, D.A. (ed.). *Studies on Fossils in Amber, with Particular Reference to the Cretaceous of New Jersey.* Backhuys Publ., Leiden: 345–354.

WICHARD, W., and WEITSCHAT, W. 1996. Wasserinsekten im Bernstein. Eine paläobiologische Studie. *Entomol. Mitt. Löbbecke Muz. + Aquazoo* **4**: 1–122.

WIEBES, J.T. 1979. Co-evolution of figs and their insect pollinators. *Annu. Rev. Ecol. Syst.* **10**: 1–12.

WIEBES, J.T. 1986. The association of figs and fig-insects. *Rev. zool. afric.* **100**(1): 63–71.

WIEDERHOLM, T., and ERIKSON, L. 1979. Subfossil chironomids as evidence of eutrophication in Ekoln Bay, Central Sweden. *Hydrobiologia* **62**(3): 195–208.

WIEGMANN, B.M., MITTER, Ch., and THOMPSON, F.C. 1993. Evolutionary origin of the Cyclorrhapha (Diptera): tests of alternative morphological hypotheses. *Cladistics* **9**(1): 41–81.

WIELAND, G.R. 1935. The Cerro Cuadrado Petrified Forest. *Carnegie Inst. Washington Publ.* **449**: 1–180.

WIGGINS, G.B., and WICHARD, W. 1989. Phylogeny of pupation in Trichoptera, with proposals on the origin and higher classification of the order. *J. North Amer. Benthol. Soc.* **8**: 260–276.

WIGHTON, D.C. 1980. New species of Tipulidae from the Paleocene of central Alberta, Cananda. *Canad. Entomol.* **112**: 621–628.

WILBY, P.R. 1993. The role of organic matrices in post-mortem phosphatization of soft-tissues. *Kaupia* **2**: 99–113.

WILDE, V., LENGTAT, K.-H., and RITZKOWSKI, S. (eds.). 1992. Die oberpliozäne Flora von Willershausen am Harz. *Ber. naturhist. Ges. Hannover* **134**: 7–115.

WILEY, E.O. 1981. *Phylogenetics: The Theory and Practice of Phylogenetic Systematics.* John Wiley & Sons, New York etc.: 439 p.

WILF, P., and LABANDEIRA, C.C. 1999. Response of plant-insect associations to Paleocene-Eocene warming. *Science* **284**: 2153–2156.

WILF, P., LABANDEIRA, C.C., KRESS, W.J., STAINES, C.L., WINDSOR, D.M., ALLEN, A.L., and JOHNSON, K.R. 2000. Timing the radiation of leaf beetles: hispines on gingers from Latest Cretaceous to Recent. *Science* **289**: 291–294.

WILLE, A. 1983. Biology of the stingless bees. *Annu. Rev. Entomol.* **28**: 41–64.

WILLE, A., and MICHENER, C.D. 1973. The nest architecture of stingless bees with special reference to those of Costa Rica (Hymenoptera, Apidae). *Rev. Biol. Trop.* **21**(suppl. 1): 1–278.

WILLEMS, H., and WÜTTKE, M. 1987. Lithogenese lakustriner Dolomite und mikrobiell induzierte Weichteil-Erhaltung bei Tetrapoden des Unter-Rotliegenden (Perm, Saar-Nahe-Becken, SW-Deutschland). *Neues Jahrb. Geol. Paläont. Abhandl.* **174**: 213–238.

WILLEMSTEIN, S.C. 1980. Pollen in Tertiary insects. *Acta Botan. Neerland.* **29**: 57–58.

WILLIAMS, C.B. 1958. *Insect Migration*, Collins, London: 235 p.

WILLIAMS, D.D. 1980. Invertebrate drift lost to the sea during low flow conditions in a small coastal stream in Western Canada. *Hydrobiologia* **75**(3): 251–254.

WILLIAMS, M.A.J. (ed.). 1994. *Plant Galls. Organisms, Interactions, Populations.* Syst. Assoc. Spec. Volume **49**. Chapman & Hall, London: 488.

WILLMANN, R. 1978. *Mecoptera (Insecta, Holometabola). Fossilium Catalogus, Animalia* **124**. W. Junk, The Hague: 139 p.

WILLMANN, R. 1981. Das Exoskelett der männlichen Genitalien der Mecoptera (Insecta). II. Die phylogenetische Beziehungen der Schnabelfliegen-Familien. *Zeitschr. Zool. Syst. EvolutionForsch.* **19**: 153–174.

WILLMANN, R. 1983. The phylogenetic system of the Mecoptera. *Syst. Entomol.* **12**: 519–524.

WILLMANN, R. 1984. Zur systematische Stellung mesozoischen und tertiärer Mecopteren einschließlich *Eoses triassica* Tindale (angeblich Lepidoptera). *Paläont. Zeitschr.* **58**: 231–246.

WILLMANN, R. 1987. Widersprüchliche Rekonstruktionen der Phylogenese am Beispiel der Ordnung Mecoptera (Schnabelfliegen; Insecta: Holometabola). *Paläont. Zeitschrift* **57**: 285–308.

WILLMANN, R. 1988. Der oligozäne Lebensraum von Sieblos/Rhön im Spiegel seines Insekten. *Beitr. Naturkunde Osthessen* **24**: 143–148.

WILLMANN, R. 1989. Evolution und Phylogenetisches System der Mecoptera (Insecta: Holometabola). *Abhandl. Senckenberg. naturforsch. Ges.* **544**: 1–153.

WILLMANN, R. 1990. Insekten aus der Fur-Formation von Dänemark (Moler, ob. Paläozän/unt. Eozän?). 1. Allgemeines. *Meyniana* **42**: 1–14.

WILLMANN, R. 1994. Raphidioidea aus dem Lias und die Phylogenie der Kamelhalsfliegen (Insecta: Holometabola) *Paläont. Zeitschrift* **68**(1/2): 167–197.

WILLMANN, R. 1999. The Upper Carboniferous *Lithoneura lameeri* (Insecta, Ephemeroptera?). *Paläont. Zeitschrift* **73**: 289–302.

WILLMANN, R., and NOVOKSCHONOV, V. 1998a. *Orthophlebia lithographica* – die erste Mecoptere aus dem Solnhofener Plattenkalk (Insecta: Mecoptera, Jura). *Neues Jahrb. Geol. Paläont. Monatshefte* (9): 529–536.

WILLMANN, R., and NOVOKSCHONOV, V. 1998b. Neue Mecopteren aus dem oberen Jura von Karatau (Kasachstan) (Insecta, Mecoptera: 'Orthophlebiidae'). *Paläont. Zeitschr.* **72**(3/4): 281–298.

WILSON, E.O. 1971. *The Insect Societies.* Belknap Press, Cambridge: 548 p.

WILSON, E.O. 1976. Which are the most prevalent ant genera? *Stud. entomol.* **19**(1–4): 187–200.

WILSON, E.O., and TAYLOR, R.W. 1964. A fossil ant colony: new evidence of social antiquity. *Psyche* **71**: 93–103.

WILSON, G.W. 1993. Initial observations of the reproductive biology and an insect pollination agent of *Bowenia serrulata* (W. Bull) Chamberlain. *Encephalartos* **36**: 13–18.

WILSON, M.V.H. 1977. New records of insect families from the freshwater Middle Eocene of British Columbia. *Canad. J. Earth Sci.* **14**: 1139–1155.

WILSON, M.V.H. 1978. Paleogene insect faunas of Western North America. *Quaest. Entomol.* **14**(1): 13–34.

WILSON, M.V.H. 1980. Eocene lake environments: depth and distance-from-shore variation in fish, insect, and plant assemblages. *Palaeogeogr., Palaeoclimatol., Palaeoecol.* **32**: 21–44.

WILSON, M.V.H. 1988a. Reconstruction of ancient lake environments using both autochthonous and allochthonous fossils. *Palaeogeogr., Palaeoclimatol., Palaeoecol.* **62**: 609–623.

WILSON, M.V.H. 1988 b. Taphonomic processes: information loss and information gain. *Geosci. Canada* **15**(2): 131–148.

WILSON, M.V.H. 1991. Insects near Eocene lakes of the Interior. In: LUDVIGSEN, R. (ed.). *Life in Stones: A Natural History of British Columbia's Fossils.* UBC Press, Vancouver: 225–233.

WILSON, M.V.H. 1996. Taphonomy of a mass-death layer of fish in the Paleocene Paskapoo Formation at Joffre Bridge. Alberta, Canada. *Canad. J. Earth Sci.* **33**: 1487–1498.

WILSON, R.S., and BRIGHT, P.L. 1973. The use of chironomid pupal exuviae for characterizing streams. *Freshwater Biol.* **3**(3): 283–302.

WING, S.L., and SUES, H.-D. 1992. Mesozoic and Early Tertiary Terrestrial ecosystems. In: BEHRENSMEYER, A.K., DAMUTH, J.D., DiMICHELE, W.A., POTTS, R., SUES, H.-D., and WING, S.L. (eds.). *Terrestrial Ecosystems through Time. Evolutionary Paleoecology of Terrestrial Plants and Animals.* Univ. Chicago Press, Chicago – London: 327–416.

WING, S.L., and TIFFNEY, B.H. 1987a. The reciprocal interaction of angiosperm evolution and tetrapod herbivory. *Rev. Palaeobot. Palynol.* **50**: 179–210.

WING, S.L., and TIFFNEY, B.H. 1987b. Interactions of angiosperms and herbivorous tetrapods through time. In: FRIIS, E.M., CHALONER, W.G., and CRANE, P.R. (eds.). *The Origins of Angiosperms and their Biological Consequences.* Cambridge Univ. Press, Cambridge: 203–224.

WITHYCOMBE, C.L. 1924. Some aspects of the biology and morphology of the Neuroptera. With special reference to immature stages and their possible phylogenetic significance. *Trans. Entomol. Soc. London* **72**: 303–412.

WITTLAKE, E.B. 1981. Fossil plant galls. In: KAISER, H.E. (ed.). *Neoplasms – Comparative Pathology of Growth in Animals, Plants, and Man.* Baltimore: 729–731.

WOLFE, J.A. 1992. Climatic, floristic, and vegetational changes near the Eocene/Oligocene boundary in North America. In: PROTHERO, D.R., and BERGGREN, W.A. (eds.). *Eocene-Oligocene Climatic and Biotic Evolution.* Princeton Univ. Press, Princeton: 419–436.

WOLFF, W.W., SPARKS, A.N., PAIR, S.D., WESTBROOK, J.K., and TRUESDALE, F.M. 1986. Radar observations and collections of insects in the Gulf of Mexico. In: DANTHANARAYANA, W. (ed.). *Insect Flight: Dispersal and Migration.* Springer, Berlin–Heidelberg–New York: 221–234.

WOOD, D.M., and BORKENT, A. 1989. Phylogeny and classification of Nematocera. In: McALPINE, J.F., and WOOD, D.M. (coordinators). *Manual of Nearctic Diptera* 3. Research Branch Agric. Canada Monogr. **3**. Ottawa: 1333–1370.

WOOTTON, R.J. 1972. The evolution of insects in fresh water ecosystems. In: CLARK, R.B., and WOOTTON, R.J. (eds.) *Essays in Hydrobiology.* Exeter Univ., Exeter: 69–85.

WOOTTON, R.J. 1988. The historical ecology of aquatic insects: an overview. *Palaeogeogr., Palaeoclimatol., Palaeoecol.* **62**: 477–492.

WOOTTON, R.J., and ELLINGTON, C.P. 1991. Biomechanics and the origin of insect flight. In: RAYNER, J.M.V., and WOOTTON, R.J. (eds.). *Biomechanics and Evolution.* Cambridge Univ. Press, Cambridge: 99–112.

WOOTTON, R.J., and ENNOS, A.R. 1989. The implications of function on the origin and homologies of the dipterous wing. *Syst. Ent.* **14**: 507–520.

WOOTTON, R.J., and KUKALOVÁ-PECK, J. 2000. Flight adaptations in Palaeozoic Palaeoptera (Insecta). *Biol. Rev.* **75**: 129–167.

WOOTTON, R.J., KUKALOVÁ-PECK, J, NEWMAN, D.J.S., and MUZÓN, J. 1998. Smart engineering in the mid-Carboniferous: how well could Palaeozoic dragonflies fly? *Science* **282**: 749–751.

WORTHY, T.H. 1984. Faunal and floral remains from F1, a cave near Waitomo. *J. R. Soc. New Zealand* **14**(4): 367–377.

WRIGHT, J.W., WILSON, L.F., and BRIGHT, J.N. 1975. Genetic variation in resistance of Scotch pine to Zimmerman pine moth. *Great Lakes Entomol.* **8**(4): 231–236.

WRIGHT, V.P. (ed.). 1986. *Palaeosols: Their Recognition and Interpretation.* Blackwell, London: 315 p.

WUNDERLICH, J. 1988. Die fossilen Spinnen im Dominikanischen Bernstein. *Beiträge zur Araneologie* **2**. Verlag Jörg Wunderlich, Straubengardt: 378 p.

WUNDERLICH, J. 1999. Two subfamilies of spiders (Araneae, Linyphiidae: Erigoninae and Anapidae: Mysmeninae) new to Dominican amber – or falsificated amber? *Estud. Mus. Cienc. Natur. Alava.* **14** (Núm. Espec. 2): 167–172.

WÜTTKE, M. 1983. "Weichteil-Erhaltung" durch lithifizierte Mikroorganismen bei mittel-eozänen Vertebraten aus den Ölschiefern der "Grube Messel" bei Darmstadt. *Senckenberg. Lethaea* **64**: 509–527.

WYGODZINSKY, P. 1961. On a surviving representative of the Lepidotrichidae (Thysanura). *Ann. Entomol. Soc. Amer.* **54**: 621–627.

WYGODZINSKY, P. 1967. On the geographical distribution of the South American Microcoryphia and Thysanura. In: DEBOUTTEVILLE, D., and RAPOPORT, E. (eds.). *Biologie de l'Amérique Australe.* **3**. CNRS, Paris: 505–524.

WYLIE, F.R., WALSH, G.L., and YULE, R.A. 1987. Insect damage to aboriginal relics at burial and rock-art sites near Carnarvon in Central Queensland. *J. Austral. Entomol. Soc.* **26**: 335–345.

YAALON, D.H. (ed.). 1971. *Paleopedology: Origin, Nature, and Dating of Paleosols.* Internat. Soil Science Society & Israel Universities Press, Jerusalem: 350 p.

YAKUBOVSKAYA, T.A. 1955. *Sarmatskaya flora Moldavskoy SSR* (The Sarmatian flora of Moldavian SSR). Trudy Botan. Inst. Akad. Nauk SSSR **1**(11). Akad. Nauk SSSR, Leningrad: 1–184. (in Russian).

YAMAMOTO, H., and SEKI, H. 1979. Impact of enrichment in a waterchestnut ecosystem at Takahamairy Bay of Lake Kasumigaura, Japan. IV. Population dynamics of secondary producers as indicated by chitin. *Water, Air, and Soil Pollution* **12**: 519–527.

YANG, Z., and HONG, Y.-C. 1980. Discovery of fresh-water triopsids from the Upper Jurassic Dabeigou Formation of Weichang, Hebei, China and its bearing on the classification of the family Triopsidae. *Acta Palaeontol. Sinica* **19**(2): 91–98 (in Chinese with English summary).

YINGST, J.Y., and RHOADS, D.C. 1978. Seafloor stability in central Long Island Sound. II. Biological interactions and their importance for seafloor erodability. In: WILEY, M. (ed.). *Estuarine Interactions*. Academic Press, New York: 245–260.

YOKOHAMA RESEARCH GROUP. 1987. Fossil insects and diatoms from the Maioka Formation at Wakitani, Yokohama. *Bull. Assoc. Kanto Quartern. Res.* **13**: 53–58 (in Japanese).

YOSHIMOTO, C.M., and GRESSITT, J.L. 1959. Trapping of air-borne insects on ships in the Pacific. Pt. 2. *Proc. Hawaiian Entomol. Soc.* **17**(1): 150–155.

YOSHIMOTO, C.M., and GRESSITT, J.L. 1960. Trapping of air-borne insects on ships in the Pacific. Pt. 3. *Pacif. Insects* **2**(2): 239–243.

YOSHIMOTO, C.M., and GRESSITT, J.L. 1961. Trapping of air-borne insects on ships in the Pacific. Pt. 4. *Pacif. Insects* **3**(4): 556–558.

YOSHIMOTO, C.M., and GRESSITT, J.L. 1963. Trapping of air-borne insects in the Pacific-Antarctic area. Pt. 2. *Pacif. Insects* **5**(4): 873–883.

YOSHIMOTO, C.M., GRESSITT, J.L., and MITCHELL, C.J. 1962. Trapping of air-borne insects in the Pacific-Antarctic area. Pt. 1. *Pacif. Insects* **4**(2): 847–858.

YURETICH, R.F. 1984. Yellowstone fossil forests: new evidence for burial in place. *Geology* **12**: 159–162.

YURTSEV, B.A. 1981. *Reliktovye stepnye kompleksy Severo-Vostochnoy Azii* (The Relict Steppe Assemblages of North-East Asia). Nauka, Novosibirsk: 168 p. (in Russian).

YUSHKIN, N.P. 1973. *Yantar' arkticheskich oblastey* (Amber of Arctic Regions). *Seriya Preprintov "Nauchnye doklady"* (Series of Preprints "Scientific Reports"). Komi Otdelenie Akad. Nauk USSR, Syktyvkar: 45 p. (in Russian).

ZABŁOCKI, J. 1960. *Pinus króli*, nowy gatunek sosny trzeciorzędowej z pokładów soli kamiennej w Wieliczce. *Stud. Soc. Sci. Torunensis*, Sect. D (Botanica) **4**(4): 43–48.

ZACHVATKIN, Yu.A. 1975. Embryology of Arthropoda. Vysashaya Shkola, Moscow 328 p.

ZAIKA, V.V. 1989. Amphibiotic insects and their role in redistribution of organic and inorganic matter between aquatic and terrestrial ecosystems. *Sovetsko-Mongol'sky eksperiment "Ubsu-Nur": Mnogostoronnee soveshchanie stran – chlenov SEV, Kyzyl, 1–10 avgusta, 1989. Tezisy dokladov* (The Soviet-Mongolian Experiment 'Ubs-Nur': Multilateral Meeting of the SEV countries, Kyzyl, Aug. 1–10, 1989. Abstracts). Pushchino: 48–50 (in Russian).

ZAJÍC, J. 2000. Vertebrate zonation of the non-marine Upper Carboniferous – Lower Permian basins of the Czech Republic. *Courier Forsch.-Inst. Senckenberg* **223**: 563–575.

ZALESSKY, G.[Yu.]M. 1935. Sur deux restes d'insectes fossiles provenant du bassin de Kousnetz et sur l'âge géologique des dépôts qui les renferment. *Bull. Soc. géol. France*. Ser. 5. **5**: 687–695.

ZALESSKY, Yu.M. 1947. On two new Permian beetles. *Doklady Akad. Nauk SSSR* **56**: 857–860 (in Russian).

ZALESSKY, Yu.M. 1953. New localities of the fossil insects in the Volga River Basin, Kazakhstan and Transbaikalia. *Doklady Akad. Nauk SSSR* **88**: 163–166.

ZALESSKY, Yu.M. 1961. Rare forms of conservation of fossil insect remains from the Miocene of Transcarpathian and other regions. *Geolog. sbornik L'vovskogo geol. obshchestva* **7–8**: 467–470 (in Russian).

ZAPFE, H. 1988. Lebensspuren. *Denkschr. Österr. Akad. Wiss., Mathem.-naturwiss. Klasse* **112**: 109–122.

ZESSIN, W. 1983. Revision der mesozoischen Familie Locustopsidae unter Berücksichtigung neuer Funde (Orthoptera, Caelifera). *Deutsche entomol. Zeitschr.* N. F. **30**: 173–237.

ZESSIN, W. 1987. Variabilität, Merkmalwandel und Phylogenie der Elcanidae im Jungpaläozoicum und Mesozoicum und die Phylogenie der Ensifera (Orthoptera, Ensifera). *Deutsche entomol. Zeitschr.* N. F. **34**: 1–76.

ZESSIN, W. 1988. Neue Saltatoria (Insecta) aus dem Oberlias Mitteleuropas. *Freib. Forsch.-Heft* C **419**: 107–121.

ZESSIN, W. 1996. Exkursion: Der Lias von Dobbertin. 3. Internat. Fachgespräch fossile Insekten in Friedrichsmoor, 14–16. Juni 1996: 31–59.

ZESSIN, W. 1997. *Thueringoedischia throsteidei* nov. gen. et nov. sp. (Insecta, Orthoptera) aus dem Unteren Rotliegenden von Thüringen. *Veröffentlichungen Naturkundemuseum Erfurt*: 172–183.

ZEUNER, F.E. 1931. Die Insektenfauna aus dem Böttinger Marmors. *Fortschr. Geol. Paläontol.* **19**: 247–406.

ZEUNER, F.E. 1938. Die Insektenfauna des Mainzer Hydrobienkalks. *Paläontol. Zeitschr.* **20**: 104–159.

ZEUNER, F.E. 1939. *Fossil Orthoptera Ensifera*. British Mus. (Natur. History), London: 321 p. + plate volume.

ZEUNER, F.E. 1941a. The fossil Acrididae (Orth. Salt.). Part I. Catantopinae. *Ann. Mag. Nat. Hist.* Ser. 11. **8**: 510–522.

ZEUNER, F.E. 1941b. The Eocene insects of the Ardtun beds, Isle of Mull, Scotland. *Ann. Mag. Nat. Hist.* Ser.11. **7**: 82–100.

ZEUNER, F.E. 1942. The fossil Acrididae (Orth. Salt.). Part II. Oedipodinae. *Ann. Mag. Natur. Hist.* Ser. 11. **9**: 128–134.

ZEUNER, F.E. 1943. On recent and fossil *Pseudonaclia* Butler (Lep., Amatidae). *Ann. Mag. Nat. Hist.* Ser. 11. **10**: 140–144.

ZEUNER, F.E. 1944. The fossil Acrididae (Orth. Salt.). Part IV. Acrididae *incertae sedis* and Addendum to Catantopinae. *Ann. Mag. Nat. Hist.* Ser. 11. **11**: 359–383.

ZEUNER, F.E. 1962. A subfossil giant dermapteron from St. Helena. *Proc. Zool. Soc. London* **138**: 651–653.

ZEUNER, F.E., and MANNING, F.J. 1976. A monograph of fossil bees (Hymenoptera: Apoidea). *Bull. Brit. Mus. (Nat. Hist.), Geol.* **27**: 149–268.

ZEZZA, F. 1974. Il Quaternario del Corridoio Interandino dell'Ecuador (Fossa di Latacunga – Ambato). *Atti Inst. Geol. Univ. Pavia* **24**: 120–130.

ZHANG, J. 1985. New data on the Mesozoic fossil insects from Laiyang in Shandong. *Geology of Shandong* **1**(2): 23–39 (in Chinese with English summary).

ZHANG, J. 1989. *Fossil Insects from Shanwang, Shandong, China*. Shandong Sci. Technol. Publ. House, Jinan: 459 p. (in Chinese, with long English summary).

ZHANG, J. 1991. Going further into Late Mesozoic mesolygaeids (Heteroptera, Insecta). *Acta Palaeontol. Sinica* **30**: 679–704.

ZHANG, J. 1992. *Congquingia rohra* gen. nov., sp. nov., – a new dragonfly from the Upper Jurassic of eastern China (Anisozygoptera: Congquingiidae fam. nov.). *Odonatologica* **21**: 97–126.

ZHANG, J. 1993. A contribution to the knowledge of insects from the Late Mesozoic in Southern Shaanxi and Henan Provinces, China. *Palaeoworld* **2**: 49–56.

ZHANG, J. 1994. Discovery of primitive fossil earwigs (Insecta) from Late Jurassic of Layang, Shandong and its significance. *Acta Palaeontol. Sinica* **33**: 229–245.

ZHANG, J., SUN, B., and ZHANG, X. 1994. *Miocene Insects and Spiders from Shanwang, Shandong*. Science Press, Beijing: 298 p. (in Chinese, with long English summary).

ZHERIKHIN, V.V. 1978. *Razvitie i smena melovykh i kainozoiskikh faunisticheskikh kompleksov (trakheinye i khelitserovye)* [Development and changes of the Cretaceous and Cenozoic faunal assemblages (Tracheata and Chelicerata)]. Trudy Paleontol. Inst. Akad. Nauk SSSR **165**. Nauka, Moscow: 198 p. (in Russian).

ZHERIKHIN, V.V. 1979. The use of palaeontological data in ecological prognostication. In: SMIRNOV, N.N. (ed.). *Ekologicheskoe prognozirovanie* (Ecological Prognostication). Nauka, Moscow: 113–132 (in Russian).

ZHERIKHIN, V.V. 1980a. Characters of insect burial. Superorder Thripidea Fallen, 1814. Insects in the terrestrial ecosystems. In: ROHDENDORF, B.B., and RASNITSYN, A.P. (eds.). *Istoricheskoe razvitie klassa nasekomykh* (Historical Development of the Class Insecta). Trudy Paleontol. Inst. Akad. Nauk SSSR **175**. Nauka, Moscow: 7–18, 69–72, 189–224 (in Russian).

ZHERIKHIN, V.V. 1980b. Class Insecta. In: SOLOVYEV, A.N., and SHIMANSKY, V.N. (eds.). *Razvitie i smena bespozvonochnykh na rubezhe mezozoya i kainozoya. Mshanki, chlenistonogie i iglokozhie* (Development and Changes of Invertebrates on the Boundary of the Mesozoic and Cenozoic. Bryozoans, Arthropods, and Echinoderms). Nauka, Moscow: 40–97 (in Russian).

ZHERIKHIN, V.V. 1985. Insects. In: RASNITSYN, A.P. (ed.). *Yurskie kontinental'nye biotsenozy Yuzhnoi Sibiri i sopredel'nykh territorii* (Jurassic Continental Biocoenoses of the Southern Siberia and Adjacent Territories). Trudy Paleontol. Inst. Akad. Nauk SSSR **213**. Nauka, Moscow: 100–131 (in Russian).

ZHERIKHIN, V.V. 1987. Biocoenotic regulation of evolution. *Paleontol. Zhurnal* (1): 3–12 (in Russian, English translation: *Paleontol. J.* **21**(1): 12–19).

ZHERIKHIN, V.V. 1989. The Oligocene seed beetles and weevils (Coleoptera: Bruchidae, Curculionidae) from Bolshaya Svetlovodnaya River (North Primorye). In: KRASSILOV, V.A., and KLIMOVA, R.S. (eds.). *Kaynozoy Dal'nego Vostoka* (The Cainozoic of the Far East). Biologo-Pochvenny Inst. Dal'nevostochnogo Otdeleniya Akad. Nauk SSSR, Vladivostok: 145–150 (in Russian).

ZHERIKHIN, V.V. 1992b. The historical changes in insect diversity. In: YURTSEV, B.A. (ed.). *Biologicheskoe raznoobrazie: podkhody k izucheniyu i sokhraneniyu* (Biological Diversity: The Approaches to Study and Conservation). Zool. Inst. Ross. Akad. Nauk, St.-Petersburg: 53–65 (in Russian).

ZHERIKHIN, V.V. 1993a. The nature and history of grass biomes. In: KARAMYSHEVA, Z.V. (ed.). *Stepi Evrazii: problemy sokhraneniya i vosstanovleniya. Pamyati akad. E.M. Lavrenko* (Steppes of Eurasia: Problems of Conservation and Restoration. To the Memory of Acad. E.M. Lavrenko). Inst. Geogr. Ross. Akad. Nauk, Botan. Inst. Ross. Akad. Nauk, Ross. Botan. Obshchestvo, St-Petersburg – Moscow: 29–49 (in Russian).

ZHERIKHIN, V.V. 1993b. Possible evolutionary effects of ecological crisis: palaeontological and contemporary data. In: KOZLOV, M.V., HAUKIOJA, E., and YARMISHKO, V.T. (eds.). *Aerial Pollution in Kola Peninsula. Proc. Internat. Workshop, April 14–16 1992, St.Petersburg*. Kola Scientific Centre, Apatity: 53–60.

ZHERIKHIN, V.V. 1993c. History of the tropical rain forest biome. *Zhurnal obshchey biol.* **54**(6): 659–666 (in Russian).

ZHERIKHIN, V.V. 1994. Genesis of grass biomes. In: ROZANOV A.YU., and SEMIKHATOV, M.A. (eds.). *Ekosistemnye perestroiki i evolyutsia biosfery* (Ecosystem Restructures and the Evolution of Biosphere) **1**. Nedra, Moscow: 132–137 (in Russian).

ZHERIKHIN, V.V. 1997a. Taphonomical characters of insects. In: OCHEV, V.G., and TVEROKHLEBOVA, G.I. (eds.). *Tafonomiya nazemnykh organizmov* (Taphonomy of Terrestrial Organisms). Izdatel'stvo Saratov. Univ., Saratov: 34–46 (in Russian).

ZHERIKHIN, V.V. 1997b. Succession pruning: a possible mechanism of biome diversification. *Rec. Queen Victoria Mus. & Art Gallery* **104**: 65–74.

ZHERIKHIN, V.V. 1998. Insects in the Eocene and Oligocene deposits of the former USSR. In: LEONOV, Yu.G. (ed.). *Geologicheskie i bioticheskie sobytiya pozdnego eotsena – rannego oligotsena na territorii byvshego SSSR. Chast' 2. Geologicheskie i bioticheskie sobytiya* (Late Eocene – Early Oligocene Geological and Biotical Events on the Territory of the Former Soviet Union. Part II. The Geological and Biotical Events). Trudy Geol. Inst. Ross. Akad. Nauk **507**. Geos, Moscow: 79–87.

ZHERIKHIN, V.V. 1999. Cladistics in palaeontology: problems and constraints. *Proc. First Internat. Palaeoentomol. Conf., Moscow, 1998*. AMBA projects Internat., Bratislava: 193–199.

ZHERIKHIN, V.V. 2000. A new genus and species of Lophioneuridae from Burmese amber (Thripida (= Thysanoptera): Lophioneurina). In: ROSS, A.J. (ed.). *The History, Geology, Age and Fauna (Mainly Insects) of Burmese amber, Myanmar*. Bull. Natur. Hist. Mus., Geol. **56**(1). London: 39–41.

ZHERIKHIN, V.V., and ESKOV, K.Yu. 1999. Mesozoic and Lower Tertiary resins in former USSR. *Estud. Mus. Cienc. Natur. Alava* **14** (Núm. Espec. 2): 119–131.

ZHERIKHIN (ZHERICHIN), V.V., and GRATSHEV, V.G. 1993. Obrieniidae, fam. nov., the oldest Mesozoic weevils (Coleoptera, Curculionidae). *Palaeontol. J.* **27**(1A): 50–69.

ZHERIKHIN, V.V. and GRATSHEV, V.G. 1995. A comparative study of the hind wing venation of the superfamily Curculionoidea, with phylogenetic implications. In: PAKALUK, J., and SLIPINSKI, S.A. (eds.). *Biology, Phylogeny and Classification of Coleoptera: Papers Celebrating the 80th Birthday of Roy A. Crowson*. Muzeum i Instytut Zoologii PAN, Warszawa: 633–777.

ZHERIKHIN, V.V., and KALUGINA, N.S. 1985. Landscapes and communities. In: RASNITSYN, A.P. (ed.) 1985. *Yurskie kontinental'nye biotsenozy Yuzhnoy Sibiri i sopredel'nykh territoriy* (Jurassic Non-marine Biocoenoses of South Siberia and adjacent territories). Trudy Paleontol. Inst. Akad. Nauk SSSR **213**. Nauka, Moscow: 137–183 (in Russian).

ZHERIKHIN, V.V., MOSTOVSKI, M.B., VRŠANSKÝ, P., BLAGODEROV, V.A., and LUKASHEVICH, E.D. 1999. The unique Lower Cretaceous locality Baissa and other contemporaneous fossil insect sites in North and West Transbaikalia. *Proc. First Palaeoentomol. Conf., Moscow 1998*. AMBA projects Internat., Bratislava: 185–192.

ZHERIKHIN, V.V., and NAZAROV, V.I. 1990. New Curculionid beetle of the genus *Trichalophus* Lec. from Yakutia and its findings in the Pleistocene of Belorussia. In: VELICHKEVICH, F.Yu. (ed.). *Novye predstaviteli iskopaemoy flory i fauny Belorussii i drugikh rayonov SSSR* (New Representatives of the Fossil Flora and Fauna of Belorussia and Other Territories of the USSR). Navuka i Tekhnika, Minsk: 92–112 (in Russian).

ZHERIKHIN, V.V., and RASNITSYN, A.P. 1980. Biocoenotic regulation of macroevolutionary processes. In: *Mikro- i makroevolyutsiya* (Micro- and Macroevolution). *Akad. Nauk Eston. SSR*, Tartu: 77–81 (in Russian).

ZHERIKHIN, V.V., and ROSS, A.J. 2000. A review of the history, geology and age of Burmese amber (Burmite). In: ROSS, A.J. (ed.). *The History, Geology, Age and Fauna (Mainly Insects) of Burmese Amber, Myanmar*. Bull. Natur. Hist. Mus., Geol. **56**(1). London: 3–10.

ZHERIKHIN, V.V., and SUKATSHEVA, I.D. 1973. On the Cretaceous insectiferous "ambers" (retinits) in the North Siberia. In: NARCHUK, E.P. (ed.). *Voprosy paleontologii nasekomykh. (Doklady na 24-m Ezhegodnom chtenii pamyati N.A. Kholodkovskogo, 1971)* [Problems of the Insect Palaeontology Lectures on the XXIV Annual Readings in Memory of N.A. Kholodkovsky (1–2 April, 1971)]. Nauka, Leningrad: 3–48 (in Russian).

ZHERIKHIN, V.V., and SUKATSHEVA, I.D. 1989. Patterns of the insect burial in resins. In: SOKOLOV, B.S. (ed.). *Osadochnaya obolochka Zemli v prostranstve i vremeni. Stratigrafiya i paleontologiya* (Sedimentary Cover of the Earth in Space and Time: Stratigraphy and Palaeontology). Moscow: 84–92 (in Russian).

ZHERIKHIN, V.V., and SUKATSHEVA, I.D. 1990. Caddis fly cases as stratigraphic tools in the Cretaceous. In: KRASSILOV, V.A. (ed.). *Kontinental'ny mel SSSR. Materialy soveshchaniya sovetskoy rabochey gruppy proekta N 245 MPGK. Vladivostok, Oktyabr' 1988* (Non-marine Cretaceous of the USSR. Materials of Conference of the Soviet working group of the Project 245 IGCP, Vladivostok, Oct. 1988). Vladivostok: 19–29 (in Russian).

ZHERIKHIN, V.V., and SUKATSHEVA, I.D. 1992. Taphonomy of inclusions in resins. In: KULYOVA, G.V., and OCHEV, V.G. (eds.). *Materialy po metodam tafonomicheskikh issledovaniy* (Contributions to the Methods of Taphonomic Investigations). Izdatel'stvo Saratov. Univ., Saratov: 74–80 (in Russian).

ZHOU, W., and CHEN, Q. 1983. On the discovery of a fossil insect from Zhejiang (Dermaptera). *Entomotaxonomia* **5**: 61–62.

ZHOU, Zh., and ZHANG, B. 1989. A sideritic *Protocupressinoxylon* with insect borings and frass from the Middle Jurassic, Henan, China. *Rev. Paleobot. Palynol.* **59**(1–4): 133–143.

ZHURAVLEVA, I.T., and ILYINA, V.I. (eds.) 1988. *Verkhniy paleozoy Angaridy* (Upper Paleozoic of Angaraland). *Trudy Inst. Geol. Geofiz. Akad. Nauk SSSR* **707**. Nauka, Novosibirsk: 265 p. (in Russian).

ZINOVYEV, E.V. 1997. *History of faunistic assemblages of the taiga zone of West Siberian Plain in the Quaternary*. Autoreferat dissert. cand. biol. nauk. Institut Ekologii Rasteniy i Zhivotnykh, Ural'skoe Otdelenie Ross. Akad. Nauk, Ekaterinburg: 17 p. (in Russian).

ZOBELL, C.E., and RITTENBERG, S.C. 1938. The occurrence and characteristics of chitinoclastics in the sea. *J. Bacteriol.* **35**: 275–287.

ZOEBELEIN, G. 1956a. Die Honigtau als Nahrung der Insekten. 1. *Zeitschr. angew. Entomol.* **38**: 361–416.

ZOEBELEIN, G. 1956b. Die Honigtau als Nahrung der Insekten. 2. *Zeitschr. angew. Entomol.* **39**: 129–167.

ZRZAVY, J., and ŠTYS, P. 1994. Origin of the crustacean schizoramous limb: A re-analysis of the duplosegmentation hypothesis. *J. evol. Biol.* **7**: 743–756.

ZWICK, P. 1973. *Insecta: Plecoptera. Phylogenetische System und Katalog. Das Tierreich* **94**. Walter de Gruyter, Berlin & New York: 465.

ZWICK, P. 1980. *Ordnung Plecoptera (Steinfliegen)*. In: Kukenthal, W. (ed.). Handbuch der Zoologie **4**(2). Walter de Gruyter, Berlin: **26**: 115 p.

ZWICK, P. 2000. Phylogenetic system and zoogeography of the Plecoptera. *Annu. Rev. Entomol.* **45**: 709–746.

General Index

Index to Taxon Names

Index to Straton Names

Index to Geographic Names

DATE DUE